XVI All-Russian Seminar with International Participation "Dynamics of Multiphase Media" (XVI DMM)

Journal of Physics: Conference Series Volume 1404

Novosibirsk, Russia
30 September - 5 October 2019

Editors:

Yevgeniy A. Bondar
Natalya N. Fedorova
Evgeniy I. Kraus
Andrey A. Sidorenko

ISBN: 978-1-7138-1311-8
ISSN: 1742-6588

Printed from e-media with permission by:

Curran Associates, Inc.
57 Morehouse Lane
Red Hook, NY 12571

Institute of Physics
Dirac House, Temple Back
Bristol BS1 6BE UK

Phone: 44 1 17 929 7481
Fax: 44 1 17 920 0979

techtracking@iop.org

Additional copies of this publication are available from:

Curran Associates, Inc.
57 Morehouse Lane
Red Hook, NY 12571 USA
Phone: 845-758-0400
Fax: 845-758-2633
Email: curran@proceedings.com
Web: www.proceedings.com

TABLE OF CONTENTS

MULTIPHASE FLOWS

COMBUSTION AND DETONATION

EXPERIMENTAL STUDIES

Preface

Dear reader, we provide you with the Proceedings of the XVI All-Russian Seminar with international participation "Dynamics of Multiphase Media" (XVI DMM) held in Novosibirsk, Russia, on September 30 - October 5, 2019. The Seminar was organized by the Khristianovich Institute of Theoretical and Applied Mechanics of the Siberian Branch of the Russian Academy of Sciences and Novosibirsk State University of Architecture and Civil Engineering (Sibstrin) under the auspice of the Russian National Committee for Theoretical and Applied Mechanics.

The main topics of the seminar were as follows:

- Problems of constructing physical and mathematical models of mechanics of reacting/inert heterogeneous media and their application to the flows of liquids and gases in porous media, wave and detonation processes in gases;
- Physical, chemical and mathematical problems of ignition and combustion of mixtures of gases and solid particles;
- Laboratory and full-scale experimental studies of the processes of filtration of liquids and gases in porous media;
- Computational and analytical methods for solving the problems of mechanics of heterogeneous media and its applications to ecology, gas/oil recovery, drying, heat and mass transfer in apparatuses of chemical technologies, etc.

The problems under consideration are characterized by a close interconnection of fundamental and applied research as well as development of new technologies. In this regard, at the seminar, special attention was paid to the practical application of promising methods to solve all of the above problems by introducing them into the design and construction of oil and gas production facilities, their explosion and fire safety.

Invited lectures by leading experts and oral presentations by participants were presented at the sessions. In conjunction with the seminar, a scientific school for young participants was organized by Novosibirsk State University of Architecture and Civil Engineering (Sibstrin) which included lectures and master classes with the demonstration of the capabilities of modern software for solving multiphase problems.

More than 150 papers were presented at the seminar sessions, and many of the authors offered their works for publishing in these Proceedings. The present volume provides comprehensive state-of-the art knowledge in the field of multiphase flows and related problem of combustion and detonation, experimental and numerical studies of homogeneous and heterogeneous flows. Taking into account the wide range of problems considered, we divided all the papers into 4 chapters.

Chapter 1 contains articles on various topics of multiphase flows, including the classical problems of the dynamics of particles and droplets in gas flows, as well as filtration flows in porous media. The works on the construction of mathematical models of mechanics of reacting / inert heterogeneous media and their application to the processes of movement of liquids and gases are presented. Various problems associated with phase transitions are also considered. Multiphase flows have wide industrial applications, and in this chapter the reader will find various examples, including cold gas-dynamic spraying, medical applications, river hydraulics, and much more.

Chapter 2 discusses issues related to the construction of new models of heterogeneous media mechanics in terms of the propagation of detonation and combustion waves in porous bodies, in rarefied and dense suspensions of reacting metal and organic particles, and in drops of various hydrocarbon fuels. The kinetic mechanisms of the combustion of hydrocarbon fuels in the capillaries and pores of the substance are considered and the effects associated with these physical processes are examined. Significant attention was paid to the issues of explosion and fire safety in industrial processes involving reactive aerosols and pyrophoric gases used in the electronic and aerospace industries.

Chapter 3 contains papers devoted to various aerophysical experimental researches. In this rather wide area there are papers discussing problems of flight dynamics and conceptual design, sonic boom simulation and attenuation, shock wave interaction, laminar-turbulent transition, vortexes and jets studies, combustion and etc. There are papers particularly discussing the experimental techniques and questions of data processing. All papers of this chapter are united by the common goal to increase the informativity of aerophysical experimental researches by combining measurements with CFD simulations.

Chapter 4 of the Proceedings contains collected works on the numerical simulation of homogeneous and heterogeneous media. The range of topics includes inert, chemically reactive and ionized flows, gas-surface interaction, high-enthaplpy and rarefied flows, flows with external forces and analysis of flow stability. Numerical techniques, methods and approaches, their program implementation in in-house codes and the use of commercially available tools are also discussed in the scope of the chapter.

All papers included in this Volume had undergone the careful evaluation by the Editorial board members and reviewers before they were selected for publication. We express our deepest gratitude to all members of the Organizing committee, invited speakers, reviewers, and participants for their valuable contribution to the successful organization of XVI DMM.

In order to make Seminar proceedings widely accessible the scientific community, we are proud having a partnership with IOP Conference Series team as one of the most reputable publishing companies with highly flexible solutions. The support and help of the IoP team is much appreciated.

XVI DMM Chair:
Vasily M. Fomin, Professor, Dr. Sc,
Full Member of Russian Academy of Sciences,
Scientific Adviser of the Khristianovich Institute of Theoretical and Applied Mechanics of the Siberian Branch of the Russian Academy of Sciences (ITAM SB RAS)

Volume Editors:
Yevgeniy A. Bondar, Ph.D., Deputy Director of ITAM SB RAS, Head of the Computational Aerodynamics Laboratory, ITAM SB RAS;
Natalya N. Fedorova, Professor, Dr. Sc., Leading Researcher of the Laboratory of Wave Processes in Fine-Dispersed Media, ITAM SB RAS;
Evgeniy I. Kraus, Ph.D., Deputy Director of ITAM SB RAS, Head of the Laboratory of Thermomechanics of New Materials and Technologies, ITAM SB RAS;
Andrey A. Sidorenko, Ph.D., Deputy Director of ITAM SB RAS, Head of the Laboratory of Physical Problems of Gas-Dynamic Flow Control, ITAM SB RAS.

1404 (2019) 011002 doi:10.1088/1742-6596/1404/1/011002

Peer review statement

All papers published in this volume of *Journal of Physics: Conference Series* have been peer reviewed through processes administered by the proceedings Editors. Reviews were conducted by expert referees to the professional and scientific standards expected of a proceedings journal published by IOP Publishing.

An experimental study of the propagation of a shock wave in a granular medium

A V Alyanov, I V Kazanin, V N Zinoviev, A S Vereshchagin and V M Fomin

Khristianovich Institute of Theoretical and Applied Mechanics SB RAS,
4/1 Institutskaya str., Novosibirsk, 630090, Russia

aljanov-ant@ya.ru

Abstract. In this work, experimental study of propagation shock waves and pressure-waves in granular medium was performed. As a granular medium are used hollow glass microspheres and composite sorbent based on them. The special experimental setup was prepared, which uses the principle of a shock tube. The setup includes a high pressure section and low pressure section with stepwise change of its internal diameter. The pressure graphics were obtained for the various tube sections for different media. The tube filling time and pressure wave velocity in the medium were estimated.

Introduction

The propagation of a shock wave and a pressure wave through selectively sorbed granular media is a pressing problem. Such problems arise in the design of facilities operating on the principle of pressure swing adsorption. In such facilities, when the elements of shut-off and control valves are quickly opened, shock waves and pressure waves can occur and propagate further through the pipelines into the adsorber with a sorbent. The parameters of gas–sorbent interaction, wave propagation velocity, adsorber filling time are needed to simulate the operation of the facilities and their design.

One of the applied problems, where it is important to take into account the parameters of interaction and propagation of a shock wave and a pressure wave in granular media, is the design of facilities for the extraction of helium from a helium-containing mixture using the membrane-sorption method. This method is developed at the Institute of Theoretical and Applied Mechanics SB RAS and combines membrane technology and short-cycle adsorption [1]. One of the distinguishing features of the proposed method is the use of hollow microspherical particles of silicate materials as membrane elements, the wall of which is selectively permeable to helium and impermeable to other components of natural gas [2]. The use of fine particles in their original form presents some technological difficulties, one of the solutions to this problem is the creation of a granular composite sorbent with microparticles as a helium-permeable component [3].

Installation description

To study the propagation of a shock wave and a pressure wave in granular media, a special experimental setup was prepared, the schematic diagram of which is shown in figure 1.

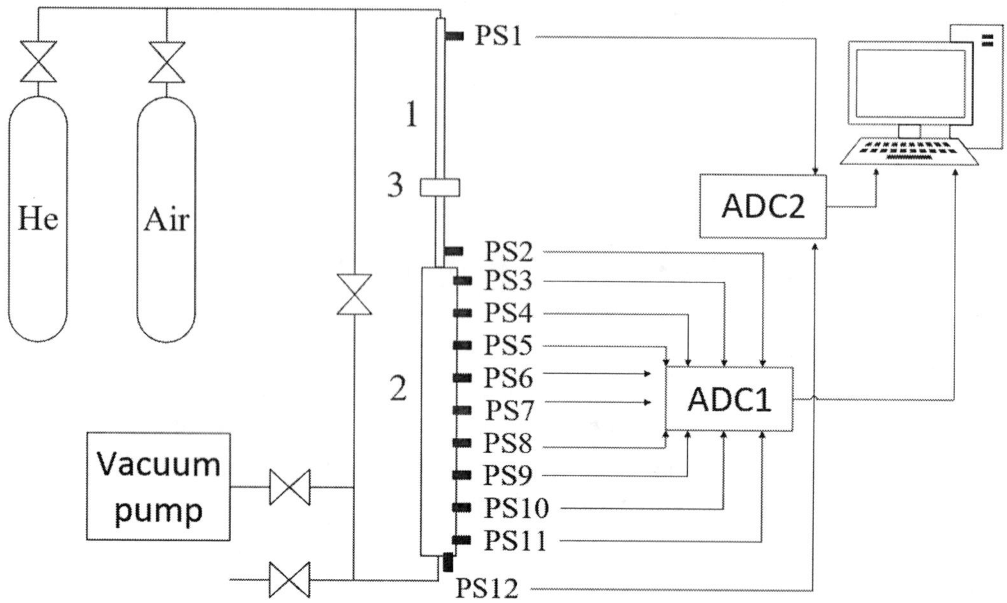

Figure 1. Schematic diagram of the setup.

The experimental setup uses the principle of a shock tube and includes a high pressure section (1), a low pressure section (2), a membrane unit (3), elements of shut-off and control valves and pressure sensors (PS). The high pressure section is a pipe made of stainless steel with length of 1120 mm and internal diameter of 25 mm. The low pressure section consists of two pipes with different diameters: the first part has the same internal diameter of 25 mm as the high pressure section and length of 160 mm, the second one has internal diameter of 50 mm and length of 793 mm. A special flange connects these parts of the low-pressure section, so that the internal diameter of the channel is changed stepwise. In the narrow part of the pressure gauge there is a pressure sensor at distance of 100 mm from the installation site of the membrane. In the wide part of the low pressure switch along its generatric, 10 pressure sensors are installed. The first one is installed at distance of 60 mm from the upper end of the wide part, followed by the sensors going with constant interval of 84 mm. At the lower end of the pressure section there is another pressure measurement sensor. To protect the pressure sensors from the ingress of solid particles of the medium, special filters made of non-woven material are installed.

Two types of materials were used as a granular medium: microspheres MS-V-2L and a composite sorbent based on microspheres.

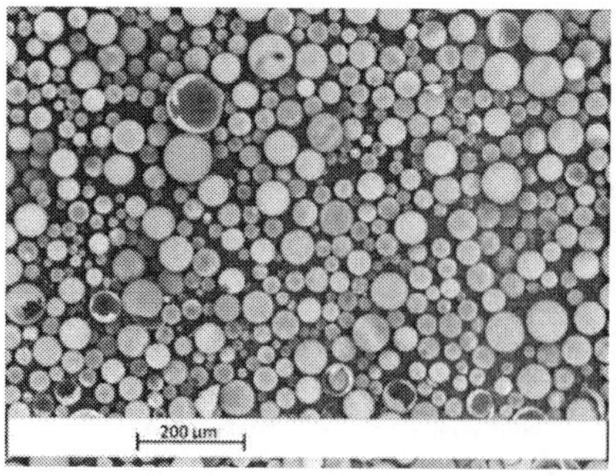

Figure 2. Microspheres MS-V-2L from an electron scanning microscope.

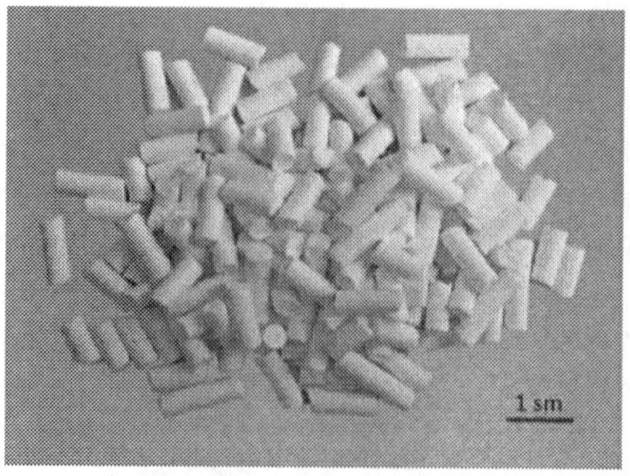

Figure 3. The composite sorbent.

Microspheres are a light granular medium in the form of a white powder, consisting of individual hollow particles of a spherical shape with size of up to 90 microns and thickness of ~ 1 micron. The microspheres MS-V-2L are made of sodium borosilicate glass at JSC "NPO Stekloplastik". Figure 2 shows the image of microspheres MS-V-2L obtained using an electron-scanning microscope. Particles have spherical form and a smooth surface. Bulk density is 0.23-0.27 g/cm^3.

A composite sorbent is made using microspheres as a helium-permeable component and a gamma alumina binder as a highly porous, strong binder framework. The content of microspheres in the sorbent under consideration is 15% by weight. The granules are cylindrical in shape. The diameter of the granule is 3 mm, the length is about 10 mm. A photograph of the composite sorbent is shown in figure. 3 The parameters of the media formed by such materials were considered in [4, 5].

The shock tube is positioned vertically to ensure uniform filling by filling the bottom of the low-pressure section. The backfill level is between the third and fourth pressure sensors. The experiments are carried out using air and helium as the working gas; the temperature in the experiments was ~ 20 C. A dried cellophane membrane is installed in the membrane block to provide brittle fracture. A needle detonated the membrane.

When conducting the experiment in air, the pressure in the low-pressure section is equal to atmospheric; in the high pressure section, it is let in by air to the pressure of 3.5 atm. During the experiment with helium, the shock tube is previously evacuated, after which the low and high pressure sections are filled with helium to values of approximately 1 atm, after which helium is additionally injected into the high pressure section with pressure of 3.5 atm. Then, the membrane is ruptured. The data from pressure sensors 2–11 are digitized using an ADC with frequency of 40 kHz, time series are 3 seconds for experiments with a composite sorbent, and 10 seconds for experiments with microspheres.

Experimental background

Previously, experiments were conducted in air and helium with an empty setup. Then, experiments were carried out with backfill of microspheres (backfill porosity 0.33) and a composite sorbent (backfill porosity 0.82). Figure 4 shows data from the pressure sensors for the empty low-pressure section.

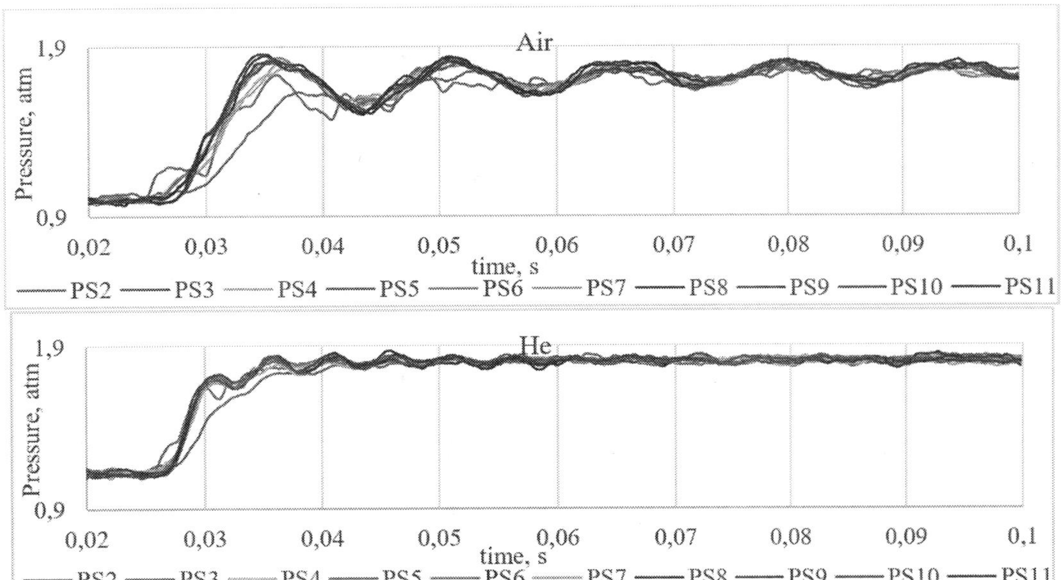

Figure 4. Graphics for empty tube.

Due to the relatively low reaction speed of pressure sensors, it is difficult to measure the speed of a shock wave in the empty tube. Near the connection of tubes of different diameters, next to pressure sensor 3, after the propagation of the shock wave, a toroidal vortex is formed, due to which the pressure measured by this sensor is lower than in the other sections of the tube. After the propagation of the shock wave, gas oscillations are observed in the tube, the period of which is shorter for helium due to its lower density and higher adiabatic index.

In the case of filling the low-pressure section with a composite sorbent, when the shock wave-pressure wave propagates, the experimental data are as follows in figure 5. As follows from the graphs, as the shock wave passes through the backfill, it is broken up and a pressure wave propagates further along the sorbent.

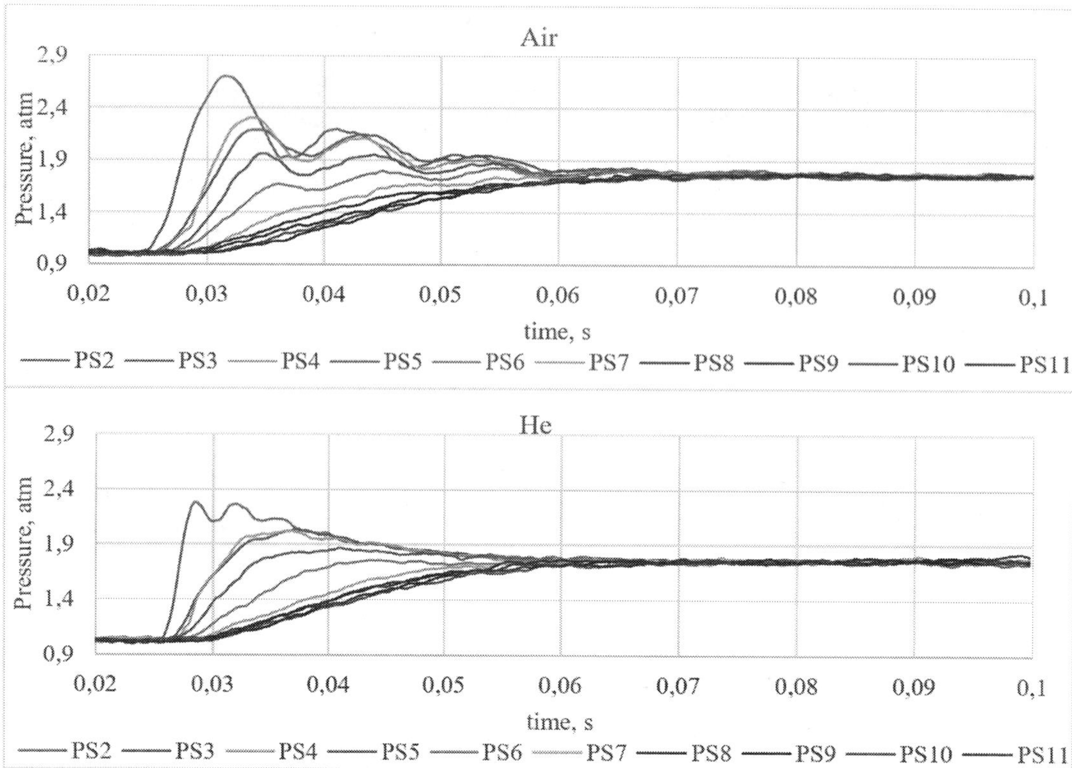

Figure 5. Graphics for composite sorbent backfill.

Figure 6 shows the pressure graphs in the tube with microspheres. At this time scale, the difference between pressure sensors 2 and 3 is not noticeable. Immediately after the collision of the shock wave with microspheres, all sensors demonstrate almost simultaneous slight stepwise increase in pressure. This phenomenon can be explained by shock wave propagation throw the system of glass microspheres, which can be regarded as a solid. Using the pressure curves from figure 6, it is possible to determine the pressure-wave velocity in the granular medium under study, which is changed from the shock wave propagation velocity up to 0.3 m/s towards the end of the tube.

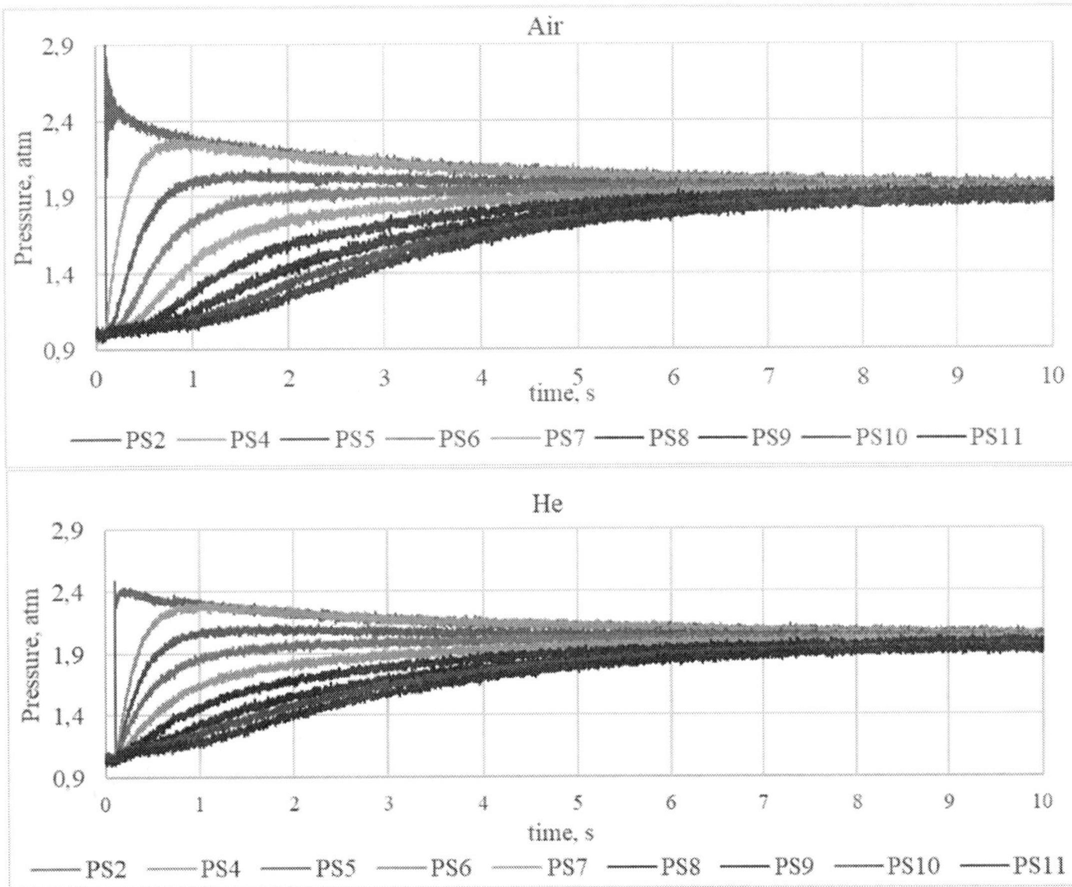

Figure 6. Graphics for microspheres backfill.

Conclusion

In this work, an experimental study of propagation of shock waves and pressure-waves in granular medium was performed. The experiments were carried out in a tube with various fillings. The pressure graphics were obtained for the various tube sections for different media. The tube filling time and the pressure wave velocity in the medium were estimated.

Acknowledgments

This work was carried out with the partial financial support of the RFBR project № 19-41-543007.

References

[1] Fomin V M, Zinovyev V N, Kazanin I V, Lebiga V A, Pak A Y, Vereshchagin A S, Fomina A F, Anshits A G, Buluchevsky E A and Lavrenov A V 2014 *R.F. Patent No. 2508156014*

[2] Zinovyev V N, Kazanin I V, Pak A Yu, Vereshchagin A S, Lebiga V A and Fomin V M 2016 *J. Engineering Physics and Thermophysics* **89** 25-37

[3] Zinovyev V N, Kazanin I V, Lebiga V A, Pak A Yu, Vereshchagin A S and Fomin V M 2016 *J. Thermophysics and Aeromechanics* **5** 771–7

[4] Kazanin I V, Zinovyev V N, Lebiga V A, Pak A Y, Fomin V M, Vereshchagin A S and Tsibulsky N G 2018 *AIP Conference Proceedings* **1939** 020029

[5] Kazanin I V, Zinoviev V N, Lebiga V A, Pak A Y, Fomin V M, Vereshchagin A S and Tsibulsky N G 2017 *AIP Conference Proceedings* **1893** 030013

Gas heat exchange at inhomogeneous combustion of solid porous media

E V Amosova[1,2] and K S Kuznetsov[3]

[1]Far Eastern Federal University, 8, Sukhanova str., Vladivostok, 690090, Russia
[2]Institute of Applied Math, Far Eastern Department of RAS,
7, Radio str., Vladivostok, 690041, Russia
[3] Novosibirsk State University, 1, Pirogova str., Novosibirsk, 630090, Russia

el_amosova@mail.ru, kuznetsovks17@gmail.com

Abstract. The authors suggest using a numerical method based upon the finite element analysis of the equations of continuous inhomogeneous media mechanics to describe thermophysical properties of transient motions of viscous gas through porous media with chemical changes and phase transitions in a solid phase. The model takes into account heat exchange between a solid frame and gas, changes in the phase volume and mass at interaction, the presence of oxidant diffusion. The software developed in FreeFem++ allows clear and vivid studying the process of gas combustion in solid porous medium with inhomogeneous combustion source at forced filtration in the course of time as well as modeling and analyzing a real experiment.

1. Introduction

Filtration gas combustion processes are widely applied in everyday routine and chemical technology. Peat fire, cigarette smoking can be modeled as the combustion of solid porous media.

The article relevance is explained by the fact that the process of porous rock inhomogeneous combustion is inseparable associated with various catastrophes both technogenic and natural. Such catastrophes can include combustion occurrence at fire explosive facilities, peat bogs and waste deposits. The study of motion structure and gas heat exchange at the combustion of solid porous rocks helps to develop understanding of the nature of these processes. This will facilitate the prevention of fires or their extinguishing.

The numerical analysis of two-dimensional gas fluids in porous media with inhomogeneous combustion sources at forced filtration and free convection was conducted in the papers [1], [2] developing a numerical method based upon the combination of explicit and implicit finite-difference schemes. According to the proposed method, the equations of power, impulse conservation and oxidant concentration are transformed into explicit finite-difference equations while the continuity equation is transformed into an implicit finite-difference equation. Further research were developed in the area of modeling gas motion through the layer of a heat accumulating material with phase transition [3]. In [4] – [6] analyzed the research of smoldering in foamed polyurethane at free convection.

The authors study a stationary porous object having lateral walls which are impenetrable and thermally non-conductive and open on the top and bottom. Cold gas can flow into a porous object and flow out through open boundaries. A solid porous substance consists of a combustible component, inert additive and solid reaction products when a solid combustible material is transformed into gas and solid products of the reaction as a result of interaction with a gaseous oxidant.

At the research of filtration gas combustion it is possible to use a volume-averaged model [7], i.e. one can use efficient (averaged in terms of volume) thermophysical properties of a porous frame and gas while cross-media interaction is described by the heat exchange factor.

A mathematical model is built upon the equations of thermal balance in solid and gaseous phases, mass transfer of gas components, continuity equations, the equations of diffusion hydrodynamics or gas filtration and state [8]. To form source terms of the equations of thermal balance and mass, the authors set down chemical kinetics equations of the considered gas components [9], [7] and state boundary conditions.

A system of differential equations, describing the thermal conductive gas dynamics in porous medium, is a linear mixed hyperbolic-parabolic equation system. In addition, to close the system one can use equations in ordinary derivatives.

In a general case this problem cannot be resolved analytically. The paper suggests a new calculation algorithm in the integrated development software FreeFem++ [10] implemented on the basis of finite-dimensional analysis allowing one to study heat exchange through porous media with chemical transformations and phase transitions in the solid phase.

2. Mathematical model

The mathematical model of viscous gas thermal conductivity at the filtration combustion of solid porous bodies is considered as the motion of two-component continuous medium consisting of solid and gaseous media within the model of multispeed continuum and interpenetrating motion of the mix components [9], [7].

The system of differential equations modeling the process is based upon a mathematical model suggested in [1], [2].

A equation system used for modeling heat transfer of non-stationary gas flows in porous media with inhomogeneous combustion sources is written as follows [1], [2]:

$$(\rho_{cf}c_{cf} + \rho_{ci}c_{ci} + \rho_{cp}c_{cp})\,\partial T_c/\partial t = -\alpha(T_c - T_g) + Q\rho_{cf0}W + \lambda_c div[(1-a_g)\nabla T_c], \quad (1)$$

$$\rho_g[\partial T_g/\partial t + (\nabla T_g \cdot \boldsymbol{v})] = \alpha(T_c - T_g) + \lambda_g div(a_g \nabla T_g), \quad (2)$$

$$\rho_g[1 + 0.5(1 - a_g)][\partial \boldsymbol{v}/\partial t + (\boldsymbol{v} \cdot \nabla)\boldsymbol{v}]$$

$$=-a_g \nabla p + \rho_g \boldsymbol{g} - a_g^2 \mu k_1^{-1} \boldsymbol{v} - (1 - \mu_p)\rho_{cf0} W \boldsymbol{v}, \quad (3)$$

$$\partial \rho_g/\partial t + (\nabla \rho_g \cdot v) + \rho_g div\, v = (1 - \mu_p)\rho_{cf0}W, \quad p = \rho_g R T_g/(a_g M), \quad (4)$$

$$\rho_g[\partial C/\partial t + (\nabla C \cdot \boldsymbol{v})] = div(\rho_g D_g \nabla C) - \mu_g \rho_{cf0} W - (1 - \mu_g)\rho_{cf0} W C, \quad (5)$$

$$W = (1 - \eta) C exp[-E/(RT_c)], \qquad \partial \eta/\partial t = W, \qquad \eta(0) = 0, \quad (6)$$

$$\rho_{cf} = (1 - \eta)\rho_{cf0}, \qquad \rho_{cp} = \mu_p \rho_{cf0} \eta, \quad (7)$$

$$a_g = a_{g0} + (a_{cf0} - a_{cpEnd})\eta, \quad (8)$$

$$D_g = D_{g0}(T_g/273)^b, \quad \mu = c_{s1} T_g^{1.5}/(c_{s1} + T_g). \quad (9)$$

where T_c is solid medium temperature, T_g is gas temperature, ρ_g is gas density, C is oxidant mass concentration, W is chemical reaction rate, η is solid combustible conversion, a_g is gas volumetric concentration, a_{g0} is gas volumetric concentration at the initial instant, a_{cf0} is fuel volumetric concentration at the initial instant, a_{cpEnd} is volumetric concentration of the reaction product condensed phase at the finite time moment, E is activation energy, D_g is gas diffusion factor, b is exponential factor in the equation for the gas diffusion factor, ρ_{cf} is fuel density, ρ_{cp} is reaction product density, $\boldsymbol{v}$ gas velocity, $\boldsymbol{g}$ is gravity acceleration.

At the inlet to the porous object (boundary Γ_1) the following values are known: gas temperature and pressure as well as oxidant mass concentration. At the outlet of the object (boundary Γ_2) the pressure is known. Additionally, the authors set the heat exchange conditions at the inlet and outlet of the porous object and limiting impermeable walls (boundary Γ_0). We can write the boundary conditions for equations (1) – (9) in the following form:

$$p = p_0(x,y),\ \ \partial T_c/\partial \boldsymbol{n} = (\beta/\lambda_c)\left(T_g - T_c\right),\ \ T_g = T_{g0},\ \ C = C_0,\ \ (x,y) \in \Gamma_1$$

$$\text{if } (\boldsymbol{v}\cdot\boldsymbol{n}) \le 0,\ (x,y) \in \Gamma_1;$$

$$p = p_1(x,y),\ \ \partial T_c/\partial \boldsymbol{n} = (\beta/\lambda_c)\left(T_g - T_c\right),\ \ \partial T_g/\partial \boldsymbol{n} = 0,\ \ \partial C/\partial \boldsymbol{n} = 0,\ \ (x,y) \in \Gamma_2$$

$$\text{if } (\boldsymbol{v}\cdot\boldsymbol{n}) > 0,\ (x,y) \in \Gamma_2;$$

$$\partial T_c/\partial \boldsymbol{n} = 0,\ \ \partial T_g/\partial \boldsymbol{n} = 0,\ \ \partial C/\partial \boldsymbol{n} = 0,\ \ (x,y) \in \Gamma_0,$$

$$(\boldsymbol{v}\cdot\boldsymbol{n}) = 0,\ (x,y) \in \Gamma_0,$$

where β is the heat removal coefficient.

Boundary conditions are obtained using the following values:

$$T_{g0} = 1,\ \ C_0 = 0.23,\ \ p_0 = 1.1,\ \ p_1 = 1,\ \ T_c = \begin{cases} 1, & \text{in the zone without burning,} \\ 2, & \text{in the combustion zone.} \end{cases}$$

3. Numerical modelling

When modeling a system with continuous time, a system with discrete time in the power equation and in the equations of oxidant concentration and continuity applies the inverse Euler method of the first order [11] while the concentration equation is transformed with the help of the direct Euler method [11].

In this case, the thermal properties of the medium depend on the temperature, in this sense all the equations included in the system are linear. Non-linearity has a non-monotonic nature.

We will study the linearized model, which is characterized by the fact that the solution at the new temporary layer is derived from the solution of the linear differential problem, the coefficients of the derivatives are taken from the previous temporary layer, while a modified Newton method is used to linearize a non-linear term in the right part of the equation, describing the reaction oxidation [12]. In addition, the continuity equation is replaced by a more regularized equation by means of adding an elliptic component with a regularization parameter.

Note that two-dimensional gas motions at inhomogeneous combustion of solid porous media were numerically studied [1], [2]. The advantage of this method is that the selection of implicit schemes allows modeling quite a long period of combustion at minimum time expenditures for software calculation [12], [12].

It seems impossible to prove the analytical convergence of the approximated solution to exact solution. The study of the approximated solution convergence was conducted both by means of comparison with the available computational data and by the experiments with the help of grid condensation and reducing time increment. The convergence analysis of the obtained solution on two computational grids, with the number of nodes on one twice more than on the other, showed that the convergence of the solution is achieved with the number of computational nodes of approximately 13912 which corresponds to h=0.02.

In the case of forced filtration, when a self-sustaining combustion occurs, the reaction wave moves up and towards the side walls, completely burning the combustible substance.

Journal of Physics: Conference Series **1404** (2019) 012002 doi:10.1088/1742-6596/1404/1/012002

Figures 1 and 2 give an example of the gas velocity field, as well as the degree of conversion of the combustible component of the solid medium inside the object some time after the occurrence of ignition initiated by the lower part of the surface of the object (presented as an ellipse).

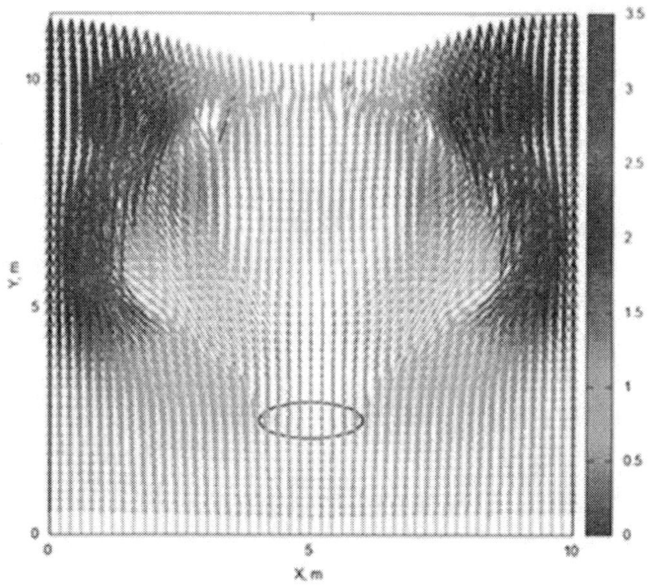

Figure 1. The gas velocity field in a porous object during forced filtration after 6 hours after the start of the ignition process.

As it is clear from Figure 1, a larger gas part flows round the hot region and keeps flowing along the edges of the porous object where no combustion has occurred yet.

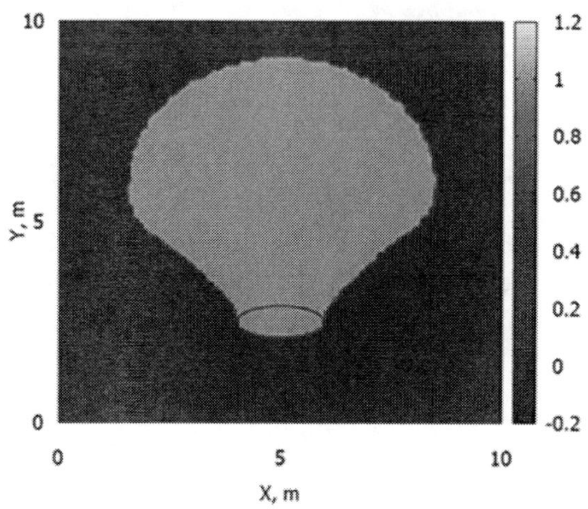

Figure 2. Degree of conversion of the combustible component of the solid medium field in a porous object during forced filtration after 6 hours after the start of the ignition process.

Increasing the contact area of the solid phase with a gaseous one occurs because of porosity growth. The mass of solid medium in a porous component reduces while the gas rate increases, and due to this the general heat exchange rate increases as well.

Figure 3 shows the dynamics of oxidant mass concentration change in the course of time - oxygen enters into the reaction with fuel in the direction of combustion propagation and, after the region burning out, oxygen enters the region again through an open boundary.

 1404 (2019) 012002 doi:10.1088/1742-6596/1404/1/012002

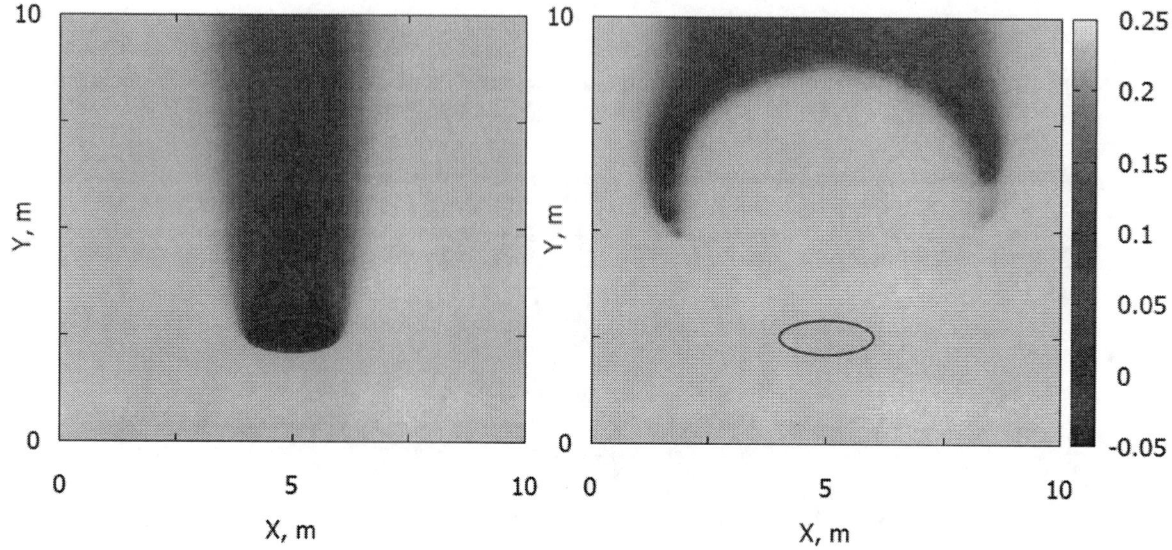

(a) (b)

Figure 3. An example of a mass concentration of an oxidizing agent in a porous object during forced filtration after 0.12 (a) and 6 (b) hours after the start of the ignition process.

Figure 4 shows that at the first gas density increases and then starts to decrease. At the initial time stage gas gives a larger part of heat energy to the porous energy which results in density growth.

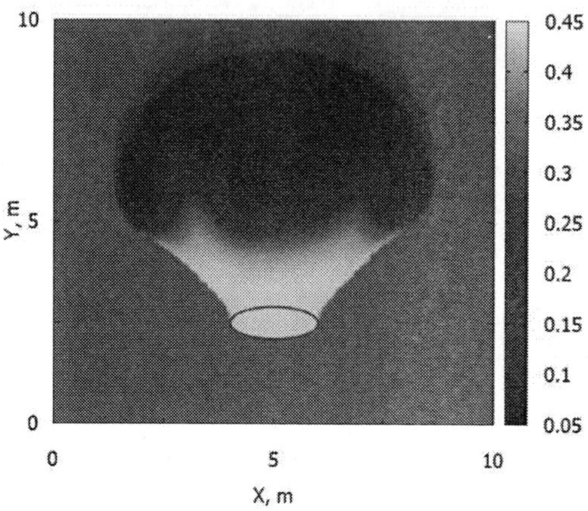

Figure 4. An example of the gas density function in a porous object during forced filtration 6 hours after the start of the ignition process.

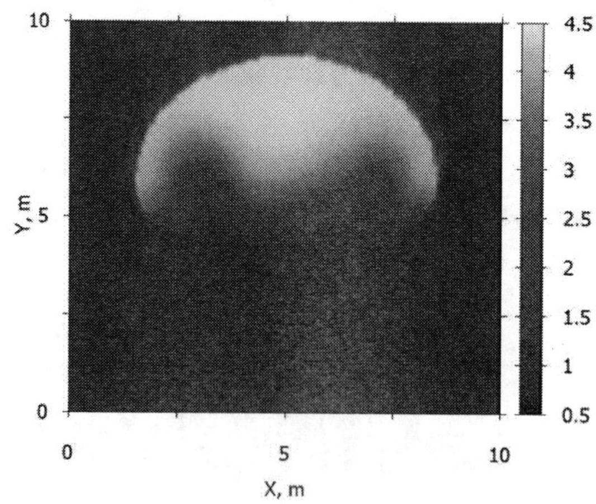

Figure 5. Example of gas temperature in a porous object during forced filtration 6 hours after the start of the ignition process.

Figures 5, 6 present the dynamics of gas temperature and pressure gradients. Temperature is rising in the combustion region and out of the boundaries during the combustion process after which the burned out region becomes cool. The change of oxidant mass concentration (Figure 3) also clearly shows the combustion process propagation. During the combustion of solid porous media, the gas dynamics inside objects is quite complex.

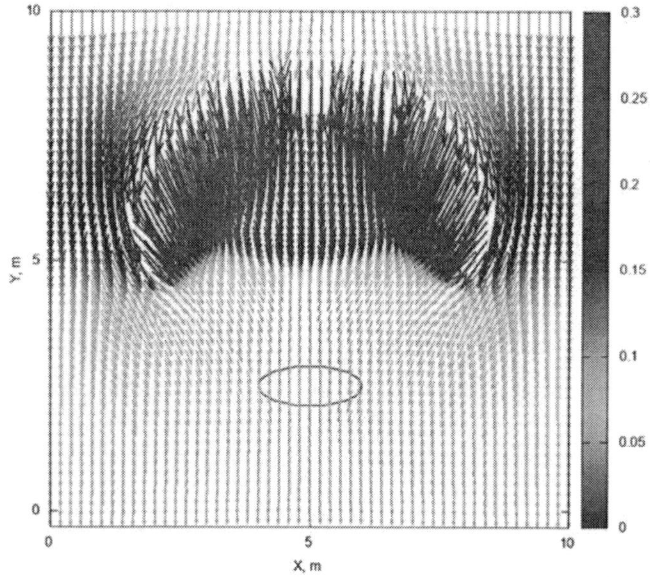

Figure 6. An example of a gas pressure gradient in a porous object during forced filtration 6 hours after the start of the ignition process.

Conclusions

To describe two-dimensional non-stationary flows of viscous thermal conductive gas at filtration combustion of solid porous media, the authors suggest a numerical algorithm. The authors managed to develop a software implementing the suggested numerical method and tested it. The prospective application of the proposed computational model and obtained results is to generate a uniform solver for diffusion reaction systems by means of the finite element method using FreeFem++ [10].

References

[1] Levin V and Lutsenko N 2018 *Reports of the Academy of Sciences* **476** 1 30
[2] Lutsenko N 2018 *Combustion Theory and Modelling.* **22** 2 359
[3] Levin V, Lutsenko N and Fetsov S 2018 *Reports of the Academy of Sciences* **479** 4 386
[4] Torero J, Fernandez-Pello A and Kitano M 1993 *Combust. Sci. Tech.* **91** 95
[5] Torero J and Fernandez-Pello 1995 *Fire Saf. J.* **24** 35
[6] Ohlemiller T, Bellan J and Rogers F 1979 *Combust. Flame* **36** 197
[7] Nigmatulin R 1978 *Basics of Inhomogeneous Media Mechanics* (Moscow: Nauka) p 336
[8] Loitsyanskiy L 2003 *Fluid Mechanics* (Moscow: Drofa) p 840
[9] Gremyachkin V 2015 *Inhomogeneous combustion of solid fuel particles* (Moscow: Publishing House of Bauman Moscow State Technical University) p 280
[10] Hecht F 2012 *J. Numer. Math.* **20** 251
[11] Kalitkin N and Koryakin P 2013 *Numerical methods in two volumes. Mathematical Physics Methods* (Moscow: Akademiya) p 304
[12] Samarsky A and Vabishchevich P 2003 *Computational Heat Transfer* (Moscow: URSS) p 784

Journal of Physics: Conference Series **1404** (2019) 012003 doi:10.1088/1742-6596/1404/1/012003

Modeling of the synthesis of composite particles based on titania and silica in flow-type plasmachemical reactor

S M Aulchenko[1,2], E V Kartaev[1] and V A Emelkin[1]

[1] Khristianovich Institute of Theoretical and Applied Mechanics SB RAS, Novosibirsk, Russia
[2] Novosibirsk State Architectural-Building University, Novosibirsk, Russia

kartayev@itam.nsc.ru

Abstract. The modeling of single-stage synthesis of composite titania-silica nanoparticles in the working zone of flow-type plasmachemical reactor has been carried out based on segregated oxidation of titanium and silicon tetrachlorides, with each precursor being preliminarily mixed with air oxygen by means of bubbling. Within the framework of the developed model of the synthesis, the data regarding core size of composite particles, their shell thickness as well as ratio of particles with and without shell have been extracted.

1. Introduction

Synthesis of nanocomposite powders of oxide ceramics is one of the most promising trends in modern technologies. These products belong to new class materials with controlled physico-chemical properties ranged depending on their applications. In particular, nanosized titania (TiO_2) powders are being widely used in industry. In many applications it is required to extremely inhibit photocatalitic activity of TiO_2 nanoparticles. For instance, it is the case when TiO_2 pigments are used to modify physico-chemical properties of paint films, plastic materials, in paper manufacturing as well as that of sun blocking agents. In these it is necessary to reduce an area of photo-active free surface of titania pigment as much as possible, with titania optical properties being unchanged. This requirement might be met by the synthesis of nanocomposite 'TiO_2 core - SiO_2 shell' particles. In these powders the greater thickness of silica amorphous layer and the lesser its microporosity, the lower photoactivity of the composite powder [1]. In [2] first time the simulation of single-stage synthesis of composite TiO_2/SiO_2 nanoparticles in the working zone of plasmachemical reactor has been performed providing segregated oxidation of titanium and silicon tetrachloride. In this work molar concentrations of reacting components fed into the reactor were chosen regardless of actual technique of mixing. The latter resulted in overpredicted concentrations of titanium and silicon tetrachloride vapors.

In this numerical study the results of simulation of aforementioned single-stage synthesis of composite TiO_2/SiO_2 nanoparticles using the chloride method have been presented based on segregated oxidation of titanium and silicon tetrachlorides taking into account variable axial location of side slit of injection of silicon tetrachloride vapors.

2. Formulation of the problem

In figure 1 the schematic of the working zone of a flow-type reactor is shown (actually, the reactor is positioned vertically). Nitrogen jet with a temperature T_1 and flow rate Q_1 is injected through the channel of the reactor to its working zone. Titanium tetrachloride vapors mixed with air (temperature

T_2, total flow rate Q_2) are supplied into the reactor through first side slit. In the mixing zone of the reactor, the synthesis of gaseous titania is followed by a formation of TiO_2 particles. While moving along the reactor, the particles grow due to the surface reaction and coagulation. In [3,4] the results of numerical and experimental studies of this stage are presented. $SiCl_4$ vapors mixed with air with temperature T_3 and flow rate Q_3 are injected via second side slit.

In this mixing zone both the homogeneous reaction of synthesis of gas-phase silica which is condensed on titania particles and heterogeneous reaction with formation of solid silica on TiO_2 particles and TiO_2/SiO_2 composite particles occur.

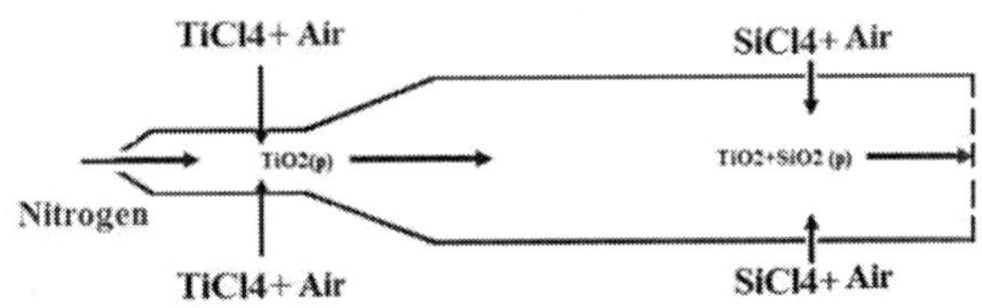

Figure 1. Schematic of the working zone of the flow-type reactor

3. Mathematical model

A flow of a viscous heat-conducted compressible gas mixture is considered. The mixture consists of following species: O_2, N_2, $TiCl_4$, $SiCl_4$, TiO_2, SiO_2, and Cl_2. The last three components are the products of following generalized gas-phase reactions of the synthesis of titania and silica with kinetic rates k_T and k_S, respectively:

$$TiCl_4 + O_2 \xrightarrow{k_T} TiO_2 + 2Cl_2 , \quad SiCl_4 + O_2 \xrightarrow{k_S} SiO_2 + 2Cl_2 .$$

A single-fluid flow is assumed. It is defined by a system of quasi-gas dynamic equations (QGD) [3] in which the external forces and the heat sources are accounted for as follows:

$$\frac{\partial \rho}{\partial t} + \mathrm{div}\,\mathbf{j} = 0 ,$$
$$\frac{\partial (\rho \mathbf{u})}{\partial t} + \mathrm{div}\,(\mathbf{j} \otimes \mathbf{u}) + \nabla p = \rho \mathbf{F} + \mathrm{div}\,\Pi , \qquad (1)$$
$$\frac{\partial (E)}{\partial t} + \mathrm{div}\,(\mathbf{j}H) + \mathrm{div}\,\mathbf{q} = (\mathbf{j} \cdot \mathbf{F}) + \mathrm{div}\,(\Pi \cdot \mathbf{u}) + Q .$$

Here all quantities are normalized: $\mathbf{u}$, ρ and p are the velocity vector, density and averaged pressure of the mixture flow, respectively. $\mathbf{F}$ is the mass force density vector, H is the total specific enthalpy, Π is the viscous stress tensor, Q is the specific heat.

The mass flux density vector $\mathbf{j}$ is determined as following:

$$\mathbf{j} = \rho \mathbf{u} - \tau \left[\mathrm{div}\,(\rho \mathbf{u} \otimes \mathbf{u}) + \nabla p - \rho \mathbf{F} \right] , \ \tau = \frac{\mathrm{M}}{\mathrm{Re\,Sc}} \frac{T}{p} .$$

where τ is the relaxation parameter, Re – Reynolds number, Sc – Schmidt number, M – Mach number, T – gas mixture normalized temperature.

The system is supplemented with continuity equations for the species in the mixture:

$$\frac{\partial \rho_i}{\partial t} + \mathrm{div}\,\mathbf{j}^i = \sum_j J^{(ji)} \qquad (2)$$

where ρ_i is normalized density of i-th species, $J^{(ji)}$ is the normalized rate of transformation of the mass of the j-th species to the i-th species in a unit volume of the mixture, $\mathbf{j}^i$ is normalized mass flux density vector of i-th species.

In addition, the equations for the volumetric concentration c_p of the solid phase should be considered for this system:

$$\frac{\partial c_p}{\partial t} + \mathrm{div}\left(c_p \mathbf{u}\right) = \sum_j J^{(jp)} \quad (3)$$

Here $J^{(jp)}$ is the normalized rate of transformation of the mass of the j-th species to the solid phase.

The right-hand sides of (2), (3) are accounted for the kinetic relations which describe variations of the concentrations of titanium tetrachloride and titanium dioxide in the gas and solid phases as well as those of silicon tetrachloride and silicon dioxide in the gas and solid phases due to the homogeneous and heterogeneous reactions and to the phase transition:

$$\frac{dC^1}{dt} = -k_{\mathrm{T}}^{\mathrm{r}} C^1 = -\left(k_{\mathrm{T}}^{\mathrm{g}} + k_{\mathrm{T}}^{\mathrm{s}} A\right) C^1, \quad \frac{dC^2}{dt} = k_{\mathrm{T}}^{\mathrm{g}} C^1 - k_{\mathrm{T}}^{\mathrm{p}} C^2,$$

$$\frac{dC^3}{dt} = k_{\mathrm{T}}^{\mathrm{s}} C^1 A + k_{\mathrm{T}}^{\mathrm{p}} C^2, \quad \frac{dC^4}{dt} = -k_{\mathrm{S}}^{\mathrm{r}} C^4 = -\left(k_{\mathrm{S}}^{\mathrm{g}} + k_{\mathrm{S}}^{\mathrm{s}} A\right) C^4, \quad (4)$$

$$\frac{dC^5}{dt} = k_{\mathrm{S}}^{\mathrm{g}} C^4 - k_{\mathrm{S}}^{\mathrm{p}} C^5, \quad \frac{dC^6}{dt} = k_{\mathrm{S}}^{\mathrm{s}} C^4 A + k_{\mathrm{S}}^{\mathrm{p}} C^5.$$

C^1, C^2, C^3, C^4, C^5, C^6 stand for corresponding mass concentrations of titanium tetrachloride and titanium dioxide in the gas phase and of titanium dioxide in the solid phase, silicon tetrachloride and silicon dioxide in the gas phase and of silicon dioxide in the solid phase. k^r_T, k^r_S are the rates of the generalized reactions of synthesis of titania and silica, respectively, k^g_T, k^g_S are the corresponding rates of the homogeneous reactions, k^s_T, k^s_S are the rates of the surface reactions, k^p_T, k^p_S – the rates of the phase transitions of titania and silica correspondingly, and A is the relative area of particles.

The additional relations including ideal gas mixture law to provide a closure of the system of equations are as follows:

$$p = \rho R_{\mathrm{m}} T \frac{m_{\mathrm{g}}}{1 - c_p}, \qquad \alpha_i = \rho_i / \rho, \qquad R_{\mathrm{m}} = R_{\mathrm{g}} \left(\sum_i \alpha_i / m_i\right).$$

In these relations R_m is the universal gas constant of the mixture, R_G is the specific gas constant, α_i is the mass fraction of the i-th species, m_g is the mass fraction of the gas, and m_i is the molar weight of the i-th species.

We supply (1)-(4) with the equation for the number of particles N per a unit volume:

$$\frac{dN}{dt} = k_{\mathrm{T}}^{\mathrm{g}} C^1 N_{\mathrm{Av}} - \frac{\beta N^2}{2}, \quad (5)$$

where N_{Av} is the Avogadro constant. Taking into account the known nuclei diameter and a total mass of particles, their number and molar concentration in an elementary volume, we can calculate the average size of particles there at an arbitrary time point. Coagulation parameter β is evaluated in accordance with [4,5]:

$$\beta = 8\pi d_{\mathrm{B}} d_{\mathrm{p}} \left[\frac{d_{\mathrm{p}}}{d_{\mathrm{p}} + g\sqrt{2}} + \frac{4 d_{\mathrm{B}} \sqrt{2}}{u_{\mathrm{p}} d_{\mathrm{p}}}\right]^{-1},$$

 1404 (2019) 012003 doi:10.1088/1742-6596/1404/1/012003

where $g = \left(\frac{1}{3d_p l_a}\right)\left[\left(d_p + l_a\right)^3 - \left(d_p + l_a\right)^{\frac{3}{2}}\right] - d_p,\ l_a = \frac{8d_B}{\pi u_p},\ d_B = \frac{3\sqrt{mkT/2\pi}}{2\rho d_p^2\left(1 + \alpha\pi / 8\right)}$.

Here d_B is diffusivity of Brownian particle, d_p is particle diameter, u_p is particle velocity, l_a is particle mean free path parameter for the coagulation process, m is mass of the molecule of carrier gas, k is Boltzmann constant, α is accommodation coefficient.

4. Boundary conditions

On the reactor walls, we imposed the no-slip conditions, the absence of the heat flux of the wall temperature, and a zero value of the normal derivative of pressure (this additional condition is caused by specific features of the QGD system). The flow rates and temperatures were specified for the jets. To calculate the pressure, density, and velocity at the inlet boundaries of the jets, we used the boundary conditions based on using the Riemann invariants for the Euler equations.

With the purpose of numerical integration of Eqs. (1)–(3), they are rewritten in the cylindrical coordinate system (the axisymmetric problem is considered) and are brought to the dimensionless form. As the main dimension parameters of the problem, the radius of the reactor channel, the velocity of sound in quiescent air at temperature of 300 K, and the density of this air are considered.

The QGD system of equations is solved numerically using a difference scheme explicit in time. The time derivatives are approximated by upwind differences with the first order of accuracy, and the spatial derivatives are presented as central differences of the second order of accuracy

5. Results of calculations

The main series of calculations whose results could be validated against experimental data was performed in the following formulation. At first, the quasi-steady flow field in the reactor, which was formed by injection of the jet from the plasma torch, was computed. After that, side jets were injected.

The reactor has the following geometry: L_R = 444 mm, d_R =32 mm, L_C = 38 mm, d_C=14 mm, L_T= 33 mm, and half angle of opening of the zone with variable cross-section α= 15°. Axial coordinates of injection of centers of side jets: z_1 = 28 mm, z_2 = 146, 192 and 292 mm. Here d_C, d_R is channel and reactor diameters, respectively. L_C, L_R, L_T is corresponding lengths of channel, reactor and zone of variable area of cross-section.

Operating parameters of the set-up: T_1=4500 K, Q_1=1 g/s; T_2=300 K, Q_2=2.5 g/s; T_3=300 K, Q_3 =1.0-8.0 g/s. The molar composition of species of first side jet: air – 99.5%, $TiCl_4$ –0.5%; the molar composition of species of second side jet: air – 99.2%, $SiCl_4$ – 0.8%.

The values of the kinetic rate constants in (4),(5) were given in compliance with [5,6] as follows:

$$k_T^r = 8.26 \cdot 10^4 \exp\left(\frac{-10\,681}{T}\right),\quad k_T^s = 4.9 \cdot 10^3 \exp\left(\frac{-8993}{T}\right),$$

$$k_T^p = 1.2 \cdot 10^{10} \exp\left(\frac{-10\,681}{T}\right), \tag{6}$$

$$k_S^r = 8.0 \cdot 10^{14} \exp\left(\frac{-400\,000}{T}\right),\quad k_S^s = 4.0 \cdot 10^{13} \exp\left(\frac{-40\,828}{T}\right),\quad k_S^p = k_T^p .$$

In figure 2 profiles of particle diameter averaged over their number $d_i = \frac{1}{N_i}\sum_j d_{ij}N_{ij}$, as well as profiles of logarithm of number $N_i = \sum_j N_{ij}$ of particles along the reactor are presented as a function of axial distance for different flow rates of mixture air+$SiCl_4$ vapors and three axial locations of the second slit side jet. Here d_{ij} and N_{ij}– size and number of particles in j-th computational cell of i-th

 1404 (2019) 012003 doi:10.1088/1742-6596/1404/1/012003

cross-section. It is evident from the figure that there are two kinds of nanoparticles at the reactor outlet: titania particles and composite particles (figure 2a); only composite particles (figure 2b – d).

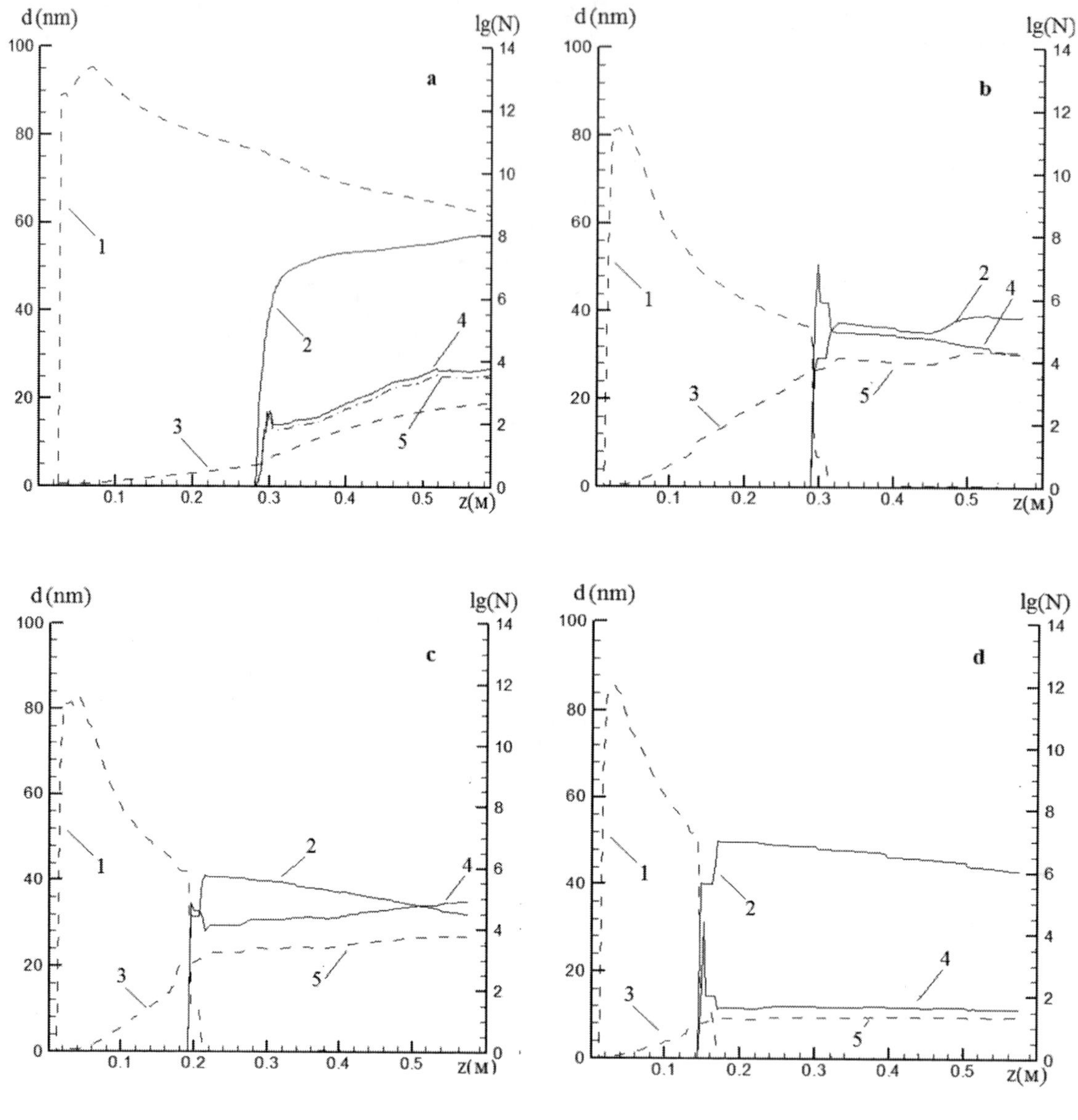

Figure 2. Distribution of mass average particle diameter (nm) and the logarithm of the number of particles along the reactor; (a) – calculation for $z_2 = 292$ mm, $Q_3 = 1.0$ g/sec; (b) – calculation for $z_2 = 292$ mm, $Q_3 = 8.0$ g/sec; (c) – calculation for $z_2 = 192$ mm, $Q_3 = 8.0$ g/sec; (d) – calculation for $z_2 = 146$ mm, $Q_3 = 8.0$ g/sec. 1 – logarithm of number of titania particles; 2 – logarithm of number of composite particles; 3 – the diameter of TiO_2 particles; 4 – the diameter of composite particles; 5 – the core diameter of composite particles.

The results of the computations show that a ratio between a number of titania particles and that of composite ones depends on a ratio between flow rates of Q_2 and Q_3 (for instance, see figure 2a and 2b) as well as locations of the jet slit through which a mixture of air and $SiCl_4$ vapors are injected (see figure 2b - d). The same parameters affect the size of composite particles and their shell thickness.

Apart of resultant higher rate of the synthesis of gaseous silica, increasing the flow rate Q_3 leads to formation of a stagnant zone nearby the reactor walls upstream of the plane of injection of mixture air+ $SiCl_4$ vapors. In this zone titania particles grow and this would result in corresponding increase of core size of composite particles. This may be explained by lower flow velocity in the vicinity of walls and consequently with a rise of residence time of particles therein, they become larger because of surface reaction and coagulation. Downstream of this zone a monotonous character of distributions of parameters of composite particles is established owing to an entrainment of titania particles from mainstream into the zone of propagation of silica vapors.

The axial displacement of the slit of injection of mixture air+$SiCl_4$ upstream results in a decrease of core size of composite particles and an increase of their number. It is caused by reducing the characteristic time of coagulation of titania particles.

6. Conclusion

Based on segregated oxidation of $TiCl_4$ and $SiCl_4$ the simulation of single-stage synthesis of composite 'TiO_2 core – SiO_2 shell' particles in working section of plasmachemical reactor has been conducted. Precursor flow rates and their fractions in gas mixtures injected thru side jet slits are in accordance with those which might be assigned to provide preliminary mixing of reagent components using bubbling procedure. These parameters will be crucial for experimental validation of applied physico-mathematical model of the synthesis of composite particles. In addition, the results of calculations will enable us to examine an extent of influence of operating parameters of the set-up on mean particle size and particle shell thickness as well.

Acknowledgments

The research was partly carried out within the framework of the Program of Fundamental Scientific Research of the state academies of sciences in 2013-2020 (project No. AAAA-A17-117030610120-2). This work was also carried out with partial financial support from the Russian Foundation for Basic Research (Grant No. 18-08-00219 a).

References

[1] El-Toni A M, Yin S and Sato T 2006 *J. Colloid Interface Sci.* **300** 123
[2] Aulchenko S M and Kartaev E V 2018 *AIP Conf. Proc.* **2027** 04002
[3] Aulchenko S M 2013 *J. Eng. Phys. Thermophys.* **86** 1027
[4] Kartaev E V, Lukashov V P, Vashenko S P, Aulchenko S M, Kovalev O B and Sergachev D V 2014 *Int. J. Chem. React. Eng.* **12** 377
[5] Kolesnikov A and Kekana J 2011 *Int. J. Chem. React. Eng.* **9** A83
[6] Park H K and Park K Y 2015 *KONA Powder Part. J.* **32** 85

Calculation of the drag coefficient of micro and nanoparticles

I A Bedarev[1,3] **D A Slastnaya**[1,2] **and V M Temerbekov**[1]

[1]Khristianovich Institute of Theoretical and Applied Mechanics, SB RAS, 4/1 Institutskaya street, Novosibirsk, 630090 Russia
[2]Novosibirsk State University, 2 Pirogova street, Novosibirsk, 630090, Russia
[3]Novosibirsk State Technical University, 20 Karl Marx av., Novosibirsk, 630073, Russia

d.slastnaia@g.nsu.ru

Abstract. A numerical simulation of the flow of air around micro or nanoparticles was carried out. The results of calculating the drag coefficient for various diameters of the free-stream Mach number are presented. For particles with a diameter of 10 and 100 μm, the influence of flow turbulence was taken into account. The calculated dependences obtained are compared with empirical formulas.

1. Introduction

When constructing a physical and mathematical model of the flow, it is necessary to accurately predict the local flow characteristics that arise when the gas medium interacts with the particle. In the interaction of shock and detonation waves with particles, the accuracy of predicting the rate relaxation of particles depends on the accuracy of setting such a flow parameter as the drag coefficient. The correctness of the results directly depends on the chosen model of the interaction of the flow and the particle at the boundary of two phases.

A number of works are devoted to the study of the interaction of flows with particles at the micro level. In [1] the problem of the flow around a system of bodies (cylinders or spheres) was solved by numerical methods. The flow regimes are described; a parametric analysis of the phenomenon under study is carried out. Based on the obtained data, a map of the modes of flow around the shock wave near the bodies is compiled, depending on the Mach number of the incoming flow and the distance between the bodies.

In [2] a numerical simulation of the interaction of a shock wave with a system of stationary particles located along the flow and perpendicular to it was carried out. The dependence of the particle drag coefficient on the Mach number of the flow behind the shock wave is found. It is shown that in the case when the particles are perpendicular to the flow, the flow regime has a negligible effect on the drag coefficient, and in the case when the particles are located along the flow their mutual influence remains at a large distance.

In article [3] a numerical simulation of the flow of a two-phase mixture of gas and submicron metal particles onto an obstacle with a needle located in front of it was carried out. The structure of the separated flow is studied and various approximation formulas (Milliken, Hendersen, etc.) are compared for the law of resistance of spherical particles.

In [4] physical and mathematical models were proposed for describing the processes of propagation, damping, and suppression of detonation in hydrogen-oxygen, methane-oxygen, and silane-air mixtures with inert micro- and nanoparticles. The resistance coefficient is specified using the empirical Cunningham formula.

The aim of this work is a numerical study of the influence of the free-stream Mach number and the Reynolds number through the particle diameter on their drag coefficient.

2. Formulation of the problem

Figure 1 presents a diagram of the simulated area. Air was chosen as the test gas at a temperature T_{st} = 300 K and an initial pressure P_{st} = 100 kPa. The particle diameter and Mach number varied: d – from 10 nm to 100 μm; M – from 1.1 to 3.

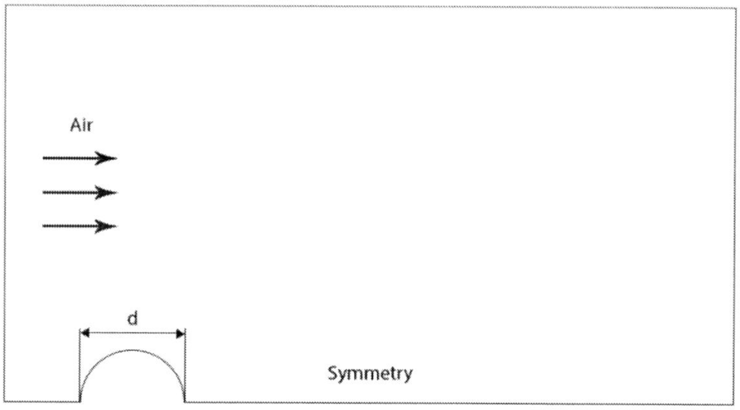

Figure 1. Scheme of the simulation.

The problem was solved in a two-dimensional axisymmetric formulation in the ANSYS Fluent calculation package. The Favre-averaged Navier-Stokes equations supplemented by the SST (k-ω) model of turbulence or the model of laminar flow, depending on the Reynolds number, were used as a mathematical model. For approximation in time, an implicit second-order scheme was used, and for approximation in space of inviscid flows, a third-order accuracy AUSM splitting scheme was used.

3. Calculations results

The drag coefficient Cd was adopted as the main simulation result. This coefficient was also calculated using the empirical Henderson formulas [5] and using the approximation formula [6] with the Cunningham correction taken from the review paper [7]. The obtained values were compared with each other.

For particles with a diameter of 10 μm and 100 μm, the calculations were performed in two versions: with a turbulence model and a laminar flow model. This was done to identify the effect of the model used on the calculation results. For these diameters, the Reynolds number is quite significant which allows us to use the turbulence model (k-ω in the SST version of the Mentor). The calculation results for a particle with the diameter of 100 μm are presented in Figure 2.

The resulting Cd values for various flow models (laminar and turbulent flow) are somewhat different. The turbulent flow model gives large values of the drag coefficient; however, this difference decreases with increasing Mach number. The numerical data are overestimated in comparison with the results of calculations using the Cunningham correction. For Henderson's formula, more complete agreement is observed as the Mach number increases.

For a particle with the diameter of 10 μm a similar trend is observed, see Figure 3. However, the difference between the values of the drag coefficient obtained from numerical calculations for laminar and turbulent flow is not so significant. For Mach numbers $M > 1.5$ this difference is negligible.

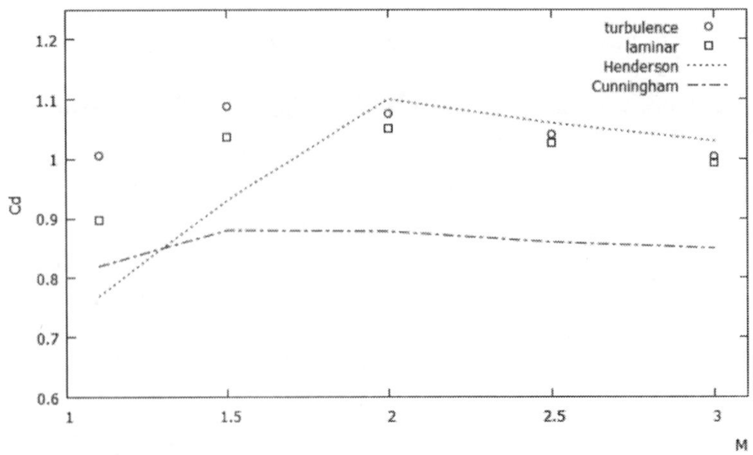

Figure 2. The dependence of the resistance coefficient on the Mach number for d = 100 μm. The results of numerical modeling and calculations by approximation dependencies.

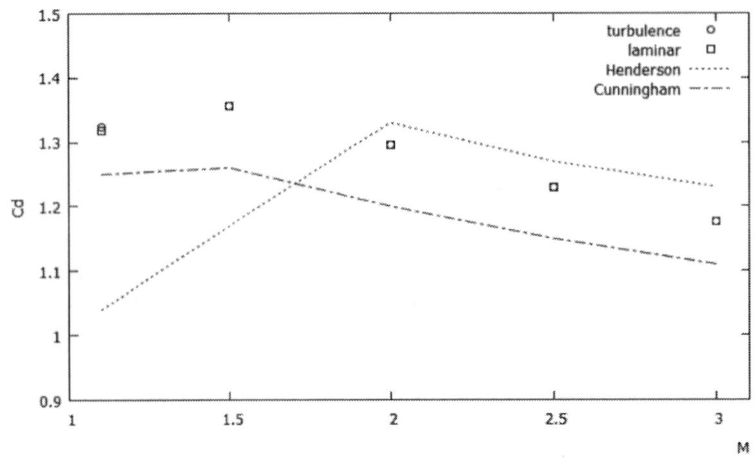

Figure 3. The dependence of the resistance coefficient on the Mach number for d = 10 μm. The results of numerical modeling and calculations by approximation dependencies.

For particles with a diameter of d = 10 nm – 1 μm calculations were carried out only using the laminar flow model, since precisely at this Reynolds number this mode is observed. Figure 4 presents the calculation results for a particle with a diameter of d = 1 μm.

As can be seen from Figure 4, obtained as a result of numerical simulation, the values of cd qualitatively repeat the approximation dependences. However, these values are overestimated in comparison with calculations using the Cunningham correction and underestimated in comparison with the curve obtained by the Henderson formula.

As for the previous cases, obtained as a result of numerical simulation, the values of the resistance coefficient for d = 100 nm qualitatively repeat the approximation dependences, see Figure 5. The numerical data show overestimated values in comparison with empirical formulas.

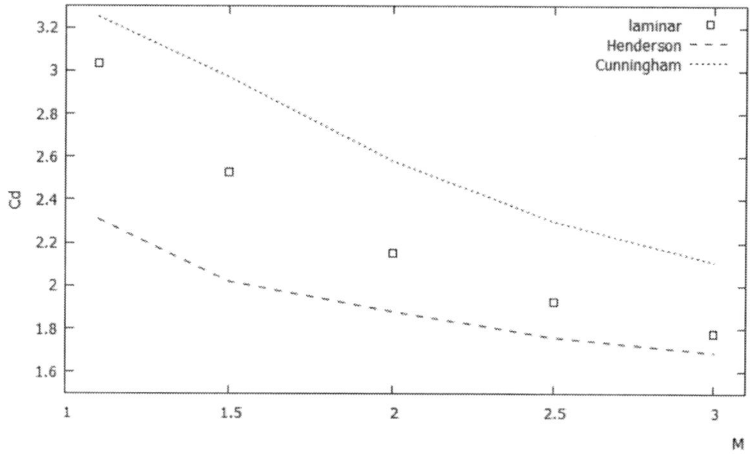

Figure 4. The dependence of the resistance coefficient on the Mach number for $d = 1$ µm. The results of numerical modeling and calculations by approximation dependencies.

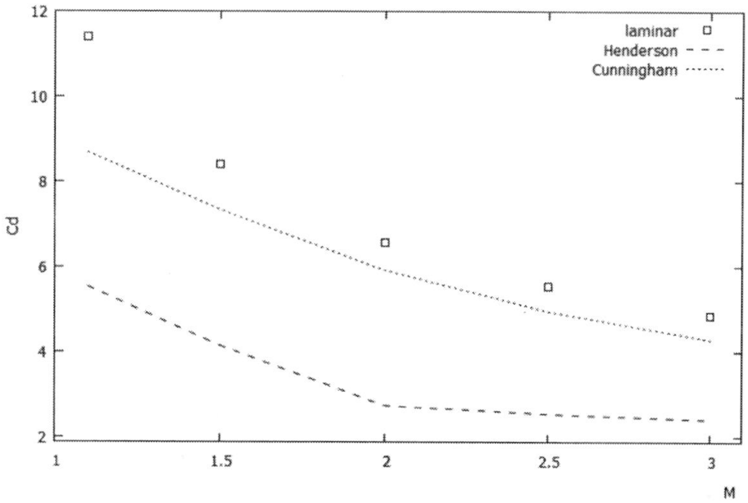

Figure 5. The dependence of the resistance coefficient on the Mach number for $d = 100$ nm. The results of numerical modeling and calculations by approximation dependencies.

For $d = 10$ nm the discrepancy between numerical and empirical data increases, see Figure 6. Such a discrepancy is associated with the realization of the free-molecular flow regime at which the Knudsen number $Kn \sim 10$ which violates the continuity hypothesis and leads to the necessity of taking into account correction factors in the boundary conditions on the particle.

4. Conclusions

Using numerical methods the dependences of the drag coefficient of particles of various diameters on the Mach number of the incoming flow are obtained.

The obtained values are compared with empirical dependences, which showed the need to take correction factors into account, especially for nanoscale particles.

Acknowledgments
This work was supported by the Russian Foundation for Basic Research (Grant No. 17-08-00634).

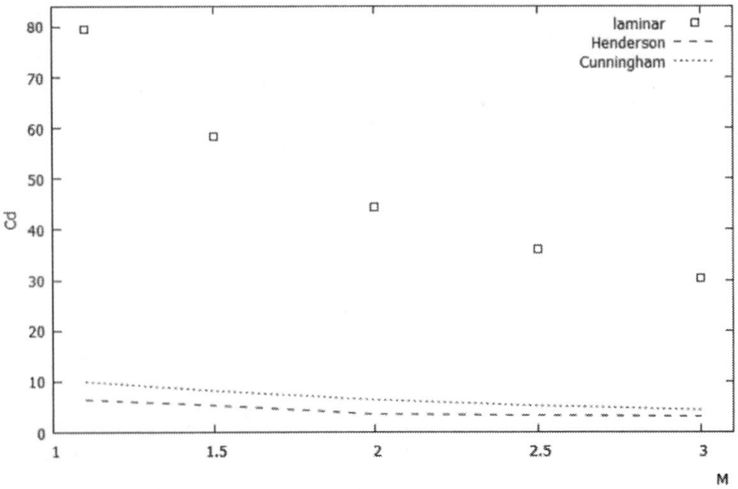

Figure 6. The dependence of the resistance coefficient on the Mach number for d = 10 nm. The results of numerical modeling and calculations by approximation dependencies.

Nomenclature

T_{st} – static temperature;
P_{st} – static pressure;
d – diameter of the particle;
M – free stream Mach number
Cd – particle drag coefficient
Kn – Knudsen number

References

[1] Bedarev I A, Fedorov A V and Fomin V M 2012 *Comb. Expl. Shock Waves* **48** 446–454
[2] Bedarev I A and Fedorov A V 2015 *J. Appl. Mech. Techn. Phys.* **56** 750–760
[3] Alkhimov A P, Bedarev I A and Fedorov A V 2013 *J. Eng. Phys. Thermoph.* **86** 905–912
[4] Tropin D A and Fedorov A V 2018 *Comb. Expl. Shock Waves* **54** 200-206
[5] Henderson C B 1976 *AIAA J* **14** 707–708
[6] Boiko V M *et al* 1995 *Dokl. Ross. Akad. Nauk* **340** 188–190
[7] Fedorov A V and Khmel T A 2019 *Comb. Expl. Shock Waves* **55** 1–17

Journal of Physics: Conference Series **1404** (2019) 012005 doi:10.1088/1742-6596/1404/1/012005

Calculation the efficiency of energy transfer from nanosecond CO_2-laser pulses to the iron target during plasma appearance

D V Bedenko and O B Kovalev

Khristianovich Institute of Theoretical and Applied Mechanics, Siberian Branch of Russian Academy of Sciences, Novosibirsk, Institutskaya str., 4/1, 630090, Russia

dmtr.bedenko@gmail.com

Abstract. The steel target evaporation and the gas-dynamic processes were calculated, thenfollowed by the plasma creation on in the metal vapor under the influence of nanosecond pulses were calculated. Mathematical model based on the Euler equations for the inviscid gas flow in conjunction statement with the radiation transfer equation in the diffusion multi-group approximation and the heat equation in solid metal was formulated. Nonlinear evaporation, depending on the energy entering to the target surface, was calculated using analytical expressions for the Knudsen layer. The structures of gas-dynamic flows with the formation of shock waves in an ambient gas (air) in the active vaporization mode and during plasma formation are were studied, also the thermal state of the target is was determined in the one-dimensional formulation. The comparison of the efficiency of energy transfer to the target is was made in both cases, : from laser radiation directly and when plasma appearsappeared, by means of thermal radiation. The significant increase in the adsorption coefficient of energy in the case of the appearance of plasma is shown.

1. Introduction

There are many works [1–5] devoted to the study of low-temperature CO_2-induced laser plasma. Previously studies were conducted for the laser radiation interaction with matter and attempts were made to use it for surface treatment: cutting, welding, cladding, heat treating. At the high radiation power density $I_{las} > 10$ MW/cm^2, the target material is actively vaporized, and at the vapor near the surface, due to low metal ionization potentials, the plasma is formed, which then propagates towards the beam. This processes has been actively studied both experimentally [2-3] and theoretically [4, 5].

The absorption of laser radiation by metals in the infrared wavelength range, such as a CO_2 laser (10.4 μm), is negligible, especially for noble metals. Therefore, the use of such lasers as the energy source for the thermal action on samples does not have high energy efficiency, although, at the same time, it has a number of other advantages. When the plasma occurs, it begins itself to actively absorb the incident radiation, and then emits the stored energy as thermal radiation. If the plasma temperature is sufficiently high (1 eV or higher), then, according to the Hagen – Rouben relationship, the absorption coefficient for photons with the corresponding wavelengths is significant, from 40% for iron to almost complete absorption [6]. As long as the plasma is near the surface, a significant fraction of the radiation from it is directed toward the target and leads to heating of the material. In this paper, we consider the effect of plasma formation on the efficiency of energy transfer from laser radiation to the target for short pulses, its ablation, and the gas-dynamic processes resulting from it.

2. Mathematical model

For the gas dynamics description of the low-temperature plasma, we write the system of equations, without taking into account the viscosity of the gas, consisting of the continuity (1), momentum (2), and energy equations (3):

$$\frac{\partial \rho}{\partial t}+\frac{\partial}{\partial x}(\rho u)=0, \tag{1}$$

$$\frac{\partial \rho u}{\partial t}+\frac{\partial}{\partial x}\left(\rho u^2\right)+\frac{\partial P}{\partial x}=0, \tag{2}$$

$$\frac{\partial}{\partial t}\left(\frac{3}{2}P+\frac{1}{2}\rho u^2\right)+\frac{\partial}{\partial x}\left(\left(\frac{3}{2}P+\frac{1}{2}\rho u^2\right)u\right)=\frac{\partial}{\partial x}(Pu)-\frac{\partial W}{\partial x}-\frac{\partial G}{\partial x}, \tag{3}$$

where ρ is the density, u is the velocity, P is the pressure, T is the temperature, W is the radiation energy flux, G is the laser radiation intensity.

For the process of mixing vapor with the surrounding air, we used the additional continuity equation per vapor volume fraction:

$$\frac{\partial \rho \gamma_v}{\partial t}+\frac{\partial \rho \gamma_v u}{\partial x}=0, \tag{4}$$

supplemented by the relations: $\rho_v=\rho\gamma_v$, $\rho_g=\rho\gamma_g$, $\gamma_v+\gamma_g=1$, where γ_v is the vapor volume fraction and γ_g is the gas volume fraction.

In addition to the equations (1-4), the equation of state in the plasma is added: $P=n_0 k_b T(1+\alpha)$, where n_0 is the initial concentration of atoms, α is the degree of ionization

The plasma is in the local thermodynamic equilibrium (LTE) if the electron concentration in it exceeds the certain critical value N_e^{cr}, determined from the condition: $N_e^{cr}\geq 1.6\cdot 10^{12} T_e^{1/2}(\Delta E)^3$, where N_e^{cr} is the critical electron concentration (in the cm^{-3}), T_e is the electrons temperature, ΔE is the highest energy level in the atom. In this case the degree of ionization are calculated by the Saha ionization equation: $\frac{n_i n_e}{n_a}=2\left(\frac{2\pi m_e k_b T}{h^2}\right)^{\frac{3}{2}}\frac{Z_i}{Z_a}e^{-\frac{I_{ion}}{k_b T}}$, where Z_a, Z_i are the statistical sums for neutral atoms and ions, n_i, n_e, n_a are the concentrations of ions, electrons and neutral atoms, respectively.

For the radiation losses description, we use the radiation transfer equation in the diffusion multigroup approximation [7]:

$$-\frac{\partial}{\partial x}\left(\frac{1}{3}\frac{c}{\chi_k(T)}\right)\frac{\partial U_k}{\partial x}+c\chi_k(T)U_k=4\chi_k(T)\sigma\left(T,h\nu_k,h\nu_{k+1}\right)T^4 \tag{5}$$

where U is the radiation energy density, χ_k is the absorption coefficient at the frequency ν_k, σ is the integral of thermal radiation in the frequency range from k to $k+1$, $k=1,2,3,\ldots\infty$. Then we find the radiation energy flux from the equation: $W^v=-\frac{1}{3}\frac{c}{\chi_v}\frac{\partial U_v}{\partial x}$.

Due to the fact that the radiation transfer equation in the form (5) includes U, which contains only the fraction of the absorbed radiation by the plasma, it is necessary to take into account the incident laser radiation separately . Accordingly, we come to Beer-Lambert-Bouguer law:

$$\frac{\partial G(x)}{\partial x}-\chi_{las}(x)G(x)=0, \tag{6}$$

where χ_{las} is the adsorption coefficient at the lasers wavelength.

Only the continuous spectrum was taken into account in the radiation absorption coefficient: $\chi_{\nu}^{\text{con}} = \chi_{\nu}^{\text{f-f}} + \chi_{\nu}^{\text{f-b}}$, where are $\chi_{\nu}^{\text{f-f}}$ is the coefficient of the free-free absorption and $\chi_{\nu}^{\text{f-b}}$ is the coefficient of the bound–free absorption [8]:

$$\chi_{\nu}^{\text{f-f}} = \frac{4}{3}\left(\frac{2\pi}{3m_e k_b T}\right)^{\frac{1}{2}} \frac{Z^2 e^6}{hcm_e \nu^3} n_i n_e \text{ and } \chi_{\nu}^{\text{f-b}} = \frac{32\pi^4}{3\sqrt{3}} \frac{e^{10} m Z^4}{h^3 c \nu^3} \frac{1}{\left(2\pi m_e k_b T\right)^{3/2}} n_i n_e \sum_{n=1}^{\infty} \frac{e^{\frac{I_{ion}}{k_b T}\frac{1}{n^2}}}{n^3}.$$

The vapor source is the target material; therefore, it is necessary to calculate its thermal state. It is assumed, that evaporation occurs from the solid phase immediately into the gaseous phase, bypassing the formation of the molten pool, since sufficiently powerful laser pulses with the peak intensity up to $I_{\max} = 86.7$ MW/cm^2 are used. In this case, to calculate the target temperature, we can simply use the heat equation:

$$c_p \rho \left(\frac{\partial T}{\partial t} - u_s(t)\frac{\partial T}{\partial x}\right) = \frac{\partial}{\partial x}\lambda \frac{\partial T}{\partial x} + \frac{1}{\delta} I_s(t) \exp\left(-\frac{1}{\delta}\cdot x\right), \tag{7}$$

where $\delta = \frac{1}{\sqrt{\pi \mu \mu_0 \lambda_e}}$ is the skin depth.

Boundary conditions. The target surface reach thermal equilibrium on the radiation with the plasma, since the speed of light c is much higher than the characteristic rates of gas-dynamic processes. The boundary condition for the radiation energy density on the target-gas boundary is: $cU = \sigma T^4$, and on the right (open) boundary, the condition for the free exit of radiation is set as: $W_{\nu} = -\frac{cU_{\nu}}{2}$. For the laser radiation incident on the target surface: $G(t) = I_0 \exp\left(-\int_0^L \chi_{las} dx\right)$, while on the open border: $G = I_0$. The pulse shape had the triangle form with the duration of 50 ns of the power density growth to the values $I_{\max} = 86.7$ MW/cm^2 followed by linear drop to zero during 250 ns.

In the case of subsonic gas outflow into the open space with atmospheric pressure on the right boundary: $P = P_{\text{atm}}$, and the density and velocity of the gas leaving the region are calculated by the method of characteristics.

For the vapor parameters determination at the target-gas boundary, the Knudsen layer model was used, namely, its analytical approximation, obtained on the solution of the Boltzmann equation for molecular gas flows [9]. As the result, the parameters of the outflowing vapor were determined. The energy conservation condition at the target-gas boundary:

$$\lambda_s \left.\frac{\partial T}{\partial x}\right|_s = AI_s - \alpha\left(T_p - T_s\right) - \varepsilon W - |u_s| \rho_s L_e - u(E(0) + \rho u(0)^2/2 + P(0)),$$

where $u(0) = \mathrm{M}\sqrt{\frac{\gamma R T_s}{\mu}}$ is the vapor velocity, $u_s \rho_s = u(0)\rho(0)$ is the mass conservation law, $E(0) = \left(\frac{n_v}{\gamma_v - 1} + \frac{n_g}{\gamma_g - 1}\right) kT$ is equation of state for the vapor, $P_s = P_{\text{atm}} \exp\left[\varepsilon/k_b\left(1/T_b - 1/T_s\right)\right]$ is the equilibrium vapor pressure at the target surface temperature T_s.

In turn, the relations for the Mach number and the temperature at the exit from the Knudsen layer:

$$M = \frac{1 - (\mathrm{P}(0)/\mathrm{P}_s)^{-0.275}}{1 - (\pi_0)^{-0.275}},$$

$$\frac{T(0)}{T_s} = \frac{\tau_0 - (\pi_0)^{-0.125} + (1 - \tau_0)(\mathrm{P}(0)/P_s)^{-0.125}}{1 - (\pi_0)^{-0.125}},$$

where $\pi_0 = 0.20742$, $\tau_0 = 0.6437$.

The thickness of the target is chosen large enough, so that an isothermal boundary condition can be used on its reverse side: $\left.\frac{\partial T}{\partial x}\right|_{\xi} = 0$.

Initial conditions. At the initial time, the ambient gas is assumed to be at rest, under normal conditions:

$$u(x)=0,\ P(x)=P_{\text{atm}},\ T(x)=T_{\text{atm}}.$$

3. Calculation results

In the process of exposure to pulsed high-power laser radiation, intense heating and active evaporation of the target material occurs. A layer of vapor appears near its surface, pushing out an ambient gas. Figure 1 (a,c) shows the main gas-dynamic parameters of the mixture, including iron vapor and ambient air incoming in the volume at the moment 100 ns. Due to the powerful evaporation process, the vapor velocity is equal to 600 m/s near the boundary, air is completely pushed out from the surface and the shock wave is formed before the vapor moving from the target. The total pressure behind the front of the shock wave is 5-6 atmospheres, and the temperature of the expiring vapor is approaching 3500 K.

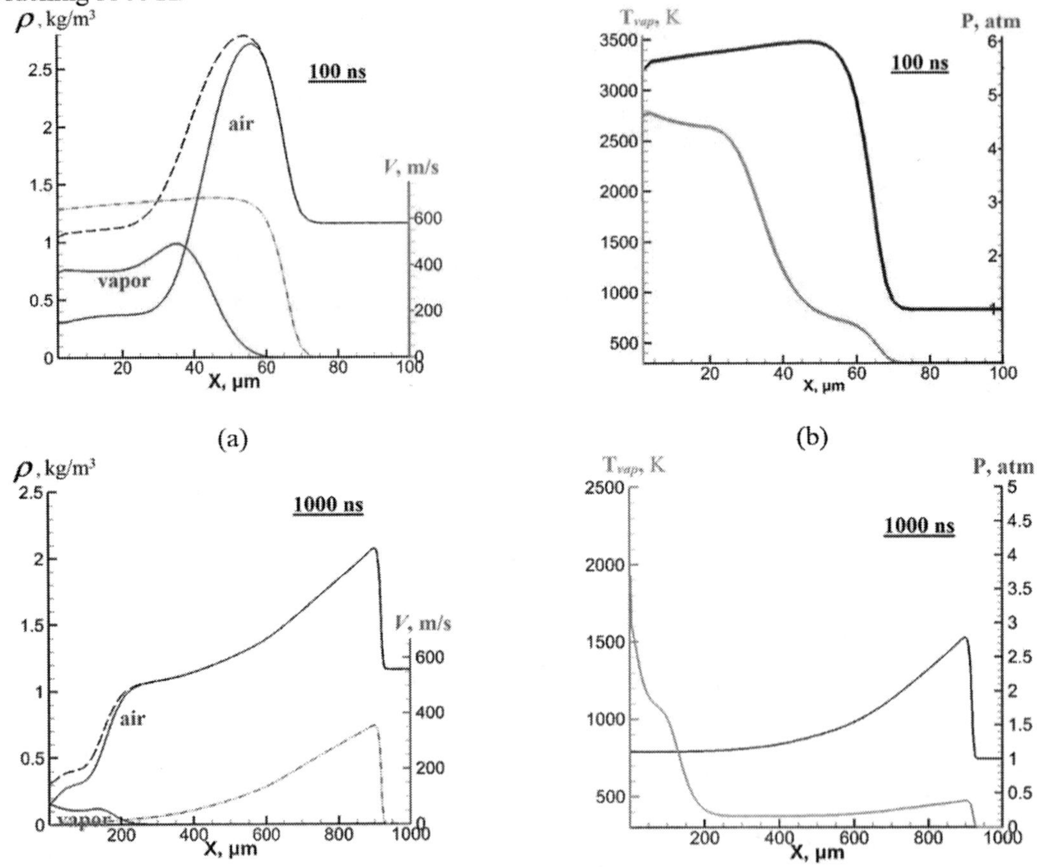

Figure 1. Main gas-dynamic parameters for vapor-air mixture at 100 ns (a,b) and 1000 ns (c,d). Density of vapor (red), density of air (blue), total density (dash), velocity of mixture (dash-dot) (a,c); temperature (orange), pressure (violet) (b,d).

Near the end of the laser pulse, at moment 250 ns, the evaporation of the target ceases, so the pressure and velocity of the shock wave rapidly decrease. At a time moment of 1000 ns, the velocity of its front decreases almost two times, and the pressure behind it drops to 2.7 atm (figure 1 (c,d)). At the same time the vapor velocity near the target is zero and it has the atmospheric pressure.

The fraction of absorbed radiation during the 300 ns pulse is about 12%. Of this energy, 84% goes deep into the target due to thermal conductivity, 14.4% is spent on heat exchange with the gas phase, 1.6% is spent on ablation (evaporation) of the metal.

With subsequent pulses near the surface, metal vapors gradually accumulate, not having enough time to completely fly away from the surface or condense on it during the time between pulses equal to 10 μs. The concentration of free electrons, which are responsible for the laser radiation absorption, increases, and conditions are created for their avalanche-like growth with quick growth of simultaneous absorption. The incident radiation is actively absorbed followed by the sharp heating and appearance of the plasma. The shock wave, due to rapid heating of the gas (vapor-air), intensifies strongly: the velocity of its front increases up to 1430 m/s, as the pressure behind it to 56 atmospheres (figure 2). The plasma temperature is 12000 K, but it is seriously limited by the powerful flow of relatively cold (4300 K) vapor, coming from the target surface. The vapor velocity near the surface is much more lower in comparison with the situation without plasma. This is due to the huge pressure in the plasma, which significantly suppresses the evaporation of the target material, up to its complete cessation.

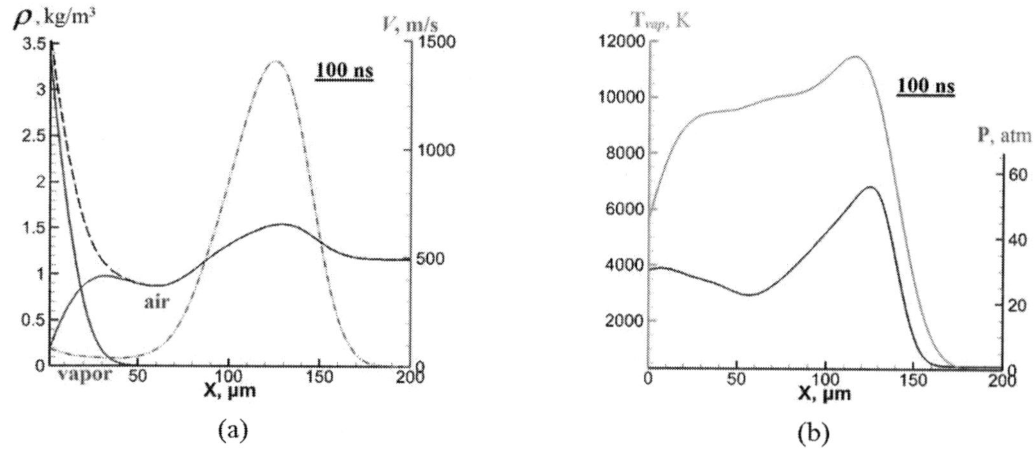

Figure 2. Main gas-dynamic parameters with plasma appearance at 100 ns. (a) density of vapor (red), density of air (blue), total density (dash), velocity of mixture (dash-dot); (b) temperature (orange), pressure (violet).

At the moment of the end of the pulse, $\tau = 300$ ns, the plasma moves up to 1 mm from the target and the quasi-one-dimensionality of the problem disappears, because we assumed that the diameter of the evaporation spot is approximately equal to the diameter of the laser beam, which equal 700 μm. At the greater distance from the target, the behavior of the plasma hasn't been studied.

When the plasma appears, a new channel for heat transfer is added due heat radiation from the plasma to the surface. When plasma forms, for a short time almost all the incident laser radiation gets absorbed by the plasma. So the amount of energy transferred directly from the laser beam to the target during 300 ns is only 11%. The fraction of energy that the target receives from plasma is 89%. In this case, heat removal from the surface occurs in the following way: thermal conductivity deep into the target equals 65.5%, 31.5% goes to heat exchange with the gas phase, 2.9% goes to evaporation.

Figure 3 shows comparison the temperatures T_s of the target surface in the modes without plasma and with its presence. The comparison is correct only up to 300 ns, when the quasi-one-dimensionality of the problem is preserved. During this time, the amount of absorbed energy increased 3.2 times from

6 mJ to 19.4 mJ, which corresponds to 12% and 38.8% of the laser pulse energy. Due to the increase in heat exchange with gas phase during plasma existence, the fraction of energy accumulated by the target increased 2.5 times from 5 mJ to 12.7 mJ.

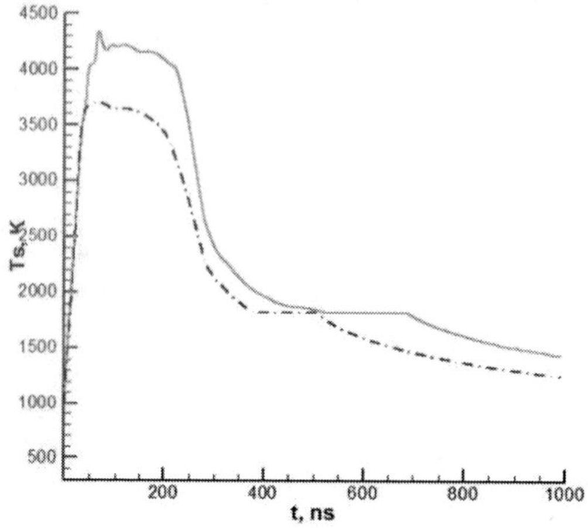

Figure 3. Comparison of the target surface temperature T_s in the different modes: without formation (dash-dot line) and with the plasma appearance (solid line).

4. Conclusions

With the first laser pulse, when the intensity is insufficient for optical breakdown, active evaporation of the material occurs. Vapors push the ambient gas from the surface and form the front of the shock wave. During the time between pulses the gas-vapor mixture reach dynamic equilibrium, however, vapor gradually accumulates in the area above the target, the electron concentration increases, and conditions sufficient for the plasma formation arise. In this case, the plasma is actively heated, the pressure behind the front of the shock wave increases by an order of magnitude, and the velocity of its front increases several times. Due to the rapid growth of the pressure above the target surface, the evaporation process can be significantly suppressed.

As long as the plasma is near the target, it transfers a significant amount of the thermal radiation into the surface due to the better absorption of short-wave radiation by the iron (from 1 μm or less). As the result, during short pulses (up to 300 ns), the total radiation absorption coefficient for the iron can increase by 3 times, and the amount of heat accumulated by the target by 2.5 times.

Acknowledgments

The work was carried out with partial financial support from the RSF, project No. 18-19-00430.

References

[1] Prokhorov A M et al 1988 *Interaction Between Laser Radiation and Metals* (Moscow: Nauka) 537 pp
[2] Fowler M C and Smith D C 1975 *J. Appl. Phys.* **46** 138–150
[3] Gus'kov K G, Raizer Yu P and Surzhikov S T 1990 *Quantum Electron* **20** 860–864
[4] Mazhukin V, Smurov I and Flamant G 1994 *J. Comput. Phys.* **112** 78–90
[5] Mazhukin V I, Nossov V V and Smurov I 2004 *J. Phys. D: Appl. Phys.* **37** 185–199
[6] Xie J, Kar A, Rothenflue J A, Latham W P 1997 *J. Laser Appl.* **9** 77-85
[7] Chetverushkin B N *Mathematical Simulation of the Problems of Radiant Gas Dynamics* 1985 (Moscow: Nauka) 304 pp
[8] Zel'dovich Ya B, Raizer Yu P *Physics of Shock Waves and High-Temperature Hydrodynamic Phenomena* 2003 (New York: Dover) 944 pp
[9] Gusarov A V, Smurov I 2005 *J. of Appl. Phys.* **97** 014307-13

Journal of Physics: Conference Series **1404** (2019) 012006 doi:10.1088/1742-6596/1404/1/012006

Deformation and acceleration-induced fragmentation of the liquid drops in the gradient air flow

V M Boiko, S V Poplavski, A A Pivovarov and I V Samsonova

Khristianovich Institute of Theoretical and Applied Mechanics SB RAS, Novosibirsk 630090, Russia

bvm@itam.nsc.ru

Abstract. The regimes of aerodynamic destruction of a falling drop of a low-viscosity liquid in a falling accelerating air flow were experimentally studied. This model formulation allows us to study the behaviour of a drop when the direction of the aerodynamic force is reversed. The experiments showed a large deformation and rather high instantaneous drop acceleration at a moderate velocity gradient. The results confirmed that a change in the sign of aerodynamic force increases the effect of Rayleigh-Taylor instability.

1. Introduction

The study of liquid drops breakup was and remains actual problem in gas dynamics. Aerodynamic atomization is used in various technologies where it is necessary to maintain high performance nozzle requirements and the fragment size dispersion of sprays. It should be noted that the variety of the gas flow velocity regimes and geometries of used channels and nozzles allows detecting transition and hybrid modes of drop breakups that are still understudied today. Also, the liquid drop breakup in the gradient flows is not sufficiently studied.

Comparing to shock wave drop breakup, where gas flow parameters are constant, there is much fewer information about drop breakup in gradient flows [1-5]. The difficulty is that the gas flow parameters can vary at characteristic lengths of velocity relaxation, deformation and breakup of a drop. Understanding of the breakup physics is important for the numerical simulations of the aerodynamic atomization of liquid drops in the low Weber number range and for different aims of the spray formation and optimization, where the time-resolved breakup features and fragment size distributions are required.

The present work aims to study the deformation and breakup of water drops falling in an accelerating falling air flow as a limiting case of the flow fast change when the direction of the aerodynamic force can be even reversed at the moderate gas velocity gradient. This problem is being solved for the first time. In addition to academic value, the results are of interest in creating industrial air strainers, large-scale systems of the dispersion and cooling of liquids, etc.

2. Experimental part

A freely falling drop enters in the vertically positioned channel with rectangular cross-section open to the atmosphere [3]. Before falling into the channel, an aerodynamic force acts from the bottom to the top; in the channel, the aerodynamic force from the falling flow is directed from above. The drop is accelerated in the air flow created by suction by a vacuum machine through the lower outlet of the channel (figure 1). Depending on critical conditions defined by the Weber criterion, it is unstable and the secondary breakup may occur in an upper or a middle part of the channel. It should be noted that some periodic drop deformations are usually observed before the channel entrance. This kind of the deformation is characteristic for the freely falling drops. The deformation periods t_k (figure 1, (a)) for water drops with different initial spherical diameters d_0 estimated from image sequences after post-processing procedures are in good agreement with experimental and predicted data [6]. The most important parameter is relative velocity ΔV between drop V_l and air flow V_{flow} in main flow direction: $\Delta V = V_{flow} - V_l$. V_{flow} in the channel is regulated by convergence angle of channel walls and by the air pumping rate (3 regimes). The velocity profiles obtained from PIV measurements. One of the V_{flow} profiles is shown in figure 1, (b). In the experiments always maintains the condition $V_{flow} > V_l$. In the gradient flows, there is an uncertainty in the start moment of interaction drop-air which do not allow to precisely measure the time characteristics of the process, unlike the shock wave disturbances. Thus, the induction time of breakup t_i is defined as the time between the visual drop deformation (the beginning) and an appearance of the first fragments during the initial drop destruction (*i.e.* the bag bursting). Experimentally observed times t are typically made non-dimensional using the characteristic interaction time: $t_0 = d_0(\rho_l/\rho_g)^{1/2}(\Delta V)^{-1}$. Here ΔV is the initial relative velocity, ρ_l/ρ_g is the liquid-to-ambient gas density ratio. For the experimental curves the non-dimensional time t/t_0 is usually calculated. For example, the water drops with $d_0 = 2.7$-2.9 mm and velocities in the entrance to the channel 2.5-2.8 m/s have $t_i \approx 14$-16 ms. The ratio $t_i \approx t_k$ is practically obtained and $t_i > t_0$, where t_0 is about 5.5-6 ms. All these times are typical for the secondary breakup regimes in the low values Weber number range ($We < 20$).

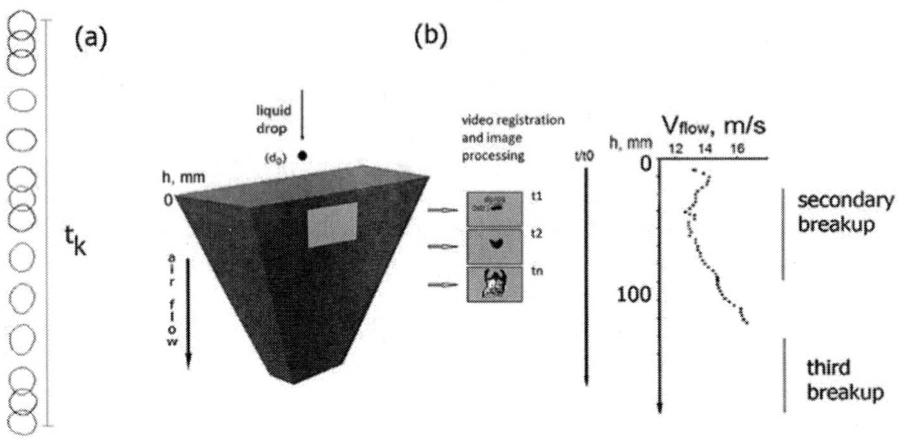

Figure 1. (a) The deformation period ($t_k \approx 28.5$ ms) of the freely falling drop with $d_0 = 4.4$ mm and $V_l \approx 2.3$ m/s; (b) Schematized configuration of breakup regime study, video registration and post-processing of the images.

The test conditions are summarized in table 1. The liquid and air properties were obtained from [7].

Table 1. Test conditions.

parameter	range
liquid	water
initial droplet diameter, d_0 (mm)	2.7 - 4.5
air velocity, V_{flow} (m/s)	12 - 32
water-air density ratio, ρ_l/ρ_g	840
Reynolds number (*Re*)	960 - 2670
Weber number (*We*)	6 - 14
Ohnesorge number (*Oh*)	0.0012-0.0024

Reynolds number range is higher to one for the regimes when the gas (air) viscosity effects on the liquid drop drag. For this range, the drag coefficient of a solid sphere is about 0.4-0.5. The Mach number is low (M ≈ 0.1).

3. Results and discussion

It is obtained that the periodic drop deformation over the channel entrance does not considerably affect the breakup mechanism. The images of the main stages of the typical secondary breakup in one process are presented on figure 2, (a).

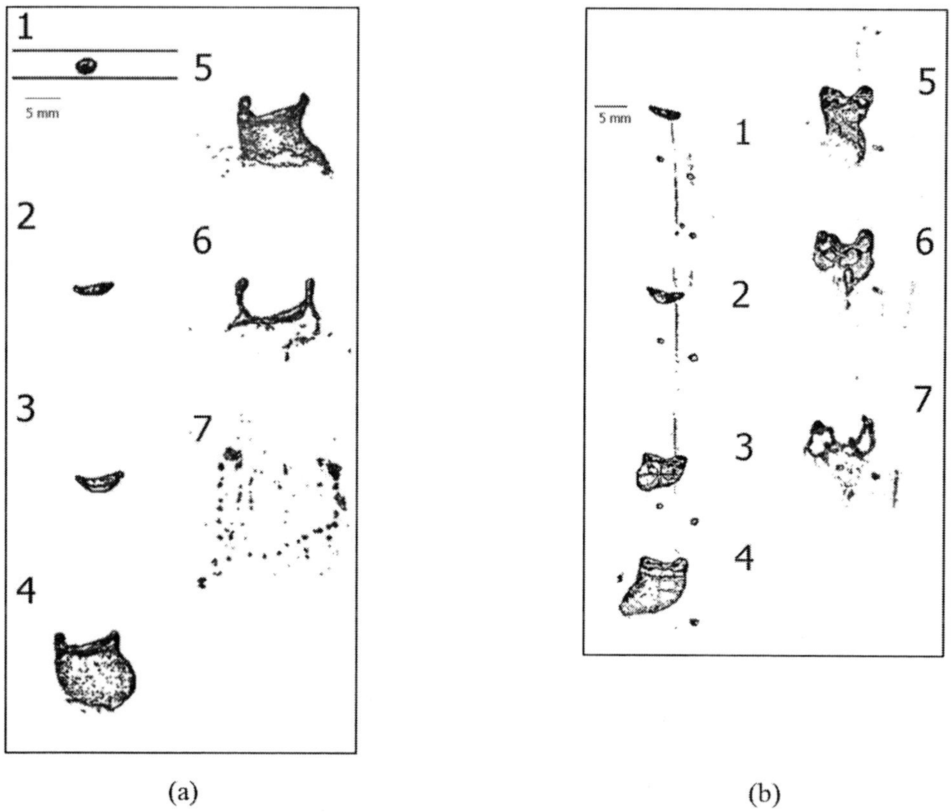

Figure 2. Drop breakup in falling accelerating air flows; the high-speed registration (Phantom, 6500 fps); (a) the secondary bag breakup: 1-3 deformation (0; 9.23; 14.46 ms), 4-7 breakup (17.69; 18.15; 19.69, 24.3 ms), d_0 = 2.9 mm, $t_0 \approx$ 5,6 ms, $We \approx$ 4-10; the third breakup on the transition bag-and-stamen mode of the fragment with $d_0 \approx$ 1.9 mm, $t_0 \approx$ 4.6 ms, $We \approx$ 3-14: 1-5 deformation (4.64; 5.87; 6.1; 6.48; 6.63 ms), 6-8 fragmentation (7.1; 7.71; 8.02 ms) (b).

Note after image filtration procedure there are no small fragments (less than 0.4 mm). The distinguishing characteristic of experiments is that there is the third breakup in the lower part of the channel (Figure 2, (b)). The third breakup (bag or transition bag-and-stamen breakup) is possible because there is the continuous air flow, and *We* numbers may exceed critical condition. After the secondary big drop breakup namely the big fragments (node-droplets and some ring-droplets) are involved in the third breakup.

It is known [8,9], that the drop deformation significantly affect the drag. In addition, to estimate the instantaneous drag coefficient, it is necessary to have experimental information about time dependence of the deformation. Besides, to estimate the instantaneous drop acceleration, experimental data about the drop velocity are required.

Accordingly, the time dependencies of the drop velocity and the deformation are given as an example (figure 3 and 4). Moreover, figure 4 represents the velocity profile of the drop along the channel length. The details of calculations and estimations are not discussed in this paper.

As shown by figure 3, the drop of a greater diameter d_0 needs more time for the deformation and the further fragmentation. In this case, inertial forces increase. It bears mentioning that the "internal flow" effects are not considerable in the range of parameters under study.

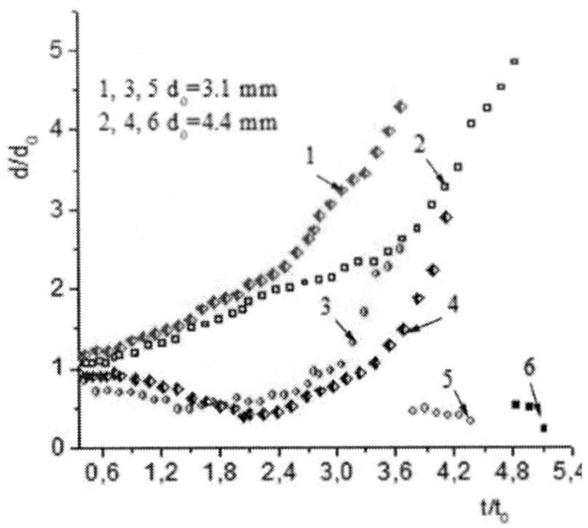

Figure 3. Deformation dynamics of drops; d_0= 3.1 mm (t_0~ 6.4ms) and d_0= 4.4 mm (t_0~ 10,2 ms): 1, 2 - expanding of the cross-stream diameter d_{cro} until the bag bursting; 3, 4 - evolution of the total stream-wise diameter d_{str} until the bag bursting; 5, 6 – d_{str} of the liquid ring until its rupture beginning.

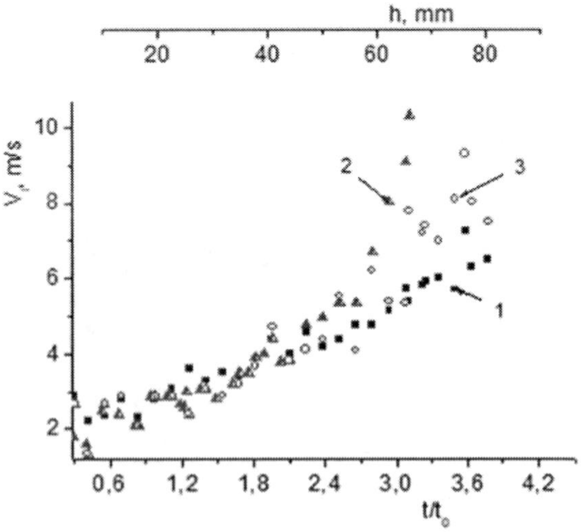

Figure 4. Dependence of the drop velocities on the channel length and dimensionless time (d_0= 2.8 mm, t_0~ 5.5 ms): velocity measurements at the forward edge (1) and back edge (3), until the liquid ring rupture beginning; at the drop bag end (2) until its bursting.

As it is mentioned above, the breakup process starts when the drop enters in the channel and is subjected to an ambient flow field. Deformation is caused by an unequal pressure distribution over the surface of the drop. There are liquid drop expanding in cross-flow direction d_{cro} and compressing in the stream-wise direction d_{str}. So, when the stage of flattened drop is obtained, which is characterized by the disk-like drop deformation, a flow separation and a wake formation begin. The difference between forward and rear pressure areas is a consequence of the flow separation. This leads to the formation of the drop bag and its growth. From the beginning up to this stage the drop velocity increases and the drop acceleration increases too. At the flattened drop stage (d_{cro} is maximal) the instantaneous drop acceleration is the biggest. Later, when the hollow bag is formed, the drop velocity change is slow and the acceleration falls down. After the bag bursting the toroidal liquid ring moves with a slow increasing of the velocity until its rupture. At this stage the pressure difference vanishes. The droplet acceleration is not very high. The results of the different point velocity measurements are shown at figure 4. There is difference between the velocities, which is associated with the bag mode. Instantaneous parameters estimations at different stages are given in table 2.

Table 2. Estimations of the instantaneous parameters for the water drop with $d_0 = 2.9$ mm.

Dimensionless time, t/t_0	Channel length, h, mm	V_{flow}, m/s	We	Drop acceleration a, m/s^2	Bond number, Bo
0.81	11.6	16	9.1	99-130	12-15
1,56	25.8	16.9	8.8	277-522	33-62
2,7	56.3	17.6	7.7	190-230	23-28
3,16	73,5	17.1	6.3	150-185	18-22

It is noteworthy that, in our experiments, "bag" breakup of droplets occurs at Weber numbers We<9, while in stationary flows this mode occurs at Weber numbers 14-50. Some unsteady effects of the ambient air flow should be noted at the fragmentation stages. For the third breakup process, these effects appear too but the quantitative data are not presented. So, more experiments are needed.

For the bag mode, the drop acceleration at the flattened drop stage can be estimated by the length of Rayleigh-Taylor instability wave. For the water drop with a diameter d_0 = 2.9 mm (for example, figure 2, (a), stage 2) and the maximal cross-stream diameter $d_{cro} \approx 5$ - 6.4 mm, $\lambda_c = d_{cro}$ in the case with two nodes. Thus, the drop acceleration $a \approx 274$ - 440 m/s^2.

4. Conclusions

In this paper, the deformation and breakup regimes of a falling drop in a falling accelerating air flow have been investigated for the first time. Such experiments models the behavior of a droplet when sudden changing the direction of aerodynamic force. The presented results show that a change in the sign of the aerodynamic force enhances the effect of Rayleigh-Taylor instability and leads to a significant decrease in the critical Weber number for this breakup mode.

Acknowledgments

The research was carried out within the framework of the Program of Fundamental Scientific Research of the state academies of sciences in 2013-2020 (project No. AAAA-A17-117030610137-0).

References

[1] Chou W H and Faeth G 1998 *Int. J. Multiphase Flow* **24** 889-912
[2] Boiko V M and Poplavski S V 2009 *Comb., Expl., Shock Waves* **45** 198–204
[3] Boiko V M, Minakov A V, Poplavski S V and Shebeleva A A 2018 *AIP Conference Proceedings* **2027** 030021
[4] Flock A K, Guildenbecher D R, Chen J, Sojka P E, Bauer H J 2012 *Int. J. Multiphase Flow* **47** 37-49
[5] Strotos G, Malgarinos I, Nikolopoulos N, Gavaises M 2016 *Int. J. Multiphase Flow* **85** 96-109
[6] Volkov R S, Zhdanova A O, Vysokomornaya O V, Kuznetsov G V and Strizhak P A 2014 *J.Engineering Physics Thermophysics* **87(6)** 1351-1361
[7] Kuchling H 1983 *Physics. Reference book* (Moscow: Mir)
[8] Guildenbecher D R, Lopez-Rivera C 2009 *Exp. Fluids* **46** 371-402
[9] Theofanous T G 2011 *Annu. Rev. Fluid Mech.* **43** 661-690

Droplet breakup in shock waves at moderate values of the Weber number

V M Boiko and S V Poplavski

Khristianovich Institute of Theoretical and Applied Mechanics SB RAS, Novosibirsk 630090, Russia

bvm@itam.nsc.ru

Abstract. The object of this study was the physical mechanisms of aerodynamic fragmentation of droplets of low-viscosity liquids in shock waves in the region of existence of the shear breakup modes. New experimental data were obtained on the time induction of the droplets breakup depending on the gas flow velocity in the poorly studied range of Weber numbers We = 100 ÷ 2200.

1. Introduction

The aerodynamic breakup of droplets in a flow behind a shock wave in the range of Weber numbers We = 100–2200 is of interest for three sections of physical gas dynamics. Firstly, protection against drip erosion of supersonic vehicles and their engines during flight in the precipitation zone [1,2]. Secondly, high-speed gas-liquid flows as applied to nozzles and burners with a high consumption of liquid fuel. Thirdly, fire and explosion safety in the oil industry [3]. In order to establish the role of droplet breakup processes in specific circumstances and conditions, it is first necessary to have reliable data on a number of determining time parameters, such as the time of gas flow exposure, the scenario and the time of deformation of the droplets, the characteristic times of the onset and end of the process. Note that in the range of Weber numbers of interest to us, there are practically no experimental data on the time of the preparatory stage of spraying (the time of breakup induction t_i, in dimensionless form $t_i/t_0 = k$, where $t_0 = d/U\left(\rho_l/\rho_g\right)^{1/2}$). Here ρ_g, ρ_l are the density of gas and liquid, U is the gas velocity; d is the diameter of the droplet; σ is the surface tension of the liquid.

In widely cited works [1, 2], the main role in droplet destruction in shock waves (SW) is attributed to the formation of a boundary layer in a liquid under the action of friction forces and its subsequent separation. The main conclusion of [2] states that the induction time is directly proportional to the diameter of the droplet and inversely proportional to the velocity of the gas flow. To confirm the correctness of this statement, experiments are given in [2] for the water drops with $d = 2.7$ mm at the Mach numbers of the shock wave Ms = 1.3, 1.5, and 1.7. These modes, according to our estimates, correspond to the Weber numbers We = 1500 , 4600, 9500 and are outside the considered range of 100 ≤ We ≤ 1000.

We managed to find only one work [3], in which a series of experiments in a shock tube was carried out to determine the induction time for liquid streams breakup over a wide range of We, including the regime We = 10^2–10^3 of interest to us. Experiments on direct measurement of the induction time for spherical water droplets was not performed, but for water streams it was shown that the dimensionless induction time $t_i/t_o = k$ = const and $k \sim 0{,}5$ only for We ≥ 1000. In the range 100 ≤

We ≤ 1000, the value of *k* is not constant and decreases from 2 to 0.5. There is no explanation of the observed phenomenon in the cited papers. Therefore, a large range of issues for the further accumulation and analysis of experimental data on droplet breakup in high-velocity flows remain relevant.

2. Experiments and discussion of the results

This paper presents the results of an experimental study of the processes of deformation and aerodynamic breakup of droplets of low-viscosity (Newtonian) liquids in shock waves in order to determine the type and characteristic time of the droplet fracture induction in the poorly studied range of Weber numbers We = 100 ÷ 2200. This range of Weber numbers We has a maximum rate deformation of the drop, which significantly affects the dynamics of the drop at an early stage of interaction with the shock wave [4].

The working gas is air at atmospheric pressure. Before each experiment, the initial temperature and atmospheric pressure in the channel of the shock tube were measured. The flow parameters were calculated according to the theory of ideal shock waves from the measured velocity of the shock wave front Vs. Velocity Vs was measured in two ways: 1 - from the shift of the shock front on two adjacent frames of shadow images; 2 - from the travel time of the shock front of a known distance (270 mm) between two pressure sensors. The relative measurement error is δVs/Vs ~ 1%. Using the measured velocity and initial parameters of the air, the Mach number of the shock wave Ms was determined and the velocity *U* and the density of the gas ρ_g were calculated.

The dynamics of the processes of interaction between the droplet and shock wave were recorded by the method of shadow visualization with high-speed filming (figure 1).

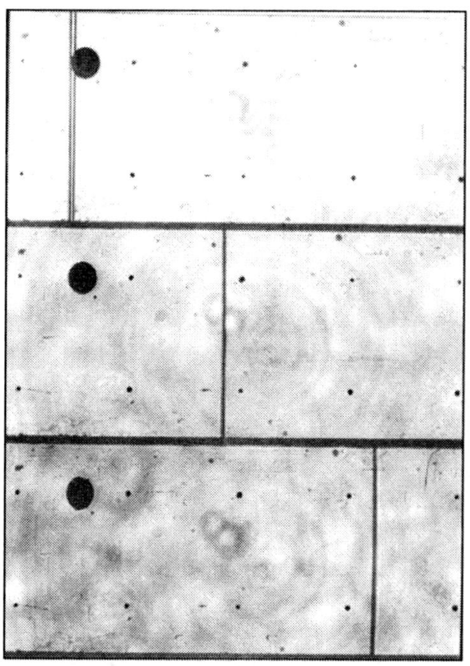

Figure 1. Three consecutive frames of high-speed filming illustrating the beginning of the process of interaction of a shock wave with a drop of water; intervals between frames Δt = 30 μs, exposure time τ ≤ 50 ns, the image size in grayscale is 9.56 Mpix.

The basic element of the diagnostic complex was the stroboscopic ruby laser light source with periodic Q-switching by the Kerr cell developed at the Institute of Theoretical and Applied Mechanics of the Siberian Branch of the Russian Academy of Sciences. The images were recorded by a high-speed camera in standby mode with rotating mirror polyhedron onto a high-resolution film 35 mm wide. The film sector for each experiment has dimensions of the 24x260 mm^2 and contains up to 20 frames. For reasons of space-time resolution optimization, the number of frames and the intervals between frames were selected taking into account the calculated duration of constant parameters behind the shock wave front ~ 600 μs in these experiments. In the present work, the following parameters of the laser strobe were set: exposure time $\tau \leq 50$ ns, interval between the frames $\Delta t = 30 \pm 0.1$ μs.

In all the original photographs (figure 1), there are two rows of points of the coordinate grid on the glass (scale 10 mm), successive positions of the shock front, position, size and shape of the droplet are clearly visible [4-6]. To obtain quantitative data, the informative area of each frame was selected and digitized (figure 2), and then a digital image was stylized with tracing of particle outlines and grid points (figure 3).

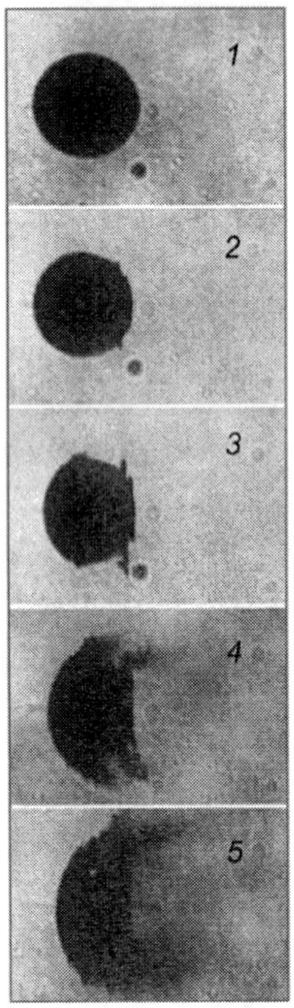

Figure 2. Selected informative area of the enlarged image of the drop - shadow image.

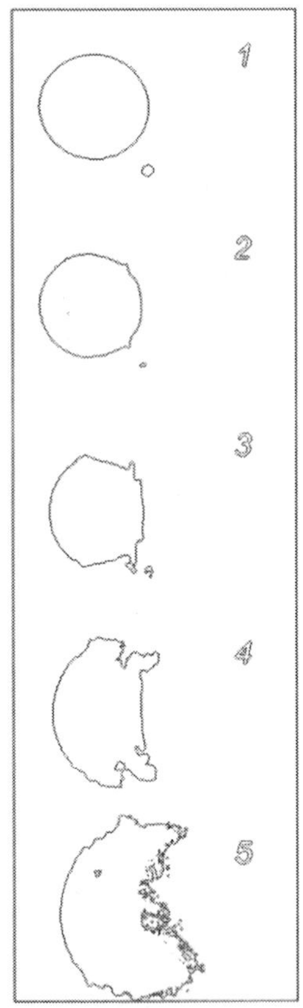

Figure 3. Selected informative area of the enlarged image of the drop - stylization.

The morphological features of the windward surface determined the type of drop destruction. The moment of manifestation of crushing products in the droplet trail was taken as the onset of breakup. The fracture induction time t_i was determined from the moment the shock wave front intersects the windward boundary of the droplet until crushing products appear as, for example, in figure 2 frame 4. The absolute accuracy of these data is limited by the half-interval between adjacent frames and in this experiment is ≈ 15 μs.

The measured delays t_i depending on the flow velocity are shown in figure 4. The data are presented in logarithmic coordinates and are approximated with a high accuracy by a straight line. The slope of line determines the powermode function of dependence of the time of the breakup beginning of the drop on the flow velocity:

$$t_i = 10^{5.4} U^{-3/2} \tag{1}$$

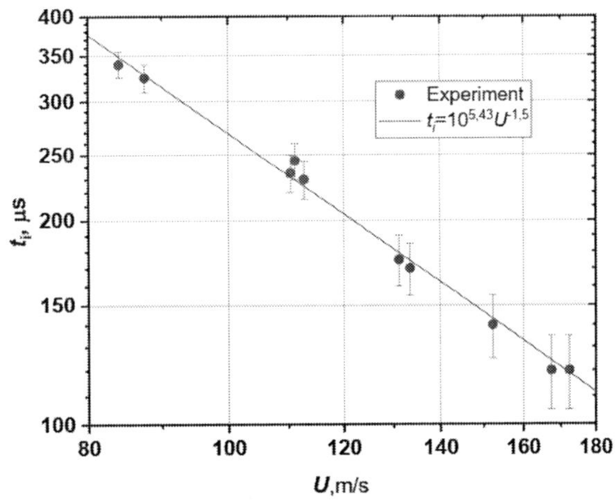

Figure 4. Dependence of the induction time of water drop breakup on the air flow velocity behind the front of a shock wave.

3. Summary

The results of an experimental study of the processes of deformation and aerodynamic fragmentation of droplets of low-viscosity (Newtonian) liquids in shock waves are presented with the aim of determining the type and characteristic time of the droplet breakup induction in the range of Weber numbers We = 100 ÷ 2200. The obtained experimental data shows that the droplets destruction induction time depends on the gas flow velocity to the degree of -3/2 instead of -1, predicted by the A. Ranger on the basis of his simplified model.

Acknowledgements
The research was carried out within the framework of the Program of Fundamental Scientific Research of the state academies of sciences in 2013-2020 (project No. AAAA-A17-117030610137-0).

References

[1] Ranger A A and Nicholls J A 1969 *AIAA J.* **7** 285–90
[2] Ranger A A 1972 *Acta Astronaut.* **17** 675–83
[3] Buzukov A A 1963 *AMTP* 154–58
[4] Boiko V M and Poplavski S V 2009 *Comb., Expl., Shock Waves* **45** 198–204
[5] Boiko V M and Poplavski S V 2012 *Comb., Expl., Shock Waves* **48** 76–82
[6] Boiko V M, Minakov A V, Poplavski S V and Shebeleva A A 2018 *AIP Conference Proceedings* **2027** 030021

On numerical modelling of gas flows through axisymmetric porous object with heterogeneous combustion sources under forced filtration

K G Borovik[1,2]

[1] Far Eastern Federal University, 8 Sukhanova street, Vladivostok 690090, Russia
[2] Institute of Automation and Control Processes, Far Eastern Branch of the RAS, 5 Radio street, Vladivostok 690041, Russia

E-mail: ksushareno@mail.ru

Abstract. The time-dependent gas flows through axisymmetric porous objects with heterogeneous combustion sources are considered. The mathematical model, which describes the process, and the numerical method are proposed. The regimes of heterogeneous combustion wave propagation in axisymmetric porous object with various location of ignition zones under forced filtration are investigated. It is shown that the gas tends to bend around the heated zone and flows through colder regions as in the case of two-dimensional plane flows.

1. Introduction

Many natural and technological processes, such as peat fires, spontaneous combustion of solid waste landfills, burning of solid fuels, associated with heterogeneous combustion in porous media. Heterogeneous combustion is one of the types of filtration combustion, which occurs when an oxidizer from the gas phase and a condensed combustible component from the solid phase interact.

Basic principles of heterogeneous combustion are briefly described in [1] for forced and free filtration. There are a huge number of papers which devoted to solid porous media combustion, in particular [2-5]. Computational modeling is one of the methods of investigation of heterogeneous combustion in porous media. A numerical model for investigating the time-dependent processes of one-dimensional combustion wave propagation in porous objects with unknown gas flow rate and gas velocity at object boundaries was proposed in [6]. In [7, 8] two-dimensional unsteady gas flows through the porous media with heterogeneous combustion sources under forced and natural convection are considered. The original numerical method, proposed in these papers, is the development of numerical algorithm for computation of the gas flow through the porous objects with heat sources when the gas pressure at object boundaries is known [9, 10].

In the present work the time-dependent gas flows though the axisymmetric porous object with heterogeneous combustion sources under forced convection are investigated. The gas velocity, the solid temperature, the gas pressure and the degree of conversion of the solid combustible component were analyzed at various gas pressure at object bottom and various locations of the ignition zone.

2. Mathematical model and numerical method

The axisymmetric porous object opened at the top and the bottom with impermeable non-heat conducting side walls is considered. A cold gas flows into the porous object through its bottom due to forced filtration, the gas flows through porous medium and flows out through object top. Mathematical model is constructed on the model of two interactive interpenetrative continua [11] and consists of the equations of energy of gas and solid phase, the equation of motion, the continuity equation, the equation of concentration, the equation of state for gas. In the energy equation of solid phase we take into account the intensity of interphase heat exchange, which is assumed to be proportional to the difference of the phase temperatures, as well as thermal conductivity and heat release, which is proportional to the rate of chemical reaction. In the energy equation of gas we take into account a thermal conductivity of gas and intensity of interphase heat exchange. The equation of motion for gas is the momentum conservation equation for porous media. The equation of motion for solid media degenerates, because the solid phase is immovable. It is shown in [12] that taking into account the temperature dependence of the gas dynamic viscosity can affect the solution of the problem of time-dependent gas flow through porous media not only quantitatively but also qualitatively, so in the present work the dynamic viscosity of gas depends on temperature according to Sutherland's formula. So, the system of equations is the following:

$$\left(\rho_{cf}c_{cf}+\rho_{ci}c_{ci}+\rho_{cp}c_{cp}\right)\frac{\partial T_c}{\partial t}=-\alpha\left(T_c-T_g\right)+Q\rho_{cf0}W+\left(1-a_g\right)\lambda_c\left(\frac{1}{r}\frac{\partial}{\partial r}\left(r\frac{\partial T_c}{\partial r}\right)+\frac{\partial^2 T_c}{\partial z^2}\right),$$

$$\rho_g c_g\left(\frac{\partial T_g}{\partial t}+v_r\frac{\partial T_g}{\partial r}+v_z\frac{\partial T_g}{\partial z}\right)=\alpha\left(T_c-T_g\right)+a_g\lambda_g\left(\frac{1}{r}\frac{\partial}{\partial r}\left(r\frac{\partial T_g}{\partial r}\right)+\frac{\partial^2 T_g}{\partial z^2}\right),$$

$$\rho_g\left(1+\chi\left(1-a_g\right)\right)\left(\frac{\partial v_r}{\partial t}+v_r\frac{\partial v_r}{\partial r}+v_z\frac{\partial v_r}{\partial z}\right)=-a_g\frac{\partial p}{\partial r}-a_g^2\frac{\mu}{k_1}v_r-\left(1-\mu_p\right)\rho_{cf0}Wv_r,$$

$$\rho_g\left(1+\chi\left(1-a_g\right)\right)\left(\frac{\partial v_z}{\partial t}+v_r\frac{\partial v_z}{\partial r}+v_z\frac{\partial v_z}{\partial z}\right)=-a_g\frac{\partial p}{\partial z}-\rho_g g-a_g^2\frac{\mu}{k_1}v_z-\left(1-\mu_p\right)\rho_{cf0}Wv_z,$$

$$\frac{\partial\rho_g}{\partial t}+\frac{\partial\left(\rho_g v_r\right)}{\partial r}+\frac{\rho_g v_r}{r}+\frac{\partial\left(\rho_g v_z\right)}{\partial z}=\left(1-\mu_p\right)\rho_{cf0}W,\quad p=\frac{\rho_g RT_g}{a_g M},$$

$$\rho_g\left(\frac{\partial C}{\partial t}+v_r\frac{\partial C}{\partial r}+v_z\frac{\partial C}{\partial z}\right)=\frac{\partial\left(\rho_g D_g\right)}{\partial r}\frac{\partial C}{\partial r}+\frac{\partial\left(\rho_g D_g\right)}{\partial z}\frac{\partial C}{\partial z}+\rho_g D_g\left(\frac{\partial^2 C}{\partial r^2}+\frac{1}{r}\frac{\partial C}{\partial r}+\frac{\partial^2 C}{\partial z^2}\right)-$$

$$-\mu_p\rho_{cf0}W-\left(1-\mu_p\right)\rho_{cf0}WC,$$

$$W=\left(1-\eta\right)Cke^{-\frac{E}{RT_c}},\quad \frac{\partial\eta}{\partial t}=W,\quad \rho_{cf}=\left(1-\eta\right)\rho_{cf0},\quad \rho_{cp}=\mu_p\rho_{cf0}\eta,$$

$$a_g=a_{g0}+\left(a_{cf0}-a_{cpEnd}\right)\eta,\quad D_g=D_{g0}\left(\frac{T_g}{273}\right)^b,\quad \mu=c_{s1}\frac{T_g^{1.5}}{c_{s2}+T_g}.$$

Here a is the porosity, b is the exponent in the expression for diffusion coefficient, C is the mass concentration of the oxidizer, c is the heat capacity, c_{s1} and c_{s2} are the constant in Sutherland's formula, D_g is the diffusion coefficient of gas, E is the activation energy, g is the gravity acceleration, k is the pre-exponential factor in the expression for the rate of reaction, p is the gas pressure, Q is the heat release of reaction, R is the universal gas constant, t is the time, T is the temperature, v_g is the gas velocity, W is the rate of chemical reaction, α is the constant determining the interphase heat transfer intensity, η is the degree of conversion of the combustible component of solid medium, λ is the thermal conductivity, μ is the dynamic viscosity of gas, ρ is the effective density, χ is the coefficient taking into account the inertial interaction of the phases in their relative motion, ∇ is the nabla operator and Δ is the Laplace operator. The subscripts are as follows: '0' denotes the initial moment, 'c'

denotes the condensed phase, '*End*' denotes the end point, '*i*' denotes the inert component, '*f*' denotes the fuel, '*g*' denotes the gas and '*p*' denotes the product.

At the inlet to the porous object the gas pressure, temperature of gas and mass concentration of oxidizer are known. At the output of the object the gas pressure is known, as the gas flows into the open space. At the inlet and outlet of the porous object and on the impermeable walls the heat exchange conditions are also known. The gas velocity and gas flow rate at the outlet of the object are unknown and have to be found from the solution of the problem. So, we can write the boundary conditions in the following form:

$$p\big|_{x\in G_1}=p_0, \quad \frac{\partial T_c}{\partial n}\Big|_{x\in G_1}=\frac{\beta H}{\lambda_c}\left(T_g\big|_{x\in G_1}-T_c\big|_{x\in G_1}\right),$$

$$T_g\big|_{x\in G_1}=T_{g0} \;\; \text{и} \;\; C\big|_{x\in G_1}=C_0, \;\; \text{если} \;\; v_g\big|_{x\in G_1}\cdot n\big|_{x\in G_1}\le 0,$$

$$\frac{\partial T_g}{\partial n}\Big|_{x\in G_1}=0 \;\; \text{и} \;\; \frac{\partial C}{\partial n}\Big|_{x\in G_1}=0, \;\; \text{если} \;\; v_g\big|_{x\in G_1}\cdot n\big|_{x\in G_1}>0,$$

$$\frac{\partial T_c}{\partial n}\Big|_{x\in G_2}=0 \;\; \text{и} \;\; \frac{\partial T_g}{\partial n}\Big|_{x\in G_2}=0, \;\; \frac{\partial C}{\partial n}\Big|_{x\in G_2}=0, \;\; v_g\big|_{x\in G_2}\cdot n\big|_{x\in G_2}=0,$$

where G_1 is the object boundary opened to an atmosphere, G_2 is the impermeable boundary of the object, $\boldsymbol{n}$ is the outward vector directed normally to G_1 or to G_2. C_0, p_0 and T_{g0} are the mass concentration of the oxidizer, the gas pressure and the temperature of gas at the inlet of the object, β is the heat removal coefficient.

A numerical method based on combination of explicit and implicit finite difference schemes was used to solve the problem. This method is a result of development of numerical method used for plane gas flows through porous media with heterogeneous combustion sources [7, 8]. According to the method the energy equations, equation of concentration and momentum conservation equation are transformed into the explicit finite difference equations. From these equations we find the temperature of solid media, gas temperature, the velocities of gas and mass concentration of oxidizer. The continuity equation is transformed into implicit finite difference equation. From this equation we find the gas pressure taking into account the equation of gas state using Thomas algorithm [13].

3. Results

We consider the axisymmetric porous object with heterogeneous combustion sources when ignition zone was located in the centre of object bottom, in the centre of object and near side walls at the object bottom. The figures 1-3 shows the field of the gas velocity, the temperature of solid medium, the gas pressure and the degree of conversion of the solid combustible component in some time after ignition when the ignition zone is located in the center of the object, in the center of the object bottom and near side walls at the object bottom respectively with the problem parameters used in the work [8]. We can see that as in the plane case, the gas tends to bend around the heated zone and prefers to flow through colder areas.

The figure 4 shows the degree of conversion of the solid combustible component at the different pressure at the object inlet when the ignition zone is located in the centre of the object bottom. We can see that the highest degree of conversion is observed at the lowest pressure.

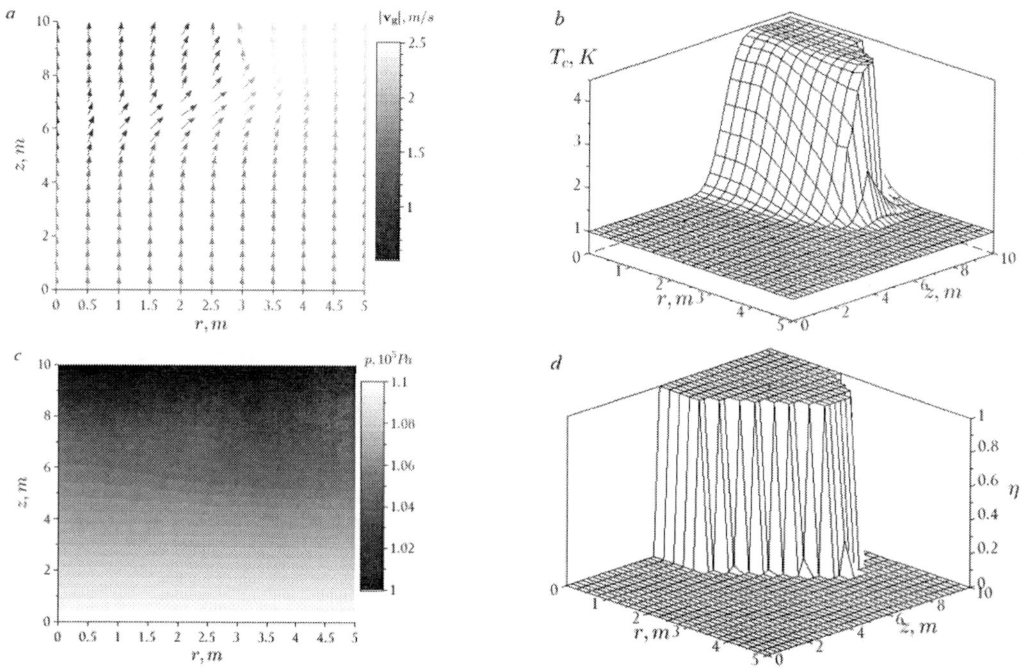

Figure 1. The field of the gas velocity (a), the temperature of solid medium (b), the gas pressure (c) and the degree of conversion of the solid combustible component (d) in 5 hours after ignition when the ignition zone is located in the center of the object.

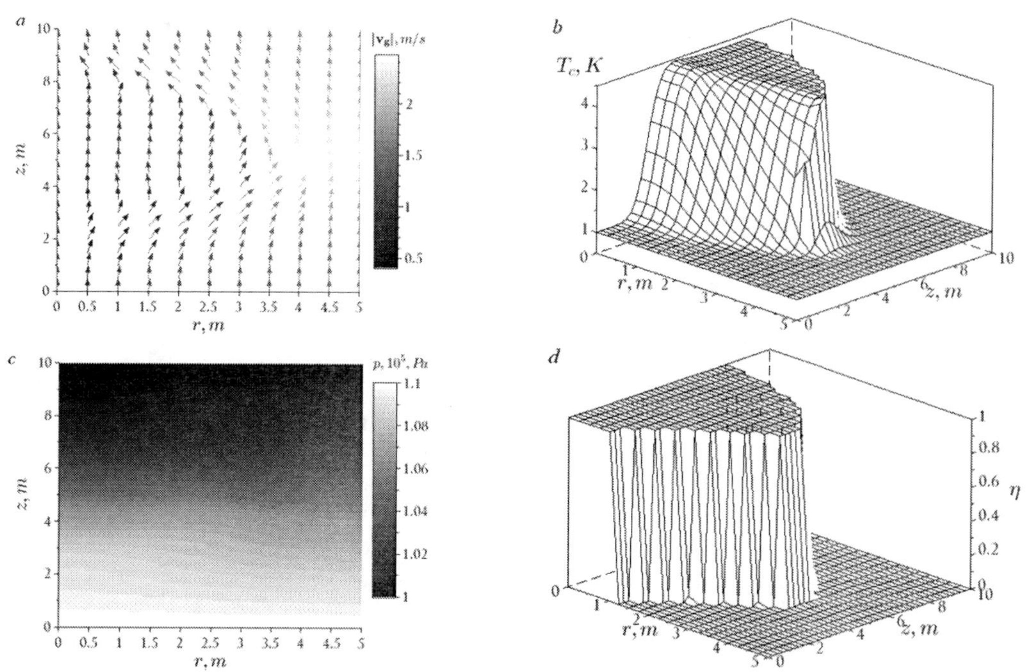

Figure 2. The field of the gas velocity (a), the temperature of solid medium (b), the gas pressure (c) and the degree of conversion of the solid combustible component (d) in 7 hours after ignition when the ignition zone is located in the center of the object bottom.

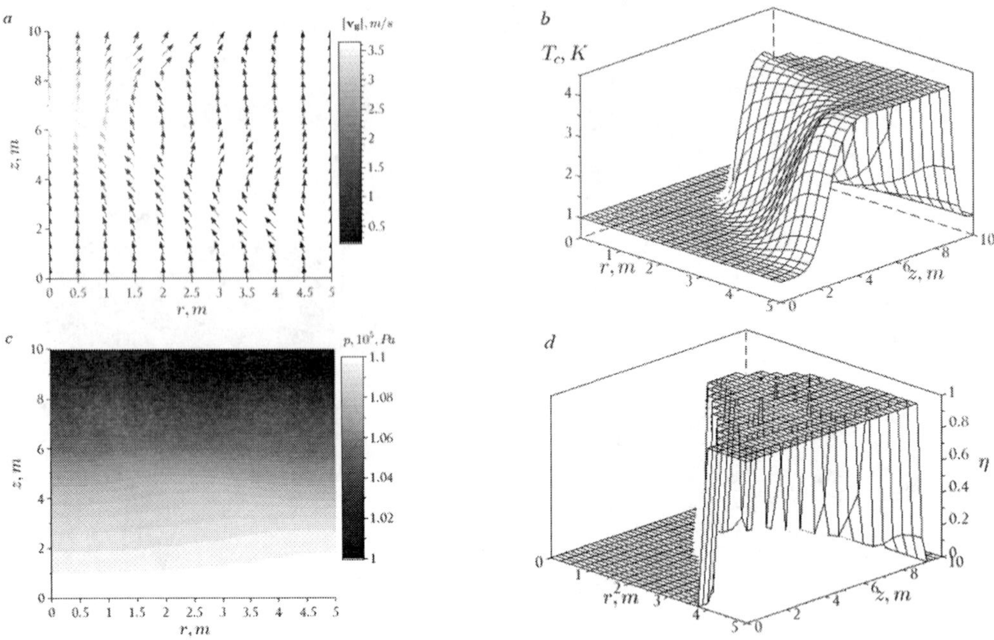

Figure 3. The field of the gas velocity (a), the temperature of solid medium (b), the gas pressure (c) and the degree of conversion of the solid combustible component (d) in 7 hours after ignition when the ignition zone is located near side walls at the object bottom.

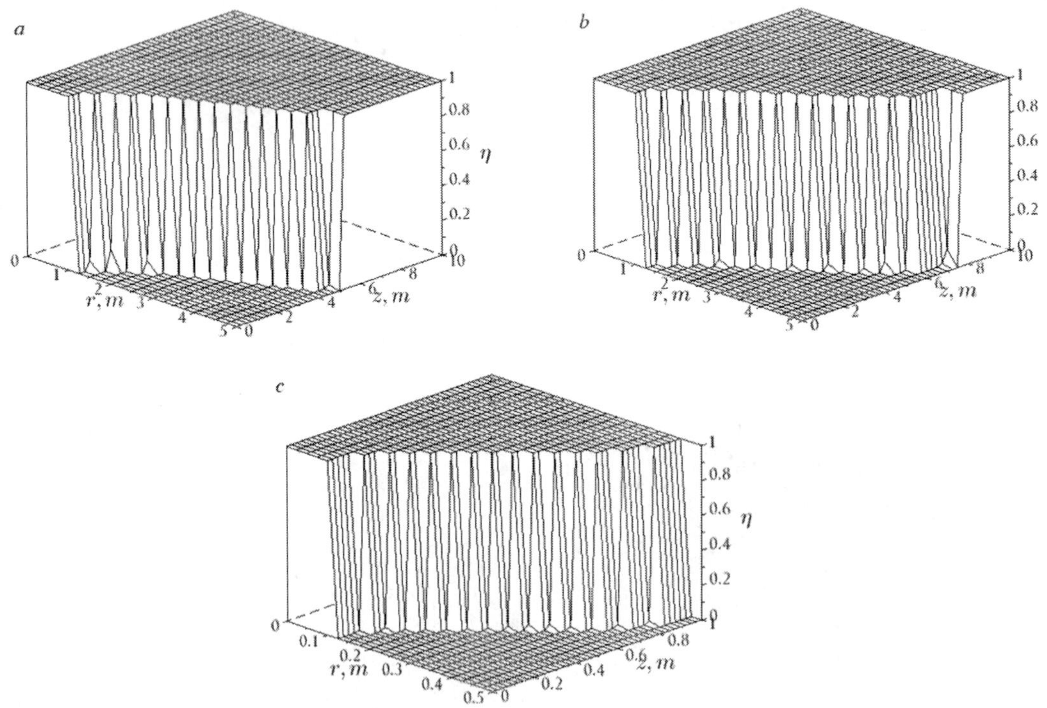

Figure 4. Degree of conversion of the solid combustible component after the end of combustion when the ignition zone is located in center of the object bottom and the pressures at the inlet is equal to 105 kPa (a), 110 kPa (b) and 120 kPa (c).

4. Conclusion

The gas flows through axisymmetric porous object with heterogeneous combustion sources under forced filtration were investigated. The various locations of ignition zone and various gas pressure at object bottom were considered. It is shown that the main features coincide with the plane case: the gas tends to go around the heated zone and flow through colder areas, the size of burn-out part of the porous object strongly depends on the location of the ignition zone and gas pressure at the object inlet.

5. Acknowledgements

The work was partially supported by the Program "Far East" 2018-2020 (project 18-5-064). The results were obtained using the equipment of Shared Resource Center "Far Eastern Computing Resource" IACP FEB RAS. The author would like to acknowledge and thank Dr. Nickolay Lutsenko for his help in the present research.

References

[1] Aldushin A P and Merzhanov A G 1988 Theory of filtration combustion: general concepts and the current state of research *Propagation of thermal waves in heterogeneous media* ed Y S Matros (Novosibirsk:Nauka Publishers) pp 9–52 (in Russian)
[2] Aldushin A P, Matkowsky B J and Schult D A 1996 *Combust. Flame* **107** 151–175
[3] Aldushin A P, Matkowsky B J and Schult D A 1997 *J. Eng. Math.* **31** 205–234
[4] Salgansky E A, Kislov V M, Glazov S V, Zholudev A F and Manelis G B 2008 *Combust. Explosion Shock Waves* **44**(3) 273–280
[5] Rabinovich O S, Malinouski A I, Kislov V M and Salgansky E A 2016 *Combust. Theory Model.* **20** 877–893
[6] Lutsenko N A 2013 *Proc. Combust. Inst.* **34** 2289–2294
[7] Levin V A and Lutsenko N A 2017 *Doklady Physics* **62** 425–429
[8] Lutsenko N A 2018 *Combus. Theory Model.* **22** 359–377
[9] Levin V A and Lutsenko N A 2010 *Math. Models Comput. Simul.* **2**(5) 635–647
[10] Levin V A and Lutsenko N A 2011 *Fluid Dyn.* **46**(5) 826–834
[11] Nigmatulin R I 1978 Principles of the mechanics of heterogeneous media (Moscow: Nauka Publishers) (in Russian)
[12] Levin V A and Lutsenko N A 2006 *J. Eng. Phys. Thermophys.* **79** 33–39
[13] Tannehill J C, Anderson D A and Pletcher R H 1997 *Computational Fluid Mechanics and Heat Transfer* (Philadelphia: Taylor and Francis)

Model of deformation of a liquid drop of zirconium dioxide after its collision with a steel porous surface

V V Bublik[1,2] and A N Cherepanov[1]

[1] Khristianovich Institute of Theoretical and Applied Mechanics of the Siberian Branch of the Russian Academy of Sciences, 630090, Novosibirsk, Russia
[2] Novosibirsk State University, 630090, Novosibirsk, Russia

E-mail: bublik@itam.nsc.ru

Abstract. Based on the integral laws of conservation of mass and energy, a mathematical numerical-analytical model of the deformation of a continuous liquid-metal drop after its collision with a flat porous surface is constructed. The model takes into account the capillary and adhesive properties of the melt, the processes of cooling a liquid drop until it solidifies. Numerical calculations are performed for the model case of collision of a continuous liquid drop of zirconia with a porous steel substrate.

1. Introduction

Investigation of the processes of droplet-substrate collision is given great attention in connection with the tasks of applying protective powder coatings to parts and mechanisms, thermal barrier coatings on turbine blades and aircraft engines, applying solders to microchips, etc. [1–12]. In this case, continuous smooth surfaces are mainly considered. One of the main problems with this is the one of the strength of the coating adhesion to the substrate. In this paper, we study the problem of the interaction of a liquid metal droplet with a porous substrate, which, in our opinion, is of interest from the perspective of increasing the strength of adhesion to the substrate, as well as for issues of dynamic impregnation of working surfaces with various materials resistant to corrosion, chemical and other aggressive influences.

The proposed model is a further development of the models of impact of a solid and hollow drop with a solid substrate [13–15]. In particular, in [14] a mathematical numerical-analytical model of the deformation of a continuous liquid-metal drop after its collision with a flat porous surface in the isothermal case is constructed. The model takes into account the capillary and adhesive properties of the melt. In this paper, this model is generalized to the nonisothermal case.

2. Brief description of the model

Let us describe briefly the model from [14]. The initial diameter of the droplet is H_0, the collision velocity of the droplet with the substrate is v_0. The initial kinetic and potential energies of the drop are equal

$$W_0 = \frac{\pi \rho H_0^3 v_0^2}{12}, \quad \Pi_0 = \sigma_{12} \pi H_0^2.$$

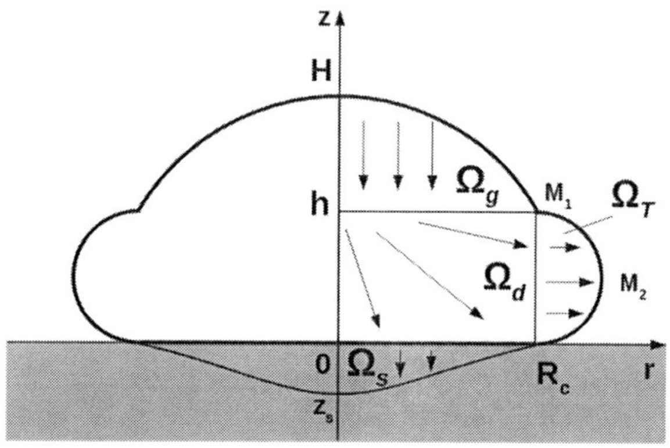

Figure 1. Drop spreading pattern.

Here, σ_{12} is the surface tension at the liquid—gas interface, and ρ is the density of the liquid.

After the collision, we divide the entire volume of the drop into four subregions (see figure 1): Ω_g is a ball segment formed by rotation around the z axis of curved triangle hM_1H; Ω_d is a disk formed by rotation of rectangle $0hM_1R_c$ around the z axis; Ω_T is a toroidal region formed by rotation of semicircle $R_cM_1M_2R_c$ around the z axis; Ω_s is the region of liquid in porous substrate.

The porosity of the substrate will be modeled in the form of vertical capillary tubes. Let $z_s \leq 0$ be the coordinate z of the maximum penetration of the melt deep into the porous substrate (we believe that the maximum penetration depth is achieved on the axis of symmetry of the drop-substrate system). The rate of fluid filtration into the porous substrate is determined from the Darcy equation:

$$\dot{z}_s = \frac{m_s r_p^2 \rho \dot{H}^2}{16\chi\mu z_s}, \tag{1}$$

where χ is the tortuosity parameter of the channel, r_p is the effective radius of the capillary, m_s is the porosity of the substrate, μ is the dynamic coefficient of viscosity of the liquid. We assume that at each moment of time the fluid filtration rate is the same in the entire region Ω_s. From the condition of constant volume of incompressible fluid, we obtain the relation connecting the value of R_c with other parameters of the drop

$$3(H+h)R_c^2 - 6m_s \int\limits_0^t \dot{z}_s R_c^2 \, d\tau + \frac{3}{2}\pi h^2 R_c + (H-h)^3 + h^3 - H_0^3 = 0.$$

In each of the subregions Ω_g, Ω_d, and Ω_T, we define our approximation of the velocity field satisfying the continuity condition:

$$u_g = 0, \quad v_g = \dot{H}, \quad u_d = (m_s\dot{z}_s - \dot{H})\frac{r}{2h}, \quad v_d = (\dot{H} - m_s\dot{z}_p)\frac{z}{h} + m_s\dot{z}_s,$$
$$u_T = (m_s\dot{z}_s - \dot{H})\frac{R_c^2}{2hr}, \quad v_T = 0.$$

The kinetic energy of a moving fluid will be determined by formula

$$W = \iiint\limits_{\Omega} \frac{u^2+v^2}{2}\rho \, d\Omega = \pi\rho \iint\limits_{\Omega_1} (u^2+v^2) r \, dr \, dz.$$

Here, Ω is the rotation figure formed by rotation around the z axis of the plane region Ω_1. A detailed conclusion and the final form of the formulas for kinetic energies in each of the subregions can be found in [14], and is not given here.

Potential energy of the drop-substrate system is

$$\Pi_\sigma = \sigma_{12}(S_g + S_T + m_s S_d) + (\sigma_{13} - \sigma_{23})((1 - m_s)S_d + S_s),$$

where σ_{13} and σ_{23} are the surface tension at the liquid—substrate and substrate—gas boundaries, respectively, $S_g = \pi(R_c^2 + (H - h)^2)$ is area of the outer surface of the spherical segment, $S_T = 2\pi h(2h + \pi R_c)$ is area of the lateral surface of the toroidal part, $S_d = \pi R_c^2$ is area of the contact spot excluding porosity, $m_s S_d$ is surface area of the liquid—gas inside the pores (for simplicity we assume that the surface of the liquid in the pores is strictly horizontal), $(1 - m_s)S_s$ is the area of the contact spot taking into account the substrate porosity, $S_s = 2(\pi H_0^3/6 - V_g - V_d - V_T)/r_p$ is the surface area of the liquid—substrate within the pores. The values σ_{12}, σ_{13}, and σ_{23} are related by formula $(\sigma_{13} - \sigma_{23})/\sigma_{12} = -\cos\theta$, where θ is the contact angle of wettability.

The work of the adhesion forces when the droplet spreads over the substrate is determined by expression

$$Q = (1 - m_s)\pi R_c^2 \sigma_{12}(1 + \cos\theta).$$

In the disk-shaped region Ω_d, the force of viscous friction acts, therefore, we take into account the work of this force

$$A_\mu = 2\pi \int_0^h dz \int_0^{R_c} F_\mu \, r \, dr = \frac{\pi\mu |m_s \dot{z}_s - \dot{H}| R_c^3}{3h}.$$

Here, μ is the dynamic coefficient of viscosity of the liquid.

The condition for energy conservation of the drop—substrate system is

$$W_g + W_d + W_T + W_p + \Pi_\sigma + Q + A_\mu = W_0 + \Pi_0. \quad (2)$$

Suppose that the parameters H and h are related by

$$h = H\left(1 - \frac{H}{H_0}\right)^n,$$

where n is the constant chosen to ensure the best agreement with the experiment.

Functions $H(t)$ and $z_s(t)$ are determined from a system of ordinary differential equations (1), (2). As initial conditions, we take

$$H(0) = H_0, \quad z_s(0) = 0. \quad (3)$$

The direct solution of the obtained Cauchy problem causes some difficulties. This is due to the fact that the system of equations is non-linear with respect to the highest derivatives. In previously developed models (with an impenetrable substrate), the highest derivative entered the equation quadratically, which made it easy to isolate the desired branch of the solution from the analysis of the physical model. In the case of a porous substrate, after eliminating $\dot{z}_s$ from equation (2) using (1) we obtain an equation that is a fourth-degree polynomial with respect to $\dot{H}$. In the general case, the roots of the fourth degree polynomial cannot be expressed in an analytical form, therefore, one must either investigate all possible solutions of the Cauchy problem (theoretically, the number of such solutions can reach four), or look for ways to unambiguously single out the highest derivative.

In this paper, we propose instead of the equation on $\dot{H}$ to take its differential consequence

$$A\ddot{H} + B = 0, \tag{4}$$

where $A = A(\dot{H}, H, z_s)$, $B = B(\dot{H}, H, z_s)$ are known functions that are not given here because of their excessive bulkyness. As the missing Cauchy condition, we take the condition for the drop velocity of the droplet and the velocity of the droplet tip at the moment of collision:

$$\dot{H}(0) = v_0. \tag{5}$$

Thus, to describe the motion of a droplet after collision with a porous substrate, it is necessary to solve the system of equations (1), (4) with Cauchy data (3), (5).

3. Temperature change simulation

We supplement the droplet spreading model with a description of the process of changing the temperature of a liquid droplet until solidification (we do not affect the solidification process itself). We introduce the effective thermophysical parameters of the region of the substrate impregnated with liquid metal:

$$\lambda_f = \lambda_l m_s + \lambda_s(1 - m_s), \qquad c_{Vf} = c_{Vl} m_s + c_{Vs}(1 - m_s),$$

where λ_l and c_{Vl} are the thermal conductivity and bulk thermal conductivity of the liquid metal; λ_s, c_{Vs} are thermal conductivity and bulk thermal conductivity of the substrate material. The drop is cooled due to the radiative heat transfer of its free surface with the surrounding medium with temperature T_a and heat transfer to the substrate with initial temperature T_{s0}. In view of the small volume of the droplet, we consider the solidification process volumetric. Then, the heat balance equation in the volume of the drop above the substrate will have the form

$$M(t) c_{Vl} \frac{dT_1}{dt} = -F_c(t) q - F_1(t) \alpha_r (T_1 - T_a),$$

where $M(t)$ is the mass of the melt in the volume of the particle above the substrate; $q = -\lambda_f(\partial T_1/\partial z)$ is substrate heat flux; T_1 is liquid metal temperature; $F_c(t) = \pi R_c(t)$ is contact spot area; $F_1(t)$ is free surface of a liquid particle; $\alpha_r = \varepsilon\sigma_0(T_1 + T_a)(T_1^2 + T_a^2)$ is radiative heat transfer coefficient; σ_0 is Stefan—Boltzmann constant; ε is emissivity.

The heat transfer equation in the region of the substrate impregnated with the melt is

$$\rho_f c_{Vf} \left(\frac{\partial T_2}{\partial t} + \dot{z}_z(t) \frac{\partial T_2}{\partial z} \right) = \lambda_f \left(\frac{\partial^2 T_2}{\partial r^2} + \frac{1}{r}\frac{\partial T_2}{\partial r} + \frac{\partial^2 T_2}{\partial z^2} \right),$$

where $T_2 = T_1 m_s + T_s(1 - m_s)$ is the mass average temperature of the substrate with the melt, T_s is substrate temperature, $\rho_f = \rho_l m_s + \rho_s(1 - m_s)$ is effective density of the substrate with liquid metal.

This equation is solved under boundary conditions

$$T_1\big|_{z=0} = T_2\big|_{z=0}, \qquad -\lambda_f \frac{\partial T_2}{\partial n}\bigg|_{\Gamma} = \frac{\lambda_s}{\phi(t,r)}(T_2 - T_{s0}),$$

where $\phi(t, r)$ is the depth of liquid penetration into the substrate, Γ is impregnation boundary. To eliminate the singularity in the denominator of the boundary conditions, we assume that $\phi(t, r) \geq r_p$, that is, we assume that at all points of the contact spot, the liquid penetrates deep into the substrate at least by a thickness of r_p .

Table 1. The dependence of the geometric dimensions of the splat on the temperature of the substrate

T_{s0}, K	t, μs	$-z_s$, μm	R_c, μm	H, μm
—	1.9	1.5	—	—
1370	1.1	1.37	47	19
970	0.95	1.29	43	24
570	0.84	1.22	41	28

4. Results of a numerical experiment

We carry out numerical calculations for the model case of collision of a continuous liquid drop of a melt of zirconium dioxide with a porous substrate. The initial diameter of the droplet is 60 μm, the speed of collision of the droplet with the substrate is 50 m/s, the capillary radius is 0.5 μm, the porosity of the substrate is 0.15, the tortuosity parameter of channel 1, the metal density is 5600 kg/m^3, the surface tension at the interface of liquid and gas is 0.43 N/m, dynamic coefficient of viscosity of the fluid is 0.021 N·s/m^2.

The time dependences of the geometrical sizes of the droplet and the region of penetration of the liquid into the substrate are obtained. Calculations show that, without taking into account solidification of the melt, the maximum radius of the contact spot can reach 550 μm in about 50–100 μs after the collision. In this case, the maximum penetration depth of the melt into the substrate of 1.5 μm is reached after about 1.9 μs. After this, only the drop spreads over the surface.

Different versions of the initial temperatures of the droplet and substrate were investigated. In the calculation, the substrate temperature was taken to be 570 K, 970 K, and 1370 K, and the droplet temperature at the moment of collision was 3300 K. The following thermophysical properties of the zirconium dioxide melt were also used: λ=3.35 W/(m·K), T_m=2960 K, c=710 J/(kg·K).

The dependence of the geometric dimensions of the splat on the initial temperature of the substrate is presented in table 1. It turned out that the crystallization temperature of the droplet in all these cases is reached before the maximum penetration of the liquid deep into the substrate. In this case, similar calculations for colliding a nickel drop with a porous steel substrate show that the crystallization of the drop begins only after the drop has completely spread.

5. Conclusion

A numerical-analytical model of the interaction of a metal drop with a porous substrate is constructed. It has been shown that for refractory materials (for example, ZrO_2), a higher substrate heating is required to obtain a greater penetration depth.

The proposed solution can be useful for assessing the parameters of the splat when forming powder coatings with dynamic impregnation of porous surfaces in order to harden them and obtain composite layers with improved properties.

Acknowledgments

The research was carried out within the framework of the Program of Fundamental Scientific Research of the state academies of sciences in 2013-2020 (project No. AAAA-A17-117030610136-3).

References

[1] Saburov V P, Eremin E N, Cherepanov A N and Minnekhanov G N 2002 *Modification of Steels and Alloys with Dispersed Inoculators* (Omsk: OmSTU Publ.) (in Russian)
[2] Moskvichev V V, Krushenko G G, Burov A E, Uskov I V and Fedorova E N 2013 *Nanopowder Technology in Mechanical Engineering* (Krasnoyarsk: Siberian Federal University Publ.) (in Russian)
[3] Orishich A M, Cherepanov A N, Shapeev V P and Pugacheva N B 2014 *Nanomodification of Welded Joints in Laser Welding of Metals and Alloys* (Novosibirsk: Sib. Branch RAS Publ) (in Russian)
[4] Popov V N and Cherepanov A N 2017 *Thermophys. Aeromech.* **24** 779–86
[5] Cherepanov A N, Mali V I, Maliutina Iu N, Orishich A M, Malikov A G and Drozdov V O 2017 *Int. J. Adv. Manufact. Technol.* **90** 3037–43
[6] Bozhanova N M, Panov I T, Manolov V K, Cherepanov A N and Lazarova R L 2018 *Thermophys. Aeromech.* **25** 759–64
[7] Drozdov V O, Cherepanov A N, Chesnokov A E and Smirnov A V 2018 *AIP Conf. Proceed.* **2053** 030010
[8] Cherepanov A, Cherepanova V, Manolov V and Yovkov L 2018 *J. Phys.: Conf. Series* **1115** 042042
[9] Cherepanova V K and Cherepanov A N 2018 *J. Phys.: Conf. Series* **1105** 012050
[10] Shapovalov V 1994 *MRS Bulletin* **19** 24–8
[11] Yanagino H, Tsumura T, Nakajima H, Hyun S-K and Nakata K 2006 *Mater. Transact.* **47** 2254–8
[12] Nakajima H 2007 *Progr. Mater. Sci.* **52** 1091–173
[13] Cherepanov A N, Solonenko O P and Bublik V V 2009 *Thermophys. Aeromech.* **15** 631–41
[14] Cherepanov A N and Bublik V V 2012 *Fizicheskaya Mezomekhanika* **15 (6)** 67–71 (in Russian)
[15] Bublik V V and Cherepanov A N 2018 *AIP Conf. Proceed.* **2053** 040010

Journal of Physics: Conference Series **1404** (2019) 012010 doi:10.1088/1742-6596/1404/1/012010

Determining the law of ice deformation

Vyacheslav Buznik[1], **Sergey Golushko**[2,3], **Evgeniya Amelina**[2], **Vasily Belyaev**[2,4], **Luka Bryndin**[2,4], **Arseniy Gorynin**[2] **and Vasily Shapeev**[2,4]

[1] All-Russia Scientific Research Institute of Aviation Materials, Moscow, Russia
[2] Novosibirsk State University, Novosibirsk, Russia
[3] Institute of Computational Technologies SB RAS, Novosibirsk, Russia
[4] Khristianovich Institute of Theoretical and Applied Mechanics SB RAS, Novosibirsk, Russia

E-mail: `va.belyaev@itam.nsc.ru`

Abstract. A mathematical model of nonlinear deformation of ice samples from distilled water is implemented. An algorithm for numerical solution of corresponding boundary-value problem is proposed and realized. Results of the numerical modeling have been compared to experimental data on the three-point bending of ice beams.

1. Introduction

Many researchers deal with the mechanics of ice in many parts of the world (see, for example, [1–11] and the references therein), since the ice structures are widely used in cold regions. However, ice is "intricate" and insufficiently studied in terms of building material because of many factors. Its physical and mechanical properties heavily depend on prehistory of the natural ice and process of preparation of the test samples, test conditions such as speed of loading and temperature, crystallographic characteristics of the ice, orientation and size of the grains, geometrical dimensions of the test samples, presence of natural or artificially introduced impurities and etc. It is also known that ice may be any one of the 18 known solid crystalline phases of water, or in an amorphous solid state at various densities. As a consequence, for example, published data on the ultimate strength of ice vary greatly [1, 2, 8, 10].

The present paper is devoted to simulating the bending of ice beams made of distilled water. Studies related to the definition of the deformation law of such samples will form the basis for the creation of mathematical models for calculating and analyzing the stress-strain state of composite materials on the basis of an ice matrix. Russia is the wealthiest country in terms of the deposits of natural ice, which is associated with the cold climate of a considerable part of its territory and adjacent northern seas. The development of country is to a large extent associated with the industrial and social development of the Arctic zone. Therefore, it is reasonable to use the ice composite materials in cold regions for constructing various buildings, roads [6], ice crossings, air strips and etc.

2. Statement of the problem and solution strategy

The problem involves determining the law of deformation based on experimental data on the three-point bending of ice samples (parallelepiped with the dimensions of $20\times50\times145$ mm^3). The tests were conducted on a Zwick/Roell Z050 all-purpose electromechanical machine (Germany) equipped with an environmental chamber (All-Russian Scientific Research Institute of Aviation

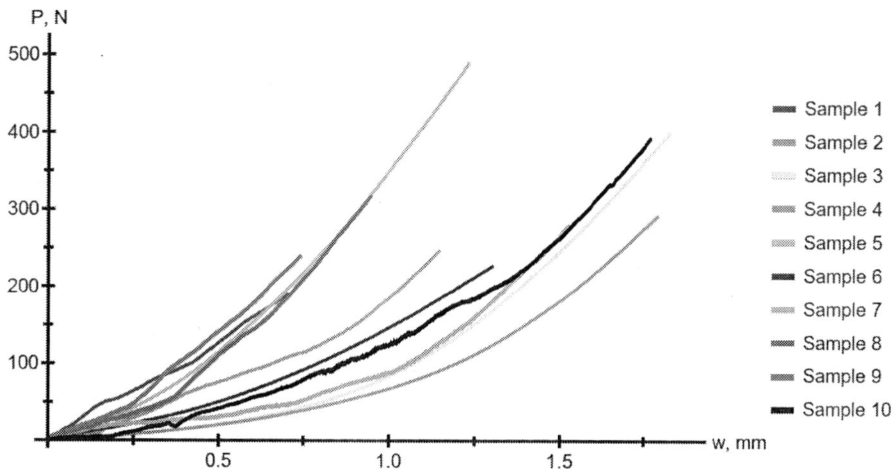

Figure 1. Experimental P-w diagram for the ice samples from distilled water

Materials). The readers are referred to [8,9] for more information about the experiments. Figure 1 shows the experimental data on the three-point bending of different ice samples (P-w diagram). Note that the experimental data range is sufficiently great, which proves the "intricacy" of ice. Additionally, at low values of loading and deformation, the experimental dependences (Fig. 1) strongly depend on the quality of the sample surface, i.e. a region of running in is present. Therefore, we are more interested in the section with sufficiently large loads, where nonlinear behavior is observed. The exact dimensions of these samples are given in Table 1.

The relative error

$$\|E_r\|_2 = \sqrt{\frac{\int_0^{P_{max}} (w_h - w_{ex})^2 dP}{\int_0^{P_{max}} w_{ex}^2 dP}}$$

was calculated after solving the direct problem for fixed values, where P_{max} denotes the maximum load before cracking (fracture of the sample), w_h is the numerical solution arising from the least squares collocation (LSC) method (section 3), and w_{ex} is deflection experimental data. Note that the coefficient a_1 should not differ much from the initial elastic modulus. Figure 2 shows the solution strategy of this problem.

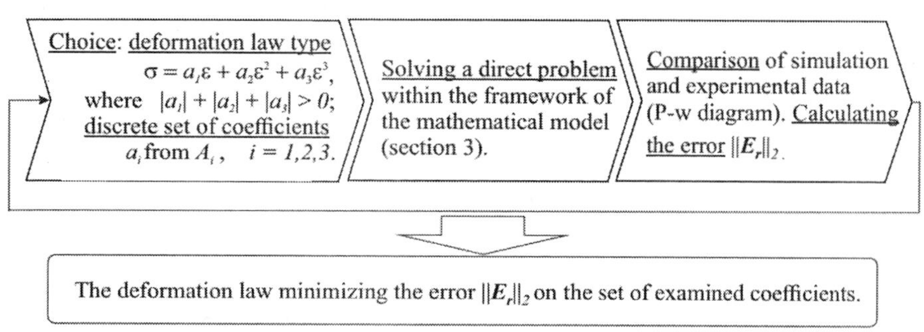

Figure 2. Solution strategy

3. Mathematical model and solving the direct problem

Let us consider the model of three-point bending of a beam with rectangular cross section $b \times 2h$ and span l between the supports used by Amelina, Golushko et al. [12, 13]. Here, $l = 125$ mm whilst the beam size $L = 145$ mm. Figure 3 shows the scheme of physical model. The left edge of the beam is hinged, while the right one is supported freely. The center of the beam bears a concentrated load. The model neglects the shape of the supports and assumes the occurring load P and support responses R_A and R_B to be concentrated. In addition, the model neglects the possible heterogeneity of the deformations in the direction normal both to the longitudinal direction and to the load direction. In this case, the beam's upper part undergoes compression strain in the longitudinal direction, the bottom part — tension strain. It is assumed that the material reacts identically to tension and compression. However, this model is capable of taking into account the difference between tensile and compressive moduli of elasticity.

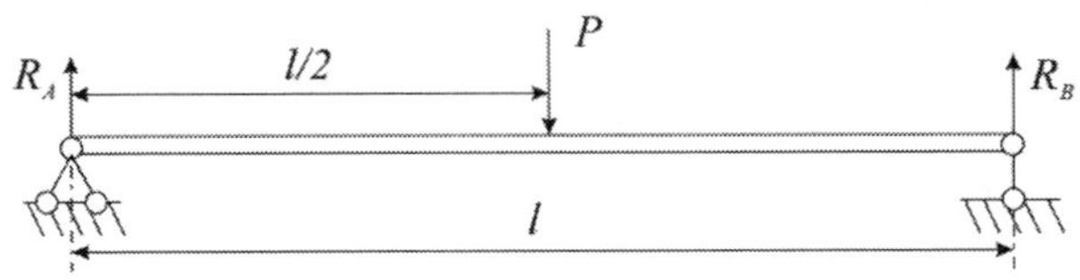

Figure 3. The scheme of physical model of three-point beam bending with rectangular cross section

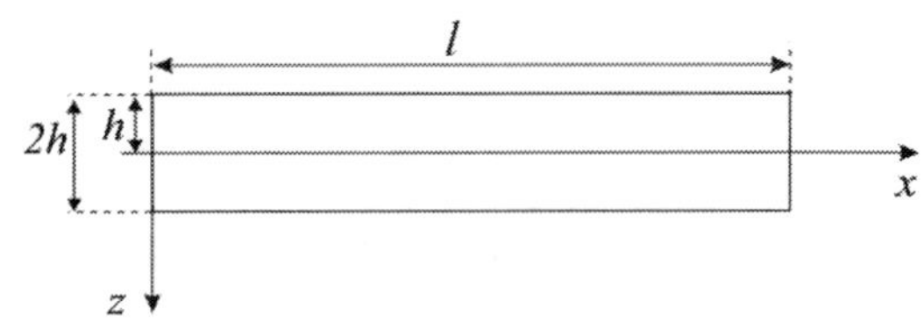

Figure 4. System of coordinates

The Euler-Bernoulli beam theory can be used for description of the equilibrium state due to the very low deformation rates (speed of tie beam equals 1 mm/min). Let the reference surface of the beam coincide with the median one $z = 0$ (Fig. 4). The corresponding equilibrium equations are written as

$$\frac{dN}{dx} = 0, \qquad \frac{dQ}{dx} = 0, \qquad \frac{dM}{dx} = Q, \tag{3.1}$$

where $Q(x)$ denotes the shear force, $M(x)$ is the bending moment, $N(x)$ is the longitudinal force.

The responses R_A and R_B are determined by ratio $R_A = R_B = P/2$. The bending moments at the support points equal zero: $M_A = M_B = 0$. The solution of (3.1) can be expressed as

$$N = 0, \qquad Q(x) = \begin{cases} P/2, & 0 \le x \le l/2, \\ -P/2, & l/2 \le x \le l, \end{cases}$$
$$M(x) = \begin{cases} Px/2, & 0 \le x \le l/2, \\ -P(x-l)/2, & l/2 \le x \le l. \end{cases} \tag{3.2}$$

Strain distribution for the beam's thickness can be obtained from the Kirchhoff-Love kinematic hypotheses:

$$\varepsilon(x, z) = e(x) + z\kappa(x), \tag{3.3}$$

$$\kappa(x) = -\frac{d^2w}{dx^2}, \tag{3.4}$$

where $\varepsilon(x, z)$ denotes the strain in the beam; $e(x)$ is the median surface strain, $\kappa(x)$ is changes in the median surface curvature, and $w(x)$ is the beam bending. Due to the fact that the deformation law for tension and compression is the same in this study, $e(x) = 0$. As mentioned above, the beam undergoes tension and compression strain. In this case for the section area

$-h \le z \le 0$, the strain will be negative, and for $0 \le z \le h$ positive. The constitutive equations can be expressed as

$$\sigma^{\pm}(x,z) = f^{\pm}(\varepsilon), \tag{3.5}$$

where the superscript "+" refers to the areas with positive strains and "−" — to the area with negative ones; $f(\varepsilon)$ denotes the relationship between the stress and strain (a linear, quadratic, or cubic function).

Bending moment M in the beam cross section is determined by equation

$$M = b\left(\int_{-h}^{0} \sigma^{-} z dz + \int_{0}^{h} \sigma^{+} z dz\right). \tag{3.6}$$

Having substituted (3.6) with relations (3.3), (3.5) and integrated it over the beam thickness, one obtains the equation to determine κ:

$$M(\kappa, x) = \begin{cases} Px/2, & 0 \le x \le l/2, \\ -P(x-l)/2, & l/2 \le x \le l. \end{cases} \tag{3.7}$$

Equation (3.7) in general case is nonlinear and the Newton method is applied to solve it. Thus, the linearized equation has the form:

$$M(\kappa_0, x) + \frac{\partial M(\kappa_0, x)}{\partial \kappa}(\kappa - \kappa_0) = M(x),$$

it can be solved for unknown value

$$\kappa = F(\kappa_0, M(x)),$$

where κ_0 is the initial approximation and $M(x)$ is determined from (3.2).

As the initial approximation at small values of the load P the solutions obtained for the linear constitutive equations $\sigma = a_1\varepsilon$ were used. Since the computation is performed with a relatively small increment of P, in case of big values of P one can use the computation results acquired at a previous step as the initial approximation for current step.

Having determined the change of median surface curvature, one can find the beam bend. It is necessary to solve the Dirichlet problem ($w(0) = w(l) = 0$) for Poisson equation (3.4) in the one-dimensional case. For that purpose, the LSC method has been applied [14, 15]. Given the limits on the length of this paper, here we only note that an approximate solution of the differential problem is reduced to solving an overdetermined system of linear algebraic equations consisting of collocation equations, matching conditions, and boundary conditions in the LSC method.

4. Comparison of the simulation and experimental data

Figures 5–7 compare data of diagram $P = P(w)$ arising from the experiment and simulation for different approximations $\sigma = \sigma(\epsilon)$. Note that $\sigma^+ = a_1\epsilon + a_2^+\epsilon^2 + a_3\epsilon^3$ and $\sigma^- = a_1\epsilon + a_2^-\epsilon^2 + a_3\epsilon^3$, where $a_2^+ = -a_2^-$ in case when material reacts identically to tension and compression. Applying the linear relationship between the stress and strain has not resulted in adequate approximation (Fig. 5). The error value is smaller in the case of a quadratic relationship compared with linear relationship. However, the application of the quadratic relationship has not resulted in a satisfactory match with experimental data for $\sigma > 150$ MPa (Fig. 6). It was found out that the part of diagram with $\sigma > 150$ MPa is poorly caught in case $\sigma = a_1\epsilon + a_2^{\pm}\epsilon^2$ in a large number

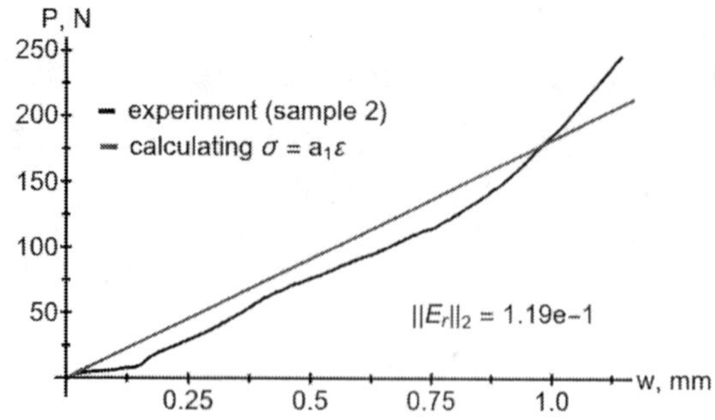

Figure 5. Linear approximation: $a_1 = 2.1e{+}2$

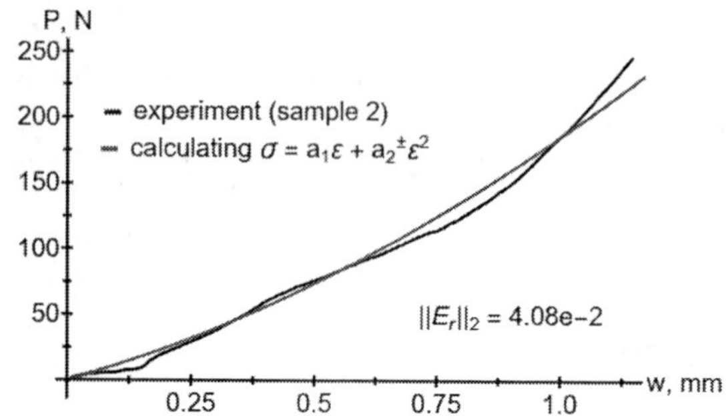

Figure 6. Quadratic approximation: $a_1 = 1.25e{+}2$, $a_2^{\pm} = \pm 2.1e{+}4$

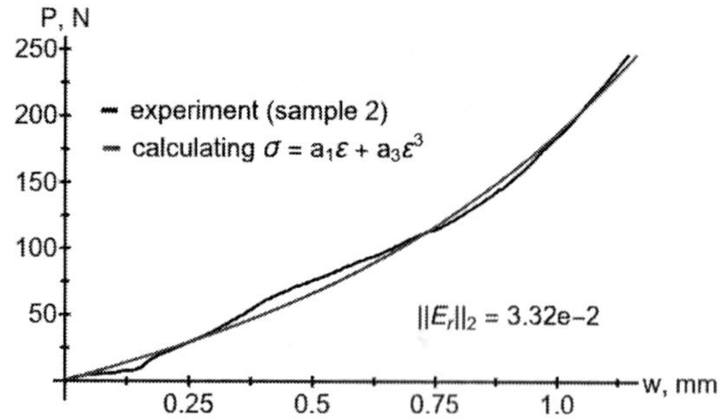

Figure 7. Cubic approximation: $a_1 = 1.3e{+}2$, $a_3 = 4.5e{+}6$

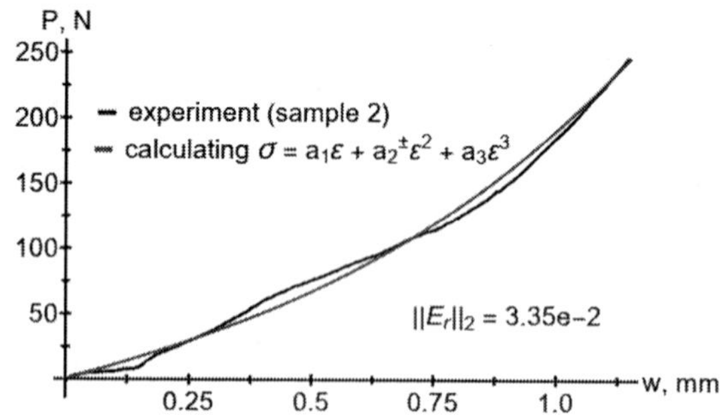

Figure 8. Cubic approximation: $a_1 = 1.3e{+}2$, $a_2^{\pm} = \pm 1.0e{+}3$, $a_3 = 4.5e{+}6$

of the numerical experiments. There is not much difference between the cubic relationship with and without ($a_2^{\pm} = 0$) the quadratic term (Fig. 7, Fig. 8).

Table 1 shows the beam dimensions in mm, coefficient values when $\sigma = a_1\epsilon + a_3\epsilon^3$, and relative errors $\|E_r\|_2$. On top of that, this table also shows the average value for these parameters. It is important to note that the average value for a_1 coincides in order with the bending modulus of

Table 1. The results of the numerical simulation when $\sigma = a_1\epsilon + a_3\epsilon^3$

Sample number	$2h{\times}b{\times}L$ (mm^3)	a_1 (MPa)	a_3 (MPa)	$\|E_r\|_2$
1	19.8×49.7×145	3.45e+2	0	5.62e-2
2	20.4×49.9×145	1.3e+2	4.5e+6	3.32e-2
3	21.6×50.5×145	3.0e+1	2.7e+6	1.63e-2
4	21.0×50.3×145	3.0e+1	2.3e+6	1.31e-2
5	20.2×50.4×145	2.1e+2	9.3e+6	3.34e-2
6	17.6×50.2×145	1.4e+2	6.8e+6	3.48e-2
7	20.6×49.05×145	2.8e+1	4.5e+6	4.2e-2
8	19.05×49.1×145	2.4e+2	1.65e+7	3.34e-2
9	23.1×48.9×145	2.0e+2	4.0e+6	4.83e-2
10	24.45×49.1×145	4.0e+1	1.4e+6	2.88e-2
Average value	**20.78×49.71×145**	**1.39e+2**	**5.2e+6**	**3.40e-2**

elasticity, which was determined by the displacement of the walking beam at the section of the diagram of loading from 1.5 to 2 MPa by the least squares method [8]. Interestingly, the linear relationship between the stress and strain is sufficient for sample 1.

5. Conclusions

The results of presented study are the following:

— the algorithm for determining the law of ice deformation based on the solution of the bending problem is proposed;

— the mathematical model for nonlinear flexural deformation of ice beams has been implemented;

— the LSC method has been proposed and realized for the numerical solution of the Dirichlet problem for the Poisson equation.

The mathematical model and numerical method developed have proved to be applicable to the problem of three-point bending of ice beams. A satisfactory match with the results of mechanical tests has been obtained. The range and average values of the coefficients in the calculated laws of deformation for ice beams have been established. It is shown that the best approximation to the bundle of experimental curves present the polynomial third-degree deformation law.

Acknowledgments

The experiments, formulation of the problem, development of algorithms for its solution, and analysis of the results was carried out within the framework of the grant from the Russian Science Foundation (project no. 18-13-00392). The software implementation of algorithms was carried out within the framework of the Program of Fundamental Scientific Research of the state academies of sciences in 2013-2020 (project No. AAAA-A17-117030610136-3).

References

[1] Pounder E R 1965 *The Physics of Ice* (Amsterdam: Elsevier) p 160

[2] Lavrov V V 1971 *Deformation and Strength of Ice* (Jerusalem: Israel Prog. for Sci. Translations) p 164

[3] Dempsey J P 2000 *Int. J. Solids. Struct.* **37** (1–2) 131–53

[4] Petrovic J J 2003 *J. Mater. Sci.* **38** (1) 1–6

[5] Marchenko A 2008 *Cold Reg. Sci. Technol.* **52** (3) 278–301

[6] Matveev S A and Sirotyk V V 2010 *Use of geosynthetic materials for reinforcing of road designs* (Khanty-Mansiysk: Dep. of Educ. and Sci. of the Khanty-Mansiysk Autonomous Okrug) p 490 (in Russian)

[7] Yakimenko O V and Sirotyuk V V 2015 *Earth's Cryosphere* **18** (1) 88–91 (in Russian)

[8] Buznik V M, Landik D N, Erasov V S, Nuzhnyi G A, Cherepanin R N, Novikov M M, Goncharova G Y, Razomasov N D, Razomasova T S and Ustyugova T G 2017 *Inorg. Mater. Appl. Res.* **8** (4) 618–25

[9] Cherepanin R N, Nuzhnyi G A, Razomasov N A, Goncharova G Y and Buznik V M 2018 *Inorg. Mater. Appl. Res.* **9** (1) 114–20

[10] Karulina A, Marchenko A, Karulin E, Sodhi D, Sakharov A and Chistyakov P 2019 *Appl. Ocean Res.* **90** Arcitle 101853, 1–15

[11] Wu Y, Liu X, Chen B, Li Q, Luo P and Pronk A 2019 *Automat. Constr.* **106** Arcitle 102862, 1–13

[12] Amelina E V, Golushko S K, Erasov V S, Idimeshev S V, Nemirovskii Y V, Semisalov B V, Yurchenko A V and Yakovlev N O 2015 *Computat. Technolog.* **20** (5) 27–52 (in Russian)

[13] Golushko S K 2019 (Optimum Composite Structures) *Mathematical modeling and numerical optimization of composite structures* (London: IntechOpen) pp 13–34

[14] Shapeev V P, Bryndin L S and Belyaev V A 2019 *Bulletin of the South Ural State University, Series: Mathematical Modelling, Programming and Computer Software* **12** (2) 140–52 (in Russian)

[15] Shapeev V P, Golushko S K, Bryndin L S and Belyaev V A 2019 *J. Phys.: Conf. Series* **1268** 012076-1–012076-7

Influence of surface active agent on the size of metal powder particles during their ball milling in a planetary mill

A E Chesnokov[1], A V Smirnov[1] and T M Vidyuk[1,2]

[1] Khristianovich Institute of Theoretical and Applied Mechanics SB RAS, 4/1 Institutskaya str., Novosibirsk, 630090, Russia
[2] Institute of Solid State Chemistry and Mechanochemistry SB RAS, 18 Kutateladze str., Novosibirsk, 630117, Russia

chae@itam.nsc.ru

Abstract. The paper presents the results of a study of the influence of surface active agent during the ball milling of metal powders in a planetary mill. It is shown that the mass fraction of the surface active agent in the powder mixture under study affects the size of the particles being treated, depending on the time of processing. The particle size distribution, the specific surface area were determined, and the surface morphology of the initial and ball milled particles was studied.

1. Introduction

One of the main tasks of solid state physics, powder metallurgy and mechanochemistry is the grinding of solids to powder with the smallest possible particle size, for example, by ball milling materials in a planetary mill. The degree of grinding depends on the properties of the material, which determine its internal structure (porosity), physical and mechanical properties, such as elasticity, ductility, and strength [1]. Depending on the properties of the material, their grinding mechanisms are realized. As shown in [2], plastic materials (Ti, Al, Ni, Cu, NiCr, Nb) require milder grinding conditions at lower temperatures and lower collision energy, while the initial particles are actively dispersed at the initial time of machining. Further, ball milling of the powder particles becomes similar to the granulation process by the rolling method, which leads to a multiple increase in the average particle size. It is possible to stop the increase in particle size and continue their grinding during mechanical processing using surfactants [3,4]. Surfactants can be liquids or solids that do not chemically interact with the mechanically processed material. The role of surfactants is associated with a modification in the mechanical properties of the material. The mechanical treatment of particles is accompanied by continuous volumetric deformation, which leads to a change in the internal structure, nucleation and the appearance of microcracks and cracks. The adsorption of surfactants on a newly formed particle surface prevents their recovery. In this case, the resulting value of the force arising during stress relaxation is directed through the volume of the material, as a result of which the particle volume is divided along the formed crack.

The purpose of the work is to experimentally study the effect of surface active agents during the ball milling of metal powders.

2. Experimental procedure

In this work, the influence of a surfactant (carbon black P803) during the mechanical treatment of powders of titanium (PTOM-2), cobalt (PK-1U) and nickel-chromium alloy (PH80N20), the fraction is less than 40 microns was studied. The histograms of volumetric and counting particle distributions were determined on particle size analyzer LS 13 320 (Beckman Coulter). The specific surface of the samples was determined by the BET method by thermal desorption of argon with an internal standard. Before the measurements, all samples were heated in a gas flow of Ar and He at temperature of 120 °C for 0.5 h. Particle morphology was determined using Evo MA15 electron microscope (Carl Zeiss). The planetary mill "Activator-2SL" has two cylinders with volume of 250 ml each with inner radius of 42.5 mm, the mass of loaded balls is 160 g for each cylinder, the acceleration of grinding bodies is 117 g, the mass of the processed material is 30 g. Maximum rotation speed of the central axis is 1045 rpm, that of the cylinders is 1550 rpm. As grinding bodies, steel balls with diameter of 5 mm were used.

3. Results and discussion

To study the effect of a surfactant on the initial particles of metal powders, powder mixtures were prepared, the compositions of which are presented in table 1.

Table 1. The composition of the powder mixture.

Powder mixture designation	The mass fraction of components
1-1	Ti
1-2	Ti – 5%C
1-3	Ti – 10%C
1-4	Ti – 20%C
2-1	NiCr
2-2	NiCr – 5%C
2-3	NiCr – 10%C
2-4	NiCr – 20%C
3-1	Co
3-2	Co – 5%C
3-3	Co – 10%C
3-4	Co – 20%C

Control samples of metal powders, as well as their mixtures (table 1) were taken after ball milling for 15, 30, 45, 60, and 90 seconds. Figure 1 shows the experimental dependences characterizing the change in the average volumetric size depending on the time of machining. From the presented data, it can be seen that an increase in the percentage of surfactant in powder mixtures from 5 to 20% during their mechanical processing, ceteris paribus, leads to a significant decrease (dispersion) of particles of metal powders and with increasing time of high-energy exposure remains approximately constant. This is due to the fact that ultrafine carbon black particles, adsorbed on a new surface formed during the bulk deformation of a metal particle, create a protective shell that prevents their further agglomeration.

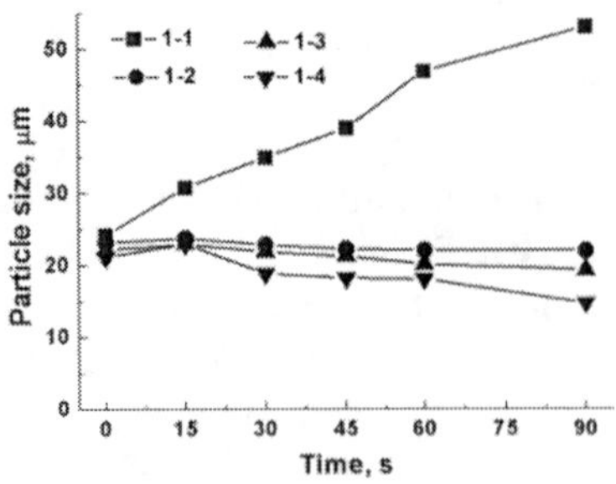

(a)

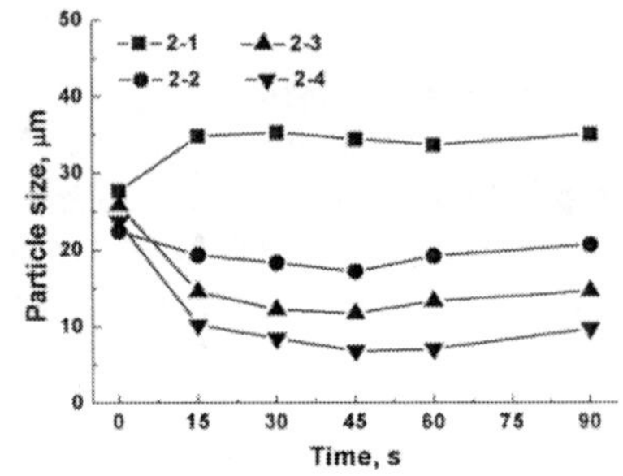

(b)

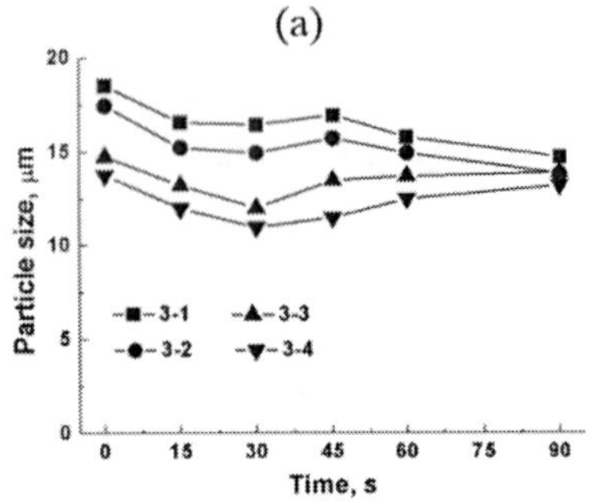

(c)

Figure 1. Change in average volumetric size depending on the time of mechanical processing of the powders: (a) titanium, (b) nickel-chromium alloy, (c) cobalt and their compositions with different mass fraction of surfactant.

Figure 2 shows the change in the values of the specific surface of metal powders and their mixtures depending on the time of machining. The ball milling of powder mixtures (table 1) is accompanied by an increase in the specific surface area (figure 2), which indicates an increase in the number of particles with sizes of one micron or less. Figure 3 shows the morphology of titanium particles, which is also typical for other materials studied, isolated from a mixture mechanically processed for 90 seconds.

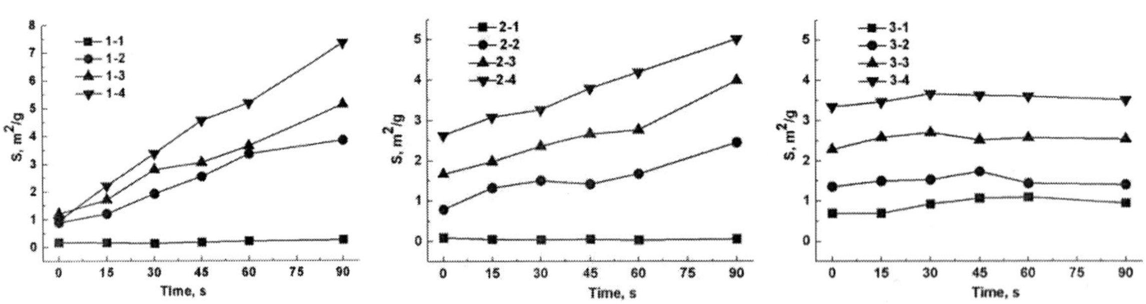

(a) (b) (c)

Figure 2. The specific surface values depending on the time of mechanical processing of the powders: (a) titanium, (b) nickel-chromium alloy, (c) cobalt and their compositions with different mass fraction of surfactant.

Figure 3. The morphology of particles of ball milled titanium within 90 seconds of the composition: (a) 1 - 1; (b) 1 - 2; (c) 1 - 3; (d) 1 – 4.

4. Conclusion

It is shown in the paper that the mass fraction of the surface active agent in the studied powder mixture affects the size of the ball milled particles, depending on the time of the processing. An increase in the content of the mass fraction of the surfactant in the powder mixture from 5 to 20% leads to a significant decrease in the volumetric average size of the metal particles, while plating them with carbon black. Thus, above mentioned features, it makes it possible to stimulate reactions of interfacial interaction of dissimilar elements, in particular, it increases the reaction rate of self-propagating high-temperature synthesis of chemical compounds, and also makes an additional contribution to the formation of a nonequilibrium structural-phase state of a cermet alloy during its synthesis.

Acknowledgments

This study was supported by the Russian Science Foundation (Project No. 19-19-00335) using equipment of the Joint Access Center «Mechanics» of ITAM SB RAS.

References

[1] Boldyrev V V 2006 *Russ. Chem.* **75** 177–89

[2] Burmeister C F and Kwade A 2013 *Chem. Soc. Rev.* **42** 7660–7

[3] Stepanchuk F N, Bilik P I, Boyko P A 1989 *The Technology of Powder Metallurgy* (Kiev: Vysshaya shkola) p 264

[4] Poluboyarov V A, Andryushkova O V, Pauli I A, Korotaeva Z A 2011 *The Influence of Mechanical Stresses on Physical and Chemical Processes in Materials* (Novosibirsk: NSTU) p 604

Impact of the rate of input of specific energy on the ball milling of aluminium in a planetary mill

A E Chesnokov[1], A V Smirnov[1] and T M Vidyuk[1,2]

[1] Khristianovich Institute of Theoretical and Applied Mechanics SB RAS, 4/1 Institutskaya str., Novosibirsk, 630090, Russia
[2] Institute of Solid State Chemistry and Mechanochemistry, SB RAS, 18 Kutateladze str., Novosibirsk, 630117, Russia

chae@itam.nsc.ru

Abstract. In the present study, the effect of the rate of the input of comparable amounts of energy into the material being processed on the particle size and on the value of specific surface area of the particles was analysed. It was shown that, irrespective of the rate of the energy input into the material, the resultant particles have similar characteristics, including their size and shape, and the value of the specific surface area. A more intense action exerted by the grinding bodies on the particles in the material being processed is accompanied by a sharp increase of temperature throughout the entire system. The heated particles plasticize, thus undergoing to a rolling-type granulation process.

1. Introduction

In powder metallurgy, one of the main processes is the mechanical processing of powders, including that in planetary-type ball mills. This process affects the size of the particles and their shape depending on the physical properties and chemical composition of the material. The aim of powder processing is to obtain particles with a minimum size, with a maximum specific surface area, and with the bulk density of the material at the minimal consumption of activation energy [1-6]. The use of high-energy attritors for powder treatment allows not only performing the mechanical processing of powders but, also, their mechanical activation [7].

Ball milling of powder mixtures is accompanied by an intense mechanical action exerted on the particles. At the initial moment of the treatment, a considerable dispersion of the initial components occurs. Then, during the mechanical processing, due to the high surface activity of ultrafine particles those particles form the so-called "layered composites" (mechanocomposites), or powder agglomerates in the case of a monophasic chemical compound [8-12]. As a result of the intense plastic deformation occurring in the formed agglomerates, the area of the inter-particle contacts many times increases, and there forms a high concentration of non-equilibrium defects and internal stresses, which process leads, for instance, to an acceleration of exothermic reactions and an increase in the SHS synthesis temperature [9, 11, 12].

When a grinding body interacts with a particle, the kinetic energy is spent on the plastic deformation of the particle proceeding with the release of a large amount of heat, up to 95% [13]. In [14], it was shown that the temperature in contacts, which arises due to the sliding friction, can reach the melting point of one of the materials. In this connection, the current studies of the influence of the energy input by grinding bodies on the material being processed, aimed at establishing the optimum regimes of material processing, are presently quite topical ones. For instance, the study of [15] was devoted to obtaining a FeNi-alloy based powder in a ball mill, where the influence of the kinetic energy of the balls on the microstructure and particle size of the powders was investigated. In [16], low- and high-efficiency regimes of ball milling of powders (Cu, Ni and Fe) were preliminarily determined. The work [17] was devoted to the study of the doping of metallic Cu and Fe particles with ceramic Al_2O_3 and SiC nanoparticles. It was shown in [18] that the final structure of the doped $Cu_{50}Zr_{43}Al_7$ alloy obtained by mechanical processing depended on the processing conditions. A low-energy planetary mill with a specific-energy input of 1440 J/g yielded a solid solution with a high copper content and, on increasing the rate of specific-energy input to 4320 J/g on a Spex ball mill (shaker), a nanostructured solid solution with a high Cu content and intermetallic compound Zr_2Al_3 formed. In [19], a study of the effect of the ball diameter on the crystal size, on the induced deformation, and on the diffusion of atoms in the system of immiscible Cu - 50%Fe alloys was performed. In [20], an analysis of the ball kinetics during the mechanical processing of copper powder was performed.

In publication [21], the processes occurring during ball milling between the grinding bodies and the material being processed were analyzed. High impact energies can be implemented using either large ball diameters or a high density of grinding bodies in combination with a high speed of their rotation [22, 23]. Very ductile materials (Ni, Zn, Cu, Al, Nb, etc.) require softer grinding conditions at lower temperatures and at lower impact energy.

The aim of the present work was to experimentally study the influence of the rate of specific-energy input into the grinded material during the ball milling of aluminum powder.

2. Experimental procedures

In the study, ball milling process ASD-1 grade aluminium powder was investigated. The histograms of the volumetric and number distributions of particles were obtained using an LS 13 320 (Beckman Coulter). Morphology of particles was determined using scanning electron microscope Evo MA15 (Carl Zeiss). The classification of powder particles in terms of their sizes was performed on a vibro-bench provided with a set of analytical sieves. The Activator-2SL planetary mill [27] had two cylinders, each having a volume of 250 ml. The inner radius of the cylinders was 42.5 mm; the mass of the loaded balls, 160 g for each cylinder; the acceleration of the grinding bodies, 117 g; and the mass of the material load to be processed, 30 g. The maximum rotational speed of the central axis was 1045 rpm, and that of the cylinders, 1550 rpm. Steel balls with a diameter of 3, 5, and 8 mm were used as the grinding bodies.

The specific power needed per unit weight of processing material in the ball mills was calculated using the following empirical formula [6]

$$N = \frac{g^{3/2}}{4\pi A}\frac{M_b}{m_p}\left(\frac{a}{g}\right)^{3/2} R^{3/2}, \tag{1}$$

In formula (1), $A = 0.03\ m$ is an empirical parameter, M_b is the mass of the balls, m_p is the mass of the loaded powder, $a = 4\pi^2 n^2 R$ is the ball acceleration, R is the inner radius of the cylinders, n is the number of revolutions per second, and g is the free-fall acceleration.

3. Results and discussion

For comparison of data concerning the input of energy by grinding bodies into the powder particles, the regimes of mechanical processing were calculated using formula (1) (see Table 1).

Table 1. Processing regimes of aluminium powder.

Processing regime of the powder	Specific energy, J/g	Parameter of ball milling 1: M_b/m_p (160/30); a = 117g	Parameter of ball milling 2: M_b/m_p (80/30); a = 117g	Parameter of ball milling 3: M_b/m_p (160/30); a = 4.58g
Regime №1	71	15 c	30 c	31 min
Regime №2	142	30 c	60 c	1 h 3 min
Regime №3	284	60 c	120 c	2 h 5 min
Regime №4	426	90 c	180 c	3 h 7 min

After mechanical processing, the powder was divided into two fractions, with the sizes of powder particles being smaller and greater than 90 µm. The separation of particles into the two fractions was due to the fact that the particles of the main fraction, with sizes less than 90 µm, are widely used ones, in particular, in gas-thermal, detonation, and cold gas-dynamic spraying [24-27]. The particles larger than 90 µm in size and shaped as deformed disks can be treated in a plasma jet to obtain a powder that consisted of spherical particles of micron and decamicron sizes [28, 29]. Such particles are currently in demand as they are widely used in additive technologies [30, 31], laser cladding [32], and in other powder-metallurgy areas.

Using the results of the study of the dispersed composition of the initial and treated aluminum particles, we have plotted the curves characterizing the changes in the mean particle size and in the main-range boundaries d_{10} to d_{90} as dependent on the input of energy into the processed material (Figure 1). The initial aluminum powder particles had a spherical shape with the mean size of the main fraction of 18 µm.

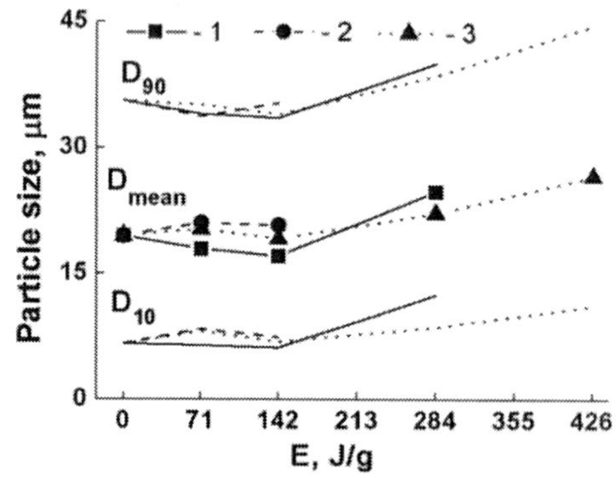

Figure 1. Size composition of the initial and mechanically processed aluminum powders. 1 – parameter of ball milling No 1; 2 – parameter of ball milling No 2; 3 – parameter of ball milling No 3.

Thus, it was shown that, irrespective of the rate of input of comparable amounts of energy into the processed material, the sizes of the particles and their specific surface areas had rather close values. The energy input intensity affects only the weight content of particles larger than 90 µm in the ball milled material. Figure 2 shows the successive changes in the surface morphology of the mechanically processed particles. It should be noted here that the behavior of aluminum particles during the mechanical processing was typical of the mechanical processing of ductile metallic materials in planetary mills irrespective of the regime of specific energy entry into the material [15, 33].

Figure 2. Evolution of the surface area in particles greater in size than 90 μm for four regimes at the parameter of ball milling No.1. Treatment duration: a) 15 s, b) 30 s, c) 60 s, d) 90 s.

At fixed mass of the grinding bodies loaded into the mill cylinder, a decrease in the ball diameter leads to an increase of the number of the balls (Table 2). Accordingly, at fixed velocity the kinetic energy of one ball decreases in proportion to its mass. Consequently, for the ball-particle interaction system a decrease in the ball diameter leads to a decrease in the degree of deformation of the particle and to a less pronounced heating of its volume. Taking into account the fact that with a decrease of ball diameter the curvature of the ball surface increases, the kinetic energy of the ball impinging on a particle is transferred to a smaller contact area with a sharp increase of local temperature occurring at the interaction site. Afterwards, sintering occurs over the hot-spot area during the interaction of powder particles. The reduction in the curvature of the surface (due to the increased ball diameter) leads to an increase of the contact area. As a result, the sintering leads to the formation of larger particles with a corresponding increase of the diameter of processed Al particles for a given operating mode of the planetary mill (figure 3).

Table 2. Characteristics of the grinding bodies.

Ball diameter, mm	Mass of one ball, g	Number of the balls for load of 160 g
3	0.11	1470
5	0.50	318
8	2.05	78

(a) (b) (c)

Figure 3. General view of the surface of the particles processed during 90 seconds at parameter No.1. Particles obtained by the interaction with the balls of diameter: a) 3 mm, b) 5 mm, c) 8 mm.

4. Conclusion

In the present study, it was shown that, irrespective of the rate of input of comparable amounts of energy into the material being processed, the particle size and the specific surface area of particles had similar values. The rate of energy input affected only the mass fraction of the particles larger than 90 µm in the mechanically processed material. The study of the influence due to the ball diameter at fixed mass of the load and at fixed rotation speed of the mill cylinders showed that the dynamics of particles and their deformations were similar. An increase in the ball diameter leads to an increase of the average diameter of the processed aluminum particles with a similar external structure.

Acknowledgments

This study was supported by the Russian Science Foundation (Project No. 19-19-00335) using equipment of the Joint Access Center «Mechanics» of ITAM SB RAS.

References

[1] Boldyrev V, Avvakumov E, Logvinenko A 1977 *Collection of Scientific Papers* (Novosibirsk: Nauka) p 3–10
[2] Butyagin P Yu 1987 *Izv. SO AN SSSR (ser. Khim. Nauk)* **17** 48–59
[3] Boldyrev V V, Pavlov S V, Poluboyarov V A and Dushkin A V 1995 *J. Inorganic Materials* **31** 1128–38
[4] Goldberg E L, Pavlov S V 1993 *Proc. of the 1st International Conference on Mechanochemistry* 1 66–70
[5] Poluboyarov V A, Pauli I A 1996 *Chemistry in Favor of Stable Development* **4** 519–23
[6] Poluboyarov V A, Korotaeva Z A, Kisilevich S N, Pankrat'ev Yu D, Sysoev V F and Andryushkova O V 1999 *J. of Physical chemistry* **73** 1094–9
[7] Suryanarayana C 2001 *Progress in Materials Science* **46** 1–184
[8] Boldyrev V V 2006 *Russ. Chem.* **75** 177–89
[9] Prokudina V K, Ratnikov V I, Borovinskaya I P and Merzhanov A G 1981 (Kiev: Naukova Dumka) p 29–34
[10] Lyakhov N Z, Talako T L, Grogorieva T F 2008 (Novosibirsk: Parallel) p 167
[11] Avvakumov E G 1988 *Mechanical Methods for Activation of Chemical Processes* (Nauka: Novosibirsk) p 305
[12] Boldyrev V V 1991 *Mechanochemical Synthesis in Inorganic Chemistry* (Novosibirsk: Nauka) ed E.G. Avvakumov p 264
[13] Taylor G I 1934 *Proc. Roy. Soc.* **145** 362–404
[14] Bowden F P, S F R and Thomas P H 1954 *Proc. Roy. Soc.* **223** 29–40
[15] Gheisari Kh, Javadpour S, Oh J T, Ghaffari M 2009 *J. of Alloys and Compounds* **472** 416–20
[16] Lebrun P, Gaffet E, Froyen L, Delaey L 1992 *Scripta Metallurgica et Materialia* **26** 1743–48

[17] Afkhama Y, Khosroshahi R, Kheirifard R, Mousavian T R, and Brabazon D 2017 *Physics of Metals and Metallography* **118** 749–58
[18] Dastanpoor E, Enayati M H 2015 Indian J. of Engineering & Materials Sciences **22** 521–26
[19] Vaezi R, Ghassemi M, Shokuhfar A 2012 *J. Theor Appl. Phys.* **6** 7
[20] Brun P L, Froyen L, Delaey L 1993 *Materials Science and Engineering* **161** 75–82
[21] Burmeister C F and Kwade A 2013 *Chem. Soc. Rev.* **42** 7660–7
[22] Broseghini M, Gelisioa L, D'Incaua M, Ricardoa A C L, Pugno N M, Scardi P 2016 *J. Eur. Ceram. Soc.* **36** 2205–12
[23] Schmidt R, Scholze M, Stolle H 2016 *Int. J. Industrial Chemistry* **7** 181–6
[24] Puzryakov A F 2007 *Theoretical Foundations of Plasma Spraying* 2nd edition (Moscow: Bauman MSTU) p 360
[25] Oksa M, Turunen E, Suhonen T, Varis T, Hannula S P 2011 *Techniques, Materials, and Applications Coatings* **1** 17–52
[26] Shorshorov M Kh, Kharlamov Yu A 1978 *Physicochemical Foundations of the Detonation-Gas Spraying of Coatings* (Moscow: Nauka) p 227
[27] Papyrin A, Kosarev V, Klinkov S, Alkhimov A and Fomin V 2007 *Cold Spray Technology* (Amsterdam: Elsevier Science) p 336
[28] Kotlyarov V I, Beshkarev V T, Kartsev V E, Ivanov V V, Gasanov A A, Yuzhakova E A, Samokhin A V, Fadeev A A, Alekseev N V, Sinayskiy M A, Tretyakov E V 2017 *Inorganic Materials: Applied Research* **8** 452–8
[29] Grigoriev A V, Razumov N G, Popovich A A and Samokhin A V 2017 *Metallurgy and Materials Science SPbPU Scientific and Technical reports* **23** 247-55
[30] Weller C, Kleer R, Piller FT 2015 *International Journal of Production Economics* **164** 43–56
[31] Herzog D, Seyda V, Wycisk E, Emmelmann C 2016 *Acta Materialia* **117** 371–92
[32] Orishich A M, Cherepanov A N, Shapeev V P, Pugacheva N B 2014 *Nanomodification of Welded Joints in Laser Welding of Alloys and Metals* (Novosibirsk: SO RAN) p 252
[33] Poluboyarov V A, Solonenko O P, Zhdanok A A, Chesnokov A E, Pauli I A 2017 *SibFU J. Engineering and Technologies* **10** 646–56

Modeling the process of river bed bottom erosion in the area of the underwater pipelines location

V V Degtyarev[1], M E Garmakova[1], M N Shumkova[1] and A V Chebotnikov[1,2]

[1] Novosibirsk State University of Architecture and Civil Engineering (Sibstrin), 630008, Novosibirsk, Russia
[2] Lavrentyev Institute of Hydrodynamics of Siberian Branch of Russian Academy of Sciences, Novosibirsk, 630090 Russia

garmakova94@mail.ru

Abstract. This paper deals with the modeling of flows and transport of sand particles during erosion of pipelines. The results of an experimental study of velocity fields and numerical CFD modeling in a three-dimensional transient formulation on the basis of Reynolds averaged Navier-Stokes equations in the ANSYS software package are presented. Modeling was carried out with different arrangement of pipelines, simulating cylinders, on a sandy bottom. The influence of the number of cylinders on the process of bottom soil reformation is analyzed.

1. Introduction

Underwater pipelines are complex engineering facilities. On the one hand, subsea oil and gas pipelines and sewer dukes when crossings navigable rivers are affected by the flow, waves and ice; on the other hand, they themselves change the speed structure of the stream, which affects the intensity and direction of channel deformations during the construction and operation of underwater facilities. Bottom erosion occurs as a result of bottom currents of various nature depending on the magnitude of the current velocity and direction, as well as the characteristics of the bottom soil. Sagging of the pipeline occurs on the diffuse part of the bottom; it leads to the generation of its free oscillations and development of resonant phenomena with a complex oscillation diagram. The task of identifying potentially dangerous, susceptible to erosion areas and their development over time is of great importance at the design and operation stages of underwater pipelines.

Due to the desire to prevent the negative consequences leading to accidents, the deformation of the channels under the pipelines has been the subject of research for many decades by both domestic [1-3] and foreign [4-6] scientists.

The classical problem of flowing around a cylinder with a fluid flow has been the subject of a large number of scientific papers. We mention here only the fundamental works [7–9]. A great amount of theoretical and experimental information has been obtained, which is crucial for solving a number of important problems, such as, for example, resistance to the movement of bodies in a liquid and the structure of the hydrodynamic wake behind a body moving in a liquid.

Until recently, the velocity field in the vicinity of the cylinder was the least studied. In theoretical papers, the case of relatively small values of the free stream Reynolds numbers was mainly considered. As a result, the development of computer technology and numerical methods made it possible to carry out direct numerical simulation of turbulent flows at high Reynolds numbers, and the

PIV-method for measuring velocity (Particle Image Velocimetry), free from the indicated drawback of the hot-wire anemometric method, became available to experimenters.

2. Experimental facility and methods

This paper presents experimental data on the horizontal and vertical components of the averaged velocity in a transverse flow around a horizontal cylinder with a turbulent incident flow with a free surface. The study is conducted in the laboratory of experimental applied hydrodynamics of the Lavrentyev Institute of Hydrodynamics of Siberian Branch of Russian Academy of Sciences.

The experimental design is shown in Figure 1. A cylinder with a diameter of $D = 0.028$ m was located perpendicular to the side walls of a rectangular channel of width B=0.2 m at various distances from its axis to the bottom z_0. A photograph of the experimental setup and PIV equipment is shown in Figure 2. Here 1 is the mirror for a shadow under the cylinder; 2 is the working channel; 3 is the flow damper; 4 is an optical system forming a light knife; 5 is the flexible mirror system; 6 is the laser and 7 is digital video camera.

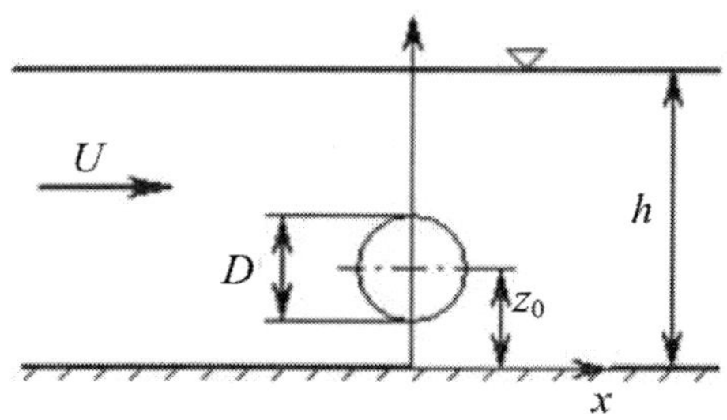

Figure 1. Experiment design.

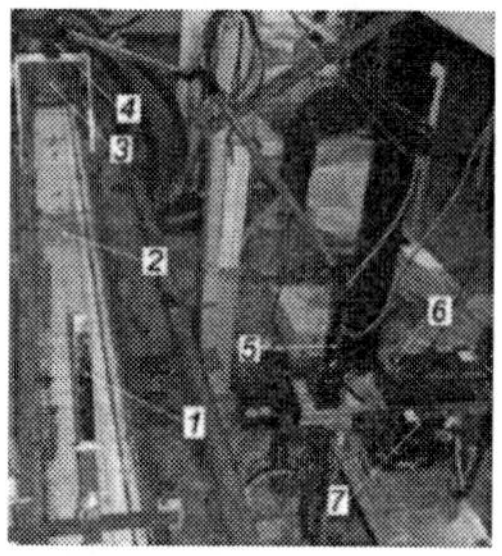

Figure 2. Photograph of the experimental setup.

In the experiments, the volume flow rate Q and the distance z_0 are changed. The values of the given parameters vary in the ranges: $0 \le z_0 \le 0.042$ m; $0.00336 \le Q \le 0.00576$ m^3/s. With the help of the free surface level controller located at the channel exit, the oncoming flow depth h is kept constant and equal to 0.12 m. The average rate U is calculated from the volume flow rate Q, channel width B and water depth h as $U = Q{\cdot}(B{\cdot}h)^{-1}$. Depth h is determined using measuring needles with a standard error of no more than 1%. Using the PIV method, implemented using Dantec Dynamics equipment, in the longitudinal plane of channel symmetry, instantaneous values of the longitudinal u and vertical w velocity components as functions of spatial variables x, z and time t are measured.

Four series of experiments are performed. In the first series, which corresponds to a completely buried pipeline, the cylinder is absent. In the second series, the cylinder is placed half of its diameter above the bottom ($z_0 = 0$). In the third series, the cylinder lays at the bottom of the channel ($z_0 = 0.5$D). In the fourth series, there is a gap between the bottom of the channel and the cylinder equal to the diameter of the cylinder ($z_0 = 1.5$D). Photographs of the cylinder in the last three series are shown in Figure 3.

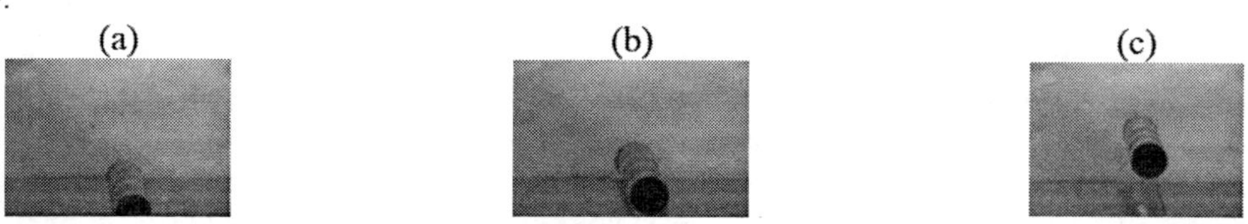

Figure 3. Photos of the cylinder at various position: cylinder rises above the bottom of the channel by half its diameter, $z_0 = 0$ (a); the cylinder lies at the bottom of the channel $z_0 = 0.5D$ (b) and a gap under the cylinder is equal to its diameter $z_0 = 1.5D$ (c).

In each series, four values of the average rate are set as U= 0.14; 0.17; 0.21 and 0.24 m/s. Speed field data are obtained in the following ranges: $-0.7 \le x^0 \le 0.5$; $\delta^0 \le z^0 \le 0.87$

3. Experimental results and discussions

Figure 4 shows the profiles of the longitudinal velocity component u^0 (z^0) in the cross section of the channel x^0 = 0 in the absence of a cylinder, and with the above three positions of its axis. For these profiles, the values of the Reynolds and Froude criteria are $2.04 \cdot 10^4$ and 0.025, respectively. In the absence of a cylinder, at the indicated Re value the velocity has turbulent profile. In the presence of a cylinder, the no-slip condition takes place on its surface, and immediately behind the cylinder a stagnant zone is formed, outside of which the speed increases significantly.

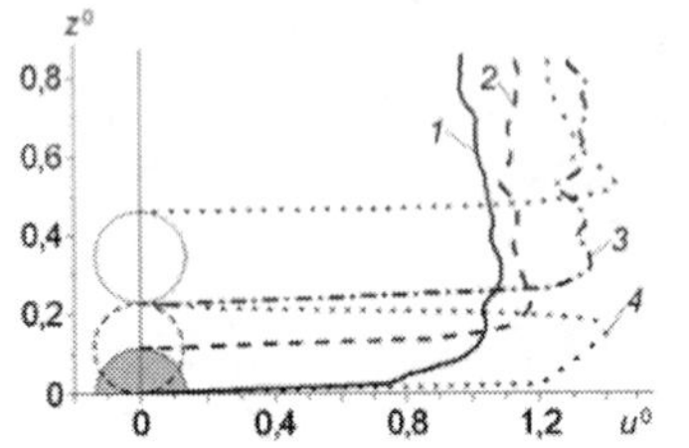

Figure 4. Longitudinal velocity profiles u^0: h = 0.12 m; U = 0.17 m/s; x^0 = 0; no cylinder (1); z_0 = 0 (2); z_0 = 0.5D (3) and z_0 = 1.5D (4).

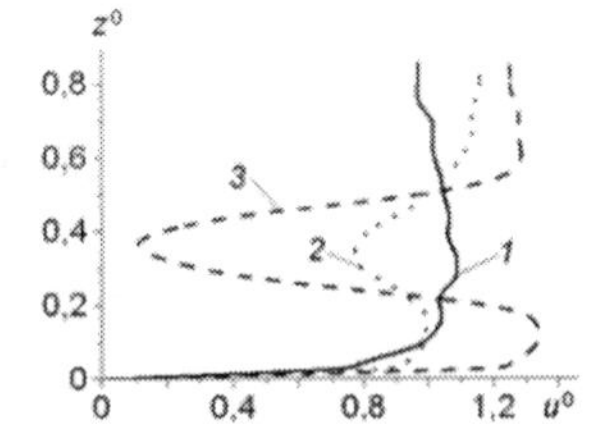

Figure 5. Longitudinal velocity profiles u_0: h = 0.12 m; U = 0.17 m/s; 1 – no cylinder, x^0 = 0 (1); z_0 = 1.5D, x_1^0 = –0.23 (2); z_0 = 1.5D, x_2^0 = 0.35 (3)

In terms of sediment transport, speed near the bottom is of greatest interest. The data given in Figure 4 show that in the case of a cylinder with a gap equal to D (profile 4 in Figure 4), the velocity at $z^0 = \delta^0$ increased by 1.7 times compared to the case when there is no cylinder in the flow (profile 1 in Figure. 4). In the case of a washable bottom, an increase in speed near the bottom leads to a more intensive sediment removal. When the cylinder rises above the bottom, a stationary vortex and a stagnant zone are formed behind the cylinder, in which sediment deposition can occur. When the gap exists under the cylinder, the vortices are shedding from either one or other sides of the cylinder. Similar picture known as the Kármán vortex street is observed in a free flow [9].

At a certain distance behind the cylinder, a turbulent hydrodynamic wake forms, the averaged speed of which is less than the speed of the incoming flow. A decrease in the averaged local velocity also occurs in front of the cylinder. The corresponding illustrations for the case of the arrangement of the cylinder with a gap equal to D are shown in Figure 6 to compare the velocity profiles at x_1^0 = - 0.23 and at x_2^0 = 0.35 with a velocity profile in the case without a cylinder.

As in cross section x^0 = 0 (see profile 4 in Figure 4), the velocities and their gradients near the bottom are greater than those in the absence of a cylinder. At a distance from the bottom $z^0 = \delta^0$, the velocity magnitude increased by 1.28 times in front of the cylinder, and by 1.74 times behind the cylinder.

The hydrodynamic pressure on the bottom and sediment transport are substantially dependent on the presence of a vertical velocity component in the flow. In the absence of a cylinder, the vertical component of the averaged velocity in a uniform flow is zero. With a cylinder in its vicinity, there are significant vertical components of the averaged velocity. The results of measuring the profiles of the longitudinal $u^0(z^0)$ and vertical $w^0(z^0)$ components of the averaged velocity for various combinations of the given parameters are presented in Figure 6. Positive values of x^0 correspond to cross sections located behind the cylinder, negative values of x^0 correspond to cross sections in front of the cylinder. The value x^0 = 0 corresponds to the cross section of the flow passing through the axis of the cylinder.

The data given in Figure 7 show that in the presence of a cylinder, the absolute value of the horizontal component of the averaged velocity exceeds 1.6 times the average consumption speed at individual points of the flow, and the vertical component reaches 50% of the average consumption speed. Noteworthy is the fact that at short distances in the wake of a cylinder u^0 may be negative (see profiles 1, 2, 3 in Figure 6 and 7). Since measurements by a hot-wire anemometer are carried out only at a sufficiently large distance behind the cylinder, the presence of negative velocities in the wake of the cylinder was not detected earlier. This flow feature in the near wake, detected by the PIV method, is due to the influence of large eddies shedding from the cylinder.

Figure 6. Profiles of the longitudinal u^0 (a) and vertical w^0 (b) velocity components: $z = 0$; $h = 0.12$ m; *1* – $x^0 = 0.35$; *2* – $x^0 = 0.23$; *3* – $x^0 = 0.175$; *4* – $x^0 = 0$; *5* – $x^0 = 0.175$; *6* – $x^0 = -0.23$; *7* – $x^0 = -0.35$
$x^0 = x/h$; $U = 0.14$ m/s.

Figure 7. Profiles of the longitudinal *u*0 (a) and vertical *w*0 (b) velocity components $U = 0.17$ m/s. Designations are the same as in Figure 6.

In the presence of a sufficiently large gap between the cylinder and the bottom of the channel, as well as between the cylinder and the free surface, the structure of the vortex motion in the near wake of the cylinder is qualitatively similar to that observed in a free flow around the cylinder [10]. In the case of free flow, instead of the Re criterion introduced above, the criterion $Re_1 = U \cdot D/\nu = D \cdot Re$ is used, in which the cylinder diameter is used as a characteristic linear dimension. In a free flow, at low Re values ($Re_1 < 1$), the flow is attached to the cylinder surface. When $Re_1 > 5$, the flow separets, and a structure with two stable vortices with the opposite direction is formed behind the cylinder. Such a structure is retained in the range $5 < Re_1 < 50$.

In the range $Re_1 > 60$, the vortices behind the cylinder are no longer stable, but they are alternately shedding from the one or the other side of the cylinder. In the range $60 < Re_1 < 5000$, at some distance behind the cylinder, there is the Karman vortex street [8]. With an increase in Re_1, starting from the lower boundary of the range of existence of the vortex street, Strouhal number $St = n \cdot D/U$, where n is the vortex shedding frequency (Hz), increases. At $Re_1 > 600$ St becomes constant and equal to 0.212.

Even in the indicated range of values of Re_1, the classical vortex strees is stable only at a limited distance behind the cylinder. With the growth of Re_1 beyond this range, destructive tendencies intensify, and the hydrodynamic wake becomes turbulent already at a small distance behind the cylinder. In our experiments, the values of Re_1 were set from the range $3900 \le Re_1 \le 6700$, i.e. either corresponded to the region of existence of the Karman classical vortex path, or only slightly exceeded the upper boundary of this region.

4. Computational problem setup and numerical methods

The process of reforming the bottom soil in the tray is considered numerically. The tray of 130 cm length has a rectangular cross-section with a width of 10 cm, and a height of 19 cm. Sand is evenly distributed at the bottom of the tray. The particle diameter (d = 0.05 mm) is chosen so as to exclude the occurrence of particle removal in the absence of a cylinder. On the sandy bottom the cylinder of diameter D = 2.5 cm is placed fixed rigidly to the side walls of the tray. This allows to simulate real conditions when the cylinder is stationary along the underwater transition path and rests on a surface formed by sand.

Two cylinder arrangements on the sand surface are considered named as Experiment 1 and Experiment 2 (Figure 8). Numerical modeling is implemented in the ANSYS Fluent software based on the Reynolds averaged Navier-Stokes equations, supplemented by the standard k-ε turbulence model and the equations of the Euler model of multiphase media taking into account the granularity of the particle phase. Details of model calculations in ANSYS are described in [11].

The identical boundary conditions are prescribed for the Experiments 1 and 2, namely, water mass flow rate at the inlet is set as Q = 1.2 kg / s and pressure is set as P=1 atm at the outlet.

Figure 8. Schemes of the experiments performed: Experiment 1 (a) and Experiment 2 (b).

5. Numerical simulation results and discussions

As a result of the calculation, the following fields were obtained: fields of velocities, pressures, and phase concentrations. Volumetric concentration iso-surfaces α = 0.01 are constructed for a visual representation of bottom soil reorganization at different points in time for the Experiment 1 (Figure 9) and Experiment 2 (Figure 10).

After 0.3 seconds from the beginning of the Experiment 1, the process of sand washing under the cylinder and formation of the bottom ridges are observed (Figure 9). After a few seconds, the ridge reaches the cylinder, and it is completely covered with sand. The process is very intense. Further, the ridge continues to move along the tray, the cylinder is again exposed and a cavity is formed under it.

Figure 9. Volumetric concentration α = 0.01 iso-surface for Experiment 1, t = 0.3 s.

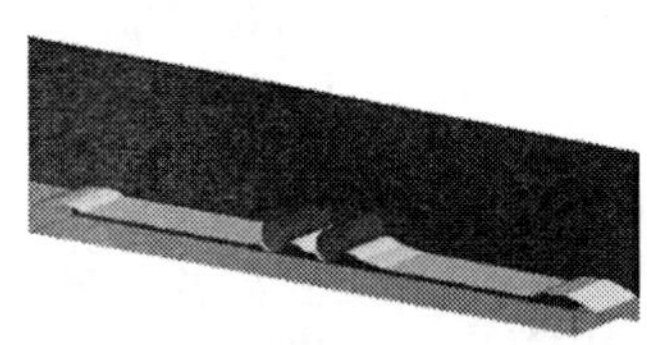

Figure 10. Volumetric concentration α = 0.01 iso-surface for Experiment 2, t = 0.1 s.

Sand erosion under the first cylinder in the Experiment 2 was recorded already after 0.1 seconds from the start of the study (Figure 10). At the tenth second, the size of the bottom ridge increases from the rear of the cylinders. The pronounced heterogeneous structure of the surface of the bottom relief confirms the general thesis that the generation of individual erosions under the cylinder is not synchronous and along the entire length of the cylinder, but locally under the influence of local small-scale disturbances. After 30 seconds from the start of the Experiment 2, the second cylinder was exposed, and under the first one, the depth of the washout funnel increased.

The following series of figures shows the formation of vortices shown by streamlines. Vortices that form in front of the pipe (Figure 11a) and between two pipelines (Figure 11b) contribute to the erosion process.

(a) (b)

Figure 11. The formation of a vortex region for Experiment 1 (a) and Experiment 2 (b).

6. Conclusions

Experimental studies of velocity fields and numerical modelling of hydro-physical processes in the area of the location of the cylinders that imitate underwater pipelines are carried out.

In the course of experimental studies using PIV equipment, it is found that the presence of a cylinder substantially changes the velocity field in a flow with a free surface. In the vicinity of the cylinder, an increase in the averaged horizontal velocity component and the appearance of an averaged vertical velocity component are recorded. Such changes should be taken into account, in particular, when analyzing the deformation of the eroded bottom. From this point of view, the most interesting experimental results are those in which there is a gap between the bottom and the cylinder. With a gap equal to the diameter of the cylinder, the longitudinal component of the averaged velocity near the bottom can 1.8 times exceeds its value in the absence of a cylinder. At the same time, the tangential stresses at the bottom increase.

Numerical experiments using ANSYS Fluent 18.2 on the basis of Reynolds averaged Navier-Stokes equations are performed. The erosion of the sandy bottom is studied at a given mass flow rate Q = 1.2 kg/s. It is found that the intensity of the reformation of the moving bottom depends on the number of cylinders located on the surface of the sand. It is revealed that when two cylinders are located on the bottom surface, the process of bottom soil reformation is accelerated.

References

[1] Degtyarev V V and Shlychkov V A 2013 *Hydraulic Engineering* **8** 21-27

[2] Degtyarev V V and Chebotnikov A V 2011 *News of Higher Educational Institutions. Construction* **10** 92-97

[3] Perezhogin Y D, Ratner A G and Spektor Y I 1998 *News of Higher Educational Institutions. Oil and Gas* **1** 81-87

[4] Mao Y 1988 Seabed scour under pipelines. *Proc. 7th Int. Conf. on Offshore Mechanics and Arctic Engineering* (Houston, TX: ASME) Vol V pp 33 – 38

[5] Shen W, Griffiths T, Zan Z and Leggoe J 2014 Shear stress amplification around subsea pipelines: Part 3. 3D study of spanning pipelines *Scour and Erosion* Proc. 7th Int. Conf. Scour Erosion (United Kingdom: CRC Press) pp 325-35

[6] Chen C and Zhang J 2009 Review on Scour Modeling below Pipelines *Pipelines 2009: Infrastructure's Hidden Assets* Ed. J J Galleher and M K Kenny pp 1019-28

[7] Strouhal V 1878 *Ann. Phys. Chem.* **5** 216-251

[8] von Karman Th 1911 *Nachrichten von der Gesellschaft der Wissenschaften zu Göttingen* 509

[9] Townsend A A 1956 *The Structure of Turbulent Shear Flow* (Cambridge Univ. Press) 315 pp

[10] Daily J W and Harleman D R F 1966 *Fluid Dynamics* (Boston, Massachusetts: Addison-Wesley Publishing Co) 454 pp

[11] Garmakova M E, Degtyarev V V, Fedorova N N and Shlychkov V A 2018 *AIP Conf. Proc.* **1939** 020037

Journal of Physics: Conference Series **1404** (2019) 012014 doi:10.1088/1742-6596/1404/1/012014

Physical model of a heterogeneous medium as applied to nuclear medicine

N V Denisova

Khristianovich Institute of Theoretical and Applied Mechanics, Siberian Branch of Russian Academy of Sciences, 4/1 Institutskaya st., Novosibirsk, 630090 Russia

denisova@itam.nsc.ru

Abstract. In nuclear medicine, a patient is administered a radiolabeled pharmaceutical which is distributed in various organs of a patient body. The patient body with distributed emitting radiopharmaceutical is a heterogeneous medium. In this work, the problem of reconstructing the image of the human brain using single-photon emission tomography (SPET) is considered. To obtain the image, the statistical MAP-ENT algorithm is applied. To develop this algorithm, it is necessary to formulate a priori information about the object to be reconstructed. The distribution of radiopharmaceuticals in the patient's body is a spatially inhomogeneous distribution of non-interacting radiopharmaceutical particles. This distribution is described using the Boltzmann model of a non-equilibrium isolated system. For such a system, one can write the entropy functional, which determines the a priori model. The algorithm was tested using the Hoffman mathematical model, which simulates the distribution of a radiopharmaceutical in the human brain. In numerical experiments, the properties of the MAP-ENT algorithm were studied under conditions close to clinical practice. The results have shown the prospects of the MAP-ENT algorithm.

1. Introduction

In nuclear medicine, namely single-photon emission computed tomography (SPECT) and positron emission tomography (PET), a patient is administered a radiolabeled pharmaceutical which is distributed in various organs of the patient body. The photons emitted by the radiopharmaceutical are collected by detectors located around the patient. Once the data have been collected, the reconstruction algorithm is applied to provide an image of radionuclide distribution within the patient body. Image reconstruction belongs to the class of inverse ill-posed problems of mathematical physics. The method of solution for this class of problems was developed by A.Tikhonov and was called as regularization [1]. This method was developed in deterministic form. At the moment, the Bayesian approach Maximum a Posteriori (MAP) is used for solving inverse ill-posed image reconstruction problems with stochastic data [2]. The MAP approach is based on statistical regularization which introduces a priori information in the probabilistic form. Within a framework of the Bayesian approach one needs to develop an *a priori* model. When solving reconstruction problems, one can propose the physical model of the object to be reconstructed and determine a priori information by using this model. We consider a cumulative model: it is assumed that the radiopharmaceutical has already accumulated in various organs and that this distribution remains constant during the imaging procedure. Radiopharmaceutical distribution in a patient body is a heterogeneous medium. In the present paper, we discuss the prior model to describe this medium in context of its applications in nuclear medicine.

2. Theory

Radiopharmaceutical distribution in a patient body is the spatially non-uniform distribution of non-interacting radiopharmaceutical particles. This distribution can be described using the Boltzmann model of non-equilibrium isolated system. We consider the distribution of particles in phase space. The entropy functional for such system can be presented as:

$$S = -\int n(p,q)\ln n(p,q)dpdq \quad (1)$$

where $n(p,q)$ is the particle density in the phase space, p and q are pulses and coordinates of the particles. We replace the variables:

$$dpdq = dq2\pi(2m)^{3/2}\varepsilon^{1/2}d\varepsilon \quad (2)$$

where ε is particle energy. Entropy functional (1) can be rewritten as:

$$S = -\sum_j \{\int n(q_j,\varepsilon)\ln n(q_j,\varepsilon)2\pi(2m)^{3/2}\varepsilon^{1/2}d\varepsilon + const\} \quad (3)$$

We assume that an equilibrium energy distribution is established in each voxel q_j. In the equilibrium state, particles have mean energy ε_0. Eq.(3) can be written:

$$S = -\alpha\sum_j n(q_j,\varepsilon_0)\ln n(q_j,\varepsilon_0)\varepsilon_0^{1/2}\Delta\varepsilon + \text{const} \quad (4)$$

$\Delta\varepsilon$ is the energy 'width' where the particle spends almost all of its time. Taking into account the normalization

$$\int \rho(\varepsilon)d\varepsilon = \rho(\varepsilon_0)\Delta\varepsilon = 1 \quad (5)$$

one obtains:

$$S = -\frac{\alpha}{\rho(\varepsilon_0)}\sum_j n(q_j)\ln n(q_j) + const\ , \quad (6)$$

where $\rho(\varepsilon_0)$ is the probability density of the energy ε_0. The mean density of emitting gamma photons $\bar{f}_j$ is assumed to be proportional to the radiopharmaceutical concentration:

$$\bar{f}_j \propto n(q_j) \quad (7)$$

The entropy-based prior (6) is rewritten as:

$$\ln P(\bar{f}_j) = -\beta\sum_j \bar{f}_j \ln \bar{f}_j \quad (8)$$

The Bayesian MAP estimation (in logarithmic form) gives the solution of the reconstruction problem:

$$\tilde{\bar{f}} = \max_{\bar{f}>0}\{\ln P(\bar{f}) + \ln P(g \mid A\bar{f})\} \quad (9)$$

Journal of Physics: Conference Series **1404** (2019) 012014 doi:10.1088/1742-6596/1404/1/012014

$P(\bar{f})$ is a priori probability density function and $P(g \mid A\bar{f})$ is the likelihood distribution of the observed data. The likelihood term $P(g \mid A\bar{f})$ is determined by the physics of data acquisition process. For applications in nuclear medicine, the likelihood distribution in logarithmic form is defined by the Poisson law:

$$\ln P(g \mid A\bar{f}) = g \ln A\bar{f} - A\bar{f} - \ln g! \tag{10}$$

Substitution of the entropy prior (8) and condition probability (10) in (9) gives the Maximum a Posteriori estimation with entropy-based prior (MAP-ENT):

$$\tilde{\bar{f}} = \underset{\bar{f} \geq 0}{\arg\max} \{-\beta \sum_{j=1}^{J} \bar{f}_j \ln \bar{f}_j + \sum_i g_i \ln \sum_j A_{ij} \bar{f}_j - \sum_j A_{ij} \bar{f}_j - \ln g_i !\} \tag{11}$$

Eq.(11) gives a solution of the reconstruction problem.

3. Numerical simulations

Numerical simulations using MAP-ENT algorithm were performed. To evaluate the method of iterative statistical regularization, the Hoffman brain phantom was used in this work. The 3D Hoffman phantom is shown in Fig.1. The image was sampled in 100*100*50 voxels. The two-dimensional (2D) projection data were generated for this phantom by using 120 views with 100*50 bins per projection. Data statistics was modeled as close as possible to the real clinical data. The method developed in [3] was used to generate realistic projection data taking into account the effect of collimator-detector response and Poisson data statistics. Fig.2a shows an example of the generated projections from one selected view.

Figure 1. The 3D Hoffman phantom. Human brain model developed for nuclear medicine applications.

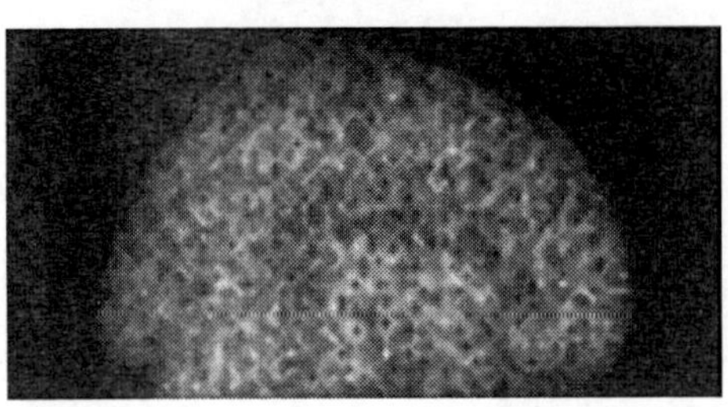

Figure 2. An example of projection data, generated ffom the Hoffman phantom.

The the MAP-ENT algorithm (11) was applied to image reconstruction using the generated projection data. The results of reconstructions are shown in Fig.3. The ‘exact’ and reconstructed images of the Hoffman brain phantom are shown in selected slices. Comparison of the ‘exact’ images of the Hoffman phantom and images reconstructed by the MAP-ENT algorithm, one can see that some fine structures are lost on the reconstructed images.

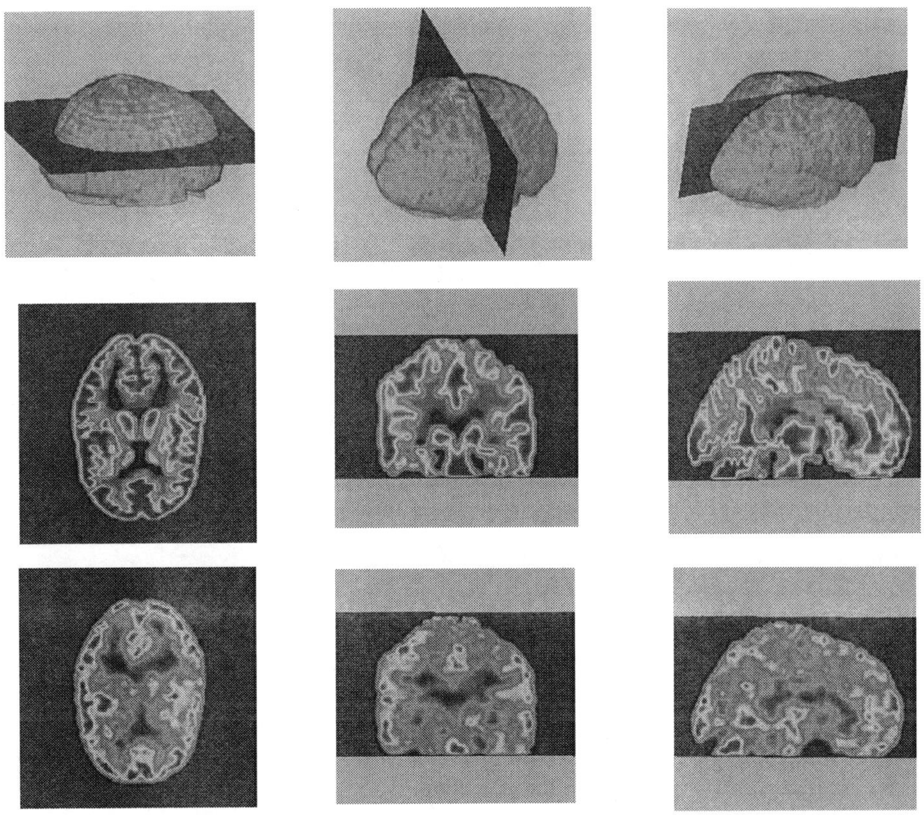

Figure 3. First row: the selected slices of the Hoffman phantom are shown; second row: images in the corresponding slices; third row: reconstructed images in the same slices.

4. Conclusion

A model of a heterogeneous environment, such as the human body, is developed for applications in nuclear medicine. Numerical studies were performed using the Hoffman phantom, imitating the structure of the human brain. The MAP-END algorithm has been applied to reconstruct this complex structure. The results obtained showed the prospects of the method and the need for its further development.

Acknowledgements

This work was partly supported by Russian Foundation for Basic Research (grant No. 19-02-00244).

References

[1] Tikhonov A N 1963 *Sov. Math. Doklady* **4** 1035-1038
[2] Turchin V F 1967 A solution of Fredholm equation in statistical ensemble of smooth functions *J. Comput. Mathem. Math. Phys*. **6** 1270
[3] Denisova N V and Terekhov I N 2016 *Biomed. Phys. Eng. Express* **3** 055015

Journal of Physics: Conference Series **1404** (2019) 012015 doi:10.1088/1742-6596/1404/1/012015

Experimental investigation of decreasing porosity and permeability of bead packing at suspension flow

S V Dimov

Institute of Thermophysics SB RAS,
1 Acad. Lavrentiev Avenue, Novosibirsk, 630090, Russia

dimov@itp.nsc.ru

Abstract. This study presents the experimental results of decreasing bead packing porosity and permeability during the flow of a suspension through the bead packing. The retention of solid particles was investigated as a function of the ratio of the particle diameter *d* to the diameter of the balls of packing *D*. The experiments were carried out for various ratios *d/D*, which varied from 0.046 to 0.109. The solids content of the inlet suspension varied from 0.001 to 0.02.

1. Introduction

Particle transport in porous media is an important process in the hydrogeological environment as well as in engineered systems. Particle deposition in porous media has been a research topic for both petroleum industry and water treatment industry. It is a complex process due to the complex nature of porous media and the properties of injected particles and fluids. The process of retaining particles of the suspension depends on the size of particles, the channels of the porous medium; physicochemical properties of particles and surface of a porous medium [1, 2, 3]. Retention leads to an increase in hydrodynamic resistance to fluid flow in a porous sample. In [1] two main mechanisms of colmatation (particle retention in a porous medium) were distinguished: mechanical filtration for large particles (diameter $d > 30$ μm), and a physicochemical filtration for small particles (diameter of about 1 μm)-for mean particles (3 μm < diameter < 30 μm), both mechanical and physicochemical phenomena intervene. Later [3, 4], a third retention mechanism of colloid deposition in saturated porous media has commonly been identified during the filtration of suspensions -straining (retention in narrow pore throat).

Bead packing and sandpack are simple models of porous medium, used by many researchers to investigate filtration of suspension, particle movement and deposition [2, 4, 6, 8-12]. Physicochemical filtration theory [1, 3-6] has been developed and is now often used to characterize colloid attachment in porous media. The influence of *d/D* for small (colloid) particle was investigated in [4-6]. Mechanisms of deposition for large particles have been determined in [8-10] in dependence of the ratio of suspension particle diameter on diameter of pack bead. They [10] found that for $d/D > 0.1$, the porous medium is irreversibly blocked and a filter cake is formed; for $d/D < 0.05$, the retention remains always low and for the intermediate values a partial blocking of the porous bed may occur (this depends on particle shape and bed porosity). Experimental studies of deposition of ~6 μm particles have been conducted in bead packing [11, 12] and in core samples [7, 13].

The aim of this work is to present the results of an experimental study on the change in porosity and permeability of monodisperse packing of balls, when a suspension of large particles (d>30 μm) of approximately the same size flows in it.

2. Experimental Equipment.

The retention of solid particles in the bead packing during a suspension flow through it under the controlled conditions was studied experimentally on the setup described in [9]. The setup (see Fig. 1) included a working section packed with glass beads, measuring graduated cylinders, differential pressure sensors, a computer, ADC, tank constant level with a mixer, inlet and outlet tubes, and webcamera.

To study particle deposition within a packed bed, a well defined granular material, i.e. spherical glass beads is packed in the column to form a porous bed with a fixed porosity. The porosity of bead packing was determined from the geometrical parameter of the working section (its length, diameter), mass of glass beads, and their density. Then the bead packing was filled under vacuum with water, followed by its displacement with a water-glycerin solution. During the flow of a water-glycerin solution, the permeability of the bead packing was determined. After preliminary measurements of permeability, the test suspension was injected into a bottom of vertical laboratory column packed with glass beads. The temperature of the suspension was monitored and kept constant during the injection process.

In the process of filtering the suspension through a porous medium, differential pressure was measured by differential strain gauges. This made it possible to determine the local change in permeability along the length of the porous medium. Preliminary calibration of the sensors showed a linear dependence of the output signal on the pressure drop across them.

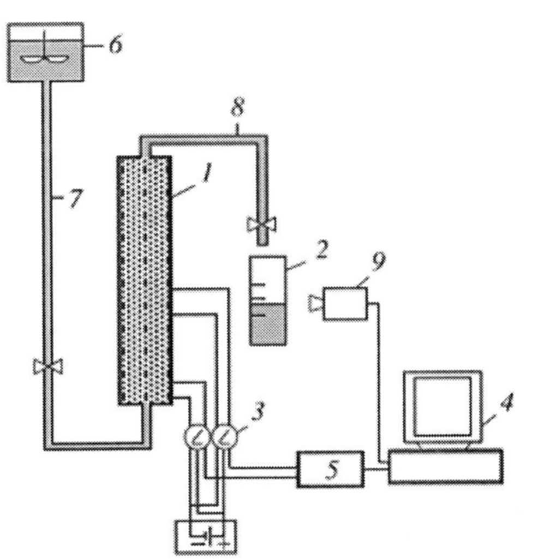

1-bead pack, 2 - measuring cylinder; 3 - differential pressure transducers; 4 - computer; 5 - analog-digital converter; 6 - constant-level tank with a mixer; 7,8 - supply and removal tubes; and 9 - web camera.

a

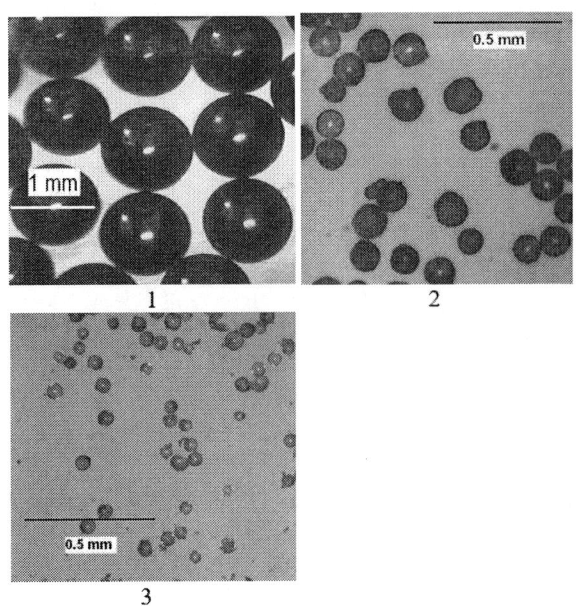

b

Figure 1. Schematic diagram of the glass pack setup–a. Pictures b show glass beads-1; aluminosilicate spheres -2 (100-140 μm), -3 (50-80 μm).

The signal from the sensors passed through the ADC to the computer. The change in the volumetric flow rate of the suspension was recorded by a video camera to fill 250 ml measuring cylinders with liquid at the outlet of the working section.

The measured flow rate and pressure drops determined the average and local permeability of the bead pack. In the experiment, the number of microspheres passed through the packing in each half-liter portion of the filtrate was determined. This was accomplished by isolating the microspheres in each portion of the outlet suspension, using the density difference between the water-glycerin solution and the aluminosilicate microspheres. After thorough washing in hot water and separation of microspheres on a fine sieve (40 μm), they were dried and weighed. After the experiment, the working section with bead packing was disassembled so that the volumes of the porous medium located between the pressure sensors were distinguished. These volumes were washed in separate containers. When washing, particles of the suspension floated in water, and balls of packing fell to the bottom. The separated suspension particles were washed, separated on a sieve, dried, and weighed.

3. Experimental Results

Figure 2 represents the average permeability of bead packing during suspension flow of 80-100 μm aluminosilicate spheres in water-glycerin solution at two pressure drops and three initial solid content. Received data show that permeability is determined by the total volume of suspension particles $V_{sus} \cdot C_{in}$. This result was obtained in [9] for local permeabilities of bead packing. Here you can see insignificant influence of initial velocity on average permeability. The maximum filtration velocity of suspension in the experiments was 0.26 cm/s. Unlike filtration of colloidal suspensions [2], the initial velocity has little effect on the change in the permeability of the bead packing. Symbols 2, 4 on figure 2 corresponds 60% increased initial filtration velocity of suspension of 80-100 μm aluminosilicate spheres.

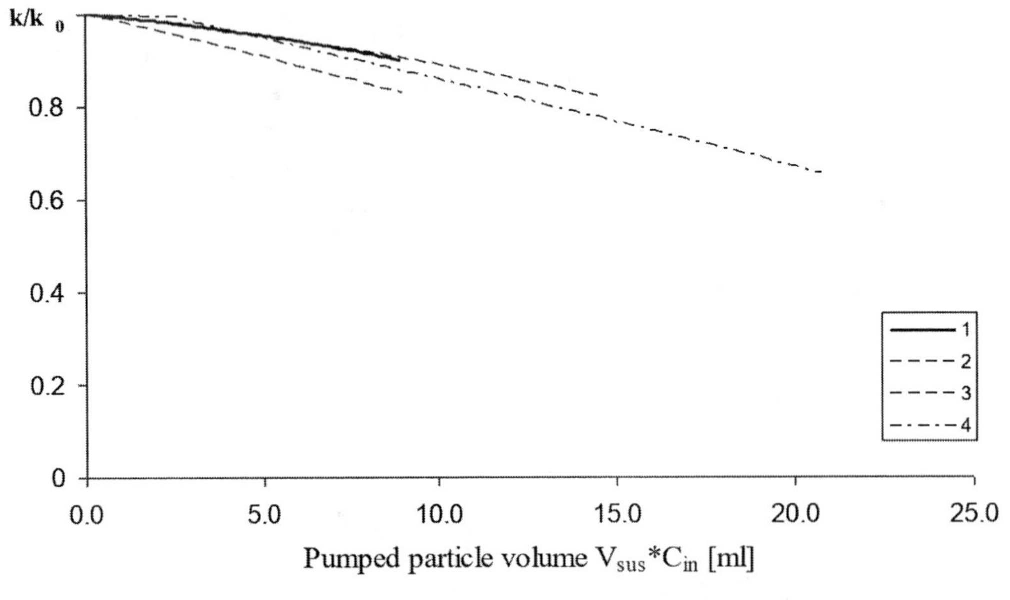

Figure 2. The normalized permeability of the bead packing depending on the volume of particles received with the suspension of 80-100 μm microspheres. Lines 1, 2 correspond to 0.3% solid content suspension, 3 -0.5%, 4 0.7%. Lines 1, 3 correspond to 14 kPa pressure drop, 2, 4 ones correspond to 22.5 kPa pressure drop.

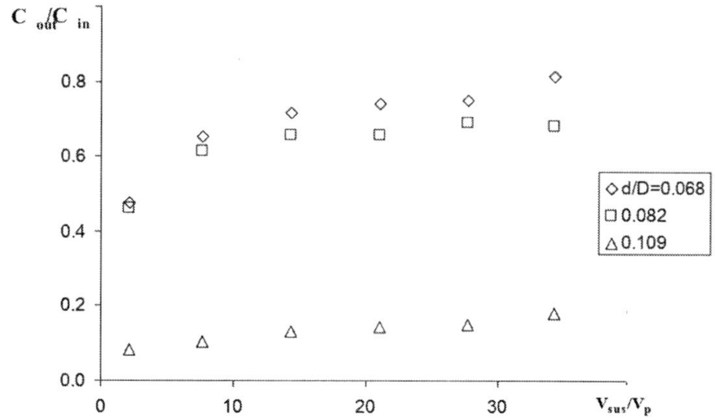

Figure 3. The normalized solid phase concentration of outlet suspension for different ratios of diameter of particle suspension to diameter of glass bead versus pumped suspension volume. Inlet solid content is 0.3%.

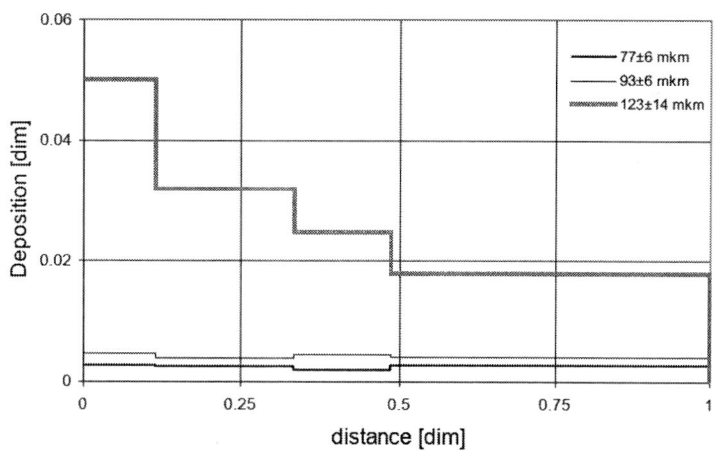

Figure 4. A histogram of the distribution of retained particles along the bead packing after flow of 0.3% suspension of aluminosilicate particles with different sizes.

The value of retained particles can be determined from data on outlet solid phase content. Figure 3 represents the outlet solid phase content during the flow suspension of different sizes 77.3 ± 6, 92.7 ± 6, 123.0 ± 12 μm. These sizes correspond to 0.068, 0.082, 0.109 ratio diameter of suspension particle to diameter of bead pack respectively. An increase in particle sizes gives in result the lower outlet content of solid particles.

Figure 4 depicts deposition profiles after 0.3 % solid content suspension flow of different size particles. The profile of maximal size particles corresponds to classical solution of deep bed filtration theory. The profiles of other retained particles do not correspond to the theory. It concerned with small quantity of deposition.

Figure 5 depicts normalized permeability of bead packing versus the value of retaining particles. The reduction of permeability is determined with the value of deposition. That parameter does not depend on the size of retained particle, as in [13], where experiments were conducted with the cores.

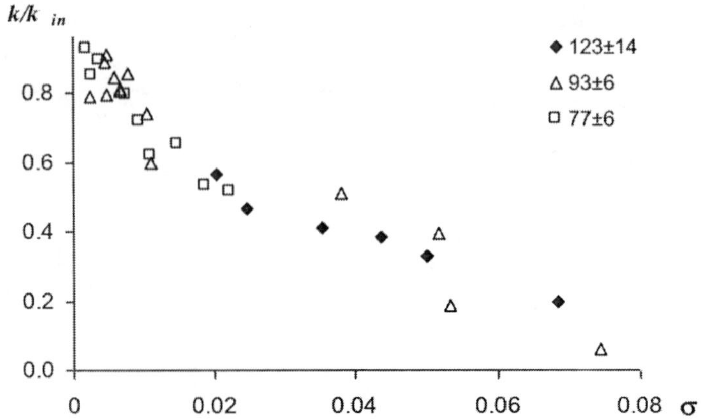

Figure 5. Normalized permeability of bead packing versus volume fraction of retained particles. Different symbols correspond to various sizes of suspension particles. Here is the volume of deposited particle per unit filter volume.

4. Summary

The suspension flow through bead packing was investigated experimentally for different ratios of the diameter of suspension particle to diameter of bead pack. It was found that the permeability reduction was uniquely correlated with the amount of particles retained within the specimen independent of the injected influent concentration, injected velocity, injected fluid volume, and particle deposition profile. Local permeability of bead pack was determined with the volume of retained particles.

Acknowledgement
This research was supported by the state assignment to the IT SB RAS (AAAA-A18-118112790037-0).

References

[1] Herzig J P, Leclerc D M and Goff P 1970 *Industrial and Engineering Chemistry* **61**(5) 8
[2] Gruesbeck C and Collins R 1982 *SPEJ* **22**(6) 847
[3] McDowell-Boyer L, Hunt J and Sitar N 1986 *Water Resourses Res.* **22** 1901
[4] Bradford S, Simunek J, Bettahar M, Genuchten M and Yates S 2006 *Wat Res Res* **42** W12S15
[5] Tufenkji N, Redman J and Elimelech M 2003 *Environ. Sci. Technol.* **37** 616
[6] Xu S, Gao B and Saiers J 2006 *Water Resources Res.* **42** W12S16
[7] Al-Abduwani F, Shirzadi A, van den Broek W and Currie P 2005 *SPEJ* **10**(2) 138
[8] Maroudas A and Eisenklam P 1965 *Chem Eng Sci* **20** 867
[9] Dimov S, Kuznetsov V, Rudyak V and Tropin N 2012 *Fluid Dynamics* **47**(2) 178
[10] Sakthivadivel R 1966 *Rep. HEL* **15-5** (Berkeley: Univ. Calif. Press.) 110 pp
[11] Moghadasi J, Muller-Steinhagen H, Jamialahmadi M and Sharif A 2004 *J Pet Sci Eng* **43** 163
[12] Ahfir N, Benamar A, Alem A and Wang H 2009 *Transp Porous Med* **76** 289
[13] Wang R and Mettananda D 2010 *Tranp Porous Med* **81** 105

Journal of Physics: Conference Series **1404** (2019) 012016 doi:10.1088/1742-6596/1404/1/012016

Gasdynamic Structure of Dispersed Flows in Two-Stage Air Purifiers

I H Enikeev

The Moscow Polytechnical University 107023 Moscow, Russia

enickeev.iX@yandex.ru

Abstract. The work deals with numerical modelling of four-phase flows in a nozzle of two-stage air purifier. The aerodynamic structure of a gas-dispersed flow is studied in a wide range of determining parameters. The flow regimes are considered in the first stage nozzle with a high dust concentration.

1. Introduction

Protecting the environment from industrial emissions is a priority mission of the world. In this regard, improving the design of dust collecting equipment is one of the urgent tasks of modern science and technology. Among the existing methods for cleaning the emissions from suspended particles, the most common is the method of dry cleaning which can be successfully used for processing dust and gas flows with a high concentration of particles.

Despite the successes achieved in the theory of multiphase flows, there is still no single approach to adequately describe the complex gas-dynamic flows in dust separators in a wide range of the determining parameters. This work is aimed at solving this problem. In this paper, gas-particle flows are considered on the basis of the interpenetrating continua (Euler-Euler) approach [1]. Within this approach, the general system of equations is considered without any simplifying assumptions about the flow structure in the region under consideration. Analytical solution is only possible to get for a narrow range of problems that do not describe the whole picture. Therefore, the general problem can only be solved numerically. In the paper, the method of large particles [2] is used for numerical simulations.

2. Problem statement

Let us consider a multiphase flow of a gas and mono dispersed solid particles in a nozzle of the first stage of a two-stage inertial separator with an input section of radius R (Figure 1).

As the primary phase, we consider the carrier gas. The secondary phase is the particle fraction supplied at the inlet section of the nozzle of the first stage. The third phase is the fraction of particles reflected by the side nozzle surface and the fourth phase is the fraction of particles reflected by the internal separator baffles located along the nozzle axis of the first and second stages. Collisions take place between particles of different fractions causing the interphase momentum exchange, so it is necessary to take into account the effective interaction force between particles from different phases. It should be noted that particle collisions can lead to randomization of particle motion, and hence the additional terms appear in the conservation law equations for particle momentum and energy. We suppose that the particle randomization effect is not significant for this type of flow, so corresponding term is not included in the governing equations.

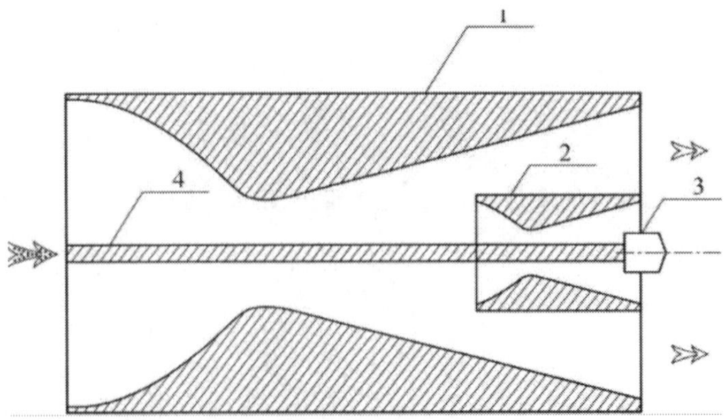

Figure 1. The operation scheme of two-stage inertial separator with a central dust extraction.
1 is the building of the first stage, 2 is the building of the second stage,
3 is the dust-collecting bag and 4 is the partition.

3. Mathematical model

The problem is considered in the cylindrical coordinate system (x, Y, φ), the direction of x-axis OX coincides with the nozzle axis. The governing equations within the framework of the interpenetrating continua (Euler-Euler approach) are as follows:

$$\frac{\partial \rho_i}{\partial t} + div\ \rho_i \vec{v}_i = 0$$

$$\frac{\partial \rho_i \vec{v}_i}{\partial t} + \nabla^k(\rho_i v_i^k \vec{v}_i) = (\delta - 1)\nabla p + \vec{f}_{ij}$$

$$\sum_{i=1}^{4}\left[\frac{\partial \rho_i E_i}{\partial t} + div(\rho_i E_i + (1-\delta)p)\vec{v}_i\right] = 0 \qquad (1)$$

$$\frac{\partial \rho_i e_i}{\partial t} + div \rho_i e_i \vec{v}_i = q_{1i} + \vec{f}_{ij}(\vec{v}_i - \vec{v}_j)$$

where $\delta = \begin{Bmatrix} 0, i=1 \\ 1, i \neq 1 \end{Bmatrix}$, the subscript i, j = 1, 2, 3, 4, $i \neq j$ refers to the parameters of the gas and the corresponding fractions of particles, respectively; $\rho_i, \vec{v}_i, e_i, E_i$ are the effective density, velocity, inner and the total energy of the i-th phase, p is the pressure in the gas, $\vec{f}_{ij}$ is the power intensity of interaction between phases, and q_{1i} is the heat transfer between gas and particles of different fractions.

As shown in [1] for $\frac{\rho_1^0}{\rho_2^0}$ <<1, the main contribution to the expression for the force interaction of the phases is made by the friction force between the gas and the particles, having the form:

$$\vec{f}_{ij} = \frac{0.75\rho_1^0 \rho_2 C_{di}|\vec{v}_1 - \vec{v}_i|(\vec{v}_1 - \vec{v}_i)}{\rho_2^0 d^2}\psi_{\alpha_i},\ C_{di} = \frac{24}{\mathrm{Re}_{1i}} + \frac{4}{\sqrt{\mathrm{Re}_{1i}}} + 0.4,$$

$$\mathrm{Re}_{1i} = \frac{\rho_1^0|\vec{v}_1 - \vec{v}_i|d}{\mu_1},\ \psi_{\alpha_i} = (1-\alpha_i)^{-2.7},\ \alpha_i = \frac{\rho_i}{\rho_i^0},$$

IOP Publishing
1404 (2019) 012016 doi:10.1088/1742-6596/1404/1/012016

$$q_{1i} = \frac{6\rho_i \lambda_1 Nu_{1i}(T_1 - T_i)}{\rho_2^0 d^2}, \quad Nu_{1i} = 2 + 0.6\,\mathrm{Re}^{\frac{1}{2}}\,\mathrm{Pr}^{\frac{1}{3}}$$

Here ρ_i^0 is the true density phase; C_{di} ,Re_{1i} , Nu_{1i} are the aerodynamic drag coefficient, Reynolds number and Nusselt number of relative flow around the particle of the i-th phase, respectively; Pr is the Prandtl number, μ_1 and d are the dynamic viscosity coefficient of gas and particle diameter. The expression for the effective interaction force between particles of different factions of $\vec{f}_{sl}$ the same, as in [3.4]:

$$\vec{f}_{sl} = \frac{k^{(f)}\rho_s \rho_l (\vec{v}_s - \vec{v}_l)\left|\vec{v}_s - \vec{v}_l\right|}{\beta^v}, \beta^v = \frac{\rho_2^0 d}{\rho_1^0 R}$$

Here the $k^{(f)}$ stands for the intensity of the force interaction between particles of different phases, an β^v is the degree of inertia of the particle. In [4] where multiphase flows in pipes are studied, the dependence of this coefficient on the phase velocity difference is defined. It was shown that $k^{(F)} \approx 0.1$ for gas velocity $v_1 \approx 10$ m/s. As closing relations for system (1), the equations of state of the phases were used:

$$p = \rho_1^0(\gamma - 1)e_1 ,e_1 = c_{v1}T_1,\ e_2 = c_2 T_2,$$

where γ is gas specific heats ratios; c_{v1},c_2 the specific heat of the gas at constant volume and specific heat of particles; T_i is the gas temperature.

To integrate the system (1), it is necessary to set boundary and initial conditions. It is assumed that the outlet boundary of the computational domain is located far enough from the . Then x ->- ∞ a flow is realized without dynamic (in speed) and thermal (in temperature) particle lag, with vertical components of gas and particle velocities equal to zero. In this case fulfillment of the flow homogeneity condition [5, 6] can be considered fair, i.e. assume that the condition is satisfied:

$$\frac{\partial v_1^{(x)}}{\partial x} = 0$$

It was also assumed that the mixture flow in this section is isentropic and isenthalpic, i.e., H_o=*const*, *S*=*const*. It was also believed that the reduced density of the second phase in the input section is a given value. In the calculations, these boundary conditions were carried into x= -2 section. On the side walls of the nozzle, the symmetry condition is fulfilled for gas guaranteeing the absence of flow through this boundary. For particles, the condition of normal reflection is set

$$v_s^{(\tau)} = v_l^{(\tau)}, v_s^{(n)} = -k^{(n)} v_l^{(n)}$$

where k^n is a reflection coefficient. In the nozzle outlet section, the relations obtained for the one-dimensional isentropic gas dynamic theory are used for the gas [7]. As the initial conditions, the parameters of the undisturbed flow in the cross section x = -2, with F_{1l}=0 and q_{1l}=0 are prescribed.

To integrate the system (1) in the curvilinear region, the finite-difference grid is constructed in the computational domain. A direct replacement leads to the appearance of irregular nodes (or calculated cells) near the boundaries of the layer region. In this case, to set the boundary conditions in the layer of irregular calculated cells in a method of large particles, fractional cells are introduced into consideration [8]. The practice of performing calculations using fractional cells has shown that this algorithm is rather cumbersome, especially in the case of low Mach numbers M_0<< 1. Therefore, it is more practical to introduce new variables $\xi = \xi(x,y)$, $\eta = \eta(x, y)$, in which the curvilinear region becomes rectangular. It was shown in [5, 6] that if, under such transformation, the Jacobean of the transformation $I = D(\xi, \eta)/D(x, y)$ exists and does not vanish at any point in the region, then the divergent form of equations (1) is preserved. Using new independent variables (x, ξ), where

$\xi = \dfrac{y - G(x)}{F(x) - G(x)}$; $F(x)$ and $G(x)$ are the equations of the upper and lower boundaries of the nozzle of the first stage of the separator, we get the rectangular computational domain N: $0 \le x \le 1, 0 \le \xi \le 1$.

The governing equations (1) in the variables (x, ξ) take the form:

$$
\begin{aligned}
&\frac{\partial \rho_i}{\partial t} + \frac{\partial \rho_i u_i}{\partial x} + \frac{\partial \rho_i U_i}{\varepsilon \partial \xi} = -\frac{\rho_i U_i^{\varepsilon}}{\varepsilon \xi} \\
&\frac{\partial \rho_i u_i}{\partial t} + \frac{\partial \rho_i u_i^2}{\partial x} + \frac{\partial \rho_i u_i U_i}{\varepsilon \partial \xi} = \frac{\delta - 1}{\varepsilon}\left(-\varepsilon \frac{\partial p}{\partial x} + \xi \varepsilon' \frac{\partial p}{\partial \xi}\right) - \frac{\rho_i u_i U_i^{\varepsilon}}{\varepsilon \xi} + f_{ij}^{x} \\
&\frac{\partial \rho_i v_i}{\partial t} + \frac{\partial \rho_i v_i u_i}{\partial x} + \frac{\partial \rho_i v_i U_i}{\varepsilon \partial \xi} = \frac{\delta - 1}{\varepsilon}\frac{\partial p}{\partial \xi} - \frac{\rho_i v_i U_i^{\varepsilon}}{\varepsilon \xi} + f_{ij}^{y} \\
&\sum_{i=}^{4}\left(\frac{\partial \rho_i E_i}{\partial t} + \frac{\partial \rho_i E_i u_i}{\partial x} + \frac{\partial \rho_i E_i U_i}{\varepsilon \partial \xi}\right) + \frac{\partial p u_1}{\partial x} + \frac{\partial p U_1}{\varepsilon \partial \xi} = -\frac{p U_1}{\varepsilon \xi} - \sum_{i=1}^{4}\frac{\rho_i E_i U_i^{\varepsilon}}{\varepsilon \xi} \qquad (2)\\
&\frac{\partial \rho_i e_i}{\partial t} + \frac{\partial \rho_i e_i u_i}{\partial x} + \frac{\partial \rho_i e_i U_i}{\varepsilon \partial \xi} = -\frac{\rho_i e_i U_i^{\varepsilon}}{\varepsilon \xi} + q_{1i} + \frac{1}{2}\vec{f}_{ij}(\vec{v}_i - \vec{v}_j) \\
&U_i = v_i - u_i \xi \varepsilon', U_i^{\varepsilon} = v_i + u_i \xi \varepsilon', \quad \varepsilon = F(x) - G(x)
\end{aligned}
$$

The system (2) can be written as:

$$
\begin{aligned}
&\frac{\partial \rho_i \varepsilon^2 \xi}{\partial t} + \frac{\partial \rho_i u_i \varepsilon^2 \xi}{\partial x} + \frac{\partial \rho_i U_i \varepsilon \xi}{\partial \xi} = 0 \\
&\frac{\partial \rho_i u_i \varepsilon^2 \xi}{\partial t} + \frac{\partial \rho_i u_i^2 \varepsilon^2 \xi}{\partial x} + \frac{\partial \rho_i u_i U_i \varepsilon \xi}{\partial \xi} = (1-\delta)\left(-\frac{\partial p \varepsilon \xi}{\partial x} + \frac{\partial p \varepsilon' \varepsilon \xi}{\partial \xi}\right) + \varepsilon^2 \xi f_{ij}^{x} \\
&\frac{\partial \rho_i v_i \varepsilon^2 \xi}{\partial t} + \frac{\partial \rho_i u_i v_i \varepsilon^2 \xi}{\partial x} + \frac{\partial \rho_i v_i U_i \varepsilon \xi}{\partial \xi} = (\delta - 1)\varepsilon \xi \frac{\partial p}{\partial \xi} + \varepsilon^2 \xi f_{ij}^{y} \qquad (3)\\
&\sum_{i=1}^{4}\left\{\frac{\partial \rho_i E_i \varepsilon^2 \xi}{\partial t} + \frac{\partial [\rho_i E_i + (1-\delta)p] u_i \varepsilon^2 \xi}{\partial x} + \frac{\partial [\rho_i E_i + (1-\delta)p] \varepsilon \xi U_i}{\partial \xi}\right\} = 0 \\
&\frac{\partial \rho_i e_i \varepsilon^2 \xi}{\partial t} + \frac{\partial \rho_i e_i u_i \varepsilon^2 \xi}{\partial x} + \frac{\partial \rho_i e_i U_i \varepsilon \xi}{\partial \xi} = \varepsilon^2 \xi [q_{1i} + 0.5 \vec{f}_{ij}(\vec{v}_i - \vec{v}_j)], \alpha = \gamma M_0^2
\end{aligned}
$$

We solve the system (3) in dimensionless variables introduced as:

$$
\bar{\rho}_i = \frac{\rho_i}{\rho_{10}^{0}}, \bar{u}_i = \frac{u_i}{u_0}, \bar{v}_i = \frac{v_i}{u_0}, \bar{E} = \frac{E_i}{u_0^2}, \bar{e}_i = \frac{e_i}{u_0^2}, \bar{t} = \frac{t}{t_0}, \bar{x} = \frac{x}{R}, \bar{\varepsilon} = \frac{\varepsilon}{R},
$$

where subscript "0" refers to the value of the corresponding parameters at the initial time, R is the characteristic length scale of the problem, in this case this is the input separator section radius.

In dimensionless variables, the system (3) has the following form (the upper line of dimensionless variables is omitted):

$$\frac{\partial\rho_i\varepsilon^2\xi}{\partial t}+\frac{\partial\rho_i u_i\varepsilon^2\xi}{\partial x}+\frac{\partial\rho_i U_i\varepsilon\xi}{\partial\xi}=0$$

$$\frac{\partial\rho_i u_i\varepsilon^2\xi}{\partial t}+\frac{\partial\rho_i u_i^2\varepsilon^2\xi}{\partial x}+\frac{\partial\rho_i u_i U_i\varepsilon\xi}{\partial\xi}=\frac{1}{\alpha}(1-\delta)(-\frac{\partial p\varepsilon^2\xi}{\partial x}+\frac{\partial p\varepsilon'\varepsilon\xi}{\partial\xi})+\varepsilon^2\xi f_{ij}^x$$

$$\frac{\partial\rho_i v_i\varepsilon^2\xi}{\partial t}+\frac{\partial\rho_i u_i v_i\varepsilon^2\xi}{\partial x}+\frac{\partial\rho_i v_i U_i\varepsilon\xi}{\partial\xi}=\frac{(\delta-1)}{\alpha}\varepsilon\xi\frac{\partial p}{\partial\xi}+\varepsilon^2\xi f_{ij}^y \qquad (4)$$

$$\sum_{i=1}^{4}\left\{\frac{\partial\rho_i E_i\varepsilon^2\xi}{\partial t}+\frac{\partial\left[\rho_i E_i+(\frac{1-\delta}{\alpha})p\right]u_i\varepsilon^2\xi}{\partial x}+\frac{\partial\left[\rho_i E_i+(\frac{1-\delta}{\alpha})p\right]U_i\varepsilon\xi}{\partial\xi}\right\}=0$$

$$\frac{\partial\rho_j e_j\varepsilon^2\xi}{\partial t}+\frac{\partial\rho_j e_j u_j\varepsilon^2\xi}{\partial x}+\frac{\partial\rho_j e_j U_j\varepsilon\xi}{\partial\xi}=\varepsilon^2\xi\left[q_{1j}+\frac{1}{2}\vec{f}_{ij}(\vec{v}_i-\vec{v}_j)\right]$$

$$\alpha=\gamma M_0^2$$

The system (4) is integrated numerically by the method of large particles using an implicit Euler scheme [8]. As it is shown in this paper, for flows in rectangular region with M_0<<1 at the Euler stage, it is advisable to use a time- implicit difference scheme for calculating pressure. In [3], a generalization of the method of large particles with implicit Euler stage is presented for the case of regions of a complex shape.

4. Results

The results of the calculations presented in Figures 2 and 3 show that since the mass concentration of dispersed phase in the input section of the separator is high $m_{20} = 0.2$, the dispersed phase not only affects the gas flow, but also significantly influences the flow of various particles fractions. So, for example, with an increase m_{20} due to an increase in the intensity of the interaction force with particles of the second phase, particles of the third phase are more intensively drifted towards the exit section of the nozzle, thereby reducing the region in which the fourth phase are located.

5. Conclusion

Based on the theory of interpenetrating continua, mathematical model and calculation method have been created that allow determining the characteristics of the gas-dispersed flows in areas of complex shapes. It was revealed that at high contents of the disperse phase in the input section of the inertial separator, particles that bounced off the side walls of the nozzle of the first stage significantly affect the flow structure. In this case, the streamline of both gas and dispersed fractions undergo significant changes due to the force interaction between particles of different phases.

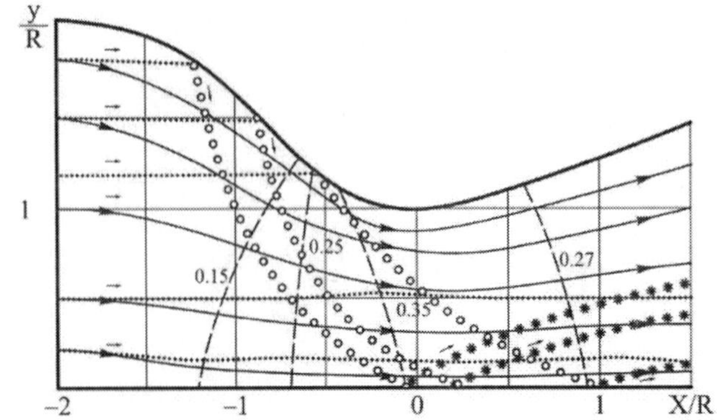

Figure 2. The flow lines gas and particles in the first stage of the first stage of the inertial separator for the case of $m_{20} = 0.2$; $M_0 = 0.05$; d= 200μm.

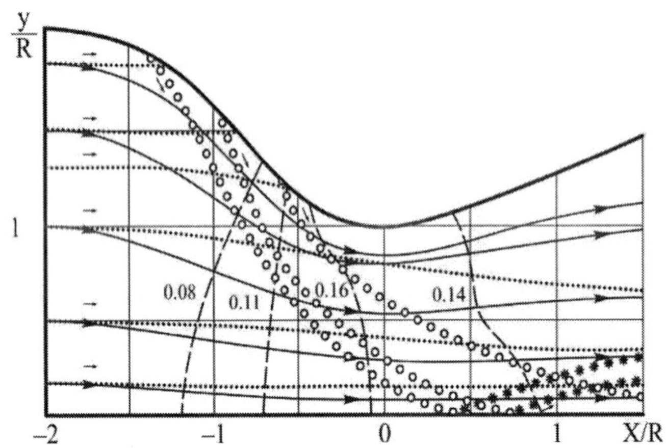

Fig.3. The gas and particles streamlines in the first stage of the inertial separator for the case m_{20} =0.5, M_0 = 0.05, d =200 μm, solid lines are for gas; dotted lines are for the second phase; circles are for the third phase; asterisks are for the fourth phase; dashed lines are the sonic lines.

References

[1] Nigmatulin R I 1987 *Dynamics of Multiphase Media* (M.: Nauka)

[2] Belotserkovsky O M and Davydov Yu M 1982 Method of Large Particles in Gas Dynamics. Computational Experiment (M.: Nauka)

[3] Enikeev I H 2009 Application of method large particles for the calculation of three-phase flow in a narrowing channel *Theor. Foundations Chem. Technol.* **43** 322 - 9

[4] Babuha T D and Schreiber A A 1972 *Interaction of Particles of Polydisperse Material in Two-Phase Flows* (Kiev: Nauk. Dumka)

[5] Rychkov A D 1976 *Mathematical Modeling of Gas-Dynamic Processes in Channels and Nozzles* (Novosibirsk: Nauka)

[6] Godunov S K, Zabrodin A V, Ivanov M Ya, Kraiko A N and Prokopov G A 1976 *Numerical Solution of Multidimensional Problems of Gas Dynamics* (M.: Nauka)

[7] Loytzyanky L G 2003 *Mechanics of Liquid and Gas* (M.: Drofa)

[8] Davydov Yu M 1981 *Differential Approximation and Representation of Difference Schemes* (M.: MIPT)

Direct numerical simulation of viscous incompressible flow with spherical particles in the flat channel

D V Esipov, V N Lapin, D S Kuranakov and D V Chirkov

Institute of Computational Technologies SB RAS, Novosibirsk, 630090, Russia

E-mail: esipov@ict.sbras.ru

Abstract. The paper deals with numerical simulation of 3D fluid flow with relatively large spherical particles in the flat channel. The simulation is performed using the previously developed SIMPLE based immersed boundary method. The collision model, describing particle-particle and particle-wall collisions, complements the numerical method. This model prevents particle overlapping and takes into account the complicated fluid flow in the space between the particles. The computed flow rates decrease significantly with increasing the particle concentration. It is shown that the value of the particle diameter has a small influence on the flow rate. The flow rate can be approximated using the value of effective viscosity computed by Maron – Peirce formula with the use of extremely high packing concentration. Particles are moving faster than the mixture, and the difference increases with increasing concentration.

1. Introduction

Flows of mixtures of viscous fluid with spherical particles often arise in technological processes. If the particle diameter is small relative to the channel width $d \ll W$, there are several appropriate approaches, in which the mixture is represented as a homogeneous fluid with effective viscosity dependent on the particles volume concentration. Specific expressions for the effective viscosity can be obtained by averaging the influence of particles and using some phenomenological assumptions [1,2,3]. Large particles significantly affect the flow field and these approaches are no longer suitable. Nevertheless, it is desirable to find an expression for effective viscosity even in this case.

In the present work, such particle-laden flows in a flat channel are computed using the direct numerical simulation approach. The mathematical model consists of the Navier – Stokes equations for a viscous incompressible fluid and the set of equations of motion and rotation for each particle. An immersed boundary method (IBM) similar to [4] is used. The fluid is considered to occupy the entire domain of the channel including the space inside the particles. The presence of particles at a given point is taken into account in the Navier – Stokes equations by adding a special force term. This force is calculated such that the no-slip condition on the immersed boundary be approximately satisfied.

2. Problem statement

The flow of a viscous, incompressible fluid in a channel with rigid particles of diameter d placed within is considered, as shown in figure 1. The fluid flow is governed by the following equations

$$\rho \frac{d\mathbf{u}}{dt} = \mu \Delta \mathbf{u} - \nabla p + \mathbf{f}, \qquad (1)$$
$$\nabla \cdot \mathbf{u} = 0,$$

where $\mathbf{u}$ is the velocity vector, p is the pressure, μ and ρ are the fluid dynamic viscosity and density respectively. Because IBM is used for the numerical solution, $\mathbf{f}$ indicates the fictitious body force that affects the fluid flow near the surface of the immersed body. The specific value for $\mathbf{f}$ is chosen in such a way that the no-slip condition on the immersed boundary is satisfied. Here, $\frac{d\mathbf{u}}{dt}$ is the total (substantial) derivative

$$\frac{d\mathbf{u}}{dt} = \frac{\partial \mathbf{u}}{\partial t} + \nabla(\mathbf{u} \otimes \mathbf{u}). \tag{2}$$

It is assumed that gravity does not affect the fluid or particles. In a certain sense, it can be considered as a zero buoyancy case.

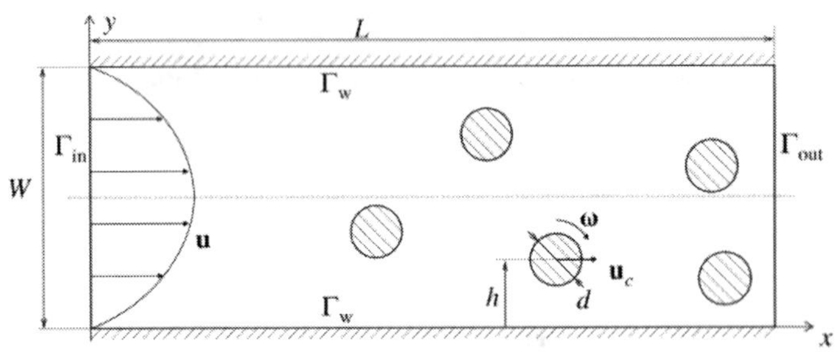

Figure 1. Problem statement for the migration of particles in the flat channel. Here, h is a distance from the boundary. z-direction is not shown for concise reasons.

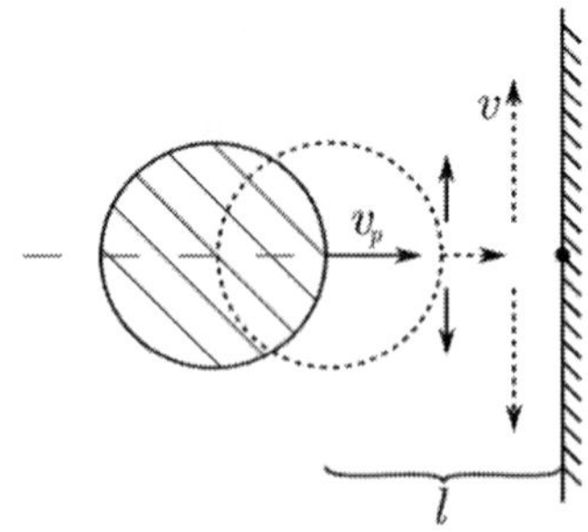

Figure 2. Motion of the particle to the wall in two successive times. The particle decelerates, while the cross-flow ahead accelerates.

Let us derive equations of the particle movement. The velocity of each point of an absolutely rigid particle $\mathbf{u}_p$ can be decomposed into a translational part and a rotational part according to

$$\mathbf{u}_p = \mathbf{u}_c + \boldsymbol{\omega} \times \mathbf{r}, \tag{3}$$

where $\mathbf{u}_c$ is the translational velocity of the particle centroid, $\boldsymbol{\omega}$ is the angular velocity of the particle, and $\mathbf{r}$ is a radius vector from the centroid to the point of the particle. The motion of one immersed particle is described by the Newton–Euler motion equations for $\mathbf{u}_c$ and $\boldsymbol{\omega}$

$$\begin{aligned} \rho_p V_p \frac{d\mathbf{u}_c}{dt} &= \int_{\Gamma_p} (\boldsymbol{\sigma} \cdot \mathbf{n}) d\Gamma + \mathbf{F}_n, \\ \rho_p V_p \frac{d^2}{10} \frac{d\boldsymbol{\omega}}{dt} &= \int_{\Gamma_p} [\mathbf{r} \times (\boldsymbol{\sigma} \cdot \mathbf{n})] d\Gamma + \mathbf{r} \times \mathbf{F}_\tau. \end{aligned} \tag{4}$$

Here, ρ_p is the density of the particle, V_p is the volume of the particle with the surface $\Gamma_p = \partial V_p$, $\mathbf{n}$ is the outward-pointing unit normal to Γ_p, and $\boldsymbol{\sigma}$ is the viscous stress tensor. In the right-hand side of equations (4) the first term stands for the force acting on the particle from the fluid flow, and the second term is the force and the moment of particles interaction with each other or with the wall when the collision occurs. The normal component $\mathbf{F}_n$ and the tangent component $\mathbf{F}_\tau$ of the force are not equal to zero only for small periods of time when the particles are in contact. The particle is assumed to be homogeneous and continuous (without cavities). The viscous stress tensor is calculated for a Newtonian fluid as follows

$$\boldsymbol{\sigma} = -\mathbf{I}p + \mu(\nabla\mathbf{u} + \nabla\mathbf{u}^T), \tag{5}$$

where $\mathbf{I}$ is a unit tensor. Evaluating $\boldsymbol{\sigma}\cdot\mathbf{n}$ from equations (1) one can rewrite (4) as

$$\begin{aligned} \rho_p V_p \frac{d\mathbf{u}_c}{dt} &= \rho \frac{d}{dt}\int_{V_p} \mathbf{u}dV - \int_{V_p} \mathbf{f}dV + \mathbf{F}_n, \\ \rho_p V_p \frac{d^2}{10}\frac{d\boldsymbol{\omega}}{dt} &= \rho \frac{d}{dt}\int_{V_p} (\mathbf{r}\times\mathbf{u})dV - \int_{V_p} (\mathbf{r}\times\mathbf{f})dV + \mathbf{r}\times\mathbf{F}_\tau. \end{aligned} \tag{6}$$

As proposed in [4] we assume that the velocity of the fluid "inside" the solid particle satisfies the rigid body motion (3). Substituting (3) to (6), the following motion equation can be obtained

$$(\rho_p - \rho)V_p \frac{d\mathbf{u}_c}{dt} = -\int_{V_p} \mathbf{f}dV + \mathbf{F}_n, \tag{7}$$

and the following rotation equation can be obtained

$$(\rho_p - \rho)\frac{d^2}{10}\cdot\frac{d\boldsymbol{\omega}}{dt} = -\int_{V_p} (\mathbf{r}\times\mathbf{f})dV + \mathbf{r}\times\mathbf{F}_\tau. \tag{8}$$

For each particle equations (7) and (8) are valid. In order to close this system of equations it is necessary to use collision model and derive expressions for $\mathbf{F}_n$ and $\mathbf{F}_\tau$, which will be discussed later.

At the inflow Γ_{in} and outflow Γ_{out} boundaries, the periodic boundary conditions are stated for the velocity vector $\mathbf{u}$. It is assumed that the fluid flow is driven by the constant pressure drop between the inlet and outlet cross sections. For the case of a pure Poiseuille flow the pressure is a linear function of the channel length, $p = p_g(x)$, and

$$\frac{\partial p_g}{\partial x} = -\frac{16}{\mathrm{Re}}. \tag{9}$$

Here, $\mathrm{Re} = \dfrac{2\rho U_{max} W}{\mu}$ is the Reynolds number of the fluid flow in the flat channel unperturbed by the particles. The pressure can be split into two parts, namely, $p(x,y,z) = p_g(x) + p_v(x,y,z)$. First part p_g is responsible for the pure Poiseuille flow through the channel, and the second one p_v is responsible for fluctuations caused by the movement of the particles. Periodic boundary conditions are used for p_v. When a particle reaches the outflow boundary, the particle segment that goes beyond the boundary is instantly translated to the inflow with conservation of its velocity and rotation. The initial condition for fluid field is taken from the undisturbed Poiseuille flow. The initial conditions for the particles are also taken from the same flow for an arbitrary position of the particle measured by the distance from the boundary h_0. At the initial time the particles are placed randomly in the channel.

The considered particle-laden flows are characterized by four main parameters: the Reynolds number Re, the relative diameter of the particles d/W, their buoyancy or the ratio of particle density to fluid density ρ_p/ρ and the volume concentration of particles c. Equations (1) are solved by the SIMPLE method on the uniform staggered grid. Equations (7) and (8) are integrated over time by the Euler method. The forces of interaction $\mathbf{F}_n$ and $\mathbf{F}_\tau$ are determined by the collision model.

3. Collision model

The simulation of the motion of many particles needs particle-particle and particle-wall collision model to overcome the overlapping of particles, as well as penetration through the channel walls Γ_w. Particle velocities can be different at the contact time, and elastic interaction should be taken into account. The elastic interaction is described by the model of soft sphere similar to the one used in [8] and taken into account by the term $\mathbf{F}_n$ in (7). We neglect the friction force $\mathbf{F}_\tau$ in (8) arising from different rotation velocities of the particles.

The modeling of interaction through a thin layer of fluid is very difficult due to the following. Let us consider a case when the rigid spherical particle moves to the wall along the perpendicular line, as shown in figure 2. As the particle approaches the wall, the cross-flow ahead occurs in the thin space between them. The velocity of this flow is higher, the closer the particle is to the wall. Consequently, the fluid pressure increases when the particle moves towards the wall. Moreover, when the particle reaches the wall at the contact point A, according to the Navier – Stokes equations, the cross-flow velocity should become infinite. Therefore, the fluid pressure becomes infinite too. A similar behavior, but with the opposite sign, should also occur when the particle moves outward from the wall.

Taking into account the use of coarse computational meshes and large time steps, together with the above-described singularity in interaction, direct numerical simulation of the collisions is unsuitable. We propose to describe the interaction through a thin layer of fluid, also using the force $\mathbf{F}_n$.

The normal component of particle's approaching velocity in the fluid is small relative to the velocity of the entire flow. In this case, the force acting on the particle far away from the obstacle can be evaluated using Stokes formula $F = 3\pi\mu d v_p$, where v_p is particle's velocity. Near the wall or particle, a change in the value of the force acting on the particle can be represented using a correction factor $\lambda(\varepsilon)$ depending on the dimensionless distance $\varepsilon = 2l/d$:

$$F = 3\pi\mu d v_p \lambda(\varepsilon). \tag{10}$$

The hydrodynamics problem with collision can be solved analytically in the Stokes stationary approximation. In this approximation, the correction factor $\lambda(\varepsilon)$ can be found as the sum of a series of exponential functions [5]. In [6] the approximations of this series are given. In the case of the particle-particle collision, its approximation is equal to

$$\lambda(\varepsilon) = \frac{1}{2\varepsilon} - \frac{9}{20}\ln\varepsilon - \frac{3}{56}\varepsilon\ln\varepsilon + 1.346, \tag{11}$$

and in the case of the particle-wall collision, its approximation is equal to

$$\lambda(\varepsilon) = \frac{1}{\varepsilon} - \frac{1}{5}\ln\varepsilon - \frac{1}{21}\varepsilon\ln\varepsilon + 0.9713. \tag{12}$$

Despite the expression on the force (10) with (11) and (12) contain non-integrable terms with $1/\varepsilon$, it is widely used in modeling of particle-laden flows [7,8,9]. On the other hand, formulae (11) and (12) are valid only for $\varepsilon < 3$. Solving model problems of particle collisions in fluids, it was found that the developed numerical method predicts correct force value only for $\varepsilon > 0.2$. Thus, the interaction for large distances $\varepsilon > 0.2$ is described by the developed numerical method, while for small distances $\varepsilon < 0.2$ the interaction is described using (11) for $\mathbf{F}_n$ in (7).

The collision model was validated on experimental data from [10]. Good agreement on the energy dissipation is obtained. For considered in the paper Stokes numbers $\mathrm{Sk} < 10$, after the collision, approximately 3% of the kinetic energy of the particle is retained.

4. Results

A flat channel of length $L = 5$, width $W = 1$ and depth $D = 2$ is considered (figure 1). In z direction the boundary conditions of symmetry are set. The flow of the mixture along the x axis is enforced by the pressure gradient (9). Calculations were carried out using the uniform Cartesian Euler grid with $250 \times 50 \times 100$ cells. The time step was taken $\tau = 0.001$. Each calculation was unsteady. Time interval $t \in [0,15]$ proved to be enough for reaching the quasi-steady state, meaning that the average characteristics of the flow are settled. Subsequent time interval $t \in [15,30]$ was computed to accurately calculate the averaged flow characteristics. Velocities were averaged over each horizontal section $y = \text{const}$ and over time.

Figure 3 shows the averaged longitudinal velocity for various particle concentrations and diameters. The velocity profile looks like the Poiseuille profile, while the mixture velocity is higher near the channel walls and lower in its center. The velocity decreases significantly with increasing the concentration of particles c. The increase of concentration leads to a formation of an almost rigid core in the center of the channel (close to the plateau in the velocity profile in figure 3 c).

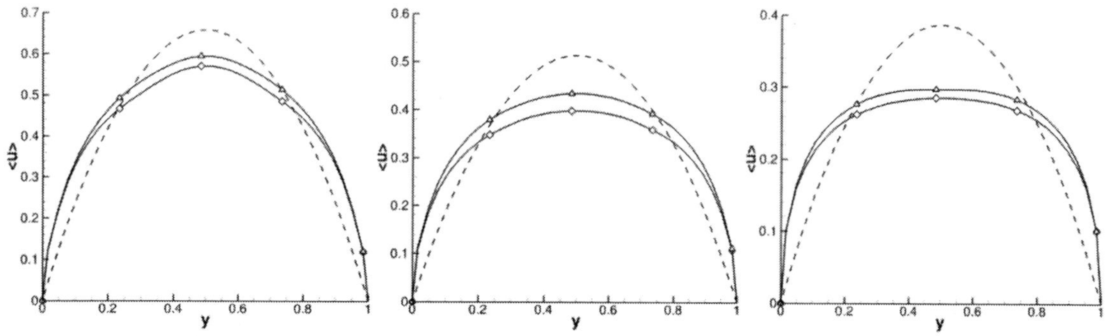

Figure 3. Distributions of the average longitudinal velocity of a mixture $<u>(y)$ of neutrally buoyant particles for $\text{Re} = 100$, various concentrations: a – $c = 0.14$, b – $c = 0.21$, c – $c = 0.28$, and various diameters: ◊ – $d / W = 0.3$, Δ – $d / W = 0.38$. The dashed line shows the Poiseuille velocity profile with the effective viscosity calculated by the formula (13).

In the field of hydraulic fracturing simulation, where similar flows arise [11,12], the Maron – Peirce formula for effective viscosity obtained experimentally in [13] and derived theoretically with some phenomenological assumptions in [3] is often used:

$$\mu_{eff} = \mu \left(1 - \frac{c}{c^*}\right)^{-2}, \tag{13}$$

where c^* is the packing concentration. For comparison the Poiseuille velocity profile with effective viscosity (13) is shown in figure 3. Here, the concentration of dense packing is taken equal to $c^* = \pi / \sqrt{18} \approx 0.74$ (the hexagonal dense packing of spheres). Nevertheless, for the diameters under consideration ($d / W = 0.3; 0.38$), such a dense packing in the flat channel is not achievable due to the presence of flat walls.

The average flow rate is obtained by integrating the average velocity over the channel width:

$$<q> = \int_0^W <u> dy. \tag{14}$$

Table 1 shows the flow rates obtained in the simulations and using formula (13). The flow rate is significantly affected by the particle concentration, while the influence of particle diameter is weaker. The flow rates obtained by the formula (13) with extremely high packing concentration are close to the ones obtained by direct numerical simulation.

Table 1. Average mixture flow rate $< q >$ for various particle concentrations and diameters.

	$d/W=0.3$	$d/W=0.38$	Formula (13)
$c=0.14$	0.43552	0.45482	0.43827
$c=0.21$	0.32222	0.34952	0.34197
$c=0.28$	0.24310	0.25463	0.25760

Table 2. Average velocity of particles $< u_p >$ for various particle concentrations and diameters.

	$d/W=0.3$	$d/W=0.38$
$c=0.14$	0.47655	0.48612
$c=0.21$	0.36923	0.38346
$c=0.28$	0.26870	0.28725

Table 2 shows the average longitudinal particle velocities obtained by direct numerical simulation. Accounting $W=1$, the average flow rate of the mixture $<q>$ equals the average velocity of the mixture. Comparing the values in tables 1 and 2, one can conclude that particles move faster about 8–12% than the mixture, and this difference increases with increasing concentration c.

5. Conclusion

Averaged velocity profiles of the fluid-particles mixture across the channel are computed by direct numerical simulation. The velocity profiles are close to the Poiseuille profile but have a more convex shape. The flow rates decrease significantly with increasing the particle concentration. The value of the particle diameter has a small influence on the flow rate. The flow rate can be approximated using the value of effective viscosity computed by Maron – Peirce formula with the use of extremely high packing concentration. In many papers devoted to the simulation of the hydraulic fracturing (e.g. [10]), the Maron – Peirce formula is used but with smaller packing concentration, which can lead to the unreliable results. Particles are moving faster than the mixture, and the difference increases with increasing concentration. The continuation of the present work would be driven by varying the value of the Reynolds number.

Acknowledgments

The authors acknowledge the financial support of this research by the Russian Science Foundation (grant No. 17-71-20139).

References

[1] Einstein A 1905 On the movement of small particles suspended in a stationary liquid demanded by the molecular-kinetic theory of heat *Annln. Phys.* **17** 549-60

[2] Batchelor G K 1972 The effect of Brownian motion on the bulk stress in a suspension of spherical particles *J. Fluid. Mech.* **83** 97-117

[3] Quemada D 1977 Rheology of concentrated disperse systems and minimum energy dissipation principle I. Viscosity-concentration relationship *Rheol. Acta.* **16** 82-94

[4] Uhlmann M 2005 An immersed boundary method with direct forcing for the simulation of particulate flows *J. Comput. Phys.* **209** 448-76

[5] Brenner H 1961 The slow motion of a sphere through a viscous fluid towards a plane surface *Chem. Eng. Sci.* **16** 242-51

[6] Dance S L and Maxey M R 2003 Incorporation of lubrication effects into the force-coupling method for particulate two-phase flow *J. Comp. Phys.* **189** 212-238

[7] Lambert A, Picano F, Breugem W-P and Brandt L 2013 Active suspensions in thin films: nutrient uptake and swimmer motion *J. Fluid Mech.* **733** 528-57

[8] Kempe T, Vowinckel B and Frohlich J 2014 On the relevance of collision modeling for interface-resolving simulations of sediment transport in open channel flow *Int. J. Multiphase Flow* **58** 214-35

[9] Ladd A J C and Verberg R 2001 Lattice-Boltzmann simulations of particle-fluid suspensions *J. Stat. Phys.* **104** 1191-251

[10] Legendre D, Zenit R, Daniel C and Guiraud P 2006 A note on the modelling of the bouncing of spherical droplets or solid spheres on a wall in viscous fluid *Chem. Eng. Sci.* **61** 3543-649

[11] Adachi J, Siebrits E, Peirce A and Desroches J 2007 Computer simulation of hydraulic fractures *Int. J. Rock. Mech. & Min. Sci.* **44** 739-57

[12] Karnakov P V, Kuranakov D S, Lapin V N, Cherny S G and Esipov D V 2018 Peculiarities of the hydraulic fracture propagation caused by pumping of proppant-fluid slurry *Thermophysics & Aeromechanics* **25** 587-603

[13] Maron M H and Pierce P E 1956 Application of ree-eyring generalized flow theory to suspensions of spherical particles *J. Colloid Sci.* **11** 80-95

Molecular-dynamic study of processes at the interface of a flat metal heterostructure during its heating

E I Golovneva, I F Golovnev, A V Utkin and V M Fomin

Khristianovich Institute of Theoretical and Applied Mechanics SB RAS, Institutskaya str., 4/1, Novosibirsk, 630090, Russia

E-mail: elena@itam.nsc.ru

Abstract. The author applies the method of heating to a flat copper-silver heterostructure. An approach has been developed for carrying out a meso-analysis: the study of the distribution of parameters such as temperature, concentration of atoms, etc. in the direction perpendicular to the interface. A parameter describing the diffusion of atoms across the interface during heating has been proposed and calculated.

1. Introduction

In recent years, interest to the molecular-dynamic modeling of alloy formation on the atomic level has been rising.

In particular, analysis of phenomena in crystals during heating is of high interest [1]. As the alloy forms, the material interface occurs. A number of works [2–6] are devoted to the investigation of the phenomena on the crystal interface (either on grain boundaries, or the substrate and thin-film coatings).

However, our paper concentrates on the analysis of the physics of the processes at alloy formation, not only on the research of thermophysical proportion of the ready system resulting from the interface melting, which is a fundamental difference from the available works of other teams. The investigation type is close to the project group [7, 8].

Thus, this work is a pilot one in the field of the development of the physical and mathematical model of alloy formation, which would permit performing on the micro-level the detailed investigation of the influence of rates of thermophysical parameters of the matrix and nanocluster materials, as well as the external controlled parameters on the physical properties of the resultant alloys.

2. Physical system, initial data assignment

We set separately a silver crystal, a copper crystal, each shaped as a rectangular parallelepiped with the atoms located in the nods of a perfect FCC lattice. The sizes of sub-systems by the contact area and the interface are determined with due regard to the fact that the size of the crystal lattice constant for copper is smaller than for silver. The copper crystal lattice constant a_{Cu} =3.615 Å; the silver crystal lattice constant a_{Ag} = 4.09 Å. We choose the sizes on the heterostructure base, regarding the previous investigation about the effect of the chosen nano-object size on the calculation results [9].

Finally, the following sizes of the heterostructure components have been chosen: n_x=30, n_y=10, n_z=5 (7046 atoms of copper) for the copper part; n_x=26, n_y=9, n_z=4 (4532 atoms of silver) for the silver part. After that, in an individual software program, we place one part above the other within the distance of a half of the copper crystal lattice.

The Verlet velocity modification [10] was used for all further calculations, the time step $\tau = 10^{-16}$ s. Atomic interaction is calculated by the embedded atom method [11] with due regard to the composition: Cu-Cu, Ag-Ag, Cu-Ag.

3. Preparation of the physical system

Now, the resultant heterostructure should be cooled, to put it in the equilibrium state. This is done by the artificial viscosity method [12]. After 1 million of computational steps, the system gets into the energy global minimum condition with a slight atomic shift about the perfect crystal structure owning to the stress occurring at the interface; the stress is caused by the constant difference in copper and silver crystal lattices (figure 1). Then, the atom coordinates and momentums in this cooled heterostructure are used as the initial data.

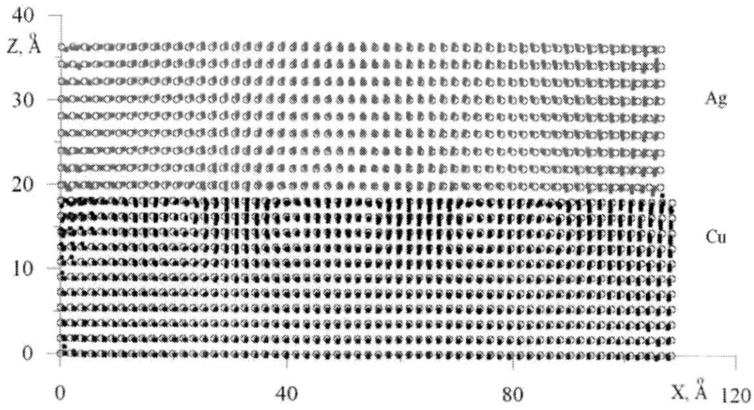

Figure 1. Outlook of the cooled Cu-Ag heterostructure.

4. Heterostructure heating

The heating procedure is carried out with the stochastic impulses method [13]. Figure 2 shows the main calculated characteristics.

The link energy is calculated as the difference of the full system energy and energies of the copper and silver sub-systems and corresponds to the energy, which is needed to split the heterostructure by the sub-systems and separate them infinitely:

$$U_b = U_t - (U_{Cu} + U_{Ag})$$

The figures below show the heated heterostructure at different temperatures (figures 3 and 4).

It is evident that at T = 1100 K (figure 4), which is close to the silver melting point, the systems begin to bend. This results from two factors: weakening bindings between atoms during the heating and difference between the constants of copper and silver crystal lattices.

 1404 (2019) 012018 doi:10.1088/1742-6596/1404/1/012018

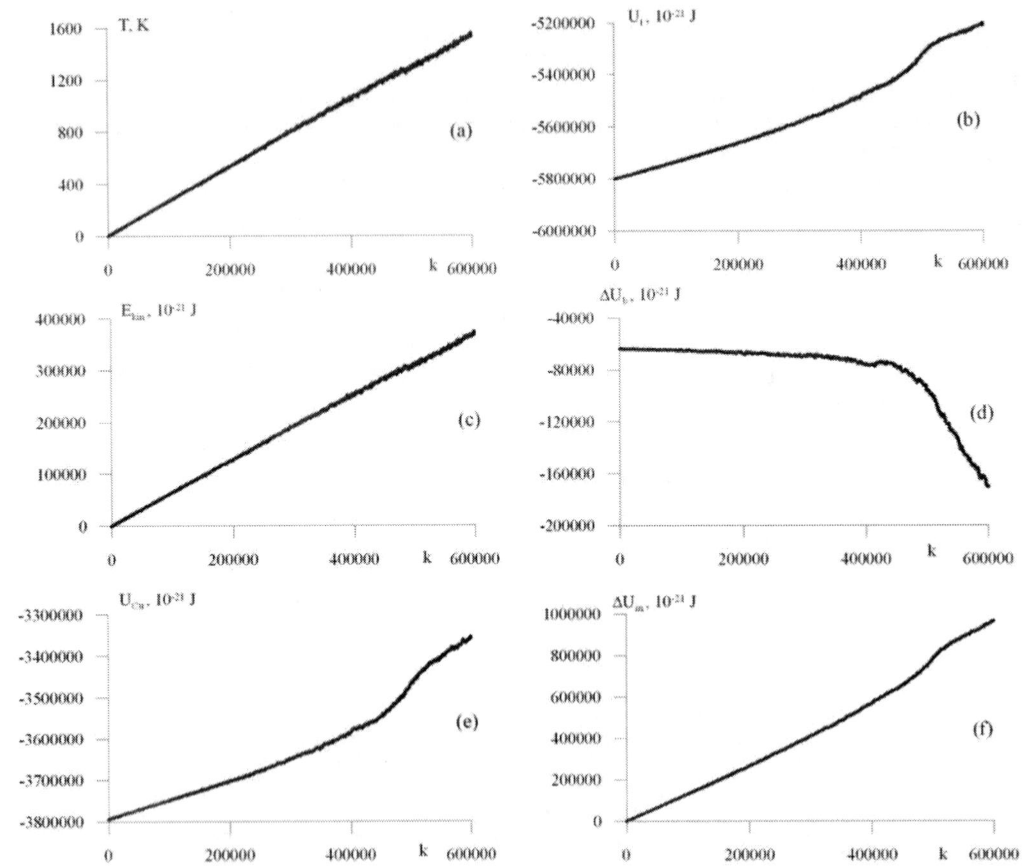

Figure 2. Energy characteristics versus the time step during the heterostructure heating. (a) the whole system temperature; (b) the heterostructure full energy; (c) the kinetic energy of heterostructure atoms (minus the energy of motion of the heterostructure mass center); (d) the varying energy of link between silver and copper parts; (e) the potential energy of the copper part; (f) the variation of the internal system energy.

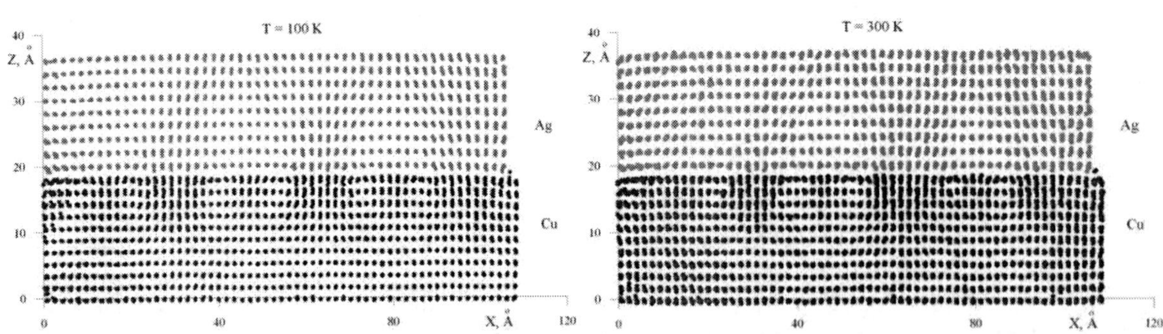

(a) (b)

Figure 3. Outlook of the heterostructure in the XZ plane at T=100 K (a) and T=300 K (b).

At 1300 K (which is close to the copper melting point and above the silver melting point), the heterostructure loses its initial rectangular-parallelepiped shape and tends to the spherical shape.

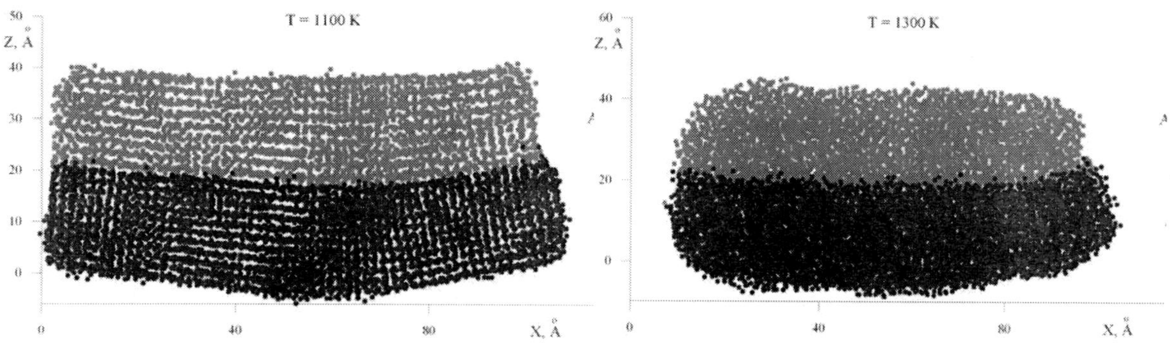

Figure 4. Outlook of the heterostructure in the XZ plane at T=1100 K (a) and T=1300 K (b).

5. Mesoanalysis

The system characteristics have been meso-analyzed at different temperatures, in order to see the dynamical changes of some parameters, which characterize the presence of diffusion through the interface. The mesoanalysis is realized via the mental splitting of the heterostructure to meso-cells by the planes parallel to the interface and perpendicular to the Z axis (in this task statement).

The coordinate Z_{int} of the copper-silver interface has been selected at the initial time instant (before heating) – and is shown by a straight grey line in the graphs below. The spatial step is equal to the half of the copper crystal lattice constant $a_{Cu}/2$. The size of the mesoanalysis domain is assigned in such a way to take the thermal dilation of metals into account.

The temperature distribution over the meso-cells along the Z axis has been studied, as well as the distribution of the relative amount of copper and silver atoms. We calculate the relative amount of atoms of the certain substance (copper or silver in this task) as the relation of this substance atoms amount in the selected meso-cell at the certain temperature to the amount of atoms of the same substance in the same meso-cell in the perfect cooled heterostructure:

$$N^i_{rel,Cu} = \frac{N^i_{mes,Cu}(T)}{N^i_{mes,Cu}(T=0)}; \qquad N^i_{rel,Ag} = \frac{N^i_{mes,Ag}(T)}{N^i_{mes,Ag}(T=0)}$$

The "beak" in figure 5 (a) results from the fact that the meso-cell partition is connected with the copper crystal lattice constant, so at such a partition, some meso-cells get much fewer atoms than the other. In future, it can be excluded via increasing the meso-cell and physical system size. The fact that the copper sub-system temperature is higher than the one of the silver sub-system also proves that during the heterostructure heating, the time interval for the system relaxation to the equilibrium state should be increased as compared to the numerical heating of homogeneous systems. It is seen that, as the system is heated, the distribution of the relative amount of atoms perpendicularly to the interface becomes smooth, the breach disappears, which indicates the beginning diffusion of copper and silver atoms through the interface (figures 5 and 6).

Journal of Physics: Conference Series **1404** (2019) 012018 doi:10.1088/1742-6596/1404/1/012018

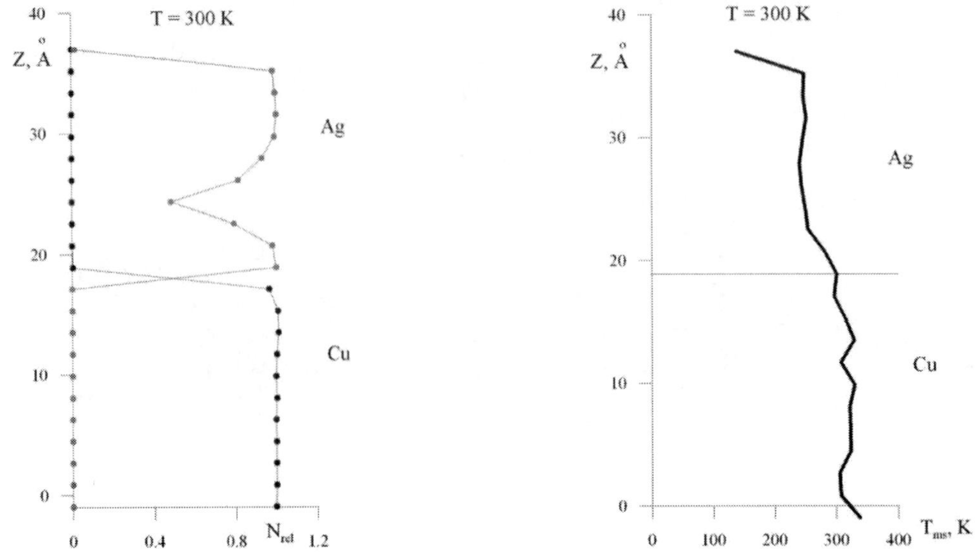

(a) (b)

Figure 5. (a) Distribution of the relative amount of Cu atoms (black dots) and Ag atoms (grey dots) perpendicularly to the interface. (b) Temperature profile at the copper-silver interface. The average heterostructure temperature T=300 K.

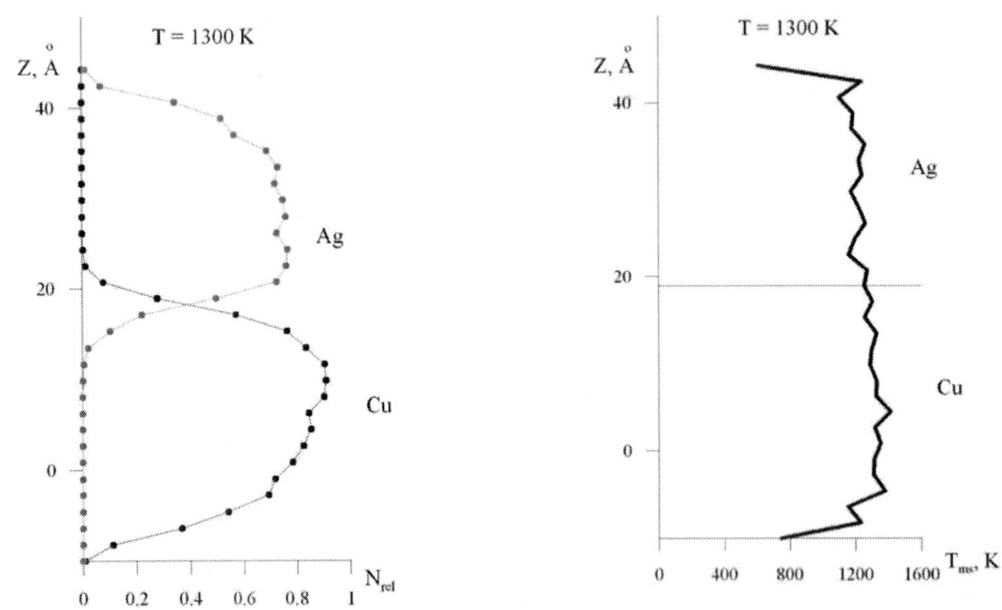

(a) (b)

Figure 6. (a) Distribution of the relative amount of Cu atoms (black dots) and Ag atoms (grey dots) perpendicularly to the interface. (b) Temperature profile at the copper-silver interface. The average heterostructure temperature T=1300 K.

6. Conclusions

To model the heating of the flat heterostructure, more time should be given for the whole system relaxation to its equilibrium state at each temperature increase step. In our case, 600 steps were chosen (10^{-16} s or 0.001 in the computational measurement units), which was enough for the uniform system consisting of one substance atoms, but this amount should be increased for the heterostructure.

It is demonstrated that the meso-cell distribution of the relative amount of atoms perpendicular to the heterostructure interface (along the Z axis) can be used as the first sign of the atom diffusion through the interface.

The peaks on the dependence of the relative amount of atoms in the computational mesoanalysis cell result from the smallness of the whole system and, hence, small sizes of the meso-cells.

It is better to use the parameter obtained in the task of the spherical heterostructure as the criterion of diffusion presence: the relation of amount of atoms of the certain substance in the meso-layer at the certain temperature to the full amount of atoms in the same meso-layer at the same temperature T. The characteristics are smoother, without dramatic drops as in the case of meso-analysis, at the diffusion in the flat heterostructure, as the relation was found in respect to the amount of atoms of the same substance at the zero temperature.

Acknowledgments

The research was carried out within the framework of the Program of Fundamental Scientific Research of the state academies of sciences in 2013-2020 (project No. AAAA-A17-117030610134-9) and with the financial support of the RFBR, grant No 18-01-00287-a.

References

[1] Kuzkin V A and Krivtsov A M 2017 *Physics of Solid State.* **59** 5 1051
[2] Korchuganov A V, Tyumentsev A N, Zolnikov K P, Litovchenko I Yu, Kryzhevich D S, Gutmanas E, Li S, Wang Z and Psakhie S G 2019 *J. Mater. Sci. Technol.* **35** 1 201
[3] Kryzhevich D S, Zolnikov K P and Korchuganov A V 2018 *Comput. Mater. Sci.* **153** 445
[4] Panin V E, Goldstein R V and Panin S V 2008 *Inter. J. of Fracture* **150** 1-2 37
[5] Psakhie S G, Moiseyenko D D, Smolin A Yu, Shilko E V, Dmitriev A I, Korostelev S Yu and Tatarintsev E M 1999 *Comp. Mater. Sci.* **16** 333
[6] Smolin A Yu, Eremina G M, Sergeyev V V, Shilko E V and Psakhie S G 2014 *Phys. Mesomech.* **17** 292
[7] Levchenko E V, Evteev A V, Belova I V and Murch G E 2010 *Comput. Mater. Sci.* **50** 331
[8] Levchenko E V, Evteev A V, Lúwisch G G, Belova I V and Murch G E 2012 *Intermetallics* **22** 193
[9] Golovnev I F, Golovneva E I and Fomin V M 2006 *Comput. Mater. Sci.* **36** 176
[10] Allen M P and Tildesley D J *Computer Simulation of Liquids*. (New York: Clarendon Press, 1987) p 385
[11] Voter A F 1993 Embedded Atom Method Potentials for Seven FCC Metals: Ni, Pd, Pt, Cu, Ag, Au, and Al *Los Alamos Unclassified Technical Report LA-UR 93-3901*
[12] Golovnev I F, Golovneva E I and Fomin V M 2003 *Phys. Mesomech.* **6** 5-6 41
[13] Golovnev I F, Golovneva E I and Utkin A V 2018 *Phys. Mesomech.* **21** 6 523

The formation of heterogeneous wear-resistant coatings by the additive technology method

A A Golyshev, A G Malikov and A M Orishich

Khristianovich Institute of Theoretical and Applied Mechanics SB RAS,
4/1 Institutskaya st., Novosibirsk, 630090, Russia

alexgol@itam.nsc.ru

Abstract. The microstructure, hardness property and wear resistance coatings of WC, Ni-Cr and Fe powders deposited by laser cladding at varying processing parameters were investigated. The results of the present study revealed the prospects of multilayer cladding of R6M5 high-speed tool steel (analogy of M2 steel) onto low-alloy steel with the use of a laser beam. Controlling thermal cycles of laser cladding, it is possible to obtain a clad coating made of high-speed steel having the structure of high-alloy austenite martensite mixture with disperse inclusions of carbides up to 10 mm thick, i.e., it is actually possible to create bimetal structures. The wear resistance of the laser-clad self-fluxing PG-10N-01 (Ni-Cr-B-Si-C) alloy increases by a factor of 5 due to additional hardening by cast tungsten carbide with spherical particles. As a result, it becomes higher than the wear resistance of high-speed steel by more than a factor of 3.

1. Introduction

Laser cladding of powders has been intensely studied recently all over the world as a method of hardening, improving the wear resistance of the material surface, creating high-strength cutting tools, recovery of worn workpieces, etc. To increase the wear resistance of clad layers, metal-composite powders based on Ni-Cr and Co-Cr mixed with carbides and borides are used [1-3].

It was demonstrated [4-6] that the characteristics of abrasion of Ni-based coatings with WC inclusions depend on the volume fraction and distribution of WC particles in resultant coatings. The basic studies are aimed at decreasing cracking in clad layers [7-11]. Researchers also study the microstructure of clad layers, stability of the process, and microhardness [12-15]. However, there are few reported investigations of mechanical properties of workpieces, their strength, abrasion and impact resistance, etc., i.e., actually the operation performance of technological products in absolute values.

In this work, we study the process of laser cladding of R6M5 high-speed tool steel powder (analog of M2 steel (USA) and HS6-5-2 (EU)), which possesses high hardness, a self-fluxing PG-10N-01 (Ni-Cr-B-Si-C) powder and a mixture of a self-fluxing Ni-Cr-B-Si-C powder with tungsten carbide (WC). Absolute values of the wear resistance of resultant clad layers are analyzed and compared. We studied laser cladding of powder mixture Ni-Cr-B-Si-C+ WC with a ceramic WC concentration of up to 45%. As a result received high-quality ceramic-metal coatings (without cracks and pores), measured the wear and microhardness of the coatings.

2. Experimental methods

Controlling the parameters of a highly focused laser beam during the laser cladding procedure allows one to vary the power density and the heating spot size within wide limits. In the present studies, the automated laser technological complex (ALTC) developed and fabricated at the Khristianovich Institute of Theoretical and Applied Mechanics of the Siberian Branch of the Russian Academy of Sciences (ITAM SB RAS) on the basis of a powerful CO_2 laser with convective cooling of test gases [16, 17] was used as process equipment. The microstructure of clad layers was studied by using an Olympus LEXT OLS3000 confocal scanning laser microscope.

3. Laser cladding of high-speed steel

A technological possibility of creating a bimetal material with a matrix made of low-alloy steel and with working surfaces made of R6M5 high-speed steel powder with particles in diameter up to 150-170 μm (its chemical composition is presented in Table 1) by means of laser cladding was studied.

In the present work, the alloying agent was the powder of R6M5 steel, which was distributed prior to laser treatment by a special dispenser-crystallizer onto the base material surface without any binder. The base material had the form of metal sheets shaped as disks 2 mm thick, which were fabricated by means of laser cutting from 65G steel (its chemical composition is presented in Table 1) after complete quenching and tempering to the hardness of 45 HRC.

Table 1. Chemical composition (percentage).

Material	C	Si	Mn	Cr	W	V	Mo	Fe	Ni	B
R6M5	0.84	0.48	0.5	3.9	6.1	1.9	4.9	Base	-	-
65G	0.62-0.7	0.17-0.37	0.9-1.2	0.25	-	-	-	Base	-	-
PG-10N-01	0.4	2.6	-	13.5	-	-	-	-	Base	2.0

Laser cladding was performed in the optimal mode: laser power of 1.0 kW. The laser beam speed was 1 m / min. The laser radiation is focused on the surface -2 mm.

Interaction of the base and clad metals mainly occurs along the boundaries of austenite grains. After cooling, the melted grains start to act as centers of growth of dendrite crystals of the clad material (figure 1). As the thickness of the powder layer subjected to laser treatment increases, the depth of melt penetration into the base material decreases. If the treated layer thickness is greater than 2 mm, there appear regions where no fusion with the base material occurs; such regions are separated from the base material by an oxide film.

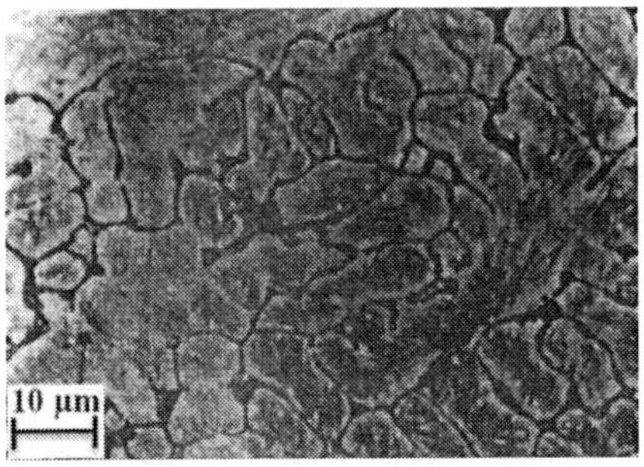

Figure 1. Metallographic analysis of single-layer cladding with R6M5 steel.

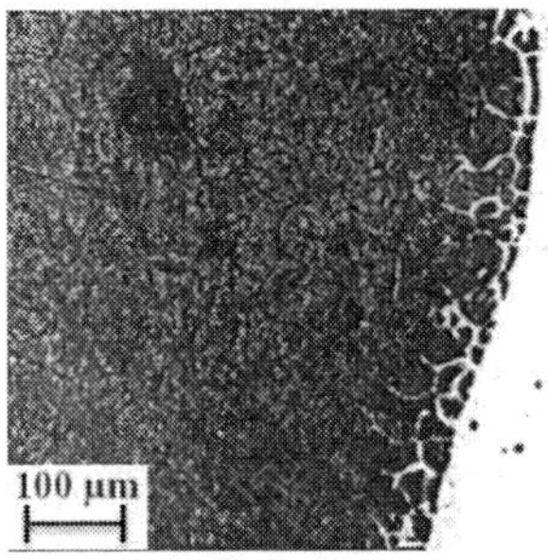

Figure 2. Infiltration of the clad R6M5 layer into the 65G base material.

The base material in the laser influence zone (LIZ) becomes structurally inhomogeneous (figure 2) as a result of incomplete quenching of 65G steel, which has a row-by-row distribution of carbides in this case. R6M5 steel used for cladding has a dendrite structure (figure 1). Cross sections of branches of high-alloy austenite dendrites accommodate martensite crystals. Carbides Fe3C; Fe2C; Fe3C4, FeC evolve in the space between these dendrites in the form of eutectic colonies and chains of individual particles, which form a continuous skeleton in some places. The microhardness of the austenite-martensite mixture is 9.48 – 10.1 GPa in the zone of cladding material fusion with the base material and 10.1 – 10.9 GPa in the middle part and on the surface.

The thickness of the resultant coating is determined by the purpose and type of cladding. If a thick coating is needed, cladding of several layers can be performed. The thickness of the second and next layers in the experiment was chosen to be approximately 80% of the depth of melting of R6M5 steel. The lower layers are located in the LIZ in the case of layer-by-layer cladding. The chosen regime of repeated laser treatment of the previously clad layer of R6M5 steel ensures favorable changes in the distribution of carbides.

Laser cladding in several passes leads to melting of the network of carbides Fe3C; Fe2C; Fe3C4, FeC separating into individual particles (figure 3 a); carbides dissolve. This, in turn, increases the level of alloying of austenite formed during subsequent crystallization.

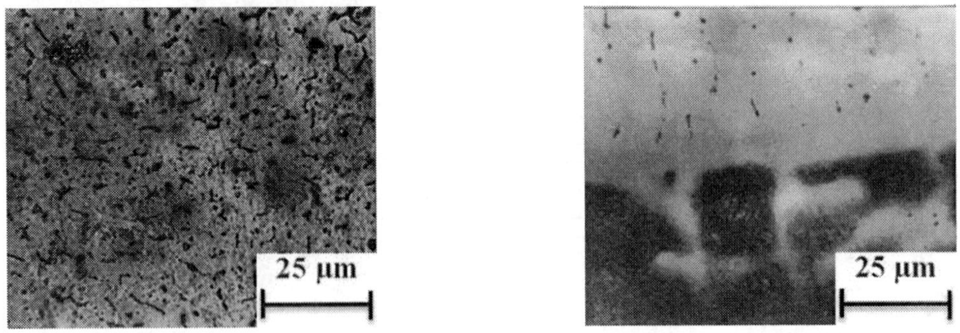

(a) (b)

Figure 3. Structure of the carbide phase of the R6M5 steel coating in the case of laser cladding in several passes (a – middle part of the coating, b- fusion zone).

The structure of the R6M5 steel coating in the front of laser beam motion and after repeated laser cladding is illustrated in figure 4. In the first case (in the front of the melted zone), a high level of etchability, microvoids, and non-metallic inclusions are observed. After re-cladding, carbide eutectic becomes separated into individual particles. The total amount of the carbide phase decreases. The level of alloying of the solid solution simultaneously increases.

The measured microhardness of the structural components of the LIZ is shown in figure 5. The microhardness of the base material with the initial structure of tempered troostite is 5 GPa. During laser treatment, those regions of the base material that are heated to temperatures below the austenization point experience tempering, and their hardness decreases to 3.5 GPa. At the LIZ edges, the austenite phase is mainly formed around carbides. These austenite re-gions with inhomogeneous concentrations turn during subsequent cooling into an austenite-martensite mixture with microhardness values ranging from 6 to 9 GPa (light areas in figure 2). On the boundary with the clad material, the LIZ temperature ensures homogenization of aus-tenite and its subsequent quenching to martensite with a microhardness value of 11 GPa.

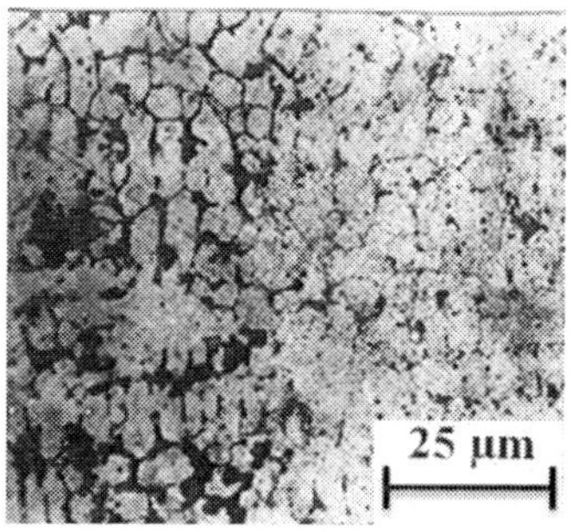

Figure 4. Carbide phase structure in the case of after repeated laser treatment of the clad R6M5 steel coating.

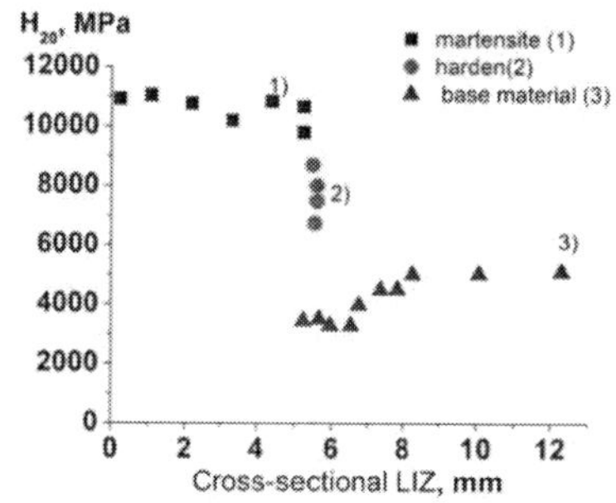

Figure 5. Cross-sectional distribution of microhardness.

Repeated quenching occurs in the base metal. The martensite hardness is 8.74 GPa. Diffusion processes in the fusion zone begin to develop (figure 3 b). There are only minor changes in the hardness of dendrite cells varying from 9.4 to 10.1 GPa.

Using layer-by-layer cladding, we managed to obtain a coating up to 8-10 mm thick with-out any macroscopic defects and with a hardness value of 63 - 64 HRC. After standard three-fold tempering at 560 °C, the hardness of the clad coating of R6M5 steel increased to 66 - 67 HRC.

4. Laser cladding of the Ni-Cr-B-Si-C alloy powder

We studied cladding of a PG-10N-01 nickel-chrome powder (its chemical composition is presented in table 1) for obtaining a wear-resistant coating, which was performed in optimal regimes at the laser power of 1.65 kW. The laser beam scanned the surface with a frequency of 62 Hz, the beam velocity was 3.2 m/s, and the scanning range was 5 mm. The workpiece feed velocity was 1.2 m/min.

After thermal treatment, the powder alloy becomes completely melted and ensures good fusion with the base material, thus, forming a coating. The resultant coating has a dendrite-cellular structure (figure 6). The finely differentiated eutectic component is identified over the dendrite cell boundaries. No aggregates of primary carboborides are visible. The solid solution has parameters that testify to an increase in the concentration of dissolved components. This fact is indirectly evidenced in this work by the increase in the solid solution microhardness by 10 – 12 % as compared to the microhardness value obtained during volumetric thermal treatment.

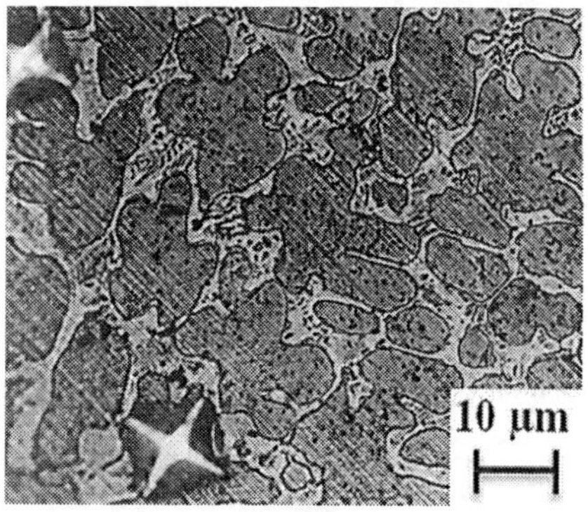

Figure 6. Dendrite-cellular structure in the case of laser treatment of the NiCr powder cladding coating

Tests with cladding of the self-fluxing nickel-chrome alloy did not reveal any embrittle-ment effect of the network of carboborides on the dendrite cell boundaries. It was found that an increase in the fraction of solid phases in the clad coating decreases the abrasion and im-pact wear resistance.

Structural studies show that this is caused by spalling of coating fragments along the brittle carbide network on the dendrite cell boundaries. However, in view of the high hardness of the coating, it can be expected to exhibit a high wear resistance value during its operation in a volume filled by abrasive particles and in the case of its dragging on a fixed abrasive material.

5. Laser cladding of the Ni-Cr-B-Si-C alloy (PG-10N-01) and WC.

Based on the data obtained, compositions of cladding materials of the self-fluxing nickel-chrome alloy (PG-10N-01) and WC were composed, with the ratio 55:45%. These compositions were fed in the form of a fine-grain powder and spheroidized particles of cast tungsten carbide.

In the case of melting of the Ni-Cr-B-Si-C alloy, the cast tungsten carbide particles remain in the solid state. The surface of the tungsten carbide particles dissolves in the melted nickel-chrome alloy. Mutual diffusion of the components occurs, and the amount of carbide eutectic and disperse carbides increases during subsequent crystallization (Figure 7). Being heavier than the melt, the cast tungsten carbide particles descend to the base material surface. The eutectic with participation of iron is formed on the interface with the base material. Simultaneously, thermoreactive dissolution of the steel base occurs and holes are formed under the cast tung-sten carbide particles, which improves adhesion between the coating and the base material (Figure 7). The hardness of the cast tungsten carbide particles is very high. The microhardness value determined by a PMT-3 tool with a 50-g load on the indenter was H50 = 32.36 GPa. For comparison, the microhardness of the base material (65G steel) quenched by laser treatment to the martensite state is H50 = 10.7 GPa.

The wear resistance of the resultant composite coating was estimated by dragging it on a fixed abrasive material in accordance with the requirements of the Russian Standard GOST 17367-71. Samples of coatings obtained by means of laser cladding of the PG-10N-01 wear-resistant alloy and of R6M5 high-speed tool steel powder were tested simultaneously. No cracks, flakes, chips, or spalls of the clad material were observed during the tests. The test results are summarized in table 2.

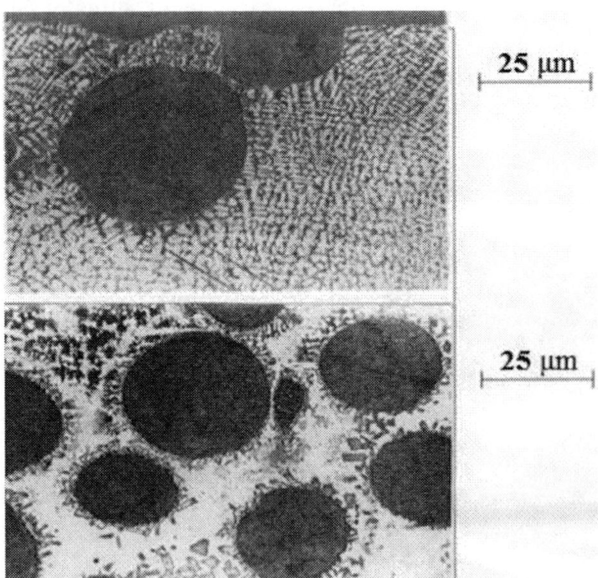

Figure 7. The structure of the deposited layer of 55% PG-10N-01 + 45% spherical relit by weight (Upper - surface after abrasive wear, lower - fusion zone).

The wear resistance of the resultant composite coating was estimated by dragging it on a fixed abrasive material in accordance with the requirements of the Russian Standard GOST 17367-71. Samples of coatings obtained by means of laser cladding of the PG-10N-01 wear-resistant alloy and of R6M5 high-speed tool steel powder were tested simultaneously. No cracks, flakes, chips, or spalls of the clad material were observed during the tests. The test results are summarized in table 2.

Table 2. The results of the wear resistance test.

Coating composition	Hardness	Vwear, 10^{-7}, kg/s
PG-10N-01	HRCexp 45 - 48	25.0
R6M5	HRCexp 66	15.6
PG-10N-01 (55%), cast WC (45%)	H_{50}=32.36 GPa	4.9

It follows from the data obtained that the wear resistance of the PG-10N-01 alloy in the case of its dragging on a fixed abrasive material increased by a factor of 5 due to additional hardening by cast tungsten carbide with spherical particles. As a result, it became higher than the wear resistance of high-speed steel by more than a factor of 3.

6. Conclusions

The results of the present study revealed the prospects of multilayer cladding of R6M5 high-speed tool steel (analog of M2 steel (USA) and HS6-5-2 steel (EU)) onto low-alloy steel with the use of a laser beam. Controlling thermal cycles of laser cladding, it is possible to obtain a clad coating made of high-speed steel having the structure of high-alloy austenite-martensite mixture with disperse inclusions of carbides up to 10 mm thick, i.e., it is actually possible to create bimetal structures.

The wear resistance of the laser-clad self-fluxing Ni-Cr-B-Si-C alloy in the case of its dragging on a fixed abrasive material increases by a factor of 5 due to additional hardening by cast tungsten carbide with spherical particles. In work [6], it was shown that the abrasive wear of Ni-Cr-B-Si-coatings with WC ceramics of 25% compared to Ni-Cr-B-Si-C without ceramics created by laser cladding according to principle "ball-on-disk" is insignificant at low sliding speed. As a result, it becomes higher than the wear resistance of high-speed steel by more than a factor of 3.

Cast tungsten carbide with spherical particles can be recommended as a component for wear-resistant coatings and clad layers for elements operating under abrasive wear conditions: drilling tools, executive devices of excavating, construction, and road-building machines, clay mixers, elements of metallurgical equipment, etc.

Acknowledgments

The research was carried out within the framework of the Program of Fundamental Scientific Research of the state academies of sciences in 2013-2020 (project No. AAAA-A17-117030610122-6).

References

[1] Zhong M, Liu W, Yao K, Goussain J C, Mayer C and Becker A 2002 *J. Surf. Coat. Technol.* **157** 128-137

[2] Sun S, Durandet Y and Brandt M 2005 *J. Surf. Coat. Technol.* **194** 225–231

[3] Paul C P, Alemohammad H, Toyserkani E, Khajepour A and Corbin S 2007 *J. Mater. Sci. Eng.* **A 464** 170-177

[4] Wu P, Zhou C Z and Tang X N 2003 *J. Surf. Coat. Technol.* **166** 84–88

[5] Dhakar B M, Dwivedi D K and Sharma S P 2012 *J. Surf. Eng.* **28** 73–80

[6] Guo C, Chen J, Zhou J, Zhao J, Wang L, Yu Y and Zhou H 2012 *J. Surf. Coat. Technol.* **206** 2064–2071

[7] Van Acker K, Vanhoyweghen D, Persoons R and Vangrunderbeek J 2005 *J. Wear* **258** 194-202

[8] Wang F, Mao H, Zhang D and Zhao X 2009 *J. Appl. Surf. Sci.* **255** 8846–8854

[9] Zhu L, Xu B, Wang H and Wang C 2012 *J. Mater. Sci.* **47** 2122–2126
[10] Kadolkar P B, Watkins T R, De Hosson J Th M, Kooi B J and Dahotre N B 2007 *J. Act. Mater.* **55** 1203-1214
[11] Lee C, Park H, Yoo J, Lee C, Woo W C and Park S 2015 *J. Appl. Surf. Sci.* **345** 286–294
[12] Abioye T E, Folkes J, Clare A T and McCartney D G 2013 *J. Surf. Eng.* **29** 647-653
[13] Xuan H F, Wang Q Y, Bai S L, Liu Z D, Sun H Gand Yan P Ch 2014 *J. Surf. Coat. Technol.* **244** 203–209
[14] Shang S, Dan Wellburn, Sun Y Z, Wang S Y, Cheng J Liang J and Liu C S 2014 *J. Surf. Coat. Technol.* **248** 46–53
[15] Fomin V M, Golyshev A A, Kosarev V F, Malikov A G, Orishich A M, Ryashin N S, Filippov A A, Shikalov V S 2017 *J. Appl. Mech. & Tech. Phys.* **5** 947-955
[16] Golyshev A A; Orishich A M and Filippov A A 2019 *J. Appl. Mech. & Tech. Phys.* **4** 758-767
[17] Tyurin A, Nagavkin S, Malikov A and Orishich A 2015 *J. Surf. Eng.* **31** 74-77

Numerical investigation of wave propagation in high-permeable cylindrical waveguide in porous medium

A A Gubaidullin[1,2], O Yu Boldyreva[1] and D N Dudko[1]

[1]Tyumen Branch of Khristianovich Institute of Theoretical and Applied Mechanics SB RAS, 74 Taymirskaya Street, Tyumen, 625026, Russia
[2]University of Tyumen, 6 Volodarskogo Street, Tyumen, 625003, Russia

A.A.Gubaidullin@yandex.ru

Abstract. The features of waves propagating in a highly permeable cylindrical waveguide in a porous medium are numerically investigated. The influence of the ratio of permeability inside the cavity and the surrounding porous medium, as well as the signal frequency on the evolution of the pressure wave inside the cavity is studied. On the basis of the calculated data, estimates of the velocity and attenuation of the signal during its propagation in the waveguide in a wide frequency range are obtained. It is shown that with the increase in the waveguide permeability, the propagation velocity increases and the attenuation of waves decreases, the depth of penetration of perturbations into the surrounding space decreases. Increasing the frequency of the initial signal leads to an increase in the velocity and attenuation of the pulse. A comparison is made with the case of wave propagation in a liquid-filled cylindrical cavity in a porous medium. The results of the study can be used in the interpretation of well logging data.

1. Introduction

It is known that the composition and structure of micro-homogeneous rocks and fluids saturating them significantly affect the velocity and damping of elastic waves propagating in such media. Therefore, by changing the signals propagating in the internal waveguide in a porous medium, it is possible to determine the properties of rocks, such as porosity, permeability, fluid saturation, etc. [1, 2]. In [3-6], waves in a liquid-filled cylindrical cavity in a porous medium containing gas bubbles or gas hydrate were studied. This work is devoted to the study of the influence of the properties of the internal waveguide and the surrounding porous medium on the peculiarities of wave propagation in such a system.

2. Governing equations

To study numerically wave propagation and attenuation in a highly permeable cylindrical waveguide in a porous medium, we use a two-velocity porous medium model [7, 8].

$$\frac{\partial \rho_f}{\partial t} + \nabla^l \left(\rho_f v_f^l\right) = 0, \qquad \frac{\partial \rho_s}{\partial t} + \nabla^l \left(\rho_s v_s^l\right) = 0, \tag{1}$$

$$\rho_f \frac{d_f v_f^k}{dt} = -\alpha_f \nabla^k p_f - F^k, \qquad \rho_s \frac{d_s v_s^k}{dt} = -\alpha_s \nabla^k p_f + \nabla^l \sigma_{s*}^{lk} + F^k, \tag{2}$$

$$F = F_m + F_\mu, \quad F_m = \frac{1}{2}\eta_m \alpha_s \alpha_f \overset{\circ}{\rho}_{f0}\left(\frac{d_f v_f}{dt} - \frac{d_s v_s}{dt}\right), \quad F_\mu = \eta_\mu \alpha_s \alpha_f \mu_f a_{s*}^{-2}\left(v_f - v_s\right), \quad (3)$$

$$\sigma_{s*}^{kl} = \alpha_s\left(\lambda_{s*}\delta^{kl}\varepsilon_s^{mm} + 2\mu_{s*}\varepsilon_s^{kl} + v_{s*}\delta^{kl}p_f\right), \quad v_{s*} = \frac{\lambda_{s*} + 2/3\mu_{s*}}{\lambda_s + 2/3\mu_s}, \quad \frac{d_s\varepsilon_s^{kl}}{dt} = \frac{1}{2}\left(\nabla^k v_s^l + \nabla^l v_s^k\right), \quad (4)$$

$$p_j - p_{j0} = K_j\left(\overset{\circ}{\rho}_j - \overset{\circ}{\rho}_{j0}\right)/\overset{\circ}{\rho}_{j0}, \quad (5)$$

$$p_{s*} = \alpha_s\left(p_s - p_f\right), \quad p_{s*} = -\frac{1}{3}\sigma_{s*}^{mm}, \quad \rho_j = \alpha_j \overset{\circ}{\rho}_j, \quad \alpha_s + \alpha_f = 1. \quad (6)$$

Here ρ_j, v_j, α_j are effective density, velocity and volume content of the solid ($j = s$) and fluid phase ($j = f$); σ_{s*}, p_f are the effective stress in the skeleton of the porous medium and the pressure in the fluid respectively. The interaction force has the form of the sum of the forces of the associated masses F_m and viscous friction F_μ. ρ_j° is true density of j - phase, a_{s*} is characteristic particle size of the skeleton, μ_f is the viscosity of the fluid, η_m, η_μ are the factors dependent on the porous medium structure. λ_{s*}, μ_{s*} are moduli of elasticity of skeleton, ε_s is deformation of the solid phase. p_s is true pressure in the solid phase, K_j are bulk moduli of elasticity for the material of j – phase; the subscript 0 refers to the undisturbed value of the variable. The permeability of a porous medium is related to the model parameters by the expression $k = \frac{\alpha_f}{\eta_\mu \alpha_s} a_{s*}^2$, which is a consequence of the first equation (2) in the filtration flow regime.

3. Calculation results

The features of waves propagating in a highly permeable cylindrical waveguide in a porous medium are numerically investigated in an axisymmetric approximation. The study was performed in the framework of a two-phase porous medium model using the Lax-Wendroff method.

The results presented below are obtained for the computational domain $0 \le z \le 2m$, $0 \le r \le r_w + 1m$, $r_w = 0.1$m. The region $r > r_w$ is filled with a main porous medium (PM1), and a part of a cylindrical cavity $z > 0.1$ m, $0 \le r \le r_w$, is filled with porous medium with a higher permeability (PM2). The region $0 \le z \le 0.1$ m, $0 \le r \le r_w$ is filled with liquid.

The perturbation of finite duration was created by a liquid source located inside the cavity at $0 \le r \le r_{00} = 0.04$ m , $0 \le z \le z_{00} = 0.04$ m. The material of the porous medium skeleton is quartz, the liquid is water. The case of open pores at the porous medium – liquid interface and at the PM1 – PM2 interface is considered.

The values of the main parameters taken in the calculations are as follows: for PM1 porosity is 0.3, characteristic grain size of the porous medium is 0.1 mm, permeability $4.3 \cdot 10^{-11}$ m^2; for PM2 porosity 0.4, characteristic grain size of the porous medium 0.2 mm, permeability $2.7 \cdot 10^{-10}$ m^2.

Fig.1 shows the change in fluid pressure on the axis of the cavity at $r = 0$, $z = 0.2$, 0.4, 0.6, 0.8, 1 m, for compression wave (carrier frequency is about 0.5 kHz) and for oscillating pulses of carrier frequency 1 kHz and 3 kHz. It can be seen that the propagation of the compression wave is characterized by a significant distortion, i.e. form-spreading and attenuation. For oscillating pulses, attenuation is also noticeable, but the waveform does not change much. Analysis of the calculated data indicates that high-frequency signals attenuate significantly stronger than low-frequency ones and propagate almost without dispersion.

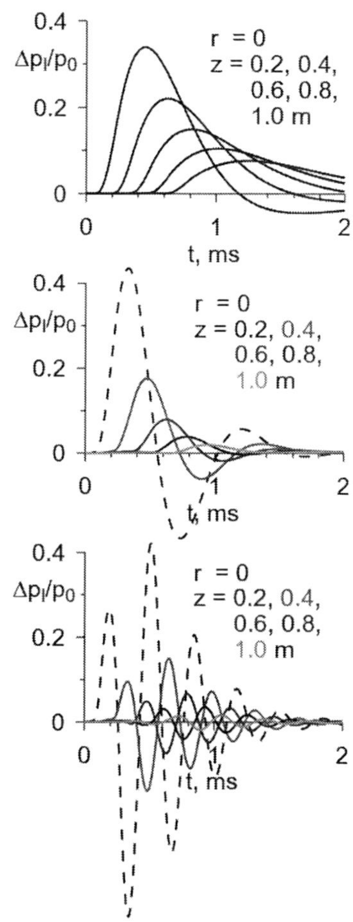

Figure 1. The pressure change in a fluid in a highly permeable cylindrical waveguide during the propagation of compression waves (carrier frequency of about 0.5 kHz, top) and the oscillating pulses of the carrier frequency 1 kHz (centre) and 3 kHz (bottom).

Based on the calculated data, estimates of the signal velocity and attenuation during its propagation in the waveguide in the considered frequency range are obtained. From the estimates it follows that an increase in the frequency of the original signal leads to an increase in the velocity and a more significant attenuation of the pulse. With increasing permeability of the waveguide, the propagation velocity increases and the attenuation of waves decreases, the depth of penetration of disturbances into the surrounding space decreases.

A comparison is made with the case of wave propagation in a liquid-filled cylindrical cavity in a porous medium. In contrast to this case, a wave in a porous waveguide contains two modes: deformation and filtration. In the limiting case, when the permeability of the porous medium inside the waveguide increases, the pressure curves coincide with the corresponding dependences for the waveguide filled with liquid.

4. Conclusions

The features of waves propagating in a highly permeable cylindrical waveguide in a porous medium are numerically investigated. The influence of the ratio of permeability inside the cavity and the surrounding porous medium, as well as the signal frequency on the evolution of the pressure wave inside the cavity is studied. In the considered frequency range with increase in the permeability of the waveguide, the propagation velocity is increased, and the wave damping decreases, and the penetration depth of disturbances into the surrounding space decreases; increase in the frequency of the original signal causes an increased velocity and increased attenuation of the pulse.

Acknowledgments
The research was carried out within the framework of the Program of Fundamental Scientific Research of the State Academies of Sciences in 2013-2020 (Project No. AAAA-A17-117030610130-1).

References
[1] White J E 1983 *Underground Sound. Application of Seismic Waves* (Amsterdam – Oxford – New York: Elsevier)
[2] Pusyrev N N 1997 *Methods and Objects of Seismic Research* (Novosibirsk: Publishing House of SB RAS)
[3] Gubaidullin A A, Boldyreva O Yu and Dudko D N 2017 *Proceedings of the Mavlyutov Institute of Mechanics* (Ufa Investigation Center, RAS) **12(1)** 9–14
[4] Gubaidullin A A, Boldyreva O Yu and Dudko D N 2018 *AIP Conf. Proc.* **1939** 020031
[5] Boldyreva O Yu, Gubaidullin A A, and Dudko D N 2018 *Physical and Mathematical Modeling. Oil, Gas, Energy* (Tyumen State University Herald) **4(4)** 210-21
[6] Boldyreva O Yu, Gubaidullin A A and Dudko D N 2019 *Physical and Mathematical Modeling. Oil, Gas, Energy* (Tyumen State University Herald) **5(1)** 111-22
[7] Nigmatulin R I 1978 *Fundamentals of Mechanics of Heterogeneous Media* (Moscow: Nauka)
[8] Nigmatulin R I 1990 *Dynamics of Multiphase Media* Part 1 (New York: Hemisphere Publ.)

Analysis of interphase heat and mass transfer in suspension plasma spraying using air

Igor Gulyaev

Khristianovich Institute of Theoretical and Applied Mechanics SB RAS,
4/1 Institutskaya str., Novosibirsk, 630090, Russian Federation

gulyaev@itam.nsc.ru

Abstract. A theoretical model of interphase interaction "suspension particle - plasma flow" is proposed. The model assumes several successive stages of the evolution of liquid material: 0) entry into the plasma stream and break-up of the liquid into droplets; 1) heating the droplet to the evaporation temperature of the carrier fluid; 2) evaporation of the liquid, accompanied by convective mixing of the material; 3) evaporation of the liquid with the formation of a dry crust; 4) heating the solid material to melting; and 5) evaporation. Based on the analysis of the intensity of the processes of interfacial heat transfer, evaporation, diffusion and filtration mass transfer, relations are obtained for evaluating at each stage of the process duration, particle size, speed and distance passed in the flow. Demonstration calculations were performed for an aqueous ZrO_2 suspension with a solid phase concentration of 20% injected into the air plasma stream at a flow velocity of 2000 m / s and a flow temperature of 5000 K.

1. Introduction

Methods using suspensions (Suspension Plasma Spraying) and precursor solutions (Liquid Precursor Plasma Spraying) of materials are actively developing in the family of thermal spraying technologies. Interest in them is caused by the possibility of forming coatings with thicknesses from units to hundreds of micrometers with a fine structure of submicron scale at a productivity of several grams per minute [1]. To date, there are a large number of works demonstrating the features and advantages of such coatings; on the other hand, the dynamics of heat and mass transfer of suspension droplets in a plasma jet remains poorly studied. The main feature of such methods is the process of mass transfer of liquid from the droplet volume (evaporation) under the action of high plasma temperatures.

In the present work, an attempt was made to outline the main stages of the interaction of suspension droplets with plasma, to obtain analytical estimates of the dynamics of the main processes of heat and mass transfer. The results of calculations of the dynamics of evaporation, heating, and acceleration of droplets of an aqueous suspension of ZrO_2 particles in an air plasma jet are presented.

2. The scenario of suspension droplets evolution in plasma

Let's consider a single spherical drop of a suspension with a given initial concentration of solid phase φ_0, which is instantly placed in a uniform plasma stream (figure 1). The flow of a particle around a particle with a high-speed gas stream causes tangential stresses on the particle surface, which lead to convective mixing of the material [2] and, as a result, the temperature profile and particle concentration in the volume become uniform. After the particle reaches the boiling point of the liquid, its evaporation begins and the volume concentration increases up to the value at which the increased viscosity of the suspension stops mixing in the drop volume. Further evaporation of the liquid leads to the formation of a dry porous crust on the surface and the gradual spread of its internal boundary to the center of the particle. The porous crust presents a significant resistance to the diffusion mass transfer of liquid vapor, so there is an increase in pressure inside the particle. The decrease in internal pressure occurs due to the filtration flow of steam. Further heating of the solid phase leads to its melting and the formation of a drop of melt of the material.

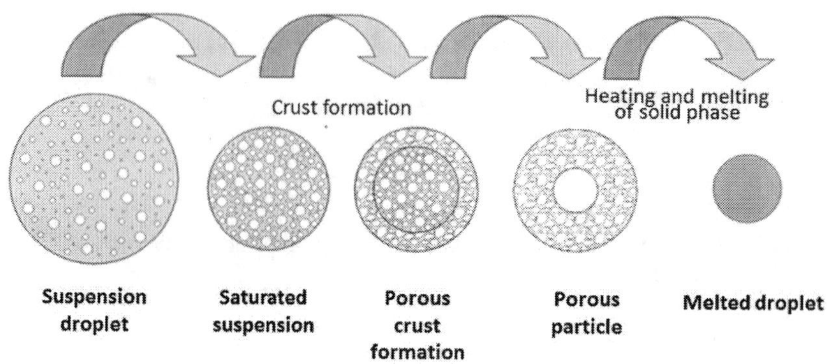

Figure 1. Schematic representation of the process of drying a suspension drop in a plasma stream.

For quantitative estimates, we will consider an aqueous suspension of submicron particles of zirconium dioxide with a concentration $\varphi_0 = 0.2$ injected into an air plasma with a temperature $T_f =$ 5000 K and a velocity $U_f =$ 2000 m/s. At all stages of particle motion in a plasma stream, the equations of one-dimensional particle motion and heat transfer are used [3].

The sizes (diameters) of droplets of suspension produced by injectors of various designs usually lie in the range of 10-30 microns [4]. After the droplet of the suspension enters the plasma stream, it breaks-up to a size $R_0 = We_{cr} / 2\rho_g U_f^2$ that corresponds to the critical Weber number $We_{cr} = 10$. The duration of the droplet spheroidization process is short and amounts to $D_p \mu / \sigma$. The concentration of the suspension at this stage does not change and remains equal to the initial value φ_0.

2.1. *Heating of droplets to a liquid evaporation temperature*
After droplet breaks-up to an equilibrium size, mixing of the material begins, caused by the tangential forces on the surface from the outside gas flow. The temperature and concentration of solid phase are equalized in volume. The drop is evenly heated to the boiling point of the liquid. The diameter of the droplet and the concentration of the suspension remain unchanged.

2.2. *Evaporation of material with convective mixing*
The evaporation of free liquid from the surface begins, the solid phase concentration increases, and the radius of the droplet decreases. All the heat flux coming from the stream is spent on evaporating the liquid. The temperature of the material remains constant, the concentration of solid phase increases, but remains uniform in volume due to mixing.

Mixing stops when the concentration of solid particles reaches a critical value of $\varphi_{fix} = 0.61$ [5], at which the viscosity becomes large enough to stop internal convection. The gel is formed. The radius of the drop decreases and is determined by the expression $R_{cr} = R_0 \sqrt[3]{\varphi_0 / \varphi_{fix}}$. The total heat flux to the particle Q remains unchanged, the evaporation rate of the material is

$$\frac{dm_{H20\_evap}}{dt} = \frac{Q}{L_{evap}} \approx 10^{-9} \text{ kg/s..}$$

The throughput of the gas boundary layer by diffusive mass transfer is higher than the liquid evaporation rate, therefore, the duration of this drying stage is limited by the heat flux to the particle.

2.3. *Evaporation without liquid mixing.*

A solid (porous) crust of radius R_{cr} is formed on the surface, the bulk density of which is equal to $\varphi_{cr} = (1 - porosity) = 0.74$. The crust has its effective coefficient of thermal conductivity $\lambda_{cr} = \lambda_{sol}\varphi_{cr}$ and diffusion coefficient of water vapor $D_{cr} \sim 10^{-6}$ m^2/s [6]. The particle size is fixed. The thickness of the dry crust is constantly growing. The "concentrated suspension - porous crust" interface moves toward the center of the particle. The speed of movement is determined by the rate of evaporation of the liquid from the suspension. Water vapor slowly passes through the forming crust, so their pressure rises sharply. The outer radius of the particle is already fixed and equal to R_{cr}, the final inner radius of the crust is determined from the mass balance (volume) of solid particles and is equal to

$$R_i = R_{cr}\sqrt{1 - \varphi_{fix} / \varphi_{cr}} .$$

Due to the fact that the thermal conductivity of the gas and the porous crust relate as $\lambda_g / \lambda_{cr} \sim 10^{-1}$ the surface temperature of the particle is much closer to the temperature of the evaporating liquid, but not to the temperature of the gas. Estimation shows that the particle surface temperature at a final crust thickness is approximately 400-500 K, which is quite far from the melting point of the solid phase.

Diffusive mass transfer of liquid vapor from the surface of the suspension through a dry crust and a boundary layer of gas into the free flow is determined by the expression:

$$\left.\frac{dm_{H_2O}}{dt}\right|_{diff} = \frac{2\pi(\rho_{v\_surf} - \rho_{v\_g})}{\dfrac{1}{D_g R_{cr} Sh} + \dfrac{\delta}{2D_{cr} R_{cr}(R_{cr} - \delta)}} = \frac{2\pi(\rho_{v\_surf} - \rho_{v\_g})}{MR_{film} + MR_{crust}} \approx 10^{-12} \text{ kg / s},$$

where $\rho_{v\_surf}, \rho_{v\_g}$ are the vapor density of the liquid on the evaporation surface and in the gas stream, D_g and D_{cr} - the vapor diffusion coefficient in the gas and the porous crust. As can be seen, the vapor diffusion mass transfer rate is 3 orders of magnitude lower than the liquid evaporation rate. In this case, the diffusion resistance of the crust MR_{crust} is approximately 3 orders of magnitude higher than the resistance of the gas boundary layer at the particle surface MR_{film}. The time for the evaporation of the entire liquid of 2 orders of magnitude is shorter than the time required for diffusion outflow of steam through the crust (at atmospheric pressure inside), therefore, an increase in gas pressure inside the cavity will occur very quickly, accompanied by a proportional increase in the rate of filtration mass transfer.

The rate of filtration flow of liquid vapor through a porous crust is proportional to the pressure gradient (differential) and is determined by Darcy's law: $v_{filt} = -\dfrac{K}{\mu_{vap}}\dfrac{\Delta P}{\Delta r}$, in which K, μ_{vap} are the crust permeability and the viscosity of the vapor. The mass transfer rate of vapor through the outer surface of a particle of a fixed radius R_{cr} calculated using the expression for the permeability of a porous medium [7] is equal to:

$$\left.\frac{dm_{H_2O}}{dt}\right|_{filt} = S_{surf}\rho_{vap}v_{filt} \approx 10^{-9} \text{ kg / s}.$$

Journal of Physics: Conference Series **1404** (2019) 012021 doi:10.1088/1742-6596/1404/1/012021

When water evaporates, its density decreases 2000 times: from 1000 to 0.560 kg / m3 (at atmospheric pressure). This leads to an instantaneous increase in pressure inside the pores of the particles and accelerate the filtration outflow of steam. Thus, the drying speed of the droplet is limited exclusively by the heat flux to the particle: all the liquid that has been evaporated will be removed from the particle due to the pressure gradient. Calculations show that the pressure value, which ensures the filtration mass transfer rate of vapors equal to the liquid evaporation rate, is approximately 3 atm.

The evaporation of free liquid from the surface begins, the solid phase concentration increases, and the radius of the droplet decreases. All the heat flux coming from the stream is spent on evaporating the liquid. The temperature of the material remains constant, the concentration of solid phase increases, but remains uniform in volume due to mixing.

2.4. *Heating the dry material to the melting temperature.*
The liquid is completely vaporized, the particle presents a porous frame. Further heating of the material leads to melting of the material (T_m = 3000K). The radius of the droplet somewhat decreases at this stage due to the fact that the solid skeleton of the main ZrO_2 material completely melts and takes the form of a dense sphere (without pores):

$$R_p = \sqrt[3]{\frac{3m_{sol}}{4\pi\rho_s}}.$$

2.5. *Heating the material to the boiling temperature.*
The entire heat flow that enters the particle is spent on heating the material to a boiling point (T_b =4573 K) and its evaporation.

3. The scenario of suspension droplets evolution in plasma

Figure 2 presents the dynamics of changes in the diameter and mass of the sprayed particle upon injection of a 20% wt ZrO_2 suspension into a one-dimensional plasma stream with a temperature of 5000 K and a velocity of 2000 m / s. As can be seen, the particle reaches a final diameter of 0.9 μm (melting of the material) at a distance of about 9 mm, while it has a velocity of 1800 m / s, close to the velocity of the gas stream. A comparison of the durations of the various stages shows that the processes of liquid evaporation during convective mixing of the suspension (stage 2) and the heating and melting of a dry particle take the longest time (stages 4 and 5).

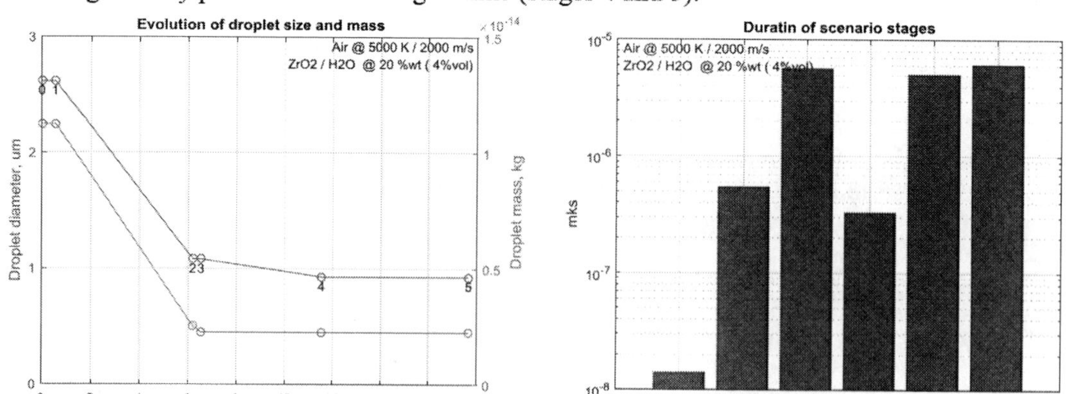

Figure 2. Evolution dynamics of 20% wt ZrO_2 suspension droplet placed in a one-dimensional air plasma stream with a temperature of 5000 K and a speed of 2000 m/s: a change in the diameter, mass, and comparison of the duration of the various stages of drying the droplet.

The developed model was used to analyze the influence of air plasma flow conditions on the characteristics of the sprayed particles. Calculations were carried out for the values of the initial concentration of the suspension $\varphi_0 = 0.01 \div 0.4$, flow rate $U_f = 500 \div 2500$m/s, and flow temperature $T_f = 3000 \div 5000$K. Typical values of the vapor pressure during drying of the porous crust are 1.5–4.0 atm, and the pressure increases with increasing flow rate, flow temperature, and suspension concentration.

An increase in the plasma flow velocity is expected to reduce the diameter of the sprayed particles. The minimum particle size values of 0.1 - 0.2 μm (for $\varphi_0 = 0.01 \div 0.05$) are achieved at the highest speed and lowest flow temperature. Plasma flow rates of less than 1500 m/s are unsuitable for suspension plasma spraying, since the distance of the complete melting of the material exceeds 50 mm. The temperature of the plasma flow weakly affects the dynamic and dimensional characteristics of the sprayed particles. An increase in the initial concentration of the suspension in the range $\varphi_0 = 0.01 \div 0.4$ leads to an increase in the size of the sprayed particles and from 0.2–0.3 μm to 1–1.5 μm, the distance of complete melting of the material from 3 mm to 20 mm.

4. Conclusion

A theoretical model of the interfacial interaction “suspension particle - plasma flow” has been developed. Based on the analysis of the intensity of the processes of interfacial heat transfer, evaporation, diffusion and filtration mass transfer, relations are obtained to estimate the duration of the stages, particle size, speed and distance traveled.

It is shown that the diffusion mechanism of mass transfer is not able to provide the vapor release rate corresponding to the intensity of the liquid evaporation. At the same time, the filtration mechanism of mass transfer allows vapor to be removed from the particle with an increase in internal pressure to 1.5–4 atm. The developed model was used to analyze the influence of air plasma flow conditions on the characteristics of the sprayed particles. The minimum particle sizes of 0.1–0.2 μm (for concentrations of 1–5%) are achieved at a high flow velocities of 2000–2500 K. Plasma flow velocities of less than 1500 m / s are not suitable for suspension plasma spraying. The temperature of the plasma flow weakly affects the dynamic and dimensional characteristics of the sprayed particles.

Acknowledgements

The research was carried out within the framework of the SB RAS program (project No. AAAA-A18-118021590026-1) and supported by the Russian Foundation for Basic Research (projects No. 18-08-01152,18-08-01475, 19-42-540008).

References

[1] Fauchais P and Montavon G 2010 *J. Therm. Spray Tech.* **19(1-2)** 226-239
[2] Solonenko O P and Gulyaev I P 2009 *Technical Physics Letters* **35(8)** 777-780
[3] Gulyaev I P and Solonenko O P 2013 *Thermophysics and Aeromechanics* **20(6)** 769-782
[4] Fauchais P, Rat V, Coudert J F, Etchart-Salas R and Montavon G 2008 *Surf. Coat. Tech.* **202** 4309–4317
[5] Ur’ev N B and Potanin A A 1992 *Flowability of Suspensions And Powders*, (Moscow: Chemistry)
[6] Nesic S and Vodnik J 1991 *Chem. Eng. Sci.* **46(2)** 527-537
[7] Tang G H, Tao W Q and He Y L 2005 *Physical Review* **E72** 056301

Calculation of rocks permeability based on periodic models of porous media

D E Igoshin[1,2] **and D Yu Legostaev**[1,2]

[1]Tyumen Branch of Khristianovich Institute of Theoretical and Applied Mechanics SB RAS, Taymyrskaya str., 74, Tyumen, 625026, Russia
[2]Tyumen State University, 6 Volodarskogo st., 625003 Tyumen, Russia

igoshinde@gmail.com

Abstract. The paper proposes a method for calculating rock permeability based on periodic cubic models of porous media, taking into account the particle size distribution of the samples. Conditional permeability is calculated for each fraction. The permeability of the sample depends on the relative position of the fractions and can be calculated as the weighted average (arithmetic, geometric or harmonic). Next, an optimization problem is solved, which allows one to choose the "weight" of a particular method of averaging permeability over fractions inside the sample for all samples of the selected formation. Results were obtained for several formations from different fields.

1. Introduction

The main filtration-capacitive characteristics of porous media are porosity and permeability (absolute and phase). In the study of rocks, an important measured quantity, in addition to porosity and permeability, is the particle size distribution. The permeability of the rock is a complex function, determined by the granulometric composition of the grains composing the rock, their shape and relative position. As a rule, core data are many points on the "porosity – permeability" diagram, with a pronounced positive correlation [1].

2. Particle size distribution of core

The particle size distribution of the core sample is a histogram of the percentage of grains for N fractions. In Figure 1 shows the particle size distribution averaged over approximately two hundred samples from the UZ formation of one of the deposits in Western Siberia ($N = 14$). Additionally, the function of the density distribution of grain sizes is given, taking into account the average size of the fractions and the width of their ranges. It can be seen that the highest distribution density is realized for grains less than 1 μm in size; for grains larger than 100 μm, the distribution density drops sharply. A practically important task is the calculation of the permeability of core samples taking into account their particle size distribution.

3. Porosity and permeability of cubic structures

In this paper, we use the approach for determining the permeability of rocks using model periodic porous media described in [1, 2] with a skeleton of spherical segments based on cubic lattice systems: primitive cubic (PC), body-centered cubic (BCC) face-centered (FCC). The model parameters in these structures are the side of the cube L and the degree of intersection of the spheres α (dimensionless). The table 1 shows the analytical expressions for the porosity of cubic structures and its limiting values.

At [4; 5] for the structures under consideration, hydrodynamic modeling of the stationary flow of a viscous incompressible fluid was carried out. Based on the numerical solution of the Navier – Stokes system of equations and the Darcy law for media formed by these structures, the permeability is determined depending on the dimensionless parameter α: $k = k(\alpha)$ and porosity m: $k = k(m)$ at $L = 10^{-5}$ *m*.

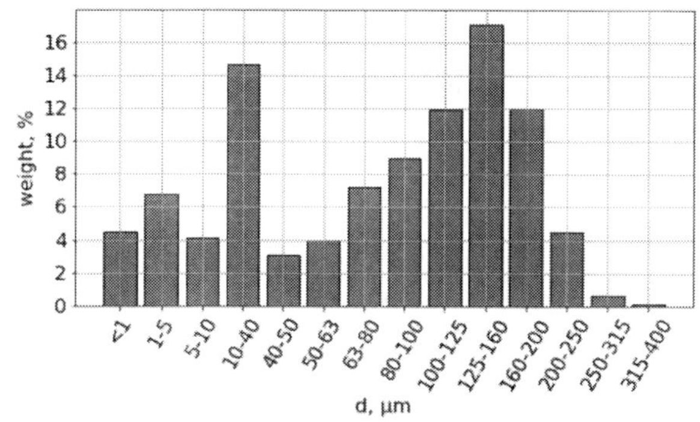

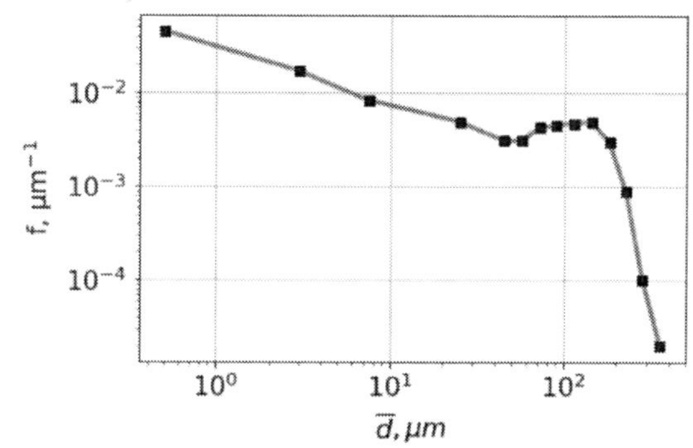

Figure 1. The average particle size distribution in the reservoir (left) and the function of the grain density distribution over the average size of the fractions (right).

Table 1. Porosity of cubic structures.

Structure	Porosity, $m = m(\alpha)$	$m(0)$	α_*	$m(\alpha_*)$	β^2
PC	$1-(\pi/12)\cdot[2-3\alpha^2(3-\alpha)]/(1-\alpha)^2$	0.4764	0.2929	0.0349	1
BCC	$1-(\pi\sqrt{3}/8)\cdot[1-2\alpha^2(3-\alpha)-1.5\alpha_2^2(3-\alpha_2)]/(1-\alpha)^2$ $\alpha_2 = 1-2(1-\alpha)/\sqrt{3} \quad at \quad \alpha > 1-\sqrt{3}/2$	0.3198	0.1835	0.0055	3/4
FCC	$1-(\pi\sqrt{2}/6)\cdot[1-3\alpha^2(3-\alpha)]/(1-\alpha)^2$	0.2595	0.134	0.0359	1/2

According to the π-theorem, the permeability of the medium can be represented in the form of products of a dimensionless function and a scalable parameter.

$$k = d^2\tilde{k}\,. \tag{1}$$

The reduced permeability $\tilde{k}$ depends only on the type of structure and does not depend on the characteristic grain size. For cubic periodic structures

$$d = \beta L/(1-\alpha), \tag{2}$$

where β value is given for each type of cubic structure in the table.

Based on system (1) – (2), we formally write down the permeability values of the i-th fraction and the values determined numerically in [3, 4]:

$$k_i = d_i^2\tilde{k}, \quad k_{num} = [\beta L/(1-\alpha)]^2\tilde{k},$$

where d_i is the average diameter of the i-th fraction, whence the conditional permeability of the i-th fraction is

$$k_i = [(1-\alpha)d_i/\beta L]^2 k_{num}. \tag{3}$$

4. Permeability at different locations of fractions

If the fractions in the rock are located in parallel, then its permeability reaches a maximum and is determined by the arithmetic mean weighted by fractions. With a random mutual arrangement of fractions, the permeability is determined by the weighted geometric mean. With a sequential (layered) arrangement of fractions, the permeability is determined by the weighted average harmonic and takes a minimum value.

Figure 2 shows the weighted arithmetic mean (WAM), weighted geometric mean (WGM) and weighted harmonic mean (WHM) permeability estimates, taking into account the particle size distribution based on the CP structure. It can be seen that the WHM is permeability estimate is significantly closer to the core data than the other two estimates, which are 1-3 orders of magnitude higher. This is due to the fact that in the rock, large grains are usually surrounded by smaller grains that gravitate towards the necks of the pores and decisively affect the permeability of the medium. Moreover, the proportion of large pores is negligible. Flow through a porous media occurs mainly through a network of channels through small fractions.

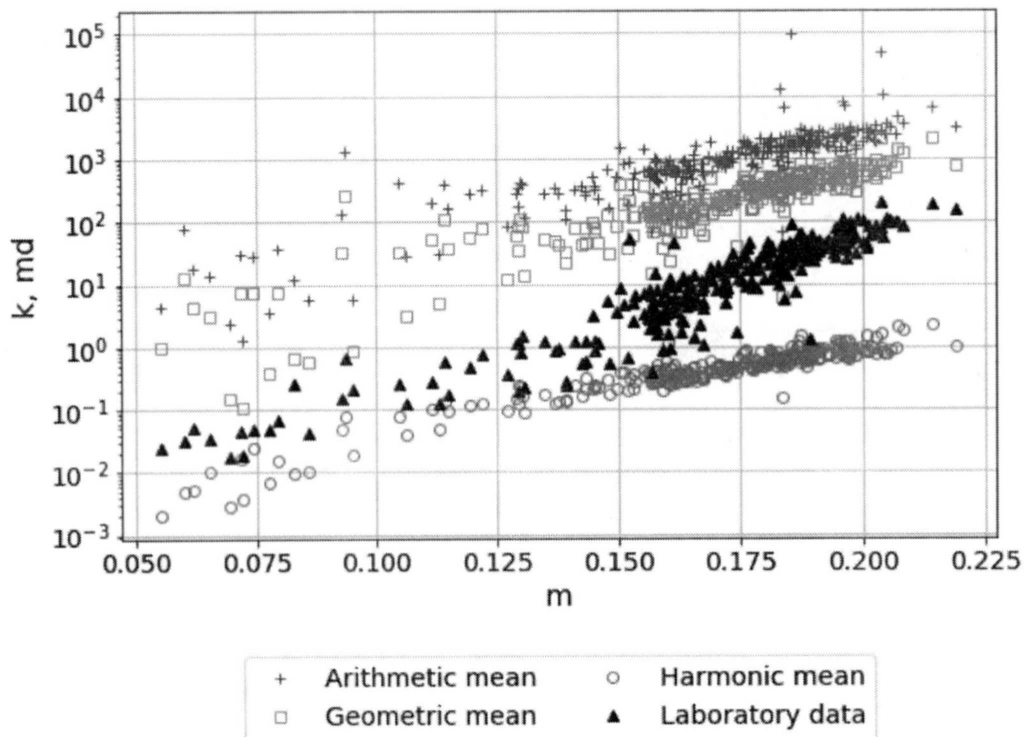

Figure 2. Porosity-permeability diagrams for various averaging methods and for laboratory data.

We calculate the effective value of core permeability based on (3) as WGM is the value between the WGM and WHM estimates for fractions:

$$k_c = \bar{k}_g^{\sigma} \bar{k}_h^{1-\sigma} \tag{4}$$

where σ is the "weight" of the permeability of the medium with a random stacking of fractions determined by the WGM — estimate, $\sigma \in [0;1]$. We find the weight σ in the form

$$\sigma = Am + B \tag{5}$$

by minimizing the objective function

$$J = \sum_{j=1}^{n} \frac{k_{f,j} - k_{c,j}}{k_{f,j}}, \tag{6}$$

indices f and c refer to actual and calculated values respectively, n is the number of core samples.

When approximating filtration-capacitive properties, the exponential dependence of permeability on porosity is often used:

$$k = k_0 \exp\left(\frac{m}{m_0}\right) \tag{7}$$

Figure 3 shows the results of the calculation of permeability for reservoir E1 and core data. It is seen that the proposed method, taking into account the particle size distribution, gives a better result than the exponential permeability estimate (7) based on the least squares method.

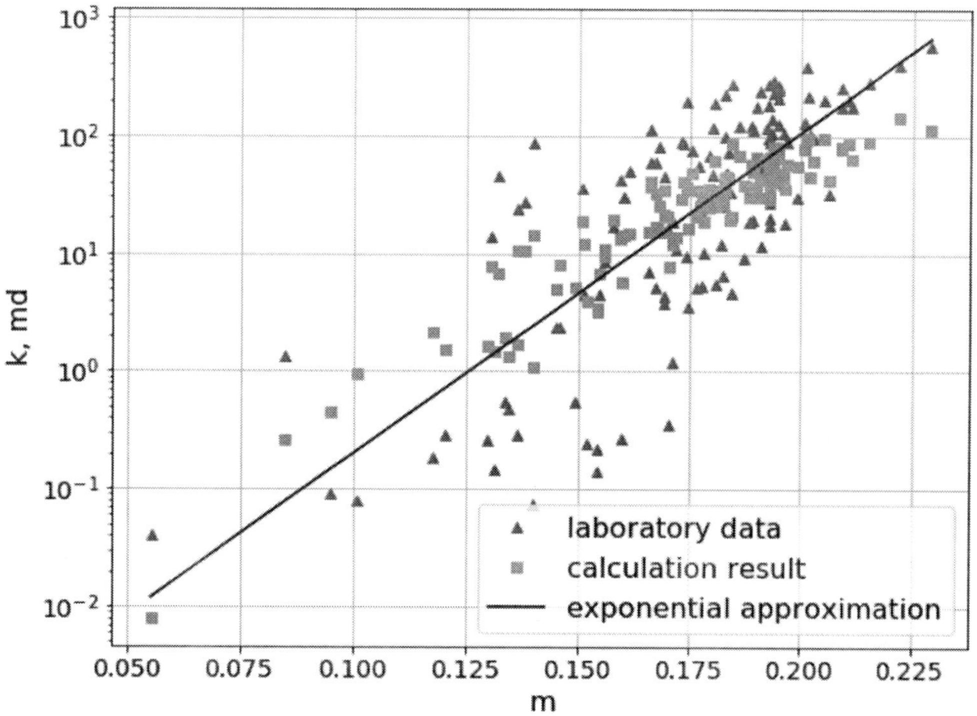

Figure 3. Porosity-permeability diagrams for various averaging methods and for laboratory data.

Table 2 shows the results of the optimization problem (4) - (6) and the parameters of the exponential dependence of permeability on porosity (7). In 6 cases out of 10, the proposed method gives the best results.

Table 2. Calculation results.

		Formation									
		E1	E2	E3	P1	P2	N	V1	MV1	MV2	MZ
Cubic	Object	1.558	1.586	1.329	0.556	0.277	0.551	1.906	0.452	1.776	0.472
	A	0.267	0.129	0.027	0.032	0.006	0.402	-0.005	1.717	-0.008	1389
	B	-0.012	0.000	-0.001	-0.001	0.002	-0.018	0.008	-0.079	0.003	-0.064
Exponen	Object	2.161	2.933	1.393	0.542	0.293	0.470	0.829	0.675	0.831	0.555
	k_0, md	3.8E-04	7.2E-04	1.7E-05	4.9E-04	1.8E-03	3.6E-04	2.1E-03	5.9E-03	6.6E-03	1.4E-02
	m_0	0.016	0.019	0.014	0.020	0.025	0.017	0.022	0.029	0.044	0.031

It can be assumed that a three-parameter model of a periodic medium [5] allows one to obtain a better approximation of permeability.

Acknowledgments
The research was carried out within the framework of the Program of Fundamental Scientific Research of the state academies of sciences in 2013-2020 (project No. AAAA-A17-117030610130-1).

References
[1] Igoshin D E, Nikonova O A and Mostovoy P Ya 2014 *Tyumen State University Herald (Physical and Mathematical Modeling. Oil, Gas, Energy)* **7** 34
[2] Igoshin D E and Nikonova O A 2015 *Tyumen State University Herald (Physical and Mathematical Modeling. Oil, Gas, Energy)* **1-2** 131
[3] Igoshin D E and Saburov R S 2015 *Tyumen State University Herald (Physical and Mathematical Modeling. Oil, Gas, Energy)* **1-1** 84
[4] Igoshin D E 2015 *Automation, Òelemechanization and Communication in Oil Industry* **12** 30
[5] Igoshin D E and Khromova N A 2016 *Tyumen State University Herald (Physical and Mathematical Modeling. Oil, Gas, Energy)* **2-3** 107

Journal of Physics: Conference Series **1404** (2019) 012023 doi:10.1088/1742-6596/1404/1/012023

Mathematical model of the submerged jet for the cases of man-made spills

S R Kildibaeva[1], I K Gimaltdinov[2,3] and G R Kildibaeva[1]

[1]Applied Informatics and Programming Department, Sterlitamak Branch of Bashkir State University, 49 Lenin avenue, Sterlitamak, Republic of Bashkortostan, Russia
[2] Physics Department, Ufa State Oil Technical University, 1 Cosmonauts street, Ufa, Republic of Bashkortostan, Russia.
[3] The Academy of Sciences of Bashkortostan, 15 Kirova street, Ufa, Republic of Bashkortostan, Russia.

E-mail: freya.13@mail.ru

Abstract. The paper considers a mathematical model of the spread of a multiphase submerged jet, the flow of which is associated with a well damage during the development of a field or pipeline during oil transport. The jet consists of oil droplets and gas bubbles, as well as water involved into the jet. The case is considered when the environmental conditions correspond to the conditions of a stable hydrate existence, because a hydrate shell begins to form and the gas bubble turns into a hydrate on the surface of the bubbles. The integral Lagrangian control volume method is used to describe the jet spread process. The thermo-physical characteristics of a multiphase submerged jet, depending on the vertical coordinate, as well as the trajectory of the jet, are obtained.

1. Introduction

Due to the increase in oil production in the deep-water fields of the world's oceans, the probability of oil spills into the ocean increases. Often such spills are man-made, occurred as a results of well, oil pipelines and drilling rigs damage. The cause of the spill may be corrosion of the pipeline, mechanical damage to the oil production structure, fire, etc. Oil spills in the Gulf of Mexico on the Deepwater Horizon platform, Ixtoc-1, as well as in the Gulf of Alaska Exxon Valdez are the prime examples indicating the need to study the leaks eliminating methods [1,2]. Over the past decade, more than 1 billion gallons of oil have been spilled into the world's oceans as a result of accidents around the world, indicating the serious scale of the problem. [3]

In this regard, there is a need to study the characteristics of the spread of the petroleum products that fall into the ocean as a result of accidental spills. Forecast migration of petroleum products will reduce the time of the leakage elimination [4,5]. To do this, it is important to study the features of the oil and associated gas spread, their interaction with the surrounding water. The study of the undercurrents effects on the oil spread in the environment is also important. To predict the spread of a multiphase submerged jet, the integral Lagrangian control volume method described in [6-8] is used. According to this method, the jet is considered as a sequence of control volumes, each characterized by the temperature, speed, density and volume content of the components included in the control volume. Knowing these parameters of the control volume, you can get similar information about the jet.

Under the conditions of stable hydrate existence on the bubbles surface, a hydrate shell is formed, turning the gas bubble into a hydrate one. The process of hydrate formation should be given sufficient attention, as it can affect the dynamics of the jet. An important role is played by the flow of the environment, which acts on the submerged jet. Under the effects of this flow, the jet can deviate, changing its trajectory, radius and concentration of substances entering the jet.

2. Basic equations

Let us assume there is some source with radius r at the bottom of the reservoir from which a mixture of oil and gas flows with a known initial temperature (T^e) and volume flow rates (Q_o^e) and (Q_g^e). The temperature (T_w) and the velocity ($\vec{V}_w$) of the surrounding water are also known. Oil spreads in the form of droplets, and gas leaves in the form of bubbles, which at the initial moment are in the center of the jet. Because of the surrounding water flow, the jet can deviate, and the bubbles leave the jet.

The integral Lagrangian control volume method (ILCVM) proposed in [7,8] is used to simulate the flow of a submerged jet. According to this method the jet (figure 1) consists of a series of control volume (CV) cylinders, characterized by the height h, radius b and spatial coordinates (x, y, z), corresponding to the center of the axial section of the cylinder (hereinafter the center of the cylinder).

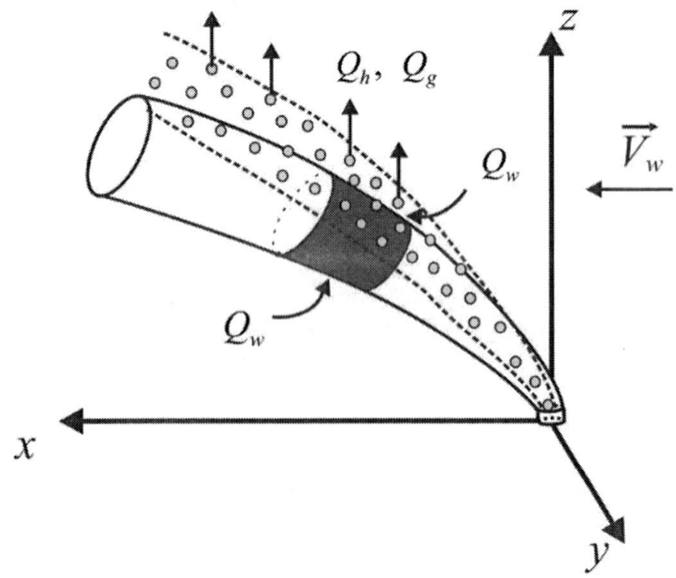

Figure 1. Scheme of a submerged jet.

The equation of mass conservation CV:

$$\frac{dM}{dt} = \rho_w Q_w - \rho_{com} Q^f, \rho = \sum_i \alpha_i \rho_i, \sum_i \alpha_i = 1, \quad (1)$$

here $M = \sum_i M_i$, ($i = o, g, h, w$) is the CV mass; $Q^f = Q_g^f + Q_h^f$ is the volume bubbles flow "leaving" the CV due to the jet curvature, α_i is the volume contents of the corresponding components in the CV, ρ_{com} is the bubble density, that can be gaseous, and gaseous with a hydrate shell or completely hydrate. Here and further the subscripts o, g, h, w refer respectively to the parameters of oil, gas, hydrate and water. The initial values of the size of the control volume correspond to the radius of the well r, where hydrocarbons come from: $h_0 = r, b_0 = r$.

Due to the formation of hydrate shells on the gas bubbles surface, the gas, hydrate and water masses change as follows:

$$\frac{dM_g}{dt} = -J_g - \rho_g Q_g^f, \quad \frac{dM}{dt} = J_h - \rho_{hh} Q_h^f, \quad (2)$$

here M_i, ρ_i (i = o, g, h, w) are the mass and density of the corresponding components in the CV, Q_w is the volume flow of the surrounding water involved in the jet, accordingly, J_w, J_g are the intensity of water and gas consumption in the hydrate formation, J_h is the intensity of gas hydrate formation.

Gas, water and hydrate flow intensities are related by the following relationships:

$$J_g = G \cdot J_h, J_w = (1-G)J_h, J_h = N4\pi a_{gh}^2 j_h, \quad (3)$$

here G is the hydrate number, N is the number of bubbles in the CV, a_{gh} is the radius of composite bubbles consisting of a gas core and a hydrate shell, j_h is the intensity of hydrate formation related to the bubble surface area.

The equations of the momentum conservation for CV formulated taking into account the action on the jet of three-dimensional flow of the environment have the following form:

$$\begin{aligned} &\frac{d}{dt}(Mu) = u_w \rho_w Q_w - u\rho_{com} Q^f, \\ &\frac{d}{dt}(M\upsilon) = \upsilon_w \rho_w Q_w - \upsilon\rho_{com} Q^f, \\ &\frac{d}{dt}\left[(M_w + M_o)w + (M_g + M_h)\cdot(w+w_b)\right] = w_w\rho_w Q_w - (w+w_b)\rho_{com}Q^f + \\ &+(\rho_w - \rho_l)\pi b^2 h(\alpha_w + \alpha_o)g + (\rho_w - \rho_{com})\pi b^2 h(\alpha_g + \alpha_h)g, \end{aligned} \quad (4)$$

here u, υ, w are the CV velocity $\vec{V} = u\vec{i} + \upsilon\vec{j} + w\vec{k}$ components, u_w, υ_w, w_w are the components of the water flow velocity $\vec{V_w} = u_w\vec{\imath} + \upsilon_w\vec{\jmath} + w_w\vec{k}$ of the ambient CV, g is the gravity acceleration, $\rho_l = \alpha_w\rho_w + \alpha_o\rho_o$ is the density of liquid component in CV.

The CV coordinates are defined by the following expressions:

$$\frac{dz}{dt} = w, \quad \frac{dx}{dt} = u, \quad \frac{dy}{dt} = v. \quad (5)$$

The equation of conservation of energy CV has the following form:

$$\frac{d}{dt}(cMT) = c_w T_w \rho_w Q_w + J_h L - J_w c_w T_{hs}, \quad c = \sum \chi_i c_i, \quad (6)$$

here c, c_w, T, T_w is the heat capacity and the temperature of CV and water, L is the heat of the phase transitions. c_i, $\chi_i = \frac{M_i}{M}$ (i = o, g, h, w) are the heat capacity and the mass content of oil, gas, hydrate and water in CV.

3. Calculation results

To analyze the thermo-physical characteristics of the submerged jet, we solve equations numerically (1)-(5). Let us consider the initial parameters of the system corresponding to the conditions of stable hydrate existence. Here and further the calculations are given for the initial parameters of the system: r = 0.25 m, $Q_o^e = 0.1$ m^3/s, $Q_g^e = 0.1$ m^3/s, $T_w = 4\,°$C, $T^e = 80\,°$C, p = 15 MPa, $\rho_w = 1030$ кg/m^3, $\rho_o = 650$ кg/m^3, $\rho_g = 99.4$ кg/m^3, $\rho_{\square} = 910$ кg/m^3, $\varphi = 90°$, $\theta = 0°$, $u_w = 1$ m/s, $\upsilon_w = 0$ m/s, $w_w = 0$ m/s, $c_o = 2090$ J/(кg K), $c_g = 2365$ J/(кg K), $c_h = 2100$ J/(кg K), $c_w = 4200$ J/(кg K), $\lambda_w = 0.58$ W/(m·K), $\mu_w = 1.57 \cdot 10^{-3}$ Pa·s, $L = 5 \cdot 10^5$ J/кg, $G = 0.12$, $D^* = 10^{-10}$ m^2/s, $g = 9.8$ m/s^2.

The trajectory of the jet is given in the figure 2a. In the graphs we see that the jet is strongly deflected under the influence of the flow of the environment. Because of the process of involving the surrounding water into the jet, the jet expands.

The dependence of the jet temperature on the vertical coordinate is given in the figure 2b. The jet temperature is calculated for two cases: the line 1 corresponds to the case when hydrate formation is limited by heat exchange, the case when the hydrate formation is limited by the gas diffusion through the hydrate shell is represented by the line 2. The jet temperature decreases because of the involvement of cold ambient water into the jet and reaches the equilibrium temperature of hydrate formation $T_{hs} = 17°$ C at the height $z = 0.5$ m. You can see also in the graph that the line 2 is located above the line 1. This is due to the exothermic process of hydrate formation, in which the heat releases heating additionally the jet.

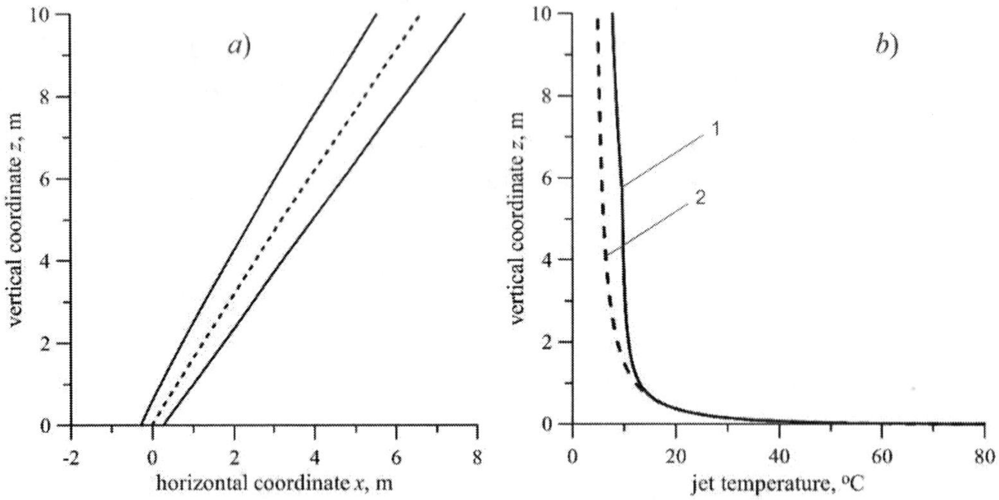

Figure 2. Jet trajectory (a), the dependence of the jet temperature on the vertical coordinate (b).

The dependence of the jet density on the vertical coordinate is presented in the figure 3a. The jet consists only of oil droplets and gas bubbles at the initial stage. Then the surrounding water begins to flow into the jet, as a result of which the jet density increases.

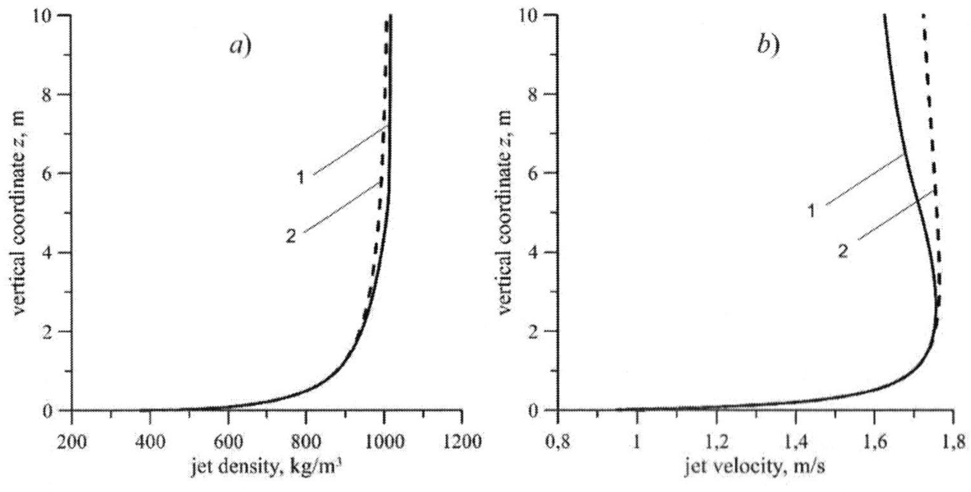

Figure 3. The dependence of jet density (a) and jet velocity (b) on vertical coordinate.

The dependence of the CV velocity on the vertical coordinate z for two cases is presented in the figure 3b. The line 1 corresponds to the case when the hydrate formation is limited by the heat exchange with the environment; the line 2 presents the limitation of the hydrate formation by gas diffusion. The jet velocity increases to a value of $z = 2.3$ м. This is due to the action of initial pulse on the jet. Then the jet velocity decreases with coordinate increasing due to the capture of the surrounding water and, as a consequence, the jet weighting.

4. Conclusion

A mathematical model of the flow of a multiphase submerged jet is considered in this paper. An integral Lagrangian control volume method is developed for the study of multiphase submerged jets taking into account the hydrate formation and the action of three-dimensional flow of the environment. The trajectory of the jet, dependences of the jet temperature, velocity and density on the vertical coordinate are obtained. As a result of numerical calculations, the dependencies of the thermo-physical parameters of the submerged jet (temperature, velocity, density) on the vertical coordinate and the jet trajectory are obtained. The influence of the three-dimensional flow of the surrounding water on the submerged jet dynamics is also analyzed.

Acknowledgments
The reported study was funded by RFBR according to the research project № 18-31-00264mol-a.

References

[1] Mason Ch F 2019 Regulating Offshore Oil and Gas Exploration: Insights from the Deepwater Horizon Experience in the Gulf of Mexico Review of environmental economics and policy vol 13 chapter 1 pp 149-54
[2] Garcia Giraldo D and Yeung R W 2019 The 2010 Oil Spill in the Gulf of Mexico: Flow-Rate Estimation Based on Satellite-Images *Analysis Journal of offshore mechanics and arctic engineering-transactions of the ASME* **141** 246-51
[3] Abdul-Hamid 2013 The OPEC annual statistical bulletin (Vienna: OPEC)
[4] Gimaltdinov I K, Kildibaeva S R 2015 About the theory of initial stage of oil accumulation in a dome-separator *Thermophysics and Aeromechanics* **22** 387-92
[5] Gimaltdinov I K and Kildibaeva S R 2018 On the theory of accumulation of hydrocarbons in a dome used to eliminate a technogenic spill at the bottom of the ocean *J. of Engineering Physics and Thermophysics* **91** 246-51
[6] Yapa P D, Wimalaratne M R, Dissanayake A L and DeGraff J A 2012 How does oil and gas behave when released in deepwater? *Journal of Hydro-environment Research* **6** 275-85
[7] Zheng L, Yapa P D and Chen F 2002 A model for simulating deepwater oil and gas blowouts – part I: theory and model formulation *J. of hydraulic research* **41** 339-51
[8] Gimaltdinov I K and Kildibaeva S R 2018 Model of a submerged jet accounting for two limiting schemes of hydrate formation *Thermophysics and Aeromechanics* **25** 75-83

Study of substrate temperature dependence upon nozzle traverse speed at cold spraying

S V Klinkov, V F Kosarev and V S Shikalov

Khristianovich Institute of Theoretical and Applied Mechanics, SB RAS, 630090, Institutskaya str., 4/1, Novosibirsk, Russia

klyiii@yandex.ru

Abstract. The paper presents the results of investigation of dependence of substrate temperature upon nozzle traverse velocity under cold spray conditions. A model is proposed for prediction of substrate temperature. Experimental and simulated data were obtained in the typical range of change in the nozzle traverse speed (1-400 mm/s) and air stagnation temperatures (200-600°C), encountered in the practice of cold spraying. The simulated data are in good agreement with experimental. An approximate analytical equation is proposed for estimation of substrate temperature during cold spraying in a wide range of gas stagnation temperatures and nozzle traverse velocities.

1. Introduction

Cold gas-dynamic spraying is a rapidly developing method, which originates from the discovery of the phenomenon of coating formation upon impact of non-molten metal particles in Khristianovich Institute of Theoretical and Applied Mechanics, Siberian Branch of Russian Academy of Sciences, in the early 1980s [1].

The presence of a critical (minimum) particle impact velocity [2], the presence of a deposition delay [3], the necessity for plastic deformation of particles, the ability to produce coatings on non-metallic materials (glass, ceramics, building materials [4-6]) etc. are the main regularities of coating formation. Cold spraying enables to obtain composite coatings [7-14]. Gas dynamics of flows in De Laval nozzles [15-19], as well as in nozzles of other geometries [20-22] are important for successful deposition because for acceleration of powder particles gases like air, nitrogen and helium are often used. At present, worldwide, further researches are being conducted and industrial technologies are being developed using cold spraying.

In practical use, a continuous coating is created by scanning the substrate surface with a two-phase jet at a given nozzle travel speed. Since the surface temperature of the substrate affects the spraying process [1, 23-31], the study of its dependence on the nozzle travel speed is an urgent task. This paper presents preliminary (without taking into account the contribution of particles) experimental and simulated data in the typical range of change in the nozzle travel speed (1-400 mm/s) and air stagnation temperatures (200-600°C), encountered in the practice of spraying.

2. Experimental

The measurements were carried out by using an assembly comprising a base to which a layer of ceramic thermal insulator is attached, on which aluminum substrate is placed sized 50 × 50 × 1 mm The substrate was pressed by 1 mm thin steel frame, also through a ceramic insulator. The frame had a rectangular cutout sized 40 × 20 mm, through which the jet impinged on the substrate. The first thermocouple (for measuring the jet temperature) was installed in front of the substrate and frame. The second and third thermocouples were placed flat to the substrate surface symmetrically relative to its middle. The distance between second and third thermocouples was 20 mm. Air flow parameters were the next: stagnation pressure 3 MPa, stagnation temperature 200, 400 and 600°C. An axisymmetric profiled supersonic nozzle with an estimated exit Mach number of 3 (critical nozzle diameter 2.8, exit nozzle diameter 6.5 mm) was used to produce supersonic jet.

3. Model description

For computation, a grid was used with a cubic cell which side was equal to the substrate thickness. We consider relatively thin substrates, the thickness of which is many times smaller than the characteristic dimensions in plane, i.e., it is assumed that the temperature gradient along the substrate thickness is negligible. In the first step, the heat fluxes are calculated through the faces of each cell (numbered i along x-axis that is directed along nozzle traveling speed, and j along y-axis that is perpendicular to the nozzle traveling speed):

$$qx_{i,j} = -\lambda h dy \frac{T_{i,j} - T_{i-1,j}}{dx},\; qy_{i,j} = -\lambda h dx \frac{T_{i,j} - T_{i,j-1}}{dy},\; qw_{i,j} = -\alpha dx dy \left(T_0 - T_{i,j}\right). \quad (1)$$

Here, $T_{i,j}$ is a substrate temperature in cell (i, j), λ is a heat conductivity of substrate material, h is substrate thickness, α is a coefficient of heat transfer between jet and substrate, T_0 is jet stagnation temperature over cell (i, j), dx is a cell size in x-direction, dy is a cell size in y-direction.

For an axisymmetric jet, the distribution of the heat transfer coefficient and the stagnation temperature along the substrate surface is taken as:

$$T_0(r) = \frac{T_{00}}{\left(1 + 15\left(\frac{r}{r_T}\right)^2\right)^{0.25}},\; \alpha(r) = \frac{\alpha_0}{\left(1 + 15\left(\frac{r}{r_\alpha}\right)^2\right)^{0.25}}. \quad (2)$$

Here, T_{00} is a stagnation temperature in nozzle pre-chamber, α_0 is a heat transfer coefficient at the jet axis, r_T is characteristic coordinate where stagnation temperature is twice lower than that at the jet axis, r_α is an analogous coordinate for heat transfer coefficient. In simulation we adopted α_0 $=7\cdot10^3$ W/K m^2, $r_T = 25$ mm, $r_\alpha = 4$ mm. The last two values are taken on the basis of previous studies [1]. As for α_0, it should be noted that its value was selected to achieve the best match of the measurement results at the jet stagnation temperature 400°C. Further, in all presented cases of calculation, it was kept constant.

In the second step of the simulation the average temperatures of the substrate cells are calculated:

$$T_{i,j}^{k+1} = T_{i,j}^{k} + \left(qx_{i,j} - qx_{i+1,j} + qy_{i,j} - qy_{i,j+1} + qw_{i,j}\right)\frac{dt}{c\rho h dx dy}. \quad (3)$$

Here, c is a heat capacity of substrate material, ρ is a density of substrate material, k is a number of time steps dt.

In the case of nozzle moving at a constant speed u_n, the radius in the distributions of stagnation temperature and heat transfer coefficient (2) for each cell is calculated by the formula:

$$r_{i,j} = \sqrt{\left(x_i - s\right)^2 + \left(y_j - L_y / 2\right)^2},\; s = u_n \cdot k \cdot dt. \quad (4)$$

Here, L_y is a width of the substrate.

4. Results and discussion

Figure 1 shows the experimental and calculated data performed for three different air stagnation temperatures. The substrate temperatures shown were taken as the average between the maximum readings of the two thermocouples. Further, the approximation of the calculated and experimental data was carried out. It is seen that in the logarithmic coordinates of the nozzle travel speed there is almost a linear relationship. Deviation from this dependence takes place at the most extreme points.

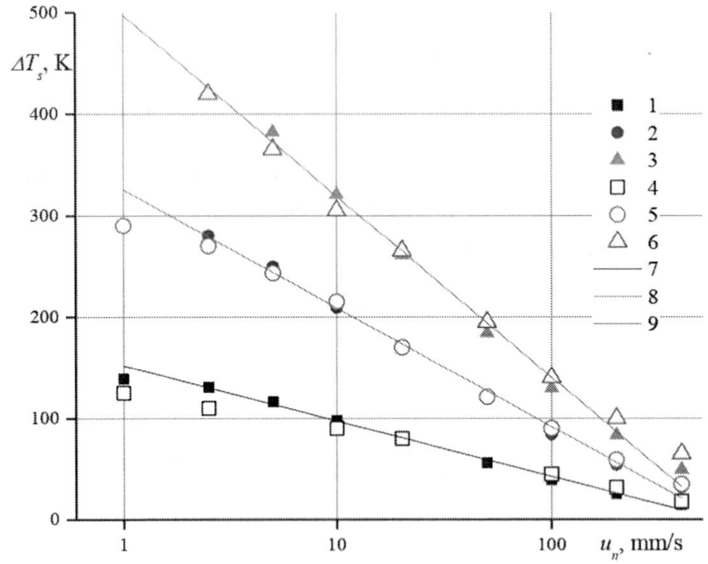

Figure 1. Dependence of the substrate temperature gain (relative to the initial temperature) on the nozzle travel speed. 1-3 simulation results, 4-6 experimental results, 7-9 approximation; 1, 4, 7 – stagnation temperature 200°C; 2, 5, 8 – 400°C; 3, 6, 9 – 600°C.

A generalization of the obtained in simulation and experimental data was also carried out, as shown in figure 2. For this generalization a reference nozzle travel speed of 10 mm/s was selected; the substrate temperature gain at this speed was selected as the base. All substrate temperatures were divided by this base temperature, so that all curves of calculations and experiments intersect at this one point in the plot.

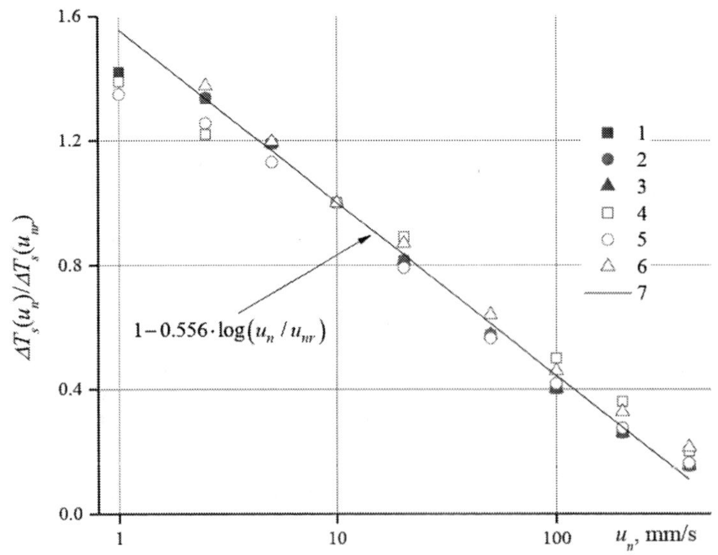

Figure 2. General dependence of the relative substrate temperature on the nozzle travel speed. 1-3 simulation results, 4-6 experimental results, 7 approximation by formula $1-0.556\cdot\lg(u_n/u_{nr})$; 1, 4 – stagnation temperature 200°C; 2, 5 – 400°C; 3, 6 – 600°C.

As it can be seen, the results are described by a single dependence with a good accuracy, at least in the studied range of stagnation temperatures and nozzle traveling speeds. The values of the base temperature are equal to 97.8, 209.5, and 321.25 C for the stagnation temperatures 200, 400, and 600°C, respectively. In the first approximation it is possible to estimate this base temperature as a half of the stagnation temperature.

The results of calculations and experiments in the stagnation temperature range of 400 and 600°C coincide with a good accuracy. At stagnation temperature 200 C there is some difference (not exceeding 30-35°C, see figure 1). Thus, for practical estimates the simple expression (5) can be used, obtained as a result of the present work:

$$\Delta T_s = \Delta T_{sr}\left(1 - 0.556 \cdot \lg\left(u_n / u_{nr}\right)\right),\ \text{where } u_{nr} = 10\ \text{mm/s},\ \Delta T_{sr} \approx 0.5 T_{00}. \tag{5}$$

These relations, however, are true only for experimental conditions: a thin (1 mm) aluminum heat-insulated substrate fixed by a frame with a slot. When using thicker substrates fixed without thermal insulation, lower substrate temperatures should be expected. Therefore, the values obtained here should be taken as the maximum possible, at least for aluminum substrates.

5. Conclusion

Dependence of substrate temperature upon nozzle traverse speed is investigated under cold spray conditions. For the first time, experimental and simulated data were obtained in a wide range of change in the nozzle traverse speed (1-400 mm/s) and air stagnation temperatures (200-600°C), typically encountered in the practice of cold spraying. The simulated data are in good agreement with experimental. It is shown that there is almost a linear relationship between substrate temperature and logarithm of nozzle traverse velocity. Moreover, the results can be described by a single generalized dependence with a good accuracy, at least in the studied range of stagnation temperatures and nozzle traveling speed. An approximate analytical equation was proposed for estimation of substrate temperature during cold spraying in a wide range of gas stagnation temperatures and nozzle traverse velocities.

Acknowledgments

The research was supported by RFBR Grant № 19-08-00538\19 using equipment of Center Mechanics of ITAM SB RAS.

References

[1] Papyrin A, Kosarev V, Klinkov S, Alkhimov A and Fomin V 2007 *Cold Spray Technology* (Amsterdam: Elsevier Science) p 336

[2] Alkhimov A P, Kosarev V F and Papyrin A N 1990 *Soviet Physics Doklady* **35** 1047–9

[3] Klinkov S V and Kosarev V F 2006, *J. Therm. Spray Technol.* **15** 364–71

[4] Shikalov V S, Klinkov S V and Kosarev V F 2016 *Proc. 18th Int. Conf. on the Methods of Aerophysical Research (ICMAR 2016)* AIP Conf. Proc. **1770**, ed V M Fomin (Mellville NY: American Institute of Physics) p 040015 https://doi.org/10.1063/1.4964084

[5] Shikalov V S, Klinkov S V and Kosarev V F 2017 *Proc. XXV Conf. on High-Energy Processes in Condensed Matter (HEPCM 2017)* AIP Conf. Proc. **1893**, ed V M Fomin (Mellville NY: American Institute of Physics) p 030038 https://doi.org/10.1063/1.5007496

[6] Kosarev V F, Klinkov S V, Melamed B M, Nepochatov Yu K, Ryashin N S and Shikalov V S 2018 *Proc. 19th Int. Conf. on the Methods of Aerophysical Research (ICMAR 2018)* AIP Conf. Proc. **2027**, ed V M Fomin (Mellville NY: American Institute of Physics) p 030047 https://doi.org/10.1063/1.5065141

[7] Lomovsky O I, Dudina D V, Ulianitsky V Yu, Zlobin S B, Kosarev V F, Klinkov S V, Korchagin M A, Know D-H, Kim J-S and Know Y-S 2007 *Mater. Sci. Forum* **534–536**, 1371–6

[8] Kim J S, Kwon Y S, Lomovsky O I, Dudina D V, Kosarev V F, Klinkov S V, Kwon D H and Smurov I 2007 *Composites Sci. Technol.* **67** 2292–6
[9] Kosarev V F, Klinkov S V, Sova A A and I. Smurov I 2008 *Surf. Coat. Technol.* **202** 5858–62
[10] Klinkov S V, Kosarev V F, Sova A A and I. Smurov I 2009 *J. Therm. Spray Technol.* **18** 944–56
[11] Sova A A, Kosarev V F, Papyrin A and I. Smurov I 2011 *J. Therm. Spray Technol.* **20**, 285–91
[12] Klinkov S V and Kosarev V F 2012 *J. Therm. Spray Technol.* **21**, 1046–53
[13] Kosarev V F, Polukhin A A, Ryashin N S, Fomin V M and Shikalov V S 2017 *Mechanics of Solids* **52**, 457–64
[14] Fomin V M, Golyshev A A, Kosarev V F, Malikov A G, Orishich A M, Ryashin N S, Filippov A A and Shikalov V S 2017 *J. Appl. Mech. Tech. Phys.* **58** 947–55
[15] Papyrin A N, Alkhimov A P, Kosarev V F and Klinkov S V 2001 *Proc. Thermal Spray 2001: New Surfaces for a New Millennium*, ed C C Berndt (Singapore: ASM International) p 423–31
[16] Alkhimov A P, Klinkov S V and Kosarev V F 2003 *J. Therm. Spray Technol.* **10** 375–81
[17] Kosarev V F, Klinkov S V, Alkhimov A P and Papyrin A N 2003 *J. Therm. Spray Technol.* **12** 265–81
[18] Kosarev V F, Klinkov S V and Papyrin A N 2007 *The Cold Spray Materials Deposition Process: Fundamentals and Applications*, ed V K Champagne (Cambridge: Woodhead Publishing) pp 178–216
[19] Klinkov S V, Kosarev V F and Zaikovskii V N 2011 *J. Therm. Spray Technol.* **20** 837–44
[20] Sova A A, Klinkov S V, Kosarev V F, Ryashin N S and Smurov I 2013 *Surf. Coat. Technol.* **220** 98–101
[21] Klinkov S V, Kosarev V F and Zaikovskii V N 2016 *Surf. Eng.* **32** 701–6
[22] Kiselev S P, Kiselev V P, Klinkov S V, Kosarev V F and Zaikovskii V N 2017 *Surf. Coat. Technol.* **313** 24–30
[23] Fukumoto M, Wada H, Tanabe K, Yamada M, Yamaguchi E, Niwa A, Sugimoto M and Izawa M 2007 *J. Therm. Spray Technol.* **16** 643-50
[24] Rech S, Trentin A, Vezzu S, Legoux J G, Irissou E and Guagliano M 2011 *J. Therm. Spray Technol.* **20** 243-51
[25] Suo X K, Yu M, Li W Y, Planche M P and Liao H L 2012 *J. Therm. Spray Technol.* **21** 1091-8
[26] Yin S, Suo X, Guo Z, Liao H and Wang X 2015 *Surf. Coat. Technol.* **268** 252-6
[27] Ernst K-R, Braeutigam J, Gaertner F and Klassen T 2013 *J. Therm. Spray Technol.* **22** 422-32
[28] Watanabe Y, Yoshida C, Atsumi K, Yamada M and Fukumoto M 2015 *J. Therm. Spray Technol.* **24** 86-91
[29] Kulmala M and Vuoristo P 2008 *Surf. Coat. Technol.* **202** 4503-8
[30] Danlos Y, Costil S, Guo X, Liao H and Coddet C 2010 *Surf. Coat. Technol.* **205** 1055-9
[31] Lupoi R, Sparkes M, Cockburn A and O'Neill W 2011 *Mater. Lett.* **65** 3205-7

Journal of Physics: Conference Series **1404** (2019) 012025 doi:10.1088/1742-6596/1404/1/012025

Study of the mechanism of attenuation of thermocapillary convection of the melt during laser powder cladding

O B Kovalev and A M Gurin

Khristianovich Institute of Theoretical and Applied Mechanics SB RAS, Institutskaya str., 4/1, Novosibirsk, 630090, Russia

kovalev@itam.nsc.ru

Abstract. The results of a numerical solution of the thermocapillary microconvection problem arising under the action of laser radiation in a liquid metal during laser surface treatment are presented. A characteristic feature of the behavior of particles of refractory metals in a multi-vortex melt pool is revealed. The influence of falling powder particles in the melt on the inhibition of thermocapillary convection is estimated. The mechanisms of slowing the melt flow associated with the addition of powder components for laser alloying or cladding is discussed. The results of three-dimensional calculations are presented, which show how these mechanisms can work.

1. Introduction

The hydrodynamics of the melt pool during laser interaction with metals has always been given great attention in the literature [1-5]. The approaches where the thermoconvection of liquid metal is described using the Marangoni condition have already become classical [2-5]. Additive laser technologies are based on the use of various powdered materials that are usually added to the melt. Powder input modes and materials can be very different: metals, alloys that melt during transport in the laser beam, as well as refractory ceramics that fall into the melt in the solid state. The interaction of powder particles (liquid or solid) with the melt pool, the effect of powder input on the thermal convection of the pool is currently insufficiently investigated. In this paper, a computational and theoretical analysis of the mechanisms of slowing the flow in the melt pool associated with the addition of powdered components used in laser alloying or cladding technologies is carried out.

2. Qualitative analysis of microconvection in a laser melt pool

The process diagram, which includes the powder input, the position of the laser beam with the coaxial nozzle and the melt pool on the substrate, is shown in figure 1. The laser beam heats the metal to the melting point. The temperature gradient that occurs on the surface of the liquid leads to the appearance of thermocapillary forces. Under the influence of these forces in the liquid there is a motion. At the liquid-gas interface, the laws of conservation of mass, momentum and energy should be considered in a strict description. In many works this is done simplistically. For example, only the mass of the powder added to the melt and the energy brought by the particles are taken into account, and the law of conservation of momentum is usually neglected [2-5]. The boundary conditions on the free surface of a liquid $z = h(t,x,y)$ can more generally be written:

$$\dot{m}_p = \xi_p \rho_p \left|\mathbf{V}_p\right| \cos\varphi, \quad (1)$$

$$\dot{m}_p \cdot (\mathbf{V}_p)_n + \mu \frac{\partial U_n}{\partial n} = \mu_g \frac{\partial (\mathbf{V}_g)_n}{\partial n} + P_g + \sigma\kappa_0, \quad (2)$$

$$\dot{m}_p \cdot (\mathbf{V}_p)_s + \mu\left(\frac{\partial U_n}{\partial s} + \frac{\partial U_s}{\partial n}\right) = \mu_g \left(\frac{\partial (\mathbf{V}_g)_n}{\partial s} + \frac{\partial (\mathbf{V}_g)_s}{\partial n}\right) + \nabla_s \sigma, \quad (3)$$

$$\lambda \left.\frac{\partial T}{\partial n}\right|_{z=h} = S_{las} + S_p - S_{conv} - S_{rad} - S_{evap}. \quad (4)$$

Here μ is the dynamic viscosity of the liquid; U_n, U_s are the projections of the velocity vector of the fluid at the normal (n) and tangential (s) directions; $\mathbf{V}_g$ is the velocity vector of the carrier gas; φ is the angle between the surface normal and the velocity vector of the particles $\mathbf{V}_p$.

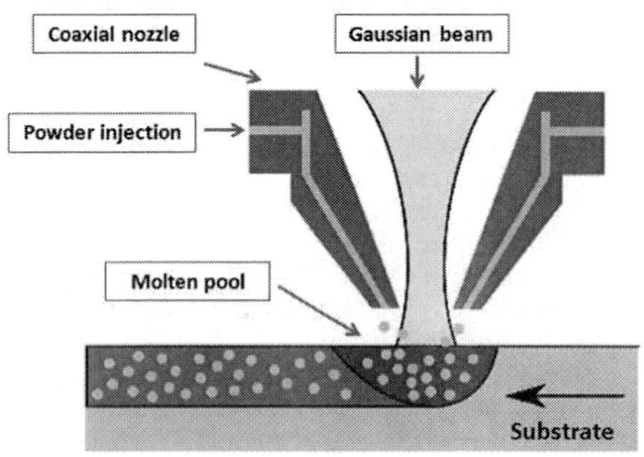

Figure 1. The scheme of the surface laser treatment process (alloying or cladding).

The first equation is the law of conservation of the mass flow of matter. The second is the equality of normal stresses taking into account the viscous stresses in the liquid, as well as the capillary pressure due to the curvature of the free surface, and the gas pressure P_g. The third is describes the balance of stresses tangent to the surface associated with the Marangoni effect, viscous stresses of the gas and the additional momentum brought by the added mass of the powder. The heat inflow (4) on the surface of the melt pool is maintained by absorbing laser radiation $S_{las} = A_{ab} I(x, y)\cos\gamma$ and energy brought by the flow of incident particles $S_p = \dot{m}_p / \rho_p \cos\phi\left(H - H_p^{\circ}\right)$. Heat losses include convective heat exchange with the surrounding environment $S_{conv} = \lambda_g Nu\left(T - T_0\right)/d$, radiation losses $S_{rad} = \varepsilon k_s\left(T^4 - T_0^4\right)$, and evaporation of part of the superheated melt $S_{evap} = L_e \dot{m}_{evap}$ [9]. The power density distribution in the beam is written using the Gauss function: $I(x, y) = 2W/\pi\omega_0^2 \exp\left(-2(x^2 + y^2)/\omega_0^2\right)$, where W is the power, ω_0 is the radius of the beam, x, y are the coordinates of the Cartesian system.

In the case where the free surface of the liquid is flat $z = 0$, the mechanism of interaction of the particles with the surface is reduced to adhesion $(\mathbf{V}_p)_n = U_n = 0$, and $(\mathbf{V}_p)_s = U_s$. In this case, equation (2) turns into an identity, and equation (3) takes the following form:

$$\dot{m}_p U_s + \mu \frac{\partial U_s}{\partial n} = \gamma_T \frac{\partial T}{\partial s}. \quad (5)$$

The equation (5) differs from the known Marangoni condition [1-5] by the appearance of an additional term $\dot{m}_p U_s$ that takes into account the mass flow of the powder on the liquid surface. In the case of a slow fluid flow, the friction stress $\tau_s = \mu \partial U_s / \partial n$ is approximated by the expression $\tau_s \approx 4\mu U_{max}/h$, [2]. Then, for flows with a small Reynolds number, the maximum velocity U_{max} of the fluid can be estimated by the formula (6) in the case of $\dot{m}_p \geq 0$.

$$U_{max} \approx \frac{(\gamma_T \Delta T/d)}{(4\mu/h + \dot{m}_p)}, \quad (6)$$

where d is the width and h is the depth of the melt pool; and the width is necessarily greater than the depth ($d > h$), and the surface temperature gradient $\partial T/\partial s$ is replaced by the ratio $\Delta T/d$, where ΔT is the temperature difference between the center of the pool surface and its periphery.

In table 1 the calculated estimates of thermal capillary convection deceleration for some low carbon steel, stainless steel SS304 and titanium alloy Ti-6Al-4V substrate materials obtained by formulas (6) with varying $\dot{m}_p$ are given. Deceleration of the flow can be caused by an increase in the viscosity of the mixture in the presence of solid dispersed components. In this case, the average velocity of the mixture begins to depend on the effective viscosity of the suspension, μ_{eff} which increases exponentially with the increase in the volume content of the solid phase f_s [6]:

$$\mu_{eff} = \mu \exp\left[\alpha f_s / (1 - \beta f_s)\right], \quad (7)$$

where α, β are empirical constants. With an increase f_s to 0.7, the effective viscosity increases several times, which can cause a significant slowdown in the flow.

Table 1. Effect of powder particle flow $\dot{m}_p$ on convection in liquid metal.

Parameter	Low carbon steel	Stainless steel SS304	Ti-6Al-4V
Surface tension gradient: γ_T, N/(m K);	-4.9×10^{-4}	-3.5×10^{-4}	-2.7×10^{-4}
Dynamic viscosity: μ, kg/(m s).	6.93×10^{-3}	4.3×10^{-3}	4.0×10^{-2}
Liquid temperature: T_m (ΔT), K	1785 (1000)	1723 (1000)	1923 (1000)
The melt pool width: d, mm	4	4	4
The melt pool depth: , h, mm	2	2	2
Max velocity: U_{max}, m/s; $\dot{m}_p = 0$	8.8	10	0.84
Max velocity: U_{max}, m/s; $\dot{m}_p = 10$ kg/(m^2 s)	5.1	4.7	0.75
Max velocity: U_{max}, m/s; $\dot{m}_p = 30$ kg/(m^2 s)	2.8	2.3	0.61

3. Statement of the problem

It is assumed that the investigated laser heating regimes cause small values of the convective velocities, so that the melt flow is considered to be laminar, and this causes a flat shape of the free surface of the liquid. Mass transfer in a fluid is described by the Navier–Stokes equations in the Boussinesq approximation (8, 9). The heat transfer equation (10), written through the enthalpy H of the melt, takes into account the processes of metal melting by the laser beam and micro-convection of the liquid phase according to [5]:

$$\nabla \cdot \mathbf{V}_l \tag{8}$$

$$\rho \frac{\partial \mathbf{V}_l}{\partial t} + \rho(\mathbf{V}_l \cdot \nabla)\mathbf{V}_l = -\nabla p + \mu \nabla^2 \mathbf{V}_l + \rho \mathbf{g} \beta_T (T - T_0) \tag{9}$$

$$\frac{\partial H}{\partial t} + (\mathbf{V}_l \cdot \nabla) H = \nabla(\lambda \nabla T)\ ,\ H = \int_{T_0}^{T} \rho c_p(T) dT \tag{10}$$

where $\mathbf{V}_l = (u, v, w)$ is the velocity of fluid and $\mathbf{g} = (0,0,g)$ is the acceleration of gravity; p, T are the pressure and temperature, ρ, λ are the density and thermal conductivity; μ is the viscosity.

The transportation method and the mechanism for adding powder particles to the molten pool are not considered in this study. Particles appear on the surface of the melt pool in locations obtained from a random number generator. The equations of conservation of energy (in the enthalpy form) and the equation of conservation of momentum during the motion of particles along their trajectories when entrained by an melt moving due to the Marangoni effect should also be added to the equations of fluid flow in the molten pool [4, 5]. Theses equations are

$$\frac{d}{dt}(m_p H_p) = 4\pi r_p^2 \left[\frac{\lambda_l}{r_p} Nu_p (T_p - T_l) \right], \tag{11}$$

$$m_p \frac{d\mathbf{V}_p}{dt} = \frac{1}{2} \rho_l C_d \pi r_p^2 |\mathbf{V}_l - \mathbf{V}_p| (\mathbf{V}_l - \mathbf{V}_p) + m_p \mathbf{g} \left(1 - \frac{\rho_l}{\rho_p}\right), \tag{12}$$

$$\frac{d\mathbf{X}_p}{dt} = \mathbf{V}_p, \tag{13}$$

$$T_p = \begin{cases} H_p / c_s, & H_p < c_s T_{mp} \\ T_{mp}, & c_s T_{mp} \le H_p \le c_s T_{mp} + H_{mp} \\ T_{mp} + (H_p - c_s T_{mp} - H_{mp}) / c_m, & H_p > c_s T_{mp} + H_{mp} \end{cases} ,H_p = \int_{T_0}^{T_p} c_p(T) dT \tag{14}$$

$$C_d(\mathrm{Re}_p) = \frac{24}{\mathrm{Re}_p}\left(1 + 0.179 \mathrm{Re}_p^{0.5} + 0.013 \mathrm{Re}_p\right), \mathrm{Re}_p \le 10^3,\ \mathrm{Re}_p = \frac{2 r_p \rho_m |\mathbf{V}_l - \mathbf{V}_p|}{\mu}, \tag{15}$$

$$Nu_p = 2 + 0.459 \mathrm{Re}_p^{0.55} \mathrm{Pr}^{0.33};\ Nu_p = (2 r_p \alpha_T)/k_m;\ \mathrm{Pr} = \mu c_m / k_m. \tag{16}$$

Here $\rho_l, \mathbf{V}_l$ are the density and velocity of the melt; $m_p = 4/3 \pi r_p^3 \rho_p$ is the particle mass; C_d is the drag coefficient; Nu_p is the Nusselt number; H_p is the enthalpy; r_p is the radius; ρ_p is the density; $\mathbf{V}_p$ is the velocity of the particle; T_{mp} is the melting temperature; and H_{mp} is the latent heat of melting.

4. Calculation results

Figures 2 and 3 show the pattern of steady-state flow in the molten pool of the steel substrate. Absorption coefficient $A_{ab} = 0.19$. The current lines and the flow velocity distribution are shown in the space occupied by the fluid. Separately, the position of the melting front, which limits the melt zone, is highlighted. For a laser beam with a power of 5 kW and a spot size of 3 mm (noted in the figure), a liquid pool with a diameter of 5 mm and a depth of about 400 microns was obtained. The behavior of the current lines indicates a multi-vortex pattern of fluid flow. In the radiation area there are two vortices symmetrically located relative to the scanning direction of the beam. The colder melt is carried from the bottom of the pool to the surface by these vortices. Within the beam spot, the melt heats up and, due to the Marangoni effect, rushes to the periphery of the pool, where a toroidal vortex is formed. Here, the heated melt is mixed with the colder liquid metal. At the back of the pool, a stream is formed, rushing to a depth. Then this flow goes up with the formation of the already mentioned two vortices transferring the melt from the bottom to the surface. Then everything repeats.

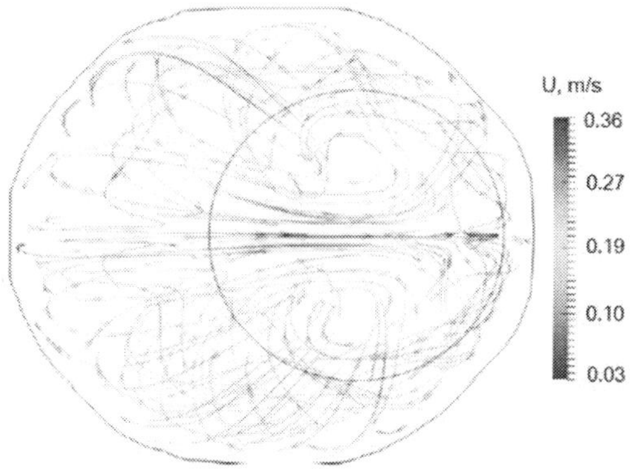

Figure 2. Velocity field along streamlines of the liquid metal, the top view in the XY plane.

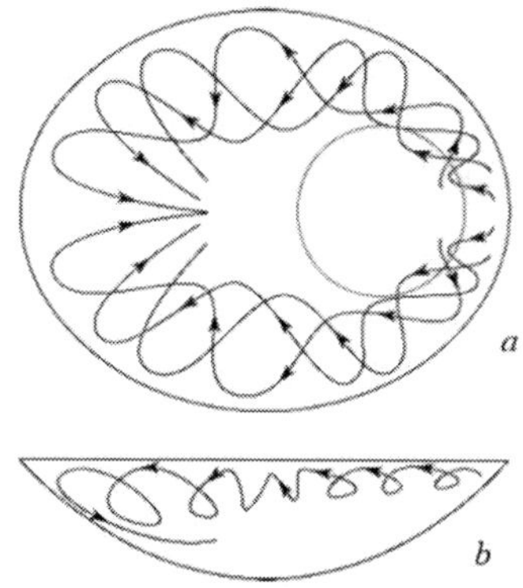

Figure 3. Flow diagram in a toroidal vortex, view in the plane: XY (a); XZ (b).

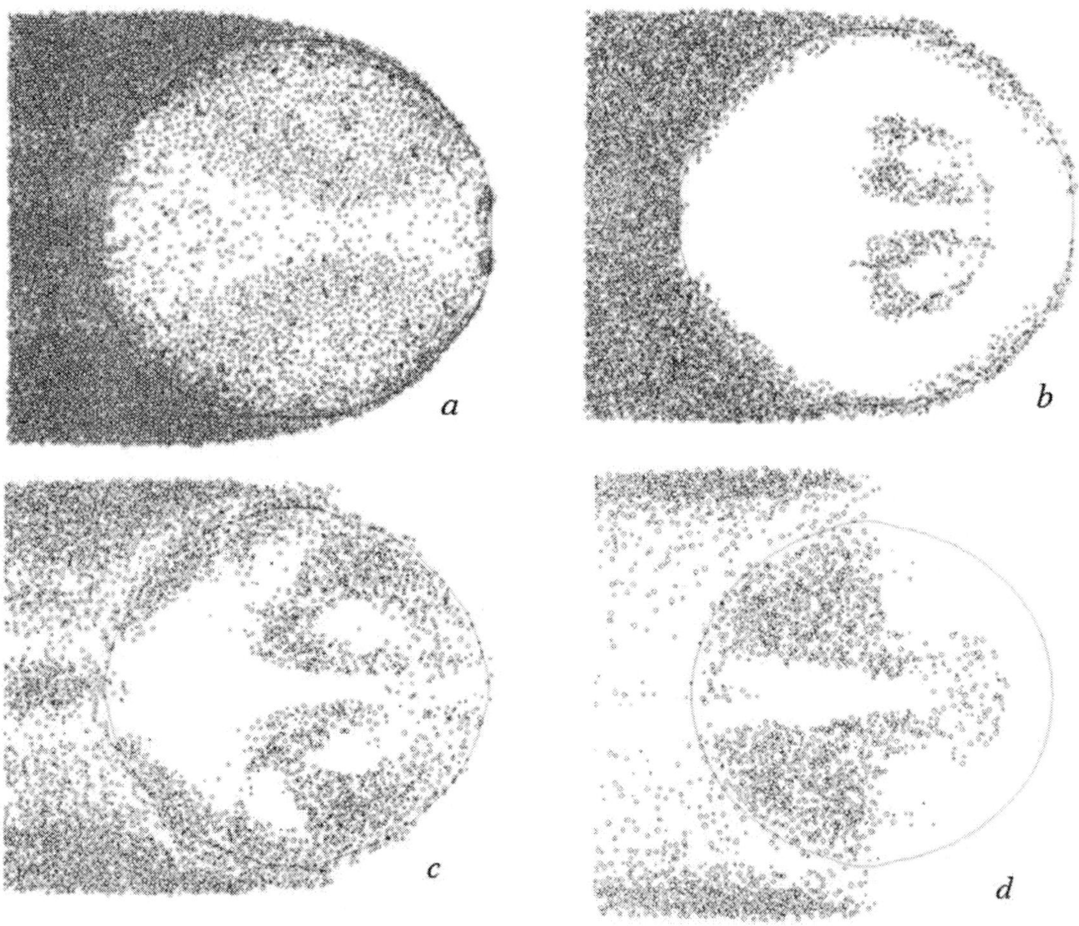

Figure 4. Horizontal sections of the molten pool with particles at a powder flow rate of 5 g/min, scanning speed 10 mm/s. Depth of sections z, microns: -10 (a), -200 (b), -300 (c), -400 (d).

The analysis of the behavior of powder particles falling on the surface of the pool with already developed thermocapillary convection is of interest. The spatial distribution of particles of refractory ceramics (TiC) in the melt pool of the steel substrate (16NCD13) is illustrated in figure 4.

The unevenness of the spatial distribution of the powder is explained by the dynamic delay of the particles with respect to the motion of the fluid carrying them. Particles accumulate most at the periphery and bottom of the pool, where the flow rate is low. Therefore, the centers of vortex flows and some regions with an increased flow velocity, which the particles overcome quickly, turn out to be free of particles.

Two series of calculations were carried out to illustrate the effect of particles on the weakening of the melt convection. First, a modified Marangoni condition (5) was used. The powder consumption varied from 0 to 50 g/min, and the effective viscosity of the melt remained constant: $\mu_{eff} = \mu$. Changes in the mass-average velocity of the melt $U(x)$ along the OX axis with variations in the flow rate of the powder G are shown in figure 5. Note that the maxima of the velocity distribution $U(x)$ decrease when the powder flow rate G>25 g/min. At the same time, the temperature of the melt in the beam spot increases, since the melt is less intensively mixed and has time to heat up more, this leads to an increase in the Marangoni force. For this reason, the convection deceleration in figure 5 is not as significant.

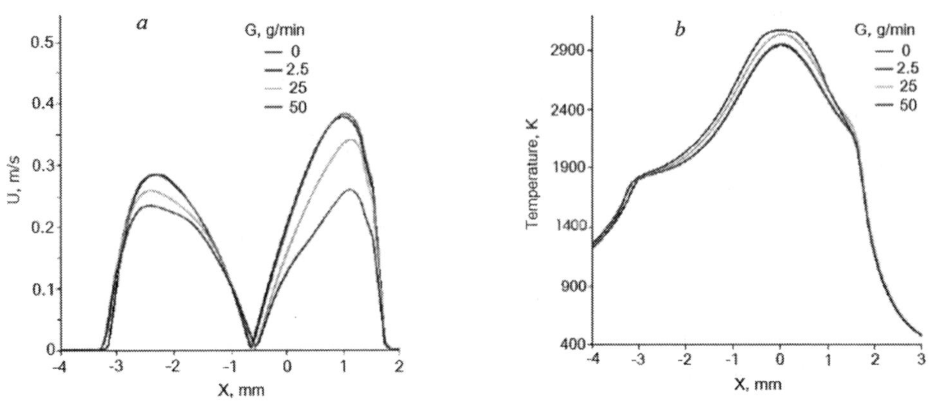

Figure 5. The effect of the modified Marangoni condition (5) on the attenuation of fluid convection (a) and temperature change (b) with varying powder flow rate G.

Further, calculations were made with variations in the flow rate of powder G, but taking into account the exponential dependence of the effective viscosity of the melt μ_{eff} on the particle content f_s according to formula (7). In this case, the classical Marangoni condition was used, which is obtained if $\dot{m}_p = 0$ in (5). Figure 6 shows the changes in the mass-average velocity $U(x)$ and temperature on the surface of the liquid. A noticeable weakening of convection is already observed at $G = 10$ g / min. It can be seen that with an increase in the flow rate of the powder added to the pool, the increase in the effective viscosity of the suspension has a decisive effect on convection.

Conclusions

A modified Marangoni condition is proposed to describe the deceleration of microconvection when microparticles of refractory powders are added to the melt. It is shown that one of the determining mechanisms of convection deceleration is to increase the effective viscosity of the mixture by increasing the content of particles fed into the pool as an alloying or depositing material.

The accumulation of particles is observed in stagnant areas of the melt pool, which increases the unevenness of the change in the effective viscosity of the mixture and further slows the flow rate. It is shown that the fluid flow regions having an increased velocity are free of particles.

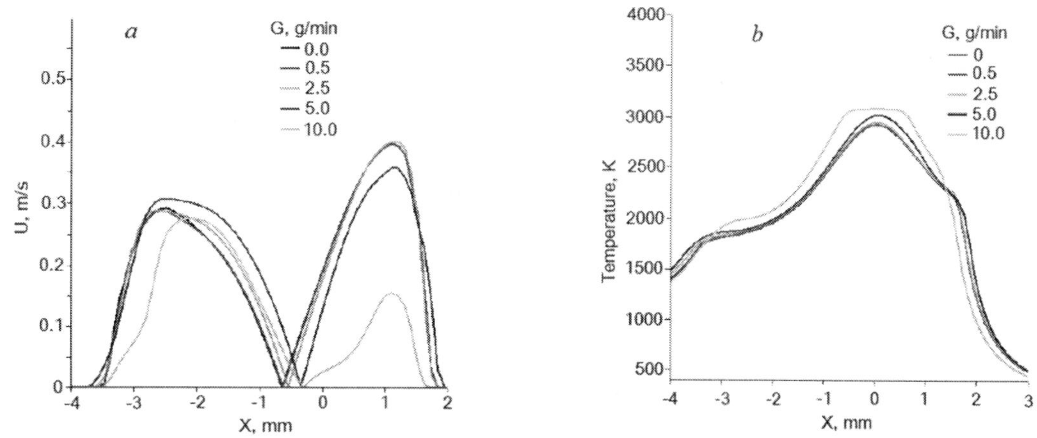

Figure 6. The effect of the effective viscosity of the suspension μ_{eff} on the attenuation of fluid convection (a) and temperature change (b) with varying powder flow rate *G*.

Acknowledgments
The authors appreciate with gratitude the financial support of the Russian Foundation for Basic Research (project No. 17-20-03197). The research was partially carried out within the Program of Fundamental Scientific Research of the state academies of sciences in 2013–2020 (projects No. AAAA-A17-117030610120-2, AAAA-A18-118021590032-2).

References
[1] Rykalin N N, Uglov A A and Kokora A N 1975 *Laser Processing of Materials* (Moscow: Mashinostroenie Press)
[2] Antonova G F, Gladush G G, Kosyrev F K, Krasyukov A G, Likhanskii V V, Loboiko A I and Sayapin V P 1998 *Quant. Electr.* **28(5)** 430
[3] Kou S and Wang Y H 1986 *Metallurgical Transactions* **A17** 2265
[4] Gurin A M and Kovalev O B 2013 *Thermophys. Aeromech.* **20** 227
[5] Kovalev O B and Gurin A M 2014 *Int. J. Heat and Mass Trans.* **68** 269
[6] Kondratiev A S and Shvydko P P 2014 *Scientific journal "Izvestiya MGTU "MAMI", Series Natural Sciences* **4(2)** 40

Numerical simulation of the high-speed interaction of a spherical impactor with a system of spaced heterogeneous plates

A E Kraus[1,2], E I Kraus[1] and I I Shabalin[1]

[1] Novosibirsk State Technical University,
20 K. Marksa av., Novosibirsk, 630073, Russia
[2] Khristianovich Institute of Theoretical and Applied Mechanics SB RAS,
4/1 Institutskaya st., Novosibirsk, 630090, Russia

kraus@itam.nsc.ru

Abstract. A 3D non-stationary problem of the interaction of space debris spherical particle and spaced protective screens is considered. To verify the 3D code, a series of calculations was performed of the impact of a spherical impactor made of ShKh15 steel at speed of 3.02 km/s with a spaced barrier consisting of three homogeneous plates, the first two made of AMg6 alloy and the third one made of steel 10. Comparison of the results of numerical calculations for hole diameters with experimental data showed (2-5)% compliance. The collisions of steel spheres with AMg6 plates were calculated at an angle to the normal. To compare the resistance of the protective system, homogeneous screens were replaced by plates of heterogeneous material AMg6+B_4C. It was shown that with the same mass characteristics, a heterogeneous system of screens had increased resistance compared to a homogeneous system.

1. Introduction

The protection of spacecraft is an important problem at present. Collision with cosmic dust or debris is a serious hazard to spacecraft.

The duration of flights increases and, thereby, increases the likelihood of meeting this danger. The issues of protection of spacecraft have arisen since the birth of the ideas of space exploration [1] and are relevant to this day. To protect the spacecraft from man-made space debris, screens are used consisting of several metal plates, usually about 0.5 mm thick, causing the disintegration of potentially dangerous objects into smaller fragments before they collide with the spacecraft's body directly [2]. All such questions come down to the study of the high-speed interaction of a striking body with spaced plates. In [3], a series of experiments was carried out and numerical simulation of the effect of debris particles weighing from 0.7 to 20.0 g on screen protection was performed. The critical speeds were obtained for aluminum and steel plates at speeds of 0.5-6.5 km/s.

In [4], a numerical simulation of the effect of long rods on layered spaced barriers was considered. The probabilistic approach and the numerical technique proposed in the work in the most complete (from the physical point of view) three-dimensional formulation allow reproducing with high accuracy the processes of breaking through multilayered and spaced barriers by high-speed rod elements. In [5–7], based on REACTOR software package, it was managed to simulate the processes of high-speed loading of complex technical objects, for example, a tubular heat exchanger, which is a part of the spacecraft thermal regime support system. The numerical solution was brought until the "end", i.e. until the reserves of kinetic energy of the incident debris particles were exhausted.

2. The purpose of the work and problem statement

The purpose of this work is to conduct a comparative analysis in the spatial case of the resistance of spaced protection systems made of homogeneous materials (two plates of aluminum alloy AMg6 and plates of steel 10) with a system of equally weighted heterogeneous material (two plates of AMg6+B_4C and a steel plate) at high speed loading.

A series of 3D non-stationary problems on the interaction of a uniform projectile with an obstacle in the form of spaced plates at angles of the meeting 0, 30, 45, 60 degrees to the normal was solved. The mathematical formulation of the problems was considered in detail in [8,9]. The calculations were performed with REACTOR software package [10,11], which implements numerical algorithms for solving balance equations using an explicit difference scheme, taking into account the deformation and fracture of materials and the fulfillment of boundary conditions.

3. Simulation of the experiment with a spaced barrier

Let a spherical impactor made of steel ShKh15 hit normally a system of spaced plates, the first two of alloy AMg6 and the third one of steel 10 at speed of 3.02 km/s. The geometry of the problem is shown in figure 1, the sphere is made of ShKh15 steel with diameter of 1.35 cm. The plates of AMg6 alloy 1.5 cm thick are located at distance of 3.0 cm from each other, and the plate of steel 10 with thickness of 1.0 cm is located at distance of 15.0 cm from the plates, as was done in the experiment in [12].

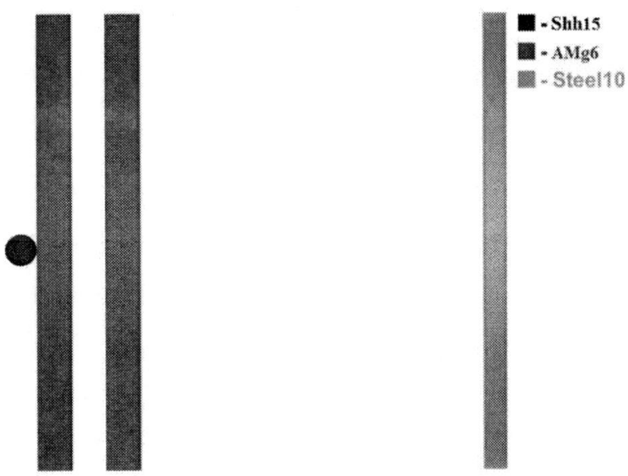

Figure 1. Geometric model of numerical experiment 1.

As a result of the calculations, the diameters of the holes were obtained; their values are given in table 1. The spherical impactor pierces both plates of AMg6, but does not pierce the third plate of steel (see figure 2). The following notation is introduced in table 1: d_1 is the diameter of the hole in the first plate, d_2 is the diameter of the hole in the second plate, h is the depth of the crater in the third plate, and δ_i is the error of the author's calculation relative to the experiment.

Table 1. Comparative data of numerical experiment 1.

	Experiment [12]	Simulation [12]	The authors' simulation	δ_i, %
d_1, mm	35	38	34	2.8
d_2, mm	40	44	40	0
h, mm	~2	3 – 4	2.5	25

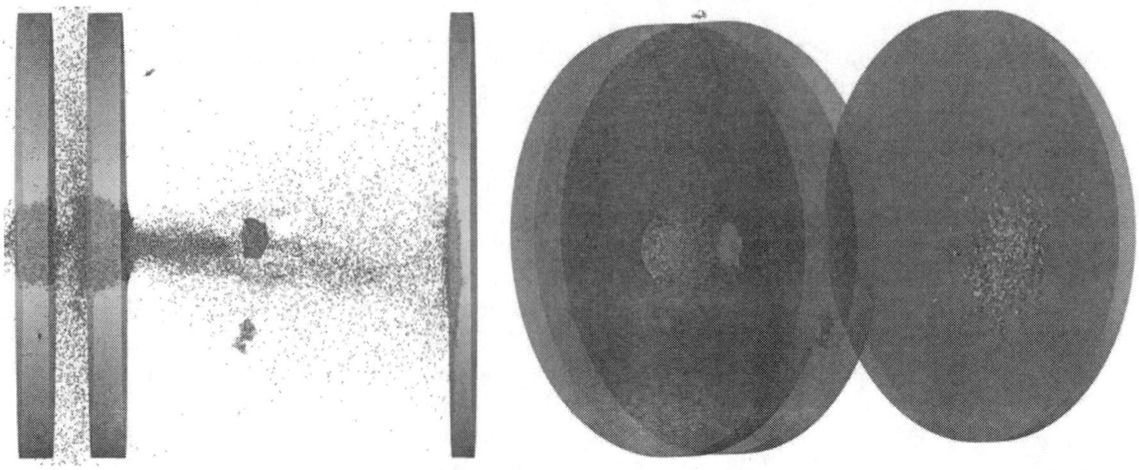

Figure 2. Visualization of the simulation results of experiment 1 with spaced plates.

Based on the series of calculations and comparison with experimental data [12], parameters were found for the equation of state of all materials [13,14], which were used below, see table 2.

Table 2. Physic-mechanical constants of the materials.

	a, km/s	S	μ	ρ, g/cm^3	G, GPa	K_s, GPa	Y, GPa
Steel 10	4.90	1.288	0.29	8.01	93.92	192.32	0.80
AMg6	5.46	1.310	0.33	2.63	26.49	73.95	0.42
ShKh15	4.90	1.288	0.29	8.01	93.92	192.32	2.00

where a is the volume velocity of sound, S is the slope of the generalized straight line in coordinates $D=a + SU$ for the shock adiabat, μ is the Poisson coefficient, ρ is the initial density, G is the shear modulus, K_s is the bulk expansion modulus, and Y is the yield strength.

In numerical experiment 2, a series of problems was solved on the interaction of a spherical impactor with a system of plates at the meeting angle of 30, 45, and 60 degrees to the normal. Figure 3 shows the geometric statement of the problem, in particular for angle of 30 degrees. It is necessary to determine the approach angle, at which the system of AMg6 plates will provide protection for the object located behind it.

 1404 (2019) 012026 doi:10.1088/1742-6596/1404/1/012026

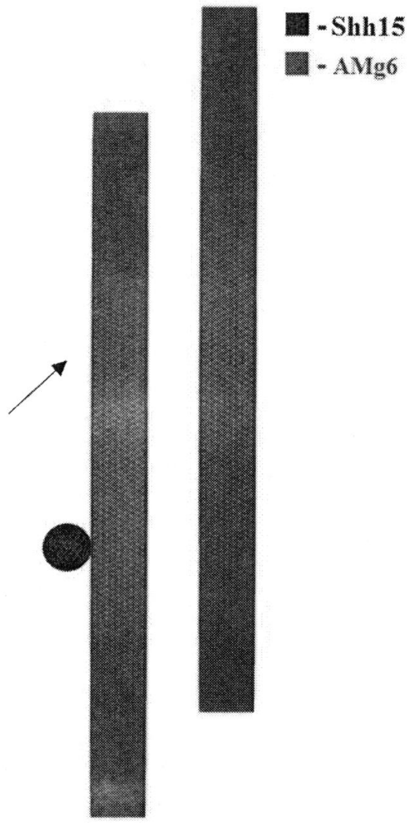

Figure 3. Geometric model with angle of 30 degrees to the normal.

Figure 4 shows the results of numerical calculations and the shape of the holes obtained in the first screen after interaction. As the angle of meeting increased relative to the normal, the size of the punched hole also increased. The largest hole in the first plate was formed at the meeting angle of 60 degree relative to the normal. After breaking through the first plate, the striker collapsed, creating a cloud of fragments with particles of the plate broken off. The cloud of fragments of both materials after interaction with the second plate left a crater on the surface of the plate. The deepest crater was formed in the second plate at the meeting angle of 30 degrees, with the smallest hole in the first plate. After that, a cloud of destroyed particles changed the ejection direction vector perpendicular to the normal direction.

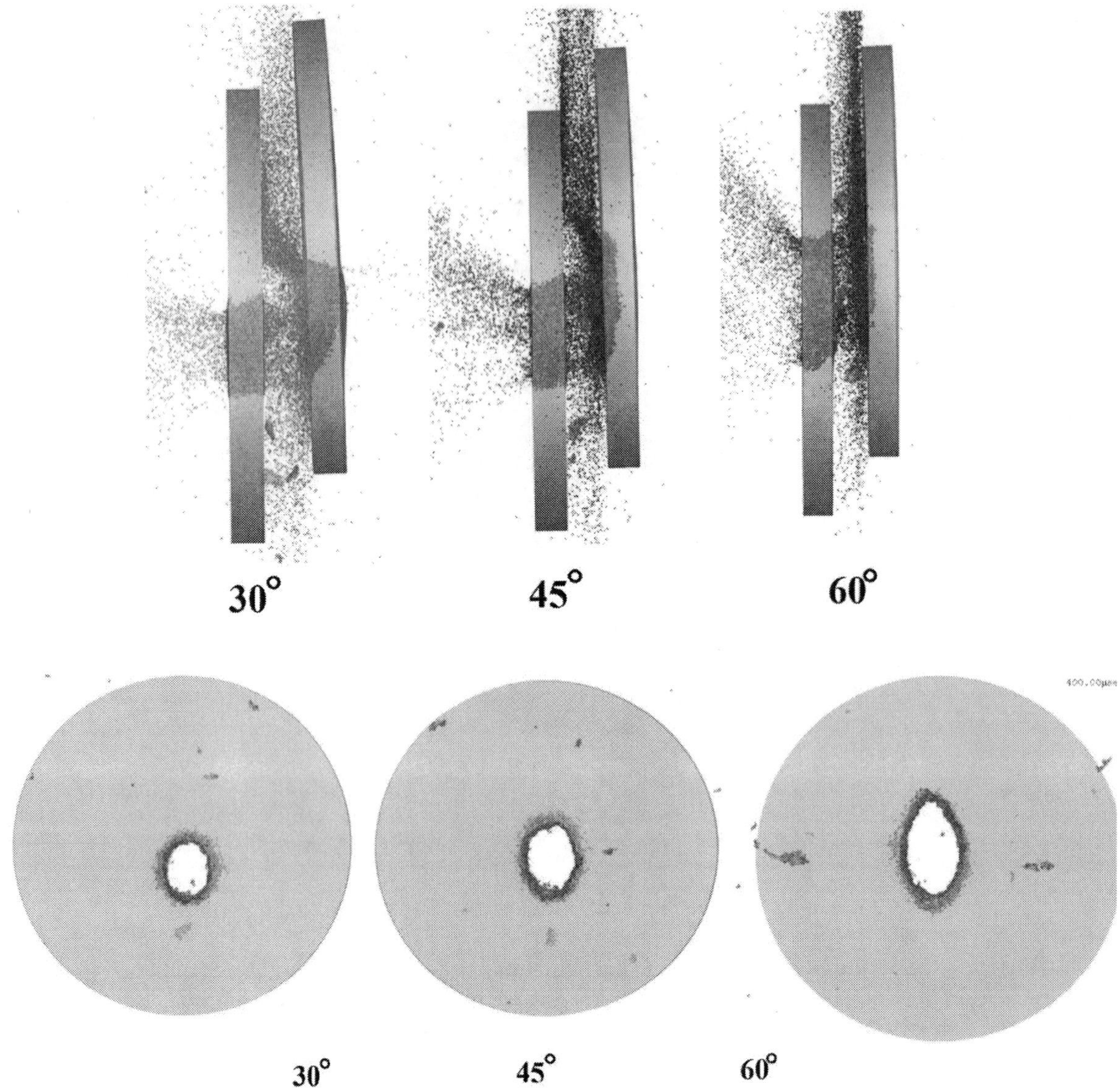

Figure 4. Visualization of the calculation results for meeting angles of 30°, 45°, 60°.

The calculation results for all meeting angles are summarized in table 3. Since the holes at an angle different from the normal form the shape of the hole close to oval, the table shows the hole width d_{n1}, the hole height d_{n2}, the crater depth h, the plate serial number n. As can be seen from the data obtained, the AMg6 screen system provides a spherical impactor ricochet upon impact at angle of 60 degrees.

Table 3. The sizes of the holes after the impact at an angle.

Angle	30	45	60
d_{11}; d_{12}, mm	26.8; 37.1	30.4; 42.0	37.3; 56.6
d_{21}; d_{22}; h_2, mm	36.9; 49.9; 140.0	33.6; 57.8; 76.0	ricochet

4. Simulation of the impact processes with a spaced heterogeneous barrier

Similarly to the formulation of numerical experiment 1, we consider the problem of the interaction of a spherical impactor made of ShKh15 steel with three spaced plates, the first two of heterogeneous materials (a mixture of AMg6 alloy and B_4C ceramics), and the third one of steel 10, a normal impact with initial velocity of 3.02 km/s. The description of the heterogeneous material was carried out according to the procedure [11,15,16], the volume fraction of ceramics in the heterogeneous barrier was 25%. The geometric formulation of the problem is presented in figure 5.

Figure 5. Geometric model with heterogeneous plates.

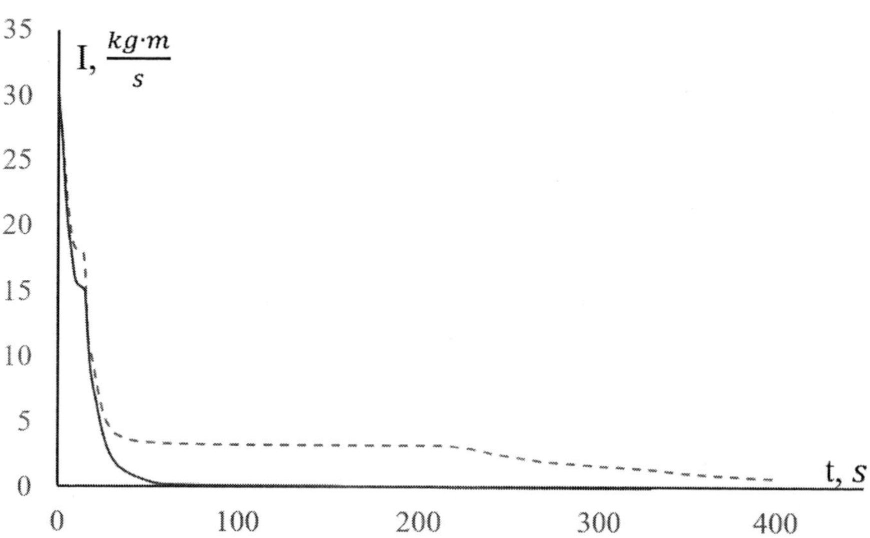

Figure 6. Impact loss of the striker. The red line corresponds to the screen of heterogeneous material, the dashed line corresponds to the screen of homogeneous material.

Comparisons of the ballistic resistance of homogeneous and heterogeneous barriers will be carried out with equal mass values. Since the density of AMg6 alloy does not differ significantly from the density of B_4C ceramics, the equivalent thickness of the heterogeneous plate, at a volume concentration of 25%, is less by 1 mm.

Table 4. The diameters of the holes in the heterogeneous plates.

d_1, mm	32
d_2, mm	37
h, mm	0

From the graphs of momentum loss of the spherical impactor (figure 6), it can be seen that the model with heterogeneous obstacle has increased ballistic resistance and, in contrast to the homogeneous obstacle of AMg6, withstands the impact.

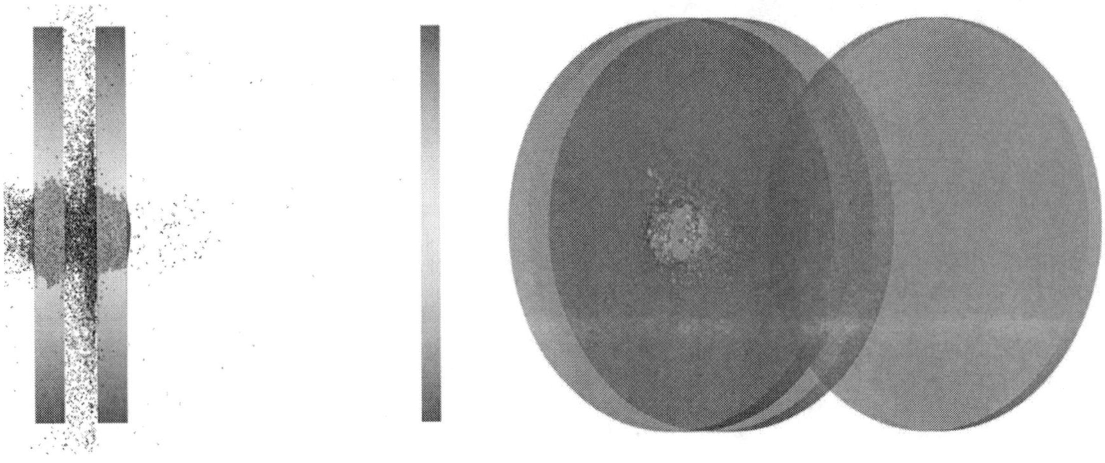

Figure 7. Visualization of the simulation results of the experiment with spaced heterogeneous plates.

The diameters of the holes in the first heterogeneous plate are smaller than in the homogeneous barrier (see figures 2 and 7). The results on the size of the holes in the heterogeneous barrier are summarized in table 4 with the designations similar to those in table 1.

5. Conclusions

- The parameters for the equation of state of the materials used in the calculation are determined.
- A screen of spaced plates made of homogeneous AMg6 material does not withstand the impact of the ShKh15 steel spherical impactor at interaction speed of 3.02 km/s.
- A screen of spaced plates made of homogeneous AMg6 material provides ricocheting of the spherical impactor at interaction speed of 3.02 km/s and the approach angle of 60 degrees.
- The screen of spaced heterogeneous AMg6+B4C plates with similar mass characteristics and ceramics volume fraction of 25% has increased ballistic resistance and withstands the impact of the ShKh15 steel striker at a similar interaction speed.

Acknowledgments
The work was supported by the Russian Foundation for Basic Research (project No. 19-08-00906).

References

[1] Whipple F L 1947 Meteorites and space travel. *Astron. J.* **52** 131
[2] Gerasimov A V., Dobritsa D B, Pashkov S V and Khristenko Y F 2016 *Cosm. Res.* **54** 118–26
[3] Bashurov V V., Bebenin G V., Belov G V., Bukharev Y N, Zhukov V I, Ioilev A G, Lapichev N V., Mikhailov A L, Smirnov G S, Fateev Y A and Schlyapnikov G P 1997 *Int. J. Impact Eng.* **20** 69–78
[4] Gerasimov A V and Pashkov S V 2017 *Procedia Engineering* **204** 284–91
[5] Fedorov M Y, Kraus E I and Shabalin I I 2018 *AIP Conf. Proc.* **2027** 030177
[6] Fedorov M Y, Kraus E I, Fomin V M and Shabalin I I 2009 *Bull. MAI* **16** 49–53
[7] Kraus E I and Shabalin I I 2016 *Res. Sci. City* 6–11
[8] Wilkins M L 1999 *Computer Simulation of Dynamic Phenomena* (Berlin, Heidelberg: Springer Berlin Heidelberg)
[9] Fomin V M and Kiselev S P 1997 Elastic-Plastic Waves in Porous Materials *High-Pressure Shock Compression of Solids IV: Response of Highly Porous Solids to Shock Loading* ed L Davison, Y Horie and M Shahinpoor (New York, NY: Springer New York) pp 205–32
[10] Kraus E I and Shabalin I I 2016 *AIP Conf. Proc.* **1770** 030092
[11] Kraus E I, Melnikov A Y, Fomin V M and Shabalin I I 2019 *J. Appl. Mech. Tech. Phys.* **60** 526–32
[12] Bashurov V V, Bukharev Yu N, Tereshin A I and Tverskov A V 2003 *Modern methods for designing and testing rocket and artillery weapons* **1** 23–33
[13] Kraus E I and Shabalin I I 2016 *J. Phys. Conf. Ser.* **774** 012009
[14] Kraus E I and Shabalin I I 2015 *J. Phys. Conf. Ser.* **653** 012085
[15] Kraus E I, Shabalin I I and Shabalin T I 2017 *AIP Conf. Proc.* **1893** 030130
[16] Kraus A E, Kraus E I and Shabalin I I 2019 *AIP Conf. Proc.* **2125** 030067

Modelling the destruction process of a thermal reactor by man-made space debris impact

A E Kraus[1,2], E I Kraus[1] and I I Shabalin[1]

[1] Novosibirsk State Technical University,
20 K. Marksa av., Novosibirsk, 630073, Russia
[2] Khristianovich Institute of Theoretical and Applied Mechanics SB RAS,
4/1 Institutskaya st., Novosibirsk, 630090, Russia

kraus@itam.nsc.ru

Abstract. The three-dimensional transient problem of the interaction of artificial space debris particles (steel bolt M12) with a spacecraft model reactor is solved. Two cases of interaction are considered, the first is the reactor filled with water, and the second is the empty reactor. The calculations show that the impact of a high-speed steel bolt on the reactor with water at a speed of 11.17 m/s leads to the reactor failure. A water-free reactor obtains catastrophic damage to the hull shell and fuel cells.

1. Introduction

Protecting a spacecraft in orbit is currently a big concern. Collision with cosmic or man-made-debris is a serious danger for spacecraft [1–3]. For example, a strike on spacecraft by a debris element with critical mass and velocity parameters damages it or even destroys it into fragments.

This article is a continuation of researches [4–6] based on the usage of the REACTOR software package [7]. The process of the interaction of a bolt, as an element of junk, with a model of a spacecraft thermal reactor is considered.

2. The work purpose and problem statement

The purpose of this work is to investigate in the 3D approximation the behavior of the thermal part of a spacecraft reactor made of 316TiSteel alloy during high-speed interaction with space debris.

The mathematical formulation of the problem, including a set of equations for balance of mass, momentum, and energy, a model of behavior of a solid deformable body, namely, equations of state and elastoplastic flow, as well as initial data and boundary conditions, was considered in detail in [8–10]. These works also contained descriptions of the explicit difference scheme and the construction of a symmetric algorithm for calculating contact boundaries.

Let a steel bolt M12 with length of 6 cm hits at a high speed a 40 cm diameter reactor containing seven rods with UO_2 in 316TiSteel casing. The calculations are performed with the REACTOR software package, which implements a numerical solver of the balance equations, taking into account the behavior of materials and the fulfillment of the boundary conditions.

3. The numerical simulation

The geometric model shown in figure 1 together with the finite-difference grid is a hollow cylinder of 40 cm diameter and a wall thickness of 0.5 cm. In the center, there is a rod of 4.9 cm diameter and a shell of 1 cm thickness. It also contains six evenly spaced rods of 2.9 cm diameter and 1 cm thick shell.

The impactor is a standard 6 cm long bolt M12 made of steel 20. The impact speed is 11.17 km/s. The impact occurs on the side of the bolt. Two cases are considered. The first one is a collision with the reactor filled with water, the second one is a collision with the empty reactor.

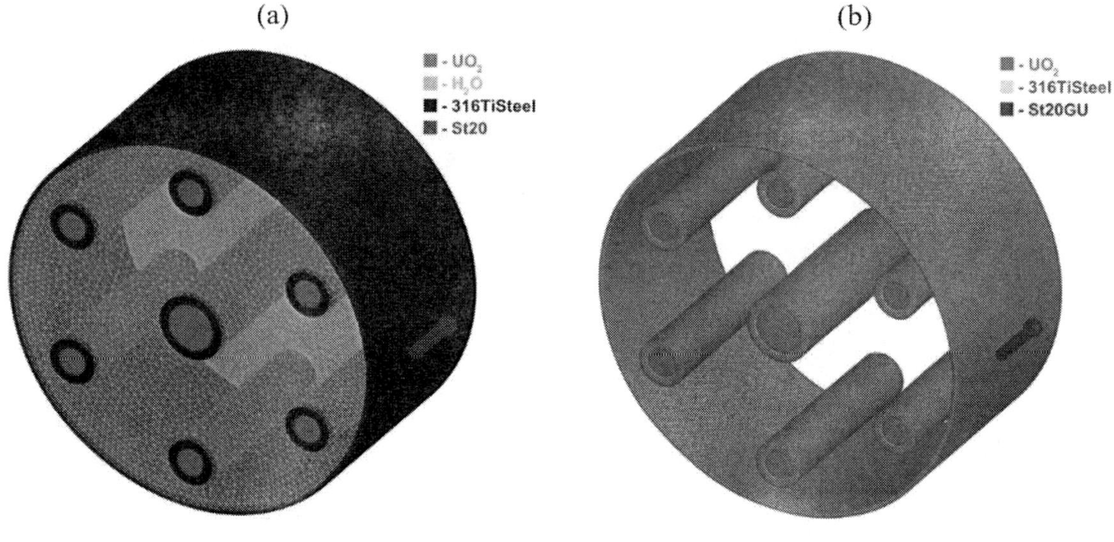

Figure 1. Geometric Model. The reactor is filled with water (a); the hollow reactor (b).

Figure 2 shows a sequence of frames illustrating the process of interaction of the bolt with the reactor in the first case.

Note that at given interaction speeds, the steel bolt is completely destroyed. The rods located near the area of interaction of the bolt with the body receive minor damage to their shells. The remaining rods are not damaged, because the fragmentation cloud is inhibited by the fluid. The reactor vessel in the area of the impact is broken and receives significant deformations. The violation of the shell tightness will lead to termination of the reactor's operating mode.

The impact of the steel bolt on the unfilled reactor leads to catastrophic consequences. Frames showing the evolution of the fragment cloud formation and their effect on the fuel cells are shown in figure 3.

As in the first case, the bolt is completely destroyed when it hits the reactor vessel. Almost all fuel rods are damaged, some of them critically, namely, the tightness of the shells of the fuel elements is broken. A cloud of fragments pierces many through holes in the reactor vessel. The damage to the reactor is catastrophic.

4. Conclusions

The results of a series of calculations showed that a model reactor filled with water, when interacting with artificial junk in the form of an M12 bolt at a speed of 11.17 km/s, received damage that put it out of action, while the unfilled reactor received catastrophic destruction.

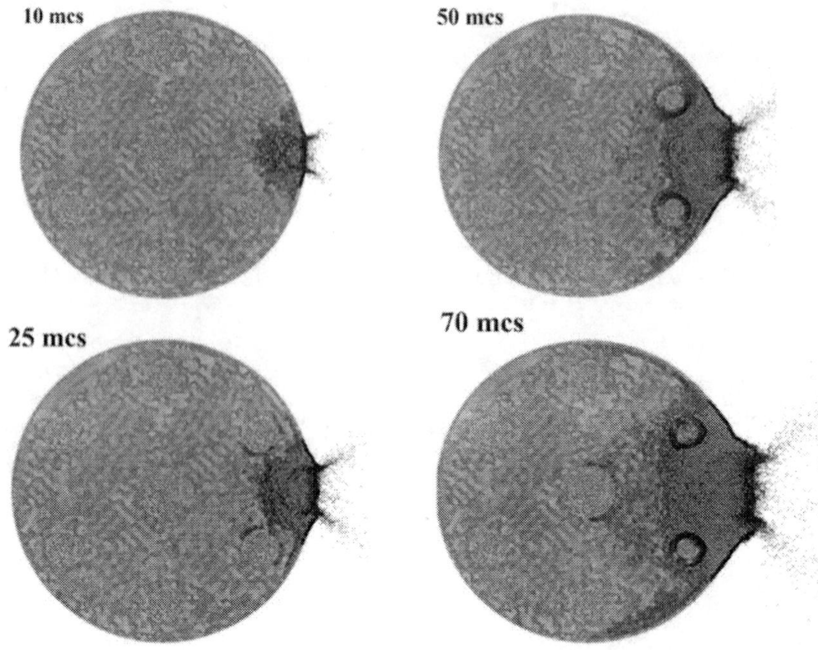

Figure 2. Visualization of the simulation in the first case at several time moments.

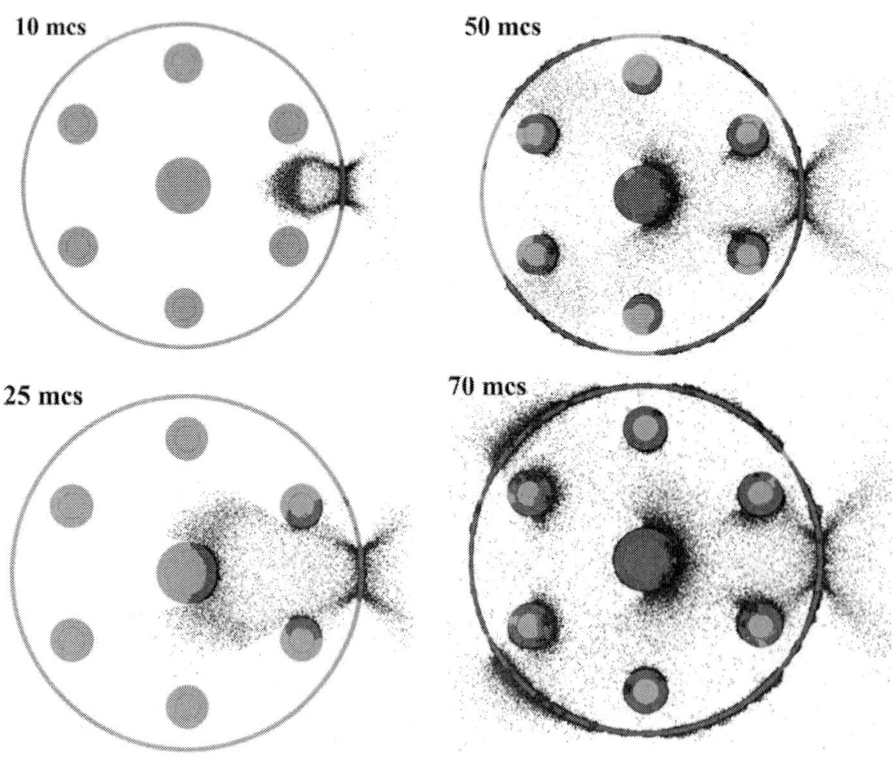

Figure 3. Visualization of the simulation in the second case at several time moments.

Acknowledgments
The research was carried out within the framework of the Program of Fundamental Scientific Research of the state academies of sciences in 2013–2020 (project No. AAAA-A17-117030610136-3).

References

[1] Smirnov N N 2002 *Space Debris: Hazard Evaluation and Mitigation* (Taylor & Francis)
[2] Johnson N L, Stansbery E, Whitlock D O, Abercromby K J and Shoots D 2008 *History of On-Orbit Satellite Fragmentations* 14th edition (Houston: NASA)
[3] Adushkina V V, Kozlova S I and Sil'nikov M V 2016 *The Impact of Rocket and Space Technology on the Environment* (Moscow: GEOS)
[4] Fedorov M Y, Kraus E I and Shabalin I I 2018 *AIP Conf. Proc.* **2027** 030177
[5] Kraus E I and Shabalin I I 2013 *Frat. ed Integrita Strutt.* **24** 138–150
[6] Kraus E I and Shabalin I I 2016 *Res. Sci. City* 6–11
[7] Kraus E I and Shabalin I I 2016 *AIP Conf. Proc.* **1770** 030092
[8] Wilkins M L 1999 *Computer Simulation of Dynamic Phenomena* (Berlin, Heidelberg: Springer)
[9] Fomin V M, Gulidov A I, Sapozhnikov G A and Shabalin I I 1999 *High-Velocity Solids Interaction* (Novosibirsk: SB RAS)
[10] Kraus E I and Shabalin I I 2017 *Math. Montisnigri* **39** 18–29

Numerical simulation of fluid flow in a saturated fractured porous media based on the linear poroelasticity model

D Yu Legostaev, A Yu Botalov and S P Rodionov

Tyumen Branch of Khristianovich Institute of Theoretical and Applied Mechanics of SB RAS, Taymirskaya Str. 74, 625026, Tyumen, Russia

E-mail: `legostaevdy@yandex.ru`

Abstract. Field development processes change the stress-strain state of the formation, which can especially affect on the state of fractured reservoir. In this paper, the coupled flow and geomechanics model was used to simulate a fractured-porous reservoir. To describe fractures in the fluid flow problem the discrete fracture model was used; the fracture was considered as an internal boundary for the geomechanics problem. The control volume method, which is widely used in computational fluid dynamics, was used for both flow and mechanics problems. Based on the developed coupled flow and geomechanics model, simulations of an fractured poroelastic system was carried out.

1. Introduction

The existence of fractures has a significant effect on the fluid flow in reservoir systems. Traditional reservoir flow simulation takes into account the dependence of the fracture permeability only on pore pressure. At the same time, laboratory studies show that the permeability of fractures depends on normal stresses [1]. The paper [2] states that shear stresses have a decisive influence on the hydrodynamic characteristics of reservoir faults: the critically-stressed-fault hypothesis is introduced, according to which mechanically active faults, determined by the Coulomb-Mohr criterion, are hydraulically conductive. Thus, to obtain a complete picture of the fluid flows in a fractured-porous medium, it is necessary to use coupled flow-geomechanical modeling.

Approaches to the coupled fluid flow and geomechanical reservoir simulation are based on the Biot theory [3, 4]. One of the methods for modeling flow in fractured-porous media is the discrete fracture model (DFM) [5]. A fracture is considered as an internal boundary for geomecanical problem, for which it is additionally necessary to set the boundary conditions that describe the mechanical contact between the fracture sides. In reservoir fluid flow simulations, the control volume method (FVM) [6] is widely known. The finite element method (FEM) is the traditional approach of continuum solid mechanics, however, recently FVM has become an alternative in solid mechanics field [7].

In this study, our aim is to explore the fractured poroelastic system. To achieve this, we have developed coupled fluid flow and geomecanical model for fractured reservoirs simulations. Developed coupled model was verified by comparison with Terzaghi and Mandel poroelasticity problems.

2. Mathematical model

2.1. Poroelsaticity

The governing equations for coupled flow flow and geomechanics come from mass and linear-momentum conservation laws. Linear momentum balance equations for poroelastic body under quasi-static condition and neglecting the body force can be written as

$$\nabla \cdot [\mu \nabla \mathbf{u} + \mu \nabla \mathbf{u}^T + \lambda \mathrm{tr}(\varepsilon)\mathbf{I} - \underbrace{b p \mathbf{I}}_{coupling}\ ] = 0 \tag{1}$$

where $\mathbf{u}$ is the displacement vector, ε is infinitesimal strain tensor, λ and μ are Lame's coefficients, b is the Biot coefficient, $\mathbf{I}$ is identity matrix.

The single-phase flow equation in a deforming porous medium is

$$\frac{1}{M}\frac{\partial p}{\partial t} + \underbrace{b\frac{\partial \varepsilon_v}{\partial t}}_{coupling} - \nabla \cdot \left(\frac{k}{\mu_f} \nabla p \right) = q \tag{2}$$

where p is fluid pressure, $\varepsilon_v = \mathrm{tr}(\varepsilon)$ is volume strain, k is rock permeability, μ_f is fluid viscosity, q is a source term, $M = \phi c_f + \frac{\phi - b}{K_s}$ is the Biot modulus, ϕ is rock porosity, c_f is fluid compressibility, K_s is is the bulk modulus of the solid grain.

The coupled poroelastic system consists of equations (1) and (2).

2.2. Fracture representation

The high aspect ratio of the fracture allows us to consider them as zero thickness inclusions for geomechanical problem. Normal σ_n and shear τ stresses acting on a fracture depend on the relative normal du_n and shear du_s displacements between fracture boundaries. For the fracture under compressive stress we adopt following linear constitutive law:

$$\begin{Bmatrix} d\sigma_n \\ d\tau \end{Bmatrix} = \begin{bmatrix} k_n & 0 \\ 0 & k_s \end{bmatrix} \begin{Bmatrix} du_n \\ du_s \end{Bmatrix} \tag{3}$$

where k_n and k_s are normal and shear stiffness respectively.

To simulate the flow in a fractured-porous medium, we used the discrete fracture model, which is applicable for modeling systems of long-scale fractures. It is assumed that the fracture has some effective hydrodynamic aperture, to describe the fluid flow inside the fracture in the equation (2), we use the permeability k_f and porosity ϕ_f corresponding to the fracture, and volumetric strains ε_v inside the fracture due to its small size is neglected. The transfer function between the matrix and the fractures determinate by pressure difference and can be written as

$$q_{f-m} = \frac{T_{f-m}}{\mu_f}(p_f - p_m) \tag{4}$$

where T_{f-m} is the transmissibility between the fracture and the matrix, p_m and p_f are pressure in matrix and fracture respectively.

3. Numerical Model

The fluid flow and geomechanical equations was discretized using the cell-centered FVM, which is up to second-order accuracy. Approaches of equations (1) and (2) discretization are discussed in detail in [7]. For the poroelasticity equations system implicit–explicit discretization was used, which leads to decoupled system of linear equations solved in the segregated manner. For solution of coupled flow-mechanics problem unconditionally stable fixed-stress split method was used [8].

Detail of using of the DFM in numerical method can be found in [5]. Using of the FVM for mechanical problem of fractured media is presented in [9].

4. Validation

Developed coupled model was verified by comparison with Terzaghi and Mandel poroelasticity problems (figure 1). Poroelastic parameters used for both problem are shown in table 1.

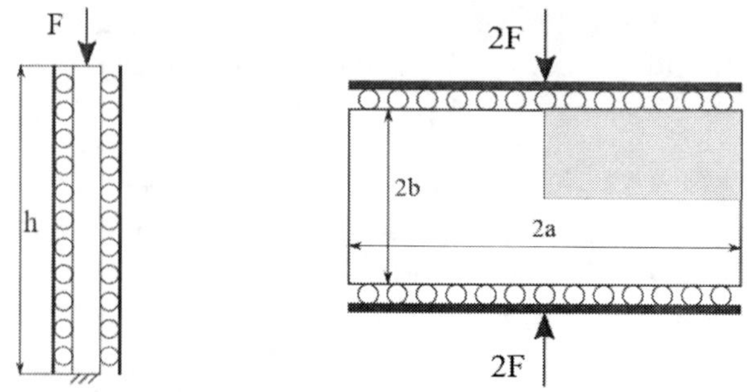

(a) Terzaghi problem (b) Mandel problem

Figure 1. Terzaghi (a) and Mandel (b) problems formulation.

Table 1. Parameters used in calculation of the Terzaghi and Mandel problems.

Parameter	Valuie	Unit
Matrix porosity	0.2	-
Matrix permaebility	10	md
Fluid viscosity	0.1	$Pa \cdot s$
Fluid compressibility	$8 \cdot 10^{-4}$	MPa^{-1}
Young's modulus	$2 \cdot 10^{4}$	MPa
Poisson's ratio	0.2	-
Biot coefficient	1	-

4.1. Terzaghi Problem

In the Terzaghi problem, a saturated column under load is considered. The column is laterally constrained, load $F = p_0$ is applied to the upper boundary. The no-flow boundary conditions are set at all boundaries except the upper one, where the pressure $p_{top} = 0$ MPa is set. The initial pore pressure $p_0 = 0.1$ MPa and initial deformations are equal to zero. The calculation domain is 10 m high (h) and 1 m thickness presented on figure 1a is divided on 99 grid blocks along the vertical-direction and on 9 in horizontal-direction.

A comparison for dimensionless pressure $\overline{p} = \frac{p}{p_0}$ and vertical displacement $\overline{v} = \frac{uK_{dr}}{hp_0}$ for dimensionless coordinate $\overline{z} = \frac{z}{h}$ at various characteristic times $\overline{t} = \frac{kt}{\mu_f\left(\frac{1}{M}+\frac{b^2}{K_{dr}}\right)h^2}$, where $K_{dr} = 2\mu + \lambda$ is the bulk modulus, is showed in figure 2.

4.2. Mandel Problem

The main feature of the Mandel problem is the Mandel-Cryer effect[10]. A saturated poroelastic medium with long of $2a = 20\,m$ and height $2b = 20\,m$ sandwiched between two rigid impermeable plates (figure 1b) was considered. The pressure at lateral boundaries of the region is equal to zero

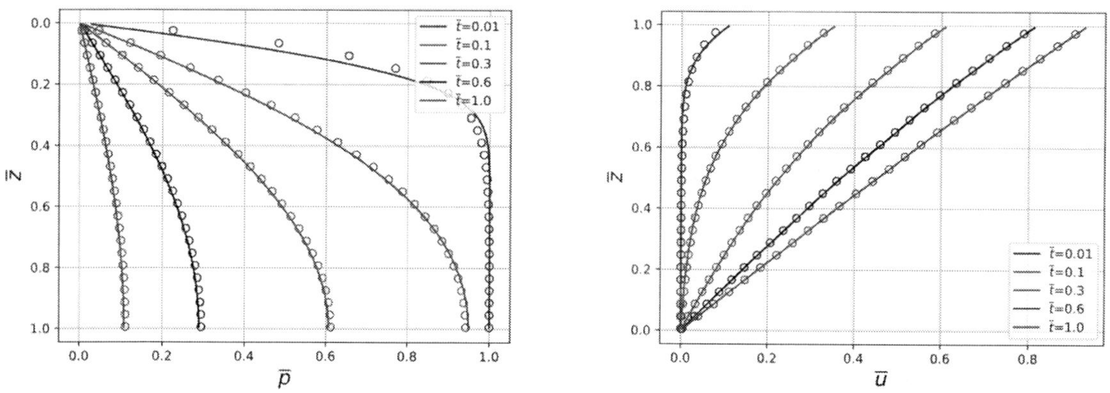

(a) Dimensionless pressure profiles (b) Dimensionless displacement profiles

Figure 2. Comparison of numerical solution (dots) for pressure (a) and vertical displacement (b) with analytical solution (lines) of Terzaghi problem.

and they are assumed traction free. At the initial time, the constant compressive force $2F = 2$ MPa immediately begins to act on the plates. The initial pore pressure p_0 and deformations are equal to zero. Because of the symmetry of this problem, the simulation was run only for one-quarter divided on 31 grid blocks in both direction.

Figure 3 shows a comparison for dimensionless pressure $\overline{p} = \frac{3ap}{FB(1+\nu_u)}$ and vertical displacement $\overline{v} = \frac{2\mu u}{F\nu_u}$ for dimensionless coordinate $\overline{x} = \frac{x}{a}$ at various characteristic times $\overline{t} = \frac{kt}{\mu_f\left(\frac{1}{M}+\frac{b^2}{K_{dr}}\right)a^2}$, where $B = \frac{bM}{K_{dr}+b^2M}$ is the Skempton's coefficient and $\nu_u = \frac{3\nu+bB(1-2\nu)}{3-bB(1-2\nu)}$ is the undrained Poisson's ratio.

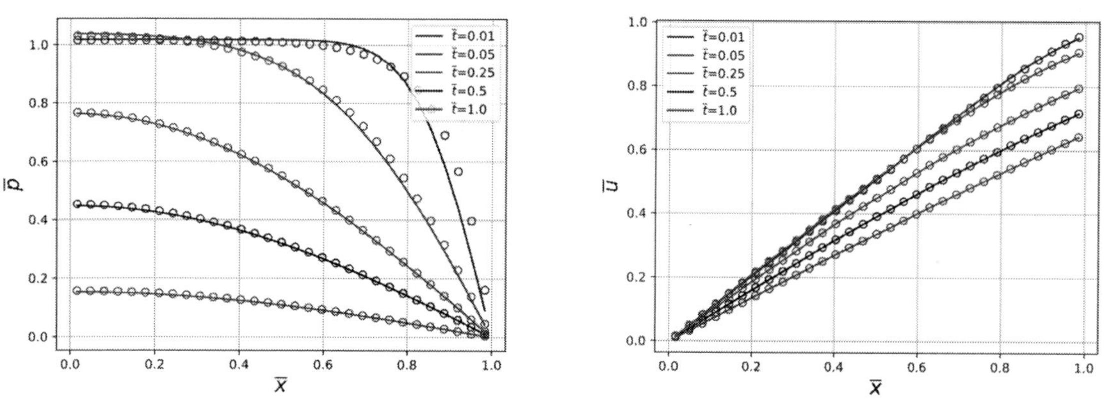

(a) Dimensionless pressure profiles (b) Dimensionless displacement profiles

Figure 3. Comparison of numerical solution (dots) for pressure (a) and horizontal displacement (b) with analytical solution (lines) of Mandel problem.

We obtain a good agreement with both Terzaghi and Mandel analytical solutions for pressure and displacement fields at all times steps.

5. Results

Based on the approaches to simulation fractured reservoirs considered above, a model for modeling fractured poroelastic media is developed. The pressure, the strain and the stress fields are considered, and special attention is paid to the fracture mechanical state. Reservoir and fracture geometry are shown in figure 4, the domain size is $500 \times 500 \times 5$ m. The fracture 300 m long was located in the center of the computational domain, three various angles of fracture inclination (30°, 45°, 60°) are investigated. An unstructured computational grid was used to describe the geometry of the fracture.

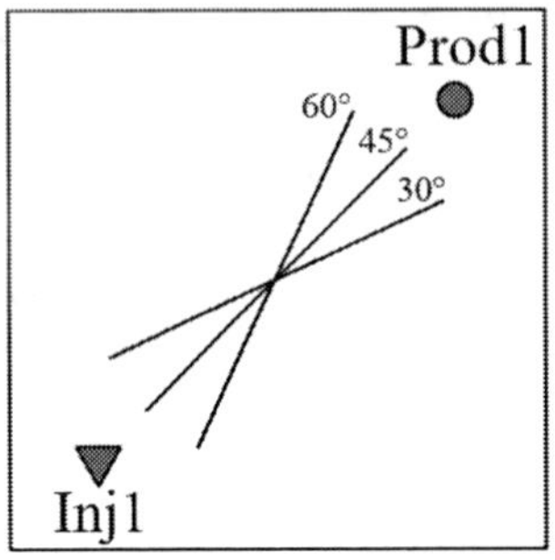

Figure 4. Reservoir geometry with different fracture orientation.

For all boundaries of the region the no-flow boundary condition was applied. For the geomechanical problem boundary the maximum stress applied vertically $\sigma_{max} = 60$ MPa and the minimum stress applied horizontally $\sigma_{max} = 40$ MPa. Production well with bottomhole pressure of $P_{prod} = 22$ MPa and injection operating with constant flow rate $Q_{inj} = 10\, m^3/day$ are located inside the computational domain. Initial strains are agree with boundary conditions and uniform initial pressure $p_{init} = 25$ MPa. Simulation time was 600 days.

The same parameters as in validation section was used, which are shown in table 1. Additionally, it is necessary to describe the properties of the fracture: $k_f = 10^7\, md$, $\phi_f = 0.5$, $k_s = 2 \cdot 10^3$ MPa and k_n is equal to matrix Young's modulus.

The figure 6 shows the distribution of pore pressure for the case of inclination of the crack at 60 degrees. The effect of a fracture on the direction of the filtration flows is observed, especially near the wells, which is well demonstrated in the figure by arrows.

Let consider the mechanical state of the investigated fracture. The increase of of the average pressure was observed for the reservoir system, which leads to the fracture effective normal stresses decrease. To estimate fracture stress state it is convenient to use the Coulomb failure function (CFF):

$$CFF = \tau - \mu\sigma_n \tag{5}$$

where μ is the coefficient of friction.

Following the critically-stressed-faults hypothesis, faults that are mechanically active ($CFF > 0$) are hydraulically alive. As shown on Mohr–Coulomb diagram (figure 5), only for fracture with 60° inclination angle are possible to reach a critical stress state. This condition is achieved after 440 days of reservoir simulation, it can serve as an indicator of the shear on a fracture plane.

6. Conclusions

A coupled fluid flow and geomechanical model for fractured reservoir has been developed using cell-centered FVM, DFM and fixed-stress split method. The model was validated against known

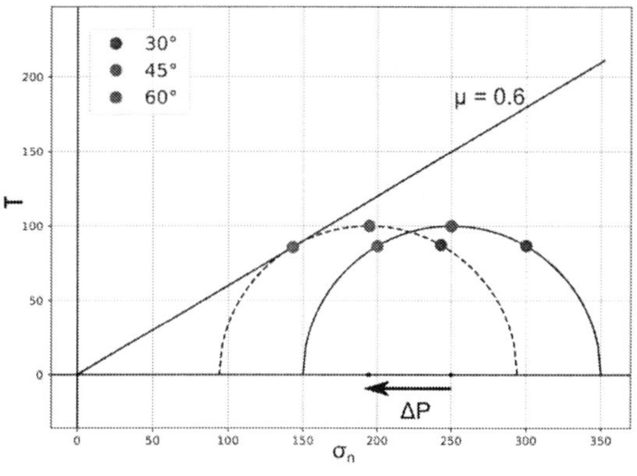

Figure 5. Mohr–Coulomb diagram represent the fracture stressed state. Solid line is initial fractures state, dashed line is fractures stressed state at 440 day.

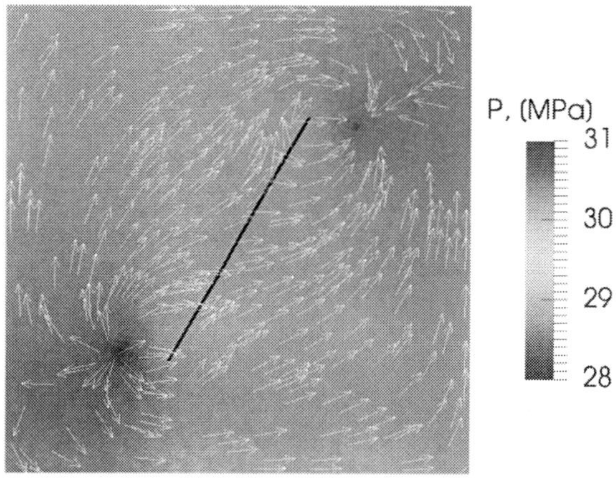

Figure 6. Pressure field for the case of a fracture 60° inclination at 300 day. The arrows indicate the direction of fluid flow in the cells. Black line represent the fracture.

poroelastic problems.

The fractures orientation has a decisive effect on their mechanical condition. Fractured reservoir modeling was carried out for various fracture orientations. In the studied reservoir only fractures with an orientation close to 60° can be critically stressed.

Acknowledgments

Acknowledgements The research was carried out within the framework of the Program of Fundamental Scientific Research of the State Academies of Sciences in 2013-2020 (Project No. AAAA-A17-117030610130-1).

References

[1] Kranz R L, Frankel A D et al. 1979 The permeability of whole and jointed Barre granite. *Int. J. Rock Mech. Min. Sci. Geomech. Abst.*, **16**, 225–234.
[2] Zoback M D 2007 *Reservoir Geomechanics* (Cambridge University Press) p 449
[3] Biot M A 1941 General theory of three-dimensional consolidation *J. Appl. Phys.* **12** 155–164.
[4] Coussy O 2004 *Poromechanics* (Chichester, England: John Wiley and Sons) p 298
[5] Karimi-Fard M, Durlofsky L J and Aziz K 2004 An efficient discrete fracture model applicable for general purpose reservoir simulator *SPE Journal* **9**(2), 227–236.
[6] Aziz K and Settari A 1979 *Petroleum Reservoir Simulation* (London: Elsevier)
[7] Tang T, Hededal O and Cardiff P 2015 On Finite Volume Method Implementation of Poro-Elasto-Plasticity Soil Model. *Int. J. Numer. Anal. Meth. Geomech.* **39** (13) 1410–1430
[8] Kim J, Tchelepi H A and Juanes R 2011 Stability, Accuracy, and Efficiency of Sequential Methods for Coupled Flow and Geomechanics. *Society of Petroleum Engineers* (doi:10.2118/119084-PA)
[9] Ucar E, Keilegavlen E, Berre I and Nordbotten J M 2016 A Finite-Volume Discretization for the Deformation of Fractured Media (arXiv:1612.06594v2)
[10] Abousleiman A, Cheng A, Detournay E and Roegiers J 1996 Mandel's problem revisited. *Geotechnique* **46** 187–195

Influence of copper on the phase composition and strength of laser-welded connections

A G Malikov[1], A M Orishich[1], E V Karpov[1,2] and N V Bulina[3]

[1] Khristianovich Institute of Theoretical and Applied Mechanics, Siberian Branch Russian Academy of Sciences, 630090, Novosibirsk, Russia
[2] Lavrentyev Institute of Hydrodynamics, Siberian Branch, Russian Academy of Sciences, 630090, Novosibirsk, Russia
[3] Institute of Solid State Chemistry and Mechanochemistry, Siberian Branch, Russian Academy of Sciences, 630128, Novosibirsk, Russia

smalik@ngs.ru, laser@itam.ncs.ru, evkarpov@mail.ru

Abstract. This paper deals with the investigation of the influence of copper on the structure-phase composition and strength of laser-welded connections of aluminum alloys of the system Al-Cu-Mg, Al-Cu-Mg-Li, Al-Cu-Li. The microstructure, mechanical characteristics and chemical composition of the welded joints have been investigated. It is founded that the laser welding results in the reduction of copper concentration in the solid solution and growth of this concentration on the dendritic boundary.

1. Introduction

Aluminum alloys are commonly used in space equipment owning to their unique properties, such as low density, high strength and elasticity module. In aircraft engineering of power components of a planer and aircraft wings, both classical aluminum alloys of the system Al-Zn-Mg-Cu, Al-Cu-Mg and alloys with lithium of the system Al-Cu-Li-X (X = Mg, Zn, Mn, Zr, Sc, Ag) are used [1, 2]. Lithium added in the system Al-Cu changes dramatically the phase composition and increases the mechanical properties of these alloys due to the proper thermal treatment (TT). This TT leads to the release of various strengthening phases $T_1(Al_2CuLi)$, $\delta'(Al_3Li)$, $\theta(Al_2Cu)$, $S(Al_2CuMg)$ from the solid solution. The phase T_1 remains in the composition the phase S in the system Al-Cu-Mg, thus we can suggest that lithium interacts with aluminum and copper in the same way as magnesium. Note that the above mentioned phases in the alloy influence on the aluminum alloy mechanical properties, their release at different TT types (quenching, annealing, artificial ageing) leads to the different composition of the solid solution. The properties of such alloys are studied carefully, the methods of improvement of their strength characteristics are developed involving different methods of thermal (quenching, artificial ageing, annealing) and deformation (elastic deformation) treatment. In order to replace the riveting technology widely used in the aircraft and space equipment and hence reduction of resultant construction weight, the technologies of welding of aluminum alloys are developed intensively (mixing welding, laser welding). Laser welding has a number of advantages. A small welded pool reduces essentially the thermal influence zone and joint distortion as compared to the conventional welding methods. High rate of the process and reproducibility reduce the consumptions and provide the high quality of the final product and maximal welding rate. The welded connections of the aluminum alloys containing copper as the major alloying element have low mechanical properties, the yield strength is 50 – 80% of the main alloy strength regarding the alloying system [3, 4]. Today, the laser welding of high-strength aluminum-lithium alloys of the system Al-Cu-Li, Al-Cu-Mg-Li is studied intensively [5, 6]. At the same time, to increase the strength of the welded joint in thermally-strengthened aluminum-lithium alloys, it is necessary to provide extra thermomechanical treatment of the welded joint [7]. This paper investigates the microstructure and phase composition of the laser welded joints of the aluminum alloys with copper as the major alloying element, in order to determine the copper effect on the mechanical characteristics of the welded joint.

2. Experimental Technique

Table 1 present the chemical composition of the studied aluminum alloys:

Table 1. Chemical composition of the aluminum alloys.

Alloy	Al	Cu	Mg	Mn	Li	Zr	Sc
D16T system Al-Cu-Mg	base	4.3	1.5	1.5			
1441 system Al-Cu-Mg-Li	base	1.6-1.9	0.7-1.1	0.01-0.04	1.7-2.0	0.02-0.26	
1461 system Al-Cu-Li	base	2.5-2.95	0.4-0.6	0.1-0.4	1.5-1.9	0.05-0.12	0.05-0.1
1469 system Al-Cu-Li	base	3.2-4.5	0.1-0.5	0.05-0.08	1.0-1.7	0.02-0.26	0.02-0.28

These alloys present the solid solution of α- Al with the alloying elements in aluminum which form different phases. Phase composition of the aluminum alloys: D16T - θ(Al_2Cu), S(Al_2CuMg); 1441- T_1(Al_2CuLi), S(Al_2CuMg), S_1 (Al_2MgLi), δ′(Al_3Li); 1461- T_1(Al_2CuLi), T_2(Al_6CuLi_3), S_1 (Al_2MgLi), δ′(Al_3Li), θ(Al_2Cu); 1469 - T_1(Al_2CuLi), T_2(Al_6CuLi_3), S_1(Al_2MgLi), δ′(Al_3Li), θ′′, θ′(Al_2Cu).

Laser beam welding was performed on the Sibir-1 automated laser technological complex, which includes a continuous-wave CO_2 laser with a power up to 8 kW developed at the Khristianovich Institute of Theoretical and Applied Mechanics of the Siberian Branch of the Russian Academy of Sciences [8]. The laser beam focuses on the alloy surface with the help of a ZnSe lens with a focal distance of 254 mm. An inert gas (helium) is used for the welded joint protection. The spectral analysis of the welded joint and base alloy is performed on a LEO 1430 VPI scanning electron microscope. The strength of the welded joints is measured at the static extension on the electro-mechanical test machine Zwick/Roell Z100. XRD patterns are recorded by a D8 Advance powder diffractometer. The laser-beam welded samples are obtained in the following welding mode: laser power 2.5 – 3.5 kW, welding speed 4 m/min, focus forced into the material 2 – 3 mm.

3. Results and Discussion

Figure 1 shows the values of the yield strength for the basic material and welded joint after the stretch tests of the studied aluminum alloys. Figure 2 shows the coefficient k (welded joint yield strength to the basic material yield strength) versus the copper concentration in the initial alloy.

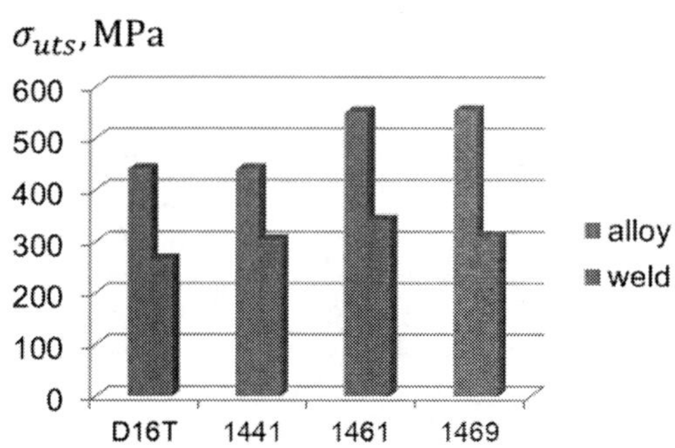

Figure 1. Mechanical characteristics.

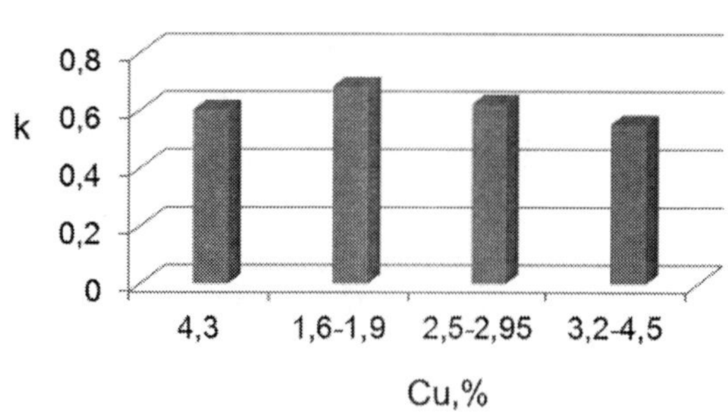

Figure 2. Coefficient k versus copper concentration in the initial alloy.

Journal of Physics: Conference Series **1404** (2019) 012029 doi:10.1088/1742-6596/1404/1/012029

As is seen from figure 2, the mechanical characteristics of the welded joints are low, and, the higher the copper concentration, the lower the coefficient k. Figure 3 presents the optical photo of the cross section microstructure of the welded joint. It is obvious that the microstructure of the welded joint differs fundamentally from the microstructure of the alloy. The laser action forms the dendritic structure in the welded connection. In the thermal influence zone, for the alloys of the system Al-Cu-Mg-Li, Al-Cu-Li there is a thin equiaxed zone of about 13 – 15 μm in width. In the limits of this zone, there are many small-size equiaxed almost-spherical grains are formed. Then, toward the melting zone center of the welded joint, we see the zone of extruded rod-like dendrites. The dendritic structures grow in the heat release direction. This structure is typical for the laser welding of the aluminum alloys containing lithium and zirconium as alloying elements [9]. In the alloy of the system Al-Cu-Mg, the thin equiaxed structure in the thermal influence zone is not observed, only the zone of extruded rod-like dendrites.

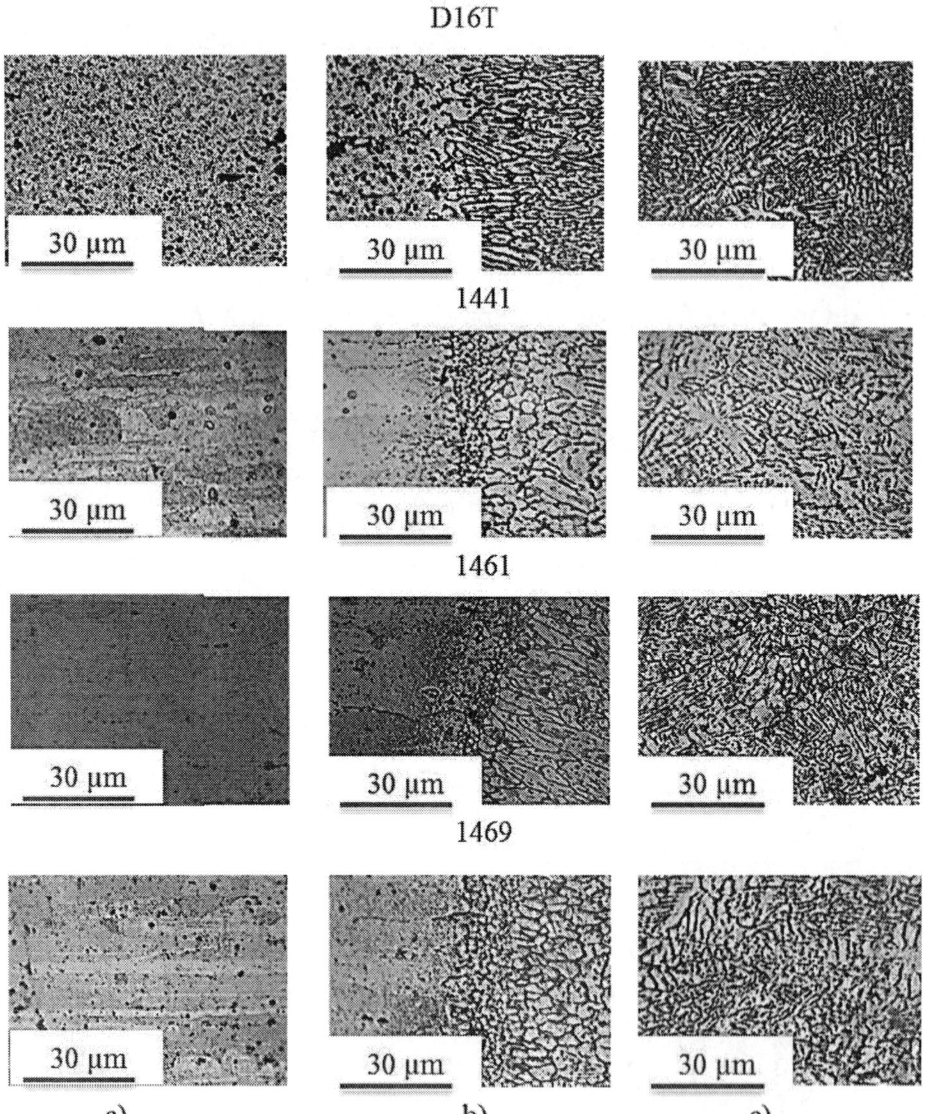

Figure 3. Optical image of the alloy cross section (a), HAZ (b), welded joint (c).

Figure 4 presents the microstructure and EDX analysis of the welded connection in the aluminum alloy (the system Al-Cu-Li and Al-Cu-Mg-Li) melting zone. The data are obtained in the electronic microscope. For the alloy 1441 of the system Al-Mg-Cu-Li, on the boundary there are both dark and light particles. For the alloys 1461 and 1469 of the system Al-Cu-Li, mainly light particles accumulate on the dendritic grain boundary. Also there are light particles observed in the welded joint, they are seen as individual agglomerates in every studied alloy. Note that in the welded joint, the light particles accumulate predominantly on the dendritic grain boundaries, which makes them very contrast.

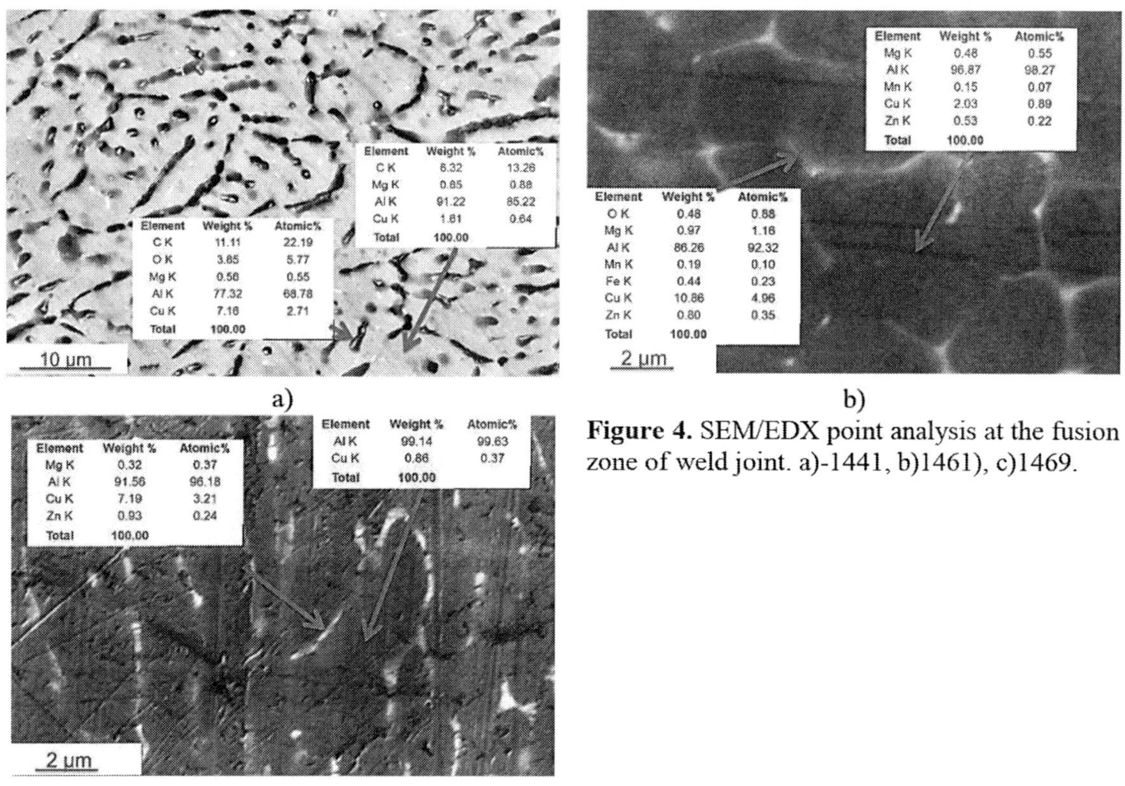

Figure 4. SEM/EDX point analysis at the fusion zone of weld joint. a)-1441, b)1461), c)1469.

Control of the chemical composition demonstrates that individual light particles, both in the initial alloys and welded joints, are essentially enriched with different alloying elements. Their concertation grows (especially in the joints) as compared to the solid solution by 10 times. For the alloy 1441 of the system Al-Mg-Cu-Li, the light particles with the characteristics size of 1 – 2 μm and the light particles combined with dark ones shows the increased copper content (7.16 weight % and 2.31 weight %, respectively) as compared to the solid solution of the welded joint (1.61%). It can be suggested that the light particles with the increase copper content form the phase T_1 (Al_2CuLi), θ′ (Al_2Cu), whereas accumulated light and dark particles over the dendritic grain boundaries from not only T_1(Al_2CuLi), θ′ (Al_2Cu), but also the phase S' (Al_2CuMg) S_1 (Al_2MgLi).

For the alloys of the system Al-Cu-Li, the alloy 1469, the measurements show that the chemical composition in the welded joint features the dramatic jump of the copper concentration up to 10% and reduction of the copper concentration in the solid solution down to 0.86%. The solid solution was depleted. It can be suggested that in the light particles mainly T_1(Al_2CuLi) and θ′ (Al_2Cu) phases form. For the welded joint of the alloy 1461, we also observe the copper concentration reduction in the solid solution down to 1.69 weight % and concentration growth on the dendritic grain boundary up to 11 weight %. Individual light particles are enriched with various alloying elements of the alloy.

Analysis of X-ray patterns (see figure 5) shows that both in the welded joint and initial alloy 1469 of the system Al-Cu-Li there are clear reflexes which can correspond to the following intermetallic phases: S_1(Al2MgLi) – the double angle 2Θ= 12.3^0 : θ(Al2Cu) –2Θ = 20.8^0 and 42^0; $T_1(Al_2CuLi)$ –2Θ = 21^0 and 42^0.

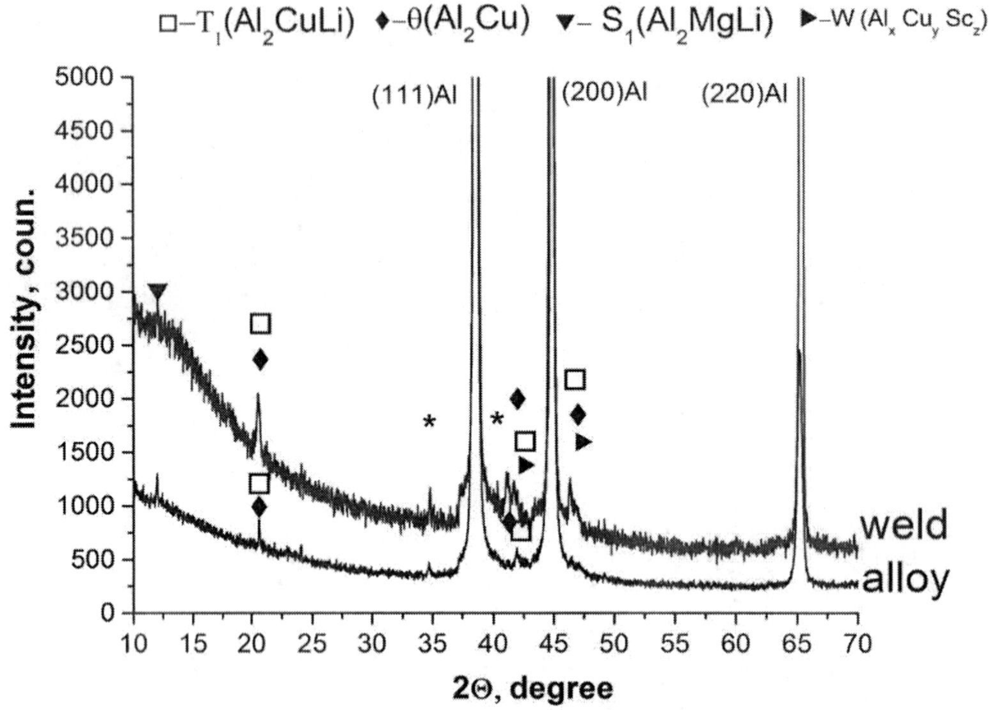

Figure 5. XRD patterns of the base alloy and welded joint.

During the laser welding of the alloy of the system Al-Cu-Mg-Li и Al-Cu-Li, copper diffused toward the dendritic grain boundary. For the alloy of the system Al-Cu-Mg-Li, the phase δ′(Al_3Li) diluted in the solid solution, and S'(Al_2CuMg) $S_1(Al_2MgLi)$ $T_1(Al_2CuLi)$ released over the dendritic grain boundaries. For the alloy of the system Al-Cu-Li, the phase δ′(Al_3Li) diluted in the solid solution, and $T_1(Al_2CuLi)$ and θ′(Al_2Cu) released over the dendritic grain boundaries.

4. Conclusions

The paper presents the analysis of the laser welding of the aluminum alloys containing copper as the main alloying element. The yield strength of the welded joint is 55 – 68% of the basic alloy yield strength. The welded joint microstructure differs fundamentally from the basic alloy microstructure and is dendritic-like. It is founded that the laser welding results in the copper diffusion toward the dendritic grain boundary with intermetallic phase formation which leads to welded joint embrittlement and worsening of the mechanical properties.

Acknowledgments
The work has been performed as the part of the project “Physical bases of the effect of Mg, Cu and their compositions of the mechanical properties of high-strength laser-welded joints of aluminum alloys”, the complex program pf fundamental researches SB RAS “Inter-disciplinary integration investigations” for 2018 – 2020. The study was conducted at the Joint Access Center «Mechanics» of ITAM SB RAS.

References

[1] Dursun T and Soutis C 2014 *Mater. Des.* **56** 862–71
[2] Abd El-Aty A, Xu Y, Guo X, Zhang S-H, Ma Y and Chen D 2018 *J. Adv. Res.* **10** 49–67
[3] Sánchez Amaya J M, Amaya-Vázquez M R and Botana F J 2013 *Handbook of Laser Welding Technologies* (Elsevier).
[4] Kashaev N, Ventzke V and Çam G 2018 *J. Manuf. Process.* **36** 571–600
[5] Han B, Tao W, Chen Y and Li H 2017 *Opt. Laser Technol.* **93** 99–108
[6] Fu B, Qin G, Meng X, Ji Y, Zou Y and Lei Z 2014 *Mater. Sci. Eng. A* **617** 1–11
[7] Malikov A, Orishich A, Golyshev A and Karpov E 2019 *J. Manuf. Process.* **41** 101–10
[8] Malikov A, Orishich A, Bulina N, Karpov E and Sharafutdinov M 2019 *Mater. Sci. Eng. A* **765** 138302
[9] Hu Y N, Wu S C and Chen L 2019 *Eng. Fract. Mech.* **208** 45–71

Journal of Physics: Conference Series **1404** (2019) 012030 doi:10.1088/1742-6596/1404/1/012030

Application of continuous symmetry groups for numerical solutions of two-phase filtration problems

Pavel Markov

CEO, MicroModel LLC, Skolkovo, Moscow 143026, Russia

E-mail: markov.pv@mail.ru

Abstract. The method of numerical solution calculations for difference schemes on the basis of continuous symmetry groups has been presented for the two-phase flow equation. This method gives an opportunity to calculate numerical solutions using transformations of a symmetry group and known particular numerical solutions what allows to achieve an acceleration of serial numerical calculations by several orders. The families of invariant difference schemes have been obtained for one of generalized Buckley-Leverett equations - the Rapoport-Leas equation. A particular nonlinear equation of multiphase flow in fractured porous media with certain coefficients from the mentioned families is analyzed numerically by the presented method.

1. Introduction

Many problems in modeling are not solved rigorously and have to be replaced by some numerical algorithms such as solving of correspondent difference equations. But for using these algorithms many times for different problems, one must be sure that this particular numerical method is still reliable. That is why there is a need to research them using different methods, which come from different modern fields of mathematics. Also, many problems in mathematical modeling have a great degree of uncertainty and different variants of optimal solutions. That is why there is a need to conduct multivariate calculations. However, this number of calculations demands fast and effective numerical calculation methods.

There are two possible ways of the calculation time decreasing: improvements of hardware and using more advanced methods for theoretical analysis of algorithms. Improvements of numerical algorithms have to be gained by using new methods for their analysis. It is the only way to increase both the speed and the effectiveness of numerical calculations in many fields. And one of these new methods of analysis is the theory of continuous symmetries for difference equations.

This article is intended to demonstrate the method of fast obtaining of numerical solutions using continuous symmetries for difference schemes when conducting a large number of numerical calculations for the Rapoport-Leas equation with some certain types of defining coefficients for relative permeability and capillary pressure.

2. Continuous symmetries of differential and difference equations

There are different types of equations, which can be analyzed using continuous symmetries: differential equations [13], difference equations [6], differential-difference equation [24] and others. There are some classical books about group analysis of differential equation such as [8], [13], [19], [21]. One of the first articles about applications of continuous groups to discrete equations was [14] where continuous symmetries were applied to discrete dynamical systems. It gave an impulse for further studying of discrete dynamical systems: [15], [18] and others. Also, specialists started to study difference equations: [3], [5], [11], [12] and others.

2.1. The method of difference scheme construction with the preservation of continuous symmetries

The method of difference schemes construction preserving continuous symmetries of an original differential equation can be found in [6]. There are two approaches: the first one uses definitions of finite-difference derivatives to prolong the action of a continuous group and the second one uses values of independent and dependent variables in points on a mesh as the basis for all calculations. The method in this article uses the second approach for further results of possible invariant difference schemes.

Invariant difference schemes for a differential equation are constructed by the following steps:

- the identification of continuous symmetries for a differential equation;
- the expression of a differential equation in terms of differential invariants;
- the construction of difference invariants for a symmetry group;
- the approximation of differential invariants via difference invariants;
- the notation of equations via approximations of differential invariants;
- the seeking of functions for the definition of a mesh.

2.2. The method of numerical solution calculations using continuous symmetries

The method of numerical solution calculations is based on the fact that a transformation from a continuous symmetry group transforms solutions into solutions. This approach is applied for differential equations [8] and can be applied for difference schemes as well [6]. The step by step description of the developed method is presented below [17]:

- the obtaining of an invariant difference scheme, i.e. a scheme with a continuous symmetry group, for a differential equation;
- the obtaining of a set of initial and boundary conditions, which can be transformed into each other by the continuous symmetry group;
- the calculating of one particular solution using an invariant difference scheme according to initial and boundary conditions, which allows to obtain a set of solutions for the gained set of initial and boundary conditions;
- the seeking of parameters for the continuous group of symmetry, which allows to transform the particular solution conditions into the set of conditions;
- the transformations of the particular numerical solution into the set of numerical solutions with conditions from the set of conditions.

Errors for numerical solutions can be controlled and decreased when solutions are generated by symmetry groups [17]. Boundary and initial conditions must be transformed one into another by using a symmetry group if we want to obtain a family of solutions. Similar problems are discussed in, for example, [7].

3. Results for the Rapoport-Leas equation

3.1. Results of group classification for the Rapoport-Leas equation

The partial differential equation of two-phase flow in one-dimensional porous media within the generalized Buckley-Leverett model [23] (also known as the Rapoport-Leas equation) is

Journal of Physics: Conference Series **1404** (2019) 012030 doi:10.1088/1742-6596/1404/1/012030

considered in this article. This equation can be presented as

$$\frac{\partial S}{\partial t}+\frac{K}{\mu_o\phi}\frac{\partial}{\partial x}\left(K_o(S)f(S)\frac{dP_c}{dS}\frac{\partial S}{\partial x}\right)+\frac{V}{\phi}\frac{df}{dS}\frac{\partial S}{\partial x}=0, \tag{1}$$

where $S(t,x)$ - water saturation, μ_w and μ_o - constant viscosities of water and oil respectively, ϕ - porosity, $P_c(S)$ - capillary pressure, $V=V_o+V_w=const$ - total flow, $f(S)$ - fractional flow rate of water, $K_w(S)$ - relative permeability for water, $K_o(S)$ - relative permeability for oil.

The equation (1) can be written in the form

$$\frac{\partial S}{\partial t}-\frac{\partial}{\partial x}\left(A(S)\frac{\partial S}{\partial x}\right)+\frac{\partial B(S)}{\partial S}\frac{\partial S}{\partial x}=0, \tag{2}$$

which is similar to the correspondent heat transfer equations. Different heat transfer equations are well studied using continuous groups of transformations [10], [20].

The equation (2) has two one-parameter groups of symmetries for arbitral functions of relative permeability and capillary pressure: the groups of translations for time and space because time and spatial variables do not appear explicitly in the equation. If the mentioned functions are defined as [16]

$$\begin{aligned}&K_w(S)=S^N,\ \ K_o(S)=(1-S^N),\ \ P_c(S)=-\tfrac{P_1}{N}\ln\left|\tfrac{S^N}{S^N-1}\right|+P_2,\\&f(S)=S^N,\ \ \mu_o=\mu_w=\mu,\end{aligned} \tag{3}$$

where P_1 and P_2 - scaling coefficients defining relations for capillary pressure, N - a dimensionless parameter fulfilling conditions $N\neq 1$ and $N\in R$, the equation (1) is transformed into the nonlinear equation below:

$$\frac{\partial S}{\partial t}-\alpha\frac{\partial}{\partial x}\left(S^{N-1}\frac{\partial S}{\partial x}\right)-\beta S^{N-1}\frac{\partial S}{\partial x}=0,\ \ \alpha=\frac{KP_1}{\mu\varphi},\ \ \beta=-\frac{VN}{\varphi}. \tag{4}$$

The equation (4) has the three-parameters group

$$\bar{t}=e^{-a_1}t+a_2,\ \ \bar{x}=x+a_3,\ \ \bar{S}=e^{\frac{a_1}{N-1}}S,\ \ a_1,a_2,a_3\in R \tag{5}$$

with the infinitesimal operators [10]

$$X_1=\frac{\partial}{\partial t},\ X_2=\frac{\partial}{\partial x},\ X_3=-t\frac{\partial}{\partial t}+\frac{S}{N-1}\frac{\partial}{\partial S}. \tag{6}$$

3.2. Families of invariant difference schemes

Differential equations of type (2) are well studied by the group analysis. This type of equations has group classifications, which can be found in [10], [20] as it is mentioned above. These classifications are used for the current section for constructions of families of invariant difference schemes for differential equations of the type (2), which represents the physical problems (1). The purpose of this section is to present differential equations (by their coefficients) and their invariant difference schemes (by their difference invariants) with coefficients, which can be found in real physical cases for two-phase flow in porous media. It means that the following results in Table 1 are not intended to present all cases from the known classifications. There are similar results for other types of heat transfer equations, which can be found in [2], [4].

In Table 1, $p_i\in\mathbb{Z}$ and $c_j\in\mathbb{R}$ are arbitral constants, operators X_1 and X_2 are from (6). The equation (4) and all further calculation results belong to the case 1 of Table 1 with coefficients

Table 1. Invariant difference schemes

No.	A	B	Operators	Difference invariants
1	$A = c_1 S^{c_2}$, $c_2 \neq -1$	$B = c_1 c_3 S^{c_2+1}$	X_1, X_2, $\bar{X}_3 = -t\frac{\partial}{\partial t} + \frac{S}{c_2}\frac{\partial}{\partial S}$	$J_1 = (t^{n+p_1} - t^{n+p_2})\times$ $\times (S_{k+p_4}^{n+p_3})^{c_2}$, $J_2 = S_{k+p_6}^{n+p_5} \Big/ S_{k+p_8}^{n+p_7}$, $J_3 = x_{k+p_9} - x_{k+p_{10}}$
2	$A = \frac{c_1}{S}$	$B = c_1 c_3 \ln S$	X_1, X_2, $\bar{X}_3 = -t\frac{\partial}{\partial t} - S\frac{\partial}{\partial S}$	$J_1 = \frac{(t^{n+p_1} - t^{n+p_2})}{S_{k+p_4}^{n+p_3}}$, $J_2 = S_{k+p_6}^{n+p_5} \Big/ S_{k+p_8}^{n+p_7}$, $J_3 = x_{k+p_9} - x_{k+p_{10}}$
3	$A = c_1 e^{c_2 S}$	$B = c_1 c_3 e^{c_2 S}$	X_1, X_2, $\bar{X}_3 = -t\frac{\partial}{\partial t} + \frac{1}{c_2}\frac{\partial}{\partial S}$	$J_1 = (t^{n+p_1} - t^{n+p_2})\times$ $\times e^{c_2 S_{k+p_4}^{n+p_3}}$, $J_2 = S_{k+p_6}^{n+p_5} - S_{k+p_8}^{n+p_7}$, $J_3 = x_{k+p_9} - x_{k+p_{10}}$

(3) where $c_1 = \alpha$, $c_2 = N - 1$, $c_3 = \beta/\alpha$. All cases from Table 1 can have certain coefficients with certain physical meanings in the field of two-phase flow in porous media and can be used for constructions of wide sets of invariant difference schemes.

Difference schemes constructed from the invariants of Table 1 should be analyzed separately for stability if these difference schemes are used for numerical calculations. Invariants from Table 1 allow to choose meshes, which have time step choice depending on saturation values in some points of meshes. Time steps can be chosen constant if we choose the mentioned values of saturation from some particular point of meshes. Spatial variable steps must be constant for all points of meshes because of the invariant J_3 of all three classes from the table.

3.3. *Numerical results of calculations*

This section is intended to apply the results of previous sections: the definition of one invariant difference scheme of the Rapoport-Leas equation, which is from the case 1 of Table 1 and can be written as

$$\begin{aligned} &S_k^{n+1} = S_k^n + \beta\frac{\Delta t}{2\Delta x}\left(S_k^n\right)^{N-1}\left(S_{k+1}^{n+1} - S_{k-1}^{n+1}\right) + \\ &+\alpha\frac{\Delta t}{\Delta x^2}\left(\left(S_k^n\right)^{N-1}\left(S_{k+1}^{n+1} - S_k^{n+1}\right) - \left(S_{k-1}^n\right)^{N-1}\left(S_k^{n+1} - S_{k-1}^{n+1}\right)\right), \\ &\Delta t = t^{n+1} - t^n = const, \quad \Delta x = x_{k+1} - x_k = const, \end{aligned} \tag{7}$$

and calculations of numerical solutions using the presented method and this invariant difference scheme. Table 2 presents parameters for the difference scheme (7), which are chosen for the sake of demonstrations of the developed method, for the sake of the presence of both convective and diffusive parts of the equation (4) and for the sake of demonstrations of practical applications for the method. The choice of initial and boundary conditions caused by the desire to check the convergence of the calculated solutions using the known accurate solution [22] for (4). Thus, these conditions are calculated using the mentioned accurate solution.

Chosen boundary and initial conditions together with the difference scheme (7) refer to the problem where drainage (oil is displacing water) of some natural fractured reservoir (fractures and porous matrix) is taking place with the constant liquid velocity. The parameters from Table 1 and functions from (3) with $N = 0.75$ can be found in the results of experiments, for example, in [1] and [9] for fractured porous media. The form of used relative permeability

Table 2. Values of parameters for numerical calculations

Parameter	Units	Values
Porosity ϕ	-	0.25
Absolute permeability K	mD	500
Viscosity μ	mPa · s	1.4
Scaling parameter P_1	atm	5
Scaling parameter P_2	atm	33
Liquid velocity V	m/s	0.00005
Parameter N	-	0.75
Length of calculation region L	m	80
Maximum time T	s	200000
Initial water saturation for left boundary S_L	-	0.2
Initial water saturation for right boundary S_R	-	0.6
Number of spatial steps N_x	pcs	40
Number of time steps N_t	pcs	200000

curves is close to X-curves what refers to fractured porous media. For all calculations of this section, water saturation and relative permeability values change between 0 and 1 but functions for relative permeability from (3) can be scaled by the changing of variables using residual saturation points. Also, relative permeability can be scaled simultaneously for water and oil what is allowed by the form of the equation (4) and functions (3).

Figure 1 shows the generation process of five numerical solutions with values of initial left boundary water saturation $S_L = 0.32, 0.28, 0.24, 0.16, 0.12$ on the basis of the calculated solution and parameters $a_1 \neq 0$, $a_2 = 0$, $a_3 = 0$. It means using the one-parameter group of time and saturation dilations from the continuous group (5). The comparison with the known particular solution of the equation (4) shows the sufficient accuracy and the convergence of the used difference scheme with the set parameters. The relative error is less than 1 percent for all points of the mesh.

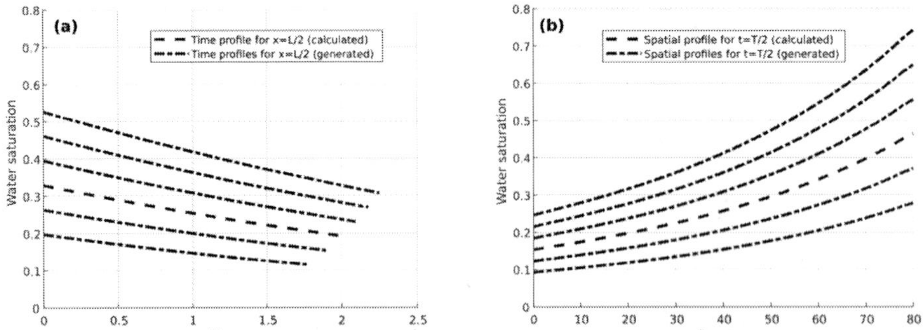

Figure 1. Time (left) and spatial (right) saturation profiles for generated solutions.

Obtained results showed that the gain in speed of calculations of the five solutions using the symmetry group in comparison with calculations of the same number of solutions using a difference scheme is several orders. In fact, the speed of the generating method for numerical solutions depends substantially on the time of calculations of the first solution where the difference scheme and some classical method for systems of linear equations are used [17].

4. Conclusions

The described method of numerical solution generations for the Rapoport-Leas equation can be applied for the increasing of the accuracy and the speed of calculations for two-phase filtration problems. The results can be used for the development and applications of simplified express techniques, which solve various problems of the development of oil and gas fields.

Unconditional advantages of the presented method can be formulated: the increasing of the calculation speed by orders because a numerical solution should be calculated only once for a family of some initial and boundary conditions; the opportunity to control accuracy for further numerical calculations because errors of generated (transformed) solutions depend on the error of the preliminary calculated numerical solution. Shortcomings of the method are connected with the following: one should narrow a family of equations using specific forms of coefficients in equations because it allows them to have continuous symmetries; it is hard to deal with arbitral initial and boundary conditions because all generated solutions must have connections for conditions by transformations of a continuous symmetry group. But these shortcomings can be neglected in conditions of high degrees of uncertainties of real initial data.

Acknowledgments

The reported study was funded by RFBR according to the research project 18-31-00436.

References

[1] Alturki A A, Maini B B, Gates I D 2013 *J. Petrol. Explor. Prod. Technol.* **3** 119
[2] Bakirova M I, Dorodnitsyn V A, Kozlov R V 1997 *J. Phys. A* **30** 8139
[3] Dorodnitsyn V 1994 *International Journal of Modern Physics C* **5** 723
[4] Dorodnitsyn V and Kozlov R 1997 The whole set of symmetry preserving discrete versions of a heat transfer equation with a source *NUMERICS-97-4* p 41
[5] Dorodnitsyn V 1999 Conservation laws for difference equations *Conf. Modern Group Analysis VII* (Nordfjordeid: MARS Publishers) p 91
[6] Dorodnitsyn V 2011 Applications of Lie Groups to Difference Equations (Boca Raton: Taylor and Francis Group) p 265
[7] Hydon P E 2005 *Journal of Mathematical Analysis and Applications* **309** 103
[8] Ibragimov N H 2009 *A Practical Course in Differential Equations and Mathematical Modelling: Classical and New Methods. Nonlinear Mathematical Models. Symmetry and Invariance Principles* (Higher Education Press, World Scientific: Singapore) p 364
[9] Izadi M, Shadizadeh S R, Moradi S 2012 *Int. J. Sci. Emerging Tech.* **3** 46
[10] Lagno V I, Spichak S V, Stogniy V I 2004 *Symmetry Analysis of Evolutional Type Equations* (Moscow, Izhevsk: Institute of computer research) p 392
[11] Levi D and Winternitz P 1991 *Physics Letters A* **152** 335
[12] Levi D and Winternitz P 1996 *J. Math. Phys.* **37** 5551
[13] Lie S 1891 *Vorlesungen uber Differentialgleichungen mit bekannten infinitesimalen Transformationen* (Leipzig: Teubner) p 568
[14] Maeda S 1987 *IMA J. Appl. Math.* **38** 129
[15] Markov P V 2013 *Rus. J. Nonlin. Dyn.* **9** 641
[16] Markov P 2017 *J. Phys.: Conf. Series* **894** p 7
[17] Markov P 2019 Numerical simulation using finite-difference schemes with continuous symmetries for processes of gas flow in porous media *Computation* **7** p 18
[18] Moritz B, Schwalm W, Uherka D 1998 *J. Phys. A* **31** 7379
[19] Olver P J 1986 *Applications of Lie Groups to Differential Equations* (Berlin: Springer) p 513
[20] Oron A and Rosenau P 1986 *Physics Letters A* **118** 172
[21] Ovsyannikov L V 2013 *Lectures on the Theory of Group Properties of Differential Equations* (Beijing: Higher Education Press Limited Company) p 140
[22] Polyanin A D and Zaitsev V F 2011 *Handbook of Nonlinear Partial Differential Equations* (Boca Raton: Chapman & Hall/CRC) p 1912
[23] Rapoport L A and Leas W J 1953 *Petroleum Transactions* **198** 139
[24] Winternitz P and Levi D 1995 *Modern Group Theoretical Methods in Physics* **18** 271

Journal of Physics: Conference Series **1404** (2019) 012031 doi:10.1088/1742-6596/1404/1/012031

Simulation of the deposition of aerosol droplets in a person's bronchial tree

A E Medvedev[1,2], **P S Gafurova**[1,2]

[1] Khristianovich Institute of Theoretical and Applied Mechanics SB RAS, 4/1 Institutskaya st., Novosibirsk, 630090 Russia
[2] Novosibirsk National Research State University, 1, Pirogova st., Novosibirsk, 630090 Russia

medvedev@itam.nsc.ru, polina_g96@mail.ru

Abstract. The work is devoted to the mathematical construction of the human bronchial tree, calculating the flow of gas in the lungs and modelling the deposition of drops of medicinal aerosols in the human bronchial tree. Analytically built a symmetrical human bronchial tree. To calculate the flow of gas in the lungs and the deposition of aerosols, the method of phasing out the flow in one branch of the bronchial tree is used. Three-dimensional numerical calculations of air flow and aerosol droplets in human lungs were carried out. It is shown that the number of past drops depends (except for the flow, size of drops) on the duration of inhalation.

1. Introduction

For the treatment of a wide range of human diseases, the inhalation (aerosol) method of drug administration is now increasingly used. This method of treatment has a number of serious advantages over other ways of administering drugs, as it acts directly at the very place of pathology (for the treatment of pulmonary diseases) or delivers the medicine directly to the blood, through the pulmonary alveoli. Knowing how and where drops or particles of medication are deposited in the respiratory system is fundamental to the correct dosage of medicines.

The human respiratory system consists of the upper (nose, nasopharynx, larynx) and lower (trachea, bronchial tree and alveoli) airways. The human bronchial tree has a complex tree-like structure. From the trachea to the alveoli, the bronchial tree has 24 branches (bifurcations). Individual bronchi in the section are not, generally speaking, circles. In pathology (e.g. asthma) there is a narrowing of the bronchi and a change in the shape of the passage section of the bronchi.

The complexity of numerical modeling of the current in the bronchial tree is determined by the geometric growth of the calculation area with a decrease in the size of the bronchus (the number of bronchus increases twice after each branching). Existing numerical models are limited to 7 generations for asymmetrical bronchial tree and 16 for symmetrical wood, while the full human bronchial tree consists of 23 generations. One-dimensional flow models are used to model the current in subsequent generations (in smaller bronchus). Existing methods of building a bronchial tree are empirical in nature and are tied to the possibilities of constructing complex geometry with a specific commercial CFD simulation package. Existing models of the bronchial tree do not describe the "star" internal structure of the bronchiol and thus do not describe the lungs normally and especially in pathology. To calculate the deposition of particles or drops of aerosol in a person's bronchial tree, it is necessary to use three-dimensional calculations, because only such calculations give a real picture of the movement of particles or droplets in the bronchi of a person.

2. Building a human bronchial tree

For computer simulation of the flow of air and deposition of aerosol droplets in the bronchial tree of a person it is necessary to build a three-dimensional model of bronchus with bifurcations. The quality of the built bronchial tree affects the accuracy and speed of calculation. The situation is aggravated by the different scale of the upper and lower bronchus, which together leads to a loss of accuracy of numerical calculations. Taking into account the non-cylindricalness of the bronchi further complicates the construction of a three-dimensional model of the bronchial tree. The analytical representation allows you to build a bronchial tree up to the alveoli. For numerical calculations, analytical formulas allow to build a grid of any complexity up to the smallest bifurcations and bronchus. The construction of the analytical model of the human bronchial tree is detailed in the works [1, 2]. Examples of analytical construction of the human star-shaped bronchial tree are shown in Figure. 1.

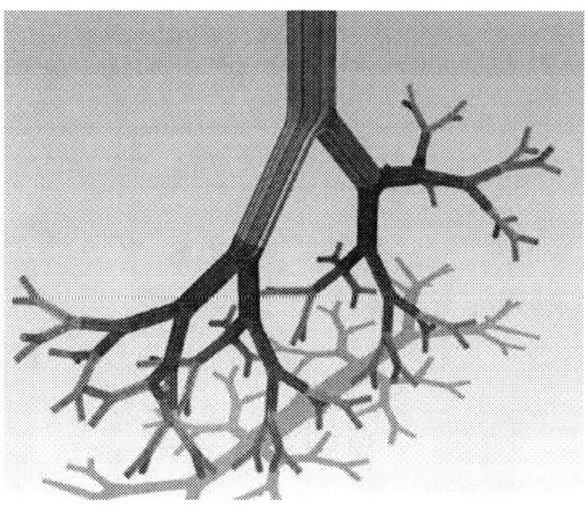

Figure 1. An example of the construction of a human star-shaped bronchial tree. Shown here is the bronchial tree of a person from 0 to 5-th bifurcation (bifurcations of one order are highlighted in one color).

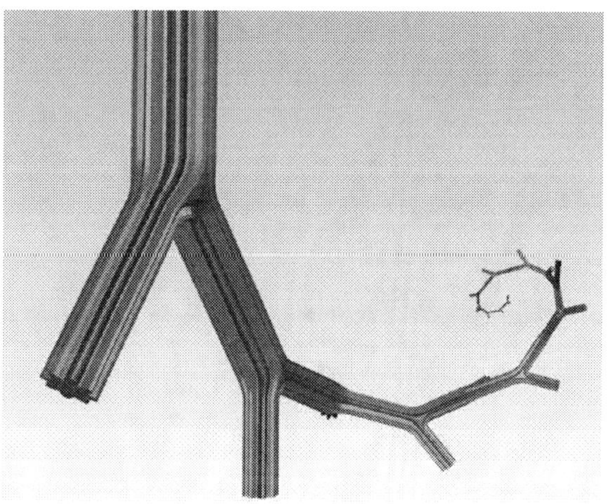

Figure 2. An example of building a separate full branch of the human star-shaped bronchial tree. Shown here is a full branch of the human bronchial tree from the 0 to the 23-rd bifurcation.

3. Calculating the flow of gas in a person's bronchial tree

Previously, a numerical method of sequential calculation of human bronchial tree was proposed [1, 2]. Due to low speeds and pressures, air compressibility is usually neglected. By construction, the bronchial tree has a symmetrical structure, so the air flow through all bronchi n-generation is the same. The calculation is carried out from n to $n+1$-st generation. At the entrance to the n generation is set a field of speeds and average pressure in the entrance section, at the exit – expense. After the calculation at the exit we will get a field of speeds and average pressure in the section. These values are transferred to n 1st generation. Thus, you can (if desired and patient) calculate the current in the entire bronchial tree of a person or at least in one branch from the 0 to the 23-rd generation. The size of the bronchi and bifurcation decreases with the increase in the number of bifurcation (the diameter of the bronchi 0 bifurcation is 18 mm, and the 23-rd is only 0.419 mm). The calculation is carried out for a separate bifurcation, so that the grid and the accuracy of the method on bifurcation do not degrade much, which allows almost no loss of accuracy of calculation up to the 23-rd bifurcation.

The results of the calculation of pressure in the bronchial tree are shown in Figure 3. On Figure 3 the value of the given pressure in the bronchial tree is shown. The given pressure is equal to $p_{\text{given}} = p_0 - p_n$ where p_0 is the pressure of inhaled air, p_n is the pressure on the way out of n bifurcation. For comparison on Figure 3 the calculation of the current in a separate branch of the bronchial tree is given. In the work [3], the calculation was carried out by iterative method for the entire branch of the bronchial tree. In the [3] calculation was carried out for the turbulent flow of gas for incomplete branch up to 14 bifurcation. It is evident that the laminar current mode gives half the value of the drop-in pressure, compared to the turbulent calculations. Perhaps this is true, because the human body over the years of evolution has developed breathing with the least energy. For the star-shaped (in pathology) section of the human bronchi (Figure 3) the given pressure above the "normal" pressure drop.

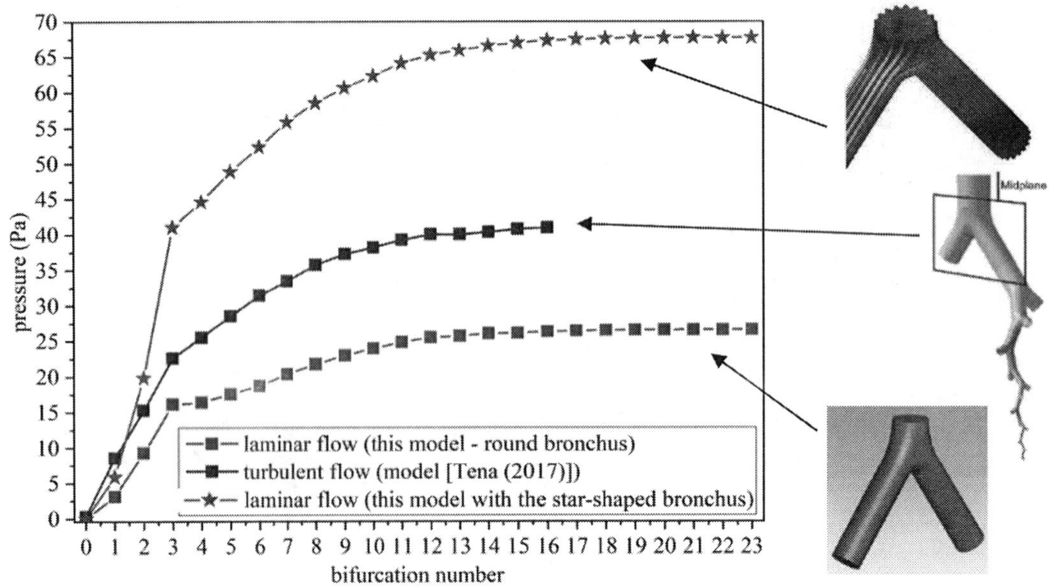

Figure 3. The results of the numerical calculation of the flow of air in the bronchial tree of a person at the consumption of 50 l/min. Calculating the given pressure depending on the bifurcation number.

4. Precipitation of aerosol drops in human bronchi

To calculate the deposition of aerosol droplets in a person's bronchial tree, a typical NE-C24 compressor ("Omron" production) is taken for example. The characteristics of the nebulizer are as follows: the size of drops 3 microns, the volume consumption of aerosols 0.3 ml/min. The typical air flow when inhaling aerosols through a nebulizer is taken equal to 50 l/min. The duration of the breath ranged from 1 to 3 seconds.

Aerosol drop equations

$$m_p \frac{d\vec{v}_p}{dt} = \vec{F}_D, \tag{1}$$

where m_p is the particle mass, $\vec{v}_p$ is the velocity of the particle, $\vec{F}_D = m_p\left(\vec{v} - \vec{v}_p\right)/\tau_p$ is the drag force, $\tau_p = \rho_p d_p^2 / 18\mu$ is the particle velocity response time, μ is the fluid viscosity, ρ_p is the particle density. These ratios are true for the slow flow of gas drops. The drops don't break apart, they don't stick together. When falling on the wall drops of aerosol stick to it.

The time of the drop's flyby through the bronchi can be estimated by the average speed of the gas flow in the bronchi. The average speed of gas flow in the bronchi is equal to

$$\tilde{w}_n = \frac{Q}{2^n R_n^2} . \tag{2}$$

Then the time of flying the drops through the n bronchi will be equal $t_n = L_n / \tilde{w}_n$. The total time of the droplets through a person's bronchial tree should be no less than time

$$t_\infty = \frac{\pi}{Q} \sum_{n=0}^{23} 2^n L_n R_n^2 . \tag{3}$$

For the flow of gas $Q = 50$ l/min, the flow of drops to the bronchi is no less then $t_\infty \approx 1.98$ s.

The calculations were carried out for one branch of the bronchial tree. The duration of the breath ranged from 1 to 3 seconds.

The results of the calculation are shown in the Figure 4. It can be seen that with the increase in the number of bifurcation (down the bronchial tree) the number of past drops of aerosol almost linearly falls. Inhale time 1 and 2 seconds is not enough for drops of aerosol to reach the alveoli. When inhaled for 1 second, aerosol drops reach only the 21-st bifurcation. When inhaled for 2 seconds, drops of aerosol reach the 22-nd bifurcation. And only with a breath duration of 3 seconds about 45% of drops reach alveoli. When you use 50 liters of air, a person needs 0.83 liters of air every second. The maximum deep breath of a person is about 2 liters.

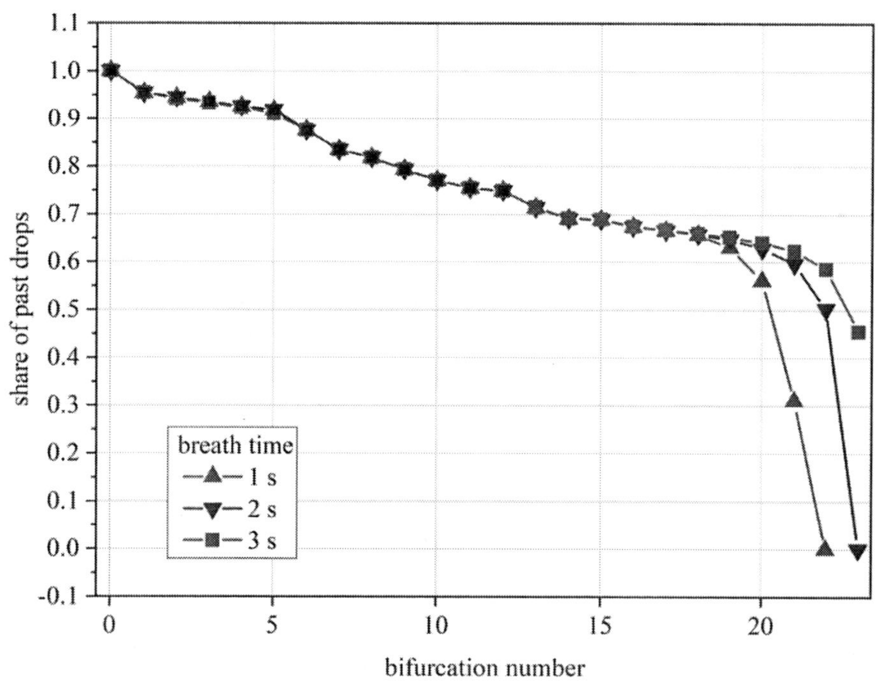

Figure 4. The proportion of past drops of aerosol depending on the number of bifurcations. The calculations for three breath durations are shown: 1, 2 and 3 seconds.

It is evident that for the initial bifurcations the nature of subsidence drops is not monotonous. Beginning with the 7-th bifurcation, the nature of subsidence is almost linear. After the 21-st bifurcation, the drops do not have time to fly to the alveoli during the breath. Therefore, the proportion of settled and under-achieving alveoli drops tend to be a unit. In the 23-rd bifurcation and in the alveoli drops during inhalation time (2 seconds) do not have time to fly. To improve the performance of the nebulizer, it is necessary to increase the breath time and/or reduce the size of aerosol droplets.

Acknowledgments
The study was supported by the Russian Foundation for Basic Research and the Subject of the Russian Federation (Novosibirsk Region) within the framework of the research project No. 19-41-540003 r_a.

References
[1] Medvedev A E, Fomin V M and Gafurova P S 2018 Simulation of the flow in the human bronchial tree *AIP Conference Proceedings* **2027** 040074
[2] Medvedev A E and Gafurova P S 2019 Three-dimensional model of the air flow in the human bronchial tree (normal and pathological cases) *AIP Conference Proceedings* **2125** 030009
[3] Tena A F *et al* 2017 Design of a numerical model of lung by means of a special boundary condition in the truncated branches *Int. J. Numer. Meth. Biomed. Eng.* **33(6)** 2830

Effect of the structure of skeleton models of cellular materials on the drag of a cylinder with a frontal gas-permeable insert in a supersonic flow

S G Mironov[1,2], S V Kirilovskiy[1,2], T S Militsina[1,3], T V Poplavskaya[1,2], I S Tsyryulnikov[1,2] and I R Valiullin[1,3]

[1]Khristianovich Institute of Theoretical and Applied Mechanics SB RAS, 4/1, Institutskaya str., Novosibirsk, 630090, Russia
[2]Novosibirsk State University, 1, Pirogova str., Novosibirsk, 630090, Russia
[3]Novosibirsk State Technical University, 20, Prospekt K. Marksa, Novosibirsk, 630073, Russia

kirilov@itam.nsc.ru

Abstract. The paper presents the results of the numerical simulation of the supersonic ($M_\infty = 7$) flow around the cylinder with the front gas-permeable porous insert at different angles of attack. The numerical simulation involves the comparative analysis of a number of skeleton models of high-porous cellular materials of the front insert: from the set of coaxial rings of different diameters, and the set of hollow spheres with different spatial resolution. The calculated data of the drag coefficient have been compared with the wind tunnel experiment results.

1. Introduction

The tasks of control of the flow around bodies which are important for practical applications can be divided into two groups: the tasks related with the inhibition of the disturbances caused by developing acoustic waves [1-5] in hypersonic boundary layers, and the tasks of control of the mean flow characteristics [6-8]. Use of gas-permeable high-porous materials in supersonic aerodynamics permits controlling the air drag for aircrafts [7-20]. Experimental and numerical results showed the high efficiency of the control of the air drag for a cylinder with a front porous insert made from cellular-porous nickel with the high porosity value. The blunt cylinder model located along the flow was chosen in order to show up the higher effect of porosity on the model drag reduction. A ring skeleton model from a high-porous cellular material (HPCM) was chosen for the conditions of the cylinder flow at the zero angle of attack [8, 14-16, 1-20].

Practical needs of supersonic aerodynamics call for experimental and numerical investigations of the flow around bodies with gas-permeable inserts under angles of attack. The transition from the axisymmetric flow task to the 3D task suggests the development of new models of the porous medium and approximation of its spatial structure to the real HPCM carcass. Today, there are no investigations in the literature devoted to the effect of the structure of gas-permeable porous materials on air drag characteristics of bodies.

This paper presents the early results of experimental and numerical investigations in this direction. The paper presents two 3D models with the different HPCM spatial structure: the ring skeleton model for the non-symmetric flow around the cylinder with the front porous insert and the skeleton model from hollow spheres. The results of these models construction and utilization are presented experimental and numerical aerodynamic coefficients of the cylinder with the front porous insert are obtained for different angles of attack, the data are compared to each other.

2. Experimental model and wind-tunnel measurements of aerodynamic forces

The model presenting a solid cylinder 14.5 mm in diameter, in which a thin tube moves with no gap (see figure 1a) is used in the experimental researches. The tube purpose is to keep the gas-permeable porous insert on the cylinder leading end-face, plus to change the insert length in the flow. The general model length L was constant (100 mm). The model photo and schematic are given in [8-10, 12, 13, 18-20, 21]. In the experiment, the porouse inserts with the porosity value $k = 0.95$ and pore cell diameter d of 1; 2, 3 and 4 mm were used.

The experiments with the model were carried out in the supersonic wind tunnel T-327B ITAM SB RAS at the Mach number $M_\infty = 7$ and unit Reynolds number $Re_1 = 1.5 \times 10^6$ m^{-1}. The aerodynamic forces measurements were performed with the aid of three-component strain gage balance. The model was fixed to the balance with a holder which permits changing the angle of attack by the model [21]. The aerodynamic forces were measured as the model angle of attack was varied within the limits $\alpha = 0$ - $15°$, and the relative length of the porous insert variation ($\Delta x/D = 0 - 2$).

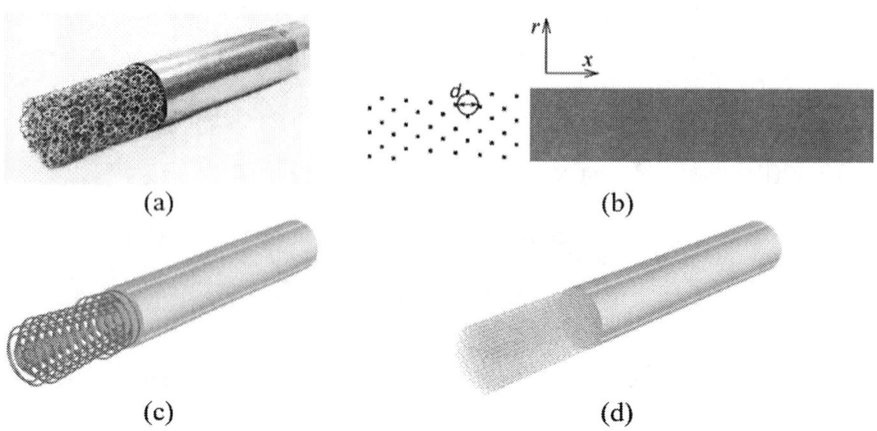

Figure 1. Experimental model of the cylinder with the front porous insert (a); cylinder model under in the numerical simulation: with the 2D-skeleton of the front porous insert (b) and with the 3D-skeleton from coaxial rings (c); with the 3D-skeleton from the set of hollow spheres (d). L = 100 mm, D = 14.5 mm, Δx = 28 mm, $k = 0.95$, $d = 2$ mm.

3. Numerical simulation of the axisymmetric flow around the cylinder with the front porous insert (2D ring skeleton model)

To simulate the flow in the front HPCM gas-permeable porous insert on the leading end-face of the cylinder flown at the zero angle of attack, the skeleton model with a simple spatial structure from coaxial rings of different diameters is suggested. In the axial cross section plane, this ring system presents a set of impermeable chequered square elements (figure 1b).

This model was previously used to construct the skeletons of the front porous insert, the pore diameters are $d = 1$ mm and 3 mm. The paper models the skeleton of the HPCM front porous insert, $d = 2$ mm. Figure 2 shows the density fields at the flow around the solid cylinder and the cylinder with the front porous insert (the length Δx = 28 mm at the Mach number M_∞= 7 and zero angle of attack. Table 1 demonstrates that the resulting computational data by the cylinder drag coefficients and the value of drag reduction coincide with the experiment within the measurement accuracy limits.

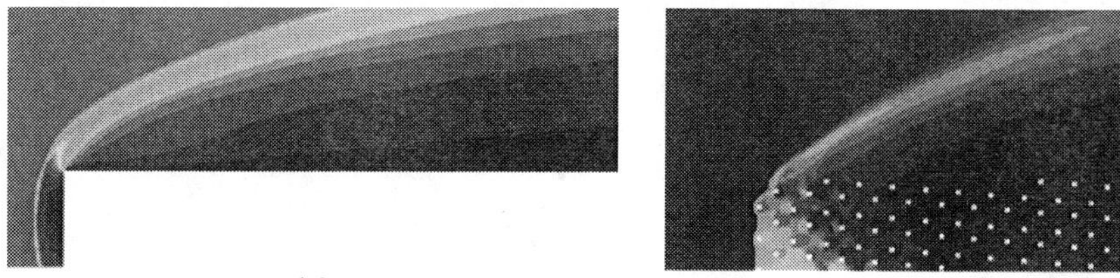

(a) (b)

Figure 2. Density fields at the flow around the cylinder D = 14.5 mm: the solid cylinder (a) and cylinder with the front porous insert Δx = 28 mm, k = 0.95, d = 2 mm (b). M_∞= 7, Re_1=1.5×10^6 1/m.

Table 1. Cylinder drag coefficients

$M_\infty = 7$ $Re_{1\infty} = 1.5 \times 10^6$ m^{-1}	C_{x0}	C_x	C_x / C_{x0}
Experiment	1.83±0.015	1.304±0.015	0.712
CFD	1.8429	1.3084	0.71

4. Numerical simulation of the flow around the cylinder with the front porous insert at the angle of attack

4.1. The solid cylinder

To determine the effect of the front porous insert in the cylinder flown by a supersonic flow at the angle of attack, the flow around the solid cylinder should be calculated. The task of the non-symmetric flow around the cylinder was solved within the framework of Reynolds-averaged 3D Navier Stockes equations, with the aid of the ANSYS Fluent package.

Figure 3a-3d show the computational density fields on the solid cylinder flown by the supersonic (M_∞ = 7) flow at different angles of attack. It is evident that at every angle of attack, the separating arched shock wave forms ahead of the leading end-face of the cylinder. As the angle of attack rises, the head shock wave presses to the cylinder surface windward, and distances from the surface leeward. The obtained calculated values of the drag coefficient (the curves) are presented in figure 4 for different angles of attack in comparison with the experimental findings (the symbols) obtained in the wind tunnel T-327B ITAM SB RAS (see Chapter 2). The data difference does not exceed 4%.

The computational code was verified also by means of comparison with the experimental data found in the wind tunnel T-313 ITAM SB RAS [22] for the solid cylinder at M_∞ = 4, Re_1 = 50 ×10^6 m^{-1} (see figure 4). It is obvious that the scatter of the calculated and experimental data does not exceed 4% either.

Figure 4 also shows that the drag coefficients rise along with the angle of attack growth. The reason is that, as the angle of attack rises, the asymmetric flow around the cylinder increases the size of the body interacting with the flow.

4.2. 3D skeleton ring model

The 2D ring HPCM skeleton model has shown good results in the calculations of the axisymmetric flow around the cylinder with the front porous insert. That is why the ring model (figure 1c) constructed in the same way as the 2D ring skeleton model has been chosen as the first 3D model. The 3D model also features only three parameters (the porosity value k, pore diameter d, and HPCM transparency length), which are easily found by direct measurements.

Figure 3e-3h presents the calculated density fields on the cylinder with the front porous insert flown by the supersonic (M_∞ = 7) flow at different angles of attack. Evident that at every angle of attack, the head shock wave attaches to the leading end-face of the model, and the angle of the head shock wave inclination toward the flow reduces. Moreover it is obvious that the intensity of the head shock wave leeward the model is much lower than it is during the flow around the solid cylinder.

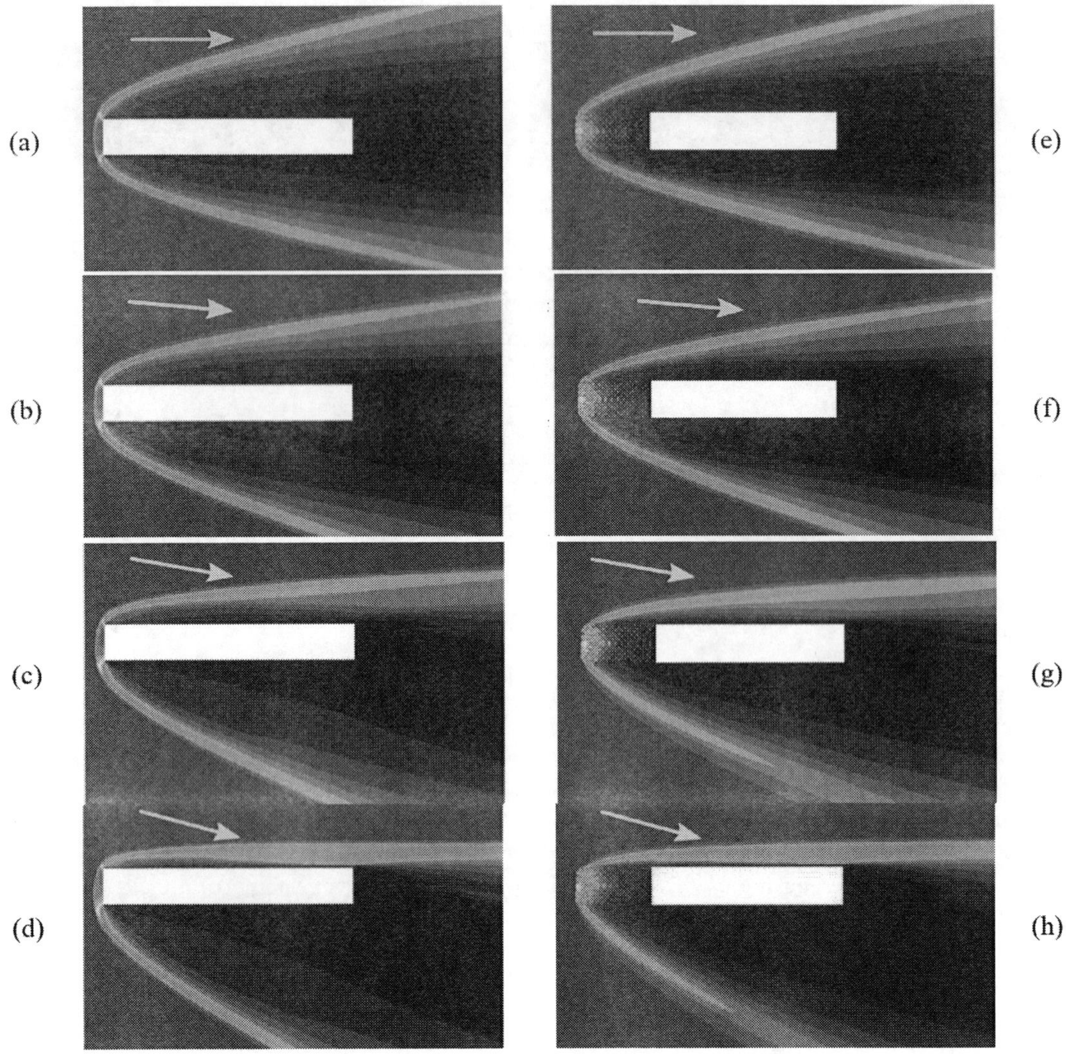

Figure 3. Calculated density fields $M_\infty = 7$, $Re_1 = 1.5\times10^6$ 1/m during the flow around the solid cylinder (a-d) and the cylinder with the front porous insert $\Delta x = 29$ mm, $k = 0.95$, $d = 2$ mm (e-h) at different angles of attack: (a,e) - $\alpha = 0$; (b, f) - $\alpha = 5°$; (c, g) - $\alpha = 10°$; (d, h) - $\alpha = 15°$.

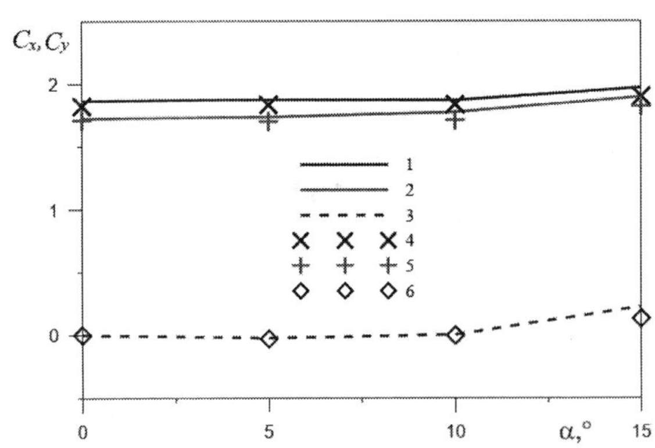

Figure 4. Drag coefficients C_x, C_y of the solid cylinder versus the angle of attack: 1-2 – calculated data C_x; 3 - calculated data C_y; 4, 6 – experimental data (T-327B); 1, 3, 4, 6 - $M_\infty = 7$, $Re_{1\infty} = 1.5\times10^6$ m^{-1}; 2, 5 - $M_\infty = 4$, $Re_{1\infty} = 50\times10^6$ m^{-1}; 5 - experimental data C_x (T-313 ITAM SB RAS) [22].

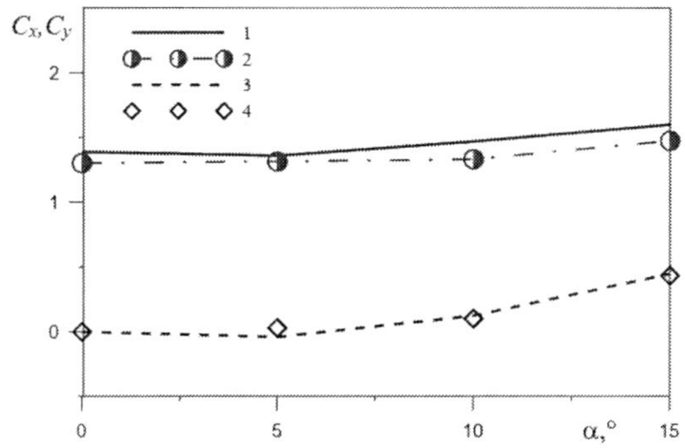

Figure 5. Drag coefficients C_x, C_y of the cylinder with the front porous insert $\Delta x = 29$ mm, $k = 0.95$, $d = 2$ mm versus the angle of attack: 1, 3 – calculated data C_x, C_y ; 2, 4 - experimental data obtained in the wind tunnel T-327B ITAM SB RAS; $M_\infty = 7$, $Re_{1\infty} = 1.5 \times 10^6$ m^{-1}.

The found calculated data of the drag coefficient values (the curves) on the cylinder with the front porous insert are shown in figure 5 for different angles of attack in comparison with the experimental data (the symbols) obtained in the wind tunnel T-327B ITAM SB RAS (see Chapter 2). The maximal difference is 10%. Here, similarly to the solid cylinder (figure 4), the drag coefficients rise along with the angle of attack growth.

4.3. 3D skeleton model from hollow spheres (spherical)

Another 3D skeleton model in appearance is more similar to the cellular structure of the HPCM; it presents the set of hollow spheres with the diameter equal to the HPCM pore diameter *d*. The sphere position is such that the sphere centers are in octahedron apexes, the octahedron rib length is *a* and $b = \sqrt{3}a/2$ (figure 6a). The gas passes through the holes in the hollow spheres. The distance between the sphere centers can be either smaller than the sphere diameter *d* – the crossing spheres (figure 6b), or bigger than *d* – the non-crossing spheres (figure 6 c). In the second case, to provide the gas passage through the spheres, the cylindrical channels of the radii R_1 and R_2, respectively, are cut along the octahedron ribs *a* and *b* (see 6a). The values of these radii correspond to the certain value of material permeability *q*. Here, *q* is the ratio of the sum of the though-holes projection areas to the full area of the material projection toward the flow. The radii R_1 and R_2 were chosen in such a way to preserve the material skeleton unbroken. Figure 6b,c demonstrates the elementary volumes of the skeleton model for the cases of the crossing and non-crossing spheres. Green indicates the edges of contact with neighboring elementary volumes, grey shows the hollow sphere boundaries, red and violet show the internal surfaces of the respective cylindrical channels.

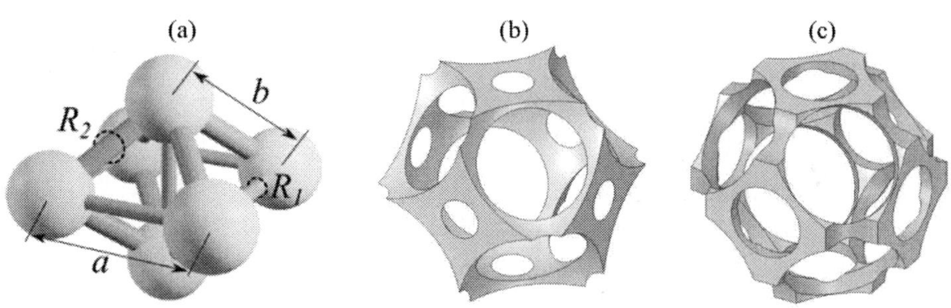

Figure 6. (a) – Scheme of the spatial position of the closely located spheres as an octahedron; (b) the elementary volume of the skeleton model from the hollow crossing spehers; (c) the elementary volume of the skeleton model from the hollow non-crossing spehers.

Figure 7 presents the projection of the skeleton model from the set of hollow spheres on the plane (y, z) for different parameters *k*, *a*, and *q*. For these variants, the calculations were carried out to clarify the influence of the structure on the drag coefficient C_x (see table 2).

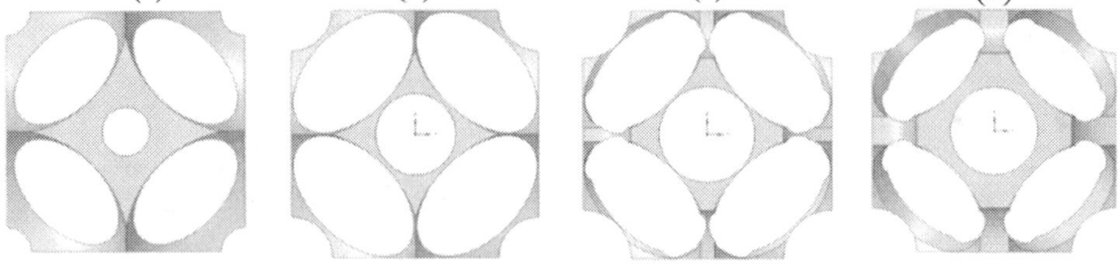

Figure 7. Projection of the skeleton model from the set of hollow spheres on the plane (y, z): model parameters are given in table 2.

According to Table 2, the coefficients C_x obtained for the used set of parameters of the spherical model turned out to be higher than the values obtained for the ring model and in the experiment. However, the wide range of variation of the coefficient C_x (from 1.55 to 1.72) permits in future, varying the spherical model parameters, having the data in agreement with the experiment. Furthermore, the variants of octahedron turn in order to increase the parameter q were beyond the consideration in the current structure of the spherical model.

Table 2. Drag coefficient C_x for different structures of the porous material d = 2 mm, Δx = 29 mm, M_∞ = 7, $Re_{1\infty}$ = 1.5 ×10^6 m^{-1}, α = 0.

Skeleton model		k	a, mm	q	C_x	C_x/C_{x0}
Ring		0.95	-	-	1.39	0.75
Spherical	Figure 7(a)	0.95	0.98	0.59	1.72	0.922
Spherical	Figure 7(b)	0.98	0.95	0.81	1.57	0.842
Spherical	Figure 7(c)	0.98	0.98	0.81	1.547	0.829
Spherical	Figure 7(d)	0.95	1.05	0.61	1.686	0.9

5. Conclusion

The effect of the structure of the high-porous cellular-materials skeleton models on the drag of the cylinder with the front gas-permeable insert has been numerically simulated in the supersonic flow. It is demonstrated that the parameters of the model porous material impose the essential influence of the drag coefficients of the flow around the cylinder with the front porous insert.

Acknowledgements

The work was supported by the Russian Foundation for Basic Research (Grant No 19-08-00113).

References

[1] Fedorov A V, Shiplyuk A N, Maslov A A et al. 2003 *J. Fluid Mech.* **479** 99-124
[2] Maslov A A, Kudryavtsev A N, Mironov S G et al. 2007 *J. Appl. Mech. Tech. Phys.* **48** 368-74
[3] Maslov A A, Mironov S G, Kudryavtsev A N et al. 2010 *J. Fluid Mech.* **650** 81-118
[4] Maslov A A, Mironov S G, Poplavskaya T V et al. 2012 *J. Appl. Mech. Tech. Phys.* **53** 162-72
[5] Tsyryulnikov I S, Maslov A A, Mironov S G et al. 2015 *Tech. Phys. Lett.* **41** 184–86
[6] Maslov A A, Mironov S G, Poplavskaya T V, et al. 1999 *European Journal of Mech. B/Fluids* **18(2)** 213-26
[7] Mironov S G, Maslov A A, Poplavskaya T V et al. 2015 *J.Appl. Mech. Tech. Phys.* **56** 549–57
[8] Maslov A A, Mironov S G, Poplavskaya T V et al. 2019 *J. Fluid Mech.* **867** 611-32
[9] Fomin V M, Mironov S G and Serdyuk K M 2009 *Tech. Phys. Lett.* **35** 117-19
[10] Bedarev I A, Mironov S G, Serdyuk K M et al. 2011 *J. Appl. Mech. Tech. Phys.* **52** 9-17
[11] Mironov S G, Maslov A A and Tsyryulnikov I S 2014 *Tech. Phys. Lett.* **40** 888-90
[12] Mironov S G, Maslov A A and Tsyryulnikov I S 2015 Method of control over supersonic aircraft

overflow. *Pat.* 2559193 C1 RU
[13] Mironov S G, Kolotilov V A and Maslov A A 2015 *Thermophysics and Aeromechanics* **22** 599-607
[14] Poplavskaya T V, Kirilovskiy S V and Mironov S G 2016 *AIP Conf. Proc.* 1770 030067
[15] Poplavskaya T V, Kirilovskiy S V and Mironov S G 2017 *J. Phys.: Conf. Series* 894 012074
[16] Poplavskaya T V, Kirilovskiy S V, Mironov S G et al. 2017 *AIP Conf. Proc.* 1893 030155
[17] Mironov S G, Poplavskaya T V and Kirilovskiy S V 2017 *AIP Conf. Proc.* 1893 030151
[18] Mironov S G, Poplavskaya T V and Kirilovskiy S V 2017 *Thermophysics and Aeromechanics* **24** 629-32
[19] Kirilovskiy S V, Maslov A A, Mironov S G et al. 2018 *Fluid Dynamics* **53** 409–16
[20] Mironov S G, Poplavskaya T V, Kirilovskiy S V et al. A A 2018 *Tech. Phys. Lett* **44** 225-8
[21] Maslov A A, Mironov S G, Poplavskaya T V et al. 2019 *J. Phys. Conf. Series* (in press)
[22] Kornilov V I and Lysenko V I 2011 *NSU Newsletter. Series: Physics* **6** 16-24

Justification of the asymptotic behavior of the filtration theory regimes by the method of conditional moment

A V Mishin[1,2]

[1] Khristianovich Institute of Theoretical and Applied Mechanics SB RAS, 4/1 Institutskaya st., Novosibirsk, 630090 Russia
[2] Novosibirsk National Research State University, 1, Pirogova st., Novosibirsk, 630090 Russia

alekseymishin1994@gmail.com

Abstract. In the work, a substantiation of the filtration modes is proposed by the method of conditional moments. The Navier-Stokes stationary equation of motion is used as initial equation. The calculations within the framework of this model include the construction of a solution of stochastic equations by the Green's function method, the subsequent conditional averaging of this solution, the direct and inverse Fourier transform in order to extract the integrals containing the structure of the medium and to obtain the desired averaged equations, respectively. At large Reynolds numbers, inertial terms dominate. When averaging them, a structural form of the quadratic filtration law is obtained. At low Reynolds numbers, the viscous terms make the main contribution, the averaging of which leads to Darcy's law.

1. Introduction

According to the description of filtration phenomena, there are many methods of various degrees of strictness. A review of many of these models, as well as the main aspects of the theory of filtration, can be found in [1]. In [2–4], emphasis was placed on mathematical calculations within the framework of spatial averaging (the mixture theory method) [2, 3, 5, 6], as a result of which it is possible to substantiate the structural dependence (Darcy's law [2, 3] by averaging the equations Stokes and Forchheimer's law [4] when averaging the Navier–Stokes equations). In these studies, the main difficulties lie both in providing evidence of the dependence of interfacial force on the difference in the velocity of the phases and obtaining coefficients of proportionality (containing information about the microstructure of the medium), binding interfacial force with the difference in the velocity of the phases. The presented approaches do not fully reflect the heterogeneity of the structure being explored. There are works aimed at determining the permeability coefficients taking into account the knowledge of the structural dependence of the Darcy law by means of simulation and analytical modeling of the framework and the flow through it [7] and there are a large number of works related to the experimental determination of the permeability coefficients, for example [8].

Currently there are a number of methods for describing heterogeneous media: the mixture theory method [2, 3, 6, 7], the Eshelby approach [9], the variational principle in the Hashin–Shtrikman interpretation [10], the self-consistent field method [11], the analysis of spatially periodic media with geometrically correct inclusions [12], Mori–Tanaka model [13], Ponte Castaneda–Willis model [14], hybrid homogenization model [15], a single-point approximation [6] and the method of conditional moments in the theory of effective modules [6, 16], the singular and generalized singular approximation [17], the hypothesis strong isotropy [18], the method of conditional moments [19]. In this paper, it is proposed to use the method of conditional moments.

2. Problem statement

The micro-inhomogeneous liquid/gas–solid system (gas is not indicated further, but we mean it when considering low velocities) is considered stochastically homogeneous and isotropic filling the entire space. Let us consider a monodisperse system, where the structural elements of the phases do not depend on the orientation, having the same size and shape. The volumetric concentration and scale of heterogeneity are assumed to be statistically homogeneous. The characteristic sizes of heterogeneity exceed the molecular sizes and at the same time are negligible with respect to the change of macro fields (statistically average fields in the system). Let us consider the material tensors of the components as isotropic ones.

Let's explore the movement of the liquid phase through the motionless skeleton. To describe the liquid phase, it is customary to use the Navier–Stokes equation. The equation of the stationary theory of deformations is usually used to describe the solid phase. The presence of liquid and solid phases should be reflected in the initial equation, which means the union of equations of state for phases. In this regard, we use the Navier–Stokes stationary equation of motion as the initial analyzed equation

$$\left(\rho v_j v_i\right)_{,j} = -p_{,i} + \left(\eta_{ij\alpha\beta} v_{\alpha,\beta}\right)_{,j} + f_i, \tag{1}$$

where f_i contains skeleton information, elastic properties of the solid phase (explanations will be given below), $\eta_{ij\alpha\beta}$ is the stochastic tensor of viscosity, ρ is the stochastic density, v_α is the stochastic velocity vector, $p_{,i}$ characterizes the derivative of pressure in the x_i direction. The stochasticity of the fields and material parameters is due to the fact that in each microdot of the medium there is one of the phases; let 1 be the liquid phase, 2 be the solid phase. We act in the skeleton reference system. Density is considered constant.

3. Solution by the method of conditional moments

Let us analyze the functional dependence of the filtration behavior at $Re \gg 1$. In this case, the contribution of the viscous terms $\left(\eta_{ij\alpha\beta} v_{\alpha,\beta}\right)_{,j}$ should be neglected () and the original equation (1) takes the form

$$\left(\rho v_j v_i\right)_{,j} = -p_{,i} + f_i. \tag{2}$$

From the stationary continuity equation $\left(\rho v_j\right)_{,j} = 0$ it follows $\left(\rho v_j v_i\right)_{,j} = \rho v_j v_{i,j}$. According to [19], we reduce equation (2) to the form

$$\rho^* V_j^* v_{i,j} = -\left(\rho v_j - \rho^* V_j^*\right) v_{i,j} - p_{,i} + f_i, \tag{3}$$

where V_j^* is the introduced auxiliary effective constant velocity of the effective medium, ρ^* is the effective constant density. Using V_j^* (also ρ^*) as a certain effective constant velocity of the effective medium, we can still consider a sufficiently strong true velocity gradient. Non-linearity remained, but moved to the right side for the possibility of applying the apparatus of the Green's function method (to the left side), conditional averaging, and the Fourier transform in the framework of the conditional moment method. The effective pulse density $\rho^* V_j^*$ should be considered proportional to the average pulse density of the liquid phase, the exact form will be determined in the process. Next, we will remove the asterisk over the effective speed and density. Let's consider a one-dimensional flow. Velocity of the liquid phase is directed along the z axis. All fields are zero at the border.

Let's write the equation for the Green's function for the left side of equation (3):

$$\rho V_z G^{(1,2)}_{,z} = \delta\left(\boldsymbol{r}^{(1)} - \boldsymbol{r}^{(2)}\right). \tag{4}$$

Performing the Fourier transformation, using the residue transform and performing the inverse Fourier transform, we find the Green's function

$$G = \frac{1}{\rho V_z}\delta(\boldsymbol{\rho})\theta(z). \tag{5}$$

Next, we construct a solution of the problem (3)

$$v_i^{(1)} = -\int\limits_{V^{(2)}} G^{(1,2)}\left(\rho^{(2)}v_j^{(2)} - \rho V_j\right) v_{i,j}^{(2)} d\boldsymbol{r}^{(2)} + \int\limits_{V^{(2)}} \left(G^{(1,2)} f_i^{(2)} - G_{,i}^{(1,2)} p^{(2)}\right) d\boldsymbol{r}^{(2)}. \tag{6}$$

The indices (1), (2) correspond to points in researched space. The pressure gradient is gone due to integration in parts, taking into account the zero pressure value at the infinitely distant boundary.

Let's apply to equation (6) the statistical averaging by the method of conditional moments over the conditional density $f\left(v_i^{(1)}, v_j^{(2)}, \rho^{(2)}, v_{i,j}^{(2)}, p_{,i}^{(2)}, f_i^{(2)} \middle| {}^{(1)}_{\nu}\right)$, which means averaging the presented fields at the corresponding points, provided that at the point (1) the component ν is found ($\nu = 1,2$).

$$\begin{gathered}\langle v_i^{(1)\nu}\rangle = -\sum_{k=1}^{2}\int\limits_{V^{(2)}} G^{(1,2)}\left(\rho_k \langle v_j^{(2)k}\rangle - \rho V_j\right)\langle v_{i,j}^{(2)k}\rangle p_{\nu k}^{(1,2)} d\boldsymbol{r}^{(2)} + \\ + \sum_{k=1}^{2}\int\limits_{V^{(2)}} \left(G^{(1,2)}\langle f_i^{(2)k}\rangle - G_{,i}^{(1,2)}\langle p^{(2)k}\rangle\right) p_{\nu k}^{(1,2)} d\boldsymbol{r}^{(2)},\end{gathered} \tag{7}$$

where $\langle v_j^{(2)k}\rangle$, $\langle f_i^{(2)k}\rangle$, $\langle p^{(2)k}\rangle$ characterize the averaged parameters for phase k at point (2). Thus, integro-differential equations are obtained for mean values (mathematical expectations) of the corresponding fields for the corresponding components at the corresponding points in space. To close the averaged equations the relation is used $\langle v_j^{(2)k} \middle| {}^{(1)}_{\nu}\rangle = \langle v_j^{(2)k}\rangle$. The function

$$p_{\nu k}^{(1,2)} = w\left({}^{(2)}_{k} \middle| {}^{(1)}_{\nu}\right) = c_k + (\delta_{\nu k} - c_k)\varphi^{(1,2)}$$

describes the probability of finding phase k at point (2), provided that phase ν is located at point (1); c_k is the volume concentration of phase k, $\varphi^{(1,2)} = e^{-\alpha|\boldsymbol{r}^{(1)} - \boldsymbol{r}^{(2)}|}$ is the correlation function for microinhomogeneous material in which the phases are stochastically uniform, α^{-1} is the scale of heterogeneity.

Further, the parameters with index 1 will be considered to belong to the liquid phase $\langle v_i^{(1)1}\rangle$ and the parameters with index 2 will be considered to be solid $\langle v_i^{(1)2}\rangle$; c_2 is the volume concentration of the solid phase ($c_1 + c_2 = 1$).

Using (7), let's find the difference $\langle v_i^{(1)1}\rangle - \langle v_i^{(1)2}\rangle$

$$\langle v_i^{(1)1}\rangle = -\int\limits_{V^{(2)}} G^{(1,2)}\varphi^{(1,2)}\left(\rho_1\langle v_j^{(2)1}\rangle - \rho V_j\right)\langle v_{i,j}^{(2)1}\rangle d\boldsymbol{r}^{(2)} + \tag{8}$$

$$+\int\limits_{V^{(2)}}\left(G^{(1,2)}\langle f_i^{(2)\prime}\rangle-G_{,i}^{(1,2)}\langle p^{(2)\prime}\rangle\right)\varphi^{(1,2)}d\boldsymbol{r}^{(2)},$$

where it is taken into account that the velocities and velocity gradients of the solid phase are zero $v_i^{(1)2}=0$, $\langle v_i^{(1)2}\rangle=0$. The difference between the corresponding averaged fields for the phases is indicated by strokes $\langle p^{(2)\prime}\rangle=\langle p^{(2)1}\rangle-\langle p^{(2)2}\rangle$. Using also (7), we find the average velocity $c_1\langle v_i^{(1)1}\rangle+c_2\langle v_i^{(1)2}\rangle$

$$c_1\langle v_i^{(1)1}\rangle=-\int\limits_{V^{(2)}}G^{(1,2)}c_1\left(\rho_1\langle v_j^{(2)1}\rangle-\rho V_j\right)\langle v_{i,j}^{(2)1}\rangle d\boldsymbol{r}^{(2)}+$$
$$+\int\limits_{V^{(2)}}\left(G^{(1,2)}\langle f_i^{(2)}\rangle-G_{,i}^{(1,2)}\langle p^{(2)}\rangle\right)d\boldsymbol{r}^{(2)}. \quad (9)$$

Let's perform the Fourier transform of equation (8) based on the convolution

$$A_i^1=-RY_i+RD_i'-R_iP', \quad (10)$$

where the corresponding introduction are of the form

$$A_i^1=\frac{1}{(2\pi)^{3/2}}\int\langle v_i^1\rangle e^{-i\boldsymbol{qr}}d\boldsymbol{r},\qquad D_i'=\frac{1}{(2\pi)^{\frac{3}{2}}}\int\langle f_i'\rangle e^{-i\boldsymbol{qr}}d\boldsymbol{r},$$
$$Y_i=\frac{1}{(2\pi)^{3/2}}\int c_1(\rho_1\langle v_j^1\rangle-\rho V_j)\langle v_{i,j}^1\rangle e^{-i\boldsymbol{qr}}d\boldsymbol{r}, \quad (11)$$
$$P'=\frac{1}{(2\pi)^{3/2}}\int\langle p'\rangle e^{-i\boldsymbol{qr}}d\boldsymbol{r},\qquad R=\int G\,\varphi e^{-i\boldsymbol{qr}}d\boldsymbol{r},\qquad R_i=\int G_{,i}\,\varphi e^{-i\boldsymbol{qr}}d\boldsymbol{r}.$$

Let us find the form of the functions R, R_i, as a series in the Fourier parameters $\boldsymbol{q}$, based on the fact that the spatial scale of the inhomogeneity α^{-1} is much smaller than the scale of variation of the macro fields q^{-1}. We restrict ourselves to terms of order q^0. Terms of a higher order are small as a consequence of $\alpha\gg q$. Expressions take a form

$$R=\frac{1}{\alpha\rho V_z},\qquad R_i=\frac{1}{\rho V_z}\delta_{iz}, \quad (12)$$

where the relation $G_{,i}=\delta(\boldsymbol{r})\delta_{iz}/(\rho V_z)$ is used by definition (4).

Now, based on the convolution, we perform the Fourier transform of equation (9)

$$c_1A_i^1=-c_1\tilde{R}Y_i+\tilde{R}D_i-\tilde{R}_iP. \quad (13)$$

The main changes with respect to (10), (11) are the integrals marked by the tilde

$$\tilde{R}=\int G\,e^{-i\boldsymbol{qr}}d\boldsymbol{r},\quad \tilde{R}_i=\int G_{,i}\,e^{-i\boldsymbol{qr}}d\boldsymbol{r}, \quad (14)$$

These integrals are considered to be the introduction of a small attenuation decrement, as is customary when working with generalized functions, and as a result, the attenuation decrement tends to zero. So, from (14) we obtain the expressions

$$\tilde{R} = \frac{1}{iq_z\rho V_z}, \qquad \tilde{R}_i = \frac{1}{\rho V_z}\delta_{iz}. \tag{15}$$

Let's substitute (12) in (10) and similarly (15) in (13) and, using the inverse Fourier transform, we obtain the expressions

$$\begin{aligned}\langle v_i^1\rangle &= -\frac{1}{\alpha\rho V_z}\big(\rho_1\langle v_j^1\rangle - \rho V_j\big)\langle v_{i,j}^1\rangle + \frac{1}{\alpha\rho V_z}\langle f_i'\rangle - \frac{1}{\rho V_z}\delta_{iz}\langle p'\rangle,\\ c_1\langle v_i^1\rangle_{,z} &= -\frac{1}{\rho V_z}c_1\big(\rho_1\langle v_j^1\rangle - \rho V_j\big)\langle v_{i,j}^1\rangle + \frac{1}{\rho V_z}\langle f_i\rangle - \frac{1}{\rho V_z}\delta_{iz}\langle p\rangle_{,z}.\end{aligned} \tag{16}$$

The first equation (16) is the main achievement of the method of conditional moments; the second equation is the averaged equation (3), which can be obtained, for example, by a generalized singular method. Taking into account $c_1\langle v_{i,j}^1\rangle = \langle v_{i,j}\rangle = \langle v_i\rangle_{,j} = c_1\langle v_i^1\rangle_{,j}$ and $\langle f_i\rangle = c_2\langle f_i^2\rangle$ the second equation takes the form

$$c_1\rho_1\langle v_z^1\rangle\langle v_z^1\rangle_{,z} = c_2\langle f_z^2\rangle - \langle p\rangle_{,z}, \tag{17}$$

in which only the z index remains, since only this direction makes sense here, as could be pointed out earlier. On the other hand, the averaged equation of the effective medium for this case should take the form

$$\rho V_z\langle v_{z,z}\rangle = c_2\langle f_z^2\rangle - \langle p\rangle_{,z}. \tag{18}$$

Equation (18) is a consequence of the fact that the coefficients in the averaged equation should be effective. Comparing expressions (17) and (18) we find that $\rho V_z = \rho_1\langle v_z^1\rangle$, $\rho = \rho_1$, $V_z = \langle v_z^1\rangle$, i.e., $\langle v_z^1\rangle_{,z} = V_{z,z} = 0$, because V_z expresses the effective constant velocity of the effective medium.

Based on the constancy of the average velocity of the liquid phase, (18) reduces to the equation

$$c_2\langle f_z^2\rangle = \langle p\rangle_{,z}, \tag{19}$$

since $c_2\langle f_z^2\rangle - \langle p\rangle_{,z} = \rho V_z\langle v_{z,z}\rangle = \rho V_z\, c_1\langle v_z^1\rangle_{,z} = \rho V_z c_1\langle V_z\rangle_{,z} = 0$. Equation (19) describes the nonlinear deformation of the skeleton under the action of an external force in the form of a pressure of the liquid phase. Under stationary deformation, without pressure of the liquid phase and bulk forces, the skeleton satisfies the equation $\langle f_i^2\rangle = 0$. Generally speaking, knowing only the skeleton equation, it is possible to show the constancy of the velocity of the liquid phase and the relation $\rho V_z = \rho_1\langle v_z^1\rangle$.

In the first equation (16), the last term vanishes $\langle p'\rangle = 0$, due to the Stokes approximation as applied to heterogeneous media (an analysis of this hypothesis is omitted in the present work, but it is proved by using the apparatus of the method of conditional moment for averaging the equation $\big(\nu p_{,i}\big)_{,i} = \Big(\nu\big(\rho v_j v_i\big)_{,j}\Big)_{,i}$, where $\nu = 1/\rho$); term $\big(\rho_1\langle v_z^1\rangle - \rho V_z\big)\langle v_{z,z}^1\rangle/\alpha\rho V_z = 0$. Then, taking into account equality for the skeleton (19), we have a finite equation that describes the filtration of the liquid phase

$$V_z = -\frac{1}{c_2\alpha\rho V_z}\langle p\rangle_{,z}. \tag{20}$$

Let us write this equation in a slightly different form

$$V^2 n_i = -\frac{1}{c_2 \alpha \rho} p_{,i}, \tag{21}$$

where n_i is the direction of the filtration velocity. Thus, in the situation $Re \gg 1$ a quadratic filtration law is obtained. For small Reynolds numbers $Re \ll 1$, the contribution of the inertial terms $\left(\rho v_j v_i\right)_{,j}$ becomes negligible and averaging by the method of conditional moments [19] of the viscous terms leads to the Darcy law in the form

$$\langle v_j^1 \rangle = -\frac{2}{3 c_2 \alpha^2 \eta^*} \langle p \rangle_{,j}, \qquad \eta^* = \frac{\eta_1}{1 - \frac{5}{2} c_2}, \tag{22}$$

where $\langle v_j^1 \rangle$ is the averaged velocity of the liquid phase, η^* is the effective shear viscosity of the carrier liquid phase in a homogeneous isotropic structure, η_1 is the shear viscosity of the liquid phase. Comparing expression (20) with Darcy's law (22), we see that the ratio of the coefficients gives the number Re

$$Re = \frac{2}{3} \frac{\rho V \alpha^{-1}}{\eta^*}. \tag{23}$$

The Reynolds number is proportional to the characteristic scale of the inhomogeneity α^{-1}, which in our case describes a homogeneous isotropic geometry of the structure. For a small scale of heterogeneity and for a large effective shear viscosity (due to the large viscosity coefficient of the liquid phase and the high concentration of the solid phase) the Reynolds number is small, the viscosity dominates, and the filtration is determined by Darcy's law. In the reverse situation $Re \gg 1$, the inertial term dominates and the filtration law is determined by expression (21).

It is known that in order to filter the liquid phase through the porous skeleton, a sufficient pressure gradient should be applied. The force of surface tension is affected (the non-Newtonian nature of the fluid). With a large number of heterogeneous surfaces, the role of surface tension increases quite strongly. As a result, the fluid that fills the pores cannot pass through the skeleton (it is part of the skeleton), which is equivalent to the condition: the effective viscosity of a heterogeneous medium should be proportional to the viscosity of the solid phase $\eta^* \sim \eta_2 \gg \eta_1$. Then, from equation (22) it follows that $\langle v_j^1 \rangle = 0$, since $\eta_1/\eta^* \to 0$.

Let's make a pressure difference Δp to the skeleton such that $\Delta p = -\tau L$, where τ is the volume force characterizing the surface effects of tension. Let δp determine the pressure difference in phases caused by capillary effects on the scale of heterogeneity, αL characterizes the number of inhomogeneities, then $\tau = \delta p \alpha$. For a spherical shell $\delta p = 2\sigma\alpha$, where σ is the surface tension coefficient. Then the critical pressure gradient takes the form $\Delta p/L = 2\sigma\alpha^2$.

References

[1] Scheidegger A E 1957 *The Physics of Flow Through Porous Media* (Toronto: Univ. Toronto Press) p 370
[2] Whitaker S 1999 *The Method of Volume Averaging* (Davis, CA: Univ. California Davis Press.) p 235
[3] Lasseux D, Quintard M and Whitaker S 1996 *Transp. Porous Media* **24** 107–37
[4] Irmay S 1958 *American Geophysical Union August IMS* **39** 702–7
[5] Nigmatulin R I 1978 *Fundamentals of the Mechanics of Heterogeneous Media* (Moscow: Nauka) p 464
[6] Khoroshun L P 2000 *Appl. Mech.* **30** 30–62
[7] Bedrikovetsky P 2008 *Transp. Porous Media* **75** 335–69

[8] Maya D, Aktasb A and Advani S G 2019 *Compos. P. A* **121** 100–14
[9] Eshelby J D 1957 *Proc. R. Soc.* **241** 376–96
[10] Hashin Z Shtrikman S 1963 *J. Mech. Phys. Solids* **11** 127–40
[11] Bruggeman D A G 1936 *Ann. Phys. Bd.* **417** 645–72
[12] Bakhvalov N S and Panasenko G P 1984 *Averaging of Processes in Periodic Media* (Moscow: Nauka) p 235 (in Russian)
[13] Mori T and Tanaka K 1973 *Act. Metall* **21** 571–4
[14] Ponte Castaneda P and Willis J R 1995 *J. Mech. Phys. Solids* **43** 1919–51
[15] Fedotov A 2018 *J. Mater. Sci.* **53** 5092–102
[16] Khoroshun L P 1987 *Appl. Mech.* **23** 100–8
[17] Shermergor T D 1977 *Elasticity Theory for Microinhomogeneous Materials* (Moscow: Nauka) p 235 (in Russian)
[18] Bolotin V V and Moskalenko V N 1968 *J. Appl. Mech. Tech. Phys.* **34** 66–72
[19] Khoroshun L P 1995 *Mech. Compos. Mater.* **31** 310–8

Journal of Physics: Conference Series **1404** (2019) 012034 doi:10.1088/1742-6596/1404/1/012034

On the issue of the solutions existence of the problem of gas hydrate dissociation in a porous medium with the formation of an extended region of phase transitions

N G Musakaev[1,2] and M K Khasanov[3]

[1]Tyumen Branch of Khristianovich Institute of Theoretical and Applied Mechanics of SB RAS, Taymirskaya Str. 74, Tyumen 625026, Russia
[2]Industrial University of Tyumen, Volodarskogo Str. 38, Tyumen 625000, Russia
[3]Sterlitamak Branch of Bashkir State University, Lenina Str. 37, Sterlitamak 453103, Russia

musakaev@ikz.ru

Abstract. Mathematical model of the gas hydrate dissociation in a porous medium with the formation of an extended region of phase transitions is proposed. In a rectilinear-parallel approximation self-similar solutions are constructed that describe the distribution of pressure, temperature, and hydrate saturation in a reservoir. Based on the obtained solutions analysis it was proved that for the problem of injecting warm gas into a hydrate saturated reservoir with piezoconductivity coefficient exceeds the thermal diffusivity coefficient, there are no solutions with an extended region of phase transitions. The existence of such solutions is possible during gas production from a gas hydrate reservoir due to a decrease in pressure at the extraction point.

1. Introduction

A significant part of the world's natural gas reserves is concentrated in gas hydrate deposits [1, 2], which are solid solutions of gas in water with a high relative mass fraction of gas in hydrate $G \geq 0.1$ [3-5]. To improve the technology of gas production from hydrate deposits, theoretical studies of the decomposition of gas hydrates in porous structures is necessary. Some features of this process are presented in [6–9]. In these works solutions are constructed for the decomposition of gas hydrates in a porous medium for the case of phase transitions on a frontal surface. Such mathematical description is adequate for a limited range of parameters characterizing the state of the "porous medium + gas hydrate" system.

In this work in a rectilinear-parallel approximation we studied the conditions under which the formation of an extended region of the gas hydrates decomposition in a porous medium is possible.

2. Problem statement and basic equations

Suppose that at the initial time moment a porous reservoir is saturated with methane and its hydrate. The pressure p_0 and temperature T_0 in the reservoir initially satisfy the thermodynamic conditions for the stable existence of gas hydrate:

$$t = 0,\ x \geq 0:\quad p = p_0,\ T = T_0.$$

We assume that the pressure p_e and temperature T_e, which correspond to the decomposition conditions of methane hydrate, are instantly established on the left reservoir boundary (in the sequel these values are kept constant):

$$t>0,\ x=0:\quad p=p_e,\ T=T_e .$$

It is known that in order to exert a thermal effect on a hydrate saturated reservoir (for example, when warm gas is injected into the reservoir [10, 11]), the following condition must be fulfilled: $p_e > p_0$, $T_e > T_0$. When the depressive effect on the reservoir is realized, the pressure at the reservoir boundary decreases, i.e. the condition is satisfied: $p_e < p_0$ [7, 12, 13].

At the thermal or depressive effects on a hydrate saturated porous medium, gas hydrate decomposes. Three regions may occur in the reservoir; the location of these regions and the composition of the fluids saturating the porous medium are shown in figure 1.

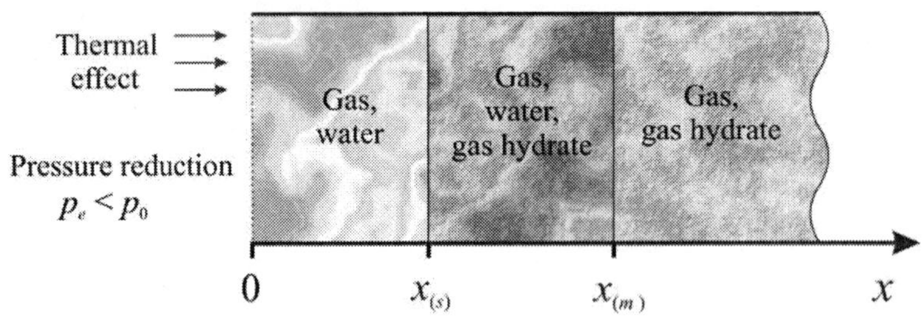

Figure 1. Scheme of the problem of the methane hydrate decomposition in a reservoir. The subscripts s and m in brackets refer to the parameters at the boundaries $x = x_{(s)}$ and $x = x_{(m)}$, respectively.

In the second region, where the process of the gas hydrate decomposition occurs, the condition of phase equilibrium (gas, water and gas hydrate) is satisfied. The values of temperature and pressure in this region are related by the condition of phase equilibrium:

$$T = T_0 + T_* \ln\left(p/p_{s0}\right), \tag{1}$$

where T_0 is the initial temperature of the system; p_{s0} is the equilibrium pressure corresponding to the temperature T_0; T_* is an empirical parameter depending on the type of gas hydrate and the phases on which it dissociates.

On the surface $x = x_{(m)}$, dividing 2 and 3 regions, the decomposition of methane hydrate begins. And this process ends on the surface $x = x_{(s)}$ (figure 1).

The system of basic assumptions and equations for the presented problem statement can be written on the basis of [9, 12, 14–16]. The dependence of phase permeability for gas k_g on gas saturation S_g is set based on the Kozeny formula: $k_g = k S_g^3$, where k is the porous medium permeability.

The gas saturation in each region (i = 1, 2, 3) will be considered a constant value. The equations of piezoconductivity and thermal diffusivity in a rectilinear-parallel approximation can be represented as:

$$\frac{\partial p}{\partial t} = \frac{k_g}{m S_g \mu_g}\frac{\partial}{\partial x}\left(p\frac{\partial p}{\partial x}\right) - \frac{\rho_h G p}{\rho_{g0}(1-\nu)}\frac{\partial S_h}{\partial t}, \tag{2}$$

$$\frac{\partial T}{\partial t} = \frac{\lambda}{\rho c}\frac{\partial}{\partial x}\left(\frac{\partial T}{\partial x}\right) + \frac{c_g k_g \rho_g}{\rho c \mu_g}\frac{\partial p}{\partial x}\frac{\partial T}{\partial x} + \frac{m\rho_h L_h}{\rho c}\frac{\partial S_h}{\partial t},$$

$$\rho c = (1-m)\rho_{sk}c_{sk} + m\sum_{j=g,w,h}\rho_j S_j c_j,\quad \lambda = (1-m)\lambda_{sk} + m\sum_{j=g,w,h} S_j\lambda_j,\quad \sum_{j=g,w,h} S_j = 1.$$

Here p is the pressure; T is the temperature; t is the time; G is the mass concentration of gas in methane hydrate; ρ_j ($j = sk, h, w, g$) is the true density of the j-th phase, S_j ($j = g, w, h$) is the pore saturation with the j-th phase; ν is the initial hydrate saturation; c_j and λ_j ($j = g, w, h$) are the specific heat and thermal conductivity of the j-th phase; μ_g is the dynamic viscosity of the gas phase; m is the porous medium porosity; L_h is the specific heat of the gas hydrate decomposition; the subscripts sk, h, w and g refer to the parameters of the porous skeleton, methane hydrate, water, and gas, respectively.

On the surface $x = x_{(s)}$ a jump in hydrate saturation occurs from $S_h^- = 0$ to $S_h^+ = S_{h(s)}$. From the conditions of the balances of mass and heat on this boundary, taking into account the Darcy law, the equation of state for gas and the condition of phase equilibrium (1), we can write:

$$k_{(1)}\frac{\partial p_{(1)}}{\partial x} - k_{(2)}\frac{\partial p_{(2)}}{\partial x} = m\mu_g S_{h(s)}\left(\frac{\rho_h G}{\rho_{g(s)}} + \frac{\rho_h(1-G)}{\rho_w} - 1\right)\dot{x}_{(s)}, \tag{3}$$

$$\lambda\frac{\partial T_{(2)}}{\partial x} - \lambda\frac{\partial T_{(1)}}{\partial x} = m\rho_h L_h S_{h(s)}\dot{x}_{(s)},$$

$$\frac{\partial T_{(2)}}{\partial x} = \frac{T_*}{p_{(s)}}\frac{\partial p_{(2)}}{\partial x},$$

where $\dot{x}_{(s)}$ is the velocity of the phase transition boundary; the regions parameters are provided with subscripts in parentheses $i = 1, 2, 3$.

On the surface $x = x_{(m)}$ the value of hydrate saturation is equal to the initial hydrate saturation ν. From the conditions of the balances of mass and heat on this boundary, taking into account the Darcy law and the phase equilibrium condition (1), we have the following equations:

$$\frac{\partial p_{(2)}}{\partial x} = \frac{\partial p_{(3)}}{\partial x}, \quad \frac{\partial T_{(2)}}{\partial x} = \frac{\partial T_{(3)}}{\partial x}, \quad \frac{\partial T_{(2)}}{\partial x} = \frac{T_*}{p_{(m)}}\frac{\partial p_{(2)}}{\partial x}. \tag{4}$$

On both interphase boundaries, the pressure and temperature will be considered continuous quantities:

$$x = x_{(s)}: \quad p_{(1)} = p_{(2)} = p_{(s)}, \quad T_{(1)} = T_{(2)} = T_{(s)},$$

$$x = x_{(m)}: \quad p_{(1)} = p_{(2)} = p_{(m)}, \quad T_{(1)} = T_{(2)} = T_{(m)}.$$

3. Self-similar solution

We introduce a self-similar variable $\xi = x\big/\sqrt{\chi^{(T)}t}$. Then the system of equations (2) based on the Leibenzon linearization method can be represented as a system of ordinary differential equations of the second order:

$$-\frac{\xi}{2}\frac{dp^2}{d\xi} = \eta\frac{d}{d\xi}\left(\frac{dp^2}{d\xi}\right) + \phi_h p^2\xi\frac{dS_h}{d\xi}, \tag{5}$$

$$-\frac{\xi}{2}\frac{dT}{d\xi} = \frac{d}{d\xi}\left(\frac{dT}{d\xi}\right) + \frac{\mathrm{Pe}}{2p_0^2}\frac{dp^2}{d\xi}\frac{dT}{d\xi} - \Delta T\frac{\xi}{2}\frac{dS_h}{d\xi},$$

$$\eta = \frac{\chi^{(p)}}{\chi^{(T)}}, \quad \chi^{(T)} = \frac{\lambda}{\rho c}, \quad \chi^{(p)} = \frac{k_g p_0}{m(1-\nu)\mu_g}, \quad \phi_h = \frac{\rho_h G}{\rho_{g0}(1-\nu)}, \quad \Delta T = \frac{m\rho_h L_h}{\rho c}, \quad \mathrm{Pe} = \frac{\rho_{g0} p_0 c_g k_g}{\mu_g \lambda}.$$

In self-similar variables relations (3) can be written as:

$$k_{(1)}\frac{dp_{(1)}^2}{d\xi}-k_{(2)}\frac{dp_{(2)}^2}{d\xi}=KS_{h(s)}\xi_{(s)}, \tag{6}$$

$$\frac{dT_{(2)}}{d\xi}-\frac{dT_{(1)}}{d\xi}=\frac{\Delta T}{2}S_{h(s)}\xi_{(s)},$$

$$\frac{dT_{(2)}}{d\xi}=\frac{T_*}{2p_{(s)}^2}\frac{dp_{(2)}^2}{d\xi},$$

where $K=m\mu_g\chi^{(T)}p_{(s)}\left(\frac{\rho_h G}{\rho_{g(s)}}+\frac{\rho_h(1-G)}{\rho_w}-1\right)$.

Similarly, in self-similar variables based on (4) we get:

$$\frac{dp_{(2)}^2}{d\xi}=\frac{dp_{(3)}^2}{d\xi},\quad \frac{dT_{(2)}}{d\xi}=\frac{dT_{(3)}}{d\xi},\quad \frac{dT_{(2)}}{d\xi}=\frac{T_*}{2p_{(m)}^2}\frac{dp_{(2)}^2}{d\xi}. \tag{7}$$

Based on (5), distributions for pressure, temperature, and hydrate saturation in each region can be found:

– in the first region ($0<\xi<\xi_{(s)}$):

$$p_{(1)}^2=p_{(s)}^2+\left(p_e^2-p_{(s)}^2\right)\int\limits_{\xi}^{\xi_{(s)}}\exp\left(-\frac{\xi^2}{4\eta_{(1)}}\right)d\xi\Bigg/\int\limits_{0}^{\xi_{(s)}}\exp\left(-\frac{\xi^2}{4\eta_{(1)}}\right)d\xi, \tag{8}$$

$$T_{(1)}=T_{(s)}+\left(T_e-T_{(s)}\right)\int\limits_{\xi}^{\xi_{(s)}}\exp\left(-\frac{\xi^2}{4}-\frac{\mathrm{Pe}_{(1)}}{2p_0^2}p_{(1)}^2\right)d\xi\Bigg/\int\limits_{0}^{\xi_{(s)}}\exp\left(-\frac{\xi^2}{4}-\frac{\mathrm{Pe}_{(1)}}{2p_0^2}p_{(1)}^2\right)d\xi,$$

$$S_{h(1)}=0,$$

– in the second region ($\xi_{(s)}<\xi<\xi_{(m)}$):

$$p_{(2)}^2=p_{(m)}^2+\left(p_{(s)}^2-p_{(m)}^2\right)\int\limits_{\xi}^{\xi_{(m)}}\exp\left(-\frac{\xi^2}{4\chi}\right)d\xi\Bigg/\int\limits_{\xi_{(s)}}^{\xi_{(m)}}\exp\left(-\frac{\xi^2}{4\chi}\right)d\xi, \tag{9}$$

$$T_{(2)}=T_0+T_*\ln\left(p_{(2)}/p_{s0}\right),$$

$$S_{h(2)}=\frac{T_*(\eta_{(2)}-1)}{2\left(\phi_h T_*+\eta_{(2)}\Delta T\right)}\ln\frac{p_{(2)}^2}{p_{(m)}^2}+\nu,$$

– in the third region ($\xi_{(m)}<\xi<\infty$):

$$p_{(3)}^2=p_0^2+\left(p_{(m)}^2-p_0^2\right)\int\limits_{\xi}^{\infty}\exp\left(-\frac{\xi^2}{4\eta_{(3)}}\right)d\xi\Bigg/\int\limits_{\xi_{(m)}}^{\infty}\exp\left(-\frac{\xi^2}{4\eta_{(3)}}\right)d\xi, \tag{10}$$

$$T_{(3)}=T_0+\left(T_{(m)}-T_0\right)\int\limits_{\xi}^{\infty}\exp\left(-\frac{\xi^2}{4}-\frac{\mathrm{Pe}_{(3)}}{2p_0^2}p_{(3)}^2\right)d\xi\Bigg/\int\limits_{\xi_{(m)}}^{\infty}\exp\left(-\frac{\xi^2}{4}-\frac{\mathrm{Pe}_{(3)}}{2p_0^2}p_{(3)}^2\right)d\xi,$$

$$S_{h(3)} = \nu ,$$

where $\chi = \dfrac{\phi_h T_* + \eta_{(2)} \Delta T}{\phi_h T_* + \Delta T}$.

From the third equation of the system (9) we obtain the value of hydrate saturation on the boundary $\xi = \xi_{(s)}$ (from the second area side):

$$S_{h(s)} = \frac{T_* (\eta_{(2)} - 1)}{2\left(\phi_h T_* + \eta_{(2)} \Delta T\right)} \ln \frac{p_{(s)}^2}{p_{(m)}^2} + \nu . \tag{11}$$

4. Analysis of the obtained solutions

For real natural reservoirs, the reservoir piezoconductivity coefficient $\chi^{(p)}$ significantly exceeds the thermal diffusivity coefficient $\chi^{(T)}$, therefore we assume that the following condition is satisfied: $\chi_{(2)}^{(p)} > \chi^{(T)}$. In this case $\eta_{(2)} = \dfrac{\chi_{(2)}^{(p)}}{\chi^{(T)}} > 1$ and therefore, the multiplier for the logarithmic function in formula (11) is a positive quantity:

$$\frac{T_* (\eta_{(2)} - 1)}{2\left(\phi_h T_* + \eta_{(2)} \Delta T\right)} > 0 .$$

Since in the initial state the porous medium was saturated only with gas and hydrate, and hydrate and water are assumed to be immovable, the hydrate saturation value in the second region cannot exceed the initial hydrate saturation value ν. In this case, from the relation (11) for the existence of a volumetric region of hydrate decomposition, the following condition must be satisfied:

$$p_{(s)} < p_{(m)} . \tag{12}$$

Substituting the solutions for the pressure distribution in the first and second regions (8) and (9) into the first equation of system (6), we obtain the following relation on the boundary $\xi = \xi_{(s)}$:

$$k_{(1)} \frac{\left(p_{(s)}^2 - p_e^2\right) \exp\left(-\dfrac{\xi^2}{4\eta_{(1)}}\right)}{\displaystyle\int_0^{\xi_{(s)}} \exp\left(-\dfrac{\xi^2}{4\eta_{(1)}}\right) d\xi} - k_{(2)} \frac{\left(p_{(m)}^2 - p_{(s)}^2\right) \exp\left(-\dfrac{\xi^2}{4\chi}\right)}{\displaystyle\int_{\xi_{(s)}}^{\xi_{(m)}} \exp\left(-\dfrac{\xi^2}{4\chi}\right) d\xi} = K S_{h(s)} \xi_{(s)} . \tag{13}$$

Similarly, substituting the solutions for the pressure distribution in the second and third regions (9) and (10) into the first equation (7), we obtain the following relation on the boundary $\xi = \xi_{(m)}$:

$$\frac{\left(p_{(m)}^2 - p_{(s)}^2\right) \exp\left(-\dfrac{\xi_{(m)}^2}{4\chi}\right)}{\displaystyle\int_{\xi_{(s)}}^{\xi_{(m)}} \exp\left(-\dfrac{\xi^2}{4\chi}\right) d\xi} = \frac{\left(p_0^2 - p_{(m)}^2\right) \exp\left(-\dfrac{\xi_{(m)}^2}{4\eta_{(3)}}\right)}{\displaystyle\int_{\xi_{(m)}}^{\infty} \exp\left(-\dfrac{\xi^2}{4\eta_{(3)}}\right) d\xi} . \tag{14}$$

In Eq. (13), the right side is a positive quantity, therefore, taking into account condition (12), we have: $p_e < p_{(s)}$. From here we can write the following inequality:

$$p_e < p_{(s)} < p_{(m)} . \tag{15}$$

Given this inequality, it can be shown that the left side of relation (14) is a positive quantity. The right side of this relation is greater than zero when the following inequality satisfied:

From here we can write the following inequality:

$$p_{(m)} < p_0 \tag{16}$$

Based on inequalities (15) and (16), we can write the necessary condition for the existence of a volume region of gas hydrate decomposition:

$$p_e < p_{(s)} < p_{(m)} < p_0 \tag{17}$$

5. Conclusions

It follows from the last inequality that the extended decomposition region of gas hydrate is realized when the pressure on the left reservoir boundary decreases, i.e. in the implementation of depressive effects on a hydrate saturated reservoir. If warm gas is injected into a reservoir with pressure p_e and temperature T_e corresponding to the decomposition conditions of methane hydrate, then the phase transition will occur only on the frontal surface.

Acknowledgments

This work was supported by the grant of the Russian Foundation for Basic Research No. 18-29-10023.

References

[1] Makogon Y F, Holditch S A and Makogon T Y 2007 *J. Petrol. Scien. Eng.* **56** 14–31
[2] Makogon Y F 2010 *J. Natural Gas Scien. Eng.* **2** 49–59
[3] Byk S Sh, Makogon Y F and Fomina V I 1980 *Gas Hydrates* (Moscow: Khimiya)
[4] Sloan E D and Koh C A 2007 *Clathrate Hydrates of Natural Gases* (Boca Raton: CRC Press)
[5] Shagapov V Sh and Nasyrova L A 1999 *High Temperature* **37** 754 – 59
[6] Tsypkin G G 2000 *Annals New York Acad. Scien.* **912** 428 – 36
[7] Tsypkin G G 2001 *Doklady Physics* **46** 806 – 9
[8] Vasil'ev V I, Popov V V and Tsypkin G G 2006 *Fluid Dynamics* **41** 599 – 605
[9] Tsypkin G G 20076 *Fluid Dynamics* **42** 798–806
[10] Musakaev N G, Khasanov M K, Borodin S L and Belskikh D S 2018 *Vestnik Tomsk Gosudarstvennogo Universiteta - Matematika i Mekhanika* **56** 88–101
[11] Musakaev N G, Borodin S L and Belskikh D S 2019 *AIP Conf. Proc.* **2125** 030114
[12] Musakaev N G, Borodin S L and Khasanov M K 2018 *Int. J. Heat Mass Transfer* **118** 455 – 61
[13] Khasanov M K, Musakaev N G and Gimaltdinov I K 2015 *J. Eng. Phys. Therm.* **88** 1052 – 61
[14] Nigmatulin R I 1991 *Dynamics of Multiphase Media* (New York: Hemisphere Publ. Corp.)
[15] Shagapov V Sh, Chiglintseva A S and Rusinov A A 2016 *Theoretical Foundations of Chemical Engineering* **50** 449–58
[16] Shagapov V Sh, Khasanov M K and Musakaev N G 2008 *J. Appl. Mech. Tech. Phys.* **49** 462 – 72

Journal of Physics: Conference Series **1404** (2019) 012035 doi:10.1088/1742-6596/1404/1/012035

Cutting transport simulation in a horizontal borehole near tool joints

E Orlik, Y Ignatenko and O Bocharov

Baker Hughes Russia,
4A Kutateladze st., Novosibirsk, 630128, Russia

evgeniy.orlik@bakerhughes.com

Abstract. In the presented study, we investigated the influence of cuttings transport in a channel with variable cross-section area to determine the pressure drop and obtain preliminary data on tool joint effect on solids deposition. The influence of boundary conditions at the channel's inlet on flow regimes was analysed. The average particle concentration over the cross-section significantly reduced due to increasing the velocity in the narrowing area on the tool joint. Behind the tool joint, particles deposition appeared due to the reverse flow, which manifested itself as a local maximum in concentration distribution along the channel. The impact of a tool joint on the pressure drop in a single-phase flow has been more pronounced if compared to one in a two-phase flow at the given conditions.

1. Introduction

Cuttings transport is one of the main issues of directional drilling, because insufficient cleaning of the borehole may result in drill-bit sticking; mud circulation loss; excessive pump loads, vibrations and torques; and reduced penetration rate. All these may lead to serious financial loses and cause sufficient accident risk [1]. Currently, there has been a significant growth in the number of studies to investigate different flow regimes and particles sedimentation in inclined annular channels for drilling muds of different rheology [2]. A wide range of the principal parameters influenced the dynamics of cuttings such as carrier and disperse phase flow rates, drill string eccentricity and its RPM, fluid and particle properties, borehole geometry and its inclination angle add additional complexity to the studies performed. The scale of the problem calls for joint theoretical and experimental investigations to be performed.

Interactions of particles with each other, wall, carrier phase, sedimentation layer, particle form and size effect and their dispersion, complex fluid rheology, unsteadiness of flow, drill string axial position changing in time and turbulence significantly complicate the theoretical description of the process. For the time being universal approaches have not been developed using computational methods. The large number of model constants and approximations require them to be tested against experimental data that are quite difficult to obtain especially if it comes to local values. Up to now, a tool joint (TJ) effect on cutting transport has never been taken into account, despite the well-known fact that joints have a significant effect on the hydrodynamic parameters and pressure drop in case of a single-phase flow [3]. This paper presents the results of our study that used numerical modeling to investigate sedimentation layer formation near a tool joint in a horizontal annular channel.

2. Problem statement and research method

Figure 1 represents the problem's geometry. The pipe's inner and outer diameters are 0.127 and 0.2159 meters respectively and the TJ's diameter is 0.1683 meters. The considered flow parameters have been selected to be close to those used while actual drilling, so that Re=300, and the particle volume concentration C =0.06 (6%). The drill string axis is displaced to be near the bottom wall in a way that the eccentricity determined relative to the TJ is equal to 0.8. In our investigation, drill string rotation was not considered and the numerical modeling was performed for an unsteady 3D problem statement based on the Reynolds-averaged Navier-Stokes equations using in-house software [4].

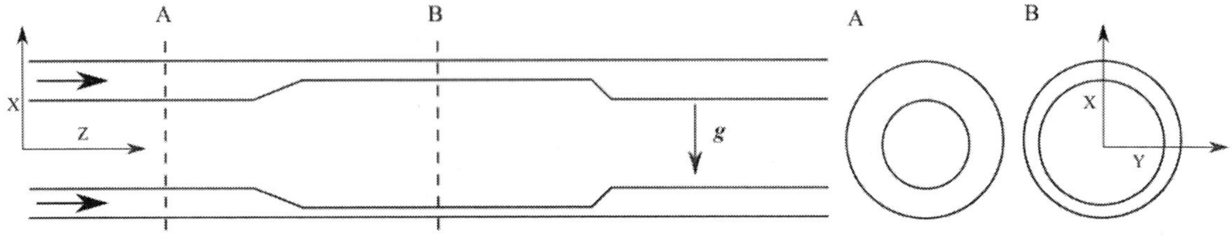

Figure 1. Scheme of TJ in annular channel.

2.1. Mathematical model

In our investigation, we applied the Eulerian method to describe a two-phase disperse flow. The model is based on the mixture approach with an algebraic relation for the disperse phase slip velocity $\boldsymbol{u}_r = \boldsymbol{u}_s - \boldsymbol{u}_f$. The mixture is considered as heterogeneous medium, where each phase occupies a part of the volume. The average values characterizing the mixture are: the mixture density $\rho = \rho_f(1-\alpha) + \rho_s\alpha$; the mixture mass averaged velocity $\boldsymbol{u} = (\rho_f\boldsymbol{u}_f(1-\alpha) + \rho_s\boldsymbol{u}_s\alpha)/\rho$, where $\boldsymbol{u} = (u, v, w)$ and α denotes the particle volume concentration. The particles are spheres with constant diameter d and density ρ_s.

The carrier is a non-Newtonian fluid of Herschel-Bulkley rheology and constant density ρ_f. The governing equations describing the mixture's motion are [5]:

$$\frac{\partial \rho}{\partial t} + \nabla \cdot (\rho \mathbf{u}) = 0 \tag{1}$$

$$\rho \frac{d\mathbf{u}}{dt} = -\nabla p + \rho \mathbf{g} + \nabla \cdot \left(\mathbf{T}_\mu + \mathbf{T}_t + \mathbf{T}_r\right) \tag{2}$$

where $\mathbf{T}_\mu$ denotes the mixture's viscous stress tensor, $\mathbf{T}_t$ is the turbulent stress tensor, $\mathbf{T}_r$ is the "diffusion" stress due to phase slip.

The equations below describe the volume particle concentration transported by the fluid and the relative particle velocity:

$$\frac{\partial \alpha}{\partial t} + \nabla \cdot (\mathrm{u}_s \alpha) = 0 \tag{3}$$

$$\rho_s \mathbf{u_r}/\tau_s = (\rho_s - \rho)(\mathbf{g} - (\mathbf{u} \cdot \nabla)\mathbf{u}) - \frac{1-\alpha}{\alpha}\nabla p_s - \frac{1}{\alpha}(\mathbf{F}_L + \mathbf{F}_t) \tag{4}$$

where τ_s denotes the particle relaxation time related to drag force, $\mathbf{F}_L$ is the lift force, $\mathbf{F}_t$ is the force generated by turbulent pulsations, $\mathbf{g}$ is the acceleration due to gravity, and p_s is the "particles" pressure.

2.2. Model Closure

The suggested model utilizes a phenomenological description of mixture flow, where the effect of the disperse phase is considered by introduction of mixture effective viscosity. The viscous stress tensor is expressed as $\mathbf{T}_\mu = 2\mu[\mathbf{S} - 1/3\,(\nabla \cdot \mathbf{u})\mathrm{I}]$. Here, $\mathbf{S} = 0.5(\nabla\mathbf{u} + \nabla\mathbf{u}^T)$ denotes the mixture's deformation rate tensor, μ is the effective viscosity coefficient, I is the unity matrix. In general, effective viscosity depends on the concentration, shape and size of particles. The literature contains many theoretical and empirical equations to describe the viscosity of a mixture that are accurate for various concentrations of particles. For our purposes, we applied an empirical expression suggested in [6].

The turbulent characteristics of the mixture flow were determined using the single-phase two-parameter low-Reynolds-number RANS k-ω SST model. Turbulent stresses are related to the deformation velocity of an averaged mixture flow as follows:

$$\mathbf{T}_t = 2\mu_t\mathbf{S} - \frac{2}{3}k_t\mathbf{I}, \quad (5)$$

where μ_t and k_t denote the turbulent viscosity and the mixture's turbulent kinetic energy. A simplified version of the model [7] was applied to model the turbulence of non-Newtonian fluid.

Particles pressure is described by the suspension balance model [8] and includes two terms related to the share stress migration and the pressure due to permanent particles contacts. The particles stresses are important mechanisms for description particles flow in highly concentrated regions. The normal pressure due to particles continuing contacts increases rapidly when the particle concentration is close to the maximum packing limit. The contribution to the particulate pressure due to permanent particle contact is taken into account, following Johnson and Jackson [9]. The particle relaxation time τ_s takes into account particle concentration through viscosity [8]. The simplest formula to describe particles diffusion due to turbulent pulsations $\mathbf{F}_t$ was derived from the gradient hypothesis [10]. The lateral lift forces on a dispersed particle borne by a shear flow are modeled by the following conventional expression with the lift coefficient C_L: $\mathbf{F}_L = C_L\rho_f\alpha\left[\mathrm{u}_r \times \left[\nabla \times \mathrm{u}_f\right]\right]$.

Our model has been verified by experimental works for both the laminar and turbulent flows of Newtonian and non-Newtonian fluids in annular pipes. More details and description of the model and model constants can be found in [5].

2.3. Input data, boundary conditions and mesh generation

In our investigation, two kinds of boundary conditions (BC) were considered as input data (figure 2). In the case of BC1 it was a steady-flow velocity profile in a channel without tool joints and particles. The profile of particle concentration was considered constant. Such a problem statement has a practical meaning as the mixture can be regarded as homogenized near drill-bit (figure 2a).

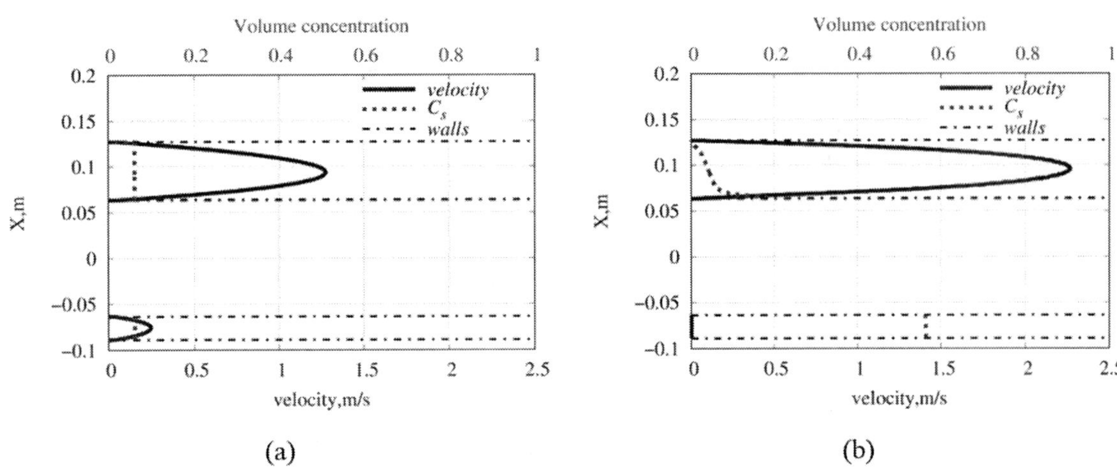

Figure 2. Boundary conditions BC1 (a) and BC2 (b).

To reduce the required computation time, we also applied the boundary conditions obtained from 2D computations for a steady flow of a mixture in a channel with constant cross-section and well-formed sedimentation layer that is noted here as BC2 (figure 2b). Preliminary calculations to verify meshes convergence have been performed. Figure 3 demonstrates that the ripples which are visible on concentration curves reduce as spatial resolution increases. The computation results demonstrate that formation of steady-state sedimentation layer downstream requires at least 100 hydraulic diameters (D_h). For that reason, in our study, two TJ positions were considered: at the distance Zj1= 0.3 and Zj2=1.3 meters from the channel's inlet (figure 3b), which is 3.4 and 14.6 D_h respectively. Comparison of the computation results against those obtained for full 3D statement confirmed that it is proper to use the BC2 conditions for relatively long channels.

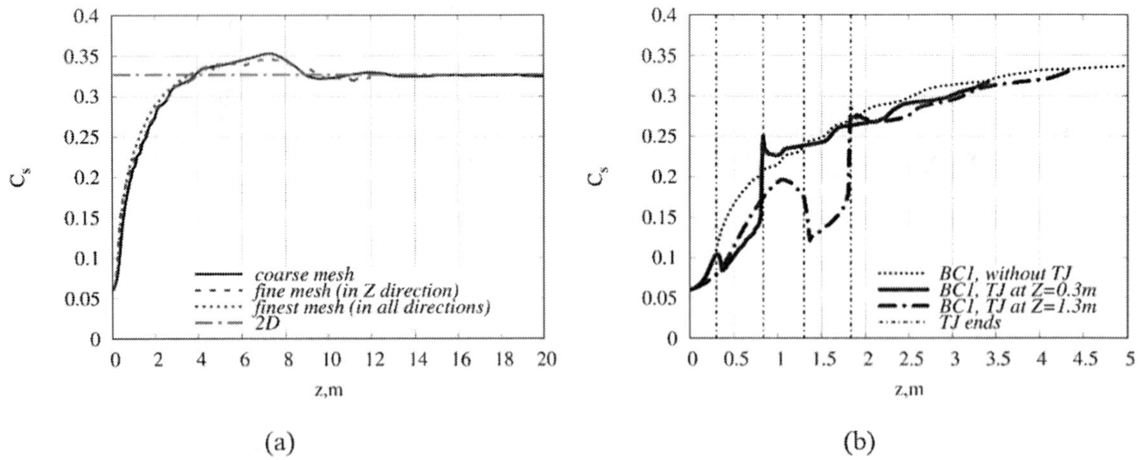

Figure 3. Distribution of mean particle concentration in section along the annular channel: (a) without TJ; (b) with TJ. BC1 boundary condition.

3. Results and discussion

The performed numerical experiments allowed us to investigate the TJ effect on the hydrodynamic parameters of the flow and on the sedimentary layer. All figures below present the results with BC2 conditions.

Figure 4 compares the velocity fields near TJ with and without cuttings transport. In the given flow conditions, a weak reverse flow is formed behind the TJ. Figure 5 shows the particle concentration distribution in different cross-sections of the channel. Sedimentation layer slightly decreases due to increasing of flow velocity on tool joint that led to reduced mean particle concentration in a section. It is noteworthy, that the zone of reduced particle concentration starts at a certain distance before the TJ (figure 6c). A local maximum in the distribution of the mean particle concentration in section (figure 6) is observed in the expansion zone of tool joint.

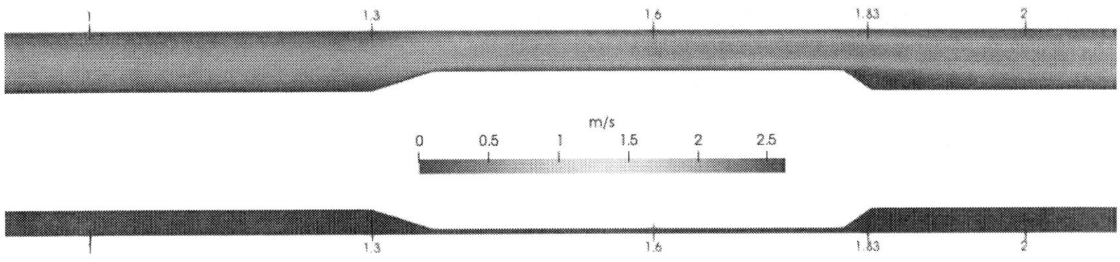

(a)

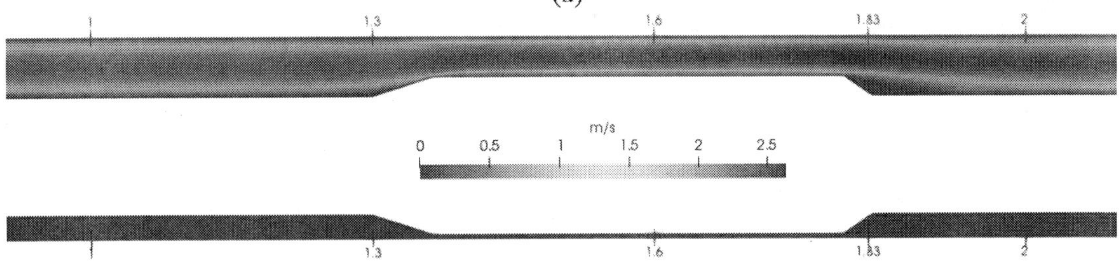

(b)

Figure 4. Velocity field in the axial section Y=0m, BC2: (a) single-phase flow, (b) two -phase flow.

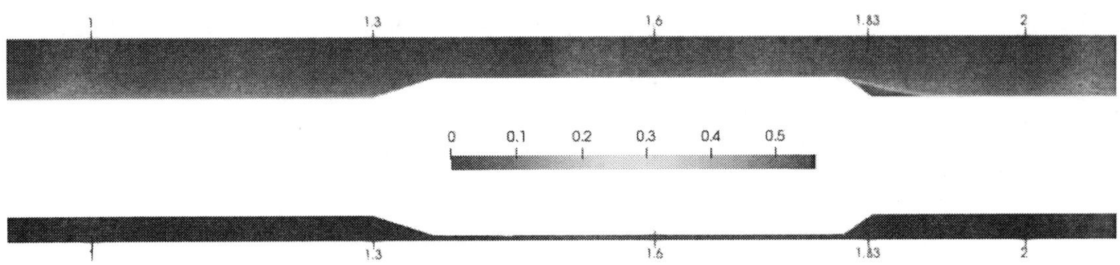

(a) Axial section Y=0m

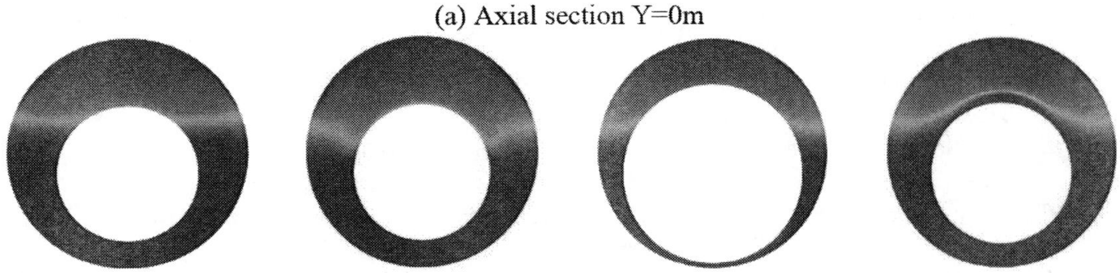

(b) Z=1m (c) Z=1.3m (d) Z=1.6m (e) Z=1.83m

Figure 5. Particle concentration distribution in sections: (a), (b), (c), (d), (e).

In the case of single-phase flow, the mean velocity at the TJ is determined by the channel's cross-section. The calculations show that the presence of cuttings affect the way the velocity changed on the TJ. In this case, the work area for the flow over the sedimentation layer was no longer proportional to the geometrical area of TJ. The calculations show that despite TJ position and boundary condition type (figures 3b and 6), sedimentation process behind the TJ is analogous to that in annulus channel without TJ.

Cuttings transport is an energy-consuming process, which is confirmed by the pressure distribution shown in figure 7. It is noteworthy, that the relative TJ effect on pressure drop for the single-phase flow is higher than for the two-phase one. While drilling, one normally uses pipes of more than 4.572 meter in length. Since the flow becomes steady behind the TJ at the distance that is much less than the mentioned 4.572 meter, it turns out to be very easy to estimate the full pressure drop in the borehole. All you need to know is how to determine the pressure drop for a single segment of a pipe.

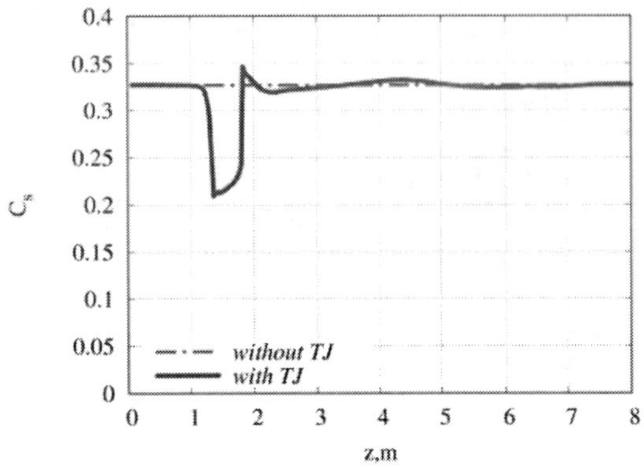

Figure 6. Mean particles concentration distribution along a channel.

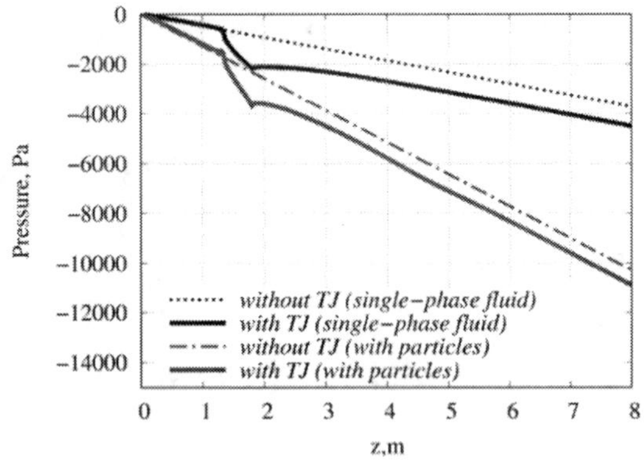

Figure 7. Pressure distribution along a channel.

4. Conclusions

In the presented study, a mathematical model and a methodology of cuttings transport numerical simulation in a channel with variable cross-section have been tested. The effect of input boundary conditions and TJ positioning on sedimentation layer formation has been estimated and the first data describing TJ effect on a sedimentation layer has been obtained. TJ effect on pressure drop for two-phase flow has been considered as well and the results obtained have been compared against those for a single-phase flow. Recommendations on determining full pressure drop in a borehole for a given drilling mode are suggested.

References

[1] Clark R and Bickham K 1994 *Proc. SPE Annual Technical Conference and Exhibition, Society of Petroleum Engineers*
[2] Ignatenko Ya, Bocharov O Gavrilov and A May R 2018 *Proc. 37th International Conference on Ocean, Offshore and Arctic Engineering*
[3] Karimi Vajargah A, Najafi Fard F, Parsi M and Hoxha B 2014 *Proc. SPE Deepwater Drilling and Completions Conference*
[4] Gavrilov A, Shebelev A and Ignatenko Y 2019 *Proc. 10th International Conference on Multiphase Flow*
[5] Ignatenko Ya, Bocharov O, Gavrilov A and May R 2019 *Proc. 38th International Conference on Ocean, Offshore and Arctic Engineering OMAE2019-95228*
[6] Cheng N-S and Law A W-K 2003 *J. Powder Technology* **129** 156–160
[7] Gavrilov A A and Rudyak V Y 2016 *J. of Non-Newtonian Fluid Mechanics* **227** 45-55
[8] Miller R M and Morris J F 2006 *J. of Non-Newtonian Fluid Mechanics* **135** 149–165
[9] Johnson P C and Jackson R 1987 *J. of Fluid Mechanics* **176** 67–93
[10] Burns A D, Frank T, Hamill I and Shi J-M 2004 *Proc. Fifth International Conference on Multiphase Flow*

Journal of Physics: Conference Series **1404** (2019) 012036 doi:10.1088/1742-6596/1404/1/012036

Models of heterogeneous medium in medicine

V I Pen'kovskiy and N K Korsakova

Lavrent'ev Institute of Hydrodynamics, SB RAS,
15 Acad. Lavrent'ev av., Novosibirsk, 630090, Russia

penkov@hydro.nsc.ru, kors@hydro.nsc.ru

Abstract. A review of models of the blood circulation in the tissues of the human brain and lower extremities is presented. The models take into account the well-known laws of physiology and theoretical mechanics. Mathematical models are based on the representation of organ tissues as a heterogeneous continuous medium consisting of three interpenetrating continua: living cells, and circulating arterial and venous blood. The results of numerical calculations of boundary value problems of the pulse blood flow in the brain of a healthy person and cases of pathologies such as arteriovenous malformation and embolism are presented. The distribution of blood pressure in the lower extremities is studied under the action of gravity and other overloads.

1. Governing equations and formulation of problems

The organism of mammals, including human body, is a complex natural system of elements in which dozens of physicochemical and electrical processes take place. Friedrich Engels defined life as "way of existence of protein bodies". A living organism of a mammal contains many other organisms, bacteria, viruses and other protozoa capable of reproduction. There are physiological laws that govern the "way of existence of protein bodies" in nature. Primary such laws are anabolism and catabolism, fission and atrophy of cells which are as well essentially living organisms. The blood circulation plays a decisive role in these processes. Nutrition and waste removal in cells are carried out by a directed blood flow in small capillaries. Violation of capillary blood flow leads to the death of cells.

The pattern of blood flow in a selected single element of organ can be presented in the scheme shown in figure 1. On the scheme, numerals 1 and 2 mark the parts of arterial and venous blood in single element respectively. Arrows show a direction of capillary flow from arterial network to venous one. The volume of organ cells is presented by light-coloured rectangles.

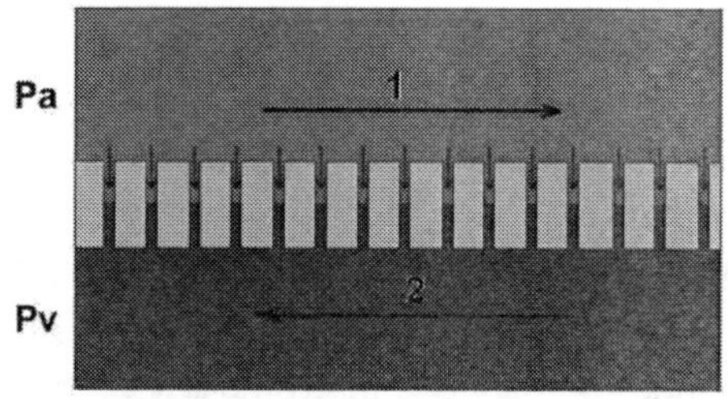

Figure 1. Blood flow scheme in single element of organ living tissue.

Journal of Physics: Conference Series **1404** (2019) 012036 doi:10.1088/1742-6596/1404/1/012036

The mathematical model of the blood flow in muscle tissue as a heterogeneous structure consisting of interpenetrating continua includes the equations of motion in continuums 1 and 2 disjoint with each other [1-3]

$$v_i = -\frac{k_i}{\mu} grad p_i \quad (i = 1{,}2)$$

and the laws of conservation of blood mass

$$\frac{\partial(m_i \rho)}{\partial t} + div(\rho v_i) + q_i = 0$$

Here v_i, p_i, k_i, μ, m_i, ρ, t, q_i are velocity, pressure, permeability of networks, fluid viscosity, porosity of continua (specific volumes of networks), fluid density, time, and values of overflows, respectively.

The equations written above are reduced to system

$$\frac{k_i}{\mu}\Delta p_i = \beta_i \frac{\partial p_i}{\partial t} + q_i \ (i = 1{,}2) \tag{1}$$

where coefficients $\beta_i = dm_i / dp_i$ describe the compressibility of vessel network. Δ is Laplace operator. The equations of the system (1) are linked to each other by cross-flows $-q_1 = q_2 = q = \frac{\alpha}{\mu}(p_1 - p_2)$, where α is specific parameter of inner exchange (capillary conduction). The system (1) contains five parameters: the conductivities of arterial and venous networks, vessel system compressibilities and capillary conduction.

The values of some of them can be estimated by statistical methods if the distribution function of the vessel sizes by their sizes is known. It is assumed that the elasticity of the vessels in the continua is identical and is actually determined by the elasticity of the cells. In addition, the total conductivity of the arterial and venous networks can be considered approximately the same.

Including the new required functions $S^{\pm}(x,t) = p_1 \pm p_2$, the system (1) can be reduced to the next system of independent equations

$$-\beta_1 \frac{\partial S^-}{\partial t} + \Delta S^- - 2\eta_1 S^- = 0 \ , \ -\beta_1 \frac{\partial S^+}{\partial t} + \Delta S^+ = 0 \ . \tag{2}$$

Applying to pulse blood flow in the cerebral cortex, the problem without initial data are solved of the form

$$z = z_w : S^{\pm} = A^{\pm} = A\cos\omega t + B^{\pm}; \quad z = z_p : \partial S^{\pm} / \partial n = 0 \, ,$$

where $B^{\pm} = B \pm p_0$, $A = (p_s - p_d)/2$, $B = (p_s + p_d)/2$, p_s is systolic pressure, p_d is diastolic pressure, p_0 is suction pressure in veins, ω is pulse regime frequency, $z_w = x_w + y_w, z_p = x_p + y_p$ are points of the Willis circle and a brain periphery respectively, $\partial / \partial n$ is derivative of normal to periphery. The last boundary condition means blood unflowing through external brain border (periphery).

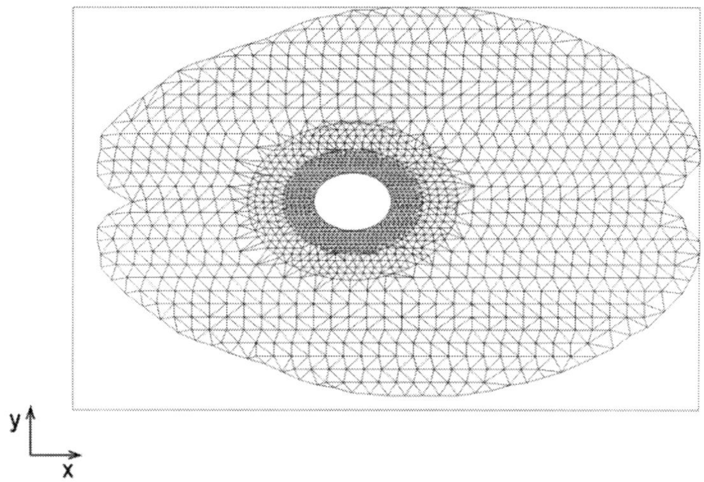

Figure. 2. Configuration of calculating domain (brain cortex).

The problem is solved with finite element method using net crowding. The calculating domain is shown in figure 2. When numerically investigating the effect of malformation or embolism in the brain on the total blood flow, the coordinates and size of the pathology zone are set. In the zone, the parameters corresponding to the nature of the pathology are specified.

In figure 3 and figure 4 are shown izolines of arterial blood pressure for cases of vessel malformation and embolism. Circles mark locations of the pathologies. Three types of pathology can be considered in a split system of equations (2). A more detailed study requires the use of a complete system of equations

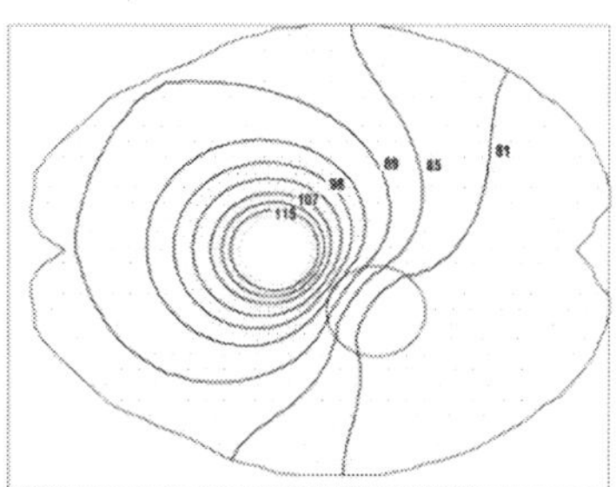

Figure 3. Malformation $\eta_1 = 1$.

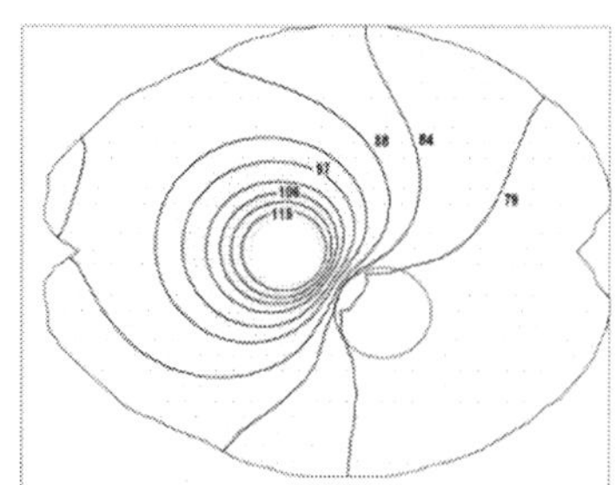

Figure 4. Vessel embolism $\eta_1 = 100$.

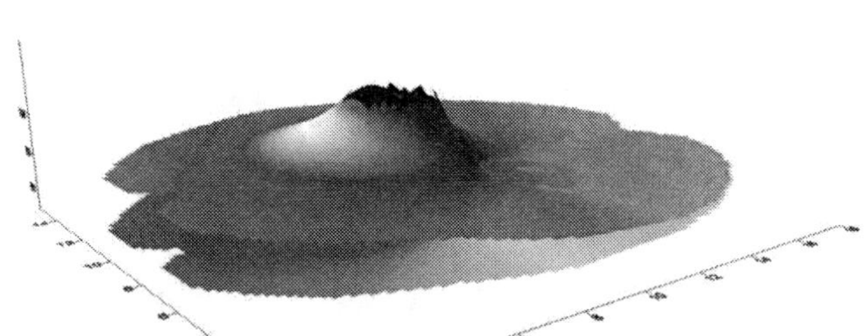

Figure 5. Pressure surfaces in the presence of embolism ($\eta_1 = 100$).

The curves correspond to the point in time at which the maximum pressure is reached in the arteries of the Willis circle.

Figure 5 presents the surfaces of pressure distribution in arterial and venous blood flow in case of embolism in circle with center point (11, 6) where parameters are taken as $\beta_1 = 0.01$; $\eta_1 = 100$.

When modeling blood flow in the lower extremities [4-5], averaged over the cross section of muscle tissue and directed along the axis z flows are considered. Let q_a be flux in distributive (arterial) network, q_v be flux in collector (venous) network. Then analog of equations (1) is the following system

$$\begin{cases} -\beta_1 \dfrac{\partial h_a}{\partial t} + \dfrac{1}{s(z)}\dfrac{\partial}{\partial z}\left(s(z)\dfrac{\partial h_a}{\partial z}\right) = \eta_1 (h_a - h_v); \\ -\beta_1 \dfrac{\partial h_v}{\partial t} + \dfrac{1}{s(z)}\dfrac{\partial}{\partial z}\left(s(z)\dfrac{\partial h_v}{\partial z}\right) = -\eta_1 (h_a - h_v). \end{cases}$$

The boundary conditions at the hip and at the foot respectively must be satisfied for this system

$$z = 0: p_a = h_a = A\cos(\omega t) + B,\ p_v = h_v = p_o; \quad z = z_p : \partial h_a / \partial z = \partial h_v / \partial z = 0.$$

Here we denote $h_a = p_a - \rho g z$; $h_v = p_v - \rho g z$; $v_a = -\partial h_a / \partial z$; $v_v = -\partial h_v / \partial z$, and $s(z)$ is a cross-section area of leg muscle bulk.

Analytical solutions are obtained for particular model cases. In the general case, the problem is solved numerically. The numerical results demonstrate that blood pressure distribution in low limbs depends essentially on a human position relative to the vector gravity and other overloads.

2. Conclusions

New mathematical models of blood circulation in human organs are considered. They take into account well-known laws of physiology, theoretical fluid mechanics and base on conception that tissues of human organ are heterogeneous continuum consisting from interpenetrative continua: living cells as well as circulating arterial, venous and capillary blood. A branched arterial network is a distribution system that passes through the capillary system into the collector system of veins. There are no intersections between systems in a healthy body. Cases of intersections or clogging of blood vessels lead to dangerous pathologies: malformation and embolism. Calculations show that pathologies significantly affect the distribution of blood pressure in the arterial and venous channels.

The mathematical models enable one to choose a treatment strategy for the human organism.

References

[1] Penkovskii V I and Korsakova N K 2014 Proc. Int. Russian-Kazakhst. Conf. in Math. model. scince-tech. ecolog. problems oil-and-gas producing industry (Atyrau) pp. 291-5

[2] Penkovskii V I and Korsakova N K 2015 Proc. 14th All-Russia Workshop in Dynamic of Multiphase Media (Novosibirsk:) pp 73-75

[3] Korsakova N, Pen'kovskij V, Shilova A and Shevchenko V 2016 *Series on Biomechanics* **30** 24

[4] Penkovskii V I and Korsakova N K 2018 *Appl. Mech. Tech. Phys.* **59** 460

[5] Pen'kovskiy V I and Korsakova N K 2018 *AIP Conference Proceedings*. **1939** 020049

Studying the process of powder transportation during laser cladding with a coaxial nozzle

T A Polyanskiy[1,2], A V Zaitsev[1,2] and G V Ermolaev[1]

[1] Khristianovich Institute of Theoretical and Applied Mechanics SB RAS, 4/1 Institutskaya st., Novosibirsk, 630090, Russia
[2] Novosibirsk State University, 1, Pirogova str., Novosibirsk, 630090, Russia

zaitsev@itam.nsc.ru

Abstract. Results of experiments and numerical modeling of gas-powder flow for laser cladding are presented. Numerical simulation is based on solving Navier-Stokes equations with Lagrange approach for particle. The main attention is paid to the process of interaction between powder particles and nozzle walls. Simple particle-wall interaction model was suggested and verified with experiments for gas-powder jet formed in the long tubes with different diameters. The model was used for simulation of powder transport in the coaxial nozzle for laser cladding.

1. Introduction

Laser cladding is additive technology that uses laser radiation for a local heating of base material and powder of metal, ceramics and metal mixture for building new details with complex 3D geometry. The technology has become widespread in industry, but several unresolved problems remain. One of the main problems is difficulty to achieve high powder usage coefficient, which directly influences the cost of the process. Other problems are associated with roughness of the cladded bead and measures to protect optical elements of the laser nozzle head from the damage of hot powder particles. Main goal of the research is construction of new cladding nozzles with high powder usage coefficient. Despite a large number of works on the modeling of a gas-powder mixture for laser cladding technology, including recently published [1,4], we found that one of the main reasons determining the shape of a powder jet is not considered qualitatively. In our previous works, it was shown [2-3] that the law of particle reflection from the nozzle walls is one of the most important. In this paper, we continued our study, specifying the laws and coefficients that describe the interaction between particles and walls and applied the model to research processes in the coaxial nozzle for laser cladding.

2. Models

Despite the fact that gas flows containing microparticles have been studied for a long time, the area of object application is so wide that there are different physical models for each specific field. Laser cladding characterized by gas and particle velocities not exceed 30 m/s. Usually particle size of powder is in range of 30 to 100 μm and dimensions of the channel are several millimeters. Therefore, particles experience multiple collisions with the walls in nozzle. The main parameter for practical application is size of powder jet spot on the substrate, it determined directly the utilization rate of powder.

The main phenomena usually considered for moving particles in the gas are the drag force, the Magnus force, the Saffman force, the effect of the attached mass. According to our research [2-3] under conditions specific to laser cladding the last three terms have a small effect and can be neglected. It means we could neglect influence particles rotation on their trajectories between collisions with walls and other particles.

But works of Sommerfeld [5] and Tsirkunov [6] showed that particle rotation may be important in the process particle-wall interaction. In our model, we tried to avoid using angular velocity with introducing simpler laws for particle-wall interaction. The model based on two restitution coefficients of the tangent and normal velocity. These coefficients are strongly empirical. They were calculated from statistical experimental data and heavily depend on nozzle and powder material. In most cases particle-particle interactions also can be neglect because of volume fraction does not exceed 10^{-4} that less then generally accepted value 0.01 [6].

In [2-3] we showed that particle backward influence on the gas could be neglected and task could be solved in two stages: 1) the gas flow pattern is calculated by solving the Navier-Stokes equations with the SST-model of turbulence, 2) the particle trajectories are calculated in the Lagrangian approach on the obtained distribution of gas velocities. We neglect direct influence of turbulence pulsation on particle motion and use averaged gas flow field because turbulent oscillations weakly affect the particle motion inside the channel (Re <500), and the gas motion outside the nozzle is not so decisive for the spot size of the powder jet on the substrate.

The particle size distribution in simulation was consistent with experimental one (Fig. 2). Volume distribution were converted to number distribution and every time particle started motion with size chosen randomly with built function.

Simulation software developed on base of open source package OpenFOAM. Additional user library with new particle interaction and injection models in order to expand functionality and replace computationally time-consuming built-in procedures was written.

3. Experimental setup and results

Experimental investigation of powder jet characteristics performed using original optical diagnostic system based on digital camera and positioning system, additionally equipped with pulse LED illumination module. Module synchronized with the camera exposition time (pulse duration 10-500 μm). That made possible to increase the resolution to 15 μm/pixel and allow to register particles with a diameter of ~20 μm or more. Frames with tracks of moving particles (about 3000 for each experiment) were recorded and processed to identify individual particle tracks on each frame, their position and velocity vector components.

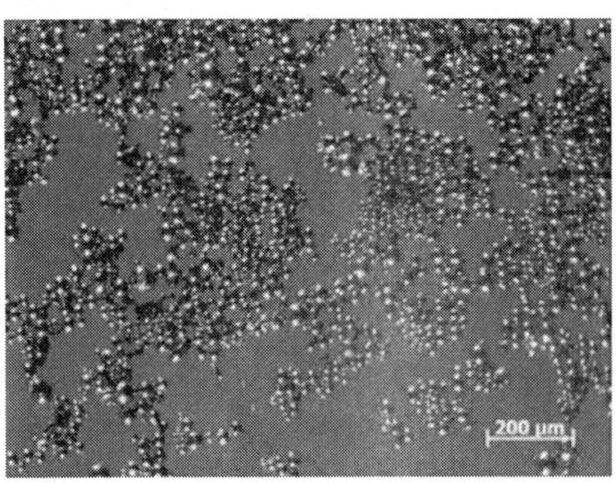

Figure 1. Powder PR-HX16CP3 particle shape,

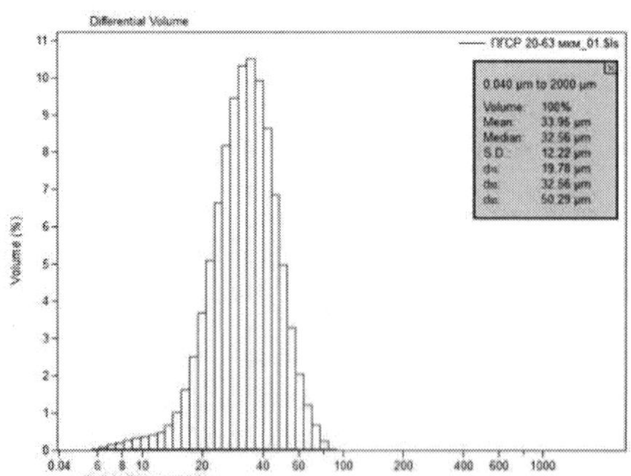

Figure 2. Powder PR-HX16CP3 particle size distribution.

All experiments were carried out with powder PR-HX16CP3 of the system (Ni (base) -16Cr-3.2Si-2.7B) characterized by particle size mostly in range 20 – 63 μm with high sphericity (figures 1 and 2).

The experiments studied the structure of gas-powder jet forming by single 100 mm long tubes of various diameters and material. The copper tubes with diameters 3, 4, 6, 8 mm and the steel tubes with diameters 2, 4, 6, 8 mm were used.

Additionally, roughness of tubes inner surface is investigated. Roughness parameters Ra and Rz were measured for all tubes on optical profiler NewView 7300. The results are presented in Table 1.

Table 1. Roughness parameters Ra and Rz for copper and steel tubes.

	Copper tubes				Steel tubes			
Inner diameter	3	4	6	8	2	4	6	8
Ra	0.83	0.72	0.45	0.39	0.79	2.14	1.48	2.7
Rz	4.1	3.5	3.1	2.5	6.6	12	8.8	12.8

Figure 3 shows how the experimental particle jet width (a) depends on the distance from the nozzle exit for 6mm steel tube and average particle velocity changes (b) after leaving 4mm copper tube. The jets had almost conical shape and nearly do not depend on speed of carrying gas. The velocity of the particles remains almost the same after leaving the nozzle for the gas speed less than 20 m/s. Similar results are obtained for other tubes. Therefore, the average angle of inclination of the particle velocities after the nozzle mainly determines the shape of the powder jet.

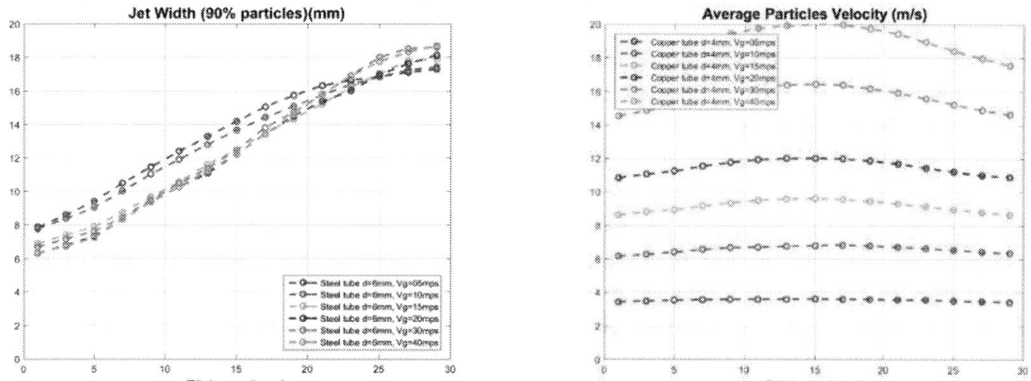

(a) (b)

Figure 3. (a) Experimental particle jet width vs distance from nozzle exit for steel tube, 6 mm; (b) Average particle velocity vs distance from nozzle exit for 4 mm copper tube. Gas speed varies from-5 to 40 m/s.

4. Numerical simulation

4.1. Long tubes experiments comparison

Numerical simulation was carried out for gas flow 10 m/s and tubes 4, 6 and 8 mm in diameter. Similarly to the experiment, in the numerical simulation, the coordinates and components of the particle velocity were tracked. Additionally, average particle concentration calculated. Model coefficients varied for to achieve better comparison with experimental results.

For copper tubes, a good agreement with the experimental data on the scattering angle of the particles was obtained with coefficients of velocity restitution about 90%. Coefficients for steel tubes lie in the range from 95% to 99%.

4.2. *Coaxial nozzle simulation*

At this stage, triple coaxial nozzle is investigated. Velocity field was calculated with the following flow rates for each gas type: shielding and transport – 10 l/min, shaping - 30 l/min. With such parameters strong gas flow from outer ring to center is forming (figure 4).

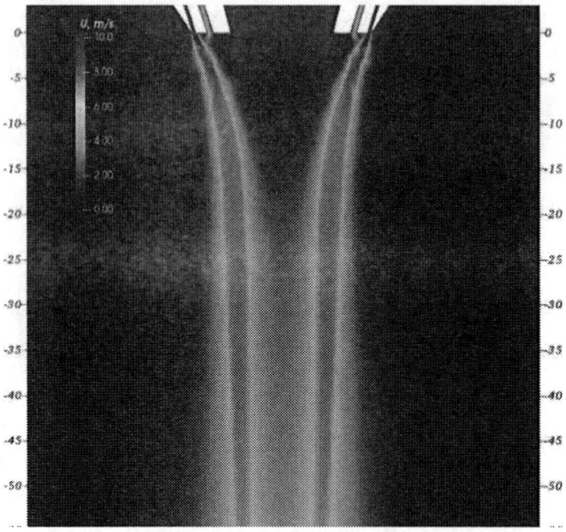

Figure 4. Gas velocity field.

Using this flow field, particle movement calculation was made with different particle-wall interaction laws.

For first case mirror reflection model was used. A large dispersion of particles in space was observed despite the presence of a gas flow directed toward the center (figure 5).

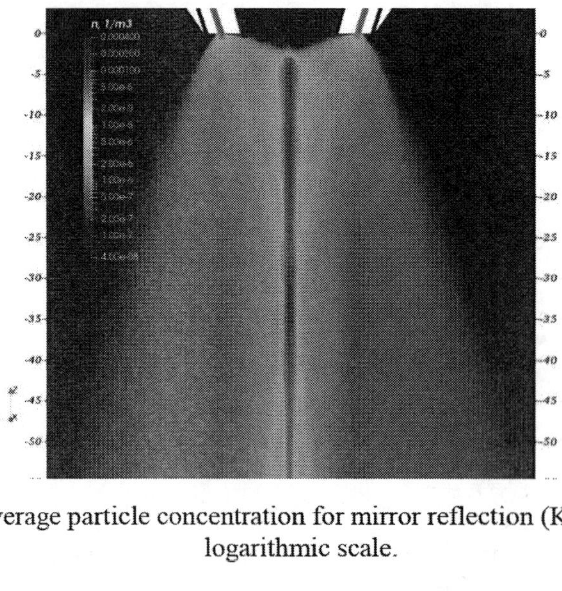

Figure 5. Average particle concentration for mirror reflection (Kn = 1, Kt = 1) in logarithmic scale.

Case with experimentally obtained coefficients gives better jet focusing (figure 6). This confirms the assumption that the shape of the powder jet under conditions typical for laser cladding, is primarily determined by process inside nozzle than gas influence on particles in free space.

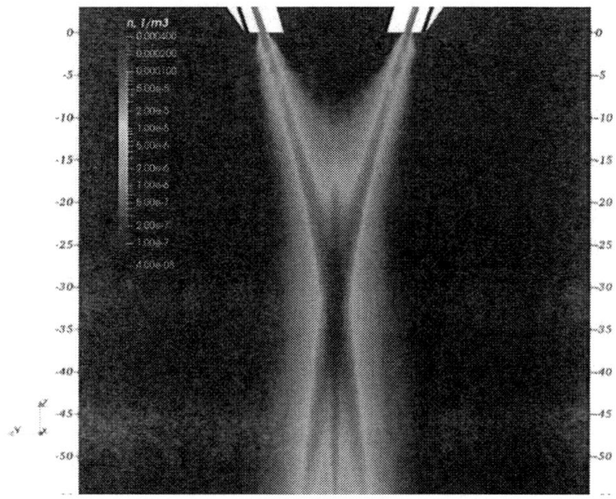

Figure 6. Average particle concentration for experimental obtained coefficients for copper tubes in logarithmic scale.

5. Conclusion

Optical diagnostics of gas-powder flows forming by long steel and copper tubes with diameters 2-8 mm was carried out with gas velocities of 5-40 m/s in typical laser-cladding conditions. According to results, particle velocity changes insignificantly after leaving tube. It means powder spot size depends on processes inside tube or nozzle due to particle-wall interaction. A model of particles collision with a wall, based on restitution coefficients of the tangent and normal velocity, is proposed and coefficients are selected by comparing numerical simulations with experimental results. Using the model and coefficients, the spatial distribution of the powder in coaxial nozzle jet is studied numerically. Simulation showed that, as in the case of tubes, the influence of the gas flow on the particle trajectory after leaving the nozzle is negligible. The use of steel nozzle walls as material with coefficients close to 100% leads to a significant expansion of the jet. Copper wall provides better focusing.

Acknowledgments

This work was partially supported by Russian Foundation for Basic Research (Project No. 18-08-00449)

Reference

[1] Zhang J, Yang L, Zhang W, Qiu J, Xiao H and Liu Y 2020 *Optics and Lasers in Engineering* **126** 105873
[2] Zaitsev A V, Gulyaev I P, Polyanskiy T A and Gurin A M 2018 *AIP Conference Proceedings* **2027** 030147
[3] Polyanskiy T A, Zaitsev A V, Gulyaev I P and Gurin A M 2018 *J. Phys.: Conf. Ser.* **1109** 012010
[4] Kovalev O B, Bedenko D V, and Zaitsev A V 2018 *Applied Mathematical Modelling* **57** 339-359
[5] Sommerfeld M 1992 *International Journal of Multiphase Flow* **18** 905–26
[6] Tsirkunov Y M, Romanyuk D A and Panfilov S V 2014 *29th Congr. Int. Counc. Aeronaut. Sci. ICAS 2014*

Journal of Physics: Conference Series **1404** (2019) 012038 doi:10.1088/1742-6596/1404/1/012038

On the near wake structure of a supersonic coaxial gas-liquid jet

S V Poplavski and A Yu Nesterov

Khristianovich Institute of Theoretical and Applied Mechanics SB RAS,
4/1 Institutskaya str., Novosibirsk 630090, Russia

s.poplav@itam.nsc.ru

Abstract. The work is devoted to a complex experimental study of the near wake of supersonic coaxial gas-liquid jets with a high concentration of the dispersed phase. At the various liquid flow rates, wave structure of a gas phase was visualized, droplet velocity profiles were obtained by the LDA method, and the spray concentration was measured. The effect of the liquid phase on the supersonic underexpanded jet was studied, in particular, a change in the general wave structure, Mach disk transformation, etc. A gas velocity profile was obtained in a two-phase core of a gas-liquid jet by the method of droplet velocity profile correction taking into account their velocity relaxation.

1. Introduction

Devices for spraying liquids - nozzles - has many technological applications: coating, medicine and pharmacology, the chemical and food industry, propulsion engineering, furnace burners, modern fire extinguishing systems, etc. Among various kinds of nozzles, there is a class of pneumatic ones; they are mainly used for burning wastes in the oil industry, but can be used in any technology that requires a finely dispersed spray with large liquid flow rate. In these nozzles, a high-speed gas flow is used to preliminarily destroy the liquid and form the spray jet. This allows one to provide dispersion (~ 10 microns) acceptable for the efficient spray combustion at a flow rate of up to several cubic meters per hour [1]. A number of issues of the optimal design of such devices remain unresolved, in particular with regard to supersonic gas regimes relevant for these purposes. However, the study of high-flow nozzles is complicated by the fact that, due to the high concentration of the spray, traditional tools and methods, such as the Pitot tube and hot-wire anemometers, are not applicable. Due to the opacity of the gas-liquid jets, visualization, PIV and other optical diagnostics are also excluded. Mathematical modeling of gas-liquid flows is limited by the volume concentration of the dispersed phase $\beta < 10^{-4}$ [2], which is two orders of magnitude less than the real spray dense of high-flow nozzles. For the development and verification of computational models of dense gas-liquid flows, reliable, and methodically verified experimental data on a wide range of physical parameters are needed. Thus, this section of the dynamics of two-phase media does not yet have a sufficient methodological base for either experimental or mathematical modeling.

Due to the complexity of the processes, it is advisable to carry out studies of the mechanisms of decay of a liquid jet and secondary destruction of droplets in a simplified formulation with model objects. One of these objects is a coaxial gas-liquid jet with a fluid supply through the nozzle exit: it has central symmetry and is available for investigation from the moment the liquid enters the gas stream. According to previous studies, the destruction of droplets in such a system occurs in the region of maximum velocity gradients, namely, in the near wake [3]. Dispersion of liquid is most effective in supersonic jets when droplets interact with a system of jumps that supposedly penetrate into a two-phase core in the near wake of a supersonic underexpanded gas jet.

A comprehensive study of the processes in the core of a supersonic gas-liquid jet, their influence on the flow structure and spray dynamics are the key research challenges. Along with this, it is necessary to develop new methodological approaches to the study of such systems. This paper presents the results of an experimental study of the structure of the near wake of a supersonic gas-liquid jet, obtained on the basis of new approaches and data on the dynamics of the gas phase and droplets.

2. Experimental setup and diagnostics

The experiments were carried out on "Gas-liquid stand" setup of ITAM SB RAS. The setup allows conducting investigations of gas-liquid jets in the range of liquid flow discharge from 10 to 5000 l/hr and the pressure in the pre-chamber can be up to 8 atm. The gas-liquid stand *1* (figure 1) comprises a nozzle unit *2*, where the nozzle was fixed with possibility of vertical movement, a receiving tank *3* for used liquid, and instruments for controlling flow parameters of gas and liquid. The atomizer constitutes convergent nozzle with input and output diameters 19 and 14 mm respectively (figure 2) with the central tube 2 mm in diameter for supply of liquid along the jet axis.

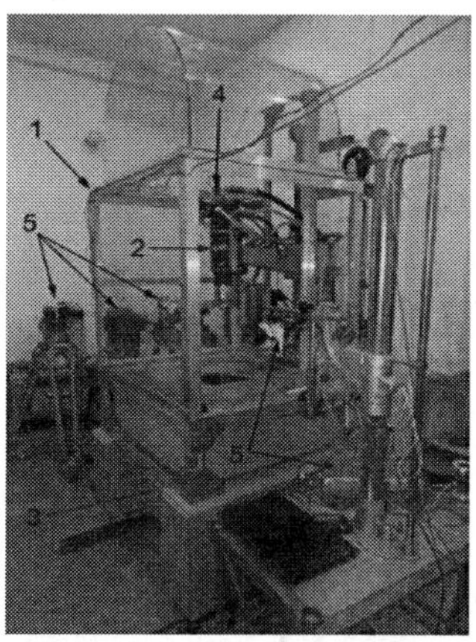

Figure 1. Gas-liquid stand: 1 – jet module, 2 – nozzle unit, 3 – receiving tank, 4 – liquid and gas channels, 5 – elements of diagnostics.

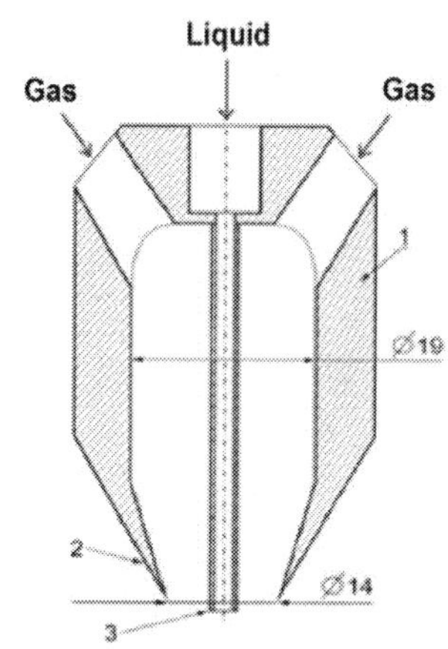

Figure 2. Sketch of the atomizer: 1 – body, 2 – outlet convergent section, 3 – the pipe for central supply of liquid.

In the present work we use the complex of optical methods for gas-liquid flow diagnostics:

- Flow visualization (shadow method, AVT visualization techniques [4]) for registration of jet structure;
- Malvern Spray Tec analyser for determining the spray dispersity;
- PIV diagnostics of gas flow velocity field;
- Laser Doppler anemometer with direct spectral analysis [5, 6] for droplets speed determination.

The experimental setup, the complex of diagnostics and the parameters of the jet in the different regimes are described in [3]. One of the important parameters of the gas-liquid jet is a concentration of the spray, it influences on the interphase heat transfer and the speed of sound in two-phase medium [7]. Note that the presence of a liquid with the volume fraction $\beta > 10^{-6}$ in the flow leads to changes in the profile of gas velocity. So the data of PIV in pure gas flow is inapplicable to two-phase medium. For determining the gas velocity $U(x)$ in two-phase core of the jet a method of data correction for velocities of liquid $V(x)$ was used to take into account the velocity relaxation. In [8] it is shown that the profile of gas speed $U(x)$ can be calculated from the liquid velocity $V(x)$ using the equation:

$$U(x) = V(x) \pm \sqrt{\lambda \cdot V(x) \cdot \left|\frac{dV(x)}{dx}\right|}, \qquad \lambda = \frac{4}{3}\frac{\rho_p}{\rho}\frac{d}{C_D}, \tag{1}$$

$$C_D(\mathrm{Re}) = \frac{24}{\mathrm{Re}} + \frac{4}{\sqrt[3]{\mathrm{Re}}} + 0.004\mathrm{Re}^{0.44} \tag{2}$$

Here λ is the velocity relaxation length of particles, the plus sign "+" is used for the case $dV/dx > 0$, and the minus sign "–" is used for the case $dV/dx < 0$, d and ρ_p are the size and the density of the particles, C_D is the particles drag coefficient assuming the characteristic density of gas $\rho \approx$ const.

In [9], this approach was used to correct the gas velocity field obtained by the PIV method in a supersonic underexpanded jet. Here, this method was applied to the droplet velocity profiles obtained by the LDA method to calculate the gas velocity profile. The correction was carried out in two stages:

- Preliminary correction with estimated value of λ gives approximate gas velocity profile $u(x)$ in two phase core;
- Refined correction performed according to Re number for the relative velocity of gas and drops $V(x)$ - $u(x)$.

3. Results and discussion

3.1. *The effect of the liquid phase on the wave structure of the jet*

The experiments were carried out for the modes Npr = 6 and 8, and the liquid discharges 20 and 300 l/h (5.6 and 83 g/s). As the liquid we use water, dispersing phase is air. The supersonic underexpanded jets with well-studied wave structure are generated on these mods without liquid [9]. In our case the flow is somewhat more complex because of the central body in the nozzle [3], but these changes are small in comparison with perturbation caused by the presence of liquid.

The flow was visualized by the shadow method without liquid and with the introduction of its small quantities, while part of the jet remains transparent, where the wave structure can be observed. Shadowgrams of the jet without liquid and at small liquid consumption are presented in figures 3 and 4.

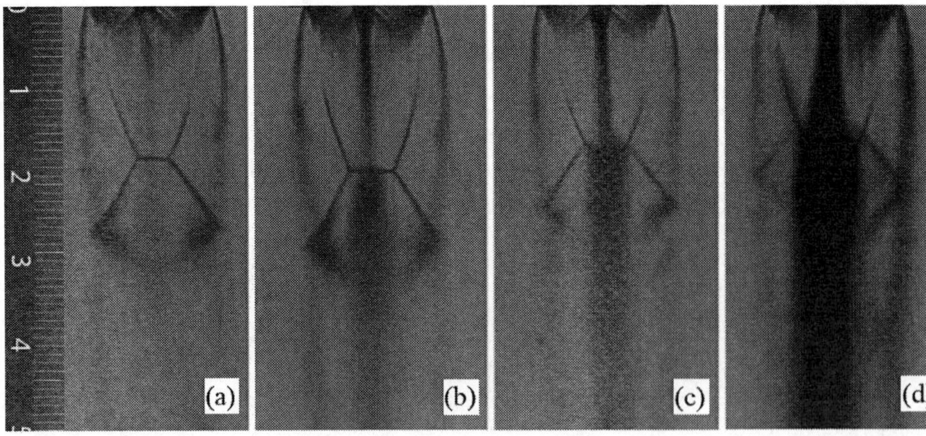

Figure 3. Shadow pictures of the jet in the Npr6 mode in the absence of liquid (a), at low liquid flow rates (b, c) and at the liquid flow rate of 20 l/h (d).

It is seen that in the presence of a liquid, the pattern of jumps changes significantly: the Mach disk bends upstream, the shift of the jump is also provided by a shift upstream triple point and deformation of the general structure. The triple point shifts in comparison to pure gas is from 18 to 16 mm from the nozzle in Npr6 mode (in figure 3c) and from 22 to 20 mm in Npr8 mode (see figure 4c). The probable position of the jump on the axis of the jet is 4–5 mm upstream. With a liquid discharge of 20 l/h, the Mach disk looks like a bow shock on a cloud of particles [10,11] (figure 3d and figure 4d) and shifts even higher. The bow shock is formed because of an increase in the particle concentration. This happens due to the deceleration of the droplets at the boundary of the first cell, and because of the recirculation zone presumably formed here [12]. Thus, a complex picture of the gas flow is observed, leading to high velocity gradients, this has a great influence on the destruction of droplets and requires a more detailed consideration.

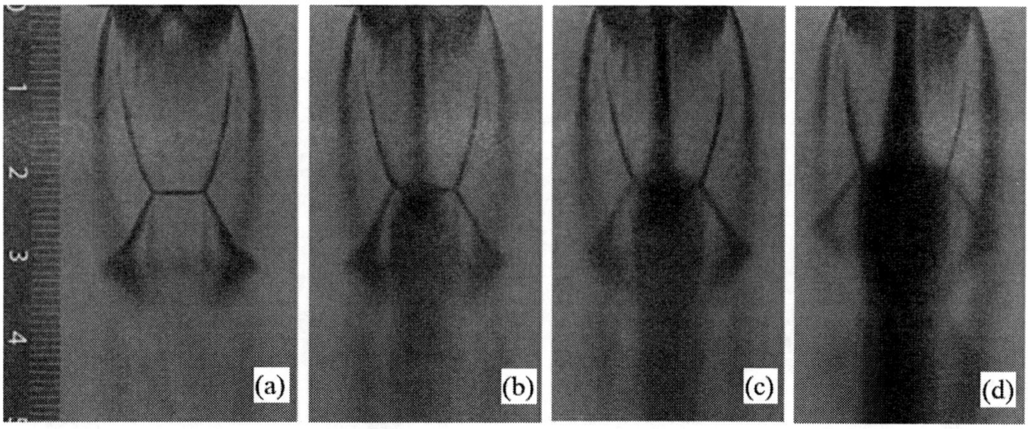

Figure 4. Shadow pictures of the jet in the Npr8 mode in the absence of liquid (a), at low liquid discharge (b, c) and at the liquid consumption of 20 l/h (d).

3.2. *The dynamics of gas-liquid jet*
The restructuring of the flow depending on the liquid discharge can also be traced by the droplets velocity profiles. Figures 5 and 6 shows these profiles along the jet axis. Their characteristic feature is a sharp drop in droplets velocity in the region from 10 to 20–25 mm, followed by a monotonic increase up to velocity relaxation in a decelerating gas flow at distances greater than 100 mm [3].

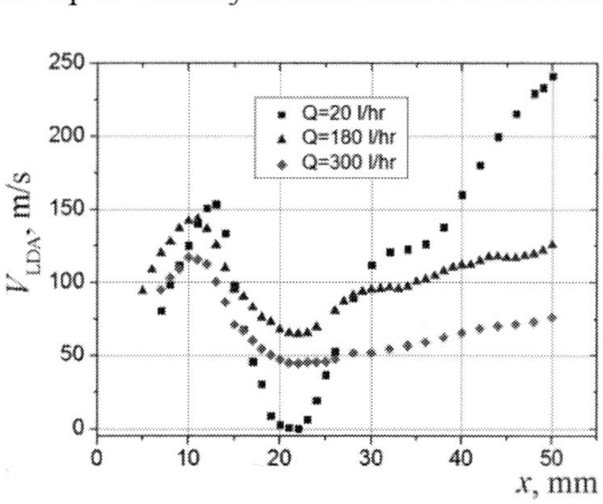

Figure 5. Profiles of the droplets velocity along the axis in the near wake of jet at various liquid consumption, Npr6.

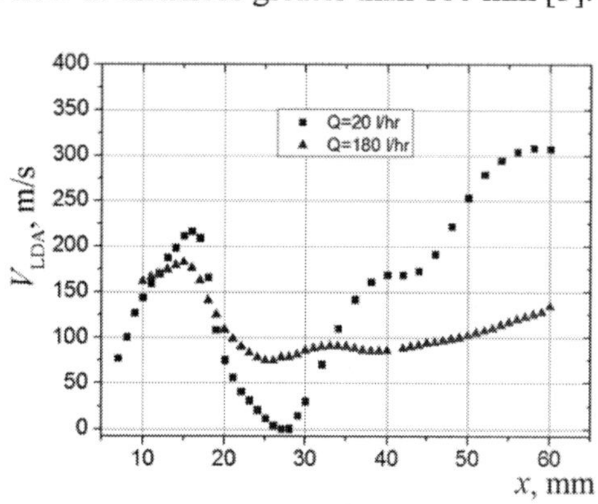

Figure 6. Profiles of the droplets velocity along the axis in the near wake of jet at various liquid consumption, Npr8.

This type of profiles is observed at all liquid consumptions; there are differences only in the quantitative characteristics and the Mach disk position. It shifts toward the nozzle with increasing concentration, which can be seen from the velocity maximum position. Thus, in this range of the parameters, the flow pattern is noticeably different from the pure gas jet.

Figures 7, 8 show the profiles of the velocity and spray concentration in the near wake, as well as the flow shadow pattern for a liquid flow rate of 20 l/h, modes Npr6 and Npr8. It can be seen that the volume concentration before the Mach disk drops to values of 3×10^{-3} in the Npr6 mode and 10^{-3} in Npr8 one; behind the jump, the concentration increases significantly with a subsequent decrease to the same values and even lower in the region of jet acceleration. In both modes, a characteristic flow pattern with the likeness of a bow shock is observed. In the Npr8 mode, the effect of the expansion of the two-phase region just behind the jump is noticeable.

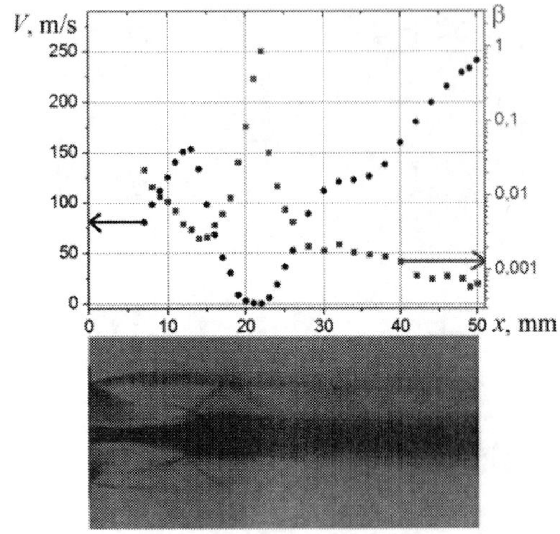

Figure 7. Droplets velocity profile, concentration and shadow flow pattern in Npr6 mode, Q = 20 l/h.

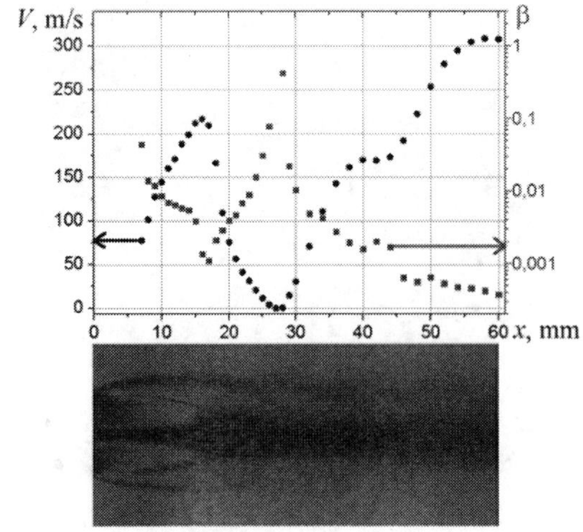

Figure 8. Droplets velocity profile, concentration and shadow flow pattern in Npr8 mode, Q = 20 l/h.

Figure 9 shows the results of the correction of the droplet velocity profile for the Npr6 mode. It is seen that the obtained velocity profile is close to the characteristic structures of the underexpanded supersonic jet of pure gas, but the velocity behind the Mach disk becomes negative. Apparently, this is due to the presence of a recirculation zone behind the Mach jump with a countercurrent near the jet axis [12], which leads to an intensive deceleration of droplets in this region.

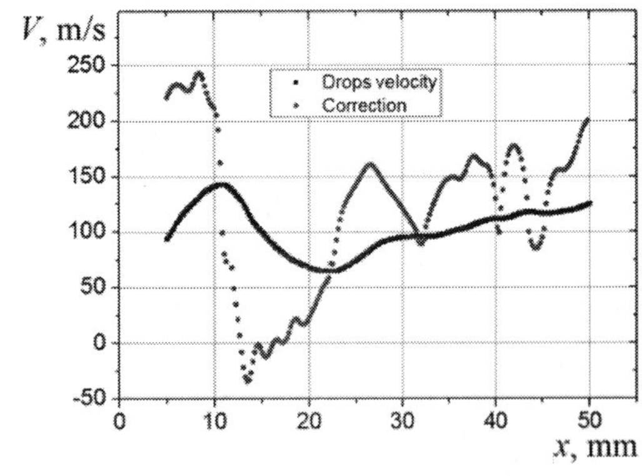

Figure 9. Experimental droplets velocity and calculated gas velocity, Npr6, Q = 180 l/h.

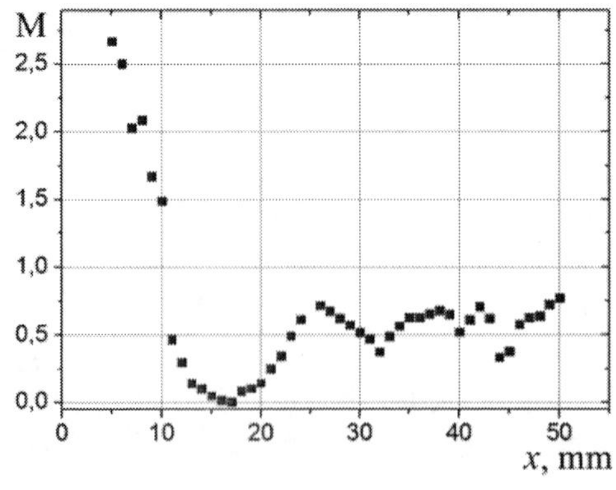

Figure 10. Mach numbers in a two-phase jet core, Npr6, Q = 180 l/h.

In figure 9 periodic velocity gradients is seen that correlate with the position of the jumps in the gas jet periphery. A significant decrease in the gas velocity in the two-phase jet core is observed in comparison with a purely gas one: the maximum velocity was ~ 240 m/s, while for a gas jet it was ~ 430 m/s. In this case, the flow remains supersonic due to a drop in the speed of sound in a two-phase medium, the graph of the Mach numbers on the axis of the jet is shown in figure 10. It can be seen that, despite a strong drop in the gas velocity, the Mach number before the jump is about 1.5-2. Due to the deceleration of the gas in the two-phase core of the jet and the growth of the speed of sound of the medium downstream, only one compression wave (Mach disk) is reproduced in the gas-liquid jet. Nevertheless, the flow retains the peroidic gradient for 5-7 calibers due to the influence of supersonic structures at the periphery.

4. Summary

The paper presents the results of a comprehensive experimental study of the near wake of a supersonic coaxial gas-liquid jet at various liquid consumptions. The change in the wave structure in the presence of a liquid is studied, and the degeneracy of the Mach disk into a conical jump is shown. The velocity and concentration profiles of droplets in two phase core of the jet are obtained. Using the correction algorithm of the droplets velocity, the gas velocity in two-phase jet core was calculated. The presence of a gas velocity jump of the "reverse step" type with a negative velocity region and periodic gradient structures downstream were observed. An assumption is made about the existence of a recirculation zone in the gas phase behind the conical shock wave.

Acknowledgements
The research was carried out within the framework of the Program of Fundamental Scientific Research of the state academies of sciences in 2013-2020 (project No. AAAA-A17-117030610137-0).

References

[1] Vitman L A, Katsnelson B D and Paleev I I 1962 *Atomization of liquid by nozzles* (Moscow, Leningrad: Gos. Energ. Izd.) p 265
[2] Varaksin A Yu 2013 *High Temperature* **51** 377–407
[3] Boiko V M, Nesterov A Yu and Poplavski S V 2019 *Termoph. Aeromech.* **26** 385–98
[4] Boiko V M, Orishich A M, Pavlov A A and Pikalov V V 2009 *Methods of Optical Diagnostics in Aerophysical Experiments* (Novosibirsk: Novosib. Gos. Univ.) p 450
[5] Boiko V M, Poplavski S V, Nesterov A U, Kondratev S V, Morozov A A and Potekhin A K 2017 *AIP Conference Proceedings* **1893** 020015
[6] Boiko V M, Nesterov A Yu and Poplavski S V 2018 *AIP Conference Proceedings* **2027** 040008
[7] Deich M E and Filippov G A 1981 *Gas Dinamics of Two-Phase Media* (Moscow: Energoizdat) p 472
[8] Boiko V M, Pivovarov A A and Poplavski S V 2013 *Comb., Expl., Shock waves* **49** 548–54
[9] Boiko V M, Zapriagaev V I, Pivovarov A A and S V Poplavski 2015 *Comb. Exp. Shock Waves* **51** 587–96
[10] Boiko V M, Papyrin A N and Poplavski S V 1993 *Comb. Expl. Shock Waves* **29** 389–94
[11] Boiko V M and Poplavski S V 2009 *Comb. Expl. Shock Waves* **45** 198–204
[12] Zapriagaev V I, Boiko V M, Kavun I N, Kiselev N P and Pivovarov A A 2016 *AIP Conference Procceedings* **1770** 030029

Study of filtration processes of a two-phase fluid in a zonal-inhomogeneous fractured-porous medium

A A Pyatkov[1], S P Rodionov[1], V P Kosyakov[1] and N G Musakaev[1, 2]

[1]Tyumen Branch of Khristianovich Institute of Theoretical and Applied Mechanics of SB RAS, Taymirskaya Str. 74, 625026, Tyumen, Russia
[2]Industrial University of Tyumen, Volodarskogo Str. 38, 625000, Tyumen, Russia

pyatkovi80@mail.ru, rodionovsp@bk.ru, hammer-rav@mail.ru, musakaev@ikz.ru

Abstract. A significant amount of proven reserves of hydrocarbons are contained in the fields to some extent have fracturing. Fractures have high conductivity, and, therefore, have a significant impact on the process of extracting oil from the reservoir. In this regard, it is necessary to investigate the filtration processes of a mixture of water and oil in fractured-porous reservoirs. In this paper, the task of selecting options for the appointment of injection and production wells in a zonal-inhomogeneous fractured-porous reservoir to the most complete extraction of oil from the reservoir. A study was also made of the influence of the intensity of water injection on the efficiency of oil recovery from a fractured-porous reservoir. The tasks were solved using our own three-dimensional two-phase hydrodynamic simulator based on the model of dual porosity-permeability.

1. Introduction

Naturally fractured reservoirs contain a significant share of the world's hydrocarbon reserves [1]. Despite the small relative volume of fractures, they have high conductivity and, therefore, can have a qualitative effect on the production of hydrocarbons [2]. From practice, there are cases of increased oil recovery with a decrease in the rate of water injection. This situation is typical of fractured-porous reservoirs. While under homogeneous reservoir conditions, a decrease in the rate of water injection does not lead to an increase in oil recovery. In addition, changing the direction of seepage fluids in fractured porous reservoirs can also lead to an increase in oil recovery.

The distribution of stresses in a real reservoir is inhomogeneous. Fractures in such a reservoir can be found both in the compaction zones and in decompaction zones. Therefore, aperture of a fracture (and hence its permeability) will depend on which of these zones it will be in. In addition, fracture aperture can be influenced by the change in pressure during non-stationary flooding. In connection with the above, the study of flow in fractured reservoirs is relevant. In particular, it is of interest to study the effect of the rate of water injection and changes in the direction of seepage fluids in a fractured-porous reservoir on the process of oil production.

The purpose of this work is to study the effect of the intensity of water injection on the efficiency of oil extraction from a fractured-porous reservoir, as well as solving the problem of selecting options for the appointment of injection and production wells in a zonal-inhomogeneous fractured-porous reservoir.

2. Mathematical model

To describe the motion of a weakly compressible fluid (oil and water) in a fractured-porous medium for each continuum, the following system of equations is used, including the equations for the conservation of phase masses and the general Darcy law [3]:

$$S_l\left(\phi\beta_l+\beta_c\right)\frac{\partial p}{\partial t}+\phi\frac{\partial S_l}{\partial t}+q_l=div(v_l)$$
$$(l=w,o)$$
$$v_l=-k\frac{k_{rl}}{\mu_l}grad(p) \qquad (1)$$
$$S_w+S_o=1$$

In the presented equations, the subscript l below indicates the values related to the l-th phase (l=w - water, l=o - oil), p is the pressure, S is the saturation, $\boldsymbol{v}$ is the velocity vector, μ_l is the viscosity, k_{rl} is the relative permeability, k is the absolute permeability, β is the fluid compressibility, ϕ is the porosity, β_c is the rock compressibility, and q is the intensity of internal sources (wells). It is assumed that the influence of gravitational and capillary forces is negligible. The flow of fluid between the continua is proportional to the difference in reservoir pressures in them [4, 5].

The dependence of the fracture permeability k_f on the pressure p was given in the form [6]:

$$k_f=k_f^0\left[1-\alpha\left(p_0-p\right)\right]^3, \qquad (2)$$

where k_f^0 is the fracture permeability at initial reservoir pressure p_0, α is the parameter of the fractured medium, which depends on the elastic properties and geometry of the fractures. The calculations were made by the method of control volume on an irregular grid.

3. The problem of selecting the options for the appointment of injection and production wells in the zonal-inhomogeneous fractured-porous reservoir

In practice, it is mainly necessary to work with reservoirs that are anisotropic in permeability. One example of such reservoirs can serve as a zonal-inhomogeneous fractured-porous reservoir. In such reservoirs it is possible to allocate a high-permeable zone and a low-permeable zone. The system of fractures can be located both in the low-permeable zone and in the high-permeable zone. For the most effective development of the field, it is necessary to solve the problem of well placement. Namely, to determine the production or injection wells are most advantageous to have in the area with fractures.

Let us consider the model of a zonal-inhomogeneous fractured-porous reservoir. Let the reservoir consist of two zones with different permeability. The system of fractures is contained in the zone with less permeability. Consider two options for well placement: injection wells are located in the zone with fractures (model 1), production wells are located in the zone with fractures (model 2). The models of a zonal-inhomogeneous fractured-porous reservoir are shown in figure 1.

The parameters of the reservoir, fluids and modes of operation of wells took the following values: the size of the reservoir - 1000×1000×5 m; porous reservoir porosity - 0.2; fractured reservoir porosity - 0.05; initial/residual water saturation - 0.2; residual oil saturation - 0.2; the pore permeability is 100 mD and 400 mD; the initial permeability of fractures is 500 mD; the initial reservoir pressure is 100 atm; water/oil viscosity ratio - 1/10; bottomhole pressure (production wells) - 50 atm; bottomhole pressure (injection wells) - 200 atm; the compressibility of water, oil and porous media is $5\cdot10^{-5}$, $5\cdot10^{-4}$ and $3\cdot10^{-5}$ 1/atm, respectively. The calculation is carried out until reaching a water cut of production wells of 0.98. The types of relative phase permeabilities for pores and fractures are shown in figure 2. Figure 3 shows the dependences of oil recovery factor and water cut on time for models 1 and 2

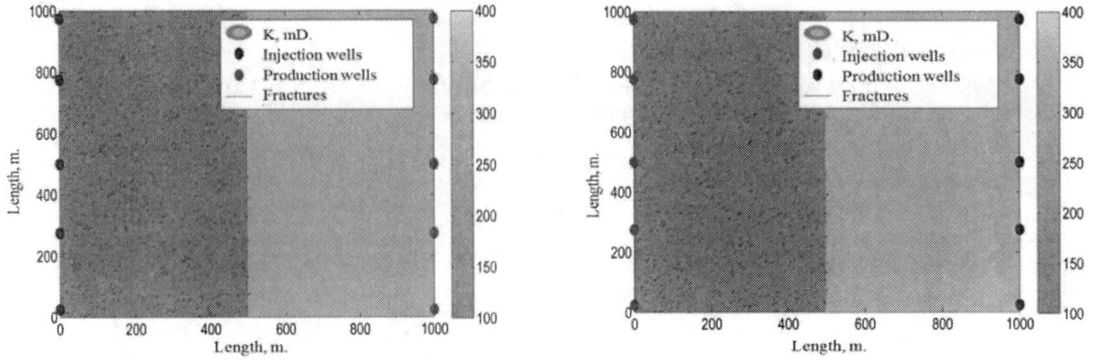

(a) (b)

Figure 1. Zonal-inhomogeneous fractured-porous reservoir models. Injection wells are located in the fractured zone (a). Production wells are located in the fractured zone (b).

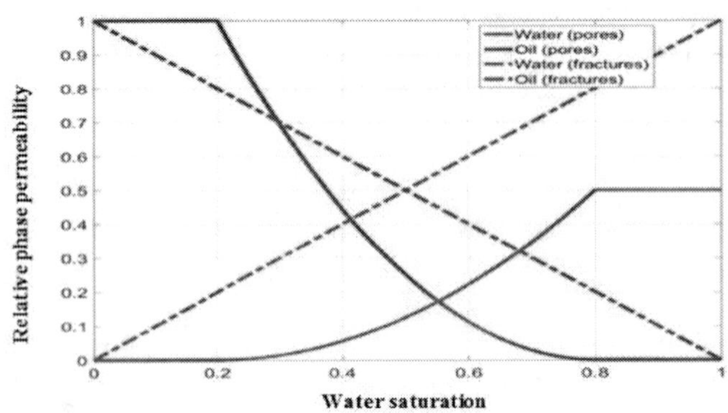

Figure 2. The types of relative phase permeabilities.

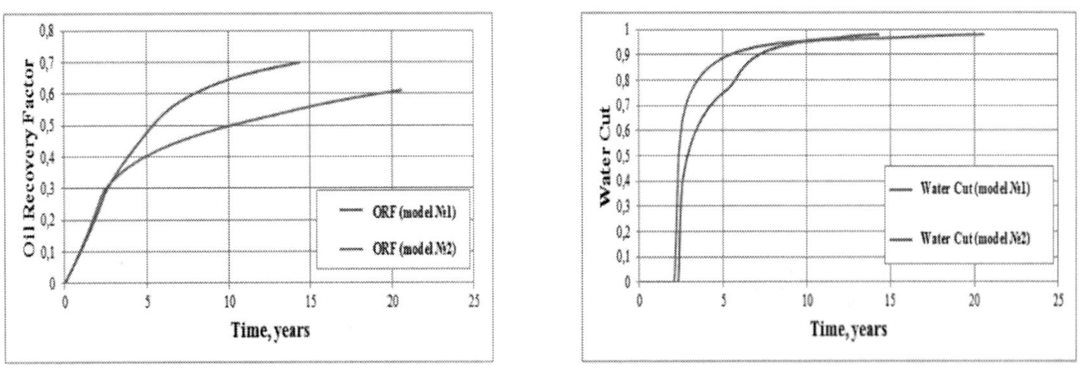

(a) (b)

Figure 3. Typical dependencies of oil recovery factor (a) and water cut (b) on time for models 1-2.

From figure 3a it can be seen that the final ORF for model 2 is higher than for model 1. In addition, the maximum water cut of producing wells is achieved faster than for the model 1. From figure 3b it can be seen that for model 1 the water cut profile is similar to the water cut profile of a medium with one porosity and one permeability. While for model 2, the water cut profile looks similar to the water cut profile for a medium with double porosity and double permeability. Thus, the model 2 is more effective – a model with producing wells located in a low-permeable zone. To explain the results, refer to figures 4a, 4b. They show the flow of oil and water between the pores and fractures, respectively.

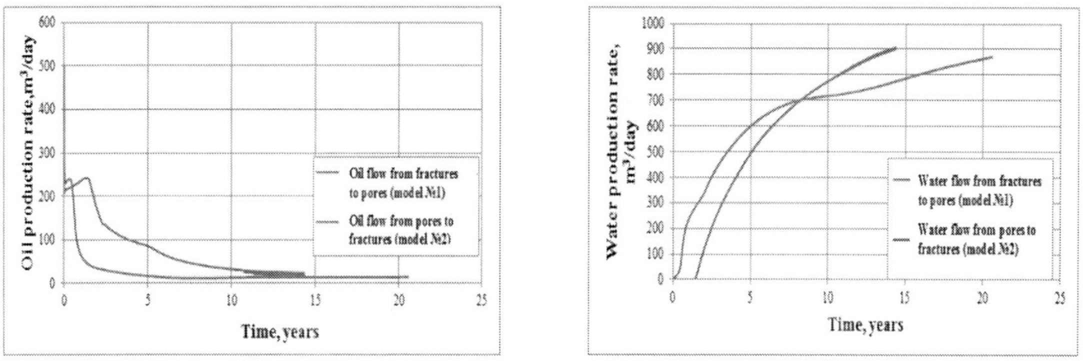

(a) (b)

Figure 4. The intensity of the exchange of water and oil between fractures and pores for models 1-2.

From figure 4a and 4b, it can be seen that for model 2 most of the development time was the flow of oil from the pores into the fractures. The result is that oil quickly reached production wells through highly conductive fractures. In model 1, oil, on the contrary, flowed from fractures into the pores. As a result, oil slowly reached production wells. Also a disadvantage of the model 1 is the following circumstance. Water injection in this model occurs in the fractured zone of the reservoir, because of which the water quickly breaks through the system of fractures into the highly permeable zone of the reservoir, increasing the water cut of producing wells. In model 2, on the contrary, the system of fractures is not prone to water flooding directly from the wells and most of the development time through it is oil is filtered.

These circumstances explain the advantage of model 2 over model 1. Thus, it is more advantageous to place production wells in a fractured zone in a zonal-inhomogeneous fractured-porous reservoir.

4. Investigation of the influence of the intensity of water injection on the efficiency of oil recovery from a fractured-porous reservoir

The study was conducted using a synthetic reservoir model developed by the row scheme of arrangement of wells. In view of the use of the double porosity and dual permeability model, we consider that the all reservoir is covered by a connected network of small fractures. When the reservoir pressure p is less than the initial pressure p_0, the fractures are closed, and their permeability becomes equal to 0.1 of the pore permeability. When the reservoir pressure p is greater than the initial pressure p_0, the permeability of fractures is determined by formula (2). The parameters of the reservoir, fluids and modes of operation of wells took the following values: the dimensions of the reservoir - $1000\times1000\times10$ m; porosity of the pore collector - 0.16; fractured reservoir porosity - 0.04; initial/residual water saturation - 0.2; residual oil saturation - 0.2; the pore permeability is 10^3 mD; the initial permeability of fractures is 10^4 mD; the initial reservoir pressure is 35 atm; water/oil viscosity ratio - 1/300; bottomhole pressure (production wells) - 30 atm; bottomhole pressure (injection wells) - 60 and 40 atm; the compressibility of water, oil and porous media is $5\cdot10^{-5}$, $5\cdot10^{-4}$ and $3\cdot10^{-5}$ 1/atm, respectively. The calculation is carried out until reaching a water cut of production wells of 0.98. The types of relative phase permeabilities for pores and fractures are shown in figure 2.

In figure 5 schematically shows a synthetic model of an oil reservoir developed using a row scheme of arrangement of wells.

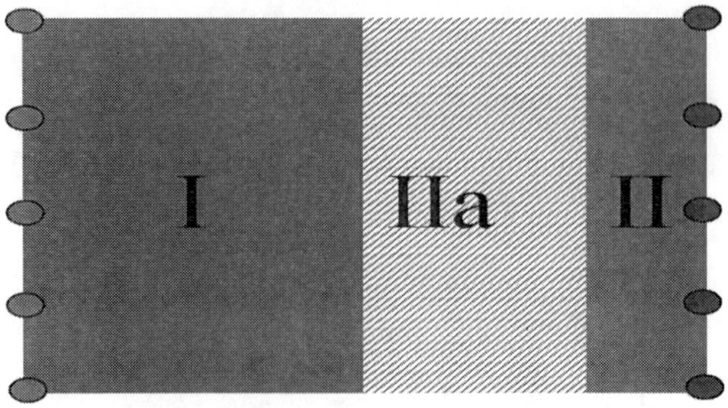

Figure 5. Synthetic model of oil reservoir.

In the case of operation of wells with constant bottomhole pressures (case 1), 2 zones are conventionally formed in the reservoir. The first zone (I) is the area in which the reservoir pressure p is higher than the initial reservoir pressure p_0. Fractures are revealed and their permeability is according to the formula (2). In this area, filtration is mainly carried out by fractures. The second zone (II) is the area in which the pressure is less than the initial reservoir, the fractures are closed. In this area, filtering is carried out both by pores and fractures. For the second version of the calculation (case 2) after 200 days after the start of the calculation, the well operation mode is changed (the bottomhole pressure in the injection wells decreases from 60 atm to 40 atm). As a result, an additional part of the porous reservoir (IIa) is involved in the development.

In figure 6 shows the dependences of the ORF and water cut on the pumped pore volume for both calculation cases.

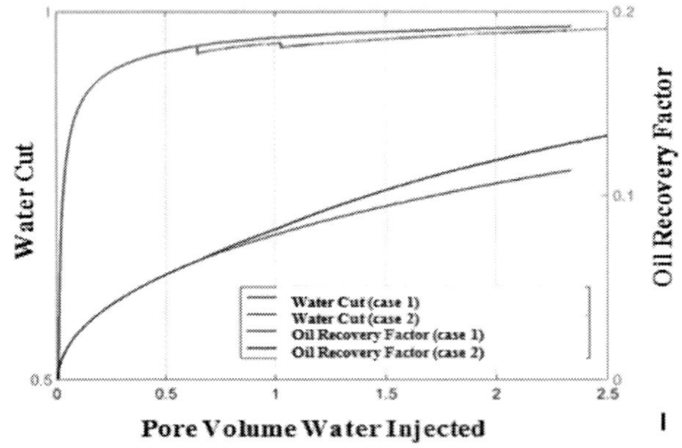

Figure 6. Typical dependencies of oil recovery factor on pore volume water injected for cases 1-2.

From figure 6, it can be seen that case 1, which implies a reduction in bottomhole pressure at injection wells, has a finite oil recovery rate higher than case 2 with a constant mode of operation of injection wells. This is explained by the fact that after reducing the bottomhole pressure at injection wells, an additional part of the porous reservoir (area IIa) is involved in the development, which helps reduce the water cut at the production wells (green and purple curves). Thus, the reduction of bottomhole pressure at injection wells, when developing a fractured porous reservoir, can lead to an increase in oil recovery.

5. Conclusion

On the example of a synthetic model of an oil field with a zonal-inhomogeneous fractured-porous type of reservoir, the problem of selecting options for the purpose of injection and production wells was solved. It is established that it is more profitable, in terms of oil recovery, to have producing wells in the fractured zone of the reservoir. This arrangement of wells allows not only to increase the final oil recovery factor, but also to reduce the time required to achieve the maximum water cut of producing wells. The result is obtained using a double porosity and double permeability model.

On the example of a synthetic model of an oil field with a fractured-porous type of reservoir, a positive effect from the point of view of oil recovery from a decrease in the rate of water injection was demonstrated. It is established that the decrease in the rate of water injection can reduce the water cut of products and increase the final oil recovery factor. The result is obtained using a double porosity and double permeability model. To assess the overall efficiency of reducing the rate of water injection, it is necessary to use economic criteria, since the increase in the volume of oil produced increases the time required to achieve the maximum water cut of producing wells.

Acknowledgments

The research was supported by Russian Science Foundation (project № 18-19-00049).

References

[1] Saidi A M 1983 *Society of Petroleum Engineers* 361
[2] Weber K J 1986 *Reservoir Characterization* 487
[3] Aziz Kh and Settari A 1979 *Petroleum reservoir simulation* (London: Applied Science Publishers)
[4] Warren J E and Root P J 1963 *Society Petroleum Engineers* **3** 245
[5] Kazemi H, Merrill L S Jr and Zeman P R 1976 *Society Petroleum Engineers* **16** 317
[6] Basniev K S, Kochina N I and Maximov M V 1993 *Underground Hydromechanics* (Moscow: Nedra)

Journal of Physics: Conference Series **1404** (2019) 012040 doi:10.1088/1742-6596/1404/1/012040

On particle incipient motion from the sedimentation layer in shear flow

A V Seryakov, O B Bocharov and Ya S Ignatenko

Novosibirsk Technology Center, Baker Hughes Russia, 4A Kutateladze Street, 630090, Novosibirsk, Russia

alexander.seryakov@bakerhughes.com

Abstract. This paper presents the results of numerical modelling for a laminar flow around a spherical particle located near a sedimentation layer. The particle being at the minimum distance from the rough layer is affected by a fluid shear flow with Reynolds numbers varying from 2 to 200. The considered fluids included Newtonian and power law ones. Depending on a shear Reynolds number, the critical Shields numbers have been determined to describe the particle's incipient motion. The obtained parameters have been compared against those for a smooth surface to demonstrate that in some cases the differences in critical stress coefficients may reach up to 70%.

1. Introduction

Cuttings transport has been one of the topical issues for drilling of inclined and horizontal wells to penetrate hydrocarbon reservoirs. Ineffective clearing of the inner annulus from cuttings particles results in ring channel clogging, oil seal, string sticking and increased bottom-hole pressure [1]. To estimate the movability of the layer of dispersed debris in a well, one uses the dependencies obtained from the analysis of a simplified model in which a spherical cuttings particle starts moving over a motionless sedimentation layer. The main integral characteristics of this motion include the drag and lift forces, and torque that depends on the Reynolds number. These characteristics are used as input parameters in different drilling support software packages to determine the mud weight and rate of used drilling fluid, penetration rate, etc. [2].

The theoretical aspect of the issue to be the problem of particle-bearing flow that determines the parameters of particle incipient motion has been considered in a number of studies [3]–[5]. All of them describing the incipient motion of a spherical particle near a smooth wall.

However, since in borehole conditions cuttings transport occurs over a rough sedimentation layer, a case of a particle-bearing flow over such a rough boundary is of sufficient interest to study. Experiments to study the flow around granular pack located on the sedimentation layer has been described in [6],[7]. The increased computation capacities during the last decades make it possible to perform 3D numerical modeling of a particle hydrodynamics over a rough sedimentation layer. Thus in [8], Lee and Balachandar present their results for a shear Newtonian particle-bearing flow over a layer of hemispheres whose radius is similar to the one of the particle. In our study, we used a similar problem statement and applied it to calculate particle incipient motion for power law fluid.

2. Problem statement

We consider a particle of diameter d that is placed over a surface composed of hemispheres of the same diameter (figure 1, a). The hemispheres are honeycombed on a substrate, and the particle is placed over a gap formed by the three hemispheres colored in green (see figure 1, b). Due to the preliminary calculations demonstrating that the flow of viscous, incompressible fluid over a rough surface is completely formed behind the fourth raw of hemispheres; the particle has been placed just there.

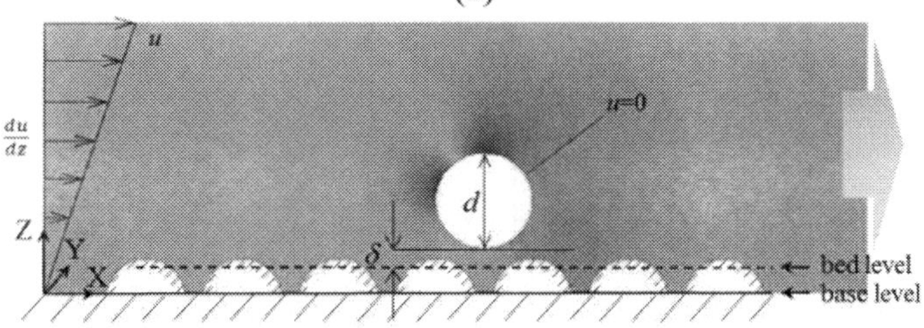

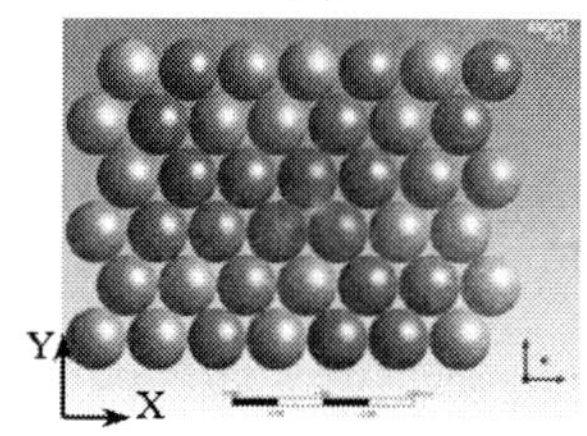

Figure 1. Problem statement for a shear flow bearing a particle over a rough surface (a). The bed level corresponds to the value δ=0, when the particle touches the two bottom hemispheres marked in green (b).

The spherical particle and sedimentation layer are flown by a shear flow moving along the x-axis. The sphere remains steady. The velocity profile $\vec{u}$ is set as input data and is characterized by its shear parameter $G = \partial u_x / \partial z$. The distance from the particle to the rough surface δ is determined based on the condition that at δ=0, the particle touches the three bottom hemispheres.

The system of hydrodynamic equations for viscous incompressible fluid in dimensionless form can be written as:

$$\begin{cases} \nabla \cdot \vec{u} = 0 \\ \dfrac{\partial \vec{u}}{\partial t} + (\vec{u} \cdot \nabla)\vec{u} = -\nabla p + \dfrac{1}{\mathrm{Re}_{\mathrm{SH}}} \nabla (2\mu \mathbf{D}) + \vec{f} \end{cases} \qquad (1)$$

where $\mathbf{D} = 1/2\left[\nabla \vec{u} + (\nabla \vec{u})^T\right]$ denotes the deformation rate tensor; $\mu = \mu(\mathbf{D})$ is the dimensionless viscosity being the function of deformation rate tensor components for non-Newtonian fluids; $\vec{f}$ is the dimensionless density of mass forces. The shear Reynolds number $\mathrm{Re}_{\mathrm{SH}}$ for power law fluid is determined as $\mathrm{Re}_{\mathrm{SH}} = \rho_f G^{2-n} d^2 / k$, where n denotes the power; ρ_f is the fluid density, d is the diameter of the particle, k is the fluid consistency index.

In our study, a laminar flow of Newtonian and power law (pseudo-plastic) fluids was considered, whose $\mathrm{Re}_{\mathrm{SH}}$ varies from 2 to 200. The geometry and mesh were constructed in ANSYS Meshing, then the prepared setup was used for calculations in the ANSYS Fluent and OpenFOAM packages.

The main integral parameters characterizing the way the flow affected the spherical particle such as the drag force F_D, lift force F_L and torque M, were calculated using corresponding surface integrals of a stress tensor over the particle. The coefficients of the drag (C_D), lift (C_L) forces and of the torque (C_M) were determined through normalization by the characteristic force F_0 affecting the particle's cross-section: $F_0 = 0.5 \cdot u_p^2 \rho_f \pi (d/2)^2$, where u_p denotes the "undisturbed" flow velocity in the particle's center that is obtained from a shear flow above the rough wall without a particle. The coefficients C_D, C_L, C_M are calculated following the formula: $C_D = F_D / F_0$, $C_L = F_L / F_0$, $C_M = 2M / (d \cdot F_0)$.

Since the main objective of our study was to determine the particle incipient motion parameters, the particle's position was considered to be as close as possible to the rough wall. Using ANSYS Meshing it became possible to make a mesh allowing the distance δ=0.005d between the particle and hemispheres. To test the correctness of the selected modeling and meshing parameters, a case of shear Newtonian flow over rough sedimentary layer with particle was calculated so that the obtained integral parameters C_D and C_L could be compared against those calculated in [8]. A satisfactory comparison was obtained when using elements of 0.005d in size on the particle's surface, inflation layers of 0.005d in thickness and cells of 0.05d in height on the hemisphere's surface. In this case, the characteristic size of the mesh was about $6 \cdot 10^6$ tetrahedral elements.

3. Newtonian fluid

The drilling support packages to calculate cuttings transport parameters, as a rule, use correlations for the forces and torques obtained for a flow around particle near smooth wall. For that reason, the effect produced by the wall's roughness on particle incipient motion in compare to the smooth wall is a matter of great interest and relevance. Newtonian fluid being the simplest one in terms of rheology and was chosen on the first step. For a flow around a sphere over a smooth wall, the modeling results for C_D, C_L, and C_M were published in [5].

The calculations for rough surface demonstrated that for the range of $\mathrm{Re}_{\mathrm{SH}}$ from 2 to 200 both the drag and lift forces increased. The increase of F_D was due to a wider area of lower pressure behind the particle. The area widened because the particle's lower part was placed in the gap between the sedimentation layer's hemispheres. The gap became a place for a stagnation region to form, so its outflow restricted by the channel between the particle and the hemisphere made the pressure behind the particle lower (figure 2).

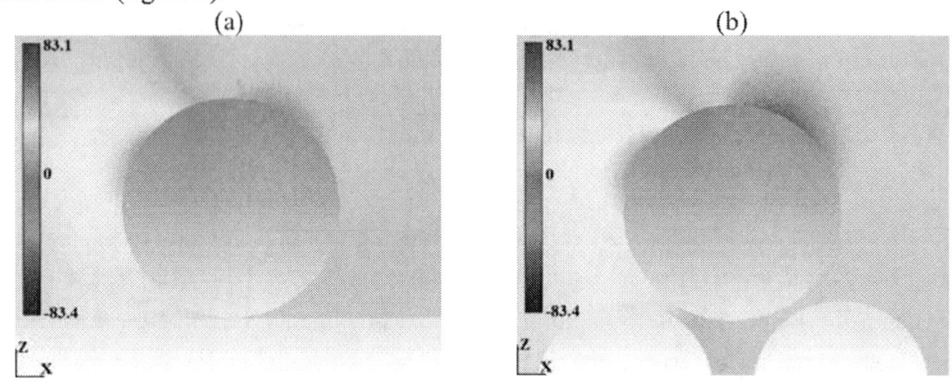

Figure 2. Distribution of the dimensionless hydrodynamic pressure p in a laminar shear flow (Newtonian fluid) around a particle, $\mathrm{Re}_{\mathrm{SH}}$ =10. The particle is placed over a smooth (a) and rough (b) surfaces.

The increased lift force F_L can be explained by forming an area of excessive pressure before the gap and that this pressure is higher than that near a smooth wall. Moreover, reducing the pressure in the suction area, whose minimum is the particle's upper boundary behind the flow also contributes to the F_L increase. The torque, in this case, increases as well because the flow does not go around the whole particle: the reduced flow in the gap doesn't counterwork the torque created at the particle's upper boundary.

As for the force coefficients, things are not so clear, because the profile of undisturbed velocity is different from a linear one: the velocity u_p over a rough wall is higher than that over a smooth one, which has also been confirmed in [8]. The dimensional specific force depends on u_p^2, so in Newtonian fluid C_D reduces, C_L increases, and C_M slightly changes if compared to the smooth wall case. In total, C_D changes up to 5%, C_L – up to 17%, and C_M – up to 7%. Since coefficients demonstrate certain invariant behavior when modeled for the rough wall, in Newtonian case the smooth wall approximations can be used to determine the forces affecting a particle in a flow.

4. Power law fluid

Analysis of the calculations for power law fluid demonstrates that the tendency to increase the torque and drag and lift forces is preserved.

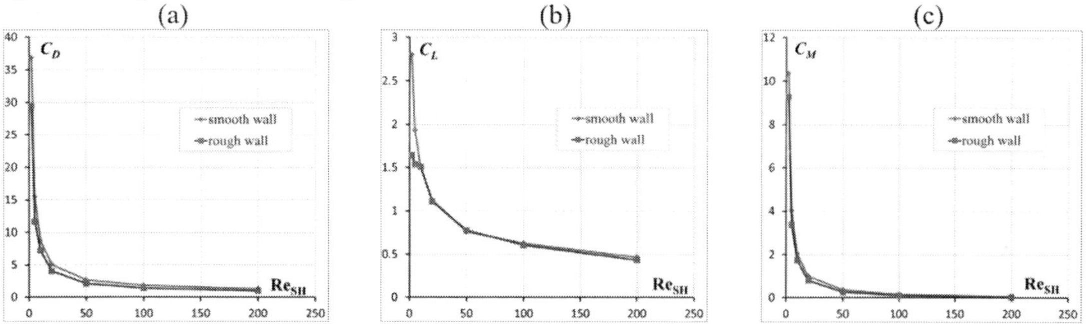

Figure 3. Drag C_D (a), lift C_L (b) and torque C_M (c) coefficients and their behaviour depending on Re_{SH} for a particle over smooth and rough surfaces in power law fluid (n=0.6).

At the same time, all three coefficients get sufficiently smaller than those calculated for the smooth wall, mainly because of the increased difference in undisturbed velocity between the flows over smooth and rough surfaces. Figure 3 demonstrates the coefficient curves calculated for power law fluid (n=0.6). As you can see, the most significant difference in the lift force (up to 40%) was registered for small Reynolds numbers (2-10). The maximum deviations of the other coefficients for the whole Re_{SH} range comprise 25 % for C_D and 35% for C_M.

5. Incipient motion conditions

The conditions that initiate the particle's motion along the sedimentation layer can be determined if we write down the balance of moments relative to point A (figure 4) where the particle touches a hemisphere being a part of the sedimentation layer [9],[10].

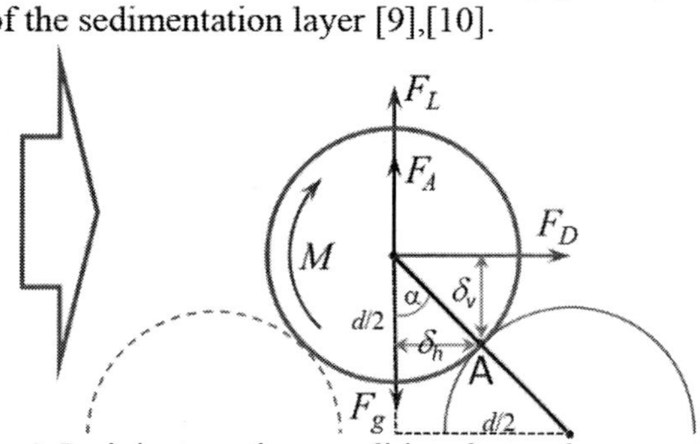

Figure 4. Incipient motion condition that makes a particle move along a rough sedimentary layer.

When the sum of these moments exceeds zero, the particle starts roll out from the gap between the hemispheres: $M + F_D\delta_v + F_L\delta_h - F_g\delta_h + F_A\delta_h \ge 0$, where $F_g = \rho_p V_p g$ denotes the gravity, ρ_p is the particle's density, $V_p = 4/3 \cdot \pi \cdot (d/2)^3$, $F_A = \rho_f V_p g$ is buoyancy force. In the case of dense layer packing and if the particle and hemispheres are of the same diameter, the angle α will be close to 60° while $\delta_v = d/4$, $\delta_h = d\sqrt{3}/4$. Having taken that into account and divided by F_0, we obtain:

$$\frac{\sqrt{3}}{16}\rho_f d^2 G^2\left(2C_M + C_D + \sqrt{3}C_L\right) \geq gd\left(\rho_p - \rho_f\right). \tag{2}$$

Now, let consider another kind of incipient motion when a particle is taken away from a rough surface by a flow. In this case, the inequality for the forces is written as $F_L - F_g + F_A \geq 0$.

After the transformations analogous to those in (2) we obtain:

$$C_L \geq \frac{16}{3}\frac{gd\left(\rho_p - \rho_f\right)}{\rho_f G^2 d^2}. \tag{3}$$

When analyzing the incipient motion conditions, it is convenient to consider the so-called Shields number τ_B that represents the ratio of the characteristic shear stress on the wall τ_{wall} and the difference between the buoyancy force F_A and the gravity F_g [9],[10]. Having defined the characteristic shear stress as G and the viscosity as μ, the following expression for τ_B can be obtained:

$$\tau_B = \frac{G\mu}{gd\left(\rho_p - \rho_f\right)}. \tag{4}$$

When transforming expressions (2), (3) for the Shields number, we obtain that the particle's rolling motion on a rough surface starts when $\tau_B \geq \tau_R$, and the particle is detached from surface when $\tau_B \geq \tau_L$, where

$$\tau_R = \frac{16}{\sqrt{3}\mathrm{Re}_{\mathrm{SH}}\left(2C_M + C_D + \sqrt{3}C_L\right)},\quad \tau_L = \frac{16}{3C_L\mathrm{Re}_{\mathrm{SH}}}. \tag{5}$$

Comparing the smooth and rough wall cases for Newtonian fluid it should be noted that the parameter τ_R changes insignificantly (up to 3-5 %) in the range of $\mathrm{Re}_{\mathrm{SH}}$ from 2 to 150, and the changes become significant (up to 17%) only at $\mathrm{Re}_{\mathrm{SH}}$=200. At the same time τ_L reduces by 9-13% for the whole range of Reynolds numbers.

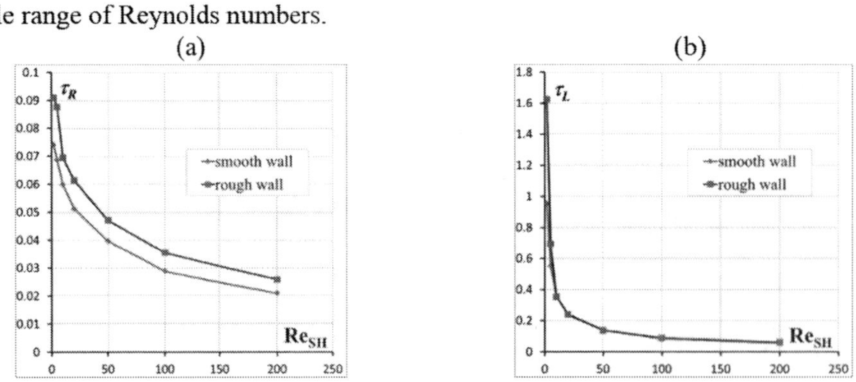

Figure 5. Comparison of the critical Shields numbers that produce the rolling motion τ_R (a) and detaching τ_L (b) of a particle from smooth and rough surfaces in a shear flow of power law fluid (n=0.6).

The critical Shields numbers for smooth and rough surfaces in a flow of power law fluid (n=0.6) can be seen in figure 5. Unlike the Newtonian case, the differences between the smooth and rough walls here are much more pronounced. For example, the parameter τ_R is higher for the rough than for the smooth wall, the difference comprises up to 27 %. Changes of the parameter τ_L reach their maximum at Re_{SH}=2 and comprise 70%. As the Reynolds number increases, the difference in τ_L between the two kinds of walls becomes smaller.

The way the coefficients τ_R, τ_L change reflects the fact that for non-Newtonian fluid it is far more difficult to make a particle move from a rough wall than from a smooth one. At the same time, the particles take-off or their mixing occurs when the parameter τ_B increases. For a power law fluid, this parameter can be written as $\tau_B = \rho_f G^2 d / [\mathrm{Re}_{SH} \cdot g(\rho_p - \rho_f)]$.

With that in mind, the particle's motion can be induce first of all by increasing of G with a simultaneous changing of the fluid's consistency k (to meet the same values of Re_{SH}). The second way to increase τ_B is the reduction of the particles density.

6. Conclusion

The obtained results can be summarized in the following highlights. Numerical simulation has been performed that describes a shear flow of Newtonian and non-Newtonian fluid around a particle on a rough sedimentary layer for Reynolds numbers ranging from 2 to 200. The key integral characteristics have determined from the simulation describing the way the particle behaves in a flow. In the engineering formulas for Newtonian fluids, devised to determine the force parameters and conditions of a particle incipient motion, approximations for a smooth wall can be used. It has been demonstrated that for non-Newtonian power law fluid near a rough wall, the drag and lift force coefficients, as well as the torque coefficient, significantly differ from those calculated for a smooth wall. For a pseudo-plastic fluid (n=0.6), critical shear stresses that characterize a particle incipient motion from a rough surface have been determined and their difference from the smooth wall case has been demonstrated.

The performed study allows one to specify the movability conditions for a cuttings layer in a pipe, which will facilitate solving the problem of annulus clearing while drilling.

Acknowledgements

The authors would like to thank Baker Hughes a GE company for the permission to publish the results of modelling.

References

[1] Nazari T, Hareland G and Azar J J 2010 *Proc. SPE Western Regional Meeting 27-29 May 2010* (Anaheim, California, US)
[2] Makovey N 1986 *Gidravlika bureniya* (Moscow: Nedra) 536 pp
[3] Krishnan G P and Leighton D T 1995 *J. Phys. Fluids* **7** 2538-45
[4] Zeng L, Najjar F, Balachandar S and Fischer P 2009 *J. Phys. of Fluids* **21** 033302
[5] Ignatenko Ya, Bocharov O and May R 2017 *OMAE2017* Proc. 36th Int. Conf. Ocean, Offshore Arctic Eng., June 25–30, 2017, Trondheim, Norway (New York: ASME) 61131
[6] Davis T R and Samad M F 1978 *J. Hydraulic Division* **104** 1171-82
[7] Watters G Z and Rao M V 1971 *J. Hydraulic Division* **97** 421-39
[8] Lee H and Balachandar S 2017 *Int. J. Multiph. Flow* **88** 116-132
[9] Lee H and Balachandar S 2012 *J. Geophysical Research* **117** F01026
[10] Ignatenko Ya, Bocharov O and May R 2017 *J. Phys.: Conference Series* **894** 012037

Numerical simulation of nucleation and solid phase growth in a nanomodified binary alloy

V G Shchukin and V N Popov

Khristianovich Institute of Theoretical and Applied Mechanics SB RAS,
4/1 Institutskaya str., Novosibirsk, 630090, Russia

shchukin@itam.nsc.ru

Abstract. A mathematical model of solidification of a binary alloy with eutectic-type phase diagram modified by refractory nanoscale particles as nucleation centers of the crystalline phase is proposed. Thermodynamic processes in the melt and solid phase, heterogeneous nucleation and growth of particles of the main (α) phase of the crystallizing material are described. The crystal phase origin occurs on the surface of nanoscale particles. The volume of the solid phase grown on separate crystallization centers determines the character of the grain structure in the solidified alloy. On the example of a binary alloy based on iron (Fe-C) numerical simulation of melt solidification in a cylindrical casting is carried out. The features of the kinetics of heterogeneous nucleation and crystallization in the melt were considered. It is determined that the undercooling magnitudes, the conditions of nucleation, the rate of crystallization and solidification time vary significantly inside the casting. It is established that as the melt cools, volume-sequential crystallization takes place. The area with the finest structure of the solidified metal is located near the wall of the mold.

1. Introduction

Among binary metal alloys whose phase diagram has eutectic, Fe-C carbon alloys are of particular interest being the main structural materials in almost all branches of modern production. However, at present, there are still urgent problems to reduce defects and increase the strength of steel castings. One way to solve these problems can be grinding the structure of the hardened metal by adding nanoscale refractory particles to the melt. Heterogeneous nucleation is possible on particles, depending on the size, shape, and wettability of their surface, as well as the consistency of their atomic lattices with the material of the nucleus. The use of this technology allows of increasing the number of crystallization centers and improving the performance properties of the metal [1-4]. Specially prepared particles of materials SiC, TiC, TiN, etc. can serve as modifiers.

In this paper, the process of solidification of a two-component Fe-C alloy modified by refractory nanoscale particles in ceramic mold is considered. The mathematical model describing thermophysical and thermodynamic phenomena in the alloy in liquid and solid state, as well as heterogeneous nucleation and crystallization of the main α-phase from the melt is formulated. Kinetics of solid phase growth and temperature regimes in solidifying casting are determined. The numerical simulation results are used to assess the adequacy of the crystallization model in the presence of nanoscale particles in a binary iron alloy.

2. Mathematical model

The solidification process of Fe-C alloy in cylindrical mold is considered. Due to the symmetry of the object, it is sufficient to limit the modeling process only to half of the central longitudinal section of the mold (figure 1).

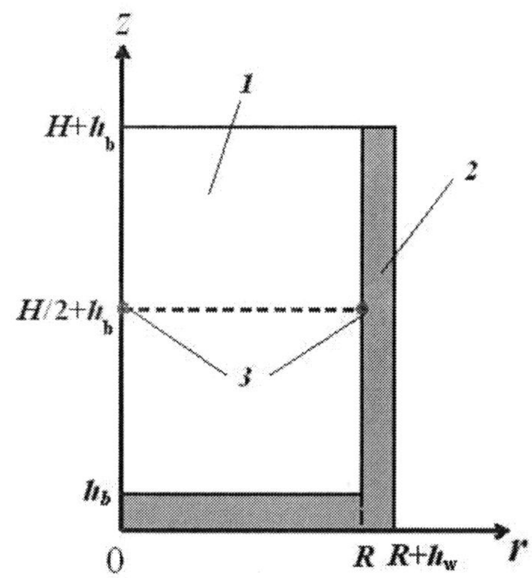

Figure 1. Schematic representation of the mold with the melt and the calculated area: *1* - melt, *2* - insulating wall and bottom, *3* - control points at half height from the mold bottom.

The height of the casting is equal to H, its radius is R, the thickness of the thermal insulation pad in the bottom of the mold is h_b, and the thickness of the side wall is h_w.

The melt is uniformly modified in the volume by refractory nanoscale particles with a characteristic size of l_p, their mass content of m_p does not exceed 0.05% and, therefore, their influence on the thermophysical characteristics of the melt can be neglected. The thermophysical properties of the metal in the liquid and solid phases are assumed to be constant and equal to the average values in the considered temperature ranges.

We introduce indices i=1 and i=2 for the liquid and solid phases of the alloy, respectively, and i=3 for the mold material. Then, taking into account the above assumptions, the heat transfer in the alloy is described by the following equation in the cylindrical coordinate system (r, z):

$$c_e\rho_e\frac{\partial T}{\partial t}=\frac{1}{r}\frac{\partial}{\partial r}\left(r\lambda_e\frac{\partial T}{\partial r}\right)+\frac{\partial}{\partial z}\left(\lambda_e\frac{\partial T}{\partial z}\right)+\rho_2\kappa\frac{\partial f_s}{\partial t},\quad 0\le r\le R,\ h_b\le z\le h_b+H, \tag{1}$$

where the local values of the coefficients are: : $c_e=c_1$, $\rho_e=\rho_1$, $\lambda_e=\lambda_1$ when $f_s=0$, $c_e=c_2$, $\rho_e=\rho_2$, $\lambda_e=\lambda_2$, when $f_s=1$ and $c_e=c_1(1-f_s)+c_2f_s$, $\rho_e=\rho_1(1-f_s)+\rho_2f_s$, $\lambda_e=\lambda_1(1-f_s)+\lambda_2f_s$ when $0<f_s<1$. Here, f_s is the fraction of the solid phase in the melt, λ, c, ρ are the thermal conductivity, heat capacity, and density, respectively, κ is the specific heat of melting.

The temperature field in the mold wall is described by equation

$$c_3\rho_3\frac{\partial T}{\partial t}=\lambda_3\left(\frac{1}{r}\frac{\partial}{\partial r}r\frac{\partial T}{\partial r}+\frac{\partial^2 T}{\partial z^2}\right),\ 0\le r\le R+h_w, 0\le z\le h_b;\ R\le r\le R+h_w, h_b\le z\le h_b+H. \tag{2}$$

Heat exchange occurs between the outer surfaces of the mold, the free surface of the melt and the environment, the heat transfer coefficients are constant. On the contact surfaces of the melt with the form, the thermal resistance, determined by the results of experiments, is taken into account [7]. The melt and mold temperatures at the initial moment are assumed to be the same and equal to a value slightly higher than the crystallization temperature of the pure metal. The model assumes that the wetting refractory nanoscale particles are uniformly distributed over the melt volume. It is shown in [8] that homogeneous nucleation in the process of solidification of such a melt is practically not observed, and the nucleation and growth of the crystalline phase occurs only on the seeds.

According to [5], under the assumption that nanoscale particles have the shape of a cube, the faces of which are characterized by the wetting angle θ, the expression for the rate of nucleation of crystals of the α-phase (i.e., iron-based solid solution) has the form:

$$I = n_p \frac{2\pi^2 D_0}{l_a^4 l_c^2} R_c^4 (1-\cos\theta)\sin^2\theta \exp[-(E+\Delta G^*)/(k_B T)] . \quad (3)$$

Here, $n_p = m_p \rho_1 / (100 \rho_p l_p^3)$ is the number of nanoscale particles per unit volume of the melt, ρ_p is the density of the particle's material, l_p is the length of the edge of nanoparticle, l_a, l_c are the interatomic distance in the melt and in the nanoparticle, respectively, k_B is the Boltzmann constant, D_0 is the pre-exponent in Arrhenius's law for self-diffusion coefficient of iron in the melt, E is the activation energy of this process. The Gibbs energy ΔG^* of formation of a critical nucleus is equal to

$$\Delta G^* = \pi \sigma_{12}^\infty R_0 (R_0 - 6\delta)(1-\cos\theta)^2 (2+\cos\theta)/3 ,$$

where R_c is the critical radius of the nucleus, determined by relation $R_c = R_0 - \delta$ [6], δ is the Tolman's parameter [9], $R_0 = 2\sigma_{12}^\infty T_{L0} / (\kappa \rho_1 \Delta T_0)$, σ_{12}^∞ is the surface tension at the interface of nucleus-melt, $\Delta T_0 = T_{L0} - T$, T_{L0} is the initial liquidus temperature.

The number of crystals of the α-phase formed during the cooling of the liquid metal, starting with the time moment t_{L0}, when the temperature fells to the value T_{L0}, is equal to

$$N(r,z,t) = \int_{t_{L0}}^{t} I(r,z,\tau)[1 - f_s(r,z,\tau)] d\tau ,$$

where f_s is the fraction of the growing solid phase determined by the relations [10]:

$$f_s(r,z,t) = 1 - \exp(-NV_s), \quad V_s(r,z,t) = (4\pi/3)(R^3 - R_{e0}^3) ,$$

where V_s is the volume of the solid phase growing on a single nanoparticle. Assuming that the solid phase growth obeys the normal mechanism and is determined by the linear dependence of the growth rate on magnitude of undercooling, $\partial R/\partial t = K_\alpha \Delta T$ [1, 11], where R is the radius of the crystal, K_α is the physical constant, $\Delta T = T_L - T$, T_L is the current liquidus temperature, we obtain

$$R(r,z,t) = R_{e0} + \int_{t_{L0}}^{t} K_\alpha \Delta T d\tau ,$$

where $R_{e0} = (3/2\pi)^{1/2} l_p$ is the radius of a spherical particle having a surface area equal to the surface of a cubic particle. To estimate the kinetic constant K_α, it is possible to use the formula [12]:

$$K_\alpha = (D \Delta H_\alpha) / (l_a k_B T_A^2) ,$$

where ΔH_α is the enthalpy of melting per atom, the diffusion coefficient in the liquid metal D is determined by the Arrhenius equation $D = D_0 \exp[-E/(k_B T)]$. Undercooling of a melt with a variable impurity content $\Delta T = T_L - T$ can be represented as

$$\Delta T = T_A - \beta C_0 / (1 - f_s)^{1-k} - T .$$

Here, the liquidus temperature T_L is approximated by the linear dependence on the concentration of the dissolved component (carbon), T_A is the melting point of the pure solvent metal (Fe), β is the modulus of the slope of the liquidus line in the Fe–C state diagram. Concentration of the dissolved component is determined by the non-equilibrium lever rule $C=C_0/(1-f_s)^{1-k}$ [11], where C_0 is the initial concentration, k is the impurity distribution coefficient.

Note that the crystallization of the α-phase from the melt occurs in the temperature range $T_{L0}\geq T\geq T_E$, where $T_{L0}=T_A-\beta C_0$, T_E is the eutectic temperature. The solidification rate $\partial f_s/\partial t$ is determined according to the equation:

$$\partial f_s/\partial t=[V_s I(1-f_s)+4\pi N K_\alpha R^2(T_L-T)](1-f_s). \quad (4)$$

The fraction of the solid phase when cooled to the eutectic temperature ($T=T_E$) is equal to $f_{s\alpha}$. After cooling the metal below the eutectic temperature, the eutectic phase solidifies. Since in binary iron-based alloys the eutectic concentration is much higher than the impurity (carbon) content and, consequently, the value of $f_{s\alpha}$ is close to one, in this model a stepwise increase to one of the fraction of the solid phase is assumed upon cooling below the T_E point.

3. Results of numerical experiments

In order to test the model, numerical simulation of solidification of Fe-0.49%C alloy in cylindrical form with introduced refractory TiN particles of 80 nm size in the amount of 0.03% by weight (bulk concentration of nanoparticles $n_p=7.65\cdot10^{17}$ m^{-3}) was carried out. To determine the initial and boundary conditions, we used the description of an experiment in which the crystallization of a cylindrical casting of a similar size was studied [7].

Numerical research of process of crystallization of the modified melt was carried out at following parameters for the mold and alloy Fe-0.49%C [7, 13-15]: c_1=680 J/(kg·K), λ_1=30 W/(m·K), ρ_1=7100 kg/m^3, c_2=580 J/(kg·K), λ_2=27 W/m·K, ρ_2=7450 kg/m^3, $\kappa=2.77\cdot10^5$ J/kg, T_0=1793 K, T_A=1803 K, T_{L0} = 1758 K, T_E=1420 K, β=91.7 K/%, C_0=0.49 wt.%, k=0.4, $K_\alpha=3.5\cdot10^{-5}$ m/(s·K), α_1=150 W/(m^2·K); density of nanoparticles TiN ρ_p=5440 kg/m^3, m_p=0.03wt.%, $l_p=8\cdot10^{-8}$ m, $l_a=2.9\cdot10^{-10}$ m, $l_c=8.23\cdot10^{-10}$ m, $D_0=1.4\cdot10^{-7}$ m^2/s, $\Delta H_a=2.58\cdot10^{-20}$ J/atom, $E=9.6\cdot10^{-20}$ J/atom, $\sigma_{12}^{\infty\infty}$=0.2 J/m^2, $k_B=1.38\cdot10^{-23}$ J/K, $\delta=1.53\cdot10^{-11}$ m, $\theta=5^0$; R=0.017 m, H=0.051 m, $h_b=10^{-2}$ m, $h_w=1.5\cdot10^{-3}$ m, c_3=1080 J/(kg·K), $\rho_3=1.65\cdot10^3$ kg/m^3; λ_3=1.28 W/(m·K), α_2=150 W/(m^2·K), $R_h=10^{-4}$ m^2·K/W, T_c=293 K.

A finite difference method was used to implement the (1)-(4) model. The calculated area was divided into $I\times K$ cells. The steps of the spatial grid h_r, h_z were chosen from the condition that the contact surfaces of the melt and the mold were located in the middle between the neighboring nodes of the grid. The temperature distribution is described by the values at the grid nodes. The solution of the algebraic system obtained by implicit approximation of the heat transfer equations was carried out by the iterative method of block sequential upper relaxation [16]. Calculations continued until the melt solidified completely.

Figure 2 shows the dynamics of temperature, undercooling of the melt, and the solid phase fraction in the melt in the center of the emerging casting (r=0) and in the point of contact with the mold's wall ($r=R$) at a distance equal to half height of the casting, $z=h_b+H/2$ (points '*3*' in figure 1).

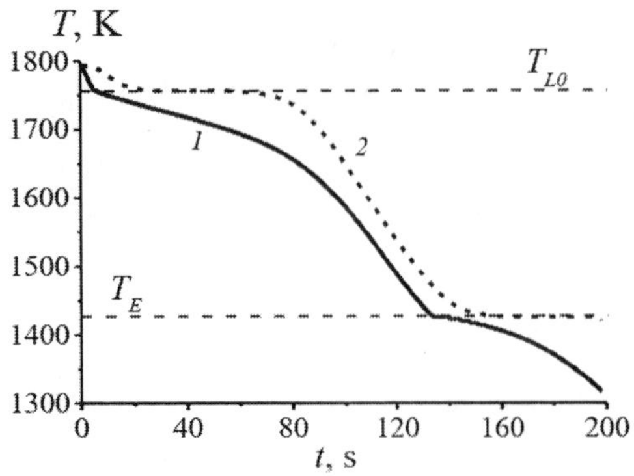

(a)

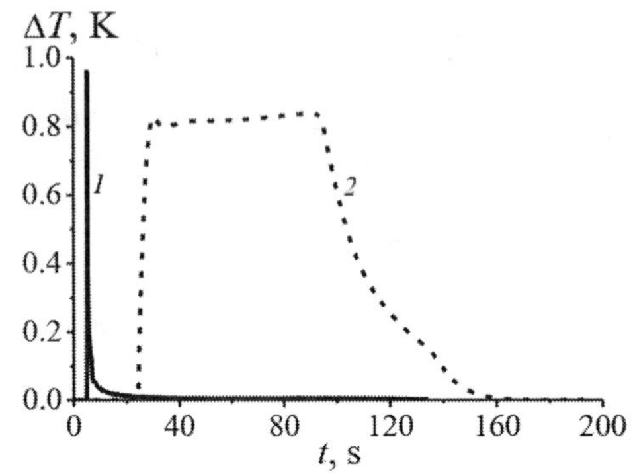

(b)

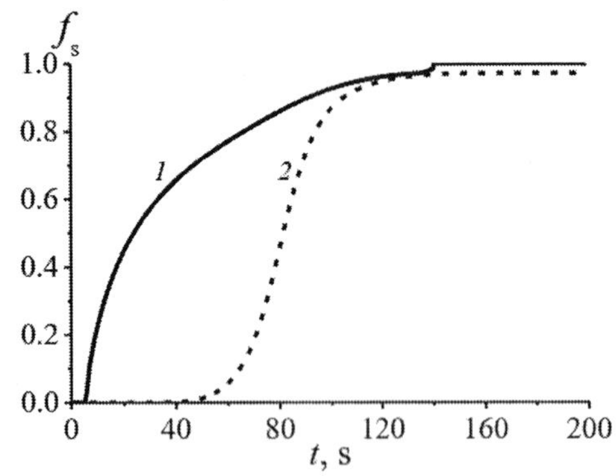

(c)

Figure 2. Time-evolution of temperature (*a*), undercooling (*b*) and fraction of solid phase (*c*) in the melt at the mold side surface (*1*) and in the center of the casting (*2*).

As follows from the simulation results, at the point of contact of the melt with the side wall (curves 1), overheating is removed in a time equal to 5 seconds (figure 2, a), and then the undercooling of the melt in a short period of time (about 0.2 s) reaches a value of 0.96 K (figure 2, b). This period of time can be considered as the stage of the melt staying in a metastable state, when the fraction of the crystalline phase in it is negligible. Then, the stage of intensive growth of the solid phase comes, in which the undercooling decreases rapidly to values less than 0.01 K due to the release of latent heat of crystallization. Crystallization of the α-phase near the wall lasts about 125 seconds, the fraction of the solid phase by the end of this period reaches fsα=0.985, and then fs increases to 1 abruptly due to eutectic crystallization of the alloy (figure 2, c).

In the centre of the casting, where heat exchange with the surrounding material is limited, the initial overheating is removed much later compared to the wall area: at t=23 s (curves *2*). The process of formation of nuclei of crystals of the α-phase with a noticeable rate begins at t >28 s, when the undercooling of the melt reaches the value of 0.8 K. This indicates a significant increase in the period of the metastable state of the crystal phase up to 20 seconds, when both the nucleation rate and the growth rate are very small.

In contrast to crystallization near the mold walls, in the central zone of the mold there is a gradual equalization of the rate of release of latent heat of crystallization and the rate of heat removal to the surrounding melt. As a result, the undercooling of the melt for quite a long time up to the time moment t=90 s is maintained at approximately a constant level of about 0.8 K. Intensive α- phase solidification in the centre of the casting occurs in the time interval $48 \le t \le 90$ s. Then, undercooling ΔT monotonically decreases due to the gradual enrichment of the melt by carbon, which leads, respectively, to a decrease in the growth rate of the solid phase. The fraction of the solid α-phase at this stage of crystallization increases to $f_{s\alpha}$=0.9 at t=120 s until the melt temperature drops to the value near eutectic point T_E. Further, crystallization process occurs at a very low rate: solidification of the alloy is completed at t=198 s (figure 2, c).

Solidification time, magnitudes of undercooling and crystallization rate vary significantly with the distance from the wall of the mold. Obviously, the conditions of nucleation in the melt also differ. Figure 3 illustrates the changes in the concentration N of nuclei and of effective size of grains as a function of the distance from the central axis of the mold in its middle section (z=H/2+h_b).

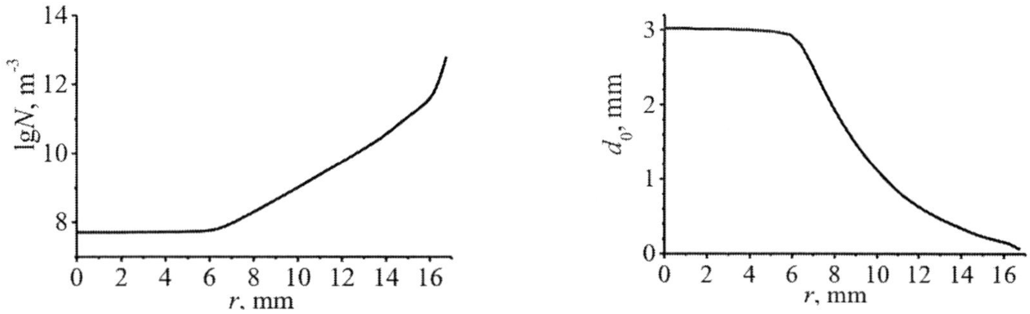

(a) (b)

Figure 3. Radial distribution of the nucleus concentration (a) and grain size (b) in the cross section of the casting.

As follows from the calculations, the values of the volume concentration of nuclei in the central region and at the boundary with the wall differ by more than five orders of magnitude: N=8·10^7 at r=0, whereas at r=R, N=10^{13} m^{-3}. The estimation of the size of the grain structure of the metal in the casting, calculated by formula d_0= $(2^{-1/2}/N)^{1/3}$ [17], derived for spherical packing of maximum density, is shown in figure 3, b. As one can see, according to the simulation results, it is possible to distinguish a central zone with radius equal to about a third of the mold radius, where the effective diameter of the grains is maximal and reaches 3 mm. Area with the smallest metal structure (d_0~100 nm) is located near the side wall of the mold, which qualitatively coincides with the available experimental data [7]. It is interesting to note that the estimate of the grain structure size near the mold wall practically coincides with the size of the edge of the cubic nanoparticles l_p=80 nm, which was taken into account in these calculations. This result may suggest that the greatest efficiency of modification of the melt by nanoparticles is achieved in the near-wall region of the mold.

4. Conclusion

A mathematical model of non-equilibrium crystallization of a binary alloy (Fe-C) with modifying refractory nanoscale particles is proposed. The numerical simulation of melt solidification in a cylindrical mold is carried out and the features of crystallization are considered. The kinetics of heterogeneous nucleation and crystallization in the melt cooling process is described.

It is determined that the conditions of nucleation and solidification differ significantly within the casting. It is established that as the melt cools, the volume-sequential solidification of the metal takes place. The calculated temperature regimes of crystallization of the alloy and the size of the structure of the hardened metal are qualitatively consistent with the known experimental data.

Acknowledgments

The research was carried out within the framework of the Program of Fundamental Scientific Research of the state academies of sciences in 2013-2020 (projects No. AAAA-A17-117030610136-3 and AAAA-A19-119051590004-5).

References

[1] Saburov V P, Eremin E N, Cherepanov A N and Minnekhanov G N 2002 *Modification of Steels and Alloys with Dispersed Inoculators* (Omsk: Omsk State Technical University Publ.) p 212
[2] Turnbul D 1953 *Acta Metallurgica* **1** 8–14
[3] Fletcher N 1958 *J. Chem. Phys.* **29** 572–576
[4] Maxwell I and Hellawell F 1975 *Acta Metallurgica* **23** 229–237
[5] Cherepanov A N, Cherepanova V K, Manolov V and Yovkov L 2018 *Journal of Physics: Conf. Series* **1115** 042042
[6] Cherepanov A N, Cherepanova V K and Manolov V 2017 *AIP Conf. Proc.* **1893** 030114
[7] Popov S, Manolov V and Cherepanov A N 2012 *Comptes rendus de l Academie Bulgare des Sciences* **65** 441–448
[8] Cherepanov A N, Popov V N and Solonenko O P 2008 *Thermophysics and Aeromechanics* **15** 483–489
[9] Tolman R 1949 *J. Chem. Phys.* **17** 333–337
[10] Kolmogorov A N 1937 *Bulletines of the Academy of Sciences of USSR. Ser. Math.* **3** 355–359
[11] Flemings M 1974 *Solidification Processing* (New York: McGraw-Hill)
[12] Christian J 2002 *The Theory of Transformations in Metals and Alloys* (Oxford, Pergamon Press)
[13] Shebzukhova M A, Shebzukhov Z A and Shebzukhov A A 2010 *Bulletin of the Russian Academy of Sciences: Physics* **74** 697–704
[14] Hollomon D and Turnball D 1956 *Nucleation during Phase Transformations Progress in Metal Physics* (Moscow: Metallurgiya)
[15] Postovalov V G, Romanov E P and Kondrat'ev V P 2009 *The Physics of Metals and Metallography* **107** 1–11
[16] Samarskij A A and Nikolaev E S 1989 *Numerical Methods for grid Equations. Volume II: Iterative Methods* (Basel-Boston-Berlin: Birkhauser Verlag)
[17] Sloane N 1984 *Scientific American* **250** 116–125

Numerical simulation of particle beam focusing in a supersonic nozzle with rectangular cross-section

A A Shershnev[1] **and A N Kudryavtsev**[1,2]

[1]Khristianovich Institute of Theoretical and Applied Mechanics, Russian Academy of Sciences, Siberian Branch, 630090 Novosibirsk, Russia
[2] Novosibirsk State University, 630090 Novosibirsk, Russia

E-mail: `antony@itam.nsc.ru`

Abstract. Particle-laden flow in 3D supersonic micronozzle with rectangular cross-section and side walls convergent/divergent in both lateral and transversal direction is studied numerically using a one-way coupled Eulerian/Lagrangian approach. The carrier gas flow is simulated numerically on the basis of the Navier-Stokes equations and then is used to calculate the particles trajectories. It is shown that a collimated beam of particles can be produced using the effect of aerodynamic focusing and the beam collimation is observed in two different ranges of particle sizes. Obtained results are consistent with previously obtained data for plane and axisymmetrical nozzles.

1. Introduction

Multiphase flows in micronozzles are of great interest for their possible industrial applications. One of the most promising applications is generation of high-speed collimated beams of microparticles which can be used in different technologies, such as Cold Gas Spraying [1, 2, 3] or relatively new material deposition method known as Collimated Aerosol Beam Direct-Write (CAB-DW) [4]. In the CAB-DW, an "aerosol gun" is employed for printing solid materials on a substrate for producing solar cells and electronic chips. It was demonstrated both numerically and experimentally [5, 6] that collimated beams can be successfully generated by subsonic (convergent) nozzles. Supersonic (convergent/divergent) nozzles could be used to create high-speed beams; however, in the supersonic part, flow streamlines are divergent, hence the possibility of aerodynamic focusing in such nozzles is questionable.

It was shown numerically [7] that for plane and axisymmetrical supersonic convergent-divergent nozzles aerodynamic focusing of particles into a collimated beam can be obtained. Moreover, it is observed for two different sizes of particles. As for 3D nozzle it was demonstrated, that when a nozzle has a constant width in the lateral direction the focusing is only feasible in the transversal direction. Apparently, the displacement of streamlines in the supersonic part of the nozzle caused by the growth of the boundary layer thickness on the lateral walls is not sufficient for beam collimation in this direction. Hopefully, introduction of subsonic nozzle part, convergent in both lateral and transversal direction will produce particle trajectories focused into a narrow beam rather than a plane sheet. Thus, the present paper is a continuation of the cited study performed to confirm that conclusions above hold for 3D nozzle with rectangular cross-section convergent/divergent in both lateral and transversal direction.

2. Problem statement and numerical techniques

Two-phase flow in a 3D nozzle with a throat width 200 μm is simulated numerically using the Eulerian-Lagrangian approach. The nozzle contour is based on the nozzles used in experiments of Rothe [8]. Unlike [8], where the author used axisymmetrical nozzles with round cross-section, in the present paper the nozzle with convergent/divergent side walls and rectangular cross-section in considered. Nozzle geometry and computational mesh are shown in figure 1a. The design Mach number of the nozzle is $M_d = 7.52$ and the Reynolds number based on the nozzle half-width and parameters in the nozzle throat is equal to $Re = 350$. The Navier-Stokes equations are used to compute the carrier monatomic gas (argon) flowfield, then Lagrangian approach is applied for calculating the particle trajectories. The stagnation temperature and the nozzle wall temperature are equal to 300 K, the stagnation pressure is 21,790 Pa. It is assumed that the gas exhausts into a chamber with an ambient pressure that is lower than the jet pressure at the nozzle exit so that the exhausting plume is underexpanded.

The particles are introduced into the flow in the nozzle pre-chamber. The flow around small micro- and nanoparticles can occur in the transitional or even the free molecular regime and strong velocity nonequilibrium between the carrier gas and particles can be observed. Therefore, the drag force acting on a particle is calculated with a formula which interpolates between continuum and free molecular expressions and takes into account particle inertia, gas compressibility, and rarefaction effects [9].

When solving the Navier-Stokes equations a shock capturing TVD scheme of 3rd order is used for calculating the convective terms. The 2nd order central differences are used to approximate the diffusive terms. Stagnation temperature, stagnation pressure and flow direction are specified on the inflow boundaries. All gasdynamic quantities are extrapolated from within the computational domain on the outflow boundary. Particle equations of motion are integrated with the 4th order Runge–Kutta scheme. Carrier phase flow parameters in particle locations are calculated using the bilinear interpolation.

3. Results and discussion

Numerical simulation of the carrier phase was performed on a mesh consisting of 295×40×40 cells. The local Mach number flowfield obtained in this computation is shown in figure 1b. As can be seen, the effective cross-section of the nozzle is reduced considerably because of the rapid growth of the boundary layer in the supersonic part of the nozzle. As a result, computed Mach number of the flow at the nozzle exit is $M = 4.5$, i.e. only 60% of the design Mach number. The flow pattern is typical for low-Reynolds number nozzle flows considered in previous study.

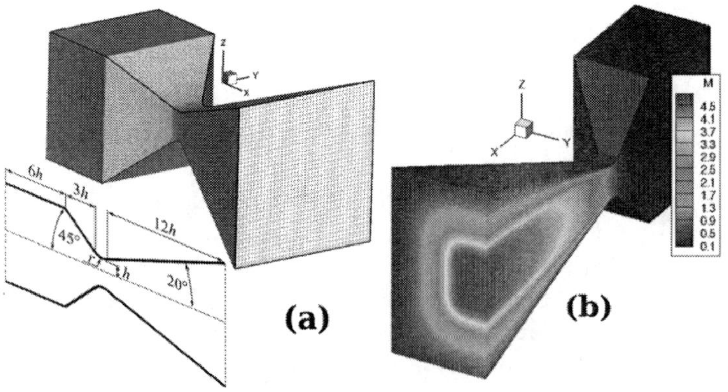

Figure 1. Computational domain sketch and grid used in carrier gas simulation (a) and local Mach number flowfield (b).

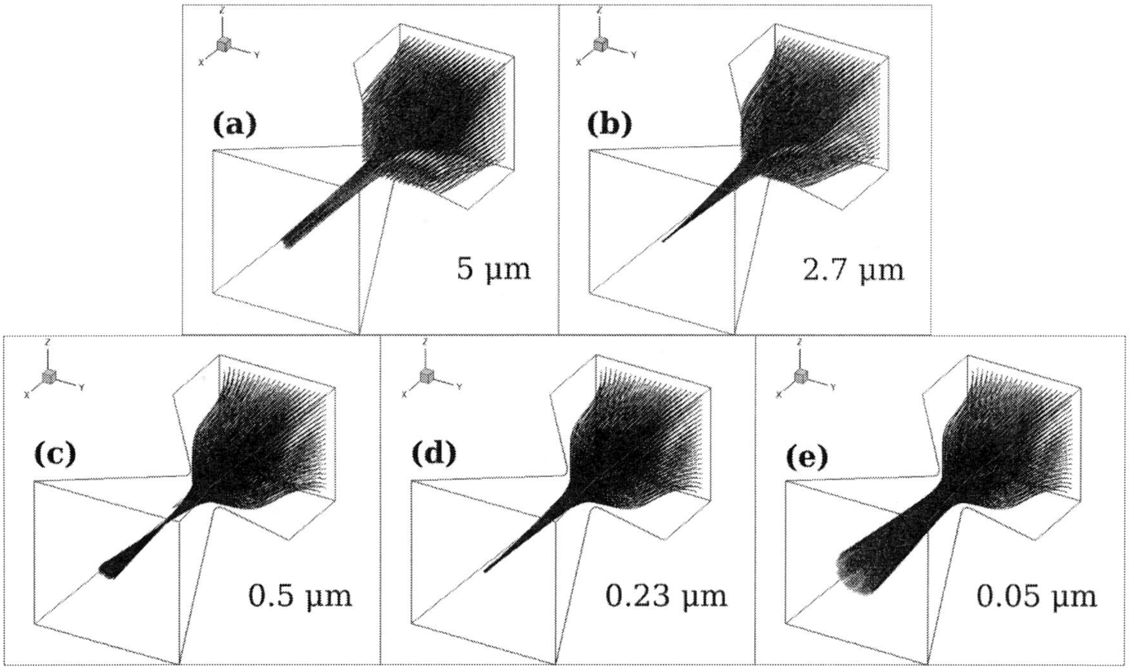

Figure 2. Trajectories of particles of different radii.

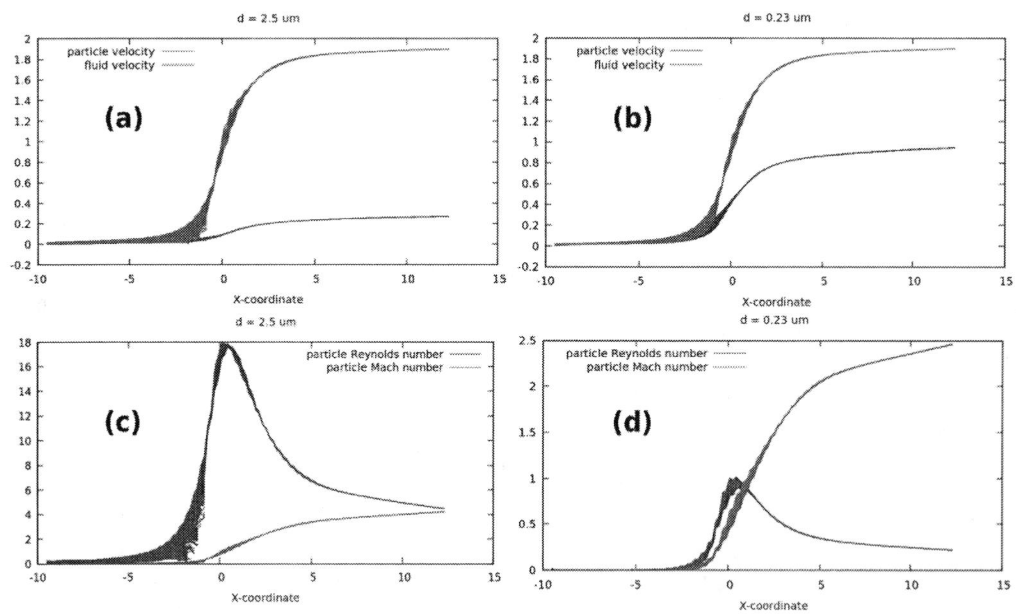

Figure 3. Distribution of the longitudinal velocity of the carrier and dispersed phases (a and b) and the particle Mach number and particle Reynolds number (c and d) along the trajectory in the cases of collimation.

The gas and particle velocities (normalized to the critical velocity) along particle trajectories are shown in figures 3a and b for the particle sizes for which aerodynamic focusing occurs. In case of large particles, the difference in velocity values is significant. Nevertheless, the particle velocity continues to grow in the supersonic part of the nozzle. This acceleration is much more pronounced for smaller particles. Evolution of the particles Reynolds and Mach numbers along the trajectories is presented in figures 3c and 3d. It can also be seen, that both parameters change quite drastically as the particle travels through nozzle. This fact emphasizes the need for formulas for the drag force approximation applicable in a wide range of flow regimes, from the creep flow to supersonic flow and from continuum to free molecular flow.

4. Conclusion

The particle-laden flow in the 3D supersonic micronozzle with a rectangular cross-section and side walls convergent/divergent in both lateral and transversal directions was investigated numerically using the one-way coupled Eulerian/Lagrangian approach. The carrier gas flow was simulated numerically on the basis of the Navier-Stokes equations and then is used to calculate the particles trajectories. It was shown that a collimated beam of particles can be produced using the effect of aerodynamic focusing and the beam collimation is observed in two different ranges of particle sizes. Obtained results are consistent with previously obtained data for plane and axisymmetrical nozzles. Further investigations are required to study the plume interaction with a transversal plate (substrate) which is of great importance for practical applications such as material deposition processes.

Acknowledgments

The work is supported by the Russian Science Foundation (grant No. 18-11-00246). This support is gratefully acknowledged.

Computational resources were kindly provided by Siberian Supercomputer Center of the Institute of Computational Mathematics and Mathematical Geophysics SB RAS (sscc.ru), Computational Center of Novosibirsk State University (nusc.nsu.ru) and Moscow State University Supercomputing Center (parallel.ru).

References

[1] Klinkov S V, Kosarev V F, Ryashin N S and Shikalov V S 2016 Experimental study of cold gas spraying through a mask. Part 1 *Thermophys. Aeromech.* **23** 735-40

[2] Klinkov S V, Kosarev, V F and Ryashin N S 2017 Experimental study of cold gas spraying through a mask. Part 2 *Thermophys. Aeromech.* **24** 213-24

[3] Klinkov S V, Kosarev V F and Shikalov V S 2019 Influence of nozzle velocity and powder feed rate on the coating mass and deposition efficiency in cold spraying *Surface and Coatings Technology* **367** 231-43

[4] Han S et al 2008 Printed silicon as diode and FET materials – preliminary results *J. Non-Crystalline Solids.* **354** 2623-6

[5] Akhatov I S, Hoey J M, Thompson D, Lufturakhmanov A, Mahmud Z, Swenson O F, Schulz D L and Osiptsov A N 2009 Aerosol flow through a micro-capillary *Proc. of the ASME 2nd Micro/Nanoscale Heat & Mass Transfer International Conference (Shanghai, China)* Paper MNHMT2009-18421

[6] Bhattacharya S, Lufturakhmanov A, Hoey J M, Swenson O F, Mahmud Z and Akhatov I S 2013 Aerosol flow through a converging-diverging micro-nozzle *Non-linear Eng.* **2** 103-12

[7] Kudryavtsev A, Shershnev A and Rybdylova O 2019 Numerical simulation of aerodynamic focusing of particles in supersonic micronozzles *Int. J. Multiph. Flow* **114** 207-18

[8] Rothe D E 1971 Electron-beam studies of viscous flow in supersonic nozzles *AIAA J.* **9** 804-11

[9] Molleson G V and Stasenko A L 2008 Acceleration of microparticles in a gasdynamic facility with high expansion of flow *High Temperature* **46** 100-7

Journal of Physics: Conference Series **1404** (2019) 012043 doi:10.1088/1742-6596/1404/1/012043

Numerical modeling of the nucleation of crystallization centers during the modification of the surface layer of a metal using laser heating

O A Shmagunov

Khristianovich Institute of Theoretical and Applied Mechanics SB RAS, 4/1 Institutskaya str., Novosibirsk, 630090, Russia

E-mail: shmag@itam.nsc.ru

Abstract. Laser surface treatment using nanoscale particles of refractory compounds is a promising way to improve the operational characteristics of machine parts. Nanoscale particles increase the number of crystallization centers in the melt that occurs on the metal surface when exposed to a laser beam and, thereby, improve the dispersion and uniformity of the crystalline grains and the quality of the surface processed. The mathematical model used in the work allows one to describe the processes of nucleation and crystallization in the melt, the kinetics of the solid phase growth. Steady-state processes of heating, phase transition, heat transfer in the melt, nucleation, and growth of the solid phase are considered. The distribution of the temperature field, the size of the crystallization zone, and the number of crystallization centers in the molten metal are estimated.

1. Introduction

A promising way to improve the operational properties of machine parts is laser processing of their surfaces with the melt modification by nanosized particles of refractory compounds (carbides, nitrides, carbonitrides, etc.) [1, 2]. This allows of increasing the number of crystallization centers, and the resulting high dispersion and uniformity of the crystalline grains contribute to improving the quality of the treated surfaces [3–6].

In this paper, using numerical simulation, we consider the nucleation of crystallization centers in molten metal of a moving substrate when its surface is exposed to laser radiation and contains modifying nanosized particles. A mathematical model is formulated that describes the thermodynamic processes in the metal, including heterogeneous nucleation and crystallization of the primary phase of the melt, which allows one to determine the kinetics of the solid phase growth. Steady-state processes are considered, including heating, phase transition and heat transfer in the molten metal, nucleation and growth of the solid phase. Based on the results of numerical experiments, the distribution of the temperature field, the size of the crystallization zone, and the number of crystallization centers that have arisen in the molten metal of the substrate are estimated.

2. Mathematical model

The effect of a laser beam with focal spot of radius r_0 and flux power density described by a Gaussian distribution on a moving substrate of mild steel is considered. The process scheme is illustrated in figure 1.

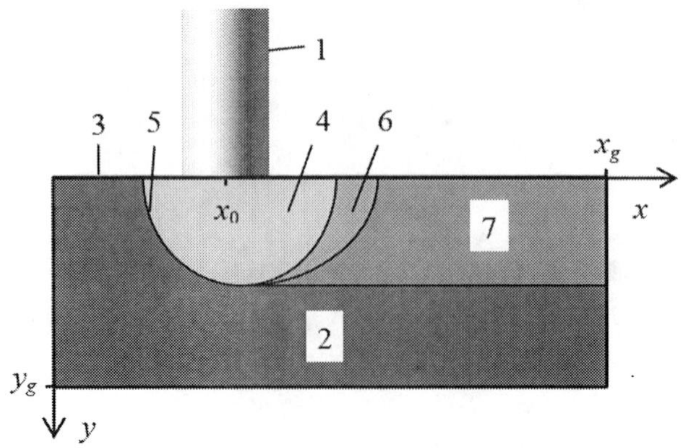

Figure 1. The processing scheme in the plane of symmetry: 1 – laser radiation energy, 2 – metal substrate and particles of modifying material 3 on its surface, 4 – melt, 5 – superheated metal boundary, 6 – two-phase zone of solidified metal, 7 – modified metal.

The plate moves with constant speed V along the x axis, therefore, the processes occurring in it under the influence of laser radiation energy (heating and melting followed by solidification of the metal) are considered in the quasi-stationary approximation. The phase transition occurs at melting temperature T_{l0} of the substrate material. The surface of the substrate is covered by a layer of modifying nanosized particles of a refractory compound. A part of the particles is assumed to penetrate from the surface of the melt and get uniformly distributed throughout the volume, contributing to the emergence of crystallization centers [4-6]. After moving the plate from the area affected by the laser beam energy, due to heat removal into unheated material and heat exchange with the environment, the melt cools and crystallizes. To protect the metal from oxidation, the surface of the processed plate is blown with an inert gas. The boundaries x_g, y_g, z_g of the region under consideration are chosen in such a way that their position does not affect the processes under study.

To simplify the problem, we assume that the thermophysical characteristics of liquid, solid, and two-phase media are the same and do not depend on temperature. Mass content M_p of the particles penetrating into the melt is small ($M_p \leq 0.05\%$), the particles are cubic with edge l_p much smaller than the characteristic size of the liquid pool, therefore, the influence of inclusions on the physical parameters of the melt can be neglected. Metal melting is considered in the approximation of the Stefan problem: it is assumed that under the considered heating conditions, the small values of convective velocities in the melt determine the flat shape of the free surface of the liquid.

The processes of heat transfer, melting and crystallization of the metal in plate are described in the quasi-stationary case by equation in Cartesian coordinate system

$$c_p \rho V \frac{\partial T}{\partial x} = \lambda\left(\frac{\partial^2 T}{\partial x^2} + \frac{\partial^2 T}{\partial y^2} + \frac{\partial^2 T}{\partial z^2}\right) + \kappa \rho V \frac{\partial f_s}{\partial x}$$

where T is the temperature; ρ is the density of the plate material; c_p is the specific heat; λ is the coefficient of thermal conductivity; κ is the specific heat of fusion; f_s is fraction of solid phase in the material ($0 \leq f_s \leq 1$).

We assume that the central point of the laser action spot of radius r_0 is located at point (x_0, 0, 0). On the surface $z = 0$ of the plate, the boundary conditions

$$\lambda(\partial T / \partial z) = q(x, y) + (\alpha_k + \alpha_r)(T - T_g)$$

Here, T_g is the temperature of the protecting gas, $\alpha_r = \varepsilon\sigma_0\left(T^2 + T_g^2\right)\left(T + T_g\right)$ is coefficient of the radiative heat transfer, ε is the reduced degree of blackness, σ_0 is the Stefan-Boltzmann constant, α_k is the convective heat transfer coefficient,

$$q(x,y)=\begin{cases} q_0 \exp\left\{-2\left[(x-x_0)^2+y^2\right]/r_0^2\right\} & \left[(x-x_0)^2+y^2\right]^{1/2}\le r_0 \\ 0, \quad \left[(x-x_0)^2+y^2\right]^{1/2}>r_0 \end{cases}$$

is power density of the laser beam with q_0 being its maximum value.

On the plane $y = 0$, $0 \le x \le x_g$, $0 \le z \le z_g$, the symmetry condition is used. The conditions at the remaining borders are as follows:

$$\lambda(\partial T/\partial x)=0,\ x=0,\ x=x_g;\ \lambda(\partial T/\partial y)=0,\ y=0;\ \lambda(\partial T/\partial z)=0,\ z=z_g$$

Fraction f_s of the solid phase is calculated according to the model proposed in [7, 8]:

$$f_s(x,y,z)=1-\exp\left\{-\frac{1}{V}\int_{x_{l0}}^{x} I(\xi,y,z)\left[1-f_s(\xi,y,z)\right]V_s(\xi,y,z)d\xi\right\},\qquad V_s(\xi)=\frac{4\pi}{3}\left(R_C+\frac{K_V}{V}\int_{\xi}^{x}\Delta T d\zeta\right)^3$$

where $I(x, y, z)$ is the nucleation rate:

$$I=\frac{M_p\rho\pi^2 D_0}{50\rho_p l_p^3 l_a^4 l_c^2}R_C^4\sin^2\theta(1-\cos\theta)\exp\left[-\frac{E+\Delta G^*}{k_B T}\right]$$

$1-f_s\ (x, y, z)$ is fraction of the liquid phase; $V_s(x, y, z)$ is the volume of growing grain; M_p is the amount of nanopowder,% by mass; ρ is the metal density; D_0 is pre-exponent in the Arrhenius law; ρ_p is the nanopowder density; l_p is the edge size of the cubic seed; l_a is the melt atom diameter; l_c is the interatomic distance of the seed material; $R_C=R_0(1-6\delta/R_0)$ is the critical radius of the solid phase nucleus, $R_0=2\sigma_{12}^{\infty}T_{l0}/(\kappa\rho\Delta T_0)$ is the initial value of the nanoseed effective radius; $\Delta T_0 = T_{l0} - T$, T_{l0} is the initial liquidus temperature; δ is the Tolman parameter; θ is the contact angle of wettability of the particle surface; E is the activation energy of diffusion of the alloying component; k_B is the Boltzmann constant; ΔG^* is the free energy of formation of the critical solid phase nucleus on a flat substrate, defined by expression

$$\Delta G^*=\frac{\pi}{3}\sigma_{12}^{\infty}R_0^2\left(1-\frac{6\delta}{R_0}\right)(1-\cos\theta)^2(2+\cos\theta)$$

$\Delta T=T_A-\beta C_0(1-f_s)^{k-1}-T$ is local supercooling of the melt ($T_{l0} = T - \beta C_0$); σ_{12}^{∞} is surface tension at the nucleus–melt interface; k is the coefficient of the alloying component distribution in the alloy; K_v is the constant of growth of the crystalline grain nucleus; x_l is coordinate on the melt–two-phase-zone surface, where the alloy temperature is equal to the equilibrium temperature T_l. The number of crystals formed in a unit volume of the solidified metal is calculated by formula

$$N=\frac{1}{V}\int_{x_{l0}}^{x} I(\xi,y,z)\left[1-f_s(\xi,y,z)\right]d\xi$$

3. Simulation results

Numerical studies were carried out with the following parameters [6, 9-11]: $r_0 = 0.001$ m, $x_0 = 0.006$ m, $x_g = 0.024$ m, $y_g = 0.012$ m, $z_g = 0.008$ m; $V = 0.005$–0.025 m/s; $T_g = 300$ K; $q_0 = (2.5$–$4.0)\cdot10^8$ W/m^2; $\sigma_0 = 5.7\cdot10^{-8}$ W/(m^2·K^4); $\alpha_k = 100$ W/(m^2·K); $\rho = 7065$ kg/m^3, $c_p = 787$ J/(kg·K), $\lambda = 27$ W/(m·K), $\kappa = 2.77\cdot10^5$ J/kg, $T_A = 1775$ K, $C_0 = 0.49\%$ mass, $k = 0.4$, $\beta_0 = 91.7$ K/%, $K_v = 3.5\cdot10^{-5}$

m/(s·K), $\sigma_{12}^{\infty} = 0.44$ J/m^2, $\theta = 5$–$15°$, $D_0 = 1.4 \cdot 10^{-7}$ m^2/s, $E = 9.6 \cdot 10^{-20}$ J, $\delta = 1.53 \cdot 10^{-11}$ m, $M_p = 0.03\%$, $\rho_p = 5440$ kg/m^3 (TiN), $l_a = 2.86 \cdot 10^{-10}$ m, $l_c = 8.23 \cdot 10\text{-}10$ m, $l_p = 8 \cdot 10^{-8}$ m.

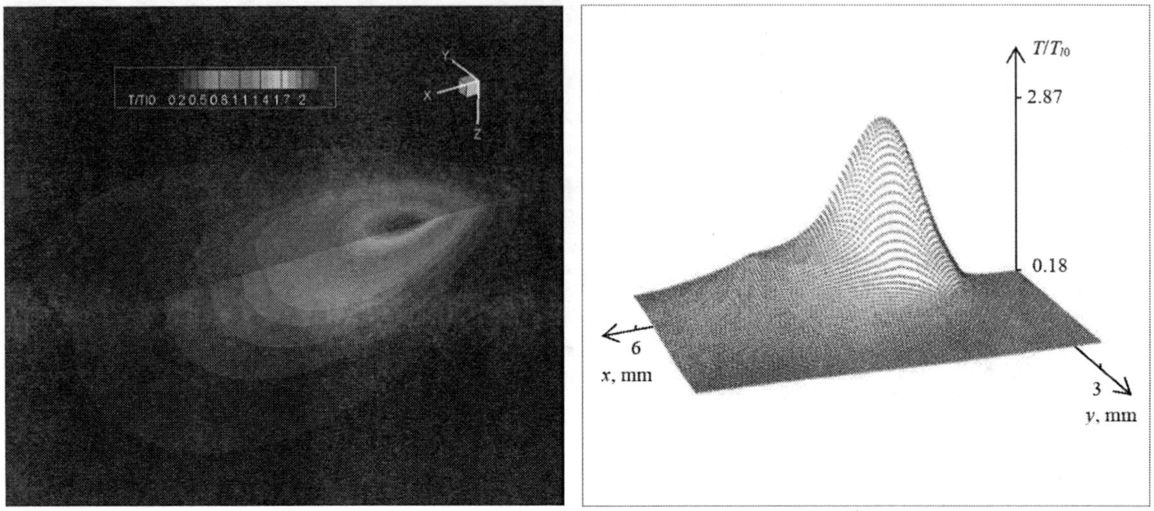

(a) (b)

Figure 2. Temperature distribution in the volume and on the plate surface at: (a) $V = 0.015$ m/s, $q_0 = 2.5 \cdot 10^8$ W/m^2; (b) $V = 0.020$ m/s, $q_0 = 3.5 \cdot 10^8$ W/m^2.

Figure 2 shows the temperature distribution inside and on the surface of the plate. Figure 2a illustrates the temperature distribution on the upper surface and plane of symmetry. Figure 2b, illustrating surface heating, shows the irregular temperature distribution due to heat generation in the two-phase crystallization zone. Only the part of the computational domain including the crystallization zone is shown.

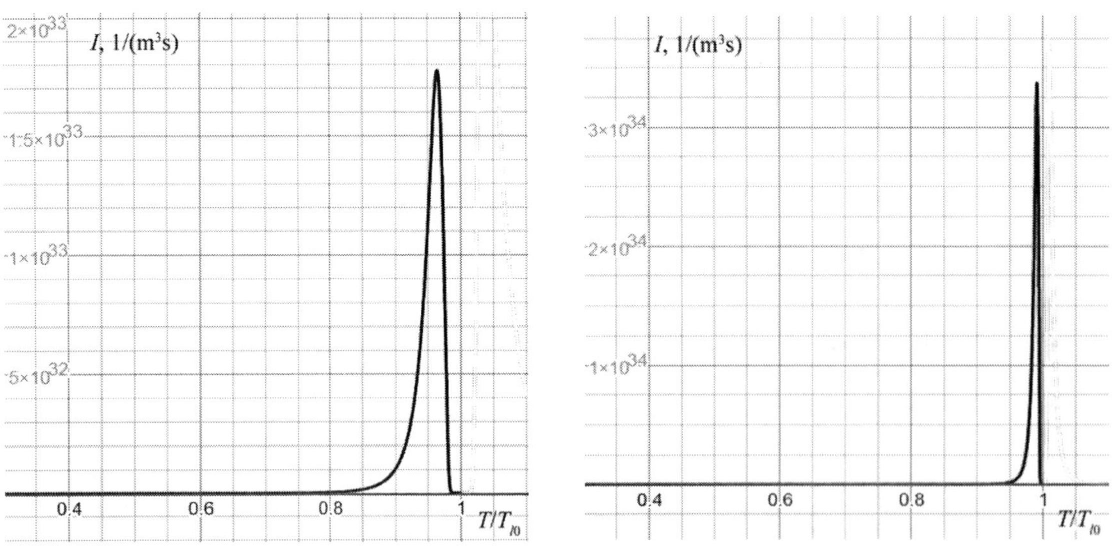

(a) (b)

Figure 3. Dependence of nucleation rate I on temperature $T = T/T_{l0}$ at wetting angles: $\theta = 10°$ (a) and $\theta = 5°$ (b).

Figure 3 illustrates the dependence of nucleation rate I on dimensionless temperature $T=T/T_{l0}$ and various values of the wettability angle θ. From the results obtained, it follows that with the emergence of supercooling, the formation of nuclei of the metal crystals begins. With an increase in supercooling, the nucleation rate increases, however, after reaching a certain maximum value, it begins to decrease and subsequently stops completely. At lower values of θ, the nuclei arise earlier, the rate of their appearance is higher and, it can be said, is explosive. Small values of θ are of the greatest interest for research, because they are more consistent with the values of supercooling at which crystallization occurs.

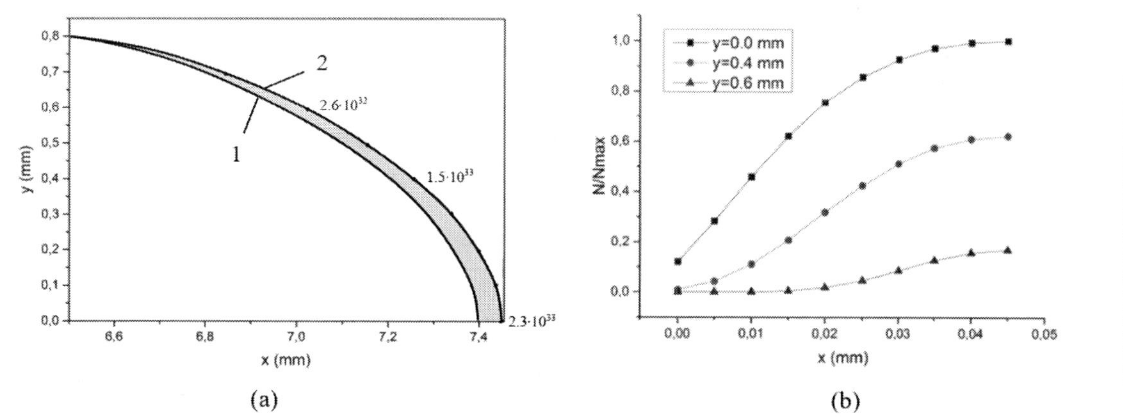

Figure 4. The number N of nuclei: (a) at the boundary (1 – beginning of the two-phase zone, 2 – end of the two-phase zone), (b) inside the two-phase zone.

Figure 4 illustrates the number N of nuclei formed at the boundary of the two-phase zone at the time of complete solidification of the melt and the patterns of their distribution inside the two-phase zone at speed $V = 0.015$ m/s and wettability angle $\theta = 10°$. A decrease in the number of nuclei (figure 4a), as well as in the rate of their nucleation, is observed inside the two-phase zone during solidification of the melt with the distance from the line of symmetry at $y = 0$ (figure 4b). This can be explained by a decrease in the melt overheating in the region of influence of the laser beam energy with increasing distance from the axis of symmetry of the system and due to an increase in the heat removal to the unheated material in the y direction. We can conclude that there are various conditions for nucleation.

Based on the results of computational experiments, it was determined that a change in velocity V affects the size and shape of the crystallization zone, but does not significantly affect the number of nuclei formed.

4. Conclusion

A model and an algorithm for its implementation are proposed, which allow one to study the processes of nucleation of crystallization centers when modifying the surface layer of a metal substrate using laser energy. The process of crystallization centers nucleation in the molten surface layer of the metal, in particular, the influence of the wettability angle of the surface of nanosized particles, is studied. The temperature fields, the sizes of the crystallization zone of the substrate, depending on the speed of its movement, as well as the distribution of the number of crystallization nuclei in the hardening and hardened metal, depending on the wettability angle of the surface of nanoscale inoculators, are determined. The study contributes to the understanding of the processes of laser surface modification of steel, which triples the microhardness of the modified layer and eliminates structural defects such as cracks and porosity [4].

Acknowledgments

The author is grateful to V.N. Popov and A.N. Cherepanov for statement of the problem and constant attention to the work.

The research was carried out within the framework of the Program of Fundamental Scientific Research of the state academies of sciences in 2013-2020 (project No. AAAA-A17-117030610136-3).

References

[1] Popov V N and Cherepanov A N 2017 Numerical evaluation of the laser-pulse modification modes of the metal surface layer in the presence of a surface-active component in the melt *Thermophys. and Aeromech.* **24** 779–86

[2] Popov V N and Shchukin V G 2019 Numerical simulation of metal surface layer modification using high-frequency induction heating *Inorganic Mater.: Appl. Res.* **10** 616–21

[3] Saburov V P, Cherepanov A N, Zhukov M F et al 1995 *Plasma Chemical Synthesis of Ultra-Dispersed Powders and its Application for Metal and Alloys Modification* (Novosibirsk: Nauka) p 344

[4] Cherepanov A N, Drozdov V N, Malikov A G, Orishich A M, Mul' D O and Shevtsova L I 2016 The effect of nanostructured powder compositions on the characteristics of the surface layer of steel during laser processing *Tyazheloe mashinostroenie* **6** 2–4

[5] Cherepanov A N and Popov V N 2015 Evaluation of influence of nano-size refractory particles modification of superalloy *Bulletin of Novosibirsk State University. Ser. Phys.* **10** 97–102

[6] Kuzmanov P M, Popov S I, Yovkov L V et al 2017 Investigation the effect of modification with nanopowders on crystallization process and microstructure of some alloys *AIP Conf. Proc.* **1893** 030104

[7] Cherepanov A, Cherepanova V, Manolov V and Yovkov L 2018 On crystallization of a metal inoculated with nanoparticles *J. of Phys.: Conf. Ser.* **1115** 042042

[8] Shiryayev A N 1992 *Selected Works of A.N. Kolmogorov. Mathematics and Its Applications* vol 26 (Dordrecht: Springer)

[9] Shebzukhova M A, Shebzukhov Z A and Shebzukhov A A 2010 The tolman parameter, self-absorption, and surface tension on flat and curved surfaces of liquid metals *Bull. of the Russ. Acad. of Sc.s: Phys.* **74** 697–704

[10] Hollomon J H and Turnbull D 1953 *Nucleation Progress in Metal Physics* vol 4 (London: Pergamon Press) pp 333–88

[11] Postovalov V G, Romanov E P and Kondrat'ev V P 2009 On the theory of mass transfer in liquid metals *The Phys. of Metals and Metallogr.* **107** 1–11

Numerical simulation of mono-disperse gravity-driven granular flow around a wedge using two-fluid model

Georgy Shoev

Khristianovich Institute of Theoretical and Applied Mechanics, 630090, 4/1 Institutskaya str., Novosibirsk, Russia

E-mail: shoev@itam.nsc.ru

Abstract. The numerical simulation of glass beads granular flow around a wedge located in a rectangular channel is performed by the Two-Fluid Model (i.e. the Euler-Euler approach) aiming to make a comparison between the numerical and the available experimental data. The channel is tilted against the horizon making the glass beads fall under the action of the gravity force. The numerical simulation shows the glass beads structure that resembles a bow shock wave around the wedge. The computed bow shock stand-off distance is compared to the data extracted from the experimental results. The details of the comparison between the numerical and the experimental data are discussed.

1. Introduction

Granular flows are of high importance for different industrial applications. Accurate prediction of a granular flow behaviour is necessary for research and development studies. The Euler-Euler or Two-Fluid Model (TFM) is one of the possible approaches that can be used for the numerical simulation of different multiphase flows. TFM was tested to describe the multiphase flow in fluidized beds, where solid particles hover due to the air jets directed against the vector of the gravity force. Usually experimental data [1, 2] show the flow parameters averaged in space and time. These averaged experimental results compare almost equally to the results of the numerical simulation based on different modifications of TFM (different drag models) even though the instantaneous flow fields can be significantly different [3].

The granular flow of solid particles and the flow of gas behave similar to a certain extent. One of such similarities is the structure that resembles the shock wave in classical gas dynamics. These shock waves in granular flows [4, 5, 6] can be observed around different bodies such as a cylinder, a square, a wedge, etc. In these cases a granular flow reaches a steady state for some period of time allowing the detailed analysis of the flow features like a detached bow shock wave, shear instabilities and a granular wake (a region of the low volume fraction of particles behind the body). These external granular flows are similar to high-speed gas flows (for example, [7, 8] and many others).

One of the classical flow problems in gas dynamics is a supersonic gas flow around a wedge. The study of this flow problem shows many important mechanisms of high-speed gas flows. Just as in classical gas dynamics, the investigation of an external granular flow around a wedge can reveal many important fundamental mechanisms, and therefore, this flow problem attracted many researchers [4, 5, 6] to study it theoretically, numerically and experimentally. The experimental results [4] for a granular flow around a wedge can be used to validate different numerical models. The important thing is that this flow reaches a steady state, hence, the necessity of averaging over time is eliminated. The aim of this study is to assess the capability of the two-fluid model implemented in the ANSYS Fluent software to describe a mono-disperse gravity-driven granular flow around a wedge for the conditions of the recently performed experiments [4].

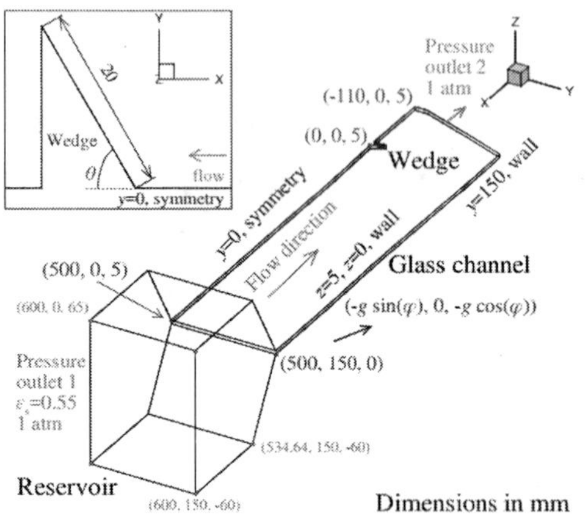

Figure 1. Computational domain and boundary conditions.

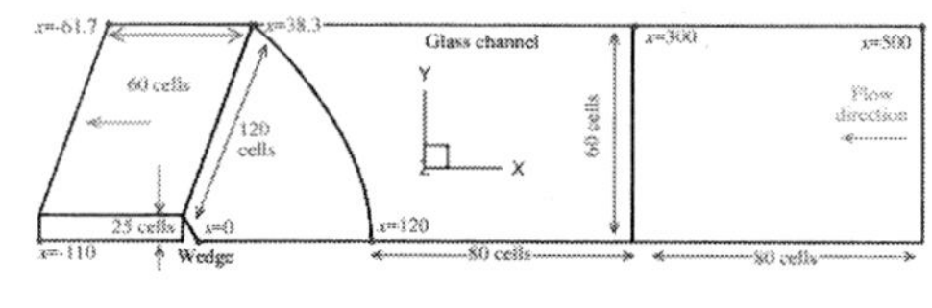

Figure 2. Sub-domains for the computational mesh.

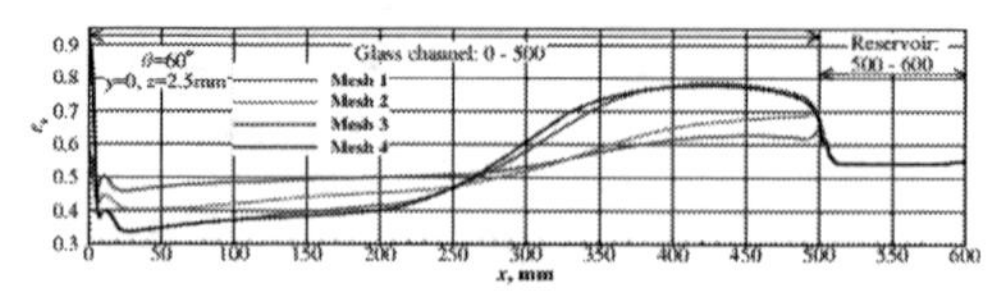

Figure 3. Volume fraction of glass beads along line (x, 0, 2.5 mm) for different computational meshes.

2. Problem formulation and numerical procedure

The current numerical setup consists of a reservoir and a fully transparent channel, where the wedge is located (figure 1). Four cases with different values of a half-angle of the wedge θ are considered, namely, 50°, 60°, 70° and 80°. The reservoir and the channel are tilted against the horizon by $\phi = 50°$ angle. At the initial moment, the computational domain has the volume fraction of the glass beads $\varepsilon_s = 0.55$. The glass beads have the diameter of $d = 125\mu m$ and the density of $\rho = 1600$ kg/m^3. The computation is performed using transient (unsteady) mode. Moving down the channel the glass beads interact with the wedge forming a steady state flow in several seconds. Convergence to the steady state was controlled by the independence of the flowfield of the glass beads volume fraction ε_s on the front ($z = 5$ mm) and the back ($z = 0$) walls and in the middle plane ($z = 2.5$ mm) of the channel. Additionally, the profile of the glass beads volume fraction along the stagnation line was monitored to make sure that the numerical solution is time independent.

The air flow is assumed to be laminar and the air is considered a single species gas with a constant density (incompressible flow). The solid particles and the air gas are considered inert media without any chemical reaction, although, the Euler-Euler approximation allows taking into account the chemical interaction between the carrier and the disperse phases [9, 10]. The energy equation is not included in the modelling. The coefficient of dynamic viscosity is assumed to be constant under normal conditions. The glass beads are considered through the Eulerian phase. The granular temperature model uses the phase property option of the flow solver. The following solver settings were used for the glass beads phase simulation: `Gidaspow` for the granular viscosity, `lun-et-al` for the granular bulk viscosity, `schaeffer` for the frictional viscosity. Angle of internal friction was set to 30°, `based-ktgf` option was used the frictional pressure, `derived` for the frictional modulus, the friction packing limit was equal to 0.63, the granular temperature used `algebraic` option, `lun-et-al` was used for the solids pressure computation, `ma-ahmadi` for the radial distribution, `derived` for the elasticity modulus. The packing limit was set to 0.99. The Gidaspow drag model [11] is used to describe phases interaction, i.e. the carrier phase affects the disperse phase and vice versa (note: one way coupling, like it was done in [12], cannot be applied). The other options for phases interaction are the lift coefficient, set to none, the restitution coefficient set to 0.97, and the surface tension coefficient set to `none`.

The computational domain contains three types of boundary conditions: pressure-outlet, wall and symmetry (figure 1). "Pressure outlet 1" is located at the top of the reservoir. During the computation the glass beads are falling down because of gravity causing reversed flow into the computational domain, hence, the volume fraction of the glass beads has to be specified and it is equal to $\varepsilon_s = 0.55$. The real size of the reservoir does not seem to be important because the mass flux into the channel seems to be independent on the quantity of the granular material in the reservoir as it was mentioned in [4]. "Pressure outlet 2" is located downstream of the wedge assuming that the flow behind this boundary does not affect the flow structure around the wedge. Both pressure-outlets use atmospheric pressure. In order to reduce the computational cost, a symmetry plane at $y = 0$ is used. The other boundaries have the type of wall, where zero shear stress is used (the inviscid wall).

The computational domain is divided into several sub-domains in order to build the computational mesh consisting of hexahedral cells. The number of computational cells on each edge of the sub-domains is given in figure 2, where (x,y) plane is shown. The computational mesh is refined toward the surface of the wedge in order to capture shock-like structures of the glass beads. The height of the cell adjacent to the wedge surface is about 0.17 mm while the farthest cell from the wedge surface in the sub-domain is about 2.9 mm. The number of cells in $z-$direction is varied because the thickness of the glass beads layer changes and can cause high gradients of the particles volume fraction. The following numbers of cells in $z-$direction inside the glass channel is used: 4 (Mesh 1), 8 (Mesh 2), 16 (Mesh 3) and 24 (Mesh 4).

Phase coupled SIMPLE algorithm is used for the pressure-velocity coupling. The spatial discretization uses `first order upwind` option for the momentum, volume, and granular temperature and `least square cell based` for gradients. Option for the transient formulation was set to `first order upwind`. Default values of the under relaxation factors are used for mesh 1 and 2, while they are two times reduced for meshes 3 and 4. Dual time stepping is used with 30 iterations per one step which is fixed at 0.001 sec.

Figure 3 shows the volume fraction of the glass beads along the line (x, 0, 2.5 mm) for different computational meshes. These profiles include the reservoir and the part of the glass channel upstream the wedge. Close to the point $x = 0$ a rapid increase of the glass beads volume fraction is observed indicating an appearing bow shock wave. As can be seen, the grid convergence is not reached, but the stand-off distance of the bow shock is not very sensitive to the spatial resolution.

3. Results

Figure 4 shows the profiles of the glass beads volume fraction along different lines (x, 0, 0), (x, 0, 2.5 mm) and (x, 0, 5 mm) ahead of the wedge with $\theta = 60°$. As we can see, the volume fraction of particles depends strongly on z-coordinate because of gravity that drags the glass beads to $z = 0$. Moreover, the volume fraction profile changes in the free-stream upstream of the bow shock, which is also caused by the influence of gravity. It is worth noting that the glass beads volume fraction exceeds the maximum packing limit (~ 0.63) on the glass channel bottom ($z = 0$) in the free-stream as well as in the close vicinity of the wedge apex. The profiles at the middle ($z = 2.5$ mm) and at the top ($z = 5$ mm) show that the glass beads volume fraction exceeds the maximum packing limit only in the bow shock or behind it close to $x = 0$. It is worth noting that the flow solver is not particularly fit for the accurate simulation of complicated gas-solid interactions. On the other hand, the flow solver has already been successfully applied to many specific flow problems [13, 14, 15, 16].

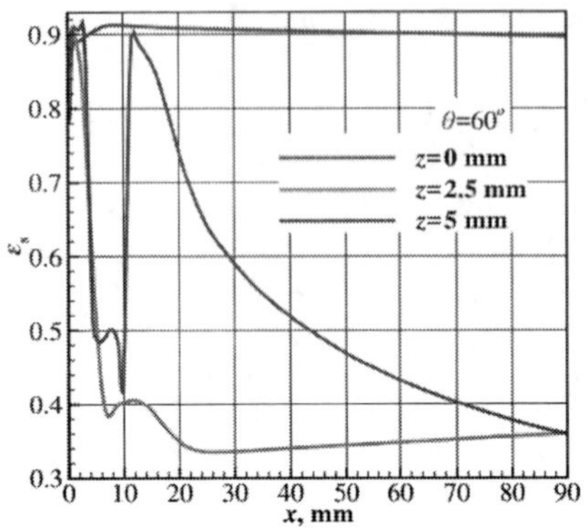

Figure 4. Volume fraction of glass beads at $y = 0$ for different z−coordinates.

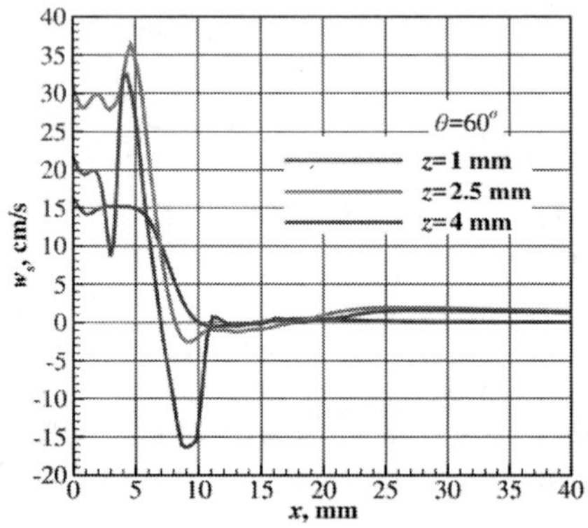

Figure 5. z-velocity (w) of glass beads at $y = 0$ for different z−coordinates.

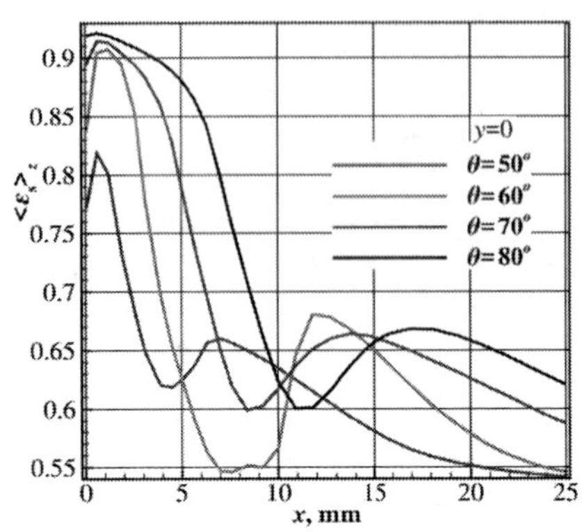

Figure 6. Average volume fraction of glass beads at $y = 0$.

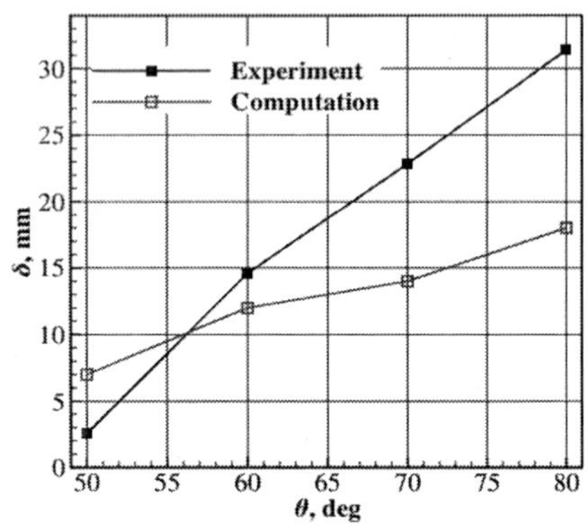

Figure 7. Shock stand-off distance: experimental data [4] and numerical data.

Figure 5 shows z-velocity component, w_s, of the glass beads. As we can see the particles at $z = 1$ mm rise in the bow shock and form a flat, plateau-like velocity profile behind it. In the middle ($z = 2.5$ mm) and close to the top ($z = 4$ mm) of the channel, the particles have more complicated behaviour of w_s that includes regions with negative and positive values.

Figure 6 shows the particles volume fraction averaged in z-direction. In a certain sense, the volume fraction averaged across the channel corresponds to the experimental measurements [4]. At the same time, the experimental shadowgraph images show only light coloured (low volume fraction) and dark coloured (high volume fraction) regions, and they do not provide information about the absolute value of the volume fraction. Therefore, the accurate comparison between experimental and numerical data is quite a challenging task. Here, the first step of such comparison is presented in figure 7 for the bow shock stand-off distance. Numerical stand-off distances are taken from the profiles of the averaged volume fraction given in figure 6. Since the free-stream value of the averaged volume fraction is not constant it is hard to identify the beginning of the bow shock. Therefore, the first peak of the averaged volume fraction of the glass beads is taken as the position to determine the numerical stand-off distance of the bow shock. Although this criterion seems to be straightforward, an agreement between the numerical and the experimental data in figure 7 is not observed. The accurate experimental identification of the averaged value of the glass beads volume fraction in the beginning of the bow shock can make such comparison more correct. On the other hand, it can be noticed that the numerically predicted averaged value of the glass beads volume fraction is much higher than the maximum packing limit (~0.63). This problem of TFM computation also needs to be properly resolved in the future in order to make a correct validation of the numerical model.

4. Conclusion

The Two-Fluid Model implemented in ANSYS Fluent can qualitatively reproduce the structure of a gravity-driven granular flow around a wedge, although the considered model does not accurately quantify some of the flow features:

- at the bottom of the channel ($z = 0$) the volume fraction of the glass beads ($\varepsilon_s \sim 0.9$) exceeds the maximum packing limit (~ 0.63);
- in the close vicinity of the wedge apex the volume fraction of the glass beads also increases higher than the maximum packing limit.

The comparison between the numerical and the experimental results shows a disagreement that can be explained by:

- inaccuracy of the model for calculating the volume fraction of the glass beads;
- impossibility of the accurate comparison of the bow shock stand-off distance in the experiment and the calculation, since the experimental data do not allow reading the absolute value of the volume fraction of the glass beads averaged through the glass channel.

Acknowledgments

The author would like to thank Ronith Stanly for the first discussion of this work, Aqib Khan for the clarification of the experimental setup. The author is especially grateful to Anna Shoeva, who helped to proof-read the paper. The research was partly carried out within the framework of the Program of Fundamental Scientific Research of the state academies of sciences in 2013-2020 (project No. AAAA-A17-117030610138-7) and Russian Foundation for Basic Research (project No. 18-38-20113). Computations are performed at Novosibirsk State University (NSU).

References

[1] Kuipers J 1990 Ph.D. thesis University of Twente
[2] Muller C, Holland D, Sederman A, Scott S, Dennis J and Gladden L 2008 *Powder Technology* **184** 241–253
[3] Stanly R and Shoev G 2018 *Chemical Engineering Science* **188** 132–149, doi:10.1016/j.ces.2018.05.030
[4] Khan A, Hankare P, Verma S, Kumar R and Kumar S 2019 *32nd International Symposium on Shock Waves*
[5] Gray J and Cui X 2007 *J. Fluid Mech.* **579** 113–136, doi:10.1017/S0022112007004843
[6] Garai P, Verma S and Kumar S 2019 *J. Vis.* **22** 729–739, doi:10.1007/s12650-019-00558-5
[7] Bondar Y, Markelov G, Gimelshein S and Ivanov M 2006 *J. Thermophys. Heat Tr.* **20** 699–709, doi:10.2514/1.18758
[8] Bountin D, Gromyko Y, Kirilovskiy S, Maslov A and Poplavskaya T 2018 *Thermophys. Aeromech.* **25** 483–496, doi:10.1134/S0869864318040029
[9] Kratova Y, Kashkovsky A and Shershnev A 2018 *AIP Conference Proceedings* **2027** 040086, doi:10.1063/1.5065360
[10] Kratova Y, Kashkovsky A and Shershnev A 2019 *Thermal Science* **23** 623–630, doi:10.2298/TSCI19S2623K
[11] Gidaspow D 1994 *Multiphase Flow and Fluidization* ed Gidaspow D (San Diego: Academic Press)
[12] Kudryavtsev A, Shershnev A and Rybdylova O 2019 *International Journal of Multiphase Flow* **114** 207–218, doi:10.1016/j.ijmultiphaseflow.2019.03.009

[13] Polivanov P, Gromyko Y, Sidorenko A and Maslov A 2017 *J. Appl. Mech. Tech. Phy.* **58** 845–852, doi:10.1134/S0021894417050108
[14] Polivanov P 2018 *Thermophys. Aeromech.* **25** 789–792, doi:10.1134/S0869864318050153
[15] Mironov S G, Poplavskaya T V and Kirilovskiy S V 2017 *Thermophys. Aeromech.* **24** 629–632, doi:10.1134/S0869864317040151
[16] Kirilovskiy S, Poplavskaya T, Tsyryulnikov I and Maslov A 2017 *Thermophys. Aeromech.* **24** 421–430, doi:10.1134/S0869864317030106

Formation of the internal structure of copper particles during their ball milling followed by spheroidization

A V Smirnov[1], A E Chesnokov[1] and T M Vidyuk[1,2]

[1] Khristianovich Institute of Theoretical and Applied Mechanics SB RAS, 4/1 Institutskaya str., Novosibirsk, 630090, Russia
[2] Institute of Solid State Chemistry and Mechanochemistry, SB RAS, 18 Kutateladze str., Novosibirsk, 630117, Russia

chae@itam.nsc.ru

Abstract. The paper presents the results of a study of the influence of high-energy effects on surface morphology and the formation of the internal structure of copper particles. It is shown that the ball milling of copper powder in a planetary mill leads to the formation of layered agglomerates consisting of many deformed ultrafine particles. It is noted that defects in the form of microcracks and micropores are observed in the volume of the agglomerated particle. Their subsequent treatment in a plasma jet leads to the formation of spherical particles. It is established that during the interaction of the molten material of a particle with distributed gas inclusions throughout the particle's volume during plasma treatment, local copper oxidation occurs. Copper oxide inclusions of predominantly rounded shape are formed, ranging in size from tens of nanometers to 7 microns, uniformly distributed over the particle volume.

1. Introduction

Metal, composite, and ceramic powders are widely used in thermal spraying and cold spraying technologies, additive technologies, powder metallurgy, etc. The size, shape, and internal structure of the initial powder particles effect on the physical, mechanical, and chemical properties of the resulting coatings. For example, in [1] it was shown that the porosity of detonation coatings when replacing the initial sprayed particles of non-spherical alumina with spherical particles decreases by half, and when using spherical hollow particles it decreases by an order of magnitude. In this case, the deposition efficiency increases and amounts to 23, 40, and 51%, respectively. Also, the maximum microhardness of the coatings is obtained by spraying a spherical hollow powder.

A preliminary high-energy effect on the initial particles, for example, during the mechanical processing of powders in planetary ball mills, leads to a change in particle size and shape depending on the physical properties and chemical composition of the material. During ball milling, agglomerated particles are formed, characterized by a layered structure, the presence in its volume of closed microcracks and micropores; surface cleaning occurs due to the destruction of oxide films and adsorbed layers, etc. [2].

Usually, as a raw material for production of spherical powder particles during plasma treatment, a powder of non-equiaxed shape is used [3, 4]. Treatment of particles in a plasma jet is accompanied by heating of the material to high temperatures, sufficient for its melting and synthesis reactions, resulting in the possibility of obtaining spherical metal-ceramic, including hollow particles [5]. In [6], a method for producing dense spherical titanium particles of decamicron size by plasma treatment of agglomerated ultrafine particles having high open porosity is shown. The addition of 10 wt. % titanium nitride nanoparticles to the initial powder leads to the formation of particles with distributed gas inclusions throughout the volume. In this paper, we study the influence of the conditions of mechanical processing of copper powder and its subsequent plasma spheroidization on the particle size distribution, surface morphology, internal structure, and phase composition of the resulting particles.

2. Experimental procedure

Copper powder PMS-1 was used in this work (figure 1 (a)). The volume distribution of particles was determined on particle size analyzer (0.04 – 2000 μm) LS 13 320 (Beckman Coulter). Particle morphology was determined using Evo MA15 electron microscope (Carl Zeiss). The classification of powder particles by size was carried out on a vibration table using a set of analytical sieves. The ball milling of the powders was carried out in Activator-2SL planetary mill, which has two cylinders with volume of 250 ml each with inner radius of 42.5 mm, the mass of loaded balls is 160 g for each cylinder, the acceleration of grinding bodies is 117 g, the mass of the processed material loaded is 30 g. Steel balls with diameter of 5 mm were used as grinding bodies. Material processing was carried out in an air atmosphere.

Spheroidization of the powder was carried out using a plasma torch with an interelectrode insert with a power of up to 50 kW. A single-sided injection of the treated powder was used immediately before the exit section of the plasma torch nozzle. The main plasma-forming gas was a mixture of argon and helium supplied at rate of 0.4 gs^{-1} and 0.2 gs^{-1}, respectively. Argon with flow rate of 0.05 gs^{-1} was used as a transporting gas for introducing the copper powder into the plasma jet from the powder feeder. The processing was carried out with the following values of the operating parameters: the feeder rate of 3.5 kgh^{-1}, the arc current of 200 A, the voltage of 160 V, the secondary gas flow supplied to protect the anode (argon) of 0.15 gs^{-1}.

3. Results and discussion

During ball milling, dispersion of the initial copper particles first occurs, which leads to an increase of their specific surface area. Then, due to the high surface activity, ultrafine particles form agglomerates having a layered structure [7-11]. During the interaction of the grinding body with the particle, kinetic energy has losses on plastic deformation of the particle with the release of a large amount of heat, up to 95% [12]. It was shown in [13] that the temperature in the contact can reach the melting temperature of materials. An increase in the machining time leads to an increase in the temperature of the entire system up to a temperature close to the melting temperature of the material processed. Particles under such conditions become plastic and begin to form agglomerates; granulation is observed by the rolling method (figure 1 (b)). Figure 2 shows the particle size distribution depending on the time of mechanical processing of the initial powder in the form of curves that characterize the change in the average particle size and the main range $d_{10} - d_{90}$.

(a)

(b)

Figure 1. Morphology of copper particles: (a) initial and (b) ball milled for 180 s.

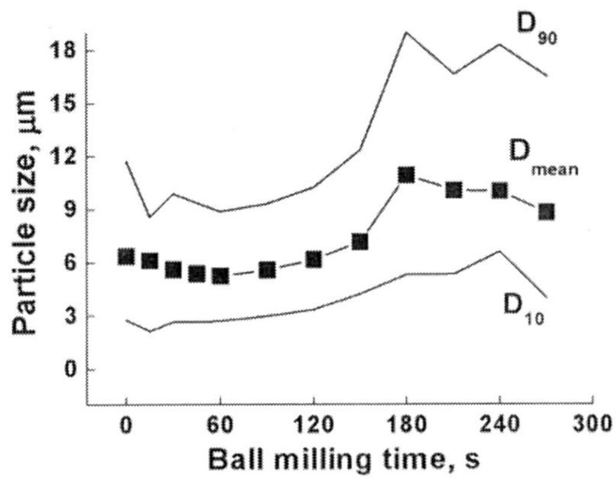

Figure 2. Volume size distribution of the initial and ball milled copper particles.

X-ray phase analysis of the ball milled powder showed the presence of peaks corresponding to the phase of only copper with a lattice parameter of 3.615 A.

Particles ball milled during 180 seconds of fraction 90 – 300 μm was used as the initial powder for plasma treatment (figure 1 (b)). As a result of treatment of the copper powder in a plasma jet, spherical particles are obtained, the morphology of which is shown in figure 3. The average particle size d_{mean} = 24 μm, d_{10} = 6, and d_{90} = 46 μm. The obtained sizes indicate the disintegration of particles in the plasma jet during processing.

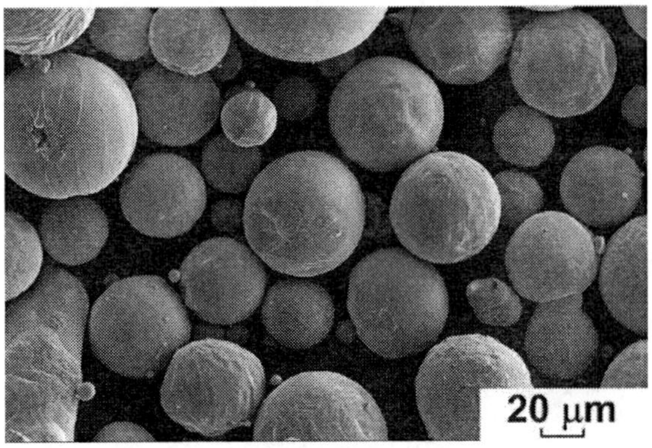

Figure 3. SEM micrographs of a general view of a spheroidized copper powder.

The initial powders consist of agglomerated particles, which are characterized by internal porosity due to the formation of a variety of randomly packed copper particles during ball milling. Their treatment in a plasma jet leads to the formation of either dense spherical particles or hollow microspheres and microspheres with distributed gas inclusions (figure 4). The realization of this or that process depends on the size of the agglomerates, their porosity, the intensity of the interphase heat exchange "agglomerated particle – plasma flow", and the time of their stay in the plasma jet [14]. As a rule, hollow spheres cannot be obtained for metal powders. This is due to the high thermal conductivity of metals, in contrast to ceramics, since to obtain hollow spheres a large temperature gradient is required over the cross section of the treated particle. Preliminary mechanical processing of the powder allows one to obtain closed porosity, which affects the final structure of the material.

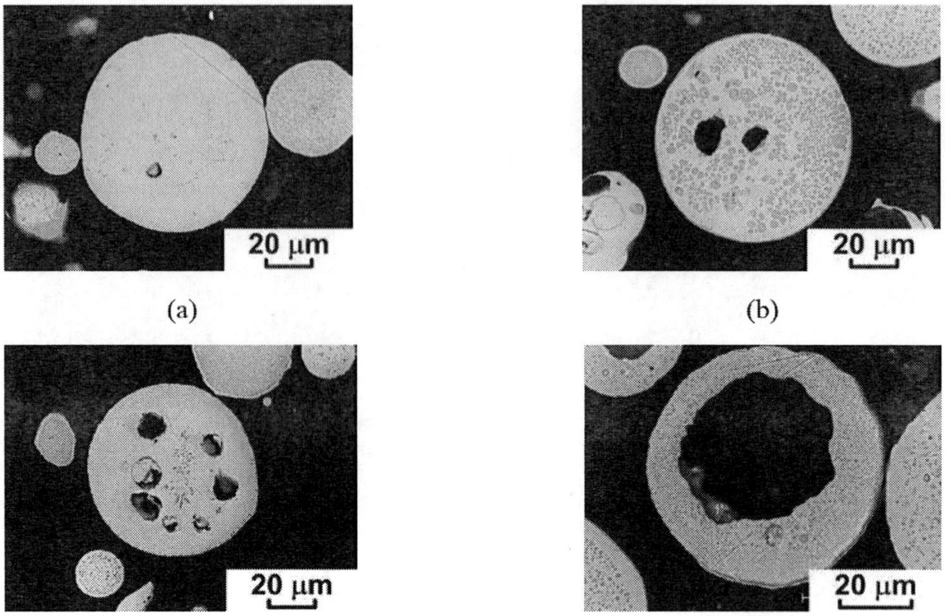

(a) (b) (c) (d)

Figure 4. A cross-section of spheroidized copper particles: (a) dense microspheres; (b, c) microspheres with distributed gas inclusions; (d) hollow microsphere.

The interaction of the molten material of the particle with localized gas volumes leads to its oxidation. Figure 4 shows the distribution of finely dispersed particles of copper oxide over the volume of the treated particle with inclusions ranging from tens of nanometers to several micrometers. X-ray phase analysis of the spheroidized powder showed the presence of peaks corresponding to the phase of copper and not more than 7 at. % of copper oxide Cu_2O.

4. Conclusion

It is shown in the work that an increase in the ball milling time leads to an increase in the mass fraction of particles larger than 90 μm in the machined material and an increase in the temperature of the entire system. Heated particles become ductile, which leads to a granulation process by the rolling method. The internal structure of the ball milled particles has a layered structure with microcracks and closed micropores. Treatment of agglomerated particles in a plasma jet allows one to obtain spheroidized particles. The internal structure of spheroidized particles contains distributed gas pores along with finely dispersed inclusions of copper oxide (with a content of not more than 7 at. %) ranging in size from tens of nanometers to several micrometers formed during plasma treatment. This preliminary ball milling of the powder allows one to produce hollow metal spheres.

Acknowledgements

This study was supported by the Russian Science Foundation (Project No. 19-19-00335) using equipment of the Joint Access Center «Mechanics» of ITAM SB RAS.

References

[1] Ulianitsky V Yu, Solonenko O P, Smirnov A V, Batraev I S 2014 *Proceed. Intern. Thermal Spray Conf.* **3878** 668–73
[2] Suryanarayana C 2001 *Progress in Materials Science* **46** 1–184
[3] Kotlyarov V I, Beshkarev V T, Kartsev V E, Ivanov V V, Gasanov A A, Yuzhakova E A, Samokhin AV, Fadeev A A, Alekseev N V, Sinayskiy M A, Tretyakov E V 2017 *Inorganic*

Materials: Applied Research **8** 452–8
[4] Grigor`ev A V, Razumov N G, Popovich A A, Samohin A V 2017 *Metallurgy and Materials Science Scientific and technical lists of the St. Petersburg State Polytechnic University* **23** 247–55
[5] Chesnokov A E, Smirnov A V 2019 *Inorganic Materials* **55** 22–7
[6] Chesnokov A E, Smirnov A V 2019 *Journal of Siberian Federal University. Engineering & Technologies* **12** 203–13
[7] Boldyrev V V 2006 *Russ. Chem.* **75** 177–89
[8] Prokudina V K, Ratnikov V I, Borovinskaya I P and Merzhanov A G 1981 (Kiev: Naukova Dumka) 29–34.
[9] Lyakhov N Z, Talako T L, Grogorieva T F 2008 (Novosibirsk: Parallel) p 167
[10] Avvakumov E G 1988 *Mechanical Methods for Activation of Chemical Processes* (Nauka: Novosibirsk) p 305
[11] Boldyrev V V 1991 *Mechanochemical Synthesis in Inorganic Chemistry* (Novosibirsk: Nauka) ed E.G. Avvakumov p 264
[12] Taylor G 1934 *Proc. Roy. Soc.* **A145** 362
[13] Bowden F 1954 *Proc. Roy. Soc.* **A223** 29
[14] Solonenko O P 2014 *Thermophys. Aeromech* **21** 735–46

Determination of relative concentrations in two-component mixtures with unknown composition by laser-induced fluorescence spectra

E O Tsibulskaya and N A Maslov

Khristianovich Institute of Theoretical and Applied Mechanics SB RAS,
4/1 Institutskaya st., Novosibirsk, 630090, Russia

euglenaria@gmail.com

Abstract. The paper is devoted to the development of a new methodology of statistical analysis of laser-induced fluorescence excitation-emission matrices. The method allows one to calculate the number of fluorescent components, their excitation and fluorescence spectra in samples under study without using any prior information about their nature. The algorithm is based on the principal component method and is modified for analysis of excitation-emission matrices. Its features were tested on two-component optical phantoms, namely, optically thin mixtures of fluorescent dyes (pyridoxine and fluorescein). The calculated excitation and fluorescence spectra of components coincided with the initial ones with good accuracy (deviation less than 5% in intensity). The weight coefficients proportional to concentrations were calculated by the new algorithm. The comparison of its ratios with those calculated from known concentrations also showed good agreement (deviation of 5%).

1. Introduction

The method of laser-induced fluorescence (LIF) has found wide application in aerophysics and combustion physics for registration of gas flows and flames [1]. In medicine, this method is used in the optical diagnosis of various diseases and the assessment of the biological tissues' state [2]. Its advantages are fluorescence's high sensitivity to changes in a substance and the possibility of non-invasive studies.

Contribution to the fluorescence spectra is determined by all fluorophores, which compose the samples. However, in condensed matter, the molecules are located close to each other. As a result, the energy levels overlap, and the lines in the fluorescence spectra expand, turning into continuous bands. Therefore, in multi-component samples, composition analysis of the samples by spectra is significantly complicated due to overlapping spectra.

The most existing studies use a single laser wavelength to excite fluorescence. However, in multi-component samples. changing the excitation wavelength leads to radical changes in the fluorescence spectra shape due to the different excitation efficiency of the components. That is, the measurement of excitation-emission matrices (sets of fluorescence spectra at different excitation wavelengths) provides additional information that can also be used in data analysis.

At the present moment, the topical problem is a development of sample diagnosis method which includes obtaining information about a number of fluorophores in the samples and its' spectral characteristics using only their statistical differences. First of all, the presented problem is important for the study of samples with a previously unknown composition, when there is no any information on fluorophores available. Also, many substances show spectral variability depending on changes in their environment (temperature, polarity, pH, etc.), which is difficult to evaluate in advance. Therefore, the new algorithm had been proposed to solve this problem and analyze the excitation-emission matrices [3]. It is based on the tensor version of the principal component analysis (PCA). The standard PCA algorithm allows one to best describe the statistical differences in the initial data. However, the calculated principal components are orthogonal, and, consequently, they have a sign variable shape. This paper investigates the ability of the proposed method to calculate the positive excitation and fluorescence spectra, and to estimate the relative concentrations of fluorophores in the sample. This algorithm was applied to the analysis of optically thin two-component samples' spectra (optical phantoms) to refine the technique. Concentrations of fluorescent substances were monitored during their creating in presented experiment.

2. Materials and methods

2.1. Experimental setup

Figure 1 shows the experimental setup for measuring of the laser-induced fluorescence spectra. Pulsed laser system with an optical parametric generator (Vibrant (HE) 355 II + UV, Opotek Inc, USA) was used to excite fluorescence. It allows of tuning the wavelength in the range 210 - 350 nm with step of 10 nm.

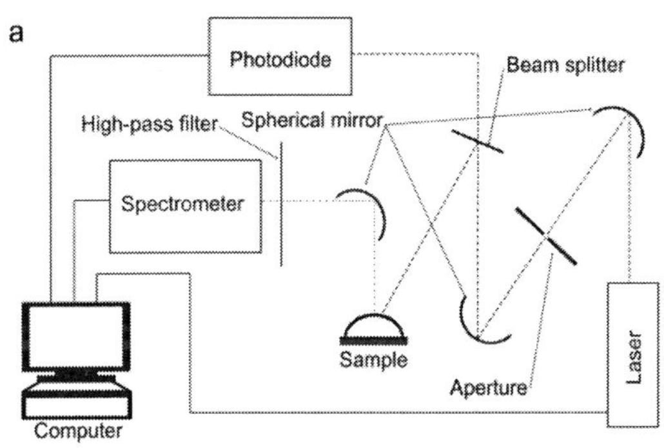

Figure 1. Scheme of experimental setup for laser-induced fluorescence spectroscopy.

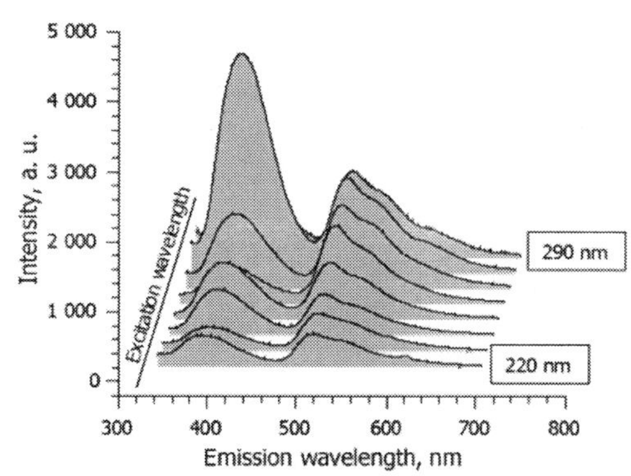

Figure 2. The excitation-emission matrix of mixture No. 2.

The system of spherical mirrors and the aperture were used to remove the scattered light from the laser system. Laser beam was divided into two beams which were sent to the examined sample and the photodiode to measure the pulse energy. Acton SpectraPro SP2300 spectrograph with a Pixis 256 CCD matrix (Princeton Instruments, United States) was used to record the fluorescence spectra. The 300 nm high pass filter was used to protect the spectrometer matrix from laser radiation.

The excitation-emission matrices were measured in 220-290 nm excitation wavelength range. The pulse energy was limited to 200 J per pulse, because the fluorescence intensity linearly depends on the excitation energy in this range. At higher energy densities, the saturation of the fluorescence signal may occur [4]. So the linear relationship of fluorescence intensity on the absorbed dose was controlled during the measurements of the spectra. Fluorescence spectra were recorded in the wavelength range 340 - 700 nm.

2.2. Modified PCA algorithm

In [3], we had proposed a modified PCA algorithm for calculating the excitation and fluorescence spectra of fluorophores from excitation-emission matrices for samples with initially unknown composition. To calculate the principal components, the tensor form of NIPALS (non-linear iterative partial least squares) algorithm [5] was used with several modifications. The total number of fluorophores was chosen based on the assumption of the samples' nature. The first approximation of principal components was calculated according with the standard algorithm. After that, components were sequentially refined over several hundred iterations using new algorithm of the minimization of initial data's functional. Unlike the standard NIPALS algorithm, iterative refinement of the principal components allows us to switch from sign variable spectra of the principal components to positive ones. Then, they can be identified using a database of known fluorophores.

3. Results and discussion

The modified PCA algorithm was tested using an optical phantoms, mixtures of two fluorescent dyes, namely, pyridoxine and fluorescein with fluorescence maxima of 395 and 514 nm, respectively. Total amount of examined samples was 3. All mixture samples had been prepared using water solutions with initial concentrations of pyridoxine 5 μg/ml and fluorescein 0.12 μg/ml. At these concentration levels, resulting mixtures were optically thin, i.e., the absorption depth of laser radiation was greater than the thickness of the sample. Series of subsequent dilutions showed that fluorescence intensity linearly depended on the concentration of the substances.

Figure 2 presents the example of excitation-emission matrix for mixture No. 2. All obtained excitation-emission matrices were analyzed using the modified PCA algorithm. After 200 iterations, the RMS deviation of the normalized calculated spectra from the initial ones was within 5% of the maximal intensity.

Figure 3 shows the principal components (PC1, PC2) been obtained as a result of PCA analysis and their comparison with the spectra of the initial fluorophores. The 1st component corresponds to pyridoxine and the 2nd component corresponds to fluorescein, and the calculated spectra repeat the initial ones rather well.

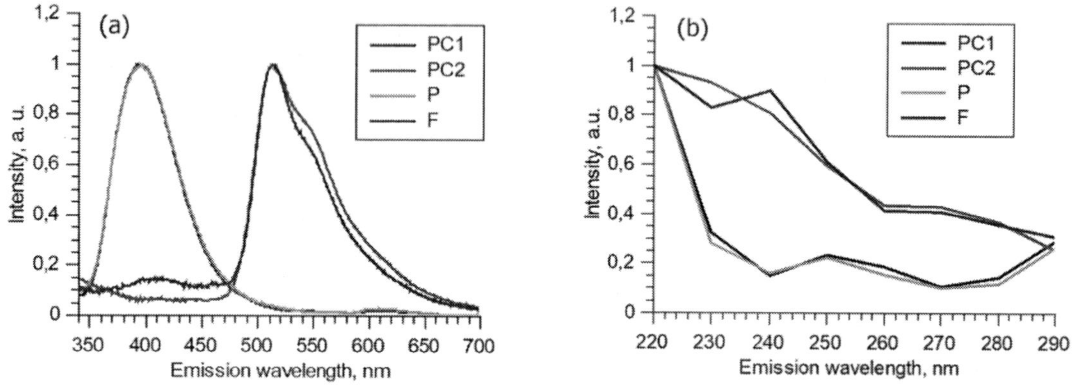

Figure 3. The principal components: (a) fluorescence spectra, (b) excitation spectra. PC1, PC2 are calculated principal components; P, F are initial substances (pyridoxine and fluorescein, respectively).

The first and the second columns of table 1 present the weight coefficients obtained as a result of PCA analysis. They determine the contributions of the corresponding components to mixtures' matrices. The ratios of the 2nd principal component's weight coefficients to 1st ones were calculated for each sample and compared with the concentrations of the initial substances in the mixtures (table 1). It can be seen that the contribution ratios of the principal components coincide with the concentration ratio with deviation of 5%.

Table 1. Calculated weight coefficients, comparison with fluorophores' concentration ratios

	Weight coefficients		Weight components' ratios normalized to unity for the first mixture	
Sample	PC1	PC2	PC2/PC1	F/P
Mixture 1	3252.64	666.83	1	1
Mixture 2	3043.38	1280.63	2.06	2
Mixture 3	6071.1	592.85	0.48	0.5

The accuracies of the reconstructed excitation and fluorescence spectra are shown in table 2. They had been calculated as the standard deviation normalized to the spectral maximum intensity and expressed as a percentage. It can be noted that the modified PCA algorithm allows us to calculate the fluorescence and excitation spectra of the initial components from matrices of two-component mixtures and to estimate the relative contributions of these components with sufficient accuracy.

Table 2. RMS deviations of calculated fluorescence and excitation spectra from the original ones

Fluorophore	Fluorescence spectra (%)	Excitation spectra (%)
Pyridoxine	1	2
Fluorescein	5	5

4. Conclusion

The present study showed that the modified PCA algorithm allows one to carry out spectral analysis of sets of optically thin two-component samples of unknown nature. It can calculate with high accuracy the fluorescence and excitation spectra of individual fluorescent components from the excitation-emission matrices. For optical phantoms the principal components' weight coefficients calculation error was within 5%. The calculated excitation and fluorescence spectra also had coincided in intensity with the initial ones with discrepancies of 5% and lower. The presented algorithm does not use any prior information about the spectra of fluorophores and can be used to study samples with initially unknown composition, like biological tissues.

Acknowledgments

The research was carried out within the framework of the Program of Fundamental Scientific Research of the state academies of sciences in 2013-2020 (project No. AAAA-A18-118021590029-2).

References

[1] Lozano A, Yip B and Hanson R K 1992 *Exp. Fluids.* **13** 369–76
[2] Razhev A M, Iskakov I A, Churkin D S, Orishich A M, Maslov N A, Tsibul'skaya E O, Lomzov A A, Ermakova O V, Trunov A N and Chernykh V V 2018 *Quantum Electron.* **48** 481-86
[3] Maslov N A and Papaeva E O 2016 *Tech. Phys. Lett.* **42** 718-21
[4] Maslov N A 2018 *J. Fluoresc.* **28** 689-93
[5] Geladi P and Kowalski B R 1986 *Anal. Chim. Acta.* **185** 1-17

Numerical modelling of plasmachemistry induced by cold atmospheric plasma jet near surface

S Vagapov[1] and I V Schweigert[1,2]

[1] Khristianovich Institute of Theoretical and Applied Mechanics SB RAS, 4/1 Institutskay str., Novosibirsk, 630090, Russia
[2] George Washington University, Washington D.C. 20052, USA

ischweig@itam.nsc.ru

Abstract. Recently the cold plasma jets are widely used for the development of anticancer therapy. The main mechanism of atmospheric plasma jet action on cells is related to the production of nitrogen and oxygen radicals in plasma-stimulated chemical reactions, followed by the diffusion of radicals into the media with cells. The increase of the plasma jet treatment efficiency depends on a detailed study of the plasma-chemical processes that occur when the plasma jet interacts with the target. In this work, a numerical study of the plasma-chemical processes for dry air is carried out, the main ways of production and loss of RONS are detected and mixture composition near the surface during plasma jet processing is estimated.

1. Introduction

Nowadays cold atmospheric plasma jets (CAPJs) are on the stage of *in vitro* and *in vivo* tests [1-3] as a perspective instrument for anti-cancer treatment. During the contact of CAPJ with ambient air large amount of various reactive oxygen and nitrogen species (RONS) are being produced near the bio target. Negative ions, atomic oxygen and nitrogen, excited states and ozone (O_3) are also important for medical applications.

The typical cold atmospheric plasma jet devices operate at 10-50 kHz-driven AC voltage of 2.5-6.5 kV amplitude and with the cylindrical and plane designs. The inert gas (helium or argon) is pumped through the plasma device with the rate of 1-5 l/min. A sequence of streamers generated with the frequency of applied voltage propagates along the flow of noble gas to the target. Intensive generation of RONS occurs in the head of the streamer due to the presence of the strong electric field and energetic electrons.

For the efficient applications of CAPJs the detailed study of plasma enhanced chemical processes is required. With appropriate plasma chemical model it is possible to detect the main trends in chemical processes, define optimal parameters of the experiment for the most intense generation of certain group of radicals, estimate their concentrations and obtain the main ways of their production and loss.

Two- or one- dimensional models of plasma chemistry require the solution of the system of 2D or 1D mass-balance equations for every species together with the solution of the balance equation of ions and electrons, the energy equation and Poisson's equation (see, for example [4]). This type of models requires an enormous computational resources so large amount of chemical reactions cannot be included into the plasma chemistry model. In this work, for the first step, we use the 0D model. In this approach all spatial derivatives in mass-balance equations are assumed to be zero and only time evolution is considered. Obviously, 0D model is not able to fully describe the plasma-chemical processes in plasma jets since the latter are an extremely heterogeneous phenomenon both in time and space. Typically, 0D models are used to describe plasma chemistry in more simple configurations, for example, for a dielectric barrier discharge (DBD) between two flat electrodes. Nevertheless, 0D model allows us to obtain qualitative and quantitative conclusions about the chemical processes that occur in the plasma and for more complex geometries. Using 0D model, it is possible to take into account a larger number of chemical reactions in order to describe the main trends in the plasma enhanced chemistry.

In this work, we study plasma chemistry near the surface with a periodical change of electron temperature that corresponds to the AC voltage behavior. In this study relative humidity of the air is assumed to be zero. Our purpose is to obtain the main ways of production and loss of RONS and estimate their concentrations near the target surface.

2. Numerical simulation

For plasma treatment only the composition of the mixture near the surface is the most important. Therefore, in this work plasma chemical processes are studied over the target surface where the gas flow rate is assumed to be zero and it is possible to neglect convection transfer of mass. For the calculation the point where the streamer touches the surface was chosen. As shown in [6]. The streamer has a toroidal shape. In the point of plasma-surface contact, the generation of RONS is the most intense because of the high density of electrons ($10^{12} - 10^{13}$ cm^{-3}) and their high temperature (3-7 eV). The initial electron density was taken from 2D fluid model simulations [6] and assumed to be $2 \cdot 10^{12}$ cm^{-3}. Initial concentrations of He, N_2, O_2 were obtained from the solution of Navier-Stokes equations ($He/\,N_2/O_2 = 0.977/0.018/0.005$).

A 2D calculation of the dynamics of streamer breakdown [6] shows that the value of the electron temperature at the surface is approximately constant for 3-5 μs after the arrival of the streamer which corresponds to a part of positive half-cycle of the voltage. Then there is a sharp drop in the intensity of ionization and excitation and during the next ~ 70 μs ionization is absent. This indicates that electron temperature at this time is very low as well as their concentration. The plasma device periodically generates streamers on the positive half-wave of alternating voltage so when the next streamer arrives, the process repeats again.

To simulate the process of streamer coming to the surface, the electron temperature is set in the form of a periodic function of time. Each period of this function is divided into 4 parts: 1) a constant maximum value of electron temperature $T_e = T_{e,max} = 3$ eV, 2) lowering T_e to gas temperature $T_{gas} = 300$ K, 3) a constant value of the electron temperature $T_e = T_{gas}$, 4) increasing of electron temperature T_e to maximum. Function of T_e and calculated evolution of electron concentration are shown in figure 1.

The evolution of concentration of species is described with the following equations

$$\frac{\partial n_\alpha}{\partial t} = \sum_{\lambda=1}^{N_\alpha} (\nu_\lambda^R - \nu_\lambda^L) k_\lambda \prod_{l=1}^{M} n_l^{\nu_l^L},$$

where n_α is number density of αth component, ν_λ^R and ν_λ^L are right- and left-handed stochastic coefficients of λth reaction, N_α is the total number of reactions where αth component participated, k_λ is the rate coefficient of λth reaction.

The kinetic scheme includes 627 chemical reactions and 38 related components. There are reactions of ionization processes (including Penning ionization), excitation and quenching of electronic states, dissociation, recombination, charge exchange, etc.

59 of all chemical reactions involve electrons. They include reactions of ionization, dissociation, attachment and excitation. The rate coefficients of these reactions were calculated from the electron energy distribution function obtained from a numerical solution of the Boltzmann equation. The Boltzmann equation was periodically solved using the scattering cross sections of the corresponding processes and the current mixture composition, using the BOLSIG+ [7] program. The reaction rate

coefficients were formed in the form of tables and linearly interpolated for the current value of the electron energy.

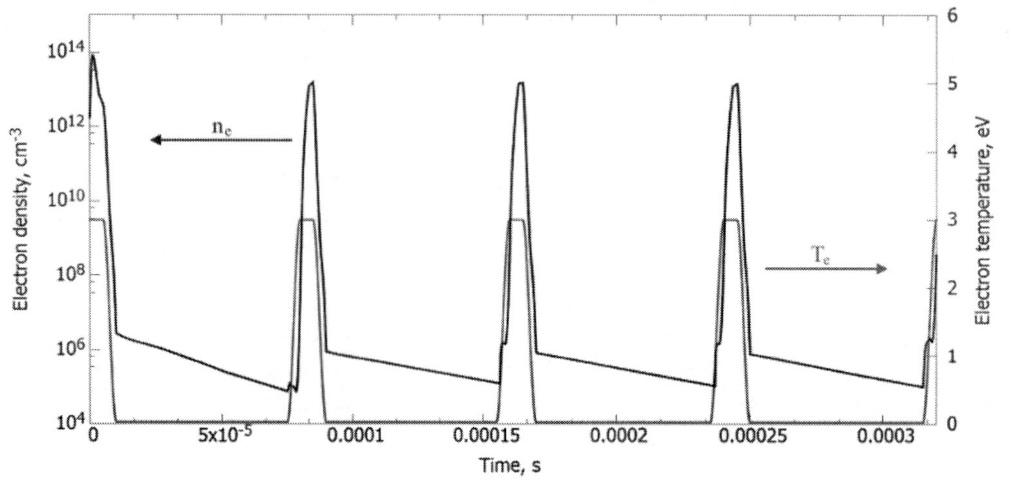

Figure 1. Function of electron temperature and calculated evolution of electrons.

3. Results

When the temperature of electrons is maximal their concentration reaches the value of $2.5 \cdot 10^{13}$ cm^{-3}. An analysis of relative contributions of the reactions showed that Penning ionization plays important role in electron production (about 55%). That are reactions of ionization of N_2, O_2 and O by excited helium:

$He^* + N_2 \rightarrow He + N_2^+ + e,\quad k = 7 \cdot 10^{-11}$ cm^3/s, (~30%)
$He^* + O_2 \rightarrow He + O_2^+ + e,\quad k = 2.6 \cdot 10^{-10}$ cm^3/s, (~12%)
$He^* + O \rightarrow He + O^+ + e,\quad k = 3.96 \cdot 10^{-10}(\frac{T_{gas}}{300})^{0.17}$ cm^3/s, (~12%)

Reaction of ionization of molecular nitrogen by electrons ($e + N_2 \rightarrow 2e + N_2^+$) contributes about one fifth in total electron production. Ionization of molecular and atomic oxygen contributes only 11% due to smaller amount of oxygen. Reaction of ionization of atomic nitrogen ($e + N \rightarrow 2e + N^+$) does not play significant role because of relatively high threshold energies of dissociation of N_2 and ionization of N (~10 eV and ~14.5 eV, respectively).

Negative ions of O and O_2 are formed during the part with maximum electron temperature by the reactions of electron attachment. As soon as the electron temperature drops to the gas temperature negative ions intensively recombine with positive ions. Concentrations of O^- and O_2^- reach the values of $3.5 \cdot 10^{11}$ cm^{-3} and $2.4 \cdot 10^{11}$ cm^{-3}, respectively.

Due to high energy of electrons dissociation processes also occur. Every time streamer comes to the surface, concentration of atomic nitrogen increases to the value of $6 \cdot 10^{14}$ cm^{-3} due to reactions of dissociation of NO and N_2:

$e + N_2 \rightarrow e + 2N,\quad f(\varepsilon)$
$e + NO \rightarrow e + N + O,\quad k = 7.4 \cdot 10^{-9}\exp(-\frac{6.5}{T_e})$ cm^3/s
$e + NO \rightarrow N + O^-,\quad f(\varepsilon)$
$N_2(a'^1\Sigma_u^-) + NO \rightarrow N_2 + N + O,\quad k = 3.6 \cdot 10^{-10}$ cm^3/s

Concentration of atomic oxygen reaches the value of $3 \cdot 10^{16}$ cm^{-3} due to dissociation of O_2 by electrons.

As soon as the electron temperature drops to the gas temperature O and N are being converted into molecules O_2 and N_2, respectively, through chain reactions:

$$e + O_2 \rightarrow O + O + e, \qquad k = 7.1 \cdot 10^{-9}\exp(-\frac{8.6}{T_e}) \text{ cm}^3\text{/s}$$
$$e + O_2 \rightarrow O(^1D) + O + e, \qquad f(\varepsilon)$$
$$O(^1D) + M \rightarrow O + M, \qquad k = 5 \cdot 10^{-12} \text{ cm}^3\text{/s}$$
$$N + NO \rightarrow N_2 + O, \qquad k = 8.2 \cdot 10^{-11}\exp(-\frac{410}{T_{gas}}) \text{ cm}^3\text{/s}$$
$$N + O_2(a^1\Delta_g) \rightarrow NO + O, \qquad k = 2.5 \cdot 10^{-10} \text{ cm}^3\text{/s.}$$

So the concentrations of O and N decrease by several orders of magnitude till the next streamer comes to the surface. Evolution of concentrations of atomic and molecular oxygen and nitrogen are illustrated in figure 2.

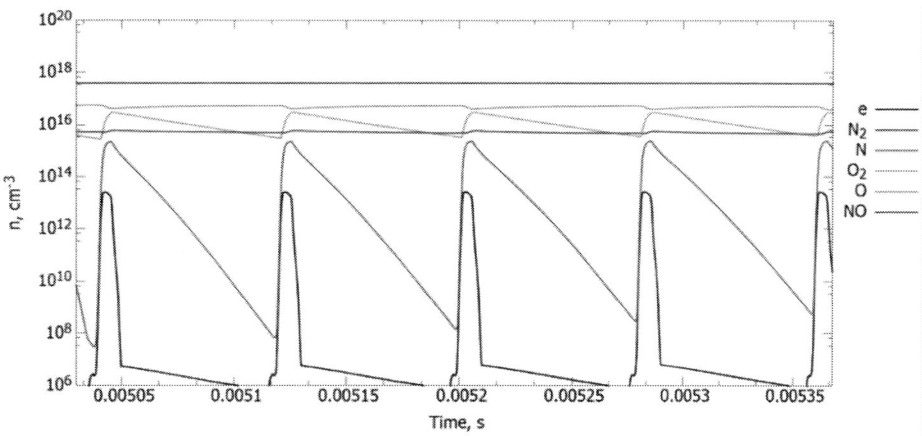

Figure 2. Calculated evolution of concentration of atomic and molecular oxygen and nitrogen, NO and electrons.

Because of high concentration of O and N in the discharge, such components as NO, O_3, NO_3, N_2O_3 and N_2O_4 are being generated through reactions of attachment of O and N with O_2, N_2 and each other. NO is very important for medical applications because of its extremely high reactivity [8]. O_3 on the other hand can cause cell's death via oxidative stress [9]. Calculated evolution of these species can be seen in figure 3.

It should be noted that concentrations of species averaged over the period vary greatly during the first ~5000 periods. As it is seen from figure 4 they gradually reach their equilibrium states.

4. Conclusion

In this work plasmachemical processes that occur during interaction of cold atmospheric plasma jets with dry air were studied. In order to simulate a sequence of streamer coming to the surface electron temperature was set as a periodical function of time. Using 0D modelling concentrations of RONS and ions near the surface were calculated. The calculation revealed that O, N, NO, O_3, NO_3, N_2O_3 and N_2O_4 are being actively generated in plasma jets. The main ways of their production and loss were determined. An analysis of relative contributions of various reactions also showed that Penning ionization reactions play crucial role in such systems.

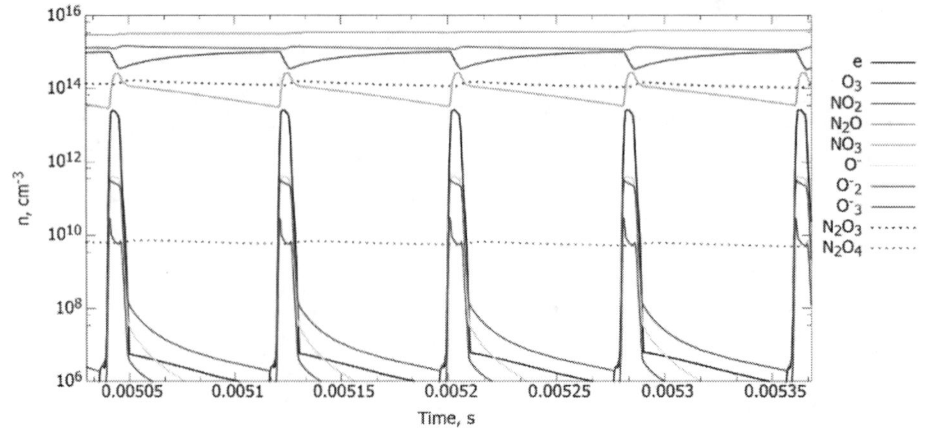

Figure 3. Calculated evolution of species concentrations.

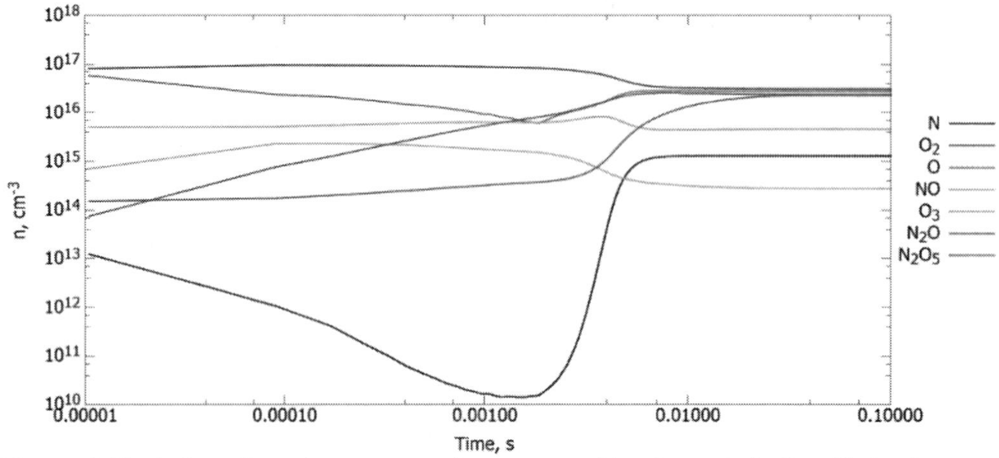

Figure 4. Evolution of species concentrations averaged by the period of applied voltage.

Acknowledgements
Authors are grateful to the Russian Science Foundation, project No. 19-19-00255, for supporting the work. One of authors, SV, thanks the RFBR, project No. 18-08-00510, for financial support for development of software for calculation of plasma chemistry.

References

[1] Gjika E, Pal-Ghosh S, Tang A, Kirschner M, Tadvalkar G, Canady J, Stepp M A and Keidar M 2018 *ACS* Appl. *Mater. Interfaces* **10** 9269–79
[2] Chen Z, Zhang S, Levchenko I, Beilis I I and Keidar M 2017 *Sci. Rep.* **7** 12163
[3] Chen Z, Simonyan H, Cheng X, Gjika E, Lin L, Canady J, Sherman J H, Young C and Keidar M 2017 *Cancers* **9** 61
[4] Schweigert I V, Vagapov S, Lin L and Keidar M 2018 *J. Phys. Conf. Ser.* **1112** 012004
[5] Breden D, Miki K and Raja L L 2012 *Plasma Sources Sci. Technol.* **21** 034011
[6] Schweigert I V, Vagapov S, Lin L and Keidar M 2019 *J. Phys.* D. **52** 29
[7] Hagelaar G J M and Pitchford L C 2005 *Plasma Sources Sci. Technol.* **14** 722–33
[8] Fridman G, Friedman G, Gutsol A, Shekhter A, Vasilets V and Fridman A 2008 *Plasma Process. Polym.* **5** 503–533
[9] Laroussi M 2009 *IEEE Trans. Plasma Sci.* 37 714–25
[10] Ariskin D A, Schweigert I V, Alexandrov A L, Bogaerts A and Peeters F M 2009 *J. Appl. Phys.* **105** 063305

Mathematical model of gas mixture adsorption by a composite adsorbent

A S Vereshchagin[1,2,3], **V N Zinovyev**[1], **I V Kazanin**[1,2], **A Yu Pak**[1], **V A Lebiga**[1,2,3] **and V M Fomin**[1,2]

[1]Khristianovich Institute of Theoretical and Applied Mechanics SB RAS, Novosibirsk, Russia
[2]Novosibirsk State University, Novosibirsk, Russia
[3]Novosibirsk Technical State University, Novosibirsk, Russia

E-mail: vereshchag@itam.nsc.ru

Abstract. Within a framework of mechanics of multiphase media a mathematical model of the dynamics of a helium and water vapor mixture flowing through a bed of composite adsorbent, based on microspheres and aluminum oxide, is derived. It is shown how to coincide the flow in porous media of adsorber and gas diffusion in a cylindrical granule of adsorbent taking into account helium adsorption by microspheres and water vapor by a porous surface of an adsorbent.

1. Introduction

Nowadays, natural gas is the only practical source of helium, and the predominant technology of helium extraction is cryogenic, which is expensive, material- and energy-consuming, and is effective only for large amounts of raw material treatment. Alternative technologies, such as membrane and absorption, are practically isolated from industrial facilities, because they cannot provide the required capacity.

Exclusively high helium selectivity of quartz membranes is known for quite a long time. The permeability factor of quartz glass for helium exceeds one for methane by seven orders ($L_{He} = 3.26 \cdot 10^{-15}$ mole m/(m^2 s Pa); $L_{CH_4} = 6.38 \cdot 10^{-22}$ mole m/(m^2 s Pa)). However, utilization of quartz or other silicate materials such as common flat or capillary-tube membranes is difficult because of the low rate of helium diffusion, complicated organization of a large mass-transfer surface, and membrane performance at high pressure difference on both membrane sides.

The authors proposed the pioneering approach for the creation of the materials selectively absorbing helium; the approach combines the membrane and sorption technologies. It is suggested that hollow glass microspheres be used as membrane elements. Due to the selective permeability of some glasses, helium diffuses into the microsphere cavity because of the difference of partial pressures of helium inside and outside the microsphere. The method is referred to as the membrane-sorption, because the microsphere shell is a membrane, and the particle itself (the microsphere) absorbs helium inside. The particle sizes enable to create the large mass-transfer surface, whereas the spherical shape provide high hydrostatic strength. The presence of many micron-size particles guarantee the large surface of the mass transfer.

The composite absorbent is created in cooperation with the Institute of Problems of Hydrocarbons Treatment SB RAS, Omsk. Synthetic glass microspheres are used as the helium-

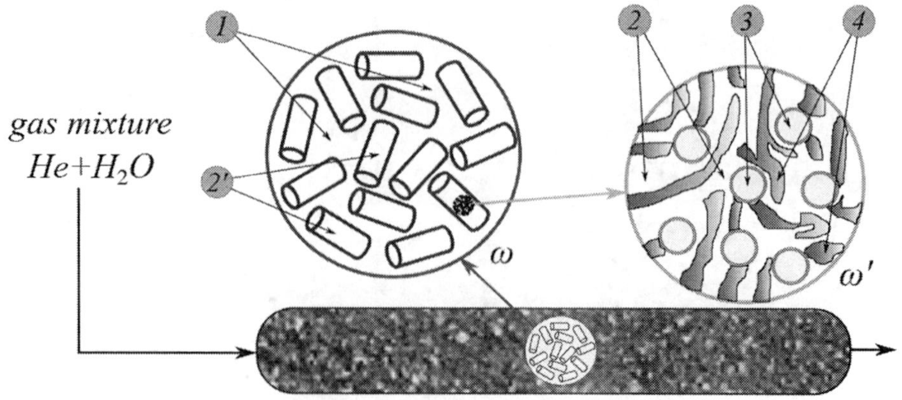

Figure 1. Structured levels of flow design: 1 – free adsorber volume; 2' – adsorber volume occupied by cylindrical granules; 2 – free space inside a granule; 3 – volume occupied by microspheres in a granule; 4 – porous binder volume in a granule.

permeable component of the composite adsorbent, and aluminum hydroxide (pseudoboehmite) works as a binder. The composite absorbent binder has a developed porosity and a big specific surface which is widely used as an absorbent for drying various gases and liquid media. Microsphere granulation into cylindrical granules from pseudoboehmite permits solving several tasks simultaneously: first, to reach the optimal granule size for feeding into an industrial absorber, second, to realize simultaneous helium occlusion and water removal from the gas mixture [1].

Gas flow through an adsorbent consists of gas filtration through the cylindrical granules, diffusion inside these granules, water vapor adsorption on a binder and helium adsorption by microspheres figure 1.

This kind of flow is described by models with double porosity with filtration through a free volume of adsorber and through a free volume of a granule. Porous structure of an adsorbent is mostly consists of mesopores with character channel diameter up to 200 nm. This fact makes the Knudsen diffusion mechanism be a driving force of flow inside the granule.

2. Mass transfer in adsorber

Suppose isothermic gas mixture flow of helium and water vapor through a bed of cylindrical granules evenly distributed in space. Mass conservation laws are averaged in investigated areas (figure 1) [2] using the method of volume averaging [3] and the interpenetrating continuum method [4]. Consider a "large" macro-volume ω in a neighborhood of a chosen point $\boldsymbol{\xi}$ in the adsorber region. Consider ω_1 to be a free volume, ω_2' – cylindrical granules volume, s_{12} – granules surface inside ω. Consider volume concentrations m_1, m_2 and a unit surface α_{12} defined by

$$m_1 = \omega_1/\omega, \quad m_2 = \omega_2'/\omega, \quad \alpha_{12} = s_{12}/\omega. \tag{1}$$

From geometric considerations it is easy to establish the following relations between the parameters

$$m_2 = 1 - m_1 = kV_p/\omega, \quad \alpha_{12} = kS_p/\omega, \tag{2}$$
$$V_p = \pi d_p^2 L_p/4, \quad S_p = \pi d_p(d_p/2 + L),$$

where V_p, S_p – volume and surface of a cylindrical granule of length L_p and diameter d_p; k – granules quantity in ω.

Further parameters m_1, k, α_{12}, L, d are considered to be constant.

Local density of each component dynamics in a free volume of adsorber is described by the following equations with boundary conditions

$$\frac{\partial \rho_{1i}}{\partial t} + \operatorname{div} \rho_{1i} \boldsymbol{v}_1 = 0, \tag{3}$$

$$\rho_{1i} \boldsymbol{v}_i \cdot \boldsymbol{n}|_{\Gamma_{12}} = -D_i \nabla \rho_{2i} \cdot \boldsymbol{n}|_{\Gamma_{12}} \quad (i = 1, 2),$$

where ρ_{1i}, ρ_{2i} – local density of helium and water vapor in a free volume of adsorber and free volume of granules ($i = 1, 2$); $\boldsymbol{v}_1$ – local gas mixture components velocity in adsorber; Γ_{12} – interphase surface in adsorber; $\boldsymbol{n}$ – outer unit normal vector to surface Γ_{12}; D_i – Knudsen diffusion coefficient of the ith phase. The last condition in (3) expresses the fact that mass flux on the surface of a cylindrical granule equals to the diffusion flux out of or into the granule.

After averaging equations (3) with boundary conditions one has

$$\frac{\partial \langle \rho_{1i} \rangle^1}{\partial t} + \operatorname{div}_{\boldsymbol{\xi}} \langle \rho_{1i} \boldsymbol{v}_1 \rangle^1 = -\alpha_{12} \langle \rho_{1i} \boldsymbol{v}_i \cdot \boldsymbol{n} \rangle_{s_{12}} \quad (i = 1, 2), \tag{4}$$

here

$$\langle \varphi(t, \boldsymbol{x}) \rangle_i (t, \boldsymbol{\xi}) = \frac{1}{\omega_i(\boldsymbol{\xi})} \int\limits_{\omega_i(\boldsymbol{\xi})} \varphi(t, \boldsymbol{x}) d\boldsymbol{x}, \quad \langle \varphi(t, \boldsymbol{x}) \rangle^i (t, \boldsymbol{\xi}) = \frac{1}{\omega} \int\limits_{\omega_i(\boldsymbol{\xi})} \varphi(t, \boldsymbol{x}) d\boldsymbol{x}, \tag{5}$$

$$\langle \varphi(t, \boldsymbol{x}) \rangle_{s_{ij}} (t, \boldsymbol{\xi}) = \frac{1}{s_{ij}(\boldsymbol{\xi})} \int\limits_{s_{ij}(\boldsymbol{\xi})} \varphi(t, \boldsymbol{x}) d\boldsymbol{x}, \tag{6}$$

where $\varphi(t, \boldsymbol{x})$ – scalar, vector or tensor function, defined in an investigated area ω_i and on an interphase boundary surface s_{ij} of ith and jth phase.

Neglecting the fluctuations of ρ_{1i} and $\boldsymbol{v}_1$ in ω_1 one gets

$$\langle \rho_{1i} \boldsymbol{v}_1 \rangle^1 \approx \langle \rho_{1i} \rangle^1 \langle \boldsymbol{v}_1 \rangle_1, \tag{7}$$

and the right hand side of (4) using the boundary condition can be estimated as

$$\alpha_{12} \langle \rho_{1i} \boldsymbol{v}_i \cdot \boldsymbol{n} \rangle_{s_{12}} \approx -k/\omega \int_{S_g} D_i \nabla \rho_{2i} \cdot \boldsymbol{n} dS, \tag{8}$$

where S_g – surface of a chosen granule, characterizing a state of all granules inside ω.

3. Mass transfer in granule

Consider gas diffusion inside cylindrical particle considering mass transfer with the free space of adsorber and interaction with inner structure of granule (helium permeation in microspheres and water vapor adsorption on binder surface).

Diffusion of local density in granule in volume 2 (figure 1) is described by the equation

$$\frac{\partial \rho_{2i}}{\partial t} = D_i \Delta \rho_{2i} \quad (i = 1, 2). \tag{9}$$

Consider the following boundary conditions for helium and water vapor

$$\rho_{21}|_{\Gamma_{12}} = \rho_{11}|_{\Gamma_{12}}, \quad -D_1 \nabla \rho_{21} \cdot \boldsymbol{n}|_{\Gamma_{23}} = \frac{C_m \beta}{d} \left(\langle p_{31} \rangle_3 - \langle p_{21} \rangle_2 \right), \quad -D_1 \nabla \rho_{21} \cdot \boldsymbol{n}|_{\Gamma_{24}} = 0, \tag{10}$$

where Γ_{12} – boundary between inner free space of a granule and a free space of adsorber; Γ_{23} – microsphere surface inside granule; Γ_{24} – surface of porous binder inside granule; C_m – permeability coefficient of microsphere shell; r, R, d – cavity radius, radius, shell thickness of microsphere; $\langle p_{31}\rangle_3$, $\langle p_{21}\rangle_2$ – averaged value of pressure inside and outside the microspheres in the neighborhood of a chosen point defined by (5).

The first condition implies the equality of local density for helium on the surface boundary, the second condition – mass flux in and out of microspheres is proportional to the difference of partial averaged pressure of helium inside and outside the microsphere; the third – helium is not adsorbed by a binder.

Helium adsorbed by microspheres is accounted by the following law

$$\frac{\partial}{\partial t}(V_m \rho_{31}) = -\frac{C_m S_m \beta}{d}\left(\langle p_{31}\rangle_3 - \langle p_{21}\rangle_2\right), \tag{11}$$

where ρ_{31} – local helium density in a chosen microsphere; $V_m = 4/3\pi r^3$ – microsphere cavity volume; $S_m = 4\pi R^2$ – microsphere surface area.

Local helium pressure in ω_2 and ω_{31} is defined by

$$p_{21} = \rho_{21} R_1 T, \quad p_{31} = \rho_{31} R_1 T,$$

where T – temperature; R_1 – helium gas constant.

For water vapor boundary conditions are

$$\rho_{22}|_{\Gamma_{12}} = \rho_{12}|_{\Gamma_{12}}, \quad -D_2 \nabla\rho_{22}\cdot\boldsymbol{n}|_{\Gamma_{23}} = 0, \quad -D_2\nabla\rho_{22}\cdot\boldsymbol{n}|_{\Gamma_{24}} = \beta_w(\langle\rho_{42}\rangle_{s_{24}} - a(\langle\rho_{22}\rangle_2, T)), \tag{12}$$

where β_w – equilibrium adsorption rate of water vapor; $\langle\rho_{22}\rangle_2$ – averaged value of local water vapor density in the neighborhood of a certain point defined by (5); $a = a(\rho, T)$ – adsorption isotherm; T – temperature.

The first condition implies the equality of local density for water vapor on the surface boundary, the second – water vapor is not adsorbed by microspheres, the third condition defines the law of water adsorbed on the binder surface.

Adsorbed mass of water vapor on the surface boundary s_{24} inside ω' is accounted by the formula

$$\frac{\partial\rho_{42}}{\partial t} = -\beta_w(\langle\rho_{42}\rangle_{s_{24}} - a(\langle\rho_{22}\rangle_2, T)).$$

4. Geometrical structure and averaged parameters of a composite particle

Main geometrical parameters of the inner composite particles structure are defined using the averaging method. Chosen volume ω' in a neighborhood of a point $\boldsymbol{\xi}$ is divided into subvolumes as follows

$$\omega' = \omega_2 \cup \omega_3 \cup \omega_4, \quad \omega_3 = \omega_{31} \cup \omega_{32},$$

where ω_2, ω_3, ω_4 – non intersecting parts of ω, defining free, occupied by microspheres and porous binder volumes; ω_{31}, ω_{32} – non intersecting parts of ω_3 defining the volume of a solid shell and cavity respectively.

Volume concentrations of phases and unit interphase surface area are defined by

$$l_2 = \omega_2/\omega', \quad l_3 = \omega_3/\omega', \quad l_{31} = \omega_{31}/\omega', \quad l_{32} = \omega_{32}/\omega', \quad l_4 = \omega_4/\omega', \tag{13}$$

$$\alpha_{23} = s_{23}/\omega', \quad \alpha_{24} = s_{24}/\omega', \tag{14}$$

where s_{23}, s_{24} – boundaries between free space and microspheres and binder inside ω'.

Assuming all microspheres are the same and evenly distributed in space it is obvious that parameters l_3, l_{31}, l_{32}, α_{23} are constant and the following relations are valid

$$l_3 = l_{31} + l_{32}, \quad l_{31} = \beta^3 l_3, \quad l_{32} = (1-\beta^3) l_3, \tag{15}$$

$$\alpha_{23} = \frac{3}{R} l_3, \tag{16}$$

where R, r – microsphere and microsphere cavity radius; $\beta = r/R$ – microsphere structure coefficient.

Assuming that porous structure is distributed evenly in space suppose l_4, α_{24} and l_2 to be constants.

5. Mass transfer in granule in terms of averaged values

After averaging diffusion equation (9) in an infinite media consisting of evenly distributed microspheres and porous binder [2], one gets the following equations

$$\frac{\partial \langle \rho_{2i} \rangle^2}{\partial t} = D_i \Delta_{\boldsymbol{\xi}} \langle \rho_{2i} \rangle^2 + D_i \operatorname{div}_{\boldsymbol{\xi}} \left[\alpha_{23} \langle \rho_{2i} \boldsymbol{n} \rangle_{s_{23}} + \alpha_{24} \langle \rho_{2i} \boldsymbol{n} \rangle_{s_{24}} \right] +$$
$$+ \alpha_{23} D_i \langle \nabla_{\boldsymbol{x}} \rho_{2i} \cdot \boldsymbol{n} \rangle_{s_{23}} + \alpha_{24} D_i \langle \nabla_{\boldsymbol{x}} \rho_{2i} \cdot \boldsymbol{n} \rangle_{s_{24}} \tag{17}$$
$$(i = 1, 2).$$

It is shown [2], that expressions under divergence operator in the second term in the right hand side of (17) being decomposed can be approximated in ω' as

$$\alpha_{23} \langle \rho_{2i} \boldsymbol{n} \rangle_{s_{23}} + \alpha_{24} \langle \rho_{2i} \boldsymbol{n} \rangle_{s_{24}} \approx - \langle \rho_{2i} \rangle_2 \nabla l_2. \tag{18}$$

These terms describe the internal structure of phase distribution inside ω'. Assuming l_2 to be constant they can be neglected. The third and the forth terms in (17) are approximated using mass transfer conditions between phases (10) and (12).

6. Mathematical model

Consider gas mixture of helium and water vapor flows through a bed of composite adsorbent consisting of cylindrical particles in a certain space area Ω (adsorber). Mathematical model of such flow is described using averaged parameters. To simplify the notation suppose the following substitutions

$$\langle \rho_{1i} \rangle^1 \to \rho_{1i}, \quad \langle \rho_{2i} \rangle^2 \to \rho_{2i}, \quad \langle \rho_{31} \rangle^3 \to \rho_{31}, \quad \langle \rho_{42} \rangle_{s_{42}} \to \rho_{42}, \quad \langle \boldsymbol{v}_1 \rangle_1 \to \boldsymbol{v}_1.$$

Parameters ρ_{1i}, $\boldsymbol{v}_1$ are functions of $(t, \boldsymbol{x})$, where $\boldsymbol{x} \in \Omega$, and $t > 0$ – time. In every point of Ω inside a macro volume ω a mass transfer flux between free volume of adsorber and granules is calculated. Thus parameters ρ_{2i} $(i = 1, 2)$, ρ_{31}, ρ_{42} are functions of $(t, \boldsymbol{x}, \boldsymbol{\xi})$, where $\boldsymbol{x} \in \Omega$, and $\xi \in G$, where G – is a set of points, restricted by a cylinder surface S_g of length L_p and diameter d_p.

Gas mixture dynamics in adsorber Ω is described by the following relations

$$\frac{\partial \rho_{1i}}{\partial t} + \operatorname{div}_{\boldsymbol{x}} \rho_{1i} \boldsymbol{v}_1 = K_{1i}, \quad K_{1i} = D_i \frac{m_1}{l_2} \frac{m_2}{V_p} \int_{S_g} \nabla_{\boldsymbol{\xi}} \rho_{2i} ds \quad (i = 1, 2). \tag{19}$$

Helium dynamics inside composite adsorbent in G is described by the following relations

$$\begin{aligned}\frac{\partial \rho_{21}}{\partial t} &= D_1 \Delta_\xi \rho_{21} + \frac{3l_3 C_m \beta}{Rd}\left(\frac{\rho_{31}}{l_{31}} - \frac{\rho_{21}}{l_2}\right),\\ \frac{\partial \rho_{31}}{\partial t} &= -\frac{3l_3 C_m \beta}{Rd}\left(\frac{\rho_{31}}{l_{31}} - \frac{\rho_{21}}{l_2}\right),\end{aligned} \tag{20}$$

with

$$\rho_{21}|_{S_g} = \rho_{11} l_2/m_1.$$

Water vapor dynamics inside composite adsorbent in G is described by the following relations

$$\begin{aligned}\frac{\partial \rho_{22}}{\partial t} &= D_2 \Delta_\xi \rho_{22} + \alpha_{24}\beta_w\left(\rho_{42} - a(\rho_{22}/l_2, T)\right),\\ \frac{\partial \rho_{42}}{\partial t} &= -\beta_w\left(\rho_{42} - a(\rho_{22}/l_2, T)\right),\end{aligned} \tag{21}$$

with

$$\rho_{22}|_{S_g} = \rho_{12} l_2/m_1.$$

7. Conclusion

In the framework of the phenomena [1], which showed the possibility of creating a technology for the separating helium from natural gas, a mathematical model is obtained which describes the dynamics of a mixture of helium and water vapor through a porous medium, consisting of cylindrical granules of a composite adsorbent capable of retaining and subsequently releasing helium and water vapor.

Acknowledgments

The work is partly carried out within the framework of the Program of Fundamental Scientific Research of the state academies of sciences in 2013-2020 (project No. AAAA-A17-117030610134-9), Complex Program of Fundamental Research SB RAS in 2018-2020 (project No. 0323-2018-0023), regional programs of Russian Foundation of Basic Research (project numbers 18-41-540012 p_a and 19-41-543007 p_mol_a).

References

[1] Zinovyev V N, Kazanin I V, Lebiga V A, Pak A Y, Vereshchagin A S and Fomin V M 2016 *Thermophys. and Aeromech.* **23** 741–746
[2] Vereshchagin A S and Fomin V M 2015 *J. of Appl. Mech. and Tech. Phys.* **56** 737–749
[3] Whitaker S 1999 *The Method of Volume Averaging* (Springer Science+Business Media Dordrecht)
[4] Nigmatulin R I 1979 *Dynamics of multiphase media. Part 1.* (Moscow: Nauka)

The effect of ball milling in a planetary mill on aluminium particles microstructure and properties of cold sprayed coatings

T M Vidyuk[1,2], A E Chesnokov[1], A V Smirnov[1] and V S Shikalov[1]

[1] Khristianovich Institute of Theoretical and Applied Mechanics, SB RAS, Institutskaya str. 4/1, Novosibirsk, 630090, Russia
[2] Institute of Solid State Chemistry and Mechanochemistry, SB RAS, 630117, Kutateladze str. 18, Novosibirsk, Russia

vidyuk@itam.nsc.ru

Abstract. Cold spraying allows one to obtain coatings without significant powder particles overheating. Hence, cold sprayed coatings have properties similar to that of sprayed particles. This paper shows results of the aluminium powder preparation for cold spraying. It is demonstrated that preliminary ball milling in the high-energy mill «Activator-2SL» has a significant impact on the particles shape and its structure. The ball milled particles have layered structure containing of a variety of the ultrafine aluminium particles.

1. Introduction

At present, coating deposition on machine parts by the thermal spraying and cold spraying becomes more significant part of technological processes in automobile, aircraft, heavy, medical and biological industries. The main function of coatings is to increase the durability of details and functional units, which work at high mechanical loads, enhanced temperatures, chemically aggressive media, etc. The coating is formed as a result of consistently collision of individual particles with the substrate. In process of spraying the basic physical parameters influence internal coating structure. These parameters include velocity, temperature, size, internal structure of particles, as well as temperature, material and condition of substrate surface [1-3]. For the methods of thermal spraying, heating above a material melting point is a distinctive feature. In contrast, the main feature of the cold spraying is ability of deposition without significant heating, which allows one to obtain coatings with properties similar to that of initial particles material [4].

As a rule, preliminary high-energy effect on initial material in a planetary mill leads to particle structure change. For example ball milling of powders in planetary mill leads to deformation energy storage throughout the volume of particles, appearing of point defects and dislocations in a crystal structure; surface cleaning as a result of destruction of oxide films and adsorbed layers that are barriers for chemical transformations, chemical reactions, etc [5]. Bearing this in mind, one could suggest that properties of coatings deposited from ball milled powder particles will be differ from physical, chemical and mechanical parameters of coatings obtained during the initial powder deposition.

The goal of this study is experimental research of the effect of aluminium particle ball milling in high-energy mill in dependence on milling time on its structure and structure of obtained cold sprayed coatings.

The following tasks have been accomplished to achieve the goal:
1. Ball-milling of powders during 15, 30, 45, 60 and 90 seconds at the constant ratio of balls weight to processed material weight and acceleration of grinding bodies;
2. Particle size composition and specific surface was determined; surface morphology and internal structure of ball milled and initial particles were investigated.
3. Cold sprayed coatings were obtained using initial and ball milled powder particles; its physical and mechanical properties were determined; its internal structure was investigated.

2. Materials and Methods

In this study the ASD-1 grade aluminium powder was used. Initial aluminium powder particles have spherical shape with a mean size of 18 μm. Figure 1 shows microphotographs of initial aluminium particles morphology and its cross section.

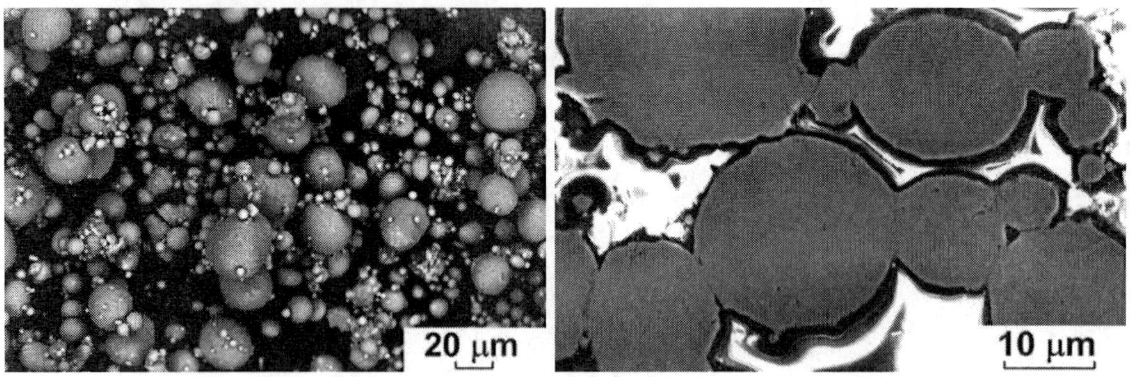

(a) (b)

Figure 1. Microphotographs of main view (a) and cross section (b) of initial aluminium particles.

Ball milling of the initial powder was carried out in the high energy mill «Activator-2SL». Its main technical details: two cylinders and each of them has volume of 250 ml and inside diameter of 42.5 mm; weight of processed balls – 160 g for each cylinder, grinding bodies acceleration -117 g, processed material shot weight – 30 g; peak velocity of centroid axis is 1045 rpm, that of cylinders – 1550 rpm. The steel balls with diameter of 5 mm were used as grinding bodies. Classification of particles by sizes was conducted with vibration table using a set of the analytical sieves. Determination of histograms of volume and counting distributions was conducted using particle size analyzer (0.04 – 2000 μm) LS 13 320 (Beckman Coulter). Powder particle morphology was investigated using scanning electron microscope Evo MA15 (Carl Zeiss). The microhardness of the coatings was measured using an automatic microhardness meter DuraScan-50. Porosity of coatings was determined using Image Analysis Software and metallographic microscope OLYMPUS GX-51. Deposition was carried out at experimental cold spray setup developed at the ITAM SB RAS, using Laval nozzle of length 145 mm and critical and exit diameters 2.8 and 6.5 mm respectively. Air was used as working and carrying gas. Pressure in the stagnation chamber was constant and it was equal to 3 MPa. Temperature in the stagnation chamber was equal to 570 K. The substrates for cold spraying were aluminium plates of 3 mm in thickness. Standoff distance was equal to 30 mm, nozzle travers speed was 25 mm/s.

3. Results and Discussion

Initial aluminium powder was ball milled during 15, 30, 45, 60, 90 and 120 seconds. After the each ball milling powder was divided into two fractions of sizes smaller and greater than 90 μm. Weight content of each fraction was determined. It was established that the increase of ball milling time leads to enhance of weight content of particles of size greater than 90 μm. Figure 2 shows the change of weight content of particles of size greater than 90 μm depending of ball milling time. The separation of the particles into two fractions is caused by the application of these powders. The particles of the main fraction with the sizes smaller than 90 μm are widely used in thermal spraying, detonation spraying and cold spraying. The particles of sizes larger than 90 μm in the shape of deformed disk can be processed in the plasma to obtain the powder consisting of micron sized spherical particles. At present, these particles are desirable for additive technologies, laser cladding and other areas of powder metallurgy.

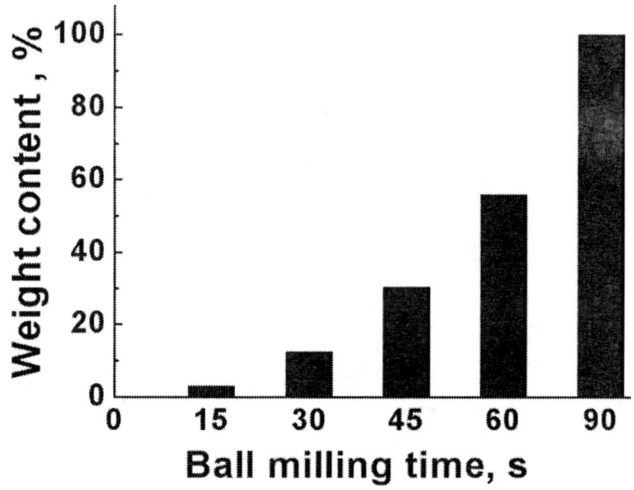

Figure 2. The change of weight content of particles of size greater than 90 μm depending on the ball milling time.

When a grinding body interacts with a particle, the kinetic energy is consumed on the plastic deformation of the particle (up to 95% [6]) with the release of a large amount of heat. As the milling time increases, the amount of energy introduced into the material increases. This leads to an increase in the heat released during particle deformation. Heat does not have time to distribute, as a result of which there is a rapid increase in the temperature of the entire system. The material is heated to a temperature close to the melting point or to its melting [7]. The particles having a temperature close to the material melting point can be plastically deformed and easy interacts with each other. Increase of their size is occurred as a result of granulation process by the method of rolling. Melting of aluminium particles is observed with a ball-milling during the 120 seconds.

Based on the research of particle size distribution for initial and ball milled aluminium particles, volume bar chart and counting bar chart of particle size distribution were plotted. Figure 3 shows the dependences of mean (d_{mean}), (d_{10}) and (d_{90}) sizes on the ball milling time of initial particles, where d_{10} – the diameter a which 10% of the sample mass is comprised of particles with a diameter less than this value; d_{mean} – the diameter of the particle that 50% of a sample mass is smaller than and 50% of a sample mass is larger than; d_{90} - the diameter a which 90% of the sample mass is comprised of particles with a diameter less than this value. It can be seen from figures 3 and 4 that the increase of mean particle size during the ball milling characterizes the formation of agglomerates, consisting of ultrafine aluminium particles. Morphology and cross section of ball milled particles is presented in figure 4.

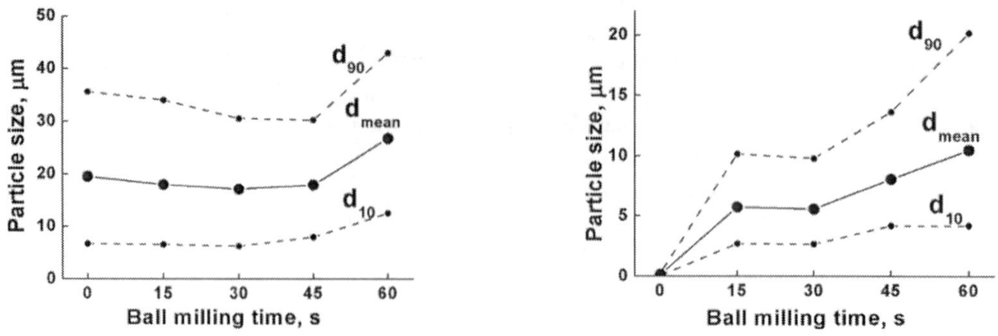

(a) (b)

Figure 3. Size distribution of the initial and ball milled aluminium powders: (a) volume and (b) counting distribution of particles.

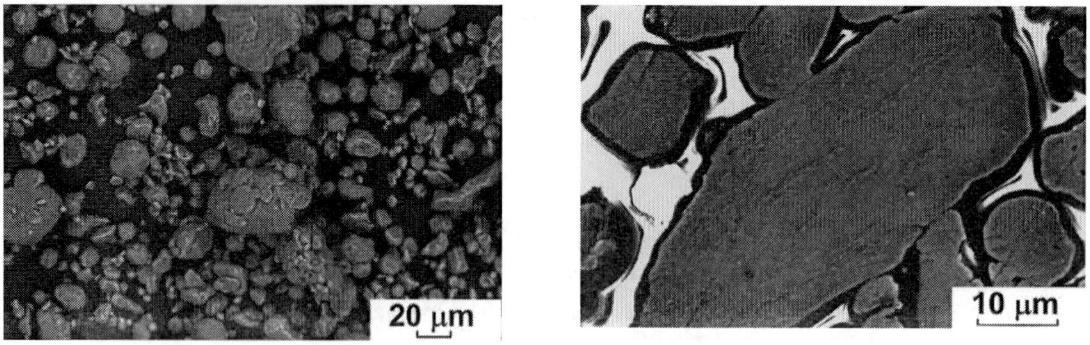

(a) (b)

Figure 4. Microphotographs of main view (a) and cross section (b) of ball milled aluminium particles.

To study the internal structure of coatings, cross sections of samples were prepared. Internal structure of coatings is presented in figure 5. The feature of the coatings is a layered structure with clearly defined boundaries formed by deformed particles, which indicates a high particle velocity before collision. It is clear that microstructure of individual particles in the coating obtained from agglomerated particles is similar to that of sprayed particles (figure 4).

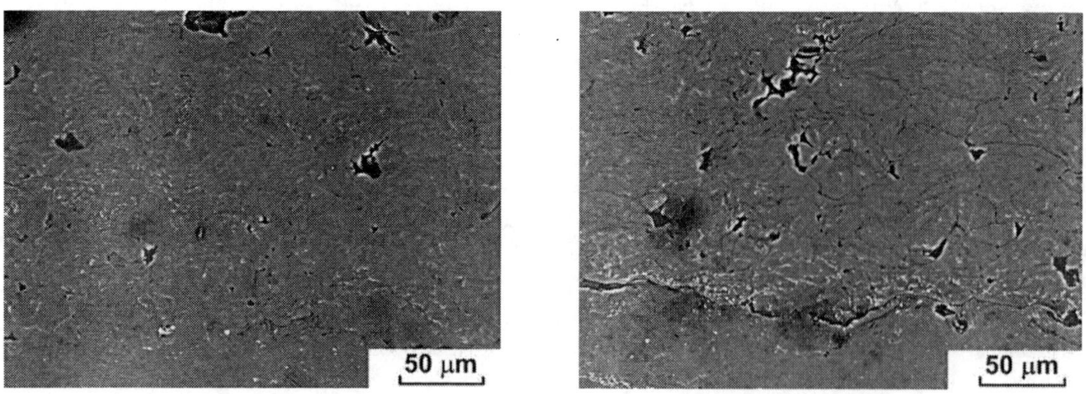

(a) (b)

Figure 5. Cross-section of coatings obtained by cold spraying of initial particles (a) and particles ball milled for 30 s (b).

Microhardness tests of coatings sprayed from both the initial powder and ball milled powder are shown in table 1 (results were averaged over 10 measurements). Table 1 shows also porosity of cold sprayed coatings. One can see that porosity and microhardness of coatings produced from ball milled and initial powders are almost the same.

Table 1. Parameters of cold sprayed metal coatings.

Coating	Porosity, %	Microhardness, $HV_{0.1}$
Initial powder	4.2	59±3.5
Ball milled powder during 30 s; sizes of particles is smaller than 90 μm	4.14	61.45±9.15

4. Conclusions

In this study experimental research of the ball milling time effect on the particle size distribution and internal microstructure of particles was conducted. It is shown that increase of ball milling time leads to enhance of system temperature (ball - particle), resulting in granulation process by the rolling method. The weight fraction of particles, which is greater than 90 μm in size, increase. When ball milling for 120 seconds the processed material reaches the melting point. It is established that microstructure of individual splats of cold sprayed coatings inherits features of initial particles microstructure.

Acknowledgements

This study was supported by the Russian Science Foundation (Project No. 19-19-00335) using equipment of the Joint Access Center «Mechanics» of ITAM SB RAS.

References

[1] Gao P-H, Li Y-G, Li C-J, Yang G-J and Li C-X 2008 *JTTEE5* **17** 742
[2] Wang H-D, Zhou Y and Shi K 2011 *Transactions of JWRI* **Special Issue** 43
[3] Ulianitsky V, Solonenko O, Smirnov A and Batraev I 2014 *ITSC* 362
[4] Klinkov S, Kosarev V and Rein M 2005 *Aerospace Science and Technology* **9** 582
[5] Avvakumov E, 1988 *Mechanical Methods of Activation of Chemical Processes* (Novosibirsk: Nauka) p 305
[6] Taylor G 1934 *Proc. Roy. Soc.* **A145** 362
[7] Bowden F 1954 *Proc. Roy. Soc.* **A223** 29

Influence of post-processing on the strength of heterogeneous materials produced by the laser welding

I E Vitoshkin[1], A G Malikov[1], A M Orishich[1] and E V Karpov[1,2]

[1] Khristianovich Institute of Theoretical and Applied Mechanics, Siberian Branch Russian Academy of Sciences, 630090, Novosibirsk, Russia
[2] Lavrentyev Institute of Hydrodynamics, Siberian Branch, Russian Academy of Sciences, 630090, Novosibirsk, Russia

igor.vitoshkin.97@mail.ru, smalik@ngs.ru, laser@itam.nsc.ru

Abstract. The paper presents the investigation of the laser welding of the alloy Al-Cu-Li and titanium alloy Ti-Al-V, as well as the analysis of the post=processing on the strength properties of welded joints. Macro- and micro-structure of the joints have been studied. The resulting heterogeneous welded materials have been post-processed. The mechanical characteristics of the welded joints have been determined before and after the thermal processing.

1. Introduction

Titanium and aluminum alloys have a number of unique properties such as high specific strength, low density (comparing to steel) and high corrosion resistance. The properties of these materials make them be commonly used in aviation and rocket technologies [1]. The need to combine the properties of titanium and aluminum alloys in one structure is often for a modern vehicle. Welding of heterogeneous materials is a challenge. Frist, the differences I the chemical and physical properties: the welded materials have different melting points, density, thermal conductivity, etc. The difference in the chemical composition may result in undesirable combinations during the welding and variation of the material properties in the near-joint zone because of the mutal diffusion of the materials. The difference in thermophysical properties causes the nonuniform heating of the materials, which in turn leads to the residual thermal stresses which can hardly be removed due to the different temperature of the thermal processing for the welded materials, as well as occurrence of brittle intermetallic compositions in the welded joint [2, 3]. Today, high-strength aluminum alloys alloyed by lithium are developed. Lithium added into the alloy decreases the alloy density an increases the strength properties owing to the strengthening phases resulting from the thermomechanical processing. These alloys are promising for the welded constructions [4].

By now, there is no technology of connection of titanium and aluminum parts by the melting welding method (for example, fixation of titanium wings to the aluminum fuselage of an aircraft) because of the intermetallic compositions forming in the melt pool which make the joint brittle and undurable. Usually this task is solved with the aid of riveting which has a number of serious disadvantages such as growing weight of the construction, occurrence of stress concentrators (which rises the specific amount of metal in the structure), complexity of sealing, low process efficiency. Rivets can hardly be smoothed to have an aerodynamic shape. At the same time, the welded connection permits providing the sealing. The structure weight is much lower without rivets, and this connection can be polished for the needed shape.

In this paper we analyze the peculiarities of the laser welding of heterogeneous materials for the aluminum alloy of the system Al-Cu-Li and titanium alloy of the system Ti-Al-V, as well as the influence of the post-processing of the strength properties of the welded connections.

2. Experimental Technique

To weld heterogeneous materials, we chose the titanium alloy VT-20 and aluminum-lithium alloy 1461 (made in Russia). VT-20 alloy is characterized by high heat resistance and welds well. The alloy is intended for the manufacture of products operating for a long time at temperatures up to 500 °C. The aluminum alloy 1461 production Kamensk Uralsky Metallurgical Works (KUMZ in Russian abbreviation) is the promising Russian high-strength welded alloy of the third generation and is involved in the parts made by PAO "Sukhoi Company". Table 1 gives the chemical composition of the studied alloys.

Table 1. Chemical composition of aluminum alloys.

Alloy	Ti	Al	Cu	Mg	Mn	Li	Zr	Sc	Mo	V
VT-20	base	5.5-7	-	-	-	-	1.5-2.5	-	0.5-2	0.8-2.5
1461	0.07	base	2.5-2.95	0.4-0.6	0.1-0.4	1.5-1.9	0.05-0.12	0.05-0.1	-	-

The titanium alloyVT-20 belongs to the titanium alloys of the system Ti–Al-V (the class by the pseudo α structure). This alloy is durable to hot cracks during the melting welding.

The aluminum-lithium alloy 1461 of the system Al-Cu-Li presents a solid solution based on Al with strengthening phases T_1 (Al_2CuLi), T_2 (Al_6CuLi_3), S_1 (Al_2MgLi), δ'(Al_3Li), θ(Al_2Cu) [5].

The laser welding (LW) was performed on the automated laser technological complex "Sibir-1". The ALTC consists of a gas-discharge CO_2-laser equipped with a SFUR-resonator [6,7]. The characteristics of the generated radiation are the following: the radiation wavelength $\lambda = 10.6$ μm; the maximal radiation power up to 8 kW; the beam quality BPP = 4.7 mm*mrad. The laser wave guide includes the following components: flat mirrors and the phase-shifting mirror, which permits transforming the flat radiation polarization into the circular polarization. The two-coordinate spanning technological table moves the welding heading in respect to the welded sheet. The values of the major table characteristics are: the working area is 1.5 x 2.5 m; the maximal motion rate is 50 m/min; the positioning accuracy is 0.1 mm.

The welding head includes the focusing ZnSe lens, the gas-dynamic nozzle protecting the welded connection from oxidation, plus the mechanism permitting the motion of the lens position along the beam axis. Figure 1 shows the photo of the laser welding process. Thermal processing of the welded connections was carried out in a tube vacuum furnace equipped with the temperature controller. The pressure in the vacuum furnace was 1 mm Hg.

Figure 1. Laser welding diagram.

The strength of the welded joints was measured at the static extension on the electro-mechanical test machine Zwick/Roell Z100.

3. Results and Discussion

3.1. Laser welding of homogeneous welded joints

At the early stage, the laser welding of the titanium alloy VT-20 and aluminum alloy 1461 was studied. The laser radiation was focused on the ally surface with the aid of the ZnSe lens, its focal distance was 254 mm. Helium, the inert gas, was used to protect the welded joint and joint core. The gas was supplied through the nozzle which was included at the angel of 45° to the welded plate, the gas flow rate was 5 l/min. prior to the test, the samples were polished with a sand paper and washed in petroleum in order to remove impurities from the surface. The titanium alloys samples (50x100x2 mm) were set on a multi-purpose table with special clamps.

Experiments gave the optimal conditions for the laser welding of titanium VT-20 sheets at which the qualitative defect-free welded joints are produced. The optimal energy conditions of the welding are: the laser radiation power is 1.2 kW, the welding rate is 1 m/min, the focus position about the upper sheet boundary is -3 mm.

Welded joints aluminum alloy 1461 without any surface defects such as hot cracks or poor fusion were obtained with the following welding mode parameters: laser beam power 3.3 kW, welding speed 4 m/min, and depth of focus in the material – 2 mm, with keyhole penetration.

Figure 2 shows an optical photograph of the cross section of the weld of a titanium and aluminum alloy, showing the difference between the microstructure in the fusion zone and the heat-affected zone (HAZ) from the base alloy

The structure of the welded joint titanium alloy is typical for the cast condition: equiaxed poliendric grains are in the center, and from them the grains elongated along the heat release toward the base. The average size of the base alloy is 5 μm, the thermal effect zone size is 20 μm.

The weld microstructure is fundamentally different from that of the aluminum alloy (see figure 2, b). The microstructure of the central fusion zone exhibits an equiaxed dendrite zone in the weld center. The heat affected zone consists of columnar dendrites. The same structure of the Al-Li alloy weld was observed in other studies.

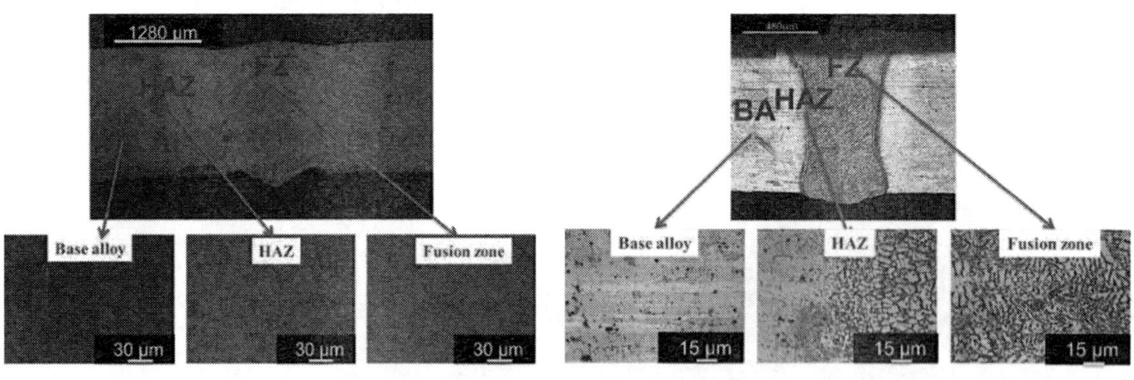

(a) (b)

Figure 2. Optical photo of the weld microstructure cross section, HAZ and base alloy. (a) weld titanium alloy, (b) alloy aluminium alloy (magnification 5x and 100x).

Table 2 contains data after mechanical testing about the ultimate tensile strength σ_{UTS}, ultimate yield strength σ_{YS}, and relative elongation δ of the alloys and welded joints.

Table 2. Mechanical properties of the alloys and welded joints.

Mode	σ_{UTS}, MPa	σ_{YS}, MPa	δ, %
Alloy VT-20	1080	1000	12
Welded joint VT-20	1050	980	5
Alloy 1461	550	470	10.1
Welded joint 1461	341	333	0.7

For a titanium alloy VT-20, the ultimate tensile strength of the weld is close to the ultimate tensile strength of the alloy. For aluminum alloy 1461, the weld the ultimate tensile strength is 0.62 of the alloy the ultimate tensile strength. In our work [6,7], it is shown that in order to increase the mechanical characteristics of the weld, it is necessary to apply post heat treatment of the weld.

3.2. Laser welding of dissimilar materials

The peculiarity of this research data is that the laser radiation is sheared in respect to the joint toward titanium for «0», «0.5» and «1» mm, where the solid titanium-based α-solution would form with the minimum intermetallic compositions. The heterogeneous materials were welded in energy modes welding titanium alloy (the laser radiation power is 1.2 kW, the welding rate is 1 m/min, the focus position about the upper sheet boundary is -3 mm).

Figure 3 shows the values of the yield strength after the stretch tests for the initial alloys and resultant heterogeneous laser-welded connections at different shearing of the laser radiation.

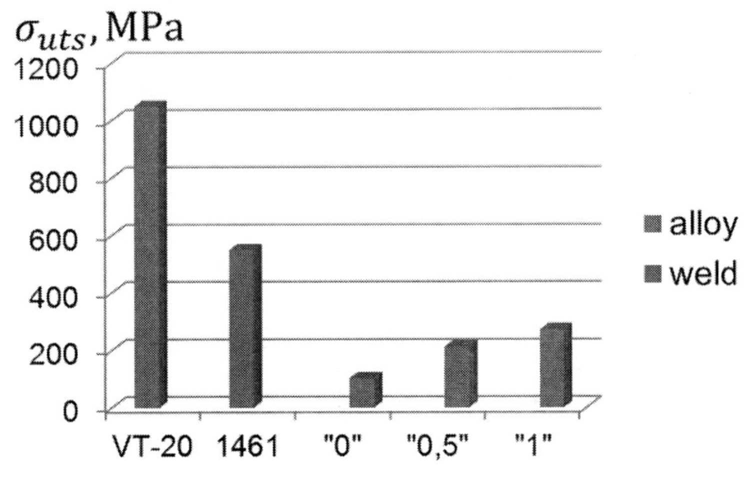

Figure 3. Mechanical characteristics of the alloy and welded joint.

As is evident from figure 3, the strength increases versus the laser radiation position, remaining low comparing the initial alloys. Figure 4 shows a general view of a high-quality heterogeneous weld.

SEM images of the weld metal microstructure for the laser beam offsets «1» are presented in figure 5.

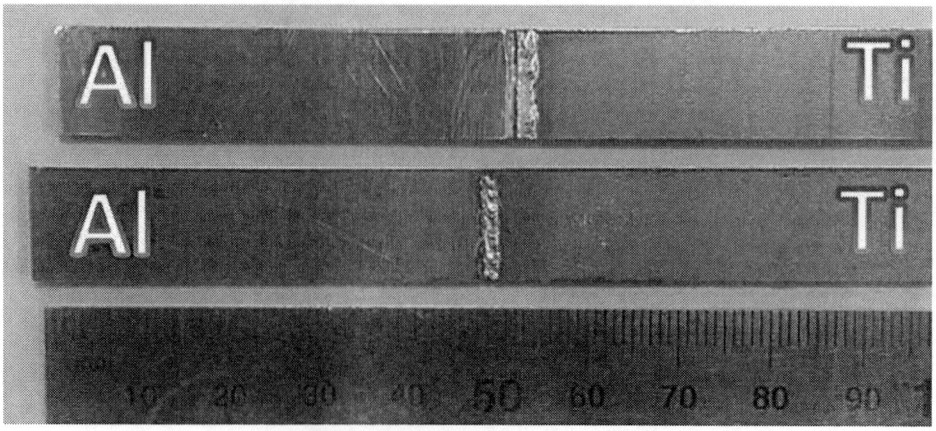

Figure 4. Photo of the laser dissimilar welded joint.

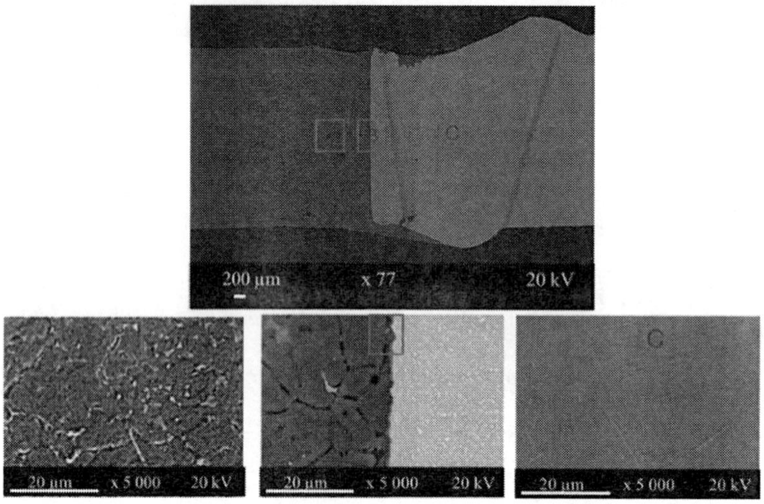

Figure 5. SEM image of cross-section of the welded joint.

Analyzing figure 5, three characteristic zones can be distinguished. The fusion zone «C» of titanium, the interface zone between aluminum and titanium «B», the fusion zone «A» of aluminum due to heat removal from molten titanium. In the interface zone there was a brazing welding.

In order to change the mechanical characteristics of the heterogeneous welded connections, the welded joints underwent the thermal processing.

This processing applied to the laser-welded connections of aluminum-lithium alloys of different systems results in the growth of the mechanical characteristics [6, 7]. In respect to the heterogeneous connections, this procedure can decrease the internal stresses remained after the welding.

The thermal processing was carried out by six modes. After the processing, the samples were cooled in air. The parameters were taken from recommendations of the references. We also preserved the control group of samples which did not undergo the thermal processing. The processed samples were tested for stretching. The test results are given in figure 6.

 1404 (2019) 012050 doi:10.1088/1742-6596/1404/1/012050

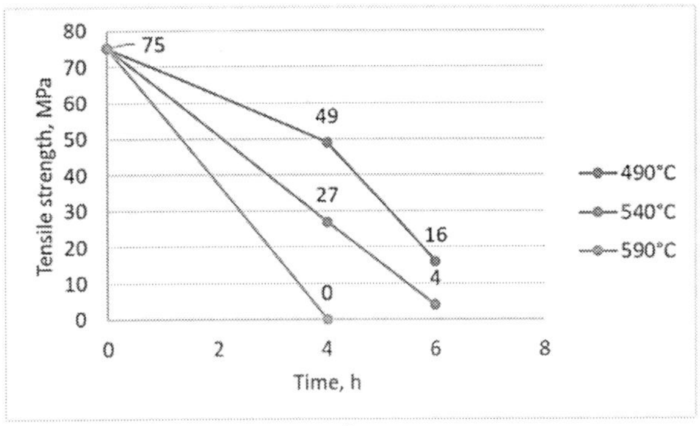

a)

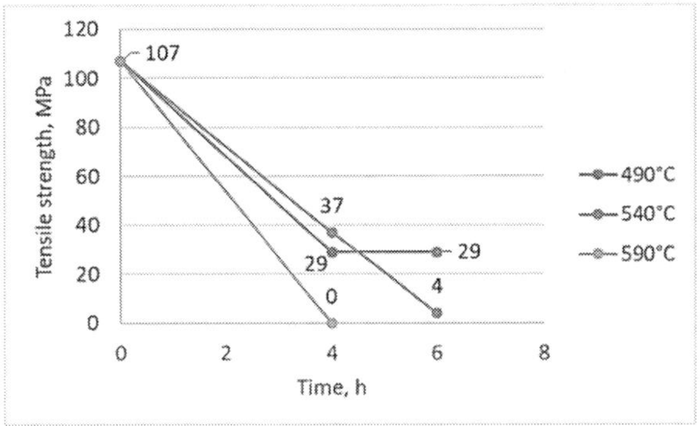

b)

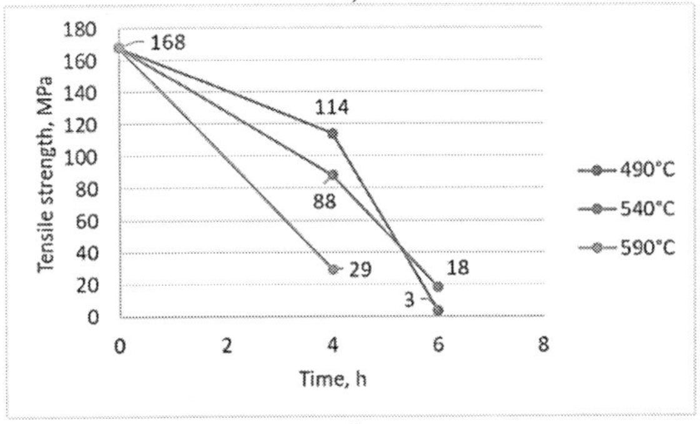

c)

Figure 6. Results of stretch tests of the samples after the thermal processing produced at different shearing without laser beam a)«0», b) «0.5», c) «1» mm.

Figure 7 shows the changed zones «B» according to figure 5 depending on the temperature and time of heat treatment.

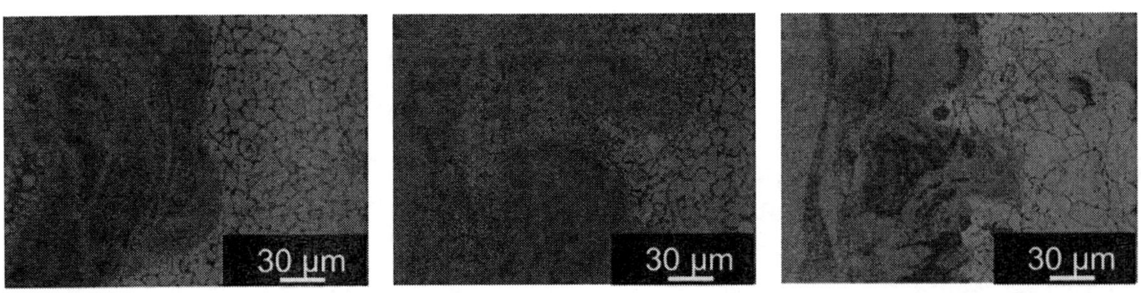

(a) (b) (c)

Figure 7. Image of cross-section of the welded joint.
T=490 °C 4 hours, b) T=490 °C 6 hours, c) T=560 °C 6 hours.

As can be seen from figure 7, the structure of the interface zone changes dramatically, with changes occurring both from the temperature at the same holding time in the furnace and from time at a constant temperature. There is a change in both the width of the interface zone and the structure. A change in the structure of agglomerates in color indicates a different chemical composition. According to the state diagram Ti-Al, various intermetallic phases (Ti_3Al, TiAl, $TiAl_2$ and $TiAl_3$) can occur, which affect the mechanical characteristics of heterogeneous laser welds based on titanium and aluminum alloy. For a more accurate determination of intermetallic phases, it is necessary to study the chemical composition of these phases using EDX analysis and transmission electron microscopy, which is planned in future studies.

4. Conclusions

The laser welding of heterogeneous materials based of titanium and aluminum alloys has been analyzed. It has been founded that the ultimate tensile strength of the produced welded connections depends on the laser radiation shear. The thermal processing of the connections produced with the sheared laser beam and without it results in the reduction of the strength. It is shown that as a result of heat treatment of the weld, the structure of the interface zone between the titanium and aluminum alloy changes its structure and size depending on both temperature and time. Apparently, as a result of heat treatment of the weld in the interface zone, intermetallic phases were formed. These intermetallic phases lead to a change in the mechanical characteristics of a heterogeneous welded joint based on titanium and aluminum alloy.

Acknowledgments

The research was carried out within the framework of the Program of Fundamental Scientific Research of the state academies of sciences in 2013-2020 (project No. AAAA-A17-117030610122-6). The study was conducted at the Joint Access Center «Mechanics» of ITAM SB RAS.

References

[1] Gloria A, Montanari R, Richetta M and Varone A 2019 *Metals (Basel)* **9** 662
[2] Zhou X, Duan J, Zhang F and Zhong S 2019 *Materials (Basel)* **12** 712
[3] Casalino G, D'Ostuni S, Guglielmi P, Leo P, Palumbo G and Piccininni A 2018 *Materials (Basel)* **11** 2337
[4] Abd El-Aty A, Xu Y, Guo X, Zhang S-H, Ma Y and Chen D 2018 *J. Adv. Res.* **10** 49–67
[5] Lukina E A, Alekseev A A, Khokhlatova L B and Oglodkov M S 2014 *Met. Sci. Heat Treat.* **55** 466–71
[6] Malikov A, Orishich A, Golyshev A and Karpov E 2019 *J. Manuf. Process.* **41** 101–10
[7] Malikov A, Orishich A, Bulina N, Karpov E and Sharafutdinov M 2019 *Mater. Sci. Eng. A* **765** 138302

Spherically symmetric stationary flows of a gas suspension

S M Voronin, A V Panov and M M Turov

Department of mathematical analysis, Chelyabinsk State University, Br. Kashirinych str. 129, Chelyabinsk, Russia

E-mail: gjd@bk.ru

Abstract. The dynamics of an isothermal gas suspension is studied. A spherically symmetric stationary submodel is derived and partially integrated. In a barochronous case, this submodel is fully integrated and results are discussed. Different modes of motion are analytically described in the barochronous case. The interpretation of the solutions is proposed. We highlight three essentially different modes of motion. In the first mode, we have a dispersion of gas suspension from the origin with a relaxation of velocities and leading to the uniform motion. In the second mode, phases focus in the origin with finite nonzero velocities and infinitely increasing densities. In the third mode, phases are interpenetrating to each other and decelerating due to the friction force, that leads to an accumulation of particles and an unlimited increase of densities in different nonzero raduises with zero velocities.

1. Introduction

Consider a system of differential equations

$$\begin{cases} \frac{\partial \rho_1}{\partial t} + \vec{u}_1 \cdot \nabla \rho_1 + \rho_1 \mathrm{div} \vec{u}_1 = 0, \\ \frac{\partial \rho_2}{\partial t} + \vec{u}_2 \cdot \nabla \rho_2 + \rho_2 \mathrm{div} \vec{u}_2 = 0, \\ \rho_1 \left(\frac{\partial \vec{u}_1}{\partial t} + \vec{u}_1 \cdot \nabla \vec{u}_1 \right) + m_1 \nabla P(\rho_1, \rho_2) = -\frac{\rho_2}{\tau}(\vec{u}_1 - \vec{u}_2), \\ \rho_2 \left(\frac{\partial \vec{u}_2}{\partial t} + \vec{u}_2 \cdot \nabla \vec{u}_2 \right) + m_2 \nabla P(\rho_1, \rho_2) = \frac{\rho_2}{\tau}(\vec{u}_1 - \vec{u}_2), \end{cases} \tag{1}$$

that describes motions of a two-phase isothermal fluid. Here $\vec{u}_1 = (u_1, v_1, w_1), \vec{u}_2 = (u_2, v_2, w_2)$ are velocity vectors of the first and the second phase, ρ_1, ρ_2 are phase densities, $P(\rho_1, \rho_2)$ is a pressure of a mixture, it is a functional parameter, $m_2 = \frac{\rho_2}{\rho_{22}}$ is a volume concentration of the second phase, ρ_{22} is an absolute density of the second phase, it is a constant parameter, $m_1 = 1 - m_2$ is a volume concentration of the first phase and τ is a time of a relaxation.

System (1) was introduced by Kh.A. Rakhmatulin [1]. There are a lot of studies of this system [2-4]. The formation and the structure of shock waves in two-phase fluids were studied, for example, in articles [5,6]. There are more complicated systems of differential equations which take to account the chaotic pressure of the particles (the second phase) [5,7]. Many exact solutions of system (1) were found in the paper [8]. In this work, we study the smooth motions of gas suspension dynamics described by system (1) with two additional constraints: spherical symmetry and stationarity.

2. Spherically symmetric stationary flows

If we focus on isotropic and stationary motions, then the solution of system (1) has a form

$$\vec{u}_i = \frac{\vec{x}}{r}\, q_i(r), \quad \rho_i = \rho_i(r), \quad r = \sqrt{x^2+y^2+z^2}, \quad i = 1, 2.$$

Here $q_i(r)$ is a radial velocity and $\rho_i(r)$ is a density of the phase on a sphere of radius r. Substitution of the last formulas to system (1) leads to a system of ordinary differential equations

$$\begin{cases} q_1\rho_1' + (2\frac{q_1}{r} + q_1')\rho_1 = 0, \\ q_2\rho_2' + (2\frac{q_2}{r} + q_2')\rho_2 = 0, \\ \rho_1 q_1 q_1' + m_1 P' = -\frac{\rho_2}{\tau}(q_1 - q_2), \\ \rho_2 q_2 q_2' + m_2 P' = \frac{\rho_2}{\tau}(q_1 - q_2), \end{cases} \tag{2}$$

with differentiation with respect to r. The first two equations are quickly integrated and we get conservation laws of a phase flow through a sphere of radius r

$$\rho_1 q_1 r^2 = c_1, \quad \rho_2 q_2 r^2 = c_2.$$

If we get $P = \frac{a^2\rho_1}{1-\frac{\rho_2}{\rho_{22}}}$ (phases are ideal), then the last two equations in (2) after the substitutions $\rho_1 = \frac{c_1}{q_1 r^2}$, $\rho_2 = \frac{c_2}{q_2 r^2}$ are reduced to a system of a form

$$A(q,r)\begin{pmatrix} q_1' \\ q_2' \end{pmatrix} = \begin{pmatrix} b_1(q,r) \\ b_2(q,r) \end{pmatrix}.$$

After multiplication on the matrix $A^{-1}(q,r)$, we get the system

$$q_1' = \frac{q_1(2a^2c_1\tau\rho_{22}(\rho_{22}r^2q_2 - c_2)q_2^2 r + c_2(q_1-q_2)(a^2c_1r^2\rho_{22} - q_1(\rho_{22}r^2q_2 - c_2)^2))}{\tau c_1(q_2(q_1^2 - a^2)(\rho_{22}r^2q_2 - c_2)^2 - a^2c_1c_2q_1)},$$

$$q_2' = \frac{2a^2c_1^2\rho_{22}\tau q_1 q_2^2 r + c_1(q_1-q_2)(\rho_{22}r^2q_2 - c_2)(-a^2\rho_{22}q_2r^2 + q_1^2(\rho_{22}r^2q_2 - c_2))}{\tau c_1(q_2(q_1^2 - a^2)(\rho_{22}r^2q_2 - c_2)^2 - a^2c_1c_2q_1)}.$$

At this moment, we have no idea how to solve this system in an analytic form. Intention to construct a phase portrait for this dynamical system leads to difficulties with singular points. Therefore, firstly, consider the simplest case when the pressure P depends on time only (barochronous flows), then the derivative P' with respect to r equals zero in (2).

3. Barochronous flows

In this case, system (2) has a form

$$\begin{cases} q_1\rho_1' + (2\frac{q_1}{r} + q_1')\rho_1 = 0, \\ q_2\rho_2' + (2\frac{q_2}{r} + q_2')\rho_2 = 0, \\ \rho_1 q_1 q_1' = -\frac{\rho_2}{\tau}(q_1 - q_2), \\ \rho_2 q_2 q_2' = \frac{\rho_2}{\tau}(q_1 - q_2). \end{cases}$$

The solution of this system is

$$\begin{gathered} \rho_1 q_1 r^2 = c_1, \quad \rho_2 q_2 r^2 = c_2, \\ c_1 q_1 + c_2 q_2 = c_3, \quad -q_2 - \frac{c_3}{d}\ln|c_3 - dq_2| = br + c_4, \end{gathered} \tag{3}$$

where $d = c_1 + c_2 \neq 0, b = \frac{c_1+c_2}{\tau c_1}$, and the constants c_1, c_2, c_3, c_4 are uniquely determined by the initial conditions

$$\rho_1(r_0) = \rho_{10}, \quad \rho_2(r_0) = \rho_{20}, \quad q_1(r_0) = q_{10}, \quad q_2(r_0) = q_{20},$$

and the equalities

$$c_1 = \rho_{10} q_{10} r_0^2, \quad c_2 = \rho_{20} q_{20} r_0^2,$$

$$c_3 = c_1 q_{10} + c_2 q_{20}, \quad c_4 = -q_{20} - \frac{c_3}{c_1 + c_2} \ln |c_3 - (c_1 + c_2) q_{20}| - \frac{c_1 + c_2}{\tau c_1} r_0.$$

The first two equalities in (3) determine a phase flow through a sphere of radius r. The third equality in (3) defines a relationship between velocities of phases, and the last equality determines a distribution of the second velocity along the radius r.

There are two different cases $c_1 c_2 > 0$ and $c_1 c_2 < 0$. Consider the first case when $c_1 c_2 > 0$. We can not explicitly represent q_2 as a function of r, so consider the function

$$r(q_2) = \frac{1}{b} \left(-q_2 - \frac{c_3}{d} \ln |c_3 - d q_2| - c_4 \right).$$

The graphs of the functions $r(q_2)$, $r(q_1)$ are presented in figures 1 and 2.

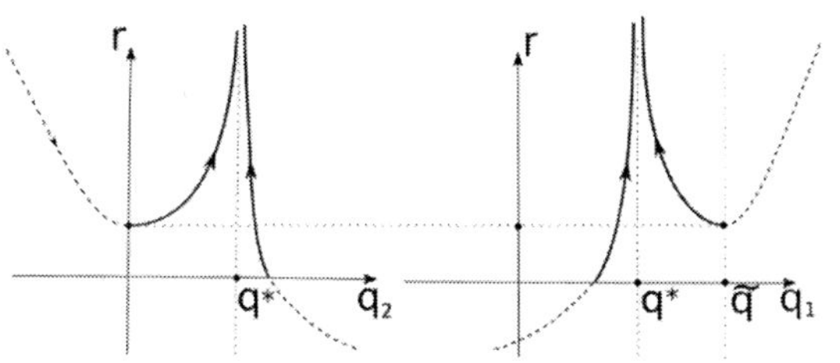

Figure 1. Case $c_1, c_2 > 0, q^* = \frac{c_3}{d}, \tilde{q} = \frac{c_3}{c_1}$.

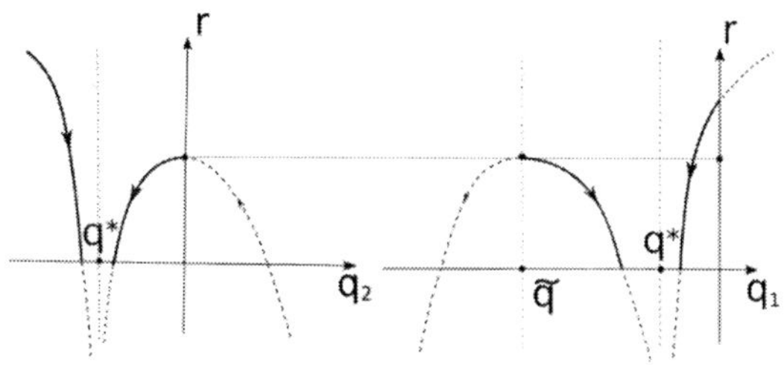

Figure 2. Case $c_1, c_2 < 0, q^* = \frac{c_3}{d}, \tilde{q} = \frac{c_3}{c_1}$.

In all figures we suppose c_4 such that $r(q_2)|_{q_2=0} > 0$. This assumption does not change a qualitative picture. A dashed line on figures indicates a prohibited mode of a phase motion. This mode is associated with the first three integrals in (3) that say velocities q_1, q_2 to be positive for all r, if they are positive for r_0. Also some dashed lines are a consequence of the condition $r > 0$. The arrows illustrate the fact that if q_1, q_2 are positive then the distance to the origin should increase.

If the initial velocities are positive, then c_1, c_2, c_3 are positive, and the dynamics is determined by figure 1. In this case, there are two possibilities. If $q_{20} > q_{10}$, then $q^* > q_{10}$ and $q^* < q_{20}$, so for all next r will be $q_2 > q_1$. When $q_{20} < q_{10}$, then $q^* < q_{10}$ and $q^* > q_{20}$, so for all next r will be $q_2 < q_1$. Thus, for positive q_{10}, q_{20}, the distance r unlimited increase, the velocities q_1, q_2 tend to q^* and the densities ρ_1, ρ_2 tend to zero. Here, we have a dispersion of a gas suspension from the origin with a relaxation of velocities and leading to the uniform motion.

If the initial velocities are negative, then c_1, c_2, c_3 are negative, and figure 2 presents the dynamics. There are also two cases. When $q_{20} > q_{10}$, then for all next $r > r_0$ will be $q_2 > q_1$. If $q_{20} < q_{10}$, then for all next $r > r_0$ will be $q_2 < q_1$. For the negative initial velocities q_{10}, q_{20}, we have the collapse of densities when $r = 0$ with some finite velocities.

Consider the second case $c_1 c_2 < 0$. If $q_{10} > 0$ and $q_{20} < 0$, then $c_1 > 0$ and $c_2 < 0$. The dynamics of phases is presented in figures 3, 4 according to $d > 0$ or $d < 0$. In figures 3 and 4 the dynamics of the first phase is determined only in the interval $(0, \tilde{q})$, because

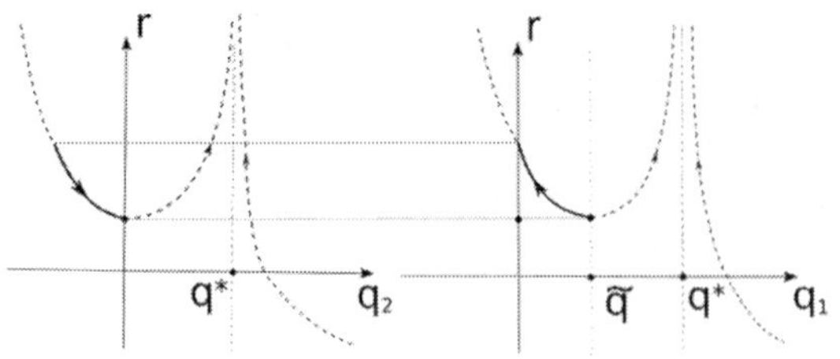

Figure 3. Case $c_1 > 0, c_2 < 0, d > 0, q^* = \frac{c_3}{d}, \widetilde{q} = \frac{c_3}{c_1}$.

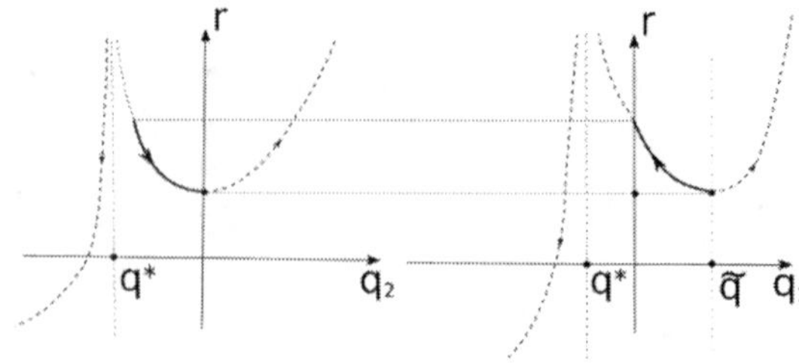

Figure 4. Case $c_1 > 0, c_2 < 0, d < 0, q^* = \frac{c_3}{d}, \widetilde{q} = \frac{c_3}{c_1}$.

$\widetilde{q} = \frac{c_3}{c_1} = q_{10} + \frac{c_2 q_{20}}{c_1} > q_{10}$. So the rest of the graph when $q_1 > \widetilde{q}$ is a dashed line. In figure 4, it is easy to see that $q_{20} > q^*$ so the rest of the graph when $q_2 < q^*$ is a dashed line. In figures 3 and 4 we see the first and the second phase, which flow to each other and, due to the friction force, they are stopping. Therefore there are two radiuses r_{1c} and r_{2c} for the first and the second phase where the particles of the phase build up and the phase density equals infinity. So, in these radiuses, there are collapses of densities with zero velocities of phases.

If $q_{10} < 0$ and $q_{20} > 0$ then $c_1 < 0$ and $c_2 > 0$. The dynamics of phases is presented in figures 5, 6 according to $d < 0$ or $d > 0$. The motion of the first phase is determined only in

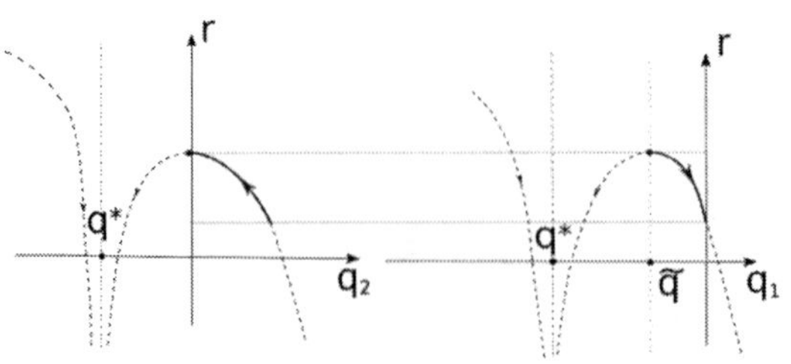

Figure 5. Case $c_1 < 0, c_2 > 0, d < 0, q^* = \frac{c_3}{d}, \widetilde{q} = \frac{c_3}{c_1}$.

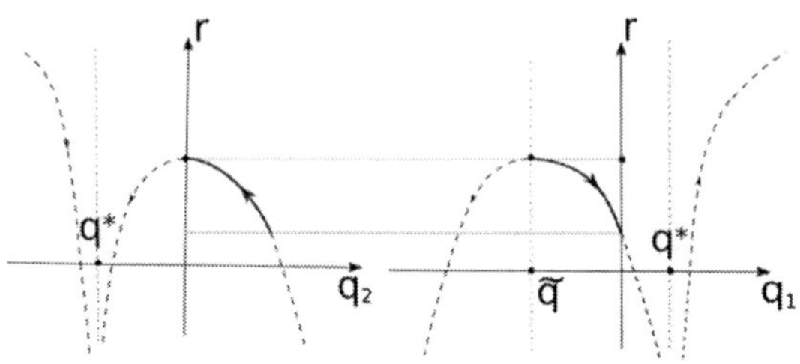

Figure 6. Case $c_1 < 0, c_2 > 0, d > 0, q^* = \frac{c_3}{d}, \widetilde{q} = \frac{c_3}{c_1}$.

the interval $(\widetilde{q}, 0)$ for the same reasons as in the previous case. Therefore, the rest of the graph when $q_1 < \widetilde{q}$ is a dashed line. In figures 5, 6 we see the same picture as in figures 3 and 4: phases are interpenetrating to each other and decelerating due to the friction force, that leads to an accumulation of particles and an unlimit increase of densities. So, in some two radiuses there are collapses of densities with zero velocities.

Note that in figures 3, 4 for q_2 and in figures 5, 6 for q_1, there may exists c_4 such that ρ_2 or ρ_1 collapses when r_{ic} equals zero, but velocities do not.

Finally, we need to see the last case $d = 0$. Here we have a solution

$$\rho_1 q_1 r^2 = c_1, \quad \rho_2 q_2 r^2 = c_2,$$
$$c_1 q_1 + c_2 q_2 = c_3, \quad \frac{q_2^2}{2} = \frac{c_3}{r c_1} r + c_5,$$

In this case we have next graphs of solutions, see figures 7 and 8. The interpretation of the graphs in figures 7 and 8 is the same one, as in the previous four cases.

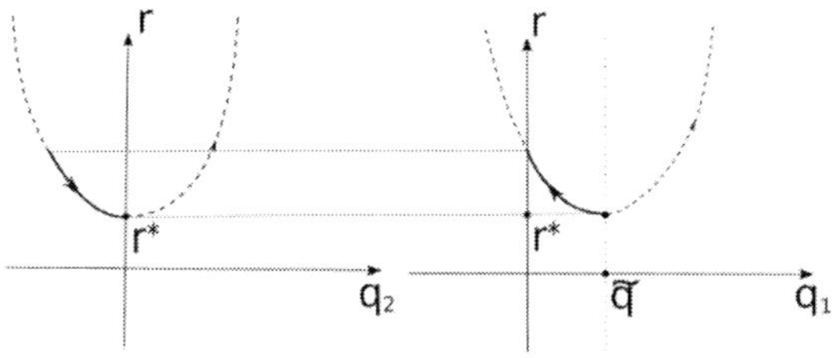

Figure 7. Case $c_1 > 0, c_2 < 0, \widetilde{q} = \frac{c_3}{c_1}, r^* = \frac{\tau c_1 c_5}{c_3}$.

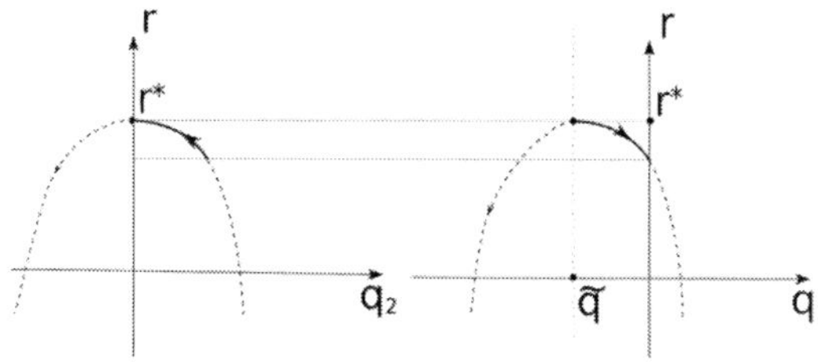

Figure 8. Case $c_1 < 0, c_2 > 0, \widetilde{q} = \frac{c_3}{c_1}, r^* = \frac{\tau c_1 c_5}{c_3}$.

Acknowledgments
The reported study was funded by RFBR according to the research projects No. 18-31-00226 (paragraph 3) and No. 18-08-00156(paragraph 2).

References
[1] Rakhmatulin Kh 1956 *J. Appl. Math. Mech.* **20** 184-95
[2] Klebanov L, Kroshilin A, Nigmatulin B and Nigmatulin R 1982 *J. Appl. Math. Mech.* **46** 66-74
[3] Kazakov Yu, Fedorov A and Fomin V 1987 *J. Appl. Mech. Tech. Phys.* **28** 773-8
[4] Yanenko N, Soloukhin R, Papyrin A and Fomin V 1980 *Supersonic Two-Phase Flow Under Conditions of Particle Velocity Disequilibrium* (Novosibirsk: Nauka)
[5] Fedorov A and Bedarev I 2018 *Math. Model. and Comp. Simul.* **10** 1-14
[6] Korobeinikov V, Markov V and Menshov I 1984 *Pr. of Math. Inst. of AS USSR* **163** 104-7
[7] Anderson K, Sundaresan S and Jackson R 1995 *J. Fluid Mech.* **303** 327-66
[8] Panov A 2019 *Int. J. Non-Lin. Mech.* **116** 140-6

Numerical investigations of dam-break flow problem

S N Yakovenko[1], E E Yakovenko[1] and K C Chang[2]

[1] Khristianovich Institute of Theoretical and Applied Mechanics, SB RAS, 4/1 Institurskaya st., Novosibirsk 630090, Russia
[2] National Cheng Kung University, Tainan 70101, Taiwan, R.O.C.

s.yakovenko@mail.ru

Abstract. A dam-break flow of water above horizontal dry bottom is studied numerically, after the comprehensive review of available literature. Computation tools include different interface resolution techniques and the continuous surface tension force model. The results are compared with the measurement data. It is shown that surface tension leads to the motion suppression and, therefore, to the acceleration decrease, providing closer agreement with laboratory experiments.

1. Introduction

One of the urgent issues of fluid dynamics of technical and natural systems is the study of evolution of the flow with a free surface near the coastline. Oncoming perturbations with huge water masses caused by natural and industrial disasters (tsunamis, tidal waves, bores produced by a sudden dam breaking, storm surf) can lead to serious consequences. In 2004 and 2011, large-scale natural disasters related to the tsunami occurred. When the waves move to the coastal zone, their height and energy increase significantly. The wave breaking near the coast is followed by run-up and run-down of large water masses, which can form a strong front and result in the coastal infrastructure damage. The velocity field reproduction at all stages of the flow evolution is a key factor for developing activities to mitigate consequences and would be useful to further understand various aspects of the impact of water masses on a coastline: land flooding, coastal erosion, wave structure interactions, sediment transport, morphology.

The water movement near the coast occurs in the zones of surf (located between the wave breaking and maximum run-down positions) and splash (between the wave run-up and run-down maxima). Study of the free surface perturbation evolution in a reservoir over horizontal and inclined bottoms for models of bore and solitary wave reflects a real picture of emergence of waves on the coast. Bore is observed in the splash zone near the coast, in an estuary (a funnel-shaped mouth of a river) and is supposed to be a moving discontinuity, where the height of the free surface grows from the level of still water to the peak of wave motion, and the horizontal velocity varies from zero to a noticeable value. Bore is usually modeled by a sudden rising of a fence (dam break), which generates a dam break flow towards a lower level of the free surface. Bore observed in tsunamis and leading to land flooding can be also viewed as a combination of incoming solitary waves with different heights and time intervals between them: the movement of bore to a region with decreasing depth generates a number of solitons.

In the present study, to resolve the free surface, the VOF concept [1] is used as in many studies, for instance, in prediction of dam-break flow [2] and Rayleigh–Taylor instability (RTI) [3]. An alternative scheme to approximate the volume fraction fluxes is based on the MUSCL approach with the QUICK interpolants and TVD limiters [4]. A dam-break flow over horizontal dry bed is chosen as a canonical test here and in many preceding studies due to available experiments data [5, 6]. The surface tension effects are inserted via the CSF model [7] used to compute RTI [3] too. The developed tools can be further applied to more general cases, e.g. to predict the dam-break bore above the sloped bottom [8].

2. Literature survey

Because of the limited accuracy of measurements, the experimental data was very scarce in early works, and simplified analytical approaches with the depth-averaged (shallow water) equations were considered. The recent development of the non-intrusive measurement tools (ADP, LDV, PIV) allowed one to obtain data for the structure of waves and bores in the channel and field observations (e.g. [8-11]).

A solitary wave moving onto an inclined shore was studied in many papers, where the free surface evolution, the breaking type, the external flow velocity field, and the maximum run-up height were considered. Recently, the velocity field evolution in the incoming wave prior to breaking, with subsequent run-up and run-down, has been studied [10, 11]. The analysis of evolution stages was carried out, data are obtained for various bottom slopes and the ratios of wave height to water depth.

The dam-break bore was studied for a number of tests, including the horizontal dry (a) and wet (b) beds, sloped bottom (c), smooth or rough surfaces, obstacles (wavebreaker, dyke, reef, prism, etc.).

In the first measurements [5] for test (a), the evolutions of free surface, (positive) surge front location x_f and water column height y_f have been obtained for five cases of square ($H \times H$) and rectangular initial water areas of depths between 57 and 115 mm. More recent experiments [6] in the larger-size channel with initial water region of depth 0.3 m, length 3.5 m also studied the evolution of the negative wave front and flow velocities. Such a flow was first computed by the VOF method in [1] to yield proper result for $x_f(t)$ with refined uniform grid of cell size $H/40$. Later numerical studies have shown that 3D effects of lateral walls can give slight improvement for $x_f(t)$ [4], as well as adding of extra turbulent viscosity in 2D runs based on RANS or LES-like techniques [12, 13]. Different volume-fraction-flux schemes and wall boundary conditions were compared in simulations [2].

For the test (b), by lifting the gate in the flume, which initially separated high and low water levels, the experiment [14] was made for weak and strong bores with turbulence generation, using the flow visualization and the water sensors to obtain the free surface distributions and to measure the wave propagation velocities. The similar studies in the water flume [15] for initial stages of dam-break flows over dry and wet beds revealed the mushroom-like jets occurring at small times which is captured in simulations by the Navier–Stokes equations [12, 16] but not by the shallow-water equations [15, 16].

For test (c), turbulence models when using with depth averaging are also not able to describe local flow features, for example, turbulence during bore breaking, due to neglect of noticeable vertical flow variations [17]. In experiments [18], bore types (strong and weak) are classified depending on the ratio of the initial water levels in front of the barrier and behind it. The effects of bore breaking interacting with a wedge of water in front of it, pushing the wedge to the shore, which is very different from the behavior described by the shallow water equations, are shown. In [17], the dam break bore, moving on the constant depth water, then over the 1:8 slope, has been studied based on the VOF and RANS methods, with the k-ε model. Free surface distributions, averaged velocity, turbulence energy during shoaling, breaking, run-up and run-down, for strong and weak bores are obtained where small-scale effects, in particular, mini-collapse during run-up and returned bore during water run-down, are found.

3. Mathematical model

The present simulation is based on the unsteady continuity, Navier–Stokes, volume fraction (f) equations in Cartesian coordinates (x, y) for incompressible fluid where x and y are the horizontal and vertical coordinates, $f = 1$ and $f = 0$ are the volume-fraction values for denser and lighter fluids, respectively. The surface tension effect is inserted as the reformulated volumetric force due to the CSF model [3, 7]. The mollified volume fraction is taken here to be equal to the volume fraction function defined numerically due to smoothing the solution by numerical scheme. One can also ignore the surface tension as in [2] and compare results. Discretization of equations is done on a staggered grid to prevent mismatch between the velocity and pressure fields. Other numerics details are given in [2, 3].

4. Computation results

The initial flowfield configuration for the problem of the rectangular water column sudden collapse is shown schematically in figure 1.

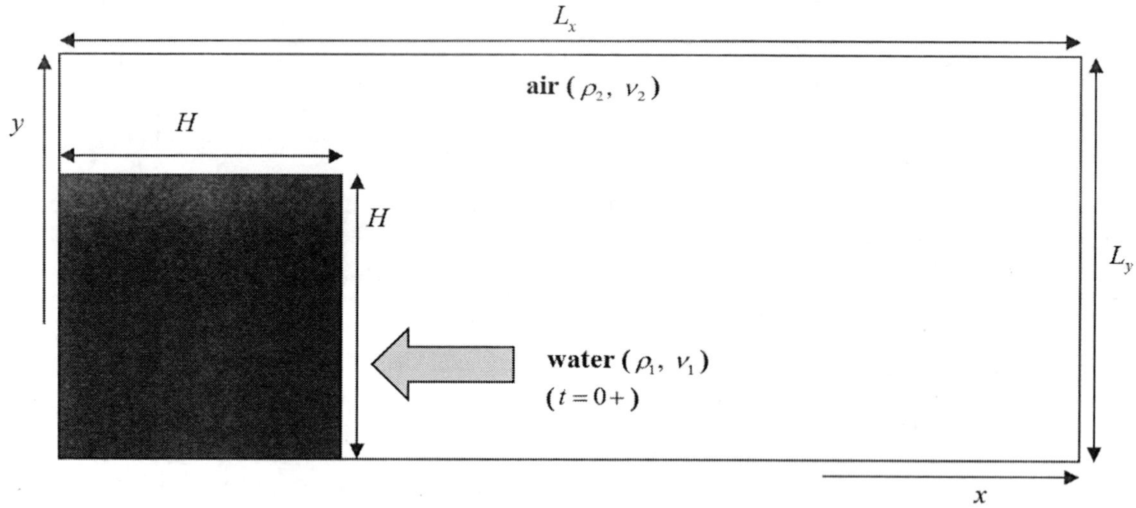

Figure 1. Scheme of dam-break problem (initial state).

Parameters of numerical simulations are as in [2, 5]: $H = 0.05715$ m, $g = 9.81$ m/s^2, $\rho_1 = 998$ kg/m^3, $\rho_2 = 1.19$ kg/m^3, $\eta_1 = 0.99 \cdot 10^{-3}$ kg/(m·s), $\eta_2 = 1.84 \cdot 10^{-5}$ kg/(m·s). The surface tension coefficient value is $\sigma = 0.072$ N/m. The reference Reynolds number $\mathrm{Re}_r = \rho_r U_r L_r / \eta_r = 43130$ is based on the reference quantities $L_r = H$, $U_r = (gH)^{1/2}$, $t_r = (H/g)^{1/2}$, $\rho_r = \rho_1$, $\eta_r = \eta_1$ used for non-dimensionalization.

The computations are performed without the CSF model (Cases 1, 2) and with the CSF model (Cases 3, 4), using two selected advection schemes described in [2]. Case 1 taken from [2] and Case 3 (both denoted as "c4minmod") correspond to the TVD MUSCL scheme with QUICK interpolants and the compressive minmod TVD limiters [4] for advection fluxes in the U, V, f equations. Case 2 taken from [2] and Case 4 (both denoted as VOF) replace the advection scheme in the f equation by the upwind-downwind donor-acceptor VOF procedure [1] while keeping the same flux expression (as in c4minmod approach) for the U and V equations. Note that both the first-order upwind scheme and the QUICK scheme without TVD limiters are not applied for flux approximation in (2)-(4). The former produces the thick unrealistic interface due to numerical diffusion, and the latter leads to significant distortion of volume-fraction contours due to spurious oscillation features of the QUICK scheme [2].

The comparison of results for different schemes (figure 2) shows as in [2] that the VOF method gives the sharper free surface whereas the c4minmod scheme produces the larger interface smearing. However, some distortion of contours of volume fraction f is seen for VOF (and for c4minmod Cases at more refined grids in [2] too). The distortion of the free surface (given by the contour of $f = 0.5$) and its wave-like behavior may be caused not only by numerical scheme drawback but also by specific physical phenomena like the developing turbulence. For the VOF scheme, more realistic reproduction of the broken dam problem evolution can be achieved using closer levels of effective interface boundaries ($f = 0.33$ and 0.67), and some shortcomings can be avoided at smaller time step [2].

For the positive surge front position and the water column height (figure 3), a small delay of the experimental values from the predicted ones at $t < 2t_r$ can evidently be explained by the measurement conditions [5]. In the experiments, the initial water column was constrained by a thin waxed paper diaphragm, and a heavy current from car batteries was used to rupture the waxed paper and thus allow motion to begin. The value $t = 0$ was taken as the time when a heater current was applied, and the water motion began with a little unfixed delay in the time interval $0 < t < 0.2t_r$ [5].

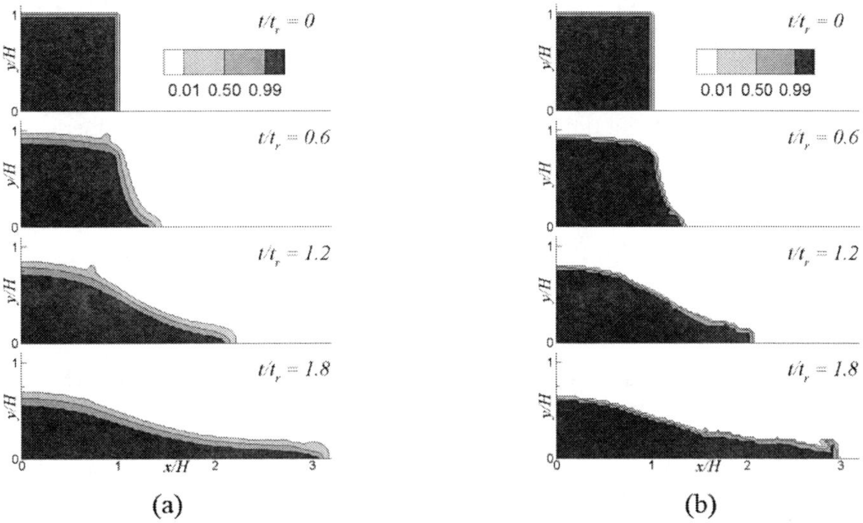

Figure 2. Contour plots of the water volume fraction obtained in computations with the CSF model for c4minmod (a) and VOF (b) advection schemes in the volume-fraction equation (Cases 3 and 4).

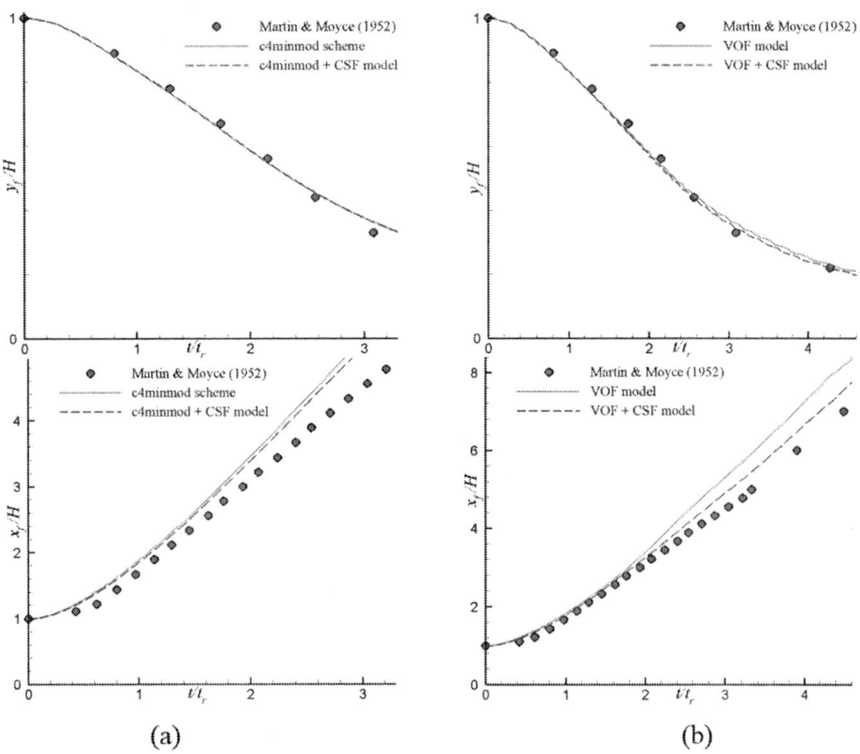

Figure 3. Surge front position x_f and column height y_f versus time with and without CSF model, for c4minmod (a) and VOF (b) schemes of convection fluxes in the volume-fraction equation.

A reason of increasing x_f over-estimation at $t > 2t_r$ (observed also in [4]) for all schemes was not evident in [2]. However, it was concluded [3] that similar actions of viscosity and surface tension effects lead to the motion suppression and, thus, to acceleration decrease, as in [4, 12, 13] due to effects of lateral walls or extra turbulent or sub-grid scale viscosities. The damping effect is clearly seen in figure 3 where values of $x_f(t)$ are smaller for both schemes and closer to experiment points when the surface tension is taken into account. Moreover, this correcting effect is stronger (figure 3b) for the VOF method giving the sharper free surface, and the smallest one for the scheme (figure 3a) producing the thicker interface. Comparison also shows approximate coincidence of water column height $y_f(t)$ for all schemes except the VOF + CSF model giving small improvements at large times.

Acknowledgments
The study is performed within the framework of the Integrated Basic Research Program of SB RAS for 2018-2020 (Project No. AAAA-A18-118021590030-8).

References
[1] Hirt C W and Nichols B D 1981 *J. Comp. Phys.* **39** 201
[2] Yakovenko S N and Chang K C 2008 *Thermophysics and Aeromechanics* **15** 169
[3] Yakovenko S N and Chang K C 2014 *Fluid Dynamics* **49** 748
[4] Kelecy F J and Pletcher R H 1997 *J. Comp. Phys.* **138** 939
[5] Martin J C and Moyce W J 1952 *Phil. Trans. Royal Soc. of London, Ser. A* **224** 312
[6] Lauber G and Hager W H 1998 *J. Hydraul. Res.* **36** 291
[7] Brackbill J U, Kothe D B and Zemach C 1992 *J. Comp. Phys.* **100** 335
[8] Lin C, Kao M J, Wong W Y, Shao Y P, Fu C F, Yuan J M and Raikar R V 2019 *European Journal of Mechanics B. Fluids* **73** 75
[9] Raubenheimer B 2002 *J. Geophys. Res.* **107** C11
[10] Lin C, Yeh P H, Kao M J, Yu M H, Hsieh S C, Chang S C, Wu T R and Tsai C P 2015 *J. Waterway, Port, Coastal, Ocean Eng.* **141** 04014038
[11] Lin C, Kao M J, Tzeng G W, Wong W Y, Yang J, Raikar R, Wu T R and Liu P L F 2015 *J. Earthquake and Tsunami* **9** 154002
[12] Shigematsu T, Liu P L F and Oda K 2004 *J. Hydraul. Res.* **42** 183
[13] Hieu P D, Katsuoshi T and Ca V T 2004 *Appl. Math. Modelling.* **28** 983
[14] Quecedo M, Pastor M, Herreros M I, Fernandez Merodo J A and Zhang Q 2005 *Comput. Methods Appl. Mech. Engng.* 194 3985
[15] Yeh H H and Mok K M 1990 *Phys. Fluids A* **2** 821
[16] Stansby P K, Chegini A and Barnes T C D 1998 *J. Fluid Mech.* **374** 407
[17] Zhang Q and Liu P L F 2008 *Coast. Eng.* **55** 1113
[18] Yeh H H, Ghazali A and Marton I 1989 *J. Fluid Mech.* **206** 563

Numerical simulation of the cellular detonation interaction with inert gas plugs

I A Bedarev[1, 2]

[1]Khristianovich Institute of Theoretical and Applied Mechanics SB RAS, 4/1 Institutskaya st., Novosibirsk, 630090 Russia
[2]Novosibirsk State Technical University, 20 K. Marksa av., Novosibirsk, 630073 Russia

bedarev@itam.nsc.ru

Abstract. The mathematical modeling of the interaction of a cellular detonation wave propagating through a hydrogen-air mixture with inert gas plugs is performed. A technology for calculating two-dimensional detonation flows in a reacting gas mixture - non-reacting gas system has been developed. The modes of cellular detonation waves interaction with inert gas plugs were studied. The geometric limits of the clouds of carbon dioxide, argon and nitrogen lead to the cellular detonation failure and isolation of the combustion zone are determined.

1. Introduction

The mathematical modeling of the interaction of a cellular detonation wave propagating through a hydrogen-air mixture with a layer of inert gas is performed in the paper. The relevance of these studies is related to issues of explosion and fire safety, namely, the problem of attenuation and suppression of detonation waves in combustible gas mixtures [1-4]. Explosions of combustible industrial gases and gas suspensions are one of the main causes of technological disasters. An example of the catastrophe is the explosion in the city of Kaohsiung, Taiwan related with a propylene leak. The explosion occurred during the transportation of propylene from the port to a chemical plant through a pipeline laid in underground utilities. One way to prevent the catastrophic consequences of these incidents is to use inert gas plugs to suppress detonation and isolate combustion zones. At the initial stage, the study in this paper proposes the study of this process in a hydrogen-air mixture with plugs of three inert gases: carbon dioxide, argon, nitrogen.

2. Mathematical model and formulation of the problem

The gas dynamics of the process was modeled by a system of Navier – Stokes equations describing the motion of a viscous compressible heat-conducting gas, taking into account the multicomponent nature of the gas mixture and chemical kinetics. The reduced kinetic mechanism is used to describe the combustion process of a mixture of a fuel and an oxidizing agent. The mechanism was verified in [5-6] by the ignition delay time, detonation wave velocity, detonation cell size, and oblique detonation wave modes. Computing technology is implemented in the frame of the ANSYS Fluent package. Previously, this computing technology was used to solve the problem of attenuation and suppression of cellular detonation waves by inert particles [7-8].

Figure 1 shows the problem formulation. At the initial moment of time, a developed cellular detonation wave (DW) is set in the right part of the channel. An inert gas plug is located in the low-

Published under licence by IOP Publishing Ltd

pressure chamber (LPC), whose left boundary is located at the coordinate $x = 0.3$ m. The plug length L varies.

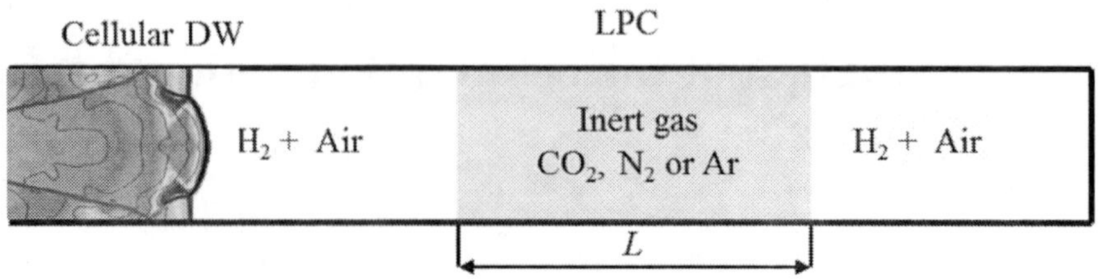

Figure 1. The problem statement. Flow pattern.

3. Calculation results

The detonation wave interaction with the plugs of three inert gases (carbon dioxide, argon, nitrogen) is compared. Depending on the type of isolating gas and the length of the plug, various flow regimes can be realized after the shock wave exits from the inert plug. Figure 2 shows the three regimes for a CO_2 plug with a length of $L = 0.2$ m (a), 0.3 m (b), and 0.4 m (c): a) detonation reinitiation; b) suppression of the detonation wave with the formation of a deflagration wave; c) suppression of the detonation wave with the combustion zone isolation. Figure 2 presents the fields of density maxima in time. For a plug with a length of $L = 0.2$ m (fig. 2a), after the shock wave leaves the inert gas layer, the detonation wave is reinitiated with the further development of the cellular structure. It is seen that at first a detonation cell with a small size is formed, but then the size of the detonation cell is enlarged to a value of 10-15 mm typical for a stoichiometric hydrogen-air mixture. The last two regimes are almost indistinguishable in the fields of maximum density.

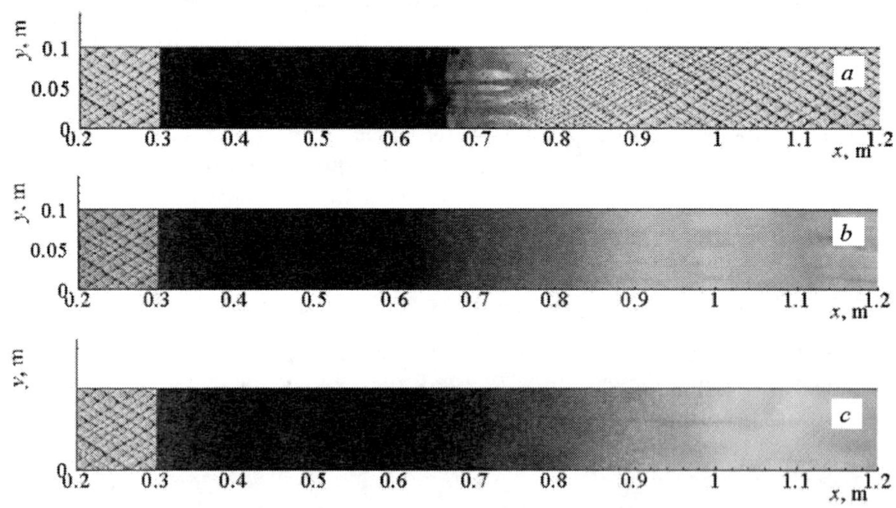

Figure 2. Fields of density maxima in time for a CO_2 plug of length $L = 0.2$ m (a), 0.3 m (b), and 0.4 m (c).

Figure 3 demonstrates the distribution of density and temperature on the symmetry axis of the channel at different times for a CO_2 plug of length $L = 0.2$ m. The density distribution shows a significant compression of the gas on the shock wave propagating in the CO_2 layer. In the figures, one can observe the rapid reinitiation of the detonation wave after the shock wave exits from the carbon dioxide layer. Figure 4 shows the temperature distributions on the symmetry axis for carbon dioxide plugs of length $L = 0.3$ m and 0.4 m. In this figure, two flow regimes for the cellular detonation failure can be observed: with the formation of a deflagration wave and with the isolation of the combustion zone. Deflagration combustion is initiated at the boundary of an inert gas layer for a plug with a length of $L = 0.3$ m (Fig. 4a).

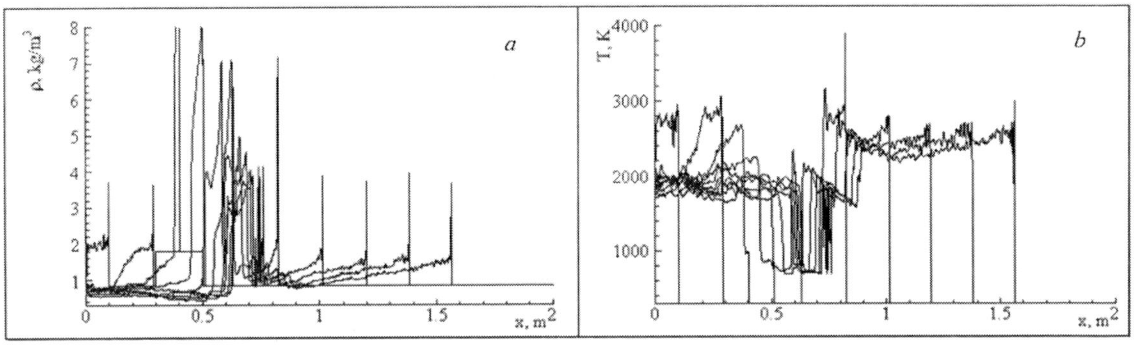

Figure 3. Density (*a*) and temperature (*b*) on the symmetry axis for a CO_2 plug of length $L = 0.2$ m.

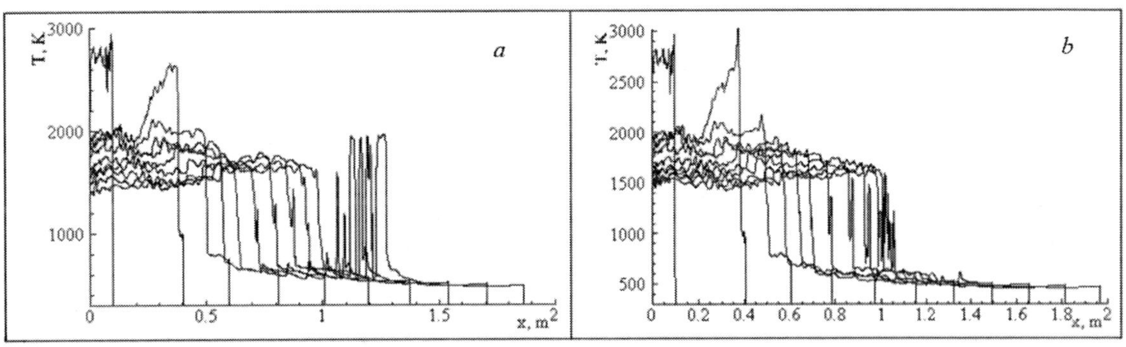

Figure 4. Temperature on the symmetry axis for CO_2 plugs of length $L = 0.3$ m (*a*) and 0.4 m (*b*).

The difference between the regimes can also be seen in the fields of temperature and mass fraction of CO_2, which shown in fig. 5. Here the calculation results at a time of 1.9 ms for clouds of 0.3 m (*a*, *b*) and 0.4 m (*c*, *d*) length are presented. It can be seen from the figure that for a cloud of inert gas with a length of 0.3 m, a combustion zone appears at its right boundary, which for a moment from 1.7 ms to 1.9 ms moves at an average velocity of about 485 m/s. Conclusions about the propagation dynamics of the combustion zone and the effect of turbulence on its behavior will allow further studies. We can also note the instability of the contact boundary of the CO_2 layer. This is a Richtmayer-Meshkov instability, which is caused by the passing of reflected shock waves through the boundary between the layer of light and heavy gas.

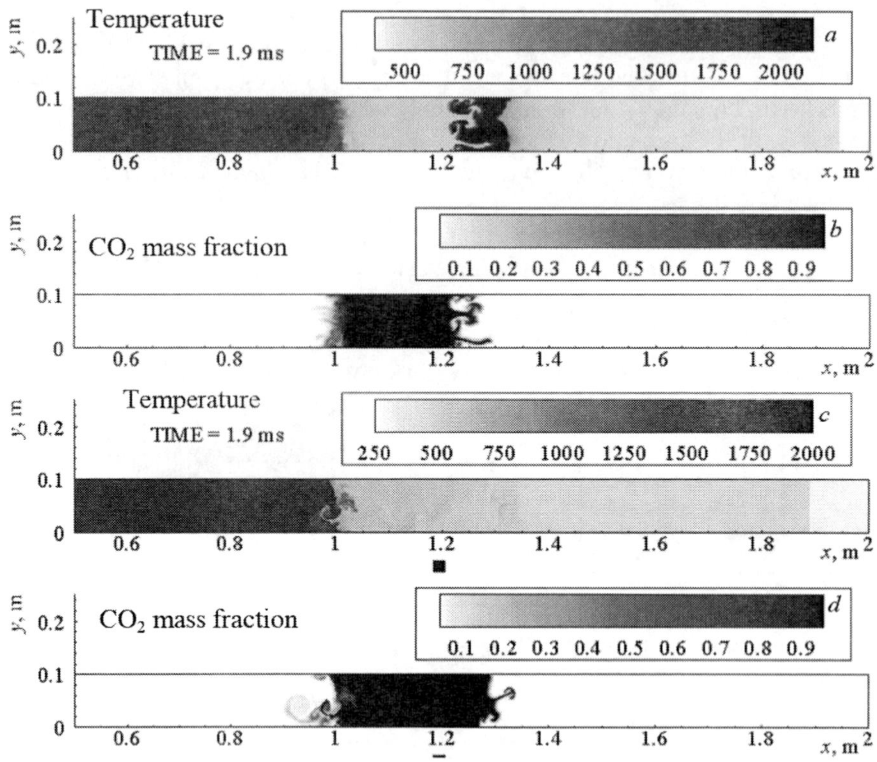

Figure 5. Fields of temperature and mass fraction of CO_2 for plugs of length $L = 0.3$ m (*a*, *b*) and 0.4 m (*c*, *d*).

Table 1 summarizes the flow regimes after the shock wave leaves the cloud for three inert gases: CO_2, N_2, Ar. The table shows that the use of carbon dioxide is more effective for detonation suppressing. In Table, the modes are indicated as follows: A – reinitiation of detonation; B - detonation wave suppression with the deflagration wave formation; C – detonation wave suppression with the combustion zone isolation.

Table 1. Interaction modes.

	$L = 0.2$ m	$L = 0.3$ m	$L = 0.4$ m	$L = 0.5$ m	$L = 0.75$ m
CO_2	A	B	C	C	C
N_2	A	A	A	B	C
Ar	A	A	B	C	C

4. Conclusions

- A technology for calculating two-dimensional detonation flows in the reacting gas mixture – inert gas system in the frame of ANSYS Fluent package has been developed.
- The modes of cellular detonation wave's interaction with inert gas plugs are studied.
- The geometrical limits of the clouds of CO_2, Ar and N_2 leading to cellular detonation failure and isolation of the combustion zone are determined.

Acknowledgments
This work was partially supported by the Program of Fundamental Scientific Research of the State Academies of Sciences for 2013-2020 (project AAAA-A17-117030610139-4).

References

[1] Tropin D A and Fedorov A V 2017 Physical and mathematical modeling of ignition, combustion and detonation of silane-hydrogen-air mixtures *Journal of Loss Prevention in the Process Industries* **49** 762-768

[2] Tropin D A and Fedorov A V 2018 Attenuation and suppression of detonation waves in reacting gas mixtures by clouds of inert micro- and nanoparticles *Combustion, Explosion and Shock Wave* **54** 200-206

[3] Tropin D A and Fedorov A V 2019 Effect of inert micro- and nanoparticles on the parameters of detonation waves in silane/hydrogen - air mixtures *Combustion, Explosion and Shock Waves* **55** 230-236

[4] Tropin D A and Fedorov A V 2019 Physical and mathematical modeling of interaction of detonation waves in mixtures of hydrogen, methane, silane, and oxidizer with clouds of inert micro- and nanoparticles *Combustion Science and Technology* **192** 275–283

[5] Bedarev I A, Rylova K V and Fedorov A V 2015 Application of detailed and reduced kinetic schemes for the description of detonation of diluted hydrogen-air mixtures *Combustion, Explosion and Shock Waves* **51** 528-539

[6] Bedarev I A, Temerbekov V M and Fedorov A V 2019 Simulating the regimes of oblique detonation waves arising at detonation initiation by a small-diameter projectile *Thermophysics and Aeromechanics* **26** 59-68

[7] Bedarev I A 2019 Micro-Level Modeling of the Detonation Wave Attenuation by inert particles *Thermal Science* **23** 439-445

[8] Bedarev I A 2018 Numerical simulation of the suppression of cellular detonation by inert particles *Journal of Physics: Conference Series* **1128** 1-6

Estimation of the initiation energy of detonation excited by a fast moving body

I A Bedarev[1,2] and V M Temerbekov[1]

[1]Khristianovich Institute of Theoretical and Applied Mechanics SB RAS, 4/1 Institutskaya st., Novosibirsk, 630090 Russia
[2]Novosibirsk State Technical University (NSTU), 20 Karl Marx avenue, Novosibirsk, 630073, Russia

tvm94@inbox.ru

Abstract. The results of numerical study of the detonation initiation by small diameter spherical supersonic projectile in reacting mixture are presented. A stoichiometric hydrogen and oxygen mixture diluted with argon was used as a reacting medium. The reduced kinetic scheme of chemical reactions was used in our study. The flow regimes and the detonation cell size obtained numerically and experimentally were compared and showed good agreement. Calculations for various sphere diameters at fixed pressure were carried out. The detonation initiation energy for the calculated data was estimated.

1. Introduction

The study of detonation initiation and stabilization by high speed body is important both from the scientific and technical points of view. This field of study covers a wide area of science connected with explosion safety, and studying the issues of initiating and re-initiating detonation.

Currently, there are a lot of works devoted to this issue, which indicates the relevance of this problem for the whole world. To model the phenomena considered it is necessary to familiarize themselves with a certain number of experimental and theoretical works devoted to detonation processes.

In work [1], an experimental study of the detonation initiation during the high-speed body passage through a combustible mixture is presented. The mixture of C_2H_2 + $2.5O_2$ and a cylindrical body with a rounded or blunt nose, propelled at a speed of about 800-1400 m/s were used. A new mechanism for the conversion of shock waves into detonation waves in near-critical regimes was observed. An analytical correlation between the aerodynamic characteristics of a high-speed body and the physicochemical parameters of a combustible mixture as a criterion for initiating gas detonation was proposed.

In [2], the initiation and stabilization of three-dimensional oblique detonation waves by spheres 4.76 mm in diameter with velocities 1.06–1.31 times higher than the Chapman – Jouguet velocity in stoichiometric mixtures of oxygen with acetylene, ethylene, or hydrogen diluted with argon were experimentally studied. The development of an oblique detonation wave was investigated. It was found that the radius of curvature needed to stabilize the oblique detonation wave, referred to the size of the detonation cell, is 8-10 for mixtures diluted by 50%, and 15-18 for mixtures diluted by 75%.

In article [3], the initiation and stabilization of detonation combustion in a supersonic flow of hydrogen-air mixtures in an axisymmetric nozzle with a central body installed in it was numerically studied. Two geometric configurations of the central body are considered: "cone – cylinder – cone" and "cylinder – cone". Modelling was carried out in a two-dimensional, axisymmetric formulation. The Euler equations for a multicomponent gas mixture taking into account chemical reactions were used as a mathematical model. It is shown that the central body can serve as an effective tool for initiating and stabilizing detonation, which provides a thrust at free-stream Mach numbers from7 to 9.

One of the most important problem of fire and explosion safety today is the assessment of the energy necessary for detonation initiation. There are a number of theoretical models that can be used to predict the initiation of detonation and evaluate the effects of an explosion. So, for example, in the work [4] a criterion for initiating detonation by a fast-flying body was proposed. It is assumed that in order to initiate gas detonation by a body flying at a high speed, it is necessary that the work of the aerodynamic drag forces per unit length of the body would be not less than the minimum initiation energy of cylindrical detonation in the mixture under given conditions (pressure, temperature).

The present work considers the detonation initiation of a hydrogen-oxygen mixture by high speed spherical particles and estimation of detonation initiation energy.

2. Problem formulation

A schematic of the simulation is shown in Figure 1. A premixed stoichiometric hydrogen–oxygen mixture diluted with argon ($2H_2 + O_2 + 3Ar$) having a velocity $V = 1.24 \pm 0.03\ V_{CJ}$, a static temperature $T_{st} = 295$ K, and an initial pressure $P_{st} = 121$kPa was supplied to the inlet of the simulation domain. The diameter of the sphere was varied $d = 3.18 \div 3.8$ mm.

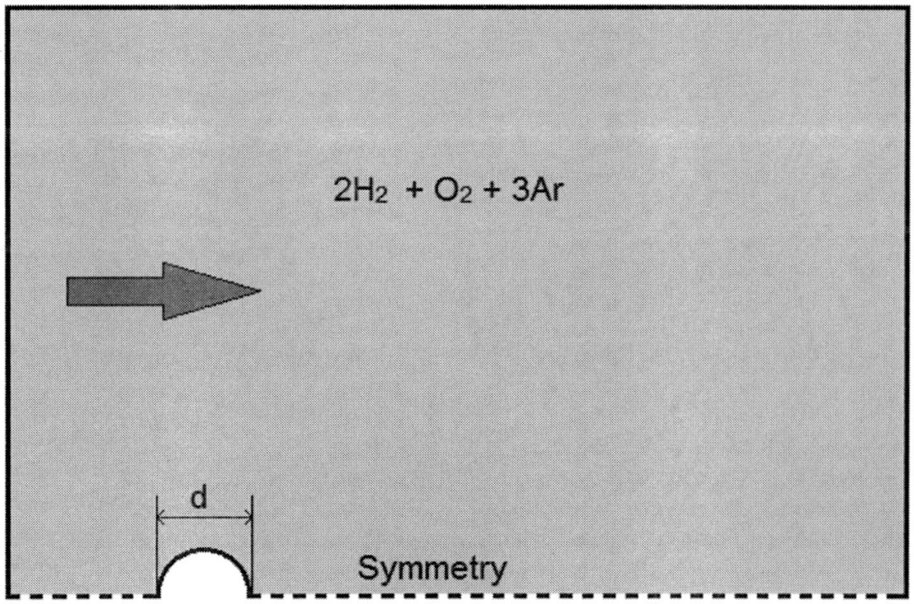

Figure 1. Schematic of the simulation.

3. Mathematical model

A Favre-averaged Navier-Stokes equations for multicomponent gas mixture supplemented by an SST modification of the *k-ω* turbulence model were used as a mathematical model. The ANSYS Fluent software package was used as a solver. An implicit second-order scheme was used for temporal approximation, and the AUSM flux vector splitting method with a second-order upwind scheme for spatial approximation. The reduced Arenius kinetic scheme including one gross reaction of hydrogen burning in air was used for the chemical kinetics simulation. In the work [5], this kinetic scheme was verified according to the experimental data on the ignition delay and the propagation velocity of the detonation wave under different conditions.

A detailed quadrangular orthogonal grid was used. For convenience, the computational grid was divided into blocks in which the grid was constructed separately. Such way allows us to avoid unnecessary condensations, reduce the total number of grid nodes without increasing the grid step.

In work [6], the mathematical model and the calculation algorithm were subjected to additional verification against experimental data [7] on the detonation cell size and detonation modes. Comparative analysis showed that the approach used allows us to model complicated detonation processes, and in addition to qualitative agreement between the calculated and experimental flow pictures we achieved a quantitative agreement between detonation cell sizes.

4. Calculations results

In previous studies [6], we numerically obtained various detonation modes observed in experiments [7]. Figure 2 shows these modes.

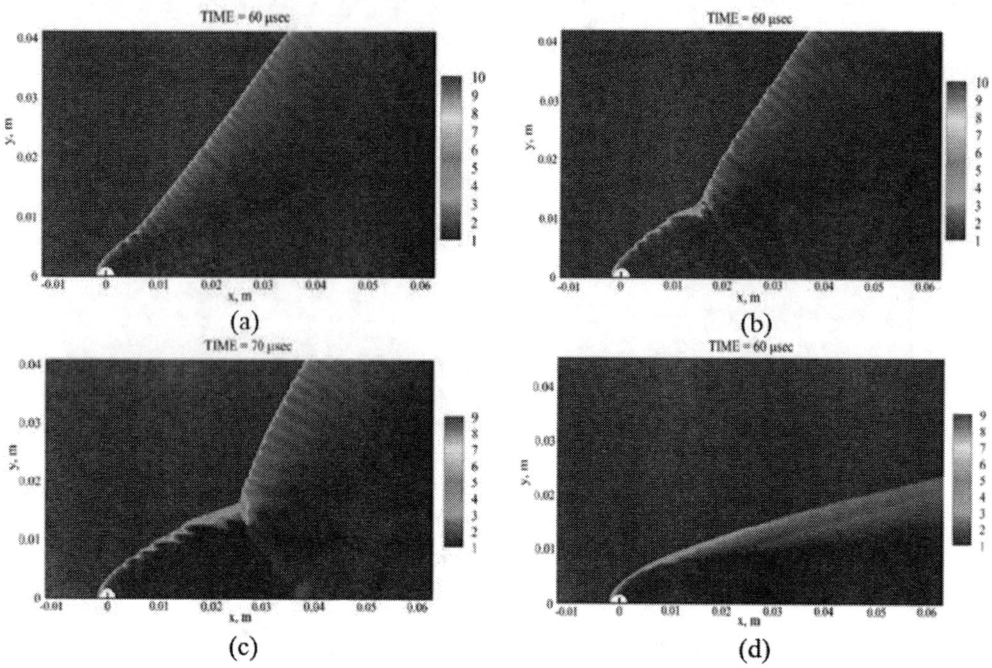

Figure 2. The results of the verification. Density fields.
(a) – Oblique detonation wave at P_{st} = 141kPa.;
(b) – Straw-hat mode with oblique detonation wave at P_{st} = 136 kPa;
(c) – Straw-hat mode with attenuated oblique detonation wave at P_{st} = 136 kPa;
(d) – Shock-induced combustion at P_{st} = 121 kPa.

For these calculations, the detonation initiation energy was estimated using the formulas from [4]. Figure 3 shows the input energy deviation (work of the aerodynamic drag forces per unit length) from the critical initiation energy of detonation under given conditions.

In this study the detonation initiation at constant pressure P_{st} = 121 kPa is considered. By increasing the diameter of the sphere starting from 3.18 mm, the same regimes were obtained as in the works [6] and [7]. The results of the simulations are shown in Figure 4.

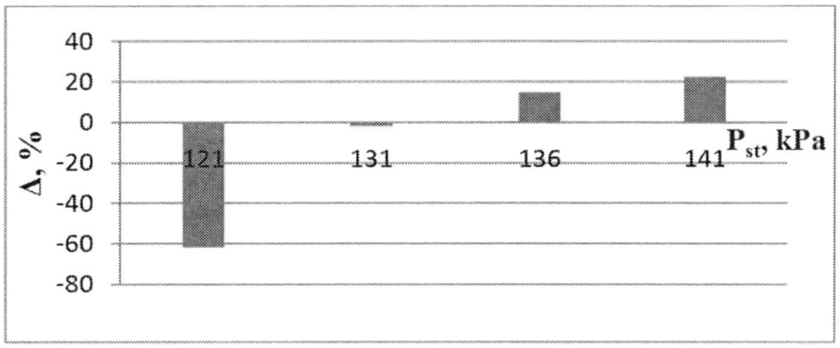

Figure 3. The input energy deviation from the critical detonation initiation energy for data [6] and [7].

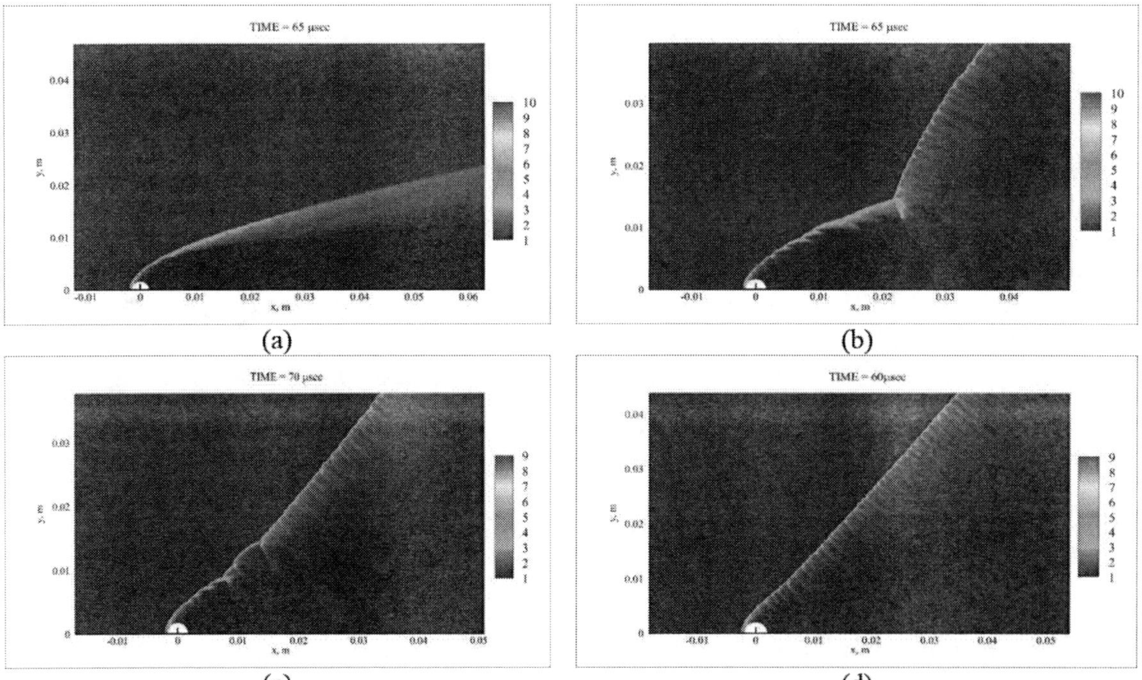

Figure 4. The results of the calculations at pressure P_{st} = 121 kPa. Density fields.
(a) – Shock-induced combustion at d = 3.18 mm;
(b) – Straw-hat mode with attenuated oblique detonation wave at d = 3.3 mm;
(c) – Straw-hat mode with oblique detonation wave at d = 3.4 mm;
(d) – Oblique detonation wave at d = 3.8 mm.

Figure 5 shows the input energy deviation (work of the aerodynamic drag forces per unit length) from the critical initiation energy of detonation at pressure P_{st} = 121 kPa for various diameter of the sphere.

For the sphere diameter d = 3.18 mm, shock-initiated combustion is carried out. For d = 3.3 mm, we can see straw-hat mode with attenuated oblique detonation. For d = 3.4 – 3.6 mm, straw-hat mode with oblique detonation wave is carried out. For sphere diameter from d = 3.8 mm we can see stabilized oblique detonation wave.

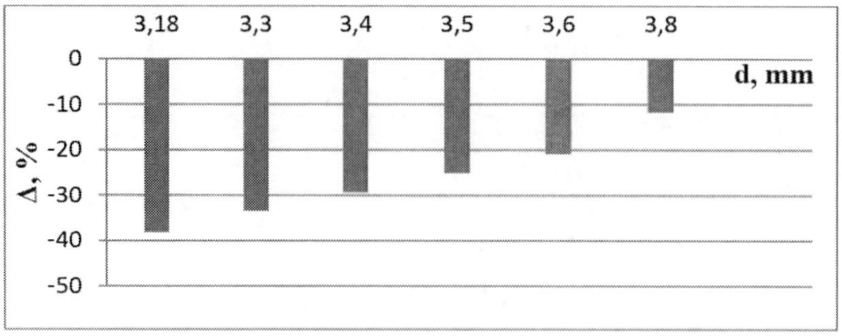

Figure 5. The input energy deviation from the critical detonation initiation energy at P_{st} = 121 kPa for various diameter of the sphere.

5. Conclusions

The problem of detonation initiation and stabilization by various diameter high speed body was solved using a previously developed mathematical model and mathematical technology for its implementation based on the ANSYS Fluent package.

The input energy deviation (work of the aerodynamic drag forces per unit length) from the critical initiation energy of detonation under given conditions was found for experimental and calculated data.

Acknowledgments

The research was carried out within the framework of the Program of Fundamental Scientific Research of the state academies of sciences in 2013-2020 (project No. AAAA-A17-117030610139-4).

Nomenclature

T_{st} - static temperature;
P_{st} - static pressure;
V_{CJ} - Chapman-Jouguet velocity;
d - diameter of the sphere;
Δ - the input energy deviation from the critical detonation initiation energy.

References

[1] Vasiljev A A 1994 *Shock Waves* **3** 321–6
[2] Maeda S, Sumiya S, Kasahara J and Matsuo A 2013 *Proc. Combus. Institute* **34** 1973–80
[3] Tunic Yu V 2015 *Proc. XI All-Russian Cong. Fund. Prob. Theoret. Appl. Mech.* (Kazan: Kazan Federal Univ. Press.) pp 3813–15
[4] Vasiljev A A 1997 *Combustion, Explosion, and Shock Waves* **33** 85-102
[5] Bedarev I A, Rylova K V and Fedorov A V 2015 *Combustion, Explosion, and Shock Waves* **51** 528–39
[6] Bedarev I A, Temerbekov V M and Fedorov A V 2019 *Thermophysics and Aeromechanics* **26** 59–68
[7] Maeda S, Sumiya S, Kasahara J and Matsuo A 2015 *Shock Waves* **25** 141–150

Effect of inhibitor addition on the Markstein length in a distorted premixed dimethyl ether/air flame

A A Chernov[1,2], A G Shmakov[1], O P Korobeinichev[1], K V Toropetsky[2], V N Moskvin[2] and V I Tatarenko[2]

[1]Voevodsky Institute of Chemical Kinetics and Combustion, Siberian Branch RAS
3 Institutskaya str., Novosibirsk, 630090, Russia
[2] Siberian State University of Geosystems and Technologies
10 Plakhotnogo str., Novosibirsk, 630032, Russia

chernov@kinetics.nsc.ru

Abstract. The paper presents the results of measuring the propagation velocity and Markstein length of pure and inhibited laminar axisymmetric dimethyl ether/air flames using the PIV method. Their dependence on equivalence ratio and the concentration of inhibitor is studied. Trimethyl phosphate is used as an inhibitor. The measurement results show a strong dependence of the Markstein length on the concentration of the inhibitor. This suggests the use of this parameter as a measure of inhibition effectiveness of turbulent flames.

1. Introduction

The main ideas that led to the modern understanding of the nature of hydrodynamic instability (or Darrieus-Landau instability (DL)) were set out by Darrieus and Landau [1, 2], who considered the flame as a structureless surface separating burned and unburned gases and moving at a constant speed with respect to the gas flow. In subsequent work, Markstein [3] took into account the effects of diffusion, thermal conductivity, and chemical reactions in the flame front on the flame speed by introducing a correction term, which was later called the Markstein length. DL instability affects the surface shape (topology) and speed of laminar and turbulent flames [4, 5]. Nonlinear theory and its numerical simulation show that hydrodynamic instability does not lead to the occurrence of a turbulent flame. Instead, a characteristic topology of the flame surface in the form of bulges and elongated cusps is formed. As the ratio of the Markstein length to the size of the flame region decreases, the curvature of the flame surface increases due to the formation of sharp peaks and a 20%–50% increase in flame speed compared to laminar flat flames [5]. Thus, the Markstein length is related to the local burning velocity and area of the distorted flame; therefore, this parameter becomes a more important indicator of the inhibitor effect on turbulent flames.

This paper presents the results of measuring the dependence of the Markstein length on the addition of inhibitor to a distorted dimethyl ether flame. Dimethyl ether (DME, CH3-O-CH3) is an environmentally friendly fuel derived from natural gas, coal or biomass. Its combustion produces 90% less nitrogen oxides than in the case of pure diesel fuel. In addition, it has a lower soot formation tendency, a high cetane number, and an almost complete absence of exhaust smoke. However, due to the physicochemical properties of dimethyl ether, its use as a fuel requires special fire and explosion safety measures, including the search for effective inhibitors, since unstable reactive peroxide compounds can form in DME/air mixtures exposed to light.

A premixed DME/air flame stabilized on a Bunsen burner with a nozzle diameter of 1 cm was studied by particle image velocimetry (PIV). The setup is shown schematically in Figure 1. A brief overview and experimental results of applying the techniques we use to study the Bunsen cone flame by PIV can be found in [6, 7]. It has been found that measurement errors are primarily due to the limited temporal and spatial resolution of PIV devices and the inertia and catalytic ability of flow tracer particles. Experimental results and theoretical estimates show that the employed method and procedures provide fairly accurate measurements of gas velocity in the flame front, tracer particles at a mass concentration > 0.02% have a slight catalytic effect, and the error in measuring the burning velocity in an optically transparent flame does not exceed 1%.

2. Experimental setup

The experiments were carried out at a temperature of the combustible mixture from 25 to 55 ^{0}C and a pressure of 748-753 mm Hg. The volumetric flow rate of the DME/air combustible mixture was selected for the corresponding stoichiometric coefficient so that the height of the flame cone was no more than 9 mm and was within the observation area of the CCD camera. Trimethyl phosphate (TMP) was added to the combustible DME/air mixture using a syringe driven by a stepper motor and an evaporator. For PIV measurements, submicron TiO_2 particles (specific surface area ~3.6 m^2/g) were introduced into the air flow at a mass concentration of no more than 0.02%. The length of the tubes supplying gas to the burner was about 1.5 m. The average size of the particles recorded by the PIV optical system at the exit of the burner nozzle was about 600 μm. Particle size and concentration were measured by an AZ-10 diffusion aero-sol spectrometer counter [8, 9].

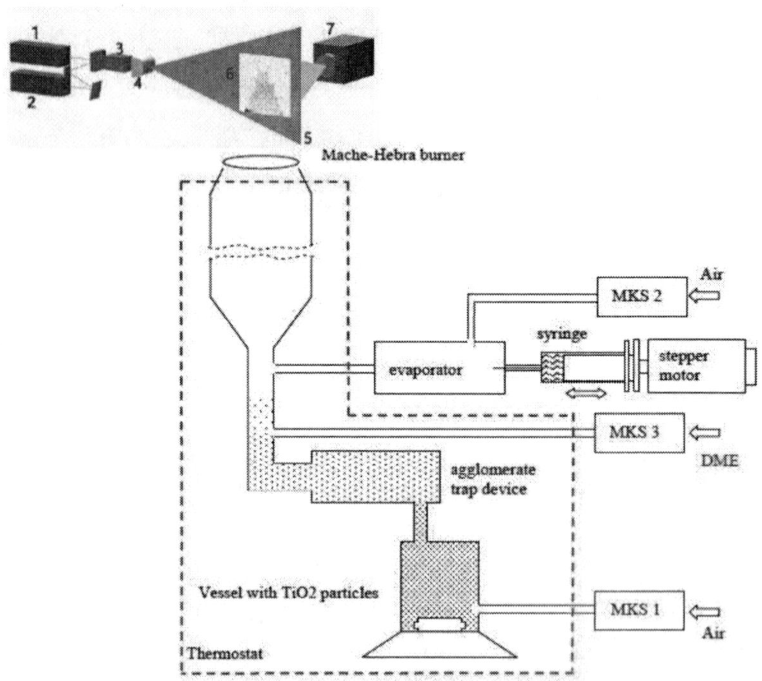

Figure 1. Experimental setup.

The system for recording and processing tracer particle image consisted of two pulsed Nd:YAG lasers (wavelength 532 nm, pulse duration 5 ns, pulse energy 45 mJ, beams collimated to the same axis), CCD cameras (1360×1024 pixels, each pixel size 4.65x4.65 μm) with the option of two-frame shooting, a Tamron SP AF 180 mm optical lens (the magnification factor was 0.34, the aperture number was maximal: f# = 32), a synchronizing processor, and a PC with Actual Flow software. The measurement plane with a laser knife thickness of 100 μm passed through the axis of symmetry of the flow of the combustible mixture. The optical resolution of the system was about 10 μm/pixel. The particle displacement in the interval between a pair of laser flashes (with a time interval of 20 μs) was calculated using an adaptive iterative cross-correlation algorithm with 50% overlapping and continuous displacement of the computational domains in the first and second frames. The cross-correlation image processing method used is described in [10]. The final size of the computational domain from which the particle group velocity was determined corresponded to 32×32 pixels (0.32×0.32 mm). From the measured tracer particle images, 500 instantaneous velocity fields were calculated, which were averaged to minimize the random measurement error. These measurements were used to determine the dependence of the local flame speed on the curvature of the flame front. The normal flame speed was obtained by linear extrapolation of these values to the zero curvature of the flame front. In the linear extrapolation, we used only values that did not undergo additional distortions in the region of the truncated flame cone, in accordance with the recommendations of [11]. The standard deviation was 0.3 (for the doped flame) to 1 cm/sec (for the undoped flame). This is due to a significant change in the transparency of the flame with increasing concentration of the inhibitor additive in the combustible mixture. The procedures for determining the line of the flame front with a maximum gas velocity gradient and its local curvature are based on recommendations from [12–16].

3. Result and discussion

For DME, these procedures were first used in [17]. The obtained data on the dependence of the burning velocity on the TMP additive concentration are in good agreement with the results of numerical simulations based on detailed kinetics and published data [18].

The dependences of the flame speed and the first Markstein length on the stoichiometric coefficient and TMP concentration in the combustible mixture are shown in Figure 2. It can be seen that a small addition of TMP has a significant effect on both the burning velocity and the Markstein length. Moreover, the effect of the addition of TMP on the Markstein length is significantly higher in absolute value than predicted by the Zel'dovich mathematical theory of combustion, but is similar in the nature of the normal flame speed dependence [18]. As this theory predicts, the higher the flame speed, the lower the proportionality coefficient in the dependence of the speed on flame front curvature — the Markstein length.

(a) (b)

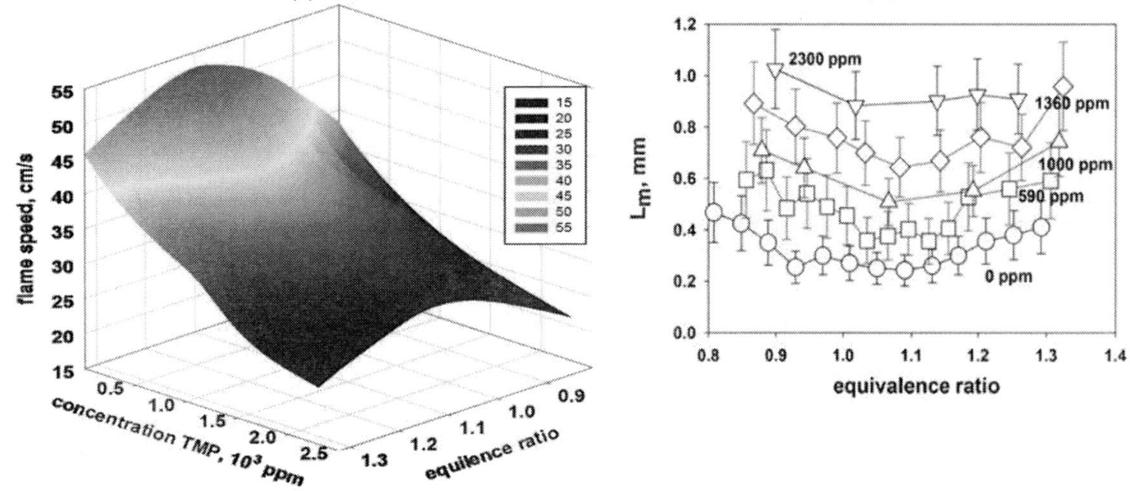

Figure 2. Results of measuring the flame propagation speed DME / air (a) and the first Markstein length (b).

4. Conclusion

We suggest that the inhibition of the turbulent flame by the addition of TMP involves a threefold effect: a decrease in the normal flame speed, a decrease in the local burning velocity, and a reduction in flame area as a result of a decrease in the number and size of characteristic topologies in the form of bulges and cavities. In this case, it becomes appropriate to use the Markstein length as an indicator of the inhibitor effect in turbulent flames.

References

[1] Darrieus G 1938 *La Technique Moderne* (Paris)
[2] Landau L 1944 *Zh. Eksp. Teor. Fiz. USSR* **14**(6) 240-244
[3] Markstein G 1964 *Non-steady Flame Propagation* (New York: Pergamon)
[4] Matalon M and Matkowsky B J 1982 *J. Fluid Mech.* **124** 239–259
[5] Patyal A and Matalon M 2018 *Combustion and Flame* **195** 128–139
[6] Korobeinichev O P, Shmakov A G, Chernov A A, Markovich D M, Dulin V M and Sharaborin D K 2014 *Combustion Explosion and Shock Waves* 50 510-517
[7] Bolshova T A, Korobeinichev O P, Toropetskii K V, Shmakov A G and Chernov A A 2016 *Combustion Explosion and Shock Waves* **52** 155-166
[8] Dubtsov S, Ovchinnikova T, Valiulin S, Chen X, Manninen H E, Aalto P P and Petaj T T 2017 *Journal of Aerosol Science* **105** 10-23
[9] Onischuk A A, Valiulin S V, Baklanov A M, Moiseenko P P and Mitrochenko V G 2018 *Journal of Aerosol Science and Technology* **52** 841-853
[10] Tokarev M P, Markovich D M and Bilsky A V 2007 *J. Comput. Technol.* **12** 109-131
[11] Fristrom R M 1965 *Physics of Fluids* **8** 273-280
[12] Choi C W and Puri I K 2001 *Combust. Flame* **126** 1640-1654
[13] Cohé C, Chauveau C, Gökalp I and Kurtulus D F 2008 *Proc. Comb. Inst.* **32** 1803-1810
[14] Gao X, Zhang Y, Adusumilli S, Seitzman J, Sun W, Ombrello T and Carter C 2015 *Combustion and Flame* **162** 3914-3924
[15] Tamadonfar P and Gulder O L 2015 *Combustion and Flame* **162** 115-128
[16] Giannakopoulos G K, Gatzoulis A, Frouzakis C E, Matalon M and Tomboulides A G 2015 *Combustion and Flame* **162** 1249-1264
[17] Knyazkov D Alyanova N Bolshova T Chernov A Shvartsberg V Shmakov A 2018 *37-th Int. Symp. Comb.* Dublin, 29 July - 3 August 2018, Book of abstracts 16657
[18] Knyazkov D A, Bolshova T A, Shvartsberg V M, Gerasimov I E, Shmakov A G and Korobeinichev O P 2019 *Proceedings of the Combustion Institute* **37** 4267-4275
[19] Zeldovich Ya B, Barenblatt G I, Librovich V B and Makhviladze G M 1985 *The Mathematical Theory of Combustion and Explosion* (Netherlands: Kluwer Academic) p 618

Filtration combustion inside a porous cylindrical pipe with axial gas flow

E P Dats[1,4], S S Minaev[1] and A I Kirdyashkin[3]
[1]Institute of Applied Mathematics FEB RAS, Radio St. 7, Vladivostok, 690041, Russia
[3] Tomsk Scientific Center SB RAS, Akademicheskiy Av. 10/4, Tomsk, 634055, Russia
[4] Vladivostok State University of Economics and Service, Gogolya st. 41, Vladivostok, 690014, Russia
E-mail: dats@dvo.ru

Abstract. The filtration gas combustion in a porous tube with injection of a combustible mixture through the end surface of a porous cylindrical tube is theoretically studied. The simulation of the filtration gas combustion was carried out in the framework of the two-temperature thermal-diffusion model. Numerical modeling allowed to estimate the range of gas flow rates at which a stable combustion regime is observed, to find the temperature distribution in the gas and the porous body, and to evaluate the radiation fluxes inside the reactor.

1. Introduction

Development of thermal energy sources based on the gas combustion with controlled temperature and power characteristics is an actual problem in power engineering. Combustion in porous media is one of the effective methods for burning gaseous fuels. Combustion here occurs inside the porous body. Heat from the combustion zone is transferred along the porous frame towards the gas flow and heats the mixture of initial reactants. Porous burners are characterized by the possibility of combustion of the lean mixtures, a broad range of adjusting, the higher specific capacities and low level of pollution [1, 4].

The theoretical research of the new type of a porous burner with axial gas flow is presented in the paper. The filtration combustion occurs in a porous tube with injection of a combustible mixture through the end surface. The samples under treatment can be placed inside a porous tube, the side surface of which is covered with quartz shells transparent for infrared radiation emitted by porous carcass. The outer surface of porous reactor is heat insulated. Radiation from the inner surface of the porous solid phase allows heating materials under treatment to high temperatures. This scheme of filtration gas combustion can be used to create chemical reactors for non-contact heating of various materials.

The aim of this study is to carry out the numerical modeling of filtrational combustion in a porous pipe: to find the temperature distribution in the gas and the porous body, the radiation fluxes inside the burner, to investigate the influence of heat loss and the geometric characteristics of the burner on the flame stabilization.

2. Governing equations. Boundary value problem statement

The porous burner is a pipe of length h with an inner radius r_1 and an outer radius r_2. The space between r_1 and r_2 is filled by a porous medium. Fresh gas mixture is injected through the end surface. Combustion occurs inside the porous body and combustion products leave the cylindrical burner from its other end surface. Scheme of the burner is presented on the Fig. 1.

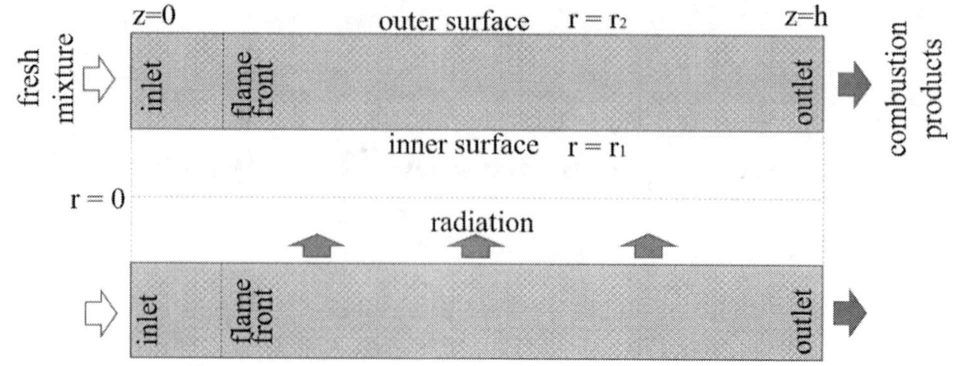

Figure 1. Scheme of the porous burner.

Simulation was conducted within the framework of a two-temperature diffusive-thermal model [2, 5, 6], consisting of the heat conduction equations for the gas and solid phases and diffusion equation for the deficient component of the mixture:

$$\begin{gathered}
\rho_g c_{pg}\left(\frac{\partial T_g}{\partial t}+u_z\frac{\partial T_g}{\partial z}\right)=\lambda_g\left(\frac{1}{r}\frac{\partial T_g}{\partial r}+\frac{\partial^2 T_g}{\partial r^2}+\frac{\partial^2 T_g}{\partial z^2}\right)+Q\cdot W(Y,T_g)-\frac{2\alpha}{d_p}(T_g-T_s)\\
\rho_s c_{ps}\frac{\partial T_s}{\partial t}=\lambda_s\left(\frac{1}{r}\frac{\partial T_s}{\partial r}+\frac{\partial^2 T_s}{\partial r^2}+\frac{\partial^2 T_s}{\partial z^2}\right)+\frac{2\alpha}{d_s}(T_g-T_s)\\
\frac{\partial Y}{\partial t}+u_z\frac{\partial Y}{\partial z}=D_c\left(\frac{1}{r}\frac{\partial Y}{\partial r}+\frac{\partial^2 Y}{\partial r^2}+\frac{\partial^2 Y}{\partial z^2}\right)-W(Y,T_g)\\
W(Y,T_g)=A\cdot Y\exp\left(-\frac{N_a}{T_g R}\right);\qquad \alpha=\frac{Nu}{d_p};\qquad d_s=\frac{d_p(1-m)}{m}
\end{gathered}\tag{1}$$

Here T_g and T_s are temperatures of gas and porous media, Y - fuel mass fraction, d_p is the average pore size, d_s is the mean space between the pores, R is the universal gas constant. Other parameters of the equations (1) are presented in the Table 1.

The set of governing equations (1) is coupled with the following boundary conditions :

$$\begin{gathered}
r=r_1:\ -\lambda_s\frac{\partial T_s}{\partial r}=\sigma_{sb}(T_s^4-T_{ref}^4),\quad \frac{\partial T_g}{\partial r}=0,\quad \frac{\partial Y}{\partial r}=0;\\
r=r_2:\ \frac{\partial T_s}{\partial r}=0,\quad \frac{\partial T_g}{\partial r}=0,\quad \frac{\partial Y}{\partial r}=0;\\
z=0:\ \frac{\partial T_s}{\partial r}=0,\quad T=T_0,\quad Y=T_0;\\
z=h:\ \frac{\partial T_s}{\partial r}=0,\quad \frac{\partial T_g}{\partial r}=0,\quad \frac{\partial Y}{\partial r}=0.
\end{gathered}\tag{2}$$

Here σ_{sb} is the Stefan-Boltzmann constant, T_{ref} - internal temperature, which sets the level of radiation heat flux. Value of T_{ref} varies to determine the effect of heat losses on the temperature distribution in the porous media. We assume that the flow velocity field is uniform and it has axial direction:

$$u_z=const,\quad u_r=0\tag{3}$$

3. Results

Calculations of the system (1) – (2) were carried out for the material constants presented on the Table 1:

Table 1. Simulation parameters

Symbol	Quantity	Value
h	length of the pipe (m)	0.16
r_2	outer radius of the pipe (m)	0.06
T_0	ambient temperature (K)	300
D_c	diffusivity (m^2/s)	$3.5 \cdot 10^{-5}$
Nu	Nusselt number	4
E_a	activation energy (J)	$125 \cdot 10^3$
m	carcass porosity	0.5
ρ_s	carcass density (kg/m^3)	2650
ρ_g,	mixture density (kg/m^3)	1.2
λ_s	carcass thermal conductivity (W/(K m))	20
λ_g	mixture thermal conductivity (W/(K m))	0.052
c_{pg}	mixture heat capacity J/(kg K)	1200
c_{ps}	carcass heat capacity (J/(kg K))	700
d_p	the average pore size (m)	0.001
Q	heat release (J/kg)	$35 \cdot 10^6$
A	pre-exponent (1/s)	10^9
Y_0	inlet fuel mass fraction	0.048

Simulation was performed by means of the finite element method. The result of the calculations is a sequence of stationary solutions of the system (1). At the first stage the problem under adiabatic conditions was solved. Due to the absence of heat loss on the inner surface, the formulation of the problem is equivalent to the one-dimensional case in which all solutions depend on z coordinate. These solutions were used as the initial conditions for two-dimensional formulation of the problem taking into account radiation on the inner surface. Solutions allowed to obtain the flow rate range at which stationary combustion mode occurs. It was shown that values of velocities have the range $1.2 < u_z < 2.8\ m/s$. The upper speed limit decreases to the value $u_z = 2.4\ m/s$ under maximum radiation flux on the inner surface ($T_{ref} = T_0$). Value $u_z = 2.4\ m/s$ was used as optimal flow velocity for further calculations.

Fig. 2 demonstrates stationary temperature distribution for gas and solid phases under condition $T_{ref} = T_0$. It was shown, that combustion occurs under super-adiabatic temperature. Maximum level of T_g corresponds to a flame front location. Calculations were carried out for the set of values $T_{ref} = 300, 600, 900, 1200, 1500, 1990\ K$. Simulation demonstrates that the flame front location weakly depends on different levels of T_{ref}. Maximum heat flux density on the inner surface takes place at $T_{ref} = 300\ K$. Temperature distribution on the inner surface for different values of inner radius r_1 is shown on the Fig. 3.

Difference between maximum values of T_s reaches about 5 percents. It means that using of small domain with high temperature of porous media is more efficiently for thin pipe. Ideal thermal insulation of outer surface provides high temperature distribution for thin-walled pipe with maximum radiation heat flux on the inner surface. The heat flux distribution is presented on the Fig. 4. Energy flux on the inner

 IOP Publishing
1404 (2019) 012056 doi:10.1088/1742-6596/1404/1/012056

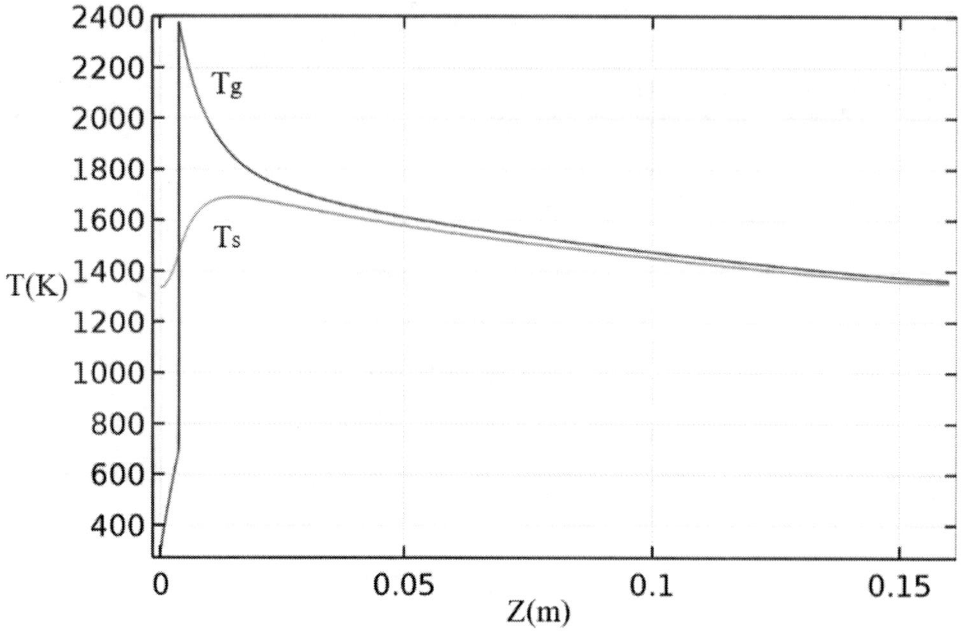

Figure 2. Temperature distribution of gas and solid body on the inner surface, $r_1 = 0.04\ m$, $T_{ref} = 300\ K$, $u_z = 2.4\ m/s$.

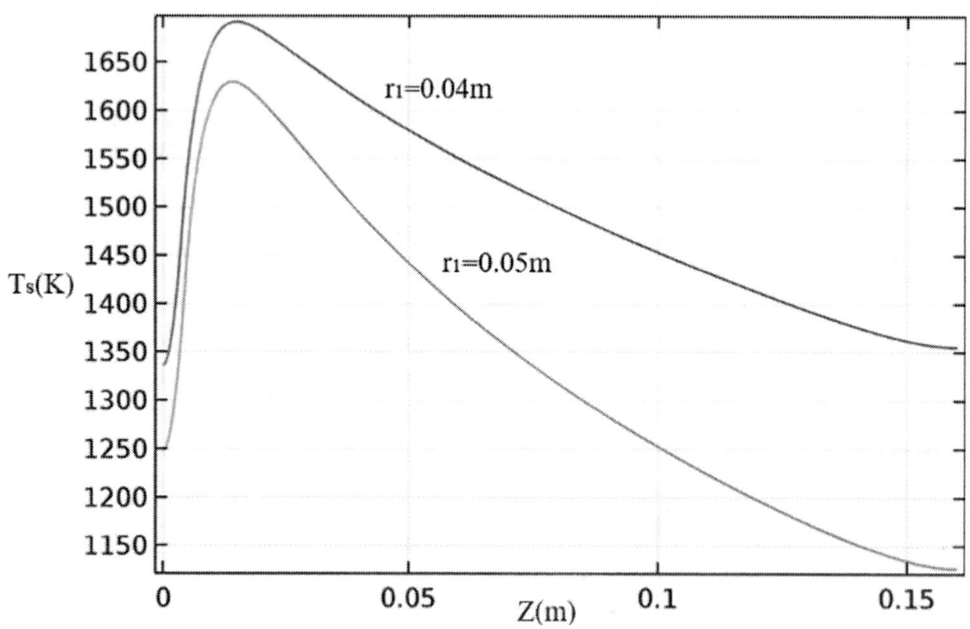

Figure 3. Temperature distribution on the inner surface for different radius r_1, $T_{ref} = 300\ K$, $u_z = 2.4\ m/s$.

surface was calculated. It was found that the energy flux for $r = 0.04$ is more than for $r = 0.05$ in the area $0 < z < h$. But we obtained the higher energy flux for a thin-walled pipe on the area $0 < z < 0.3h$.

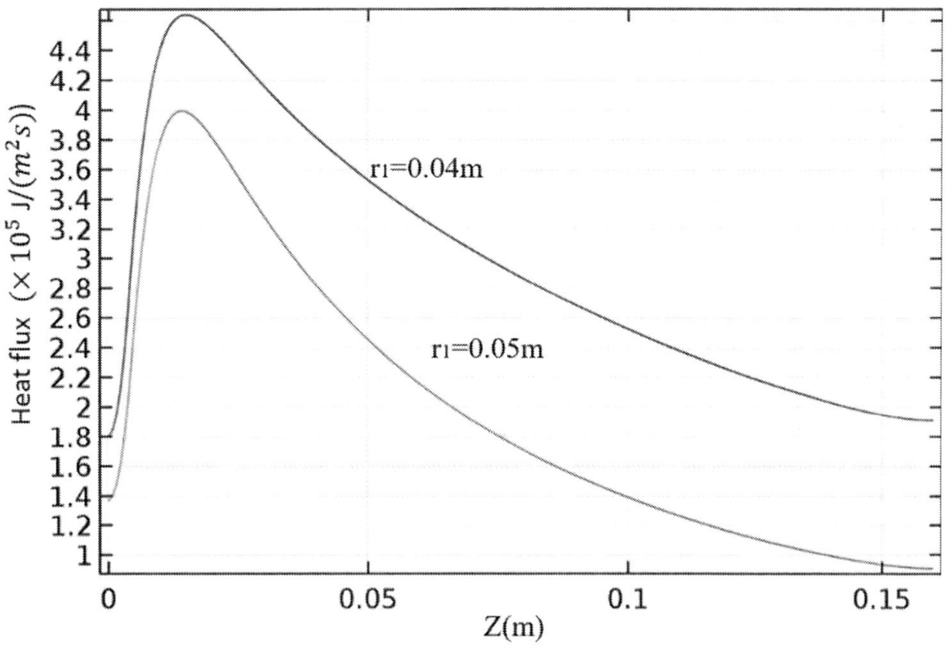

Figure 4. Axial temperature distribution of solid body on the inner surface for different radius r_1, $T_{ref} = 300\ K$, $u_z = 2.4\ m/s$.

4. Conclusion

The filtration gas combustion in a porous tube with injection of a combustible mixture through the end surface of a porous cylindrical tube is theoretically studied. The simulation of the filtration gas combustion was carried out in the framework of the two-temperature thermal-diffusion model. The influence of gas glow, heat loss, and the geometric characteristics of the reactor on the flame stabilization were studied. It was shown, that steady combustion can take place with maximum radiation loss in the inner surface. High levels of temperature distribution allow us to conclude that a burner of this type is a reactor of high efficiency. Further research is related to studying the effect of external heat losses on the temperature and heat flux formation.

Acknowledgments

The work was supported by the Russian Foundation for Basic Research (project No. №18-48-700037 r_a).

References

[1] Dobrego K, Hall M and Ellzey J 1996 Combustion of hydrocarbon fuels within porous inert media *Progress in Energy and Combustion Science* **22** 121–145

[2] Kennedi L, Fridman A and Saveliev A 1996 Superadiabatic combustion in porous media: Wave propagation, instabilities, new type of chemical reactor *International Journal of Fluid Mechanics Research* **22** 1–26

[3] Howell J, Zhdanok S and Futko S 2000 Effect of porous media transparency on spherical and cylindrical filtrational combustion heaters performance *Int. J. Heat Mass Transfer* **43** 3469–3480

[4] Fursenko R, Maznoy A, Odintsov E, Kirdyashkin A, Minaev S and Sudarshan K 2016 Temperature and radiative characteristics of cylindrical porous Ni-Al burners *Int. J. Heat Mass Transfer* **98** 277–284

[5] Palesskii F, Fursenko R and Minaev S 2014 Modeling of filtration combustion of gases in a cylindrical porous burner with allowance for radiative heat transfer *Combustion, Explosion, and Shock Waves* **50** 625–631

[6] Dats E, Miroshnichenko T, Kirdyashkin A and Maznoy A 2018 Influence of heat conductivity and pore size of porous materials on the efficiency of cylindrical radiative burners with filtrational gas combustion *Journal of Physics: Conf. Series* **1115** (2018) 042023

Propagation of shock waves formed by explosion of the air-propylene mixture in a T-pipeline

Fedorova N N[1, 2], **Fomin P A**[1,4,5] **and Valger S A** [1,3]

[1] Khristianovich Institute of Theoretical and Applied Mechanics SB RAS, Institutskaya 4/1, Novosibirsk, 630090, Russia
[2]National Research Novosibirsk State University, Pirogova str, 1, Novosibirsk, 630090, Russia
[3]Novosibirsk State University of Architecture and Civil Engineering, Leningradskaya st. 113, Novosibirsk, 630008, Russia
[4]Lavrentyev Institute of Hydrodynamics SB RAS, Lavrentyev av. 15, Novosibirsk, 630090, Russia
[5]Siberian State University of Geosystems and Technologies Plakhotnogo str., 10, Novosibirsk 630108 Russia

nfed@itam.nsc.ru

Abstract. The paper deals with mathematical modelling of the process of an explosion of a cloud of reactive gases and its impact on the environment. The research may be relevant to ensure fire safety and to prevent explosion hazards at industrial production facilities that use reactive gas mixtures as a working medium and at high-pressure industrial pipelines used for their transportation. The goal of the study is to develop and test a computational technology for numerical modelling of clouds of reactive gas mixture expansions and loads evaluation. To describe the chemical processes in the explosion, a simplified mathematical model is proposed which based on the assumption of frozen chemical reactions. Using the model, a set of User Defined Functions is implemented in ANSYS Fluent and applied for solving a problem of a "propylene-air" cloud explosion inside the T-shape tube with two closed and one open ends.

1. Introduction

Among the most important tasks of ensuring explosion safety of industrial and living environments is calculation of parameters of a blast wave generated during an explosion of a cloud of a combustible mixture in a space of complex geometry. Formation of such clouds occurs during accidental leaks of explosive liquids or gases and their mixing with air. As an example, we could refer to the disaster that took place in Kaohsiung, Taiwan in 2014 [1]. In this accident, there was leakage of liquid propylene from an underground high pressure pipeline transporting propylene from the harbour to industrial units and passing through the centre of the populous city. The vaporized hydrocarbon mixed with air and the explosive mixture cloud spread along the underground drainage system. The ensuing explosion claimed dozens of lives and led to extensive damage to residential and public buildings.

In order to describe processes of gas-air mixing and mixture ignition as well as to predict the intensity of a specific explosion and its impact on the environment, CFD simulation may be used. The paper presents the results of calculations of shock-wave flows formed as a result of an explosion in a T-shape channel with two closed ends. The calculations were carried out in the ANSYS Fluent for two problems. First, testing of the simulation method is performed on the problem of an explosion of a spherical charge of a condensed explosive. The statement of the test problem and the initial conditions are selected in accordance with the experiment [2]. Comparisons of the results obtained using viscous and inviscid models show a good agreement with experimental data. In the second part, an air-propylene mixture cloud explosion is simulated for the same channel. The simplified simulation method is used; it is based on the hypothesis of frozen chemical reaction [3]. Under these assumptions, the gaseous products of the explosion modeled as inert mixture with the varying specific heat ratio. Based on the simulation data analysis, the structure of the unsteady process is studied, the shock wave and the hot supersonic jet coming out from the channel into open space is simulated, and the excess pressure and momentum are estimated.

2. Problem setup

The geometry under investigation is a T-shaped tube with two of its outlets closed; the third outlet is adjacent to the open space. The geometry and dimensions of the channel are shown in Fig. 1, a, where H=0.868 m, W=0.7 m, D= 0.168 m, $L1$=0.360 m, $L2$=1.280 m, $L3$=1.416 m. Two problems are simulated for this geometry. In the first problem, the charge of a condensed explosive of a spherical shape with the mass of M = 185 g, which corresponds to a 500 kg in full scale, is placed in the tube outlet section adjoining an open air region. Figure 1 also shows the position of the three sensors P1, P2, and P3, in which the static pressure was measured in the experiment [2]. In the second problem, at the initial moment the part of the channel is occupied with a stoichiometric air-propylene mixture and the area-averaged pressure on the closed ends was monitored.

The calculations are performed in ANSYS Fluent. For the first problem several mathematical models are applied, namely inviscid gas dynamics (Model 1), Navier-Stokes equations (Model 2) and Reynolds averaged Navier-Stokes equations complemented with the k-ω SST turbulence model (Model 3). For the second problem computations, only Model 1 is used. All models are closed with ideal gas equation of state (EoS).

The 3D calculation region, consisting of a T-shaped channel of circular cross section and an external prismatic air region, is constructed in the ANSYS Design Modeler taking into account the symmetry of the problem in the transverse direction. The origin of the global coordinate system is in the center of the sphere, which is the explosive charge. The direction of the x axis coincides with the channel axis. The calculation grids for both the problems are created in the ANSYS Meshing. The first problem grid (Fig. 1, b) is unstructured with a total cell number of approximately 10^6. It has a refinement in the charge region and includes 10 inflation layers located near the channel walls. The structured grid for the second problem (Fig. 1, c) is constructed with the sweep method and contains about 1.6×10^6 cells. Since this problem is only computed as inviscid, no inflation layers are created.

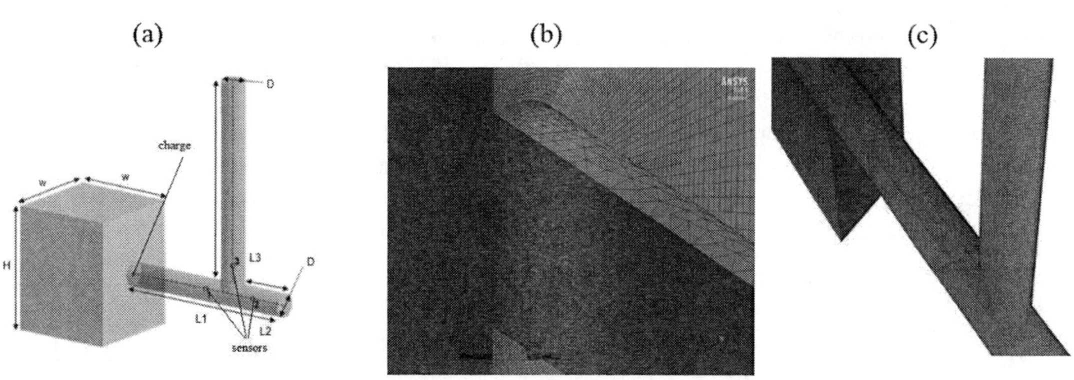

Figure 1. Geometry of the problem (a) and the structure of computational grids used in the first (b) and second (c) problems.

At the boundary of the air region, pressure-outlet boundary conditions are specified. In the simulations with Model 2 and 3, adiabatic thermal conditions on the channel walls are used. At the initial time, the pipe and air space are filled with still air under normal atmospheric conditions. In the regions of condensed explosive and propylene-air mixtures, special conditions are set described in the next section.

3. Mathematical Model and Computational Method

3.1. *Initial stage of condensed explosive detonation*

When solving the first problem, an important issue is describing the initial stage of development of the detonation process of a condensed explosive. Various approaches are available. In ANSYS Autodyn, the behavior of condensed explosive material at high pressures and temperatures is modeled on the basis of gas dynamics equations, supplemented by the John-Wilkins-Lee EoS. Alternative methods include setting the detonation parameters at the boundary of a certain spherical region using empirical functions and the "balloon" method. In [2], it was proposed to add a source term S to the energy equation that describes the energy release from the detonation as

$$S = E_r \cdot step\left(\frac{-(t-t_0)(t-t_1)}{t_c^{\,2}}\right)$$

where t is the current time [s], $t_0 = 0$ is the moment of detonation start [s], t_1 is the moment of detonation end [s], $t_c = 1$ s is the dimensionless parameter, E_r is the energy release rate, or the explosion power [kg·m^2· s^{-1}].

The values of the parameters t_1 and E_r can be calculated on the basis of the available experimental data. The charge radius r=0.0141 [m] is calculated through the volume: $V = M/\rho$=1.171·10^{-5} [m^3]. The energy release rate is calculated as $E_r = \Delta H{\cdot}M\ /\ (t_1 - t_0)$ = 54.351·10^9 [kg·m^2·s^{-1}], ΔH is the specific energy; M is the mass of the charge. The end of detonation is equal to the time the detonation wave travels through the charge: $t_1 = r\ /\ U$ = 1.65726E-06 [s], where U is the detonation velocity. Table 1 presents the parameters of the plastid explosive used in the computations.

Table 1. Physical properties of the plastid explosive.

Parameter	Specific energy ΔH [KJ/kg]	Density ρ [kg/m^3]	Detonation velocity U [m/s]
Value	4870	1580	8500

The source term works during the detonation period. As a result, high values of temperature T = 5900 [K], density ρ = 1870 [kg/m^3] and pressure P = 3.1 [GPa] are reached in the charge region.

3.2. *Modeling gas cloud expansion*

In the second problem, it is assumed that the explosion of the stoichiometric propylene-air mixture happens instantly. As a result, the reaction products are formed which are modeled as a special "gas" [3] with a molar weight μ^* = 28.31 [g/mol].

Following [3], several approaches maybe used to describe the cloud expansion scenarios. The first is based on the assumption of finite rates of the chemical reactions taking place in the mixture. Depending on the composition of the mixture, a system of ordinary differential equations describing the kinetics of combustion can include hundreds of equations, which solution complicates their stiffness. The integration of such a system when solving multidimensional problems for real geometries requires significant computing time.

In addition, there are two limiting scenarios that do not require massive computations. It is assumed for both the scenarios that at the first moment in time a gas in a cloud jumps into a state of chemical equilibrium (instantaneous explosion). The first scenario assumes that all the chemical reactions are frozen, and the cloud is expanded as a chemically inert gas with a heat capacity ratio depended on temperature. The second limiting case is based on the assumption of infinite rates of the chemical reactions. At each moment of time, the gaseous products of the explosion are in a state of chemical equilibrium, which shifts as pressure and temperature change. In this case, as the explosion products expand and the temperature decreases, the reaction products recombine and the thermal effect of the chemical reaction is maximal. Blast wave parameters calculated taking into account the detailed kinetics will fit between the corresponding parameters calculated within the framework of the limiting approximations.

A detailed description of the approaches can be found in [3]. In [4], the equilibrium approach is used to calculate the problem of the expansion of a cloud of a reacting mixture in a semi-open space. In this work, the hypothesis of "frozen" reactions is used, in which the molar composition of the mixture does not change $\mu = \mu^*$. Internal energy U, specific heats c_p and c_V and heat capacity ratio γ are computed by the formulas [3]:

$$U(T)=\left[\frac{3}{4}\left(\frac{\mu^*}{\mu_a}+1\right)+\frac{3}{2}\left(\frac{\mu^*}{\mu_a}-1\right)\frac{\theta/T}{\exp(\theta/T)-1}\right]\frac{RT}{\mu^*},$$

$$\gamma(T)=\left[\frac{3}{4}\left(\frac{\mu^*}{\mu_a}+1\right)+\frac{3}{2}\left(\frac{\mu^*}{\mu_a}-1\right)\frac{\theta/T}{\exp(\theta/T)-1}\right]\frac{R}{\mu^*},$$

$$c_V=\left.\frac{\partial U}{\partial T}\right|_{V=const}=\gamma+\frac{3}{2}\left(\frac{\mu^*}{\mu_a}-1\right)\frac{RT}{\mu^*}\frac{\theta/T^2}{\exp(\theta/T)-1}\left(-1+\frac{\theta}{T}\frac{\theta/T}{\exp(\theta/T)-1}\right),$$

$$c_p=c_V+\frac{R}{\mu^*},\quad H=U+\frac{RT}{\mu^*}$$

where μ_a=12.73 [g/mol] is a molar mass of a gas in atomic state, θ =2500 [K] is the effective temperature of excitation of vibrational degrees of freedom of molecules and H is a total enthalpy. Using these relations, a set of User Defined Functions is implemented in ANSYS Fluent and applied for solving the second problem.

Computations are performed using a density-based solver and AUSM scheme of the third order of approximation by space variables. Both implicit and explicit time-integration methods are used during one computation. For the first time period characterized by very high gradients of the flow, parameters are computed using the implicit scheme of the second order with time step of approximately 0.1 μs and 10 – 15 internal iteration. After 2 ms of flowtime, the pressure peaks are weakened, and since that moment the explicit time integration is used that is less computationally expensive.

4. Results and Discussions

4.1. First problem

As a result of the calculations, pressure records are obtained at the monitors $P1$ - $P3$ (see Fig. 1), as well as the fields of gas-dynamic parameters at various time moments, on the basis of which the structure of the transient flow in the channel and the shock wave output were studied and hot supersonic stream from the channel into the open space. The impact on the channel walls is due to the passage of the primary, most powerful shock wave and two waves of lower amplitude reflected from the closed ends of the channel, as well as the rarefaction wave.

In Figure 2, fields of static pressure, static temperature and Mach number fields for the first problem at the time t=0.5 ms and 1 ms computed using Model 1 are shown. At t=0.5 ms, the primary shock wave approached the region where the first pressure sensor is located. The pressure distribution shows the presence of a sufficiently extended high-pressure region with P_{max} = 2.23 MPa and a straight front, behind which a supersonic M =1.6 flow is formed with a temperature exceeding 2200 K. A spherical shock wave (P_{max} = 0.4 MPa) first enters the free space, followed by a supersonic (M = 2.4) hot jet (T = 1400 K).

The pressure field for time t= 1 ms (Fig. 2, d) shows that the primary shock approaches the closed pipe end face and is reflected from it, resulting in the formation of a region of high pressure

($P_{max} = 3.8$ MPa) and high temperature ($T_{max} = 2000$ K). The shock wave also entered the second branch of the pipe; however, the pressure and temperature parameters are much lower ($P_{max} = 0.8$ MPA, $T_{max} = 1300$ K).

A quantitative comparison of the data on the maximum overpressure (Table 2) and the time of arrival of the shock waves at the monitor points (Table 3) shows good agreement between the experimental data [2] and the results obtained in this work using various models. It should be noted that the most significant effect of viscosity is manifested on the secondary pressure peaks caused by the action of waves reflected from the closed channel ends. The results obtained with Models 1-3 differ by no more than 10%, so the second problem is simulated on the basis of a more economical Model 1.

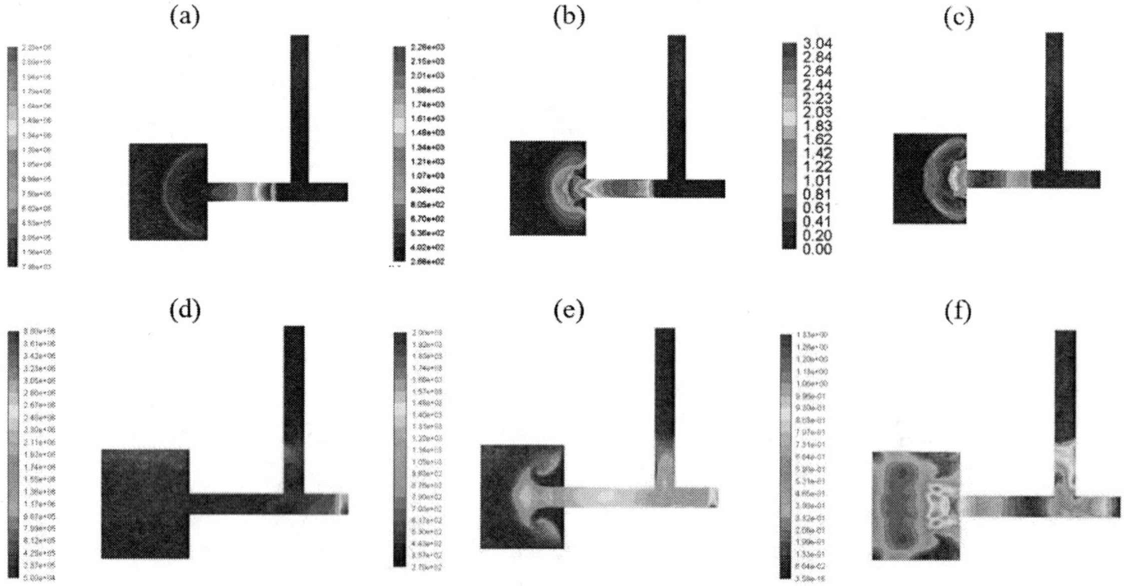

Figure 2. Computed static pressure [Pa] (a, d), static temperature [K] (b, e) and Mach number (c, f) fields in the symmetry plane for the first problem at the time t=0.5 ms (a, b, c) and 1 ms (d, e, f).

Table 2. Comparison of experimental and computed overpressure values [kPa] in monitor points.

Monitor	Exp	Model 1	Model 2	Model 3	Comp [2]
Point 1	1850	1737	1732	1686	2009
Point 2	1050	1162	1064	1056	1580
Point 3	465	520	496	490	465

Table 3. Comparison of experimental and computed shock wave arrival times [ms] in monitor points.

Monitor	Exp	Model 1	Model 2	Model 3	Comp [2]
Point 1	0.3	0.355	0.37	0.37	0.25
Point 2	0.44	0.595	0.622	0.62	0.46
Point 3	0.43	0.605	0.66	0.66	0.39

4.2. Second problem

At the next stage, the explosion of a cloud of air-propylene mixture in the same channel was simulated. It is supposed that the explosion of the stoichiometric propylene-air mixture $C_3H_6 + 4.5O_2 + 16.92N_2$ happens instantly. As a result, the reaction products are formed which are modeled as a special "gas" [3] with a molar weight $\mu^* = 28.31$ [g/mol] under the following conditions:

static pressure: P^*=0.962 [GPa] and temperature T^* = 2739 [K]. Applying the ideal gas EoS gives the initial "gas" density value of ρ^* = 1.196 [kg/m^3].

At the initial moment, the stoichiometric air-propylene mixture ("gas") occupies some part of the channel (Fig. 3, a). After the explosion causing a sharp increase in pressure, the gaseous explosion products are expanding creating a shock wave in front of them. A rarefaction wave goes inside the gas cloud. As a result, a complex gas-dynamic flow is formed with transient pressure and temperature distributions. Fig. 3 presents the dynamics of mass fraction of "gas" at various time moments. In Fig. 4, static pressure fields are shown. Fig. 5 presents the area-averaged pressure values recorded during the computations at the short (SE) and long (LE) channel ends.

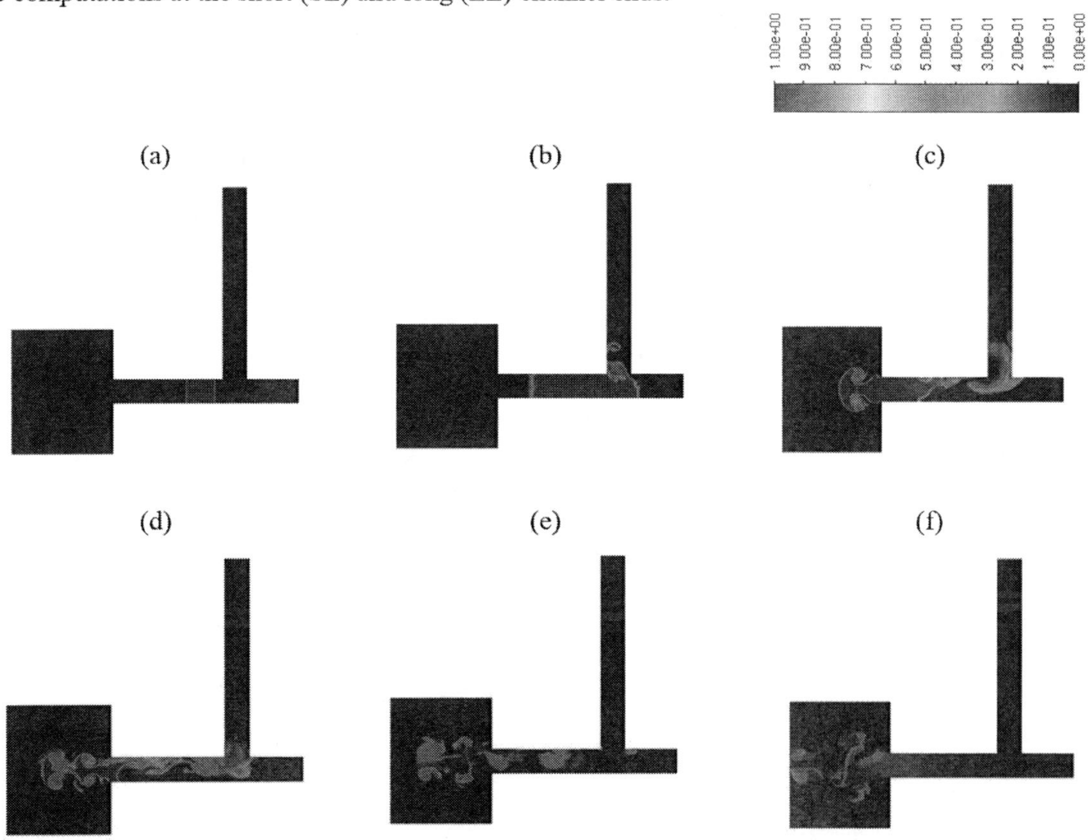

Figure 3. Computed "gas" mass fraction fields in the symmetry plane for the second problem at the time t=0 (a), 1 (b), 4 (c), 6 (d), 8 (e) and 12 (f) ms.

5. Conclusions

In the paper, a computational technology is developed for numerical simulations of clouds of reactive mixture expansions and loads evaluation. The technology is tested on the experimental data [2]. Good agreement of experimental data and simulation result is achieved. Viscous and turbulence effects are evaluated. A simplified mathematical model is proposed to describe the chemical processes in the explosion which is based on the assumption of frozen chemical reactions. Using the model, a set of User Defined Functions is implemented in ANSYS Fluent and used to simulate a problem of a propylene-air cloud explosion inside a T-shape tube with one open and two closed ends. Based on the data analysis, the structure of the unsteady flow in the channel, the exit of the shock wave (SW) and the hot supersonic jet from the channel into the open space were studied, and the excess pressure and impulses were estimated.

IOP Publishing
1404 (2019) 012057 doi:10.1088/1742-6596/1404/1/012057

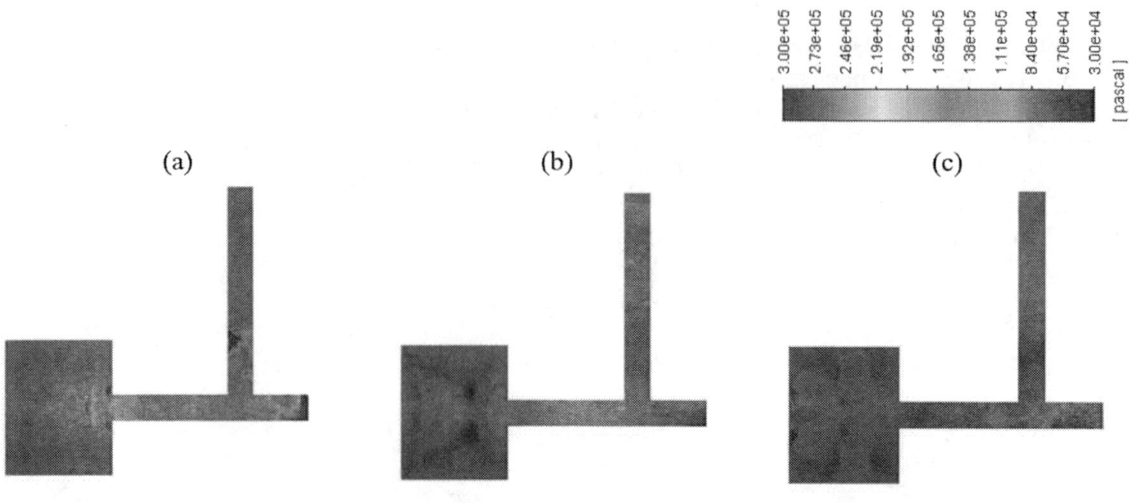

Figure 4. Computed static pressure fields in the symmetry plane for the second problem at the time t=1 (a), 2.5 (b) and 6 (c) ms.

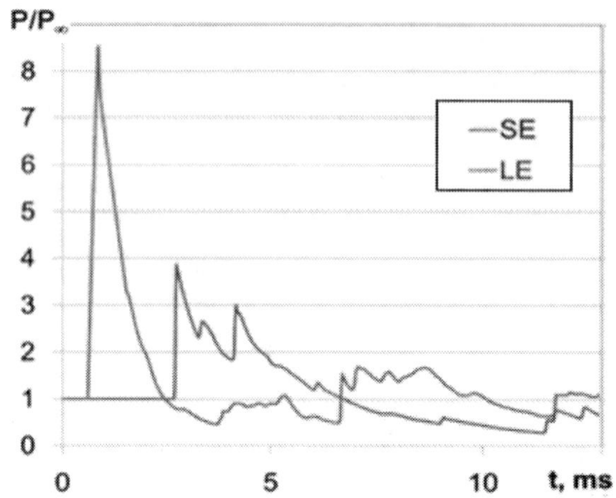

Figure 5. Computed area-averaged static pressure history recorded at the short (SE) and long (LE) closed channel ends.

Acknowledgments
The study is supported by the Russian Foundation for Basic Research project 18-58-53031.

Reference

[1] Yang H N, Chen J H, Chiu H J, Kao T J, Tsai H Y and Chen J R 2016 *J. Loss Preven. Proc. Indust.***41**(5) 107-20
[2] Rigas F and Sklavounos S 2005 *J. Hazard Mater* **121** 23-30
[3] Nikolaev Yu A and Fomin P A 1982 *Combustion, Explosion and Shock Waves* 18(1) 53-8
[4] Valger S A, Fomin P A and Fedorova N N 2018 *IOP Conf. Ser.: Materials Science Engineering* **456** 012031

Initiation of Richtmyer–Meshkov instability by a detonation wave

A V Kashkovsky, Yu V Kratova and A A Shershnev

Khristianovich Institute of Theoretical and Applied Mechanics, 4/1 Institutskaya st., Novosibirsk, 630090, Russian Federation

E-mail: sasa@itam.nsc.ru

Abstract. As part of the HyCFS-R numerical code development, a module for the simulation of multiphase chemically reacting flows based on the Euler–Euler approach was implemented. The code was verified on simple 0D and 1D problems, such as constant volume explosion and 1D shock tube problem. For more comprehensive testing the numerical simulation of the Richtmyer–Meshkov instability in reactive heterogeneous media was performed. The comparison against the reference solution revealed that numerical method implemented in the HyCFS-R either introduce too much dissipation and suppress the detonation wave front instabilities or lead to numerical artifacts in the Richtmyer–Meshkov instability flow pattern.

1. Introduction

HyCFS-R is a code currently being developed at Khristianovich Institute of Theoretical and Applied Mechanics to simulate a wide range of compressible flows. A number of numerical schemes were implemented for spatial and temporal approximation, as well as various models for calculating the rates of chemical reactions. As part of this work, a module for modeling multiphase chemically reacting flows was implemented. The code was mainly tested on the external flow, namely high-speed flows around bodies. When adding new physical models, verification is complicated by the need to check the effects of the interaction of various options and solvers with each other. It turned out that modeling a Richtmyer–Meshkov instability (RMI) in a reacting heterogeneous medium can serve as a fairly good comprehensive test.

2. HyCFS-R numerical code

The HyCFS-R [1, 2] code solves 3D unsteady Euler and Navier-Stokes equations in general curvilinear coordinates. The convective terms are approximated using the shock-capturing MUSCL TVD or the 5th order weighted essentially non-oscillatory (WENO) scheme. In the TVD scheme, the variables are reconstructed on the faces of the cells from the values in the cell centers with 2nd or 3rd order of accuracy. After that the limiter **minmod** is applied. Numerical fluxes are calculated using approximate solvers for the Riemann problem: HLLE (Harten–Lax–van Leer–Einfeldt) solver or hybrid AUSM–van Leer solver. The explicit time advancement schemes, the so-called Runge–Kutta TVD schemes of 2nd and 3rd orders of accuracy, are used.

The module for multiphase reacting flows [3] uses the description of the dispersed phase based on the Euler approach. The model for the oxygen laden with aluminum particles from the works of Fedorov and Khmel [5–9] is used at the current stage of the HyCFS-R development. The basic verification of the implemented model was performed using simple 0D and 1D problems, including the constant volume explosion and 1D shock tube. As a part of more comprehensive code testing it was decided to simulate numerically the Richtmyer-Meshkov instability in a chemically reactive heterogeneous media. The HyCFS-R was verified by comparing the computation results against the reference solution obtained using the Fortran code developed in the cited above papers of Fedorov, Khmel and Kratova.

3. Richtmyer-Meshkov instability

Richtmyer-Meshkov instability occurs when a boundary between two fluids with different densities is impulsively accelerated by the shock wave (SW). In this case, small amplitude perturbations begin to increase. Such instability is found in hydrodynamics, astrophysics, nuclear engineering, etc. Therefore, RMI research is of both fundamental and practical interest [11, 12]. Usually in the RMI numerical simulation the fluids interface is artificially disturbed in a shape of a sine wave with an amplitude smaller than the wavelength. A plane SW propagates through a lighter medium, interacts with the curved interface and leads to the increase in distance between the minimum and the maximum of the sine wave. In the detonation wave (DW), the cellular structure is formed, so the DW front is always curved. But will an RMI emerge if the curved DW front hits a plane interface?

To test this hypothesis, a numerical experiment was conducted. A plane 2D channel with a length of L = 1.4 m and a width of H = 0.06 m with specular walls (see figure 1) is filled with oxygen (ρ = 1.28 kg/m^3, T = 300 K, P = 1 atm). A cloud of 1 μm aluminum particles (ρ = 1.56 kg/m^3) is placed in the middle section of the channel. The combustion of these particles ensures the formation of the DW cellular structure. The detonation is initiated by a Mach 5 SW, propagating through the oxygen layer in the left section of the channel. The interaction of the SW with a cloud of particles results in their heating, subsequent ignition and transition to detonation. To accelerate the formation of detonation cells, the boundary of the particle cloud has a sine curve shape (amplitude ~7% of the wavelength). This interface disturbance also leads to the appearance of RMI on the left boundary. The DW can also be initiated by creating a small spot of gas with the temperature much higher than the ignition temperature of particles [10], but in this case the transition to detonation develops more slowly, and the computational domain needs to be much longer.

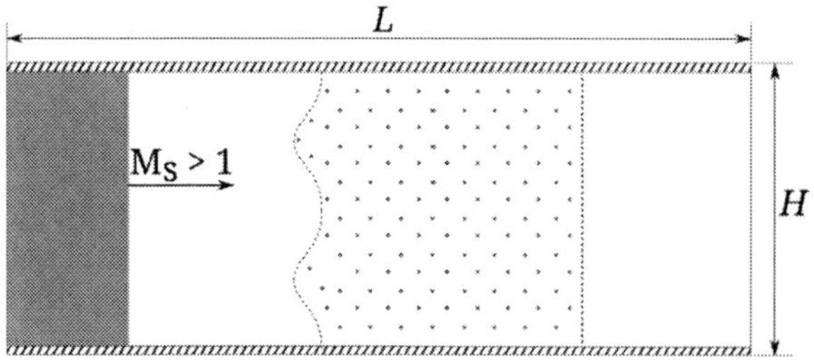

Figure 1. Computational domain geometry and the initial conditions set-up.

Figure 2 shows the density flowfields of carrier and dispersed phases at different time moments. At time $t = 0$ the shock wave approaches a cloud of particles. The particles combustion process starting at $t = 0.1$ ms leads to the formation of detonation cells seen at $t = 0.2$ ms. By the time DW exits the particle laden section, the pronounced cellular structure of DW is observed. At the same time, the combustion products at the left boundary demonstrate the classical pattern of the RMI development.

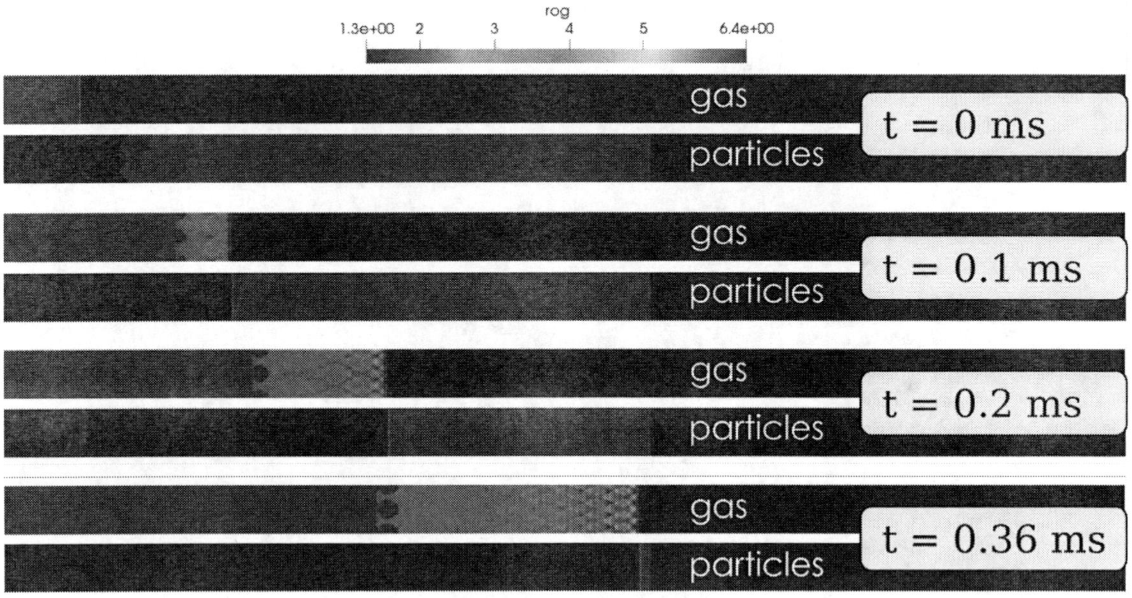

Figure 2. Formation of cellular structure of DW.

The DW, exiting from a cloud of particles, generates an SW in the carrier phase (see Fig. 3). It can be seen that an RMI also emerges at the interface, slowly growing in time, and the number of disturbances coincides with the number of detonation cells.

It should be noted that RMI emerges during the transition of SW from a lighter gas to a denser one. In this case, the density of the combustion products compressed by DW, is more than 4 times higher than the density of the gas phase. But the speed of DW is about 6 times higher than the speed of sound. It indicates that the emergence of RMI in this case depends not only on the densities of the media, but also on the speed of sound and speeds of DW and SW in these media The emergence of RMI during the transition of DW into an inert medium was observed earlier. So in [13,14], RMI-specific structures are observed when DW enters a cloud of neutral particles.

The data were extracted from the lines $Y = const$ to measure the positions of the peak and the bubble RMI. The difference in these positions illustrates the rate of RMI development. Figure 4 shows the RMI development for both left and right boundary cases. For the left RMI, the distance at first corresponds to initial disturbance amplitude (0.004 m), and after the shock-interface interaction increases non-linearly, tends to a constant value. Because the right RMI is initiated by detonation cells, peaks and bubbles positions oscillate on DW front. As a result, the size of the RMI varies within the size of the detonation cell. After emergence of RMI, a non-linear increase in size is also observed, which tends to a constant value. Oscillations in the profiles are caused by acoustic waves.

The numerical simulation were repeated using the HyCFS-R code. Comparison with the results discussed above showed that the correct reproduction of RMI development pattern strongly depends on the choice of the Riemann problem solver and MUSCL reconstruction. In the calculations presented above, the 2nd order TVD scheme with the Row solver was used. It is a well-known fact, that the Roe solver introduces rather small numerical dissipation to the flow. As a result, it is less robust than other solvers but does not suppress the development of various disturbances as much. On the other hand, HLLE and AUSM–van Leer solvers

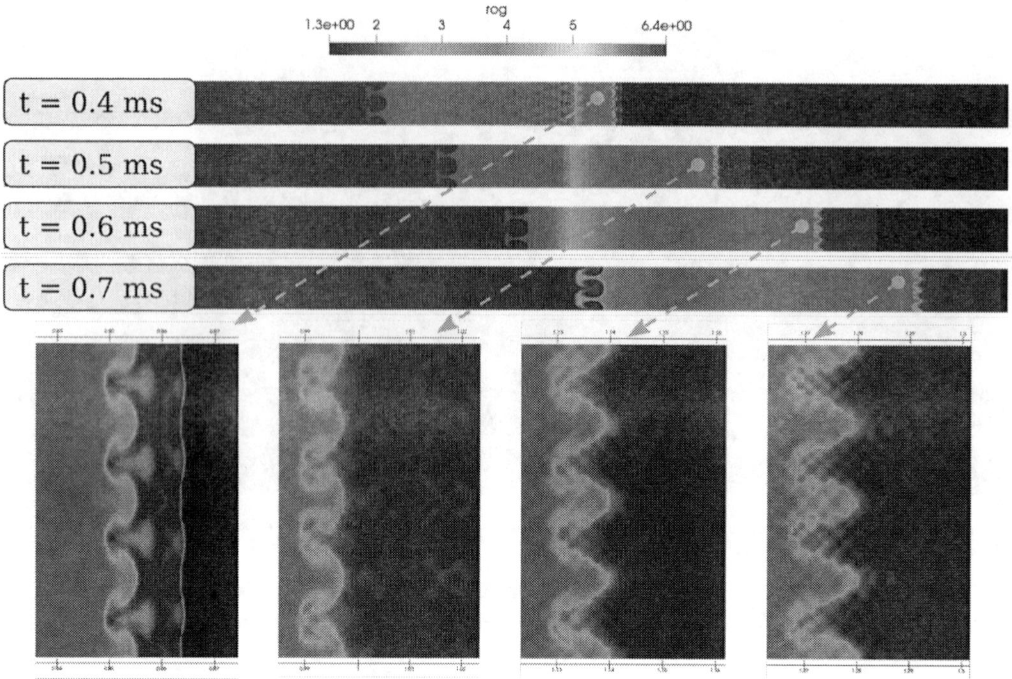

Figure 3. Carrier density flowfields at different time moments.

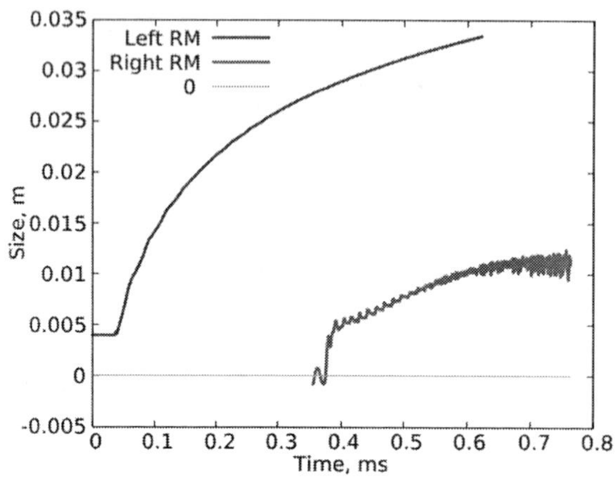

Figure 4. Distance between the bubble and the spike as function of time.

introduce more artificial viscosity to the flow which makes them more suitable for external aerodynamics problems, containing strong shocks and large flow gradients. However, the RMI problem indicated that the combination of approximation methods and solvers in the HyCFS-R code could not correctly resolve the entire flow. In figure 5 the temperature field of the carrier phase obtained in calculations with the 2nd order MUSCL reconstruction just before the end of the particle combustion process. It can be seen that the left RMI is resolved quite correctly, at the same time, the structure of the detonation cells is rather smeared by the numerical dissipation.

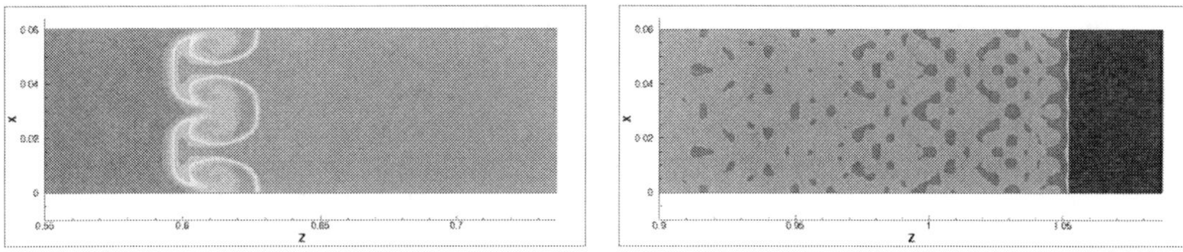

(a)

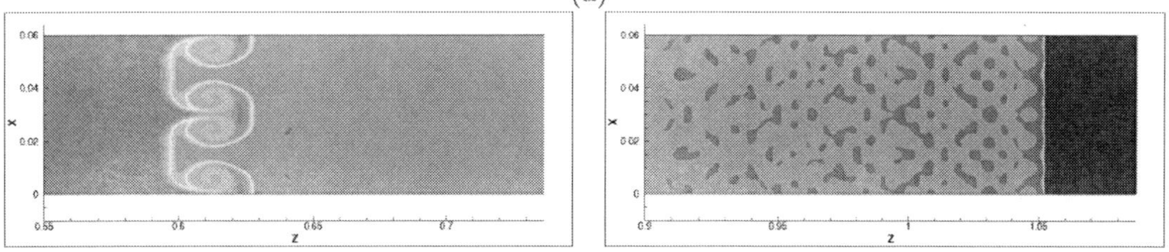

(b)

Figure 5. Carrier phase temperature flowfield in HyCFS-R computations using HLLE (a) and hybrid AUSM-van Leer Riemann solvers. 2nd order MUSCL reconstruction.

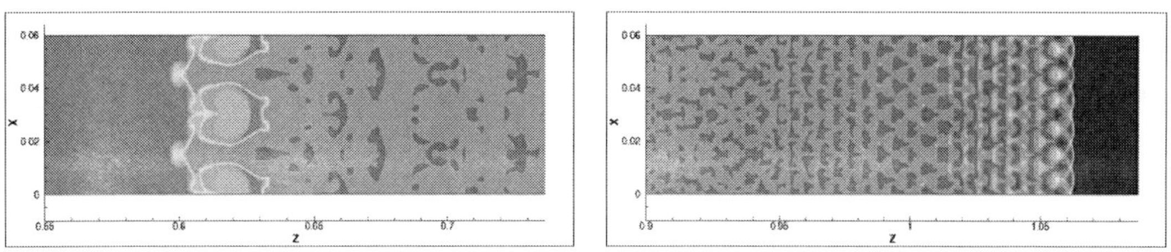

(a)

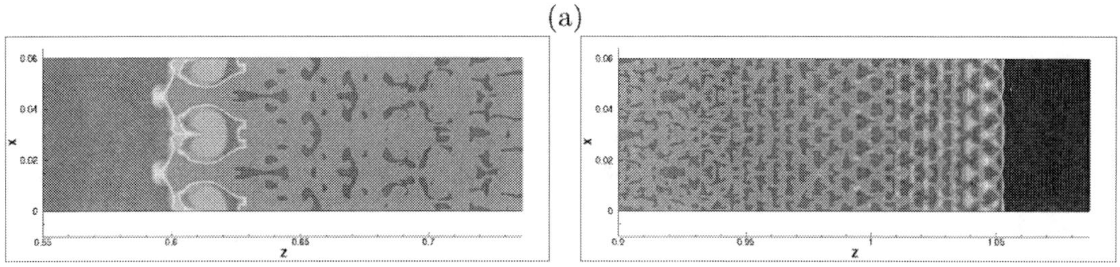

(b)

Figure 6. Carrier phase temperature flowfield in HyCFS-R computations using HLLE (a) and hybrid AUSM–van Leer Riemann solvers. 3rd order MUSCL reconstruction.

In figure 6 the same flowfields are obtained with the 3rd order MUSCL reconstruction. The structure of detonation cells was well resolved, but the certain numerical artifacts distorting the shape of the left RMI can be seen. Obviously, numerical viscosity has major impact on this flow on this flow. To solve such a problem, the solver and the numerical scheme need to treat different scales of flow inhomogeneities with equal accuracy. In the present case a more detailed study of both the flow and the HyCFS-R solvers is required. The latter ones need to be compared with other numerical schemes to locate the source of numerical artifacts.

4. Conclusion

As part of the HyCFS-R numerical code development, a module for the simulation of multiphase chemically reacting flows based on the Euler-Euler approach was implemented. The code was verified on simple 0D and 1D problems, such as constant volume explosion and 1D shock tube problem. For more comprehensive testing the numerical simulation of the Richtmyer–Meshkov instability in reactive heterogeneous media was performed. The comparison against the reference solution revealed that numerical method implemented in the HyCFS-R either introduce too much dissipation and suppress DW front instabilities or lead to numerical artifacts in the RMI flow pattern. Thus, it can be concluded that further improvement of the numerical techniques is required. Future work will include implementation of TVD schemes of 4th and 5th order by Yamamoto and Daiguji, employment of other schemes for the dispersed phase advection, and 3D calculations.

Acknowledgments

This work was supported by Russian Foundation for Basic Research grant No. 18-08-01442. Numerical code development was partly carried out within the framework of the Program of Fundamental Scientific Research of the state academies of sciences in 2013-2020 (project No. AAAA-A17-117030610138-7). Computational resources were kindly provided by Siberian Supercomputer Center of the Institute of Computational Mathematics and Mathematical Geophysics SB RAS (sscc.ru) and Computational Center of Novosibirsk State University (nusc.nsu.ru)

References

[1] Shershnev A A , Kudryavtsev A N, Kashkovsky A V and Khotyanovsky D V 2016 HyCFS, a high-resolution shock capturing code for numerical simulation on hybrid computational clusters *AIP Conf. Proc.* **1770** 030076
[2] Kudryavtsev A N, Kashkovsky A V, Borisov S P and Shershnev A A 2017 A numerical code for the simulation of non-equilibrium chemically reacting flows on hybrid CPU-GPU clusters, *AIP Conf. Proc.* **1893** 030054
[3] Kratova Yu V, Kashkovsky A V and Shershnev A A 2019 Numerical simulation of shock wave propagation in 2-D channels with obstacles filled with chemically reacting gas suspensions *Thermal Science* **23** 623-30
[4] Collins J P, Ferguson R E, Chien K Y, Kuhl A L, Krispin J and Glaz H M 2012 Simulation of shock-induced dusty gas flows using various models *AIAA 94-2309*
[5] Fedorov A V, Khmel T A, Fomin V M 1999 Non-equilibrium model of steady detonations in aluminum particles-oxygen suspensions *Shock Waves* **9** 313-8.
[6] Strauss W A 1968 Investigation of the detonation of aluminum powder oxygen mixtures *AIAA J.* **6** 1753-61
[7] Dreizin E L 1999 On the mechanism of asymmetric aluminum particle combustion *Combustion and Flame* **17** 841-50
[8] Fedorov A V, Khmel T A 2005 Numerical simulation of formation of cellular heterogeneous detonation of aluminum particles in oxygen *Combust. Expl. Shock Waves* **41** 435-48
[9] Fedorov A V, Khmel T A 2008 Formation and degeneration of cellular detonation in bidisperse gas suspensions of aluminum particles *Combust. Expl. Shock Waves* **44** 343-53
[10] Kratova Yu V, Khmel T A, Fedorov A V 2016 Axisymmetric expanding heterogeneous detonation in gas suspensions of aluminum particles *Combust. Expl. Shock Waves* **52** 74-84
[11] Zhou Ye 2017 Rayleigh-Taylor and Richtmyer-Meshkov instability induced flow, turbulence, and mixing I *Physics Reports* **720-722** 1-136
[12] Zhou Ye 2017 Rayleigh-Taylor and Richtmyer-Meshkov instability induced flow, turbulence, and mixing II *Physics Reports* **723-725** 1-160
[13] Fedorov A V, Kratova Yu V 2015 Analysis of the influence of inert particles on the propagation of a cellular heterogeneous detonation *Shock Waves* **25** 255-65
[14] Fedorov A V, Kratova Yu V 2015 Influence of non-reactive particle cloud on heterogeneous detonation propagation *J. Loss Prevent. Proc.* **36** 404-15

Propagation of cellular detonation in a gas suspension in the presence of a concentration gradient

T A Khmel[1] and P A Fomin[1,2]

[1]Khristianovich Institute of Theoretical and Applied Mechanics SB RAS, 4/1 Institutskaya st., 630090, Novosibirsk, Russia
[2]Lavrent'ev Institute of Hydrodynamics of the SB RAS, 15 Lavrent'ev av., 630090, Novosibirsk, Russia

khmel@itam.nsc.ru

Abstract. Models for the description of detonation flows in inhomogeneous gas particle mixtures are presented. The physical and mathematical model of heterogeneous detonation of aluminum suspensions is based on approaches to the mechanics of heterogeneous media and semi-empirical laws of ignition and combustion of particles. The model is extended for suspensions that are inhomogeneous in particle concentrations. This suggests the solution of an additional equation for the spatial distribution of initial concentrations. The dependence of the integral heat release of chemical reactions and the fraction of unburned particles on the initial composition is determined from the solution of the stationary problem of the structure of the detonation wave and its agreement with empirical data on the rate of normal detonation. The model for describing detonation in mixtures of reacting gases with inert particles is based on an analysis of the state of chemical equilibrium. The problem of the initiation and development of cellular detonation in a flat channel in the presence of a longitudinal or transverse gradient of the concentration of aluminium particles is considered. The limits of changes in concentrations from poor to super stoichiometric compositions are considered for 1 μm particles. When detonation propagates along a channel with a longitudinal positive (negative) concentration gradient, the cellular structures do not change, but the average and peak values of pressure and phase density increase (decrease). With a transverse concentration gradient, cell-like structures with a curved front, asymmetric and strongly elongated cells form in the channel. The results are consistent with studies of similar gas detonation flows.

1. Introduction

Detonation studies are related both to problems of explosion and fire safety, and to the development of detonation technologies. In detonation engines, a gaseous medium or dispersed suspension can be used as a working mixture. One way to control processes is to organize and use various kinds of heterogeneities. The study of gas and heterogeneous detonation, both theoretically and experimentally, was carried out, as a rule, for mixtures with a uniform concentration. The problems of suppression of detonation by dispersed clouds were considered in [1, 2], in [3] the features of cellular detonation in a two-fuel mixture were described. The homogeneity of the composition of gas mixtures is achieved quite quickly; nevertheless, the question of the detonation characteristics in inhomogeneous mixtures was considered in [4–7]. For two-phase media, these questions were considered in the problems of lifting, ignition, and detonation combustion of dust layers [8-10].

One of the problems is the description of the combustion of powders in the absence of an oxidizing agent. Combustion of aluminum can proceed both with the formation of oxide Al_2O_3 and with the formation of suboxides AlO, AlO_2, Al_2O. The review [11] discusses models of detonation combustion, in which the formation of suboxides is taken into account in various ways. The issues of modeling the deflagration combustion of aluminum with an oxidizer deficiency are discussed in [12]. In detonation combustion, the total heat release and detonation velocity are determined by the equilibrium composition of the detonation products. Some data of thermodynamic calculations of the equilibrium composition of aluminum detonation products are given in [13]. In [14], the dependence of the detonation velocity of aluminum suspensions in oxygen on the initial concentration of particles is presented, which will be used to construct a model of detonation combustion of aluminum in a gas suspension inhomogeneous in concentration.

2. Physical and mathematical model of detonation in non-uniform suspensions of reactive particles in gas oxidiser

Two-dimensional detonation flows are described within the framework of the mechanics of interpenetrating continua by the Euler equations for the gas phase and particles. The system is closed by the equations of state and relations describing the interphase exchange of masses, momentum and energy.

$$\frac{\partial\rho_i}{\partial t}+\frac{\partial\rho_i u_i}{\partial x}+\frac{\partial\rho_i v_i}{\partial y}=-(-1)^i J$$

$$\frac{\partial\rho_i u_i}{\partial t}+\frac{\partial\left[\rho_i u_i^2+(2-i)p\right]}{\partial x}+\frac{\partial\left[\rho_i u_i v_i\right]}{\partial y}=(-1)^i(f_x-Ju_2) \tag{1}$$

$$\frac{\partial\rho_i v_i}{\partial t}+\frac{\partial\left[\rho_i u_i v_i\right]}{\partial x}+\frac{\partial\left[\rho_i v_i^2+(2-i)p\right]}{\partial y}=(-1)^i(f_y-Jv_2),$$

$$\frac{\partial\rho_i E_i}{\partial t}+\frac{\partial\rho_i u_i\left[E_i+(2-i)p/\rho_i\right]}{\partial x}+\frac{\partial\rho_i v_i\left[E_i+(2-i)p/\rho_i\right]}{\partial y}=(-1)^i(q+f_x u_2+f_y v_2-JE_2),$$

$$p=\rho_{11}RT_1,\quad E_i=u_i^2/2+c_{v,i}T_i+(i-1)Q.$$

Here p is pressure, $\rho_i, u_i, v_i, T_i, E_i, c_{v,i}$ are mean density, longitudinal and transversal velocity components, temperature, total energy per mass unit and heat capacity of the i-th phase (1, 2 indicate gas and particles, respectively). Mass concentrations of gas and particles are defined as $\xi_i=\rho_i/\rho$, $\rho=\sum_i\rho_i$, $\rho_i=\rho_{ii}m_i$, m_i is the volume concentration of the i-th phase.

Closure relations for the exchange of mass, momentum, and energy of micro- and nano-sized particles are presented in [15]. Particle combustion is described in the framework of the simplest model of one-stage kinetics, taking into account incomplete particle burning (the presence of residual concentration of unburned particles) and the transitional regime from diffusion-limited to kinetically-controlled combustion for particles with a diameter of less than 1 micron [16]. The dependence of the characteristic burning time of aluminum particles in the submicron range from 135 nm to 1 μm is taken as:

$$J=\frac{\rho}{\tau_\xi}(\xi-\xi_k)\exp(-E_a/RT_1)\ \text{at}\ T_2\geq T_{ign},\ \xi>\xi_k.\quad J=0,\ \text{at}\ T_2<T_{ign}\ \text{or}\ \xi\leq\xi_k.$$

$$\tau_\xi=\tau_0(d/d_0)^{0.3}\exp(E_a/RT_1)(p/p_*)^{-m} \tag{2}$$

where $m = -0.25\ln d[\mu m]$, $E_a = 0.5\{E_{micro}(2 + \ln d[\mu m] - E_{nano}\ln d[\mu m]\}$, and E_{micro} is the activation energy for micron particles (32 kJ/mol [11]), E_{nano} is the activation energy for nanoparticles (60 kJ/mol [17]), p is pressure, and d is the particle diameter.

The heat release of chemical reactions Q in a homogeneous suspension can be determined from empirical data or thermodynamic calculations of the normal detonation velocity, which for the stoichiometric composition of a mixture of aluminum and oxygen is about 1.6 km/s. The relationship between the detonation velocity and integral heat release is determined from the problem of the structure of a stationary wave. The solution corresponding to the Chapman-Jouguet regime has either an internal sound point (in terms of the frozen speed of sound) or a final equilibrium state, sound in terms of the equilibrium speed of sound [17]. In the pre-stoichiometric interval, according to experimental data [18], it is assumed that $\xi_k = 0.1\xi_0$. Approximating dependence of the parameter Q on the initial concentration in the form $Q(\xi_0) = 20.3\exp(-3.17\xi_0)$ provides agreement with the data of [14] at $0.2 < \xi_0 < 0.55$. For high load conditions (super stoichiometry), an additional determining parameter is ξ_k. The values of the residual concentration at a value of Q=3.52 MJ/kg, corresponding to stoichiometry, are approximated by a function $\xi_k = 0.1\xi_0 + 3.3(\xi_0 - 0.55)(\xi_0 - 0.4)$, approaching 1 at $\xi_0 \to 1$. In an inhomogeneous suspension, the initial concentration distribution is determined by a function $\xi_0(x, y)$. Accordingly, the thermal effect of the reaction and the fraction of unburnt particles at each point in space vary and depend on what was the initial concentration of the volume of particles that arrived at this point in space. To do this, the function $\zeta(x, y)$ is introduced as follows. The initial distribution corresponds to the concentration distribution $\zeta_0(x, y) = \xi_0(x, y)$, and as the mixture moves, it is determined from the equation $D\zeta / Dt_2 = 0$. With (1), it is equivalent to

$$\frac{\partial \rho_2 \zeta}{\partial t} + \frac{\partial \rho_2 u_2 \zeta}{\partial x} + \frac{\partial \rho_2 v_2 \zeta}{\partial y} = -J\zeta \quad . \tag{3}$$

Parameters $Q(\zeta)$ and $\xi_k(\zeta)$ are defined by the above functions.

The statement of the problem of shock-wave initiation, formation and propagation of cellular detonation in a flat channel is similar to [15]. The numerical technique is based on the conservative flux-splitting schemes: the TVD scheme by Harten for gas and the Gentry-Martin-Daly scheme for particles. The numerical method has been tested earlier and applied for 2-D numerical simulations of the shock wave and detonation flows. The step of the finite-difference grid was determined according to the scale of the relaxation zones, varied in the test calculations.

3. Description of detonation flows of inhomogeneous gas suspensions of inert particles based on an analysis of the equilibrium of the gas chemical composition

It is assumed that incompressible particles are in mechanical and thermal equilibrium with gas, and their volume fraction is small. The influence of partial pressure and pressure of saturated vapors of the condensed phase is also neglected. The density and internal energy of a two-phase mixture are calculated by the relations:

$$\rho_* = \rho + \rho_C,\ \rho = (1-\alpha)\rho_*\ \ \rho_C = \alpha\rho_*,\ U^* = (1-\alpha)U + \alpha U^C \ , \tag{4}$$

where U^*, U, U^C are internal energy of two-phase mixture, gas and particles, α is mass fraction of particles in two-phase mixture, ρ, ρ_C, ρ_* are density of a gas, volume density of particles and gas-particles mixture.

According to the model [19-21], molar mass μ and internal energy of a gas U can be calculated via the formulas:

$$\mu(\rho,T)=\frac{B\mu_{\min}-2\mu_{\max}+\sqrt{D}}{2(A^*-1)}\,, \quad (5)$$

$$B=\frac{AT^{3/4}}{4K_+}(1-e^{-\theta/T})^{3/2}e^{-E/RT}\frac{\mu_{\max}^2}{\mu_{\min}\rho}\,,\ D=(B\mu_{\min}-2\mu_{\max})^2+4(B-1)\mu_{\max}^2\,,$$

$$U(T,\mu)=[\frac{3}{4}(\frac{\mu}{\mu_a}+1)+\frac{3}{2}(\frac{\mu}{\mu_a}-1)\frac{\theta/T}{e^{\theta/T}-1}]\frac{RT}{\mu}+E(\frac{1}{\mu}-\frac{1}{\mu_{\min}})\,.$$

Here T is the temperature, R is the universal gas constant, μ_a, $\mu_{\min}$, $\mu_{\max}$, K_+, A, θ, E are constants. From the first law of thermodynamics, the equation of state of an ideal gas, the adiabatic condition, and (4) - (6) the following formula follows for calculating the speed of sound c^* and the adiabatic index γ^* of a two-phase mixture:

$$c_*^2=\frac{dP}{d\rho_*}=\gamma_*\frac{P}{\rho_*}\,,\ \gamma_*=\frac{d\ln P}{d\ln\rho_*}=\frac{d\ln P}{d\ln\rho}=1-\frac{\rho}{\mu}\mu_\rho+\frac{\rho}{T}(1-\frac{T}{\mu}\mu_T)\frac{dT}{d\rho}\,,$$

$$\frac{dT}{d\rho}=\frac{1}{1-\alpha}\frac{dT}{d\rho_*}=-\frac{(1-\alpha)(U_\mu\mu_\rho-RT/\rho\mu)}{(1-\alpha)(U_T+U_\mu\mu_T)+\alpha U_T^C}\,.$$

Algebraic formulas for calculating $U_\mu, U_T, \mu_\rho, \mu_T$ can be obtained from the equation of state of an ideal gas and (4), (5).

In the presence of a spatial gradient α the chemical equilibrium model under consideration remains unchanged. In the presence of a spatial gradient of the chemical composition of the gas phase, two options for using the model are possible. In the first version, distribution of the space of constants $\mu_a, \mu_{\min}, \mu_{\max}$ is calculated, while the remaining model constants do not change. In the second version, the spatial distribution of all model constants is calculated. The first method for calculating thermodynamic parameters is easier to implement, while the second method is more accurate.

4. Results of calculations

4.1. The propagation of detonation waves in aluminium - oxygen mixtures with longitudinal concentration gradients

The concentration distribution along the channel was set by linear functions: decreasing $\xi_0(x)=0.55-0.3125x$ and increasing $\xi_0(x)=0.3+0.3125x$.

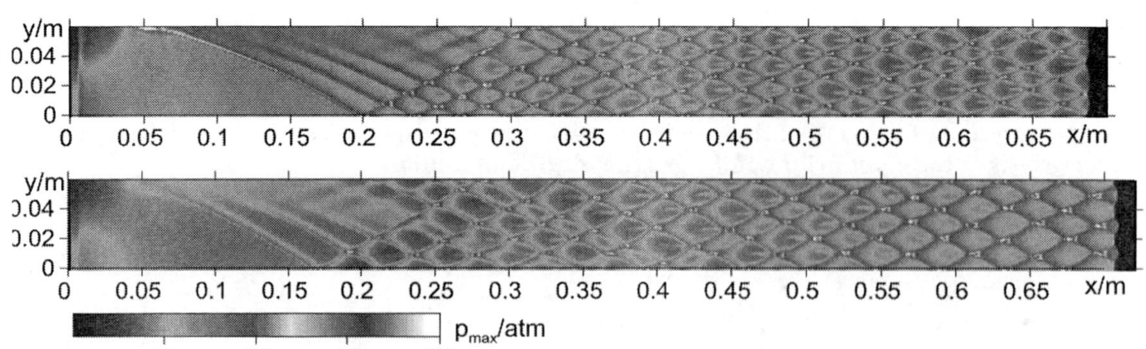

Figure 1. Maximal pressure fields at cellular detonation propagation in mixtures with longitudinal concentration gradients: negative (top), positive (bottom), $\Delta t = 0.04$ ms.

Figure 1 shows the patterns of maximum cellular detonation pressures. It can be seen that the structure of the cells at the positive and negative concentration gradients is the same, their size remains practically unchanged. In this case, the average and peak pressures along the channel vary in accordance with the initial loading of particles. Their values in areas of higher initial concentrations are higher than in areas of reduced initial particle concentration.

Figure 2 shows the instantaneous pressure profiles on the channel wall ($y = 0$) with a step of 0.04 ms that characterize cellular detonation flows. Here, as in Fig. 1, the maximum peak pressure values (due to the reflection of the transverse wave from the channel wall) also correspond to regions of higher values of the initial particle concentrations.

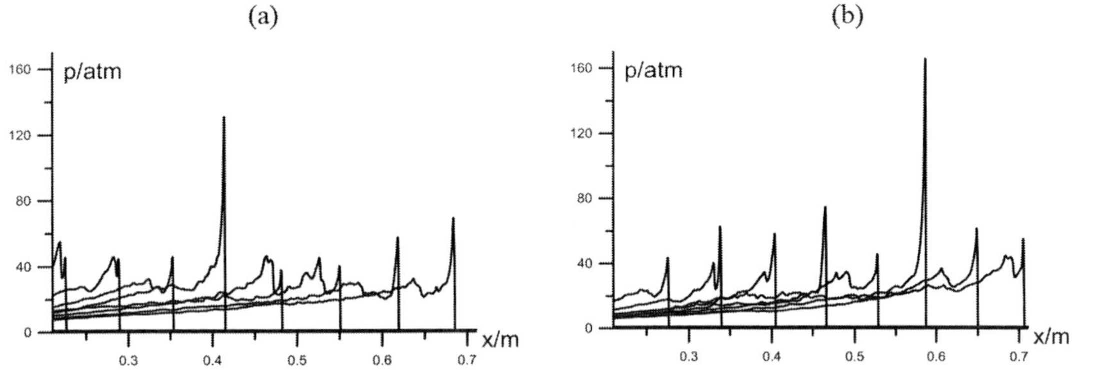

Figure 2. Pressure distributions on the wall (y=0) at cellular detonation propagation in mixtures with longitudinal concentration gradients: negative (a), positive (b), $\Delta t = 0.04$ ms.

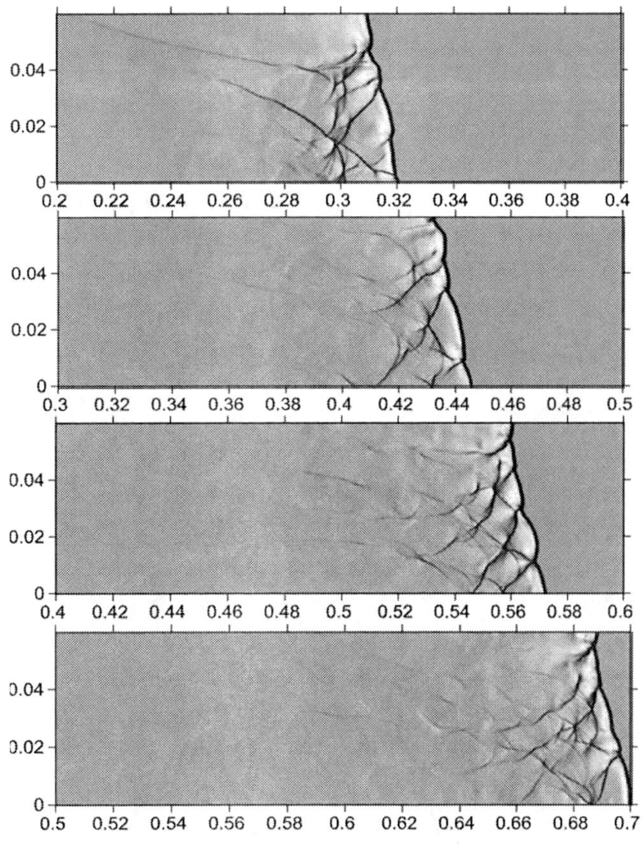

Figure 3. Pressure gradient fields at cellular detonation propagation in mixture with transversal concentration gradient, $\Delta t = 0.08$ ms.

4.2. The propagation of a detonation wave in a mixture with transversal concentration gradient

The concentration distribution across the channel was specified by a linear function $\xi_0(y) = 0.3 + 12.5y$. Figure 3 presents pictures of the field of pressure gradients (some analog of Schlieren images) with a time step of 0.08 ms, which reveal a strong curvature of the shape of the leading edge (thick lines) and the presence of many transverse waves. From Fig. 3, it can also be seen that the average propagation velocity of the curved front is constant. Figure 4 shows the field of maximal pressure history of cellular detonation. One can see that the system of transverse waves is irregular, and the distance between them in the lower part of the channel is greater than in the upper part. The result is a strongly elongated and asymmetric form of detonation cells. Moreover, the average and peak pressures in the upper part of the channel are much higher than in the lower part, which corresponds to the patterns revealed with longitudinal concentration gradients (Fig. 1). Asymmetric cells were previously modeled in inhomogeneous gas media in [22].

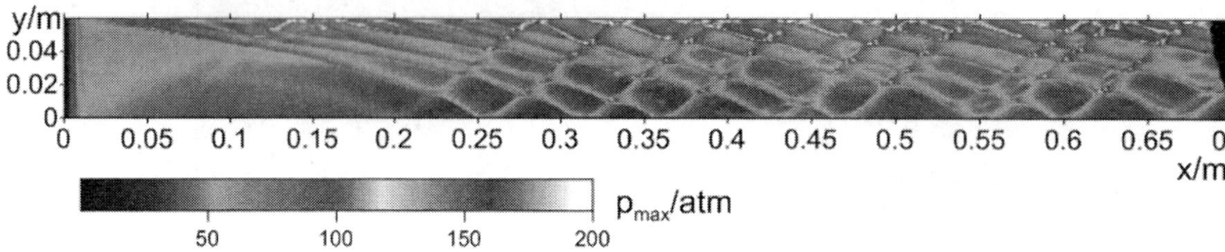

Figure 4. Cellular detonation in the mixture with transversal concentration gradient: maximal pressure histories.

5. Conclusions

The paper presents the models of reduced kinetics for the analysis of detonation flows in inhomogeneous gas particle suspensions. The models are based on an analysis of the equilibrium state or the use of the known dependence of the detonation velocity on the initial loading of reacting particles. Examples of calculating the propagation of cellular detonation in a gas suspension of aluminum particles in a channel with longitudinal and transverse particle concentration gradients are presented. Regular cellular structures are formed in channels with longitudinal concentration gradients; the cell size along the channel remains almost unchanged. In a channel with a transverse concentration gradient, irregular cells of an asymmetric and elongated shape with an inclined detonation front are formed. Peak and average pressure values are higher in areas of higher initial concentrations. The results are consistent with studies of similar gas detonation flows in [22].

Acknowledgements

The research was funded by RFBR and NSFC according to the research project № 18-58-53031) and supported by the Program of Fundamental Scientific Research of the State Academies of Sciences for 2013-2020 (project AAAA-A17-117030610139-4).

References

[1] Tropin D A and Fedorov A V 2018 *Comb. Explosion Shock Waves* **54** 200
[2] Tropin D A and Fedorov A V 2019 *Comb. Explosion Shock Waves* **55** 230
[3] Trotsyuk A V and Fomin P A 2019 *Comb. Explosion Shock Waves* **55** 384
[4] Li J, Mi X and Higgins A J 2015 *Proc. Comb. Inst.* **35** 2025
[5] Chengken Q, Peng D, Hao Yu and Zheng C 2017 *Proc. Comb. Inst.* **36** 3633
[6] Prokhorov E S 2017 *Comb. Expl. Shock Waves* **53** 705
[7] Boulal S, Vidal P and Zitoun R 2016 *Combustion and Flame* **172** 222
[8] Smirnov N N *et al* 2015 *Acta Astronautica* **115** 94
[9] Smirnov N N *et al* 2007 *Proc. Comb. Inst.* **31** 2195

[10] Semenov I, Utkin P and Markov V 2013 *J. Loss Prevention in the Process Industries* **26** 380
[11] Fedorov A V and Khmel T A 2019 *Comb. Explosion Shock Waves* **55** 1
[12] Kryukov A Yu and Malinin V I 2018 *Comb. Explosion Shock Waves* **54** 35
[13] Vasil'ev V M *et al* 1980 *Comb. Explosion Shock Waves* **16** 355
[14] Veyssiere B, Khasainov B A and Briand A 2008 *Shock Waves* **18** 307
[15] Khmel T A 2019 *Comb. Explosion Shock Waves* **55** 580
[16] Sundaram D S, Yang V and Zarko V E 2015 *Comb. Explosion Shock Waves* **51** 173
[17] Khmel T A and Fedorov A V 2018 *Comb. Explosion Shock Waves* **54** 189
[18] Strauss W A 1968 *AIAA J.* **6** 1753
[19] Nikolaev Yu A and Fomin P A 1982 *Comb. Explosion Shock Waves* **18** 53
[20] Nikolaev Yu A and Fomin P A 1983 *Comb. Explosion Shock Waves* **19** 737
[21] Nikolaev Yu A and Zak D V 1988 *Comb. Explosion Shock Waves* **24** 461
[22] Fomin P A, Kash T, Trotsyuk A V and Hieronymus H 2001 *Proc. 5th Int. Symp. Exp. Comp. Aerotherm. Inter. Flows* vol. 2, ed. Piotr Doerffer (Gdansk: IFFM Publ.) p 935

Study of propagation of heterogeneous detonation in a layer of aluminium particles

T A Khmel and S A Lavruk

Khristianovich Institute of Theoretical and Applied Mechanics SB RAS, Institutskaya 4/1, 630090, Novosibirsk, Russia

E-mail: lavruks@mail.ru

Abstract. The propagation of heterogeneous detonation along a channel partially filled with a reacting mixture is considered using numerical simulation methods. Two different regimes of propagation are observed. Flow patterns and fields of maximum pressure are analyzed. It is found that decrease of the cloud thickness leads to decreasing velocity of detonation propagation.

1. Introduction

The problems of lifting, ignition and detonation of reacting dust remain relevant, due to the ongoing explosions in industry and coal mines [1, 2]. Investigation of dust lifting after interaction with shock wave was studied as theoretically [3–7] as experimentally [8–10]. Mainly researches were made for coal dust while detonation initiation in aluminum powders was investigated poorly [4]. Data as for flow fields as for critical regimes of detonation propagation in aluminum dust layer, is not enough. In this study the interaction of the shock wave with a heterogeneous mixture of aluminum particles in channel with next detonation propagation is investigated. The main goal is to investigate the influence of the thickness of the reacting layer on critical conditions and flow fields.

2. Physical and numerical model

Main equations follows from laws of conservation of mass, momentum and energy for each phase similar to [11, 12]. The system is closed by the equation of state of an ideal gas; particles are assumed to be incompressible. Relations for the processes of mass transfer between components (evaporation, condensation, combustion), momentum exchange (resistance forces) and heat transfer between gas and particles are presented in [13]. The model of aluminum detonation burning is used the same that is described in [12, 14, 15]. Aluminum burning is described within reduced Arrhenius-type chemical kinetic with taking into account incomplete combustion of the particles and ignition temperature criterion. Transition from diffusion to kinetic combustion of the aluminum macro- and nanoparticles is also taken into account [16–18], unlike to [4].

We consider a channel partially filled with pre-mixed stoichiometric mixture of aluminum particles in oxygen (figure 1). Shock wave with Mach number M=5 propagates in the channel. During the interaction of the shock wave with the particles cloud the aluminum particle ignites and the detonation initiates. The combination of the detonation-like structure in the layer and a shock wave in the gas over the layer propagates along the channel. Structure of the flow fields behind propagating combined detonation / shock wave is analyzed by the methods of mathematical and numerical modelling.

Boundary conditions on the top and bottom channel walls correspond to impermeability and thermal insulation. Taking into account that the velocity on the left boundary is negative (a rarefaction wave is specified as the initial conditions), some soft boundary conditions were accepted on the left boundary of computational domain in the form of $\frac{du_1}{dx} = 0$, $\frac{d\rho_1}{dx} = 0$, $\frac{dT_1}{dx} = 0$, $v_1=0$.

The numerical technique is based on the conservative flux-splitting schemes: the TVD scheme by Harten for gas and the Gentry-Martin-Daly scheme for particles. The numerical method has been tested earlier and applied for *2*-D numerical simulations of the shock wave and detonation flows in [4, 11, 12]

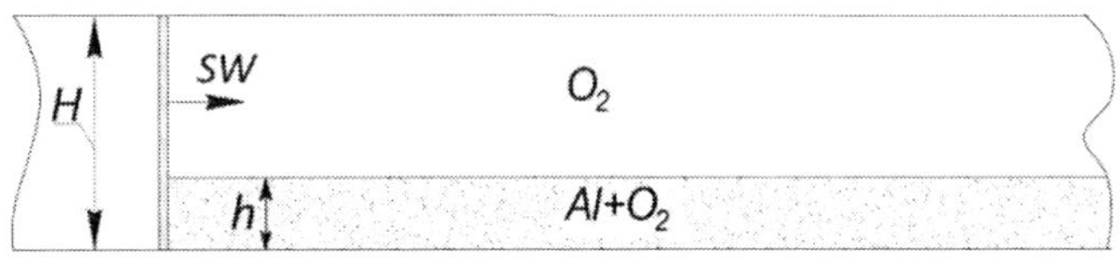

Figure 1. Scheme of simulating area.

3. Results of calculations

Numerical results for the channel height H=20 cm and particle diameter d=3.5 μm are presented in figures 2 – 5. For the width of the reacting area h=10 cm there is supercritical regime of detonation propagation. This regime of propagation is characterized by the lack of detonation failure. The detonation front expands with constant velocity that can be observed in maximum pressure history fields (figure 2). In area where the dust layer is situated there is a cellular structure with one cell on the reacting layer. The pressure values in the transverse waves are about 70 atm, that correlates with the results obtained earlier in [11]. Pressure values in the triple points reach 110 – 120 atm. It should be pointed out that these values are lower in comparison with the case when the mixture fills whole channel. Outside of the reacting layer the pressure values are less than 20 – 30 atm and only in the points of transverse waves collisions or reflections from upper wall they are about 40 atm. The average propagation velocity of combination of the detonation and shock wave in the channel in this regime is about 1.5 km/s. This value is close to the Chapman-Jouguet velocity in the stoichiometric aluminum-oxygen mixture.

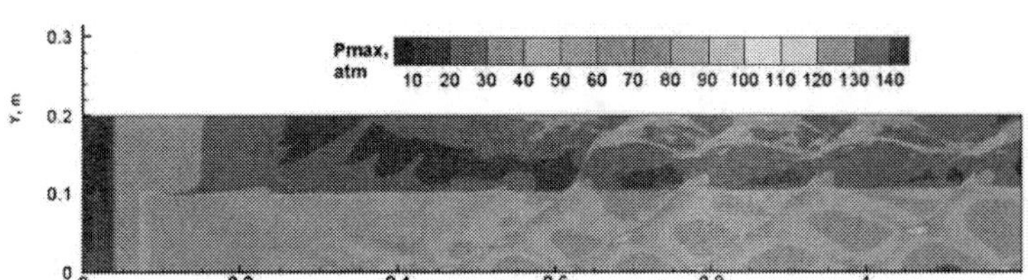

Figure 2. Maximum pressure history, h=10 cm, H=20 cm, d=3.5 μm, t=0.80 ms.

The schlieren picture (figure 3a) and the field of mean density of the discrete phase (figure 3b) at time moment t=0.45 ms demonstrate full burnout of aluminum particles in the area behind the front of the detonation wave. From the schlieren pictures it can be seen that detonation wave begin to overtake shock wave that propagate in upper part of the channel (figure 3a). The Kelvin-Helmholtz instability develops on the on the contact surface behind the shock wave. One can see the vortex structures forming in this area (figure 3a, figure 3b: 0.4<x<0.6, 0.1<y<0.18).

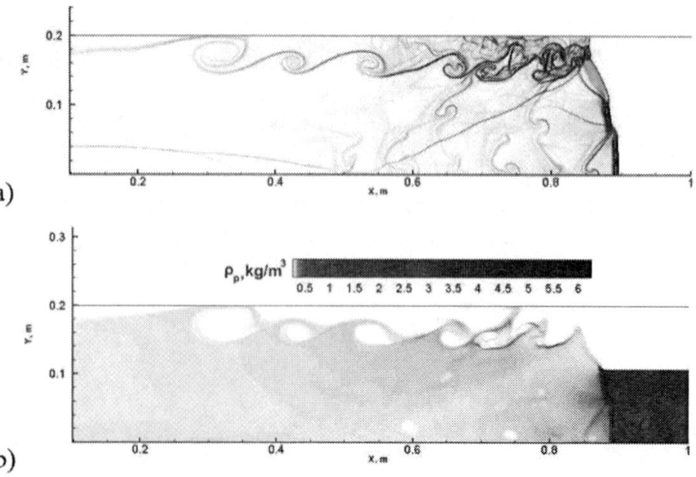

(a)

(b)

Figure 3. Flow fields h=10 cm, H=20 cm, d=3.5 μm, t=0.45 ms. (a) schlieren pictures of the flow , (b) particles density.

With thickness of the reacting layer decreased to h=5 cm the regime of the detonation propagation is close to critical. This regime is characterized by partial detonation failure in the reacting layer and following re-initiation in the transverse waves. The area with unburned particles behind the detonation wave can be seen in the flowfields of the particle mean densities (figure 4). In whole the vortex structures behind the detonation front are similar to those for h=10 cm (figure 3b). The reaction products and unburned particles form complicated vortex structures behind the shock wave (figure 4: $0.8<x<1$, $0.08<y<0.2$).

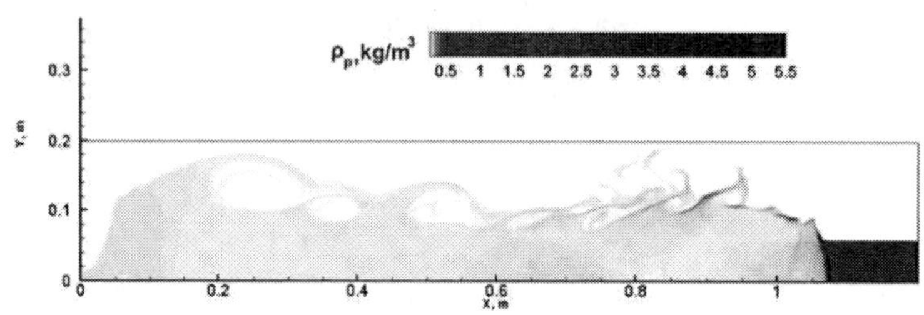

Figure 4. Particles density, h=5 cm, H=20 cm, d=3.5 μm, t=0.72 ms.

From the analysis of the maximum pressure fields (figure 5), it can be concluded that in this case (h=5 cm) the detonation in the reacting mixture initiates and propagates from the left part of the channel. Then the detonation decays and the maximum pressure decreases to 30 – 35 atm in this area. The detonation front stays flat. Subsequently, the transverse wave initiation and propagation occur, where maximum pressure exceeds the Chapman–Jouguet value (55 atm). After going out of the reacting mixture the transverse waves propagate upward in the channel and reflect from the upper wall. Also the reflection of transverse waves occurs on contact surface that enforce detonation in reacting layer (figure 5). One can see the re-initiation of detonation in the plane of symmetry where pressure rises to 110 atm (x= 0.95, y=0). In the end of the computational domain the detonation fails and the pressure does not exceed 40 atm. In this regime the average propagation velocity is about 1.3 km/s, that's less than the Chapman – Jouguet velocity (1.56 km/s).

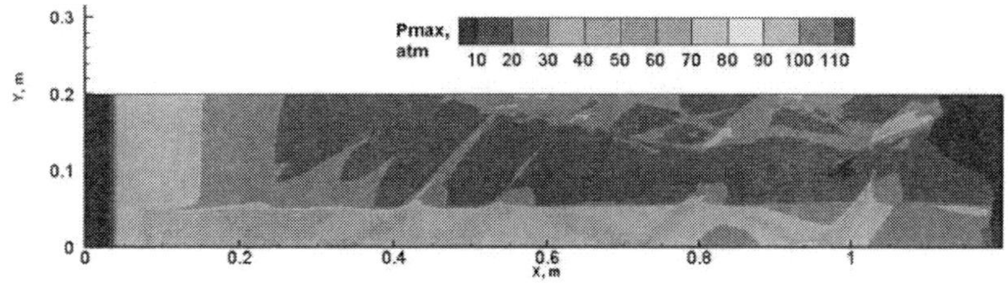

Figure 5. Maximum pressure history h=5 cm, H=20 cm, d=3.5 μm, t=0.72 ms.

4. Conclusions

Two different propagation regimes were obtained by numerical simulation of heterogeneous detonation in a channel partially filled with a reacting mixture. Based on the analysis of flow patterns and the maximum pressure field, it was found that with a decrease of cloud thickness, a decrease in the velocity of detonation propagation occurs. In the considered cases, complete detonation failure was not observed.

Acknowledgments

The research was carried out within the framework of the Program of Fundamental Scientific Research of the state academies of sciences in 2013-2020 (project No. AAAA-A17-117030610139-4).

References

[1] Li G, Yang H X, Yuan C M and Eckhoff R K 2016 *J. Loss Prev. Process Ind.* **39** 121–30
[2] Taveau J, Hochgreb S, Lemkowitz S and Roekaerts D 2018 *J. Loss Prev. Process Ind.* **51** 84–93
[3] Semenov I, Utkin P and Markov V 2013 *J. Loss Prev. Process Ind.* **26** 380–6
[4] Fedorov A V, Fomin V M and Khmel T A 2006 *Combust. Explos. Shock Waves* **42** 735–45
[5] Houim R W and Oran E S 2015 *Proc. Combust. Inst.* **35** 2083–90
[6] Zydak P and Klemens R 2007 *J. Loss Prev. Process Ind.* **20** 417–26
[7] Semenov I, Frolov S, Markov V and Utkin P 2007 S *J. Loss Prev. Process Ind.* **20** 366–74
[8] Klemens R, Zydak P, Kaluzny M, Litwin D and Wolanski P 2006 *J. Loss Prev. Process Ind.* **19** 200–9
[9] Lebecki K, Cybulski K, Śliz J, Dyduch Z and Wolański P 1995 *Shock Waves* **5** 109–14
[10] Li Y-C, Harbaugh A S, Alexander C G, Kauffman C W and Sichel M 1995 *Shock Waves* **5** 249–58
[11] Fedorov A V, Khmel' T A and Lavruk S A 2017 *Combust. Explos. Shock Waves* **53** 585–95
[12] Fedorov A V, Khmel' T A and Lavruk S A 2018 *Combust.Explos. Shock Waves* **54** 72–81
[13] Fedorov A V 1992 *Combust. Explos. Shock Waves* **28** 277–86
[14] Khmel T A and Fedorov A V. 2018 *Combust. Explos. Shock Waves* **54** 189–99
[15] Fedorov A V 2018 *Combust. Explos. Shock Waves* **54** 345–9
[16] Sundaram D S, Yang V and Zarko V E 2015 *Combust. Explos. Shock Waves* **51** 173–96
[17] Fedorov A V and Khmel T A 2019 *Combust. Explos. Shock Waves* **55** 1–17
[18] Khmel T 2019 *Combust. Explos. Shock Waves* **55** 580–8

Journal of Physics: Conference Series **1404** (2019) 012061 doi:10.1088/1742-6596/1404/1/012061

Combustion of round hydrogen microjet in a concurrent flow

V V Kozlov[1,2], G R Grek[1], M V Litvinenko[1,2], Yu A Litvinenko[1], A S Tambovtsev[1,2] and A G Shmakov[3,4]

[1]Khristianovich Institute of Theoretical and Applied Mechanics SB RAS,
4/1, Institutskaya str., Novosibirsk, 630090, Russia
[2]Novosibirsk State University,
1, Pirogova str., Novosibirsk, 630090, Russia
[3]Voevodsky Institute of Chemical Kinetics and Combustion SB RAS,
3, Institutskaya str., Novosibirsk, 630090, Russia
[4]Siberian State University of Geosystem and Technology,
10, Plakhotnogo str., Novosibirsk, 630018, Russia

kozlov@itam.nsc.ru

Abstract. Experimental data on the diffusion combustion of round hydrogen microjet in concurrent coaxial flows are presented. The effects on combustion of the concurrent flow of air and that of air premixed with nanopowder of TiO_2 are of interest. The hydrogen microjet is emanated from a round micronozzle which is surrounded by a coaxial slit to produce the concurrent flow. Combustion events found in the present conditions are similar to those observed in the previous studies on diffusion combustion of hydrogen microjets at subsonic and supersonic velocities. In a subsonic range, the so-called "bottleneck-flame region" is generated close to the nozzle exit, while in high-speed conditions the flame separates from the nozzle. At increasing velocity of both the hydrogen microjet and the concurrent flow, the "bottleneck-flame region" is still found and the combustion becomes more intense. The "bottleneck-flame region" is suppressed at the microjet velocity approaching transonic values.

1. Introduction

Experimental data on the diffusion combustion of round hydrogen microjets [1–5] testified to a variety of combustion events at increasing velocity of the microjet emanating from the nozzles with the exit diameter ranging from 0.25 to 1.00 mm. The following main stages of the combustion were suggested. 1 – combustion of the microjet with an extended laminar flame at $U_0 \leq 150$ m/s. 2 – generation of the laminar "bottleneck-flame region" close to the nozzle exit with the following turbulization of the microjet overcoming the narrow layer of density gradient at $U_0 > 150$ m/s. 3 – separation of turbulent flame from the "bottleneck-flame region" at $U_0 > 200$ m/s. 4 – termination of combustion in the turbulent section of microjet, however, the laminar portion of microjet is still burning. The last option applies to flow velocity up to transonic values. In our previous experiments [5, 6] we did not manage to examine the supersonic combustion because of the nozzle chocking at $U_0 > 331$ m/s. 5 – combustion of the microjet is completely terminated at $U_0 \approx 331$ m/s. Similar scenarios were found at the diffusion combustion of plane hydrogen microjets [2, 7]. Note that the nozzle chocking takes place at the velocity of hydrogen close to the speed of sound in air, that is, $U_0 \approx 331$ m/s. Combustion of both round and plane microjets is stabilized by the generation of "bottleneck-flame region".

In our previous experiments we failed to achieve supersonic velocities of the hydrogen microjet at preservation of its diffusion combustion. One expects that the nozzle chocking was due to the generation of "bottleneck-flame region" resulting in the nozzle heating and preventing the flame lift-off from its exit. A feature of the supersonic jet flow is the generation of supersonic cells in conditions of the flame lift-off from the nozzle which was demonstrated in Refs. [6, 8]. Some experimental and numerical data on the combustion of round hydrogen microjets at subsonic and supersonic velocity are presented in Refs. [9–11].

One more issue is the effect of additives to the hydrogen microjet on its combustion. Experimental results on the diffusion combustion of hydrogen premixed with methane, helium, or nitrogen in round microjets are presented in Ref. [12]. As is shown, the diffusion combustion of gas mixtures is featured with the origination of the "bottleneck-flame region" similarly to that at the diffusion combustion of pure hydrogen microjet.

The microjet combustion of hydrogen/methane mixture appears more stable than that of pure methane so that the combustion of hydrocarbon can be stabilized in a wide range of flow velocity due to hydrogen additive. Also, it was shown that to make more stable the diffusion combustion of microjets of hydrogen premixed with methane, helium, or nitrogen at flow velocity getting higher, the portion of hydrogen should be increased. Then, the focus of Ref. [13] was the microjet combustion of hydrogen premixed with oxygen. It was found that the oxygen additive makes the velocity range of burning microjet smaller as compared to the diffusion combustion of pure hydrogen. At increasing velocity of the microjet, the spherically-shaped "bottleneck-flame region" of laminar microjet turned into a cylindrical configuration. Note that in the above experimental studies the gas mixtures were generated upstream of the nozzle section.

The objective of the present work is experimental investigation of the diffusion combustion of round hydrogen microjet interacting with concurrent flows of air and that of air premixed with nanopowder. The combustion is examined at subsonic and supersonic flow velocities. The present results are compared with our previous experimental data on combustion of hydrogen microjets.

2. Experimental arrangement

Experimental set-up is shown in figure 1. To generate a burning microjet, the nozzle section of the test facility was supplied with hydrogen from a vessel at 100 – atm pressure; a concurrent air flow emanated from a coaxial slit through injection of compressed air as is shown in figure 1. In conditions of the concurrent air flow seeded with nanoparticles, the micronozzle was designed as a small-sized port for injection hydrogen in the centre of the nozzle section surrounded by a coaxial slit to produce the concurrent flow.

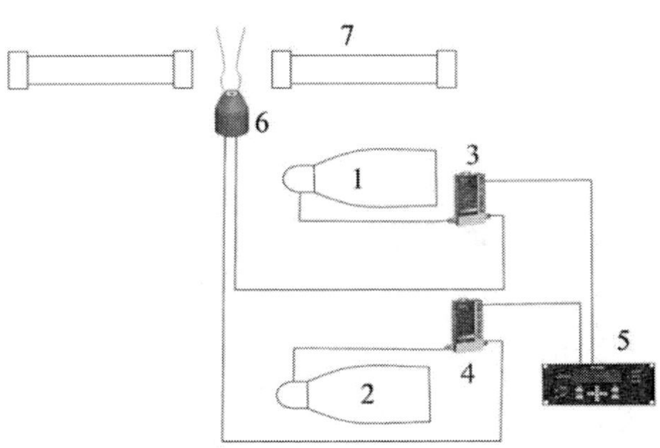

Figure 1. Experimental sketch: 1- hydrogen, 2 – air flow, 3, 4 – flow-meter valve, 5 – flow-meter controller, 6 – nozzle section, 7 – schlieren device; cross-section of the nozzle is shown in the bottom.

Journal of Physics: Conference Series **1404** (2019) 012061 doi:10.1088/1742-6596/1404/1/012061

Diffusion combustion of the hydrogen microjet, occurring with and without the concurrent flow, was recorded by a digital camera and examined at acquisition of its shadow images using a schlieren method and a digital camera providing a resolution of 24 Mpix. The volume flow rate Q (sm^3/s) was controlled by precision flow meters from MKS Instruments (USA) with an accuracy of 0.7 %. Microjet velocity was calculated as $U = Q/S$ where S stands for the cross section of the round nozzle of hydrogen microjet or that of the coaxial slit generating the concurrent flow. Note that the above formula is not appropriate in supersonic conditions as far as it does not take into account a set of parameters, particularly, gas compressibility. Thus, we also used another way to determine the microjet velocity through the pressure difference $\Delta P = P_{inlet} - P_{atm}$ at the inlet of micronozzle (P_{inlet}) and its exit (P_{atm}). If so, the microjet velocity makes $\boldsymbol{U} = \mathbf{2} \cdot \left(\frac{\Delta P}{\rho}\right)^{1/2}$ where ρ is density of hydrogen. In what follows, the crossection of round micronozzle generating the hydrogen microjet interacting with the concurrent air flow is calculated as

$$S_1 = \frac{\pi d^2}{4} = \frac{3.14 \cdot 0.1^2}{4} = 0.078\ cm^2$$

and that of the coaxial slit for air injection is given by the formula

$$S_2 = \pi r_1^2 - \pi r_2^2 = 0.0772\ cm^2.$$

In conditions of the concurrent flow of air with nanopowder, the crossection S1 of the central port of micronozzle generating the hydrogen microjet makes

$$S_1 = \frac{\pi d^2}{4} = \frac{3.14 \cdot 0.11^2}{4} = 0.095\ cm^2$$

and that of the coaxial slit S2 for injection of concurrent flow is measured as

$$S_2 = \pi r_1^2 - \pi r_2^2 = 3.14 \cdot 0.17^2 - 3.14 \cdot 0.14^2 = 0.0292\ cm^2.$$

3. Experimental results

3.1. Combustion of round hydrogen microjet in a concurrent air flow in subsonic conditions

Shadow images of the microjet combustion at variation of the hydrogen velocity (U1, m/s) and of the coaxial air microjet (U2, m/s) are presented in figure 2. In the absence of concurrent air flow (figure 2a) one can observe the origination of "bottleneck-flame region", the phenomenon which we discussed in

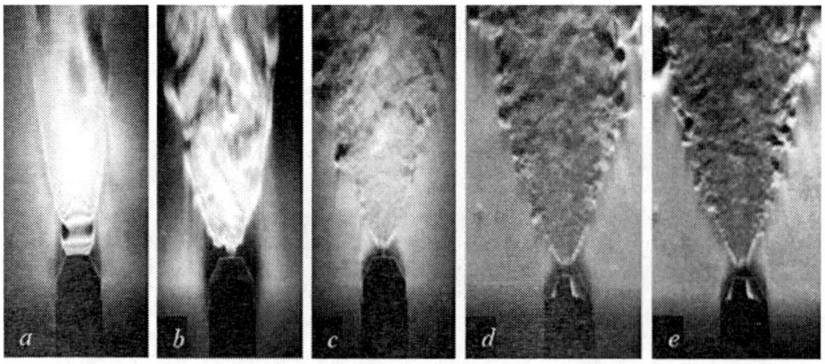

Figure 2. Shadow images of the microjet combustion at variation of the hydrogen velocity (U_1, m/s) and of the coaxial air microjet (U_2, m/s) as *a*) U_1 = 130, U_2 = 0 m/s; *b*) U_1 = 130, U_2 = 17,5 m/s, *c*) U_1=130, U_2 = 35m/s, *d*) U_1 = 204, U_2 = 18 m/s; *e*) U_1 = 204, U_2=26 m/s

our previous studies [1–5]. The "bottleneck-flame region" is still found in conditions of the hydrogen microjet interacting with the air microjet (figures. 2b, c, d and e); however, the above spherically-shaped flow region turns into a cylindrical one. At increasing velocity of the concurrent air flow, the "bottleneck-flame region" shrinks similarly to that at acceleration of the hydrogen microjet which was observed in Refs. [1–5]. Also the turbulent combustion is intensified downstream of the "bottleneck-flame region" at the speed of axial air jet getting higher.

Thus, we conclude that the main features of subsonic diffusion combustion of a hydrogen microjet surrounded by a concurrent air jet are similar to those we observed earlier in hydrogen microjets [1–5] excepting some details such as the deformation of "bottleneck-flame region" and the intensification of turbulent combustion at increasing velocity of the coaxial air jet, etc.

3.2. Supersonic air microjets with and without the concurrent air flow in the absence of combustion
Flow patterns of the interacting air microjets are given in figure 3. As in the previous sections, U1 and U2 stand for the velocity of microjets emanating from the round nozzle and the coaxial slit, respectively. In cases a–d, the value of U1 keeps constant as 340 m/s at U2 varying from 0 to 400 m/s. Oppositely, in conditions of e–h, the velocity U2 = 340 m/s is one and the same while U1 ranging from 0 to 400 m/s. In this way, the interaction of two microjets is made clear at supersonic velocities of the round (a–d) and the concurrent (e–h) jet flows.

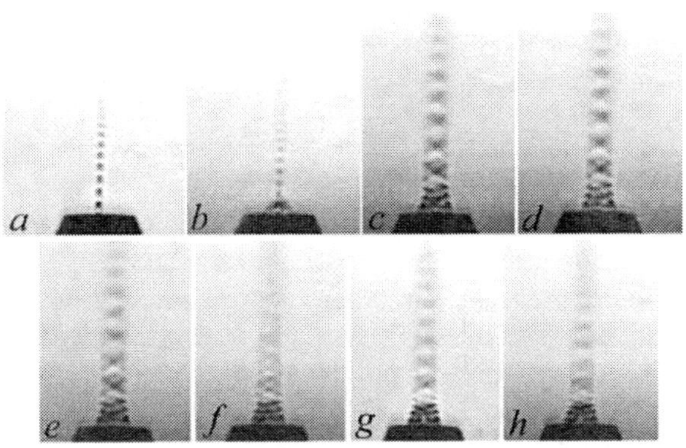

Figure 3. Shadow images of the round and concurrent microjets at U1= 340m/s, U2=0,260, 340, and 400 m/s (a–d), and U2=340m/s, U1 = 0, 280, 340, and 400 m/s (e–h)

4. Diffusion combustion of round hydrogen microjet in the concurrent flow of air with nanopowder in high-speed conditions

Shadow images of the diffusion combustion of round hydrogen microjet in the concurrent flow of air with nanopowder presented in figure 4. In particular, illustrates the interaction of both jets at their combustion visualized in orange. Here the subscripts "1" and "2" stand for the hydrogen microjet and the concurrent flow, respectively.

The flame separates from the nozzle exit so that the laminar combustion is not sustained (figure 4a and b). At reduction of flow velocity, one can observe laminar flame (figure 4c and d). Then, the flame turns necked at the border of concurrent flow with distinct radiation by the heated nanoparticles (figure 4e and f).

Thus, the present data on diffusion combustion of the high-speed round hydrogen microjet affected by the coaxial concurrent air flow with nanopowder support the scenarios of combustion which we observed earlier in a wide range of flow velocity [1–7]. Particularly, in the case of flame separation from the nozzle exit the combustion becomes turbulent in the absence of laminar section of the microjet (figure 4a and b). If so, the nozzle is not heated and the stability of flame goes down at increasing flow velocity. At diminution of the velocity of hydrogen microjet, the flame attaches to the nozzle (figure 4c and d). In such conditions, the temperature of nanoparticles which are injected through the coaxial slit grows up to 1500 °C and even higher so that the burning mixture radiates in yellow. At further reduction of flow velocity down to 50–60 m/s (figure 4e and f), the laminar section of microjet combustion originates close to the nozzle exit; in this case, the radiation of nanoparticles is not so pronounced, compare to figure 4c.

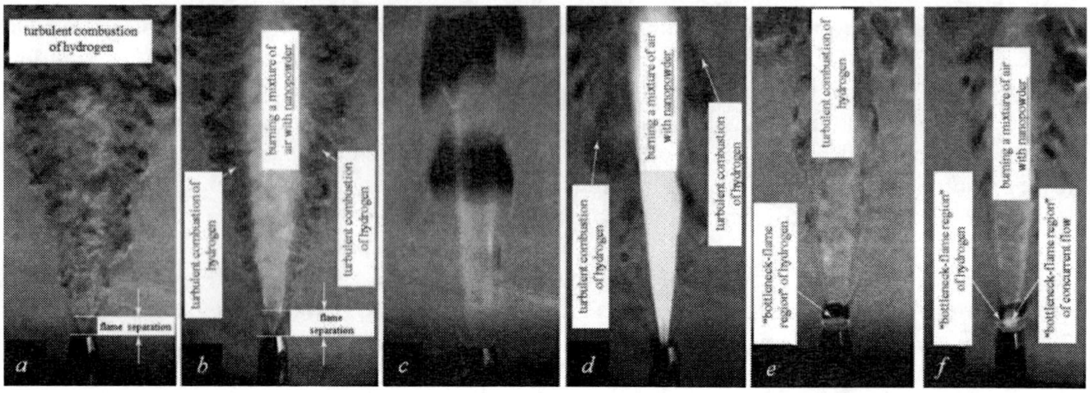

Figure 4. Shadow images of the burning hydrogen microjet in the absence of concurrent flow *a)* U_1= 210 m/s; *c)*U_1= 150 m/s; *e)* U_1= 50 m/s and at their interaction *b)* U_1=210 m/s, U_2= 6 m/s; *d)* U_1=175 m/s, U_2= 6 m/s; *f)* U_1=60 m/s, U_2= 4.5 m/s

5. Conclusion

In the present paper we reported our recent experimental results on the combustion of round hydrogen microjet interfering with concurrent coaxial flows. Two options were considered including the concurrent flow of air and that of air seeded with nanoparticles. In the first case, the flame configuration is distinguished by the origination of "bottleneck-flame region" close to the nozzle exit that is in agreement with our previous observations on the subsonic diffusion combustion of hydrogen microjets. At variation of flow velocity, the spherically-shaped "bottleneck-flame region" turns into a cylindrical one. In conditions of supersonic combustion, the supersonic cells are generated both in the hydrogen microjet and in the concurrent air flow. The diffusion combustion of round hydrogen microjet affected by the supersonic concurrent air jet occurs with the flame detached from the nozzle.

In the other case of concurrent air flow premixed with nanopowder, the experimental data are also obtained at subsonic and transonic flow velocities. As is found, the scenarios of diffusion combustion in conditions of the concurrent flow are similar to those of the isolated hydrogen microjet. The origination of "bottleneck-flame region" at the nozzle exit is observed as well as the flame lift-off at increasing velocity of hydrogen. At flow velocity getting higher, the diffusion combustion is intensified. At the generation of "bottleneck-flame region", the flame becomes deformed both in the center of microjet and in the concurrent flow. One more observation is that heating of nanoparticles is not uniform over the microjet cross section. As is seen in figure 4b, the nanoparticles emanating from the coaxial slit are heated most of all in central area of hydrogen flow. The effect is so strong that they become radiating visible light. At the same time, flow temperature of the burning concurrent flow is lower so that the radiation goes down. Finally, when the hydrogen velocity approaches the transonic range of air, the laminar combustion is terminated and the flame becomes completely turbulent.

Acknowledgements
The present work was supported by the RSF (Project number 16-19-10330) and was partly carried out within the framework of the Program of Fundamental Scientific Research of the state academies of sciences in 2013-2020 (project No. AAAA - 22.6.3).

References

[1] Shmakov A G, Grek G R, Kozlov V V, Korobeinichev O P and Litvinenko Yu A 2015 *Vestnik NSU Series: Physics* **10** 27-41
[2] Litvinenko Yu A, Grek G R, Kozlov V V, Korobeinichev O P and Shmakov A G 2015 *Vestnik NSU Series: Physics* **10** 52-66
[3] Grek G R, Katasonov M M, Kozlov V V and Litvinenko Yu 2015 *Vestnik NSU Series: Physics* **10** 42-51
[4] Kozlov V V, Grek G R, Korobeinichev O P, Litvinenko Yu A and Shmakov A G 2016 *Doklady Physics* **61** 457-62
[5] Shmakov A G, Grek G R, Kozlov V V, Kozlov G V and Litvinenko Yu A 2017 *Siberian Journal of Physics* **12** 28-45
[6] Kozlov V V, Grek G R, Kozlov G V, Litvinenko Yu A and Shmakov A G 2019 *International Journal of Hydrogen Energy* **44** 457-68
[7] Kozlov V V, Grek G R, Korobeinichev O P, Litvinenko Yu A and Shmakov A G 2016 *International Journal of Hydrogen Energy* **41** 20240-9
[8] Kozlov V V, Grek G R, Kozlov G V, Litvinenko Yu A and Shmakov A G 2017 *Siberian Journal of Physics* **12** 62-73
[9] Kalghatgi G T 1984 *Combust. Sci. Technol.* **41** 14-29
[10] Annushkin Yu M and Sverdlov E D 2008 *Investigation of the stability of diffusion flooding flame at subsonic and supersonic off-design nozzle at flowing of gaseous fuels. Chemical Physics Hydrogen combustion and explosion parameters* (Moscow: Fizmatlit Publ.) pp 53-63
[11] Shentsov V, Sakatsume R, Makarov D, Takeno K and Molkov V 2016 Lift-off and blow-out of under-expanded jets: experiments versus simulations *8th International Seminar on Fire and Explosion* (Hefei, China)
[12] Shmakov A G, Grek G R, Kozlov V V, Litvinenko Yu A and Korobeinichev O P 2016 *Vestnik NSU Series: Physics* **11** 56-76
[13] Shmakov A G, Vikhorev V V, Grek G R, Kozlov V V, Kozlov G V and Litvinenko Yu A 2018 *Siberian Journal of Physics* **13** 54-67

Investigation of the heterogeneous detonation suppression by clouds of inert particles and droplet suspension

S A Lavruk and D A Tropin

Khristianovich Institute of Theoretical and Applied Mechanics SD RAS, Institutskaya 4/1, 630090, Novosibirsk, Russia

lavruks@mail.ru

Abstract. Using the methods of mechanics of heterogeneous media, we studied the process of suppressing heterogeneous detonation by clouds of inert particles and using a droplet suspension. The main flow regimes obtained as a result of modeling are described.

1. Introduction

Issues of explosion and fire safety of industrial production attract great attention. So, when using homogeneous and heterogeneous reacting media in production, accidental explosions may occur, that lead to great destruction. Therefore, the attenuation and suppression of explosive and detonation waves is one of the priority areas of research in this area. One of the ways to weaken or suppress detonation is to create clouds of solid inert particles or a suspension of inert liquid droplets in the flow field. In [1–6], the effect of chemically inert solid particles on the parameters of detonation waves in reacting gas mixtures was shown. Various regimes of detonation propagation during interaction with a cloud of inert particles were obtained: detonation propagation in the same regime, partial decrease in the velocity of detonation propagation, partial detonation failure, followed by re-initiation of the detonation wave, and complete detonation failure with separation of the shock front and the combustion wave. Galloping mode of detonation propagation was also obtained, in which a detonation propagation intensity decreases in a cloud of inert particles, next re-initiation followed by a decrease in the detonation wave intensity and repetition of the cycle [5,7]. In [3], a comparison of the one-dimensional and two-dimensional approaches to the description of the suppression of gas detonation is given. The suppression of detonation by droplet suspension was considered in [8–12]. In [8], a review of the available experimental work on the suppression of explosions of gas mixtures (hydrogen - air, methane - air, etc.) using water fog, as well as water sprays, was carried out. Numerical modeling of detonation suppression processes was considered in [10–12]. In the calculations, the authors considered the relations on the detonation wave, and investigated how the evaporation of liquid droplets behind the front of the detonation wave occurs.

In [13–15], the investigation of suppressing heterogeneous detonation in a flat channel by clouds of inert particles was considered. In [13,14], critical conditions were obtained for suppressing detonation with allowance for diffusion combustion of aluminum particles. In [15], the model took into account the collision between particles and the refinement of the flow patterns behind the front of the shock/detonation wave was obtained.

Previously, a diffusion-controlled combustion model was used to describe the heterogeneous detonation of aluminum particles. However, in [16] it was shown that, in the range of diameters of aluminum particles close to 3.5 μm, a transition occurs from the diffusion to the kinetic-controlled combustion. In this work, we studied the interaction of detonation waves in gas-suspended aluminum particles with a diameter of less than 1 μm in air by clouds of inert particles of aluminum oxide and a suspension of inert liquid water droplets, taking into account the transition regime of combustion from diffusion to kinetic.

2. Physical and numerical model

The mathematical model of the detonation of aluminum particles in oxygen is based on the concepts of multi-velocity multi-temperature continua. In a two-dimensional unsteady flow, the basic equations describing detonation processes in monodisperse gas-suspended aluminum particles are presented in [13]. In this work, we use a mathematical model of a bi-dispersed heterogeneous medium, one fraction of which is reactive (index $i=2$), and the second is inert ($i=3$). A model of the combustion of aluminum particles taking into account the transient combustion regime was presented in [17].

With using water droplets as an inert phase, it is necessary to take into account their evaporation and crushing during interaction with a shock or detonation wave [10,12]. The breakage of droplets is described by the formula for detachment of the boundary layer [12]:

$$\frac{dm_{br}}{dt} = \frac{3\sqrt{2}\rho_3}{d_3 \mathrm{Re}^{1/2}}\left(\frac{\rho_{11}}{\rho_{33}}\right)^{\frac{1}{3}}\left(\frac{\nu_3}{\nu_1}\right)^{\frac{1}{3}}(u_3 - u_1).$$

And the evaporation of drops is described by the formula [10,18]:

$$\frac{dm_{ev}}{dt} = -(4\pi)^{\frac{2}{3}} 3^{\frac{1}{3}} \frac{1}{8} \rho_{33}^{\frac{2}{3}} m_{ev}^{\frac{1}{3}} K_v \left(1 + 0.27\,\mathrm{Re}^{0.5}\right),$$

where ρ_{33} is water density, ν_1, ν_3 are gas and liquid viscosity, u_1, u_3 are gas and liquid velocity, K_v is evaporation constant, Re is Reynolds number. More detailed information on the model of droplet breaking is presented in the works [10,18].

The interaction scheme of the detonation wave with a cloud of inert phase is shown in Fig. 1. A cellular detonation wave propagates through the channel and interacts with a cloud of inert phase that consist of aluminum oxide particles or water droplets and reacting mixture of aluminum in oxygen. The cloud occupies the entire width of the channel.

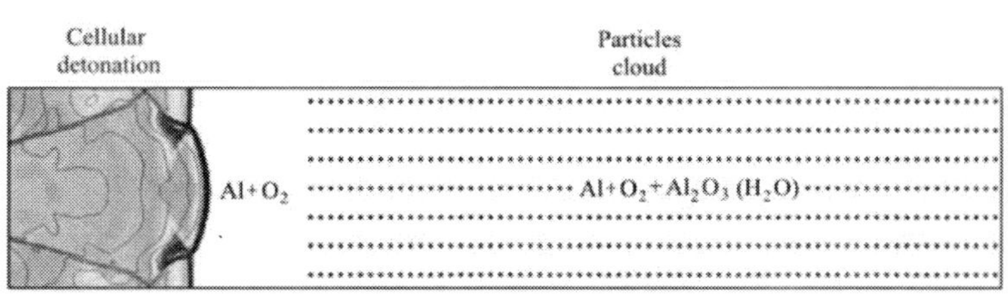

Figure 1. Scheme of computational area.

3. Estimation of times of thermal and velocity relaxation of alumina particles and water droplets

Thermal and velocity relaxation times of an individual particle and a droplet behind the detonation wave front are estimated. The method of such estimation is described in detail in [4]. In fig. 2 shows the dependences of times of velocity and thermal relaxation of water droplets and aluminum oxide particles on droplet/particle size. It is seen that times of velocity and thermal relaxation of water droplets are lower than the corresponding times for alumina particles with the same droplet and particle sizes. Thus, liquid droplets will accelerate and warm up in the gas stream behind the detonation wave front faster than aluminum oxide particles. Thus, rates of impulse and energy take off from the reacting mixture will be higher in the case of the interaction of the detonation wave with liquid droplets. It can be assumed that the efficiency of detonation waves attenuation by liquid droplets will be higher than by solid particles.

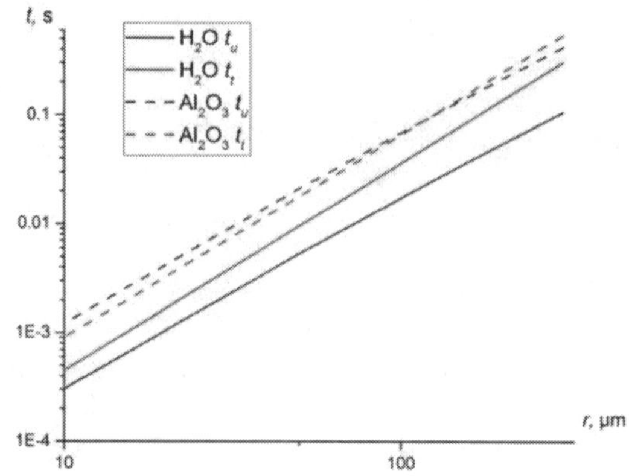

Figure 2. Dependences of times of velocity (blue curves) and thermal (red curves) relaxation of water droplets (solid lines) and aluminum oxide particles (dashed lines).

4. Suppression by inert particles

After the interaction of a detonation wave with clouds of inert particles, various regimes of detonation propagation were obtained. At a volume concentration of inert particles of $m_3=10^{-4}$, regardless of their diameter, detonation combustion is continue to propagate with the same velocity, the cellular structure remains unchanged (Fig. 3). The maximum pressure patterns show that the cellular structure does not change when the detonation wave enters the cloud of inert particles. For this case a cloud of inert particles is located at a point $x=0.3$ m. The pressure value at triple points also remains in the same range as before a cloud of inert particles (about 220 atm).

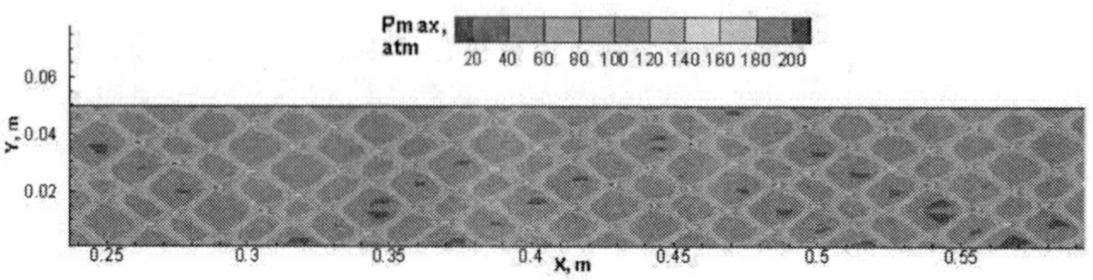

Figure 3. History of maximum pressure d_2=1 µm, d_3=1 µm, $m_3 = 1\cdot10^{-4}$.

With an increase in the concentration of inert particles m_3 to $5\cdot10^{-4}$, the flow is rearranged, the cellular structure changes and the sizes of detonation cells increase. This is clearly seen in the schlieren images (Fig. 4). The inert cloud is at the point $x = 0.1$ m. Before interaction with a cloud of inert particles or at the initial stage of interaction, many transverse waves propagate along the leading front (Fig. 4a). When propagating through the cloud, the cellular structure is rearranged and the number of transverse waves decreases to 2–4 per channel width (Fig. 4b). Figure 4c shows that in some cases there is a partial separation of the combustion front and the shock front, and re-initiation occurs in transverse waves.

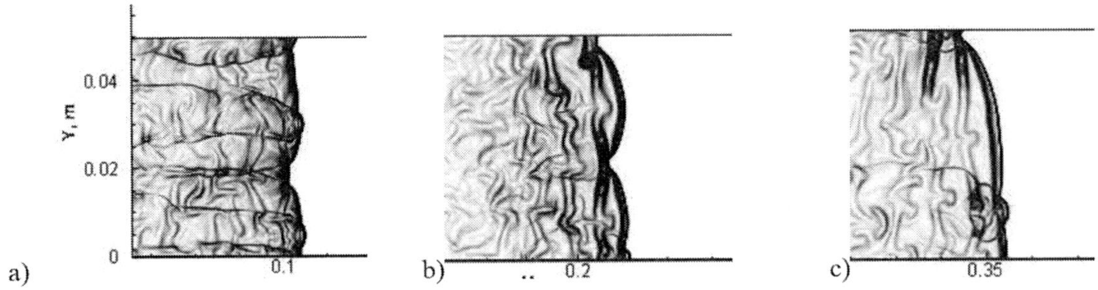

Figure 4. Detonation propagation d_2=200 nm, d_3= 5 μm, $m_3=5\cdot10^{-4}$ at moment , a) 0.04ms, b) 0.2 ms, c) 0.3 ms.

In the pictures of maximum pressure for this case shown in Fig. 3 there is a decrease in pressure values at triple points, as well as a rearrangement of the cellular structure (Fig. 5). The pressure at triple points drops from 250 atm to 170-180 atm.

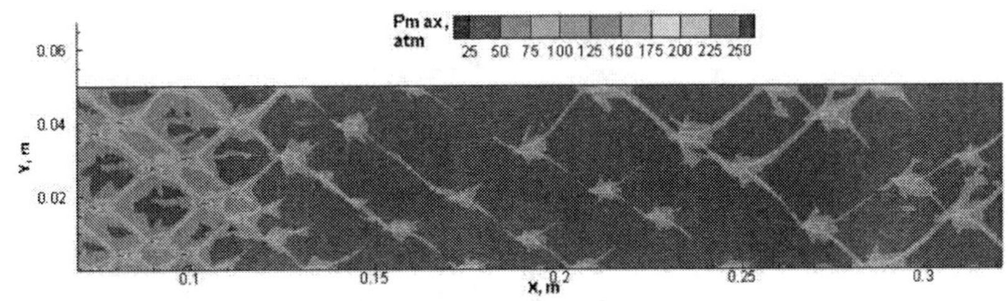

Figure 5. History of maximum pressure d_2=200 nm, d_3= 5 μm, $m_3=5\cdot10^{-4}$.

Detonation failure in gas suspensions of aluminum submicron particles is observed with an increase in the concentration of inert particles to $m_3=10^{-3}$ and higher. As the result the failure of the cellular structure that can be seen from the history of maximum pressure (Fig. 6a). Splitting of the detonation wave into a frozen shock the wave and the combustion front lagging behind can be seen on the schlieren image (Fig. 6b). Behind the combustion front, the so-called "bundles" begin to form, which were obtained in [14].

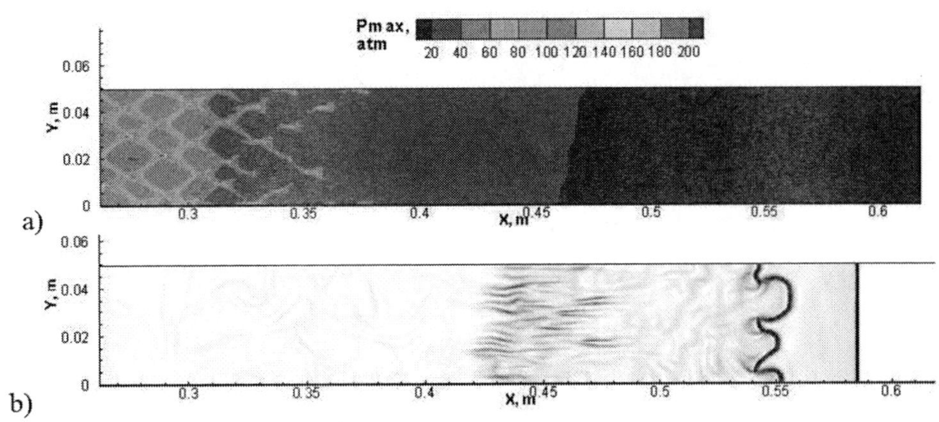

Figure 6. Flow fields d_2=200 nm, d_3=1 μm, $m_3=1\cdot10^{-3}$. a) History of maximum pressure, b) schlieren images.

5. Conclusions

Three different propagation modes are obtained by numerical simulation of heterogeneous detonation in a channel filled with a mixture of reacting and inert particles. Based on the analysis of the flow patterns and the maximum pressure field, it was found that with an increase in the volume concentration of the inert phase, the velocity of detonation propagation decreases and a complete or partial detonation failure can be observed.

Acknowledgment

The work was partly supported by the Russian Foundation for Basic Research № 18-08-00156-a.

References

[1] Fedorov A V. and Tropin D A 2011 *Combust. Explos. Shock Waves* **47** 464–72
[2] Tropin D A and Fedorov A V. 2014 *Combust. Sci. Technol.* **186** 1690–8
[3] Bedarev I A and Tropin D A 2018 *AIP Conf. Proc.* **2027**
[4] Bedarev I 2019 *Therm. Sci.* **23** 439–45
[5] Tropin D and Fedorov A 2018 *Combust. Sci. Technol.* **00** 1–9
[6] Tropin D A and Fedorov A V. 2019 *Combust. Explos. Shock Waves* **55** 230–6
[7] Tropin D A and Fedorov A V. 2018 *Combust. Explos. Shock Waves* **54** 200–6
[8] Mawhinney J R and Darwin R 2000 *Halon Options Tech. Work. Conf.* 215–26
[9] Thomas G O 2000 *Process Saf. Environ. Prot.* **78** 339–54
[10] Nikolaev Y A and Fomin P A 1984 *Combust. Explos. Shock Waves* **20** 447–55
[11] Frolov S M and Gel'fand B E 1992 *Combust. Explos. Shock Waves* **27** 756–63
[12] Belikov V V, Belikova G V, Goloviznin V|M, Semenov V N, Sarodubtseva L P and Fokin A L 1995 *High Temperature* **33** 452–7
[13] Fedorov A V. and Kratova Y V. 2013 *Combust. Explos. Shock Waves* **49** 335–47
[14] Kratova Y V. and Fedorov A V. 2014 *Combust. Explos. Shock Waves* **50** 183–91
[15] Khmel' T A and Fedorov A V. 2014 *Combust. Explos. Shock Waves* **50** 196–207
[16] Sundaram D S, Yang V and Zarko V E 2015 *Combust. Explos. Shock Waves* **51** 173–96
[17] Khmel T A and Fedorov A V. 2018 *Combust. Explos. Shock Waves* **54** 189–99
[18] Borisov A A, Gel'fand B E, Gubin S A, Kogarko S M and Podgrebenkov A L 1973 *J. Appl. Mech. Tech. Phys.* **11** 741–6

Combustion of a coal dust suspension in a pipe during the swirl

K M Moiseeva, A Yu Krainov and Yu A Tarasenko

Tomsk State University, 36 Lenina st., Tomsk, 36634050, Russia

moiseeva_km@t-sk.ru

Abstract. This manuscript presents the research on the combustion problem of a coal dust suspension in a pipe under the swirl flow condition. The physico-mathematical model is based on the approaches of the mechanics of multiphase reacting media by R. I. Nigmatulin. The solution to the problem is performed in a two-dimensional axisymmetric formulation. Godunov's scheme is used to solve the equation system of the model. The paper provides the results of the parametric analysis of the problem and data on the influence of the flow swirl on the gas temperature fields in the pipe at the initial stage of combustion. The presence of a significantly swirl flow leads to the incomplete combustion of coal dust in the pipe.

1. Introduction

Swirl combustion is currently one of the most important engineering problems in combustion physics. There is a hypothesis about increasing the combustion efficiency of the reacting gas mixtures in the swirl combustion chambers [1], as well as improving the efficiency of the fuel burning devices [2–5]. The combustion of the gas suspensions in vortex burners is the object of thorough investigation for a number of authors. The swirling is believed to lead to the heat transfer intensification and, consequently, to the increase in efficiency of the dispersed fuel combustion. The Reynolds equations [3] or the Navier-Stokes equations [1], which take into account the turbulent sources, are usually used to simulate the swirling flows. The choice of the turbulence model depends on the combustion conditions.

In the present study, it is proposed to model the swirl flow in a cylindrical pipe by taking into account the angular component of the flow velocity. The aim of the research is to determine the influence of the angular component on the flow and combustion characteristics of a coal dust gas suspension in the pipe.

2. Mathematical model and solution method

We have solved the combustion problem of a coal dust suspension in the air in a two-dimensional axisymmetric formulation. The system of equations describing the motion of the suspension takes into account the axial and radial change of the flow velocity and it is supposed that the derivative of the flow velocity due to angular component is equal to zero. It is assumed that before the process begins, the pipe is filled with a cold air and there are no particles of coal dust. The right end of the pipe is open, whilst the left end is the inlet of the coal dust suspension heated to a high temperature. The incoming coal dust-air flow is swirling. The diffusion and thermal conductivity processes in the gas phase are neglected. The heated particles of coal dust are known to be capable of chemical reaction. The first-order heterogeneous oxygen reaction proceeds on the particle surface. The rate of the reaction is limited by the mass transfer coefficient β [7]. The model takes into account the interaction between particles and gas phase.

The physical-mathematical formulation of the problem is based on the approaches of the mechanics of multiphase reacting media by R.I. Nigmatulin [6] and on the previously developed physical-mathematical model of the coal dust suspension combustion in the air [7]. Under the formulated assumptions, the model includes the following conservation law equations for the gas and particles:

$$\frac{\partial r\rho_g}{\partial t}+\frac{\partial r\rho_g u_g}{\partial x}+\frac{\partial r\rho_g v_g}{\partial r}=rG, \tag{1}$$

$$\frac{\partial r\rho_g u_g}{\partial t}+\frac{\partial r\left(p_g+\rho_g u_g^2\right)}{\partial x}+\frac{\partial r\rho_g u_g v_g}{\partial r}=-r\tau_x+rGu_k \tag{2}$$

$$\frac{\partial r\rho_g v_g}{\partial t}+\frac{\partial r\rho_g u_g v_g}{\partial x}+\frac{\partial r\left(p_g+\rho_g v_g^2\right)}{\partial r}=-r\tau_r+rGv_k+p_g+\rho w_g^2, \tag{3}$$

$$\frac{\partial r\rho_g w_g}{\partial t}+\frac{\partial r\rho_g w_g u_g}{\partial x}+\frac{\partial r\rho_g w_g v_g}{\partial r}=-r\tau_\phi+rGw_k-\rho_g w_g v_g, \tag{4}$$

$$\frac{\partial r\rho_{O2}}{\partial t}+\frac{\partial r\rho_{O2}u_g}{\partial x}+\frac{\partial r\rho_{O2}v_g}{\partial r}=-r\alpha_r G, \tag{5}$$

$$\begin{aligned}\frac{\partial r\varepsilon_g}{\partial t}+\frac{\partial ru_g\left(\varepsilon_g+p_g\right)}{\partial x}+\frac{\partial rv_g\left(\varepsilon_g+p_g\right)}{\partial r}&=-r\left[u_k\tau_x+v_k\tau_r+w_k\tau_\phi+n_k\alpha_k S_k\left(T_g-T_k\right)\right]+\\&+rGc_kT_k+0.5rG\left(u_k^2+v_k^2+w_k^2\right),\end{aligned} \tag{6}$$

$$\frac{\partial r\rho_k}{\partial t}+\frac{\partial r\rho_k u_k}{\partial x}+\frac{\partial r\rho_k v_k}{\partial r}=-rG, \tag{7}$$

$$\frac{\partial r\rho_k u_k}{\partial t}+\frac{\partial r\rho_k u_k^2}{\partial x}+\frac{\partial r\rho_k u_k v_k}{\partial r}=r\tau_x-rGu_k, \tag{8}$$

$$\frac{\partial r\rho_k v_k}{\partial t}+\frac{\partial r\rho_k u_k v_k}{\partial x}+\frac{\partial r\rho_k v_k^2}{\partial r}=r\tau_r-rGv_k+\rho w_k^2, \tag{9}$$

$$\frac{\partial r\rho_k w_k}{\partial t}+\frac{\partial r\rho_k w_k u_k}{\partial x}+\frac{\partial r\rho_k w_k v_k}{\partial r}=r\tau_\phi-rGw_k-\rho_k w_k v_k, \tag{10}$$

$$\begin{aligned}\frac{\partial r\varepsilon_k}{\partial t}+\frac{\partial ru_k\varepsilon_k}{\partial x}+\frac{\partial rv_k\varepsilon_k}{\partial r}&=rQG+\\&+r\left[u_k\tau_x+v_k\tau_r+w_k\tau_\phi+n_k\alpha_k S_k\left(T_g-T_k\right)-Gc_kT_k-0.5G\left(u_k^2+v_k^2+w_k^2\right)\right],\end{aligned} \tag{11}$$

$$\frac{\partial rn_k}{\partial t}+\frac{\partial rn_k u_k}{\partial x}+\frac{\partial rn_k v_k}{\partial r}=0, \tag{12}$$

and the initial conditions:

$$\begin{aligned}&u_g(x,r,0)=v_g(x,r,0)=w_g(x,r,0)=u_k(x,r,0)=v_k(x,r,0)=w_k(x,r,0)=0,\\&n_k(x,r,0)=\rho_k(x,r,0)=T_k(x,r,0)=0,\ T_g(x,r,0)=T_b,\ \rho_g(x,r,0)=\rho_{gb},\ \rho_{O2}(x,r,0)=a_{O2b}\rho_{gb},\end{aligned} \tag{13}$$

The boundary conditions at the axis $r = 0$ are:

$$\frac{\partial u_g(x,0,t)}{\partial r}=\frac{\partial u_k(x,0,t)}{\partial r}=\frac{\partial v_g(x,0,t)}{\partial r}=\frac{\partial v_k(x,0,t)}{\partial r}=\frac{\partial w_g(x,0,t)}{\partial r}=\frac{\partial w_k(x,0,t)}{\partial r}=0,$$
$$\frac{\partial n_k(x,0,t)}{\partial r}=\frac{\partial \rho_k(x,0,t)}{\partial r}=\frac{\partial T_k(x,0,t)}{\partial r}=\frac{\partial T_g(x,0,t)}{\partial r}=\frac{\partial \rho_g(x,0,t)}{\partial r}=\frac{\partial \rho_{O2}(x,0,t)}{\partial r}=0. \quad (14)$$

The following initial values are set at the inlet ($x = 0$): the enthalpy, the gas flow rate, the gas and particle velocity along the radial and tangential directions, the oxidizer mass concentration in the gas, particle mass flow rate and the number of particles.

$$H_k(0,r,t)=c_kT_{st}+\frac{u_b^2}{2}+\frac{w_g^2(0,r,t)}{2},\ H_g(0,r,t)=c_pT_{st}+\frac{u_b^2}{2}+\frac{w_g^2(0,r,t)}{2},$$
$$G_g(0,r,t)=u_b\rho_{gb},\ G_k(0,r,t)=\Delta_k u_b\rho_{gb}, \quad (15)$$
$$v_g(0,r,t)=v_k(0,r,t)=0,\ w_g(0,r,t)=w_k(0,r,t)=A_z(r/R),\ n_k=V_k\rho_k^0/(\Delta_k\rho_{gb}),\ a_{O2}(0,r,t)=a_{O2b}.$$

The boundary condition at the outlet of the pipe ($x = L_x$) is the atmospheric pressure.

In (1)–(15) the following notations are used: $\varepsilon_g=p_g/(\gamma-1)+0.5\rho_g\left(u_g^2+v_g^2+w_g^2\right)$ is the total energy of the gas, $\varepsilon_k=c_kT_k\rho_k+0.5\rho_k\left(u_k^2+v_k^2+w_k^2\right)$ is the total energy of the particles, $\alpha_k=Nu_k\lambda_g/(2r_k)$ is the gas-particle heat exchange coefficient, $\gamma=c_p/c_v$ is the heat capacity ratio, $\tau_{x,r,\phi}$ is the friction force along axial, radial and tangential directions, ρ is the density, u, v, w are the axial, radial and tangential velocity components, t is the time, r is the coordinate, p is the pressure, T is the temperature, A_z is the maximum tangential velocity, Δ_k is the particle volume fraction in the whole mixture, $r_k=\sqrt[3]{3\rho_k/(4\pi n_k\rho_k^0)}$ is the particle radius, Q is the heat of the reaction; k_0 is the chemical rate constant, $\alpha_r=\mu_{O2}\nu_{O2}/\mu_{gl}\nu_{gl}$ is the oxygen discharge ratio in the reaction with the coal dust particles, μ_{O2}, μ_C are the molar masses of oxygen and carbon; ν_{O2}, ν_C are the stoichiometric coefficients of the reactions; a_{O2} is the mass fraction of the oxidizer in the gas phase, $H_{k,g}$ is the enthalpy of the gas and particles, $G_{k,g}$ is the gas and particle flow rate. The following subscripts are used: b for the initial parameters, k for the particles parameters, g for the gas parameters, st for parameters at the inlet, $O2$ for oxygen.

The mass change rate of the particles during combustion is calculated from the expression: $G=n_kS_k\rho_{O2}\beta_mk_0e^{-E_a/R_uT_k}/\left[\beta_m+k_0e^{-E_a/R_uT_k}\right]$, where $\beta_m=\lambda_g(T_g)Nu_D/(c_g\rho_gr_g)$ is the mass-transfer coefficient. The friction force along the axial direction is equal to $\tau_{tr,x}=n_kF_{tr,x}$ [7], where $F_{tr,x}=C_{rx}S_m\rho_g\left(u_g-u_k\right)\left|u_g-u_k\right|/2$ is the interaction force of a single particle with gas, is the friction coefficient, $\mathrm{Re}_x=2\rho_gr_k\left|u_g-u_k\right|/\eta$ is the Reynolds number, S_m is the mid-sectional area, η is the dynamic viscosity of the gas. The components of the friction force along the radial and tangential directions were determined in a similar way.

The solution method of the problem is the same as in the study [7] and is based on the algorithms described in [8, 9]. The spatial grid steps along the axial and radial directions were set equal to 10^{-3} m, the time step was calculated from the Courant–Friedrichs–Lewy condition [8].

3. Results

We have carried out the numerical investigation on the features of the combustion front motion. We have varied the size and mass concentration of the particles and the constant A_z, which characterizes the flow swirl at the inlet. The aim of the parametric calculations is to determine the influence of the swirl on the combustion front motion in the coal dust suspension in the air and on the degree of particle burn-out. The thermo-physical and kinetic parameters used in the calculations has been taken from [7]. The calculation results are presented in Figures 1 - 6.

Figure 1 shows the initial stage of the combustion in case of unswirled flow, whilst Figures 2-3 presents a swirling flows according to the law $u_\varphi(0,r,t) = A_z(r/R)$, where R is the radius of the pipe and A_z is the maximum velocity along the angular component. The calculations have been performed for the suspension with the particle radius $r_k = 10^{-6} m$; the mass concentration of the particles supplied to the tube was equal to 5% of the supplied gas mass $\rho_{kb} = 0.05\rho_g$; the axial feed rate was $u_{xb} = 2 m/s$. The particles are assumed to move together with the gas, the velocity lag of the particles is slight.

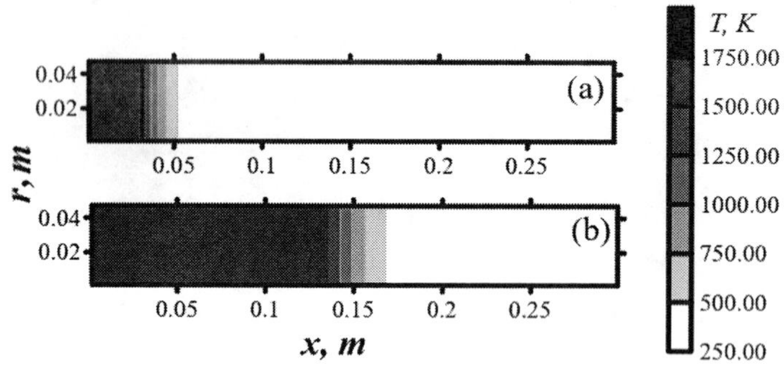

Figure 1. Particle temperatures at time $t = 0.005$ *s* (a), 0.015 s (b); $r_p = 10^{-6}$ *m*, $A_z = 0$ *m/s*, $u_{xb} = 2$ *m/s*, $\rho_{kb} = 0.05\ \rho_{gb}$.

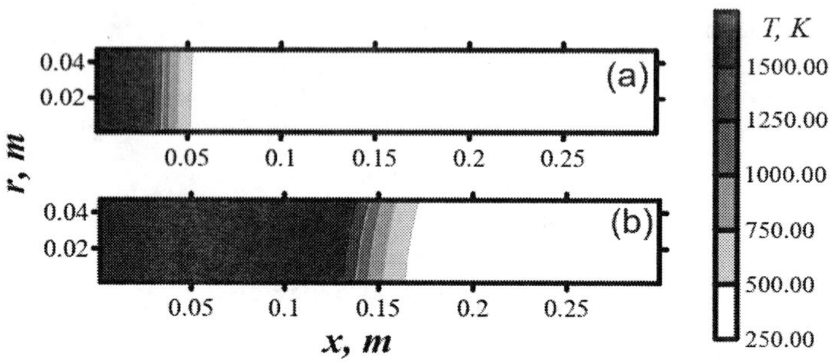

Figure 2. Particle temperatures at time $t = 0.005$ *s* (a), 0.015 s (b $r_k = 10^{-6}$ *m*, $A_z = 1$ *m/s*, $u_{xb} = 2$ *m/s*, $\rho_{kb} = 0.05\ \rho_{gb}$.

The effect of the combustion front inhomogeneity, arising due to the angular component of the flow velocity, can lead to the significant changes in the nature of the combustion wave propagation through the pipe. This effect can influence the combustion completeness of the suspension.

For the cases shown in Figures 3-4, the contours of the mass concentration of particles along the channel at the time $t = 0.015$ s have the form shown in Figure 4. According to Figure 4, there is a region of unburned particles, located around the inlet and near the wall.

In the case of the unswirled flow, the combustion front motion can be described by the one-dimensional formulation. Whilst a swirled flow provides the inhomogeneous temperature distribution along the radius of the pipe (Figures 2, 3). In Figures 2-3, the maximum swirl velocities at the inlet are 1 and 10 m/s, respectively. The presence of a swirl leads to a bending of the combustion front, with a subsequent increase in the angular component of the gas velocity at the inlet, which shifts the combustion front to the side surface of the tube. Therefore, there is a significant heterogeneity of the temperature field.

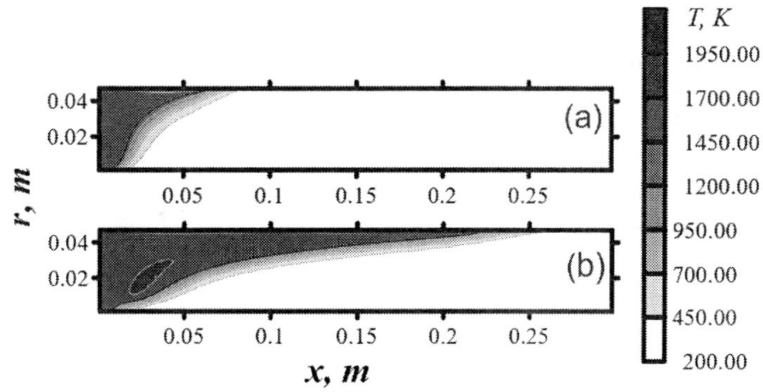

Figure 3. Particle temperatures at time $t = 0.005$ s (a), 0.015 s (b); $r_k = 10^{-6}$ m, $A_z = 10$ m/s, $u_{xb} = 2$ m/s, $\rho_{kb} = 0.05\,\rho_{gb}$

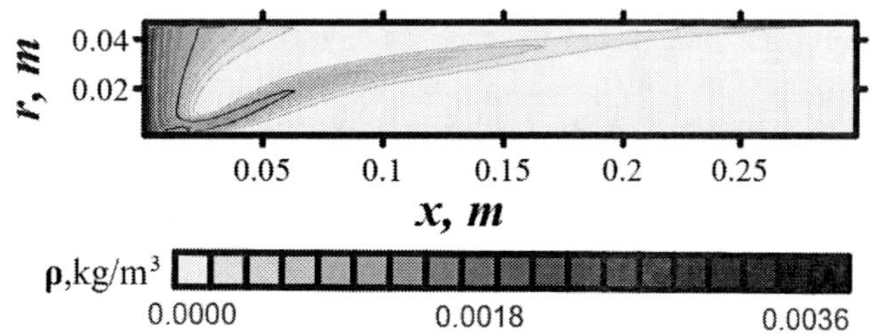

Figure 4. Particles mass concentration at time $t = 0.015$ s; $r_k = 10^{-6}$ m, $A_z = 10$ *m/s*, $u_{xb} = 2$ *m/s*, $\rho_{kb} = 0.05\,\rho_{gb}$

The conducted calculations shows that the swirling affects the temperature distribution in the pipe, therefore the temperature heterogeneity occurs during the steady state combustion. The steady-state temperature distribution is almost uniform along the pipe at the relatively low swirl speed ($A_z <= 0.5$ m/s). At high speeds, after the establishing the steady combustion regime, the fields of the gas temperature and particle mass concentration have the form shown in Figures 5-6.

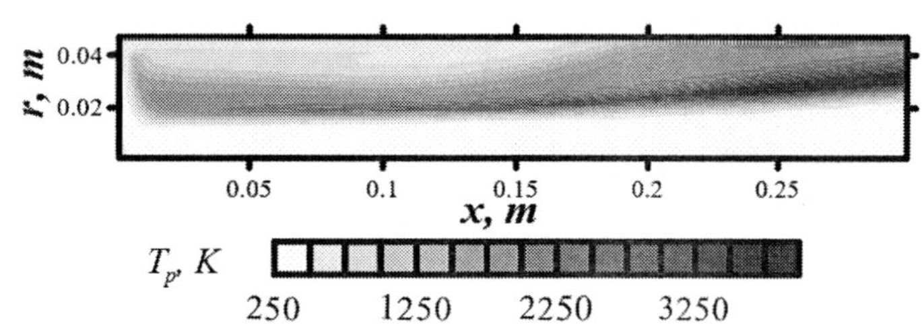

Figure 5. Particles temperature at time $t = 0.1$ *s*; $r_k = 10^{-6}$ m, $A_z = 10$ m/s, $u_{xb} = 2$ m/s, $\rho_{kb} = 0.05\,\rho_{gb}$

According to Figures 5-6, a region of the unburned particles occurs near the pipe wall and there are no particles on the axis of symmetry. The maximum gas temperature is observed at a distance from the inlet and pipe wall. There are no unburned particles at the outlet of the pipe. Thus, the final combustion of the particles takes place at the outlet region of the pipe.

We also have conducted the calculations with large particles, which have a major velocity lag, due to greater inertia. In this case, the effect of the swirling is weaker, since large particles are less carried away by the swirled flow. Thus, swirling is of great importance in the case of a finely-divided coal dust combustion suspended in the air.

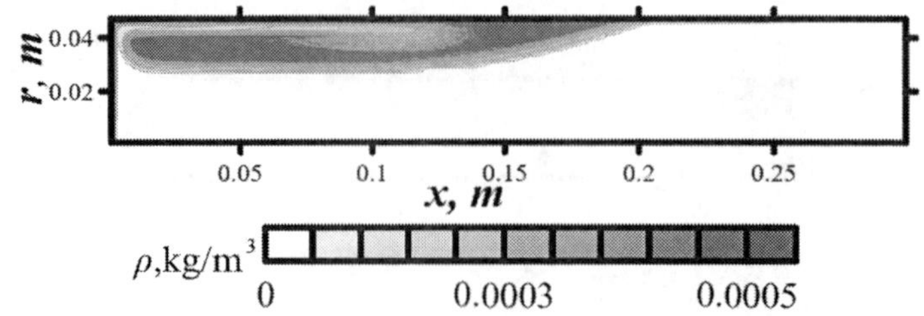

Figure 6. Particles mass concentration at time $t = 0.1$ s; $r_k = 10^{-6}$ m, $A_z = 10$ m/s, $u_{xb} = 2$ m/s, $\rho_{kb} = 0.05\ \rho_g$.

4. Conclusion

We have developed a physical-mathematical model of coal dust combustion in the swirled airflow in a pipe. The model is axisymmetric and the swirling is taken into account by the tangential component of the velocity vector. Using the model, we have conducted a numerical study on the influence of the swirling speed on the gas temperature field inside the pipe and on the completeness of coal dust combustion. The obtained data shows that for the small particles, which is carried away by a swirling flow due to the low inertia, the presence of the swirling leads to the formation of a region with unburned particles at the inlet of the pipe and near the wall. The indicated features of the coal dust combustion during flow swirling can effect significantly on the flame propagation regimes in the suspension.

References

[1] Piralishvili Sh A, Markovich D M, Lobasov A S and Vereshchagin I M 2016 *J Eng Phys Thermophys* **89** 1339-46
[2] Mikhailov A S, Piralishvili S A, Evdokimov O A, Stepanov E G and Spesivtseva N S 2018 *J Eng Phys Thermophys* **91** 925-32
[3] Matvienko O V and Bubenchikov A M 2016 *J Eng Phys Thermophys* **89** 127-34
[4] Kuznetsov V A, Dekterev A A, Sentyabov A V and Chernetskiy M Y 2016 *J Siberian Federal University. Engineering & Technologies* **9(1)** 15-23
[5] Gavrilov A A, Dekterev A A and Sentyabov AV 2015 *Fluid Dynam* **50** 471-82
[6] Nigmatulin R I 1990 *Dynamics of Multiphase Media* (CRC Press; Rev Sub edition)
[7] Moiseeva K M and Krainov A Y 2018 *Combust. Explos. Shock Waves* **54** 179-88
[8] Godunov S K, Zabrodin A V, Ivanov M Ya, Kraiko A N and Prokopov G P 1976 *Numerical Solution of Multi-Dimensional Problems of Gas Dynamics* (Moscow: Nauka)
[9] Kraiko A N 1979 *J Appl Math Mech* **43** 539-49

Effect of a smooth temperature gradient on the combustion of a preliminary mixed methane-air mixture

A A Ponomareva[1,2], S N Mokrin[1,2], G V Uriupin[2] and S S Minaev[1]

[1] Institute of Applied Mathematics FEB RAS, Vladivostok, Russia
[2] Far Eastern Federal university, Vladivostok, Russia

ap_k@inbox.ru

Abstract. In this work, the ranges of different combustion modes of a methane-air mixture were studied in dependence on the flow velocity under conditions of a smooth temperature gradient of the channel. To carry out the experiments, a laboratory setup was designed and constructed to obtain a smooth temperature gradient in a quartz tube, and experiments were performed to determine the temperature gradient. The boundaries of different combustion regimes were determined for the mixtures with different fuel-air ratio coefficients (0.8, 1, 1.2), moreover, transient modes between main regimes were found.

1. Introduction

Obtaining data on the oxidation kinetics of various fuels is important for the development of new combustion technologies. Characteristics of fuel combustion were previously studied in systems such as a perfect mixing reactor [1], a flow reactor [2, 3], shock tubes [4-6] and special "rapid compression machines" [7–10]. Currently, a microflow reactor is widely used to verify the kinetic models of combustion of various fuels [11–13]. The reactor is usually a quartz tube with an inner diameter less than critical one for a combustible mixture calculated from ambient temperature. A quartz tube is heated by an external heat source, which forms a stationary temperature gradient along the gas flow in the tube. The mixture of fuel and oxidizer is fed into the tube from the cold end. It is moving into the microchannel and rapidly heating up from the walls due to the small diameter of the tube. When a certain temperature is reached the mixture ignites and stabilizes at a certain place in the quartz tube depending on the flow rate (one of the modes). With the continuous heating of tube's walls it is possible to sustain combustion in a tube having an internal diameter smaller than the ordinary quenching diameter, even in a case of lean gas-air mixture [11-15]. The results of [13, 14] showed that, despite the small diameter of the reactor, reactions on the surface of the quartz tube can be neglected.

During the study of a flame behavior inside such reactors three combustion modes were revealed: a stable flame for high flow velocities, an unstable combustion regime with periodic ignition and extinction (FREI) in the intermediate flow rate range, and a weak flame at low flow velocities [11-15]. It was also found that the temperature gradient has a strong effect on flame behavior [13, 15-17].

Numerical simulation of the behavior of flames in a heated microchannel with the wall temperature smoothly increasing in the downstream direction demonstrated that at very small values of the temperature gradient and at a low gas-air ratio ($\phi \leq 0.8$) flame stabilization is possible in the entire velocity range [16]. Moreover, as the theoretical investigation shown the smooth temperature gradient could enlarge the distance between ignition points and extinction point in FREI regimes up to several centimeters.

In this work, we studied the behavior of a pre-mixed methane-air mixture in a narrow-heated channel in the speed range from 5 to 150 cm/s in a setup with a new type of reactor of the conical spiral form.

2. Experimental details

A new type of microreactor was done using a quartz tube ~ 100 cm long with an inner diameter of 2 mm and an outer diameter of 4 mm. The tube was twisted into a conical spiral. The diameter of the base of the spiral is 80 mm, the height is 60 mm. The geometric parameters of the spiral were selected so that the turns of the spiral did not overlap each other in the frontal and lateral projections. A flat cylindrical burner was installed at the base of the spiral, operating on a pre-mixed methane-air mixture. Thus, it was possible to construct a compact microreactor with a very smooth temperature gradient, in which the maximum temperature is reached at the top of the spiral.

The temperature distribution (Fig. 1) inside the reactor was studied using a NEC TH9100WB thermal imager, which was placed on the top of the same axis with a spiral tube and the center of an external cylindrical burner. The measurements of the thermal imager were corrected according to the data obtained using a K-type thermocouple. The tip of the thermocouple was in a direct contact with the heated wall and was hold in this position for several minutes in order to obtain an average temperature.

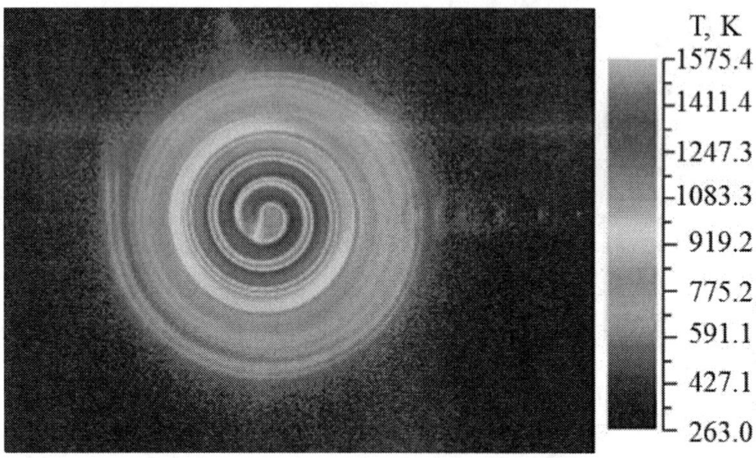

Figure 1. The temperature distribution in a spiral-type microflow reactor obtained using a thermal imager.

A study of the flame behavior in a heated microchannel with the wall temperature smoothly increasing in the downstream direction was performed in the velocity range from 5 to 150 cm/s. The flow inside the tube was laminar under atmospheric pressure. The position of the flame front was recorded using a Nikon D7200 digital camera. The camera captured the image from the mirror, which was installed at an angle above the reactor in order to display a top view. This recording method allows performing an easier estimation of the flame position using a polar coordinate grid. From the images the angle of the flame position was determined in polar coordinates centered at the beginning of the spiral (maximum temperature). Then, using the common-used formula for a conical spiral (1), the distance to the point with the maximum temperature (that was chosen as the zero point) was determined.

$$L(\eta) = \frac{1}{2\beta}\left(\eta(a^2+\eta^2)^{\frac{1}{2}} - r_1(a^2+{r_1}^2)^{\frac{1}{2}}\right) + a^2 \ln\left(\frac{\eta+(a^2+\eta^2)^{\frac{1}{2}}}{r_1+(a^2+{r_1}^2)^{\frac{1}{2}}}\right) \quad (1)$$

where $a^2 = \alpha^2 + \beta^2$; $\beta = \frac{r_2 - r_1}{2\pi}$; $\alpha = \frac{h_1}{2\pi}$;
r_1 and r_2 are the radii of the first and second turns, η is the radius of the spiral for the calculated point; h_1 is the distance between the turns along the vertical axis.

3. Results and Discussion

In the experiment, the temperature of the tube varies from 500 to 1350K, and the length of the working area was no more than 85 cm. The temperature gradient was calculated by processing images obtained by the thermal imager. As can be seen from Figure 2, the temperature change has an almost linear dependence on the length of the tube of about 50 cm. As further experiments showed, this particular segment of the tube is the main reaction zone where all visible effects can be seen. In the further section of the tube, the temperature change is even smoother.

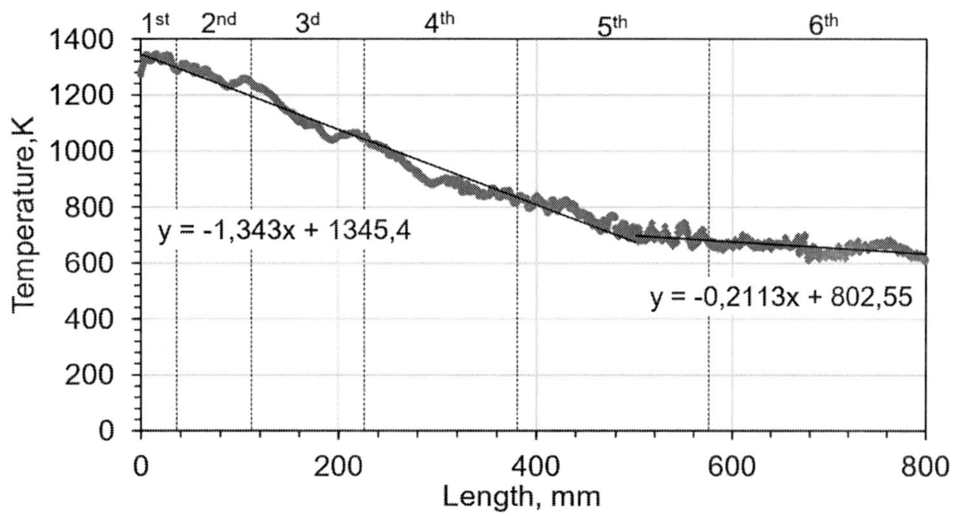

Figure 2. Experimental obtained data of temperature change of spiral microflow reactor (the dashed lines point out of the ends of marked turns).

When studying a stoichiometric methane-air mixture, the boundaries of two main combustion modes were found: a stable flame was observed at flow velocities from 39 to 150 cm/s (the A zone on fig. 3), the FREI mode was from 9 to 37 cm/s (the B zone on fig. 3). The detection of weak flame at low flow velocities was not included in this study. To visualize a weak flame, it is planned to use special methods using optical filters. That is the main goal of future experiments. We also observed transient effects between the main modes. At a speed of 38 cm/s, the flame ignites at the upstream, slowly moves downstream (as in stable mode until the flame stabilizes), and then begins to oscillate (as in FREI mode). This effect was previously observed and described in papers [15, 18]. At the velocity of 8 cm/s, the flame oscillates several times (as in FREI mode), disappears, and for some time there is no visible flame (perhaps there is a weak flame that we can see without filters). The effects have been repeated continuously and shown a temporary dependence. It should be noted that due to the high luminosity of the first two turns of the spiral and the insufficient recording speed of the used camera, difficulties arose in determining the ignition points in the FREI modes. In Fig. 3. The first visible flame positions that have appeared on the recorded videos are indicated. Thus, the ignition zone is marked on the graph.

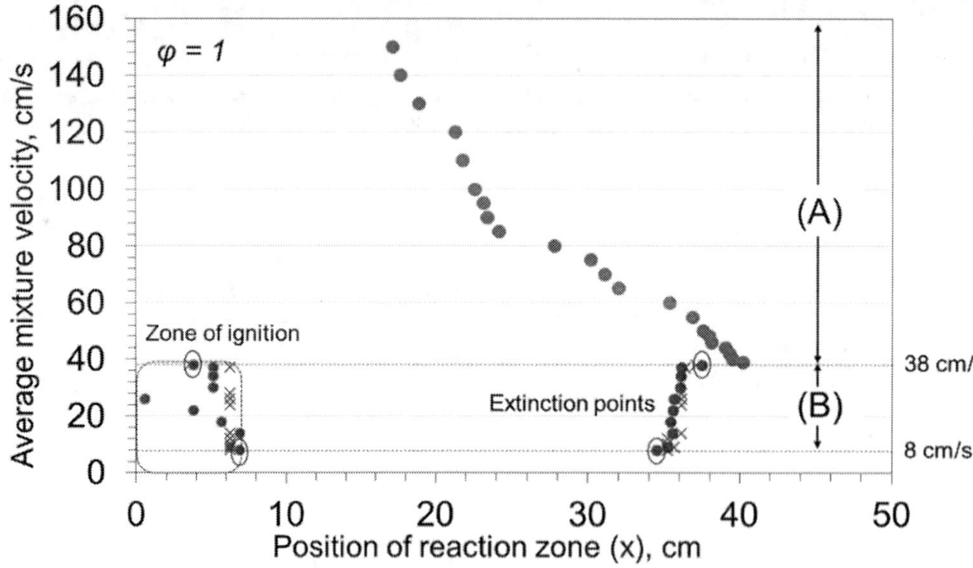

Figure 3. The position of the reaction zone depending on the velocity of the mixture (experimental results); ovals mark the extreme positions of the flame during transient effects.

As experiments for the mixture with φ=1.2 showed, the lower boundaries of the main combustion regimes shifted in the direction of higher velocities: a stable flame was observed at flow velocities from 41 to 150 cm/s, the FREI mode was from 15 to 39 cm/s. The transient effects like as for the mixture with φ=1 were also detected at 40 cm/s and 14 cm/s, respectively. In addition, the distance between the ignition and extinction points also changed significantly and became almost two times less (from 28 cm (see Fig. 3) to 13 cm).

Investigation of the flame behavior in case of methane-air lean mixture with φ=0.8 demonstrated that the stable flame regime had the same boundaries as for the stoichiometric mixture. On the other hand, the FREI regime became to be wider, up to 6 cm/s. The distance between the ignition and extinction points varies from 32 cm (at higher velocities) to 20 cm (at the boundary between the FREI and a weak flame regimes). Concerning to the transient effects, we could observe only one of them: between stable flame regime and FREI regime at 38 cm/s. It was not possible to detect the transient effect at low velocities probably due to the large measurement interval (velocities were changing with the step of 1 cm/s).

4. Concluding Remarks

New construction of microreactor with smooth temperature gradient was developed. The spiral shape of microtube allowed us to achieve almost linear temperature gradient in temperature range from 500 to 1300 K. Thus, the resolution and sensitivity of experimental setup became higher. As a result, we found the features of flame behaviour for different methane-air mixture as well as the transition effects between stable flame and FREI regimes and between FREI regime and weak flame.

References

[1] Dagaut P *et al* 2009 *Proc. Combust. Inst.* **32** 229
[2] Mueller M A *et al* 1999 *Int. J. Chem. Kinet.* **31** 113
[3] Fischer S L *et al* 2000 *Int. J. Chem. Kinet.* **32** 713

[4] Gauthier B M *et al* 2004 *Combust.* Flame **139** 300
[5] Chaos M and Dryer F L 2010 *Int. J. Chem. Kinet.* **42** 143
[6] Burcat A *et al* 1971 *Combust. Flame* **16** 29
[7] Livengood J C and Wu P C 1955 *Symp. Combust.* **5** 347
[8] Griffiths D F *et al* 1993 *Combust. Flame* **95** 291
[9] Minetti R *et al* 1994 *Combust. Flame* **96** 201
[10] Kizaki Y *et al* 2015 *Proc. Combust. Inst.* **35** 3389
[11] Yamamoto A *et al* 2011 *Proc. Combust. Inst.* **33** 3259
[12] Hori M *et al* 2012 *Combust. Flame* **159** 959
[13] Suzuki S *et al* 2013 *Proc. Combust. Inst.* **34** 3411
[14] Oshibe H *et al* 2010 *Combust. Flame* **157** 1572
[15] Di Stazio A *et al* 2016 *Exp. Therm. Fluid Sci.* **73** 79
[16] Mazurok D B *et al* 2014 *Combust. Explo. Shock+* **50** 25
[17] Grajetzki P *et al* 2018 *Comb. Sci. and Tech.* **190** 1950
[18] Tsuboi Y et al 2007 *Proc. ASME Int. Mechanical Engineering Congress and Exposition* (Seattle, Washington) IMECE2007-43339 p 155

Journal of Physics: Conference Series **1404** (2019) 012065 doi:10.1088/1742-6596/1404/1/012065

Three-dimensional numerical simulation of continuous spin detonation in hydrogen-oxygen and hydrogen-air mixtures using OpenFOAM package

A I Rybnikov, E V Simonov, A M Gurin, A V Trilis and A N Samsonov

Lavrentyev Institute of Hydrodynamics SB RAS,
15 Lavrentyev av., Novosibirsk, 630090, Russia

alexey_rybnikov@ngs.ru

Abstract. A mathematical model of continuous spin detonation for hydrogen-oxygen and hydrogen-air mixtures is developed within the framework of a nonstationary gas-dynamic formulation. An OpenFOAM package is used to carry out numerical calculations in an annular cylindrical combustor in order to compare the main solution parameters with experimental data for geometric dimensions of combustors 40 mm and 306 mm in diameter corresponding to the experiments for hydrogen-oxygen and hydrogen-air mixtures. Regimes with a single continuously rotating transversal detonation wave are obtained for given fuel mixtures. It is shown that the solution significantly depends on the coordinate along the radius. The results for the hydrogen-oxygen mixture are in good agreement with an experiment in terms of a wave velocity and satisfactory agreement relative to a front height and the number of waves (a 1.5- to 2-fold difference). The results for the hydrogen-air mixture are in satisfactory agreement with an experiment for the wave velocity and front height and significantly differ from it on a minimal flow rate of the mixture, in which the existence of continuous spin detonation is possible in the combustor.

1. Introduction

Continuous spin detonation (CSD) is currently under experimental and numerical studies based on a scheme proposed by B.V. Voitsekhovskii in 1960s in [1, 2]. Detonation regimes with CSD are described in detail in annular combustors (ACs), such as a liquid propellant engine (LPE) [3] and flow-type combustors [4, 5] for a series of fuel-oxygen and fuel-air mixtures.

The purpose of this study is to perform the three-dimensional nonstationary formulation of a closed mathematical model of CSD for hydrogen-oxygen and hydrogen-air mixtures, use the OpenFOAM package (www.openfoam.org) for developing a universal computational algorithm for three-dimensional simulation of the CSD of the above-mentioned mixtures in a wide range of original parameters for different types and sizes of ACs, numerically investigate the combustion dynamics of the hydrogen-oxidizer mixtures in transversal detonation waves (TDWs) in two annular ACs of the LPE type for two kinds of oxidizers (oxygen and air), and verify the constructed mathematical model by comparing the obtained results with experimental data [3, 4].

2. Formulation of the problem

2.1. Mathematical model

Within the framework of this model, a perfectly stirred mixture of hydrogen and oxidizer is injected into an AC with a specific flow rate g_{in} via Laval micronozzles uniformly distributed over the entrance of the AC. The latter is a space bounded by two cylinders with an annular gap of width Δ. The AC length is L_{ch} and its outer diameter is D_{ch}. The mixture is burned, and detonation products are released into the space with backpressure $P_{backpressure}$ and flow rate g_{out}. The gas-dynamic flow of the reacting hydrogen-oxidizer mixture in the solution domain is described by a three-dimensional system of equations of nonstationary gas dynamics with chemical transformations. At an initial instance, the energy required for detonation initiation is released in the small part of the AC.

$$\begin{gathered}\frac{\partial\rho}{\partial t}+\mathrm{div}\rho\vec{U}=0\\ \frac{\partial\rho\vec{U}}{\partial t}+\mathrm{div}\rho\vec{U}\vec{U}+\mathrm{grad}P=0\\ \frac{\partial\rho E}{\partial t}+\mathrm{div}\rho E\vec{U}=0\\ \frac{\partial\rho\mu}{\partial t}+\mathrm{div}\rho\mu\vec{U}=F_\mu\\ \frac{\partial\rho I}{\partial t}+\mathrm{div}\rho I\vec{U}=F_I\end{gathered}\tag{1}$$

Here t is the time, ρ is the density, $\vec{U}$ is the velocity vector, P is the pressure, $E = U_p + |\vec{U}|^2/2$, U_p is the total internal energy of the gas, μ is the current molecular mass of the mixture, and I is the fraction of the chemical induction period.

The energy release is described by a two-stage kinetics model [6] using the Nikolaev chemical kinetics [7].

The first stage is the induction stage ($0 < I < 1$): $F_I = -\frac{1}{t_{ind}}$, $F_\mu = 0$. Here t_{ind} is the induction time for the corresponding mixture. Similarly to [3, 5], the induction time is described by the expression:

$$t_{ind}=\frac{K_{ind}}{[C]}\exp\left(\frac{E_{ind}}{RT}\right)\tag{2}$$

Here $[C]=\frac{\rho v_{O_2}}{\sigma}\cdot 1000$ is the oxygen concentration in [mol/m^3], E_{ind} = 17.15 kcal/mol is the activation energy, $K_{ind} = 5.38\cdot10^{-8}$ (mol·s/m^3) is the pre-exponent, R is the universal gas constant, v_{O2} is the number of moles of the oxygen, $\sigma=\sum\mu_i v_i$, μ_i denotes the molar masses of the mixture components (H_2, O_2, N_2, Ar), and v_i is the number of moles of the components.

The second stage is the energy release [7] (I = 0):

$$F_\mu=\frac{4K_+\rho^2}{\mu}\left(1-\frac{\mu}{\mu_{max}}\right)^2-4K_+K_-\rho\left(\frac{\mu}{\mu_{min}}-1\right)\left(\frac{T}{T_0}\right)^{\frac{\beta}{2}}\left(1-\exp\left(-\frac{\theta}{T}\right)\right)^{\beta}\exp\left(-\frac{E_d}{RT}\right),$$

$$U_p=U_{th}+U_{ch},\ U_{th}=\frac{RT}{\mu}A(\mu,T),\ U_{ch}=E_d\left(\frac{1}{\mu}-\frac{1}{\mu_{min}}\right),\tag{3}$$

IOP Publishing
1404 (2019) 012065 doi:10.1088/1742-6596/1404/1/012065

$$A(\mu,T)=\frac{3}{4}\left(\frac{\mu}{\mu_a}+1\right)+\frac{3}{2}\left(\frac{\mu}{\mu_a}-1\right)\frac{\theta/T}{\exp(\theta/T)-1}$$

Here μ_{min}, μ_{max}, and μ_a are the molar masses of gas in ultimately dissociated, ultimately recombined, and atomic states, respectively; θ is the effective temperature of excitation of vibrational degrees of freedom of molecules; E_d is the mean dissociation energy of the reaction products.

Constants of the model: μ_{H2} = 2 kg/kmol, μ_{O2} = 32 kg/kmol, μ_{N2} = 28 kg/kmol, μ_{Ar} = 40 kg/kmol, ν_{H2} = 2, ν_{O2} = 1, $R = 8.3144\cdot10^3$ J/(kmol·K), $K_+ = 6\cdot10^8$ $m^6/(kmol^2\cdot s)$, E_d = 109.68 kcal/mol, β = 1.5, θ = 3500 K. For hydrogen-oxygen mixture: K_- = 1769 $kmol/m^3$, μ_{min} = 6 kg/kmol, μ_{max} = 18 kg/kmol, μ_a = 6 kg/kmol, ν_{N2} = 0, ν_{Ar} = 0. For hydrogen-air mixture: K_- = 3500 $kmol/m^3$, μ_{min} = 14.54 kg/kmol, μ_{max} = 24.62 kg/kmol, μ_a = 10.53 kg/kmol, ν_{N2} = 3.7275, ν_{Ar} = 0.0445.

2.2 Boundary conditions.

The condition of the mixture inflow from the Laval micronozzles is specified at the combustor entrance, and the backpressure $P_{backpressure}$ is given at the combustor exit. Boundary parameters are calculated similarly to [3, 5].

2.3 Initial conditions.

A motionless combustible mixture is set as initial data in the AC.

3. Numerical results

The problem is solved numerically for the stoichiometric hydrogen-oxygen and hydrogen-air mixtures for the corresponding geometric dimensions of the AC and the parameters in the mixture injection system. The solution domain is covered by a stationary grid with uniform cell steps along all axes. The number of cells is 192 000 for the hydrogen-oxygen mixture and 240 000 for the hydrogen-air mixture. The Godunov finite-difference scheme is applied in the calculations with the help of the OpenFOAM package.

3.1 Geometric parameters

For the case where oxygen is used as an oxidizer, the geometric parameters of the AC are taken from [3] (D_{ch} = 4 cm, Δ = 0.5 cm, and L_{ch} = 4 cm); for air, the parameters are taken from [4] (D_{ch} = 30.6 cm, Δ = 2.3 cm, and L_{ch} = 66.5 cm).

3.2 Injection system parameters

The flow rate at the combustor entrance is set in accordance with the experiments [3, 4]: $g_{in} \approx 46$ $kg/s/m^2$ and $g_{in} \approx 75$ $kg/s/m^2$ for the calculations with oxygen and air, respectively. At the combustor exit, the backpressure is set as $P_{backpressure}$ = 0.2 atm for oxygen and 1 atm for air.

3.3 Detonation initiation

At the initial time (t = 0), the energy sufficient for detonation initiation is released in one section of the domain. Thus, a TDW begins propagating over the AC. In order to ensure that the TDW propagates only in a single direction (counterclockwise), the solution domain is temporarily blocked by a vertical solid wall.

3.4 Calculations in the hydrogen-oxygen mixture

The calculations of the initiation and propagation of the TDW for the stoichiometric hydrogen-oxygen mixture show that, for $g_{in} \approx 46$ $kg/s/m^2$, by the time $t \approx 8.2$ msec after the initiation, the CSD with a single TDW becomes quasystationary and self-supporting with a velocity $D \approx 2.07$ km/s and a front height of $\approx$ 3.2 cm. Figure 1 illustrates a three-dimensional temperature distribution (K) in the combustor (a) and horizontal (b) (at a distance of 1 cm from the entrance) and vertical (c) (at a distance of 1 cm behind the TDW) cross sections of temperatures in the AC. Clearly, the TDW moves around the annular space of the combustor in a counterclockwise direction and occupies most of the AC by height. An oblique shock wave (tail) moves downstream from the TDW. Figure 1.b clearly shows the dependence of the solution on radius: the injected mixture inflows nonuniformly, filling up the space closer to the outer diameter of the combustor earlier than to the inner diameter. Directly behind the TDW (figure 1.c), the mixture stops inflowing, while the height of the layer of the fresh mixture at a distance of half the combustor perimeter is about half the AC height.

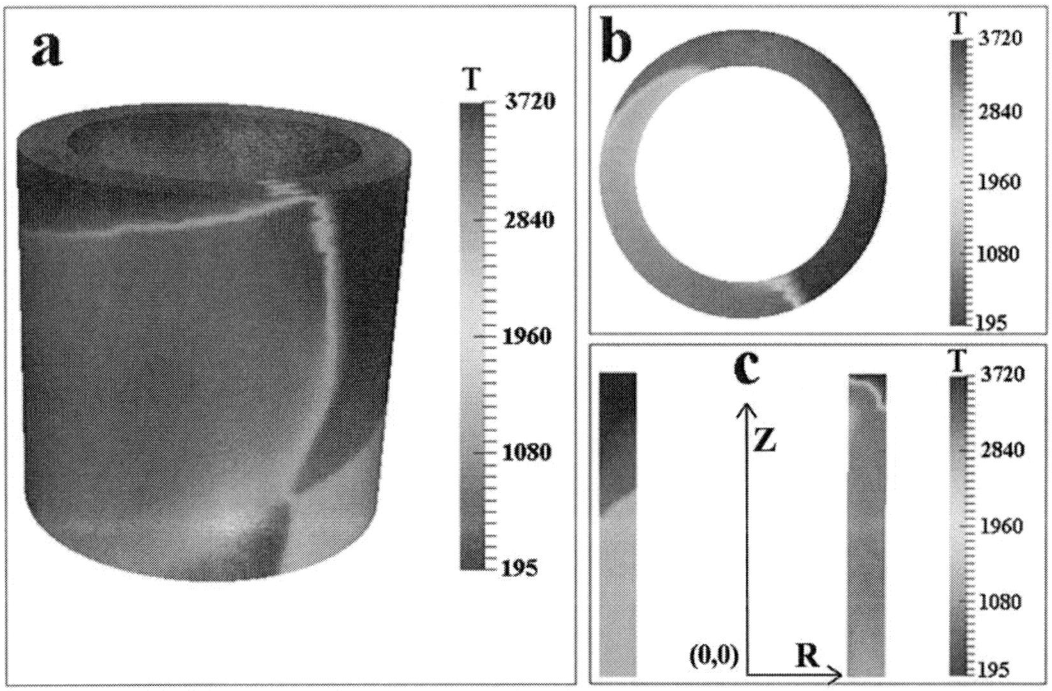

Figure 1. Calculated temperature distribution (K) in the AC for $g_{in} \approx 46$ kg/s/m^2 at the time $t \approx 8.2$ msec.

The minimal flow rate g_{min} of the existence of CSD is determined by conducting calculations with decreasing flow rate of the mixture injection g_{in}. It is shown that, as the flow rate drops below $g_{in} \approx 40$ kg/s/m^2, the CSD is stopped. This means that the solution with a single TDW, obtained for $g_{in} \approx 46$ kg/s/m^2, is physically admissible because it is located in the vicinity of g_{min} from the standpoint of the flow rate.

3.5 Calculations in the hydrogen-air mixture

Similarly to the calculations from the previous part of the study, the CSD of the hydrogen-air mixture is performed. Figure 2 shows the pressure-time dependence at a point $R = 15$ cm, $Z = 64$ cm. Periodic pressure pulsations associated with the arrival of the TDW at this point are visible. Knowing the time interval Δt between the pulsations and the AC perimeter $= D_{ch} \cdot \pi$, one can calculate the detonation velocity $D \approx 1.57$ km/s. With time, the flow rate g_{out} at the AC exit is compared with $g_{in} \approx 75$ kg/s/m^2, which indicates that the TDW becomes quasistationary. The height of the TDW front is ≈ 14 cm. The structure of the TDW for the hydrogen-air mixture is similar to that obtained for the hydrogen-oxygen mixture. As in the calculations with oxygen, the injected mixture flows nonuniformly along the radius.

IOP Publishing
1404 (2019) 012065 doi:10.1088/1742-6596/1404/1/012065

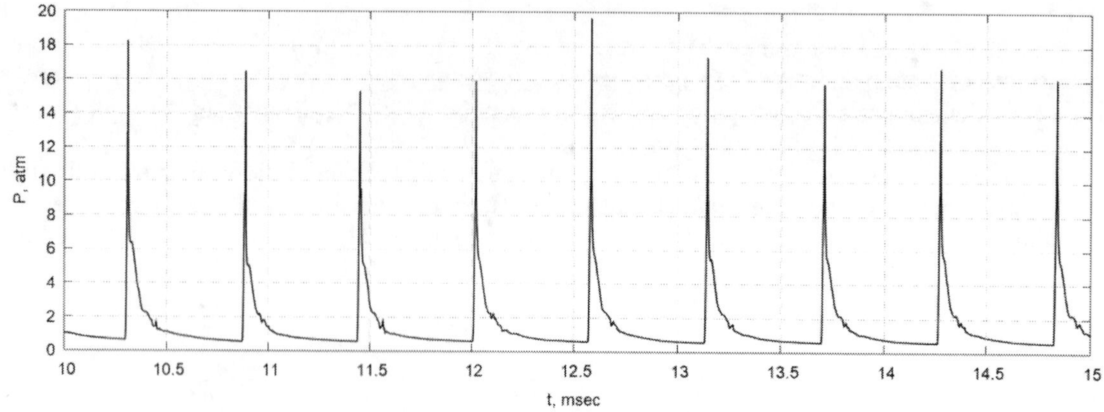

Figure 2. Pressure P versus the time t.

Figure 3 and 4 below illustrate the dependences of the velocity and rotation frequency of the TDW on the specific flow rate of the mixture (green points). Red points indicate a disruption of the TDW and the end of the CSD regime. Clearly, the velocity drops along with the flow rate, which can be related to a decrease in the height of the inflowing layer, thereby reducing the energy release behind the TDW front.

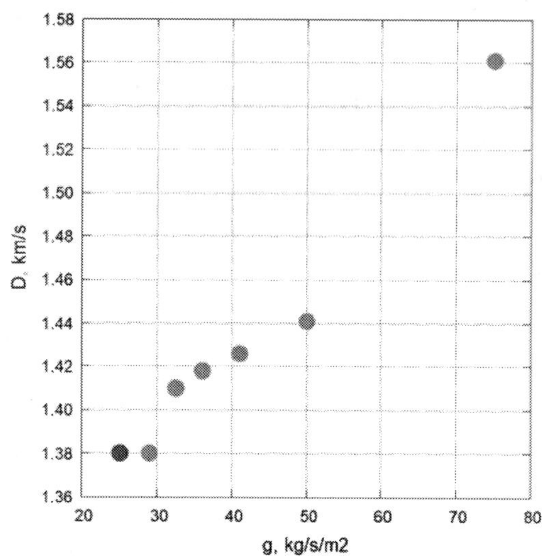

Figure 3. TDW velocity D (km/s) versus the flow rate (kg/s/m^2).

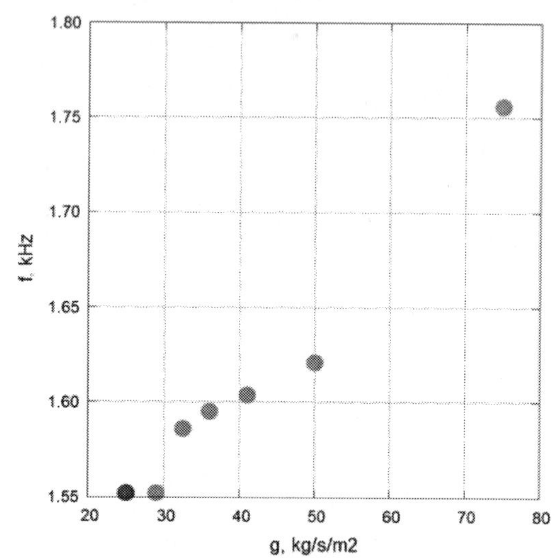

Figure 4. TDW rotation frequency f (kHz) versus the flow rate (kg/s/m^2).

4. Results and Discussion

The calculation results for the hydrogen-oxygen mixture are in good agreement with the experiments on the TDW velocity [3] and in satisfactory agreement in terms of front height and the number of waves (in the experiment, with this flow rate of the mixture, a two-wave regime with a front height lower than 2-2.5 times is implemented). Satisfactory agreement is obtained for the hydrogen-air mixture in terms of the TDW velocity (the estimated velocity is 25% higher than that in the experiments) and front height (40% lower than the experimental one). There is a significant difference from the experiment in the minimum flow rate g_{min} of the hydrogen-air mixture, with which CSD can exist in the combustor. In the experiment [4], $g_{min} \approx 46$ kg/s/m^2; in this calculation, $g_{min} \approx 25$-30 kg/s/m^2, indicating an approximately 1.5-fold difference from the experimental data. A significant decrease in g_{min} relative to the experiments is presumably related to a simplified mixture injection system in the presented mathematical model.

The analysis shows that three-dimensional effects strongly influence the inflow of the fresh mixture, which induces the formation of a complex shape of the contact boundary between the detonation products and the cold mixture. The fact that the fresh mixture nonuniformly fills the space along the radius can reduce the energy release behind the TDW front, thereby causing a decrease in the TDW propagation velocity in comparison with the quasi-three-dimensional calculation [5]. Thus, the constructed mathematical model and the algorithm for solving the three-dimensional problem allow describing the CSD in combustors of the LPE type and analyze three-dimensional effects that are not observed in two-dimensional calculations.

5. Conclusions

A closed mathematical model of continuous spin detonation is described in a three-dimensional nonstationary formulation for hydrogen-oxygen and hydrogen-air mixtures. The OpenFOAM package is used for these mixtures for the combustor sizes and parameters in the mixture injection system corresponding to the experiments to numerically obtain the solution with a single transverse detonation wave. Comparison with the experimental data is performed. For the hydrogen-oxygen mixture, there is good agreement with the experiment in terms of the wave velocity and satisfactory agreement in terms of the number of waves and the front height. For the hydrogen-air mixture, there is satisfactory agreement related to the wave velocity and front height and the 1.5-fold difference in the minimum flow rate g_{min} of the existence of the continuous spin detonation. For both mixtures, the significant three-dimensional effects that influence the processes in the combustor are revealed.

Acknowledgements

The study was financially supported by the Russian Foundation for Basic Research and the Novosibirsk Region within the framework of Grant No. 19-41-543009 and partially funded as part of the budget project of the Lavrentyev Institute of Hydrodynamics SB RAS No. III.22.2.1.

References

[1] Voitsekhovskii B V 1959 Steady detonation *Dokl. Akad. Nauk SSSR* **129** 1254-1256
[2] Bykovskii F A and Zhdan S A 2013 *Continuous Spin Detonation* (Novosibirsk: Izd. Sib. Otd. Ross. Akad. Nauk) p 423
[3] Bykovskii F A, Zhdan S A and Vedernikov E F 2008 Continuous spin detonation of hydrogen-oxygen mixtures. 1. Annular cylindrical combustors *Combust., Expl., Shock Waves* **44** 150–162
[4] Bykovskii F A, Zhdan S A and Vedernikov E F 2006 Continuous spin detonation of fuel-air mixtures *Combust., Expl., Shock Waves* **42** 463–471
[5] Zhdan S A, Rybnikov A I and Simonov E V 2018 Calculation of continuous spin detonation in a hydrogen-air mixture in an annular combustor *AIP Conference Proceedings* **2027** 030007
[6] Levin V A and Korobeinikov V P 1969 Strong explosion in a combustible gas mixture *Fluid Dynamics* **4** 30-32
[7] Nikolaev Yu A and Zak D V 1988 Agreement of models of chemical reactions in gases with the second law of thermodynamics *Combust., Expl., Shock Waves* **24** 461–463

Experimental study of the low-melting hydrocarbons regression rate in the air flow

A N Shiplyuk[1], V I Zvegintsev[1], S M Frolov[2], D A Vnuchkov[1], S V Lukashevich[1] and D G Nalivaychenko[1]

[1] Khristianovich Institute of Theoretical and Applied Mechanics SB RAS, 4/1 Institutskay st., Novosibirsk 630090, Russia
[2] Semenov Federal Research Center of Chemical Physics, Moscow 119991, Russia

vnuchkov@itam.nsc.ru

Abstract. Gasification of organic and inorganic materials in a high-temperature gas flow is a promising technology for various industrial applications like chemical industry, waste processing, rocket and air-breathing propulsion, etc. However, the characteristics of the gasification process of low-melting hydrocarbon materials remain insufficiently studied so far. In this work, experiments on the gasification of cylindrical polypropylene (PP) samples by the airflow passing through multiple longitudinal channels 3 to 4 mm in diameter and 200 mm long are performed. The temperature, pressure, and velocity of the airflow in the experiments range from 300 to 1500 K, 0.35 to 1.38 MPa, and 17 to 100 m/s. As expected, oxygen available in the airflow interacts with PP and creates an additional heat source for PP gasification, thus, enhancing the gasification process. At an airflow temperature of 1480 K, the maximum mass flow rate of gasification products is shown to attain 17.7 g/s. The minimum ratio of the mass flow rates of the airflow and gasification products reaches the value of 2. In tests with airflow temperature of 300 K, the maximum mass flow rate of gasification products is 8.5 g/s. The minimum ratio of the mass flow rates of airflow and gasification products is 5 in these experiments.

1. Introduction

The concept of solid-fuel ramjets is currently associated mainly with the scheme of a ducted ramjet engine or a Throttleable Ducted Rocket (TDR) [1-2]. A mandatory element of the TDR is a gas generator (GG) aimed to gasify a solid fuel with a minimum amount (10–20%) of an oxidizer. Such an amount of oxidizer allows gasification of solid fuel due to its partial combustion in the GG and feeding the resulting gaseous products to the main ramjet combustion chamber, where they are completely burned in a high-speed flow of atmospheric air.

In the recent scientific literature, another relevant direction is discussed actively, that is the development of a low-temperature GG for gasification of low-melting fuels (LMF) like polymers (polyethylene, polypropylene (PP), polystyrene, polybutadiene, etc.) with a melting point of about 100 °C [3-7]. In the cited references, the design of a two-chamber GG with a sequential arrangement of a conventional solid propellant and LMF is studied numerically. In this design, burning of solid propellant generates a flow of high-temperature combustion products, which passes through the porous layer of LMF and gasifies it. In contrast to LMF gasification by a flow of combustion products, the use of air as a carrier gas could be considered more beneficial due to the possibility of obtaining an additional heat source for LMF gasification by partial burning of the LMF material.

Journal of Physics: Conference Series **1404** (2019) 012066 doi:10.1088/1742-6596/1404/1/012066

The objective of this work is to study experimentally the gasification of PP in airflow. The main issues addressed are the ways of arranging PP combustion in airflow and the influence of the airflow temperature and mass flow rate on the yield of gasification products. These objective and issues are the novel distinctive features of this research.

2. Description of the experimental setup

Figure 1 shows the design of the experimental laboratory-scale GG. At the left end of the GG there is combustion chamber 3 and spark plug 13 of the fire heater operating on combustion of the stoichiometric hydrogen–oxygen mixture. In mixing chamber 5, the carrier gas (atmospheric air) is heated to the required temperature due to mixing with the combustion products. Thereafter, the heated carrier gas is fed through leveling pre-chamber 6 and profiled nozzle 7 to LMF test sample 8. Polypropylene of the PP H030 GP/1 brand is used as an LMF. After passing through the LMF test sample, the resultant gas flow enters small cylindrical channel 9, which ends with throttling insert 10 with orifice diameter of 6.5 mm. The use of a throttling insert allows maintaining the subsonic flow velocity in the GG. The exhaust gases expel from the GG to a vacuum tank (not shown in figure 1).

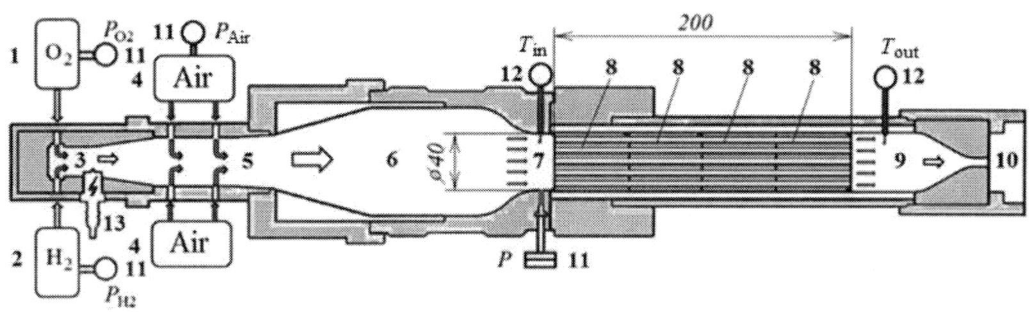

Figure 1. Scheme of laboratory experimental GG.

Two series of experiments are performed with the GG. The results of these series are presented below.

3. The results of the first series of experiments

In the first series of experiments, the airflow entering the GG is heated by combustion of hydrogen–oxygen mixture in a fire heater. The PP test sample is assembled from 4 cylindrical elements with diameter of 40 mm and length of 50 mm (figure 2). Each element has 19 evenly spaced longitudinal channels 3mm in diameter. The total length of the test sample is 200 mm.

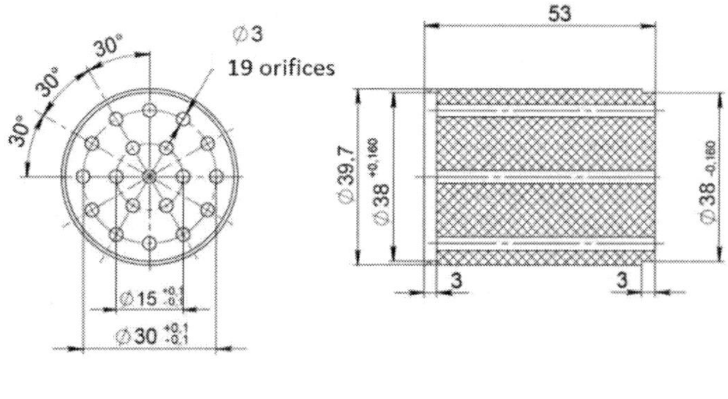

Figure 2. Test samples PP in the first series of experiments.

Journal of Physics: Conference Series **1404** (2019) 012066 doi:10.1088/1742-6596/1404/1/012066

The experimental procedure is as follows. The command of a synchronization system triggers the data acquisition system and opens the valves of oxygen and air supply. In 200 ms, it triggers the spark plug and opens the valve of hydrogen supply. The mass flow of hydrogen in the experiments ranges from 0.2 g/s to 0.7 g/s, depending on the desired temperature of the carrier gas. In 300 ms, the spark plug is turned off, implying that the hydrogen–oxygen mixture is ignited, and its combustion is established. The airflow entering the GG is then heated to a preset temperature and directed to the channels in the LMF test sample to heat and gasify the sample material. The mass flow rates of hydrogen and oxygen are chosen so that the mass fraction of oxygen in the carrier gas entering the LMF test sample and consisting of steam, oxygen, and nitrogen is close to 23%, the same as in pure air. Within the LMF test sample, chemical reactions between oxygen present in the heated carrier gas and PP gasification products lead to generation of additional heat for PP gasification. The operation of the GG is terminated in 2300 ms when the synchronization system closes the valves of hydrogen, oxygen, and air supply. The data acquisition system is turned off in 3000 ms.

Table 1 shows the measured mass flow rates of hydrogen (G_{H2}), oxygen (G_{O2}), and air (G_{Air}) supplied to the GG in some representative experiments. Also presented are the total mass flow rate of the carrier gas (G_{in}), the mass flow rate of oxygen ($G_{O2,in}$), and the mass fraction of oxygen ($Y_{O2,in}$) at the inlet of the test sample. In all the experiments, the total mass flow rate of the carrier gas entering the PP test sample (G_{in}) is seen to remain constant at a level of 35–36 g/s, and the mass fraction of oxygen in the carrier gas is ~23%.

Table 1. Results of Experiments on PP Gasification by Heated Air.

No.	Test No	G_{H2}, g/s	G_{O2}, g/s	G_{Air}, g/s	G_{in}, g/s	$G_{O2,in}$, g/s	$Y_{O2,in}$, %
1	34	0.29	3.02	32.79	36.10	8.24	22.83
2	32	0.36	3.74	31.51	35.60	8.15	22.88
3	30	0.42	4.61	30.37	35.40	8.24	23.26
4	23	0.53	5.50	29.07	35.10	7.95	22.64
5	25	0.60	6.45	28.65	35.70	8.24	23.08
6	27	0.71	7.57	27.92	36.20	8.31	22.96

Table 2 shows the time-averaged (over time $t_g \approx$ 1.0–1.8 s) characteristics of the gasification process. Increasing of the carrier gas temperature T_{in} from 900 to 1480 K is seen to increase the yield of PP gasification products (G_{PP}) from 4.7 to 17.7 g/s, whereas the ratio of the mass flow rate of the carrier gas to that of PP gasification products (G_{in}/G_{PP}) decreases from 2.9 to 1.8.

Table 2. Time-averaged characteristics of PP gasification by heated air.

No.	Test No.	t_g, sec	T_{in}, K	T_{out}, K	P_{in}, kPa	G_{PP}, g/s	G_{in}/G_{PP}
1	34	1.065	880	729	723	4.7	7.7
2	32	1.512	1015	1008	1075	12.4	2.9
3	30	1.698	1106	1100	1204	14.6	2.4
4	23	1.814	1272	1054	1196	15.2	2.3
5	25	1.836	1348	1119	1295	16.4	2.2
6	27	1.842	1483	1091	1380	17.7	2.1

Figure 3 shows typical records of pressure (P_{in}) in the GG and temperatures upstream (T_{in}) and downstream (T_{out}) of the LMF test sample in experiments with PP gasification by hot steam-air mixture.

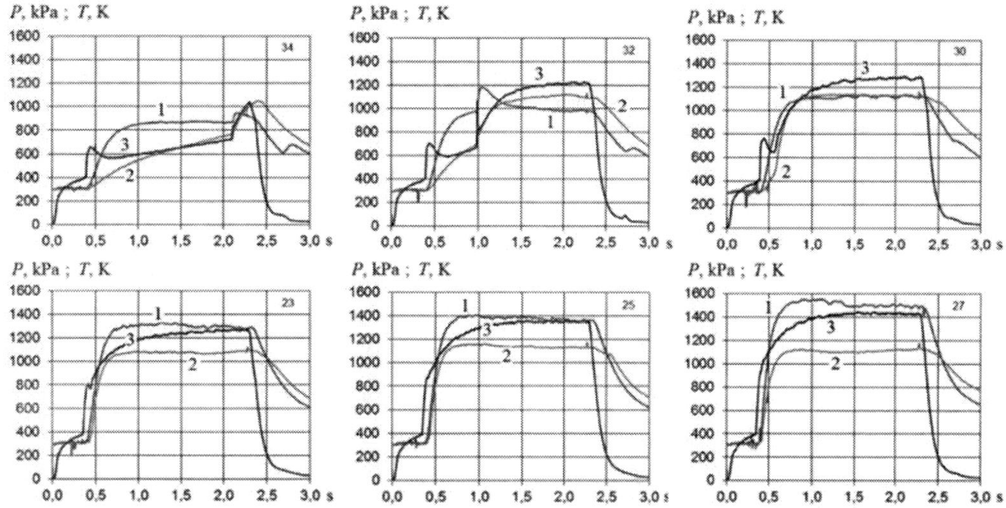

Figure 3. Typical records of temperature and pressure during PP gasification by heated air.
1 - T_{in}, 2 - T_{out}, 3 - P_{in}.

4. The results of the second series of experiments

In the second series of experiments, unheated air (at room temperature) is used as the carrier gas. Unlike the first series, PP test samples are assembled from 10 cylindrical elements with diameter of 40 mm and length of 20 mm (figure 4). Each element has 19 evenly spaced longitudinal channels 4 mm in diameter. All other details of the experimental setup are the same as in the first series of tests.

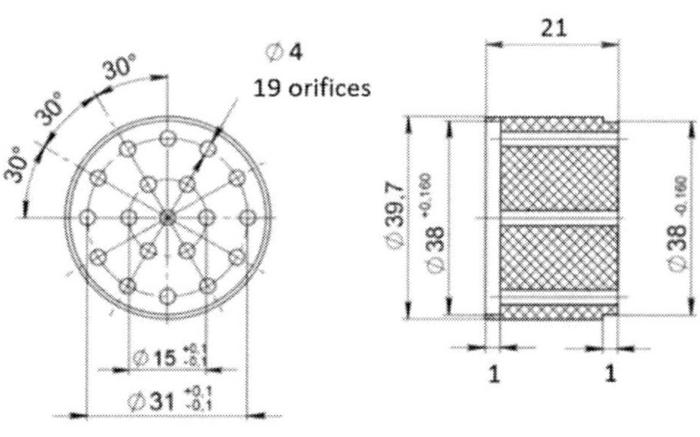

Figure 4. Test samples PP in the second series of experiments.

The experimental procedure is as follows. The command of the synchronization system turns on the data acquisition system and opens the oxygen and air supply valves. In 200 ms, the system starts triggering the spark plug and in 300 ms it opens the hydrogen valve to start combustion in the fire heater. After 300 ms of operation, the spark plug is turned off while oxygen and hydrogen supply to the GG is continued to support combustion. After 800 ms of operation (ignition stage), the oxygen and hydrogen valves are closed and only the air of room temperature is supplied to the GG with the already preheated PP test sample. Special experiments indicate that the average mass loss of the PP test sample at the ignition stage is 6.8 g. Thereafter, due to the interaction of oxygen available in the airflow with the preheated walls of the PP test sample, the combustion of PP in the sample channels is established, which provides heat for gasification of the sample material. The exhaust gas expelling from the GG through the throttling insert is mainly composed of the PP combustion and gasification products, nitrogen and residual oxygen. The operation of GG is terminated in 5300 ms by closing the valve of air supply.

Figure 5 shows typical pressure and temperature records in the experiments of the second series. As seen, immediately after the termination of the ignition stage, the temperature of the carrier gas at the inlet of the PP test sample decreases to 300 K, while the outlet temperature remains at the level of 1000–1100 K. This is the indication of the self-sustained combustion process in the test sample.

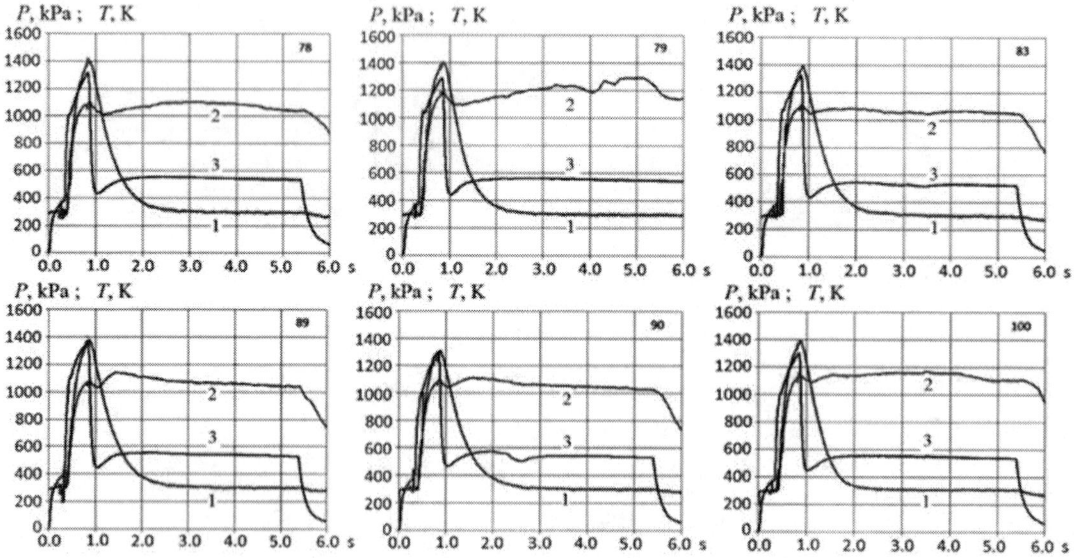

Figure 5. Typical temperature and pressure records during PP gasification by unheated air.
1 - T_{in}, 2 - T_{out}, 3 - P_{in}

Table 3. Time-averaged test results of PP gasification with cold air.

No.	Test No.	G_{Air}, g/s	T_{in}, K	T_{out}, K	P_{in}, kPa	Δm, g	G_{PP}, g/s	G_{in}/G_{PP}
1	78	27.5	304	1 080	546	21.33	4.68	5.88
2	79	27.5	300	1 189	538	34.75	4.06	6.78
3	83	27.5	310	1 061	529	20.53	4.50	6.11
4	89	27.5	310	1 065	539	22.47	4.93	5.58
5	90	27.5	307	1 055	535	23.52	5.16	5.33
6	100	27.5	310	1 142	546	18.41	4.04	6.81
7	103	27.5	305	1 085	523	16.79	3.68	7.47
8	105	33.0	333	1 066	691	28.77	6.31	5.23
9	107	27.5	304	1 031	545	24.41	5.35	5.14
10	108	33.0	307	1 107	695	29.45	6.46	5.11
11	109	42.0	298	1 227	903	32.75	7.18	5.85
12	110	42.0	297	1 200	887	38.71	8.49	4.95

Table 3 shows the time-averaged (over $t_g \approx$ 2.0–5.3 s) characteristics of the gasification process. The yield of PP gasification products (G_{PP}) is seen to attain 4–5 g/s, whereas the ratio of the mass flow rate of the carrier gas to that of PP gasification products (G_{in}/G_{PP}) ranges from 5 to 7.

5. Conclusion

The laboratory-scale gas generator aimed at gasification of a low-melting combustible (polypropylene) in a carrier gas is designed, fabricated, and tested. Two series of experiments on gasification of polypropylene samples in a carrier gas with different temperatures are performed. In the first series, the carrier gas is represented by the airflow continuously (during the entire test fire) heated by combustion of the fuel-lean hydrogen–oxygen mixture. Chemical reactions between oxygen available in the heated carrier gas and the gasification products provide additional heat for polypropylene gasification. In the second series, the carrier gas is represented by a pure airflow. Initially, the airflow is heated by combustion of fuel-lean hydrogen–oxygen mixture to ignite a polypropylene sample. After ignition, the supply of hydrogen and oxygen is terminated, and the polypropylene sample is gasified solely due to the heat generated by continuous combustion of the gasification products in the airflow. In both series, the polypropylene samples are assembled from several (4 to 10) cylindrical elements with multiple (19) longitudinal channels 3 or 4 mm in diameter. The main issue addressed in this study is the effect of the carrier gas mass flow rate and temperature on the overall yield of polypropylene gasification products.

In the experiments of the first series, the air temperature, pressure, and velocity of the carrier gas are ranged from 700 to 1500 K, 0.57 to 1.38 MPa, and 60 to 100 m/s, respectively. The maximum mass flow rate of gasification products is 17.7 g/s. The minimum ratio of mass flow rates of air and gasification products is 2.

In the experiments of the second series, the airflow temperature is fixed at 300 K, whereas pressure and velocity of the airflow are ranged from 0.35 to 1.0 MPa and from 17 to 19 m/s, respectively. The maximum mass flow rate of gasification products is 8.5 g/s. The minimum ratio of mass flow rates of air and gasification products is 5.

Acknowledgements

The research was carried out within the framework of the Program of Fundamental Scientific Research for the state academies of sciences in 2013-2020 (project No. AAAA-A17-117030610121-9, project No. AAAA-A17-117030610121-9) and was partly supported by the Russian Foundation for Basic Research (grant 18-08-00076a).

References

[1] Alexandrov V N, Bytskevich V M, Verkholomov V K *et al.* 2006 *Integrated Ramjet Engines on Solid Fuels* (Moskow: Academic Book Press) p 332

[2] Obnosov B V, Sorokin V A, Yanovsky L S *et al.* 2012 *Design and Engineering of Combined Solid Propellant Rocket Engines*. (Moskow: Publishing House of MSTU) p 279

[3] Averkov I S, Arefyev K Yu, Baykov A V and Yanovsky L S 2017 *Thermophys. & Aeromech.* **24(1)** 149–160

[4] Salgansky E A, Kislov V M, Glazov S V, Zholudev A F and Manelis G B 2010 *Journal of Combustion & Explosion Physics* **46(5)** 42–47

[5] Alexandrov V Yu, Arefiev K Yu, Ilchenko M A and Ananyan M V 2015 *Science and education. MSTU named after N.E. Bauman. Electron. journal* **08** 75

[6] Salgansky E A, Lutsenko N A, Levin V A and Yanovskiy L S 2019 *Aerospace Science and Technology* **84** 31

[7] Levin V A, Lutsenko N A, Salgansky E A and Yanovskiy L S A 2018 *Doklady Physics* **63(9)** 375–379

 1404 (2019) 012067 doi:10.1088/1742-6596/1404/1/012067

An experimental study of the combustion of paraffin and ceresin with the addition of metal-organic compounds in an oxygen stream

A N Shiplyuk[1], S V Lukashevich[1], V I Simagina[2], O V Netskina[2] and O V Komova[2]

[1] Khristianovich Institute of Theoretical and Applied Mechanics SB RAS, 4/1 Institutskaya str., Novosibirsk, 630090, Russia
[2] Boreskov Institute of Catalysis SB RAS, 5 Lavrentiev Ave., Novosibirsk, 630090, Russia

lukashevich@itam.nsc.ru

Abstract. Experimental studies of the combustion of experimental samples of ceresin and paraffin with the addition of metal-organic compounds in a gaseous oxygen flow were performed. The tests were carried out at a pressure of 8.5 to 10.5 bar and an oxygen flow velocity of 10.5 m/s. An increase in specific impulse of 10% was obtained for samples based on ceresin and from 6 to 18% for samples based on paraffin.

1. Introduction

The hybrid engine facilities (HRE) working on solid fuel (SF) and liquid or gaseous oxidizer [1-6] are a promising direction in the developing rocket-engine technology to provide the long-term high-speed flight. It is predicted that hybrid engine systems are going to find wide application, including the control in orbital flight on upper stages, tactical and strategical rocket systems. Hybrid rocket engines combine the advantages of SPRE and LPE and, at the same time, do not have many disadvantages. The specific energy-mass characteristics of the hybrid rocket engines put them in between the LPE on cryogenic fuels and SPRE. At the same time, the hybrid rocket engines demonstrate higher values of the specific pulse than SPRE owning to the fuels with the bigger energy store. The most important advantage of the hybrid engine is the wide range of draught module control and the possibility of multiple engine ignition and cutoff during one flight. Utilization of thermoplastic materials as the major component of the fuel for the hybrid engines enables to use modern additive technologies for the development and production of charges for these engines, which essentially decreases the cost of the development and production of these engine charges. All these advantages make the hybrid engines very promising in for the rocket equipment in the XXI century.

Creation of the energy-intensive oxygen-free solid fuels with high combustion rate and provision of their high combustion efficiency in the gaseous oxidizer flow is a key challenge in the HRE development. In this work we investigate the effect of energy-intensive additives in the form of metal-organic compounds on the thrust characteristic of hybrid combustion chamber on paraffin and ceresin with oxygen as an oxidizing agent.

2. Experimental setup for investigation of solid fuel combustion in the gaseous oxidizer flow

The experimental setup of ITAM SB RAS was used to perform experimental investigations on the definition of the specific impulse of hybrid axisymmetric combustion chamber. The setup schematic is shown in figure 1. The setup consists of a flame heater 1, prechamber 2, subsonic nozzle 3, elliptical cowl 4, solid fuel sample 5, pylon 6, combustion chamber 7, output sonic nozzle 8 and spark plug. The outlook of the setup is shown in figure 2. The prechamber and input nozzle are purposed to create a uniform oxidizer flow with the specified pressure and velocity at the combustion chamber (CC) entrance. The internal diameter of the combustion chamber is 40 mm the length is 400 mm. Output sonic nozzle diameter is 9 mm. In the combustion chamber, on the pylon, there is the sample 5 of the studied solid fuel made as a cylinder of 21 mm in diameter and 40 mm in length. Ahead of the sample there is the elliptical cowl to form the uniform flow around the sample. To initiate the combustion process, the flame heater is used; some hydrogen (3 g/s) is supplied into the prechamber for the 150 ms flow temperature jump at the nozzle exit up to 2000 to 2200 K. Then the flame heater is cutoff, and the gaseous oxygen of room temperature is supplied to the combustion chamber entrance, and then the self-sustaining combustion of solid fuel takes place in the oxidizer flow. Combustion and carry-away of the solid-fuel sample mass occur only on the external surface of the sample. The typical time of combustion is 1.25 second, which is sufficient to determine the variation of the sample mass, the sample size variation is insignificant and makes almost no effect on the flow parameters on the sample surface. The total pressure necessary for calculating the specific impulse is measured in the output section of the combustion chamber.

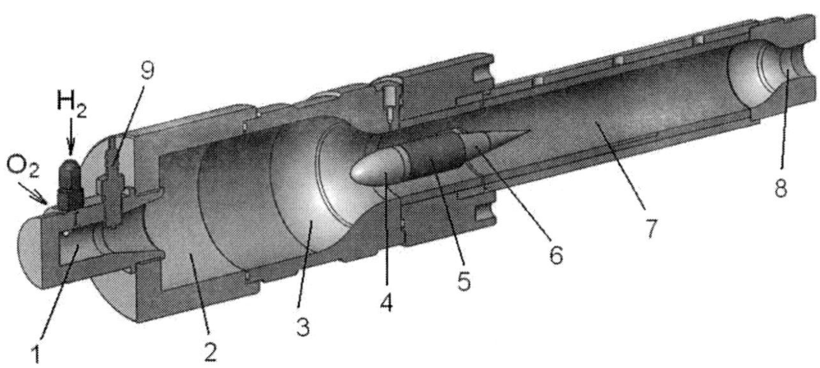

Figure 1. Scheme of experimental setup. 1 – flame heater, 2 – prechamber, 3 – subsonic nozzle, 4– elliptical cowl, 5 – solid fuel sample, 6 – pylon, 7 – combustion chamber, 8 – output sonic nozzle, 9 – spark plug.

3. Method of experimental result processing

The average mass flow rate of the sample within one regime is defined by the formula

$$Q_{sf} = \Delta m/t_{comb} \tag{1}$$

where Δm is the measurement of the sample mass within the regime, t_{ccomb} is the sample combustion time. The variation of the sample mass Δm is determined by means of the control weighting of the sample before and after the start on the laboratory balance, its accuracy is 0.02 g. The sample combustion time is preset by the oxidizer supply time in the combustion chamber.

To evaluate the efficiency of solid fuel according to the results of experiments, the specific impulse of the output sound nozzle I_{sn} is calculated by the formula

$$I_{sn} = J_{sn}/(Qg) \tag{2}$$

where: J_{sn} is the impulse of the output sound nozzle; Q, g/s is the total mass flow rate of oxidizer and solid fuel; g is the gravity acceleration

To calculate the impuls of the output sound nozzle, the expression of the jet impuls in terms of the Mach number and pressure in the critical section of the nozzle is used

$$J_{sn} = S_{cs}P_{cs}M_{cs}^{2}\gamma_{cs} + S_{cs}(P_{cs} - P_{atm}) \tag{3}$$

where: S_{cs} is the area of the critical section of the sonic nozzle, P_{atm} is the pressure in the chamber into which the outflow occurs, P_{cs} is the static pressure in the critical section of the sonic nozzle, M_{cs} is the Mach number in the critical section of the sonic nozzle in these experiments, it was assumed that $M_{cs} = 1$, γ_{cs} is the adiabatic index determined using the "Terra" thermodynamic calculation program for the theoretical composition of the gas flowing out of the nozzle. The error in determining the specific impulse of a sound nozzle by this method does not exceed 3%.

4. Results of solid fuel combustion tests in the oxygen flow

The tests were carried out at a pressure of 8.5 to 10.5 bar and an oxygen flow velocity of 10.5 m / s. In the experiments, the burning rate of solid fuel samples and the specific impulse of the output sound nozzle were determined. The experimental compositions are based on paraffin and ceresin with the addition of the following compounds: $Ni(AG)_2(NO_3)_2$, LaFeCuGly [7], $NaBH_4$ [8]. The mass fraction of $Ni(AG)_2(NO_3)_2$ and LaFeCuGly in all samples was 5%, $NaBH_4$ 10%. Additionally, a paraffin-based sample with the addition of a small amount of solid oxidizer was tested. The collective results of the tests are given in table 1.

Table 1. Test results[a].

Sample	Run	$P_{0\ hot}$, kPa	Q, g/s	I, s
Ceresin	45	858	5.21	113
Ceresin+$Ni(AG)_2(NO_3)_2$	46	944	4.97	125
Ceresin+LaFeCuGly	47	939	5.00	125
Paraffin	40	894	> 15.92	< 100.6
Paraffin+$Ni(AG)_2(NO_3)_2$	41	1061	> 16.10	< 119
Paraffin+LaFeCuGly+oxidiser	42	1018	> 17.84	< 111.6
Paraffin+$NaBH_4$	43	1008	> 17.26	< 111.5
Paraffin+$NaBH_4$+LaFeCuGly	44	1053	> 17.61	< 115.8

[a] Gly – Glycine; AG – Aminoguanidine nitrate

Figures 2 and 3 show the time dependences of the total pressure in the output section of the combustion chamber for samples based on ceresin and paraffin, respectively. A rapid increase in pressure at t = 0.3 s corresponds to the initiating hydrogen flame at time t = 0.45 s, the hydrogen supply ceases and the combustion of the sample in the oxygen flow begins. The maximum pressure level is reached in the time interval from 0.8 to 1.55 s. It can be seen that for all samples with additives, the pressure level during combustion is higher than for pure ceresin and paraffin. The specific impulse calculation results (table 1) show that the addition of $Ni(AG)_2(NO_3)_2$ and LaFeCuGly to ceresin increases the specific impulse by 10%.

In experiments with paraffin, it was not possible to accurately measure the mass flow rate of the sample due to samples completely burned out during the run. From our previous studies, it is known that the maximum increase in the burning rate of paraffin samples was 5% for samples with the addition of $Ni(AG)_2(NO_3)_2$. Thus, if we assume that the burning rates of paraffin based samples in these experiments vary no more than 5%, we obtain the following increase in specific impulse for paraffin based samples: from 13 to 18% for the sample with $Ni(AG)_2(NO_3)_2$, from 6 to 11% for samples with LaFeCuGly + oxidiser and $NaBH_4$; and from 10 to 15% for the sample with $NaBH_4$ + LaFeCuGly.

 1404 (2019) 012067 doi:10.1088/1742-6596/1404/1/012067

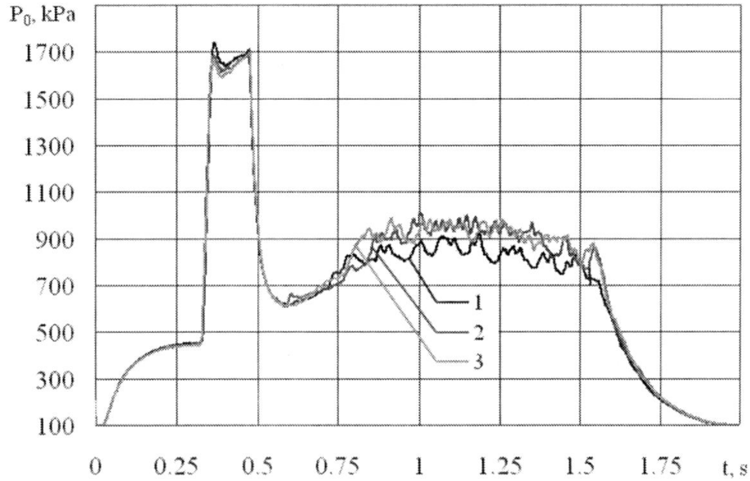

Figure 2. Combustion chamber pressure. 1 – Ceresin; 2 – Ceresin+Ni(AG)$_2$(NO$_3$)$_2$; 3 – Ceresin+LaFeCuGly.

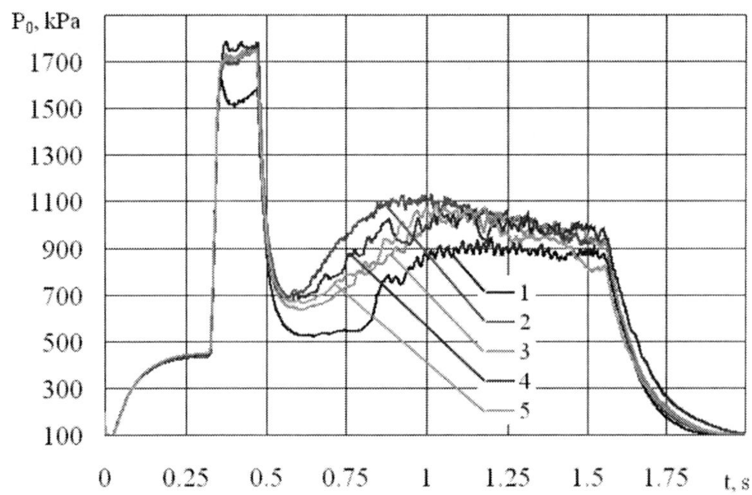

Figure 3. Combustion chamber pressure. 1 – Paraffin; 2 – Paraffin+Ni(AG)$_2$(NO$_3$)$_2$; 3 – Paraffin+LaFeCuGly+oxidiser; 4 – Paraffin+NaBH$_4$; 5 – Paraffin+NaBH$_4$+LaFeCuGly.

5. Conclusion

Experimental studies of the combustion of experimental samples of ceresin and paraffin with the addition of metal-organic compounds in a gaseous oxygen flow were performed. The tests were carried out at a pressure of 8.5 to10.5 bar and an oxygen flow velocity of 10.5 m/s. An increase in specific impulse of 10% was obtained for samples based on ceresin and from 6 to 18% for samples based on paraffin.

Acknowledgments

This work was supported by the Program no. 56 of the Presidium of the Russian Academy of Sciences Fundamental Bases of Breakthrough Technologies for the Benefit of National Security".

Journal of Physics: Conference Series **1404** (2019) 012067 doi:10.1088/1742-6596/1404/1/012067

References

[1] Potapkin A V and Lee T S 2004 *Explosion, and Shock Waves* **40(4)** 386-92
[2] Arkhipov V A, Zhukov A S, Zolotorev N N, Perfil'and K G and Bondarchuk S S 2017 *Combustion, Explosion, and Shock Waves* **53(6)** 634-40
[3] Arkhipov V A, Zharova I K, Zhukov A S, Kozlov E A, Zarko V E, Aksenenko D D and Kurbatov A V 2016 *Combustion, Explosion, and Shock Waves* **52(5)** 497-513
[4] Risha G, Boyer E, Wehrman R, Evans B and Kuo K 2003 *39th AIAA/ASME/SAE/ASEE joint propulsion conference and exhibit* **4593**
[5] Nakagawa I and Hikone S 2011 *J. Propul. Power* **27(6)** 1276-79
[6] Dun C, Gustafson G, Edwards J, Dunbrack T and Johansen C 2018 *Aerosp. Sci. Technol.* **72** 371-79
[7] Komova O V, Simagina V I, Mukha S A, Netskina O V, Odegova G V, Bulavchenko O A, Ishchenko A V and Pochtar' A A 2016 *Advanced Powder Technology* **27(2)** 496-503
[8] Netskina O V, Ozerova A M, Komova O V, Odegova G V, Simagina V I 2015 *Catalysis Today* **245** 86-92

Journal of Physics: Conference Series **1404** (2019) 012068 doi:10.1088/1742-6596/1404/1/012068

Experimental study of the low-melting hydrocarbons regression rate in the inert gas flow

A N Shiplyuk[1], V I Zvegintsev[1], S M Frolov[2], D A Vnuchkov[1], S V Lukashevich[1] and D G Nalivaychenko[1]

[1] Khristianovich Institute of Theoretical and Applied Mechanics SB RAS, 4/1 Institutskaya str., Novosibirsk 630090, Russia
[2] Semenov Federal Research Center of Chemical Physics, Moscow 119991, Russia

vnuchkov@itam.nsc.ru

Abstract. Gasification of organic and inorganic materials in a high-temperature gas flow is a promising technology for various industrial applications like chemical industry, waste processing, rocket and air-breathing propulsion, etc. However, the characteristics of the gasification process of low-melting hydrocarbon materials remain insufficiently studied so far. The present authors propose a methodology for investigating the gasification of combustible materials for ramjets. Experiments on the gasification of cylindrical polypropylene (PP) samples by the high-temperature inert carrier gas (nitrogen) passing through multiple longitudinal channels of 3 mm in diameter and 200 mm long are performed. The flow of the carrier gas is heated by combustion of hydrogen–oxygen mixture. The temperature, pressure, and velocity of the carrier gas in the experiments range from 900 to 1700 K, 0.35 to 1.0 MPa, and 80 to 130 m/s, respectively. The yield of PP gasification products is shown to increase linearly with the carrier gas temperature. At the temperature of 1700 K, the yield of PP gasification products in the test facility attains 8.0 g/s, whereas the ratio of the mass flow rates of the carrier gas and gasification products reaches the value of 4.5.

1. Introduction

The concept of solid-fuel ramjets is currently associated mainly with the scheme of a ducted ramjet engine or a Throttleable Ducted Rocket (TDR) [1-2]. A mandatory element of the TDR is a gas generator (GG) aimed to gasify a solid fuel with a minimum amount (10–20%) of an oxidizer. Such an amount of oxidizer allows gasification of solid fuel due to its partial combustion in the GG and feeding the resulting gaseous products to the main ramjet combustion chamber, where they are completely burned in a high-speed flow of atmospheric air.

In the recent scientific literature, another relevant direction is discussed actively, that is the development of a low-temperature GG for gasification of low-melting fuels (LMF) like polymers (polyethylene, polypropylene, polystyrene, polybutadiene, etc.) with melting point of about 100 °C [3-7]. In the cited references, the design of a two-chamber GG with a sequential arrangement of a conventional solid propellant and LMF is studied numerically. In this design, burning of solid propellant generates a flow of high-temperature combustion products, which passes through the porous layer of LMF and gasifies it.

The objective of this work is to study experimentally the gasification of PP in a flow of heated inert gas (nitrogen). The main issue addressed is the influence of the carrier gas temperature and mass flow rate on the yield of gasification products. These objective and issues are the novel distinctive features of this research.

2. Description of the experimental setup

Figure 1 shows the design of the experimental small-sized GG. At the left end of the GG there is combustion chamber 3 of the fire heater, operating on combustion of the stoichiometric hydrogen–oxygen mixture. In mixing chamber 5, the carrier gas (nitrogen) is heated to the required temperature due to mixing with the combustion products. Thereafter, the heated carrier gas is fed through leveling pre-chamber 6 and profiled nozzle 7 to LMF test sample 8. Polypropylene of the PP H030 GP/1 brand is used as an LMF. Each test sample is assembled from 4 cylindrical elements with diameter of 40 mm and length of 50 mm (figure 2). Each element has 19 evenly spaced longitudinal channels 3 mm in diameter. The total length of the test sample is 200 mm. After passing through the last element of the test sample, the resultant gas flow enters small cylindrical channel 9, which ends with throttling insert 10 with orifice diameter of 6.5 mm. The use of a throttling insert allows maintaining the subsonic flow velocity in the GG. The exhaust gases expel from the GG to a vacuum tank (not shown in figure. 1).

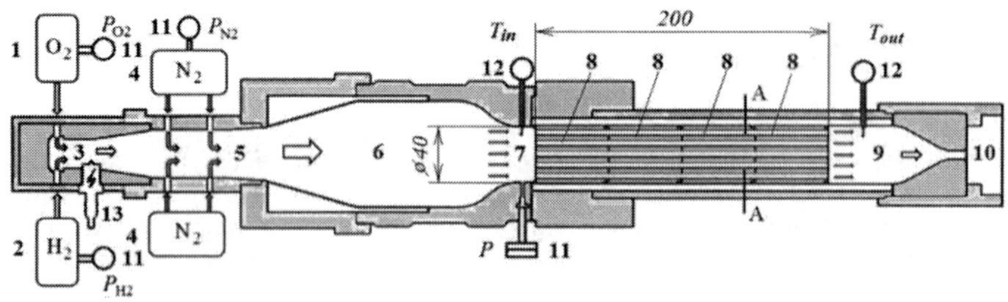

Figure 1. Scheme of a small-sized experimental gas generator.

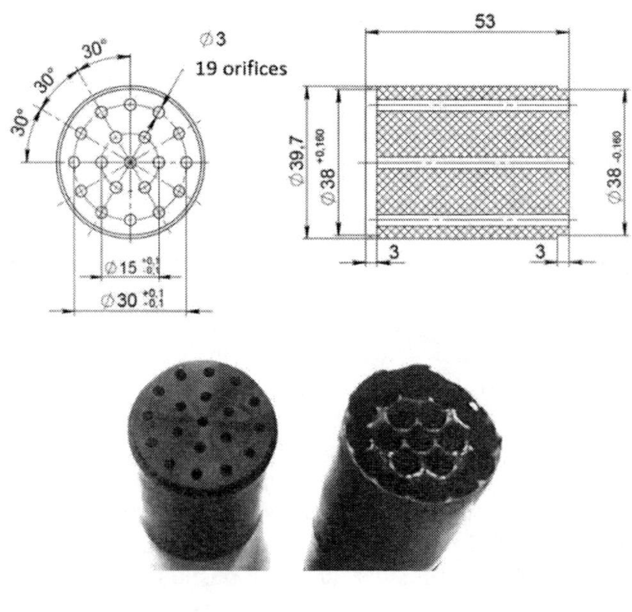

Figure 2. Total view of the test samples. (a) before burning; (b) after burning.

The experiments are performed as follows. The command of a synchronization system triggers the data acquisition system and opens the valves of oxygen and nitrogen supply. In 200 ms, it triggers the spark plug and opens the valve of hydrogen supply. The mass flow rate of hydrogen in experiments ranges from 0.3 g/s to 0.9 g/s depending on the required temperature of the carrier gas. In 300 ms, the spark plug is turned off, implying that the hydrogen–oxygen mixture is ignited, and its combustion is established. The mass flow rates of hydrogen and oxygen are maintained at the stoichiometric ratio. Measurements show that combustion of the hydrogen–oxygen mixture in the fire heater is complete. The carrier gas (nitrogen) is then heated to a preset temperature and directed to the channels in the LMF test sample to heat and gasify the sample material. As a result of experiment, a mixture of nitrogen, steam, and PP gasification products is formed at the outlet of the GG. In 2300 ms, the synchronization system closes the valves of oxygen, nitrogen, and hydrogen supply to terminate the experiment. The data acquisition system is turned off in 3000 ms.

3. Experiment Results

Table 1 shows the measured mass flow rates of hydrogen (G_{H2}), oxygen (G_{O2}), and nitrogen (G_{N2}) supplied to the GG in some representative experiments. In all the experiments, the total mass flow rate of the carrier gas (G_{in}) is seen to remain constant at a level of 35–36 g/s. The gas velocity in the channels of the LMF test sample is 120–130 m/s at the sample inlet and 90 m/s at its outlet.

Table 1. The Initial conditions of experiments on PP gasification by high-temperature nitrogen.

Test No.	G_{H2}, g/s	G_{O2}, g/s	G_{N2}, g/s	$G_{in} = G_{H2} + G_{O2} + G_{N2}$, g/s
14	0.30	2.38	33.20	35.9
8	0.39	3.05	32.20	35.6
12	0.48	3.80	31.00	35.3
10	0.59	4.70	30.50	35.8
16	0.72	5.64	28.60	35.0
20	0.82	6.51	28.20	35.5
18	0.94	7.54	27.50	36.0

Figure 3 shows typical records of pressure and temperature in experiments with PP gasification. Increasing of hydrogen mass flow rate is seen to increase the carrier gas temperature upstream of the LMF test sample (T_{in}) from 900 to 1716 K. The temperature of the gas flow at the outlet of the LMF test sample (T_{out}) is seen to remain relatively low and nearly constant (ranging from 800 to 900 K). During operation of the GG, the pressure in it (P) increases by about 15% ranging from 7 to 10 bar.

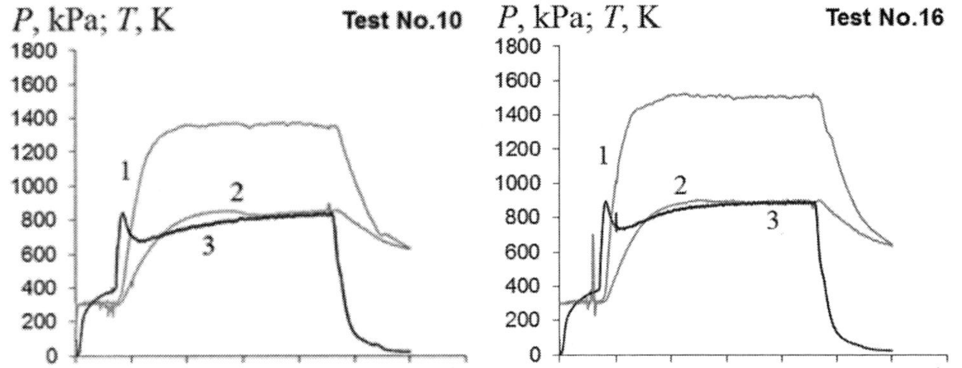

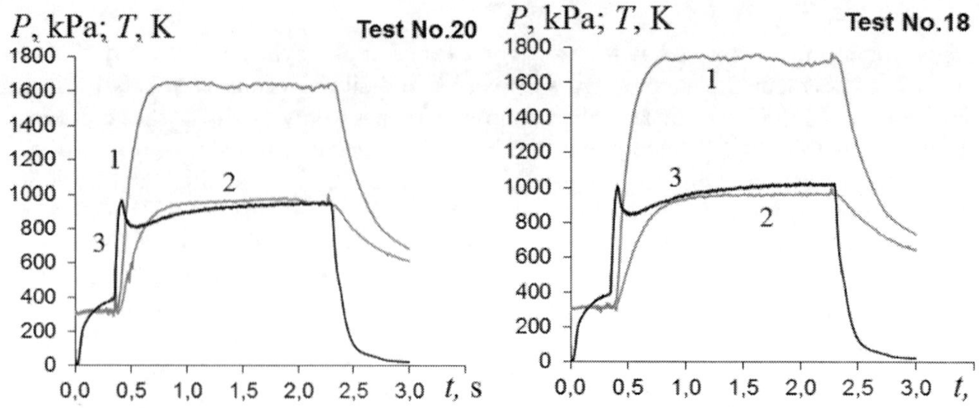

Figure 3. Typical temperature and pressure records in experiments with PP gasification. (1 - T_{in}, 2 - T_{out}, 3 – P)

Table 2 shows the time-averaged (over $t_g \approx 1.7$ s) characteristics of the gasification process. In particular, table 2 shows the yields of gasification products (G_{PP}) and the ratios of the mass flow rates of the carrier gas to those of PP gasification products (G_{in}/G_{PP}) in different experiments. The maximum yield of gasification products is 8 g/s at the carrier gas temperature of 1716 K. The minimum value of G_{in} / G_{PP} is 4.5 at the carrier gas temperature of 1716 K.

Table 2. Time-averaged test results of PP gasification with a high-temperature nitrogen.

TestNo.	t_g, s	T_{in}, K	T_{out}, K	P_{in}, kPa	G_{PP}, g/s	G_{in} / G_{PP}
4	1.059	909	650	622	0.13	271.41
8	1.507	1050	729	666	1.27	27.97
12	1.678	1183	825	709	2.42	14.58
10	1.666	1348	816	786	4.08	8.77
16	1.712	1494	865	855	5.64	6.20
20	1.788	1620	935	910	6.49	5.48
18	1.772	1716	925	976	8.00	4.50

Figure 4 shows the dependence of the mass flow rate of PP gasification products on the temperature of the carrier gas at the LMF test sample (T_{in}). The yield of gasification products is seen to increase linearly with the temperature of the carrier gas.

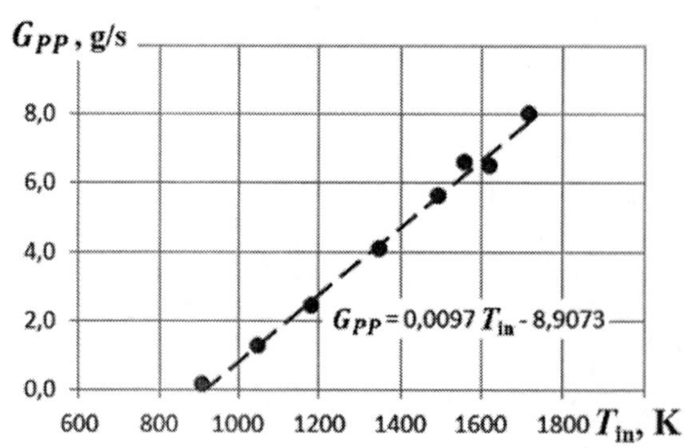

Figure 4. Yield of PP gasification products vs. the carrier gas temperature.

4. Conclusion

An experimental facility is developed and experiments are performed to study the gasification of low-melting fuel (polypropylene) by a flow of high-temperature inert carrier gas (nitrogen) through a test sample with 19 longitudinal channels 3 mm in diameter and 200 mm long. The carrier gas temperature, pressure, and velocity in the experiments range from 900 to 1700 K, 0.35 to 1.0 MPa, and 80 to 130 m/s, respectively. The yield of gasification products is shown to increase linearly with the temperature of the carrier gas. At the carrier gas temperature of ~1700 K, the yield of gasification products in the facility attains 8.0 g/s, and the ratio of mass flow rates of the carrier gas and gasification products reaches the value of 4.5.

Acknowledgements

The research was carried out within the framework of the Program of Fundamental Scientific Research for the state academies of sciences in 2013-2020 (project No. AAAA-A17-117030610121-9), (project No. AAAA-A17-117030610121-9) and was partly supported by the Russian Foundation for Basic Research (grant 18-08-00076a).

References

[1] Alexandrov V N, Bytskevich V M, Verkholomov V K *et al.* 2006 *Integrated Ramjet Engines on Solid Fuels* (Moskow: Academic Book Press) p 332

[2] Obnosov B V, Sorokin V A, Yanovsky L S *et al.* 2012 *Design and Engineering of Combined Solid Propellant Rocket Engines* (Moskow: Publishing House of MSTU) p 279

[3] Averkov I S, Arefyev K Yu, Baykov A V and Yanovsky L S 2017 *Thermophys. & Aeromech.* **24** 149–160

[4] Salgansky E A, Kislov V M, Glazov S V, Zholudev A F and Manelis G B 2010 *Journal of Combustion & Explosion Physics* **46(5)** 42–47

[5] Alexandrov V Yu, Arefiev K Yu, Ilchenko M A and Ananyan M V 2015 *Science and education. MSTU named after N.E. Bauman. Electron. journal* **08** 75

[6] Salgansky E A, Lutsenko N A, Levin V A and Yanovskiy L S 2019 *Aerospace Science and Technology* **84** 31

[7] Levin V A, Lutsenko N A, Salgansky E A and Yanovskiy L S A 2018 *Doklady Physics* **63**(9) 375–379

Calculation of continuous spin detonation of a hydrogen-oxygen mixture in an annular combustor with oxygen ejection

E V Simonov and S A Zhdan

Lavrentyev Institute of Hydrodynamics SB RAS,
15 Lavrentyev av., Novosibirsk, 630090, Russia

simon_nsk@mail.ru

Abstract. A closed mathematical model of continuous spin detonation in a H_2-O_2 mixture with oxygen ejection from the environment has been developed. The model takes into account the inverse effect of pulsation processes in the combustor on oxygen ejection. A numerical study was performed for a 10 cm diameter ramjet annular combustor with geometric dimensions corresponding to experiments. One-wave modes of continuous spin detonation were calculated and the flow structure was analyzed with variation in the specific hydrogen flow rate from 3 to 3.75 kg/(s·m^2). Decreasing the flow rate of ejected hydrogen during continuous detonation in the combustor is found to lead to a monotonic decrease in the detonation velocity, wave front height, average static pressure, and equivalence ratio, but to an increase in the oxygen ejection coefficient. The results are validated against experimental data.

1. Introduction

The results of studies of continuous spin detonation (CSD) according to Voitsekhovskii's scheme are presented in [1, 2]. CSD of a hydrogen-oxygen mixture with oxidizer ejection was first obtained experimentally in [3, 4], and the calculations of CSD were performed in [5] using the Abramovich classical ejector flow model [6] at the combustor entrance. The objectives of this study were to develop a closed mathematical model of CSD in the O_2 ejection mode without using the simplifying assumption of [5], to numerically study the CSD dynamics of a H_2-O_2 mixture in the O_2 ejection mode in a ramjet annular combustor with a diameter of 10 cm, and to compare the results with experiments [4].

2. Formulation of the problem

2.1. Mathematical model

Mathematical modeling of the initiation and propagation of CSD in a H_2-O_2 mixture in the O_2 ejection mode was carried out in a ramjet annular combustor (diameter d_c, length L, expansion of the annular combustor channel from δ to Δ and then to Δ_{exit}) and an attached entrance section in the form of an annular converging nozzle (length L_0, contraction of the channel from Δ_0 to δ), which was initially filled with oxygen. For experimental annular combustors [3, 4], the following inequalities hold: $\delta < \Delta << d_c/2$, $\Delta_0 << d_c/2$, and $\Delta_{exit} << d_c/2$. Therefore, by expanding the annular region into a rectangular solution domain $\Omega_0 \cup \Omega$ (see figure 1), we can formulate a quasi-three-dimensional mathematical problem in a similar manner as was done in [2].

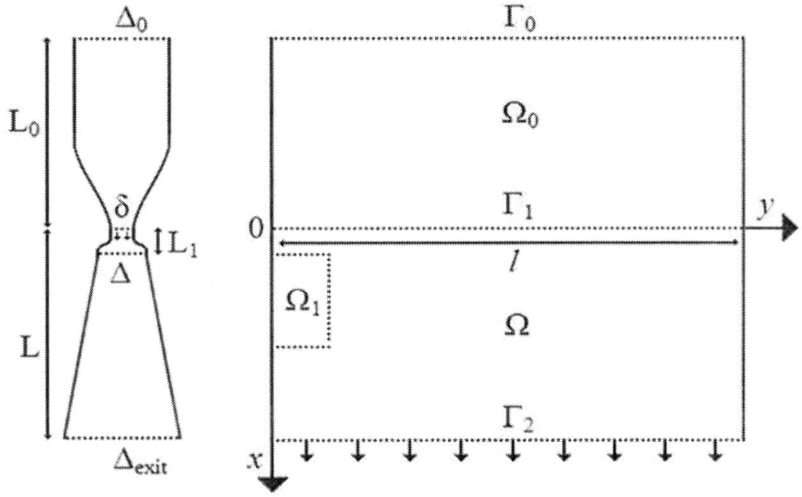

Figure 1. Solution domain of the periodic problem $\Omega_0 \cup \Omega$.

The H_2-O_2 mixture flow in the computational domain $\Omega_0 \cup \Omega$ is described by the following system of equations of non-stationary gas dynamics with chemical reactions:

$$\begin{gathered} \rho_t + S^{-1}(\rho u S)_x + (\rho v)_y = f_1^{H2}, \\ (\rho u)_t + S^{-1}(\rho u^2 S)_x + (\rho u v)_y + p_x = f_2^{H2}, \\ (\rho v)_t + S^{-1}(\rho u v S)_x + (\rho v^2)_y + p_y = 0, \\ (\rho E)_t + S^{-1}[\rho u(E + \frac{p}{\rho})S]_x + [\rho v(E + \frac{p}{\rho})]_y = f_4^{H2}, \\ (\rho Y)_t + S^{-1}(\rho u Y S)_x + (\rho v Y)_y = \rho f_5, \\ (\rho \mu)_t + S^{-1}(\rho u \mu S)_x + (\rho v \mu)_y = \rho f_6 + f_6^{H2}, \\ (\rho z)_t + S^{-1}(\rho u z S)_x + (\rho v z)_y = 0. \end{gathered} \tag{1}$$

Here t is time, x and y are the spatial variables of the orthogonal coordinate system, ρ is the density; u and v are the velocity components, p is the pressure, $E = U + (u^2 + v^2)/2$, $U(T, \mu)$ is the total internal energy of the gas, T is the temperature, μ is the current molar mass of the mixture, Y is the fraction of the chemical induction period, z is the mass fraction of oxygen, $S = S(x)$ is channel cross-sectional area in the domain Ω_0 and the combustor, which varies along the x coordinate as

$$S(x) = \begin{cases} \Delta_0 \cdot l, \ -L_0 < x < -L_0/2 \\ [(\Delta_0 - \delta)\sin(-\pi x/L_0) + \delta] \cdot l, \ -L_0/2 < x < 0 \\ \delta \cdot l, \ 0 < x < L_1/2 \\ [(\Delta - \delta)\sin(-\pi x/L_1 - \pi/2) + \delta] \cdot l, \ L_1/2 < x < L_1 \\ [\Delta + (\Delta_{exit} - \Delta)\cdot(x - L_1)] \cdot l, L_1 < x < L \end{cases} \tag{2}$$

The quantities f_i^{H2} (i = 1, 2, 4, 6) describe the injection of H_2 through the supply system located in the initial part of the domain Ω ($0 < x < L_1/2$). The supply of H_2 was carried out through a set of Laval micro nozzles uniformly distributed around the combustor circumference with total throat area per unit length S_{cr}. Under the assumption of instantaneous mixing of the components over the combustor cross section, the right sides f_i^{H2} in system (1), expressing the mass, momentum, and energy fluxes of H_2 in the combustor, are uniquely determined using the formulas of [2].

As in [7, 5], the quantities f_5 and f_6 describe the energy release within the framework of the two-stage kinetic model. The induction stage ($0 < Y \leq 1$, $f_5 = -1/t_{ind}$, $f_6 = 0$), where energy release is absent; the induction period was calculated using the White formula [8]:

$$t_{ind} = \frac{K_a}{\rho}\left(\frac{\mu_{O2}\mu_{H2}}{z(1-z)}\right)^{1/2} \exp\left(\frac{\varepsilon_a}{RT}\right), \tag{3}$$

where $\varepsilon_a = 18.1$ kcal/mol is the activation energy, $K_a = 4.17 \cdot 10^{-11}$ (mol·s/l) is the pre-exponential factor, R is the universal gas constant, μ_{O2} is the molar mass of oxygen, and μ_{H2} is the molar mass of hydrogen. Chemical transformation stage ($Y = 0$, $f_5 = 0$, $f_6 \neq 0$), where the rate of energy release is determined by the rate of chemical reactions [7]:

$$f_6 = 4K_+[W_1(\mu)\rho^2 - W_2(T,\mu)\rho],\ W_1(\mu) = \frac{(1-\frac{\mu}{\mu_{max}})^2}{\mu}, \tag{4}$$

$$W_2(T,\mu) = K_-(\frac{\mu}{\mu_{min}} - 1)\left(\frac{T}{T_0}\right)^{\frac{\beta}{2}}\left[1 - \exp(\frac{-\theta}{T})\right]^{\beta} \exp(-\frac{E_d}{RT}),$$

where K_+ is the generalized recombination rate constant, $K_-(z)$ is the equilibrium constant, T_0 is the initial temperature of the mixture, $\beta(z) = 1 + \sigma_{max}/(\mu_{max}/\mu_{min} - 1)$, $\sigma_{max}(z)$ is the molar fraction of triatomic molecules in the ultimately recombined state, $\mu_{min}(z)$ and $\mu_{max}(z)$ are the molar masses of the gas in the ultimately dissociated and ultimately recombined states. The quantity $K_-(z)$ was calculated in accordance with the equation of chemical equilibrium for the corresponding composition, and the dependence of the remaining quantities on z are described in detail in [2].

System (1) was supplemented by the equations of state

$$p = \rho RT/\mu,\ U = U_{th} + U_{ch}, \tag{5}$$

where U_{th} and U_{ch} are the thermodynamic and chemical components of the internal energy, respectively, which were determined in a similar manner as in [7].

System (1–5) is closed and completely determines the unsteady motion of the H_2-O_2 reaction mixture with variable heat release in the reaction zone behind the detonation wave.

2.2 Boundary conditions

a) at the nozzle entrance (boundary Γ_0: $x = -L_0$; $0 \leq y \leq l$) and at the combustor exit (boundary Γ_2: $x = L$; $0 \leq y \leq l$), the boundary fluxes of mass, momentum, and energy due to the interaction of the gas with the environment at pressure $p = p_a$ were determined using the method of [9].

b) on the left and right boundaries of the domains Ω and Ω_0, the condition of the periodicity of the solution (with period l) was used:

$$F(x, 0, t) = F(x, l, t),\quad -L_0 \leq x \leq L. \tag{6}$$

2.3 Initial constants of the model

The following initial constants of the model were specified (1)-(6):

$$\mu_{H2} = 2 \text{ kg/ kmol},\ \mu_{O2} = 32 \text{ kg/kmol},\ \gamma = 1.4,$$
$$R = 8.3144 \cdot 10^3 \text{ J/(kmol·K)},\ E_d = 110 \text{ kcal/mol}, \tag{7}$$
$$K_+ = 6 \cdot 10^8 \text{ m}^6/(\text{kmol}^2\cdot\text{s}),\ T_0 = 300 \text{ K},\ p_0 = 1.013 \cdot 10^5 \text{ Ps}$$

The following geometric parameters were used:

$$L_0 = 10 \text{ cm},\ L_1 = 1 \text{ cm},\ L = 10 \text{ cm},\ S_\delta/S_\Delta = \delta/\Delta = 0.338,$$
$$S_{exit}/S_\Delta = \Delta_{exit}/\Delta = 3.36,\ S_{\Delta 0}/S_\Delta = \Delta_0/\Delta = 3. \tag{8}$$

2.4 Governing parameters of the model

For fixed geometric dimensions of the domains Ω_0 and Ω and the relations for S(x), the solution of the non-stationary CSD problem (1–8) depends on the governing parameters:

$$p_{H2}^*,\ T_{H2}^*,\ S_{cr},\ p_a,\ l. \quad (9)$$

Here p_{H2}^* and T_{H2}^* are the pressure and stagnation temperature in the H_2 supply system, p_a is the ambient pressure, and l is the period of the problem along the y axis.

3. Calculation results

In the computational domain (-10 cm $< x <$ 10 cm, $0 < y < l$), the following two-dimensional non-stationary periodic problem with a period $l = \pi(d_c - \Delta) = 29.845$ cm and parameters in the H_2 supply system: stagnation pressure $p^*_{H2}/p_0 = 3.133$, stagnation temperature $T^*_{H2}/T_0 = 1$, initial specific hydrogen flow rate $g_{H2} = G_{H2}/S_\Delta = 3.75$ kg/(s $\cdot$m^2), which corresponds to experiments [4], was solved. Here $x = 0$ is the coordinate of the combustor entrance, $S_\Delta = \pi(d_c - \Delta)\cdot\Delta = 14.9$ cm^2 is the cross-sectional area at a distance of 1 cm from the combustor entrance. At the combustor end, a counterpressure $p_a/p_0 = 1$ was specified. Oxygen at rest in the domain Ω_0 and the stoichiometric H_2-O_2 mixture at rest in the domain Ω were used as initial parameters. First, over $\approx$ 1 ms, we calculated the purging of the combustor: supplying H_2 to the initial part of Ω led to O_2 ejection from Ω_0, resulting in the formation of a detonable mixture in the combustor.

The calculations of the initiation of a transverse detonation wave (TDW) and the dynamics of its formation show that by the time t $\approx$ 11.2 ms after the initiation, self-sustained CSD of the TDW is established. The time dependence of the dimensionless pressure $P = p/p_0$ at a fixed point $x = 0.5$ cm, $y = 0$ and the average combustor pressure are presented in figure 2. One can see periodic oscillations with a period $\Delta t \approx 0.127$ ms, which corresponds to a rotation frequency of the TDW $f \approx 7.87$ kHz and a detonation velocity D $\approx$ 2.35 km/s. The height of the TDW front is h $\approx$ 1.8 cm.

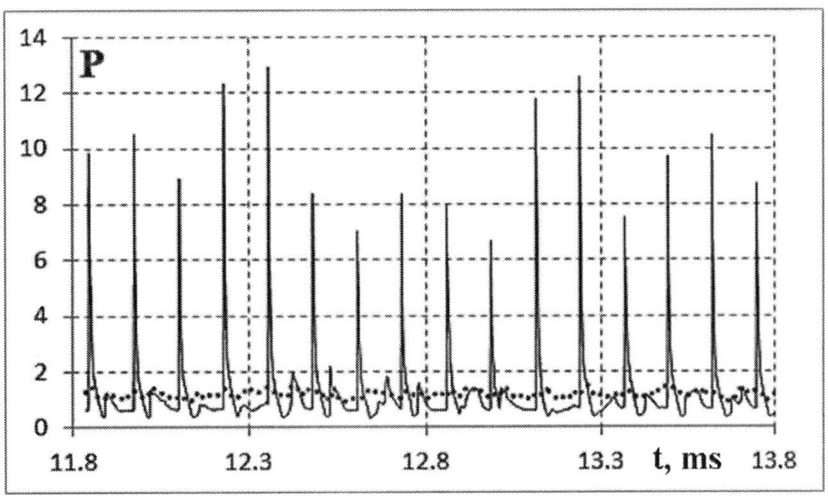

Figure 2. Dependence of the dimensionless pressure $P = p/p_0$ on time t (ms) at the point with coordinates ($x = 0.5$ cm, $y = 0$) (solid line) and the mean pressure (dashed line) at a distance $x = 0.5$ cm from the combustor edge.

The distribution of the dimensionless temperature T/T_0 (a) and the equivalence ratio $\phi = 8(1/z - 1)$ (b) in the computational domain at the time $t = 11.2$ ms are presented in figure 3. It is evident from the temperature field that the TDW front moves from left to right along the triangle of the incoming cold mixture formed by mixing of the ejected hydrogen and oxygen ejected from the environment (x < 0). An oblique shock wave (train) moves down to the left. Figure 3.b clearly shows a light zone of an elevated H_2 content in the mixture ($\phi > 1.5$) behind the TDW front. Note that photo records [4] also showed a low-temperature zone in the TDW structure, presumably due to the inflow of low-temperature hydrogen at high pressure.

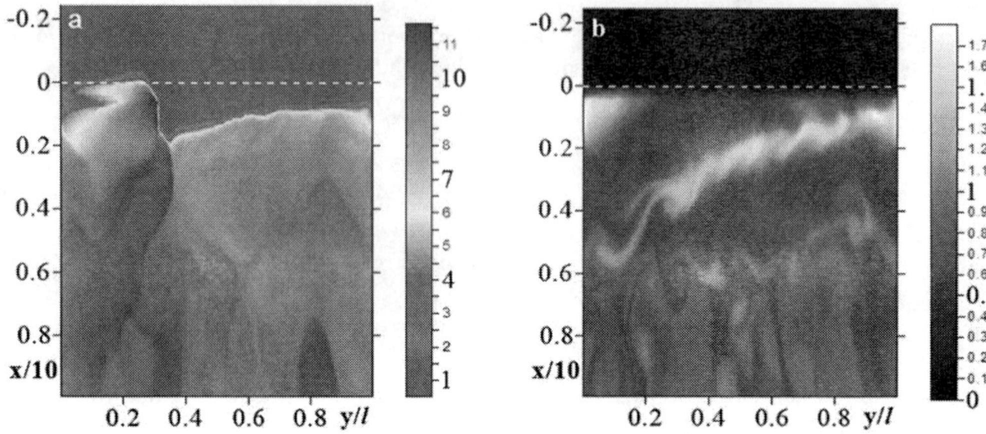

Figure 3. Calculated two-dimensional structure of CSD in the H_2-O_2 mixture in the O_2 ejection mode for $g_{H2} = 3.75$ kg/(s·m^2): (a) – temperature field T/T_0; (b) – equivalence ratio ϕ.

Comparison with experiments for a flow rate $g_{H2} = 3.75$ kg/(s·m^2) shows complete agreement on the number of waves in the combustor (one TDW in the combustor), good agreement on the composition of the mixture: the calculated equivalence ratio $<\phi>$ of the mixture in the combustor is 5% higher than the experimental estimates. Note that the calculated TDW velocity $D \approx 2.35$ km/s was 1/3 higher than that in experiments [4]. This may be due to the ideal model of mixing of the components of the mixture.

We performed calculations of CSD in the O_2 ejection mode for the specific hydrogen flow rate varied in the range $g_{H2} = 3 - 3.75$ kg/(s·m^2). Some results are presented in table 1. Here $<p>/p_a$ is the average static pressure at a distance of 5 mm from the combustor entrance and $n = G_{O2}/G_{H2}$ is the oxygen ejection coefficient.

Table 1. Calculated parameters of CSD in the H_2-O_2 mixture with oxygen ejection.

g_{H2}, kg/(s·m^2)	$<p>/p_a$	h, cm	D, km/s	$<\phi>$	n
3.75	1.19	1.8	2.35	0.88	9.09
3.5	1.19	1.75	2.31	0.83	9.64
3.25	1.18	1.7	2.28	0.76	10.53
3	1.17	1.6	2.19	0.69	11.6

It follows from the calculations that decreasing the specific hydrogen flow rate g_{H2} leads to a monotonic decrease in the velocity and rotation frequency of the TDW, the front height h, and the average static pressure $<p>/p_a$. There is also a noticeable drop in the parameter $<\phi>$, which directly depends on g_{H2}. The obtained numerical results agree with the experimental data of [4], where a decrease in g_{H2} from 3.75 to 2.25 kg/(s·m^2) also led to a decrease in TDW velocity (from 1.76 to 1.57 km/s) and equivalence ratio (0.84–0.47).

We also calculated the average specific impulse per unit mass of fuel $I_{sp,f} = (<p_0> - p_a)\cdot S_{exit}/(g_{H2}\cdot g)$. Here $<p_0> = 1/l\int_0^l (p(L,y,t)+\rho u^2(L,y,t))dy$ is the total pressure at the combustor exit, S_{exit} is the combustor exit area, and g is the acceleration due to gravity.

For all calculation variants listed in table 1, the specific impulse changed slightly and was approximately $I_{sp,f} \approx 1300\pm100$ s. For $g_{H2} < 3$ kg/(s·m^2) in the combustor with a specified period l = 29.845 cm, there was a disruption and attenuation of the TDW with the subsequent establishment of the one-dimensional (without energy release) classical mode of O_2 ejection from the environment due to H_2 supply to the chamber entrance.

4. Conclusions

A quasi-three-dimensional non-stationary mathematical model of continuous spin detonation in a hydrogen-oxygen mixture in the mode of oxygen ejection from the surrounding space was developed. Numerical simulation was performed for geometric dimensions of the combustor corresponding to the experiment. At a specific hydrogen flow rate g_{H2} = 3.75 kg/(s·m^2), a periodic solution with one transverse detonation wave moving at a velocity D ≈ 2.35 km/s was obtained. The structure of the gas-dynamic flow was analyzed. The results show agreement with the experiment for the number of waves in the combustor and the composition of the mixture and satisfactory agreement for the velocity and rotation frequency of the detonation wave. Monotonic dependence of the main parameters of the solution on g_{H2} was found by varying g_{H2} = 3–3.75 kg/(s·m^2) in calculations, which agrees with experimental data. It is shown that in continuous spin detonation of the hydrogen-oxygen mixture in the oxygen ejection mode, the calculated specific impulses per unit mass of fuel are approximately equal to $I_{sp,f} \approx 1300\pm100$ s.

References

[1] Bykovskii F A, Zhdan S A and Vedernikov E F 2006 Continuous Spin Detonations *J. Propul. Power* **22** 1204–16
[2] Bykovskii F A and Zhdan S A 2013 *Continuous Spin Detonation* (Novosibirsk: Izd. Sib. Otd. Ross. Akad. Nauk) p 423
[3] Bykovskii F A, Zhdan S A and Vedernikov E F 2009 Continuous detonation in the regime of nonstationary ejection of the oxidizer *Dokl. Akad. Nauk* **424** 40–2
[4] Bykovskii F A, Zhdan S A and Vedernikov E F 2010 Continuous detonation in the regime of self-oscilatory ejection of the oxidizer. 1. Oxygen as a oxidizer *Combust. Expl. Shock Waves* **46** 344–51
[5] Simonov E V 2017 Modeling of continuous detonation of hydrogen-oxygen mixture in the annular chamber in ejection regime due to the classical model of gas ejector *Sib. J. Pure and Appl. Math.* **17** 55–72
[6] Abramovich G N 1991 *Applied Gas Dynamics. Part 1* (Nauka, Moscow) p 888
[7] Nikolaev Yu A and Zak D V 1988 Agreement of models of chemical reactions in gases with the second law of thermodynamics *Combust. Expl. Shock Waves* **24** 461–63
[8] White D R 1967 Density induction times in very lean mixtures of D2, H2, C2H2, and C2H4 with O2 *Eleventh Symposium (International) on Combustion* (Berkeley, California, August 1966) (Pittsburgh, Pa: Comb. Inst.) pp 147–154
[9] Godunov S K *et al* 1976 *Numerical Solution of Multidimensional Problems of Gas Dynamics* (Moscow: Nauka) p 400

Kerosene combustion in a supersonic flow

P K Tretyakov[1,2]

[1]Khristianovich Institute of Theoretical and Applied Mechanics SB RAS,
4/1 Institutskaya st., Novosibirsk, 630090, Russia
[2]Novosibirsk State Technical University,
20 K.Marksa av., Novosibirsk, 630092, Russia

paveltr@itam.nsc.ru

Abstract. A new approach is proposed to the initiation of kerosene combustion in a supersonic airflow with the stagnation temperature below 2000 K, which consists of a pulsed-periodic gas-dynamic action on the flow. The experiments have shown that it is possible to implement a quasi-steady pulsed combustion mode by choosing the parameters of pulses: energy, duration, and frequency.

1. Introduction

It is well known that in order to arrange for the kerosene combustion in a supersonic airflow certain conditions must be created that ensure mixing with air in the range of concentration limits for flame propagation, a temperature sufficient to initiate ignition, and the ratios of typical times (for flow and combustion) for a stable process. For the ignition in supersonic flow, a high level of the stagnation temperature is required (for numbers $M = 2.0–2.5$) $\geq$ 2000K (see [1]). To reduce it, the use of chemically active agents ([2]) is considered, the handling of which requires special safety measures.

2. Experiment arrangement

A new approach is proposed in the work to the kerosene combustion initiation in a supersonic airflow with the stagnation temperature below 2000K, which consists of a pulsed-periodic gas-dynamic action on the flow.

Figure 1 shows the diagram of a duct, in which the flow parameters at the inlet after the nozzle are close to the conditions of the flight of a flying vehicle with hypersonic velocity. The experiment conditions: the Mach number $M = 2.2$; the total pressure $P_0 = 0.71$ MPa; the total temperature $T_0 = 1680$ K. The kerosene was fed through an injector on the duct axis. The excess air factor was chosen in such a way that at a complete combustion, the flow mean velocity in the duct remained supersonic. The length of the duct part with a constant cross section was subject to change. A special valve using the air and operating according to a given program was applied as a generator of pulses.

Figure 2 shows the package of gas-dynamic pulses, which is created by the generator. The pulses have constant frequency ($f = 52$ Hz), duration (10 ms), and a variable intensity, which is characterized by the air pressure in the generator (MPa). In the cyclogram, the digits identify the regions (1–3), which are typical of different modes of the combustion process. Two combustion modes were observed in the process of experiment.

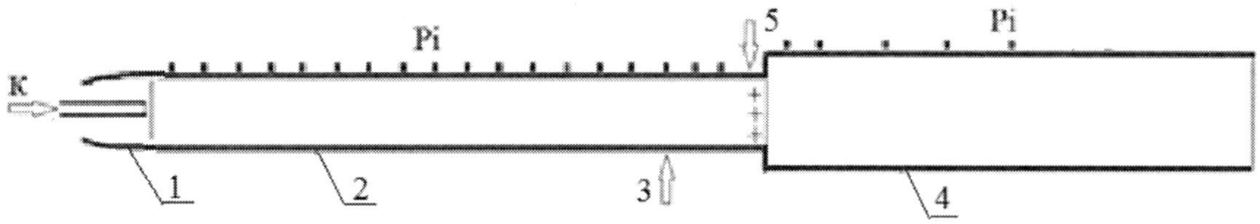

Figure 1. 1 – the nozzle; 2 – the duct of a constant cross section 50 mm in diameter, which passes to a duct with a sudden expansion 90 mm in diameter (see position 4); K is the kerosene supply along the duct axis via the injector; 3 – the introduction of gas-dynamic pulses; 5 – the location of a possible additional supply of the combustible gas; Pi is the pressure on the duct wall.

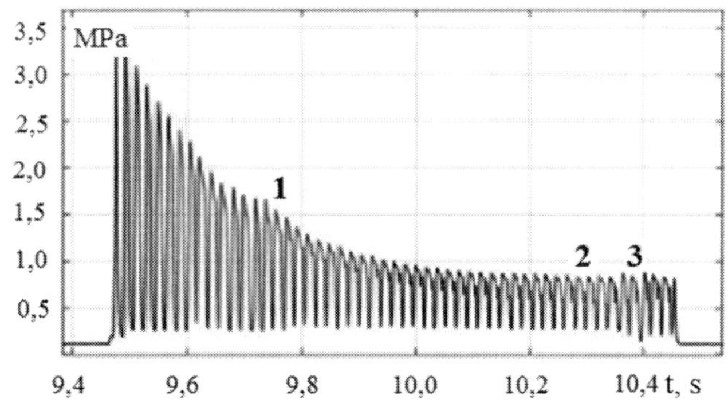

Figure 2. Cyclogram of the packages of pulses.

Shown in figure 3 relative pressure distributions along the duct length reflect the changing nature of the process from the beginning of the initiation of combustion of kerosene at a high intensity of pulses at the beginning of the package (1) to a gradual reduction of the pressure level (2) and the failure of combustion (3). During some period of time (9.6–9.9 s), there was a pulsating combustion (1). At a reduction of the intensity (10.2–10.4 s), the combustion deteriorated (2), and the flame blowout occurred (3).

Experiments have shown that the selection of parameters of gas-dynamic pulses may exercise quasi-stationary pulsating mode of combustion.

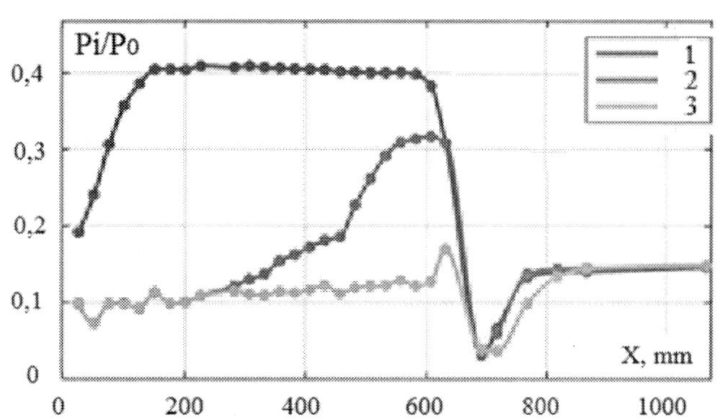

Figure 3. Distribution of relative pressure along the length of the duct.

3. Proof of the combustion effectiveness

A special experiment was made to confirm the feasibility of intensive combustion of kerosene in part of the combustion chamber of constant cross section. The essence of the experiment was as follows. The amount of kerosene was supplied on the basis that when it is completely burned in the part of the duct of constant cross section, the average flow velocity at the outlet of it will be close to the speed of sound (Mach number ≈ 1). It is assumed that it is possible to retain this mode when the supply of the package of pulses is switched off if the combustion process in the expanding part of the channel is organized in such a way that an appropriate backpressure is created. Settings at the nozzle exit are M=2.2; total pressure P_0 = 0.70 MPa; total temperature T_0 = 1680 K. Ahead of the entrance into the expanding part of the duct, ethylene was fed through 32 holes of 2 mm diameter, the combustion of which was stabilized by the recirculation zone behind the sudden expansion. The cyclogram of the mode is shown in figure 4.

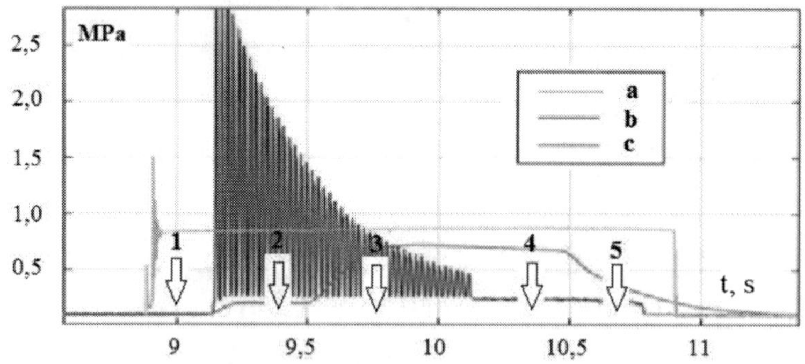

Figure 4. Cyclogram of the experiment. a, b, c – the supply in time of the kerosene, package of pulses, and ethylene after the steadying of the flow exhaustion from the nozzle; time intervals within which the pressure distribution on the wall was recorded. The pressure is measured along the ordinate axis in corresponding devices before the supply to the duct.

Parameters of the package of gas-dynamic pulses: frequency (f = 45 Hz) and duration (12 ms). The coefficient of excess fuel for kerosene φ = 0.17, and ethylene varied within φ = 0.5–0.33.

The results of the experiment are shown in figure 5. From the figure, it follows that in the absence of exposure, the combustion of kerosene is of low intensity, or there is no combustion. The introduction of a package of pulses (modes 2, 3) leads to a noticeable increase in pressure, which indicates the intensive course of the combustion process, and the combustion begins near the injection site of kerosene. At a switch-off of the package of pulses (modes 4, 5), the kerosene combustion process remains intense, the beginning of the combustion is shifted downstream, and the length is approximately four diameters of the duct of constant cross section. Upon termination of the feeding of ethylene, the combustion of kerosene in the duct of constant cross section stops. According to the previously proposed (see [3,4]) quasi-one-dimensional method of estimating the combustion completeness for combustion in a pseudo-shock, the values of the Mach number and the combustion completeness were determined. The Mach number at the end of the section of the duct of constant cross section M≈0.9, and the combustion completeness is close to 1.

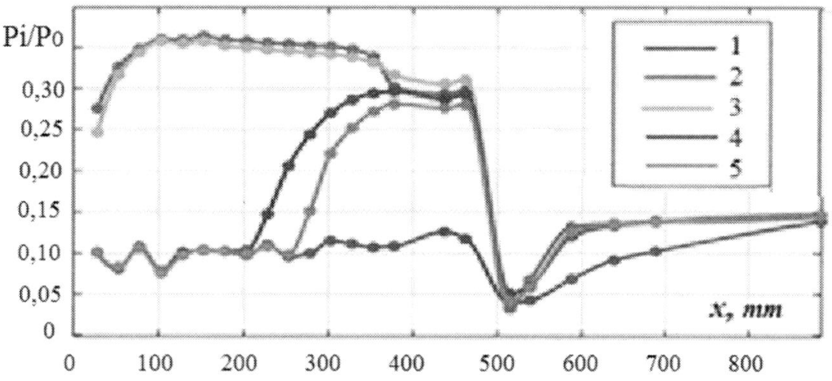

Figure 5. Distribution of relative pressure along the length of the duct for different combustion modes: 1 – kerosene supply prior the introduction of the package of pulses (see figure 4); 2, 3 – combustion modes at the action of the package of pulses and at the beginning (3) of the ethylene supply; 4, 5 – the combustion in the absence of pulses and with the ethylene combustion (5 – at a reduction of the ethylene flow rate).

4. Conclusion

The conducted study shows that it is possible to control the combustion process in a supersonic flow by means of a package of gas-dynamic pulses of variable intensity with the implementation of a pulsating combustion mode. The intense combustion mode was experimentally realized in the part of the duct of constant cross section after switching off the external pulsed-periodic gas-dynamic action when creating at the end of it the corresponding backpressure by organizing the ethylene combustion in the expanding part of the duct.

Acknowledgments

The work was partially carried out within the framework of the program of fundamental scientific research of the state academies of sciences for 2013–2020 (project AAAA17-117030610135-6) and with the support of the RFBR (project No. 17-08-00183).

References

[1] Mestre A and Viand L 1964 *Combustion supersonique dans un canal cylindrique. Supersonic Flow Chemical Processes and Radiative Transfer* ed D B Olfe and V Zakkay (Pergamon Press) pp 93–111

[2] Morris N A, Morgan R G, Paull A and Stalker R J 1987 Silane as an Ignition Aid in Scramjets *AIAA Paper* 87-1636.

[3] Tretyakov P K 1993 *Combustion, Explosion, and Shock Waves* **29** 320-6

[4] Tretyakov P K 1993 *Combustion, Explosion, and Shock Waves* **29** 694-8

 IOP Publishing
1404 (2019) 012071 doi:10.1088/1742-6596/1404/1/012071

Shock wave and detonation flow regimes in gas suspension of reacting gases and inert particles

D A Tropin and E S Bochenkov

Khristianovich Institute of Theoretical and Applied Mechanics SB RAS,
4/1 Institutskaya str., Novosibirsk, 630090, Russia

d.a.tropin@itam.nsc.ru

Abstract. In this research, the interaction of shock waves (SW) with clouds of chemically inert solid particles of silica with a diameter of 100 μm in a hydrogen-air mixture is considered. Depending on the initial volume concentration of particles, three shock wave and detonation flow regimes are realized: 1. attenuation of the SW by a cloud of inert particles, 2. initiation of detonation behind the front of the SW reflected from the rigid wall, 3. initiation of detonation in a cloud of inert particles.

1. Introduction

In the modern world, great attention is paid to the study of the behavior of reacting gas mixtures, since they can be dangerous to humans. For example, they can self-ignite, followed by the initiation of detonation. The danger of reacting gas mixtures in industrial production in which gas mixtures are used as working substances should be highlighted. Therefore, it is worth developing a theory of explosion and fire safety.

2. Formulation of the problem

Studies on the suppression of detonation waves by clouds of inert particles have already been investigated in researches [1], [2]. A shock wave comes out from a cloud of inert particles after the detonation wave was suppressed by the cloud of particles. This shock wave is dangerous because it can transform into a detonation wave due to geometric obstacle (it's shown in figure 1). Figure 1 shows the pressure distribution in the shock tube. The gas parameters in the high-pressure chamber at the initial time: p = 100 atm, T = 3000 K, as a gas O_2 is used. The gas and particle parameters in the low-pressure chamber at the initial time: p = 1 atm, T = 300 K, gas mixture – hydrogen and air, coordinates of a cloud of inert particles 0.5 – 0.75 m, the volume concentration of particles 2×10^{-3}, particle diameter – 100 μm. The bold line in figure 1 shows the distribution of pressure in a shock tube after a SW exits from the cloud of inert particles. Next the shock wave reaches the rigid wall, it is reflected from the rigid wall with the subsequent initiation of detonation. Therefore, it is important to study shock wave attenuation. One of the ways to attenuate shock waves is considered to be the addition of a cloud of inert particles in gas mixtures, but there is currently little research on the interaction of shock waves in reactive gas mixtures with clouds of inert particles. Therefore, the interaction between a shock wave and clouds of inert particles should be studied.

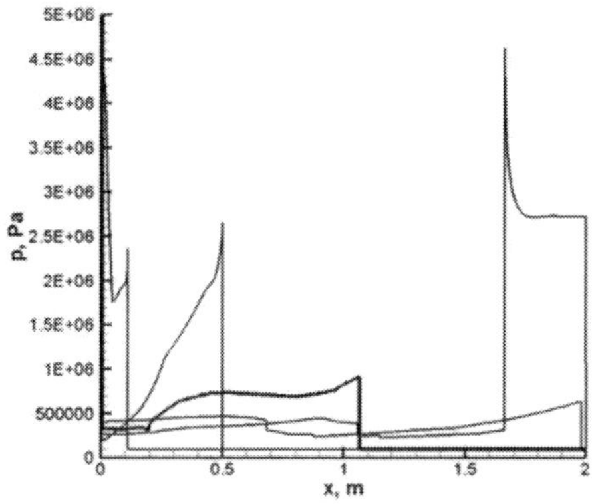

Figure 1. Pressure distribution in the shock tube in the presence of a shock wave.

In this study, the interaction of SW in a hydrogen-air mixture with clouds of chemically inert solid particles of SiO_2 with a diameter of 100 μm is considered and the resulting shock wave and detonation flow regimes are determined. The mathematical model of the mechanics of reacting gas mixtures (hydrogen and air) and inert particles is a system of ordinary differential equations of the mechanics of heterogeneous media and has the form presented in [3]. The detailed kinetics model [4], which takes into account 38 reactions of 8 components, will be used to describe chemical reactions in reacting gas mixtures.

3. Calculations

Initial conditions in the shock tube for all calculations: gas mixture – hydrogen and air; the gas parameters before the accident shock wave front: p – 1 atm, T – 300 K; the velocity of the incident shock wave front is 1120 m/s (the maximum velocity at which self-ignition of the gas mixture doesn't occur behind the front of the incident shock wave).

The first emerging regime is the regime of SW attenuation by a cloud of chemically inert particles (figure 2). A cloud of inert particles with a volume concentration of 5×10^{-3} is located between 0.49 m (the coordinate where the rarefaction wave catches up with the shock wave) and 0.6 m. The shock wave is attenuated, when it passes through the cloud of inert particles, because of interaction between the shock wave and the cloud of inert particles. Ignition with subsequent initiation of detonation isn't observed both behind the front of a passing shock wave and after its reflection from a rigid wall (bold line in figure 2).

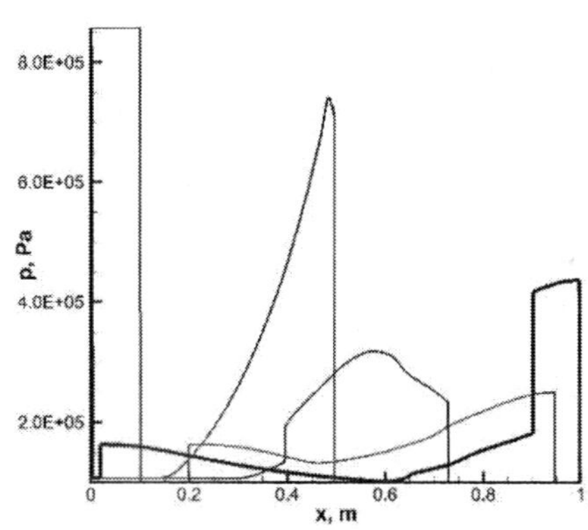

Figure 2. Pressure distribution in the shock tube in the presence of a shock wave with a rarefaction wave. The regime of SW attenuation by a cloud of inert particles.

The second regime is the initiation of detonation behind the front of the shock wave reflected from the rigid wall (figure 3). A cloud of inert particles with a volume concentration of particles 10^{-4} is located near a rigid wall in the interval from 0.35 m to 0.5 m. In the interaction of the SW with a cloud of chemically inert solid particles, part of the shock wave passes into the cloud, and part is reflected from the cloud of particles. The shock wave as it passes through a cloud of particles is weakened by it. The bold line in figure 3 shows the moment of reflection of a shock wave from a rigid wall. Further, behind the front of the reflected shock wave, the mixture ignites, followed by the initiation of detonation.

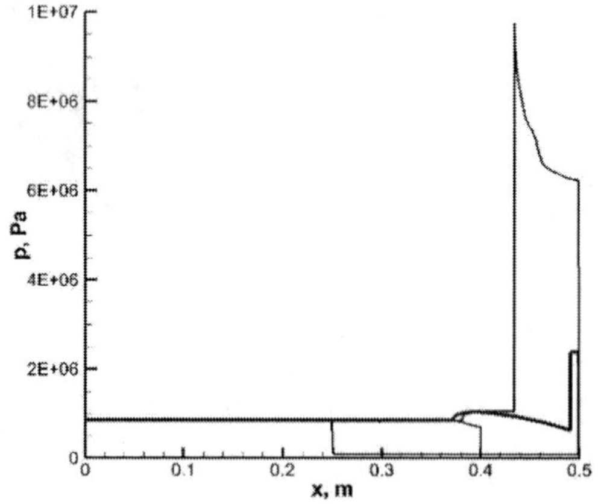

Figure 3. The pressure distribution in the shock tube with a cloud of inert particles with a volume concentration of 10^{-3}. The regime of initiation of detonation behind the front of the SW reflected from the rigid wall.

With an increase in the initial volume concentration of particles to 2.5×10^{-3}, the third regime is realized — initiation of detonation in a cloud of inert particles (figure 4). In this case, the interaction of the shock wave with a cloud of particles and a rigid wall ignition of the mixture isn't observed behind the fronts of the transient and reflected shock waves. Ignition of the mixture with subsequent initiation of detonation occurs in the cloud of inert particles (thick line in figure 4(a)). Figure 4 (b) shows the temperature distribution in the shock tube. Figure 4(b) shows the cause of the initiation of detonation in a cloud of inert particles. The thick line shows the temperature distribution at the moment of gas ignition in a cloud of inert solid particles. It can be seen from the figure that the temperature of the gas in the cloud has increased and this temperature is enough for the self-ignition of the gas. The gas temperature increases due to friction between the gas and the cloud of inert particles.

4. Conclusion

The interaction between SW and clouds of inert particles of SiO_2 with a diameter of 100 μm in a hydrogen-air mixture is considered. As a result of research, three shock-wave and detonation flow regimes were obtained, which is depend on the initial volume concentration of particles. First regime — attenuation of the SW by a cloud of inert particles, second regime — initiation of detonation behind the front of the SW reflected from the rigid wall, third regime — initiation of detonation in a cloud of inert particles.

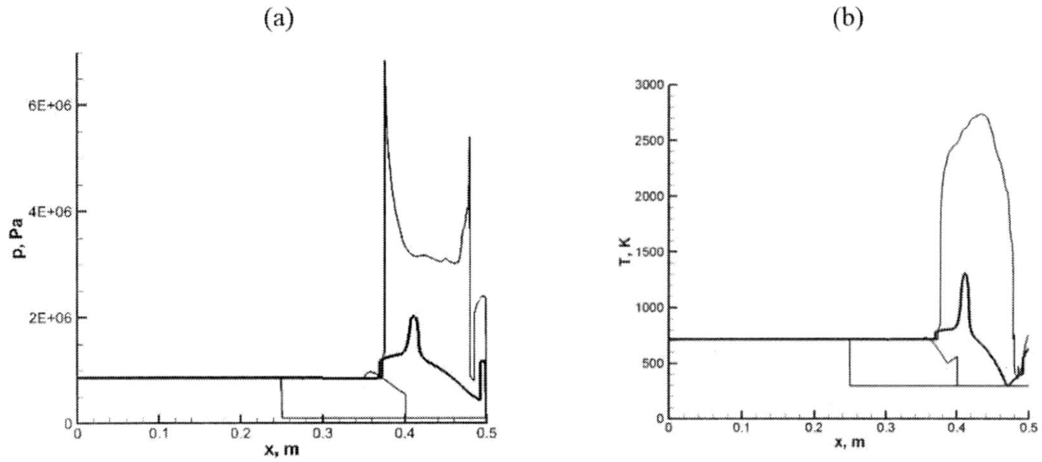

Figure 4. The pressure distribution in the shock tube with a cloud of inert particles with a volume concentration of 2.5×10^{-3}. The regime of initiation of detonation in a cloud of inert particles (a). The temperature distribution in the shock tube with a cloud of inert particles with a volume concentration of 2.5×10^{-3}. The regime of initiation of detonation in a cloud of inert particles (b).

Acknowledgments
The work was supported by Russian Science Foundation (project № 19-79-00096).

References

[1] Tropin D A and Fedorov A V 2019 Effect of inert micro- and nanoparticles on the parameters of detonation waves in silane/hydrogen - air mixtures *Comb. Expl. Shock Waves* **55** 230-6

[2] Fomin P A and Chen J R 2009 Effect of chemically inert particles on parameters and suppression of detonation in gases *Comb. Expl. Shock Waves* **45** 303-13

[3] Fedorov A V and Tropin D A 2013 Modeling of detonation wave propagation through a cloud of particles in a two-velocity two-temperature formulation *Comb. Expl. Shock Waves* **49** 178-87

[4] Bedarev I A and Fedorov A V 2006 Comparative analysis of three mathematical models of hydrogen ignition *Comb. Expl. Shock Waves* **42** 19-26

Journal of Physics: Conference Series **1404** (2019) 012072 doi:10.1088/1742-6596/1404/1/012072

Flame stretching by electric field

A V Tupikin[1] and V V Zamashchikov[2]

[1] Khristianovich Institute of Theoretical and Applied Mechanics SB RAS, 4/1 Institutskaya, Novosibirsk, 630090, Russia
[2] Voevodsky Institute of Chemical Kinetics and Combustion SB RAS, 3 Institutskaya st., Novosibirsk, 630090, Russia

tupikin@itam.nsc.ru, albor@kinetics.nsc.ru

Abstract. The paper considers the issues related to the study of the mechanisms of controlling the combustion of hydrocarbon fuels by electric fields. It is shown that influence of weak electric field on combustion in laminar mode leads to a change in the degree of flame stretching (Karlovitz number). For the Bunsen flame and reversed flame, estimates of changes in values of the degree of flame stretching when imposing an external constant electric field on flame are made.

1. Introduction

Nowadays, the ways to control burning aiming at achieving maximal energy saving at minimal level of hazardous substances in the combustion products is being actively studied. One of the perspective directions is associated with electric current impact on flame, which is illustrated by a number of works in this field [1-7]. The research covers a wide range of issues from controlling flame acoustical disturbances [1] to controlling burning in large scale combustion chambers [3]. Also, the works consider both combustion of poor homogenous mixtures [4] and diffusion flames [7]. The sensibility of hydrocarbon fuels combustion to the impact of electric fields is associated with the presence of chemoionization reactions [8]. Depending on hydrodynamic conditions and electrical field intensity, various mechanisms can play leading role: heat extraction from current passage (ohmic heat), changes in chemical kinetics or electro-hydrodynamic impact (ionic wind). This work is concerned with weak electric fields, where E[B/cm]/P[тор]<2 (E is the electric field intensity, P is the medium pressure) and electric power heats insignificantly the mixture in the pre-flame zone (IU << GCpT0, where I is the flowing current, U is the applied voltage, G, Cp, and T0 are the mixture bulk flow, its heat capacity at constant volume, and temperature). Paper [9] shows that the impact of weak electric field on flame is localized in the area of chemical reactions, and the physical mechanism is probably linked to flame extension. Putting it another way, the electrical force changes pressure distribution on the flame front, which results in changes in the heat and mass transfer balance causing stretching or contraction of the flame front. With regards to the assumption mentioned above, this work analyzes the experimental data on the field influence on laminar flame. Two variants are considered: Bunsen flame and reversed flame.

Journal of Physics: Conference Series **1404** (2019) 012072 doi:10.1088/1742-6596/1404/1/012072

2. Flame stretch factor

A general expression for flame stretch factor was suggested by Williams [10], it is written as $\gamma = \dfrac{d(\ln A)}{dt}$, where γ is the stretch factor, A is small flame element surface area. To determine the value of γ it is necessary to follow the evolution of area of selected flame element in time. The approach suggested in [11] was used for determining the flame stretch factor, where an element of the flame surface was defined in vector form as $\vec{r} = \vec{r}(p,q,t)$, where p and q are generalized coordinates. Subsequently, the area of small surface element of thin infinite flame will be $\Delta A(t) = \left(\dfrac{\partial \vec{r}}{\partial p} \times \dfrac{\partial \vec{r}}{\partial q}\right)\vec{n}\Delta p \Delta q$ ($\vec{n}$ is vector of normal to the surface at a given point). In such a case, the stretch factor γ is a sum of two summands. The first one is resulting from changing the tangential component of mixture velocity along flame surface $\vec{v}_t$, the second one is proportional to the normal component of flame surface movement velocity and is equal to 0 in stationary case. Then, the ratio for stretch factor will be $\gamma = \dfrac{\left|\dfrac{\partial \vec{r}}{\partial p}\right|\left(\dfrac{\partial \vec{v}_t}{\partial q} \cdot \vec{e}_q\right) + \left|\dfrac{\partial \vec{r}}{\partial q}\right|\left(\dfrac{\partial \vec{v}_t}{\partial p} \vec{e}_p\right)}{\left|\dfrac{\partial \vec{r}}{\partial p} \times \dfrac{\partial \vec{r}}{\partial q}\right|}$, where $\vec{e}_q$ and $\vec{e}_p$ are unitary vectors in the plane tangent to the flame surface.

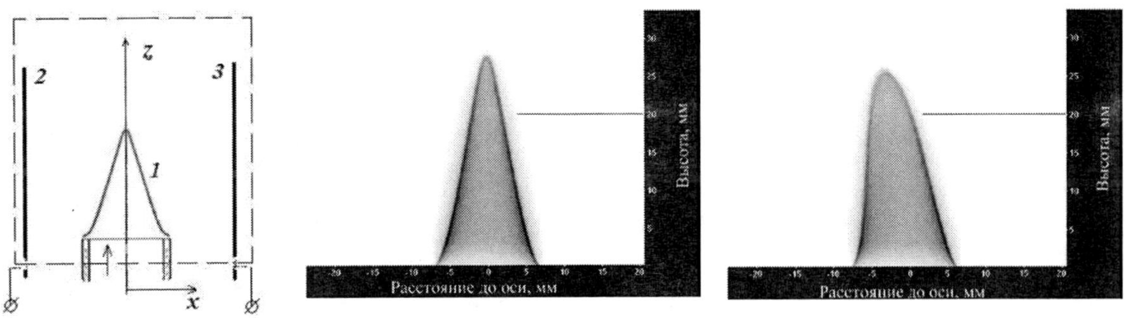

(a) (b) (c)

Figure 1. Bunsen flame in electrical field: (a) scheme of experiment; (b) and (c) spectrozonal photography (without field and in electrical field, respectively).

First, let us consider the Bunsen flame placed between two plane-parallel plates, where the voltage was applied (see figure 1). The experiments were conducted under normal conditions. The average consumption mixture velocity u_{av} = 1.25 m/s, the fuel excess ratio φ = 0.83, the electrical field intensity $E \approx 1$kV/cm ($E/P \approx 1.3$). Current intensity in the electrical circuit did not surpass 1 mA under the voltage of 3 kV. In the absence of field, the Bunsen flame is a cylindrically symmetric surface, where the mixture flow velocity $u(x)$ is directed along the axis z (let us direct the axis z along the axis of symmetry, x is a coordinate perpendicular to z). For the generalized coordinates, $p=x$ and $q=z$ are taken. In such a case, γ can be represented as a sum of three summands:

$$\gamma = \frac{u\dfrac{d^2Z}{dx^2}}{\left[1+\left(\dfrac{dZ}{dx}\right)^2\right]^2} + \frac{u\dfrac{dZ}{dx}}{x\left[1+\left(\dfrac{dZ}{dx}\right)^2\right]} + \frac{\dfrac{du}{dx}\dfrac{dZ}{dx}}{\left[1+\left(\dfrac{dZ}{dx}\right)^2\right]}$$

($Z(\rho)$ is equation of the flame surface), two of them are associated with the flame deformation, and one is associated with the flame stretching.

When switching the electrical field, the axisymmetry is disturbed, but the plane (passing though the axis of burner and perpendicular to the electrodes) is the plane of flame symmetry. $p=x$ and $q=z$ are accepted as generalized coordinates. The surface of flame is defined as $Z(\vec{r}) = x\vec{i} + y\vec{j} + z\vec{k}$, but it does not depend on the component y because of the symmetry. In such a case the stretch factor can be represented as:

$$\gamma = \frac{u_z \dfrac{d^2 z}{dx^2}}{\left[1+\left(\dfrac{dz}{dx}\right)^2\right]^2} + \frac{\dfrac{du_z}{dx}\dfrac{dz}{dx}}{1+\left(\dfrac{dz}{dx}\right)^2} - \frac{u_x \dfrac{dz}{dx}\dfrac{d^2 z}{dx^2}}{\left[1+\left(\dfrac{dz}{dx}\right)^2\right]^2} + \frac{\dfrac{du_x}{dx}}{1+\left(\dfrac{dz}{dx}\right)^2}$$

Here, as in the previous case, there are two groups of summands associated with flame front deformation and stretching. Thus, two phenomena can be observed separately when electrical field affects combustion: flame deformation and stretching.

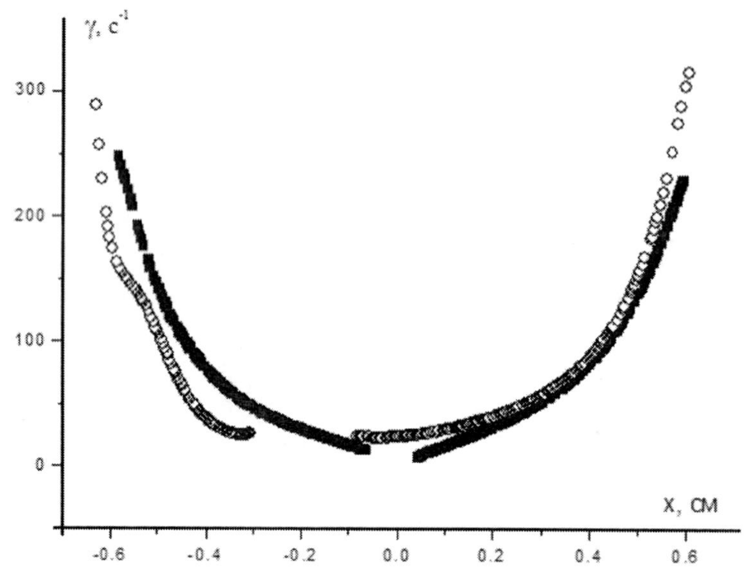

Figure 2. Flame stretch factor: square markers – without field, round markers – with field.

As for Bunsen flame, the main cause of deformation is expansion in gradient flow, and to determine the efficiency of electrical field affecting flame, we can introduce comparison of the corresponding components. Figure 2 pictures the dependences of stretch factor on the coordinate x in the case with electrical field and without it. A decrease in stretch factor can be noticed from the cathode side as affected by electrical field, while its influence from the anode side is insignificant.

In case of the reversed flame, when activating the electrical field, the symmetry remained (see figure 3). The experiments were performed at average velocity u_{av}=1.2 m/s of mixture efflux into the atmosphere, the equivalence ratio φ = 0.75. The voltage was varying from 0 to 4 kV at maximal current in the circuit I=250 mkA. The intensity of electric field was not superior to E=1.4 kV/cm ($E/P\approx 1.8$), approaching the limit value. In [12], the stretch factor is shown in dimensionless form (one of the representations of Karlovitz number): $K = \frac{d(\ln A)}{dt} \cdot \frac{\delta_0}{u_n}$, where u_n is normal velocity of burning, and δ_0 is thickness of pre-flame zone (heat penetration zone).

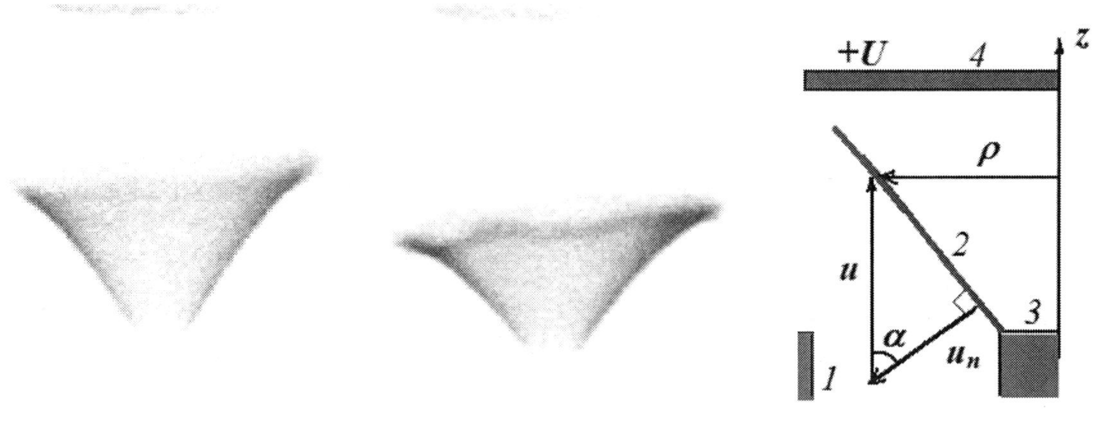

(a) (b) (c)

Figure 3. Flame shape: multispectral survey ((a) without field, (b) with field), and flow diagram (c).

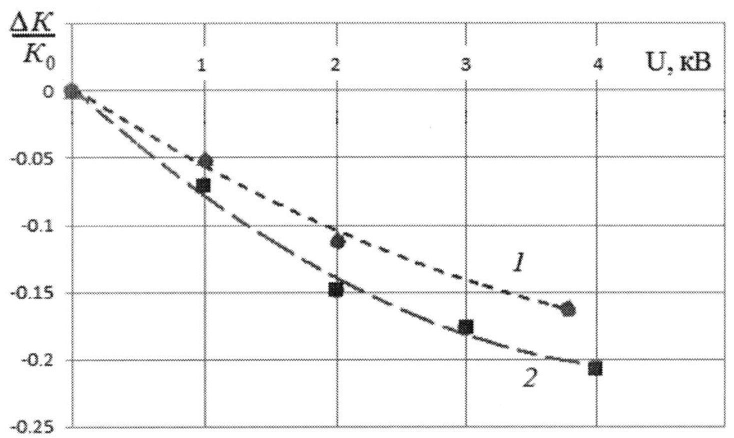

Figure 4. Ratios of criterion relative variation:
the round markers – propane; the square markers – methane.

Thus, in the case represented, using the equation from [12] for flame with axial symmetry, one can study relative variation of Karlovitz number as a function of applied voltage $\Delta K/K_0 = f(U)$, where K_0 is the value of criterion in the case without electrical field. With field intensity increasing, the rate of change $\Delta K/K_0$ (see figure 4) lowers, i.e. there is a value limit of the effect produced by the electrical field.

3. Conclusions

Consequently, the experimental data with pre-mixed fuel-air mixtures show that the electrical field impact occurs locally at the front of combustion and results in changing the degree of flame stretching. Therewith, the results of electrical field impact on combustion can be considered as two independent phenomena: flame deformation and flame stretching. Note also that the electrical field influence on Karlovitz number is limited both by the conditions (*E*/*P*<2) and by the effect magnitude.

Acknowledgments
The research was (partly) carried out within the framework of the Program of Fundamental Scientific Research of the state academies of sciences in 2013-2020 (projects No. AAAA-A17-117030610135-6 and AAA-A17-117112 00018-4).

References

[1] Volkov E N, Kornilov and de Goey L P H 2013 *Proceedings of the Combustion Institute.* **34(1)** 955
[2] Murphy D C, Sanchez–Sanz M S and Fernandez-Pello C 2014 *IOP Conf. Series: Journal of Physics: Conf. Series* **557** 012076
[3] Barmina I, Cipijs A, Valdmanis Ja, Valdmanis R, Zake M 2015 *Engineering for rural development Jelgava* 516
[4] Jianfeng Fang, Xiaomin Wu, Hao Duan, Chao Li and Zhongquan Gao 2015 *Energies* **8** 2587
[5] Tupikin A V, Tretyakov P K, Denisova N V et.al. 2016 *Combustion, Explosion, and Shock Waves* **52(2)** 167
[6] Paulo Roberto Salvador, Kunning Gabriel Xu 2017 *AIAA Paper* **2017-1585** Grapevine (TX)
[7] Tupikin A V, Tretyakov P K and Venediktov V S 2017 *Combustion, Explosion, and Shock Waves* **53(1)**32
[8] Colcote H F 1957 *Combustion and Flame* **1** 385
[9] Tretyakov P K, Tupikin A V, Denisova N V et. al. 2012 *Combustion, Explosion, and Shock Waves* **48(2)** 130
[10] Williams F A 1975 AGARD Conference Proceedings **164**.
[11] Chung S H and Law C K 1984 *Combustion and Flame* **55** 123
[12] Strehlow R A and Savage L D 1978 *Combustion and Flame* **31** 209

 1404 (2019) 012073 doi:10.1088/1742-6596/1404/1/012073

Modelling of mixing and ignition of a cold hydrogen jet in a supersonic hot air flow

O S Vankova[1] and N N Fedorova[1,2]

[1] Khristianovich Institute of Theoretical and Applied Mechanics, Siberian Division of Russian Academy of Sciences, 4/1 Institutskay st., Novosibirsk, 630090, Russia
[2]National Research Novosibirsk State University, 1 Pirogova st, Novosibirsk, 630090, Russia

vankova@itam.nsc.ru

Abstract. The paper presents the results of numerical studies of self-ignition and supersonic combustion of cold hydrogen jet supplied to the M=2 vitiated air flow. The calculations are carried out using the ANSYS Fluent on the basis of the full Favre-averaged Navier-Stokes equations supplemented by the κ–ω SST turbulence model and several kinetic mechanisms including 1, 16 and 38 reactions. The results of calculations of mixing of a cold hydrogen jet and a hot inert mixture are in good qualitative and quantitative agreement with experimental data. The results obtained for the reactive flow show that in the calculation the combustion layer is closer to the channel bottom wall than in the experiment, and the temperature of the reaction product is by 10 % lower.

1. Introduction

Turbulent combustion is one of the most important issues in systems such as rockets, internal combustion engines, industrial burners and furnaces [1]. Therefore, the study of the phenomenon is a key task for the development and improvement of practical systems in terms of increasing efficiency, reducing fuel consumption and the formation of pollutants.

After validation on verified experimental data and available theoretical formulas, simulation methods allow one to better understand impact of various physical mechanisms participating in combustion, to analyze the stability of the process and to predict the formation of pollutants. The structure of the turbulent flame in the combustion chambers is controlled by turbulence, chemical kinetics, acoustics, and radiative heat transfer. The choice of the turbulence model, the kinetic mechanism, and the description of the Turbulence and Combustion Interaction (TCI) are the key points in the numerical simulation of supersonic combustion processes [1][2]-[4].

The usage of models based on detailed kinetic mechanisms requires huge computer resources, especially in the case of a large number of chemical reactions. The flamelet concept based on the assumption that the chemical reaction times are shorter than the turbulent time scales help to significantly reduce the necessary computer resources. This approach allows chemical reactions to be calculated independently of the main flow calculations using special tables containing the values of chemical parameters for various values of scalar quantities.

Most of works on supersonic combustion uses simplified / reduced chemical mechanisms and the explicit transfer of chemical components involved in the process. In the framework of this approach, closure is necessary for the ensemble-averaged source term in the transport equations of chemical

 1404 (2019) 012073 doi:10.1088/1742-6596/1404/1/012073

components. This can be achieved using a simple but less accurate model based on using the Arrhenius law for averages, as well as using the Eddy Dissipation Concept (EDC) or algebraic or transport equations for the probability density function (PDF). Due to the strong nonlinearity of the source term and a wide range of scales due to combustion, the transport equations are rigid and therefore difficult to solve.

At present, hydrogen is considered as one of fuels of the future. In order to control hydrogen self-ignition and stable combustion, a full understanding of these processes is vital. The study of hydrogen combustion in supersonic flows presents great difficulties, both theoretically and experimentally. A theoretical study is complicated by the fact that the laws of combustion at supersonic speeds are determined by the intensity of the processes of turbulent exchange, the rates of chemical reactions in the stream, and the influence of gas-dynamic effects accompanied by heat generation. Moreover, each of these factors can have a significant impact on the processes of mixing and combustion [5].

The purpose of the research is to present a numerical study of the processes of mixing, self-ignition and combustion of cold hydrogen jet supplied into a supersonic (M=2) flow of hot vitiated air. We use a laminar combustion model that does not take into account subgrid fluctuations in the composition and temperature of the mixture. The simplicity of this approach explains its application for obtaining the first approximation for further more complex models.

As a test case, the problem is chosen of mixing and ignition of a cold hydrogen jet supplied into the supersonic flow (M=2) in an expanding channel. The case was experimentally investigated in [6] both for a non-reactive flow with the inert main flow gas and a reactive flow in which the hot vitiated air was supplied to the channel. These experimental data were used for mathematical model validation in [7] where a good agreement was obtained between the experimental and computed profiles of composition and Pitot pressure.

2. Problem setup and methods

The problem is computed with the same geometry as in the experiments [6] (Figure 1 (a)). The length of the combustion chamber is L = 0.406 m, its height at the inlet is H_{inlet} = 0.089 m and at the outlet H_{outlet} = 0.1048 m. The hydrogen is injected parallel to the channel bottom wall at a distance of 0.2 m from the channel inlet. The height of the slot for hydrogen injection is h = 0.004 m, the lip thickness is h_j = 0.00076 m. The scheme of 2D computation domain is presented in Figure 1 (b) where the positions of inlet, outlet sections and the origin of the coordinate system are shown. In the computations, the direction of x-axis coincides with the channel bottom wall.

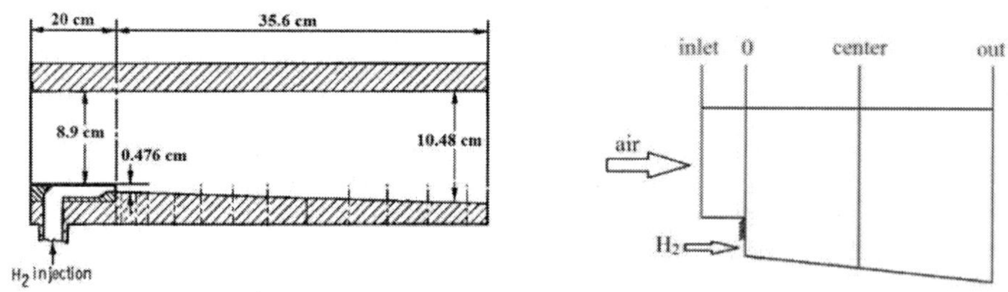

(a) (b)

Figure 1. The experimental scheme [6] (a) and computation domain (b).

The initial data for the free flow and hydrogen jet are given in Table 1. Experimental data set [6] includes two runs, the first for the inert mixture with a low oxygen concentration, and the second for the reactive mixture with a high oxygen concentration.

Mathematical modeling of flows in the combustion chamber is carried out using the ANSYS Fluent software on the basis of the Favre-averaged Navier-Stokes equations of viscous heat-conducting gas, supplemented with the k–ω SST turbulence model and the conservation law equations for multispecies mixture. A density –based solver and the AUSM scheme [8] of third order approximation are applied.

To simulate the combustion process of a hydrogen-air mixture the detailed kinetic schemes including 1, 16, and 38 reactions [[9], [10]] are used. Previously, in [11], 12] these kinetic schemes were verified by the authors using the experimental data [7].

Table 1. Initial data.

	Mach number	Static temperature, K	Velocity, m/s	Static Pressure, MPa	Species mass fraction H_2	O_2	N_2	H_2O
Jet	1	254	1216	0.1	1	0	0	0
Inert flow	2.44	1270	1764	0.1	0.0005	0.0509	0.7324	0.2162
Reactive mixture	2.44	1270	1764	0.1	0	0.258	0.486	0.256

Structured computational grids are constructed in ANSYS Meshing using sizing instruments. The coarse, medium and fine grids consist of about $2\cdot10^5$, $3\cdot10^5$ and $5\cdot10^5$ cells, respectively. For the "cold" case, a grid convergence study has been performed to ensure the results of computations on various grids differ by no more than 5 percent. Most of the computational results presented below are obtained using the medium grid. The y^+ values do not exceed 30 for the upper channel wall. For the bottom wall, it lays in a range of $1 \div 5$ that allows resolving the laminar sublayer of the turbulent boundary layer.

At the inlets, the static pressure, static temperature, Mach number and species mass fraction are prescribed in accordance with data of Table 1. At the outlet, static pressure and temperature are applied corresponding to standard atmospheric conditions. Since the main part of the flow at the outlet is supersonic, this conditions works only a very narrow subsonic part of the boundary layer. At the walls, the no-slip conditions for velocity are applied, and for temperature, cold wall condition Tw=300K is applied in accordance to experiments [6].

3. Results and discussion

The solution of the problem is implemented in several stages. First, the flow of inert/reactive mixture is simulated in a plane channel without hydrogen injection. The profiles of gas-dynamic and turbulent parameters are obtained having the boundary layer thickness equal to that in the experiments. These profiles are recorded and set at the inlet when computing flows in the combustion chamber.

At the second stage, the "cold" flow is simulated in which the mixing of hydrogen and in the inert gas mixture is studied. At the third stage, the "hot" flow is computed taking into account a hydrogen ignition in the wet air supersonic flow.

3.1. "Cold" case results

Figure 2 shows the profiles of dimensionless total temperature $T0/T_{ref}$, T_{ref} =2380K, in the $x = 0$ cm cross-section. The graph shows that the thickness of the boundary layer on the channel walls is approximately equal to $\delta \approx 1$ cm, and a good agreement between the calculated curve and the experimental points has been obtained.

In Figure 3, the computed distributions of static pressure are presented on the upper and lower walls of the combustion chamber for the "cold" case. The graph shows that at the inlet to the channel, the pressure does not change significantly. At $x = 0$, the pressure drop occurs on the lower wall which is due to the jet. Then the pressure rises sharply up to the level of $P_{st} \approx 95000$ Pa and remains practically constant till the outlet. Within $1 < x < 7$, small pressure drops are visible on the bottom wall. On the upper wall, in the region $13 < x < 18$, the pressure drop occurs due to action rarefaction wave originated at the injection slit, after which pressure are recovered and remain approximately constant $P_{st} \approx 88000$ Pa till the exit.

Figure 4 shows the calculated static pressure field for the "cold" case. The picture shows that two rarefaction waves emanate form on the lip. The first wave propagates downstream, then falls on the upper wall and reflects from it. Namely this wave causes the pressure fall that is seen in Figure 3. The second rarefaction wave falls and reflects from the bottom wall and the contact surface formed by the upper jet boundary. These results in a formation of a reflected compression wave expanding to the upper wall and a series of weak expansion /compression waves that can be seen in the transonic near-wall regions near the bottom wall (Fig. 4 (b)).

Figure 5 presents the computed profiles of the molar fraction of the mixture components (H2, N2, and H2O) in the cross-section x = 35.6 cm together with experimental points. The graph shows good qualitative and quantitative coordination of data.

Thus, as a result of calculations of the "cold" flow, it is found that the mathematical model quite accurately describes the process of mixing a cold hydrogen stream with a supersonic external stream of moist hot air.

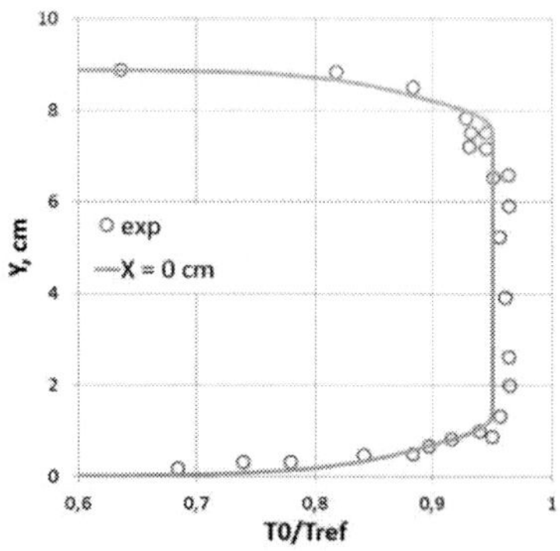

Figure 2. Profile of dimensionless total temperature in the x = 0 cross-section.

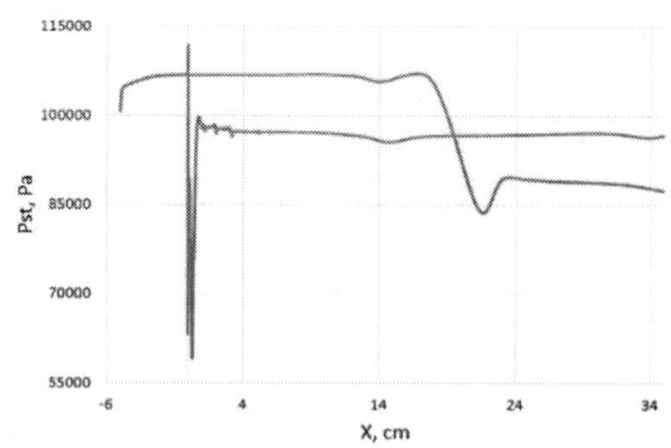

Figure 3. Computed static pressure distributions along the top (red line) and bottom (blue line) walls.

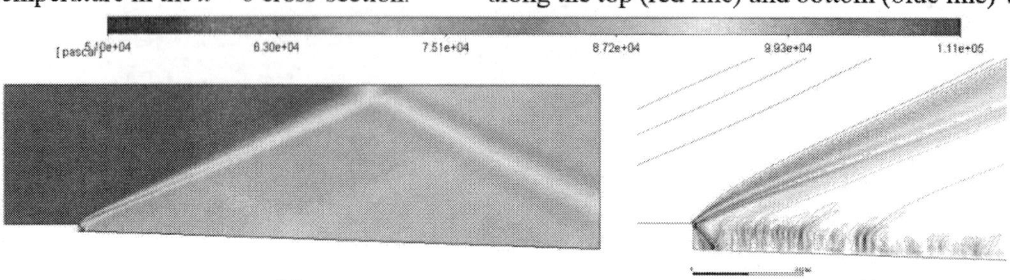

(a) (b)

Figure 4. Static pressure field in the whole computation domain (a) and in the vicinity of the jet injection slit (b).

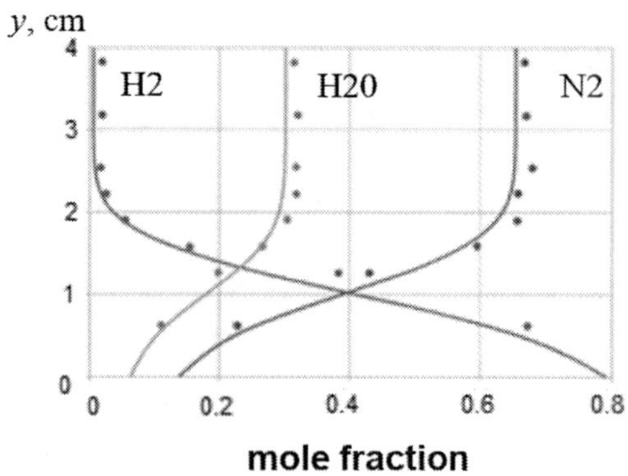

Figure 5. Computed and experimental mole fraction profiles of the mixture components in the cross-section x = 35.6 cm.

3.2. "Hot" case results

After getting the cold flow field, the blocks of chemical kinetics containing 1, 16 and 38 reactions for hydrogen combustion in air are activated in the computation model, and the "hot" flow is computed taking into account a hydrogen ignition in the wet air supersonic flow.

Figure 6 shows the calculated static temperature in this case. The figure shows that almost constant temperature is maintained in the most part of the channel except the region of the weak expansion and compression waves generated at the injection slit. The formation of a mixing layer at the periphery of the cold jet can be seen. A significant increase in temperature up to $T_{st} \approx 1500$ K testifies on the hydrogen self-ignition. At the same time, an area with a low temperature and high hydrogen concentration remains near the wall till the exit section. In Figure 7, the profiles of dimensionless temperature $T0/T_{ref}$ are presented at the $x = 0$ cm and $x = 35.6$ cm cross-sections. The calculated data are in a rather good qualitative and quantitative agreement with the experiment except the combustion region $1 < y < 3$ cm in which the total temperature is 10% underpredicted.

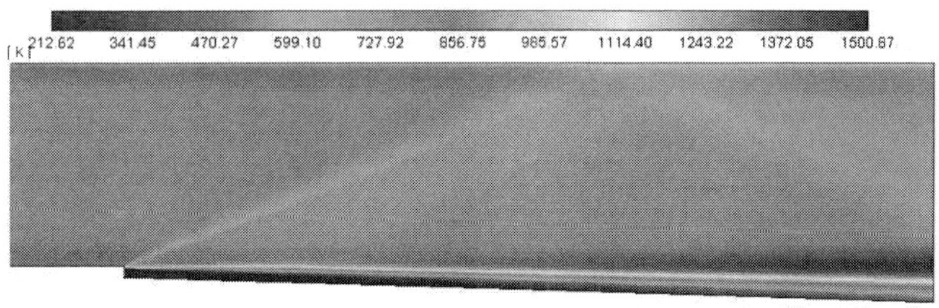

Figure 6. Computed static temperature field in "hot" case.

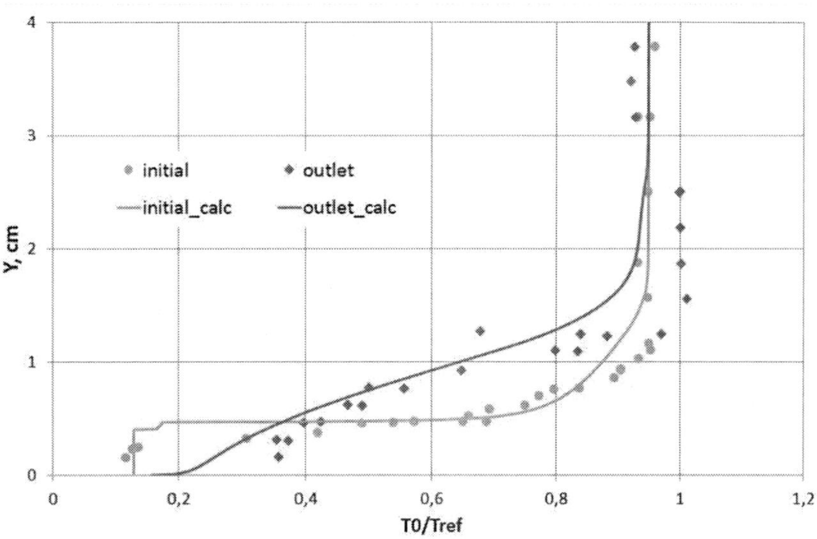

Figure 7. Profiles of dimensionless total temperature)
at the inlet ($x = 0$ cm) and outlet ($x = 35.6$ cm) of the combustion chamber.

Figure 8 shows the profiles of the molar fraction of the mixture components in the section X = 35.6 cm computed with kinetic schemes consisting of 1, 16 and 38 reactions indicated by solid, dotted, and dashed lines, respectively. The calculated curves for the H2, O2 and N2 components qualitatively coincide with the experimental points, but the maximum peak of the H2O molar concentration profile is shifted to the low channel wall in the computations comparing to the experimental data. It should be noted that the peak concentration of the main reaction product is well predicted by kinetic schemes containing 1 and 16 chemical reactions, while the most detailed scheme with 38 reactions, underestimates the maximum value of H20 concentration by 20%.

4. Conclusion
The processes of mixing, ignition, and combustion of a cold hydrogen jet supplied parallel to the main hot wet supersonic air flow in are simulated using ANSYS Fluent. The calculated data are compared with the experimental data [6]. The results of calculations of mixing of a cold hydrogen jet and a hot inert mixture are in good qualitative and quantitative agreement with experimental data.

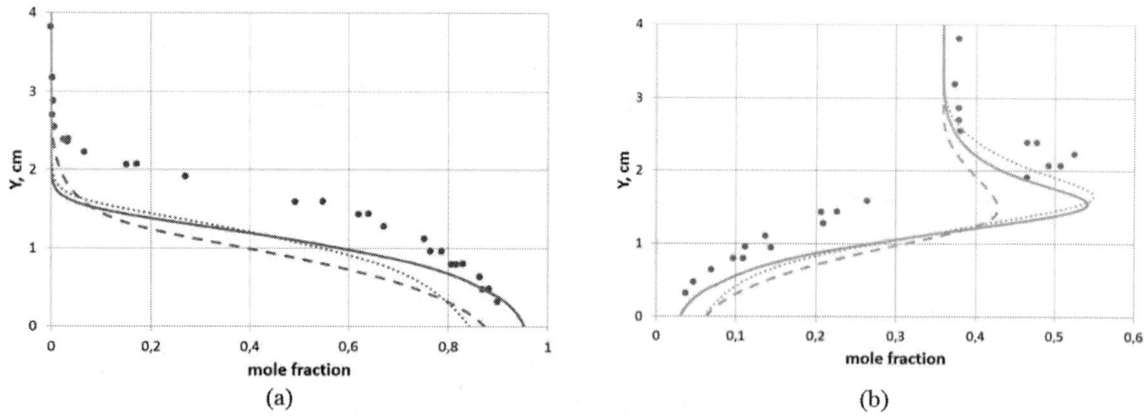

Figure 8. Profiles of the H2 (a) and H2O (b) mole fraction for the "hot" case at $x = 35.6$ cm.

The results obtained for the reactive flow show that in the calculation the combustion layer is closer to the lower wall than that in the experiment. Differences in the calculated and experimental data can be due to insufficient grid resolution of wave structures in the flow, three-dimensional effects, and also due to neglecting the processes of interaction of turbulence and combustion. All these factors will be taken into account in future work.

5. Acknowledgments
The research was carried out within the framework of the Program of Fundamental Scientific Research of the state academies of sciences in 2013-2020 (project No. AAAA-A17-117030610139-4) under the partial support of the Grant №17-08-01158 A by the Russian Foundation for Basic Research.

References
[1] Poinsot T, Candel S and Trouv´e A 1995 *Prog. Energy Comb. Scien.* **21** 531–76
[2] Gonzalez-Jueza E D, Kerstein A R, Ranjanc R and Menonc S 2017 *Prog. Energy Comb. Scien.* 60 26-67
[3] Kong D, Eckhoff R K and Alfert F 1995 *J. Hazard. Mater.* **40** 69 – 84
[4] Kong S C and Reitz R D 2002 *Trans. Amer. Soc. Mech. Eng.* **124** 702 - 7
[5] Curran E T, Heiser W H and Pratt D T 1996 *Ann. Rev. Fluid Mech.* **28** 323-60
[6] Burrows M C and Kurkov A P 1973 *Analytical and Experimental Study of Supersonic Combustion of Hydrogen in a Vitiated Airstream* NASA TM X-2828 (34 pp
[7] Evans J S and Schexnayder C J Jr 1980 *AIAA Jour.* **18** 188-93
[8] Liou M S and Steffen C J Jr 1993 *J. Comp. Phys.* **107** 23–39
[9] Tien J H and Stalker R J 2002 *Combus. Flame.* **130** 329–48
[10] Ladeinde F 2009 A Critical Review of Scramjet Combustion Simulation *AIAA Paper 2009-127* 13 pp
[11] Bedarev I A and Fedorov A V 2006 *Combust. Expl. Shock Waves* **42** 19–26
[12] Fedorov A V, Fedorova N N, Vankova O S and Tropin D 2018 *AIP Conf. Proc.* **1939** 020019

Journal of Physics: Conference Series **1404** (2019) 012074 doi:10.1088/1742-6596/1404/1/012074

The correlation of the pulsations of flow in the settling chamber with the pulsations of supersonic flow

L V Afanasev[1,2], V L Kocharin[1], A A Yatskikh[1,2], A D Kosinov[1,2], N V Semionov[1], Yu G Yermolaev[1,2] and G L Kolosov[1]

[1]Khristianovich Institute of Theoretical and Applied Mechanics SB RAS, 4/1, Institutskaya str., Novosibirsk, 630090, Russia
[2]Novosibirsk State University, 1, Pirogova str., Novosibirsk, 630090, Russia

l.afanasev@itam.nsc.ru

Abstract. The relationship of the pulsations of flow in the settling chamber with the pulsations of supersonic flow in the test section of the T-325 wind tunnel of ITAM SB RAS was investigated. Experimental data on the correlation between flow pulsations in different sections of wind tunnel and efficiency of turbulence reducing system were obtained. The system of three-channel hot-wire measurements in subsonic and supersonic flows was tested.

1. Introduction

Experimental studies of the development of disturbances in supersonic boundary layers play an important role in the creation and verification of theoretical models of the laminar-turbulent transition. There are known two methods for studying the laminar-turbulent transition of a supersonic boundary layer: under natural perturbations [1-3] and with controlled perturbations [4]. The main problem of studying the laminar-turbulent transition of the supersonic boundary layer under natural perturbations is the problem of the sources of these perturbations. It is known that acoustic disturbances are always present in a supersonic flow. The main source of acoustic disturbances is supersonic turbulent boundary layer. But some part of acoustic disturbances can come from settling chamber. If the level of acoustic pulsations in the test section is reduced by polishing the surface of the walls and nozzle inserts, then the proportion of acoustic pulsations percolate into the test section from the settling chamber may increase. This case corresponds to the operating conditions of the T-325 at the present time, but so far there have been no detailed measurements of the relationship of pulsations in the settling chamber and the test section of the T-325.

2. Experimental research of the relationship of the pulsations of flow in the settling chamber with the pulsations of supersonic flow

The experiment was performed in the low-noise supersonic wind tunnel T-325 of the Khristianovich Institute of Theoretical and Applied Mechanics SB RAS. In the test section flow Mach number was $M = 2$ and unit Reynolds number was $Re_1 = (11.0 \pm 0.1) \times 10^6\ m^{-1}$.

The scheme of the experiment is shown in figure 1. During the work flow pulsations were measured downstream of the guide vanes (1), downstream of the honeycomb and turbulence reducing screens (2) and in supersonic free flow of test section T-325 (3). Hot-wire probes 1 and 2 were located 150 mm from the inner surface of the wind tunnel: probe 1 was located at the top of the installation, probe 2 was located to the right downstream in the side wall. The wire of probe was placed horizontally.

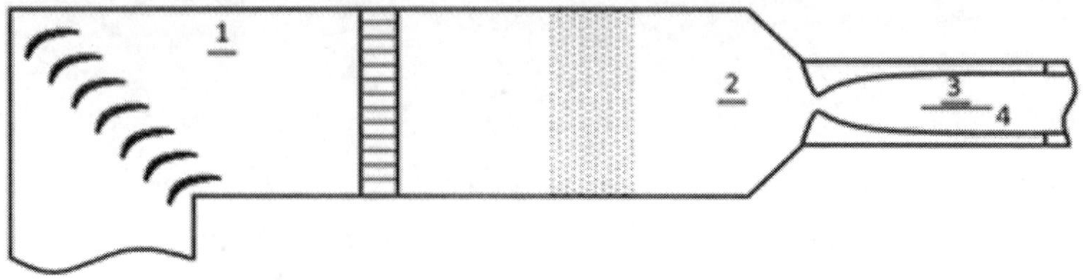

Figure 1. The scheme of the experiment: 1 – the hot-wire probe after guide vanes; 2 – the probe after honeycomb and turbulence reducing screens; 3 – the probe in the test section; 4 – flat plate.

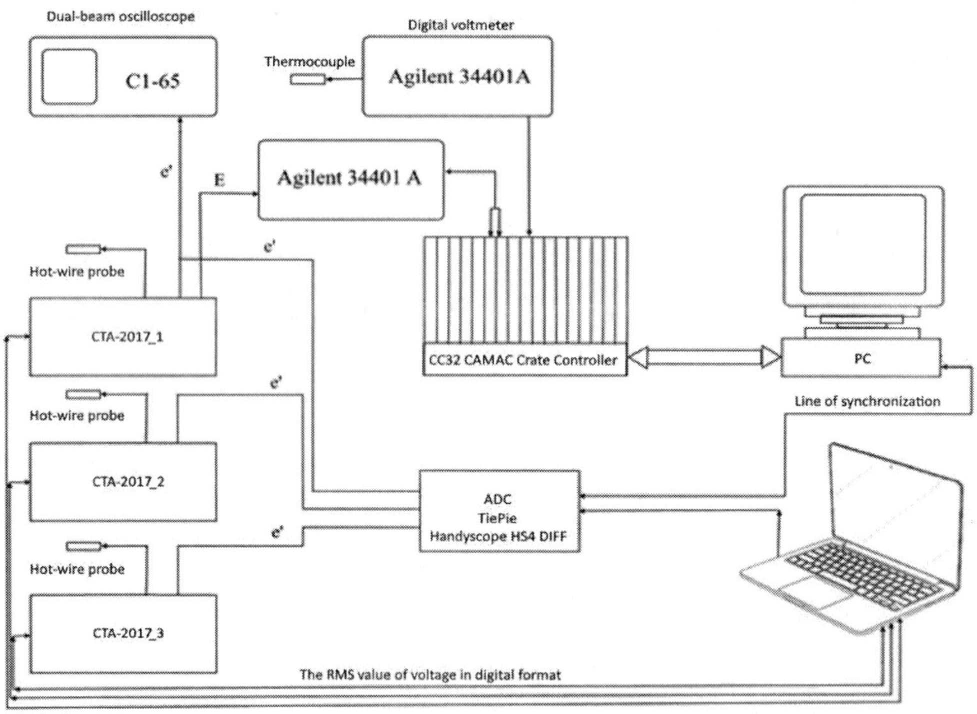

Figure 2. Scheme of the measuring System.

Three CTA-2017 series constant-temperature hot-wire anemometers were used to measure pulsations and mean flow characteristics [5]. Hot-wire probes had a tungsten wire 10 microns thick and 1.7 mm long. Measurements were carried out at an overheat equal to 0.7. The choice was dictated by the fact that the hot-wire probe is susceptible to mass flow fluctuations at this overheating and practically insensitive to changes in the total temperature [6]. The measuring system is presented in figure 2. The mean voltage was measured using an internal ADC with a 24-bit capacity. The pulsation signal was digitized using the TiePie Handyscope HS4 DIFF ADC with a resolution of 14 bits and a sampling rate of 195.3125 kHz, the implementation length of 131 072 points, which corresponds to 0.67 s. Oscillograms of the pulsating signal e'(t) and the flow velocity in the settling chamber T-325 were obtained with the help of hot-wire. Also, the RMS voltage fluctuations and frequency spectrum of signals are considered.

Journal of Physics: Conference Series **1404** (2019) 012074 doi:10.1088/1742-6596/1404/1/012074

Evaluation of the correlation of electrical signals was performed using the following form:

$$Corr(\tau) = \frac{\int f(t)g(f-\tau)dt}{\sqrt{\int f^2(t)dt * \int g^2(t-\tau)dt}} \tag{1}$$

Using the data obtained from the ADC, the correlation coefficient depending on the time shift τ between the signals was obtained by numerical calculation using the Formula (1). For example, figure 3 shows a graphic of the correlation coefficient between signals from the settling chamber. An ideal high-pass filter was used to estimate the correlation of different frequencies. It should be noted that the fact that the devices have separate power supplies, and for this reason the mutual influence of the channels caused by the devices, is negligible.

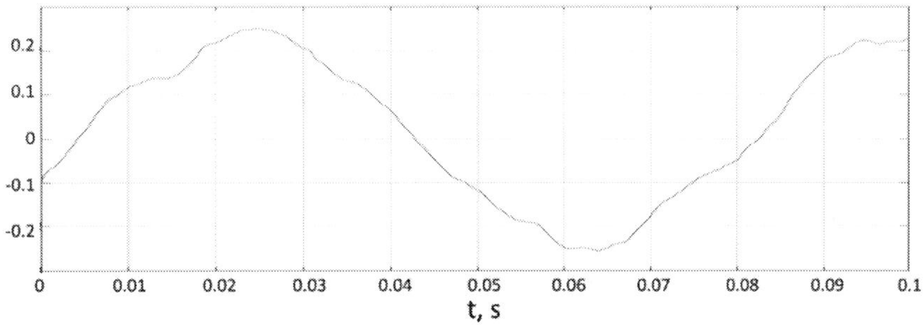

Figure 3. The correlation function of signals of the settling chamber downstream of the guide vanes and downstream of the honeycomb and screens.

3. Results

Figure 4 shows the frequency spectrum of the signal from the probe located in the settling chamber. It can be seen that the signal has a damped low-frequency component with a frequency of about 15 Hz. It is also found that this component is not stationary (see figures 4-6), its presence was observed in about half of the obtained waveforms.

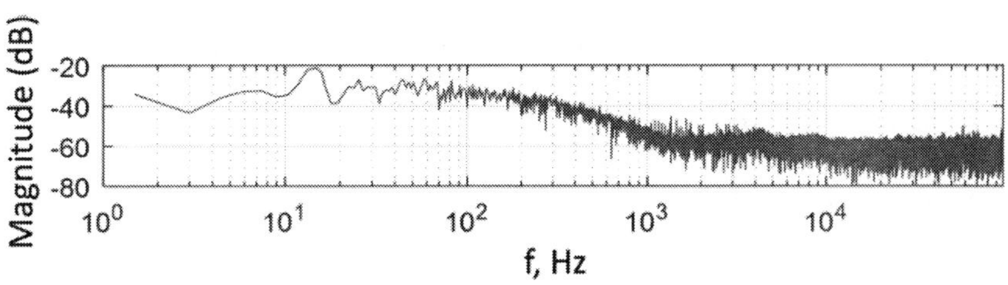

Figure 4. The frequency spectrum of signal from the probe installed in a settling chamber.

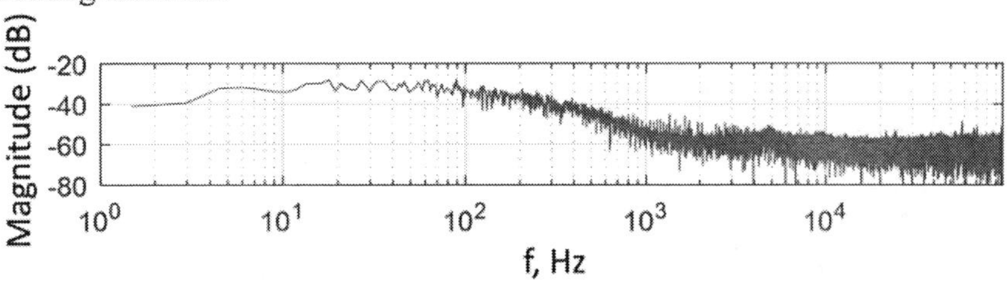

Figure 5. The signal spectrum of the probe in the settling chamber obtained a few seconds after the previous measurement.

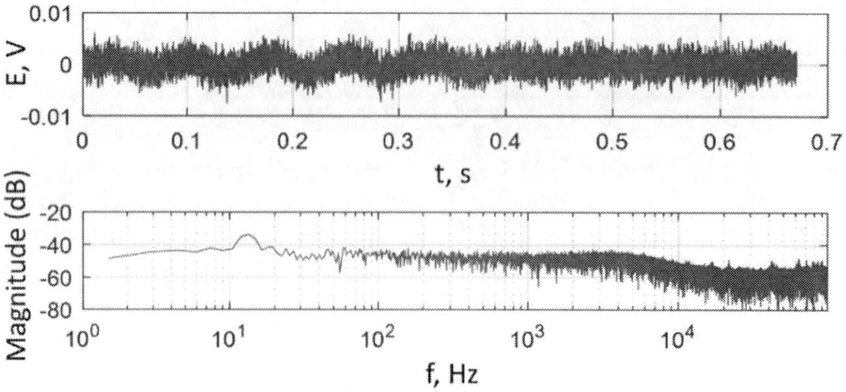

Figure 6. The signal and its spectrum from the probe in the free flow of the test section T-325 before filtration.

The root-mean-square pulsation of voltage, velocity and mass flow are presented in table 1. The flow fluctuation in the settling chamber decreased from 12.1% to 1.2% of the flow velocity after the flow passed the honeycomb and the screens. It can be seen that the pulsation of the flow has decreased by almost an order of magnitude. The flow velocity measurement by the hot-wire probe showed that the flow velocity after the turbulence reducing system equal (4.03 ± 0.02) m/s.

Table 1. RMS voltage and mass flow pulsations.

Location of hot-wire probe	<e'>, %	<u'>, %	<m'>, %
Downstream of the guide vanes	3,0	12,1	
Downstream of the honeycomb and the turbulence reducing screens	0,3	1,2	
Free flow of test section	0,02		0,09

Table 2 shows the maximum values of the correlation coefficients when applied to the initial signals of the ideal high pass filter. The value 0 Hz, 100 Hz and 500 Hz in the first line of table 2 are cutoff frequencies of high pass filter. A value of 0 indicates no filter. The decrease in correlation with increasing frequency of the high pass filter is the expected result due to the predominance of low-frequency signals in the spectra (see figures 4-6).

Table 2. The value of the maximum correlation of the frequency of the High Pass Filter.

Frequency of the High Pass Filter, Hz	0	100	500
$< Max(Corr(\tau)) >$, 1-2	0,15	0,1	0,07
$< Max(Corr(\tau)) >$, 2-3	0,09	0,08	0,08
$< Max(Corr(\tau)) >$, 1-3	0,07	0,06	0,03

1 – The probe downstream of the guide vanes.
2 – the probe downstream of the honeycomb and turbulence reducing screens.
3 – the probe in the test section.

Thus, the results show that the correlation is present between the flow pulsations in the settling chamber after the rotary blades and downstream of the detubularization system. There is a weak correlation between pulsations of flow downstream of the turbulence reducing screens and pulsations of free flow of the test section. Note that in both cases, the correlation is weak and is due to low frequencies.

4. Conclusion

A three-channel hot-wire system for synchronous measurements in subsonic flow in the settling chamber and the supersonic flow in test section of the T-325 wind tunnel has been developed. An experimental study of the correlation of the flow pulsations in the settling chamber flow and the pulsations of the free supersonic flow was carried out. The low level of correlation between the settling chamber flow pulsations and the free supersonic flow pulsations (about 0.09-0.03) was revealed. This may mean that the main part of the disturbances in the test section comes from the near-wall boundary layers. The level of pulsations in the settling chamber and the test section of the T-325 was measured. Also, the flow rate in the settling chamber after the turbulence reducing system was measured. It was obtained that downstream of honeycomb and turbulence reducing screens the flow pulsations are reduced by an order of magnitude.

Acknowledgement

The research was carried out within the framework of the Program of Fundamental Scientific Research of the state academies of sciences in 2013-2020 (project No. AAAA-A17-117030610125-7). Experiments were carried out using the equipment of the Joint Access Center "Mechanics".

References

[1] Gaponov S A, Yermolaev Y G, Kosinov A D, Lysenko V I, Semionov N V and Smorodsky B V 2010 *Thermophysics and Aeromechanics* **17** 259-68

[2] Gaponov S A, Ermolaev Y G, Kosinov A D, Lysenko V I, Semenov N V and Smorodsky B V 2012 *Thermophysics and Aeromechanics* **19** 555-60

[3] Yermolaev Y G, Kosinov A D and Semionov N V 2014 *Journal of Applied Mechanics and Technical Physics* **55** 764-72

[4] Yatskikh A A, Ermolaev Y G, Kosinov A D and Semionov N V 2017 *Journal of Visualization* **20** 549-57

[5] Yatskikh A A, Kosinov A D, Semionov N V, Smorodsky B V, Ermolaev Yu G and Kolosov G L 2018 *AIP Conference Proceedings* **2027** 040041

[6] Kosinov A D, Semionov N V and Yermolaev Yu G 1999 Disturbances in test section of T-325 supersonic wind tunnel *Preprint* 6-99

Study of the possibility of steady horizontal flight of the dual aircraft platform with the wind shear

A M Akimov[1,2] and P A Polivanov[1,2]

[1]Khristianovich Institute of Theoretical and Applied Mechanics SB RAS, 4/1, Institutskaya str., Novosibirsk, 630090, Russia
[2]Novosibirsk State Technical University, 20, Prospekt K. Marksa, Novosibirsk, 630073, Russia

polivanov@itam.nsc.ru

Abstract. In the first part of paper accurate and approximate solution for the equation of motion of steady horizontal flight for dual-aircraft platform (DAP) was given. The two gliders (sailplanes) fly at different heights and connected by a tether. To perform a horizontal flight only the effect of wind shear is used. Minimum of platform velocity to cross wind shear ratio was founded. In the second part of paper overview airflow for troposphere and stratosphere of Earth was given. Wind velocity profile was investigated for some random places. Opportunity of flight without engine was approved. However it is possible at specific conditional only.

Nomenclature

- D – aerodynamic drag force;
- L – aerodynamic lift force;
- V_∞ – free stream velocity, gliders airspeed;
- W – wind speed;
- ΔW – shear of wind speed;
- θ – pitch, angle between velocity and velocity projection on horizontal plane;
- γ – roll, angle between lateral axis and horizontal plane at zero yaw angle;
- ψ – yaw, angle between velocity projection on horizontal plane and vertical plane.

1. Introduction

The dual aircraft platform (DAP) is a version of the atmospheric satellite. Two gliders connected by a long tether and located at different heights are able to perceive the difference in wind speed and direction in altitude. Such an aircraft configuration can the support horizontal flight or even gain the altitude using only the kinetic wind energy. This technical solution was first proposed and patented in 2012 by William Englblom [1, 2].

Figure 1 shows the scheme of forces acting on the DAP. In this paper, the characteristics of the upper glider are indicated by the index “S” (the sailplane). The lower glider is called the “anchor” because it serves to keep the set direction of the DAP. Its characteristics are indicated by the index “A”. The motion equations are written in the inertial frame associated with the Earth. The presence of a cross wind shear on the altitude ΔW allows creating the roll angles γ_S, γ_A and yaw angle ψ which turns the “sailplane” aerodynamic lift force vector in the DAP movement direction. It creates the effective thrust.

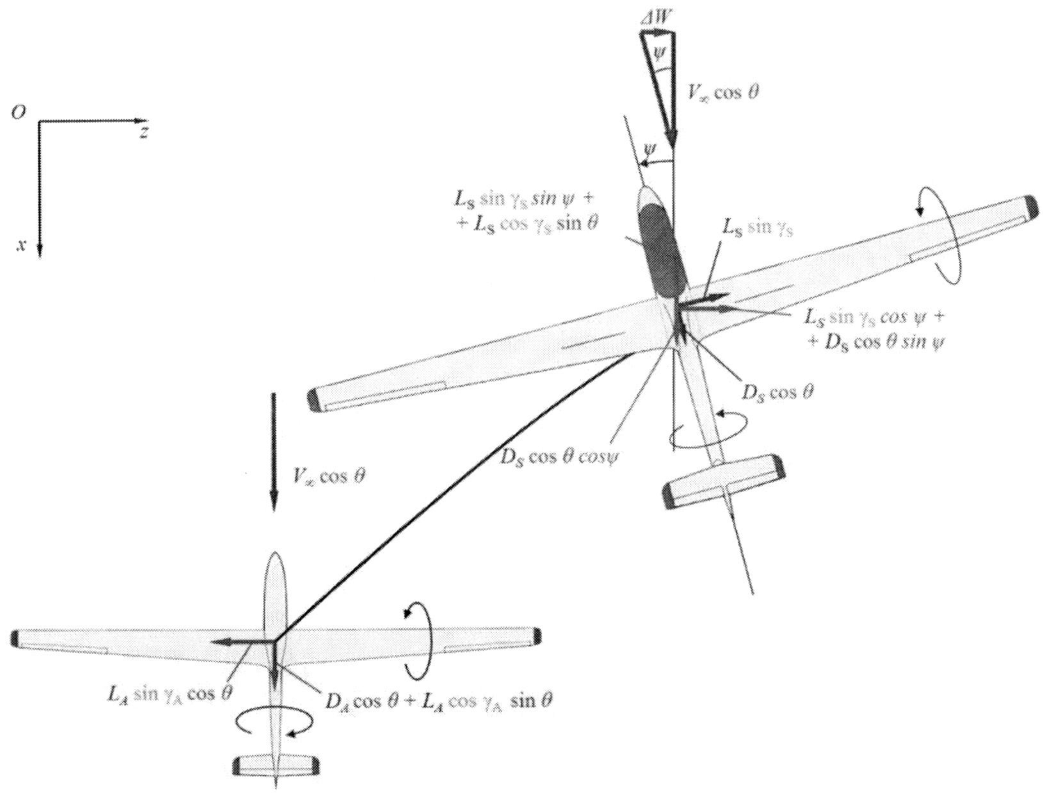

Figure 1. Top view of a two-tier system of gliding and the forces acting on it in the horizontal plane.

2. The horizontal flight conditional

The wind shear was considered to be small enough compared to the DAP flight speed. No slip angle. The projections of the forces on the axis of the earth's coordinate system for steady motion in the vertical plane are written as:

$$\begin{cases} L_S \sin\theta\cos\gamma_S + L_S \sin\gamma_S \sin\psi - D_S \cos\theta\cos\psi - L_A \sin\theta\cos\gamma_A - D_A \cos\theta = 0, \\ L_S \cos\theta\cos\gamma_S - L_A \cos\theta\cos\gamma_A + D_S \sin\theta + D_A \sin\theta - G_S - G_A = 0, \\ L_S \sin\gamma_S \cos\psi - L_S \sin\theta\cos\gamma_S - D_S \cos\theta\sin\psi - L_A \sin\gamma_A = 0. \end{cases}$$

The following assumptions and designations are made:

$$\frac{L_S}{D_S} = K_S, \ \frac{L_A}{D_A} = K_A, \ K_S = K_A = K = const;$$

$$\frac{L_A}{L_S} = k, \ \frac{D_A}{L_S} = \frac{1}{K}\frac{L_A}{L_S} = \frac{1}{K}k;$$

$$\bar{L} = \frac{L_S}{G_S + G_A}.$$

The result is:

$$\begin{cases} tg\theta = \dfrac{1}{K}\dfrac{\cos\psi + k}{\cos\gamma_S - k\cos\gamma_A} - \dfrac{\sin\gamma_S}{\cos\gamma_S - k\cos\gamma_A}\sin\psi, \\ \bar{L} = \dfrac{\cos\theta}{\cos\gamma_S - k\cos\gamma_A}, \\ k = \dfrac{\sin\gamma_S}{\sin\gamma_A}\cos\psi + \dfrac{1}{K}\dfrac{\cos\theta}{\sin\gamma_A}\sin\psi. \end{cases} \quad (1)$$

The first equation of the system (1) can be represented as:

$$tg\theta = \bar{P}_{eff} - \frac{1}{K_{eff}},$$

where

$$K_{eff} = K\frac{\cos\gamma_S - k\cos\gamma_A}{\cos\psi + k}, \; \bar{P}_{eff} = \frac{\sin\gamma_S}{\cos\gamma_S - k\cos\gamma_A}\sin\psi.$$

This is the classic equation for a steady plane flight in a vertical plane. The effective lift-to-drag ratio K_{eff} of the system is determined by the roll angles of the gliders for small yaw angles, because the vertical component of the lift force decreases. The effective thrust P_{eff} is generated due to the roll and yaw angles and does not exist when these angles equal to zero. The characteristics γ_S, K_{eff}, and P_{eff} are maximal when $\gamma_A = 90°$.

The condition under which the horizontal flight DAP is possible, is the following:

$$\psi_{min} = \frac{kK - \sqrt{(kK)^2 - 4(1+k)}}{2}.$$

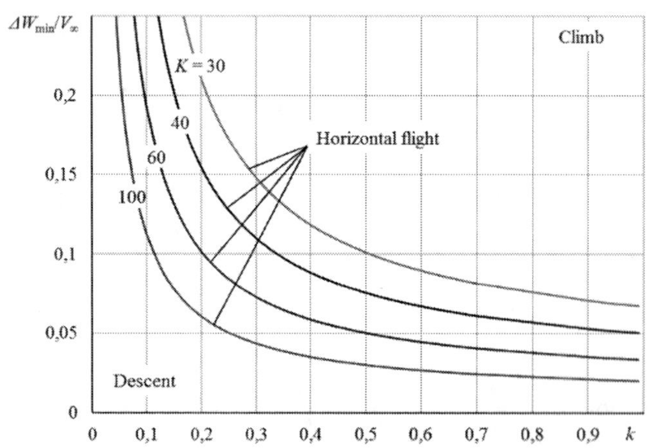

Figure 2. The minimal ratio of the wind shear to horizontal flight speed of the sailplane-anchor system required for the horizontal flight.

The yaw angle ψmin is the minimal angle between the speed of the DAP V_∞ and the total sailplane airspeed when flying without slipping. The tangent of this angle characterizes the minimal rate $\Delta W/V_\infty$. The fraction ΔW is 7 ... 8% at V_∞ for gliders with the lift-to-drag ratio 40 and for parameter $k = 0.4...0.6$. The wind shear should be 2.0...2.2 m/s at a DAP speed of 100 km/h for the horizontal flight. The required wind shear is provided due to the location of the gliders at different heights. Therefore the tether length is one of the values determining the wind shear amount. For example, for the case of difference

 IOP Publishing
1404 (2019) 012075 doi:10.1088/1742-6596/1404/1/012075

in height between gliders of about 500 m, to perform the horizontal flight, the wind gradient should be at least 4 m/s per 1 km of height for the above conditions (figure 2).

3. The continuous flight problem

The problem of ensuring a long flight is associated with heterogeneity and non-stationarity of the Earth atmosphere. Two cases are considered. The first case: the wind speed has a gradient close to zero. In this case the wind speeds can vary from 0 to 30...40 m/s. The wind gradient is not enough to create the effective thrust. The DAP descents and also drifts by the wind. In this case, the minimal speed should be equal in magnitude and opposite in the direction of the wind speed. Otherwise the DAP will drift at high speed from the given flight zone. The horizontal speed determines the vertical speed. High wind speeds with a small gradient can cause a significant reduction in altitude.

The second case: in middle latitudes of 30 ... 50°, a jet stream generates at the border of the troposphere and stratosphere (8-12 km) [3]. It creates a significant speed gradient. The example of a west wind velocity profile of 35S at a random point (approximately 177E) according to [4] is shown in figure 3. The speed was averaged over a period of 2 hours. The presence of a constant velocity gradient of more than 10 m/s per 1 km at the altitudes from 4 to 9 km allows the DAP to fly at high horizontal speeds. For example, with a tether length of about 2 km, the wind shear will be up to 20 m/s, and the horizontal flight speed can exceed 200 m/s (720 km/h), theoretically. However the lateral wind speed reaches 90 m/s (324 km/h). Even taking into account the use of 20 m/s to create the effective thrust, the DAP will drift to the east 60 m every second. This is true if the flight is carried out along the meridian. When flying along the fixed latitude, in this case, the cross wind speed (from north to south) is constant and approximately equal to 10 m/s, and has the close-to-zero gradient. This means that the DAP must fly against the wind at a speed of at least 90 m/s (324 km/h), while decreasing by 160 m/min (at K = 40).

The DAP must be equipped with a power plant to correct its trajectory in adverse weather conditions, to make a controlled flight. Since the wind can significantly change the direction in space and time, this can lead to a rapid loss of the DAP thrust.

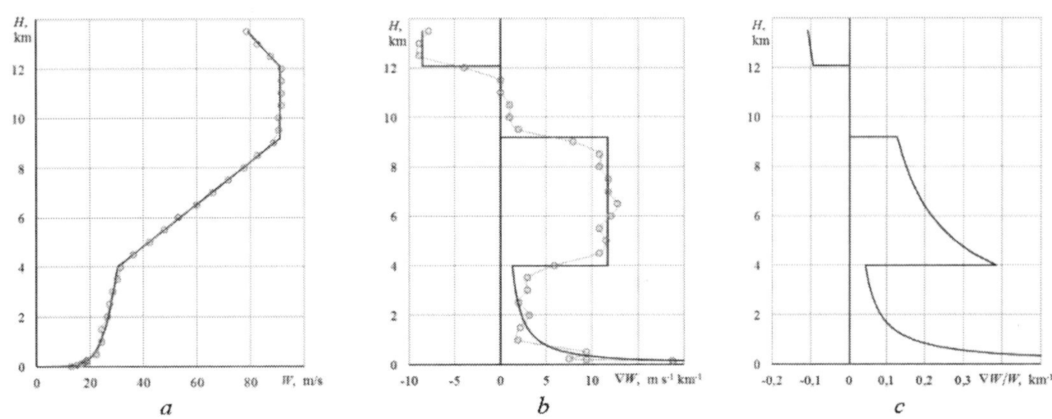

Figure 3. Profile of the west wind average velocity at the latitude of the jet stream: *a* – characteristic velocity profile; *b* – velocity gradient profile; *c* – ratio of the local velocity gradient to the local wind speed. Markers are the data of the GFS weather model [4], solid line is the approximation.

4. Conclusion

The steady horizontal flight due only to the wind shear is possible for the DAP. The minimal required wind shear is approximately 10% of the horizontal flight speed of the DAP with the lift-to-drag ratio of 40.

The results of the work showed the existence of the local wind conditions that can allow a long flight in the Earth atmosphere. However, the wind parameters unsteadiness and anisotropy do not allow the DAP moving in any direction without height loss or displacement in the horizontal plane.

The most likely scenario for using this type of aircraft is the temporary use a transverse wind gradient to save the energy and recharge batteries.

Acknowledgments
The reported study was funded by RFBR and Novosibirsk Oblast according to the research project No 19-48-540017 r_a.

References

[1] Engblom W 2012 Dual-Aircraft Atmospheric Platform *U.S. Patent* 13/414,451
[2] Engblom W 2012 *AIAA Paper* 2012-3203
[3] https://www.britannica.com/science/climate-meteorology/Jet-streams
[4] https://www.windy.com/

Flow features in a hypersonic module with a compact cross-section

S A Akinin[1,2] and A V Starov[1]

[1]Khristianovich Institute of Theoretical and Applied Mechanics SB RAS,
4/1, Institutskaya str., Novosibirsk, 630090, Russia
[2]Novosibirsk State University,
1, Pirogova str., Novosibirsk, 630090, Russia

digetex100@gmail.com

Abstract. A model of a hypersonic air intake based on the principle of compression along the directions of the captured jet converging in space is studied. Such an air intake allows a high compression ratio and a compact section of the internal channel, which simplifies the heat protection of the air intake and the combustion chamber. Experimental studies were carried out in a pulsed wind tunnel IT-302M for Mach number M = 6 - 8 in the range of angles of attack from 0° to 8°. Numerical simulation has been performed based on solving the full averaged Navier-Stokes equations, supplemented *k-w* turbulence model. The pressure distributions on the compression surfaces and in the air intake channel are obtained, the coefficients of the total pressure recovery and flow rate, the Mach number in the throat of the air intake are determined. The features of the flow structure for three variants of the air intake in the area of external compression are established: with side walls, side walls with slots and without side walls. The influence of the boundary layer bleed and lateral spreading on the launch and integral characteristics of a hypersonic air intake with a compact cross-section of the inner channel is determined. Satisfactory agreement between the calculated and experimental data is shown.

1. Introduction

Recently, more attention has been paid to the development and the possibility of using three-dimensional air intakes, providing higher thrust and aerodynamic efficiency of the aircraft compared to traditional ones. In this regard, in the development of hypersonic jet engines, interest has arisen in using the configuration of air intakes in which the jet is compressed in directions converging to the center, which allows a compact cross-sectional shape of the entrance to the channel (combustion chamber). Reducing the surface area of the channel simplifies the heat protection of the air intake and combustion chamber [1]. Air intakes of this type (figure 1) were called convergent [2].

A well-known example of such air intakes is the convergent air intakes developed at the Institute of Theoretical and Applied Mechanics of the Siberian Branch of the Russian Academy of Sciences [3] based on sector cutting from internal supersonic flows converging to the axis of symmetry, in particular, compression flows in conical funnels [4].

However, there are some of difficulties arise associated with the launch of a convergent air intake due to the complex structure of the three-dimensional flow. The issues of ensuring the launch of such types of air intakes with a high degree of compression are not fully understood. The difficulty lies in the fact that with this configuration of the air intake, lateral spreading is difficult due to the flow draining to the center of symmetry and the associated thickening of the boundary layer at the entrance to the throat

[3]. Therefore, it becomes extremely urgent to control the flow and bleed of the boundary layer to ensure the implementation of the inflowing and start the three-dimensional air intake. The boundary layer usually bleeds through slots or openings on the compression surfaces.

In the framework of this work, complex computational and experimental studies of a model of a hypersonic air intake with a compact cross section of the internal channel were carried out with the following objectives:

- obtaining flow parameters on the convergent surface of external compression of a given contour;
- determination of the integral characteristics of the air intake at Mach numbers M= 6-8 in the range of angles of attack from 0 ° to 8 °;
- determination of the influence of the effect of the boundary layer on the start-up and integral characteristics of the air intake by applying the bleed of the boundary layer through slots and lateral spreading on the external compression surface.

2. Models and experiment conditions

In [5], sector cuttings from the axisymmetric flow 180 °, 90 °, and 70 ° were studied. It has been found that better supersonic spreading in front of the throat is achieved using a 90 ° sector cut. At initial external part, the external compression was formed on the basis of a cut of the internal conical section.

2.1. The experimental model of the air intake

For research, 3 variants of a convergent air intake were chosen, the schemes and main dimensions of which are shown in figure 1.

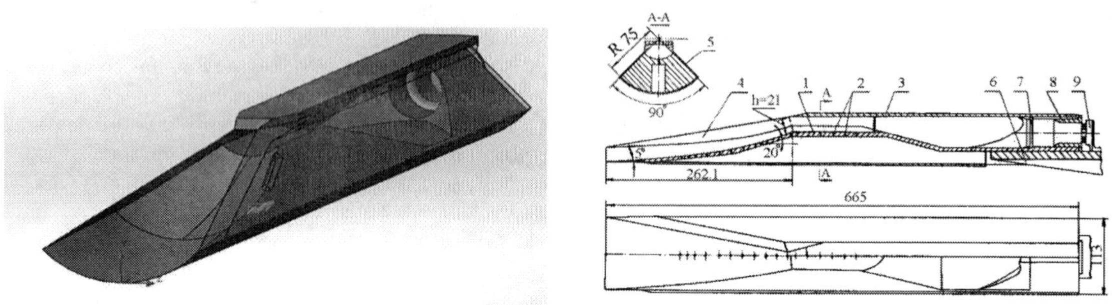

Figure 1. Convergent intake. 1-central body, 2-static pressure probes, 3-cowl, 4, 5-sidewall, 6-sting, 7-metal mesh, 8-nozzle, 9-total pressure rake. All dimensions in mm.

The design of the model allowed the installation on the body of the external compression section of the side walls of various shapes to study the influence of lateral spreading and the possibility of controlling the boundary layer. In the lateral walls there were slots of a rectangular shape with a height of 4 mm to drain the boundary layer. It is also possible to test the model without side walls in the area of external compression. The experimental model had an external contour surface 262.09 mm long, calculated on the Mach number 6.3. With this Mach number, the initial shock wave should come to the front edge of the cowl. The inner channel had a trapezoidal cross-sectional shape 100 mm long with 90° cut. The throat height of the model was 19.5 mm.

2.2. The experimental setup and measurement system

Tests of the convergent air intake model were carried out in a pulsed wind tunnel IT-302M for Mach number M = 6 - 8 in the range of angles of attack from 0° to 8°. The flow parameters during the tests varied in the range of total pressures P_0 = 42-294 kg/sm^2 and total temperature T_0 = 1600-1900K.

During the tests, measurements were made: the distribution of static pressure on the compression surfaces, the total and static pressure in the throat, the total and static pressure at the output of the sonic flowmeter nozzle. Based on the results of these measurements, the flow rate and the total pressure recovery coefficients and the Mach number in the throat were determined. The Schlieren visualization of the flow around the model entrance was carried out in all experiments. Visualization of the flow and measurement made it possible to determine the conditions for starting the air intake and the implementation of the calculated flow in the throat and channel of the model.

Due to the short start-up time (~ 100ms) in pulsed wind tunnel IT-302M, changing the angle of attack of the model during the experiment is impossible. Therefore, the experimental technique consisted of a discrete change in the angle of attack of the model before each experiment.

3. Numerical model, boundary conditions, calculation method and grid

Numerical calculations were performed using the commercial software product ANSYS Fluent based on the complete Reynolds averaged Navier-Stokes equations supplemented by the *k-w*/SST turbulence model. A finite-volume difference scheme based on the solution of equations for density (Density-Based) was used. The numerical calculations were considered in a stationary setting.

Given the symmetry of the model, the computational domain included only half of the model, which was limited only by the area of external compression and the internal channel (throat) in order to optimize and reduce the calculation time (figure 2).

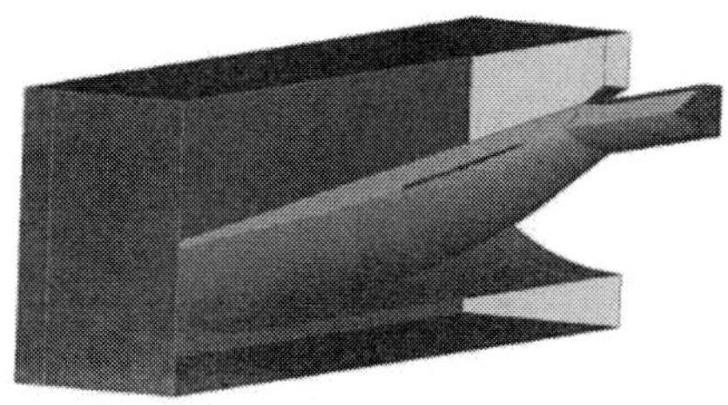

Figure 2. Computational domain.

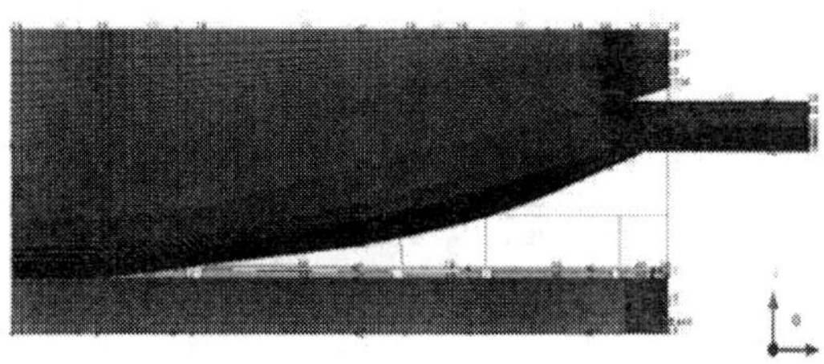

Figure 3. Computational grid.

In the computational domain, a regular rectangular grid was constructed with thickening of nodes towards solid surfaces and to areas with an increased pressure gradient. Due to the complex three-dimensional curved surfaces, the computational grid for each configuration contained a different number of nodes in three directions. In most calculations, the grid contained 170–250 nodes in the x-direction, 150–220 in the y-direction, and 50–100 in the z-direction. As an example in figure 3 show computational grid in the plane of symmetry for the configuration of the model without side walls. On the surface of the model, the slip condition for the velocity and the conditions for the constancy of the wall temperature $T_w = 300K$ were used.

4. Results and discussion

The pressure distribution along the model for various variants is shown in figure 4 (a), which shows data for models without side walls and with side walls in which there were slots for the boundary layer bleed. The pressure is significantly greater increase at utilization of variant with side walls than without their installation. This effect is weakened as the Mach number decreases. The experimental data are consistent with the calculation before entering the channel, especially for $X \leq 225$ mm. In the cross section of the entrance to the model channel with the walls, the calculated pressure is much higher than the experimental values. Slots in the side wall for drainage of the boundary layer do not lead to a noticeable decrease in pressure, which is a confirmation of the tendency for a decrease in the effect of the boundary layer on pressure increase with increasing Mach number. This conclusion is also confirmed by Schlieren visualization of the flow (figure 5).

The distribution of the static pressure P_z across the model was calculated. The calculations showed (figure 4 (b)) that, in the configuration without side walls, the decrease of pressure to the side edges reaches 70% in the initial (X = 165 mm) section of external compression from the maximum pressure in the plane of symmetry. This value varies depending on the Mach number and decreases downstream to 30% at the channel inlet (X = 258 mm).

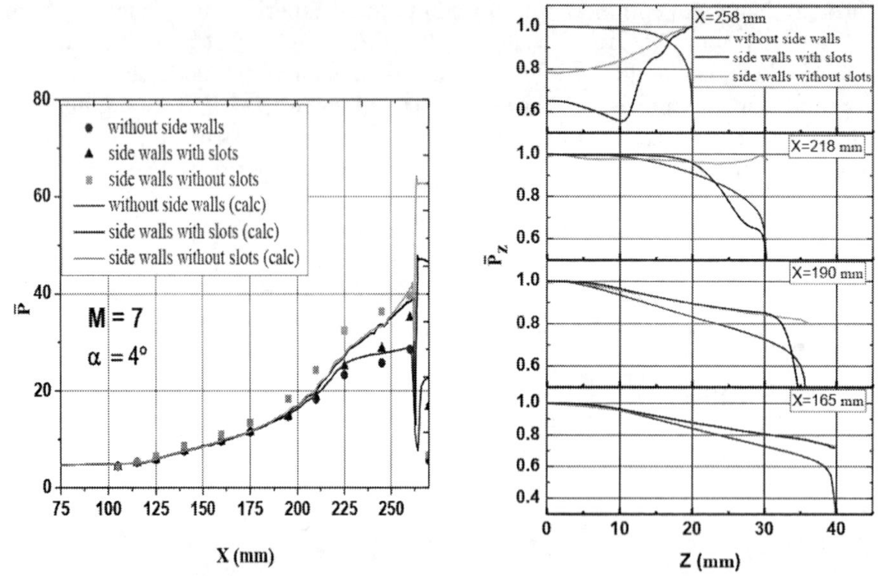

(a) (b)

Figure 4. Static pressure distribution on the convergent inlet surface (a) in the symmetry plane, (b) in Z planes for Mach number M=7 and angle attack α=4°.

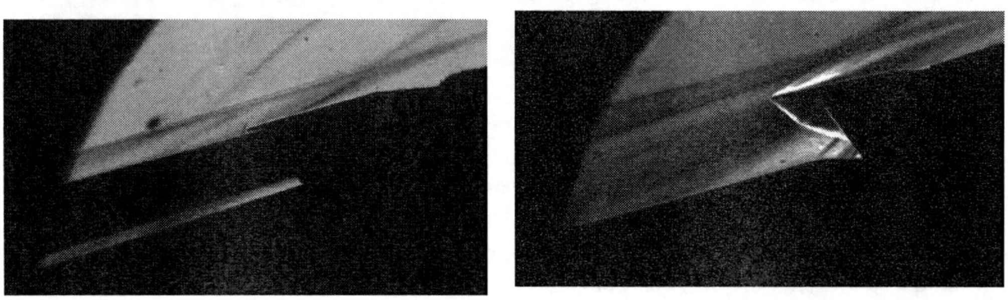

(a) (b)

Figure 5. Started air intake for the configuration of models (a) with side walls and slots and (b) without side wall for Mach number M = 7, α = 4 °.

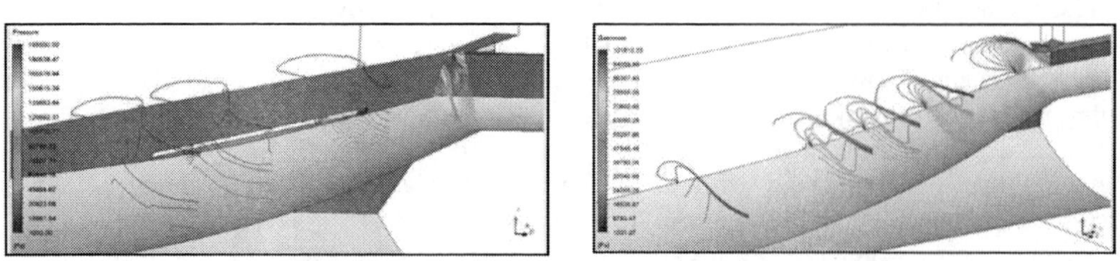

(a) (b)

Figure 6. Static pressure fields in Z planes for the configuration of models (a) with side walls and slots and (b) without side wall for Mach number M = 7, α = 4 °.

The installation of the side walls leads to a decrease in spreading, as follows from the same figure. The absence of drainage of the boundary layer even leads to some increase in pressure near the lateral side walls due to the complex interaction in the angular configuration. The data obtained suggest the non-uniformity of the pressure field across the channel width at the inlet of the engine can be ensured at an acceptable level by selecting the appropriate slots for the boundary layer bleed. The structure of the three-dimensional flow around the external compression surface is shown in figure 6. Downstream of the leading edge, the shock wave changes shape from concave in the initial section to convex in the final section. This form of the shock wave near the leading edge remains concave over the entire range of Mach numbers.

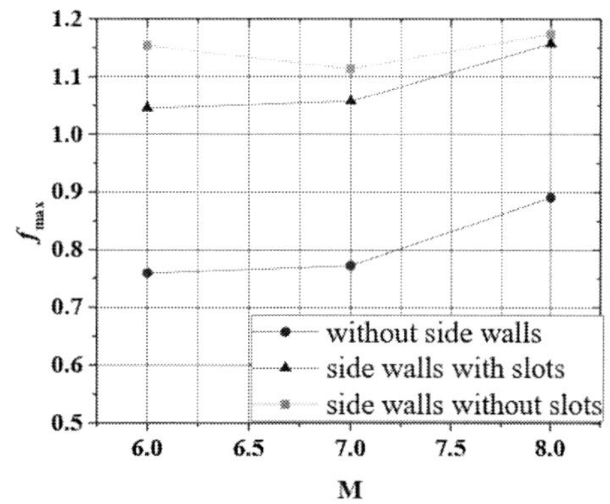

Figure 7. Mass flow rate coefficient.

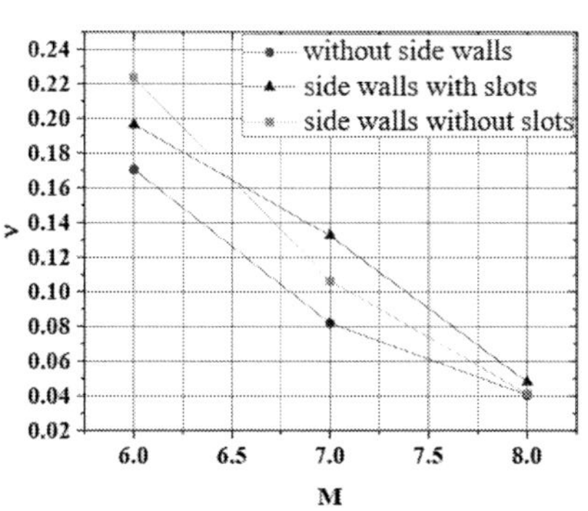

Figure 8. Total pressure recovery coefficient.

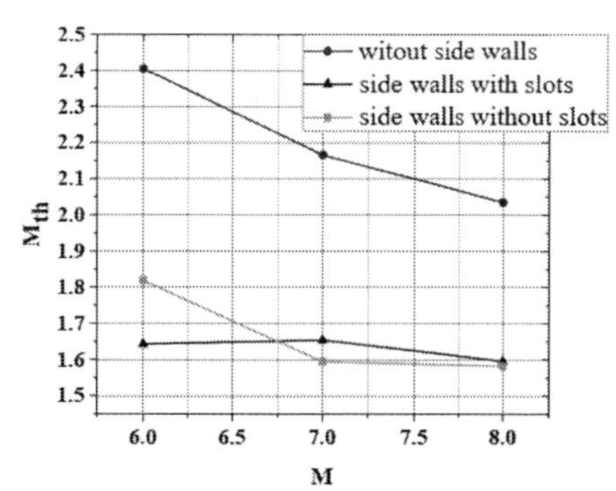

Figure 9. Mach number at the combustor entrance.

The of flow rate *f* and total pressure recovery *v* coefficients are determined by measuring total and static pressure at the outlet of air intake. The use of a large number of pressure receivers makes it possible to obtain an accuracy of measuring air flow not worse than 3% under conditions of uneven supersonic flow.

With the Mach number M = 8, the flow rate coefficients with the installed side walls are almost equal to each other (figure 7). The flow rate coefficients for the convergent air intake without side walls were less due to intensive lateral spreading. An increase of attack angle leads to a proportional increase of flow rate coefficient and ensures a supersonic flow in the throat even without regulation and control of the boundary layer over the entire range of Mach numbers.

The total pressure recovery coefficients and Mach numbers in throat (M_{th}) based on the measured total and static pressures in the throat of the model, were determined (figure 8 and 9). It can be seen that over the entire range of Mach numbers of the oncoming flow, the velocity in the throat of the model remains supersonic, even when a thick boundary layer arrives at the model input. It can be seen that the total pressure recovery coefficient with slots was lower at M = 6 than in the configuration without slots, but at high Mach numbers M = 8, the pressure recovery coefficient for all three variants becomes the same.

Comparative calculation and experimental studies of convergent air intakes of various variants should allow a more correct comparison of the characteristics of such air intakes. Calculation and experimental studies of a hypersonic air intake with a compact section of the internal channel allowed:

- To obtain flow parameters and characteristics of the boundary layer on the compression surface and in the channel of the convergent air intake.
- Determine the integral characteristics of the air intake for Mach numbers M = 6, 7 and 8 and attack angles $\alpha = 0°$, 4° and 8°.
- Determine the influence of the boundary layer bleed and lateral spreading.
- To establish that the maximum flow coefficient is achieved in the variant of the model with side walls without slots ($f = 1.17$ at M = 8) with the maximum total pressure recovery coefficient ($v = 0.22$ at M = 6). The average Mach number in the throat M_{th} for all variants of the model was supersonic and ranged from $M_{th} = 1.58$ (side walls without slots) to $M_t = 2.4$ (without side walls).
- To establish that the shock start in the impulse wind tunnel made it possible to obtain the air intake starting in the entire range of Mach numbers and angles of attack under study, including the version of the model with side walls without slots.

Acknowledgments
The research was carried out within the framework of the Program of Fundamet Scientific Research of the state academies of sciences in 2013-2020 (project No. AAAA-A17-117030610126-4).

References

[1] Mazhul I I and Scherbik D V 2001 *Thermophysics and Aeromechanics* **8** 219–35
[2] Goldfeld M A and Starov A V 2019 *Siberian Journal of Physics* **14** 51–62
[3] Gutov B I and Zatoloka V V 1973 *Aerophysical Research* **2** 64–7
[4] Blokhin A M et al. 1972 *Aerodynamic research* **1** 105–8
[5] Walker S, Tang M, Hamilton B A, Morris S and Mamplata C 2008 *AIAA Paper* 2008-2544 11

Innovative projects of the development of high-speed transport in Russia

V M Fomin, V Y Malov and D G Nalivaychenko

Khristianovich Institute of Theoretical and Applied Mechanics
4/1, Institutskaya str., Novosibirsk, 630090 Russia

denis@itam.nsc.ru

Abstract. Continuous development of transport systems at the state level is an urgent task, the success of which will largely determine the dynamics of improving the quality of life and trade and economic efficiency of regions, cities and the state as a whole. To make a qualitative transition to a fundamentally new concept of transport is possible only based on scientifically sound approaches and fundamental analysis of all aspects of the creation and operation of the transport system.

1. Introduction

The modern World is characterized by a tendency to "Accelerate changes", and changes in almost everything: in new types of goods and services, in the media, in the Internet. But perhaps the most important changes relate to the growth in the speed of both the delivery of goods and services to the final consumer, and in the speed of movement of people themselves. It is difficult, if not impossible, to accurately predict the new types of needs that may arise in 10, especially in 20 and 30 years. But the need to increase the speed of movement of material bodies will certainly increase. Aviation, of course, partially solves these problems, but it also has a physical limit on the volume of transportation of both goods and passengers.

For Russia, as the largest power by the territory, it is land transport that has always been the basis for the implementation of the tasks of maintaining the unity of economic, political and social spaces.

Traditional modes of transport, road and rail, in its modern form, in the very near future will not meet the challenges of qualitative acceleration of the movement of goods and people. In Russia, rail transport is one of the leaders of passenger and freight transport. The development and modernization of rail transport is on the way to creating an extensive network of high-speed railway. However, it is already obvious that for the used technology of movement "wheel-rail", there are problems of two technological limits. First, it limits the dynamics of acceleration and braking, which depend on the adhesion of the wheel with the rail and the reliability of current collection at speeds above 350 km/h. Secondly, with the growth of speed, the power costs for overcoming the aerodynamic resistance to the movement of the vehicle grow.

The most effective way to simultaneously overcome two limits can be the implementation of vacuum-levitation transport, the principle of which is a fruitful symbiosis of two ideas, the concept of transport on magnetic suspension in an artificially created vacuum inside a sealed overpass.

The emergence of fundamentally new types of land transport, including, perhaps, vacuum magneto-levitation transport is a matter of the near future. Russia needs to be at the forefront of both scientific and technological developments here. It is necessary to once again become world leaders in the field of innovative solutions in the field of transport and implement the idea of an Asia-Europe transport bridge using domestic technologies. The latter is sure to cause a multiplier effect in many other sectors of knowledge and economy. Reducing the time of movement of people is sure to increase the possibility of social control over the vast, but still sparsely populated regions of the country.

At present, the idea of vacuum-levitation transport is actively discussed all over the world in popular science literature, at international conferences dedicated to the transport of the future and trends in the development of rail transport. A number of foreign companies (ET3, Hyperloop Transportation Technologies Inc., Hyperloop One) actively propagate the idea of vacuum-levitation transport (VLT), offering its implementation at the expense of investors. In China, at the end of September 2019, a new state program was approved with a goal to achieve a speed of 600÷1000 km/h on land rail transport. Obviously it will be a magnetic levitation transport for which China has the necessary technology and economic resources.

Despite the bold beliefs of the representatives of the above organizations, the analysis of the literature review allows us to judge the insufficient or complete lack of scientific and technical justification of the basic aspects of VLT. A qualitative transition to a fundamentally new concept of transport is only possible based on scientifically sound approaches and fundamental analysis of all aspects of the creation and operation of the transport system.

2. Problem setup

One of the most important aspects of this approach is the task of identifying the needs of the national economy, the need to create completely new industries, materials, and the intensification of the use of new logistics schemes in remote and sparsely populated areas of the country. For this purpose, the development of theoretical and methodological schemes of a comprehensive assessment of options for creating new forms of transport systems in their unity with the transport system of the country as a whole, ensuring the satisfaction of the needs for interregional industrial and social relations, as well as in creating favorable conditions for entering world markets. The solution of this task will require improvement of the calculation tools that means study of compatibility of models of development of the whole country, models of transport systems and models that reflect the interests of individual industrial and social structures. As a result, it is expected to create a user-friendly system of calculations, with the conclusion of the results of decisions on a cartographic basis, which dramatically increases the susceptibility of information.

The proposed method is based on an optimization inter-sectoral interregional model, supplemented by a block of direct links, which makes it possible to estimate the load of individual sections of the transport network. This model is supplemented by a group of models for the formation of a support transport network with the allocation of separate sections for new vacuum-levitation transport and possible transshipment points. The economic approach in market economy conditions should remain in demand, because only such a vision of the future development of the country's economy is able to evaluate the options that meet the requirements of maintaining social control over vast but sparsely populated territories of the Asian part of Russia.

In the practice of world economic and geographical sciences, certain components of spatial development problems, taking into account the transport factor, have been successfully solved for a long time. These are the tasks of inventory optimization, optimization of transportation, logistics, and development of cities and agglomerations.

In this regard, names must be noted of U Aizard, A Lesh, L Kantorovich, A G Granberg, L Mani, S Tarkhov and the program TACIS. The approach covered in this paper, as a continuation of the work of A. G. Granberg, is that the criterion of optimality when choosing options in the proposed approach is the indicator of final consumption, proportionally divided by regions of the country in accordance with living conditions and population. The fundamental differences between our approach and the profit indicator of a separate entity of economic relations, a separate transport and logistics company or GDP indicator for a particular region. In our proposed approach, the transport system is represented by all modes of transport interacting with each other complementary and / or interchangeable. The models are designed by us taking into account the interaction of different transport modes and possible and necessary transshipment supported by the appropriate software protected by a patent. Spatial intersectoral models are also original and have no analogues in other countries and research teams.

3. Mathematical model

The basis for the study was the optimization model- of the inter-sectoral balance of production with a restriction on the production labor resource [1]. In this case, the basic model of interindustrial balance is combined with the restriction on the production resource and the condition of maximizing the final consumption (z) is added. To simplify the task, it was decided do not use dual variables and capacity constraints in the model. Taking into account the accepted assumptions the formal generalized model looks as follows:

$$z \to \max!$$
$$(E - A) \cdot X - az \geq Q$$
$$f \cdot X \leq C$$
$$X \geq 0$$

The initial data for the model are the matrix A, the row vector of coefficients *f*, the vector-columns of the consumption structure α with the condition that the sum of these coefficients is equal to one, and *Q* - vector "other" needs of the national economy (defense, balance of exports-imports, etc.), as well as labor restriction *C*. Let's say a few words about a matrix *A*, since it is key to the model. Matrix *A* contains the coefficients of direct costs of products of the *i*-th industry for the production of a unit of products of the *j*-th industry. Data for matrix *A*, coefficients *f*, *α*, *Q* and *C* were taken from previous studies. The row and column “Vacuum Levitation Transport” were added to the matrix *A*. The introduction of this column into the optimization model will help to understand how the final consumption will change when the transport system is put into operation. Also, the corresponding parameters were added to the arrays *f*, *α* and *Q*. The main difficulty was the selection of parameters for the above data sets, since the technical documentation does not mention the materials and methods of manufacturing parts for the pipe and capsule of the transport system. Therefore, strong assumptions were made about the similarity of the rows and columns “VLT” and "AVIA", except for the indicators of capital investments, which were significantly increased. It was decided that the coefficients *f*, *α* and *Q* in the respective categories will be similar, but with minor changes. Not much more manpower is needed to build a transport system VLT than for conventional transport. It is assumed that VLT occupies a place in the structure of consumption the same place as the industry "AVIA", and the state for its own needs (vector-column *Q*) will need the services of this new type of transport as much as other types of transport as extremely promising.

In the absence of other cost indicators, the indicators of the VLT variant were generally based on estimates of “Hyperloop " indicators.

4. Results and discussions

The results of preliminary calculations using the proposed approach for the integration of VLT into the prospective 2035 inter-industry balance have shown the following:

- The growth rate of final consumption in the country during the implementation of the VLT project by 2035 is slightly reduced: by 0.5-0.9 points, which is about 400-700 billion rubles of final consumption annually;
- At the same time accelerated development is required by machine building and construction complexes;

- Metallurgy (ferrous and nonferrous) industries are growing at a faster pace. The accelerated growth of the energy sector is explained not only by the needs of VLT, but also by the growth of the metallurgical complex, which is the main consumer of electricity;
- The growth rates of the industries focused on the final consumption of the population are slightly reduced: agriculture, light and food, services, etc.;
- The industries related to scientific and design studies, which are required in many related (with VLT) industries, receive a noticeable acceleration

If we conduct another computations with a model in which we assume the implementation of this project at an accelerated pace not for 20, but for 10 years, the annual losses in terms of final consumption will amount to more than 1.2 trillion rub or 0.12 points reduction in its growth rate.

The structure of export-import supplies has a significant impact on the final consumption index. It is desirable to change it in the direction of reducing exports, especially products of extractive industries.

The use of the model with a more detailed nomenclature of industries and / or with a spatial component could give us more accurate data to which industry and in which region it is useful to activate the corresponding production activity.

Acknowledgments
The research was carried out within the framework of the RAS Presidium Program in 2018-2020 (project No. AAAA-A18-118021590034-6).

References
[1] Granberg A G 2003 *Fundamentals of Regional Economy: Textbook for Universities* (M.: Higher School of Economics) p 495

Gas dynamics of the body flight in the channel

V M Fomin and D G Nalivaychenko

Khristianovich Institute of Theoretical and Applied Mechanics SB RAS,
4/1, Institutskaya str., Novosibirsk, 630090, Russia

denis@itam.nsc.ru

Abstract. The gas dynamics of the body high-speed flight in a channel is studied by the numerical simulation method. It is shown that in the considered formulation of the problem, it is possible to implement one of the two versions of the gas-dynamic flow pattern: the body movement with a forward straight shock wave ("unstart") and the supersonic current near the body ("start"). The behaviour of aerodynamic and thermodynamic loads on the body as it accelerates to high speeds is studied.

1. Introduction

Interest in the problem of high-speed motion of a body in a long channel arose against the background of public attention to the idea of a promising vacuum-levitation transport system [1]. It is believed that the vehicle of such a transport system relying on the magnetic levitation cushion flies at a high speed in a sealed pipeline. The pressure in the pipeline is significantly below the atmospheric one. There is a significant gap for air flow between the vehicle walls and the pipeline.

The literature review revealed a significant gap in knowledge about the gas dynamics of the high-speed flight in the channel. The exception is the problem of the piston (projectile) in the pipe, analytical solutions and numerous numerical studies which result from the development of a variety of throwing systems for military, industrial and scientific purposes.

Comparison with the work of air intake devices of power plants of high-speed aircraft gives the good results for the analysis of the problem stated. The geometric similarity of axisymmetric air intake devices with the body flying in the channel is evident (figure 1). The channel walls play the role of an infinite feedwell.

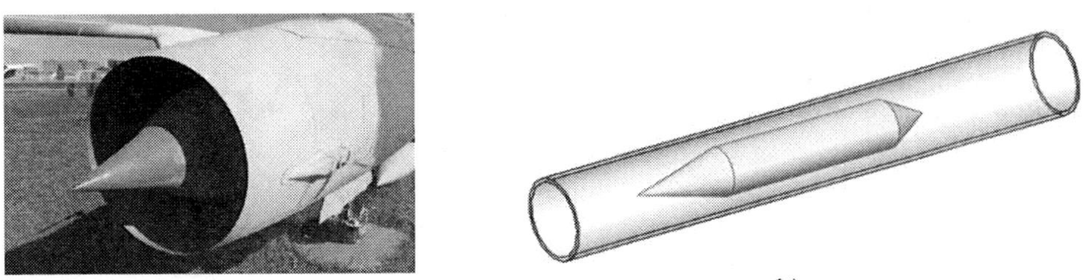

a) b)

Figure 1. a) Air intake device of the high-speed aircraft, b) The body in the channel.

The air intake of an air-jet engine is designed to take a portion of air from the incoming flow, brake this portion with minimal losses of total pressure and supply air to the combustion chamber. The most important characteristics of the air intake are the air flow coefficient (ϕ) and the compression ratio ($F_{neck}/F_{overpass}$) of the captured flow. The flow coefficient characterizes the ratio of the actual air flow passing through the engine to the maximum possible at a given flight speed and the air intake size. The high flow compression ratio ensures the efficiency of the fuel combustion process in the engine combustion chamber.

Often, in a real flight, not the air entering the air intake can be accepted ($\phi \leq 1$), herewith at the entrance before the feedwell, a direct shock wave forms, and part of the air is forced out into the atmosphere. This mode is characterized as an air intake "unstart" (see figure 2 [2]). Generally, the fixed geometry air intake operates only in a narrow range of design flight parameters.

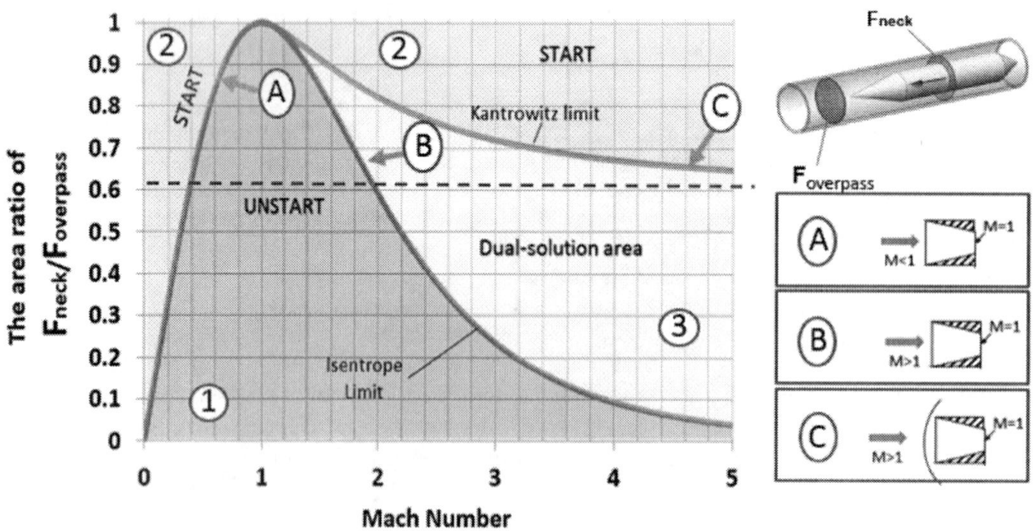

Figure 2. Start-no-start diagram of flow compression limits within the channel.

Figure 2 shows the "start-no-start" diagram widely used by developers of air intake devices for the high-speed aircraft. Here, the areas of all possible air intake operation modes are delineated by the known limits – the isentropic compression limit and the Kantrowitz limit (with the forming direct shock wave). For the fixed geometry, the region "1" defines the conditions under which the gas-dynamic "start" of the channel (air intake) never realizes. Area "2" defines the conditions when the channel "start" occurs spontaneously. Area "3" defines the conditions under which the channel can be "started", but this requires certain additional activities. Area "3" is for developers of high-speed aircraft the field of implementation of their scientific and technical ideas to create effective air intake devices.

The diagram (figure 2) is useful for analyzing the problem of uniformly accelerated motion of the body in the channel and comparing it with the work of the air intake. The horizontal line drawn at the level of the selected geometric compression ratio (figure 3) permits tracing every gas-dynamic regime in the path of the body as it accelerates.

Similarly, in the case of a body flying through the channel, all the air on its way must pass from the bow to the stern as efficiently as possible, with minimal losses of the total pressure. However the significant difference is evident. For the problem of body flight in the channel, as the "body + channel" system is unable to swallow completely all the incoming air, it obviously leads to the appearance of a leading shock wave in front of the body. This is a consequence of the "feedwell" infinity and the inability to discharge the excess air into the surrounding space. Numerical modeling helps to trace how the leading shock wave forms, as well as the gas-dynamic flow pattern develops during the body acceleration.

2. Numerical simulation of the body acceleration in the channel

This paper presents a series of numerical simulations of acceleration of axisymmetric simple geometry bodies (figure 1) from rest to high supersonic speeds. At a constant diameter of the channel ∅D=100 mm, acceleration of bodies of different sizes (∅d =20mm (4%-blocking channel), ∅d=30 mm (9%), ∅d =40 mm (16%), ∅d =50 mm (25%), ∅d=60 mm (36%) and ∅d=70 mm (49%), see figure 3) was calculated.

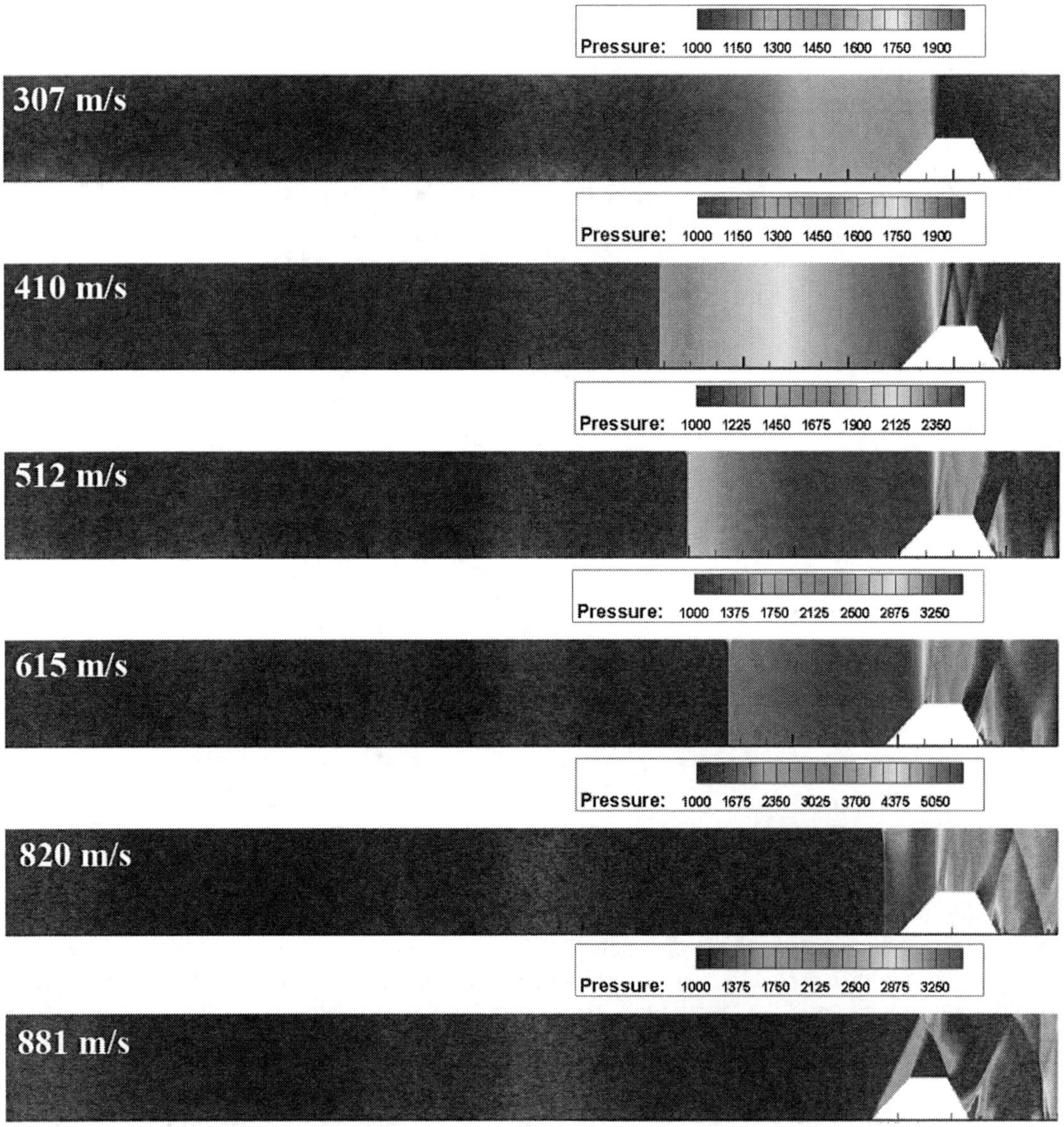

Figure 3. Stages of body acceleration in the channel. Pressure fields.

The numerical simulation is performed with the aid of the software product Fluent. The unsteady Navier-Stokes equations for the laminar flow were solved in the axisymmetric statement. Air was used as the working gas. The pressure in the channel was assumed to be P_∞=1000 Pa. Adhesion conditions and adiabatic temperature conditions were set on the model and channel surface. Free borders (pressure outlet) were set on the left and on the right of the calculated area. In the computational domain, a structured grid is constructed, with decreasing cell size of near the channel and model walls. The minimum cell size is 0.2 x 0.2 mm. To set the motion of the model, we used a software module (User defined function) which defines the dependence of the speed on time. The model movement was given by a rectilinear uniformly accelerated along the axis from right to left, with an acceleration of $a = 20500$ m/s^2. The time step was $\Delta t = 10^{-6}$ s. The computational grid was rebuilt at each time step - the entire grid moved with the model, the cells disappeared on the left border and were added on the right. The solution at every time instant was accepted as agreed if discrepancies were below 10^{-7}.

In the diagram of the isentropic compression limit and the Kantrowitz limit (figure 4), dotted lines indicate the geometric characteristics of the models for which the calculations were performed.

Figure 3 presents the characteristic pressure fields near the body (r=20mm) at different times as it accelerates, for one of the calculation options. The diagrams show the stages of the shock wave formation in front of the body, catching up of the wave front to the body and the shock wave swallowing with in the supersonic flow regime.

For every case, the body acceleration begins in the favorable, from the gas-dynamic point of view, "start" area of the channel (area "2", see figure 2). Despite this, the calculation results show that not all the air encountered in the body path is swallowed by it. A bit of air is pushed forward by the body (see figure 4), creating a zone of high pressure which later develops into a shock wave (figure 5). This is probably due to the imperfection of the aerodynamic body shape which does not provide its smooth and continuous flow.

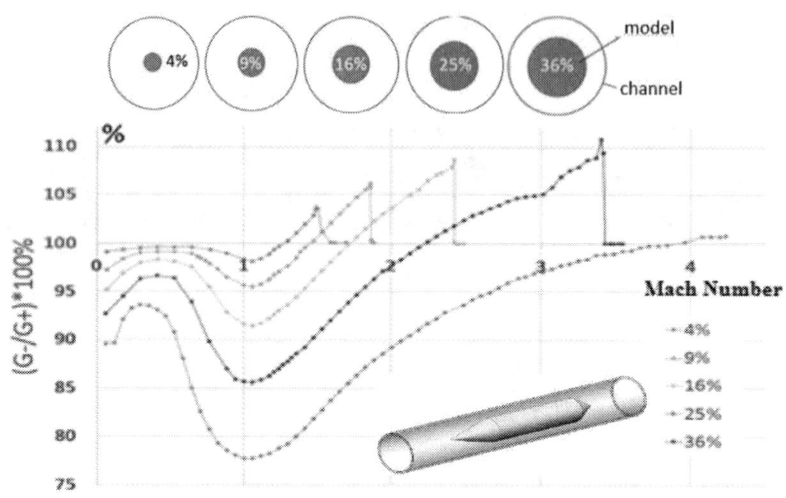

Figure 4. Proportion of air ingested.

Figure 5 shows how the pressure profile changes and the shock wave front forms in front of the body (r = 35 mm) as it accelerates in the channel.

Figure 4, for different cases of calculation, shows the relationship which in percentage terms demonstrates what part of the "incoming" air (G+) flows from the bow to the stern of the body (G-) as it accelerates. It is clearly seen on these graphs due to what and how the zone of high pressure forms in front of the body.

The lowest air flow through the duct between the body and the channel, for all calculated cases, coincides with the Mach number close to M=1, which is in good agreement with the theory of the isentropic compression limit. In the supersonic region of flight, the difference in flow $\Delta G = (G+)-(G-)$: incoming (G+) and ingested (G-) decreases.

At some point, the mass of the incoming gas becomes less than the mass of the "ingested" $\Delta G<0$. It results in the gradual disappearance congestion of the air in front of the body, leading shock wave weakening and reaching its "ingestion" moment. A favorable regime of supersonic flow is formed (see figure 4, the moment of the curve falling to the mark of 100%).

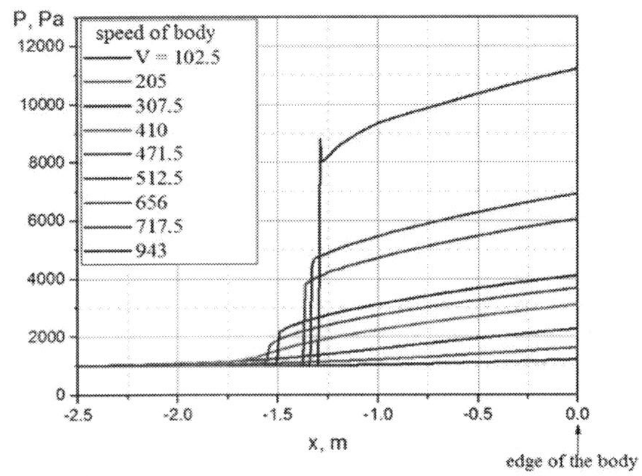

Figure 5. Pressure profile in front of the body (∅d=70 mm) as it accelerates.

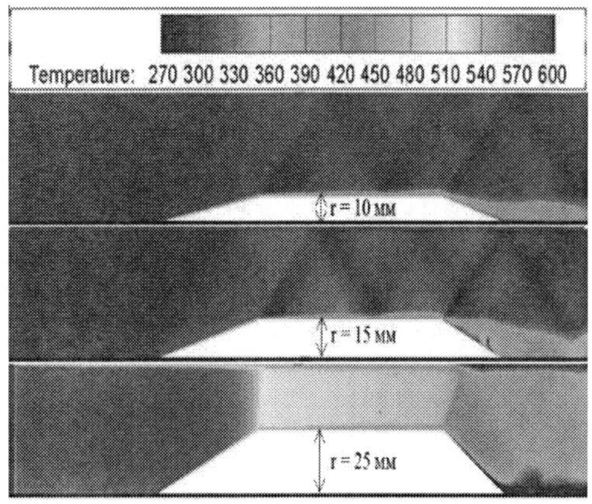

Figure 6. Temperature distribution (in Kelvins) around the body that flies at the speed of U = 820 m/s.

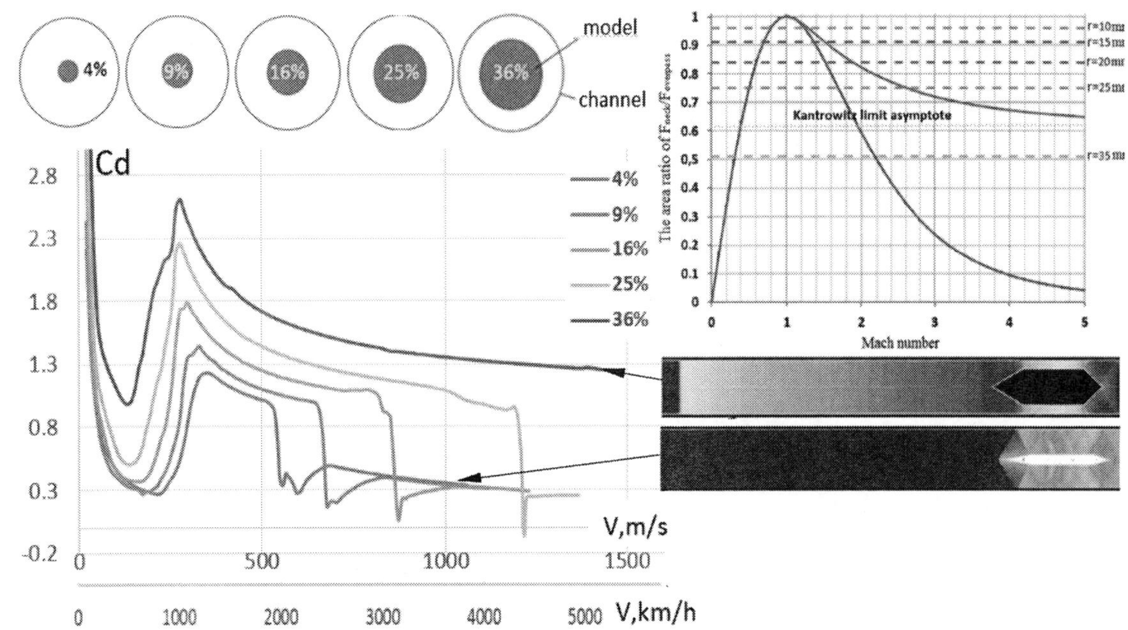

Figure 7. Dependences of aerodynamic drag coefficients of different size bodies as they accelerate in the channel.

Comparing the moments of the channel "launch" and the supersonic flow onset around the body (see figure 4) on the diagram of the isentropic compression and Kantrowitz limits (figure 2), one sees that the body moving to the favorable supersonic region "2" continues to move in the high pressure zone, behind the shock wave for quite a long time. This is due to the presence of an air plug in front of the body and the time to absorb it, according to figure 4. In this case, the body experiences high aerodynamic (figure 7) and thermal loads (figure 6).

The greater the transverse size of the body, the greater the exit section length to the supersonic flow regime. Obviously, the reduction of this segment promises significant energy benefits for the

Journal of Physics: Conference Series **1404** (2019) 012078 doi:10.1088/1742-6596/1404/1/012078

implementation of the acceleration process. The solution of the problem of Kantrowitz limit overcoming, in this formulation, is the task of further research for the authors.

Acknowledgments
The research was carried out within the framework of the complex Program of Fundamental Scientific SB RAS in 2018-2020 (project No. AAAA-A18-118021590031-5).

References
[1] Fomin V M, Zvegintsev V I, Nalivaichenko D G and Terent'ev Y A 2016 *Transportation Systems and Technology* **2** 18
[2] Zvegintsev V I 2017 *Thermophysics and Aeromechanics* **6** 829

Journal of Physics: Conference Series **1404** (2019) 012079 doi:10.1088/1742-6596/1404/1/012079

Influence of surface sublimation on boundary layer stability at high Mach number

S A Gaponov, A N Semenov and B V Smorodsky

Khristianovich Institute of Theoretical and Applied Mechanics SB RAS,
4/1, Institutskaya str., Novosibirsk, 630090, Russia

gaponov@itam.nsc.ru

Abstract. Results of theoretical investigation of the linear stability properties of Mach 6 boundary layer on a flat-plate with sublimation coating are presented. Naphthalene ($C_{10}H_8$) was chosen as a substance for such a coating. This is an organic compound capable to pure sublimation (phase transition from the solid state directly to gas, bypassing the melting stage) at moderate temperatures. Calculations performed on the basis of binary-mixture boundary-layer equations in the approximation of local self-similarity revealed that with increasing flow stagnation temperature the adiabatic surface temperature also increases that causes growth of mass flow rate of naphthalene evaporation. This leads to an increase in the density of the binary mixture (air and sublimation vapors) near the wall. The performed linear stability computations show that such a modification of the boundary layer profiles causes a reduction in the local linear amplification rates of perturbations of the first mode, while disturbances of the second mode are destabilized by the surface sublimation.

1. Introduction

The necessity to study the boundary layer under conditions of mass ablation from a surface is known for a long time. The reason for that is a practical need for thermal protection systems of hypervelocity vehicles (specifically reentry apparatus) by usage of destructible heat-shielding coatings. The importance of such studies is emphasized in the monograph [1]. As it was mentioned there, the thermal protection of the very first reentry spacecraft was greatly exaggerated due to the lack of knowledge of a number of important problems of aerothermodynamics at that time. Boundary-layer laminar-turbulent transition is one of most important problems in this row.

Various issues of transition are extensively investigated around the world. It is generally accepted that, at least in conditions of low external noise, the turbulence onset is determined by the boundary layer instability. The foundations of the theory of stability of compressible boundary layers were laid back in the early fifties [2]. Subsequent studies of the stability of supersonic boundary layers are reviewed in [3-5]. All these studies concerned the boundary layers of a single-component gas.

Stability of boundary layers with chemical reactions was investigated for the first time in [6], where stability of boundary layers of a nonequilibrium dissociating gas was analyzed. The results of these investigations were later summarized in a book [7]. Stability of hypersonic boundary layers with chemical reactions of a more general kind were studied e.g. in [8-10].

Stability and laminar-turbulent transition under conditions of ablation have hardly been studied. At present, only papers [11, 12] are known for the stability of the sharp cone hypersonic boundary layer under conditions of mass ablation from the surface. They have considered only two-dimensional (2D) perturbations at very high Mach numbers of $M = 16$ and 20. The results of those computations have never been compared with experiment. Therefore, the expansion of the theoretical studies initiated in those papers onto the development of three-dimensional (3D) perturbations at moderate Mach numbers seems important, regarding especially possibilities of experimental verification of theory.

High-temperature ablation, which is specific for the conditions of a real reentry flight in the atmosphere, is quite difficult and expensive to reproduce in laboratory conditions. However, sublimation under conditions of moderate temperatures, which could be in principle realized in modern wind tunnels, allows to study physical processes of ablation under simplified conditions, in the absence of chemical reactions and other high-temperature gas-dynamic phenomena.

Surface sublimation supplies boundary-layer flow with a foreign gas (sublimation material vapor) which is injected at a certain mass flow rate. In [13, 14] it was shown that injection of heavy foreign gas (with molecular weight greater than for the air) into the near-wall sub-layer of the supersonic boundary layer through the permeable (porous) surface of the model has favorable influence on the boundary layer stability. Theoretical findings of these papers have later been confirmed by stability and transition experiments performed in continuously running supersonic wind tunnel at Mach number 2 [15, 16]. However, permeable surfaces have natural inherent roughness [17], which to a certain degree diminishes useful influence of the heavy gas injection, because roughness normally contribute to the boundary layer destabilization. That is why extension of the theory [13, 14] to investigate application of smooth sublimation coatings with negligible roughness, which would provide injection of heavy foreign vapors into the boundary layer, seems logic and looks promising.

In [18] boundary value problem for self-similar boundary-layer equations on a sublimation surface has been deduced. Computations of base-flow properties of supersonic Mach=2 flat plate boundary layer have been performed. Linear stability computations of such a sublimation surface boundary layer at Mach 2 has been performed in [19]. It was shown that sublimation leads to noticeable stabilization of the first mode perturbations. Theory developed in [18, 19] will be used in this paper to study influence of surface sublimation on linear stability of hypersonic Mach 6 flat-plate boundary layer in relation to first and second instability modes.

2. Base-flow computations

The flat-plate model is considered, which is immersed into a flow of compressible gas. It is assumed that the model surface is coated by the chemical substance, which is able to sublimate at moderate temperatures (that is capable to phase transition from the solid state directly to vapor bypassing the melting stage). Thus, the vapor of the sublimation material, which is actually a foreign gas with molecular mass m_1, penetrates from the model surface into the boundary layer of the primary gas with a molecular mass m_2. The sublimation process is assumed to be slow enough. In this case the shape of the model surface could be treated as unchanged. Then the boundary layer flow can not be considered as a single-component flow anymore, but becomes rather binary-mixture boundary layer. Time-dependent dynamics of such a binary mixture of viscous heat-conducting compressible gases is described by the system of partial differential equations which can be found, e.g., in [7]. From these general equations the system describing 2D stationary supersonic flat-plate boundary-layer of binary gas mixture in the absence of any chemical reactions have been deduced in the approximation of local self-similarity [16]. The equations taking into account sublimation vapor diffusion across the boundary layer and corresponding boundary conditions on the adiabatic sublimation surface have been formulated in [18]. It was shown there that the most important parameter influencing the boundary layer properties is the sublimant injection factor $f_w = \dfrac{G_w}{\rho_e U_e}\mathrm{Re}$, where G_w is mass flow rate of sublimation substance vapors from the surface, ρ is the mixture density, U is streamwise velocity. Subscript *e* stands for the values taken at the boundary layer outer edge. $\mathrm{Re} = \rho_e U_e \delta / \mu_e = \sqrt{\mathrm{Re}_1 x}$ - Reynolds number based on the Blasius length scale $\delta = \sqrt{\mu_e x / \rho_e U_e}$; μ - viscosity, x – streamwise coordinate, Re_1 – unit Reynolds number. Mass flow rate of the vapor from the surface G_w was determined by Knudsen-Langmuir and Clapeyron-Clausius equations [20] (also [18, 19]).

The viscosity and thermal conductivity coefficients for the components of the mixture, as well as foreign gas diffusion coefficient, were calculated on the basis of kinetic theory using Lennard-Jones potential [21]. The viscosity and thermal conductivity coefficients for the mixture were calculated using semi-empirical Wilke's formula. The Eucken-type correction was used for calculation of thermal conductivity of polyatomic gases. In the present paper the boundary-value problem for the binary-mixture boundary-layer equations has been integrated numerically by means of the fourth-order Runge-Kutta numerical algorithm. Shooting technique and nested Newton-Raphson iterations have also been used to satisfy the boundary conditions. Boundary-layer equations, boundary conditions and numerical technique used in the present investigation were presented in a more detail in [18].

3. Linear stability computations

Linear stability theory (LST) for binary-mixture boundary-layer flow has been developed and presented earlier, e.g. in [13,14]. It was applied for the supersonic Mach=2 flat-plate sublimation boundary layer in [19]. It was shown in these papers that after linearization of equations of motion of viscous heat conducting binary gas mixture for wavy disturbances $q(x,y,z,t)=\tilde{q}(y)\exp i(\alpha x+\beta z-\omega t)$ one can deduce a tenth-order system of ordinary differential equations. Here (α,β) are streamwise and spanwise wavenumbers; $\omega=\alpha C=\omega^*\delta/U_e$ – nondimensional frequency, f – dimensional frequency in Hertz. The sought vector $\tilde{q}(y)=(\tilde{u},\tilde{u}',\tilde{v},\tilde{w},\tilde{w}',\tilde{p},\tilde{h},\tilde{h}',\tilde{c},\tilde{c}')^T$ is composed of fluctuations of the three velocity components, pressure, enthalpy and foreign gas (sublimation vapor) concentration, and their derivatives with respect to y, denoted by the prime. In the spatial stability analysis it is assumed that ω and β are real numbers, while $\alpha=\alpha_r+i\alpha_i$ is complex. Imaginary part of the streamwise wavenumber $(-\alpha_i>0)$ correspond to perturbation spatial amplification rate. System for the sought vector $\tilde{q}(y)$ together with ten homogeneous boundary conditions [19] for disturbances at the sublimation surface and at the boundary layer outer edge constitute the eigenvalue problem. This problem in the present investigation was numerically integrated by means of the method of orthonormalizations [7]. The streamwise wavenumber α was found as an eigenvalue of the problem with maximal $-\alpha_i$. Additional details on stability equations and used numerical procedures can also be found in [16].

4. Results

Boundary layer computations in the present paper have been performed for the flow of air over the flat-plate at hypersonic Mach number M=6. It was assumed that the model surface was coated with a layer of substance capable to pure sublimation at moderate temperatures, in the absence of other complication phenomena, such as chemical reactions, dissociation and ionization. Such property has, for example, dry ice, camphor, iodine. In this paper we consider a supersonic boundary layer above the surface coated with a layer of naphthalene ($C_{10}H_8$). It is a chemical compound with molecular mass $m_1=128.17$ that is 4.4 times heavier than air. Other thermo-physical properties of $C_{10}H_8$ vapors required for computations are given in [19]. The choice of naphthalene for our investigations, in addition to its favorable thermo-physical properties, is explained by the fact that $C_{10}H_8$ is an easily accessible and inexpensive hydrocarbon.

In all presented further computational results it was assumed that the value of stagnation pressure $P_0=10$ bar, which is in the operational range of hypersonic wind tunnel Transit-M of ITAM [22]. This value of P_0 was kept constant, while stagnation temperature T_0 was varied. In the present paper we studied theoretically the supersonic boundary layer on thermally insulated, naphthalene coated, flat-plate with increasing mass flow rate G_w of sublimation vapors. Such an increase of surface material evaporation can be obtained by increasing sublimation coating temperature T_w, that was achieved by raising flow stagnation temperature T_0. The computed with adiabatic wall conditions cases are tabulated in table 1, where columns are: case number (used in subsequent figures as curve numbers), flow stagnation temperature T_0, surface temperature T_w (adiabatic wall temperature, obtained as numerical solution for a given value of T_0) and corresponding flow unit Reynolds number Re_1. Case *0* presents result of computations in the absence of sublimation and serves as a reference point for our investigation.

Table 1. The computed with adiabatic wall conditions cases.

No	T_0, K	T_w, K	Re_1, mio/m
0	350	296	13.7
1	378	300	12.3
2	432	310	10.1
3	482	315	8.53
4	563	320	6.72

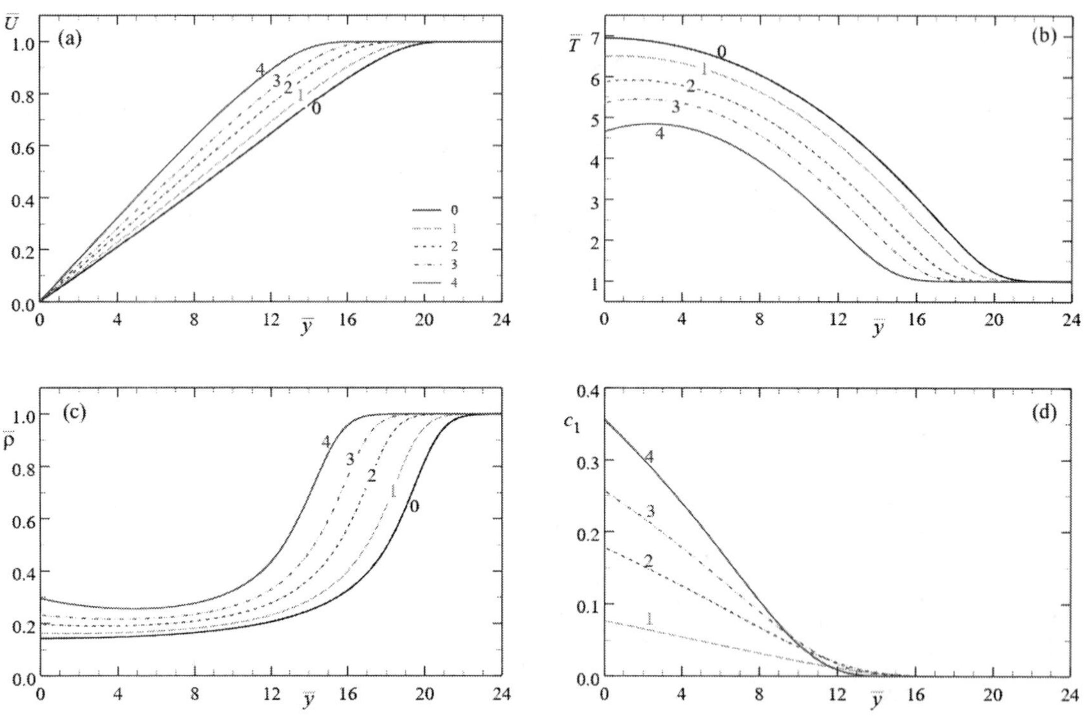

Figure 1. Binary mixture (*Air+C₁₀H₈* vapour) boundary layer profiles: dimensionless velocity $\bar{U}=\bar{U}(\bar{y})$ (a), temperature $\bar{T}=\bar{T}(\bar{y})$ (b), density $\bar{\rho}=\bar{\rho}(\bar{y})$ (c), and foreign gas concentration $c_1=c_1(\bar{y})$ (d) versus dimensionless normal coordinate $\bar{y}=y/\delta$; for a number of values of flow stagnation temperature T_0 (curves *0-4* are numbered in accordance with table 1).

Consider the base-flow properties. Figure 1 shows computed boundary layer profiles: nondimensional velocity $\bar{U}=\bar{U}(\bar{y})$ (figure 1a), temperature $\bar{T}=\bar{T}(\bar{y})$ (figure 1b), density $\bar{\rho}=\bar{\rho}(\bar{y})$

(figure 1c), and $C_{10}H_8$ vapor concentration $c_1 = c_1(\bar{y})$ (figure 1d); for various values of the stagnation temperature T_0. The overbar denotes dimensionless values normalized by corresponding value at the boundary layer outer edge (e.g. $\bar{U} = U/U_e$); $\bar{y} = y/\delta$. One can see that increase of T_0 leads to insignificant reduction of the boundary layer thickness in nondimensional representation, that is seen in streamwise velocity profiles (figure 1a). Temperature profiles $\bar{T} = \bar{T}(\bar{y})$ given at figure 1b demonstrate noticeable decrease of the dimensionless temperature near the wall $\bar{T}(0)$ with increasing T_0, despite that dimensional wall temperature is increasing (table 1). Density profiles (figure 1c) demonstrate much variation: with increasing T_0 from 350 K to 563 K the value of the mixture density at the wall $\bar{\rho}_w = \bar{\rho}(\bar{y} = 0)$ increases in two times: from $\bar{\rho}_w \approx 0.15$ to 0.3. Such an increase of ρ_w according to our previous findings with heavy gas injection through permeable wall [13-16], has a favorable influence on boundary layer stability with respect to first mode perturbations. At figure 1d one can see that growth of the mass flow of the sublimation vapors with increasing T_0 leads to an increase of foreign gas concentration at the wall $c_{1,w} = c_1(0)$ and reaches the value of $c_{1,w} \approx 35\%$ for the case *4*. However, $c_1(\bar{y})$ decreases rapidly with distance from the surface and at the boundary layer outer edge, determined by the velocity profiles ($16 < y_e < 20$, figure 1a), c_1 becomes negligibly small. So, foreign gas injection due to naphthalene coating sublimation can be considered as weak in the presented range of parameters, because sublimant does not go beyond the boundary layer bounds.

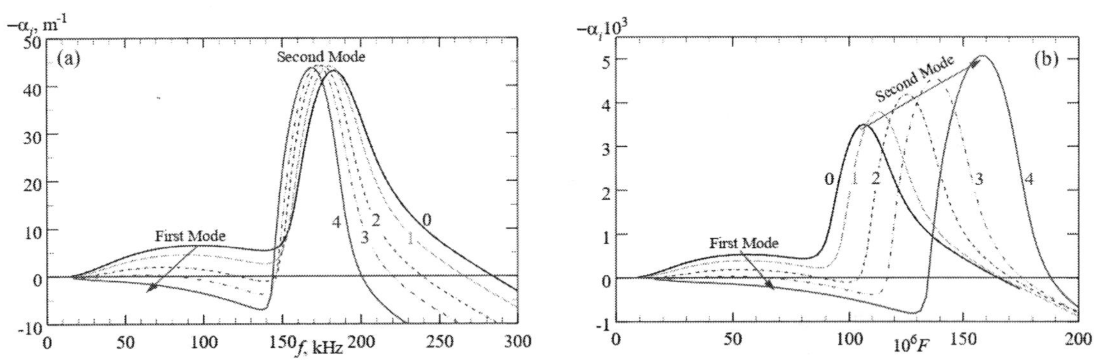

Figure 2. Dimensional (a) and dimensionless (b) presentation of spatial amplification rates $-\alpha_i$ of 2D perturbations versus frequency f (a) and reduced frequency $F = 2\pi f \mu_e / \rho_e U_e^2$ (b); for various values of flow stagnation temperature T_0 (curves *0-4*, according to table 1), $x = 90$ mm.

Figure 2a presents results of linear stability computations. It shows spatial amplification rates of 2D ($\beta = 0$) perturbations versus frequency for various values of flow stagnation temperature T_0 at $x = 90$ mm. Numbers near different curves on this plot correspond to case numbers in table 1. The black curve *0* depicts boundary layer stability results computed at lowest stagnation temperature, without sublimation. One can see that the first instability mode occupies the range of $20 < f < 150$ kHz, with maximal amplification rate of $-\alpha_{i,\max} \approx 7$ m^{-1}. Much more unstable second mode is located in the range $150 < f < 280$ kHz with $-\alpha_{i,\max} \approx 43$ m^{-1}. Increase of stagnation temperature leads to a monotonous reduction of amplification rates in the low-frequency range $20 < f < 150$ kHz. On can see a complete stabilization of 2D first mode on the sublimation surface in the boundary layer with stagnation temperature $T_0 = 563$ K (case *4*, table 1), since amplification rate of all frequencies in this range becomes negative $(-\alpha_i < 0)$. That means, linear 2D first-mode perturbations are stable and their amplitude decreases downstream.

At the same time influence of surface sublimation on perturbations of second mode is not so evident. However, one should keep in mind that increase of T_0 not only intensifies the evaporation of $C_{10}H_8$ from the surface, but also leads to a decrease of the Reynolds number (table 1), and these two factors act in opposite directions in relation to second mode. Figure 2b presents similar plot like figure 2a only in nondimensional representation, that allows to eliminate reduction of the Reynolds number. As a result one can see that intensification of surface sublimation causes increase of second mode amplification rates, while the reduced frequency corresponding to maximal amplification perturbation also increases. The latter is explained by the mentioned above thinning out of the boundary layer (figure 1a,b) in the result of increase of the binary-mixture density near the wall (figure 1c). Red arrows at figure 2a,b demonstrate how surface sublimation influences the first and second instability modes in hypersonic Mach 6 boundary layer.

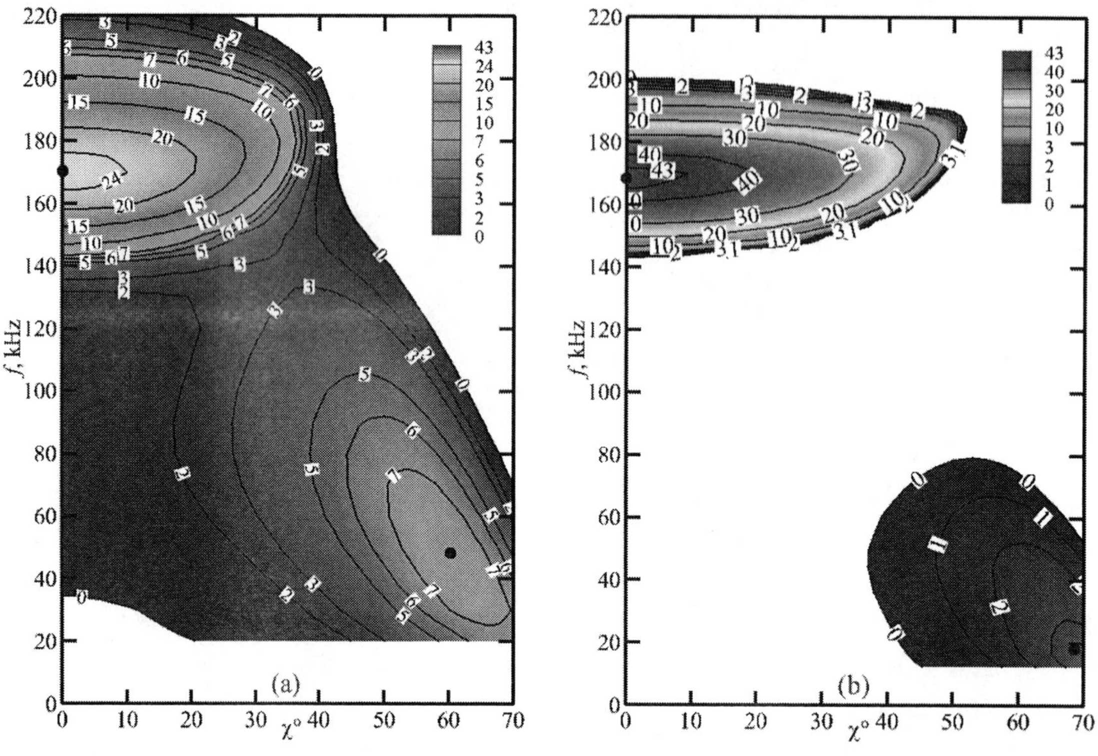

Figure 3. Stability diagrams of hypersonic boundary layer in relation to 3D perturbations: spatial amplification rate contours $-\alpha_i = -\alpha_i(\chi, f)$ [m^{-1}] on adiabatic surface at $T_0 = 563$ K, $x = 90$ mm in the absence of sublimation (a); and on naphthalene coating (case *4*, table 1) (b).

Figure 2 shows results for 2D disturbances only. Now we present an overview of the local linear stability properties of hypersonic binary-mixture boundary-layer on the sublimation surface. Figures 3a,b demonstrate stability diagrams as contour lines of spatial amplification rates on the plane: angle of wave vector orientation $\chi = \arctan(\beta/\alpha_r)$ – frequency f [kHz]. Computations have been performed for streamwise station $x = 90$ mm and stagnation temperature $T_0 = 563$ K (case *4*, table 1). Color-filled area shows linear instability region $(-\alpha_i > 0)$, where amplitude of propagating disturbances grows downstream. Figure 3a shows stability diagram for the boundary layer in the absence of surface sublimation, while figure 3b presents similar diagram for the plate with naphthalene coating. Color schemes of stability diagrams figures 4a,b are taken the same to make the comparison of the computed results easier.

One can see that without sublimation (figure 3a) the instability domain can be provisionally separated on first ($10 < f < 130$ kHz, $0 \le \chi < 75^\circ$) and second ($140 < f < 220$ kHz, $0 \le \chi < 40^\circ$) instability mode areas which merge in one instability region. The second mode has maximal amplification $-\alpha_{i,\max} \approx 25$ m^{-1} at $\chi = 0$, $f_{\max} = 170$ kHz. This is a 2D perturbation. While the first mode has much lower growth rate maximum $-\alpha_{i,\max} \approx 8$ m^{-1} at $\chi \approx 60^\circ$, $f_{\max} \approx 50$ kHz. That is the most unstable first mode disturbance is 3D, oblique wave. Local growth rate maxima of first and second instability modes are depicted by small circles on stability diagrams.

Comparing figures 3a and 3b one can reveal that under the influence of surface sublimation the instability domain is separated into two regions for the first and second modes. The second mode instability domain occupies approximately the same place ($140 < f < 200$ kHz, $0 \le \chi < 50^\circ$) however the maximal growth rate is increased in about 75% to $-\alpha_{i,\max} \approx 44$ m^{-1} at $\chi = 0$, $f_{\max} = 170$ kHz. While the first mode instability region is considerably shrunk (to $8 < f < 80$ kHz, $40^\circ < \chi < 75^\circ$) under the influence of surface sublimation. Maximal amplification of $-\alpha_{i,\max} \approx 3.2$ m^{-1} has now a 3D wave with $f_{\max} \approx 18$ kHz and $\chi_{\max} \approx 68^\circ$. It is also worth to note that 2D perturbations ($\chi \approx 0$) of the first mode (which are normally precede the amplification of the second mode) are completely stabilized by the surface sublimation.

Thus, the presented modification of the boundary layer stability diagram (figure 4a → figure 4b) under the influence of naphthalene surface sublimation in adiabatic wall conditions means stabilization of low-frequency first mode perturbations. At the same time, higher frequency disturbances of the second mode, which are normally dominating hypersonic boundary layer transition, are destabilized under the influence of surface sublimation.

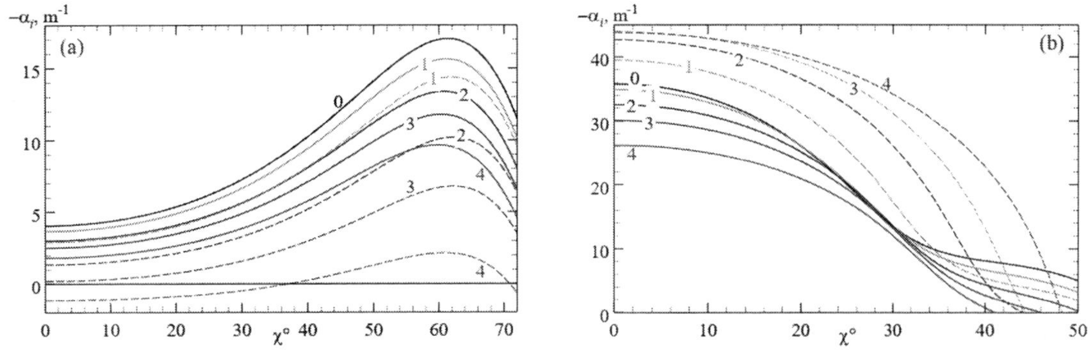

Figure 4. Spatial amplification rates $-\alpha_i$ [m^{-1}] for 3D perturbations of the (a) first mode with $f = 50$ kHz and (b) second mode with $f = 170$ kHz, versus angle χ; for various values of T_0, $x = 90$ mm. Dashed lines – on $C_{10}H_8$ sublimation coating; solid lines – in the absence of sublimation.

Figures 4a,b demonstrate comparison of spatial amplification rates of 3D disturbances of the selected frequencies $f = 50$ and 170 kHz (with maximal amplification for the first and second modes, figure 3a) versus wave-vector orientation angle χ for various values of T_0 computed on the sublimation surface (dashed lines) and in the absence of sublimation (solid lines). One can see that at the lowest considered value of the flow stagnation temperature, there is no noticeable influence of sublimation since solid and dashed lines *0* merge (figures 4a,b).

First mode perturbation (figure 4a) has the largest amplification rate $-\alpha_{i,max} \approx 17$ m^{-1} at $\chi \approx 60^\circ$. One can see that increase of the stagnation temperature causes monotonous reduction of growth rates for all $0 \le \chi < 72^\circ$. Comparing pairs of solid and dashed curves of the same color and numbers (that means, LST-computed in the flow with the same T_0) at figure 4a, one can clearly see consecutive stabilization of first mode perturbations by means of naphthalene coating sublimation. Increase of stagnation temperature from $T_0 = 350$ K (case *0*, table 1) to $T_0 = 563$ K (case *4*) in the absence of sublimation (solid lines *1-4*) causes reduction of the maximal growth rate to $-\alpha_{i,max} \approx 10$ m^{-1}, i.e. in two times. At the same time, application of naphthalene coating allows to reduce maximal growth rate till $-\alpha_{i,max} \approx 2$ m^{-1} (dashed line, case *4*, figure 4a) that is five times smaller. So, the first mode undergoes considerable stabilization by the surface sublimation, at least in conditions of figure 4a.

Growth of T_0 (and therefore reduction of Re_1, table 1) leads to the reduction of growth rates of the second mode (solid lines *0→4*, figure 4b) from $-\alpha_{i,max} \approx 36$ m^{-1} at $\chi = 0^\circ$ to $-\alpha_{i,max} \approx 26$ m^{-1}. At the same time, on naphthalene sublimation coating increase of T_0 causes noticeable destabilization of the second mode, up to $-\alpha_{i,max} \approx 44$ m^{-1} (dashed line, case *4*, figure 4b).

One can see, that local growth rates of the second mode in conditions of performed in this paper investigation are at least two times larger in comparison with the growth rates of the first mode. However, the length of amplification of individual frequencies of the second mode in the downstream direction is usually much shorter than for the first mode, which has much longer way of amplification. Therefore, for a final conclusion about stabilizing-or-destabilizing influence of the sublimation coating on the Mach 6 flat-plate boundary layer it is necessary to perform further LST-computations with application of e^N-method to get more complete information about development of the first and second instability modes in the sublimation boundary layer.

5. Conclusions

Computations of the base-flow and its linear stability have been performed for Mach=6 boundary-layer over flat-plate model with naphthalene ($C_{10}H_8$) coating. It has been found that heavy foreign gas injection into the near-wall sub-layer of the boundary layer due to evaporation of the surface material leads to an increase of the binary-mixture (air + naphthalene vapor) boundary-layer density near the sublimation wall. Linear stability computations of such a modified base-flow revealed a possibility of considerable reduction of local amplification rates of 3D disturbances of the first-mode, when the model surface is heated enough by raising flow stagnation temperature. Computations also show that naphthalene sublimation leads to destabilization of 2D second mode. Further LST-calculations with application of e^N-method are desirable to quantify possibilities of such naphthalene coating usage for Mach 6 boundary layer stability and transition control.

Acknowledgments

This research was carried out within the framework of the Program of Fundamental Scientific Research of Russian state academies of sciences in 2013-2020 (project AAAA-A17-117030610125-7, No 0323-2018-0009) and was also supported by Russian Foundation for Basic Research (project 18-01-00070a).

References

[1] Tirsky G A 2011 *Hypersonic Aerodynamics and Heat-Mass-Transfer of Reentry Space Vehicles and Planetary Probes* (Moscow: FML) (in Russian)
[2] Lin C C 1954 *The Theory of Hydrodynamic Stability* (Cambridge: CUP)
[3] Mack L M 1969 *Boundary layer stability theory* Document 900-277 (Rev.A., Pasadena)
[4] Gaponov S A and Maslov A A 1980 *Development of Disturbances in Compressible Flow*

(Novosibirsk: Nauka) (in Russian)
[5] Zhigulev V N and Tumin A M 1987 *Onset of Turbulence. Dynamical Theory of Excitation and Development of Instabilities in Boundary Layers* (Novosibirsk: Nauka) (in Russian)
[6] Petrov G V 1974 *Combustion, Explosion and Shock Waves* **10** 719–21
[7] Gaponov S A and Petrov G V 2013 *Stability of the Boundary Layer with Nonequilibrium Gas Dissociation* (Novosibirsk: Nauka) (in Russian)
[8] Malik M R and Anderson E C 1991 *Phys. Fluids* A. **3** 803–21
[9] Chang C L, Vinh H and Malik M R 1997 *AIAA Paper* 1997-2012
[10] Johnson H B, Seipp T G, Candler G 1998 *Phys. Fluids* **10** 2676–85
[11] Mortensen C and Zhong X 2014 *AIAA J.* **52** 1632–52
[12] Mortensen C and Zhong X 2016 *AIAA J.* **54** 976–96
[13] Gaponov S A and Smorodsky B V 2016 *AIP Conference Proceedings* **1770** 030047 https://doi.org/10.1063/1.4963989
[14] Gaponov S A and Smorodsky B V 2017 *AIP Conference Proceedings* **1893** 030087 https://doi.org/10.1063/1.5007545
[15] Lysenko V I, Smorodsky B V, Ermolaev Y G and Kosinov A D 2018 *Thermophys. Aeromech.* **25** 183–90 https://doi.org/10.1134/S0869864318020038
[16] Lysenko V I, Gaponov S A, Smorodsky B V, Yermolaev Y G and Kosinov A D 2019 *Phys. Fluids* **31** 104103 https://doi.org/10.1063/1.5112145
[17] Lysenko V I, Gaponov S A, Smorodsky B V, Yermolaev Y G, Kosinov A D and Semionov N V 2016 *JFM* **798** 751–73 https://doi.org/10.1017/jfm.2016.347
[18] Gaponov S A and Smorodsky B V 2019 *Siberian Journal of Physics* **14** 25–39 (in Russian) https://doi.org/10.25205/2541-9447-2019-14-1-25-39
[19] Gaponov S A and Smorodsky B V 2019 *AIP Conference Proceedings* **2125** 030103 https://doi.org/10.1063/1.5117485
[20] Polezhaev Y V and Yurevich F B 1976 *Thermal Protection* (Moscow: Energiya) (in Russian)
[21] Hirschfelder J O, Curtiss C F and Bird R B 1954 *Molecular Theory of Gases and Liquids* (NY: Wiley)
[22] Bountin D, Maslov A and Gromyko Y 2018 *Phys. Fluids* **30** 054103 https://doi.org/10.1063/1.5024025

Effect of stationary longitudinal structures on the growth of Tollmin-Schlichting waves

S A Gaponov and N M Terehkova

Khristianovich Institute of Theoretical and Applied Mechanics SB RAS,
4/1, Institutskaya str., Novosibirsk, 630090, Russia

gaponov@itam.nsc.ru

Abstract. The paper considers the possibility of increasing of the nonstationary wave (similar to Tollmin-Schlichting (T-S) wave) growth by striped stationary structures of eigen-problem of stability of the supersonic boundary layer stability. The problem is solved within the framework of three-wave resonance interaction in local-parallel approximation. In this work the pumping wave is the stationary almost striped structures. Studies have shown that even in a stable area under the influence of longitudinal structures there is an increase of T-S waves. At sufficiently high amplitudes of stationary perturbations their effect on T-W waves is noticeable even in unstable area.

1. Introduction

It is known that the interaction of the external longitudinal vorticity with the boundary layer leads to the formation of longitudinal structures in the boundary layer. For the first time longitudinal structures were investigated in [1]. It was found that the amplitude of the boundary layer thickness, periodically changing in z-direction with the wave number β, increased linearly downstream under the influence of longitudinal external vorticity and was proportional to β. About the interaction of longitudinal or localized structures with T-S waves can be found in [2].

In supersonic flows along with longitudinal stationary structures in the boundary layer, there are stationary disturbances, excited by Mach waves which were explored in [3]. Stationary disturbances as the limit of non-stationary waves propagating down the stream were detected in [4]. There properties of three types of perturbations were investigated in detail: 1) moderately fading disturbances (similar to T-S waves), 2) fading downstream perturbations with a small longitudinal wave number, 3) strongly growing waves, propagated upstream. In the paper first of all special attention is paid to the three-dimensional perturbations of the second type, the structure of which is similar to the longitudinal structures. Their parameters, obtained in a locally parallel approximation, are close to stationary perturbations studied using parabolized stability equations.

The paper considers the amplification(deceleration) possibility of the growth of T-S waves by the striated stationary structures of the eigen-task of the supersonic boundary layer stability. Taking into account that parameters of recent disturbances are similar to parameters of striated structures excited by external turbulence, it is possible to use results on effect of eigen stationary disturbances on T-S waves in a prediction of an external turbulence influence on a laminar-turbulent transition of a boundary layer. In addition, this data can be used to controlT-S waves growth, and thus the transition location.

The problem is solved within the framework of three-wave resonance interaction in a locally parallel approximation. In most of papers researches were based on Craik's theory [5] for three waves, one of which (with the largest increment) had a frequency ω, and the other two — ω/2. However in the boundary layer, especially at supersonic speeds, may be in a quasi-resonance interaction with the main wave waves with other frequencies. In addition, low-frequency disturbances can play the role of the main wave (pumping waves). As an example, it is possible to refer on the paper [6] where the pumping wave had the frequency much less than the frequency of T-S waves. In the paper the pumping wave is the stationary almost striped structure.

2. Formulation of a task

Studies have been carried out for a boundary layer on a flat plate based on the theory of three-wave interaction. The Interaction of a stationary wave with an amplitude of V and two oblique non-stationary waves which are symmetric on the relation to each other with amplitudes of G_1 and G_{-1} which parameters are proportional to $q(y)\exp(i(\theta + i\int \alpha^i dx)$, where $\theta = \beta z + \int \alpha^r dx - \omega t$, is considered. Here x, z - coordinates in the direction of the main stream and normal to it, t— time, α^r and β — wave vector projection values **k** on x and z, ω— a frequency of a moving wave. The directions of wave vectors are shown in the figure 1.

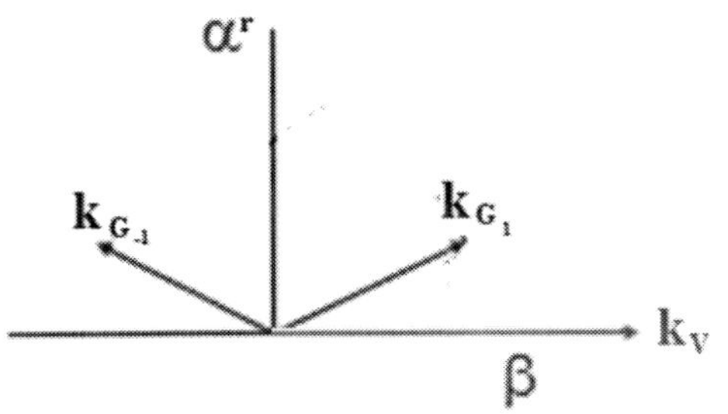

Figure 1. Directions of wave vectors of interacting resonant waves.

In accordance with Craik's theory amplitude equations are used in a non-size form [6]:

$$dV / d\xi = -\bar{\alpha}_V^i V + S_{G_1,G_{-1}}^V G_1 G_{-1} \exp(i\Delta),$$
$$dG_1 / d\xi = -\bar{\alpha}_{G_1}^i V + S_{V,G_{-1}^*}^{G_1} V G_{-1}^* \exp(i\Delta),$$
$$dG_{-1} / d\xi = -\bar{\alpha}_{G_{-1}}^i V + S_{V,G_1^*}^{G_1} V G_1^* \exp(i\Delta).$$

Here $-\bar{\alpha}_j^i = -\delta\bar{\alpha}_j^i$ — degree of linear growth (fading) of the corresponding waves, $S_{k,l}^j$ — interaction coefficients, $d/d\xi = 1/2 \cdot d/d\,\mathrm{Re}$, $\bar{\alpha}_j = \delta\alpha_j$ — dimensionless linear problem eigenvalues, Re — Reynolds number constructed by the thickness of the boundary layer $\delta = \sqrt{x\nu_e / u_e}$, ν_e, u_e — Kinematic viscosity and velocity at a boundary layer edge, x — distance to the beginning of the plate, $\Delta = \int_{x_0}^{x} \alpha_V^r dx'$ —detuning. Asterisk indicates complex conjugated values. Interaction coefficients $S_{k,l}^j$ are determined by the ratios:

$$S_{k,l}^j = \int_0^\infty (\mathbf{Z}_j^+ \cdot \mathbf{M}_{k,l}^j) dY / \int_0^\infty (\mathbf{Z}_j^+ \cdot \partial L(\mathbf{Z}_j) / \partial \alpha_j) dY,$$

where L— linear differential operator of stability equations and $\mathbf{Z}_j$ — vector-function of solving of the equations system $L(\alpha_j)\mathbf{Z}_j = 0$ with homogeneous boundary conditions, $\mathbf{Z}_j^+$ — vector-function of solving equations of conjugate operator L^+. As dimensionless amplitude the relation of amplitude of velocity perturbation in Y direction on boundary layer edge to the velocity of the running flow (to speed on boundary layer edge) is taken. Dann-Lin equations [7] were used to construct the solution and find eigenvalues $\bar{\alpha}_j$ in the paper. Detailed information on the three-wave interaction for the supersonic boundary layer can be found in the [8]. All results were obtained with Mach number M = 2.

3. Results

The text figure 2 shows the dependence of the increments of stationary (a) and non-stationary (b) waves on the Reynolds number, which show that the increments of stationary waves of the first family are negative and much smaller than the increments of the second family. Increments of the second family are small on an absolute value and therefore amplitudes corresponding to them decrease slowly on longitudinal coordinate (on a Reynolds number).

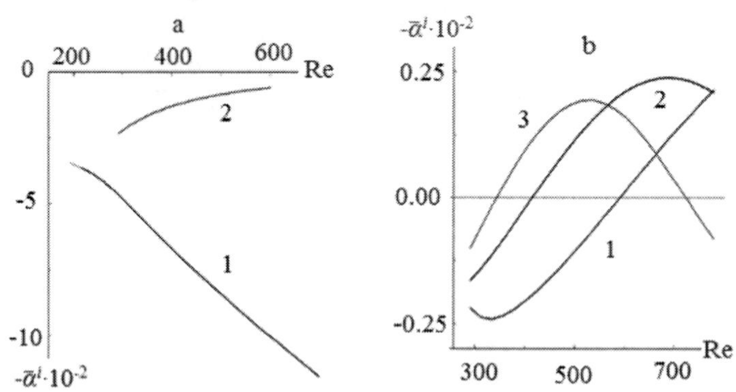

Figure 2. Increments of stationary waves (a) of the first (1) and second family (2) with $b = \bar{\beta} \cdot 10^{-3} / \mathrm{Re} = 0.15$ and moving waves (b) with b = 0.075 and $F = \omega v_e / u_e^2$ = (0.192, 0.384, 0.576) $\cdot 10^4$ (1-3)

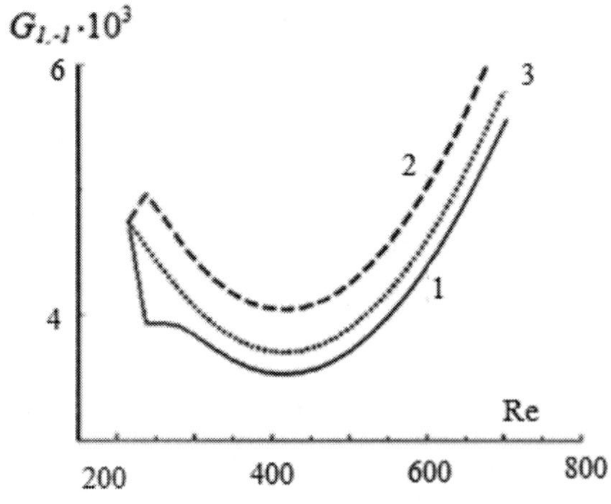

Figure 3. Amplitudes of moving waves at frequency F= 0.384 10^{-4} and b = 0.075 (1), b = -0.075 (2) in the absence of an initial phase shift at the initial amplitude of the stationary wave of the first family V =1.5%, linear amplitude (in the absence of a stationary perturbation) (3).

One of the tasks of this work was to study the possibility of amplifying traveling waves like T-S waves under the influence of stationary waves. Therefore a little of the traveling waves amplitude at the beginning of the calculation area in comparison with amplitude of a stationary wave (similar to longitudinal structure) was assumed.

 IOP Publishing
1404 (2019) 012080 doi:10.1088/1742-6596/1404/1/012080

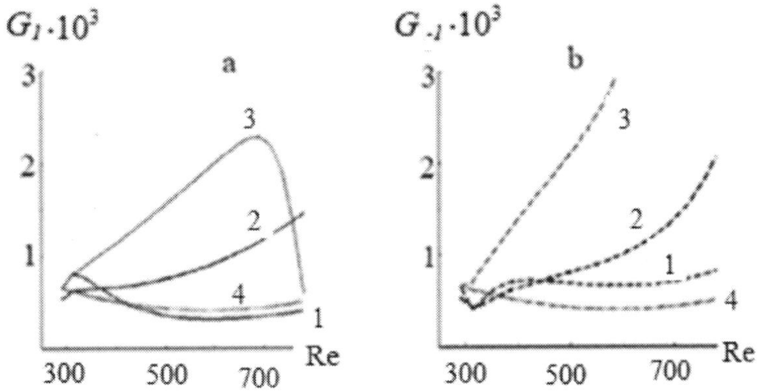

Figure 4. Amplitudes of moving waves of the spatial orientation β=0.075 (a) and negative β=-0.075 (b) with frequencies F= (0.192, 0.384, 576)$\cdot 10^{-4}$) (1-3) (Initial $G_{1,-1}$= 0.05%, initial phase shift equal to -π/4) at the initial amplitude of the stationary wave of the second family V=3.5%. linear amplitude (4).

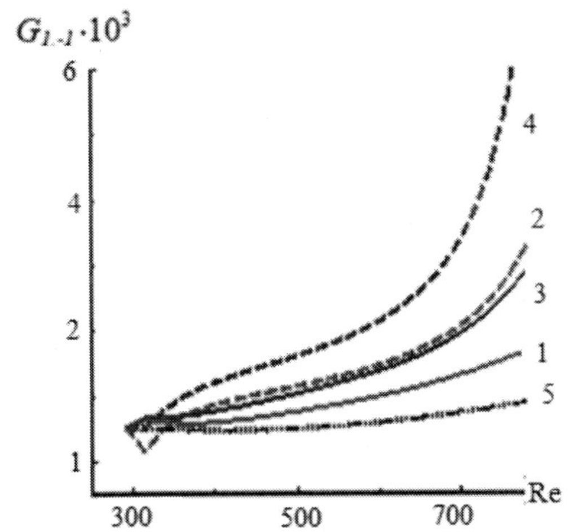

Figure 5. Dependances of $G_{1,-1}$ at F = 0.384 $\cdot 10^{-4}$ on Reynolds number (initial $G_{1,-1}$ = 0.05%, V=3.5%, initial phases: -π/2 (1-2) and - π/4 (3-4). linear amplitude (5).

From the amplitude equations it is visible that at an intensive reduction of stationary wave amplitude a fading (increase) of traveling waves with increase of longitudinal coordinate tends at the linear law with growth rate α_G^i.This is well demonstrated in figure 3. It is evident that, if to normalize the dependencies on the appropriate minimums, all the lines will merge.

In the case of the stationary wave of the second family, the damping of which is weak, its effect on the amplification of the moving waves is significant, figure 4. It should be noted that the amplitudes of G_1 and G_{-1} is different at the same frequency, that is, they depend on the sign of the wave number β. From this data it can be seen that stationary perturbations can lead to an intense of non-stationary waves significantly exceeding linear growth. Other interesting fact consists that with increase in frequency of non-stationary perturbations their growth rate of amplitudes increases. Intensity of the nonlinear process directly depends on initial values of stationary waves amplitudes. Also the growth rate of amplitudes of non-stationary disturbances dependences on ratios of their initial phases, figure 5.

In figure 6 dependences of moving waves amplitudes at F=0.384 $\cdot 10^{-4}$ on Reynolds number are shown. It is visible that with increase of the initial amplitude of a stationary wave leads to an increase of the non-stationary perturbations growth rate.

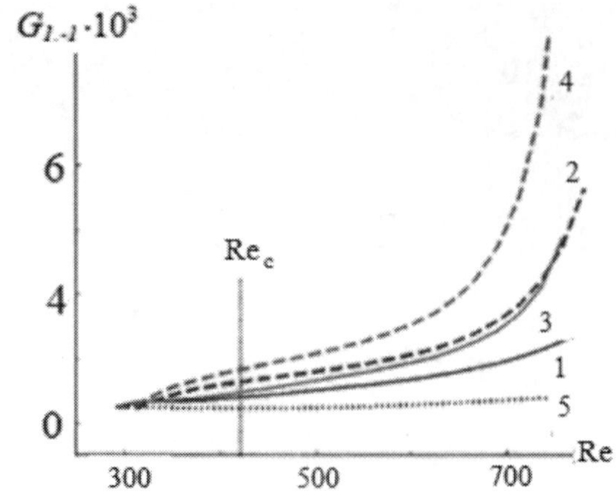

Figure 6. Dependances of moving waves amplitudes at F=0.384 $\cdot 10^{-4}$ on Reynolds number (initial $G_{1,-1}$ =0.05%; V=3.5% (1-2) and 4.5% (3-4). linear amplitude (5).

4. Summary

The results of the studies show that under the influence of stationary waves there is an active increase of non-stationary (running) waves, several times higher than the linear growth.

Growth of the moving waves amplitudes dependences on its initial phase shift, the initial amplitude of the stationary wave.

With increase of the frequency of non-stationary perturbations their growth amplitudes rate amplitudes increases.

Growth of the moving waves amplitudes dependences on its spatial orientation (positive or negative wave number β.

Intense oscillation initiated by moving waves, in turn, can destroy the original stationary wave, and lead to the beginning of a laminar-turbulent transition or the turbulent spots formation.

Acknowledgments

The research was carried out within the framework of the Program of Fundamental Scientific Research of the state academies of sciences in 2013-2020 (project AAAA-A17- 117030610125-7, No 0323-2018-0009).

References

[1] Crow S C 1966 *J. Fluid Mech.* **24** 153–64
[2] Boiko A V, Dovgal A V, Grek G R and Kozlov V V 2012 *Physics of transitional shear flows: instability and laminar-turbulent transition in incompressible near-wall shear layers* (London: Springer) DOI: 10.1007/978-94-007-2498-3
[3] Gaponov S A 2014 *International Journal of Aeroacoustics* **13** 85-111
[4] Gaponov S A and Terehkova N M 2002 *Aeromechanics and gas dynamics* **4** 35-42 (in Russian)
[5] Craik A D D 1971 *J. Fluid Mech.* **50** 393-413
[6] Gaponov S A, Maslennikova I I and Tyushin V Yu 1999 *Journal of Applied Mechanics and Technical Physics* **40** 865-70
[7] Dunn D W and Lin C C 1955 *J. Aeronaut. Sci.* **22** 455-77
[8] Gaponov S A and Maslennikova I I 1997 *Thermophysics and Aeromechanics* **4** 1-9

Correction of inertance of temperature sensing devices in high-speed flow

M A Goldfeld and V V Pickalov

Khristianovich Institute of Theoretical and Applied Mechanics SB RAS, 4/1, Institutskaya str., Novosibirsk, 630090, Russia

goldfeld@itam.nsc.ru

Abstract. The influence of the thermocouple calibration method on the reconstruction of the real temperature in impulse wind tunnel with operation duration of up to 100 ms is presented. It is shown that in an impulse set-up allow obtaining a stepwise thermal load with long temperature plateau which enough to determination of instrumental function, necessary for temperature reconstruction. To this aim, the developed technique, based on the application of the solution of the convolution integral equation, was used. It was included the information on the instrumental function, which was obtained from the calibration measurements. In the deconvolution procedure, the signal was first smoothed by splines, the convolution kernel was extracted by the differentiation of smoothed function, and regularization was performed by Kaczmarz algorithm with relaxation parameter. A comparison is made of the temperature reconstruction in the regular mode of hot shot wind tunnel with two different methods for measuring the response of thermocouples to the temperature step: measurement behind the shock wave front and rapid immersion in aluminum melt. It is found that the use of the instrumental function obtained in a wind tunnel allows one to more reliably determine the position of the temperature maximum in real experiment, which corresponds to the physical structure of the flow.

1. Introduction

The development and research of hypersonic vehicle and scramjet engines as a propulsion for them has been carried out for many years. As a result of these studies, fundamental results were obtained and demonstration flights were carried out. They confirmed the possibility of implementing hypersonic flight in an atmosphere with an air-jet engine [1]. The basis for successful development future hypersonic technologies, along with numerical simulation, is to obtain reliable experimental data on the characteristics of the vehicle and its elements. At the initial stage of research, supersonic and hypersonic periodic-duration wind tunnels with an operation time of up to 10 minutes were used. Set up of this type made allowed simulating with reasonable accuracy the flow around a hypersonic aircraft and performances its propulsion [2].

To simulate real heat transfer processes, a higher level of heat fluxes and temperatures is required. Therefore, for these purposes short-time mode wind tunnels are used, such as shock and pulsed wind tunnels, expansion tubes, which allow raising the temperature level up to 5000 K [2, 3]. The peculiarity of such set up is that the operation time lies in the range of 2-3 ms up to 100 ms. Despite the intensive development of optical methods of thermal measurements, the main measurement techniques remain thermocouples of various types [4] due to their ease of application and reliability. At using thermocouple gauges in short duration set, it is necessary to solve the inertia (time lag of the response) problem of such gages.

There are various approaches to solving this problem. A brief review of possible approaches is given in [5]. In this study, as in a number of other works [6], is indicated, the most attractive method is [5], based on the use of the integral convolution equation to describe the signal and its reconstruction. To implement this method, it is necessary to obtain the instrumental function of the measuring device that is to obtain the time dependence of the thermocouple response on a stepwise thermal load. Previously, the technique of quickly immersing temperature gages in an aluminum molt with a constant temperature was used for this purpose, which allowed obtaining an extended branch with a constant temperature. As a result, the maximum temperature was reconstructed, but the position of the temperature maxima was not determined correctly. A possible reason could be that when calibrating thermocouples, the temperature level was significantly lower than the actual temperature in the impulse wind tunnel and the heat transfer conditions differed from the conditions of real measurements in an aerodynamic experiment.

Therefore, the aim of the present work was to determine the influence of the thermocouple calibration conditions and the accuracy of calculation of the instrumental function on the error of temperature reconstruction under conditions of high-speed processes in highly enthalpy wind tunnels. In addition, the aim was also to verify the applicability of the developed deconvolution method for the new thermocouples calibration.

2. Experiment technique

To perform thermocouples calibration, a high-enthalpy hot-shot wind tunnel IT-302M was used, which at Mach numbers from 5 to 10 has flow parameters close to flight conditions [7]. The peculiarity of this wind tunnel is that in regular operation with a constant Mach number the flow parameters (pressure, temperature, and Reynolds number) are significantly reduced. However, in a special mode of operation with the multiplier, the flow parameters can be kept constant for 80-120 ms. This property of the wind tunnel was used to calibrate the thermocouples and obtain the response of the thermocouple to a step temperature impact. Such dependences were obtained for the temperature range 1100–1620 K at Mach numbers 6 and 8. In the last case, it was possible to obtain a temperature plateau with duration of approximately 120 ms. To determine the temperature, we used the data of temperature measurement in first prechamber of impulse wind tunnel taking into account heat losses [7] and special measures were applied to ensure constant temperature along the plateau. Based on the obtained calibration dependence, an instrumental function was obtained, which was used to reconstruction the temperature, measured during the regular operation of the wind tunnel. During calibration and testing, it was used the same thermocouples, which were used at the previous stage of study, described in [5].

Figure 1 demonstrates the form of the temperature step, which is implemented in the hot shot wind tunnel at nominal temperature of 1500K. It should be noted two features of the initial dependence (line 1), which was obtained because of measurements in the first prechamber.

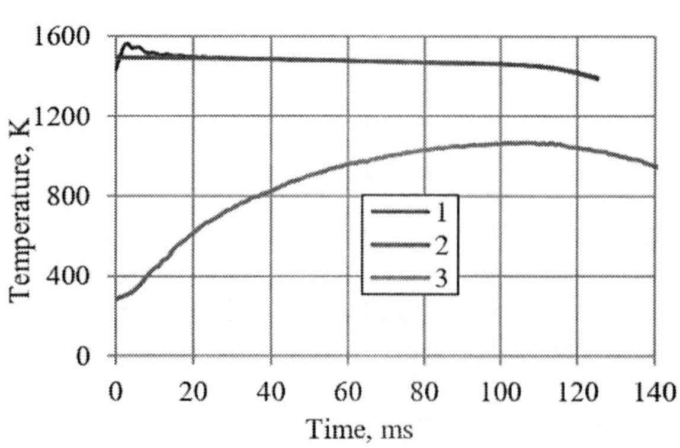

Figure 1. Calibration dependency of chromel – alumel thermocouple with junction size of 50 micron.

The first of them consists in the existence of a small in time region (6-8 ms) of temperature increase, the appearance of which is caused by wave processes during a discharge in the first prechamber and is limited by the duration of the filling of the second prechamber. The maximum temperature rise relative to the average level was not more than 1.6%. Second feature is the decrease in temperature at the end of calibration process after the 110th millisecond because of the pressure drop in the second prechamber at the end of the wind tunnel operation mode. These features are a consequence of the technological properties of the wind tunnel of this type, and they cannot be adjusted during the experiment. An analysis of the time temperature dependence indicates that the temperature pulsation in flow realized on the level of 3-5%. This is a consequence of the influence of the discharge in the first chamber and wave processes at the second prechamber filling. Since this form of the temperature step does not allow one to obtain an instrumental function suitable for the correct temperature recovery, the initial function was smoothed out. This made it possible to obtain temperature dependence with a constant temperature over the entire time range (line 2 in figure 1). The error in temperature deviation from the mean constant value did not exceed ∓0.9% in the time interval between 2 and 110 milliseconds. The response of a thermocouple with a diameter of 50 microns to a step temperature effect (line 3) shows that the maximum temperature is reached only at the end of the operation mode of the wind tunnel. It was also revealed that the pulsations of the temperature, measured by means of the thermocouple, remain at about 3-5% that can create difficulties at determining the instrumental function.

3. Signal processing and method of deconvolution

Due to the inertia of thermocouples, the measured temperature depends not only on the current flow temperature, but also on its previous temperatures. Therefore, a special approach was taken to interpret the measured signal, namely, the convolution integral equation [8] was used to describe the relationship between the true temperature of the thermocouple and the measured one [7]. The problems of usage of the instrument hardware function are usually reduced to solving similar convolution equations [9]. Signal processing and solving these equations is called the deconvolution problem [10-12]. After discretization, the integral equation was written in the form of a system of linear algebraic equations (SLAE). Details of the used algorithm are described in the authors' paper [5]. In summary, its essence is as follows. The unknown temperature distribution *T(t)* is restored from the solution of the integral equation (actually, we need to resolve the temperature differences *E-E(0)* and *T-T(0)*, which we will denote by *E* and *T* below):

$$E(t) = \int_0^t K(t-\tau)T(\tau)d\tau \qquad (1)$$

where *K(t)* is the instrumentation function (in optics it is called the point spread function, PSF), and *E(t)* is the measured value of the thermocouple temperature after subtracting its initial temperature E(0). After discretization, this equation reduces to the matrix equation *E* = *KT*, the solution of which is found by the Kachmarz iterative method [13]. In this case, the next (*k* + *1*) iteration of the restored temperature distribution *T* is found as follows:

$$T_n^{k+1} = T_n^k + \lambda K_{mn}(E_m - (KT^k)_m)/\sum_{j=1}^{N}|K_{mj}|^2 . \qquad (2)$$

Here λ is a relaxation parameter, the matrix *K* has the dimension *M* x *N*, the rows of the matrix are handled according to the rule *m* = *k* (*mod M*) + *1*, the stop criterion is the nondecreasing residual norm of the original equation (1).

To suppress instabilities inherent in the deconvolution procedure, smoothing splines and median filtering were used for experimentally measured signals, as well as regularization of the obtained SLAE. Figure 2 shows the thermocouple signal received in an impulse wind tunnel, here the temperature is shown in Kelvin and starts from a value of 274 K. The derivative of temperature is given in relative units, starting from zero.

An example of testing the obtained experimental PSF (curve 5) in a numerical experiment with a known distribution of model temperature (curve 3) is depicted in figure 3. After solving the direct problem for these functions, the inverse problem (curve 1) was solved by the authors' algorithm. For a random noise of 1%, the reconstruction error (RMS) was only 0.45%.

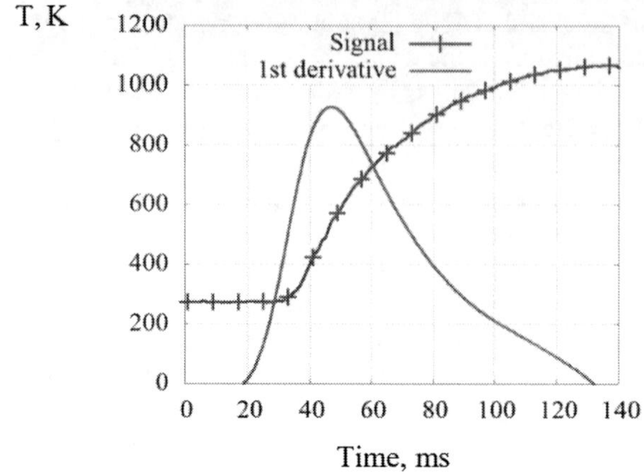

Figure 2. The part of the calibration signal of the thermocouple (K), together with its derivative representing the device function.

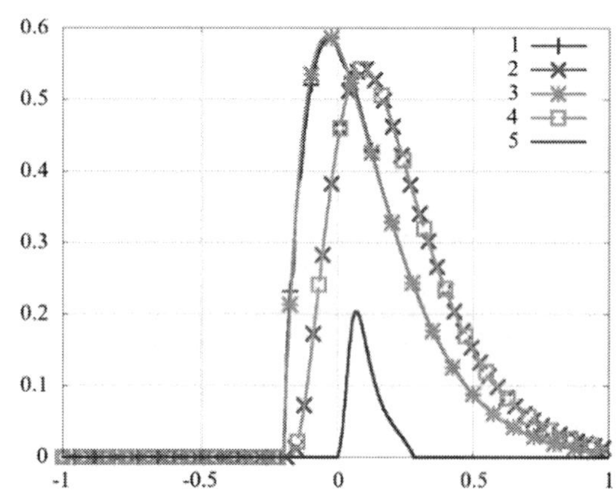

Figure 3. Verification the properties of a new instrumental function in a numerical experiment; random noise 1%, reconstruction error RMS = 0.45%. Here 1 is the reconstruction (solution of the inverse problem), 2 is the solution of the direct problem, the measured signal, 3 is the exact solution, 4 is the calculation of the direct problem for the solution found (pseudo signal), 5 is the PSF.

Figure 4 shows an example of the restoration of a calibration signal with this PSF, and the initial hypothesis assumed a stepwise behavior of the temperature curve towards the end of its rise. Although the initial portion on curve (1) shows the obvious influence of the Gibbs effect when processing high-gradient signals, the range of this curve from 30 to 100 ms can be considered conditionally constant on average.

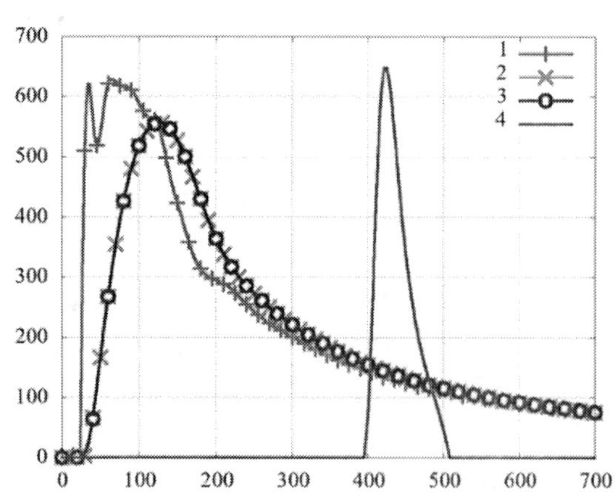

Figure 4. Restoration of the calibration temperature step at the initial measurement part. Here 1 is the reconstruction (solution of the inverse problem), 2 is the measured signal, 3 is the calculation of the direct problem for the solution found (pseudo signal), 4 is the PSF.

Note that the found hardware calibration function (figure 1) turned out to be somewhat wider than when calibrating a thermocouple in an aluminum melt [5], apparently due to weaker heat transfer in the impulse wind tunnel. This circumstance complicates the deconvolution procedure [14] and requires further research on the algorithm for processing such a signal.

4. Summary

Studies in the impulse wind tunnel have shown that, in order to obtain an instrumental function, calibration of thermocouples under conditions close to the measurement conditions is preferable. To realization of the temperature step and obtain the instrumental function with acceptable accuracy, an impulse wind tunnel with high flow parameters can be used.

In the framework of the proposed signal processing scheme and temperature reconstruction model, based on the deconvolution method, it was revealed that the use of the instrumentation function obtained in a pulsed wind tunnel makes it possible to more accurately determine the position of the temperature maximum at the real experiments, which corresponds with the physical structure of the flow.

An analysis of the simulation results indicates that the approaches based on processing signals using smoothing splines, median filtering, and an iterative deconvolution algorithm with regularization are universal and do not depend on the scheme and calibration conditions of thermocouples.

Acknowledgments

The work was partially completed within the framework of the Program of Fundamental Scientific Research of the State Academies of Sciences for 2013-2020 (project AAAA-A17-117030610126-4). The study was conducted at the Joint Access Center "Mechanics" of ITAM SB RAS.

References

[1] Segal C 2009 The Scramjet Engine: Processes and Characteristics (Cambridge: Cambridge Univ. Press)

[2] Kraetzig B, Buttsworth D R, Zande F and Löhle S 2015 *J of Thermophysics and Heat Transfer* **29** 37–46

[3] Holden M S, Wadham T P and Candler G V 2004 Experimental studies in the lens shock tunnel and expansion tunnel to examine real-gas effects in hypervelocity flows *AIAA Paper* 2004-0916

[4] Lee H J, Lee B J, Jeung I S, Kim S L and Kin I 2010 *Int. J. Aeronaut Space Sci.* **11** 193–200

[5] Goldfeld M A and Pickalov V V 2017 *Applied Thermal Engineering* **113** 731–38

[6] Frankel J I and Keyhani M 2014 *J. Thermophys. Heat Transf.* **28** 203–17

[7] Goldfeld M A., Maslov A A, Starov A V, Shumskii V V and Yaroslavtsev M I 2016 *AIP Conference Proceedings* **1770** 030020

[8] Riad S 1986 *Proceedings of the IEEE* vol 74 pp 82–85

[9] Mee D 2003 *J Shock Waves* **12** 443–55

[10] Alifanov O M 1994 *Inverse Heat Transfer Problems* (Berlin: Springer)

[11] Mueller J L and Siltanen S 2012 *Linear and nonlinear inverse problems with practical applications* (Philadelphia, SIAM)

[12] Stark H 1987 *Image recovery: Theory and application* (Orlando, FL: Academic Press)

[13] Byrne C L 2014 *Iterative Optimization in Inverse Problems* (Boca Raton: CRC Press)

[14] Jansson P A 2014 *Deconvolution of Images and Spectra: Second Edition* (New York: Dover Publications)

The influence of dual stream jet nozzle internal parts on the three-dimensional flow structure in a supersonic jet

N P Kiselev, V I Zapryagaev and S G Kundasev

Khristianovich Institute of Theoretical and Applied Mechanics SB RAS, 4/1 Institutskaya str., Novosibirsk, 630090, Russia

nkiselev@itam.nsc.ru

Abstract. The work is devoted to the experimental study of a high-speed jet exhausting from a model dual stream jet nozzle, performed using non-contact (shadow visualization) and probe (total-pressure pneumatic receiver) methods for measuring gasdynamic quantities. Azimuthal non-uniformity of the pressure distribution, whose value in the external duct is much higher than that in the internal duct, is revealed. The cause for the formation of the spatial flow structure is related to the supporting pylons inside the nozzle contour and to the occurrence of a transonic flow regime in the external duct.

1. Introduction

The study of the structure of high-speed gas jets exhausting from the nozzles of jet engines is an important problem connected with the problems of reducing the level of the jet emitted acoustic noise and minimizing their impact on aircraft elements. Gasdynamic parameters of the jet substantially depend on the design of the nozzles used for forming the jet flow. Currently, the number of numerical studies of the structure of high-speed jets discharged from single and dual stream jet nozzles has substantially increased. In these studies, in-house numerical codes (such as Zeus in TsAGI) or commercial software packages (such as ANSYS, FlowVision, etc.) are used [1]. In [2], calculations and experiments for a dual stream jet nozzle with blades disposed inside the ducts at various angles to the incident flow were performed. The flows of cold and hot jets exhausted from the nozzle were studied both numerically, using numerical methods for solving the Navier-Stokes equations with the use of Reynolds-averaged equations, and experimentally, by measuring the velocity fields and acoustic characteristics.

In the present study, we carried out an experimental study of the spatial structure of the steady flow of a supersonic jet exhausting from an axisymmetric dual stream jet nozzle, taking into account the real features of the internal structure of the nozzle. The experiments were carried out for supersonic mode of the jet flow. The experimental data include results of shadow visualization of the flow and results of probing the jet flow with a total-pressure pneumatic receiver in radial and azimuthal directions. The purpose of the work was an experimental study of the 3D structure of the flow of a high-speed jet ejected out of a model dual stream jet nozzle and identification of reasons causing the formation of a substantial azimuthal non-uniformity of the flow in the dual-stream jet.

2. Experimental instrumentation

The experiments on studying the 3D flow structure were carried out using the jet module of the T-326 blowdown wind tunnel of ITAM SB RAS, whose description is given in [3]. In the experiment, a total-pressure probe (Pitot tube with an internal diameter of 0.4 mm) was used for probing the flow. The probe was moved along the three coordinates, X, Y, and Z, using a three-component traversing gear installed in the Eiffel chamber over the nozzle. The range of movement was 0.2 meters along each coordinate.

The flow of the supersonic jets exhausted from the convergent dual stream jet nozzle was visualized using the schlieren method (shadow visualization). In the experiments, a shaped axisymmetric convergent dual stream jet nozzle with a central body was used. The nozzle attached to the settling chamber of the wind tunnel and a diagram of the nozzle with the adopted coordinate system are shown in figures 1a and 1b, respectively.

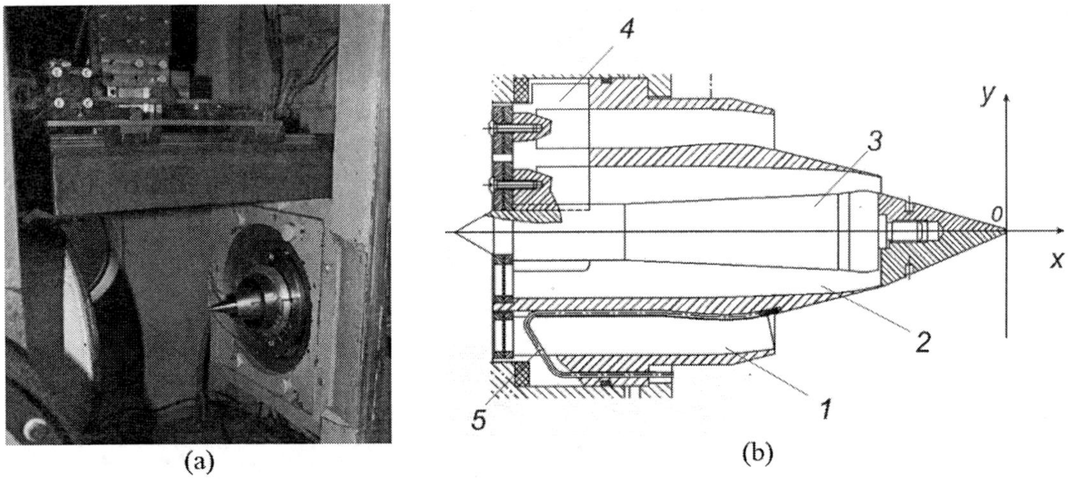

Figure 1. Photo of the model nozzle installed in the test section of the T-326 wind tunnel (a), the nozzle contours and the adopted coordinate system (b), *1* – external contour, *2* – internal contour, *3* – central body, *4* – supporting pylons, *5* – Pitot tubes.

The nozzle was prepared for collecting the experimental data intended for testing results of numerical studies [1]. The geometric shape of the internal contours of the nozzle was chosen using numerical simulation data.

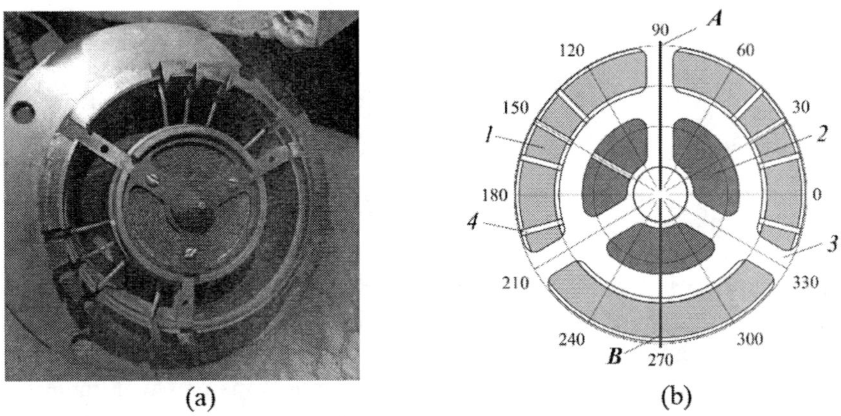

Figure 2. Dual stream jet nozzle, rear view, *1*, *2* - external and internal contours, *3* - pylons, *4* – Pitot tubes, *A* - probing direction along the pylon, *B* - between pylons.

The design feature of the nozzle was that at the nozzle exit there were structural elements obstructing the inlet channels of the contours, namely, supporting pylons *3* (3 items) and Pitot tubes *4* for measuring the wall pressure and the pressure at the inlet to the external and internal contours (see items *1* and *2* in figure 2b). The obstruction of the internal channels with pylons and Pitot tubes for both nozzle contours reduces the internal passage area by 17%, both in the internal and external circuits of the nozzle.

The cold air jet exhausted from the convergent dual stream jet nozzle into ambient space. The experiment was carried out in steady gasdynamic regime of the jet flow with pressure ratio $NPR = P_0/P_c = 2.25$ (P_0 and P_c are the pressures in the pre-chamber and in the test section of the experimental facility). The jet Mach number calculated by the formula for isentropic flow was $M_j = 1.14$. The stagnation temperature in the settling and the working chamber of the facility varied within the range $T_0 = T_c = 282$-300 K.

3. Results

Visualization of the jet flow exhausting from the dual stream jet nozzle was performed over the initial length of the flow using a shadow device and a digital camera. Visualization data taken from a supersonic jet at two exposures corresponding to the mean and instantaneous patterns of the flow are shown in figure 3.

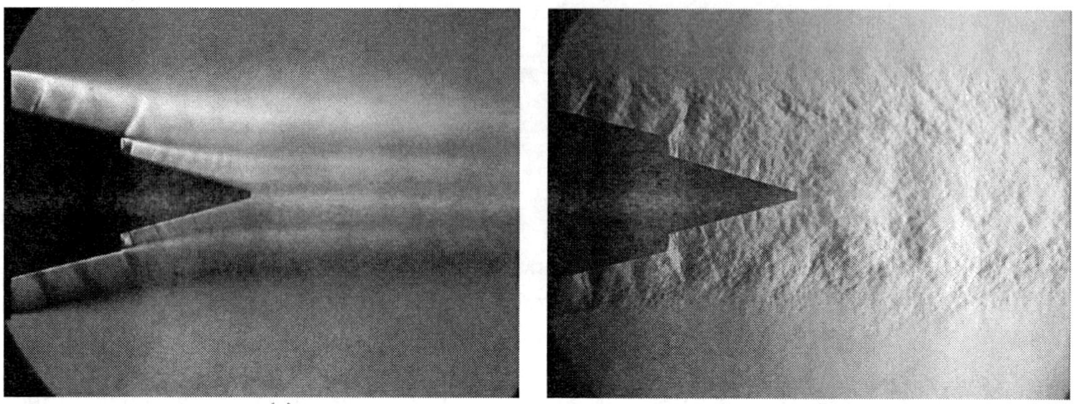

(a) (b)

Figure 3. Schlieren photograph of a dual-stream jet at NPR = 2.25 and M_j = 1.14, time of exposure 10 ms (a) and 4 μs (b).

In this regime, a slightly underexpanded supersonic jet is formed. In addition to the boundaries of the jets exhausting from the shaped nozzle and apart from the wake produced by the central body, the photographs show the shock-wave structure of the flow at the nozzle exit, both in the inner and outer jet. Compression shocks formed at the nozzle exit and accompanying the ejection of the supersonic jet are distinctly seen (Fig. 3, a). Possible reason for the dissipation of compression shocks in the flow region behind the central body is the high turbulence level in the jet stream. At low exposures (figure 3, right), substantial fluctuations of flux density are registered.

Using the traversing gear of the data acquisition system, the radial and azimuthal distributions of total pressure, $P_t(r,\varphi)$, measured by a Pitot tube in characteristic cross sections of the jet flow were obtained. The measurements were carried out in two directions from the flow centerline corresponding to the section in front of the pylon and in between adjacent pylons, as shown in figure 2b (lines *A* and *B*).

The radial distributions are shown for the cross-section $x/R_a = 2.32$ (figure 4). The profiles have a similar appearance involving features due to the gas-dynamic structure of the flow. At the centerline, $r/R_a \approx 0$, a pressure minimum due to the presence of the separated flow in the wake behind the cone of the central body is registered. Two regions can be distinguished in the graph, the flow regions for

$r/R_a = 0 - 0.38$ and $r/R_a = 0.38 - 1.2$, corresponding to the jet exhausting from the internal and external contour of the nozzle. The local pressure minimum in the region $r/R_a \approx 0.38$ corresponds to the flow in the mixing zone of the inner and outer jets.

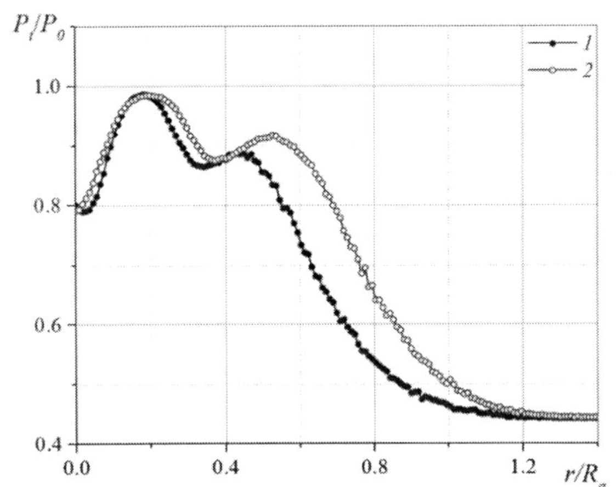

Figure 4. Radial distributions of measured total pressure in section $x/R_a = 2.32$ in between pylons and in front of one of the pylons (curves *1* and *2*, respectively).

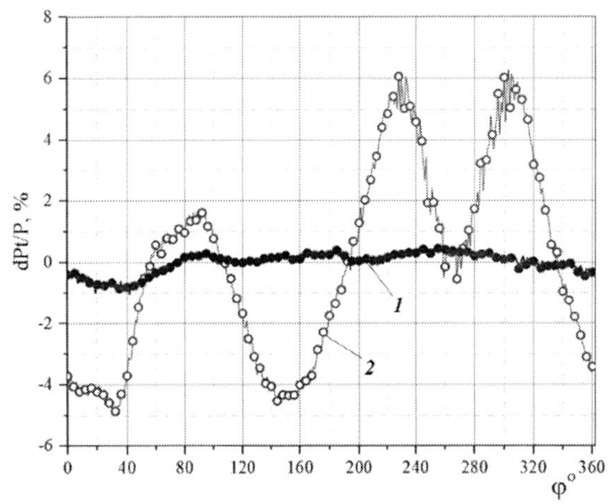

Figure 5. Azimuthal distributions of pressure in section $x/R_a = 2.32$, curve *1* - $r/R_a = 0.17$, curve *2*- $r/R_a = 0.43$.

One can see a difference between the profiles in the region $r/R_a > 0.4$ occupied by the outer jet. The pressure maximum observed at radius $r/R_a = 0.53$ in front of the pylon (curve *2*) is shifted towards the periphery with respect to the maximum in the section between pylons, $r/R_a = 0.45$ (curve *1*). The position of the maxima corresponds to the internal boundary of the mixing layer; therefore, it can be concluded that in the cross-section in front of the pylon the jet becomes narrower than in the region between pylons, since the external boundary of the jet observed at radius $r/R_a \approx 1.2$ remains roughly unchanged.

The characteristic measured azimuthal profiles of mean relative pressure measured by the Pitot tube are shown for the supersonic regime with NPR = 2.25 in the jet cross-section $x/R_a = 2.32$ in Fig. 5. The measurements were performed at values of radial coordinate corresponding to the jet streams discharged from the internal and external contours. Here, lines *1* and *2* refer respectively to the inner and outer jets. The graphs show a substantial azimuthal non-uniformity of both jet flows. The maximum amplitude of non-uniformities (from minimum to maximum) with a range of 11% is registered in the outer jet, and the minimum one, in the inner jet, with amplitude of 1%.

The spatial non-uniformity of the flow is caused by the presence of vortex structures formed as a result of disturbances generated by the internal design elements of the nozzle. A similar flow structure was observed for a supersonic jet exhausted from a nozzle with chevrons located at the nozzle exit [4].

The geometric Mach number calculated by the formulas for isentropic flow, without taking into account the blockage of flow passage area in the external and internal contour ("clean channel") was $M \approx 0.6$ and 0.24, respectively. In the presence of supporting pylons and Pitot tubes, the flow in the external contour was accelerated to a transonic speed (Mach number $M \approx 0.9$), whereas the flow in the inner channel, with allowance for the blockage, was accelerated slightly, to $M \approx 0.3$.

4. Conclusion

An experimental study of the 3D structure of a high-speed jet exhausting from an axisymmetric dual stream jet nozzle with a central body was performed in supersonic regime of the gas-dynamic flow in both nozzle contours at NPR = 2.25 and $M_j = 1.14$.

Shadow visualization of the flow showed the presence of shock waves, wakes of the cone of the central body and the nozzle edge in the interface region of the external and internal jets. The results of probing the flow field of the jet have demonstrated the presence of a substantial azimuthal non-uniformity of the pressure distribution registered in the outer supersonic jet and related with the presence of intense disturbances formed as a result of the interaction of nozzle elements (pylons and Pitot tubes) with the transonic flow.

Acknowledgments
The study was performed as part of the Program of Fundamental Scientific Research of the State Academies of Sciences for the years 2013-2020 (project No. AAAA-A17-117030610137-0).

References

[1] Vlasenko V, Bosniakov S, Mikhailov S, Morozov A and Troshin A 2010 *Progress in Aerospace Sciences* **46** 141-97
[2] Juntao Xiong, Preben Nielsen, Feng Liu, and Dimitri Papamoschou 2010 *AIAA J.* **48** 2249-62
[3] Zapryagaev V I, Kiselev N P and Pivovarov A A 2015 *Fluid Dynamics* **50** 87-97
[4] Zapryagaev V I, Kavun I N and Kiselev N P 2010 *J. Appl. Mech. Tech. Physics* **51** 202-10

The experimental study of the weak shock wave action on the boundary layer of the sweep flat plate

V L Kocharin[1], A D Kosinov[1,2], A A Yatskikh[1,2], L V Afanasev[1,2], Yu G Ermolaev[1,2] and N V Semionov[1]

[1]Khristianovich Institute of Theoretical and Applied Mechanics SB RAS, 4/1, Institutskaya str., Novosibirsk, 630090, Russia
[2]Novosibirsk State University, 1, Pirogova str., Novosibirsk, 630090, Russia

kocharin1208@gmail.com

Abstract. The weak shock waves impact on the flow in the boundary layer of the sweep flat plate at Mach number 2 is studied experimentally. The model was installed at a zero angle of attack. Shock waves in the oncoming flow were generated by a two-dimensional roughness on the side wall of the wind tunnel test section. The data were obtained with a two-dimensional roughness on the wall of the wind tunnel and with a smooth wall at sweep angles (35 and 40 degrees) of the leading edge of the flat blunt plate. The measurements were carried out with a constant temperature anemometer.

1. Introduction

The turbulence developing in high-speed boundary layers is one of the least studied fundamental problems of fluid and gas mechanics. The study of the turbulence generated in supersonic boundary layers which are realized near the aircraft surface is necessary for the development of promising high-speed aircrafts. The process of laminar-turbulent transition in supersonic boundary layers significantly depends on the perturbations level in the free flow [1]. A turbulent boundary layer emitting acoustic pulsations into the free flow is one of the main sources of disturbances in supersonic wind tunnels [2-4]. In addition to acoustic pulsations, quasi-stationary perturbations, such as weak shock waves generated by roughness in the walls of the test section of the wind tunnel can impact on the flow around the models installed in wind tunnels.

A number of experimental [5–12] and numerical [13–14] works are devoted to analysis of the weak shock waves impact on the flow in model boundary layers. The case of an incident weak shock wave at a sweep leading edge at a small angle, when the wave “catches up” the model, is of fundamental and practical interest. In practice, this situation can occur when weak shock waves from the fuselage or other elements of the aircraft get onto a wing with a supersonic leading edge. This paper is devoted to the experimental study of the weak shock waves effect on the flow in the boundary layer of the flat plate at the Mach number 2 and the leading edge sweep angles $\chi = 35^\circ$ and $\chi = 40^\circ$.

2. Experimental setup and data processing

The experiments were performed in the low-noise supersonic wind tunnel T-325 of ITAM SB RAS at the Mach number $M = 2$ and the unit Reynolds number $Re_1 = (8 \pm 0.1)\times10^6\text{m}^{-1}$. In the experiments, the model rotation mechanism was used; it allowed changing the sweep angle of the model leading edge during the experiment. The flat plate model had a blunting radius of the leading edge $r = 2.5$ mm and was installed in the test section of the wind tunnel at the zero angle of attack.

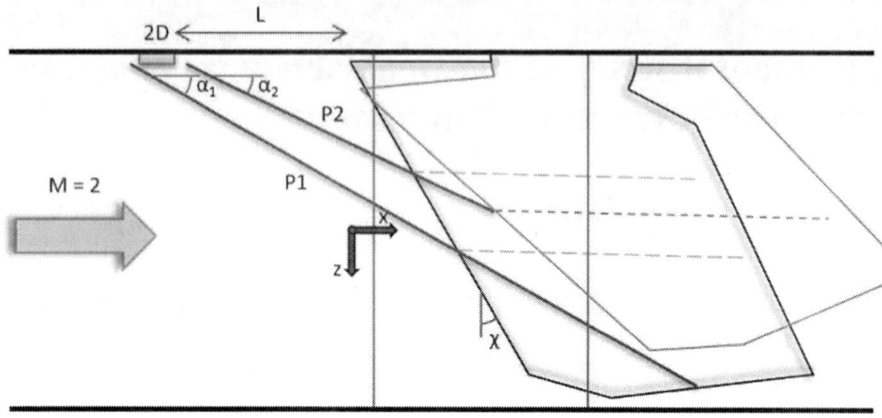

Figure 1. Experiment schematic.

The side wall surface roughness in the wind tunnel test section was used to generate the weak shock waves. An adhesive PVC tape was used as the two-dimensional roughness with a width (downstream) of 7 mm, a length (across the stream) of about 150 mm and a thickness of 0.13 mm. The schematic of the experiments is shown in figure 1, where 2D is a two-dimensional roughness, P is the trailing edge of a weak shock wave from the roughness, and L is the distance from the ledge to the leading edge of the plate. The distance $L = 100$ mm was chosen so that the trailing edge of a weak shock wave from a two-dimensional roughness get on the leading edge of the plate for all sweep angles. The measurements were performed along the z coordinate perpendicular to the oncoming flow at the fixed value of the transverse coordinate $y \approx 0.5$ mm. To measure the pulsations and characteristics of the mean flow, the constant temperature hot-wire anemometer was used. The experimental setup, experimental equipment, and methods for measuring and processing experimental data are described in details in [9, 15–18].

3. Results

The measurements in the oncoming flow ahead of the leading edge of the model were performed. Figure 2 shows the distributions of the mass flow rate pulsations $<m'>$ and the mean mass flow rate ρU depending on the transverse coordinate z for two cases: with a sticker on the side wall of the test section ADT T-325 and with a smooth wall. In the first case, a perturbation in the form of an N-wave is detected in the oncoming flow. The area $z > -12$ mm corresponds to an unperturbed oncoming flow. At $z = -12 \div -15$ mm, a significant transverse gradient of the mean flow is observed, which is accompanied by an increase in the pulsations distribution $<m'>$. This area corresponds to the wave P1. In the area $z = -15 \div -28$ mm, a decrease in the mean mass flow rate is observed, which occurs unevenly in the area $z = -20 \div -22$ mm, the mean mass flow rate is constant and is accompanied by a change in the pulsations level. At $z = -28 \div -36$ mm, a sharp change in the mean mass flow corresponds to the wave P2 and is accompanied by a sharp change in the distribution of the level of pulsations of the oncoming flow. The distance between the waves P1 and P2 in the z-axis direction is approximately 16 mm. For the case of a smooth wall of the test section, in the oncoming flow the level of the mean mass flow rate and pulsations remains constant in the entire measured cross section.

Figure 3 shows the distributions of the root-mean-square pulsations and the mean mass flow rate normalized to the free flow depending on the transverse coordinate z in the boundary layer of the sweep plate for the sweep angle of the leading edge $\chi = 35°$. For the case without two-dimensional roughness on the wall of the test section, the mean mass flow rate decreases monotonically with a decrease in the transverse coordinate z. In the case with 2d roughness, the flow pattern in the boundary layer changes over the entire measured cross section. The red lines in the figure indicate the longitudinal vortex generation region after the interaction of the wave from a two-dimensional roughness with the leading edge of the plate. In this part of the boundary layer, when the incident wave generated from the recess edge of the two-dimensional roughness, the mean mass flow rate changes along with the rising level of pulsations.

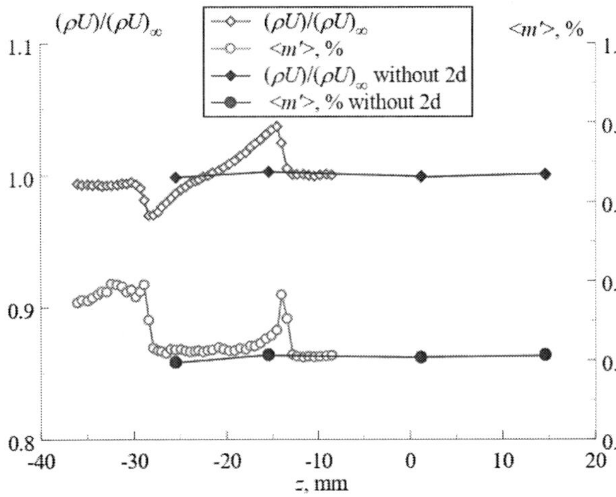

Figure 2. The mean mass flow rate and pulsations in the oncoming flow as functions of the transverse coordinate *z*.

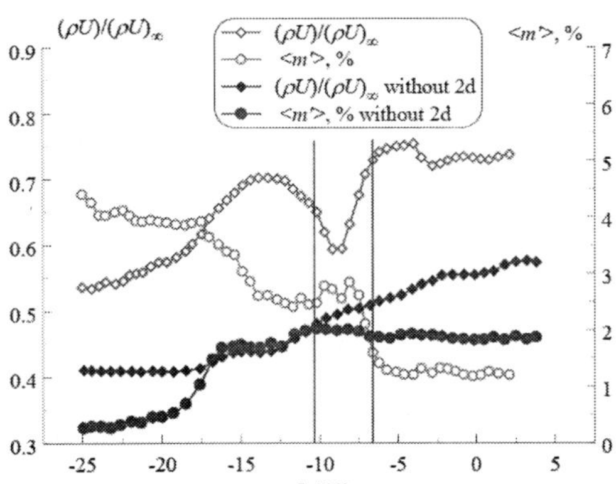

Figure 3. The mean mass flow rate and pulsations in the boundary layer as functions of the transverse coordinate z at the leading edge sweep angle $\chi = 35^{\circ}$.

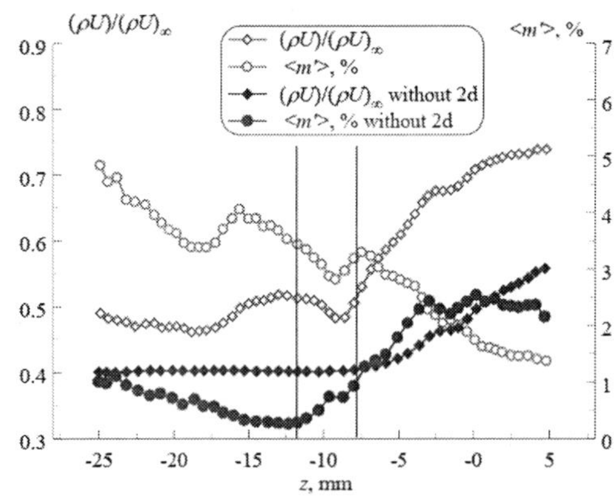

Figure 4. The mean mass flow rate and pulsations in the boundary layer as functions of the transverse coordinate z at the leading edge sweep angle $\chi = 40^{\circ}$.

The similar flow pattern in the boundary layer is observed at $\chi = 40^{\circ}$ (figure 4). In the case of 2D roughness on the wall of the test section of the wind tunnel, the flow pattern in the boundary layer changes over the entire measured cross section. In the area of longitudinal vortex generation after the interaction of the wave from the two-dimensional roughness with the leading edge of the plate, the mean mass flow rate and the level of pulsations change dramatically.

4. Conclusion

An experimental study of the impact of weak shock waves (which are introduced into the oncoming flow by roughness on the wall of the test section of the wind tunnel T-325) on the flow in the boundary layer of a blunt flat plate with the leading edge sweep angles $\chi = 35^{\circ}$ and $\chi = 40^{\circ}$ was carried out at the Mach number 2.

The perturbation in the form of an N-wave was detected by the measurements in the oncoming flow ahead of the leading edge of the model in the case of a roughness on the wall. The distance between the waves in the z axis direction was approximately 16 mm. For the case of a smooth wall of the test section, in the oncoming flow, the level of the mean mass flow rate and pulsations remains constant in the entire measured cross section.

At the sweep angles of the leading edge $\chi = 35°$ and $\chi = 40°$, the generation of a longitudinal vortex was detected in the boundary layer after the interaction of a wave from the two-dimensional roughness with the leading edge of the plate, accompanied by a change in the level of the mean mass flow rate and pulsations.

Acknowledgments
The research was carried out within the framework of the Program of Fundamental Scientific Research of the state academies of sciences in 2013-2020 (project No. AAAA-A17-117030610125-7). The study was conducted at the Joint Access Center "Mechanics" of ITAM SB RAS.

References

[1] Gaponov S A and Maslov A A 1980 *Development of Disturbances in Compressible Flows* (Nauka, Novosibirsk)
[2] Kendall J M 1975 *AIAA J.* **13** 290-9
[3] Laufer J 1961 *J. Aerospase Sci.* **28** 685-92
[4] Gaponov S A and Semenov A N 2018 *Fluid Dynamics* **53** 795-804
[5] Kocharin V L, Kosinov A D, Yermolaev Yu G and Semionov N V 2019 *EPJ Web of Conf.* **196** 00018
[6] Kosinov A D, Yatskikh A A, Yermolaev Yu G, Semionov N V, Kolosov G L and Piterimova M V 2017 *AIP Conf. Proc.* **1893** 030072
[7] Kocharin V L, Semionov N V, Kosinov A D, Yermolaev Yu G and Yatskikh A A 2018 *AIP Confe. Proc.* **2027** 040026
[8] Kocharin V L, Afanasev L V, Kosinov A D, Yatskikh A A, Semionov N V and Yermolaev Yu G 2019 *AIP Conf. Proc.* **2125** 030104
[9] Ermolaev Yu G, Kosinov A D, Kocharin V L, Semenov N V and Yatskikh A A 2019 *Fluid Dynamics* **54** 257-63
[10] Kocharin V L, Yatskikh A A, Kosinov A D, Yermolaev Yu G and Semionov N V 2019 *Siberian J. of Physics* **14** 46-55 (in Russian)
[11] Vaganov A V , Kosinov A D , Noev A Yu , Radchenko V N and Skuratov A S 2018 *AIP Conf. Proc.* **2027** 030112
[12] Yermolaev Yu G, Yatskih A A, Kosinov A D, Semionov N V, Kolosov G L and Panina A V 2016 *AIP Conf. Proc.* **1770** 020012
[13] Din Q H, Egorov I V and Fedorov A V 2017 *Uch. Zap. TsAGI* **48** 10-9
[14] Din Q H, Egorov I V and Fedorov A V 2018 *Fluid Dynamics* **53** 690-701
[15] Kosinov A D, Semionov N V, Ermolaev Yu G, Kolosov G L, Yatskikh A A and Kocharin V L 2018 *AIP Conf. Proc.* **2027** 040087
[16] Semionov N V, Yermolaev Yu G, Kocharin V L, Kosinov A D, Semenov A N, Smorodsky B V and Yatskikh A A 2018 *AIP Conf. Proc.* **2027** 030156
[17] Yatskikh A A, Ermolaev Y G, Kosinov A D and Semionov N V 2019 *Tech. Phys. Lett.* **45** 242-5
[18] Kosinov A D, Semionov N V and Yermolaev Yu G 1999 Disturbances in test section of T-325 supersonic wind tunnel *Preprint* 6-99

Evolution of artificial disturbances in swept wing supersonic boundary layer

G L Kolosov[1], A V Panina[1], A D Kosinov[1,2], V S Kosorygin[1] and A A Yatskikh[1,2]

[1]Khristianovich Institute of Theoretical and Applied Mechanics SB RAS,
4/1, Institutskaya str., Novosibirsk, 630090, Russia
[2]Novosibirsk State University,
1, Pirogova str., Novosibirsk, 630090, Russia

kolosov@itam.nsc.ru

Abstract. Experimental data on the evolution of artificial controlled travelling disturbances in a three-dimensional supersonic boundary layer on a 45° swept wing at Mach number 2.0 is presented. Artificial disturbances were introduced in the boundary layer using a periodical glow discharge with a frequency of 20 kHz. Pulsations of the boundary layer were measured with a constant-temperature hot-wire anemometer and the probe moved parallel and not parallel to the leading edge of the model. In both cases spatial-temporal and spectral-wave characteristics of the wave train development were obtained and quantitatively compared with each other. This paper discusses the suitability of the method for measuring fluctuations fields when the probe is moving not parallel to the leading edge of the model.

1. Introduction

The problem of predicting the laminar-turbulent transition onset in three-dimensional (3D) boundary layers has been studied for more than one decade [1, 2]. The practical significance is obvious, since the same boundary layers are realized on the wings of aircraft. But the task is complicated by the fact that in the case of 3D boundary layer, additional factors appear that affect the laminar-turbulent transition process. Thus, a distinctive feature of the boundary layer of the swept wing is the presence of a crossflow, which is responsible for the emergence of new types of instabilities. As a rule, four main types of flow instability are distinguished in 3D boundary layer: attachment line instability on the leading edge, stationary crossflow instability, cross-flow instability to the travelling disturbances, instability of the Tollmien-Schlichting waves. What type of instability will be the key and determine the laminar-turbulent transition strongly depends on environmental conditions, such as Mach number, surface quality, etc.

As for the supersonic boundary layer, it is somewhat more complicated than the case of an incompressible boundary layer. Differences between incompressible and compressible flows are already observed at the linear stage of development of unstable disturbances. Several studies were devoted to the evolution of traveling pulsations in 3D supersonic boundary layer [3-5], where it was found that the laminar-turbulent transition is determined by this type of disturbance. To excite controlled travelling pulsations in a supersonic boundary layer, a periodic discharge in the chamber is used at the ITAM SB RAS [6-8]. This approach made it possible to study in detail the linear development of a wave train on a swept wing (sweep angle of 45 degrees) with a Mach number of 2.0 for disturbance frequencies of 10 and 20 kHz [9, 10].

In the above works, measurements with the hot-wire probe were performed parallel to the leading edge of the model. Since in previous experiments [10] it was found that the maximum of the pulsations for all waves is approximately in one place along the boundary layer, it is of interest to measure the wave characteristics in the coordinate system associated with the incident flow. In this approach, the probe does not move parallel to the leading edge of the wing, but perpendicular to the incident flow – as in the case of a flat plate. This approach has its advantages and disadvantages. Of the benefits – a simplification in the processing of experimental data, namely in the Fourier transform and the estimation of the longitudinal wave numbers. On the minus side – the movement of the probe in experiment occurs along a curved path with a change in the normal coordinate, since the upper surface of the wing model is not flat, but convex. The results of the experiments at Mach 2 and their comparison, performed in two approaches – when the probe is moving in parallel and not parallel to the leading edge, are presented below.

2. Experimental setup and data processing

The experiments were performed in a T-325 low-noise supersonic wind tunnel of ITAM SB RAS, at Mach number $M = 2.0$ and unit Reynolds number $Re_1 = 5\times10^6\ m^{-1}$. Disturbances in the flow were measured with a constant-temperature hot-wire anemometer. A tungsten wire with a diameter of 10 microns and a length of 1.7 mm was used. Using a coordinate device, the hot-wire anemometer probe could move in all three directions in space.

A swept-wing model was used in the experiments (see figure 1). The model had a sweep angle of 45° and a slightly blunted leading edge of radius 0.2 mm. The swept wing had a 2.6% lens-shape airfoil on the upper side and a flat surface on the bottom side, with a maximum thickness of 12 mm. The curvature radius of the upper side of the model was approximately 4000 mm. Inside the model, a source of localized artificial disturbances was located, similar to that described in [6]. Controlled pulsations from the periodic discharge were introduced into the boundary layer through the hole with a diameter of 0.4 mm on the upper surface of the model. The point source was located at a distance of $x_0 = 56.6 \pm 0.3$ mm from the leading edge of the model. The discharge was ignited at a frequency of 20 kHz and introduced disturbances into the boundary layer at the same frequency, while the measurements were synchronized with the operation of the discharge. In each position of the probe, time traces were recorded containing the AC component of the signal. Two coordinate systems associated with the flow and with the leading edge of the wing are shown in figure 1. Note that the position $x = 0$ was chosen at the source aperture.

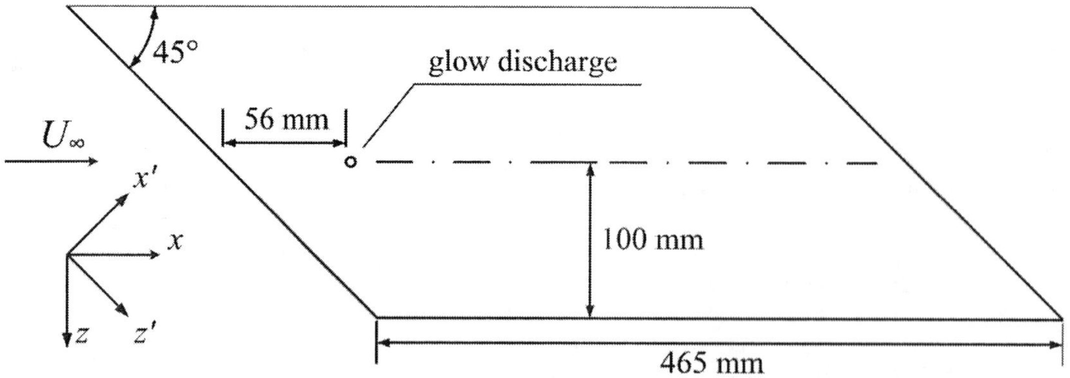

Figure 1. A swept wing model and coordinate systems used.

As already mentioned, this paper presents the results of experiments in two approaches: when the hot-wire probe moves parallel to the leading edge of the model and not parallel. Approximate trajectories of movement of the probe are shown in figure 2. The value x_s corresponds to the distance from the discharge to the measuring section along the central line of symmetry of the model.

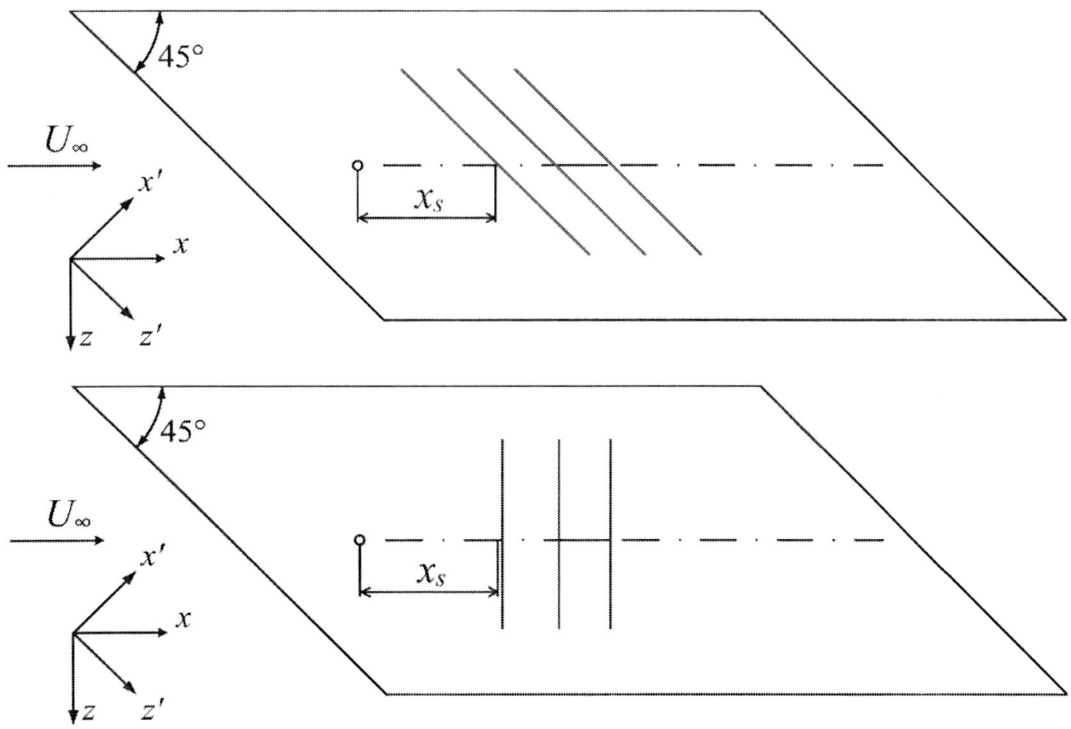

Figure 2. The trajectory of movement of the hot-wire probe: parallel to the leading edge (up), not parallel to the leading edge (down).

The hot-wire anemometer measures the pulsation of the mass flow in the longitudinal direction $m(x,z,t)$ [10]. After that, the frequency-wave disturbance spectra were calculated using the discrete Fourier transform (see formula (1) in [10]).

To obtain α_r, we used the estimation:

$$\alpha_r = \Delta\Phi_{f\beta}/\Delta x_s, \tag{1}$$

where Δx_s is the distance between the measured sections in the x-direction. Since with a different motion of the probe, after the Fourier transform, the wave and phase spectra $A_{f\beta}$, $\Phi_{f\beta}$ and $A_{f\beta'}$, $\Phi_{f\beta'}$ are obtained (in the case of parallel and not parallel to the leading edge motion of the probe, respectively), for the comparison of the dispersion relations $\alpha_r(\beta)$ the connection of two coordinate systems was used:

$$\beta = \beta'\cos(45°)\text{-}\alpha_r'\sin(45°), \tag{2}$$

where 45° is the wing sweep angle.

3. Results

Measurements of the controlled disturbance evolution were performed at three different x_s-coordinate values for the case of movement of the probe parallel to the leading edge of the model, and at five x_s-coordinate for the case of not parallel movement. These measurements were carried out in a layer of maximum fluctuations, which corresponds to the condition $y/\delta \approx 0.6$, where δ is the thickness of the boundary layer. Note, that in the first case, the measurements in each section took place at a constant value of the y-coordinate, whereas in the second case it was necessary to change the y-coordinate when the probe moving perpendicular to the oncoming flow. In both experiments, the source of controlled disturbances introduced pulsations at frequency $f = 20$ kHz in the supersonic boundary layer. The evolution of disturbances at this frequency downstream is discussed below.

The amplitude distributions of disturbances at frequency 20 kHz in the spanwise direction are shown in figure 3. We can note the increase in the wave train amplitude downstream, the expansion of its boundaries, and the shift of the center of the wave packet to the region of positive values of the z'- and z-coordinate. Asymmetry is also observed with respect to the central line where discharge is ($z' = 0$ mm). This fact is associated with the presence of the crossflow in the boundary layer of the swept wing, which is not observed in the amplitude distributions for the case of a flat plate [6]. On the whole, the obtained picture of the spatial development of the wave train corresponds to the results of [9, 10].

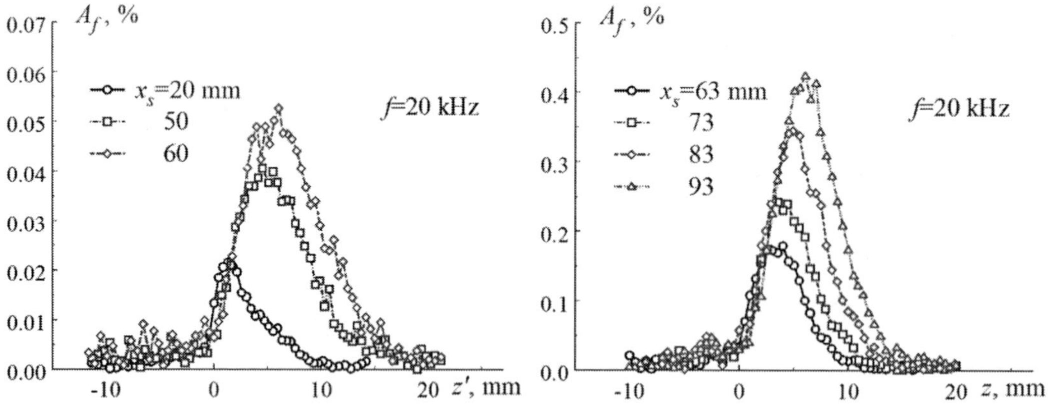

Figure 3. Amplitude distributions of disturbances in the spanwise direction when hot-wire probe moves parallel (left) and not parallel (right) to the leading edge.

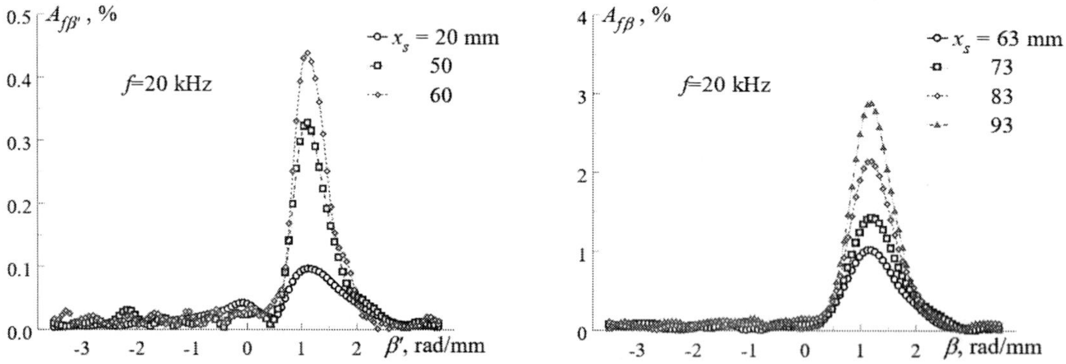

Figure 4. Amplitude β-spectra of disturbances when hot-wire probe moves parallel (left) and not parallel (right) to the leading edge.

Figure 4 shows the amplitude β-spectra of the controlled disturbances at frequency 20 kHz. Again, as in figure 3, there is an asymmetry of the β-spectra for the boundary layer of the swept wing (in contrast to the case with a flat plate). The amplitude growth is observed in a wide range of spanwise wavenumbers, but the amplitude maximum is located at $\beta' = 1.11$ rad/mm for the case of parallel moving of the hot-wire probe and at $\beta = 1.17$ rad/mm for the case of not parallel moving. This values belong to the linear peak in downstream wave train development and correspond to β'-spectra in [9, 10].

Having amplitude and phase β-spectra, it is possible to evaluate other wave characteristics of the development of unstable disturbances, namely, the wave number of the longitudinal component of the wave vector α_r and the angles of inclination of the wave vectors. The estimated results for α_r for both cases of hot-wire probe movement are shown in figure 5 (left). The dependences $\alpha_r(\beta)$ are seen to be close to the linear ones and numerically coincide with each other and with estimates in [9, 10]. The directions of the wave vectors for two cases (see figure 5 (right)) also show the correspondence of the wave characteristics in the measurements parallel and not parallel to the leading edge. Note that other wave characteristics, such as the phase velocities, the wavelengths, and the amplification rates (increments) also coincide in two different approaches.

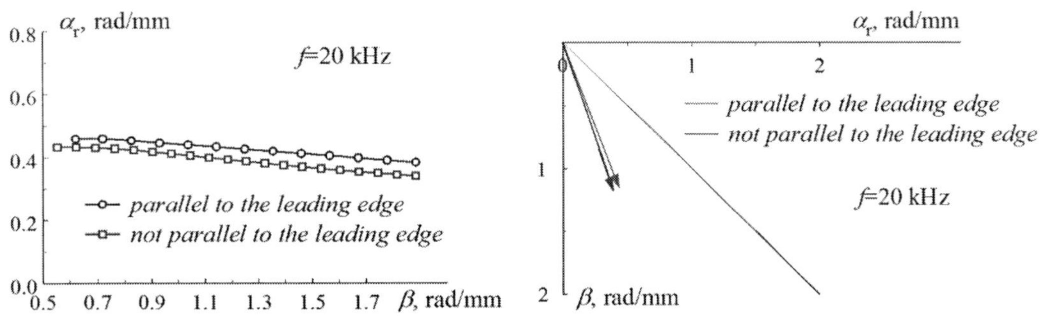

Figure 5. Dispersion relations $\alpha_r(\beta)$ (left) and directions of wave vectors (right).

4. Conclusion

The wave characteristics of the development of artificial traveling disturbances with a frequency of 20 kHz in a supersonic boundary layer on a swept wing with a Mach number of 2.0 are experimentally obtained. Two approaches were tested in the movement of the hot-wire anemometer probe: parallel and not parallel to the leading edge of the model. In both cases, it was shown that the distributions of spatial amplitudes and β-spectra are asymmetric. This fact is associated with the presence of the crossflow in 3D boundary layer. It was found that the amplitude maximum is located at $\beta' = 1.11$ rad/mm for the case of parallel moving of the hot-wire probe and at $\beta = 1.17$ rad/mm for the case of not parallel moving. Comparison of the results of two approaches to the movement of the probe shows the equivalence of data processing methods for determining the wave characteristics of controlled disturbances, which makes it possible to measure the fields of controlled disturbances in the boundary layer of the swept wing in the traditional way for flat plate models. Also, the obtained dispersion dependences $\alpha_r(\beta)$ will allow one to check the condition of wave phase synchronism to reveal the mechanism of nonlinear interaction.

Acknowledgments
This work was supported by the Grant of the President of the Russian Federation (MK-2491.2019.1). The study was conducted at the Joint Access Center "Mechanics" of ITAM SB RAS.

References

[1] Bippes H 1999 *Prog. Aerospace Sci.* **35** 363–412
[2] Saric W S, Reed H L and White E B 2003 *Annu. Rev. Fluid Mech.* **35** 413–40
[3] Cattafesta III L N, Iyer V, Masad J A, King R A and Dagenhart J R 1995 *AIAA J.* **33** 2032-37
[4] Li F and Choudhari M 2011 *Theor. Comput. Fluid Dyn.* **25** 65–84
[5] Yermolaev Y G, Kosinov A D and Semionov N V 2014 *Journal of Applied Mechanics and Technical Physics* **55** 764–72
[6] Kosinov A D, Maslov A A and Shevelkov S G 1990 *J. Fluid Mech.* **219** 621–33
[7] Yermolaev Y G, Kosinov A D, Semionov N V, Tagaev S N and Semisynov A I 2009 *FMA '09:*

Proceedings of the 7th IASME / WSEAS International Conference on Fluid Mechanics and Aerodynamics p 89
[8] Lysenko V I, Gaponov S A, Smorodsky B V, Yermolaev Y G, Kosinov A D and Semionov N V 2016 *J. Fluid Mech.* **798** 751–73
[9] Ermolaev Y G, Kolosov G L, Kosinov A D and Semenov N V 2014 *Fluid Dynamics* **49** 188–97
[10] Kosinov A D, Kolosov G L, Semionov N V and Yermolaev N V 2016 *Phys. Fluids* **28** 064101

Journal of Physics: Conference Series **1404** (2019) 012085 doi:10.1088/1742-6596/1404/1/012085

Effect of a uncertainties of flow parameters on the separation zone at supersonic speeds

A I Kutepova, P A Polivanov and A A Sidorenko

Khristianovich Institute of Theoretical and Applied Mechanics SB RAS, 4/1, Institutskaya str., Novosibirsk, 630090, Russia

a.kutepova@g.nsu.ru

Abstract. The CFD calculations were performed basing on RANS approach with the Menter transition model [1]. The shock wave / boundary layer interaction (SWBLI) developed on a flat plate was considered. The oblique incident shock wave was generated by a wedge which was placed over the plate. The sensitivity of the separation zone to various inflow and geometrics parameters was studied. Verification of the calculations was carried out on the basis of experiments performed in the T-325 wind tunnel [2] for an angle of attack of wedge $\beta = 4°$.

1. Introduction

Comparison of experimental data with calculation results is complex task. A qualitative comparison of experimental and numerical data on graphs is usual way, which can be founded in a large number of scientific papers. There is no standard technique for comparing experimental and numerical data, which leads to a wide range of methods [3]. In addition, it is necessary to take into account that the experiment in a wind tunnel is accompanied by uncertainties in the geometry of the model and test section, errors in determining the incoming flow parameters and etc. For example in paper [4] it was shown that numerical simulation of experimental model with supporting devices allowed to get better agreement between the experiment and simulation. In paper [5] the effect of the geometry shape uncertainties of the experimental model on the coefficients of lift force and moment is estimated. As a result the obtained confidence interval made it possible to establish acceptable agreement between the experimental and numerical data. The effect of errors and uncertainties of the angle of attack of the wing and the incoming Mach number on the aerodynamic coefficients was founded in paper [6] by performing RANS calculations. These calculations made it possible to quantitatively find the effect of various experimental errors and inaccuracies on the resulting confidence interval. In addition these calculations provide a better understanding of what needs to be improved in the experiment. For example in paper [7] numerical simulation are used to improve the gas flow meter based on ultrasonic sensor. The aim of this work is to determine the effect of deviations of the model geometry and incoming parameters on the problem of SWBLI at low supersonic Mach number.

2. Formulation of problem

To study the effect experimental uncertainties the stationary 2D RANS calculations were performed for various β angles of the wedge (figure 1) and several Mach numbers in the range from 1.37 to 1.5. The geometry of the computational domain corresponds to the experimental setting for the laminar case [1]. The regular hexahedral computational grid was used to numerical simulation. The total number of the mesh cells was approximately 113k .

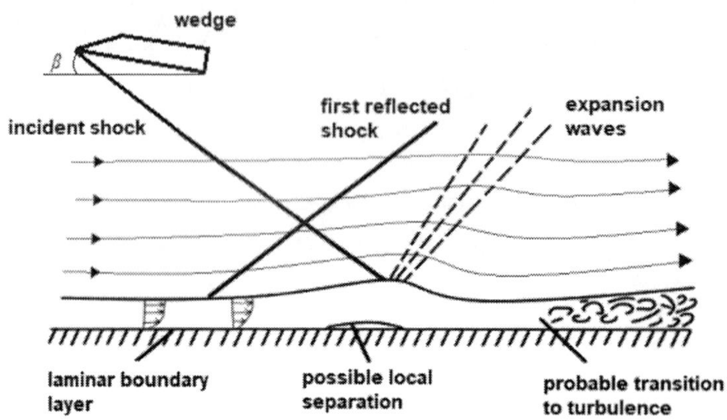

Figure 1. Geometry of the computational domain

The problem was simulated numerically by ANSYS Fluent software. The steady Navier-Stokes equations were used. The implicit second-order scheme in time and space was used together with the AUSM method of splitting the convective fluxes. The RANS calculations were performed using the Menter k-w-γ-Re_θ transition model and the k-w SST turbulent model.

3. Calculation result

Figure 2a and figure 2b shows the dependences of the displacement thickness of the boundary layer along longitudinal coordinate for various wedge angles β. The peak of the δ^* approximately corresponds to the points of intersection of the incident shock wave with the plate. For transition case (figure 2a) with a decrease the β, this point shifts downstream. Wherein the beginning of the interaction zone defined as the beginning of intensive growth δ^* also shifts downstream. As a result the length of the interaction zone weakly depends on the wedge angle. From the distribution of the shape factor shown in figure 3a, we can conclude that for angles β=2-4° laminar-turbulent transition (LTT) occurs immediately behind the incident shock wave. A comparison of the calculation and experimental data shows that for small angles, the transition in the experiment occurs later.

Figure 2b shows the results for a turbulent case. From figure it can be seen that for β = 1-2° the displacement thickness does not decrease along longitude coordinate. This means the absence of a rarefaction wave for these flow cases, which is generated due to a separated flow. This means that for the for β = 1-2° the separation zone is absent or very weak.

Figure 3b shows the longitudinal distribution of the shape factor of the boundary layer for the turbulent case. It can be seen that after passing through the zone of SWBLI the turbulent boundary layer quickly recovers to almost equilibrium state.

An example of a comparison of experimental and numerical data for two total pressure is shown in the figure 4. It can be seen that the qualitative behavior of the curves coincides, but there is a quantitative difference. The maximum δ^* is located upstream in the experiment which is possibly caused by a lower value of the Mach number in comparison with the calculated one. The decrease in the Mach number M is associated with the presence of irregularities on the wall of nozzle in the experiment. Another reason may be three-dimensional effects that are not taken into account in two-dimensional simulation.

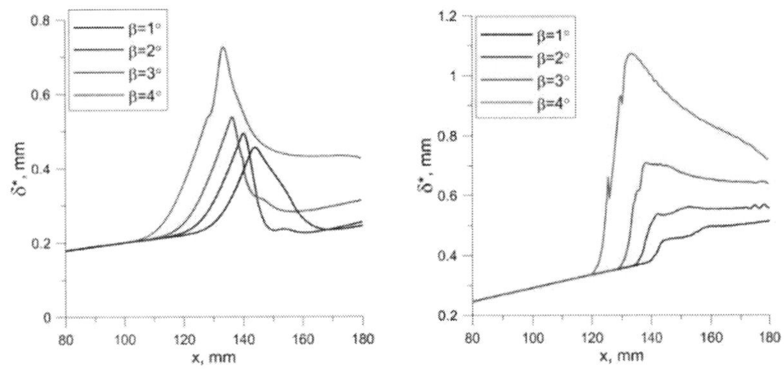

a) b)

Figure 2. Distribution of the displacement thickness along the longitudinal coordinate (a - Transition SST and b - k-w SST)

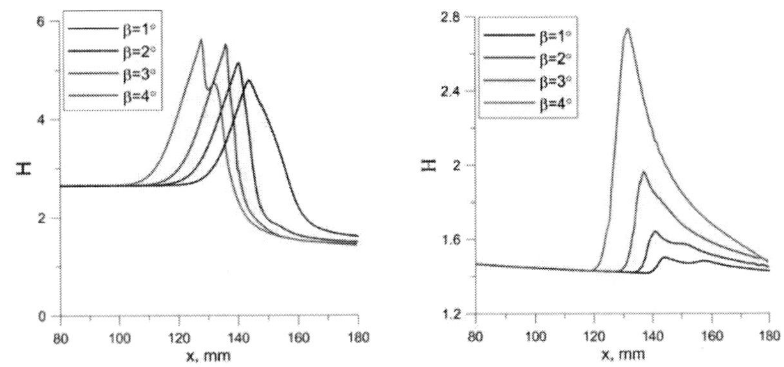

a) b)

Figure 3. Distribution of the shape factor along the longitudinal coordinate (a – Transition SST and b – k-w SST)

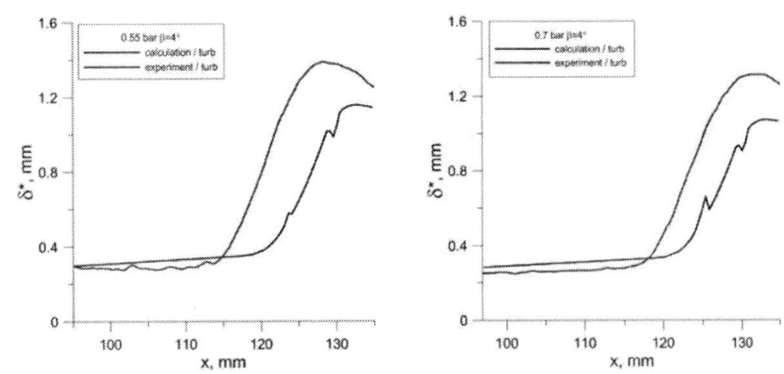

a) b)

Figure 4. Comparison of the experimental and numerical distribution of the displacement thickness along the longitudinal coordinate (a – P_0 = 55 kPa, b – P_0=70 kPa)

Calculations for different Mach numbers (figure 5) were carried out in the range from 1.39 to 1.5 for two RANS turbulence models to explain the role of the uncertainty of incoming Much number. It can be seen that a small change in the value of the Mach number significantly affects the parameters of the flow downstream. The influence of the Mach number on the size of the laminar (figure 5a) and turbulent (figure 5b) separation zones is weak especially for the transition case.

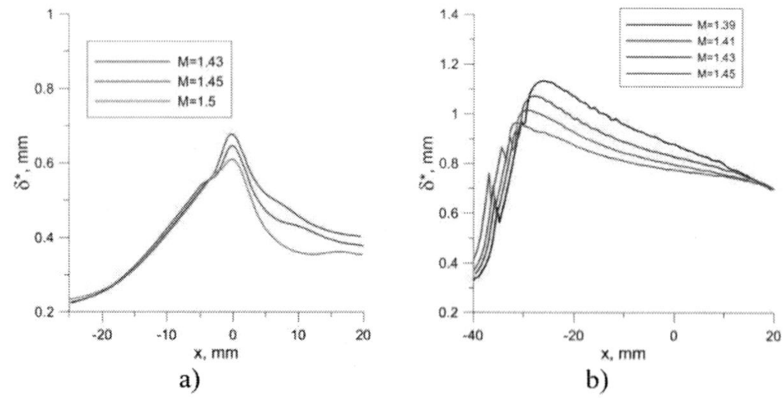

Figure 5. Distribution of the displacement thickness along the longitudinal coordinate for different Mach numbers, $\beta = 4°$, (a – Transition SST and b – k-w SST)

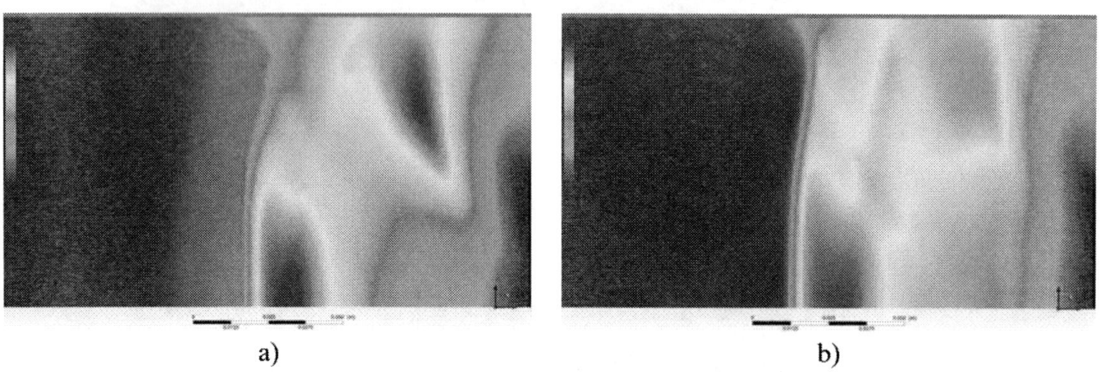

Figure 6. Pressure distribution on the plate model obtained by three-dimensional modeling ($\beta = 4°$) for a – transition and b – turbulent cases

3D RANS calculations were performed using the Transition SST and k-w SST models to study the effect of three-dimensionality of experimental geometry. The computation geometry corresponded to the actual experimental configuration. The computational domain included a supersonic nozzle, test section with a plate and wedge. The distribution of surface pressure on the plate is shown in figure 6. The lower edge of the field corresponds to the symmetry plane and the upper – to the side wall of the test section.

Figure 6 shows that near the plane of symmetry there is a SWBLI zone of quasi-two-dimensional flow, which occupies about 40% of the channel width. It can be seen that near the side walls there are no strong separated flows, which is a consequence of the gaps between the wedge and the side wall. A comparison of two-dimensional and three-dimensional calculations is shown in figure 7. The main difference was founded in the wake. The most likely reason of this is the effect of rarefaction waves propagating from the side walls.

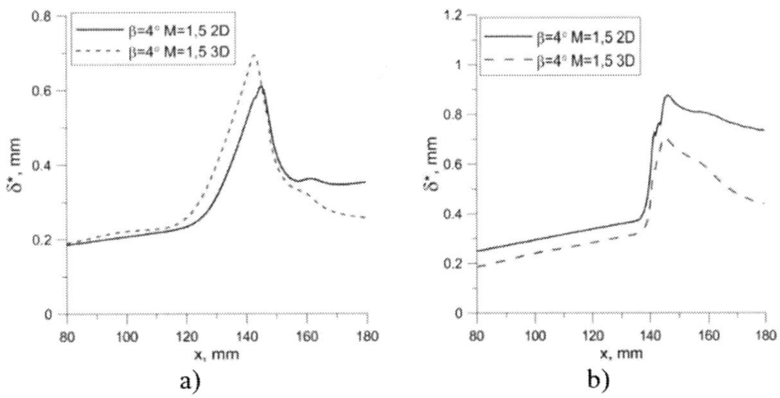

Figure 7. Comparison of distributions of displacement thicknesses for 2D and 3D simulation (a – Transition SST and b – k-w SST)

4. Conclusion

As a result of the calculations it was shown that even a small change in the parameters (for example the Mach number) has a significant effect on the flow characteristics. The results obtained in this work will be used to find quantitative estimates of the effect of experimental uncertainties on the confidence interval.

Acknowledgments
The work was supported by grant of RSF 18-19-00547.

References

[1] Menter F R, Langtry R B, Likki S R, Suzen Y B, Huang P G and Völker S 2004 *Part I: Model Formulation. J. Turbomach* **128** 413-22
[2] Polivanov P A, Sidorenko A A and Maslov A A 2015 The influence of the laminar-turbulent transition on the interaction between the shock wave and boundary layer at a low supersonic Mach number *Technical Physics Letters* **41** 933-7
[3] Oberkampf W L and Barone M F 2006 Measures of agreement between computation and experiment: Validation metrics *Journal of Computational Physics* **217** 5-36
[4] Melber-Wilkending S, Heidebrecht A and Wichmann G 2007 CFD validation by wind tunnel measurements: uncertainty assessment by numerical simulation of complete wind tunnel flows *NATO-RTO Applied Vehicle Technology (AVT 147) Symposium on Computational Uncertainty in Military Vehicle Design (Greece)*
[5] Boon D J, Dwight R P , Sterenborg J J H M, and Bijl H 2012 Reducing uncertainties in a wind-tunnel experiment using Bayesian updating *53rd AIAA/ASME/ASCE/AHS/ASC Structures, Structural Dynamics and Materials Conference (Honolulu, Hawaii)* AIAA 2012-1856
[6] Xu Lincheng, Wang Gang, Wu Jie and Ye Zhengyin 2014 Uncertainty Analysis of Airfoil Wind Tunnel Experiment Using CFD and Automatic Differentiation *Eighth International Conference on Computational Fluid Dynamics (China)* ICCFD8-0286 pp 1-12
[7] Belligoli Z, Dwight R P, Kok G J P and Lucas P 2017 A Bayesian study of uncertainty in ultrasonic flow meters under non-ideal flow conditions *Metrologia* **54** 584-98

Journal of Physics: Conference Series **1404** (2019) 012086 doi:10.1088/1742-6596/1404/1/012086

Investigation of jets mixing in the channel of the hypersonic rocket-ramjet engine model

A V Lokotko

Khristianovich Institute of Theoretical and Applied Mechanics SB RAS, 4/1, Institutskaya str., Novosibirsk, 630090, Russia

lok@itam.nsc.ru

Abstract. A technique of investigating jet mixing in the channel of a rocket-ramjet engine model in the conditions of an external supersonic airflow is proposed. Rocket exhaust jets are modeled with heated high-pressure air. Numerical computation of the mixing ratio (V. I. Kopchenov, Central Institute of Aviation Motors (CIAM)) yielded satisfactory qualitative agreement with the experimental results.

1. Introduction

There has been a number of recent studies on constructing a conceptual design of a transatmospheric vehicle propulsion system. One of the promising schemes considered in this research is the hypersonic ram-rocket engine [1]. It is a compound propulsion system that has a number of significant advantages due to combining elements of both a rocket engine and a jet engine. The main advantage is the possibility of obtaining a high specific impulse without increasing the mass while the design remains relatively simple. As the Mach number and the altitude increase, the engine operates first as a regular rocket-ramjet, then as a ramjet with subsonic channel speed (supersonic combustion ramjet, or scramjet), and finally as a ramjet with supersonic speed (hypersonic combustion ramjet, or scramjet). In the last two modes, the rocket nozzles operate as fuel injectors, without implementing the impulse. In vacuum, the propulsion system operates purely as a rocket engine. One of the main objectives in the development of a propulsion system is to achieve a high combustion efficiency at an air-fuel ratio close to stoichiometric, with minimal losses of total pressure in the airflow duct [2]. This requires a high air-fuel mixing efficiency. At the same time, it is known [3,4] that at high Mach numbers turbulent exchange processes slow down and traditional fuel supply schemes (for example, jet feed in the combustion chamber) are rendered inefficient. Two ways of increasing the mixing intensity can be outlined in a most general formulation: by generating long-lived three-dimensional vortex structures, or by increasing the initial turbulence level (through the choice of the nozzle shape, wedges, shock wave inducers, etc.).

The possibility of generating streamwise vorticity in rocket nozzles by distorting the axisymmetric nozzle shape is additionally considered [5, 6]. All of that leads to the concept of an integral air inlet with spatial flow compression.

This paper presents the results of the so called cold blowdowns of the combustion chamber [7, 8] with the air inlet in the external airflow, during which the most significant gasdynamic flow features in the channel are determined. The applicability of the results of such studies to combustion chambers relies on the fact that the structure of a microdiffusion flame follows the structure of the turbulent airflow in which the combustion occurs [9].

The process of jet mixing in rocket engines in a cocurrent air flow has been studied by heating the jets. It is known that the diffusion processes of impurities and heat (the so called transportable substances) are very close qualitatively and follow the same patterns [9]. Characteristics of turbulent transfer can be determined through the intensity of impurity diffusion.

The main purpose of in-depth study of the gasdynamic structure of the channel flow and of the mixing process is to verify numerical flow solution models.

2. Description of the model and experimental procedure

The experiments were conducted in the T-313 supersonic wind tunnel of the Institute of Theoretical and Applied Mechanics, Siberian Branch of the Russian Academy of Sciences with the test section size of 0.6 x 0.6 x 2 m. The flow parameters were fixed: $M_\infty = 4.03$, $Re = 54$ mln m^{-1}, $T_{0p} = 270 - 280$ K, stagnation pressure = 1.046 MPa, static pressure = 6.655 kPa. The tests were carried out in two modes: passive flow (no jet discharge) mode and jet injection mode.

The model (figure 1) is a rectangular channel equipped with a supersonic air inlet. The air inlet consists of side walls *1*, a lower compression wedge *2* with a 6° angle, an upper cowl *3*, and a wedge-shaped pylon *5* located on the longitudinal symmetry axis. The incoming air flow is compressed on the edges of the pylon, so the air inlet serves as a spatial compression device. Nozzles that mimic rocket engines nozzles with an exhaust Mach number of 2.94 are placed at the rear end of the pylon. The test medium of rocket engine jets – cold or heated (up to ≈ 200 °C) high-pressure air – was fed into the pylon. The air is heated in a high pressure gas generator that operates on the basis of burning hydrogen [10].

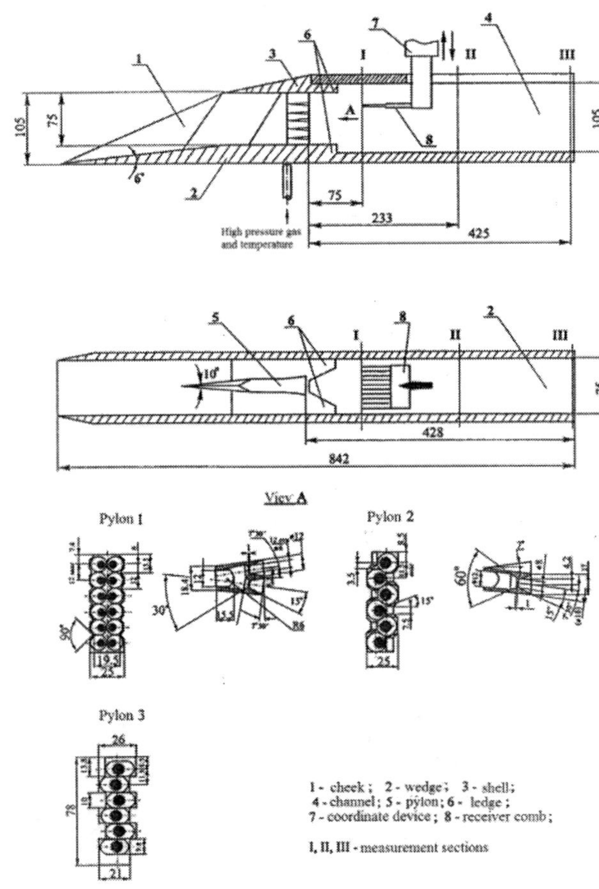

Figure 1. Model outline

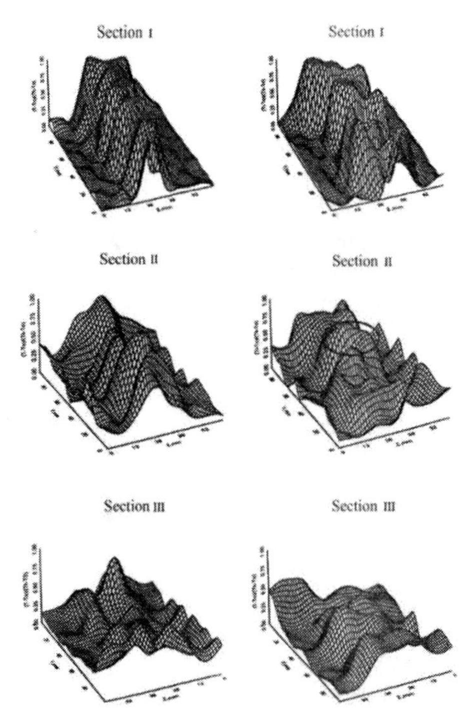

Figure 2. Relative excess temperature

At the channel entrance, after the air inlet, there are steps 6 that receive the impulse that occurs at the diffusion and heating of the incoming airflow by the rocket engine jets.

The model was installed in the test section of the wind tunnel on a high prop, which ensured its position in the core of the unperturbed flow.

The following model parameters were variable:

- the dispersal degree of the high-pressure flow, varied by injecting either twelve or six jets while the other parameters (the relative rate, the Mach number, and the exhaust pressure ratio) remained approximately the same – pylons 1 and 2, respectively;
- the structure of the flow in the model channel, varied by increasing its turbulence in the deep indentations during the installation of pylons 2 and 3;
- the structure of the injector flow, varied by using nozzles with an elliptical exhaust area (pylon 3). Pylon 2 had six round conical nozzles, and pylon 3 reproduced its geometric and rate characteristics as closely as possible, changing only one parameter – jet configuration.

The study of the gasdynamic parameter fields was performed using rake probes of total pressure P_0, static pressure P_{st}, and temperatures T_0 that scanned the flow field across the channel using a traversing probe. The traversing probe was adjusted in the longitudinal direction on installation. The rakes contained: ten P_0 and T_0 probes set in 7 mm increments; six P_{st} probes set in 12.6 mm increments. Chromel-alumel (ChA) thermocouples were used. As a primary analog-to-digital converter for temperature measurement, the digital voltmeter Shch-1516 was used. The accuracy of the rakes' movement along the height of the airflow duct is 0.1 mm. The flow parameters in each cross-section were calculated based on the measurement results for 160 points.

The total pressure in the gas generator at the exhaust of cold jets was about 3.0 MPa, for hot jets - 1.6 MPa, the pressure ratio of the exhaust $n = p_c/p_g$ (where p_c is the pressure at the nozzle section, p_g is the pressure at the air inlet throat section) varied from ~ 1.2 at the top of the channel up to ~ 6.4 at the bottom for cold jets, and, accordingly, from 0.3 to 1.6 for hot jets. The uneven nature of the pressure ratio was determined by the change in pressure along the height of the air inlet throat.

3. Measurement validation

The validity of the results depends on the following factors: maintaining the equity of the measured rate in three cross sections along channel length *G*; adherence to the impulse balance as measured in cross sections of the mixing chamber and calculated from the parameters of high-pressure and low-pressure flows within the mixing chamber *I*; stability of the averaged temperature T_0, (table 1), calculated as an example for pylon 2 during the hot jet tests. Evidently, the maximum deviation of rate δ from the average value is 0.91%. Rates are absolute values measured in different tests, so they indirectly characterize the reproducibility of maintaining the parameters both in the pre-chamber of the wind tunnel and in the gas generator.

Table 1. Values of rates *G*, impulses *I*, and temperatures T_0 (the average rate is 2.204 kg/s; in brackets - deviation δ (%))

Cross section	*G*, kg/s	*I*, N	T_0, K
I	2.201 (0.14)	1569.0	334.8
II	2.186 (0.82)	1531.7	338.4
III	2.224 (0.91)	1524.9	337.3

Calculated values: impulse, disregarding losses in the air inlet, $I = 1603.2$ N; average flow temperature in the mixing channel after gas generator nozzles:
$T_{0m} = r_{ai}\, T_{0p} + r_{gg}\, T_{gg} = 0.7896 \cdot 291.6 + 0.2115 \cdot 485 = 332.8$ K, here r_{ai} and r_{gg} are proportions of gas rate respectively through the air inlet and through the nozzle of the gas generator, T_{0p} and T_{gg} are

proportions of temperature of the incoming flow diffusion (in the pre-chamber of the wind tunnel) and of the combustion products in the gas generator. Evidently, the received measurements are very valid.

A calculation of turbulent Prandtl numbers can also serve as indication of measurement validity. It is known [3,12] that for turbulent flows the transfer of momentum, heat, and matter is proportional to the gradients of the averaged values of velocity, temperature, and impurity concentration, *i.e.* $\tau = \mu_t(\partial U/\partial y)$, $Q = c_p a_t(\partial T/\partial y)$, $G = \rho D_c(\partial c/\partial y)$. The coefficients in these equations are proportional to the gradient of the averaged velocity: $\mu_t = \rho l_u^2 |\partial U/\partial y|$, $a_t = l_t^2 |\partial U/\partial y|$, $D_c = \rho l_c^2 |\partial U/\partial y|$, where *lu, lt, lc* are the mixing path lengths in terms of speed, temperature, and concentration. Suppose that the derivative of the velocity with respect to the coordinate can be replaced by the corresponding differences or, according to Prandtl, present these coefficients depending on the difference in the characteristic velocities: $v_t = k\delta|U_m - U_\delta|$, where: U_m; U_δ; δ - respectively, the maximum velocity in the stream, the velocity of the satellite stream and the width of the mixing zone. Then the turbulent Prandtl and Schmidt numbers can be represented [3] in the form $Pr_t = l_u/l_t$; $Sc_t = l_u/l_c$.

Following [3, 11], the length of the mixing path can be expressed as the ratio of the maximum difference of a parameter (for example, the velocity in a given cross section ΔU_m) to the maximum gradient of the parameter $|\partial U/\partial y|_{max}$:

$$l_u = \Delta U_m (|\partial U/\partial y|_{max})^{-1} \tag{1}$$

In our case the maximum difference for velocity will be characterized by the maximum difference in velocity coefficient $\Delta\lambda_m = |\lambda_{max} - \lambda_{min}|$, for temperature – by the value: $\Delta T_{0m} = T_k|T_{0max}/T_k - T_{0min}/T_k|$. Next, we make the following assumption: the values of standard deviations σ are taken as analogues of the maximum gradients of corresponding parameters. Then the formula (1) can be represented as $l_u = \Delta U_{max}/\sigma U$. Following that assumption, we calculate the lengths of the mixing paths: $l_u \sim \Delta\lambda_{max}/\sigma\lambda$, $l_{T0} \sim \Delta T_{0max}/\sigma T_0$. The results of calculating the Pr_t numbers are presented in table 2. It is evident that the obtained Pr_t numbers are close, within some limits, to the known values for turbulent jet flows: $Pr_t = 0.5 - 0.8$. The patterns of Pr_t number changes with the development of the jet (a relative decrease in its value due to an increase of the mixing path by temperature) are also consistent with the data [3,13]. The values of the parameter differences were determined through the results of spline interpolation.

Table 2. Turbulent Prandtl numbers

Pylon type	Cross sections		
	I	II	III
1	1.15	0.52	0.32
2	0.69	0.60	0.48
3	1.14	0.49	0.56

4. Experimental results

Three-dimensional distributions of the relative excess temperature for pylons 1 and 2 are presented in figure 2, and contours of the temperature distribution in three cross-sections along the length of the channel are presented in figure 3. The figures demonstrate the process of jet ensemble expansion. Characteristic features are identified: jet axes, determined by Mach numbers maxima, (square markers) and temperature (round markers) are some cases not congruent. For pylon 3, with elliptical cross-section of the nozzle, a bifurcation of the jets and the appearance of two maxima is observed. That can contribute to improving the mixing process.

An analysis of flat distribution patterns of relative temperature (figure 3) allowed to construct jet trajectories in the channel (figure 4).

A significant confluence of the jets is observed for pylon No 1 in the second cross section. That feature obviously does not improve the mixing process.

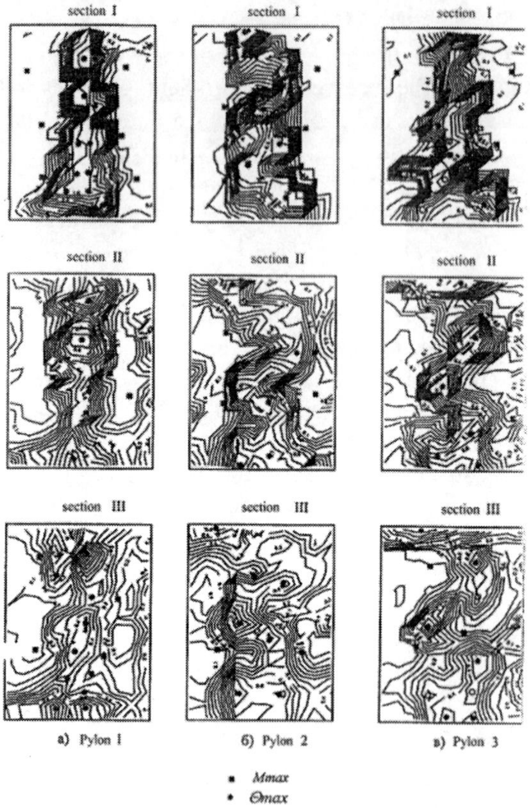

Figure 3. Contours of relative excess temperature

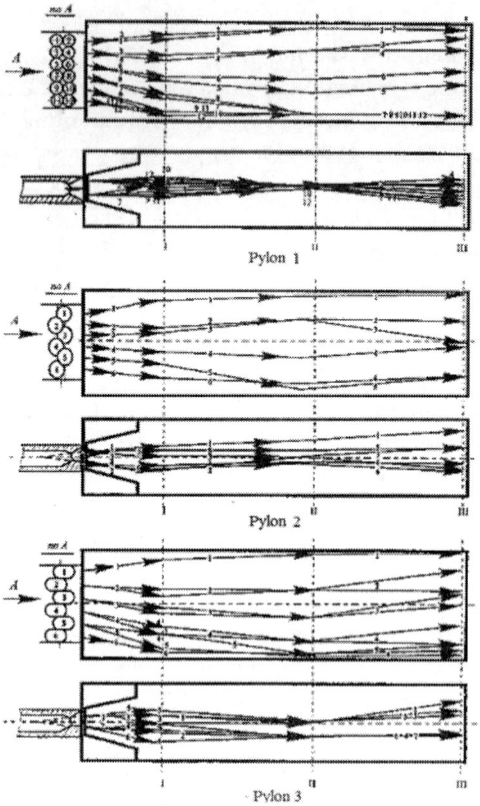

Figure 4. Rocket trajectories in the model channel

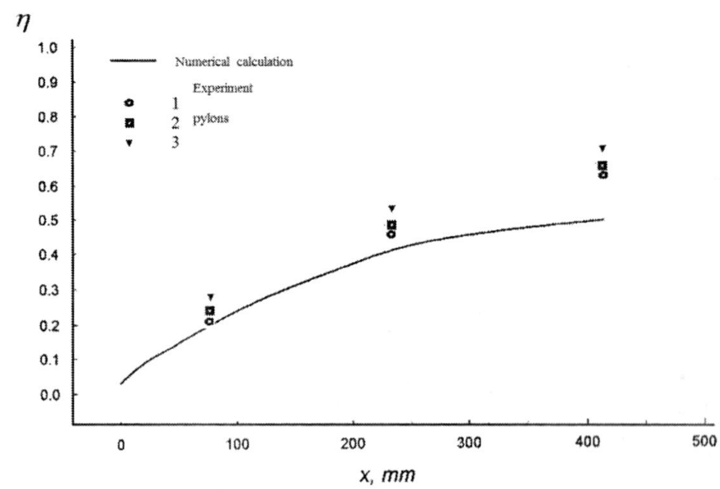

Figure 5. Results of calculation and experimental determination of the displacement coefficient in the channel of the RPD model.

The dependence $\eta = 1 - \sigma_t/T_{0m}$ where σ_{tt} is the standard deviation of the temperature in the cross section, and T_{0m} is the defined above temperature of the hot and cold flows mixture averaged over the relative areas, was taken as the mixing coefficient. As σ_T tends to zero (ideal mixing), η tends to 1, which determines the physical essence of this coefficient. Figure 5 presents the values of η for the three types of pylons, obtained in experiments and through numerical computation by a group of Central Institute of Aviation Motors (CIAM) employees under the leadership of V. I. Kopchenov.

5. Conclusions

A technique of studying jet mixing in the channel of a ram-rocket model has been developed. Numerical computation results (V. I. Kopchenov, CIAM) were shown to be in satisfactory qualitative agreement with the experimental results. The structure of the external and internal flows was demonstrated to have a significant influence on the mixing intensity.

Acknowledgments

The author expresses gratitude to L.V. Gogish, Eng.Sc.D., A.S. Rudakov, Eng.Sc.D for setting research objectives, and A.M. Kharitonov, Eng.Sc.D., for administrative support of the study. The study was conducted at the Joint Access Center "Mechanics" of ITAM SB RAS.

References

[1] Fry R 2004 *J. Propulsion and Power* **20** 27-58
[2] Lapin M S, Malyutina E V and Shilov Yu Ya 1979 *Tech. information* **9** 7-36
[3] Dubov N A, Lavrukhin G N and Lomakina M P 1983 *Central Institute of Aerohydrodynamics Review* **629**
[4] Yagodkin V I, Furletov V I, Sviridenkov A A et al 1993 *Central Institute of Aviation Motors Report* No 005-13/75 (Moscow) 36 p
[5] Drummond J P, Carpenter M H, Riggins D W and Adams M S 1989 *AIAA Paper* **2794**
[6] Kumar A, Bushnell D M and Hussaini M Y 1989 *J. Propulsion and Power* **5** 514-21
[7] Zuev V S and Makaron V S 1971 *Theory of ramjet and rocket ramjet engines* (Moscow: Mashinostroenie)
[8] Yagodkin V I 1987 *Separate flows in combustion chambers: collected papers* (Moscow)
[9] Tyulpanov R S 1970 *Some issues of turbulent combustion of two-phase air-fuel mixtures* (Novosibirsk)
[10] Lokotko A V 2016 *AIP Conf. Proc.* **1770** 040001
[11] Abramovich G N, Girshovich T A, Krashennikov S Yu and Sekundov A N 1984 *Theory of Tubulent Jets* (Moscow: Nauka)
[12] Hintze I O 1963 T*urbulence: mechanism and theory* (Moscow: Gos. izd. phys.-mat. literatury)
[13] Prudnikov A G 1956 *On turbulent combustion of a homogeneous mixture* (Moscow: Oborongiz)

Influence of individual roughnesses nozzle edge on the length of the supersonic section of an underexpanded microjet

S G Mironov and V M Aniskin

Khristianovich Institute of Theoretical and Applied Mechanics SB RAS, 4/1, Institutskaya str., Novosibirsk, 630090, Russia

mironov@itam.nsc.ru

Abstract. The paper presents the results of the experimental investigation of the influence of a small group of roughnesses uniformly located on the edge of the convergent nozzle on the length of the supersonic core of an underexpanded air microjet. The tests are carried out on the low-pressure test bench with the aid of the Pitot tube. The set of nozzles with the diameter of 2 mm, with the smooth output edge and the edge of the similar shape roughnesses (from 2 to 4) is used. The underexpanded microjets flowing out from the nozzles of the diameter 10.6; 16.1; 21.4, and 34.8 μm are simulated by the outflow Reynolds number. It is shown that the presence of roughnesses can reduce essentially the supersonic section length and eliminate the previously found effect of microjet "relaminarization", however the reduction value dubiously depends on the amount of the roughnesses on the nozzle edge.

1. Introduction

The influence of nozzle edge roughness on the measurement results is a serious challenge in the research of gas dynamics and stability of gas microjets. In particular, analysis of the roughness effect on the length of the supersonic core of axisymmetric underexpanded microjets is of interest. Since the length of the supersonic section depends on the intensity of the jet flow and ambient gas mixing, the supersonic section length should be influenced by the disturbances created by the roughness elements on the nozzle edge. Among them, the lengthwise stationary vortex structures are the best studied; they can gain in the underexpanded jets because of the positive curvature of the jet flow boundary [1, 2]. The vortex structures generated by the roughness can form initial running disturbances amplifying in the jet flow shear layer and resulting in the flow tubulisation and reduction of the length of the jet supersonic section. Note that most investigations in this field are made for completely turbulent macrojets.

The important peculiarity of the microjets is the moderate outflow Reynolds numbers which rarely exceed 5,000. Only the laminar and transition modes of the jet flow lie in this range. Moreover, we fail to produce the perfectly smooth surface of the output edge during the manufacture of the real convergent micronozzle. There are always relatively large humps and cavities of uncontrolled size and amount (see, for instance, [3, 4]). Because of small sizes of the micronozzles, it is difficult to connect the measured disturbances of the jet flow field with specific roughnesses on the micronozzle edge.

This paper presents the measurements of the length of the supersonic core of the axisymmetric underexpanded air jets flowing put from the smooth convergent nozzle with the roughnesses of the same size and shape uniformly imposed on the output edge – but the quantity of the roughnesses varies from two to four. Such an approach permits understanding the influence of a single roughness on the jet

IOP Publishing
1404 (2019) 012087 doi:10.1088/1742-6596/1404/1/012087

stability and mixing process intensity, plus finding the possible mutual influence of the disturbances in the jet flow from each individual roughness.

2. Experimental equipment and measurement technique

The experiments are carried out on the jet low-pressure test bench described in [5]. The test bench permits independently maintaining the jet outflow Reynolds number Re_d and jet underexpansion degree n. The experiments involve the convergent axisymmetrice nozzle with the diameter 2 mm. [5] previously showed the possibility to simulate the micronozzles by the millimeter-diameter nozzles by Re_d. In particular it was demonstrated that the dependencies of the supersonic core length on the underexpansion degree value n, obtained for the jets flowing out from the micron and millimeter nozzles, coincide in quantity as the outflow Reynolds numbers Re_d are equal. The framework of this approach, in the experiments the outflow of the underexpanded air jets from the nozzles diameter 10.6; 16.1; 21.4 and 34.8 µm was simulated in the underexpansion degree range from 1.1 to 6 and outflow Reynolds numbers from 800 to 5000.

Figure 1 presents the scheme of the nozzle itself and nozzle plenum chamber. There was a possibility to rotate the nozzle about the lengthwise axis. The nozzle conicity of 60° provides the minimal effect of acoustic wave reflection from the nozzle end face and discrete tone appearance. The output edge of one nozzles was absolutely smooth (nozzle No.1), whereas the output edge of the other nozzles were covered with lengthwise N-shaped scratch marks with the height and deepening h = 170 µm, width 300 µm and length 1 mm. There were two scratch marks on the edge (nozzle No. 2), three (nozzle No. 3), four (nozzle No. 4), and they were located uniformly over the nozzle output cross-section circle.

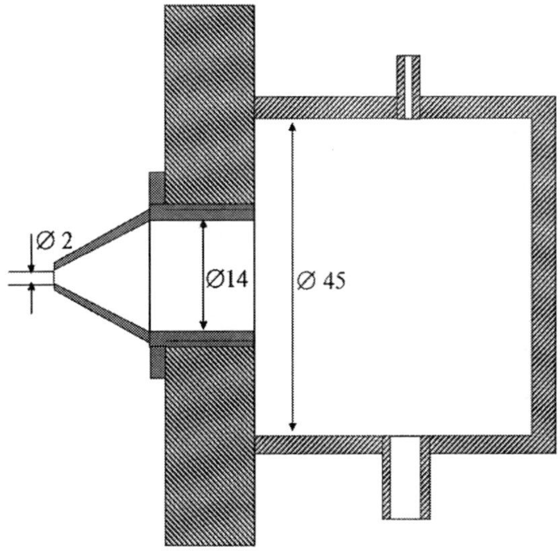

Figure 1. The scheme of nozzle and nozzle plenum chamber. There are characteristic sizes of the nozzle and chamber, air supply tubes and tubes for the manometer connection.

The measurements in the jet flow were carried out with the Pitot tube, its inner diameter 0.17 mm and outer diameter 0.4 mm. During the experiments, the Pitot pressure P_0' was measured. The tube was connected with a miniature differential pressure gage TDM4-IV1, which forms the Pitot probe. Characteristic pressure relaxation time in the probe was 0.3 s. The probe was installed on the 3-component pointing device, its position alignment accuracy was ± 0.1 mm. The main measurements were performed as the probe moved along the jet axis. Additive measurements were carried out for the Pitot pressure distributions in a number of cross sections of the jet flowing from the nozzle No. 2. The length of the supersonic section of model microjets was found as the distance from the nozzle cut to the point in the axis where the pressure P_0' corresponds to the value of this the pressure for the Mach number 1. The technique of determination of the supersonic section length with the Pitot tube is described in detail in [6].

3. Results and discussion

Figure 2 shows the results of measurement of the length of the supersonic section of *Lc* air jets rated for the nozzle diameter *d*. The graph (a) corresponds to the model jet flowing out from the nozzle with the diameter (a) — d = 10.6 µm; (b) – 16.1 µm; (c) – 21.4 µm; (d) – 34.8 µm. The solid, dashed and dash-dot curves on the graphs show the supersonic core length L_c/d versus the jet underexpansion degree n, obtained for turbulent macrojets, respectively in [7], [8] and [9].

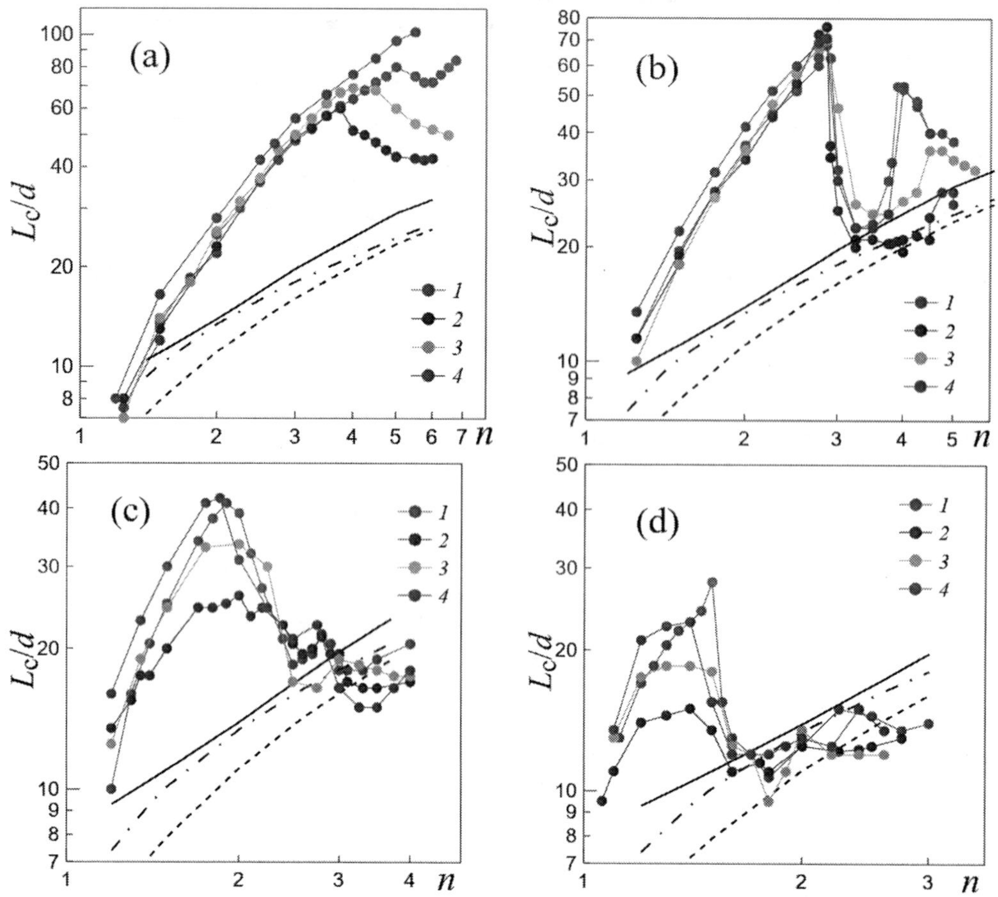

Figure 2. Dependencies of the rated supersonic core length L_c/d of the model air microjets on the jet underexpansion degree n for the nozzles with different roughness. *1* – nozzle No. 1; *2* — No. 2; *3* — No. 3; *4* — No. 4. (a) — nozzle diameter d = 10.6 µm; (b) – 16.1 µm; (c) – 21.4 µm; (d) – 34.8 µm.

The graphs show that, as a certain underexpansion degree is reached, the length of the supersonic core of the model jets falls fast to the level typical for the turbulent macrojets. The short-time jump of the supersonic core length after the fall (see figure 2*b*) results from the jet "relaminarization" effect [4, 10]. The presence of the roughness on the output nozzle edge reduces the value L_c/d and eliminates the "relaminarization" effect. The effect of the supersonic core length reduction appears better for the model nozzles of a bigger diameter. The degree of L_c/d reduction dubiously depends on the roughness amount. The strongest reduction effect is observed for the nozzle No. 2 with two roughnesses. For the nozzle No. 4 with four roughnesses this effect is weak or absent.

Evident that the roughness influence on the jet flow should depend on the relation of the roughness

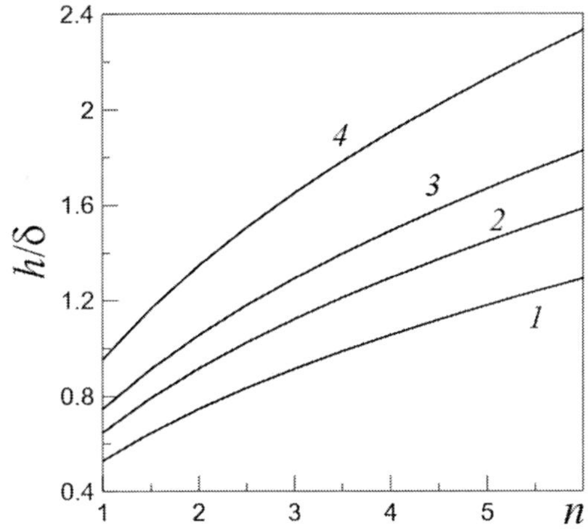

Figure 3. The roughness height/depth value h to the boundary layer thickness δ on the output nozzle edge versus the underexpansion degree value n for four diameters of the model micronozzles. *1* – d = 10.6 μm; *2* – 16.1 μm; *3* – 21.4 μm; *4* – 34.8 μm.

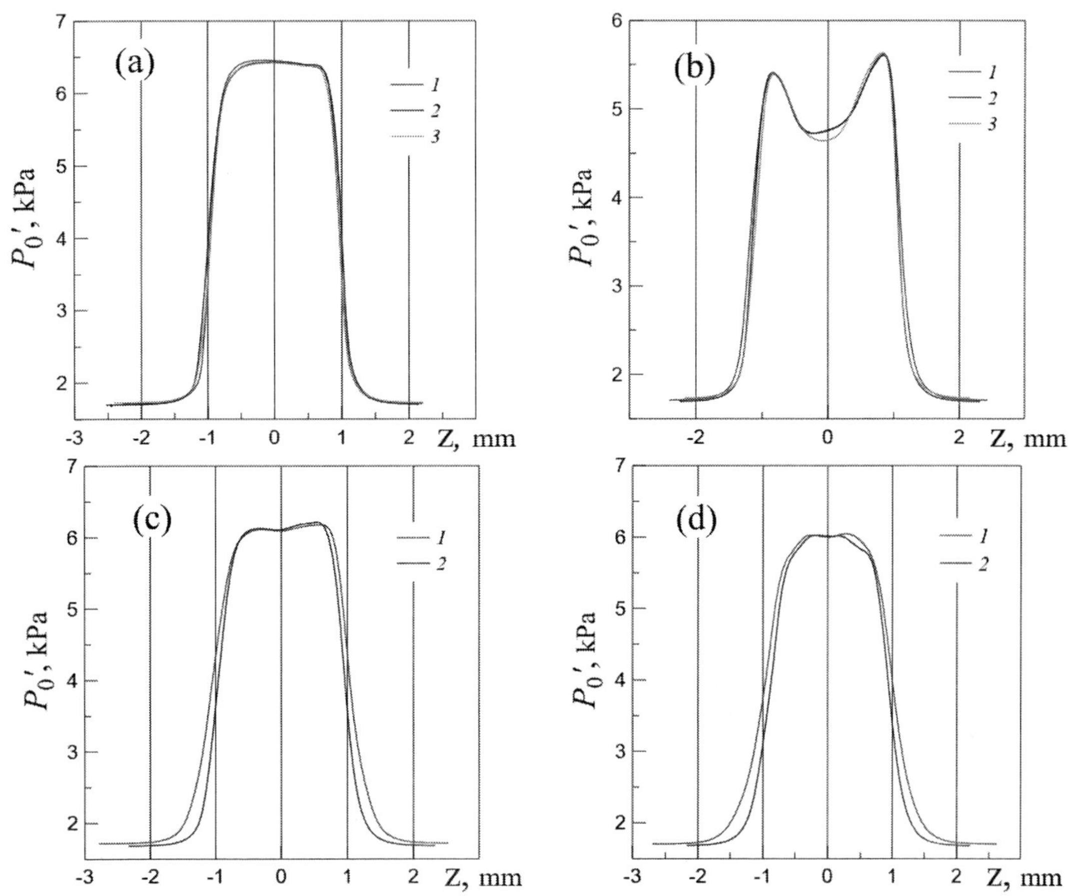

Figure 4. The pressure transversal distributions P_0' in several cross sections of the underexpanded microjet (n = 2, the model nozzle diameter d = 34.8 μm) flowing out from the nozzle No. 2. (a) – dimensionless distance from the nozzle x/d = 0.3; (b) – 0.8; (c) – 1.75; (d) – 3.2. The curve *1* is the measurement along the center line passing through the roughness; *2* – along the center line perpendicular to the line *1*. The curve *3* is obtained for the smooth nozzle No. 1.

height and depth to the boundary layer thickness on the nozzle edge. To gather the data about the boundary layer thickness on the nozzle edge, the numerical simulation of the flow in the nozzle settling chamber and in the nozzle conic contraction was made with the ANSYS Fluent package (see the scheme in figure 1). Figure 3 presents the calculated dependencies of the ratio of the roughness hump/cavity characteristic size h to the mixing layer thickness δ on the nozzle edge versus the underexpansion degree value n for four diameters of the model micronozzles. It is seen that the roughness size is comparable or exceeds the boundary layer thickness on the nozzle edge within the whole measurement area n, and the roughnesses should influence more intensively on the mean flow in the jet for the model nozzles of the bigger diameter.

To determine the roughness influence on the mean flow in the jet, we tried to measure the value of the pressure transversal distribution deformation $P_0{}'$, created by two roughnesses in the nozzle No. 2 during the simulation of the microjet flowing out from the nozzle with the diameter 34.8 μm. The Pitot tube was used to perform the measurements across the jet axis along the center line passing through the nozzle roughness, as well as along the center line in the perpendicular direction where there were no roughnesses. Additive measurements were carried out across the axis of the jet flowing out from the nozzle without roughness (nozzle No. 1). The measurements were performed in the same jet cross section by the lengthwise axis x/d. Figure 4 shows the examples of these measurements. It is evident that on the graphs there are no significant distortions of the mean flow field, except for the visible jet expansion along the direction binding two roughnesses on the nozzle No. 2 edge.

One reason of the non-monotony behavior of the supersonic core length vs the roughnesses amount on the output nozzle edge may be the interaction between the flow disturbances created by the neighboring roughnesses. The N-shaped roughness along with the flow can create one lengthwise vortex instead of two counter-rotating vortices. Then the closely located vortices can inhibit each other as the amount of the roughnesses rises on the nozzle edge. This is probably the explanation of the essential difference between the supersonic core lengths outflowing from the nozzles No. 1 and No. 4.

The other possible reason of the non-monotony influence of the roughness amount on the supersonic section length can be the inhibition of the convective instability of the shear layer of the jet lengthwise vortices created by the bigger amount of roughnesses. Inhibition of the convective instability prevents the appearance of the global jet flow instability, generation of the jet discrete tone, jet transition in the turbulent flow mode and reduction of the supersonic core length. The definite answer to the question about the possible reasons of the non-monotony influence of the roughness amount on the supersonic core length can be found from the numerical simulation results.

4. Conclusion

The paper presents the results of the experimental investigation of the effect of a small group of the roughnesses uniformly located on the convergent nozzle edge, on the length of the supersonic core of the model underexpanded air microjets.

The non-monotony influence of the amount of roughnesses on the length of the supersonic core of the model microjets. As the roughness amount rises from two to four, the supersonic core length rises and reaches the values obtained for the smooth nozzle. Moreover, the "relaminarization" effect recovers in the model microjet. Suggestions about two possible reasons of such an effect of the roughness amount are proposed:

- mutual inhibition of the closely located lengthwise vortices created by the neighboring roughnesses;
- inhibition of the convective instability of the mixing layer of the jet lengthwise vortices.

Acknowledgment

This study was funded by the Russian Science Foundation (Grant No. 17-19-01157).

References

[1] Zapryagayev V I and Solotchin A V 1991 *J. Appl. Mech. Tech. Phys.* **32** 503-507

[2] Zapryagayev V I, Mironov S G and Solotchin A V 1993 *J. Appl. Mech. Tech. Phys.* **34** 634-40
[3] Aniskin V M, Mironov S G and Maslov A A 2013 *Microfluid Nanofluid* **14** 605–14
[4] Aniskin V M, Mironov S G, Maslov A A et al. 1015 *Microfluid Nanofluid* **19** 621–34
[5] Aniskin V M, Maslov A A and Mironov S G 2018 *Fluid Dynamics* **53** 1-8
[6] Aniskin V M, Mironov S G , Korotayeva T A et al. 2019 *Micromachines* **10** 235-48
[7] Shirie J W and Seubold J G 1967 *AIAA J.* **5** 2062-63
[8] Pogorelov V I 1977 *Sov.Phys. Tech. Phys.* **22** 444-49
[9] Nescheret P A and Lentsov I A 1988 *Izvestia Vysshikh Uchebnykh Zawedeniy, Energetika* **10** 84-86
[10] Aniskin V M, Maslov A A and Mironov S G 2013 *Tech. Phys. Lett.* **39** 734–36

Flame propagation features in still wood dust suspension in the air

K M Moiseeva, A Yu Krainov and S K Chentsova

Tomsk State University, 634050, Russia, Tomsk, Lenina 36

E-mail: moiseeva_km@t-sk.ru

Abstract. This paper presents the numerical investigation on combustion of a still wood dust suspension on the air. The aim of the research is to determine the flame propagation features through the suspension under changing the size and mass concentration of the dust particles. The manuscript also provides data on parametric analysis of the problem. The conducted simulation has shown that the shape of the flame depends on the size, the particle mass concentration and the content of volatile components in the particles

1. Introduction

Fire and explosion risk at the industrial enterprises is one of the urgent problems for the modern physics of combustion and explosion. A potential threat at coal mining and wood processing enterprises is organic dust, which is a byproduct of the production. Depending on the environmental conditions and dust characteristics, potentially hazardous situations may arise at the plants.

By analogy with coal dust, the governing factors of the wood dust burning rate are the composition and concentration of the dust in the air. The composition is determined by the size and mass concentration of the dust particles, the content of volatile components in the particles, and the moisture content of the dust. The concentration of the dust in the air can be distributed heterogeneously, therefore it is necessary to solve the combustion problem of the wood dust in a two-dimensional formulation.

The aim of this study is to define the effect of the wood dust composition on the propagation velocity of the combustion front in a still wood dust-air suspension.

2. Mathematical model and solution method

The mathematical formulation of the problem is based on study [1]. The problem is solved in the 2D formulation. Wood dust particles are assumed to be immobile and irregularly distributed along the vertical coordinate. This situation may arise, for example, during the wood processing with the absence of ventilation in a working space. Wood dust can contain a fraction of volatile components. The particle size, mass concentration of the dust and the content of volatile components are set by the values r_k, $\rho_k(y)$ и V^C, respectively. The distribution of the particle mass concentration obeys the law, $\rho_k(y) = m_{dust}\left(1+\sin\left(\pi y / L_y\right)\right)$, $0 \le y \le L_y$, where m_{dust} corresponds to the minimum particle mass concentration along the y axis for the given interval. The particle distributions along the x direction are uniform. The model takes into account the drying of wood dust and the emission of volatile components with subsequent combustion of the particle and volatile components.

It is assumed that all the processes related to the wood dust drying, such as emission and combustion of the volatile components, combustion of the dust coke residue, do not occur until the wood dust reaches the temperature of 150 °C. The model takes into account the consumption of the oxidizer in two reactions: heterogeneous on the surface of the particles and homogeneous in the gas. The volatile components from particles evolve according to the first-order Arrhenius law. The released volatiles transform into the gas and are capable of chemical reaction with the oxidizer. The rate of a homogeneous reaction between gaseous volatiles and oxygen is defined by the second-order Arrhenius law. The first-order heterogeneous oxygen reaction proceeds on the particle surface. The rate of the heterogeneous reaction takes into account mass transfer [1]. The diffusion and thermal conductivity coefficients for the gas depend on temperature [1]. The temperature inside the particle is assumed to be uniform. The physico-mathematical formulation of the problem under the made assumptions can be written as the set of equations.

The heat transfer equation for the gas is

$$c_g\rho_g\frac{\partial T_g}{\partial t}=\frac{\partial}{\partial x}\left(\lambda(T_g)\frac{\partial T_g}{\partial x}\right)+\frac{\partial}{\partial y}\left(\lambda(T_g)\frac{\partial T_g}{\partial y}\right)+\alpha_k n_k S_k\left(T_k-T_g\right)+c_k T_k\left(G_2+G_3\right)+ \\ +Q_1\rho_1\rho_2 k_{01}\exp\left(-E_1/RT_g\right), \tag{1}$$

the energy equation for the particles is

$$c_k\rho_k\frac{\partial T_k}{\partial t}=-\alpha_k n_k S_k\left(T_k-T_g\right)-c_k T_k\left(G_2+G_3\right)+Q_2G_2-Q_3G_3, \tag{2}$$

the oxidizer mass-conservation equation is

$$\frac{\partial\rho_1}{\partial t}=\frac{\partial}{\partial x}\left(D(T_g)\rho_g\frac{\partial a_1}{\partial x}\right)+\frac{\partial}{\partial y}\left(D(T_g)\rho_g\frac{\partial a_1}{\partial y}\right)-\alpha_2G_2-\alpha_1\rho_1\rho_2k_{01}\exp\left(E_1/RT_g\right), \tag{3}$$

the volatile mass-conservation equation in the gas is

$$\frac{\partial\rho_2}{\partial t}=\frac{\partial}{\partial x}\left(D(T_g)\rho_g\frac{\partial a_2}{\partial x}\right)+\frac{\partial}{\partial y}\left(D(T_g)\rho_g\frac{\partial a_2}{\partial y}\right)+G_3-\rho_1\rho_2k_{01}\exp\left(E_1/RT_g\right), \tag{4}$$

the particle mass-conservation equation is

$$\frac{d\rho_k}{dt}=-G_2-G_3, \tag{5}$$

the volatile mass-conservation equation in the particles is

$$\frac{d\rho_l}{dt}=-G_3, \tag{6}$$

the mixture mass-conservation equation is

$$\frac{d}{dt}\left(\rho_g+n_k\rho_k\right)=0. \tag{7}$$

The initial conditions are

$$T_g(0,x,y)=T_0+T_{hot}\exp(-x/L_{hot}),\ T_k(0,x,y)=T_0,\ \rho_1(0,x,y)=\rho_{1b},\ \rho_2(0,x,y)=0, \\ \rho_g(0,x,y)=\rho_{gb},\ r_k(0,x,y)=r_{kb},\ \rho_k(0,x,y)=m_{dust}\left(1+\sin\left(\pi y/L_y\right)\right), \\ \rho_l(0,x,y)=\rho_k^0V_kV_C\left(1+\sin\left(\pi y/L_y\right)\right),\ n_k(0,x,y)=\rho_k/V_k\rho_k^0. \tag{8}$$

The boundary conditions are

$$T_g(t,0,y)=T_b+T_{hot},\ \frac{\rho_1(t,0,y)}{\partial x}=\frac{\rho_2(t,0,y)}{\partial x}=0,\ \frac{T_g(t,L_x,y)}{\partial x}=\frac{\rho_1(t,L_x,y)}{\partial x}=\frac{\rho_2(t,L_x,y)}{\partial x}=0,$$
$$\frac{T_g(t,x,0)}{\partial y}=\frac{\rho_1(t,x,0)}{\partial y}=\frac{\rho_2(t,x,0)}{\partial y}=0,\ \frac{T_g(t,x,L_y)}{\partial y}=\frac{\rho_1(t,x,L_y)}{\partial y}=\frac{\rho_2(t,x,L_y)}{\partial y}=0. \quad (9)$$

The following notation is used in equations (1) - (9): ρ_g is the gas density, ρ_k is the distributed particle mass per unit volume, ρ_l is the distributed mass of the volatiles in the particles, ρ_1 is the oxygen partial density, ρ_2 is the volatiles density in the gas, t is the time, x, y are the longitudinal and transverse coordinates, $r_k=\left(3(\rho_k-\rho_l)/4\pi n_k\rho_k^0(1-V_c)\right)^{1/3}$ is the particle radius, Q is the thermal effect of the reaction, k_0 is the constant of the chemical reaction rate, T is the temperature, E is the activation energy, R is the universal gas constant, c_p is the specific heat of gas at a constant pressure, c_k is the specific heat of coal dust particles, λ_g is the thermal conductivity coefficient of the gas, D_g is the gas diffusion coefficient, $\alpha_k=Nu_k\lambda_g/2r_k$ is the heat transfer coefficient between the gas and the particles, S_k, V_k are the surface area and volume of the particles, respectively. The rate of the particle mass change during combustion is defined by the equation $G_2=n_kS_kj_2\rho_{O2}$, where $j_2=\beta_m k_{02}\exp(-E_2/R_uT_k)/\left[\beta_m+k_{02}\exp(-E_2/R_uT_k)\right]$ is the rate of the heterogeneous reaction on the particles, $\beta_m=\lambda_g(T)Nu_D/(c_g\rho_g r_k)$ is the mass transfer coefficient for the particles [2]. The particles mass change rate during the volatile evolving is determined as $G_3=n_kV_kj_3$, where $j_3=\rho_{kl}k_{03}\exp(-E_3/RT_k)$ is the rate of the heterogeneous gasification reaction of the volatile components, V^C is the percentage of the volatile components in the particle. Indices: 0 is for the initial values of the state parameters, k – particle parameters, g – gas parameters, 1, 2, 3 – indices for the kinetic parameters of the homogeneous reaction in the gas, for the heterogeneous reaction on the particle surface, and for the homogeneous gasification reaction of the volatile components, respectively.

The problem is solved by the longitudinal-transverse sweep method [3]. The reliability of the calculations is verified by solving the model problems. According to the test calculation of the adiabatic temperature of the particle combustion, the calculation error does not exceed 3%. The grid step is equal to $h_x=3\cdot10^{-5}$ m, $h_y=3\cdot10^{-5}$ m. The Courant number is equal to 1. The calculations are performed for the area within the coordinates $L_y=0.02$ m, $L_x=0.06$ m.

3. Results

The kinetic parameters of the volatile evolving and combustion of the coke residue has been borrowed from the candidate thesis [4]. The kinetic constants of the volatile burning rate correspond to [1]. Volatile components released during the combustion of coal and wood dust are a mixture of light gases such as methane and hydrogen and combustible resins. The study [1], according to the monograph [5], uses the parameters corresponding to the burning of resins to generalize the combustion reaction of volatile components. The present model implies the same assumption.

Figures 1 – 3 show the results of the numerical simulation of the wood dust combustion in the air. The gas temperature flowfields at different time moments corresponding to the combustion front position along x-axis $x_f=0.2\ L_x$ (a), $0.4\ L_x$ (b), $0.6\ L_x$ (c), $0.8\ L_x$ (d) are shown in figures 1-2. The coordinate of the combustion front has been determined for the longitudinal coordinate $y=L_y/2$. The radius of the particles in the wood-air suspension is $3\cdot10^{-5}$ m, the mass concentration $m_{dust}=0.1$ kg/m^3 (figure 1) and $m_{dust}=0.15$ kg/m^3 (figure 2), and the mass fraction of volatile components in the particles is $V^C=0.1$.

According to figures 1 – 2, the combustion front for the selected air suspension compositions moves with almost the same speed and same nature of the temperature distribution along the area. The highest temperature is observed on the left border of the area and on the side surfaces. The combustion processes presented on figures 1-2 proceed with a lack of oxidizer.

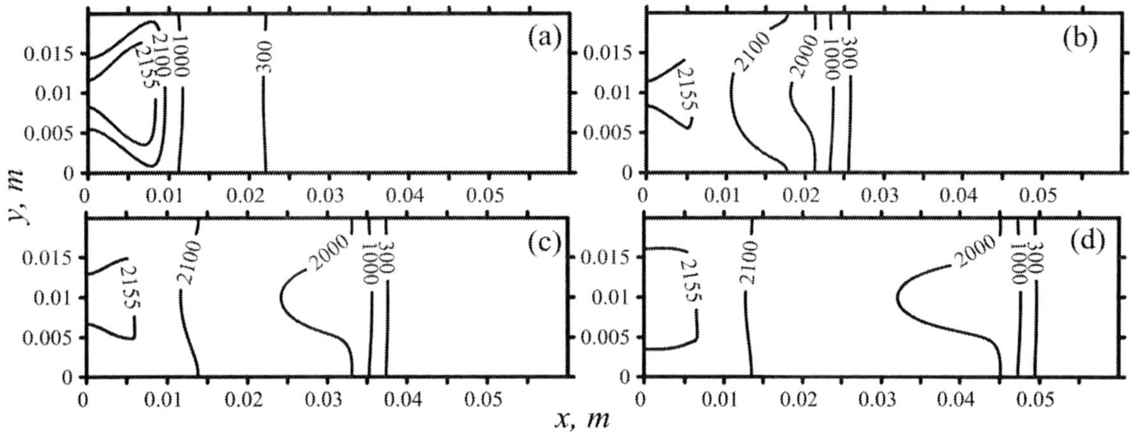

Figure 1. Gas temperature in the time moments $t = 0.142$ s (a), $t = 0.367$ s (b), $t = 0.601$ s (c), $t = 0.836$ s (d), $m_{dust} = 0.1$ kg/m^3, $V^C = 0.1$, $r_k = 3 \cdot 10^{-5}$ m.

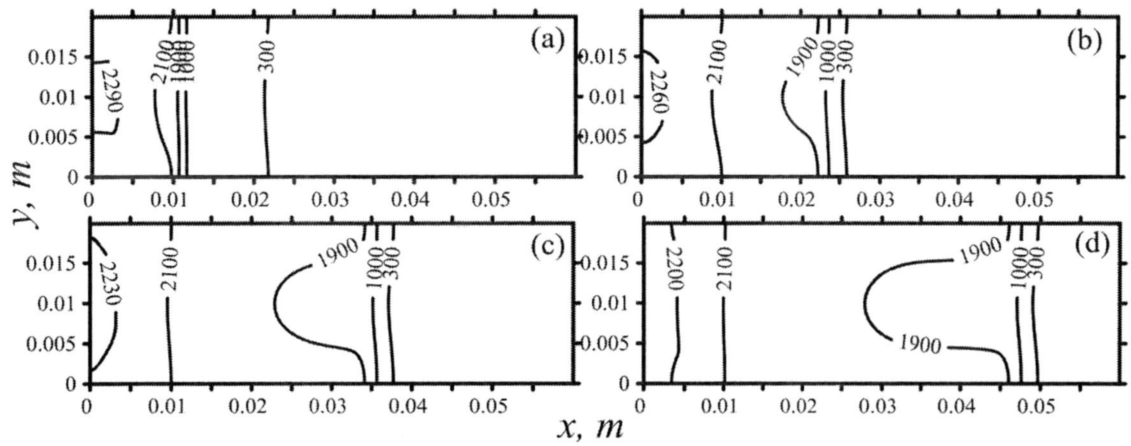

Figure 2. Gas temperature in the time moments $t = 0.142$ s (a), $t = 0.369$ s (b), $t = 0.607$ s (c), $t = 0.845$ s (d), $m_{dust} = 0.15$ kg/m^3, $V^C = 0.1$, $r_k = 3 \cdot 10^{-5}$ m.

The calculation presented in figure 3 corresponds to the case of oxidizer-rich combustion, and for the suspension with the particle mass concentration of $m_{dust} = 0.075$ kg/m^3. Figure 3 is plotted at the time moments when the coordinate of the combustion front is equal to $x_f = 0.1\ L_x$ (a), $0.2\ L_x$ (b), $0.3\ L_x$ (c), $0.4\ L_x$ (d), $0.5\ L_x$ (e), $0.6\ L_x$ (f). With the excess of oxidizing agent, the maximum gas temperature is noticed in the center, the temperature at the side is lower. The velocity of the combustion front is slightly lower than the velocity shown in figures 1 – 2.

The calculation results show that for the suspension with the particle radius of $3 \cdot 10^{-5} m$ and for the mass concentration of wood dust $m_{dust} = 0.075$ - 0.15 kg/m^3, the flame propagation velocity along the x axis is nearly the same. With a lack of oxidizer, the highest gas temperature is observed in the area of the lowest particle concentration. In the case of an oxidant-rich mixture, the maximum gas temperature is inside the region of the maximum particle concentration.

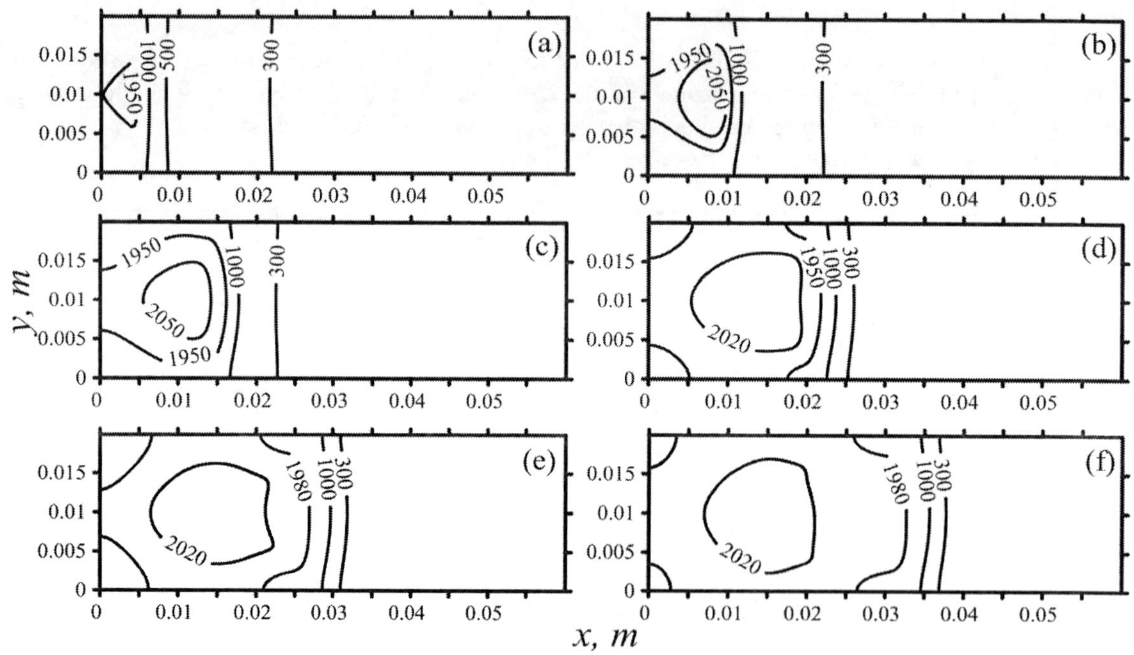

Figure 3. Gas temperature in the time moments $t = 0.05$ s (a), $t = 0.149$ *s* (b), $t = 0.262$ s (c), $t = 0.38$ s (d), $t = 0.5$ s (e), $t = 0.621$ s (f), $m_{dust} = 0.075$ kg/m^3, $V^C = 0.1$, $r_k = 3 \cdot 10^{-5}$ m.

The study [1] claims that for coal dust suspension with a high content of volatile components in the particles, the flame front moves asymmetrically. In this work, we have calculated the problem (1) – (9) for a mixture with the particle radius of 10^{-6} m, with the mass concentration of 0.1 kg/m^3 and the mass fraction of volatile components in the particles of 0.1 and 0.4. The calculation results are presented in figures 4 – 5.

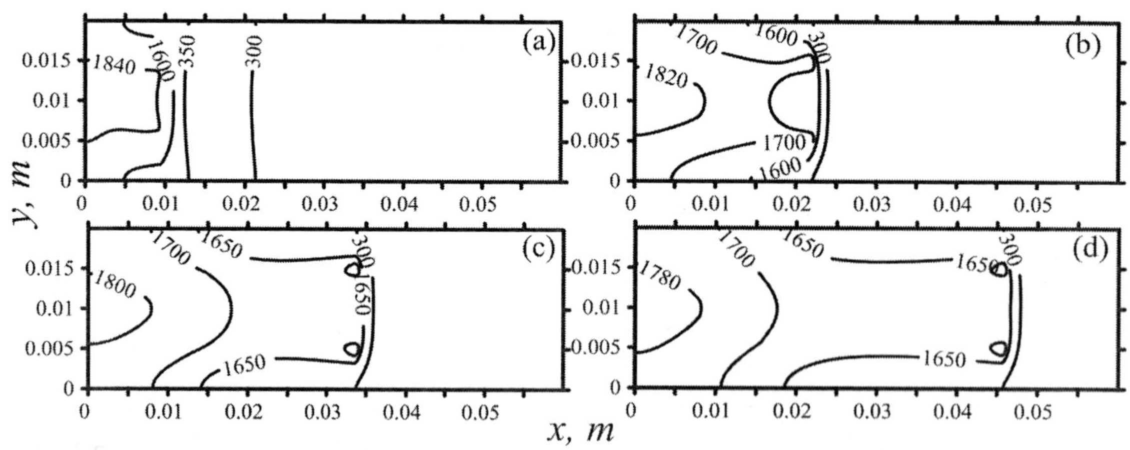

Figure 4. Gas temperature in the time moments $t = 0.04$ s (a), $t = 0.122$ s (b), $t = 0.21$ s (c), $t = 0.298$ s (d), $m_{dust} = 0.1$ kg/m^3, $V^C = 0.4$, $r_k = 10^{-6}$ m.

IOP Publishing
1404 (2019) 012088 doi:10.1088/1742-6596/1404/1/012088

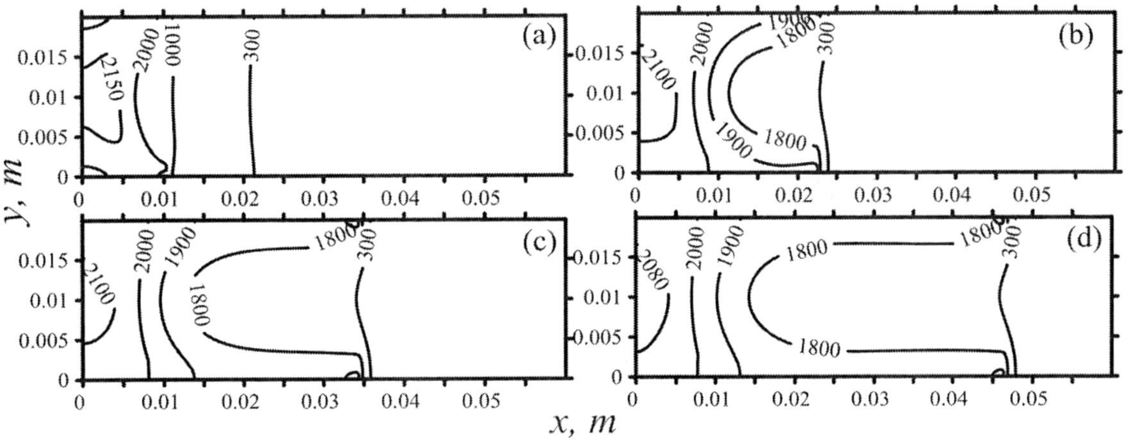

Figure 5. Gas temperature in the time moments t = 0.034 s (a), t = 0.101 s (b), t = 0.172 s (c), t = 0.244 s (d), $m_{dust} = 0.1\ kg/m^3$, $V^C = 0.1$, $r_k = 10^{-6} m$

According to figure 4, the combustion front has a prolate form with a maximum achieved on axis $y = L_y/2$, when there is a high content of volatile components in the particles. With a decrease in the content of volatile components, the shape of the combustion front changes and the maximum temperature is observed along $y = L_y$ and $y = 0$. These facts are explained by the distribution law of the particle mass concentration, since the value $y = L_y/2$ corresponds to the maximum of the law function, whilst the axes $y = L_y$ and $y = 0$ correspond to the minimum of the function. Therefore, the maximum temperature is observed in the area with the high concentration of the reactive gas components. With a low concentration of volatile components, higher temperature is achieved in the areas where the ratio between fuel and oxidizer is close to stoichiometric.

From a comparison of figures 1 and 5, we can conclude that the particle size affects the pattern of flame propagation in the suspension. With the same percentage of volatile components in the particles and with the same particle mass concentration, the suspension with small particles has higher burning rate.

4. Conclusion

We have solved the problem of the combustion front propagation through a still wood dust air suspension with an inhomogeneous dust distribution. It has been shown that the shape of the flame depends on the size, the particle mass concentration and the content of volatile components in the particles. We have distinguished the patterns of the flame along the y axis. The possibility of flame formation along the $y = L_y/2$ has been shown.

Acknowledgments
This work was supported by the Russian Foundation for Basic Research under Grant №19-48-703006.

References

[1] Moiseeva KM and Krainov A Y 2018 *Computer Research and Modeling* **10** 221-30
[2] Frank-Kamenetskiy D A 1987 *Diffusion and Heat Transfer in Chemical Kinetics* (Moscow: Nauka)
[3] Samarsky A A 2001 *The theory of difference schemes* (New York: Basel. Marcel Dekker)
[4] Golubev V A 2014 *Justification and improvement of methods of using a plant waste as source of energy PhD Thesis* (Barnaul) p 160
[5] Boyko E A 2011 *Reactivity of energetic coal* (Krasnoyarsk: Siberian Federal University)

Stability of the laminar boundary layer on the surface of the Mach-6 contoured nozzle

S O Morozov[1,2] and A N Shiplyuk[1,3]

[1]Khristianovich Institute of Theoretical and Applied Mechanics SB RAS, 4/1, Institutskaya str., Novosibirsk, 630090, Russia
[2]Novosibirsk State University, 1, Pirogova str., Novosibirsk, 630090, Russia
[3]Novosibirsk State Technical University, 20, Prospekt K. Marksa, Novosibirsk, 630073, Russia

morozov@itam.nsc.ru

Abstract. The stability of the laminar boundary layer on the surface of the Mach 6 contoured nozzle designed for short-duration wind tunnel Transit-M is investigated. The numerical simulation of the laminar flow inside the nozzle is performed using Ansys Fluent software. The obtained boundary layers profiles on the nozzle surface are used to calculate the growth rate of the first and second Mack mode and Görtler vortices in the framework of the linear stability theory. It has been shown that, for the Mach 6 nozzle, the Görtler vortices grow most rapidly, the first Mack mode grows slower, and the second mode is insignificant.

1. Introduction

Investigations of the stability and laminar-turbulent transition of the boundary layer in wind tunnels are important for understanding the fundamental nature of the transition, as well as for the validation of theoretical and numerical models. For this purpose, wind tunnels must create a uniform flow with a low noise level. As for the supersonic and hypersonic wind tunnels, the flow uniformity is achieved by special contouring of a nozzle surface [1]. The noise in the flow is created due to design future (throttles, joints of parts, measurement equipment and others) and turbulent boundary layer on the nozzle wall [2]. Special honeycombs and screens are used to decrease the fluctuations level just before the convergent section of a nozzle. At the low noise level of a flow and small roughness of a nozzle surface, the laminar to turbulent boundary layer on the nozzle wall takes place due to growing Görtler vortices or/and first and second Mack modes. The linear stability theory predicts well the growth rate of the small disturbances in the laminar boundary layers [3]. In this work, the linear stability theory is used to determinate the most unstable disturbances of the boundary layer on the surface of the Mach 6 nozzle designed for short-duration wind tunnel Transit-M. Using this nozzle, many important experimental data were obtained on the stability, laminar-turbulent transition, and stability control of hypersonic boundary layers (for example, in recent works [4–7]). It was shown that the level of pressure pulsations is below 3%, and the level of pulsations of the heat flux is below 8% in the range of free stream unit Reynolds numbers $Re_1 = (4–24) \times 10^6$ m^{-1} at the nozzle exit [8]. Therefore, the research of the causes and methods of reducing high-level flow fluctuations for a certain nozzle is an urgent task.

2. Calculations method

2.1. Mean flow

The calculations are performed for the Mach 6 nozzle designed for the Transit-M hot-shot wind tunnel based at ITAM SB RAS. The mean flow in the nozzle is obtained by the direct numerical simulation of axisymmetric Navier-Stokes equations with the Ansys Fluent software. The Mach 6 nozzle has an exit diameter 0.3 m and nozzle expanding section length of 1.46 m. Figure 1 shows the nozzle geometry and contours of Mach number. The boundary condition on the inlet of the convergent section of the nozzle is the *pressure inlet* with the stagnation temperature T_0 = 380 K and different cases of the stagnation pressure P_0 = 1.5 MPa and 3.75 MPa, which corresponds to the unit Reynolds number at the nozzle exit $Re_{1\infty}$ = 19.4×10^6 m^{-1} and 48.1×10^6 m^{-1}. The exit from the nozzle is the *pressure outlet* with the static pressure P = 0 and static temperature T = 300. The influence of the outlet condition on the boundary layer is checked by the nonphysical change of the wall shear stress, and the affected zone (x > 1.42) is not used for further research. The nozzle surface is the *wall* with no slip condition and temperature T_w = 295 K.

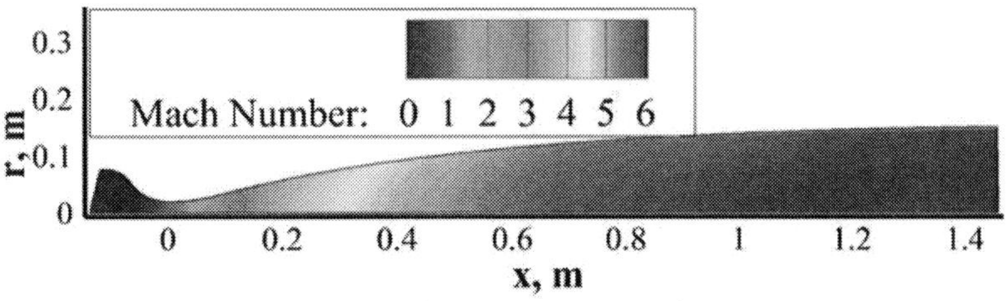

Figure 1. The nozzle geometry and Mach number contours.

The working gas is air with the perfect-gas law and the laminar viscosity. The viscosity and the thermal-conductivity are determined by the three coefficients Sutherland law and the kinetic theory, respectively.

The structured mesh with quadrilaterals elements is used. The size of the mesh elements decreases toward the surface in the normal direction and to the nozzle throat in the axis direction. The total grid size is 1700×200.

2.2. Stability

The linear stability theory is used to analyze the stability of the laminar boundary layer on the nozzle expanding section surface. The Görtler vortices and first and second Mack modes are significant for the supersonic and hypersonic boundary layers on the concave surfaces. Calculations of the Mack modes involve the linearized Navier-Stokes equations for propagation of disturbances from [9]. The growth rates of Mack modes are found for the fixed dimension transverse wave number β and frequency f. The system of equations for the Görtler vortices is taken from [10]. The Görtler vortices with only f = 0 Hz are taken into account because the steady Görtler vortices grow faster than the unsteady ones [11]. The Görtler vortices growth rates are found for the fixed number of waves around the nozzle periphery:

$$n = \frac{2\pi r}{\lambda}, \quad (1)$$

where r and λ are the local nozzle radius and spanwise wavelength, respectively.

The collocation method is used to solve the systems of equations for the Mack modes and Görtler vortices similarly to [9]. The amplitudes of the disturbances on the wall and on the upper bound are set to zero. The upper bound height y_{max} is set as twenty thicknesses of the boundary layer. The mesh points for stability equations defined by the equation:

$$k = \cos\left(\frac{\pi j}{L}\right), j = 0,1, \ldots L,$$

$$y = C_1 \frac{1+k}{C_2-k},$$

where L is total number of mesh points, y_δ is the boundary layer thickness, and $C_1 = y_\delta y_{max}/(y_{max} - 2y_\delta)$, $C_2 = 1 + 2C_1/y_{max}$. So, the half of the total numbers of collocation points is concentrated in the boundary layer.

3. Results

The stability analysis of the laminar boundary layer on the Mach 6 nozzle surface is performed for the nozzle exit unit Reynolds numbers $Re_{1\infty} = 19.4\times10^6$ m^{-1} and 48.1×10^6 m^{-1}. The example of the boundary layer profile is show in figure 2. The mesh nodes number in the boundary layer is from 10 to 80. Then these profiles are interpolated to the mesh points for stability equations with the cubic interpolation.

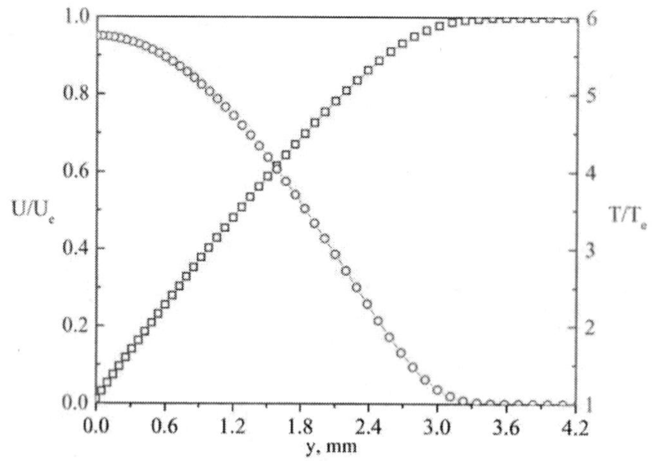

Figure 2. Boundary layer profile at x = 1 m for the $Re_{1\infty} = 19.4\times10^6$ m^{-1}.

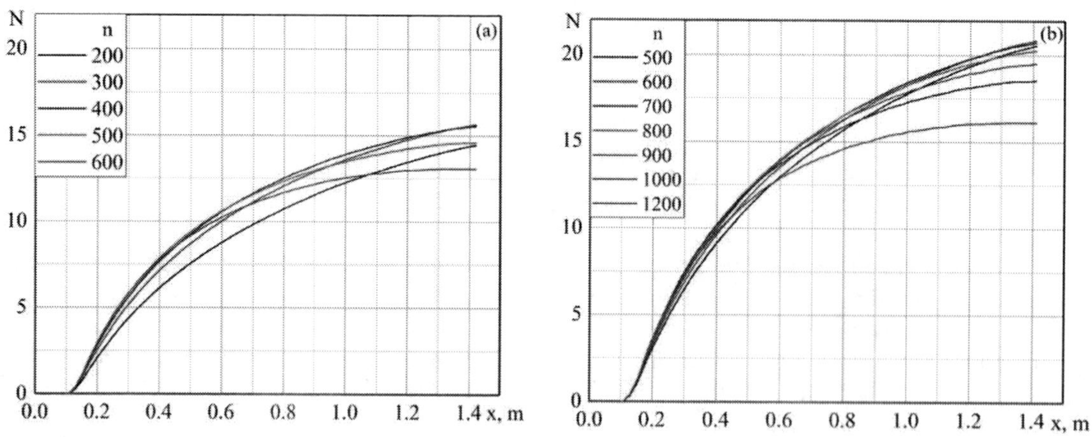

Figure 3. N-factor of the Görtler vortices along the nozzle surface for $Re_{1\infty} = 19.4\times10^6$ m^{-1} (a) and for $Re_{1\infty} = 48.1\times10^6$ m^{-1} (b).

To show the results of the stability calculations, the N-factor is obtained as

$$N = \int_{x_0}^{x} -\alpha_i(x)dx$$

 IOP Publishing
1404 (2019) 012089 doi:10.1088/1742-6596/1404/1/012089

where $-\alpha_i$ is the grow rate of disturbances, x is the coordinate along the nozzle surface, x_0 is neutral point of disturbances.

Görtler vortices N-factors are shown in figure 3 for $Re_{1\infty} = 19.4\times10^6$ m^{-1} (a) and for $Re_{1\infty} = 48.1\times10^6$ m^{-1} (b). Evident that the Görtler vortices grow fast, and the higher is unit Reynolds number, the faster they grow. The number of waves around the nozzle periphery n of the most unstable Görtler vortices increases with the increase of the unit Reynolds number.

First Mack mode disturbances are shown in figure 4 for $Re_{1\infty} = 19.4\times10^6$ m^{-1} (a) and for $Re_{1\infty} = 48.1\times10^6$ m^{-1} (b). The growth of the N-factor of disturbances starts from $x \approx 0.2$ m. The Mack first mode grows much slower than the Görtler vortices. The maximum N-factor for the first mode is $N_{max} = 5.3$ whereas for the Görtler vortices is $N_{max} = 15.6$ at the unit Reynolds number $Re_{1\infty} = 19.4\times10^6$ m^{-1}. The transverse wave number β and the frequency f of the most unstable Mack first mode increases as the unit Reynold number increases.

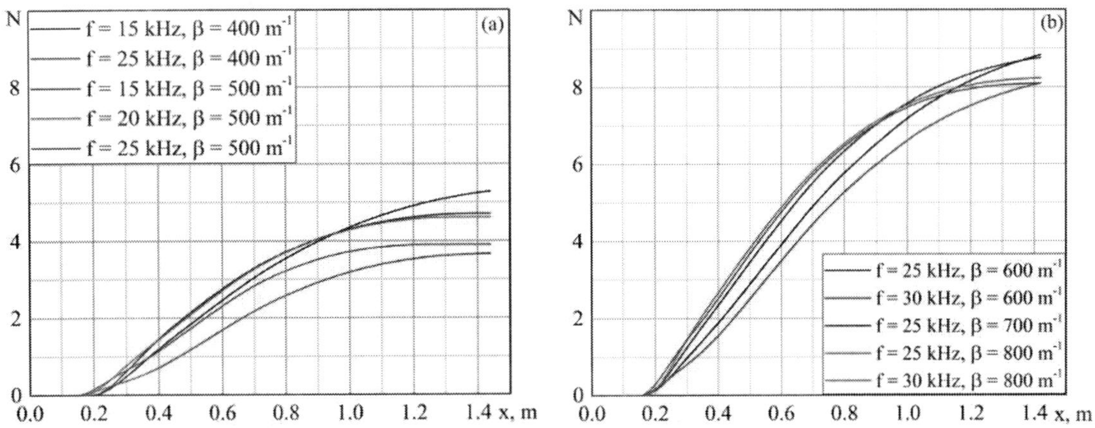

Figure 4. N-factor of first Mack mode disturbances along the nozzle surface for $Re_{1\infty} = 19.4\times10^6$ m^{-1} (a) and for $Re_{1\infty} = 48.1\times10^6$ m^{-1} (b).

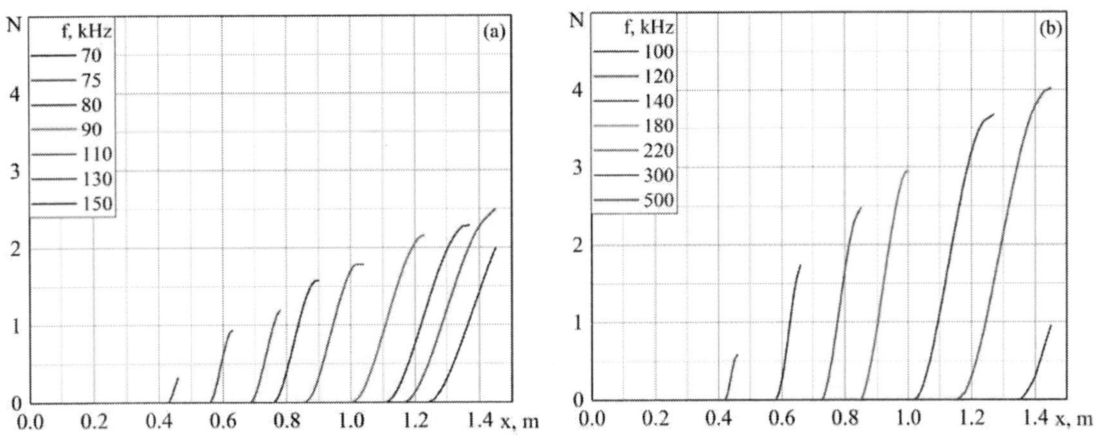

Figure 5. N-factor of second Mack mode disturbances along the nozzle surface for $Re_{1\infty} = 19.4\times10^6$ m^{-1} (a) and for $Re_{1\infty} = 48.1\times10^6$ m^{-1} (b).

The second Mack mode instability is typical for the hypersonic boundary layers with Mach number on the edge above 5. For this reason, the second mode (figure 5) starts to grow from the $x > 0.4$ were $M > 5$. Evident that the amplification of the second mode disturbances is negligible compared to the amplification the Görtler vortices and thirst Mack mode. The frequency of the most unstable second mode disturbances increases along with unit Reynolds number.

4. Conclusions

The numerical study of the stability of the boundary layer on the Mach-6 contoured nozzle surface is carried out in the framework of the linear stability theory. The calculation is carried out for the nozzle designed for the wind tunnel Transit-M of ITAM SB RAS. Numerical simulation of the steady laminar flow in the nozzle is performed using the Ansys fluent software. N-factor of the Görtler vortices and first and second Mack mode are obtained.

It is shown that for this nozzle, the most unstable are Görtler vortices are for unit Reynolds numbers at the nozzle exit $Re_{1\infty} = 19.4\times10^6$ m^{-1} and 48.1×10^6 m^{-1}. The first Mack mode grows slower than the Goertler vortices, and the first mode is negligible for this nozzle.

Acknowledgments
The reported study was funded by RFBR according to the research project No 19-08-00794.

References

[1] Aulchenko S M, Galkin V M, Zvegintsev V I and Shiplyuk A N 2010 *J. Appl. Mech. and Tech. Phys.* **51** 218-25
[2] Schneider S P 2008 J. *Spacecraft and Rockets* **45** 641-64
[3] Boiko A V, Kirilovskiy S V, Maslov A A and Poplavskaya T V 2015 *J. Appl. Mech. and Tech. Phys.* **56** 761-76
[4] Lukashevich S V, Morozov S O and Shiplyuk A N 2018 *Exp. in Fluids* **59** 130
[5] Morozov S O, Lukashevich S V, Soudakov V G and Shiplyuk A N 2018 *Thermophys. and Aeromech.* **25** 793–800
[6] Lukashevich S V, Morozov S O and Shiplyuk A N 2016 *J. Appl. Mech. and Tech. Phys.* **57** 873-8
[7] Bountin D A, Gromyko Yu V, Kirilovskiy S V, Maslov A A and Poplavskaya T V 2018 *Thermophys. and Aeromech.* **25** 483–96
[8] Gromyko Y V, Polivanov P A, Sidorenko A A, Bountin D A and Maslov A A 2013 *Thermophys. and Aeromech.* **20** 481-93
[9] Boiko A V, Demyanko K V and Nechepurenko Y M 2017 *Russ. J. Num. Anal. and Math. Mod.* **32** 1-12
[10] Spall R E and Malik M R 1989 *Phys. Fluids* A **1** 1822-35
[11] Boiko A V, Ivanov A V, Kachanov Y S and Mischenko D A 2010 *Europ. J. Mech. B/Fluid* **29** 61-83

Journal of Physics: Conference Series **1404** (2019) 012090 doi:10.1088/1742-6596/1404/1/012090

Gas-dynamic factors controlling the level of the sonic boom generated by two bodies in a supersonic flow

D Yu Moskvichev and A V Potapkin

[1]Khristianovich Institute of Theoretical and Applied Mechanics SB RAS, 4/1, Institutskaya str., Novosibirsk, 630090, Russia

di_mail@ngs.ru

Abstract. The results of calculation of the sonic boom generated in a supersonic air flow by two bodies (disk and slender body of revolution) are presented. The bodies are arranged one behind the other. The slender body is aerodynamically shaded by the disk. The free-stream Mach number is 2. The calculations are performed by a combined "phantom bodies" method. The sonic boom level can be reduced by more than 12% via the disk position and size changing, as compared to the sonic boom generated by the slender body. the gas-dynamic factors affecting the sonic boom level are described on the base of the calculation results.

1. Introduction

The sonic boom problem comprises a study of the dynamics of weak shock waves at large distances from the flight trajectory and ways to reduce the sonic boom level on the ground surface. The studies of the sonic boom problem have already been reviewed in [1, 2].

Currently, researchers develop different methods to reduce the sonic boom level from a supersonic airplane, which requires the knowledge of gas-dynamic factors reducing the sonic boom level from a body in a supersonic flow. One of the active methods to affect the sonic boom is to change the oncoming flow parameters using aerodynamic surfaces in front of the body [3, 4].

This paper describes the gas-dynamic factors changing the sonic boom level from a slender body of revolution located in an aerodynamic shadow of another body. The aerodynamic shadow is formed by a disk located in front of the slender body perpendicular to the oncoming supersonic flow.

2. Calculation method

The slender body with the disk located perpendicularly to the oncoming air flow is illustrated in figure 1. The Mach number of the oncoming flow is $M_0 = 2$. The surface of the slender body of revolution in dimensional values is specified by the equation $r = 0.2{\cdot}(x - x^2/L)$, $0 \le x \le L$, where L is the body length, x is the longitudinal coordinate counted from the tip of the body downstream, and r is the radial coordinate. The distance from the disk to the tip of the body is $x_d/L = 0.025$. The disk thickness is $h_x/L = 0.000125$.

The calculation of the near field of the flow is carried out using an ANSYS Fluent software package with the help of a viscous heat-conducting gas model. Computational parameters: $M_0 = 2$, $L = 40$ m, $p_0 = 5474.87$ Pa, and $T_0 = 216.65$ K. Here, p_0 is the static pressure in the oncoming flow, and T_0 is the static temperature in the oncoming flow. The calculations are carried out in the Cartesian and cylindrical coordinate systems. The three-dimensional flow of the near field is calculated using the Cartesian coordinate system $\{x,y,z\}$: x is the longitudinal coordinate counted from the tip of the body downstream, y is the vertical coordinate, $\{xy\}$ is the symmetry plane, and z is the lateral distance from the symmetry plane. The near field of the flow having axial symmetry is determined using the cylindrical coordinate system $\{r,x\}$: r is the radial coordinate, and x is the longitudinal coordinate counted from the tip of the body downstream.

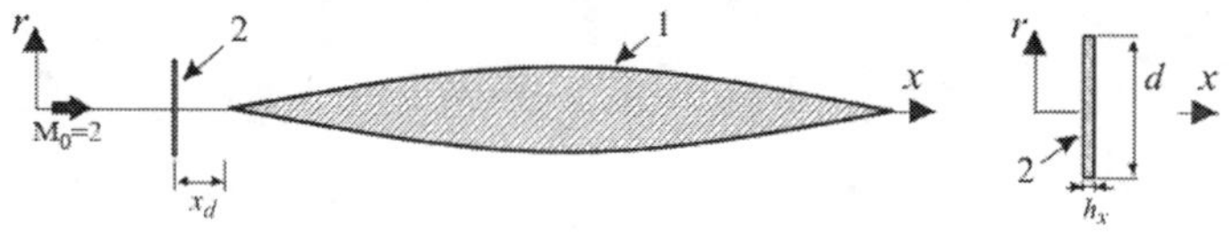

Figure 1. Body with the disk: *1* – slender body of revolution; *2* – disk.

The near-field parameters are used as the initial data for calculating the far field, which is carried out via the "phantom bodies" method [5]. The overall schematic of the method is shown in figure 2.

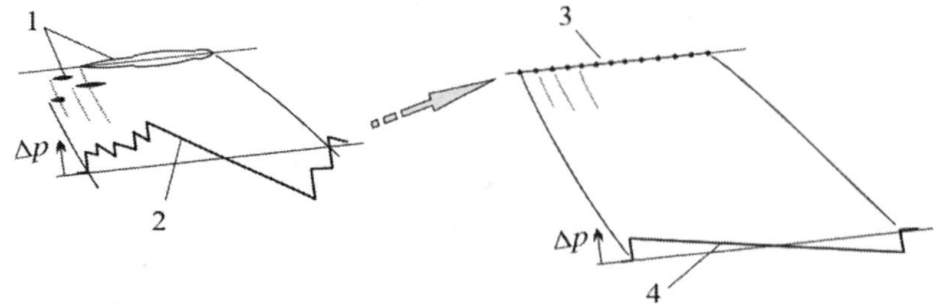

Figure 2. Overall schematic of the "phantom bodies" method: *1* – shock wave sources; *2* – excess pressure profile in the near field; *3* – phantom body; *4* – excess pressure profile in the far field.

The near-field parameters (pressure profile *2*) are used to construct the "phantom body". The phantom body is given by a discrete sequence of points *3* on a flight trajectory in which the values of the Whitham function and the longitudinal coordinates of the points are calculated. The excess pressure profile in the far field *4* is constructed using the analytical solutions of Landau [6], Whitham [7], and Rao [8] in the coordinate system with the axial symmetry. The initial data in the 3D flow calculations are chosen on the azimuthal plane $\{\varphi = \text{const}\}$ in the cylindrical coordinate system $\{r,\varphi,x\}$, as shown in figure 3.

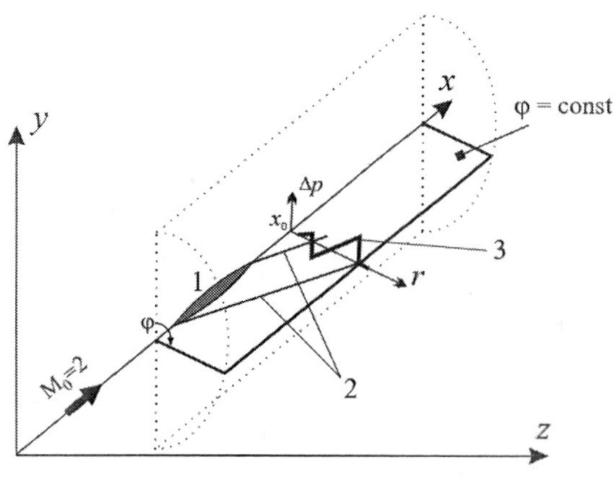

Figure 3. Initial data choice: *1* – slender body; *2* – shock waves; *3* – pressure profile in the near field.

In this case, the corresponding phantom body is constructed in each plane $\{\varphi = \text{const}\}$, and the far field is calculated independently in each azimuthal plane in the cylindrical coordinate system $\{r,x\}$ with the axial symmetry. The axis of symmetry is the x axis. The spatial pattern of the flow is obtained by calculations in several azimuthal planes. The potential of this method is indicated in [9] and validated in [5] in solving spatial problems.

The "phantom bodies" method was tested in [4, 5]. The test calculations of the dynamics of shock waves from the slender body of revolution (the Mach number of the oncoming flow was $M_0 = 2.4$) and blunt bodies ($M_0 = 1.8$) were performed. For the slender pointed body, the calculation results were compared with analytical solutions and the results of calculations carried out using other methods. The possibility of calculating the sonic boom from blunt bodies with a departing shock wave was studied using the example of the sphere problem in a supersonic gas flow. The test material was the results of experimental studies of sonic booms with spheres of different diameters in ballistic ranges and wind tunnels [10]. The test calculations showed that the "phantom bodies" method could be applied to calculate the shock wave parameters in the far field.

3. Results and discussion

Figure 4 presents the diagram of the flow near the body with the disk and the results of calculating the far field parameters. The denotations are as follows: Δp is the excess pressure and d is the disk diameter.

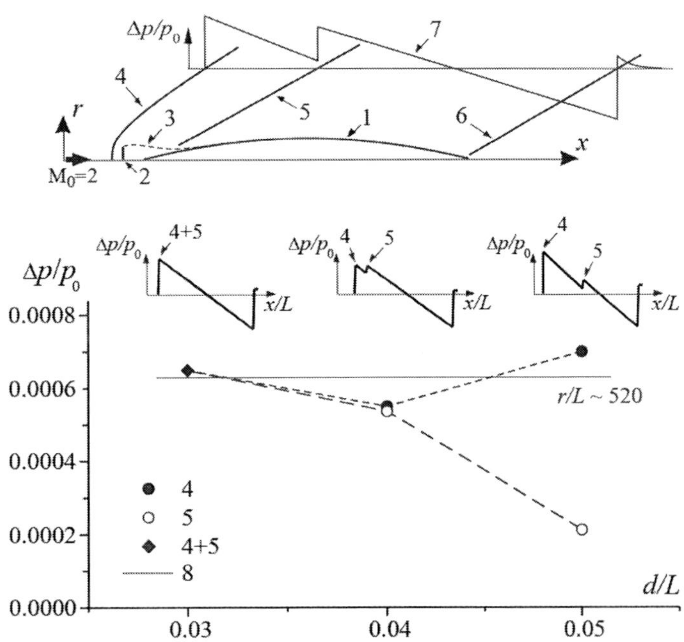

Figure 4. Body with the disk in the flow and the relative shock wave intensity: *1* – slender body of revolution; *2* – disk; *3* – sonic line; *4* – shock wave from the disk; *5* – shock wave from the body (barrel shock); *4+5* – shock wave formed after the shock waves from both the disk and the body become combined; *6* – tail shock wave; *7* – relative excess pressure profile; *8* – relative intensity of the bow shock wave from the slender body with no disk.

The calculation results related to the relative intensity of shock waves for the body with disks of different diameter and the corresponding relative excess pressure profiles behind the shock waves are shown for the distance $r/L \sim 520$. For comparison, the figure illustrates the relative intensity of the bow shock wave from the slender body without disk. The "+" sign denotes the shock wave formed after the shock waves are combined. Clearly, the disk size and the size of the aerodynamic shadow affect the sonic boom level in the far field. For the disk with a diameter $d = 0.03L$, the shock wave from the body catches up with the shock wave from the disk, combines with it, and the sonic boom in the far field becomes stronger than that from the original slender body. In the near field, the intensities of the shock waves from the disk and the body are smaller than those from the bow shock wave from the slender body with no disk, which allows one to preliminarily predict a decrease in the sonic boom level in the far field. Using the near-field data to predict the sonic boom level in the far field without precise calculations is often unreasonable and yields erroneous results. The disk with the diameter $d = 0.04L$ is in such a position relative to the body and creates such an aerodynamic shadow that the shock waves from the disk and the body do not combine within the distance of $r/L \sim 520$, thereby reducing the sonic boom. As the disk diameter increases up to $d = 0.05L$, the aerodynamic shadow becomes larger, but the intensity of the shock wave from the disk increases, too. The intensity of the shock wave from the body decreases, but the sonic boom level is determined by the shock wave from the disk and exceeds the sonic boom level from the original body.

Figure 5 shows the calculation results for the body with the disk $d = 0.04L$.

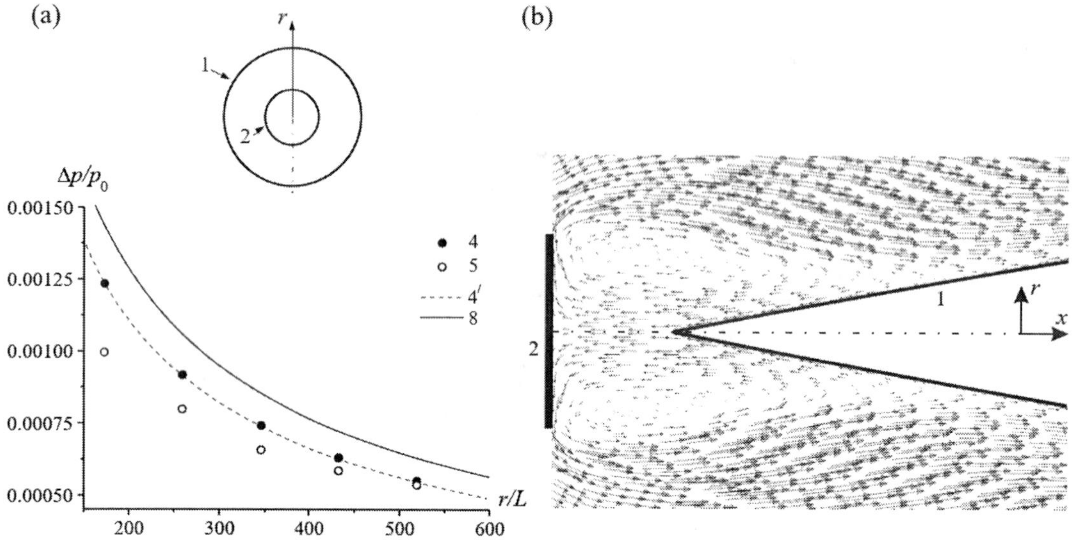

Figure 5. Body with the disk $d/L = 0.04$ (axial symmetry). Relative intensity of shock waves (a) and the flow velocity vector field (b): *1* – slender body of revolution; *2* – disk; *4* – relative intensity of the shock wave from the disk; *5* – relative intensity of the barrel shock; *4′* – relative intensity of the shock wave from the disk with no body; *8* – relative intensity of the bow shock wave from the slender body with no disk.

Figure 5(a) illustrates the frontal image of the body with the disk and the calculations of the relative intensity of shock waves at different distances r/L from the axis of symmetry x. The flow velocity vector field is shown in the plane $\{rx\}$ in figure 5(b). An aerodynamic shadow with a subsonic reverse flow is formed behind the disk. A barrel shock is formed in the outer flow above the region of the aerodynamic shadow (see figure 4). The barrel shock propagates through the gas which parameters are given by a rarefaction wave following the shock wave from the disk. The barrel shock relative intensity at all distances r/L is smaller than the bow shock wave relative intensity from the body with no disk. The relative intensity of the shock wave from the disk coincides with that of the shock wave from the disk with no body. As is shown by the calculations, a decrease in the sonic boom level in the far field can reach a value of 12 percent and larger. At the distances $r/L > 520$, the barrel shock catches up with the shock wave from the disk, and they combine.

Figure 6 shows the calculation results for the body with the disk $d = 0.04L$ that is displaced in a negative direction along the y axis at a distance of $0.005L$. Figure 6(a) shows the frontal image of the body with the displaced disk and the calculations of the relative shock wave intensity at different distances r/L from the x axis in azimuthal planes $\{\varphi = 0°\}$ and $\{\varphi = 180°\}$. Figure 6(b) shows the flow velocity vector field in the plane $\{xy\}$.

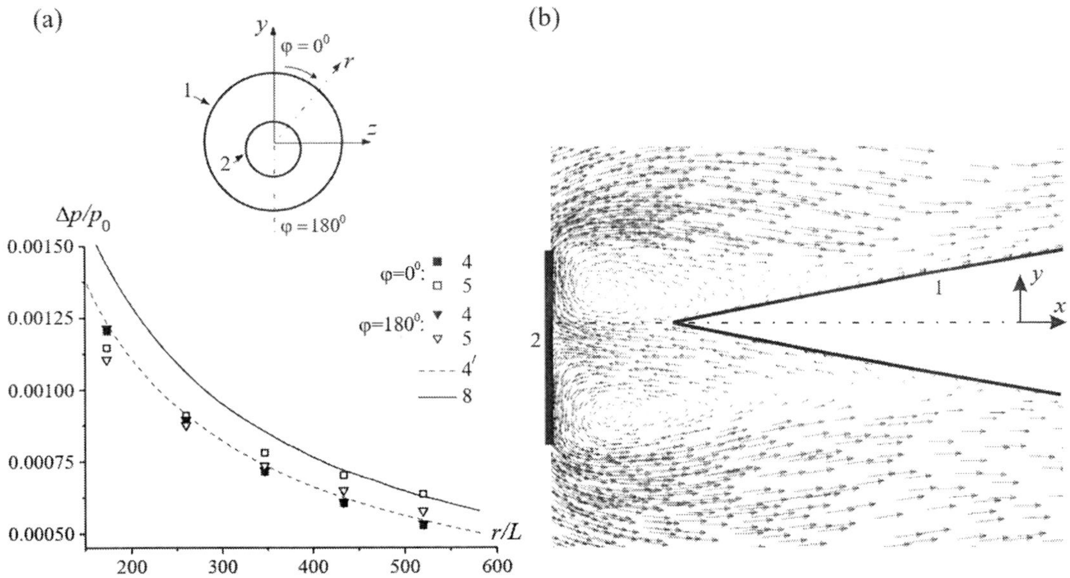

Figure 6. Body with the disk $d/L = 0.04$ (flow with no axial symmetry). The relative shock wave intensity (a) and the flow velocity vector field (b): *1* – slender body of revolution; *2* – disk; *4* - relative intensity of the shock wave from the disk for different azimuthal planes; *5* – relative intensity of the barrel shock for different azimuthal planes; *4′* – relative intensity of the shock wave from the disk with no body; *8* – relative intensity of the bow shock wave from the slender body with no disk.

Thus, the disk displacement reduces the aerodynamic shadow size in the top section of the space on the body surface. In the bottom section of the space, the size becomes larger as compared with the axisymmetric case. One could expect that the sonic boom level varies in accordance with the changes in the aerodynamic shadow size, i.e., expect the sonic boom level decrease in the bottom section of the space and an increase in the sonic boom level in the top section of the space. However, the calculations of the flow field near the body show that there are azimuthal flows, and as a result the pressure on the top and bottom surfaces of the body becomes higher than the pressure obtained for the axisymmetric case. The changes in the layout increase the sonic boom level because the intensity of the barrel shock that follows the shock wave from the disk becomes larger than that in the axisymmetric case.

4. Conclusion

Thus, it can be concluded on the basis of the calculations that there are dimensions and positions of barriers in front of the body that reduce the sonic boom at large distances from the body by forming an aerodynamic shadow.

The results show that the displacement of blunt-nosed bodies in the layout does not ensure the sonic boom level reduction in the chosen azimuthal plane. The sonic boom can be amplified due to the development of azimuthal flows.

There are two gas-dynamic factors determining the sonic boom level from two bodies in a supersonic flow. The first one is the aerodynamic shadow behind the first body with the second body placed within it. The second factor is the azimuthal flows forming near the body surface in the case of an asymmetric flow.

Acknowledgments
The research was carried out within the framework of the Program of Fundamental Scientific Research of the state academies of sciences in 2013-2020 (project No. AAAA–A17-117030610126-4).

References
[1] Benson L R 2013 *Quieting the Boom: the Shaped Sonic Boom Demonstrator and the Quest for Quiet Supersonic Flight* (NASA)
[2] Sun Y and Smith H 2017 *Prog. Aerosp. Sci.* **90** 12-38
[3] Potapkin A V and Moskvichev D Yu 2017 *AIP Conference Proceedings* **1893** 030156
[4] Potapkin A V and Moskvichev D Yu 2018 *Shock Waves* **28** 1239-49
[5] Potapkin A V and Moskvichev D Yu 2011 *J. Appl. Mech. Tech. Phys.* **52** 169-77
[6] Landau L D 1945 *Prikl. Matem. Mekh.* **9** 286-92
[7] Whitham G B 1952 *Commun. Pure Appl. Math.* **5** 301-48
[8] Rao P S 1956 *Aeronaut. Q.* **7** 21–44
[9] Whitham G B 1974 *Linear and Nonlinear Waves* (New York: John Wiley)
[10] Krasilschikov A P and Guriashkin L P 2007 *Experimental Investigations of the Hypersonic Flow Around Bodies of Revolution* (Moscow: Fizmatlit)

Determination of the heat transfer coefficient during the leakage of a subsonic axisymmetric air mini jet onto a plate

K A Mukhin[1,2], V M Aniskin[1] and N A Maslov[1]

[1]Khristianovich Institute of Theoretical and Applied Mechanics SB RAS, 4/1, Institutskaya str., Novosibirsk, 630090, Russia
[2]Novosibirsk State University, 1, Pirogova str., Novosibirsk, 630090, Russia

muhinkos@mail.ru

Abstract. In the present article, the heat transfer coefficient was determined experimentally when a subsonic jet flows onto a plate. The jet escaped from axisymmetric nozzles with a diameter of d = 290, 540, and 980 µm. The experiments were carried out at the speeds of V = 5, 20, 40, 60, 80 and 100 m/s in the Reynolds numbers range of 100 – 7000. The Nusselt number distribution was obtained for different distances between the nozzle and cooled surface.

1. Introduction

There are a lot of papers the heat transfer coefficient determining when a subsonic jet flows onto a heated surface [1, 2]. However, the number of papers where mini- and microjets were used, are extremely small. In [3], the authors used a jet escaping from the stainless tube with an external diameter of D = 1.6 mm and an internal d = 125 µm. The channel length was 25 mm; the Reynolds numbers range varied from 690 to 1770. The authors used a stainless plate 50 µm thick coated with a black paint from below. A thermal imaging microscope was used to record the surface temperature.

The Nusselt number is the main characteristic of the heat transfer process when the jet flows onto a plate. The paper [4] describes several general approaches for the Nusselt number determination. A metal foil heated by the electric current is exposed to a gas jet. The heat flux is set by the electric current, the surface temperature is measured by glued thermocouples, thermal imagers or liquid crystals. In every case, the foil thickness becomes a decisive factor for microscales due to the lateral heat flux in the foil with the high thermal conductivity. The manufacture of thin foils means great technical difficulties for the thermocouple method. Thermal imaging is also possible only if the foil is coated with the quite thick layer of a thermally emitting paint.

The plate surface is underestimated by the thermal imager when using plastics. The reason is the plastic transparency for infrared radiation, due to which the measured temperature is averaged in the layer with a thickness equal to the depth of infrared radiation penetration into the plastic. The scale of the problems considered requires the temperature averaging in the layer of max 10 microns thickness. The method of surface temperature distribution measuring based on a thermosensitive fluorescent coating was developed for this purpose.

The authors compared two panoramic methods to study the heat transfer process between a jet and a plate. The first method suggests that the cooled surface was made of polyether ether ketone (commercial name is PEEK) with the relatively low thermal conductivity in order to avoid the lateral heat transfer inside the material. This plastic has the high heat dissipation and is suitable for direct thermal imaging measurements. The second approach involves the thermosensitive fluorescent coating based on rhodamine B, applied to a stainless plate. Data obtained by these methods are compared with the results of numerical simulation.

The surface of the stainless plate of 15 μm thick was coated with a thin layer of the fluorescent paint; its fluorescence intensity depends on its temperature. Rhodamine B is a convenient fluorophore that is already used in aerodynamic measurements. The coating included rhodamine B and an optically transparent polymer – the polyurethane varnish. Under the influence of the exciting radiation, the resulting coating fluoresced with a maximum intensity at a wavelength of 600 nm. To measure the temperature distribution on the surface of the plate, a polyurethane varnish coating with a rhodamine concentration of 30 mg/ml was used. In this case, the penetration depth of the exciting radiation was about 2 μm, and the layer thickness (about 50 μm) did not matter.

The experiments to study the heat transfer of a microjet plate involving the fluorescent coatings based on rhodamine B, enabled to measure the surface temperature distributions with much simpler technical means with the higher spatial resolution. Therefore, we have chosen this very method of heat transfer coefficient determining. It is necessary to know the surface temperature T_w and the adiabatic temperature of the surface T_{aw}, as well as the heat flux q, which was determined by the input electric power to determine the heat transfer coefficient:

$$q = UI \quad \alpha = \frac{q}{T_w - T_{wa}}.$$

The Nusselt number was defined in the standard way as $Nu = \alpha\frac{d}{\lambda}$ where α is the heat transfer coefficient, d is the diameter of the nozzle, and λ is the air thermal conductivity coefficient.

2. Experiments

Figure 1 shows the nozzle diagram, their internal geometry and photos of the nozzle exit. Table 1 presents the parameters at which the experiments were performed.

The prechamber was a cylinder with a conical narrowing to the nozzle exit. There was a hole for measuring pressure in the prechamber.

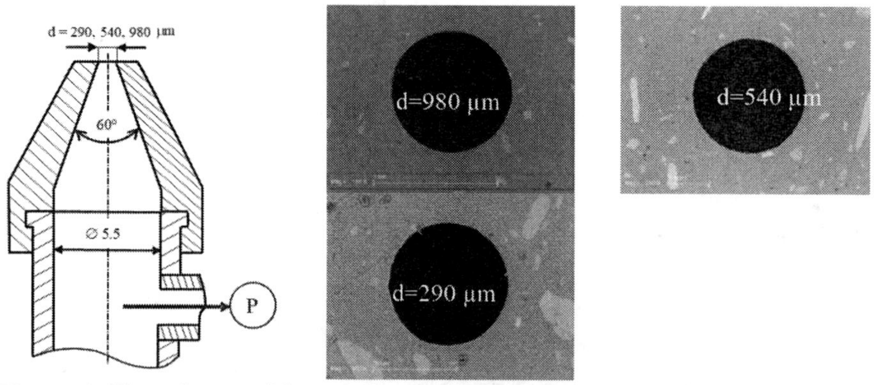

Figure 1. The scheme of the nozzles, its internal geometry and SEM images of the nozzle exit.

Table 1. Experimental parameters.

d (μm)	Re	h/d	V (m/s)
980	340-7000	0.5-10	5-100
540	187-3700	1-18	5-100
290	100-2000	1.7-30	5-100

The prechamber was a cylinder with a conical narrowing to the nozzle exit. There was a hole for measuring pressure in the prechamber.

Figure 2 shows the experimental scheme and a photo of the test bench.

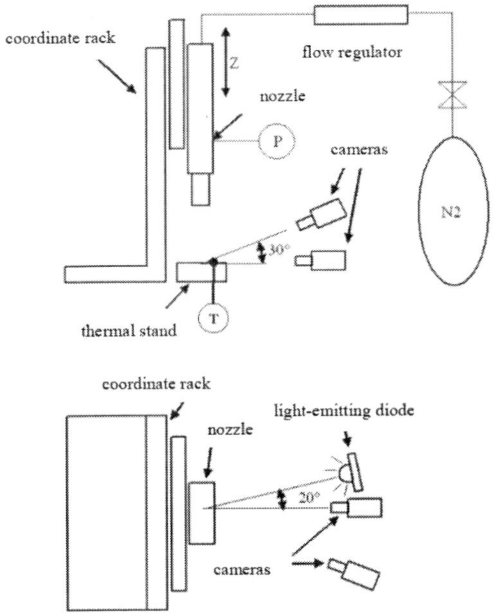

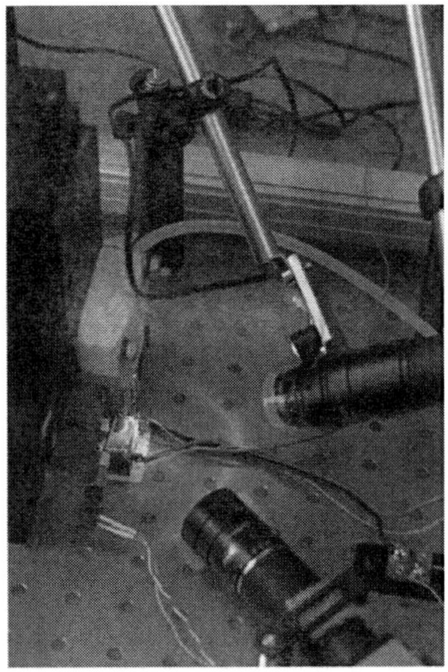

Figure 2. The experimental scheme and a test bench photo.

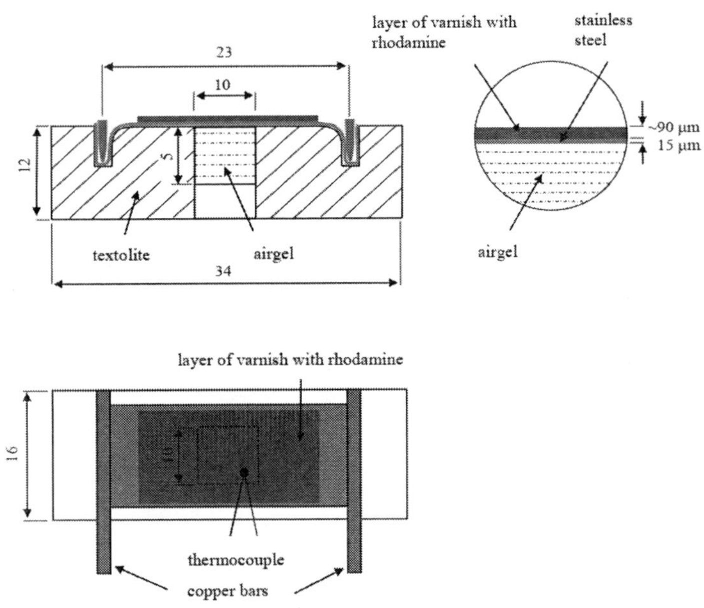

Figure 3. The thermal test bench scheme.

Nitrogen was used as the working gas. The gas flow rate was set with a flow controller. The nozzle was mounted on a coordinate rack which permitted the nozzle to be moved in the vertical direction. This enabled to change the distance between the nozzle and the plate. Two cameras were used in the experiments: one recorded the fluorescence intensity of the coating of the heat stand, the second recorded the distance between the nozzle and the surface of the heat stand. To excite the heat stand surface fluorescence, a blue LED located next to the first camera was used.

The heat stand (figure 3) was a textolite parallelepiped with a rectangular hole in the center. The stainless steel foil with a thickness of 15 μm was fixed with copper bars in the grooves of the textolite. An electric current was supplied to the copper busbars. A layer of varnish with rhodamine was applied on a stainless foil, the layer thickness of about 90 μm. The bottom plate was insulated with an airgel. The airgel was chosen as a heat insulator, for the material has a very low thermal conductivity coefficient $\lambda = 0.016$ W/(mK).

As an example, figure 4 shows the distribution patterns of the adiabatic temperature, the temperature of the heated plate, and the Nusselt number for the nozzles of 540 and 290 μm. The good axisymmetric distribution of the adiabatic temperature, the heated plate temperature, and the Nusselt number are clearly seen at a small distance from the heated surface. Then, as the distance increases, the distribution becomes not axisymmetric.

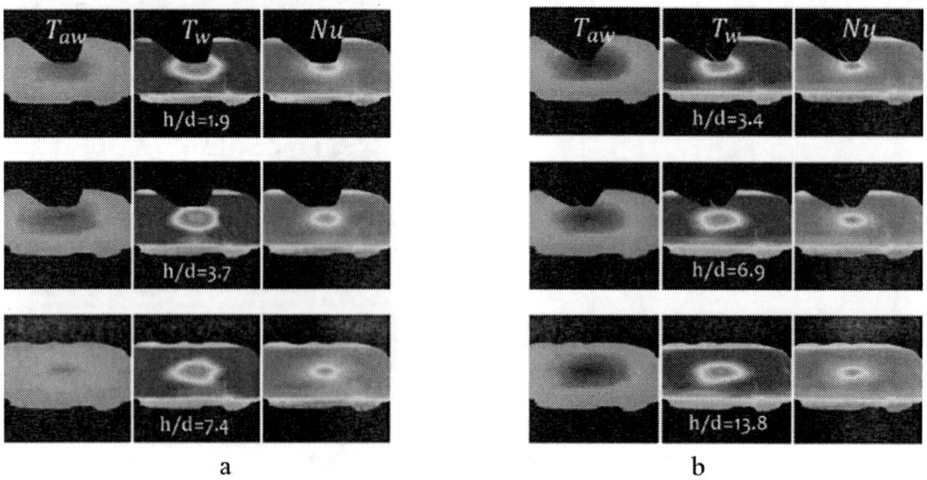

Figure 4. The distribution of adiabatic temperature, the heated plate temperature, and the Nusselt number for the nozzles with a diameter d = 540 μm, Re = 2220 (a) and d = 540 μm, Re = 1200 (b)

Figure 5 shows the Nusselt number distribution at different distances between the cooled surface and nozzle with a diameter of 980 μm at V = 60 m/s.

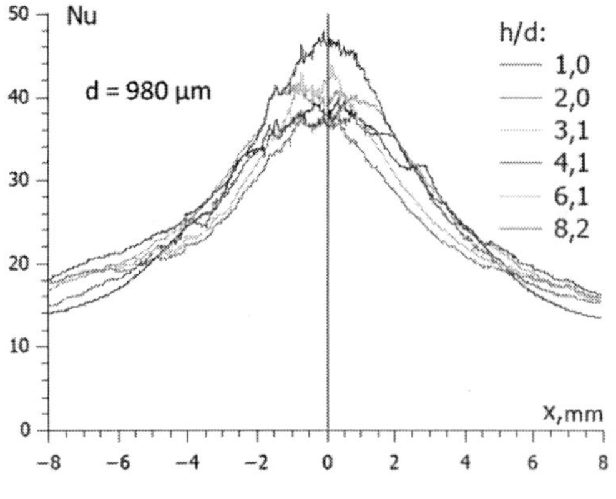

Figure 5. The Nusselt numer distribution for the 980 μm nozzle.

The Nusselt number at the frontal point ranges from 37 to 46 and reaches its maximum value at a height of h/d = 4.1.

The Nusselt number distribution at the frontal point when the jet escapes from the 540 μm nozzle is almost the same and equals to 17-19 (see figure 6).

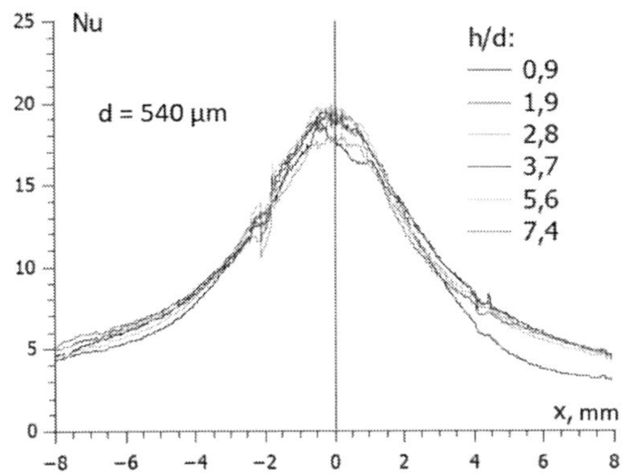

Figure 6. The Nusselt number distribution for the 540 μm nozzle.

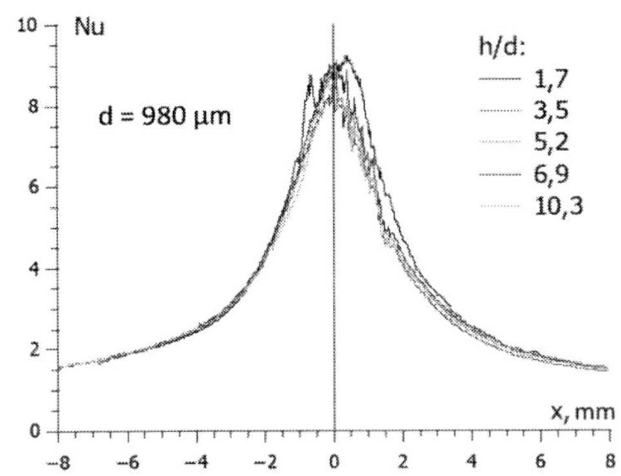

Figure 7. The Nusselt number distribution over the plate for the 290 μm nozzle.

The Nusselt number distribution at the frontal point in the case of the smallest nozzle (290 μm in diameter) is of the similar nature (figure 7). Here, the Nusselt number reaches 8 – 8.5.

3. Conclusion

The Nusselt numbers decrease significantly as the nozzle diameter decreases.

For the nozzles with a diameter of 540 and 290 μm, the Nusselt number distribution is practically independent of the location of the nozzle above the heated surface in the range of distances x/d from 0.9 to 10.

Acknowledgments

The research was partly carried out within the framework of the Program of Fundamental Scientific Research of the state academies of sciences in 2013-2020 (project No. 22.6.1).

References

[1] Dyban E P and Mazur A I 1982 *Convective heat transfer during jet flow around bodies* (Kiev: Naukova Dumka)
[2] Colucci D W and Viskanta R 1996 *Exp. Therm. and Fluid Science* **13** 71-80
[3] Patil V A and Narayanan V 2005 *Microscale Thermophysical Engineering* **9** 183-97
[4] Maslov N A, Aniskin V M, Korotaeva T A and Tsibulskaya E O 2019 *AIP Conference Proceedings* **2125** 030002

Features of the interaction of localized free-stream disturbances with the straight and swept wing boundary layer

M M Katasonov, V V Kozlov, A M Pavlenko and I A Sadovskii

Khristianovich Institute of Theoretical and Applied Mechanics SB RAS,
4/1, Institutskaya str., Novosibirsk, 630090, Russia

mikhail@itam.nsc.ru

Abstract. Under the conditions of a model experiment, the features of the appearance and spatial development the boundary layer disturbances – wave packets (precursors) and longitudinal localized structures were studied. Localized disturbances were created artificially upon local exposure of an external source to the boundary layer of the straight and swept wing with a low turbulence level of incoming flow. Wave packets formed near the fronts of longitudinal localized structures. The studies were carried out in a subsonic low-turbulent wind tunnel. Flow perturbations were recorded using a constant-temperature hot-wire anemometer.

1. Introduction

Laminar-turbulent transition in the boundary layer at a moderate or elevated level of free-stream turbulence is associated with disturbances created by external turbulence. The effect of free stream disturbances on the boundary layer leads to the formation of longitudinal localized structures or streaks, which are areas with an excess and deficit of longitudinal velocity [1, 2]. Longitudinal structures provide the basic conditions for the development of high-frequency wave disturbances, such as secondary instability and wave packets [3]. Under favorable conditions, wave disturbances turn into turbulent spots. According to recent experimental studies [4–6], a localized effect on the boundary layer can lead to the appearance of wave packets along with banded structures.

In this work, under controlled conditions, we study the occurrence and spatial development of perturbations of the boundary layer – wave packets (precursors) and longitudinal banded structures. Wave packets are formed near the fronts of longitudinal localized structures due to a sharp local change in the longitudinal flow velocity inside the boundary layer.

2. Experimental technique

The studies were carried out in a subsonic low-turbulent wind tunnel MT-324 of ITAM SB RAS. The free-stream velocity was $U_\infty = 6.8$ m/s. The freestream turbulence level Tu was 0.18% U_∞. Localized disturbances were created artificially by local exposure of the source to the boundary layer under a low turbulence level of incoming flow. The source of disturbances was located in the incoming flow near the model, see figure 1. The source of disturbances was a tube with an inner diameter of 2.5 mm through which air flowed out with a short pulse of 200 or 300 ms duration. Pulsed air blowing was provided by an electromagnetic valve. Pulses were repeated every 0.5 seconds. The method of introducing controlled perturbations was used, in which the moment of introducing disturbances and the moment of their registration always coincides, which makes possible a detailed study of the characteristics of the

 1404 (2019) 012092 doi:10.1088/1742-6596/1404/1/012092

boundary layer disturbances. It also ensures the preservation of phase information and allows tracking the dynamics of the development of the created disturbances at all stages of their development. Disturbances in the incoming flow and in the boundary layer were recorded using a constant-temperature hot-wire anemometer. In the paper an isocontours of velocity fluctuations used, representing the so-called hot-wire visualization of disturbances.

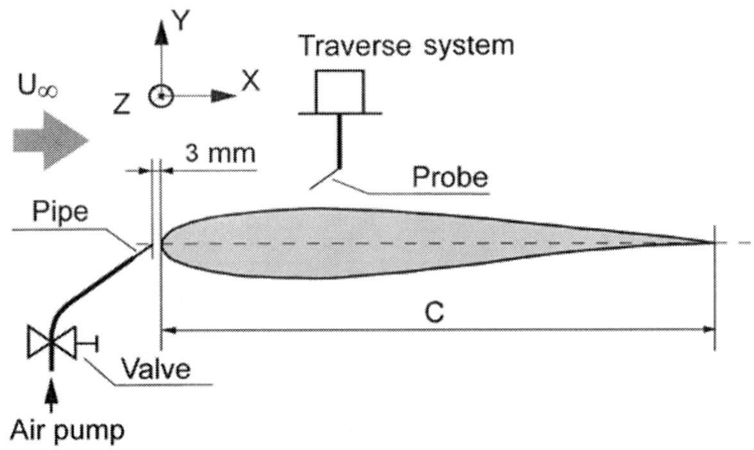

Figure 1. Scheme of experiment.

3. Results

3.1. Straight wing

Initially, the experiment was carried out on a straight wing model with a chord C1 = 290 mm, installed in the closed wind tunnel test section at zero angle of attack. Pulse air blown out under a small excess pressure from the tube and interact with the boundary layer of the model in the region of the leading edge, which has a large curvature radius about 7 mm. Figure 2 presents the pictures of hot-wire anemometric visualization of this process. The cross sections of disturbances are shown in the region of their middle in the Y-Z plane for different longitudinal coordinates X. The first cross section at X/C1 = -0.04 shows a disturbance in the incoming flow. The transverse size of the air jet flowing out of the tube remains almost unchanged (about 2.7 mm), while slightly compressing in the vertical direction, apparently due to interaction with the associated main flow. Faced with the leading edge of the model, the jet core generates a disturbance in the boundary layer, which has a much larger transverse dimension. Only the red region with excess of velocity has a transverse dimension in 4.6 times larger than the initial disturbance (12.5 and 2.7 mm, respectively). In this case, regions with a velocity defect (blue color) appear symmetrically on both sides. The disturbance generated on the model (including all regions of the defect and velocity excess) has a transverse dimension in 10 times larger than the boundary layer thickness δ. Here, the δ can be estimated from the height of the disturbance (along the Y coordinate). Moving downstream, the disturbance slightly diffuses, in proportion to the increase of the boundary layer thickness, which was noted in a number of previous studies [2, 3], when disturbances were introduced from the model wall [4, 5]. In [2], at a similar incoming flow velocity, the transverse size of the disturbance in the boundary layer approximately corresponded to the initial one, however, the radius of the leading edge in this experiment was much smaller, 0.5–0.7 mm. It is likely that the air jet in the present experiment, interacting with a strongly blunted leading edge, behaves like an impact jet, which is not observed in [2], where the leading edge is quite sharp.

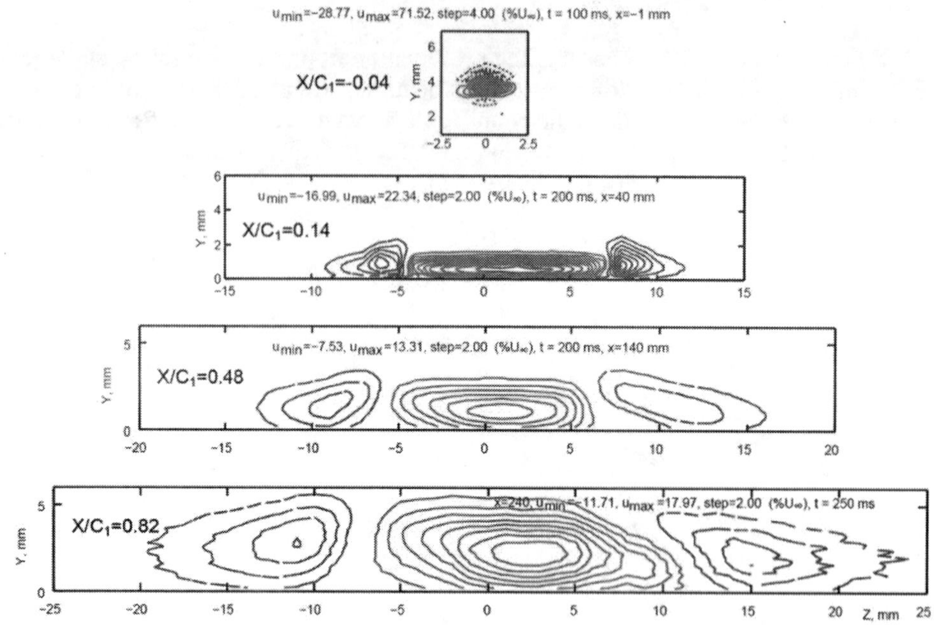

Figure 2. Isocontours of the longitudinal component of velocity fluctuations in different positions along the chord C1 of the straight wing in Y-Z plane in the region of the middle of the longitudinal localized disturbance.

Below is the evolution of the disturbance in the Z-t plane. Figure 3 shows typical visualization patterns for various coordinates along the flow direction. At X/C1 = 0.76 and further downstream, the formation and growth of the wave packet near the leading edge of the longitudinal localized disturbance are seen.

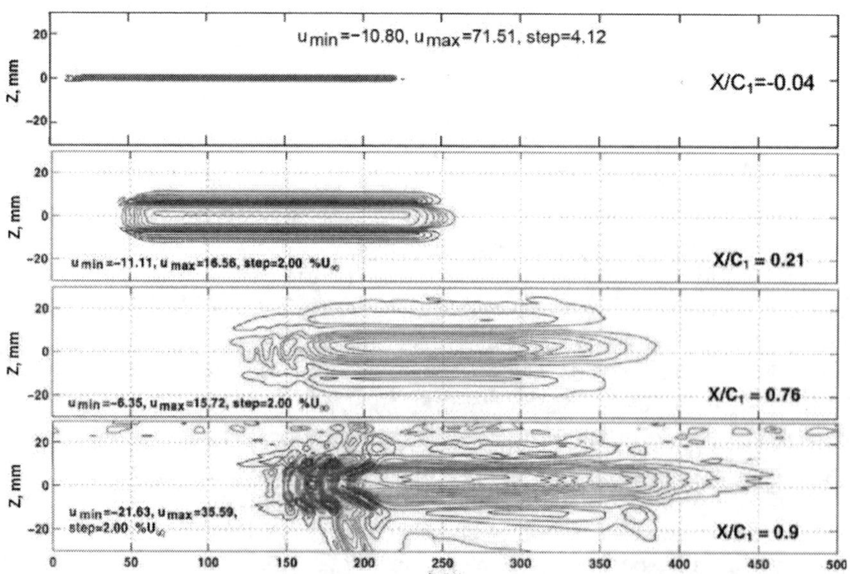

Figure 3. Isocontours of the longitudinal velocity component fluctuations in different positions along the chord C1 of the straight wing in Z-t plane.

With a low Reynolds number, in our case ReC1 = 131000, flow separation occurs in the rear part of the wing. As was shown earlier, the region of the unfavorable pressure gradient, and in particular the flow separation, strongly destabilize the wave packet, which in the present experiment manifests itself in the region near the leading front of the localized disturbance. In the present experiment, the development of the wave packet led to the appearance of the Λ-structure. A similar result was observed upon introducing perturbations from the model surface [4].

3.2. Swept wing

Further experiment was carried out on a swept wing model of with a sweep angle of 45 degrees and a chord of C2 = 410 mm, also set at zero angle of attack. Figure 4 shows the visualization of the interaction of the disturbance generated by the tube with the boundary layer of the swept wing. Near the model nose at X/C2 = 0.02, the disturbance shifts upward from the axis of symmetry by 10 mm. Moreover, the internal structure of the disturbance differs from the case of a straight wing. There are only two areas - defect and exceeds of velocity. The total disturbance width turned out to be smaller than for the straight wing (7.5 and 12.5 mm). Due to the influence of the secondary flow directed along the leading edge of the wing, the perturbation becomes asymmetric. In the process of evolution downstream, the disturbance shifts first down and then up. Its width also increases in proportion to the increase of the boundary layer thickness. Figure 5 shows the disturbance cross section along the vertical coordinate Y at X/C2 = 0.61. The velocity excess region is located above the velocity defect region, which means that, unlike the straight wing, see figure 2, the secondary flow spins the disturbance in main flow direction. In the region of an unfavorable pressure gradient, at X/C2 = 0.85, due to the secondary flow, new region is formed with exceed of the velocity. Here, as in the case of a straight wing, a high-frequency disturbance (wave packet) comes out. The symmetry of the wave packet is also broken by the influence of the secondary flow, which was previously noted in [7].

The general view of the movement of longitudinal localized disturbance along the wing, made on the basis of real-scale visualization patterns (figure 4), shows its S-shaped trajectory, see figure 6, which coincides with basic ideas about the flow structure on a swept wing [8].

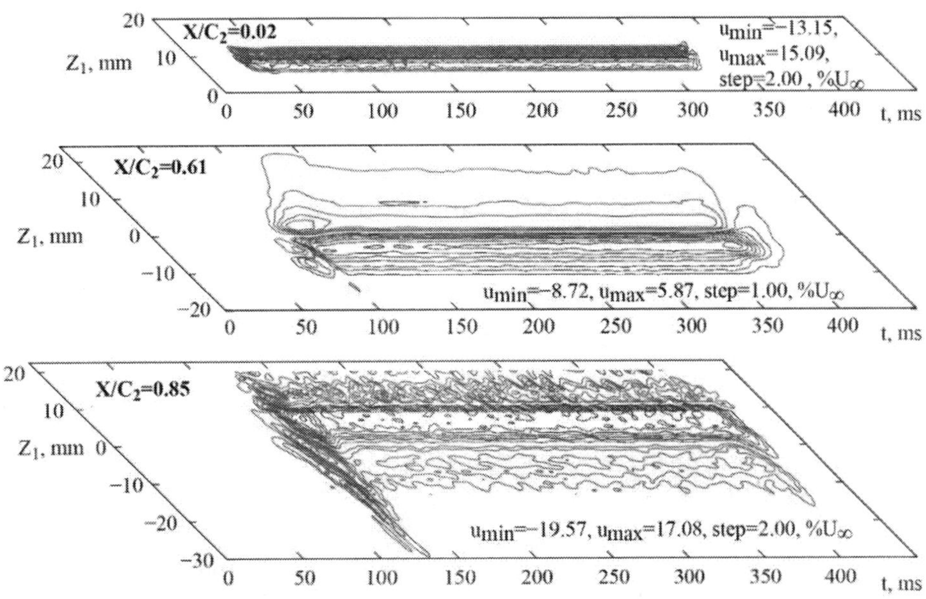

Figure 4. Isocontours of the longitudinal velocity component fluctuations in different positions along the chord C2 of the swept wing in Z-t plane.

A study of the dynamics of the downstream development of the generated disturbances on the straight and swept wings showed that the amplitude of the longitudinal localized disturbances damps. Moreover, their main characteristics coincide with the boundary layer disturbances generated by increased or high external turbulence [3]. The amplitude grow of the wave packets, which are also excited by the interaction of disturbances from the incoming flow and the boundary layer of the model, occurs in the region of an unfavorable pressure gradient and flow separation.

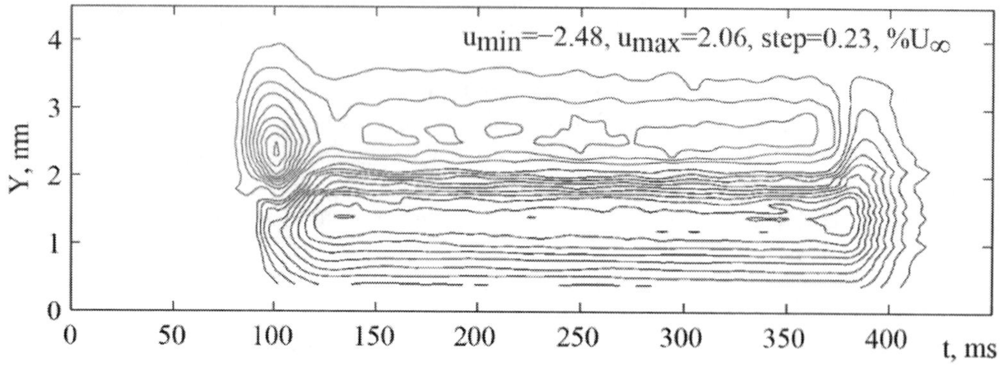

Figure 5. Isocontours of the longitudinal velocity component fluctuations in different positions along the chord C2 of the swept wing in Y-t plane, X/C2=0.61, Z1=1 mm.

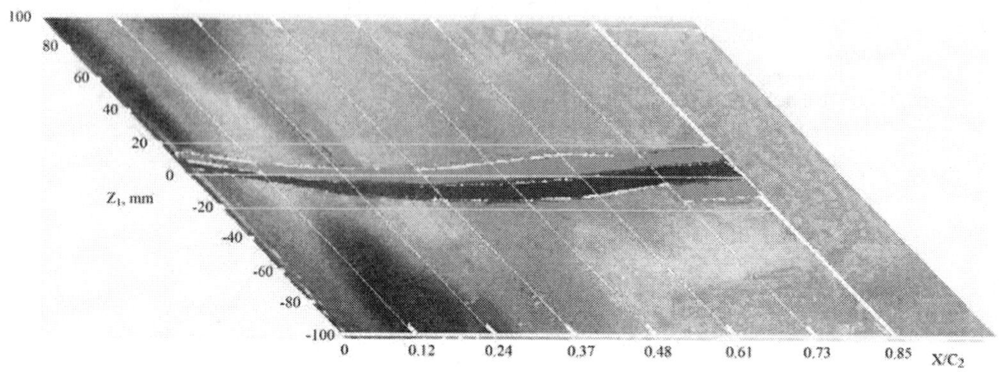

Figure 6. Downstream evolution of the regions of the defect (blue) and velocity exceed (red color) of the longitudinal localized disturbance in the swept wing boundary layer.

4. Conclusions

It was found that longitudinal localized disturbances are arose in all considered cases. These disturbances were classified as longitudinal localized structures, in their characteristics identical to disturbances arising in the boundary layer under the influence of an increased level of free-stream turbulence. It was noted that the intensity of the longitudinal localized structures decreases in the downstream direction. High-frequency wave packets also appear by the interaction of the artificial external flow disturbances with the boundary layer. Wave packets rapidly increase in the presence of an unfavorable pressure gradient and in the flow separation area. During the interaction of free stream disturbances with a thin boundary layer in the region of the model nose, the disturbance expands in the transverse direction. The presence of a secondary flow on a swept wing influences strongly the structure of both longitudinal localized structures and wave packets. The longitudinal structures twist in the direction of flow, and their motion path takes an S-shape. Wave packets become asymmetrical.

Acknowledgments
This work was supported by the project of Russian Science Foundation 16-19-10330 and was partly carried out within the framework of the Program of Fundamental Scientific Research of the state academies of sciences in 2013-2020 (project No. AAAA-A17-117030610128-8).

References
[1] Morkovin M V 1984 *NASA-CP-2386*
[2] Westin K J A, Bakchinov A A, Kozlov V and Alfredsson P H 1998 *Eur J Mech-B/Fluids* **17** 823-46
[3] Boiko A V, Grek G R, Dovgal A V and Kozlov V V 2002 *The Origin of Turbulence in Near-Wall Flows* (Berlin: Springer)
[4] Gorev V N and Katasonov M M 2004 *Thermophysics and Aeromechanics* **11** 391-403
[5] Chernorai V G, Spiridonov A N, Katasonov M M and Kozlov V V 2001 *J. Appl. Mech. Tech. Phys.* **42** 765-72
[6] Matsubara M, Takaichi K and Kenchi T 2009 *Proc. IUTAM Symp Laminar-Turbulent Transition Ed Schlatter P, Henningson D S* (Sweden: Springer-Verlag) pp 277-82
[7] Gorev V N, Katasonov M M and Kozlov V V 2007 *Fluid Dynamics* **42** 732-39
[8] Schlichting H and Gersten K 2000 *Boundary layer theory* (Berlin Heidelberg New York: Springer)

An experimental study of the influence of the type of turbulent wake on the flow around models of wings of various shapes

A M Pavlenko, B Yu Zanin, M M Katasonov and A V Bykov

Khristianovich Institute of Theoretical and Applied Mechanics SB RAS,
4/1, Institutskaya str., Novosibirsk, 630090, Russia

pavlyenko@gmail.com

Abstract. This paper presents the results of experimental studies aimed at studying the influence of the type of turbulent wake on the flow around models of various shapes. The patterns of visualization of flow structure for each flow regime were obtained on all models. The structure of the turbulent wake in space for two sources of disturbances is investigated. A comparison was made between the results. It was found that, depending on the type of turbulent wake, the separation region can significantly decrease in size.

1. Introduction

It is well known that during the cruise flight mode, local separation zones at subsonic speeds may form on the surface of the wings of aircraft [1-2]. With an increase in the angle of attack, the local separation region shifts toward the leading edge until it disappears completely. Moreover, near the trailing edge of the wing, a separation of the turbulent boundary layer begins to form. A further increase in the angle of attack to critical values leads to a separation of the flow from the leading edge, the formation of large-scale vortices and a return flow. The presence of stationary disturbance sources on the surface of the wings can lead to a change in the structure of the separated flow, up to its complete elimination, depending on the flow regime [3-6]. A new area of research is the study of the influence of external disturbances on the structure of separated flows [7]. The results of experimental studies presented below are the result of the continuation of a whole range of work on the study of the flow around an aircraft when it enters a turbulent wake. Due to the rapid development of small unmanned aircraft, the relevance of these studies is not in doubt.

2. Experimental setup

The experiments were carried out in two closed-type subsonic wind tunnels MT-324 and T-324 ITAM SB RAS Novosibirsk (Russia). The degree of free-stream turbulence of the MT-324 wind tunnel was 0.4% and T-324 was 0.04%. The study was carried out on models of wings with a straight and arrow-shaped leading edge, the dimensions of which are shown in figure 1. The span of the straight wing was 200 mm and the chord was 100 mm. Three types of hairy thread with a diameter of 0.5, 1.5 and 3 mm and three smooth fishing lines with a diameter of 0.8, 2 and 3 mm were used as a source of external disturbances. One specific thread (fishing line) was pulled upstream in front of the model and created a turbulent wake behind itself, which ran onto the wing model. The distance between the thread (fishing line) and the model was 760 mm and did not change in all experiments.

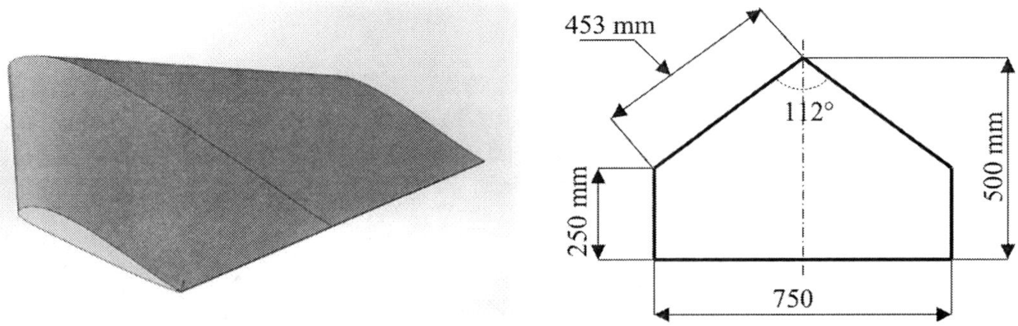

(a) (b)

Figure 1. The flying wing model (a) and geometrical dimensions of it (b).

Soot-oil visualization was used as the main research method. A mixture of titanium dioxide and kerosene was applied to the surface of the model. After drying of kerosene, limiting streamlines appeared on the wing. To obtain quantitative data on the structure of the flow behind the thread and fishing line, the method of hot-wire anemometry was used.

3. Results

3.1. Investigation of the effect of a turbulent wake on the flow around a model with an arrow-shaped leading edge

The first series of experiments was aimed at studying the flow around a flying wing model when it enters a turbulent wake at a supercritical angle of attack. These studies have focused on the stall flow regime. The angle of attack was set at $\alpha = 18°$. The free-stream velocity was $U_\infty = 22.5$ m/s. For given parameters, a stall flow regime is realized on the wing over the entire surface of the wing (see figure 2a). The direction of flow in photos is from top to bottom. The flow breaks from the leading edge, and two large-scale against the rotating vortices are formed. Then a thread (fishing line) was pulled to generate a turbulent wake 20 mm below the level of the leading edge at a zero angle of attack. When a wing enters a turbulent wake generated by a fishing line with a diameter of 0.8 mm, a stall flow regime is preserved on the model surface (see figure 2b). An increase in fishing line diameter to 2 mm led to a change in flow around. One large-scale vortex has substantially decreased in size and its focus has shifted in the left lateral edge (see figure 2c). The area of the attached flow in the left part of the wing increased. An increase in the diameter of the fishing line to 3 mm led to similar results in the case of 2 mm fishing line. It is worth noting that the right large-scale vortex has slightly decreased in size (see figure 2d). The wake of the hairy with a diameter of 0.5 mm affected the flow around the model in the same way as a fishing line with a diameter of 3 mm (see figure 2e).

The most interesting result was obtained under the influence of a hairy thread 1.5 mm in diameter (see figures 2f and 2g). This experiment was repeated several times and it was found that under the same conditions on the wing two types of flow are possible. The first type of flow is a significant increase in the region of the attached flow with a shift of the focal points of the vortices to the lateral edges of the model (see figure 2f). The second type of flow is a similar result in the case of hairy thread with a diameter of 0.5 mm (see figure 2g). A turbulent wake behind a hairy filament with a diameter of 3 mm led to a stable flow regime with the formation of small-sized vortices at the lateral edges (see figure 2h).

Then the thread (fishing line) was lowered 40 mm below the level of the leading edge at a zero angle of attack. Again experiments were conducted with three types of fishing line and three types of hairy thread. As an example, figure 3a shows the effect of the trace of a hairy thread 1.5 mm in diameter.

 1404 (2019) 012093 doi:10.1088/1742-6596/1404/1/012093

On the wing, the region of the attached flow significantly increased, and small-sized eddies formed near the lateral edges. This result was obtained for all 6 types of turbulent wake sources.

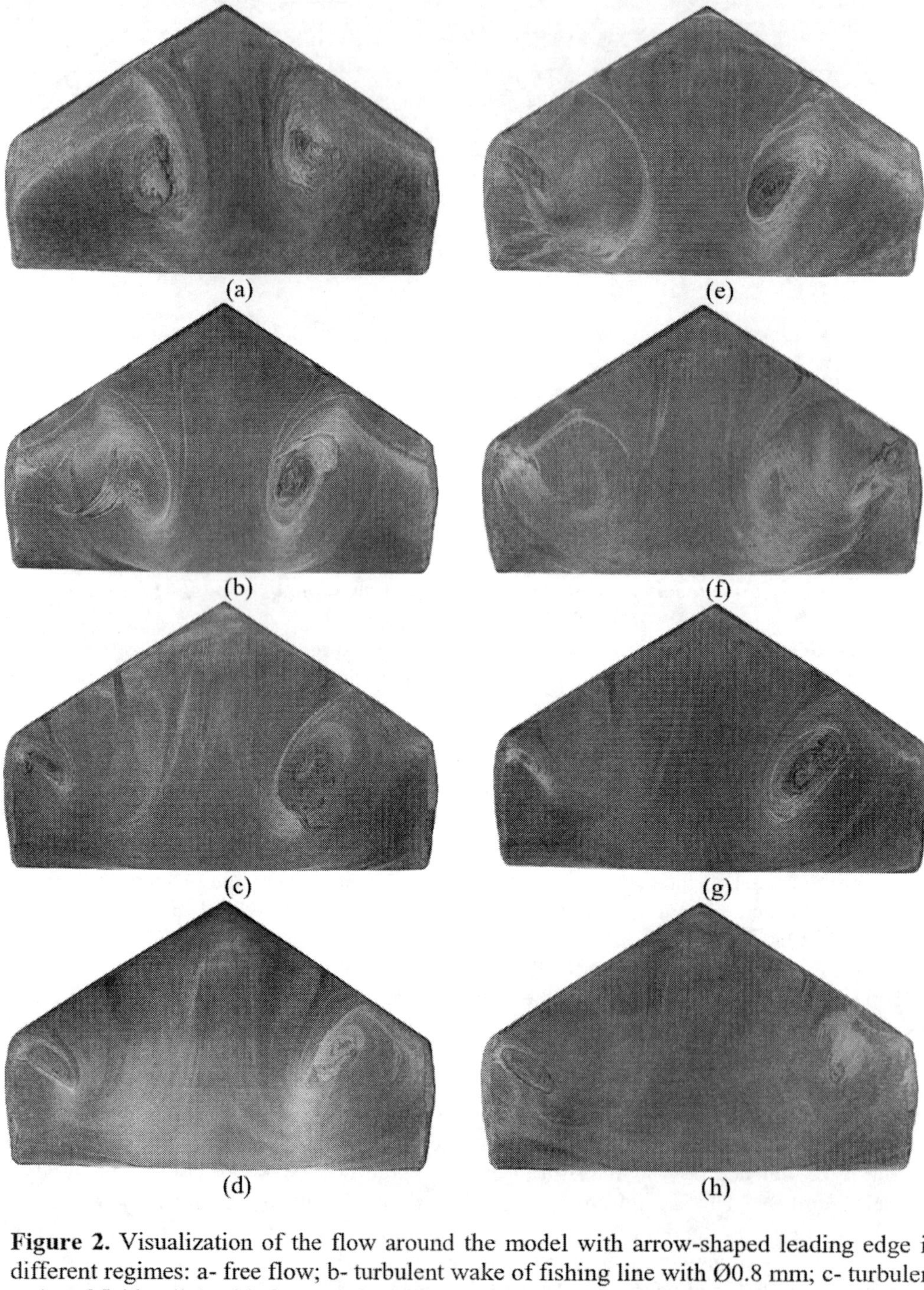

Figure 2. Visualization of the flow around the model with arrow-shaped leading edge in different regimes: a- free flow; b- turbulent wake of fishing line with Ø0.8 mm; c- turbulent wake of fishing line with Ø2 mm; d- turbulent wake of fishing line with Ø3 mm; e- turbulent wake of hairy thread with Ø0.5 mm; f and g- turbulent wake of hairy thread with Ø1.5 mm; h- turbulent wake of hairy thread with Ø3 mm.

The installation of turbulators in the form of cones on the surface of the model led to a change in the flow structure (see figure 3b). The height of the cone was 12 mm, the diameter of the base was 8 mm. The cones were installed 100 mm from the side edges. Near the surface, there are two regions of return flow and two large-scale vortices. This flow pattern was observed both during free-flow flow and when a wing enters a turbulent wake.

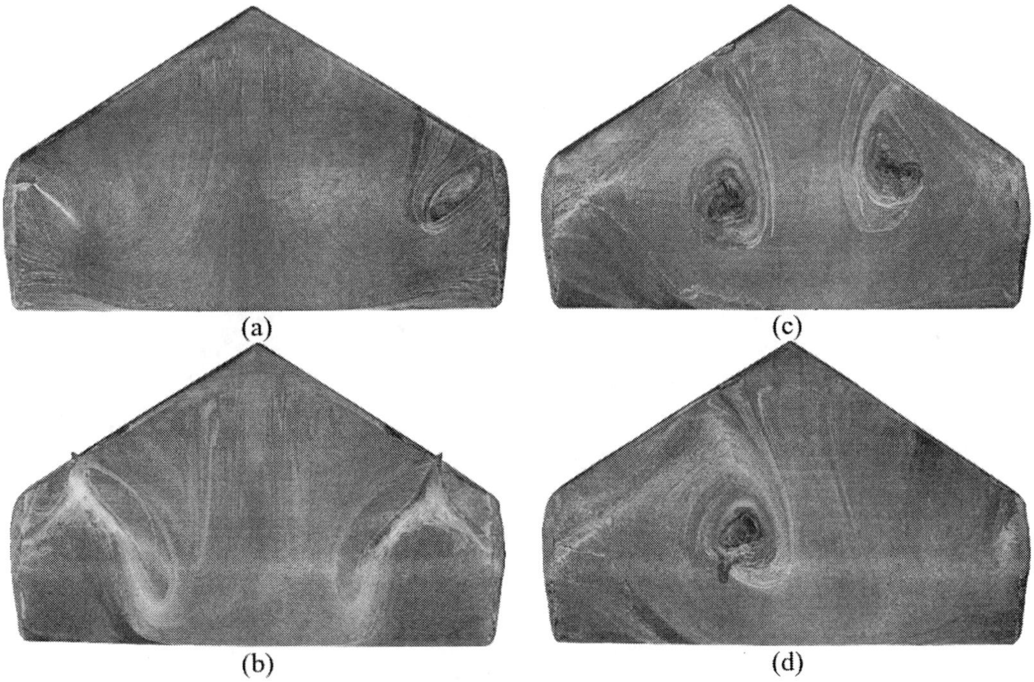

Figure 3. Visualization of the flow around the model with arrow-shaped leading edge in different regimes: a- turbulent wake of hairy thread with Ø1.5 mm; b- turbulent wake of hairy thread with Ø1.5 mm and two cones; c- turbulent wake of hairy thread with Ø1.5 mm at an angle of 27 degrees; d- turbulent wake of hairy thread with Ø1.5 mm at an angle of 90 degrees.

The effect of a turbulent wake behind a hairy thread 1.5 mm in diameter, set at an angle to the horizontal plane of the wing at 27 degrees, did not lead to a change in the flow structure (see figure 3c). As in the case of a free flow around a wing, there is a global stall with a pair of large-scale vortices on the wing.

Then the same thread was installed on ¼ of the span of the model at an angle of 90 degrees to the horizontal plane of the wing. This led to the full attachment of the flow on the right side of the model (see figure 3d).

3.2. Investigation of the effect of a turbulent wake on the flow around a model with an straight leading edge

The second series of experiments was devoted to the study of the influence of the type of turbulent wake on the flow around the straight wing model at different yaw angles (see figure 4). The model was installed at an angle of attack of $\alpha = 12°$. The yaw angle varied from 0 to 30 degrees. The free-stream velocity was $U_\infty = 13$ m/s.

The classical separation of the flow from the leading edge with the formation of a return flow and a pair of large-scale vortices is observed in a free flow (see figure 4a). An increase in the yaw angle to 15 degrees led to a shift of the local separation zone towards the left side edge (see figure 4b). It should

 1404 (2019) 012093 doi:10.1088/1742-6596/1404/1/012093

be noted that large-scale vortices disappeared. The local separation region decreases significantly when the yaw angle reaches 30 degrees (see figure 4c).

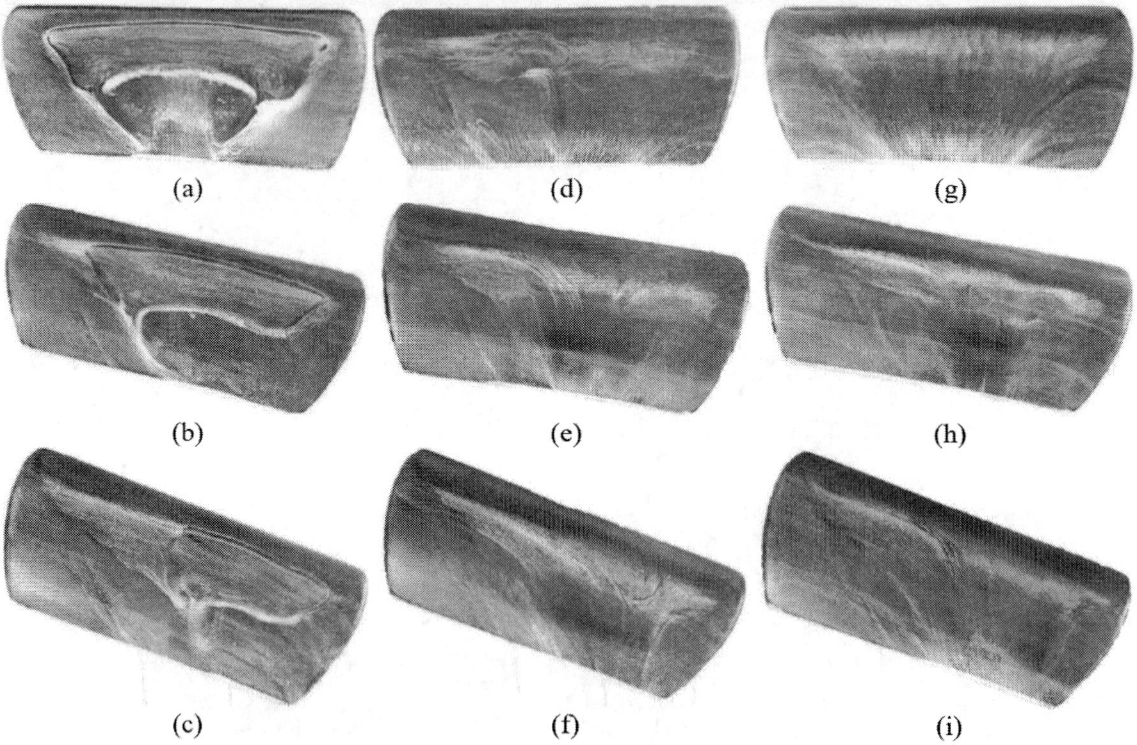

Figure 4. Visualization of the flow around the model with straight leading edge in different regimes: a- free flow, yaw angle χ=0°; b- free flow, yaw angle χ=15°; c- free flow, yaw angle χ=30°; d- turbulent wake of fishing line, yaw angle χ=0°; e- turbulent wake of fishing line, yaw angle χ=15°; f- turbulent wake of fishing line, yaw angle χ=30°; g- turbulent wake of hairy thread, yaw angle χ=0°; h- turbulent wake of hairy thread, yaw angle χ=15°; i- turbulent wake of hairy thread, yaw angle χ=30°.

Then a fishing line with a diameter of 0.8 mm was installed in front of the model. Experiments were conducted at yaw angles of 0, 15, and 30 degrees, which gave results on the influence of the turbulent wake around the model. In all three cases, the attached flow prevails on the wing with a significant decrease in the local separation region (see figures 4d, 4e, and 4f).

The maximum effect from the influence of the turbulent wake was observed from a hairy thread 1.5 mm in diameter, which was installed instead of the fishing line (see figures 4g, 4h and 4i). At zero yaw angle, we managed to maximally attach the flow (see figure 4g). The local separation regions disappeared.

3.3. Investigation of the structure of the flow behind hairy thread and smooth fishing line

The third series of experiments was devoted to obtaining quantitative data on the flow structure behind a hairy thread 1.5 mm in diameter and a smooth fishing line with a diameter of 0.8 mm. The method of hot-wire measurements was used. The free-stream velocity was $U\infty$ = 13 m/s. The measurements were carried out across the thread (fishing line) in the Y coordinate, along the thread (fishing line) in the Z coordinate, and depending on the distance between the sensor and the thread (fishing line) at x = 95, 350 and 540 mm (see figures 5 and 6).

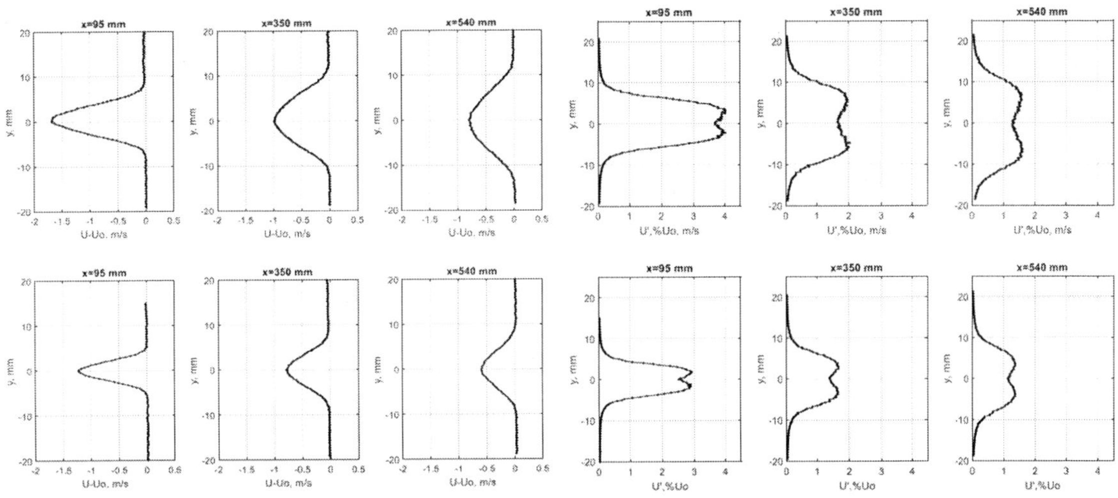

(a) (b)

Figure 5. Hot-wire measurements of turbulent wake depending on the longitudinal coordinate x. a- profiles of average velocity along the Y axis behind the thread (top) and fishing line (bottom). The absolute value. b- the amplitude of the velocity pulsations along the Y axis behind the thread (top) and fishing line (bottom).

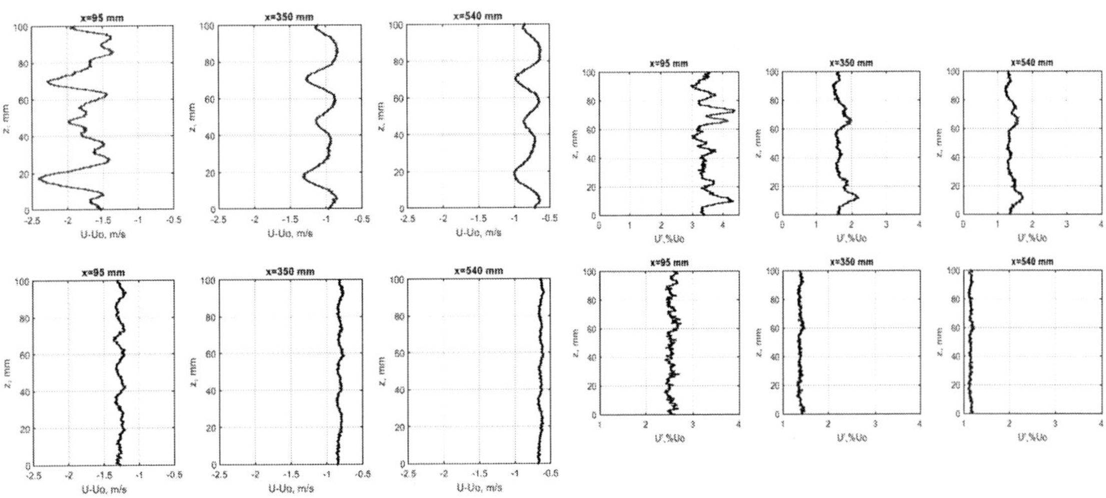

(a) (b)

Figure 6. Hot-wire measurements of turbulent wake depending on the longitudinal coordinate x. a- profiles of average velocity along the Z axis behind the thread (top) and fishing line (bottom). The absolute value. b- the amplitude of the velocity pulsations along the Z axis behind the thread (top) and fishing line (bottom).

It was found that the width of the trace of the thread is about 25 mm at a distance of 540 mm, which is 5 mm more than the trace of the fishing line under the same conditions. The average amplitude of the velocity pulsations at x = 95 mm is 1% higher in the wake of the thread compared to the trace of the fishing line, but moving downstream the amplitude of the velocity pulsations is leveled. A distinctive feature is the presence of longitudinal structures in the wake of the thread. In the trace of the fishing line, longitudinal structures are not observed.

4. Conclusions

Research of the influence of the type of turbulent wake on the structure of the flow around wing models in the detached mode at supercritical angles of attack were conducted. Photos of visualization of the limiting streamlines near the surface of the models for each case are obtained. Quantitative data were obtained on the structure of the trace of hairy thread and fishing line. A comparative analysis of the results are obtained. It was found that under certain conditions on the flying wing, two types of flow are possible when it enters into a turbulent wake. The influence of a turbulent wake can lead to either a complete flow attachment or a partial one. Partial flow attachment implies the presence of one large-scale vortex on half of the wing and attached flow around the other half of the wing. Such a flow regime can lead to dramatic consequences during the flight of an aircraft, due to the difference in the lifting force at different halves of the model in the presence of a stall and an attached flow. It was found that there is a certain region in space relative to the wing model, by installing a turbulent wake source in it, it is possible to achieve a stable and detached flow near the surface of the wing model.

Acknowledgments

This work was supported by the project of Russian Science Foundation No 18-79-00189. The authors are grateful to the Joint Access Center “Mechanics” of ITAM SB RAS for the provided equipment.

References

[1] Boiko A V, Dovgal A V, Zanin B Yu and Kozlov V V 1996 *Thermophysics and Aeromechanics* **3** 1-13
[2] Traub L W and Cooper E 2008 *Journal of aircraft* **45** 1322-33
[3] Pavlenko A M, Zanin B Y, Katasonov M M and Zverkov I D 2010 *Thermophysics and Aeromechanics* **17** 15-20
[4] Zanin B Yu, Kozlov V V and Pavlenko A M 2012 *Fluid dynamics* **47** 133-40
[5] Pavlenko A M, Zanin B Y and Katasonov M M 2015 *Vestnik of NSU* **10** 19-25 In Russian
[6] Boiko A V, Dovgal A V, Zanin B Yu, Kozlov V V, Lushin V N and Syzrantsev V V 1995 *Thermophysics and Aeromechanics* **2** 37-45 In Russian
[7] Pavlenko A M, Zanin B Y and Katasonov M M 2018 *AIP Conference Proceedings* vol 2027 edited by V.M. Fomin NY American Institute of Physics Melville 030060 DOI: 10.1063/1.5065278

The controlled periodic impact on the longitudinal vortex in the boundary layer at Mach 2

M V Piterimova[1,2], A D Kosinov[1,2], A V Panina[1], N V Semionov[1], Yu G Ermolaev[1] and A A Yatskikh[1,2]

[1]Khristianovich Institute of Theoretical and Applied Mechanics SB RAS, 4/1, Institutskaya str., Novosibirsk, 630090, Russia
[2]Novosibirsk State University, 1, Pirogova str., Novosibirsk, 630090, Russia

marina.piterimova@yandex.ru

Abstract. Experimental data on the development of controlled disturbances in the boundary layer with a longitudinal vortex are analysed. It is found experimentally that the mean flow nonuniformity can vary from 0.7 to 0.9 V in terms of the CTA mean voltage or 30–40% in ρU. Such flow nonuniformity complicates the estimation of the wave spectra, where normalization on the mean voltage of the anemometer is required. The applicability of the standard processing procedure is investigated. Comparative analysis of the data shows that both normalization types considered in the work can be applicable. However, in the case of flow nonuniformity in the transverse direction, it is preferable to use the normalization on the median value over the mean voltage spanwise distribution.

1. Introduction

The laminar-turbulent transition phenomenon is of great importance in aerohydrodynamics. Its study is necessary for the modern aviation technology development. At supersonic flow velocities, perturbations in the form of Mach waves can be emitted from the surfaces of aircrafts. This process affects downstream wings, leading edges, and other parts of the aircrafts. Experimental studies of such an effect are described in [1-5]. As a result it was found that a two-dimensional roughness on the wall of the test section generates a disturbance in the form of N-wave in space. When it interacts with the flow near the model, the mass flow pulsations level increases significantly. Also a mean flow distortion is observed and what is more, the stationary longitudinal vortices are generated in the boundary layer. In [6, 7] the N-wave influence on the perturbations development in the boundary layer was studied. It is shown that stationary longitudinal vortices do not affect the beginning of the laminar-turbulent transition, however, at the same time, its end is shifted downstream.

The experimental data techniques processing used previously [8, 9], was developed for the boundary layer homogeneous in the transverse direction. Since the measurements are performed at y = const, the mean flow inhomogeneity in the transverse direction is fixed in the form of E(z) dependence. However, for the Fourier processing, the scaled (dimensionless) output signal of the hot-wire anemometer is used, and the modulated nature of E(z) dependence, in accordance with the Fourier transform properties, will affect the wave spectra estimation results by β.

The aim of this work is the study of experimental data techniques processing applicability for the case of an inhomogeneous flow in the boundary layer.

2. Experimental set up and data processing

The experiments are carried out in a T-325 ITAM SB RAS supersonic wind tunnel at Mach number M = 2. The detailed description of the used equipment and experiment set up is given in [6, 10, 11]. In this study, the model of a flat plate with a sharp leading edge was installed in the wind tunnel test section at the zero attack angle. Figure 1 shows the experiment scheme.

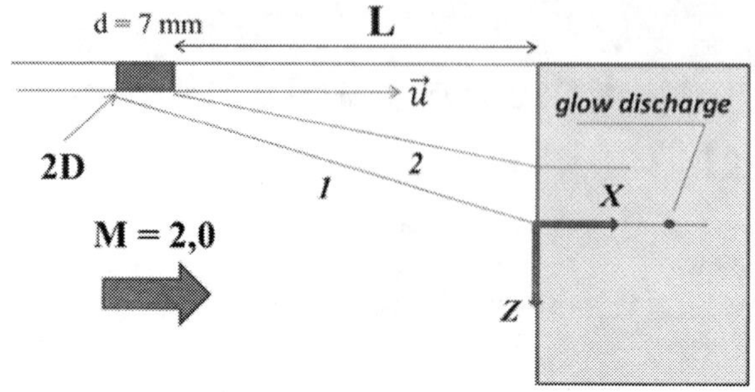

Figure 1. Experiment with 2D-irregularity on the test section wall, the discharge is on

The processing of the experimental data involves the collection of the frequency-wave spectra of the controlled disturbances [12]. The frequency-wave spectra of the controlled (periodic) disturbances were determined by the discrete Fourier transform (DFT) in the following form:

$$\tilde{A}(x_k,\beta,f_l)=\frac{\sqrt{2}}{\tilde{\delta}\cdot T\cdot Q}\sum_{n=1}\sum_{j=1}A(x_k,z_j,t_n)\cdot e^{-i(\beta\cdot z_j-\omega_l\cdot t_n)}\Delta z_j\Delta t\,, \tag{1}$$

where $\tilde{\delta}=1$ mm is the boundary layer thickness scale; T is the implementation length over time; Q is the sensitivity coefficient to mass flow pulsations; $A(x_k,z_j,t_n)$ is the instantaneous value of dimensionless pulsations of the hot-wire anemometer output signal; Δz_j is the analog-to-digital converter time sampling step.

The amplitude and phase of the disturbances were calculated after the DFT using the formulas:

$$A(x_k,\beta,f_l)=\mathrm{mod}\left[\tilde{A}(x_k,\beta,f_l)\right],$$
$$\Phi(x_k,\beta,f_l)=arctg\left\{-\frac{\mathrm{Im}\left[\tilde{A}(x_k,\beta,f_l)\right]}{\mathrm{Re}\left[\tilde{A}(x_k,\beta,f_l)\right]}\right\}. \tag{2}$$

3. Results

For this problem statement, the experimental data processing procedure allows for three different normalizations for the instantaneous value of the pulsation signal of the hot-wire anemometer $A(x_k,z_j,t_n)=e'(x_k,z_j,t_n)/E_m(x_k)$:

- to local measured voltage value $E_1=E(x_k,z_j)$,
- E_2 mean voltage (mean(E)) [13],
- E_3 median voltage value (median(E)) [13].

Table 1 shows the average and median voltage values for the cross sections x = 60 - 100 mm and the difference between these values Δ in percent. Evident that these values are different for each x. Although the difference is small, nevertheless, it is necessary to study the applicability of normalization to mean(E) and median(E).

Journal of Physics: Conference Series **1404** (2019) 012094 doi:10.1088/1742-6596/1404/1/012094

Table 1. Evolution of the mean, median voltage value and their difference downstream.

x, mm	mean(E), V	median(E), V	Δ, %
60	8.51	8.52	0.1
70	8.47	8.49	0.2
80	8.51	8.47	0.5
90	8.71	8.67	0.5
100	8.64	8.61	0.3

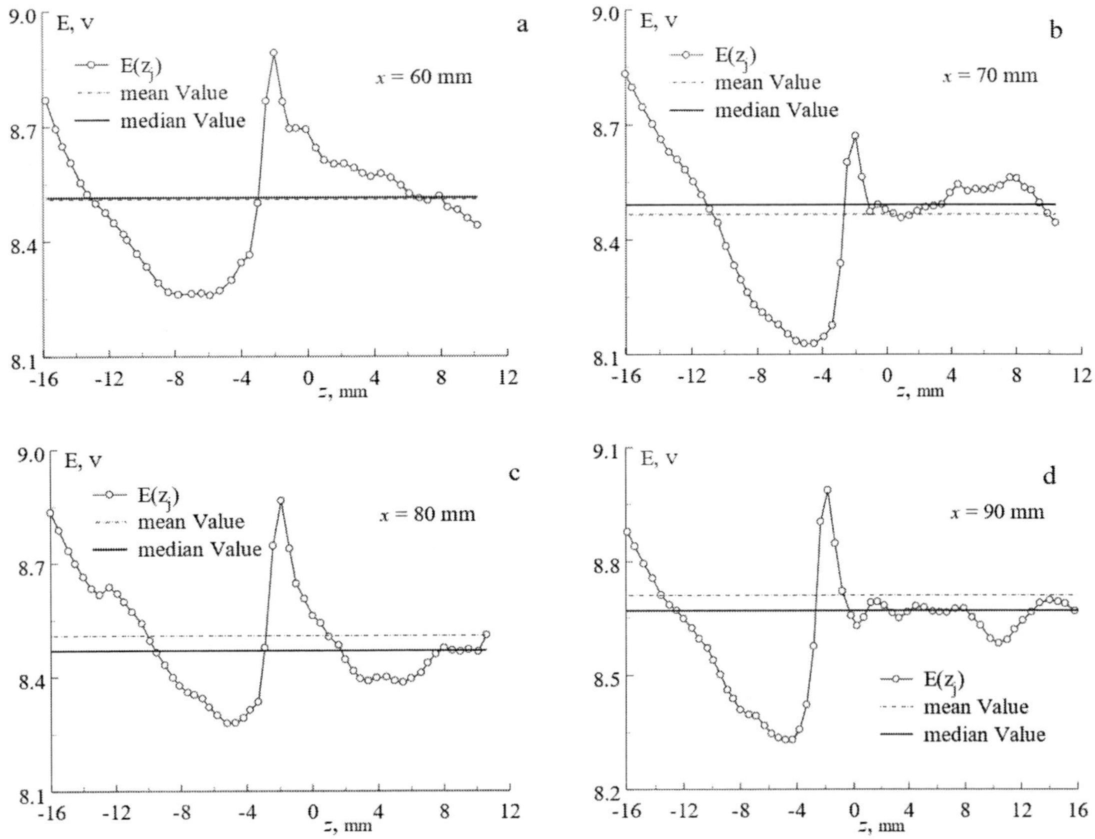

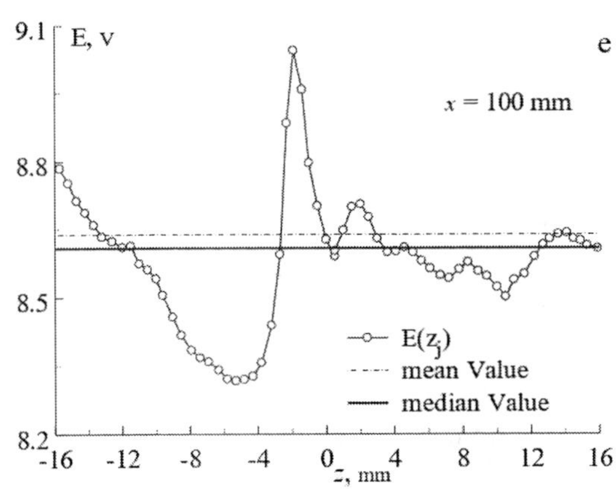

Figure 2. Mean mass flow defect downstream evolution (longitudinal vortex): a) x = 60 mm; b) x = 70 mm; c) x = 80 mm; d) x = 90 mm; e) x = 100 mm

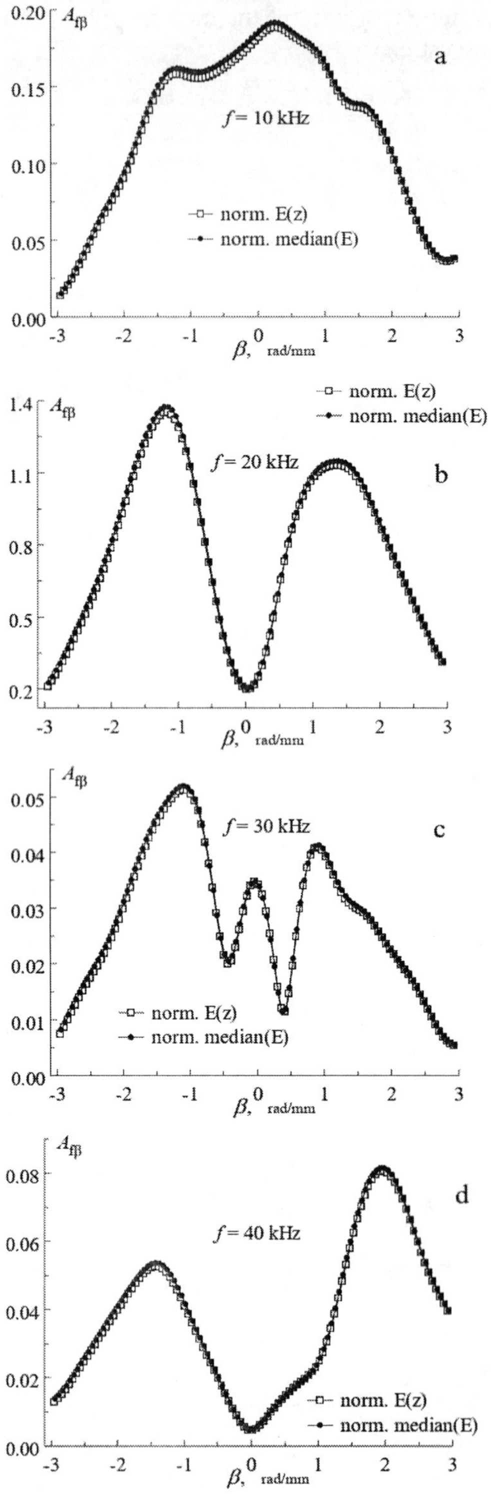

Figure 3. Amplitude β-spectra for the cross section x = 60 mm:
a) f = 10 kHz; b) f = 20 kHz; c) f = 30 kHz; d) f = 40 kHz.

 IOP Publishing
1404 (2019) 012094 doi:10.1088/1742-6596/1404/1/012094

In figure 2, the curve E(z) displays the voltage values measured by the hot-wire anemometer sensor. The area $z > 0$ corresponds to the undisturbed flow. In the interval from z = - 4 mm to z = 0, the stationary longitudinal vortex generated by a weak shock wave from 2D - irregularity can be identified.

From the voltage distributions a) - e) shown in figure 2 one can see that in this case it is preferable to use the normalization on the median voltage value (mean(E)), because it lies closer to the values of E(z) in the unperturbed flow region than the average voltage (mean(E)).

Further, in the work, the instantaneous value of the pulsation hot-wire anemometer signal was dimensionless by the median voltage value. Figures 3 and 4 show the amplitude and phase β-spectra, received via various normalizations. The data for the case of normalization on the values of E(z) were taken from [14].

Comparison of these results shows that the difference between the spectra is insignificant, and both types of normalization are applicable.

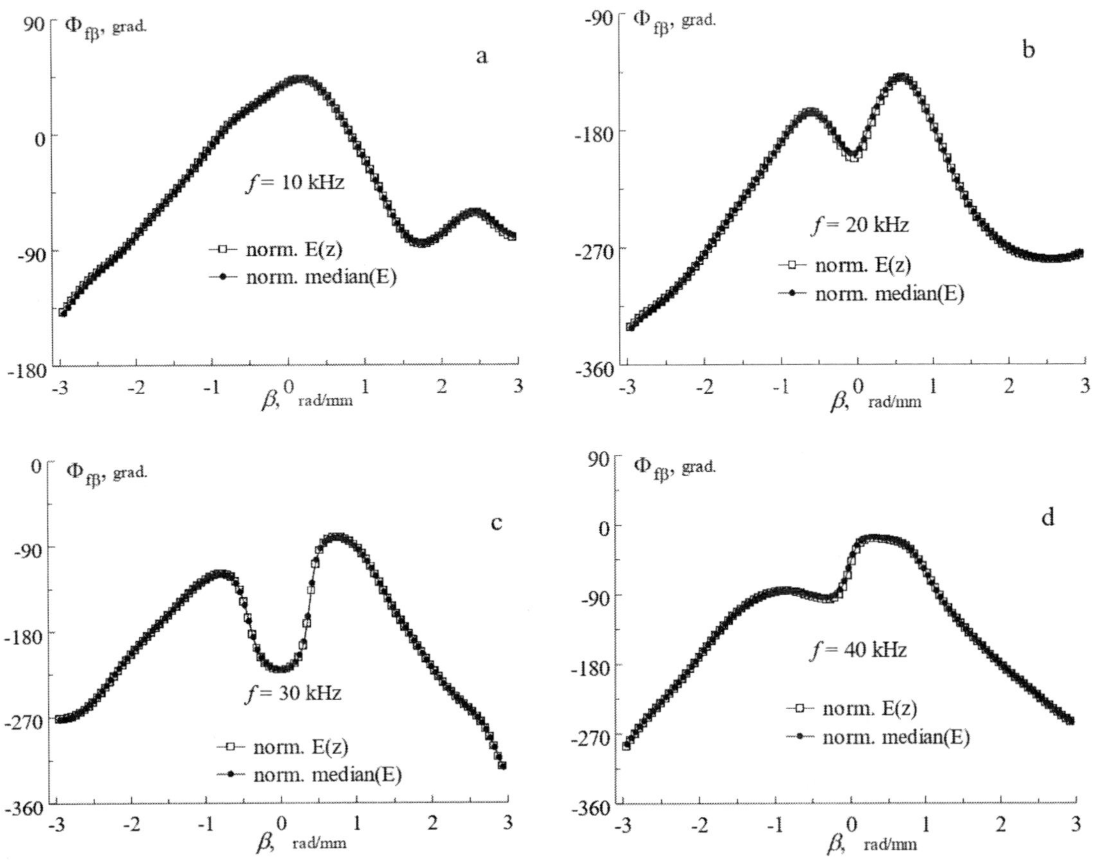

Figure 4. Phase β-spectra for the cross section x = 60 mm: a) f = 10 kHz; b) f = 20 kHz; c) f = 30 kHz; d) f = 40 kHz

4. Conclusion

Analysis of experimental data [14] on the development of controlled disturbances in the boundary layer with a longitudinal vortex is performed. The measurements in [14] are conducted at a fixed height of the hot-wire above the model surface. The choice of this height is based on the maximum location of the disturbance profile in the boundary layer for each value of the x coordinate varying from 60 to 100 mm at the same value of the z coordinate approximately equal to -16 mm. It is found under the experimental conditions that the nonuniformity of the mean flow can vary from 0.7 to 0.9 V in terms of the CTA mean voltage or 30–40% in terms of the mean mass flow ρU. Such flow nonuniformity complicates the estimation of the wave spectra, where normalization on the mean voltage of the anemometer is required. The purpose of this work was to obtain the amplitude and phase β-spectra using various normalizations.

The applicability of the standard processing procedure is investigated. Comparative analysis of the data shows that both normalization types considered in the work can be applicable. However, in the case of flow nonuniformity in the transverse direction, it is preferable to use the normalization on the median value over the mean voltage spanwise distribution.

Acknowledgments
The research was carried out within the framework of the Program of Fundamental Scientific Research of the state academies of sciences in 2013-2020 (project No. AAAA-A17-117030610125-7). The study was conducted at the Joint Access Center "Mechanics" of ITAM SB RAS.

References

[1] Kosinov A D, Yatskikh A A, Yermolaev Yu G, Semionov N V, Kolosov G L and Piterimova M V 2017 *AIP Conference Proceedings* **1893** 030072 DOI: 10.1063/1.5007530
[2] Vaganov A V, Ermolaev Yu G, Kolosov G L, Kosinov A D, Panina A V and Semionov N V 2016 *Thermophysics and Aeromechanics* **23** 43 DOI: 10.1134/S0869864316010054
[3] Kosinov A D, Semionov N V, Yatskikh A A, Yermolaev Yu G and Piterimova M V 2018 *Siberian physics journal* **13** 16 DOI: 10.25205/2541-9447-2018-13-3-16-23 (in Russian)
[4] Kocharin V L, Semionov N V, Kosinov A D, Yermolaev Yu G and Yatskikh A A 2018 *AIP Conference Proceedings* **2027** 040026 DOI: 10.1063/1.5065300
[5] Ermolaev Yu G, Kosinov A D, Kocharin V L, Semenov N V and Yatskikh A A 2019 *Fluid Dynamics* **54** 257 DOI: 10.1134/S0015462819020058
[6] Egorov I V, Fedorov A V and Dihn Q H 2018 *TsAGI Science Journal* **49** 495
[7] Dihn Q H, Egorov I V and Fedorov A V 2018 *Fluid Dynamics* **53**
[8] Kosinov A D, Yermolaev Y G, Nikolaev N N, Semionov N V and Semisynov A I 2007 *13th International Conference on the Methods of Aerophysical Research (ICMAR-2007)* vol 5 (Novosibirsk: Parallel) p 81
[9] Yermolaev Y G, Kosinov A D, Semionov N V and Semisynov A I 2007 *13th International Conference on the Methods of Aerophysical Research (ICMAR-2007)* vol 4 (Novosibirsk: Parallel) p 198
[10] Kosinov A D, Semionov N V, Ermolaev Yu G, Kolosov G L, Yatskikh A A and Kocharin V L 2018 *AIP Conference Proceedings* **2027** 040087 DOI: 10.1063/1.5065361
[11] Semionov N V, Yermolaev Yu G, Kocharin V L, Kosinov A D, Semenov A N, Smorodsky B V and Yatskikh A A 2018 *AIP Conference Proceedings* **2027** 030156 DOI: 10.1063/1.5065250
[12] Kosinov A D, Panina A V, Semionov N V, Yermolaev Yu G, Gaponov S A 2013 *Proceedings of the 11th International Conference on Fluid Mechanics & Aerodynamics (FMA '13)* p 19.
[13] Chernova N I 2014 *Mathematical statistics* (Novosibirsk: NSU).
[14] Panina A V, Kosinov A D, Semionov N V and Yermolaev Yu G 2018 *AIP Conference Proceedings* **2027** 040077 DOI: 10.1063/1.5065351

Journal of Physics: Conference Series **1404** (2019) 012095 doi:10.1088/1742-6596/1404/1/012095

Quadcopter propeller characteristics in the oblique flow

P A Polivanov and A A Sidorenko

Khristianovich Institute of Theoretical and Applied Mechanics SB RAS, 4/1, Institutskaya str., Novosibirsk, 630090, Russia

polivanov@itam.nsc.ru

Abstract. The paper analyses the effect of the yaw angle and free-stream velocity on the stationary and unsteady characteristics of a propeller with a fixed geometry. Propeller characteristics are calculated in two ways: the first uses the 3D RANS modelling, the second involves the simplified analytical model. Comparison of the two calculation methods showed a qualitative coincidence of the data. The effect of the yaw angle on the propeller characteristics is investigated for a wide range of parameters. It is found that the dynamic loads generated by the propeller are significant and should be taken into account when designing quadcopter propellers.

1. Introduction

Modeling of flight dynamics is an important stage of the design of small unmanned aerial vehicles (UAV) [1]. This is important for the correct prediction of the aircraft response to internal control signals and external effects. Thrust characteristics of the propellers make an important part of the quadcopter mathematical model [2]. The flight of the quadcopter is carried out mainly in the oblique flow, when the propeller rotation axis and the external flow vector are at a significant angle. It changes the aerodynamic characteristics of the propeller as compared to the fixed-wing aircraft [3]. The helicopter rotor aerodynamics is also not suitable for a quadcopter mathematical model, since the quadcopter propellers do not have a cyclic blade control. Studies of the fixed-wing or convertiplane propellers at the yaw angles do not show all necessary ranges of characteristic for quadcopters [4]. All these factors mean the need to gather information about the flow around the rotating quadcopter propeller. Experimental data on the aerodynamic characteristics of the quadcopters are obtained to validate the numerical analysis in recent studies [5]. This paper numerically investigates the flow characteristics of a two-blade propeller 5040 (diameter 5 inches, figure 1) for various rotational speeds and external flow conditions.

2. Calculation setup

Two ways were used to study the propeller 5040characteristics. In the first one, the simulation was performed by the Finite Volume Method in the Ansys Fluent software. Calculations were made on the unstructured tetragonal grid with a prismatic mesh near the boundary layer. In the simulation, the RANS based on the Spalart–Allmaras turbulence model was used. The value of the $y+$ on the wall was approximately 15. The numerical domain size 20 times exceeded the propeller diameter. The computational mesh was approximately 5 million cells. The numerical simulation was performed at different propeller rotation speeds, different incoming flow velocities and different yaw angles θ (figure 1). Figure 2 shows the example of the pressure on the propeller surface found in RANS calculations.

The second way involved a simplified method assuming the local two-dimensionality of the flow near the blade airfoil. The linear approximation of the lifting force dependence on the angle of attack and the quadratic law for drag force was used in the airfoil local characteristics calculation, the separation flow being ignored. It permitted calculating the propeller lift and drag forces versus the radial coordinate and azimuthal angle β of the blade position.

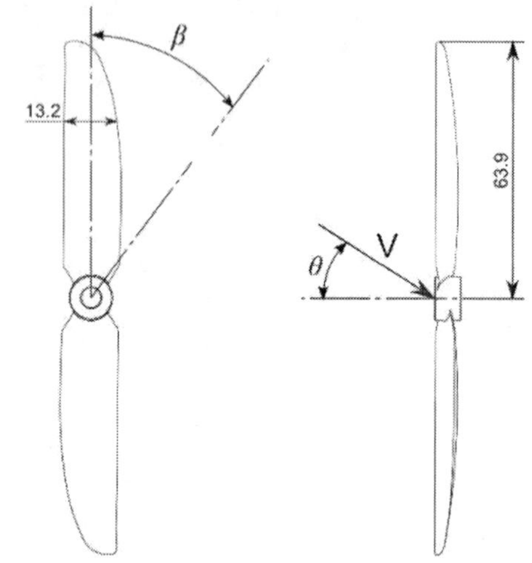

Figure 1. Sketch of propeller 5040.

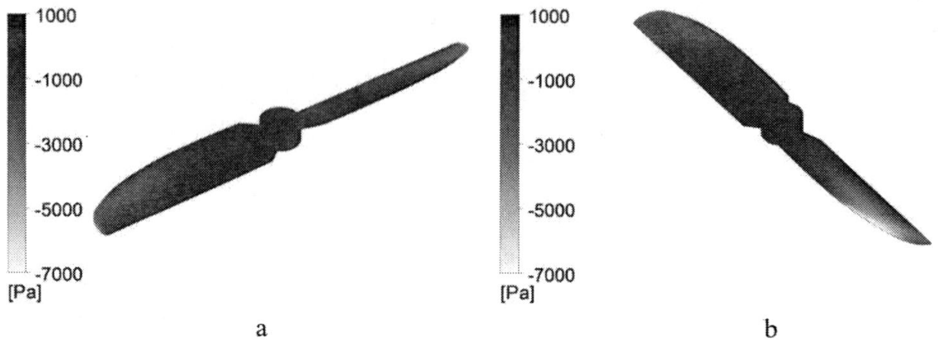

Figure 2. Pressure distribution on the propeller wall(a – θ= 0°, b – θ= 90°) at the incoming flow velocity of 10 m/s.

3. Calculation results

Figure 3 shows an example of RANS calculation. The figure presents the dependence of the force perpendicular to the propeller rotation plane (F_y) during one rotation cycle. At the zero angle θ, the flow parameters are independent on the azimuthal angle, therefore the force oscillations are zero. The increase of the angle θ goes along with the increasing average force and force oscillations. The average force increase results from the quadratic dependence of the lifting force on the incoming velocity. Therefore the force growth on the part of the propeller moving against the flow significantly exceeds the lift loss on the part of the propeller moving with the flow.

The rising oncoming flow velocity leads to the lift force drop due to the angle of attack drop on the blade airfoil. Figure 3b shows the moment arm M_x which reaches the values comparable to the propeller size. These data demonstrate that the lateral unsteady forces adversely affecting of the motor shaft occur on the quadcopter propellers.

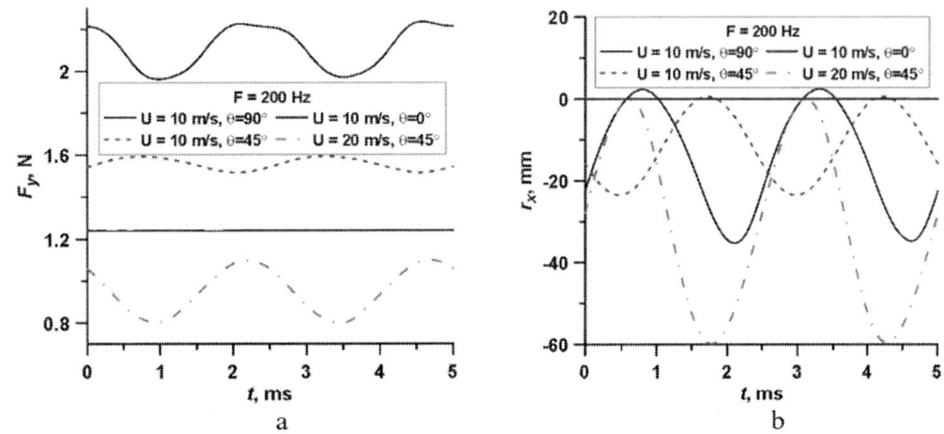

Figure 3. Propeller thrust (a) and longitudinal moment arm distance (b) vs time for different test cases.

Figure 4a shows the data similar to the ones in figure 4a but obtained using the simplified model. The qualitative data agreement is evident. Figure 4b compares the two calculation ways (RANS and the simplifiedone) for the forces F_y during the propeller rotation period at the frequency of 400 Hz and $\theta = 90°$). Good agreement between the results is clearly visible. For the simplified calculation method, the force oscillations approach to the sinusoidal law. The RANS calculation results demonstrate the non-sinusoidal force law or similar to the sinusoidal law, which depends on the separation effect on the blade.

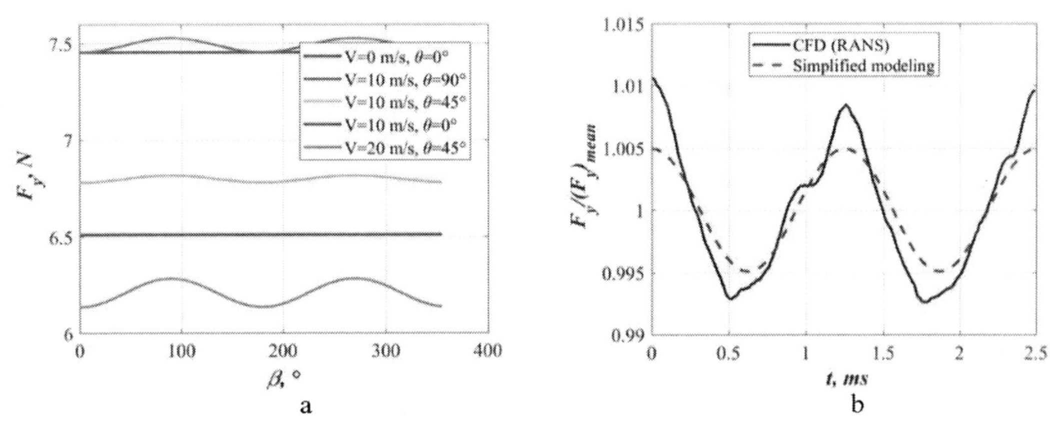

Figure 4. Distribution of propeller thrust during one rotation cycle (a – different test cases, b – comparison of CFD simulation with simplified model)

Figure 5 presents the comparison of the average force F_y and the root-mean-square (RMS) pulsation of the force F_y founded by two calculation methods for different flow regimes (1 – U = 10 m/s, θ = 0°; 2 – U = 10 m/s, θ = 90°; 3 – U = 10 m/s, θ = 45°; 4 – U = 20 m/s, θ = 0°; 1 – U = 20 m/s, θ = 45°). These data show that the mean force difference does not exceed 5%. This suggests that the simplified model can be used for the analysis of the quadcopter flight dynamics. But the simplified model cannot be used to analyze the unsteady forces, since this model underestimates the pulsations level by approximately two times.

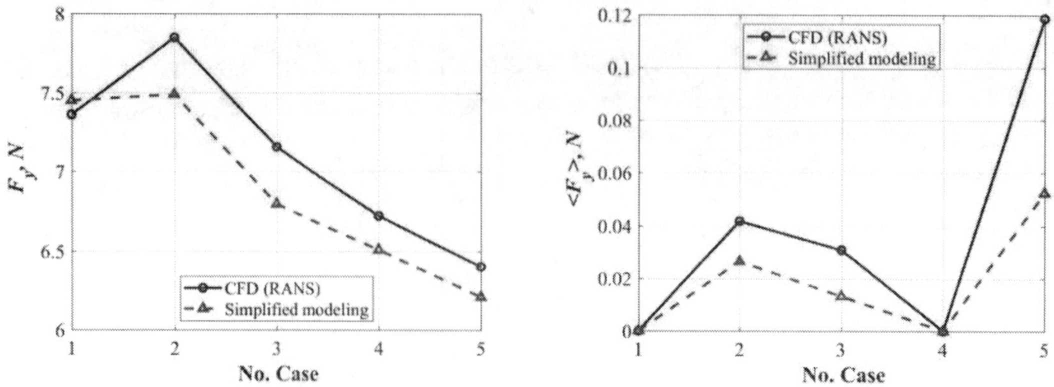

a b

Figure 5. a) Average thrust and b) pulsations of the thrust depending on the flow regime

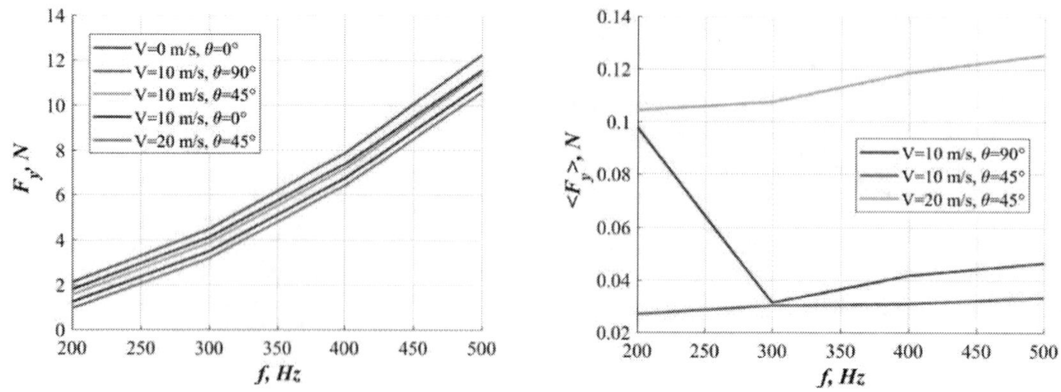

a b

Figure 6. Distribution of a) average propeller thrust and b) rms pulsation of propeller thrust vs frequency of rotation (CFD simulation)

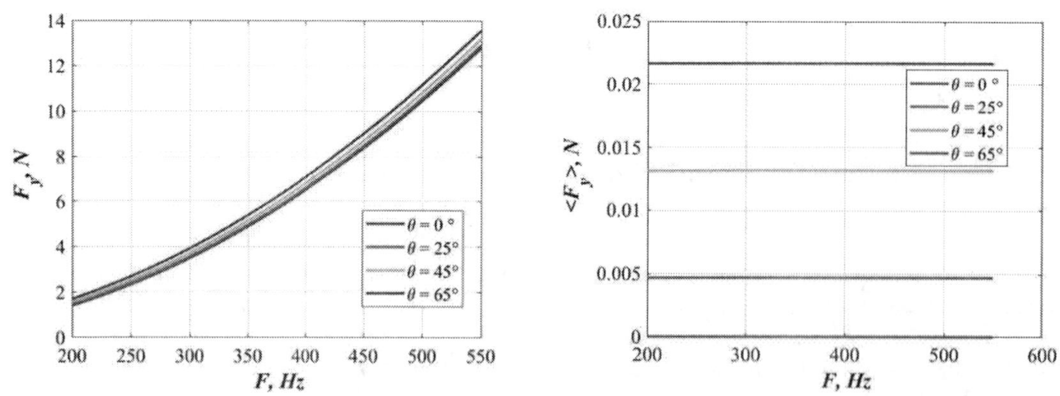

a b

Figure 7. Distribution of a) average propeller thrust and b) rms pulsation of propeller thrust vs frequency of rotation (Simplified model, U = 10 m/s)

Figures 6 and 7 show the force F_y and the RMS pulsations versus the propeller rotation speed. It is evident how the absolute effect of the incoming flow velocity and angle θ on the mean force F_y changes slightly with the increasing frequency. But the mean force grows along with the increasing frequency, therefore the relative effect of the flow parameters weakens.

The analysis of pulsations for the simplified model showed a weak frequency effect on the RMS force pulsation (figure 7b). Most of the data obtained by the RANS simulation showed the similar dependence, but the weak pulsation growth along with the frequency was founded. For the case $U = 10$ m/s, $\theta = 90°$, $F = 200$ Hz, the dramatic pulsation growth was detected as a result of the strong separated flow occurrence.

4. Conclusion

The aerodynamic characteristics of the quadcopter propeller 5040, as well as the values of forces and moments at the external oblique flow, were numerically obtained. It is shown that, during the quadcopter horizontal flight, the significant pulsations of forces and moments occur on the propellers. Unsteady forces and moments can lead to the significant reduction in the lifetime of the propulsion system.

Acknowledgments

The reported study was funded by RFBR and Novosibirsk Oblast according to the research project No 19-48-540017 r_a.

References

[1] Selig M S 2010 Modeling Full-Envelope Aerodynamics of Small UAVs in Realtime *AIAA Paper* 2010-*7635*

[2] Polivanov P A and Sidorenko A A 2018 *AIP Conference Proceedings* **2027** 040071

[3] Molter C and Cheng P W 2017 *International Micro Air Vehicle Conference and Flight Competition (IMAV)*

[4] Golovkin M A and Tarasov N N 2009 *TsAGI Science Journal* **40** 437–52

[5] Russell C, Willink G, Theodore C, Jung J, Glasner B 2018 Wind Tunnel and Hover Performance Test Results for Multicopter UAS Vehicles *NASA/TM* 2018–219758

Journal of Physics: Conference Series **1404** (2019) 012096 doi:10.1088/1742-6596/1404/1/012096

Studying of the resonator depth influence on amplitude-frequency characteristics operating flow in the two-channel system

A V Primakov[1] and A A Zhilin[1,2]

[1]Khristianovich Institute of Theoretical and Applied Mechanics SB RAS, 4/1 Institutskaya st., Novosibirsk, 630090 Russia
[2]Siberian State University of Water Transport, 33 Shetinkina st., Novosibirsk, 630099 Russia

lab20@itam.nsc.ru

Abstract. This paper is devoted to an experimental study of the characteristics of the operating flow in a two-channel system of an acoustic-convective dryer (ACD), developed at the Institute of Theoretical and Applied Mechanics (ITAM) SB RAS. A parametric study of cylindrical resonator depth and pressure in settling chamber influence on amplitude-frequency characteristics (AFC) of the formed flow is carried out. The influence of pressure in settling chamber and geometry of resonant cavity on AFC are considered. A tendency toward a decrease in Hartmann effect existence region with a decrease in cavity depth was found. A satisfactory agreement was obtained between the frequencies registered in the experiments and calculated using the Helmholtz formula for the natural frequency of resonant cavity.

1. Introduction

The idea of using high-intensity sound in drying of porous materials has led to development of acoustic-convective dryer (ACD) at the Institute of Theoretical and Applied Mechanics (ITAM) SB RAS. This device operating is based on the principles of a Hartmann-type gas-jet sound emitter operation (Hartmann whistle). Recently, increased interest in such applications has been shown in various applied fields. The study of physical processes occurring in the ACD two-channel system under the influence of a sound generator has a high priority in the development of acoustic drying technology, and also entails a number of fundamental discoveries related to the properties of deep resonators.

The use of ACD for drying porous materials allows one to avoid thermal effects on the drained material [1], which is a significant advantage for most materials: biological [2, 3], chemical [4–6], pharmaceutical and food [7–9]. An important advantage of the proposed drying technology is its higher intensity in comparison with the thermo-convective method [4]. At the same time, the structure of original material is not destroyed and its useful properties are preserved in the finished product. The relatively low energy consumption and simplicity of design makes the ACD attractive in economic terms [10].

A series of experimental [11] and numerical [12–15] studies is devoted to understanding the nature of the mechanisms occurring in the ACD system. The effects discovered during these studies require further research, including a series of physical and computational experiments.

2. Experimental setup

The experiments were carried out at ACD, the schematic diagram of which is shown in figure 1. The main elements of the setup are a settling chamber (1) subsonic narrowing conical nozzle (2) and a coaxially mounted resonant cavity of cylindrical shape (3). Elements (2) and (3) constitute a single structural element of the system — the primary channel. The nozzle and resonator are placed inside the secondary square section channel (4). The design described above forms the ACD two-channel system. Using a adjusting piston (5), the depth of the resonating cavity *l* was varied, and the maximum value of the depth is 31.5 caliber of the resonator. In this experiment, the cavity depth varied in the range from 23 to 31.5 with a constant nozzle diameter (0.5). The static pressure in the settling chamber varied in the range of 0-10 atm. The operating principle of the ACD consists in the periodic change of two processes: 1) filling an underexpanded supersonic jet from the nozzle into the resonating cavity, 2) outflow from the resonator of the jet reflected off the piston [16]. In the interaction of the jets flowing from the nozzle and from the resonate cavity, a seal region is generated, which is a source of sound fluctuations [17]. The formed acoustic-convective flow enters the operating section (4) and is recorded using the LH-610 highly sensitive piezoelectric sensor (6). The uncertainty in measurement of acoustic pressure is within ± 3%.

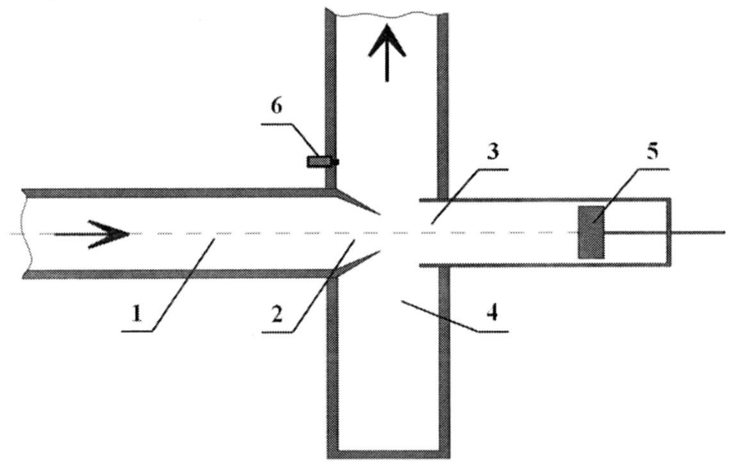

Figure 1. Schematic diagram of ACD.

3. The effect of pressure in the settling chamber on the AFC of the working flow

A series of experiments on recording the amplitude-frequency characteristics of the flow when pressure in the settling chamber changes, showed a number of patterns presented in figure 2. Firstly, depending on pressure in settling chamber, three formation stages of a high-intensity acoustic signal can be distinguished: generation and increase in the intensity of acoustic flow at low pressures up to 4 atm (figure 2.a); steady high-intensity mode of a operating stream (figure 2.b); attenuation of acoustic fluctuations in operating flow at high pressures of more than 9 atm (figure 2.c). Secondly, at low and high pressures, first harmonic dominates in spectrum, and second harmonic prevails in the interval between them. Of most interest is the high-intensity sounding mode, since nonlinear physical processes are realized in this mode.

At the second stage, at the high-intensity operation mode of the ACD, characteristic maxima in the AFC are clearly distinguished, which are the harmonic sequence of the main predicted frequency. The dominant second harmonic has a frequency of 250 Hz and an amplitude of 130 to 140 dB, while the amplitude of the first harmonic at 125 Hz is from 127 to 132 dB. Peaks in the spectrum with a frequency of 375, 500, 625, 750, 875, 1000 Hz are visible in figure 2b, which corresponds to 3–8 harmonics, but their intensity is much lower. The amplitudes of the 5th and 7th harmonics are low and almost comparable to background noise.

 1404 (2019) 012096 doi:10.1088/1742-6596/1404/1/012096

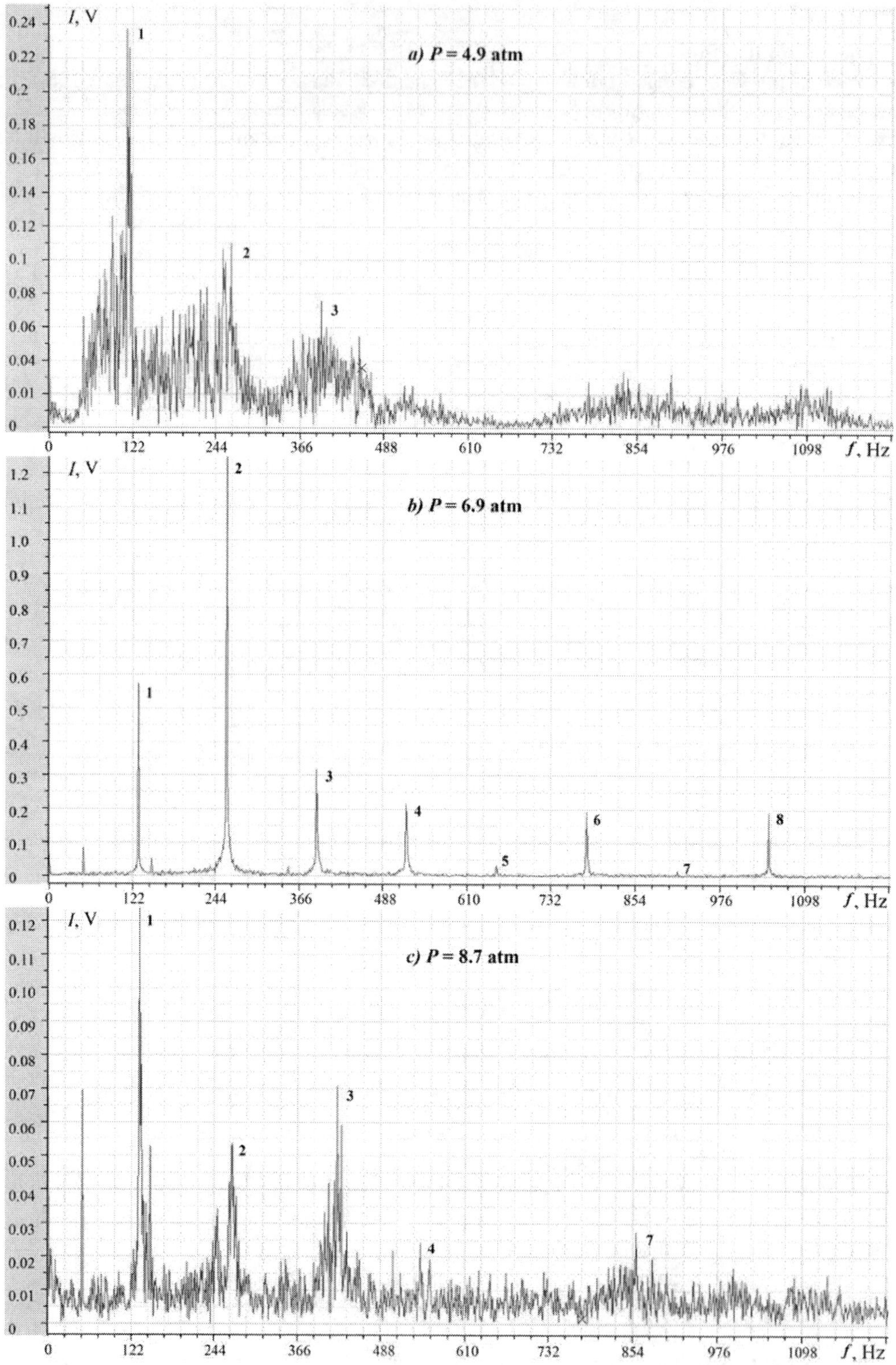

Figure 2. Characteristic acoustic emission spectra of deep resonators at a) 4.9; b) 6.9; c) 8.7 atm. In the settling chamber.

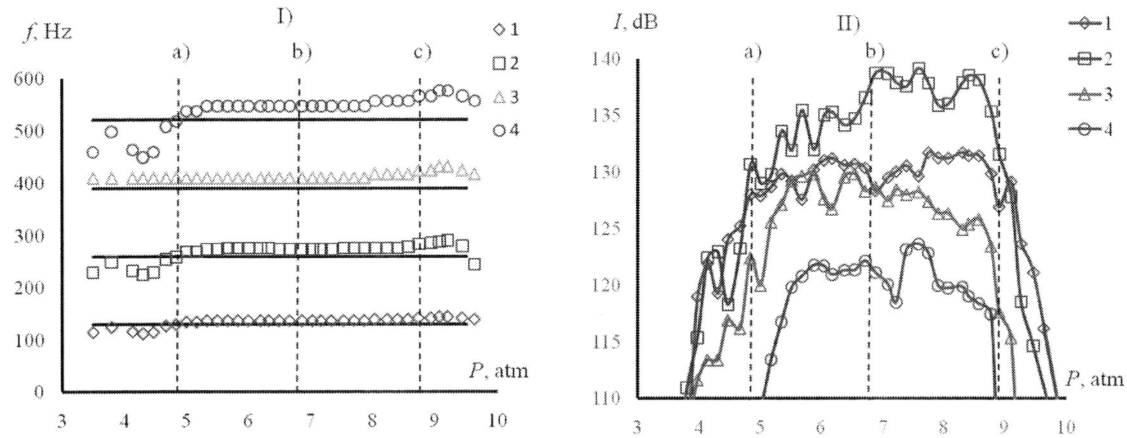

Figure 3. Comparison of the frequency (I) and intensity (II) of characteristic acoustic emission harmonics at different pressures in the settling chamber (black lines are the calculated frequency values of the corresponding harmonics).

The transition to the second harmonic is accompanied by a sharp increase in the frequency and intensity of all harmonics and persists until the end of the high-intensity operation mode, with a sharp decrease in all tones (figure 3). There is a slight increase in the oscillation frequency with increasing pressure in settling chamber (figure 3.I). At high pressure for 2-3 harmonics, the frequency rises before attenuation. Consider the change in intensity on the example of 3 and 4 harmonics, it can be seen that amplitude of high-frequency harmonics is much smaller than first and second in entire range of existence of the acoustic mode. The variation of harmonic intensity depending on pressure in settling chamber is nonmonotonic. In the interval between a) and c) the value of the fundamental harmonic goes almost constant, while the dynamics of the second, third, and fourth harmonics has a nonmonotonic character with a maximum.

4. The effect of cavity depth on AFC

In this experiment, the effect of resonator depth on frequency and intensity of generated acoustic-convective flow was studied. An analysis of the data showed that a decrease in cavity depth l leads to a shift in the fundamental frequency upward. In particular, maximum intensity is realized for second harmonic, at $l = 31.5$ and a frequency of 245 Hz, and for $l = 23.5$ at 310 Hz (figure 4.I). It is noted that the depth of resonator affects the intensity of generated sound. Larger depths correspond to greater intensity: as the depth of the resonant cavity decreases from 31.5 to 23.5, the intensity increases by 10 dB (figure 4.II).

High-intensity sound is observed at certain values of static pressure in settling chamber (4 - 10 atm.), called the region of existence of the Hartmann effect [18]. In this experiment, a tendency is observed to narrow the region of existence of acoustic emission with decreasing cavity depth. Thus, for the deepest cavities, reaching the mode of high-intensity oscillations (transition to the second harmonic) is observed in the pressure range from 4.4 to 9.7 atm, when high-intensity tones are observed for resonators with a smaller depth at 5.8 - 8.3 atm (figure 4.I).

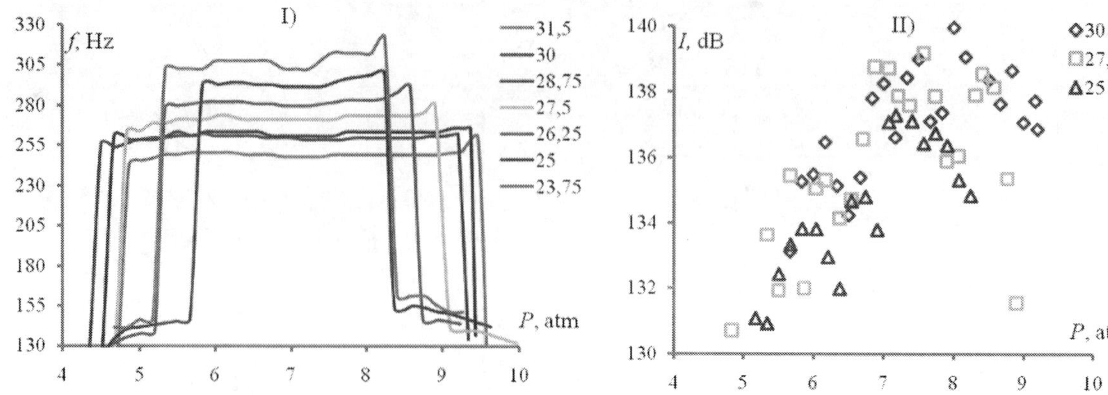

Figure 4. Comparison of the frequency (I) and intensity (II) of acoustic emission harmonics for different cavity depth values l.

5. Verification

To verify the obtained experimental data, the basic frequency formula of the Helmholtz cylindrical resonator was used $f = c/[4(l-0.3d_r)]$, where c is the speed of sound of the working gas. The frequencies determined by this formula are in good agreement with the experimental data, and their change depending on the length of the resonator is shown in figure 5. Both the experimental data and the calculated values monotonically increase with decreasing resonator depth. The increase in the discrepancy between the curves is clearly noticeable, which was also observed in [11].

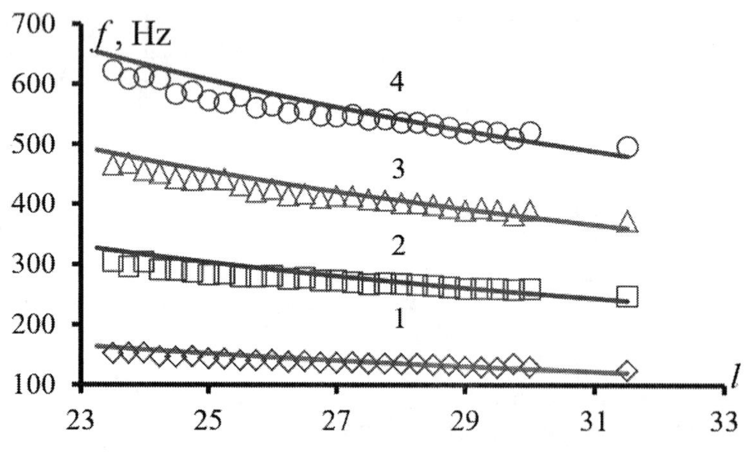

Figure 5. Comparison of the natural frequencies obtained in the experiment (points) with the calculated (lines)/

6. Conclusion

As a result of an experimental study of deep resonators of an acoustic-convective dryer, it was shown that the generated acoustic flow has a series of harmonics with a dominant second harmonic at a frequency of 250 Hz, and its evolution with increasing pressure in the settling chamber of ACD is nonmonotonic.

The dependence of flow amplitude-frequency characteristics on cavity depth was established. It was found that with decreasing cavity depth, the frequency of flow increases and the intensity decreases. It has been experimentally shown that a decrease in cavity depth leads to a narrowing of the Hartmann effect existence region.

Acknowledgments
The research was carried out within the framework of the Program of Fundamental Scientific Research of the state academies of sciences in 2013-2020 (project No. AAAA-A17-117030610139-4).

References
[1] Gosteev Yu A, Korobeinikov Yu G, Fedorov A V and Fomin V M 2005 *J. Appl. Mekh. Tech. Phys.* **46** 711–6
[2] Zhilin A A, Fedorov A V and *Grebenshchikov* D M 2018 *Foods and Raw Materials* **6** 370 – 8
[3] Zhilin A A 2019 *AIP Conf. Proc.* **2125** 030085
[4] Korobeinikov Yu G, Fedorov A V, Buluchevskii E A and Lavrenov A V 2009 *J. Eng. Phys. Thermophys.* **82** 246–50
[5] Fedorov A V, Zhilin A A and Korobeinikov Yu G 2011 **84** 965–74
[6] Zhilin A A and Fedorov A V 2017 *J. Eng. Phys. Thermophys.* **90** 1412–26
[7] Zhilin A A and Fedorov A V 2016 *J. Eng. Phys. Thermophys.* **89** 323–33
[8] Yu G Korobeinikov, G V Trubacheev, Fedorov A V, Chu K M, Zheong D M and Kim Yu I 2008 *J. Eng. Phys. Thermophys.* **81** 676–9
[9] Zhilin A A and Fedorov A V 2014 *J. Eng. Phys. Thermophys.* **87** 908–16
[10] Korobeynikov Yu G, Nazarov A A and Fedorov A 2004 Energy Costs for Drying Wood by Acoustic Method *Woodworking Industry* No 4 6–7
[11] Zhilin A A and Golubev E A 2018 *AIP Conf Proc* **1939** 020016
[12] Fedorov A V, Fedorchenko I A, An S B, Lee J H and Choo K M 2010 *J. Eng. Phys. Thermophys.* **83** 72–82
[13] Fedorchenko I A and Fedorov A V 2013 *J. Eng. Phys. Thermophys.* **86** 731–4
[14] Kravchenko A S, Zhilin A A and Fedorova N N 2018 AIP *Conf. Proc.* **1939** 020018
[15] Kravchenko A S and Zhilin A A 2019 *J. Phys.: Conf. Ser.* **1268** 012037
[16] Borisov Yu G 1967 Gas-jet sound emitters of Hartmann type Density matrix theory of coherent ultrafast dynamics *Physics and Technology of Powerful Ultrasound* ed L D Rozenberg, book 1 (Moscow: Nauka Publ.) pp 7–110
[17] V P Kurkin 1961 *Akust. Zh.* **7** 442–5
[18] Glaznev V N and Korobeinikov Yu G 2001 *J. Appl. Mekh. Tech. Phys.* **42** 616–20

Experimental study of the laminar-turbulent transition on models of wings with subsonic and supersonic leading edge at M = 2

N V Semionov[1], Yu G Yermolaev[1,2], A D Kosinov[1,2], V L Kocharin[1], A V Panina[1], A N Semenov[1,3], S A Shipul[1,2] and A A Yatskikh[1,2]

[1]Khristianovich Institute of Theoretical and Applied Mechanics SB RAS,
4/1, Institutskaya str., Novosibirsk, 630090, Russia
[2]Novosibirsk State University,
1, Pirogova str., Novosibirsk, 630090, Russia
[3]Institute of Geology and Mineralogy, SB RAS,
3, Ac. Koptyuga ave., Novosibirsk, 630090, Russia

semion@itam.nsc.ru

Abstract. The paper is devoted to an experimental study of laminar-turbulent transition in supersonic boundary layers on swept wings at Mach 2. The experiments were fulfilled in the low nose supersonic wind tunnel T-325 of Khristianovich Institute of Theoretical and Applied Mechanics on the models with supersonic or subsonic leading edges. The transition location was determined using a constant temperature hot-wire anemometer. while experiments for the case of a supersonic leading edge are a continuation of previous studies, then for the case of a subsonic leading edge, the first data have been obtained that show the possibility of carrying out studies on the development of unstable perturbations. It was obtained that transition take place parallel to the leading edge for both models. The destabilizing effect of the surface curvature of the model on the position of the laminar-turbulent transition for the case of a supersonic leading edge was confirmed. This experimental data is the room for further investigation.

1. Introduction

The study of the laminar-turbulent transition on a swept wing at supersonic flow rates is of practical interest for the development of high-speed aircraft.

There are only a few experiments in which the position of a laminar-turbulent transition was measured on three-dimensional swept wing models in supersonic flow. Supersonic flow around a swept wing is classified according to the Mach number along the normal to the leading edge. The leading edge is called subsonic if Mach number along the normal is $M < 1$; sound if $M = 1$, and supersonic if $M > 1$. Experiments on delta wing with 74° swept angle were fulfilled at Mach number range from 2.8 to 5.3 in the ballistic facilities [1]. So experiments were made for the cases of subsonic and supersonic leading edge (depend from Mach number). The effects of Mach number, temperature ratio, unit Reynolds number, leading-edge diameter, and angle of attack were investigated. The transition was measured by the optical method and so with low accuracy. It was obtained that for the subsonic leading edge $Re_{tr}=3\times10^6$ at M=2.78, $Re_{tr}=3.3\times10^6$ at M=3.35, for the supersonic leading edge $Re_{tr}=4.8\times10^6$ at M=4.1 and $Re_{tr}\approx(3\div4)\times10^6$ at M=5.3. It is noted in [1] that a high error did not allow to reveal the dependence of Re_{tr} in the range of angles of attack from -3° to 4°. At the same time the obtained values of Re_{tr} in the wind tunnel experiments [2] were approximately 3 times smaller than in [1] in the range of Mach numbers from 2 to 4. The effect of swept angle on laminar-turbulent transition in supersonic boundary layers on flat plate was studied in [2]. It was obtained that at all supersonic Mach numbers, increases in sweep angle reduced the transition Reynolds number. In [3], the results of measurements on a model of a delta wing with a swept angle χ=77.1° (subsonic leading edge) at M=3.5 are presented. It was noted that the transition was observed parallel to the lateral edge of the model and depend from unit Reynold number.

Transition measurements using hot-film sensors and an infrared camera were performed in [4] on a wing model with a subsonic leading edge at M=2 (sweep angles of the inner and outer parts of the wing were 66° and 61.2°, respectively). An effect of angle of attack was studied in two sections along the wing span in the range of angles from −1.5° to 5.5°. It was obtained monotonic increase of transition Reynolds number $Re_{tr}\approx0,6\times10^6$ to $Re_{tr}\approx1,3\times10^6$ with increasing angle of attack at 70% semispanwise probe station. Contrariwise at the 30% semispanwise probe station the longest laminar region was detected at α=1.7°, and a slight deviation in the angle of attack from 1.7° moved the transition location significantly upstream. But at zero angle of attack transitional Reynolds number was approximately 0.8×10^6 in both cases. It was shown that unit Reynolds number had no effect on the transition location [4].

The results of flight measurements of Re_{tr} on a swept wing at M=2 are presented in [5]. The leading edge of the wing had the same two swept angles as in [4]. The Reynolds numbers of the transition are not given in the work, and from the presented data we can only say that the transition occurs near the leading edge.

The influence of flow parameters on the position of the laminar-turbulent transition, such as the Mach number [6], unit Reynolds number [7], the influence of external perturbations [8] and small angles of attack [9], was studied at the Institute of Theoretical and Applied Mechanics of the Siberian Branch of the Russian Academy of Sciences on the model of a wing with a supersonic leading edge.

It was shown that with an increase in the Mach number, Re_{tr} decreases [6], with an increase in the unit Reynolds number, the Reynolds number of the transition also increases [7], and with an increase in the angle of attack, Re_{tr} monotonically increases [9]. At a high noise level caused by an increase in perturbations, the effect of flow parameters on the laminar-turbulent transition is practically not fixed [8].

In the course of studies on the influence of various flow parameters on the laminar-turbulent transition in the supersonic boundary layer of the swept wing, it was revealed that in all these cases, Re_{tr} varies from 0.7×10^6 to 2.4×10^6. The higher the noise level, the more Re_{tr} tends to a value of 0.7×10^6. This paper is a continuation of our experimental research on the wing model with supersonic leading edge [6-9] and represents the direction of future studies.

2. Setting up experiments

The experiments were performed in a low-noise supersonic wind tunnel T-325 ITAM SB RAS at Mach number M=2. The level of disturbances in the test section of the wind tunnel does not exceed 0.2% [10]. First model is a symmetrical wing with a 45° sweep angle, a 3 percent-thick circular-arc airfoil. The model length is 0.4 m, its width is 0.2 m, and the maximum thickness is 12 mm. Scheme of the model is presented in [11]. Photo of another model with swept angle χ=72° and lenticular profile is presented in figure 1. Swept angles χ=45° and χ=72° corresponds to a supersonic and subsonic leading edge at M=2. The experiments were performed at zero angle of attack. Perturbations in the flow were recorded by a constant-temperature hot-wire anemometer (CTA). The experimental data were obtained in two ways. First, a change in the position of the coordinate with a hot-wire anemometer along the model (the x coordinate changes) with a constant mode of operation of the wind tunnel (Re_1=const). The second method is to change the operating mode of the wind tunnel when the hot-wire probe is installed in a certain section (x=const, Re_1 changes). These methods are interchangeable, if a complex geometry model is used; method two is more convenient for obtaining data.

The flow characteristics were measured using an automated data acquisition system [12]. Using an Agilent 34401A digital voltmeter, the DC component of the voltage was measured from the output of the hot-wire anemometer E. The ripple signal in the diagonal of the hot-wire anemometer bridge was digitized by a 12-bit analog-to-digital converter (ADC) and then recorded into a computer. The experimental data were processed using the fast Fourier transform, and the power spectra were determined from complete waveforms.

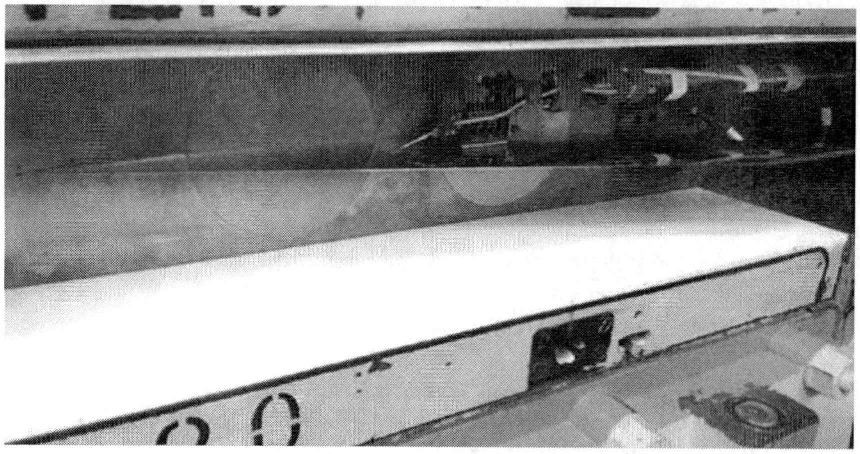

Figure 1. Photo of wing model with χ=72° in the test section of T-325.

3. Results

As noted above, studies of the influence of flow parameters on the development of disturbances and on the position of the laminar-turbulent transition were performed on a wing model with a supersonic leading edge [6 - 9, 11]. But some questions remained open, for example, how does the transition occur along the wingspan.

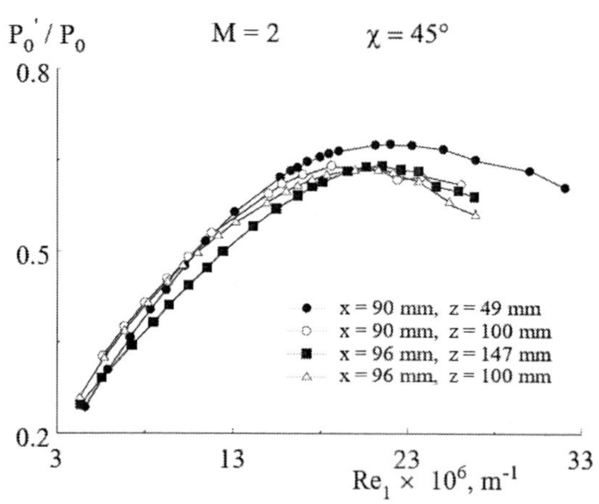

Figure 2. Transition measurement on a swept wing with a supersonic leading edge.

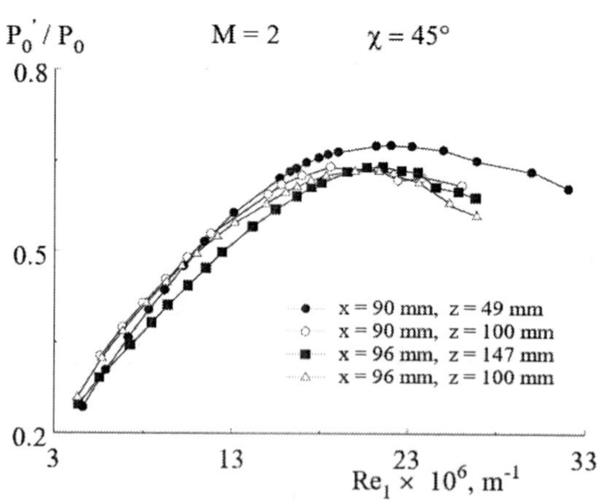

Figure 3. Transition measurement on a swept wing with a subsonic leading edge.

The measurement of the transition position on the wing with a swept angle of 45° is presented in figure 2. These measurements were carried out using a Pitot tube. The Pitot probe were installed in a fixed position along two coordinates x and z, which are indicated on the graph, and during the experiment, the unit Reynolds number (Re_1) changed. The measured dependences of the normalized pressure are close in form, and the obtained estimates of the transition Reynolds number have a value of $(2 \pm 0.1)\times10^6$.

The results of measuring the position of the laminar-turbulent transition on a swept wing with a subsonic leading edge (the sweep angle is 72°) are shown in figure 3. Measurements on this model were carried out both by a hot-wire anemometer sensor and Pitot probe. Measurements with a hot-wire anemometer were made only at one point: x=110 mm, z=-35 mm. The hot-wire anemometer sensor was installed in the layer just below the ripple maximum along the vertical coordinate y, and measurements were made with a variation of the unit Reynolds number. Hot-wire anemometer transition measurement data show that the emergence of turbulence on a swept wing with a subsonic leading edge occurs at $(1.5\pm0.1)\times10^6$ Re_{tr}. Pitot probe measurements give close estimates of the transition position.

The dependence of the laminar-turbulent transition on the radius of curvature of the surface is shown in figure 4. A flat plate has an infinite radius of curvature, and a profile of 7.8% is about 0.9 m. Transition measurements using a hot-wire anemometer were performed only for a model with a relative profile thickness of 3%. All other results were obtained using measurements with a Pitot probe. For a flat plate, the transition occurs at $Re_{tr}=1.5\times10^6$, for a model with a relative profile thickness of 3% $Re_{tr}=2.1\times10^6$, for 5% - $Re_{tr}=1\times10^6$, for 7.8% - $Re_{tr}=0.9\times10^6$. Measurements were taken at different points. We can conclude that with an increase in the relative thickness of the profile, the unit Reynolds number decreases monotonically. In order to confirm or refute this conclusion, measurements are needed that were made at one point with different models of the swept wing.

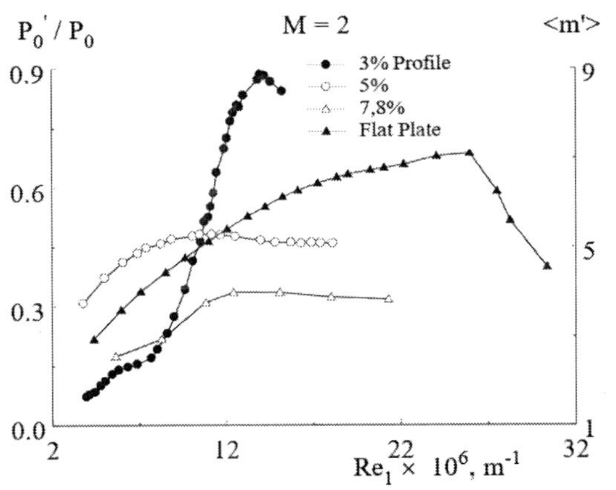

Figure 4. The effect of surface curvature of the model on the laminar-turbulent transition.

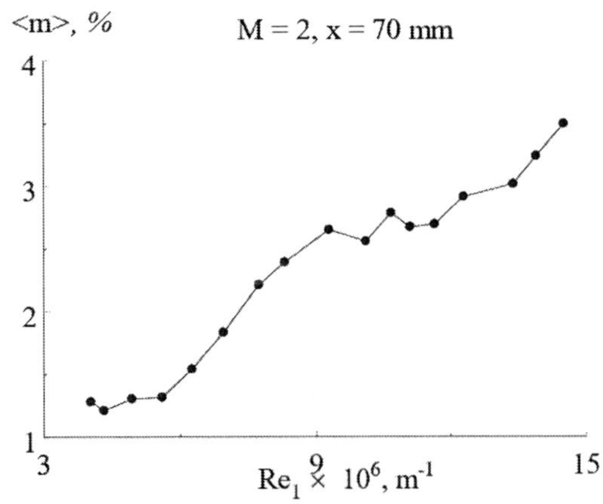

Figure 5. Growth curve of mass flow pulsations.

The curve of the increase in pulsations of the mass flow is shown in figure 5. The measurements were carried out at a distance of 70 mm from the leading edge of the wing with a sweep angle of 72°. The measurements were made by a hot-wire anemometer. The transition from laminar to turbulent flow was not achieved, since at an increased high-pressure head ($Re_1>15\times10^6$ m^{-1}) the thread of the hot-wire anemometer sensor was torn. Considering that in the model of a swept wing with a subsonic leading edge, it occurs at $Re_{tr}=(1.5\div1.6)\times10^6$ m^{-1}, then at x=70 mm the transition should occur at $Re_1=(22\div24)\times10^6$ m^{-1}. However, from these data we can conclude that the onset of turbulence occurs at $Re_1>5.6\times10^6$ m^{-1}, which is slightly less than in the case of transition measurements on the same model at x=110 mm (see figure 3).

It was obtained that transition take place parallel to the leading edge for wings with subsonic and supersonic leading edge. The destabilizing effect of the surface curvature of the model on the position of the laminar-turbulent transition for the case of a supersonic leading edge was confirmed. This experimental data is the room for further investigation.

Acknowledgments
The study was conducted at the Joint Access Centre "Mechanics" of Khristianovich Institute of Theoretical and Applied Mechanics SB RAS. The work is supported by Russian Foundation for Basic Research (grant number 17-19-01289).

References
[1] Chapman G T 1961 Transition of the laminar boundary layer on a delta wing with 74 deg. sweep in free flights at Mach numbers from 2.8 to 5.3 *NASA TN* D-1066
[2] Jillie D W and Hopkins E J 1961 Effects of Mach number, leading-edge bluntness, and swept on boundary layer transition on a flat plate *NASA TN* D-1071
[3] Cattafesta L N III, Iyer V, Masad J A, King R A and Dagenhart J R 1995 *AIAA J* **33** 2032
[4] Sugiura H, Yoshida K, Tokugawa N, Takagi S and Nishizawa A 2002 *J. of Aircraft* **39** 996
[5] Kwak D-Y, Tokugawa N and Yoshida K 2007 *WEHSFF conference* (Moscow)
[6] Semionov N V, Kosinov A D and Yermolaev Yu G 2011 *Journal of Physics: Conference Series* **318** 032018
[7] Ermolaev Y G, Kosinov A D, Semenov A N, Semionov N V and Yatskikh A A 2018 *Thermophysics and Aeromechanics* **25**, 659
[8] Dryasov A D, Yermolaev Yu G, Kosinov A D, Semionov N V and Semionov A N 2016 *Vestnik NSU. Physics* **11** 16 (in Russian)
[9] Yermolaev Yu G, Kosinov A D, Kosorygin V S, Semionov N V, Semionov A N, Smorodsky B V and Yatskikh A A 2017 *Siberian Physics J* **12** 35 DOI: 10.25205/2541-9447-2017-12-3-35-40 (in Russian)
[10] Kosinov A D, Semionov N V and Yermolaev Yu G 1999 Disturbances in test section of T-325 supersonic wind tunnel *Preprint* 6-99
[11] Kosinov A D abd Semionov N V 2019 *AIP Conference Proceedings* 2125 030105 DOI: 10.1063/1.5117487

On determining the vortex core center in wing wake at M = 2

A M Shevchenko, A S Shmakov and V V Kulikov

Khristianovich Institute of Theoretical and Applied Mechanics SB RAS, 4/1, Institutskaya str., Novosibirsk, 630090, Russia

shevch@itam.nsc.ru

Abstract. The paper presents the processing of the results of numerical simulation of the vortex flow in the wing wake, the purpose of which is to search for the vortex core center. It is based on the results of numerical simulation of the vortex flow in the wake behind of the rectangular wing, performed at the Keldysh Institute of Applied Mathematics RAS. The center of the vortex core was determined in several ways: by the minima of the total and static pressure, the axial Mach number, mass flow rate, the maximum of axial vorticity, and as the center of mass of the axial vorticity distribution. A comparison of the results of determining the position of the vortex core center showed good quantitative agreement. As a result dependencies of vortex core center coordinates on longitudinal coordinate were obtained.

1. Introduction

The study of vortex flows is one of the urgent tasks of modern aerodynamics. The relevance of investigation of free vortex wake caused by potential application study results for solving the following tasks: decreased visibility of supersonic aircraft; air traffic security; the effect of the vortex wake on ground facilities. It is well known that flowfield wing wake is accompanied by a pressure and axial velocity loss in the vortex core center, circumferential velocity maxima at vortex core boundary [1].

As a result of numerical modeling, we get a complete set of gas-dynamic parameters (P, ρ, U, V, W). In contrast to numerical simulation, a limited set of gas-dynamic parameters can be obtained in the experiment. It can be either data on pressure and Mach number, or data on mass flow rate. And these data are not enough to close the system of gas-dynamic equations. 5-hole pressure probe data [2, 3] and hot-wire anemometer data [7–10] were presented in earlier studies. Recently a possibility to extract integral density data from flow visualization results was showed [11–13].

To validate the results of numerical simulation, it is necessary to establish whether it is possible to use experimental mass flow rate data to determine the center of the vortex core. Thus, the main purpose of this work is to compare information on the position of the vortex core center, determined by various parameters obtained in the numerical simulation.

2. Results

As the source material for the search for the center of the vortex core, we used the results of wing wake numerical simulation performed at Keldysh Institute of Applied Mathematics RAS [4–6].

Rectangular half-wing with sharp edges with a chord of $b = 30$ mm and semi-span of $l = 95$ mm was used as a vortex wake generator. Numerical simulation performed at $M = 2$ and wing angle of attack of $\alpha = 10°$. In detail computational technique was reported in [6].

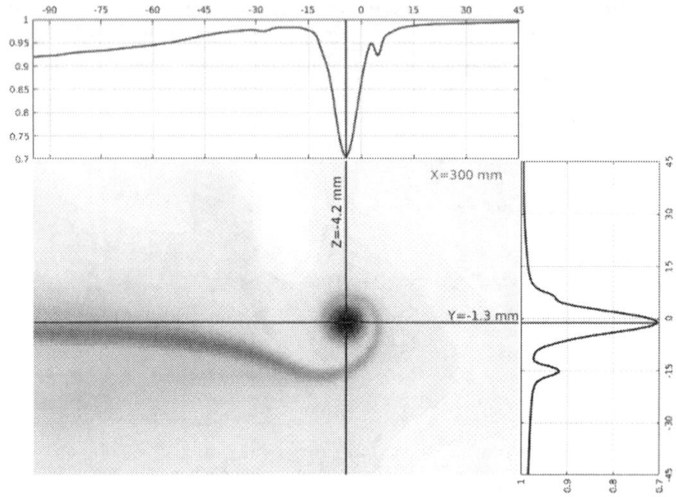

Figure 1. Mass flow rate field at $X/b = 10$.

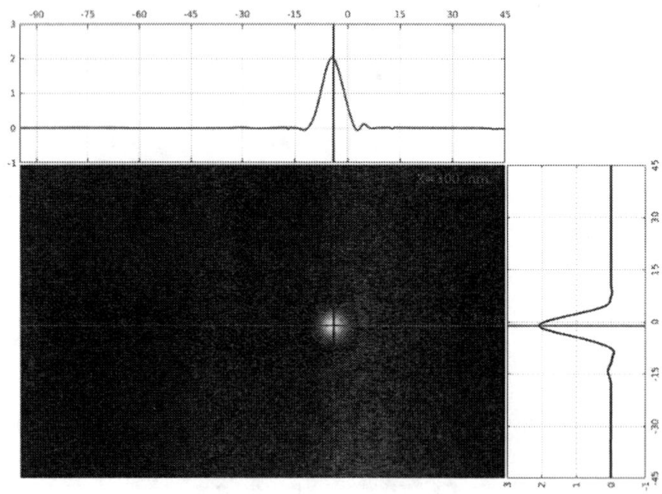

Figure 2. Axial vorticity distribution at $X/b = 10$.

The numerical data were approximated to a regular grid with a step of 0.1 mm in a ranges of $Y \in [-45, 45]$ mm and $Z \in [-95, 45]$ mm and in a wide range of longitudinal coordinate $X \in [15, 900]$ mm. As a result, for each cross section of $X = const$, a field of gas-dynamic parameters was obtained.

Figure 1 shows typical example of mass flow rate field with plots showing the distribution of this parameter across the center of the vortex core. The figure demonstrates a technique of vortex core center detection on a minima of mass flow rate.

Figure 2 shows typical example of axial vorticity field with plots showing the distribution of this parameter across the center of the vortex core. The figure demonstrates a technique of vortex core center detection on a maxima of axial vorticity.

Vortex core center coordinates were found in several ways:

- as coordinates of minima of static pressure (P), Pitot pressure (P_t), axial Mach number (M_X), mass flow rate (ρU),
- as coordinates of maxima of axial vorticity (Ω_X)

$$\Omega_X = \frac{\partial W}{\partial y} - \frac{\partial V}{\partial z}, \tag{1}$$

- as 'center of gravity' of axial vorticity (Ω_X) field according the following equations

$$Y_c = \frac{\int y\Omega_X dydz}{\int \Omega_X dydz}; \qquad Z_c = \frac{\int z\Omega_X dydz}{\int \Omega_X dydz}. \tag{2}$$

Figures 3 and 4 shows dependencies of vortex core coordinates on longitudinal coordinate. It is clearly seen quite good agreement in coordinates found by different ways.

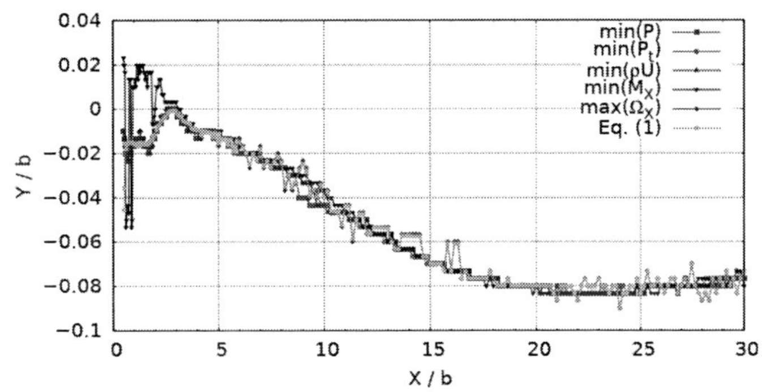

Figure 3. Y coordinate of vortex core center.

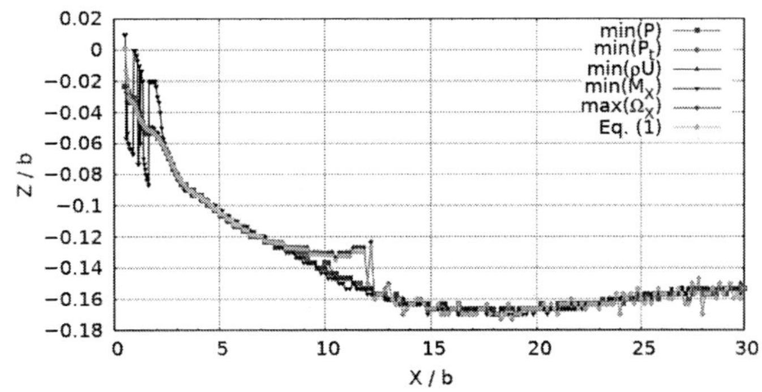

Figure 4. Z coordinate of vortex core center.

Errors in determining the center of the vortex core at small values of the longitudinal coordinate X/b are explained by the fact that the vortex core in the near sections has not yet been formed.

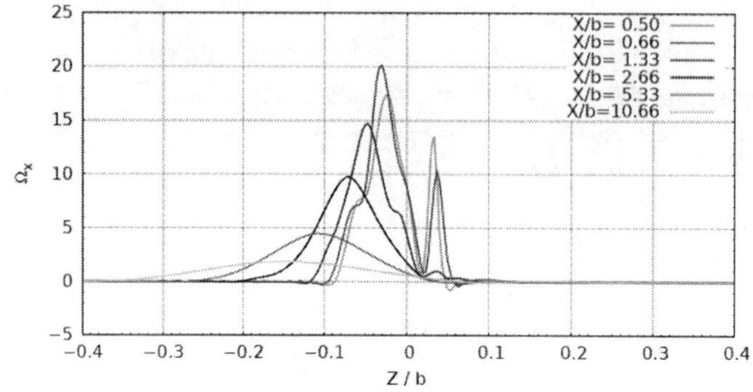

Figure 5. Longitudinal vorticity profile along Z axis

Figure 5 shows axial vorticity distribution across the vortex core at diferent distances far downstream from the wing. The plots clearly indicate two peaks of the same order in two nearest cross-sections ($X/b = 0.5$ and 0.66). Ω_X-distributions at $X/b > 1$ unlike the near wake demonstrate typical plots with one peak located at the center of the vortex core.

3. Conclusions

Based on the analysis of the results of numerical simulation and its comparison with experimental data the position of the center of the vortex core in the wing wake at M = 2 was determined. It was found that the values of the center coordinates determined by different values are close to each other. In the near sections, the vortex core is not formed, which confirmed by the presence of several peaks of the same order in distribution of longitudinal vorticity.

Acknowledgments

The work was carried out with the partial financial support of the Russian Foundation for Basic Research (Grant No. 19-01-00765).

References

[1] Brodetsky M D and Shevchenko A M 1998 *Thermophysics and Aeromechanics* **5** 281
[2] Shevchenko A, Kavun I, Pavlov A and Zapryagaev V 2005 *Proc. European Conf. for Aerospace Sci.* (Moscow, Russia, July, 4-7)
[3] Shevchenko A M, Lutsky A E, Chernoguzov A S and Polkova K Y 2007 *Proc. XIII Int. Conf. Methods Aerophysical Res.* (Novosibirsk: Parallel Publ.) pp 215–220
[4] Borisov V E, Davydov A A, Konstantinovskaya T V, Lutsky A E, Shevchenko A M and Shmakov A S 2018 *AIP Conference Proceedings* **2027** 030120
[5] Borisov V E, Konstantinovskaya T V, Davydov A A, Lutsky A E, Shevchenko A M and Shmakov A S 2018 *Journal of Physics: Conference Series* **1250** 012037
[6] Borisov V E, Davydov A A, Kostantinovskaya T V, Lutsky A E, Shevchenko A M and Shmakov A S 2018 Simulation of supersonic flow in the wake behind a wing at M=2-4 *KIAM Preprints* **50** pp 1-19
[7] Shmakov A S, Shevchenko A M, Yatskikh A A and Yermolaev Yu G 2016 *AIP Conference Proceedings* **1770** 030019
[8] Shmakov A S and Shevchenko A M 2017 *AIP Conference Proceedings* **1893** 030089
[9] Shmakov A S and Shevchenko A M 2018 *AIP Conference Proceedings* **2027** 030103
[10] Shmakov A S and Shevchenko A M 2019 *AIP Conference Proceedings* **2125** 030108
[11] Shevchenko A M, Golubev M P, Pavlov A A, Pavlov Al A, Khotyanovsky D V and Shmakov A S 2017 *Technical Physics Letters* **43** 473
[12] Pavlov Al A, Shevchenko A M, Shmakov A S and Pavlov A A 2017 *AIP Conference Proceedings* **1893** 030079
[13] Shmakov A S, Pavlov Al A, Golubev M P and Shevchenko A M 2018 *AIP Conference Proceedings* **2027** 040065

Experimental research of the supersonic core length in the supersonic jet flowing from the rectangular micro-nozzles with different aspect ratios

I V Timofeev and V M Aniskin

Khristianovich Institute of Theoretical and Applied Mechanics SB RAS, 4/1, Institutskaya str., Novosibirsk, 630090, Russia

ivan.timofeev.koi@gmail.com

Abstract. The supersonic core length of the air microjets flowing from rectangular micro-nozzles with different aspect ratios has been experimentally studied. The height and width of the nozzles varied from 119 to 135 μm and from 927 to 2185 μm, respectively.

1. Introduction

Subsonic and supersonic micro-jets can be used for the macro-flow control [1-5], to reduce the noise of macro-jets, as well as in jet cooling systems. There is evidence of attempts to use synthetic mini-jets (with zero integral flow rate) to suppress pulsations in laminar and turbulent flows. Supersonic micro-jets re promising to be used in the aerospace field to protect the surfaces streamlined by high-temperature flow and suppress plasma formations around a high-speed aircraft and lander spacecraft [6, 7]. Small-size supersonic jets have also been used in reactive satellite orientation systems with a mass below 10 kg, the so-called nanosatellites [8, 9]. One of the technological applications of jets is gas mixing and protection of surfaces from chemically aggressive or high-temperature media. The jet penetrating power and the intensity of the mixing processes are the main jet characteristics in this case.

At the expiration of the supersonic jet, a complex shock-wave structure forms (figure 1). The supersonic core length is the distance from the nozzle exit to the point on the jet axis where the flow rate becomes equal to the local sonic speed. In other words, it is the supersonic part of the jet.

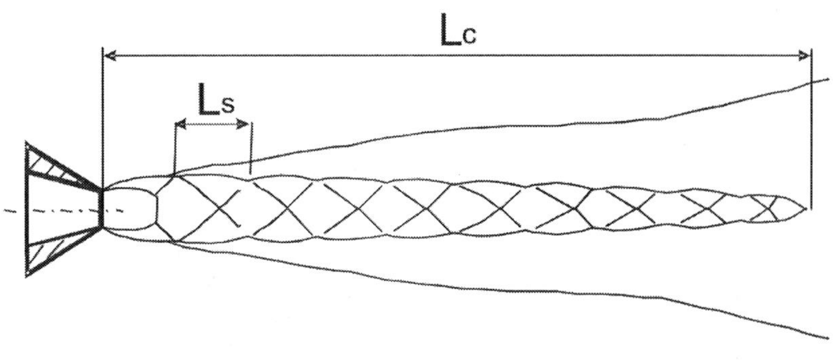

Figure 1. Scheme of the supersonic underexpanded jet. Ls is the shock sell size, Lc is the supersonic core length.

Previously, the gas-dynamic parameters of supersonic air micro-jets escaping from the circular and rectangular micro-nozzles with a large aspect ratio (AR> 40) were investigated. The dimensions of the first shock wave structure, the average shock sell size and the supersonic core length of such jets were determined. It has been shown that, if the supersonic microjet escapes from the nozzle below 50 μm, a high supersonic core length forms. The supersonic core length of the microjets is 6-8 times longer than the one for macrojets [11, 12]. The long supersonic core length from the microjets results from the laminar flow in the mixing layer.

The object of this study is the supersonic microjets escaping from the rectangular micron-sized nozzles with different aspect ratios. The nozzle parameters are shown in Table 1; the nozzle exits are shown in figure 2.

Table 1. The nozzle parameters

h, μm	w, μm	AR
121	927	7.7
119	1470	12.3
135	2185	16.2

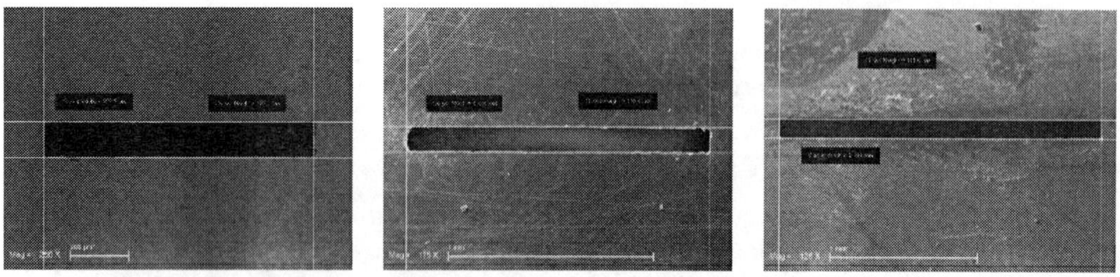

Figure 2. SEM images of the nozzles.

2. Experimental technique

The supersonic underexpanded microjets are studied by the Pitot tube. The total pressure distribution is measured along the jet axis line. Figure 3 presents the scheme of the experiment. The room-temperature air is used as the process fluid, and the supersonic air jet is escaping into atmosphere. The glass Pitot tube with the outer diameter of 40 μm and inner diameter of 20 μm is installed in the holder that can be moved in three mutually perpendicular directions with a Narishige NT 88E micromanipulator, and be positioned in space within an accuracy of ±1 μm. The Pitot tube position is monitored by the Nikon SMZ1500 stereoscopic microscope.

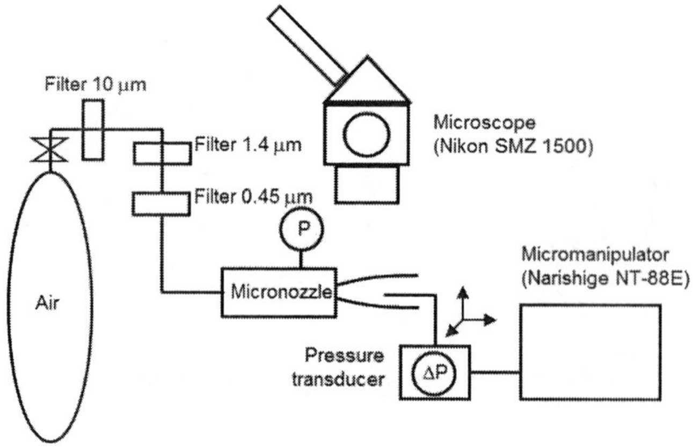

Figure 3. The scheme of the experiment.

The Pitot tube is located at a certain distance from the nozzle exit. The pressure in the nozzle pre-chamber rises gradually, and the pressure in the Pitot tube is measured. Figure 4 shows as an example the graph of the Pitot tube pressure versus the jet pressure ratio when the Pitot tube is located at a distance of 8 calibers from the nozzle exit. The red line corresponds to the pressure of 1.89 atm. According to isentropic formulas, this pressure corresponds to M=1. As a result, the jet pressure ratio is determined, which corresponds to the supersonic core length specified by the position of the Pitot tube.

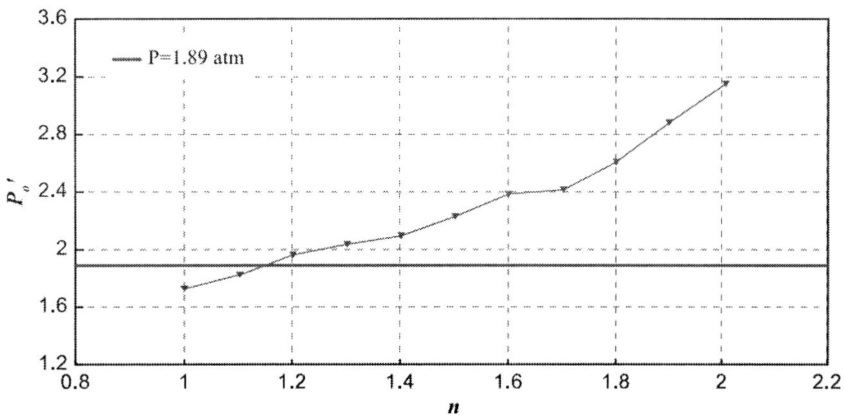

Figure 4. Graph of the Pitot tube pressure growth at the fixed distance with changing jet pressure ratio.

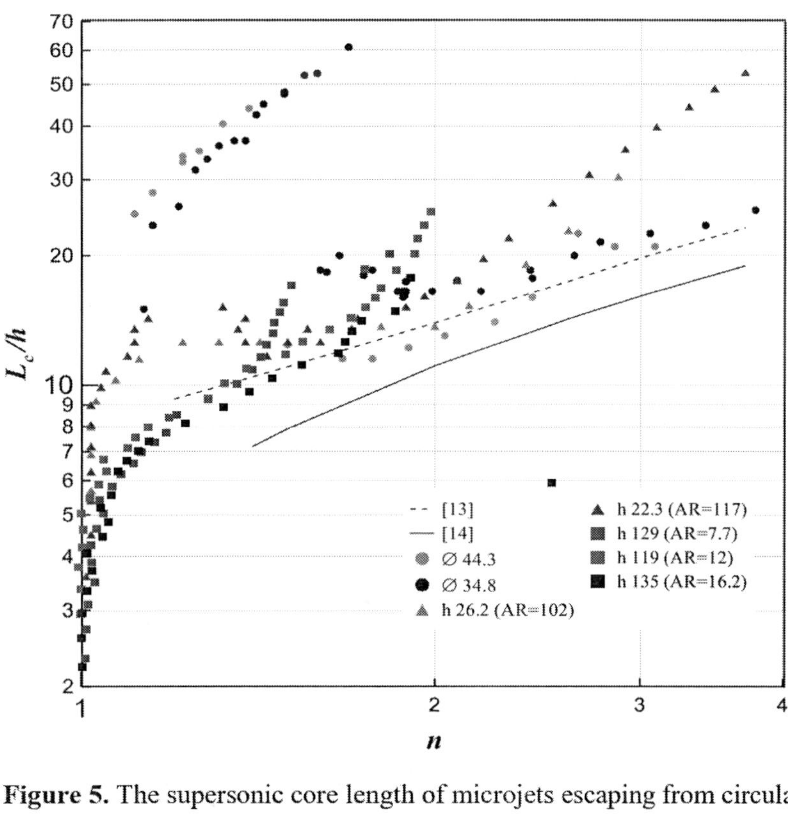

Figure 5. The supersonic core length of microjets escaping from circular and rectangular micro-nozzles with different aspect ratios.

Figure 5 shows the results of measurements of the supersonic core length.

It is evident from the graph that the supersonic core lenght of microjets escaping from rectangular micronozzles with AR≤16.2 does not differ significantly from the one of the jets escaping from flat nozzles with a large aspect ratio at low JPR values. However, later the difference is growing. The lower the aspect ratio, the earlier growth begins.

3. Results

The supersonic core length of microjets escaping from the rectangular nozzles generally coincides with the range of jets escaping from flat nozzles. At some jet pressure ratio, the supersonic core length of jets escaping from rectangular nozzles exceeds the range of the flat jets.

Acknowlegments
This work was supported by the Russian Science Foundation (Grant No. 17-19-01157).

References

[1] Tabeling P 2005 *Introduction to microfluids* (Oxford: Oxford University Press)
[2] Alvi F S, Shih C, Elavarasan R, Garg G and Krothapalli A 2003 *AIAA J.* **41** 1347-55
[3] Lou H, Alvi F S and Shih C. 2006 *AIAA J.* **44** 58-66.
[4] Choi J J, Annaswamy A M, Lou H and Alvi F S 2006 *Exp. Fluids* **41** 841-55
[5] Tanney J W 1970 *Prog. Aeron. Sci.* **10** 401-510
[6] Parmentier E M, Wray K L and Weiss R F 1970 *Proceedings the reentry plasma sheath and its effect on space vehicle electromagnetic systems* vol 1 (NASA Langley Research Center) pp 579-616
[7] Akey N D 1970 *Proceedings the reentry plasma sheath and its effect on space vehicle electromagnetic systems* vol 1 (NASA Langley Research Center) pp 25-6
[8] Bayt R and Breuer K 2001 *AIAA Paper* 2001-0721
[9] Zilić A, Hitt D L and Alexeenko A A 2007 *AIAA Paper* 2007-3984
[10] Kandlicar S G and Grande W J 2003 *Heat Transfer Engineering* **24** 3-17
[11] Aniskin V M, Mironov S G and Maslov A A 2013 *Microfluidics and Nanofluidics* **14** 605-14
[12] Aniskin V M., Mironov S G and Maslov A A. 2013 *Tech. Phys. Let.* **39** 734–6
[13] Shirie J W and Seubold J G 1967 *AIAA J.* **5** 2062-4.
[14] Pogorelov V I 1977 *Sov Tech. Phys. Let.* **47** 444-5

Low-frequency waves produced by a package of laser plasma clouds in a magnetized background

V N Tishchenko[1], A G Berezutsky[1], E L Boyarintsev[1], Yu P Zakharov[1], I B Miroshnichenko [1,2], V G Posukh[1], A G Ponomarenko[1], A A Chibranov[1,2] and I F Shaikhislamov[1]

[1]Institute of Laser Physics SB RAS, Akademika Lavrentieva Avenue, 15B, Novosibirsk, Russia
[2]Novosibirsk State Technical University, Karl Marx Avenue, 20, Novosibirsk, Russia

E-mail: tvn25@ngs.ru

Abstract. It was shown for the first time that in a laboratory experiment a train of laser plasma clouds makes it possible to increase the length of the whistler waves generated in the power tube of a magnetized medium. The intensity of waves is orders of magnitude higher than the level achieved by known methods.

1. Introduction

In laboratory experiments and simulations, it was shown that a packet of laser plasma clouds forms Alfven and slow magnetosonic waves in a magnetized medium (hereinafter referred to as the background). Such waves transfer 50% of the clouds energy in the power tube in the form of a stream of compressed, rotating plasma [1, 2]. High efficiency energy injection is achieved by fulfilling a set of dimensionless conditions ("resonance") relating the parameters of clouds and background. The large length of the flows is the result of the mechanism of merging of waves (MMW) generated by individual clouds. Previously, MMW was proposed by the authors for gases, where periodic packets generated infra- and ultrasound.

The present work shows for the first time the possibility of pumping whistlers created by a packet of laser plasma clouds into the power tube of a magnetic field in the background. The record intensity, characterized by the ratio of the amplitude of the magnetic field of the whistlers to the external field $B_W \sim 0.1 \div 0.2$, is achieved when the clouds "resonate" with the background, when only the whistlers are excited. With the simultaneous generation of whistlers and the Alfven wave, the B_W value decreases by factor of ~ 5, which is associated with an additional channel for the removal of clouds energy [3, 4, 5]. The amplitude of whistlers excited by exposure to laboratory plasma with modulated radio emission is much lower [6, 7].

2. The theoretical basis of the experiment

Whistlers are formed at the stage of explosive expansion of laser plasma clouds as a result of interaction with the magnetic field and background plasma by the Lorentz force. In this case, torsion deformation of the magnetic field lines occurs, the electrons acquire azimuthal rotation around the axis of symmetry passing through the center of the cloud and oriented along the unperturbed magnetic field B_0 of the background. Whistlers contain azimuthal B_φ and radial

B_r magnetic field components, an electric field, and a longitudinal current. The hodogram of B_φ and B_r corresponds to the right-handed polarization of the magnetic field vector. The whistler propagates along B_0 with a speed that is approximately equal to the group velocity $V_g/c = 2\Omega\sqrt{1-\Omega_{pe}^2/(\Omega^2-\Omega)}/(2\Omega+\Omega_{pe}^2/(1-\Omega)^2)$, where $c = 3\cdot10^{10}$ cm/s is the speed of light in vacuum, $\Omega = f/f_{ce}$ is the ratio of the whistler frequency to the cyclotron frequency of the electrons, $f_{ce} = eB_0/cm_e \approx 2.8\cdot10^6 B_0$, $\Omega_{pe} = f_{pe}/f_{ce}$, $f_{pe} = 9\cdot10^3\cdot\sqrt{n_0}$ is the plasma frequency of the electrons, n_0 cm^{-3} is the background plasma density. The experimentally measured velocity of low-frequency whistlers $V_g \sim (4\div6)\cdot10^7$ cm/s depends on n_0 and B_0. Whistlers correspond to the frequency range of $F_1 < f < F_2 \ll f_{pe}$ [7], where $F_1 = (f_{ci}\cdot f_{ce})^{1/2} = 6.54\cdot10^4\cdot B_0\cdot\sqrt{Z_0/m_0}$, $F_2 = f_{ce}$, $f_{ci} = 1.53\cdot10^3\cdot Z_0\cdot B_0/m_0$ is the cyclotron frequency of the background ions. A laser plasma clouds creates only whistlers if the ratio of the ion-plasma length to the dynamic radius R_d of the cloud is:

$$L_{pi} = c/(2\pi f_{pi} R_d) = 3.61\cdot10^4 Z_0^{-1}\sqrt{m_0/n_0}\sqrt[3]{B_0^2(1+\beta_0)/Q} > 0.6$$

$R_d = (8\pi\cdot Q/B_0^2(1+\beta_0))^{1/3} \approx 630(Q/B_0^3(1+\beta_0))^{1/3}$, $f_{pi} = 210\cdot Z_0\cdot\sqrt{n_0/m_0}$, $\beta_0 \sim 0.1\div1$ is ratio of the background plasma pressure to pressure B_0. When values of L_{pi} are $0.05-0.25$, the Alfven wave is formed [1, 2]. If L_{pi} is $0.25-0.4$, form the Alfven wave and weak whistlers [3, 4, 5].

From the mention above relations, the laser plasma clouds with energy $Q \sim 20$ J used in the experiment create a whistler in the background with heavy ions, for example argon, with magnetic fields $B_0 = 50-200$ G in the range of plasma density $n_0 \approx (1\div5)\cdot10^{12}$ cm^{-3}. The whistler frequency is close to $\sim F_1/5$ and, when using two consecutive clouds, it decreases by about two times. Clouds are created at a fixed point upon irradiation of a point target with two laser pulses with delay τ, which can be estimated from the characteristic time of clouds expansion $\tau = R_d/V_0 \sim 5\div7$ μs, where $V_0 \sim 1.5\cdot10^7$ cm/s is the initial thermal expansion velocity of the laser plasma. In the experiment, when the delay time τ of the laser pulses is varied, the following whistler generation modes are possible: at $\tau \gg \tau_d$, two successive whistlers; $\tau \ll \tau_d$ the length is approximately the same as from one cloud; at $\tau = \tau_d$, the length of the whistler wave increases by a factor of 2. From the calculations it follows that the train of periodic clouds generates a single whistler wave, the length of which linearly depends on the number of clouds used. The whistler group velocity $V_g \approx (3\div6)\cdot10^7$ is much higher than the Alfven wave velocity. One of the main characteristics of the interaction of plasma clouds with the background is that the Alfven Mach number under the experimental conditions is anomalously large: $M_A = V_0/C_A \approx 10$.

3. Experiment setup

The experimental design at the KI-1 facility is shown in figure 1. In a cylindrical vacuum chamber (length 5 m, diameter 1.2 m, pressure $\sim 2\cdot10^{-6}$ torr), an axial magnetic field $B_0 \sim 50\div200$ G and a longitudinal stream of argon plasma (background) were created with speed $\sim 1.2\cdot10^6$ cm/s, temperature $T_0 \sim 10$ eV, $\beta_0 \sim 0.5$. The radiation from CO_2 lasers created one or two clouds with energies $Q \sim 20$ J upon irradiation of a polyethylene target (diameter ~ 25 mm) located on the camera axis. Each laser pulse was previously divided into two equal parts in energy, which allowed the formation of a plasma expansion, which ensured the azimuthal symmetry of the whistler. The sensors measured the plasma concentration, longitudinal current, electric field, and magnetic field components azimuthal B_φ, radial B_r, and longitudinal B_Z. We varied B_0, n_0, and the corresponding L_{pi} values, as well as the delay of the second pulse relative to the first laser pulse.

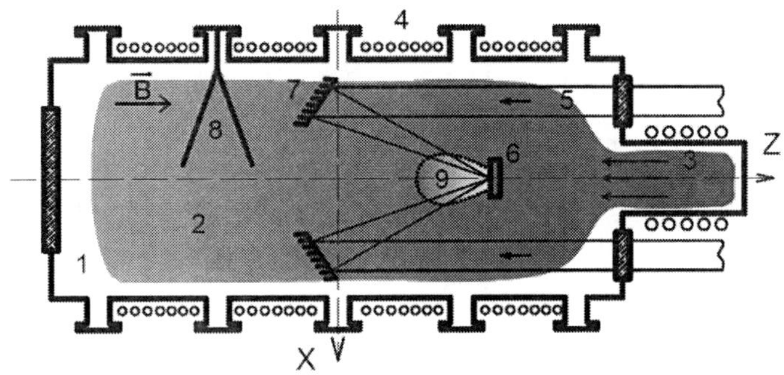

Figure 1. The experimental setup of KI-1 facility. 1 - vacuum chamber, 2 - background plasma flow created by the θ-pinch (3), 4 - solenoid creating an external magnetic field parallel to the axis of the chamber, 5 - emission of a CO_2 laser, 6 - target, 7 - focusing lenses, 8 - measuring probes placed in various sections of the chamber, 9 - a flow of plasma clouds.

4. Experimental results

Figures 2 and 3 show the azimuthal magnetic fields of whistlers generated by one and two clouds of laser plasma in a magnetized background, in which the ion-inertial length $L_{pi} = 1.3$. The hodograms characterizing the polarization of the magnetic field vector are presented in the insets. Hereinafter, the time $t = 0$ corresponds to the moment of irradiation of the target with the first laser pulse, Z is the distance from the target to the sensors. As can be seen from the hodogram, the magnetic field vector has the right-handed polarization, which is one of the properties of whistler waves. As can be seen from figure 4, the spectrum of whistlers created by two clouds of laser plasma is shifted to the region of lower frequencies in comparison with the spectrum of whistlers from a single cloud. An increase in the number of clouds, as follows from the calculations, is accompanied by a decrease in the frequency of the resulting wave. The amplitude of the whistler depends on L_{pi}. Thus, when changing the magnetic field by 2 times, the value of L_{pi} being fixed, the maximum value of B_φ practically does not change. With a decrease to $L_{pi} \sim 0.5$, due to an increase in the background plasma concentration, B_W decreases by factor of 4, which is consistent with experiments [3, 4, 5], where a single cloud of laser plasma generated, simultaneously with the whistlers, an Alfven wave in helium and hydrogen plasma at $L_{pi} \sim 0.25 \div 0.38$.

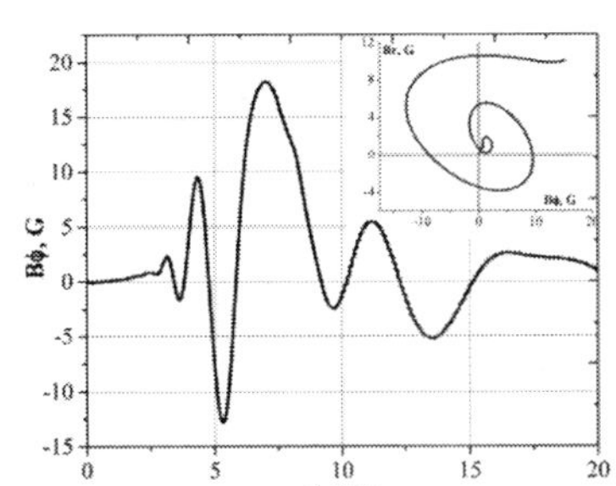

Figure 2. The time variation of B_φ for whistlers created by one plasma cloud.

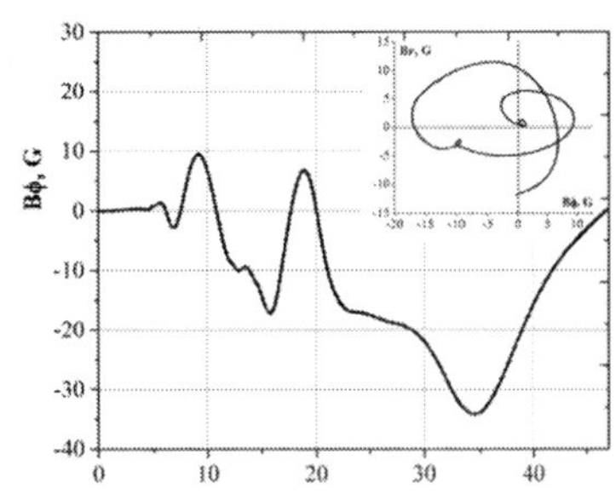

Figure 3. The time variation of B_φ for whistlers created by two plasma clouds.

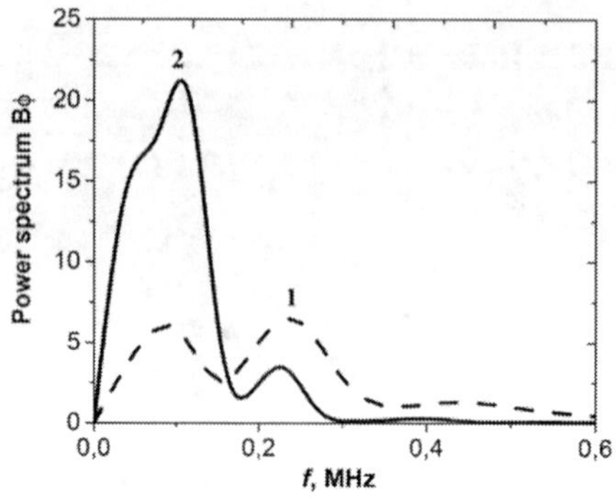

Figure 4. Spectrum of whistlers generated by one (1) and two (2) plasma clouds

5. Conclusion

The clouds of laser plasma, resonantly interacting with magnetized background plasma, generate intense whistlers in the power tube, the length of which linearly depends on the number of clouds. The amplitude of the whistlers reaches record values of ~ 0.2 relative to the external field.

Acknowledgments

This work was carried out as part of the state assignment of the Ministry of Science and Higher Education of the Russian Federation (topic No. AAAA-A17-117021750017-0). The experiment was justified using calculation modeling of Moscow State University, SSCC SB RAS, NSU and the Interdepartmental Supercomputer Center of the Russian Academy of Sciences. The reported research was funded by Russian Foundation for Basic Research and the government of the region of the Russian Federation, grant No. 18-32-00029 / 19 and 18-42-543019 and the Basic Research Program of the Russian Academy of Sciences No. 56.

References

[1] Tishchenko V N, Zakharov Y P, Shaikhislamov I F, Berezutskii A G, Boyarintsev E, Melekhov A V, Ponomarenko A G, Posukh V G and Prokopov P A 2016 *JETP letters* **104** 293–296
[2] Tishchenko V N, Berezutskiy A, Boyarintsev E, Zakharov Y P, Melekhov A V, Miroshnichenko I B, Ponomarenko A G, Posukh V G and Shaikhislamov I F 2017 *Quantum Electronics* **47** 849
[3] Vshivkov V, Dudnikova G, Zakharov Y and Orishich A 1989 *Physics of space and laboratory plasma*
[4] Prokopov P, Zakharov Y P, Tishchenko V, Boyarintsev E, Melekhov A, Ponomarenko A, Posukh V and Shaikhislamov I 2016 *Solar-Terrestrial Physics* **2** 19–33
[5] Heuer P, Weidl M, Dorst R, Schaeffer D, Bondarenko A, Tripathi S, Van Compernolle B, Vincena S, Constantin C, Niemann C *et al.* 2018 *Physics of Plasmas* **25** 032104
[6] Gushchin M E, Korobkov S V, Kostrov A V and Strikovskii A V 2008 *JETP letters* **88** 720–724
[7] Gushchin M E, Korobkov S V, Kostrov A V, Odzerikho D, Priver S and Strikovskii A V 2010 *JETP letters* **92** 85–90

Effect of side walls on the supersonic ramp flow structure

L P Trubitsyna, V I Zapryagaev and I N Kavun

Khristianovich Institute of Theoretical and Applied Mechanics SB RAS, 4/1, Institutskaya str., Novosibirsk, 630090, Russia

lukeria.trubitsyna@gmail.com

Abstract. The reattachment zone structure of a supersonic separated ramp flow was experimentally studied at $M_\infty = 6$. Several ramp model configurations with and without side walls were considered. It was found that the geometric size of streamwise vortices in the reattachment zone increases due to the absence of transverse flow in the separation region. Moreover, flow separation regions were detected on the model side walls which interact with the ramp-induced separation region, forming corner vortex structures.

1. Introduction

The features of a supersonic separation flow (such as streamwise vortex structures in the reattachment zone or shock wave-boundary layer interactions) can lead to heavy local thermal and power loads on the aircraft surface. Review papers [1] and [2] consider the main types of separated flows, their features, numerical characteristics, and the physical formation mechanism. Streamwise wall vortices arising in the reattachment zone of a supersonic separated flow are described in [3]. Review paper [4] examines the structure and gas-dynamic characteristics of such vortices.

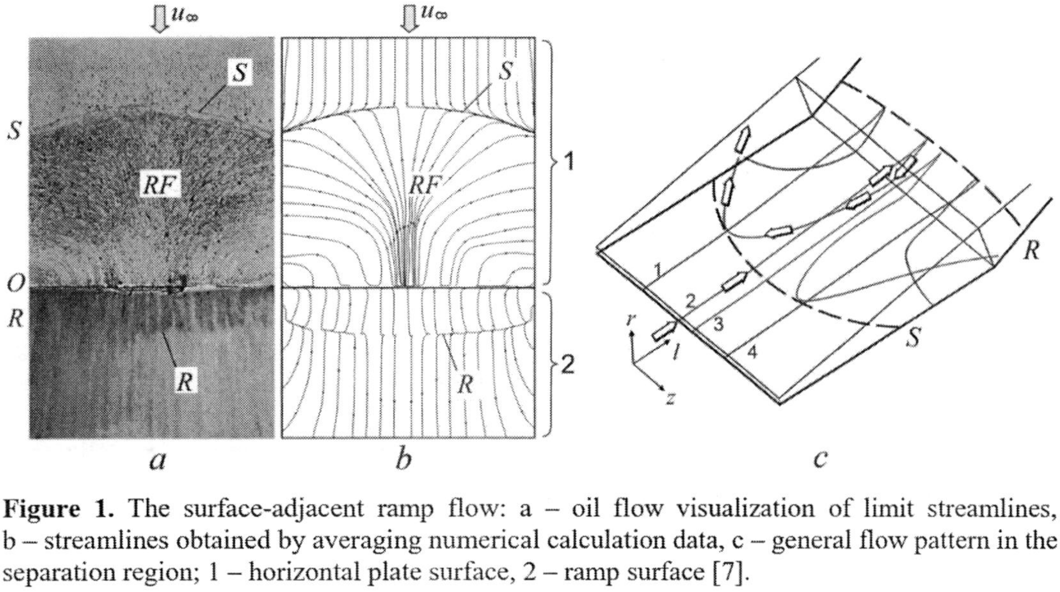

Figure 1. The surface-adjacent ramp flow: a – oil flow visualization of limit streamlines, b – streamlines obtained by averaging numerical calculation data, c – general flow pattern in the separation region; 1 – horizontal plate surface, 2 – ramp surface [7].

The reattachment zone structure of a separated flow at hypersonic freestream velocity ($M_\infty = 6 - 8$) was studied in [5] – [7] for a ramp with no side walls. The flow in the separation zone was established to have a three-dimensional mass-flow nature without a single separating streamline surface between the separation line and the reattachment line (figure 1). In this case, the gas continuously flows into the separation region and continuously flows out of it through the side edges of the model (figure 1, (c)). An intensive flow is established inside the reverse flow region both towards the flow and in the transverse direction from the midline to the side edges of the model.

The authors of [9] consider the supersonic separation flow on a similar ramp model with installed side walls at the Mach number of $M_\infty = 6$. This work shows the existence of corner streamwise wall vortices and analyses the heat flux distribution on the model surface which was not done in [5] – [7]. The flow structure was not however compared to the case of a ramp model without side walls. Meanwhile, the flow structure in the reattachment zone can change in presence or absence of the lateral flow, and the magnitude and spatial distribution of the gas-dynamic parameters in it can vary significantly, therefore, analysis of this effect is an important fundamental problem.

2. Experimental techniques

The reattachment zone flow for a ramp model with installed side walls was studied experimentally. Three ramp models were studied, all sizes were the same except for the ramp angle φ which was 20°, 30° and 40°. The horizontal plate length was L = 50 mm, the model was equal to the horizontal plate length, and the curvature radius of the sharp leading edge was R = 5-10 μm. Freestream Mach number was $M_\infty = 6$, and Reynolds number calculated by the length of the horizontal plate was ReL = 6.1· 105. The experiments were carried out in T-326 hypersonic blowdown wind tunnel in the Khristianovich Institute of Theoretical and Applied Mechanics in Novosibirsk.

The experiment included surface oil flow visualization, Schlieren visualization of the flow using IAB-451 shadowgraph instrument, and surface temperature visualization using a Seek Thermal Compact thermal imager. In addition, the Pitot pressure distribution was measured behind the reattachment line above the ramp surface using a Pitot probe (figure 2). The angle between the probe axis and the model surface is 5–10 degrees so that the probe mount frame does not disturb the near-wall flow. The height and width of the probe tip are 0.2 and 1.1 mm, respectively; the measuring hole has a height of 0.07 mm and a width of 0.8 mm. The probe can be positioned with an error of max ± 20 μm along any of the coordinate axes by a 3D traverse gear. The Pitot pressure is measured in the transversal direction along the Z-axis over the entire model width at several distances l from the plate-ramp joint O.

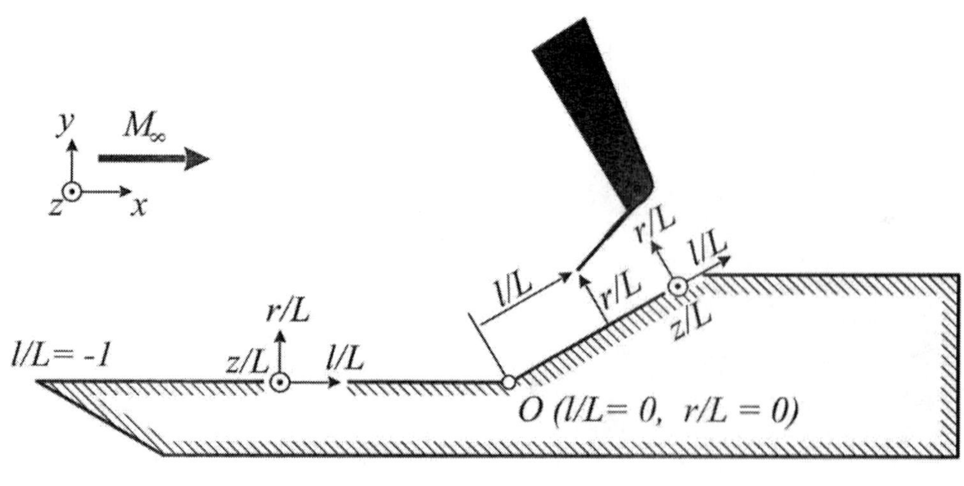

Figure 2. Experimental setup.

Studying the flow by independent methods (since visualization is a non-contact technique and probing is a contact technique) enabled to increase the overall validity of the study and mutually complement the results obtained by these methods.

3. Results and discussion

3.1. Surface flow visualization

Figure 3 presents the results of the surface oil flow visualization for the 30 °ramp model without side walls (a) and with them (b). Evident that for the model with side walls the separation line S moves toward the leading edge and the reattachment line R – toward the ramp end. The period of streamwise vortex structures in the reattachment zone increases (from 2–3.5 mm for the ramp without walls to 4.5-10 mm for the ramp with walls).

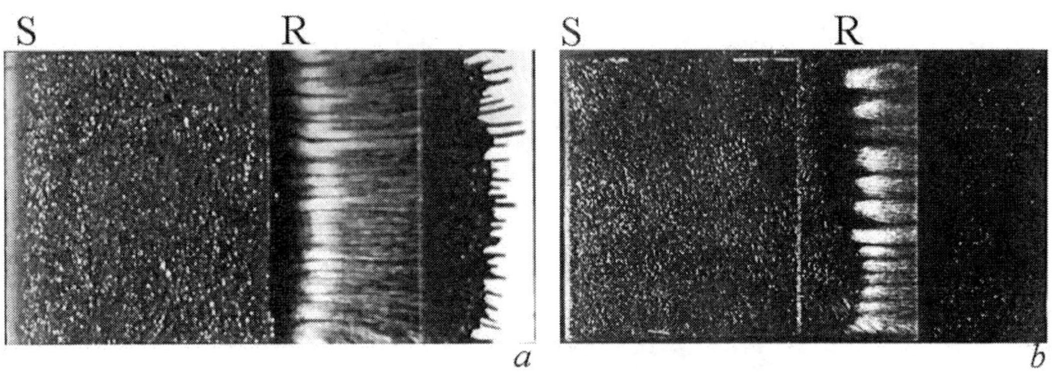

Figure 3. Surface oil flow visualization for the 30°ramp model: a) without side walls; b) with side walls.

Figure 4 presents the temperature distribution for the 30 °ramp model without side walls (a) and with them (b). The temperature distribution pattern qualitatively agrees with the data of the surface oil flow visualization.

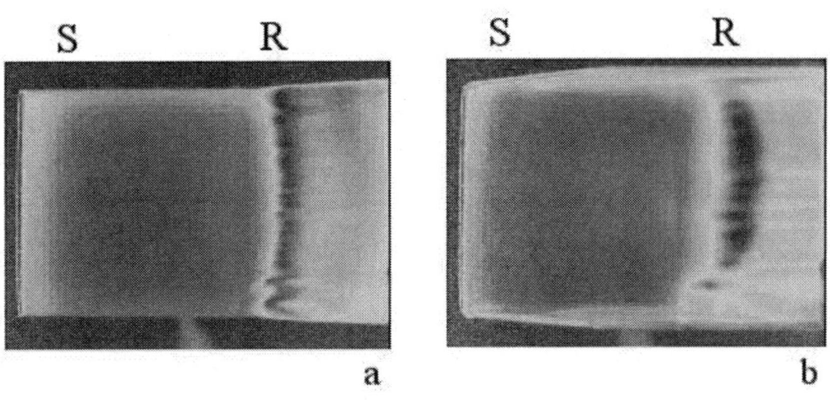

Figure 4. Temperature distribution for the 30°ramp model: a) without side walls; b) with side walls.

3.2. Flow structure on the model side walls

Figure 5 shows the results of the surface oil flow visualization and Schlieren visualization for the 30° ramp model. To carry out the Schlieren photography, the model was equipped with transparent glass walls. Evident that, in addition to the main separation region S formed by the ramp, symmetrical separation regions V appear on the side walls of the model (the dark region in figure 5, (b), marked by a dashed line, shows the absence of gas flow along the wall, which indicates a reverse flow zone). The streamlines on the model surface at the boundaries of the separation regions V (figure 5, (c)) show the interaction between them and the main separation region S with the forming corner streamwise vortices described in [8].

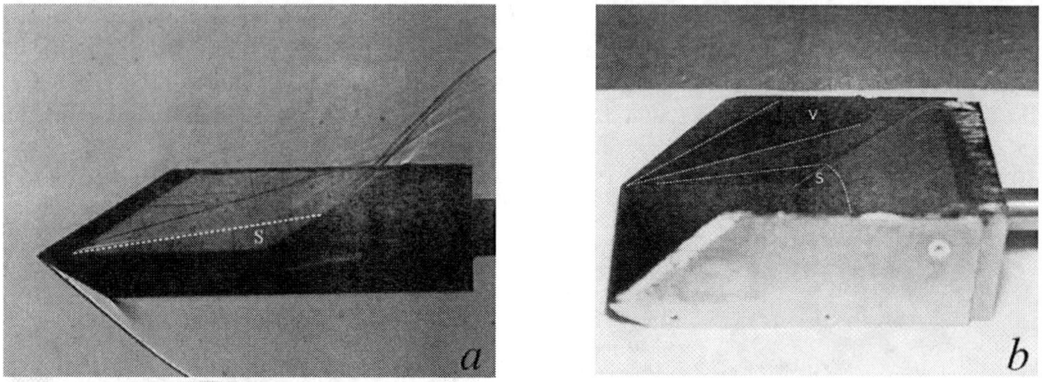

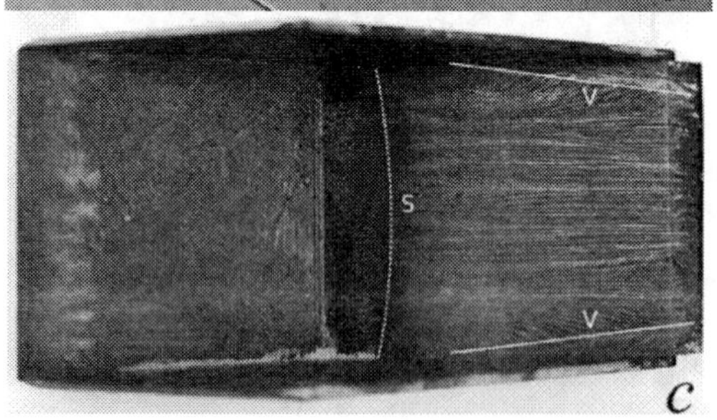

Figure 5. Flow structure visualization for a 30° ramp model with side walls: a) Schlieren flow visualization; b) surface oil-flow visualization, side view; c) surface oil-flow visualization, top view. S - ramp–induced separation zone, V - corner separation zone

The similar flow pattern is observed for the 40° ramp model (figure 6). The boundaries of the separation regions V have a larger inclination angle compared to figure 5, which may indicate an increase in the corner interaction intensity.

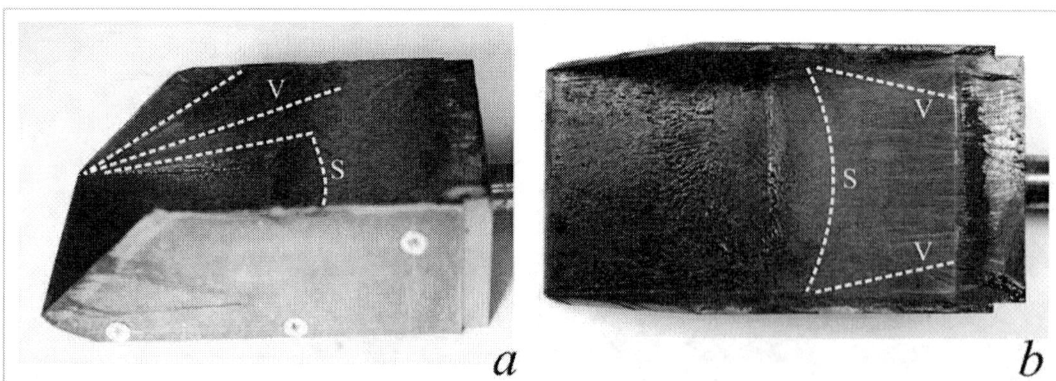

Figure 6. Flow structure visualization for a 40° ramp model with side walls: a) surface oil flow visualization, side view; b) surface oil flow visualization, top view.

3.3. Transversal Pitot pressure distribution behind the reattachment line

The Pitot pressure was measured behind the reattachment line in the transverse direction. Figure 7 shows the Pitot pressure distribution for models with a ramp angle of $\varphi = 20°$ and 30°. The geometric size and the period of streamwise vortex structures increase in the case of the model with walls which is confirmed by the visualization data.

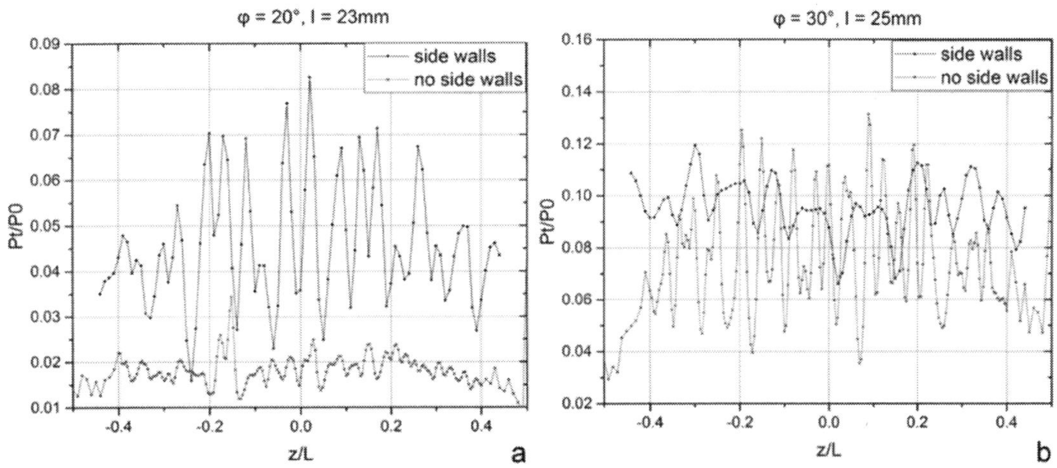

Figure 7. Transversal Pitot pressure distribution for the ramp model with and without side walls: a) 20° ramp model measured 23 mm from the plate ramp joint; b) 30° ramp model measured 25 mm from the plate ramp joint

4. Conclusions

The comprehensive experimental study of the reattachment zone structure was conducted in a supersonic separated ramp flow in the presence of side walls. Due to the absence of the lateral flow, the geometric (transverse) dimensions of the streamwise vortex structures behind the reattachment line increase. Moreover, the flow separation regions were detected on the model side walls which interact with the ramp-induced separation region, forming the corner vortex structures.

Acknowledgements

This work was partially supported by the Program of Fundamental Scientific Research of the state academies of sciences in 2013-2020 (project No. AAAA-A17-117030610137-0) and by the Russian Foundation for Basic Research (project No. 19-31-90035).

The study was conducted at the Joint Access Center "Mechanics" of ITAM SB RAS.

References

[1] Chang P K 2014 *Separation of flow* (Elsevier)

[2] Babinsky H and Harvey J K 2011 *Shock wave-boundary-layer interactions* vol 32 (Cambridge: Cambridge University Press.)

[3] Ginoux J J 1971 *AIAA Journal* **9** 759-60

[4] Borovoy V Ya 1983 *Techenie gaza i teploobmen v zonah vzaimodejstvija udarnyh voln s pogranichnym sloem* (Moscow: Mashinostroenie) (in Russian)

[5] Zapryagaev V I, Kavun I N and Lipatov I I 2014 *Fluid Dynamics* **49** 819-26

[6] Zapryagaev V I and Kavun I N 2016 *TsAGI Science Journal* **47** 27-36

[7] Kavun I N, Lipatov I I and Zapryagaev V I 2019 *International Journal of Heat and Mass Transfer* **129** 997-1009

[8] Chuvakhov P V, Borovoy V Y, Egorov I V, Radchenko V N, Olivier H and Roghelia A 2017 *Journal of Applied Mechanics and Technical Physics* **58** 975-89

Journal of Physics: Conference Series **1404** (2019) 012102 doi:10.1088/1742-6596/1404/1/012102

"N-wave" propagation in supersonic flow at flow past the flat plate with sharp edge

Yu G Yermolaev[1,2], Yu I Chudinova[1,3], A D Kosinov[1,2], N V Semionov[1]

[1]Khristianovich Institute of Theoretical and Applied Mechanics SB RAS,
4/1, Institutskaya str., Novosibirsk, 630090, Russia
[2]Novosibirsk State University,
1, Pirogova str., Novosibirsk, 630090, Russia
[3]Novosibirsk State Technical University,
20, Prospekt K. Marksa, Novosibirsk, 630073, Russia

yermol@itam.nsc.ru

Abstract. The paper discusses the results of an experimental study of the propagation of the "N-wave" in a supersonic flow and its interaction with the bow shock wave, which is formed at flowing around a flat plate with a sharp leading edge. It was found that the "N-wave" does not change its direction when interacting with the bow shock. However, this interaction leads to a reduction of the "N-wave". Estimates of the propagation angles of the "N-wave" in a supersonic flow are also obtained.

1. Introduction
The problem of passing weak pressure shock through the bow shock wave during the flow around aerodynamic bodies and their impact on the boundary layer is still relevant. So, a small disturbance of the oncoming flow can cause high-intensity low-frequency oscillations in the shear layer [1, 2], which usually leads to a local increase in heat fluxes. The models of a delta wing and a flat plate with blunt leading edges were used in previous experiments, but the case of a sharp leading edge, where an attached bow shock is realized, remains open. This work is devoted to the study of the development and interaction of the "N-wave" with the bow shock when flowing around the flat plate with a sharp leading edge.

2. Experimental setup
The experiments were carried out at the Joint Access Center "Mechanics" ITAM SB RAS in a low-noise supersonic wind tunnel T-325 at Mach number of 2.5. The experimental setup is described in detail in [3]. The flat plate with a sharp leading edge located at a zero angle of attack was used in experiments. The thickness of the leading edge did not exceed 0.1 mm. The "N-wave" was generated by a sticker mounted on the side wall of the wind tunnel test section. The sticker was a scotch tape 14 mm wide and 130-135 μm thick. The experimental scheme is shown in figure 1. The constant temperature hot-wire anemometer was used to measure the mean and fluctuation characteristics of a supersonic flow. The analog pulsed signal of the hot-wire was digitized by a 12-bit analog-to-digital converter with a sampling time τ=1.33 μs. The waveforms had length of 65536 ADC samples and four measurements were performed at each point in space.

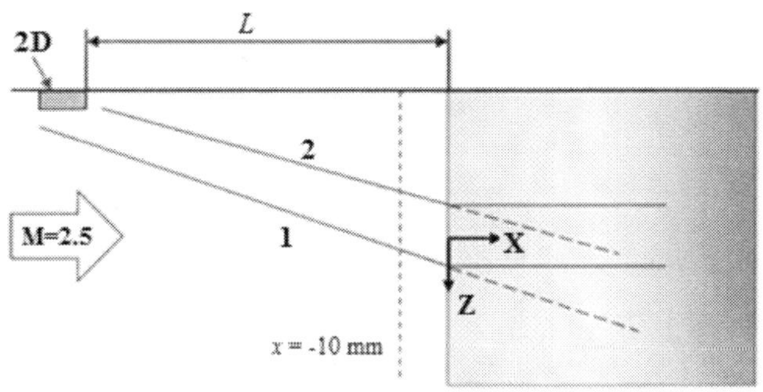

Figure 1. Scheme of the experiment.

3. Results

First, we consider the results of measurements in the oncoming flow. The measurements were carried out in the transverse direction *z* at a distance of 10 mm in front of the plate at a unit Reynolds number $Re_1=8\times10^6$ 1/m. In these experiments, the sticker was located at the distance of L=225.5 mm in front of the plate. The experimental data of the mean flow and integral pulsations are shown in figure 2. From the distribution ρU(*z*) it can be seen that the radiation from the scotch tape is reminiscent of the well-known "N-wave". It consists of two weak shock waves, which, apparently, arise in the vicinity of front lug and trailing scarp of the sticker. Weak shock waves locally distort the mean flow by 3-4% and increase the level of pulsations several times relative to the undisturbed flow. We note that the highest level of rms pulsations is observed in the regions of the maximum gradient of the mean flow. The maximum in the integral distributions of pulsations at z=9.6 mm corresponds to a weak shock wave from the front lug of the sticker (shock wave 1), and the maximum at z=-7.4 mm corresponds to a weak shock wave from the trailing scarp (shock wave 2). The stream region at z>10 mm is undisturbed and corresponds to the oncoming flow.

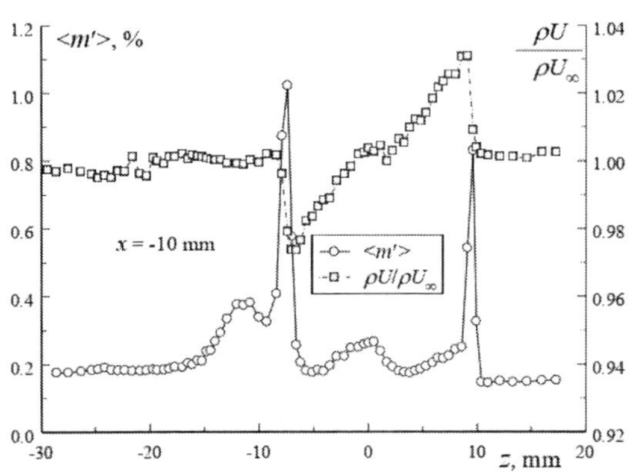

Figure 2. Distributions of the mean mass flow rate and rms pulsations in the transverse direction of *z* at *x*=-10 mm

Analysis of the amplitude-frequency spectra provides more complete information. The frequency spectra of pulsations in the propagation region of weak shock waves are shown in figure 3: graphs (a) and (b) correspond to the leading and trailing edges of the "N-wave". Each graph shows amplitude spectra in the region of the passage of a weak shock wave, where the rms pulsations are maximal. Also, the amplitude spectrum in the region of the undisturbed oncoming flow (z≈13.3 mm) is shown for comparison. Spectral analysis data show that a weak shock wave causes a growth of fluctuations in a wide frequency range. Disturbances are amplified with frequencies up to 20-30 kHz. However, the largest increase of disturbances in the oncoming flow occurs for low-frequency oscillations with a frequency of less than 1 kHz, which is consistent with [3].

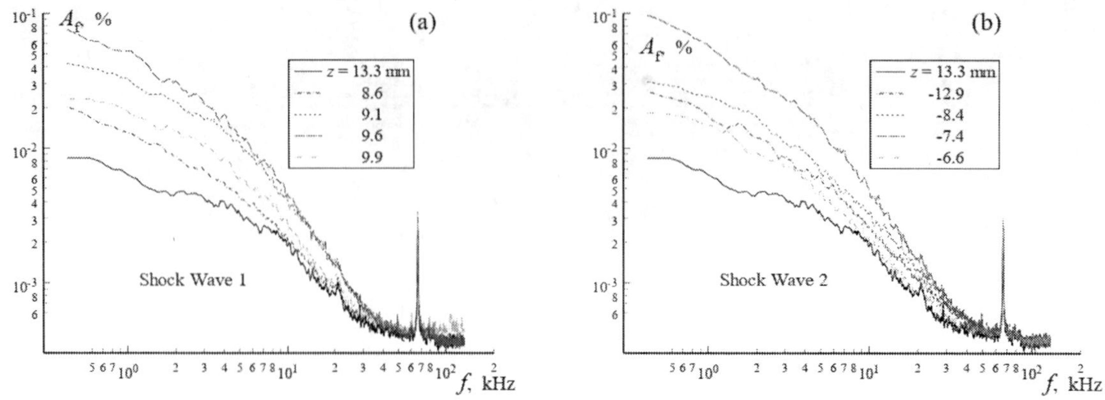

Figure 3. The amplitude-frequency spectra of disturbances in the vicinity of the fronts of the "N-wave": (a) the leading edge; (b) - trailing edge.

Next, consider the result of the passage "N-wave" through the bow shock wave originating from the leading edge of the plate. "N-wave" partially dissipates at interaction with the bow shock. This is seen in figure 4 (a), where the data on the mean flow and rms pulsations in the free flow above the plate at x = 60 mm are represented. Above the plate, the "N-wave" locally distorts the mean flow by 2% relative to the oncoming flow. This is 1.5-2 times less than before the bow shock wave.

Also, the attenuation of the "N-wave" can be seen by the dependence of the integral pulsations in the transverse direction. This is especially markedly for a weak shock wave, which makes up the leading edge of the "N-wave". So, the level of integral pulsations in the vicinity of shock wave 1 in front of the plate is 1% , and above the plate, about 0.3%. For shock wave 2, which originates from the rear edge of the sticker, the situation is less critical, but there is still a decrease in integral pulsations. Note that the level of rms pulsations in the undisturbed flow in front of the plate and above it is the same. The results of measurements inside the supersonic boundary layer at x = 60 mm are shown in figure 4(b).

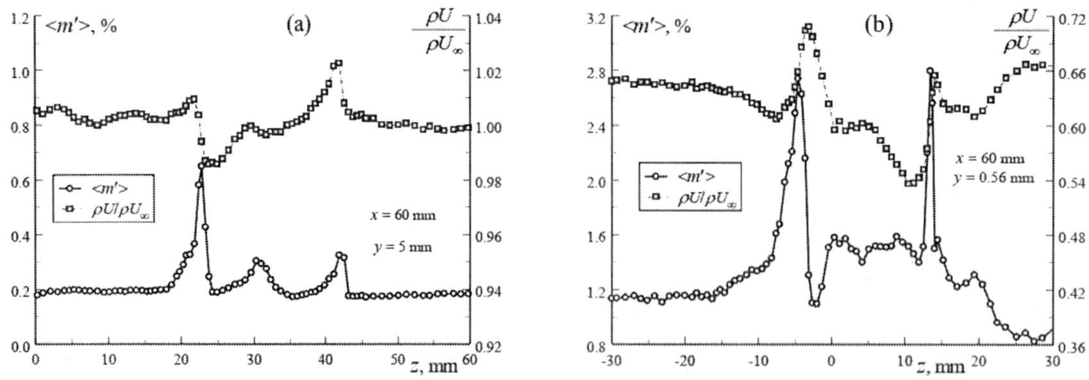

Figure 4. Distributions of the mean mass flow rate and integral pulsations in the spanwise direction at x = 60 mm: (a) in free flow; (b) inside the boundary layer.

The amplitude-frequency spectra of disturbances in the area of passage of weak shock waves, measured in a free supersonic flow above a plate at x = 60 mm, are shown in figure 5. Shock wave 1 in the free flow above the plate causes a growth in fluctuations in the frequency range up to 7 kHz, which is much smaller than in the case in front of the plate. If it compare the frequency spectra measured at x = -10 and 60 mm, it can come to the conclusion about the attenuation of the intensity of the "N-wave" when it passes through the bow shock from the plate.

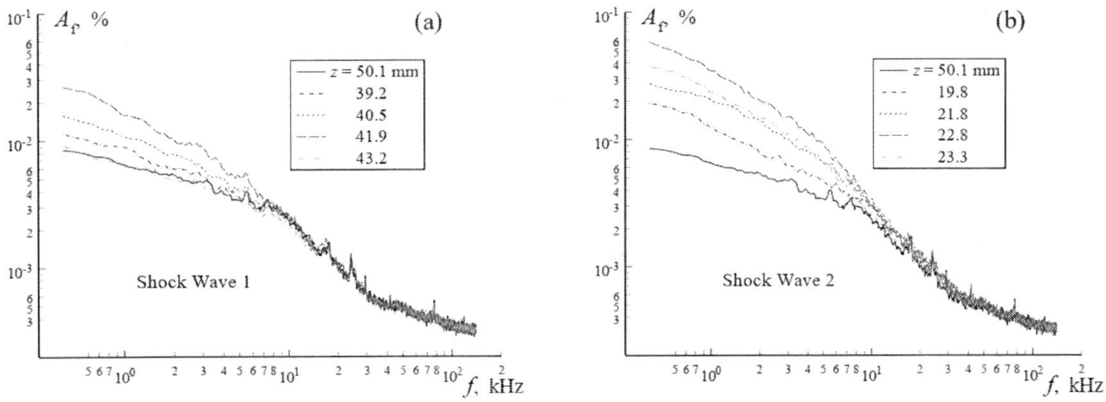

Figure 5. The amplitude-frequency spectra of disturbances in the vicinity of the fronts of "N-wave" above the plate at x=60 mm and y=5 mm

In paper [4] studied the propagation of the "N-wave" and its interaction with the bow shock which is formed at flowing around a plate. The flat plate in these experiments had a large blunting radius of the leading edge. Estimates of the propagation angle of the leading edge of the "N-wave" in oncoming flow and its change during the passage of the bow shock were made in [4] by using optical visualization of the flow. The results of these estimates show that the "N-wave" deviates by about 1.5 degrees when interacting with the detached bow shock wave. Therefore, in our case, at the passage of the attached bow shock which is formed at flowing around a sharp flat plate, the "N-wave" practically does not change its direction. Also, in consideration of the fact that, disturbances from the fronts of "N-wave" propagate parallel to the oncoming flow in the boundary layer from the leading edge of the model [5], it can be estimate the propagation angles of the "N-wave" in the oncoming flow with a Mach number of 2.5.

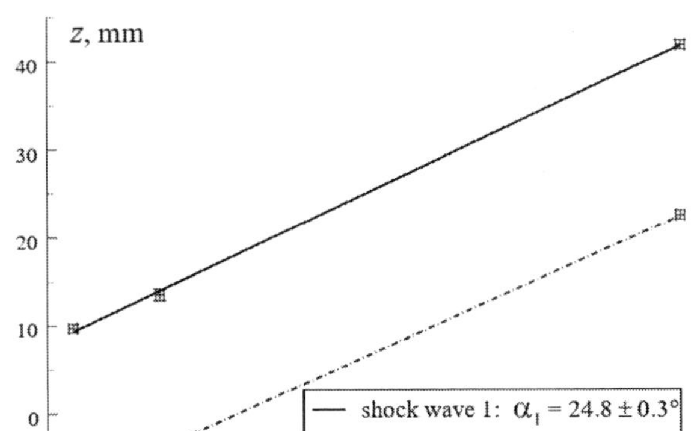

Figure 6. The propagation angles estimation of the fronts of the "N-wave" in a supersonic flow at M = 2.5

The results of estimates of the propagation angles of weak shock waves formed during the flow around the front and rear edges of the sticker are shown in figure 6. The basic error in determining the angles is associated with the initial position of the hot-wire sensor. Estimates show that the trailing edge of the "N-wave" propagates in a supersonic flow at a Mach angle equal 23.6 degrees at M = 2.5. The propagation angle of the leading edge of the "N-wave" in oncoming flow is 24.8 degrees, which coincides with the results of [4]. Approximation of the data on the side wall of the wind tunnel test section shows that the weak shock waves that make up the "N-wave" originate in the vicinity of front lug and trailing scarp of the sticker.

4. Conclusion

Experimental studies of the propagation of the "N-wave" in a supersonic flow and its interaction with the bow shock formed during the flow around a flat plate with a sharp leading edge have been performed. The measurements in an oncoming flow, free stream above the plate, and in the boundary layer were carried out. Inside the boundary layer, disturbances caused by the "N-wave" propagated parallel to the oncoming flow. Above the plate in free flow the "N-wave" does not change its direction after passing through the bow shock. Estimates show that the pressure wave generated by the trailing scarp of the sticker propagates at the Mach angle. The propagation angle of the wave generated by the front lug of the sticker is larger than the Mach angle and coincides with the results of the optical experiment [4]. The interaction of the "N-wave" with the bow shock leads to its attenuation.

Acknowledgments

The research was carried out within the framework of the Program of Fundamental Scientific Research of the state academies of sciences in 2013-2020 (project No. AAAA-A17-117030610125-7).

References

[1] Vaganov A V, Ermolaev Yu G, Kosinov A D, Semionov N V and Shalaev V I 2013 *Proc. of the Moscow Institute Physics and Technology* **5** 164–73

[2] Yermolaev Yu G, Yatskih A A, Kosinov A D, Semionov N V, Kolosov G L and Panina A V 2016 *AIP Conf. Proc.* **1770** 020012

[3] Kosinov A D, Kolosov G L, Panina A V and Semionov N V 2015 *Book of abstracts 11th ERCOFTAC SIG33 Workshop Progress in Transition Modelling and Control* (St Helier, Jersey) pp 1–6

[4] Kosinov A D, Golubev M P and Pavlov Al A 2017 *Siberian journal of Physics* **12** 20 (in Russian)

[5] Vaganov A V, Ermolaev Yu G, Kolosov G L, Kosinov A D, Panina A V, Semionov N V and Yatskikh A A 2016 *Thermophysics and Aeromechanics* **23** 43-8

Effect of Mach and Reynolds numbers on the parameters of the high-pressure layer in the supersonic separated flow near a ramp

V I Zapryagaev, I N Kavun and L P Trubitsyna

Khristianovich Institute of Theoretical and Applied Mechanics SB RAS, 4/1 Institutskaya str, Novosibirsk 630090, Russia

E-mail: zapryagaev@itam.nsc.ru

Abstract. Results of a study of the structure of a supersonic separated flow formed in the vicinity of a compression ramp are reported. At high supersonic speeds, a thin high-pressure layer may form in the re-attachment zone over the inclined wall of the ramp. The gas-dynamic quantities in such a high-pressure layer can noticeably differ from the flow quantities in the surrounding stream. The influence of Mach number (in the range of M = 3-8) and Reynolds number (laminar or turbulent mode of separation) on the flow structure is studied, and the occurrence possibility of the high-pressure layer is analyzed.

1. Introduction

A supersonic separated flow in the vicinity of a compression ramp can be realized with the formation of a thin layer in re-attachment zone (figure 1, the layer is designated as HTL) whose flow quantities considerably differ in magnitude from the quantities in the surrounding stream [1 – 3]. For example, it was found that the total pressure in the thin layer with laminar separated flow at M = 6 could reach 0.95 of the total-pressure value in the free stream. This thin layer arises directly near the re-attachment line and is located over the boundary layer formed by the inclined surface of the ramp.

Also, figure 1 shows: the compression shocks C_1 (shock due to the leading edge of the model), C_2 (shock due to the flow separation process), and C_3 (shock due to the flow re-attachment process); the shear layer SL detached from the model; and the zone of reverse flow RF. The line of flow separation is designated as S, and the line of flow re-attachment, as R.

The absence in the literature of a description of such flows points to the necessity of studying the conditions under which the formation of the high-pressure layer of interest becomes possible. In [4], it was shown that the high-pressure layer could form at M = 6 on a model whose range of ramp angles was $\varphi = 20 - 50°$.

In the present work, we analyze the impact of Mach number (in the range of M = 3 – 8), Reynolds number ($Re = 0.3 \cdot 10^6 - 5.4 \cdot 10^6$), and the flow regime (laminar or turbulent) both on the structure of the separated flow and on the presence or absence of the high-pressure layer in the system under study.

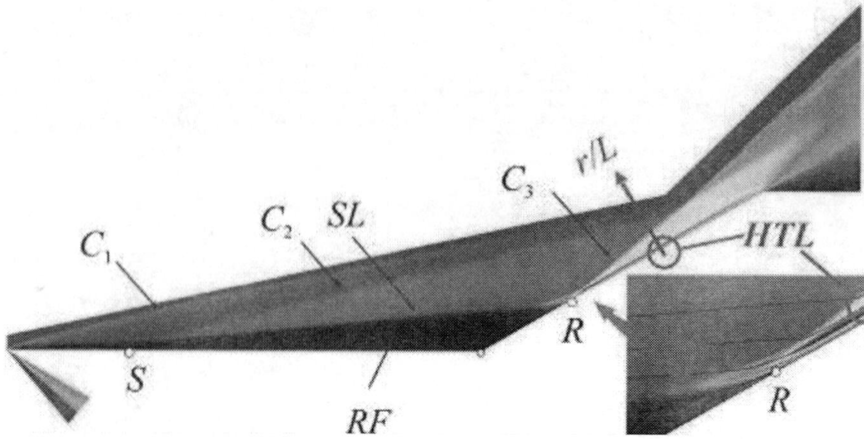

Figure 1. High-pressure layer (HTL) formed in the vicinity of the compression ramp.

2. Experimental procedure

The experiment was carried out in the T-326 and T-333 wind tunnels of ITAM SB RAS. Both Schlieren visualization of the flow and its probing with a total-pressure probe on the surface of the model were performed.

The influence due to Mach and Reynolds numbers was studied using a model ramp (figure 2) with ramp angle $\varphi = 30°$. The width of the model was equal to the distance L from the leading edge of the model to the line of joint of the horizontal and inclined surfaces of the ramp ($L = 50$ mm). The leading edge of the model was sharp, with the radius of its curvature being equal to $R \sim 5$ - 10 μm.

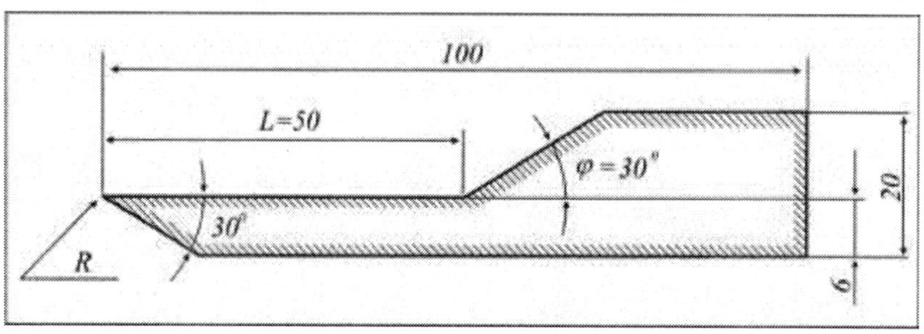

Figure 2. Schematic of the model in the symmetry plane.

The Mach number varied in the range M = 3 – 8. The Reynolds number based on the length of the horizontal part of the model L varied in the range $Re = 0.3\cdot10^6 - 5.4\cdot10^6$, which values have allowed us to realize both laminar and turbulent modes of separation. For achieving a high Reynolds number, part of the experiments was performed without heating the air in the wind-tunnel prechamber. In the latter case, for Mach numbers M = 5 and 6 a flow with air condensation in the wind-tunnel test section established. The regimes of flow separation at M = 6 and $Re = 1.9\cdot10^6$–$5.4\cdot10^6$ can be transitional ones from laminar to turbulent mode. The flow at M = 6 and $Re = 5.4\cdot10^6$ without heating (or, in other words, with air condensation) was close in its structure to the turbulent flow.

The numerical calculations were performed using the Fluent software package. The calculated data were verified by means of their comparison with experimental data.

Three-dimensional Navier–Stokes equations or turbulent Reynolds equations in stationary statement were solved using a differential two-parameter k-ω SST model of turbulence. The solution of Navier–Stokes equations has allowed us to simulate the laminar separated flow past the model at Mach numbers M = 6 – 8 ($Re = 0.6\cdot10^6 - 5.3\cdot10^6$), and the solution of Reynolds equations, at Mach numbers M = 3 – 6 (Re = $0.6\cdot10^6 - 2.4\cdot10^6$). The working medium was perfect gas (air), whose heat capacity was assumed constant. The thermodynamic state of the gas was described using the Mendeleev–Clapeyron equation. The molecular viscosity was calculated using the Sutherland formula, and the thermal conductivity, using the Aiken formula. Possible condensation of air was disregarded.

The geometry of the compression ramp in the calculations was the same as that of the experimental model. In the calculations, the walls of the model were assumed non-heat-conducting.

At the inlet and lateral boundaries of the computational domain, the Mach number, the static pressure, and the static temperature were specified. At the outlet boundary, reference values of static pressure and stagnation temperature were posed. For a turbulent flow, the value of turbulence intensity (I = 0.5%) and the ratio between the coefficients of turbulent and molecular viscosity ($\mu_t/\mu = 1$) were set.

In various cases, the computational domain contained a total of 5 to 9 million cells, all of these cells being structured into blocks. Near the wall, the grid was made condensing for adequate resolution of the near-wall flow (the boundary layer in front of the separation line and the shear flow behind the re-attachment zone). The problem was solved by the relaxation method with second-order accuracy in approximating the derivatives in conservation equations and with first-order accuracy of such approximation in transport equations used for calculating the turbulence parameters.

3. Results

Because of the absence of lateral walls on the model, which would prevent the downward gas flow through the lateral boundaries of the zone of reverse flow, the separated flow is a three-dimensional one. Figure 3, (a) (at the top) shows a schlieren photograph of the turbulent separated flow with M = 3. The calculated distribution of limiting streamlines drawn from the shear-stress vectors on the ramp surface is shown below. A similar picture is observed in the case of laminar separation occurring at M = 6 (figure 3, (b)). Evidently, there exists a good qualitative agreement between the calculated and experimental data, allowing one to use numerical results for analyzing the structure of the flow in the re-attachment zone.

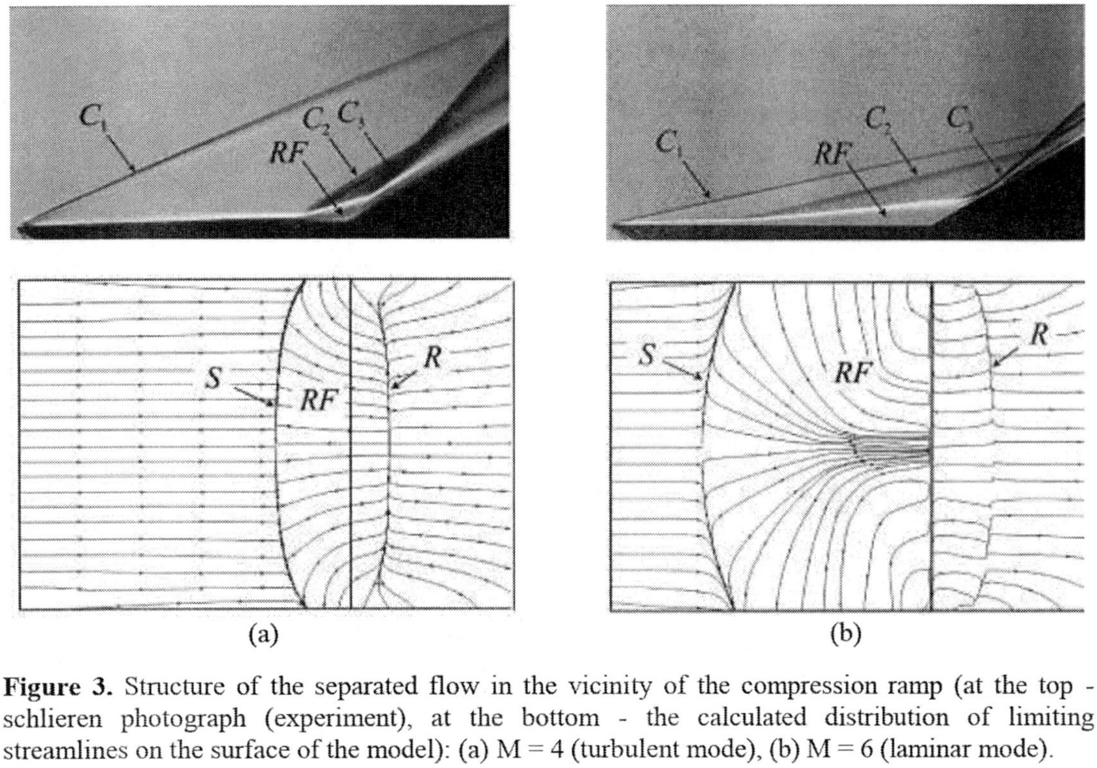

Figure 3. Structure of the separated flow in the vicinity of the compression ramp (at the top - schlieren photograph (experiment), at the bottom - the calculated distribution of limiting streamlines on the surface of the model): (a) M = 4 (turbulent mode), (b) M = 6 (laminar mode).

The presence of curvilinear lines of flow separation and re-attachment points to a complex 3D structure of the flow in the separation region, in which the gas in the zone of reverse flow moves not only against the free stream but, also, toward the lateral boundaries of the model.

Figure 4 shows schlieren photographs of the turbulent separated flow in the range of Mach numbers M = 3–6 and Reynolds numbers Re = $3.1\cdot10^6$–$5.4\cdot10^6$. At M = 5 and M = 6, the experiment was conducted without heating the air, which led to the condensation of the nitrogen and oxygen contained in the air, and the flow regime was not strictly turbulent.

Figure 5, (a) shows a diagram of the separated flow in the plane of symmetry for the range of Mach numbers M = 3 – 6 drawn according to the calculated data (turbulent flow regime) and the results of schlieren visualization of the flow in the vicinity of the model. Figure 5, (b) shows the calculated distribution of the total pressure $p_0/p_{0\infty}$ in the section l/L along the normal coordinate r/L for the cases illustrated by figure 5, (a). Here, l is the longitudinal coordinate parallel to the surface of the model, r is the coordinate normal to the surface, p_0 is the total pressure in the section under consideration, and $p_{0\infty}$ is the total pressure in the free stream.

Evidently, with increase in the Mach number the angle of inclination of all shocks to the horizontal plane decreases in value, the losses of the total pressure behind the C_3 compression shock show an increase, and the height r_{CF} of the fan of compression waves over the surface of the model shows a decrease. At Mach number M = 3, the losses of total pressure in the flow having passed the separation shock C_2 and the compression fan CF are relatively low, and the height of the fan r_{CF} almost reaches the point of interaction of the separation shock C_2 and re-attachment shock C_3. As a result, the whole gas stream between the boundary layer BL and the slipstream SS remains a high-pressure one. As the Mach number increases from M = 4 to M = 5, the losses of total pressure in the region between BL and SS grow in value, and there appears a distinct maximum of total pressure p_{0max}.

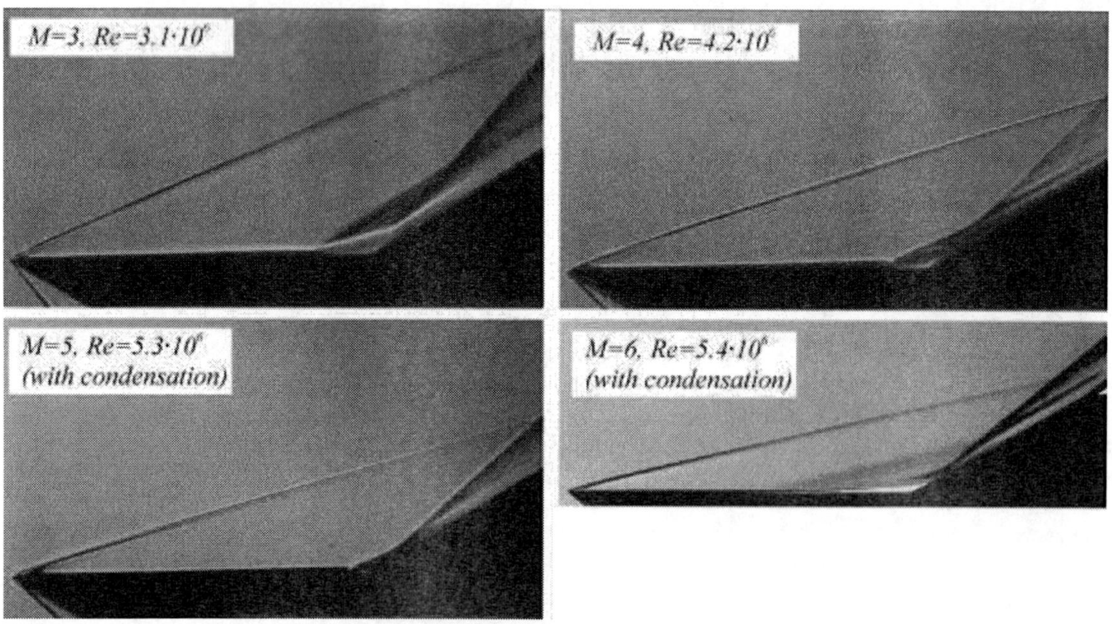

Figure 4. Schlieren photograph of the separated flow at M = 3 – 6.

At M = 6, the CF is pressed to the model surface, and the losses behind the C_3 shock become even more pronounced. As a result, a thin layer of high-pressure gas — a high-pressure layer — forms near the wall of the model ramp over the boundary layer. The obtained data suggest that a high-pressure layer can be formed at Mach numbers M > 5.

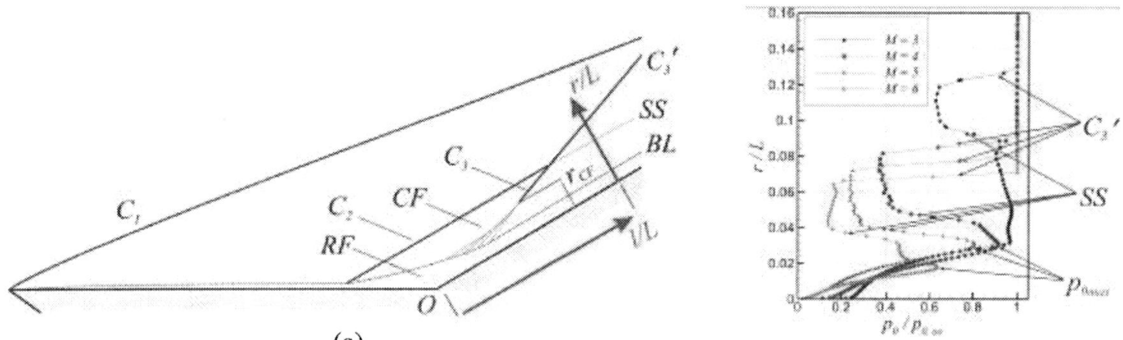

(a) (b)

Figure 5. Schematic of the separated flow (a) and the profile of total pressure (b) in the cross-sections $l/L = 0.4$ (M = 3, 4), $l/L = 0.45$ (M = 5), and $l/L = 0.5$ (M = 6).

The influence due to Reynolds number in the laminar regime of the separated flow for the case of M = 6 - 8 can be analyzed considering figures 6 and 7. Figure 6 shows schlieren photographs of the separated laminar flow at M = 6 in the range of $Re = 0.7{\cdot}10^6 - 2.9{\cdot}10^6$ and M = 8, Re = $0.3{\cdot}10^6$. A photograph taken at Re = $2.5{\cdot}10^6$ without heating of air in the test section of the facility with condensed nitrogen and oxygen in the free stream is given to illustrate the effect of condensation on the regime of the separated flow. It is seen that the general structure of the flow in all photographs is similar, and the effect due to condensation is of minor importance.

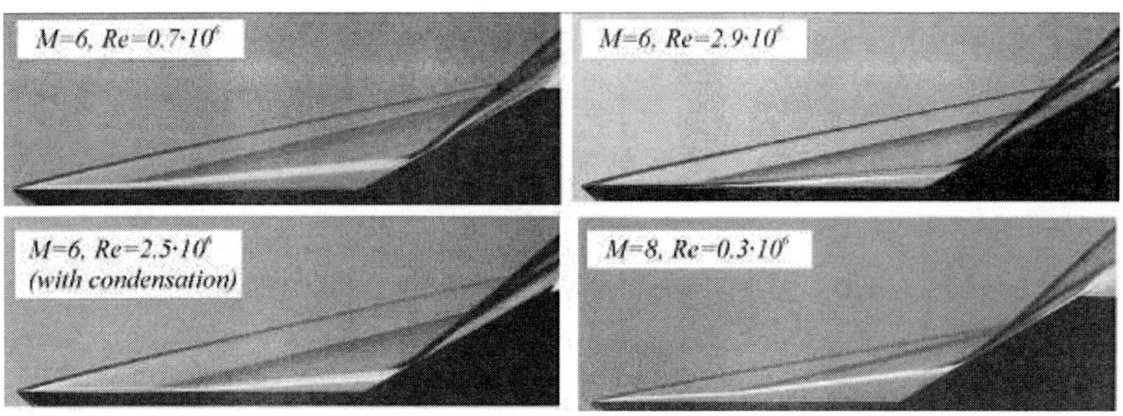

Figure 6. Schlieren photograph of the separated flow at M = 6 – 8.

Figure 7 shows the measured Pitot-pressure profiles (the Pitot pressure is the total pressure in subsonic flow and the total pressure behind a normal shock in supersonic flow) in the cross-sections $l/L = 0.35$ (for M = 6, figure 7, (a)) and $l/L = 0.3$ (for M = 8, figure 7, (b)). Local maxima (designated as HTL) indicating the presence of a wall layer with a high total pressure are distinctly seen.

Two profiles obtained in different experiments are shown in figure 7, (b). The difference in the maxima of total pressure in the HTLs illustrates the effect due to the near-wall longitudinal vortexes of the Görtler type on the parameters of the high-pressure layer.

The results show that in the case of laminar separated flow with M = 6 and M = 8 the high-pressure layer can be identified throughout the entire range of Reynolds numbers studied ($0.3{\cdot}10^6 - 2.9{\cdot}10^6$). The height of the layer over the surface of the model and its intensity both decrease slightly with increasing the Reynolds number. The effect due to air condensation on the formation of the layer is weak.

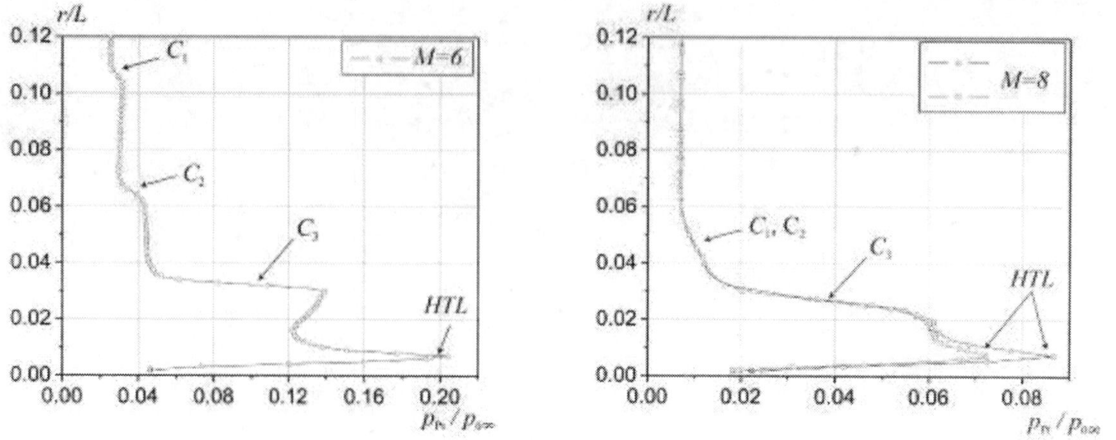

(a) (b)

Figure 7. Profiles of Pitot pressure in the cross-sections $l/L = 0.35$ (M = 6) and $l/L = 0.3$ (M = 8).

4. Conclusion

The reported results of the performed study of the gas flow structure in the re-attachment zone of the 3D supersonic separated flow formed in the vicinity of a ramp made it possible to identify conditions for the occurrence of the high-pressure layer.

A necessary condition is a high velocity of the free stream (M > 5). The impact of the regime of the flow (laminar or turbulent) and the impact of air condensation on the formation process of the layer are less pronounced.

A high-pressure layer was found to exist on the model with a ramp angle of 30° in the range of Mach and Reynolds numbers $M = 6 - 8$ and $Re = 0.3 \cdot 10^6 - 2.9 \cdot 10^6$ for the laminar regime, and $M = 6$, $Re = 5.4 \cdot 10^6$ for the turbulent regime. The absence of this layer at $M = 3 - 5$ and $Re = 3.1 \cdot 10^6 - 5.3 \cdot 10^6$ is shown.

Acknowledgments

The study was performed as part of the Program of Fundamental Scientific Research of the State Academies of Sciences for the years 2013-2020 (project No. AAAA-A17-117030610137-0). The study was conducted at the Joint Access Center “Mechanics” of ITAM SB RAS.

References

[1] Zapryagaev V I, Kavun I N and Lipatov I I 2014 *Fluid Dynamics* **49** 819 – 26
[2] Zapryagaev V I and Kavun I N 2016 *TsAGI Science J.* **47** 267 –81
[3] Kavun I N, Lipatov I I and Zapryagaev V I 2019 *International J. of Heat and Mass Transfer* **129** 997–1009
[4] Volkov K N, Zapryagaev V I, Emelyanov V N, Gubanov D A, Kavun I N, Kiselev N P, Teterina I V and Yakovchuk M S 2018 *Visualization of Physical and Mathematical Modeling Data in Gas Dynamics*, ed. Volkov K N, Emelyanov V N (Moscow: Phismatlit) p 360

Dynamics of drying of fibrous material by the acoustic-convective method

A A Zhilin[1,2]

[1] Khristianovich Institute of Theoretical and Applied Mechanics SB RAS, 4/1 Institutskaya st., Novosibirsk, 630090 Russia
[2] Siberian State University of Water Transport, 33 Shetinkina st., Novosibirsk, 630099 Russia

lab20@itam.nsc.ru

Abstract. The drying of a cellulosic fibrous material by acoustic-convective and thermal-convective methods under different conditions has been studied. Using the thermal-vacuum drying method, the weight of an absolutely dry sample of the starting material was determined and used to calculate the initial absolute and relative moisture contents, which were 12.38% and 10.96%, respectively. The dynamics of capillary impregnation and sorption wetting of samples was determined. The limiting values of the absolute and relative moisture content of cellulose fibers were found to be 862.16% and 89.47%, respectively. The kinetic curves of moisture extraction obtained in acoustic-convective and thermal-convective drying under different conditions were constructed, analyzed, and compared. It is shown that the acoustic-convective extraction of moisture has an advantage over the thermal-convective method.

1. Introduction

In recent years, projects for the construction of environmentally responsible low-rise housing have been actively developed in many regions of Russia. As a rule, cellulosic fibrous materials are used as heat-insulating materials for this class of housing. Spring floods and heavy rains lead to significant wetting of some buildings, as result of which the heat-insulating material absorbs a large amount of moisture and ceases to perform its heat-insulating function. Therefore, in order for fibrous heat-insulating material to preserve useful properties, it should have minimum moisture content. In addition, the removal of excess moisture from heat-insulating materials significantly reduces the load on the load-bearing structures of buildings.

Traditionally, excess moisture is removed from heat-insulating materials using a thermal-convective approach based on the supply of heat to the material to be dried [1, 2]. Under prolonged heating, wet material begins to actively evaporate excess moisture. The spent warm moist air is exhausted into the environment, resulting in large heat losses and hence high costs per unit of dried material. In this paper, it is proposed to dry fibrous heat-insulating material using the novel approach developed at the ITAM SB RAS. The approach is based on the exposure of materials being dried to high-intensity acoustic-convective flow. The proposed technology provides a significant intensification of moisture extraction from a wide range of biological [3-7], inorganic [8-12], etc. materials. One of the main advantages of this technology is the absence of heating of the dried material [13], i.e. drying proceeds at room temperature.

2. Material

We studied cellulose wool, a heat-insulating fibrous material containing 80% cellulose and 20% additives that reduce flammability and protect against fungal microorganisms and bacteria. Before starting the study, it was required to determine the initial moisture content of the material selected for the study. For this, several control groups of three samples each were prepared from the studied material. The samples in each group were equivalent in weight, with the weight of the second sample being two times the weight of the first sample, and the weight of the third sample four times the weight of the first. The groups of samples were placed in a vacuum oven at a working temperature of 100 °C. The samples were removed from the oven at predetermined time intervals to perform measurements and determine the current moisture content. The experiment was continued until the current moisture content of samples with higher initial weight became comparable with the current moisture content of samples with lower initial weight. Moisture content deviations between three samples did not exceed one percent. As a result, the total drying time was about 40 h.

The results obtained in the experiment were processed and averaged within each group. Subsequently, the obtained data were analyzed and averaged between the groups. As a result, the studies of the drying of cellulosic fibrous samples showed that the absolute initial moisture content of the studied material was $W = 12.38\%$ and their relative initial moisture content $w = 10.96\%$.

Next, the limiting moisture content of the material studied was determined. This was done using two wetting methods: capillary impregnation and sorption wetting in two modes.

In the study of the dynamics of capillary impregnation, all samples were placed in a bath of water for predetermined time intervals and then removed and weighed. The dynamics of capillary impregnation of the samples is presented in Figure 1, which shows that all samples absorb moisture very rapidly and reach the final equilibrium value in just three to four seconds. This is due to the high porosity of the cellulosic fibrous material; i.e., its large surface area through which moisture is absorbed. At subsequent times, the moisture content of the test material remained almost unchanged. As a result, the limiting final moisture content averaged over all test samples was 862.16 (89.47)%.

For the experiments to study the wetting dynamics of the fibrous material by sorption, we prepared an experimental setup consisting of a bath with water, above which test samples were placed without touching water. The assembled setup was placed in a sealed polyethylene bag to maintain constant parameters. Due to the tightness, the moisture content was maintained at 100% and the temperature at 19 °C. During the sorption wetting experiment, the samples were sequentially, at a predetermined time interval, briefly removed from the setup and subjected to weighing.

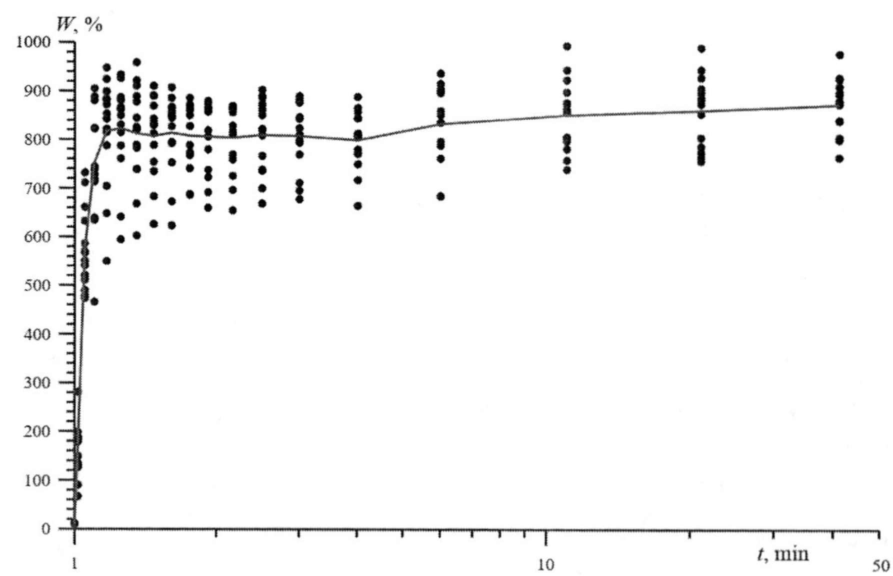

Figure 1. Change in the absolute moisture content of samples during capillary impregnation.

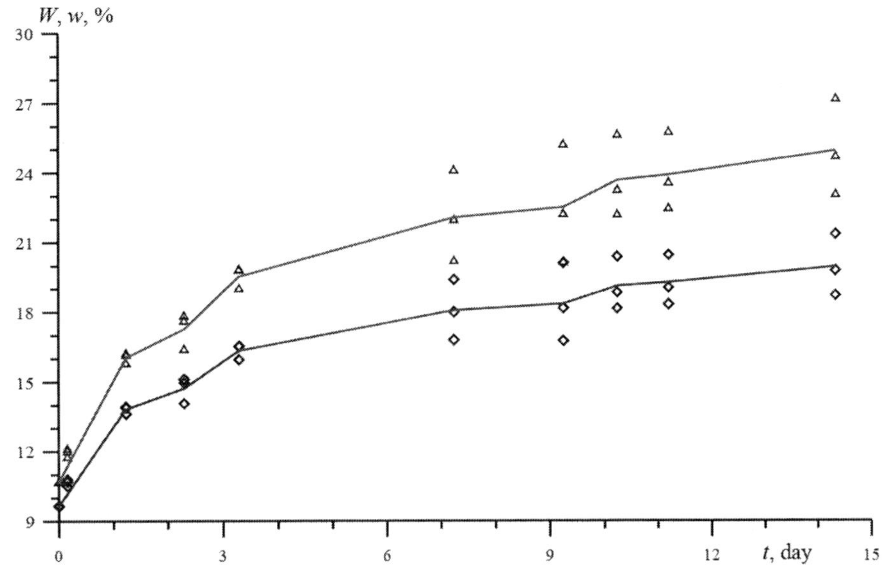

Figure 2. Dynamics of the absolute (Δ) and relative (◊) moisture content during sorption wetting.

The dynamics of sorption wetting of the samples is shown in Figure 2. After the first three days of wetting, the moisture content of the samples increased quite rapidly: by 9.14 (6.90)% for the samples of the first group, by 9.12 (6.88)% for the second group, and by 8.31 (6.31)% for the third group; i.e., on average, by 8.86 (6.69)% for the three groups. After the next four days of continuous sorption wetting, the moisture content of the test samples increased by 4.24 (2.85)%, 2.14 (1.47)%, and 1.19 (0.83)%, respectively, for each group of samples. The average value over four days was 2.53 (1.72)%. The general wetting dynamics over the next six days with periodic removal of samples from the setup and weighing is characterized by a significant slowdown and a gradual approach to asymptotic behavior. The increase in moisture content for the samples of each group was 3.07 (1.95)%, 2.71 (1.78)% and 2.84 (1.92)%. After averaging over the groups, we have 2.97 (1.88)%.

Thus, after 15 days, the average moisture content of the samples increased by 14.26 (10.29)% and amounted to 26.67 (21.25)% in absolute value.

3. Acoustic-convective drying

Experiments on acoustic-convective drying of the fibrous material were carried out on the acoustic-convective dryer (ACD) of the ITAM SB RAS. The dryer operates as a Hartmann gas-jet generator [14]. The flow pattern occurring in the dryer duct consisting of two perpendicular channels was obtained numerically in [15, 16]. The ACD was started without the material to be dried. After the startup period, the drying flow parameters were recorded. Next, containers with prepared samples were loaded. During the experiments, the static pressure in the ACD settling chamber was kept constant and equal to 6.5 atm. In the study, five ACD modes were used, by analogy with [12]. Switching between modes was performed by changing the cavity depth [17]. The modes had the following characteristics. In Mode 1, the frequency was f = 123 Hz and the intensity I = 161 dB; in Mode 2, f = 285 Hz and I = 178 dB; in Mode 3, f = 848 Hz and I = 171 dB; in Mode 4, f = 1.7 kHz and I = 168 dB; in Mode 5, a resonant frequency was absent, i.e., convective flow occurred.

Figure 3a shows the drying dynamics of cellulosic fibrous samples in different acoustic-convective flow modes. After 30-min drying in ACD in Mode 1, the absolute moisture content of the samples decreased by 241.23%, and the relative moisture content by 20.10%. During the same time of drying in Mode 2 with a frequency of 285 Hz and an intensity of 178 dB, the absolute moisture content of the fibrous material samples decreased by 262.10% and the relative moisture content by 23.86%. In the next Mode 3 with drying flow at a frequency of 848 Hz and an intensity of 171 dB, there was a significant increase in rate of moisture extraction: after 30-min drying, the absolute moisture content decreased by a record-breaking value 346.52% and the relative moisture content by 50.45%. Mode 4 with a resonant flow frequency of 1.7 kHz and an intensity of 168 dB turned out to be slightly less efficient: the loss of the absolute moisture was 327.43% and that of the relative moisture was 41.99%. The results of ACD cellulose in background Mode 5 showed a decrease in absolute moisture content by 239.12% and in relative moisture content by 19.76%.

Thus, the moisture extraction by drying the material at a flow frequency of 848 Hz is twice as high as that at a frequency of 285 Hz and two and a half times higher than that in the case of no distinct resonant frequency in the drying flow. It is also seen that the drying dynamics of the test material in acoustic-convective flow with operating parameters corresponding to Mode 1 and background mode 5 is almost identical.

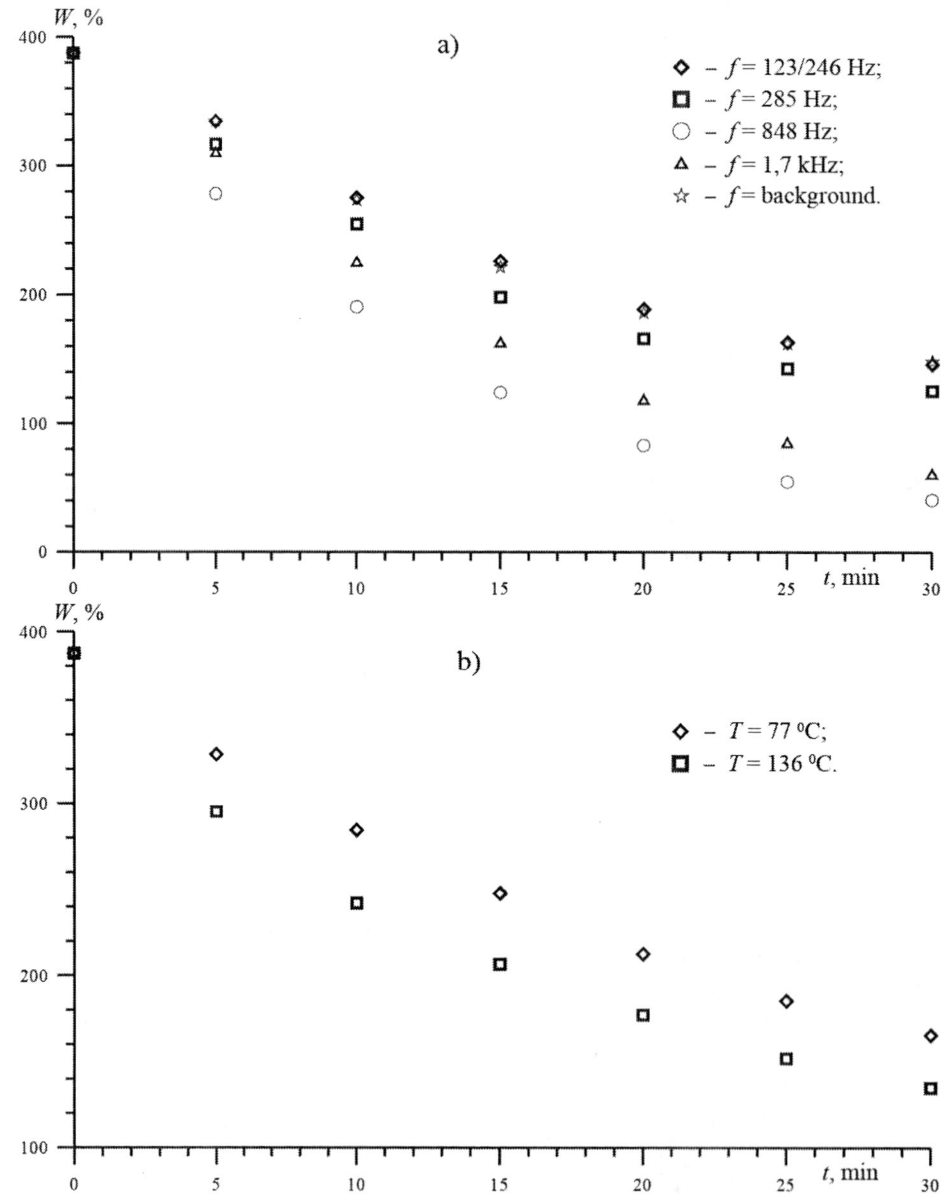

Figure 3. Dynamics of moisture extraction from the cellulosic fibrous material in different modes of acoustic-convective (a) and thermal-convective (b) drying.

4. Thermal-convective drying

To compare the proposed drying method with the traditional thermal-convective method, we designed a test bench which allowed us to perform experiments on drying the fibrous material in two thermal-convective flow modes: Mode 1 at a temperature of 77 °C and Mode 2 at 136 °C.

Pre-wetted cellulose samples were placed in the center of the thermal-convective flow. At intervals of five min, the samples were removed from the thermal flow and weighed. The duration of the experiment on thermal-convective drying (TCD) in the first and second mode was 30 min

The processed weighing results for both thermal-convective drying modes are presented in Figure 3b. It can be seen that after 30-min drying of samples at a thermal flow temperature of 77 °C, the absolute moisture content decreased by 221.94% and the relative moisture content by 17.15%. After the same time of TCD at a flow temperature of 136 °C, the absolute moisture content decreased by 252.55% and the relative moisture content by 22.06%. Thus, a twofold increase in the temperature of the drying flow leads to a slight acceleration of drying by only one-quarter. It should be noted that a final moisture content of 165.49 (62.33)% of the material dried at a flow temperature of 77 °C was reached in 30 min and at a temperature of 136 °C in about 22 min. Practice shows that, to achieve even lower moisture content, the differences in drying times at different temperatures become more significant.

5. Discussions

Let us compare the experimental results on the dynamics of drying of the cellulosic fibrous material using the acoustic-convective and thermal-convective methods in different modes. Figure 4 shows the results on moisture extraction from fibrous samples in the first 30 min of drying.

Consider in detail the result of drying the test samples. It can be seen that the slowest process is TCD in the first mode at a temperature of 77 °C, where the moisture content decreased by 221.94 (17.15)% in 30 min. The background and first modes of ACD at a flow temperature of 14.45 °C showed almost the same results: the moisture content was reduced by 239.12 (19.76)% and 241.23 (20.10)%, respectively. A slightly better result was obtained in the second mode of TCD at a flow temperature of 136 °C; here the moisture content decreased by 252.55 (22.06)%. A close result was obtained in Mode 2 of ACD at a flow temperature of 14.45 °C, in particular, the moisture content decreased by 262.10 (23.86)%. The fourth mode of ACD, in which the moisture loss was 327.43 (41.99)% was faster. Finally, the fastest drying of the cellulosic fibrous material was observed in Mode 3 of ACD with a resonant frequency of 848 Hz and a temperature of 14.45 °C, where the moisture loss was 346.52% (50.45%) after 30-min operation of the dryer.

Thus, appropriately choosing the resonant frequency of the drying flow correctly provides a significant (almost twofold, as in this case) increase in the efficiency of acoustic-convective drying of the fibrous material at the same energy consumption.

6. Conclusions

Using thermal-vacuum drying, the absolute and relative initial moisture content of the fibrous material was determined to be 12.38% and 10.96%, respectively.

It was shown that during acoustic-convective drying of the fibrous material at room temperature of 14.45 °C, an appropriate choice of a resonant frequency led to a significant increase in the rate of moisture extraction.

In the studies of the dynamics in different ACD modes, it was determined that the optimal drying mode was Mode 3 with a frequency of 848 Hz. It was found that a two-fold (from 77 °C to 136 °C) increase in the temperature of the thermal-convective flow drying the fibrous material led to a one-quarter decrease in the drying time.

Comparison of the acoustic convective and thermal-convective drying methods showed that after 30-min drying in the optimal ACD mode at room temperature, the moisture loss from samples is twice as much as that in thermal-convective drying at a flow temperature of 136 °C.

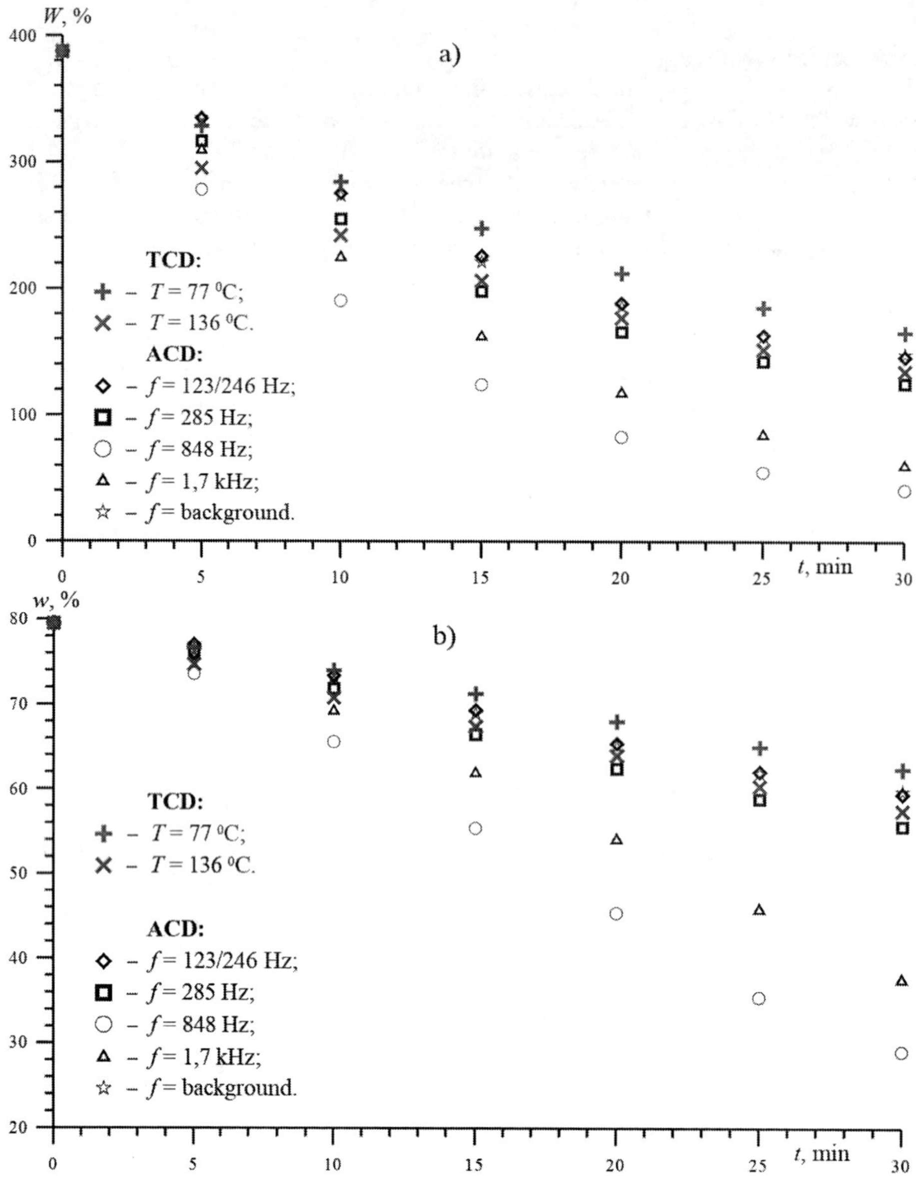

Figure 4. Comparison of the changes in the absolute (a) and relative (b) moisture content in cellulose in different modes and methods of drying.

Acknowledgments
The research was carried out within the framework of the Program of Fundamental Scientific Research of the state academies of sciences in 2013-2020 (project No. AAAA-A17-117030610139-4).

References

[1] Lykov A V 1968 *The Theory of Drying* (Moscow: Energia Publ.) p 472
[2] Lykov A V 1956 *Heat and Mass Transfer in Drying Processes* (Moscow: State Power Engineering Publ.) p 464
[3] Zhilin A A and Fedorov A V 2016 *J. Eng. Phys. Thermophys.* **89** 323–33
[4] Korobeinikov Yu G, Trubacheev G V, Fedorov A V, Chu K M, Zheong D M and Kim Yu I 2008 *J. Eng. Phys. Thermophys.* **81** 676–9
[5] Fedorov A V and Zhilin A A 2014 *J. Appl. Mech. Tech. Phys.* **55** 1016–9

[6] Zhilin A A and Fedorov A V 2014 *J. Eng. Phys. Thermophys.* **87** 879–86
[7] Zhilin A A, Fedorov A V and Grebenshchikov D M 2018 Dynamics of acousto-convective drying of sunflower cake compared with drying by a traditional thermo-convective method *Foods and Raw Materials* **6** 370–8
[8] Korobeinikov Yu G, Fedorov A V, Buluchevskii E A and Lavrenov A V 2009 *J. Eng. Phys. Thermophys.* **82** 246–50
[9] Zhilin A A and Fedorov A V 2011 *J. Eng. Phys. Thermophys.* **84** 965–74
[10] Korobeinikov Yu G and Fedorov A V 2003 *J. Eng. Phys. Thermophys.* **76** 6–9
[11] Zhilin A A and Fedorov A V 2017 *J. Eng. Phys. Thermophys.* **90** 1412–26
[12] Zhilin A A 2019 *AIP Conference Proceedings.* **2125** 030085
[13] Gosteev Yu A, Korobeinikov Yu G, Fedorov A V and Fomin V M 2005 *J. Appl. Mech. Tech. Phys.* **46** 711–6
[14] Borisov Yu G 1967 Gas-jet sound emitters of Hartmann type Density matrix theory of coherent ultrafast dynamics *Physics and Technology of Powerful Ultrasound* ed L D Rozenberg (Moscow: Nauka Publ.) book 1: Sources of powerful ultrasound pp 7–110
[15] Kravchenko A S, Zhilin A A and Fedorova N N 2018 *AIP Conference Proceedings.* **1939** 020018
[16] Kravchenko A S and Zhilin A A 2019 *J. Phys.: Conf. Ser.* **1268** 012037
[17] Zhilin A A and Golubev E A 2018 *AIP Conference Proceedings.* **1939** 020016

Ballistic facility for heavy bodies acceleration

V I Zvegintsev and A Yu Melnikov

Khristianovich Institute of Theoretical and Applied Mechanics SB RAS, 4/1, Institutskaya Str. Novosibirsk, 630090, Russia

zvegin@itam.nsc.ru, melnikov@itam.nsc.ru

Abstract. This paper discusses the issues of creating a ballistic facility for use in laboratory conditions and achieving throwing speeds of up to 2000 m/s for massive objects. The design of the facility was proposed and the results of the analytical calculation of the achieved speeds were presented. The verification of the analytical calculation and the results of experiments on the prototype ballistic facility was done.

1. Introduction

At present, aeroballistic facilities are widely used in scientific research, both in the field of collision of solids and in aerophysics problems. The current state of methods for accelerating bodies to hypersonic speeds and the possibility of using various propelling devices (gas, light-gas and powder guns, explosive, explosive-cumulative and rail-accelerated accelerators) in aeroballistic experiments are discussed in reviews [1–6]. The main direction of development of traditional acceleration tools is to increase the speed by reducing the mass of accelerated bodies. In most existing aeroballistics facilities, the mass of objects is several grams [7]. In this paper, we consider the possibility of increasing the mass of accelerated objects. To this end, the task was set to create a laboratory ballistic facility for high-speed throwing massive bodies weighing up to 1 kg.

2. Mathematic model of a single-stage ballistic facility

As the main scheme for the facility, the one of a single-stage ballistic gun was chosen, in which the projectile is accelerated only due to the pressure of the pushing gas. The peculiarity of the scheme under consideration is the use of a barrel of large elongation (up to L/D = 500), which makes it possible to accelerate rather heavy bodies in comparison with the known light-gas guns. The use of compressed gas simplifies the design of the facility and allows its use in laboratory conditions.

The design model of a single-stage ballistic facility is shown in figure 1. Pressure receiver 3 contains a specific mass m_{03} of compressed gas with parameters P_{03} and T_{03}. Barrel 2 is attached to pressure receiver 3, in which accelerated projectile 1 is placed. After being released from the holding device, projectile 1 begins to move along the barrel with increasing speed W_3 under the influence of the compressed gas coming from the receiver. Compressed gas flows from tank 3 into space 2 behind the projectile. For simplicity, it is assumed that there is no air in front of the projectile. Projectile 1 is affected by static pressure P_3, which depends on stagnation pressure P_{03} and on velocity $W_3$of the projectile. As the projectile moves, the volume-average gas-dynamic parameters P_{03} and T_{03} decrease. The velocity of the projectile depends on the rate of change of pressure in the pressure receiver, on the barrel length, and on the mass of the projectile.

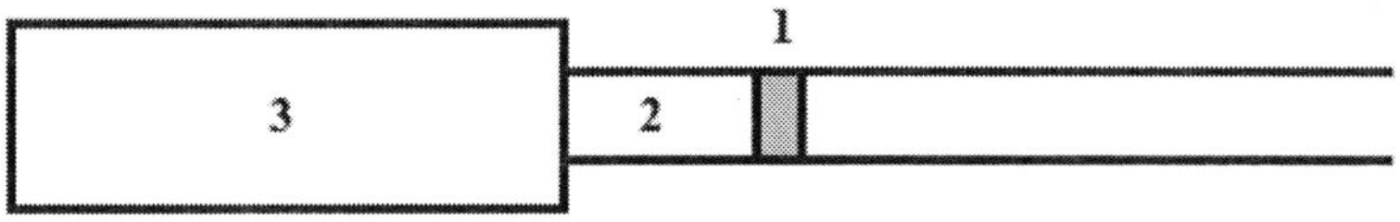

Figure 1. Calculation model of a single-stage ballistic facility.

In accordance with the methodology of [8], a one-dimensional quasistationary calculation model was created, which considers the flow of a perfect inviscid gas with constant thermophysical properties from tank 3 to tank 2. The change in parameters in tank 3 is described by the relations for the adiabatic process of gas outflow through a constant section hole.

$$P_{03}(t) = P_{03}(0) \cdot (1 + B \cdot t)^{-\frac{2\gamma}{\gamma-1}}$$

$$T_{03}(t) = T_{03}(0) \cdot (1 + B \cdot t)^{-2},$$

where:

$$B = \frac{\gamma-1}{2} \cdot \mathrm{R_\Gamma} \cdot \mu \cdot \mathrm{m} \cdot \frac{\mathrm{F}}{\mathrm{V}} \cdot \sqrt{T_{03}(0)} \quad ; \ \mathrm{m} = \sqrt{\frac{\gamma}{\mathrm{R_\Gamma}} \cdot \left(\frac{2}{\gamma+1}\right)^{\frac{\gamma+1}{\gamma-1}}}$$

The gas parameters in volume 3 are considered as stagnation parameters for moving gas in volume 2 acting on accelerated projectile 1.

$$P_{02} = P_{03}$$

$$T_{02} = T_{03}$$

The static pressure acting on the projectile is determined by stagnation pressure P_{02} and the velocity (Mach number) of the projectile.

$$P_2 = P_{02} / \left(1 + \frac{\gamma-1}{2} \cdot M_1^2\right)^{\frac{\gamma}{\gamma-1}}$$

To determine the Mach number M_1, the relative velocity of the projectile in relation to the maximum possible speed at a given stagnation temperature T_{02} is determined.

$$La_1 = W_1 / \sqrt{2 \cdot C_P \cdot T_{02}}$$

Static temperature corresponding to this speed

$$T_2 = T_{02} \cdot (1 - La_1^2)$$

Mach number corresponding to the velocity of the projectile at this static temperature:

$$M_1 = W_1 / \sqrt{\gamma \cdot R_\Gamma \cdot T_2}$$

The force acting on the projectile

$$R_1 = P_2 \cdot F$$

The acceleration

$$A_1 = R_1 / m_1$$

Projectile speed gain

$$dW_1 = A_1 \cdot dt$$

Changing the position of the projectile

Journal of Physics: Conference Series **1404** (2019) 012105 doi:10.1088/1742-6596/1404/1/012105

$$dX_1 = W_1 \cdot dt + A_1 \cdot dt \cdot dt / 2$$

Unsteady equations are solved by the Euler method with constant step *dt* in time. The calculation continues from the beginning of the projectile moving to its leaving from the barrel ($X_1 = L$). Tables 1 and 2 show the results of calculating the achieved velocities of the projectiles using air and helium as the working gas.

Table 1. Working gas is air, the volume of the gas receiver is 0.14 m^3, T = 600 K, the barrel length is 30 m, and the diameter is 50 mm.

Projectile weight, kg	120 bar	140 bar	160 bar	180 bar	200 bar
0.1	928.17	938.54	946.79	954.00	961.12
0.2	872.37	886.31	897.58	907.05	914.92
0.3	831.63	847.99	861.09	872.37	882.05
0.4	798.23	816.82	831.63	844.30	854.72
0.5	769.94	789.87	806.09	819.89	831.63

Table 2. Working gas is helium, the volume of the gas receiver is 0.14 m^3, T = 600 K, the barrel length is 30 m, and the diameter is 50 mm.

Projectile weight, kg	120 bar	140 bar	160 bar	180 bar	200 bar
0.1	1832.88	1885.29	1928.65	1964.07	1994.92
0.2	1551.58	1619.19	1675.94	1723.34	1764.31
0.3	1361.29	1435.02	1497.86	1551.58	1597.81
0.4	1219.96	1296.40	1361.29	1417.79	1467.65
0.5	1110.55	1186.20	1252.27	1309.63	1361.29

It can be seen that the maximum possible speed for a heavy projectile weighing 0.5 kg using heated air is 832 m/s; and when using heated helium, it is 1361 m/s.

3. The results of experiments on a prototype ballistic facility

To test the basic structural elements, a model (prototype) of a ballistic facility was developed and manufactured (see figure 2).

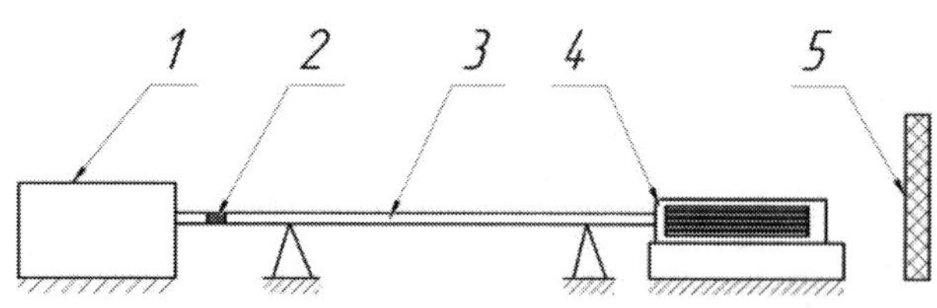

Figure 2. Scheme of the prototype ballistic facility.
1 - pressure receiver, 2 - projectile, 3 - barrel, 4 - muzzle attachments, 5 - target.

The prototype of ballistic facility consists of pressure receiver 1, into which compressed gas is supplied. A pallet with projectile 2 is installed in ballistic barrel 3 and secured using a special holding device. The ballistic barrel is made of a honed (Ra = 0.3) pipe with inner diameter of 50 mm and length of 8 to 21.5 m. At the end of the barrel, muzzle nozzle 4 is installed, on which longitudinal grooves are made to divert the gas stream. The pressure receiver is a cylinder (figure 3) with volume of 0.04 m^3, equipped with a push gas supply line and a manometer to control the initial pressure.

Figure 3. Pressure receiver.

Figure 4. Thrown projectile and pan.

The pallet is made in the form of a cylinder, at the end of which a rubber cuff is installed, which carries out the sealing of the barrel. The material of the pallet PTFE was chosen due to the low slip coefficient. The projectile is placed in the pallet (figure 4). Various options for separating the projectile and the pallet were considered. The main method of separation was chosen aerodynamic. The pallet consists of two or four identical parts, not fastened together. When a projectile with a pallet leaves the barrel, the pallet is divided into its components due to aerodynamic forces. The accelerator acceleration speed is determined using target frames mounted on a section of the barrel, with a known distance between them. The time interval between the operations of the target frames was determined using the high-speed ADC LA-2USB with frequency discretization of 500 kHz. Using the prototype of a ballistic facility, test experiments were performed to accelerate heavy bodies. For the experiments, pallets made of PTFE were used, in which metal cylinders h×d = 15×15 mm were installed. The receiver was filled with gas with a required pressure and then a shot was fired. A series of experiments for cylinders h×d = 15×15 mm and barrel with length of L = 8 m was made. The total mass of the pallet and cylinder was 130 g. Mass of the cylinders was 20.8 g. Air was used as the pushing gas. The receiver was 40 dm^3. The results of a series of experiments are shown in tables 3 and 4.

Table 3. The results of experiments using air.

№ No.	m, kg	P_{rec}, MPa	U_{exp}, m/s	U_{theor}, m/s	Δ, %
1	0.13	0.5	218	236	7.6
2	0.13	1	290	323	10.2
3	0.13	1.5	340	380	10.5
4	0.13	2	397	447	11.2
5	0.13	3	430	493	12.8
6	0.13	4	470	523	10.1
7	0.13	4	430	523	17.8
8	0.13	4	450	523	14.0
9	0.13	4	500	523	4.4
10	0.13	4	485	523	7.3
11	0.13	5	490	550	10.9

 1404 (2019) 012105 doi:10.1088/1742-6596/1404/1/012105

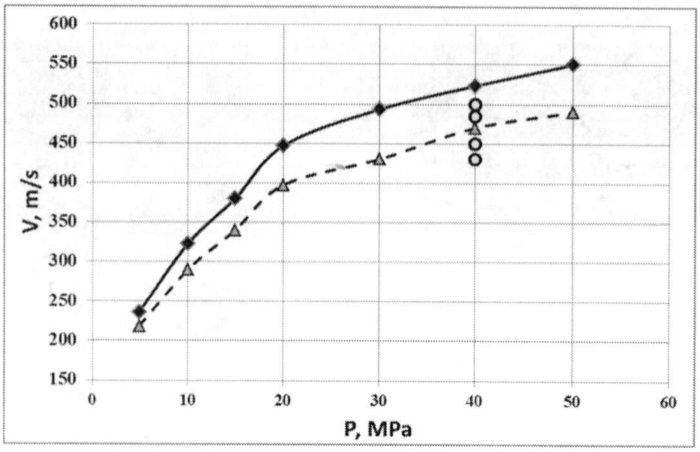

Figure 5. Dependence of speed on pressure in the receiver (barrel length 8 m, diameter 0.05 m).

The solid line in the graph (figure 5) shows the dependence of the speed obtained in the theoretical calculation on the pressure in the receiver. The dashed lines show the experimental velocity values at various initial pressures in the receiver. It is seen that the experimentally obtained velocity values are consistent with the calculated data. The average deviation of the values is 10%. The resulting difference is caused by the absence in the theoretical model of the friction forces of the pallet when moving along the barrel of the installation. Several experiments (No. 6-10), performed at pressure in the receiver of 4 MPa, show a spread of the obtained velocities, i.e. repeatability of experiments. Table 4 shows the results of experiments using helium as a pushing gas and different barrel lengths (table 4). In experiment No. 2, steel rod h×d = 300×10 mm with a mass of 323 g was used.

Table 4. The results of experiments using helium.

№ No.	m, kg	P_{rec}, MPa	U_{exp}, m/s	U_{theor}, m/s	Δ, %	L, m
1	0.130	5	700	727	3.7	8
2	0.323	4	420	490	14.3	21.5

It is seen that the use of helium can significantly increase the speed of throwing heavy projectiles.

Conclusion

The theoretical and practical study of the possibility of creating a single-stage ballistic facility for throwing projectiles weighing from 0.1 to 0.5 kg at a speed of 600 to 1300 m/s has been performed. A mathematical model has been developed that describes the acceleration of propelled projectile in a single-stage gas ballistic facility. Parametric calculations of various acceleration options for projectiles weighing from 0.1 to 0.5 kg were performed. When using compressed air, it is possible to obtain a throwing speed of projectile up to 832 m/s with weighing 0.5 kg. When using helium, it is possible to obtain a throwing speed of up to 1361 m/s. A prototype of a ballistic facility with the barrel diameter of 50 mm and length of up to 21.5 m was created. Trial experiments were carried out, which confirmed the correctness of the main decisions in the created installation. When using air without heating, the acceleration speed of 500 m/s was obtained. When using helium, the acceleration speed of 700 m/s was obtained.

Acknowledgments

This work was carried out within the framework of the Program of Fundamental Scientific Research for the Russian state academies of sciences in 2013-2020 (project No. AAAA-A17-117030610121-9) and was partly supported within the framework of the Program of Basic Research of the Presidium of the Russian Academy of Sciences entitled "Basic Fundamentals of Breakthrough Technologies in the Interests of National Security".

References

[1] Cable L 1973 Super Throwing Accelerators *High-speed shock phenomena* ed. V N Nikolaevsky(Moscow: Mir) pp 13–28
[2] Leconte K 1971 *Physics of fast processes* vol 2 ed. O N Zlatina (Moscow: Mir) p 252
[3] Pilyugin B N Leontiev P E 1999 Possibilities of increasing the speed of throwing bodies in ballistic facility (Institute of Mechanics of Moscow State University Preprint No. 52–99)
[4] Titov V M Shvetsov G L 1986 Acceleration of particles to high speeds *Dynamics of a Continuous Medium* vol 78 (Novosibirsk: Institute of Hydrodynamics SB RAS USSR) pp 128-136
[5] Khvostov I I and Shirmanov P M 1973 *ONTI TSAGI* **425** 60
[6] Ostashev V E Lebedev E F Fortov V E 1993 *TVT* **2(31)** 313–320
[7] Golubyatnikov A N Leontiev N E Pilyugin N N 2003 **2(2)** 97–125
[8] Zvegintsev V I 2014 *Gas-dynamic facility of short-term action Part 1 Facilities for scientificresearch* (Novosibirsk: Parallel)

Simulation of linearly extended streamer with planar geometry at atmospheric pressure

A L Alexandrov, I V Schweigert and Dm E Zakrevsky

Khristianovich Institute of Theoretical and Applied Mechanics SB RAS,
4/1 Institutskaya str., Novosibirsk 630090, Russia

a_alex@itam.nsc.ru

Abstract. The cold atmospheric plasma (CAP) jet is used in different applications, for example, in medicine. Typical plasma devices have the cylindrical symmetry. In this work, the new type of device with planar geometry of CAP jet is considered. The initiation of two-dimensional streamer and its propagation from linearly extended electrode to the treated surface is considered with the help of 2D fluid discharge model. Differently to the case of an axially symmetric plasma jet, the streamer head has a plate shape and the ionization wave propagates with considerably larger velocity.

1. Introduction

Nowadays the cold atmospheric plasma (CAP) jet is widely used for the healthcare, such as wound healing, sterilization and even anticancer therapy (see for example, [1]). Low power atmospheric plasma operates at alternative voltage with frequency of tens kHz and amplitude of few kV, without dangerous electric current or thermal damage to the treated tissue and producing chemical species and electrical field necessary for medical treatment and drug delivery. Usually the CAP jet is formed in cylindrical tubes [2, 3]. It was known that the plasma jet is initiated by a streamer, whose leading front propagates at velocities of 10^6–10^7 cm/s. The streamer originates on the powered electrode inside the dielectric tube and moves through a stream of inert gas to the treated tissue, placed on the grounded electrode. The treatment effect is achieved by presence of high electric field, various chemical species, produced by plasmochemical reactions in atmospheric air and delivered to the tissue by plasma jet. In work [2], simulation of streamer propagation in a cylindrical geometry, with axial symmetry, have been performed using the 2D fluid model of electric discharge. The structure of streamer head, ionization rate and electric field distribution were appropriately described by the model. Recently a new geometry of the CAP jet for medical applications was suggested, where a linearly extended powered electrode in a planar dielectric gap are used instead of cylindrical tube, so the plasma jet has a planar geometry. The aim of this work is to consider the initiation and propagation of streamer in such planar geometry, by means of the 2D fluid plasma model, similar to proposed in [2].

2. The physical model and simulation algorithm

The general view of planar CAP jet device is shown in figure 1. The linear powered electrode (on the top of photo) is placed in 2 mm wide and 25 mm depth gap between dielectric walls. A sinusoidal voltage (amplitude 3-7 kV and frequency 10-30 kHz) is applied on the powered electrode and the grounded electrode is 50 mm apart from it.

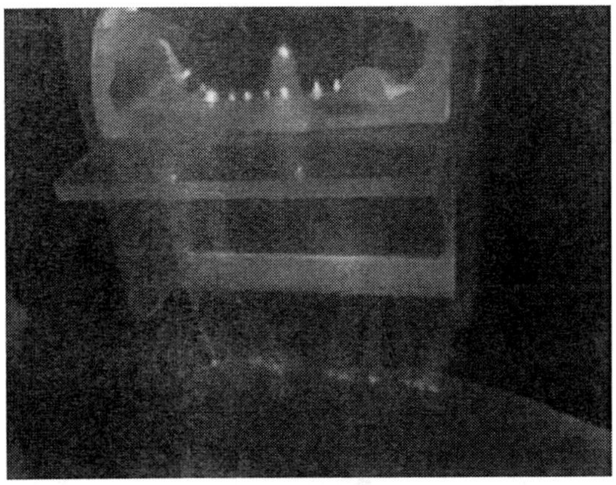

Figure 1. The ignition of CAP jet with planar geometry.

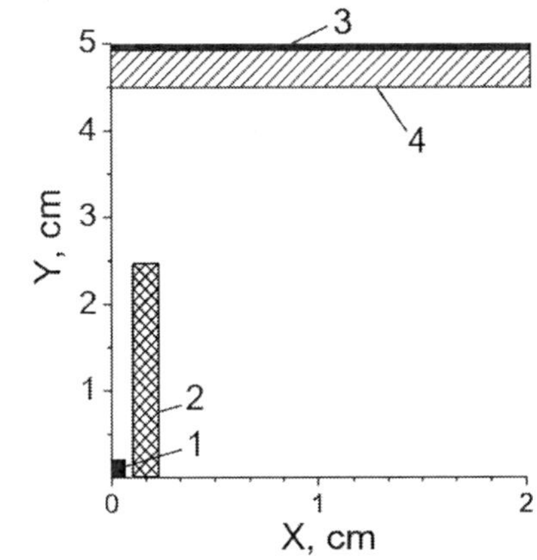

Figure 2. The scheme of simulation domain: 1 – powered electrode, 2 – dielectric wall of the gap, 3 – grounded electrode, 4 – water or PBS layer.

The streamer ignites during the positive phase of voltage and propagates towards the grounded electrodes for time less than one microsecond.

To simulate the streamer propagation, the fluid discharge model [2, 4] is applied, which includes three continuity equations for electrons, ions and mean electron energy:

$$\frac{\partial n_e}{\partial t} = div(\mu_e n_e \vec{E} + \vec{\nabla}(D_e n_e)) + \alpha_i n_e N_g - \beta n_e n_i ,$$

$$\frac{\partial n_i}{\partial t} = div(\mu_i n_i \vec{E} + \vec{\nabla}(D_i n_i)) + \alpha_i n_e N_g - \beta n_e n_i ,$$

$$\frac{\partial (n_e U_e)}{\partial t} = div(\mu_u n_e U_e \vec{E} + \vec{\nabla}(D_u n_e U_e)) + e\vec{E}(\mu_e n_e \vec{E} + \vec{\nabla}(D_e n_e)) - v_i n_e N_g - \beta n_i n_e U_e$$

and the Poisson equation for the electric potential is also solved:

$$\Delta\varphi = 4\pi e(n_e - n_i), \quad \vec{E} = -\nabla\varphi .$$

Here n_e, n_i, μ_e, μ_i, D_e, D_i are the densities, mobilities and diffusivities of electrons and ions respectively, U_e, μ_u, D_u are the mean electron energy with its effective mobility and diffusivity, E and φ are the electrical field vector and potential, N_g is the gas density, β is the electron-ion recombination coefficient. The transport coefficients together with ionization coefficient α_i and electron unelastic collision rate coefficient rate v_i are obtained by solving the Boltzmann equation for local E/N_g value. The detaled description of the model is published in [2]. Also the photoionization process in the region of streamer head is added to the model, as described in [2].

The simulation domain for planar-geometry streamer is shown in figure 2. The domain is symmetrical on Y-axis and extended in Z direction. The grounded electrode is covered by 5 mm thick layer of water of phoshate-buffered saline (PBS) layer, denoting the treated biological specimen. For streamer simulation, a positive voltage is applied to powered electrode and zero voltage to the grounded one. A pre-ionized area with plasma density of 10^{10} cm^{-3} is specified near the Y-axis in order to accelerate the calculations. The continuity equations are integrated with time step of 10^{-11} s together with solution of the Poisson equation. The spatial calculation grid needs to be fine and has 0.002 cm step in both dimensions, although gradually extends in X-direction outside the dielectric gap.

Journal of Physics: Conference Series **1404** (2019) 012106 doi:10.1088/1742-6596/1404/1/012106

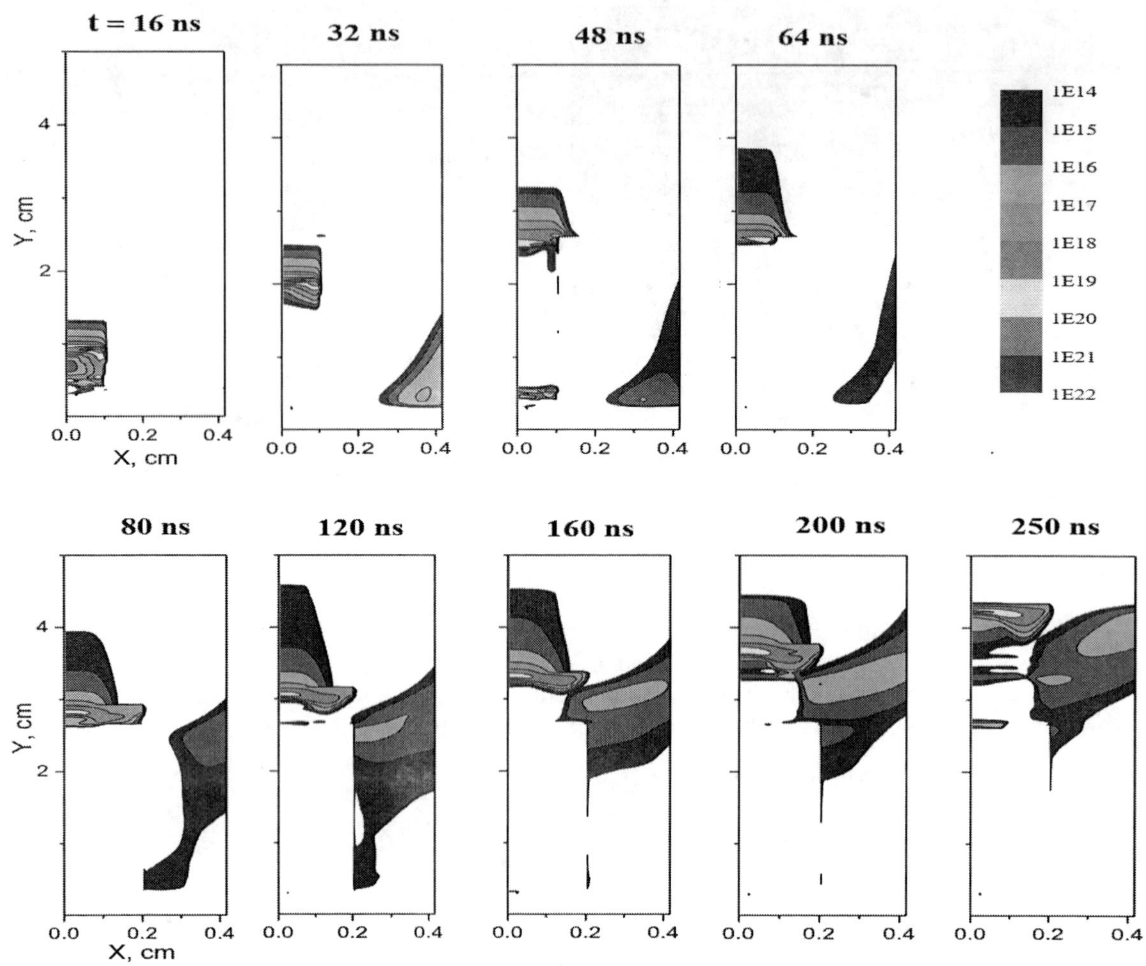

Figure 3. The configuration of ionization wave during 16–250 nanoseconds of streamer motion. The plots show contour maps of ionization rate $\alpha_i n_e N_g$ in cm^3/s for different time moments.

3. Results and discussion

The streamer initiates as a wave of ionization moving from the powered electrodes inside the dielectric gap. The propagation of ionization wave is shown in figure 3. The velocity of planar streamer head is extremely large inside the gap and achieves 10^8 cm/s at the first 40 ns of streamer propagation. After this, when the ionization wave reach the escape from the gap, its velocity strongly decreases (see the moments 48-60 ns of streamer motion in figure 3) and after this the ionization wave propagates in the atmospere with more spatially extended structure (see moments 80-250 ns). The wave velocity in the atmoshere is less than 10^7 cm/s. The spatial region of large electrical field and strong ionization becomes wider in both directions, but their magnitudes diminish. It is more clearly seen in figure 4, where the electrical potential φ and ionization rate are plotted along the Y-axis. The motion of streamer significantly decelerates at the exit from the gap for a few nanoseconds, but then goes on with slower velocity, and also smaller electric field and ionization rate magnitudes in the streamer head. Note also that the maximal gradients of φ and magnitudes of ionization rate are shifted from the Y-axis, thus not laying in the symmetry plane. This is similar to the results of axially symmetric model, where these maximal magnitudes were also aside the axis of symmetry, thus streamer head having a toroidal form [2]. In the gap, the streamer head fills all space between walls. After escaping the gap, the head firstly remains 0,2 cm in width, but tends to extend when reaching the grounded surface (see figure 3, $t = 250$ ns).

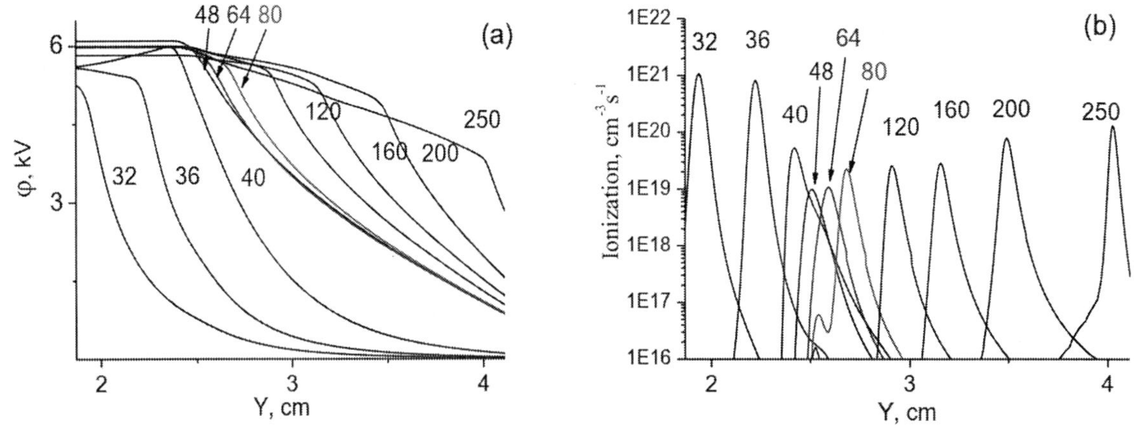

Figure 4. The profiles of electric field potential (a) and ionization rate (b) for the different time moments of of streamer propagation, 32-250 ns, plotted along the *Y*-axis.

The main difference in simulation results compared with axially symmetric streamer is much larger velocity of streamer head propagation, which in [2] was 10^7 cm/s in the dielectric tube and 10^6 cm/s in the atmosphere. The presented simulations in planar geometry shows velocity $7 \cdot 10^7$ cm/s in the dielectric gap and $8 \cdot 10^6$ cm/s in the atmosphere. This difference may be due to linear extension of large electric field region in *Z*-direction.

The maximal ionization rate in the streamer head has the similar order of magnitude as for axially symmetric case, exceeding 10^{21} cm^{-3}/s in gap. It diminishes to 10^{19} cm^{-3}/s when crossing the gap exit and then gradually rises to 10^{20} cm^{-3}/s after departure from the gap (figure 4b).

In conclusion, the 2D fluid model simulations of planar streamer propagation in application to CAP treatment of biological tissues have been performed. The structure of two-dimensional, linearly extended streamer was considered, its evolution during propagation through linear gap and extension into atmosphere was studied. The main difference with streamer simulations in axially symmetric geometry is the much larger velocity of propagation.

Acknowlegements
Authors are grateful to the Russian Science Foundation, project No. 19-19-00255, for supporting the work.

References

[1] Keidar M, Walk R, Shashurin A, Srinivasan P, Sandler A, Dasgupta S, Ravi R, Guerrero-Preston R and Trink B 2011 *Br. J. Cancer* **105** 1295
[2] Schweigert I V, Vagapov S, Lin L and Keidar M 2019 *J. Phys. D: Appl. Phys.* **52** 295201
[3] Schweigert I V, Vagapov S, Lin L and Keidar M 2018 *J. of Physics: Conf. Series.* **1112** 012004
[4] Alexandrov A L and Schweigert I V 2010 *J. Exp. Theor. Phys.* **110** 498

Journal of Physics: Conference Series **1404** (2019) 012107 doi:10.1088/1742-6596/1404/1/012107

Comparison of nonequilibrium dissociation models in the direct simulation Monte Carlo method

A Ye Galeyev, A V Kashkovsky, A A Shevyrin and Ye A Bondar

Khristianovich Institute of Theoretical and Applied Mechanics,
4/1 Institutskaya, 630090 Novosibirsk, Russia

galart32@gmail.com

Abstract. Various chemical dissociation reactions models (Total collision energy model – TCE, Quantum Kinetic model – QK and Extended vibrational bias model – EVB) were implemented in the direct simulation Monte Carlo code SMILE++. It was shown that the dissociation models considered predict the same reaction rate over a wide range of temperatures in thermal equilibrium case. The analysis for the nonequilibrium case was carried out on the basis of dissociating nitrogen gas flow modelling around the wedge for the Hornung and Smith experiments conditions. All dissociation models proposed predict various results both in the standoff distance of the bow shock wave and in the temperature values behind it. It was generally shown, that taking into account the vibrational favoring (during dissociation) results in decreasing the reaction rate, which can be manifested in increasing the shock wave standoff distance.

1. Introduction

Nowadays, the direct simulation Monte Carlo (DSMC) [1] method proposed by Graeme Bird in the early sixties is the main tool for studying aerodynamics at high-altitudes above about 80 km, when the free mean path of gas particles becomes comparable with the linear size of space vehicle. The underlying ideas of the method are that, firstly, the rarefied gas flow is considered as a set of $10^5 – 10^{10}$ modelling particles (each of them represents a large number of real gas molecules), and secondly, continuous motion and collisions of gas particles are divided into two consecutive stages. Flow fields of different parameters are obtained by averaging of molecular properties in cells of the computational domain.

It is important to note that the thermal nonequilibrium that occurs in the viscous layer behind the bow shock wave when a spacecraft enters the upper atmospheric layers complicates the gas flow structure and has a strong effect on the chemical reactions rates, including dissociation, which primarily determines the degree of thermal loads exerted on a space vehicle. Thus, correct modelling of chemical reactions in the DSMC method [2] is an essential objective for high-altitude aerodynamics calculations, since nonequilibrium chemical reactions have a significant effect on aerothermodynamic characteristics of the spacecraft. It should be noted, that in the DSMC method chemical reactions are implemented during the binary collision stage and it is necessary to know the reaction probability as a function of colliding particles parameters.

One of the most widely used software systems for rarefied gas flows calculation using the DSMC method is SMILE (Statistical Modeling In Low-density Environment) [3]. It is a software tool developed in the laboratory of computational aerodynamics of the Institute of Theoretical and Applied

Mechanics of the Siberian Branch of the Russian Academy of Sciences. Its more modern and flexible version is the SMILE++ [4, 5] code, which is being actively developed at the moment.

The main objectives of this work are implementation and validation of existing molecular models of chemical dissociation reactions in the SMILE++ software system, as well as their analysis in the case of nonequilibrium supersonic flow around a wedge.

2. Dissociation reaction models

In this work, the following chemical dissociation reactions ($A_1A_2 + R \rightarrow A_1 + A_2 + R$) models were considered:

1. Total Collision Energy (TCE) model [6]. In this model the probability of chemical dissociation reaction depends on the total energy of colliding particles:

$$P_d(E_c) = C\frac{(E_c - E_d)^{x+\frac{\xi_c}{2}+1}}{E_c^{\frac{\xi_c}{2}-1}}, \tag{1}$$

where $E_c = E_{tr} + E_{int}$ is the total energy of the colliding pair, E_{tr} is the relative translational energy, E_{int} is the internal energy of the colliding particles equal to the sum of rotational and vibrational energies of both molecules, E_d is the reaction threshold energy, ξ_c is the sum of the translational and internal degrees of freedom of the colliding pair; the constants C, x are related to the constants in the Arrhenius law.

The TCE model is widely used in practice, since the Arrhenius law constants are known for a large number of chemical reactions. However, a significant drawback of this model is that it does not take into account the favoring of the vibrational energy over the translational energy in the dissociation process.

2. Quantum Kinetic (QK) model [7, 8]. In this model dissociation always occurs when the total energy of the colliding particles exceeds the molecule dissociation energy, i.e. when the following condition is true:

$$\left\lfloor\frac{E_c}{k\theta_v}\right\rfloor > \frac{\theta_d}{\theta_v}, \tag{2}$$

where θ_v, θ_d are characteristic vibration and dissociation temperatures; $E_c = E_{tr} + E_v$ is the sum of the relative translational energy E_{tr} of the colliding pair and the vibrational energy E_v of the particle that may dissociate, k is the Boltzmann constant and square brackets denote truncation.

The main advantage of the QK model is the absence of any empirical coefficients. Thus, the model can be used in situations when there is no experimental data necessary to determine the parameters of other models.

3. Extended vibrational bias (EVB) model [9]. In this model the probability of dissociation strongly depends on the vibrational level of a molecule, and it can be represented as follows:

$$P_d = A_d\left(1 - \frac{E_d - E_v}{E_{tr} + E_r}\right)\exp\left(\lambda\left(\frac{E_v}{E_d} - 1\right)\right), \tag{3}$$

where E_r, E_v are rotational and vibrational energies of the dissociating molecule, A_d is the calibration constant, λ is the degree of vibrational favoring.

A number of studies noted [9, 10] that the EVB model provides better agreement with known equilibrium and nonequilibrium reaction rates than other widely used models due to the fact that it takes into account the vibrational level (vibrational favoring) of molecules. The calibration constants values for various chemical reactions are given in [10].

3. Numerical results

To perform the comparative analysis of physico-chemical models the QK and EVB dissociation models were implemented in the SMILE++ software system (the TCE model had been implemented before in this system [11] as the basic chemical model and, therefore, did not require any addition coding). Verification results and analysis of the models implemented in the SMILE++ software system for the homogeneous gas case as in [2] are presented below. To verify the correctness of the chemical dissociation reactions models implementation in the program, isothermal reservoir case calculations for a temperature range of 8000 – 20000 K were performed and the chemical dissociation rates under thermal equilibrium conditions were obtained for the following reactions: $N_2 + N \rightarrow N + N + N$ and $N_2 + N_2 \rightarrow N_2 + N + N$.

The verification results are shown in figure 1. It can be seen that an excellent agreement was obtained between the reaction rates of the implemented dissociation models with the Arrhenius law, which confirms the correctness of the dissociation models description under thermal equilibrium condition.

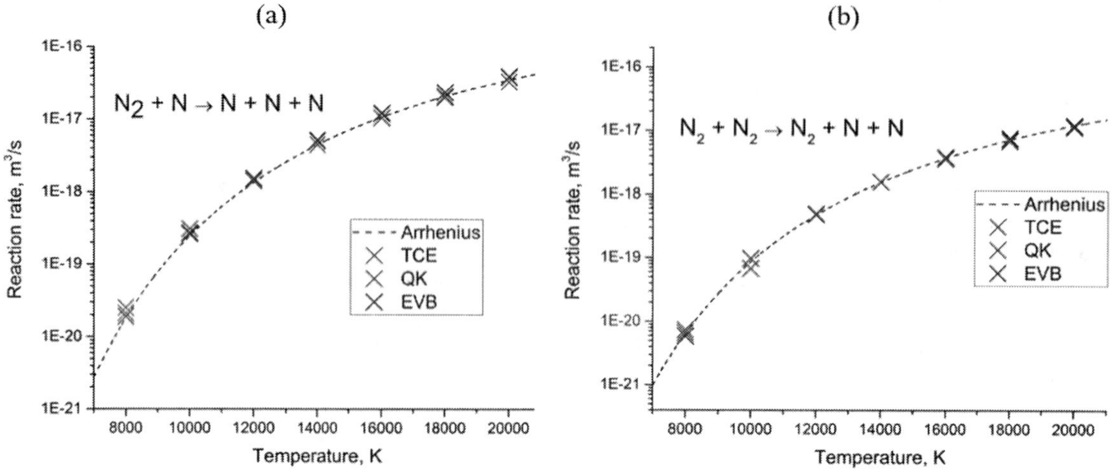

Figure 1. Dissociation reactions rates for $N_2 + N \rightarrow N + N + N$ (a) and $N_2 + N_2 \rightarrow N_2 + N + N$ (b). Symbols – calculated rates, lines –Arrhenius rates.

Table 1. Freestream conditions for the supersonic flow around the wedge.

Parameter	Value
T_∞, K	1100
$V_\infty, m/s$	5500
$\rho_\infty, kg/m^3$	0.0026
p_∞, Pa	940
Kn_∞	6.52×10^{-4}
Wedge length, m	0.051
Wedge angle, °	62.5
Composition, mol/kg	$N_2 - 31.9$; $N - 7.7$

The physico-chemical dissociation models were also verified on the problem of nonequilibrium high-enthalpy supersonic flow about a wedge for the Hornung and Smith experiments conditions [12] with $N_2 + N \rightarrow N + N + N$ and $N_2 + N_2 \rightarrow N_2 + N + N$ chemical reactions. The gas flow parameters were set similarly to the parameters given in the [13] and are shown in table 1. Elastic collisions were calculated using the Variable Soft Sphere (VSS) model [14]. The Larsen-Borgnakke (LB) model [15] was used for inelastic collisions modelling and the TCE, QK and EVB models were considered for dissociation modelling. The general flow structure obtained with all three models is demonstrated by the flowfields of translational temperature in figure 2. Numerical results for various chemical models differ both in temperature flowfields and in standoff distance of the shock wave (both QK and EVB models predict equal shock standoff distance which is more than 10% lower for the TCE model).

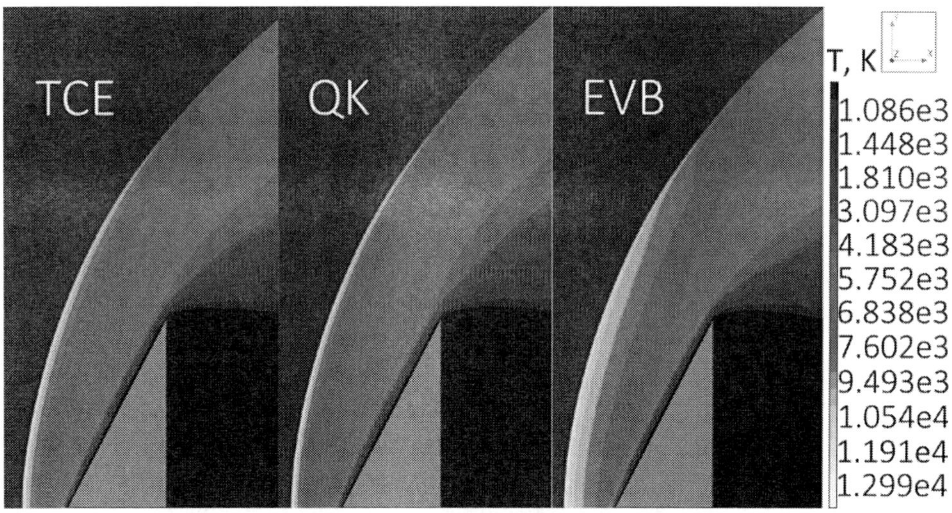

Figure 2. Translational temperature field.

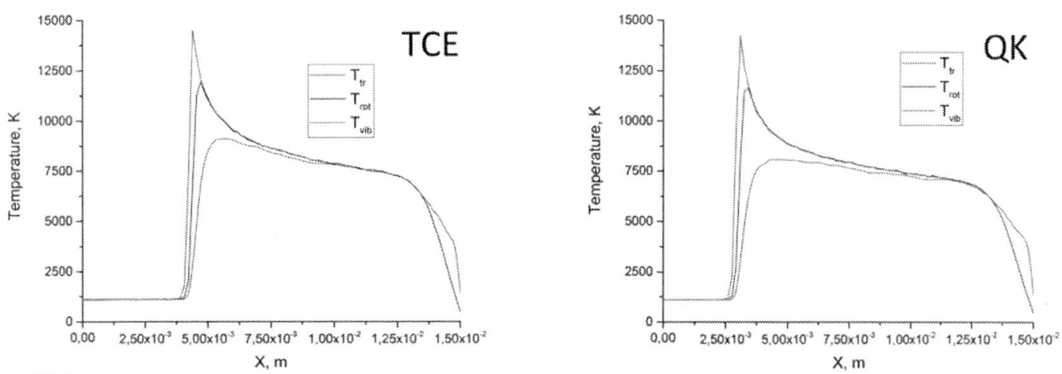

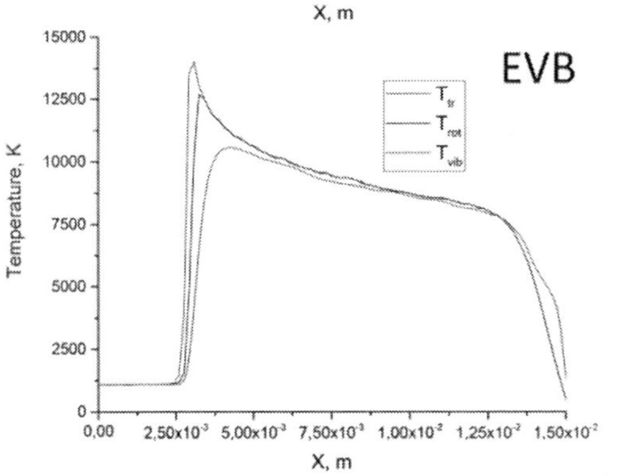

Figure 3. Translational (T_{tr}), rotational (T_{rot}) and vibrational (T_{vib}) temperatures along the stagnation streamline for the considered dissociation models.

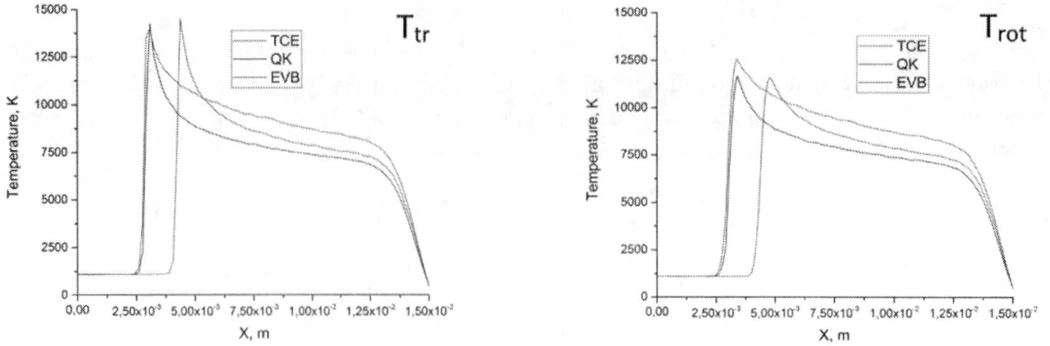

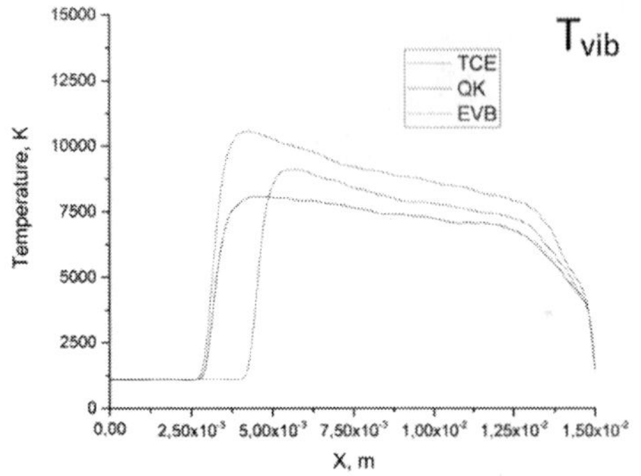

Figure 4 Translational (T_{tr}), rotational (T_{rot}) and vibrational (T_{vib}) temperatures of the nitrogen gas mixture along the stagnation streamline.

Translational, rotational and vibrational temperatures along the stagnation streamline for all considered dissociation models are shown in figure 3. The flow direction coincides with the X-axis direction, and the stagnation point was at 1.5×10^{-3} m from the origin. From these plots we can conclude, that chemical dissociation models predict the same picture from the qualitative point of view. But, from the quantitative perspective, the results obtained by different models are different.

It can be seen from the figure 3 that the translational, rotational and vibrational temperatures differ significantly from each other due to the strong thermal nonequilibrium in the shock layer. Also, it should be noted that at a distance of approximately 6×10^{-3} m from the wedge a thermal equilibrium is reached. However, when approaching the wedge an area with the vibrational nonequilibrium appears again.

Figure 4 shows the comparison of temperatures of different modes obtained with different models. The maximum gas flow temperature in the QK model is lowest among three models. The highest temperature is observed for the EVB model. It is interesting to note that the maximum value of the translational temperature is about 14000 K and is nearly identical for all three models.

4. Conclusion

The implementation in the SMILE++ software system and verification of existing molecular models of dissociation were carried out. In the process of a qualitative analysis of dissociation models, it was revealed, that while in the isothermal equilibrium case, the considered models predict the same dissociation rate, the results for the nonequilibrium flow over the wedge are substantially different both in terms of bow shock standoff distance and the values of gas-dynamic parameters, such as temperatures of different modes. To access the accuracy of the models validation studies should be performed for available ground-based or in-flight experimental data as in [13, 16-23].

Acknowledgements

The work was supported by the Russian Science Foundation (Grant No. 17-19-01375). This support is gratefully acknowledged. The authors would like to thank S. N. Tatarnikova for careful reading of the manuscript.

References

[1] Bird G A 1994 *Molecular Gas Dynamics and the Direct Simulation of Gas Flows* (Oxford, England, UK: Clarendon)
[2] Bondar Ye A, Gimelshein N E, Gimelshein S F, Ivanov M S and Wysong I J 2005 *Rarefied Gas Dynamics* (*AIP Conf. Proc.* vol. 762) ed M Capitelli (Melville, NY: American Institute of Physics) pp 607–13
[3] Ivanov M S, Kashkovsky A V, Gimelshein S F, Markelov G N, Alexeenko A A, Bondar Ye A, Zhukova G A, Nikiforov S B and Vashchenkov P V 2006 *Proc. 25th Int. Symp. on Rarefied Gas Dynamics (Saint-Petersburg)* pp 539–44
[4] Kashkovsky A V, Bondar Ye A, Zhukova G A, Ivanov M S and Gimelshein S F 2005 *Rarefied Gas Dynamics* (*AIP Conf. Proc.* vol 762) ed M Capitelli (Melville, NY: American Institute of Physics) pp 538-588
[5] Ivanov M S, Kashkovsky A V, Vashchenkov P V and Bondar Ye A 2011 *27th Int. Symp. on Rarefied Gas Dynamics* (*AIP Conf. Proc.* vol 1333) ed D A Levin, I J Wysong and A L Garcia (Melville, NY: American Institute of Physics) pp 211–18
[6] Bird G A 1981 *Progr. Astro. and Aero.* **74** 239–55
[7] Bird G A 1994 *Progr. Astro. and Aero.* **159** 185–96
[8] Bird G A 2011 *Phys. Fluids* **23** 106101
[9] Wadsworth D C and Wysong I J 1997 *Phys. Fluids* **9** 3873–84
[10] Gimelshein S F and Wysong I J 2017 *Phys. Fluids* **29** 067106
[11] Gimelshein S F, Gimelshein N E, Levin D A, Ivanov M S and Markelov G N 2002 *AIAA Paper* 2002-2759
[12] Hornung H G and Smith G H 1979 *J. Fluid Mech.* **93** 225–39
[13] Bondar Ye A, Markelov G N, Gimelshein S F and Ivanov M S 2006 *J. Thermophysics and Heat Transfer* **20** 699-709
[14] Koura K and Matsumoto H 1991 *Phys. Fluids A: Fluid Dynamics* **3** 2459–65
[15] Borgnakke C and Larsen P 1975 *J. Comp. Phys.* **18** 405–20
[16] Wysong I J, Gimelshein S F, Bondar Ye A and Ivanov M S 2014 *Phys. Fluids* **26**(4) 043101
[17] Shevyrin A A, Bondar Ye A, Kalashnikov S T, Khlybov V I and Degtyar V G 2016 *High Temp.* **54** 383
[18] Bird G A 1989 *AIAA Paper* **89** 1882
[19] Boyd I D 2007 *Phys. Fluids* **19** 096102
[20] Bird G A 2012 *28th Int. Symp. on Rarefied Gas Dynamics* (*AIP Conf. Proc.* vol. 1501) pp 595-600
[21] Gimelshein S F, Wysong I J, Bondar Ye A and Ivanov M S 2012 *28th Int. Symp. on Rarefied Gas Dynamics* (*AIP Conf. Proc.* vol. 1501) pp 637–44
[22] Boyd I D 2015 *J. Spacecr. Rockets* **52** 38
[23] Sebastiao I B, Kulakhmetov M and Alexeenko A 2017 *Phys. Fluids* **29** 017102

Starting flow in an impulse wind tunnel with a throttled prechamber

Yu P Gounko and I N Kavun

Khristianovich Institute of Theoretical and Applied Mechanics SB RAS,
4/1 Institutskaya st., Novosibirsk 630090, Russia

gounko@itam.nsc.ru

Abstract. The unsteady axisymmetric flow forming in the process of starting an impulse wind tunnel equipped with a throttled prechamber was numerically simulated. A regime with the operating flow Mach number M = 8 was under consideration. Numerical simulation on the basis of Reynolds-averaged Navier–Stokes code and the k-ω SST turbulence model was performed. The starting process was simulated as a sudden rupture of a diaphragm separating the prechamber volume with high pressure and temperature from low pressure part of the facility including a nozzle, an operating section and an exhaust vacuum tank.. The diaphragm rupture initiates a gasdynamic process associated with the Riemann problem when the primary shock wave arises which moves through the nozzle to the operating section. Computed data show another "inverse" shock wave emerging in the nozzle. A starting "shock-wave package" forms between these shock waves, it includes a flow portion of high temperature and pressure. The specificity is discussed of identified the characteristics and flow patterns forming during the process of moving the shock-wave package.

1. Introduction

As noted in [1, 2], hot-shot wind tunnels have been developed and used from the years 1960th through 1970th for experimental investigations of gasdynamic problems of hypersonic flying vehicles propelled by air-breathing engines and designed to fly with Mach numbers M > 4-5. These are impulse high-enthalpy facilities of short duration which operating principle is a virtually "instantaneous" input of energy by an electric discharge to the working gas in a prechamber separated by a diaphragm from the other parts of the facility a nozzle, subsequent an operating section and an exhaust vacuum tank. Prior to the wind-tunnel start up, the working gas is feed into the prechamber volume under a high pressure and the vacuum tank is pumped out to a very low pressure. As soon as the energy is input, the diaphragm ruptures and the working gas begins to outflow through the nozzle to the operating section where a tested model is installed. An operating pseudo-steady flow occurs in the wind tunnel after a certain starting process.

Simple hot-shot wind tunnels were used in earlier years. They were characterized by a virtually unchanged operating flow Mach number but gradually decreasing pressure and temperature. An aerodynamic facility IT-301 of such a kind was introduce into operation at ITAM in 1970 [1-3] and demounted at the present time. Improved hot-shot wind tunnels of different arrangement were developed over time. In particular, a hot-shot wind tunnel IT-302M was developed at ITAM in 1980 [2,3], and it is operated nowadays. It has two operating regimes, a one provides stabilizing the pressure in the prechamber by a moving two-stage piston, and another is a regime without pressure stabilization when the operating gas flows out from the capacity of a fixed volume. And besides, its prechamber is throttled for increasing the operational regime duration.

The initial phase of forming the flow at starting impulse wind tunnels is largely unexplored. At the same time, the starting process of impulse wind tunnels proper affects the starting process of inlets or ducted bodies tested in them as studied at a numerical simulation of these processes as applied to an impulse wind tunnel of a type IT-301 [4]. The unsteady process of the flow evolution in the IT-302M was numerically simulated in [5], where a porous insert was used to model a prechamber throttling. Flow patterns of this process are presented in [5] but their wave structure was not analyzed. The same unsteady process of the flow evolution in an impulse wind tunnel which had a simplified configuration equivalent to IT-302M was numerically simulated in the present work. Throttling of the prechamber was provided by a grate with round slots equivalent to holes of the actual throttling insert of IT-302M. The regime of flowing the operating gas from the capacity of a fixed volume was considered. The specificity of identified characteristics and patterns of the starting flow is discussed.

2. Configuration of the considered impulse wind tunnel

A schematized configuration of the impulse wind tunnel accepted for the numerical simulation is presented in Figure 1.

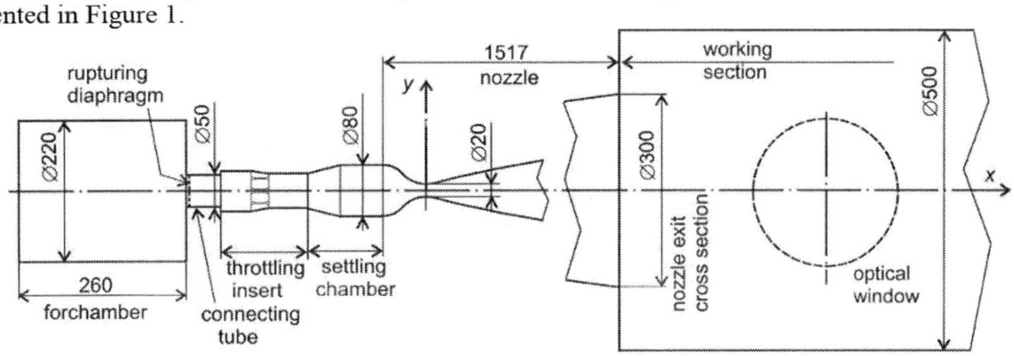

Figure 1. Scheme of an impulse wind tunnel accepted for numerical computations (sizes in mm).

The facility incorporates a prechamber of 0.22 m diameter, 0.26 m length and 0.00988 m^3 volume which in an initial time moment is filled with the heated and high pressured working gas (air). The facility includes following components: a short tube of 0.05 m diameter and 0.054 m length at entrance of which a rupturing diaphragm is assumed to be mounted; an insert of 0.134 m length with a throttling grate; a second settling chamber of 0.08 m maximal diameter and 0.16 m length. An actual throttling grate of IT-302M consists of 7 holes of 0.006 m diameter, one is central and other 6 are positioned along a circle. For the considered simplified configuration of the impulse wind tunnel, 6 holes were replaced by a round slot of the same total area. The nozzle consists of a converging section of 0.067 m length with a throat of 0.02 m diameter and also a diverging contoured section of 1.45 m length with an exit cross section of 0.3 m diameter. Geometrical parameters of the nozzle correspond to a design nominal operating flow Mach number $M = 8$ at the nozzle exit. The nozzle exit cross section was connected with an operating section of the facility of 0.5 m diameter.

3. Conditions of numerical computations

Numerical computations of the unsteady axisymmetric turbulent flow forming in the considered impulse wind tunnel were carried out on the basis a Reynolds-averaged Navier–Stokes equations and the *k-ω* SST turbulence model. The cylindrical coordinate system was accepted in computations. The origin of its reference *Oxy*-plane was positioned in the critical cross section of the nozzle throat and the *x*-axis was directed downstream along the wind tunnel axis. The computational grid was designed with uniform spacing in the radial directions. The number of radial cells in the pre-nozzle and nozzle channels was equal to 100, the $4.2 \cdot 10^3$ cells were distributed along the axis of this domain. The time step was 0.1 μs with 200 iterations per step.

As for the impulse wind tunnel accepted for computations, initial parameters of the working gas (air) in the prechamber corresponding to a moment just before the diaphragm rupture were preset as $P_{ch} = 3.5 \cdot 10^7$ Pa, $T_{ch} = 890$ K; the initial pressure and temperature of the air in the vacuum tank and in the wind tunnel duct were equal to $P_v = 215$ Pa, $T_v = 290$ K. The heat capacity, viscosity, and thermal conductivity of the air were set depending on the temperature. The boundary layer was not resolved; near-wall functions for its determining were used. The walls of the wind tunnel were assumed to be adiabatic.

4. Initial flow corresponding to the Riemann problem

The well-known one-dimensional gasdynamic description of the process arising at the sudden rupture of a thin diaphragm is based on the solution of the Riemann problem for the perfect gas. The rupture of the diaphragm generates a high-speed flow which spreads into the low-pressure part of the tube initiating a primary shock wave propagating in the same direction. The high-pressure gas expands through an expansion wave which propagates into the high-pressure part of the tube. The expanded gas is separated from the compressed gas by a contact discontinuity, which can be regarded as a fictitious membrane traveling into the low-pressure part of the tube at a constant speed. The flow Mach number and temperature vary in a jump on this discontinuity but the velocity and pressure are the same across this surface. The initial data for the Riemann problem were at the rapturing diaphragm position $x = -0.366$ m.

The numerical solution of the Riemann problem occurs in the tube connecting the prechamber and the throttling insert. Note as for the considered axisymmetric flow, the shock waves or other gasdynamic discontinuities occurring in it have the shape of the surfaces of revolution; and in a longitudinal Oxy-plane, the flow picture is analogous to that for a 2D flow.

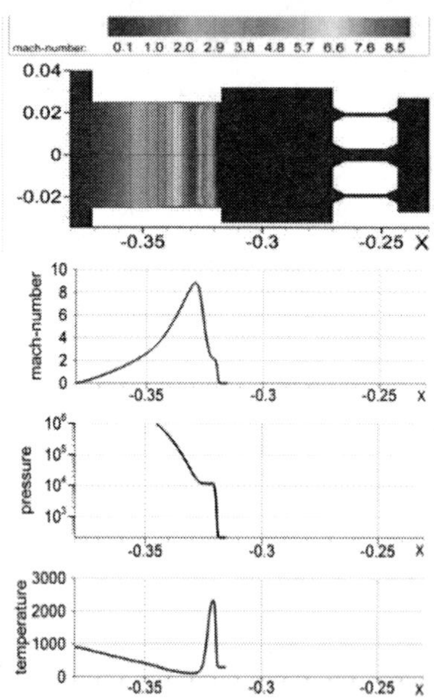

Figure 2. Flow pattern (Mach number contours) and axial distributions of flow parameters at $\tau = 23$ μs.

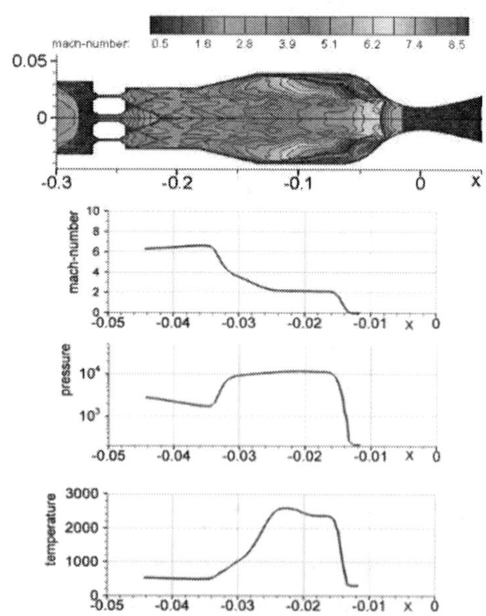

Figure 3. Flow pattern and axial distributions of flow parameters in the throttling insert and settling chamber at $\tau = 150$ μs.

A flow pattern obtained at $\tau = 23$ µs is shown in Figure 2. The flowfield is pictured as the Mach number contours, distributions of Mach number, static pressure and temperature along the flow axis are also depicted. The primary shock wave has reached a position $x \approx -0.319$ m, the flow Mach number after this shock wave is $M_1 \approx 2$. An interval $\Delta x = -0.323...-0.327$ m with constant pressure corresponds to a "smeared" contact discontinuity. An interval $x < -0.33$ m corresponds to an expansion wave.

5. Flow in the throttling insert and settling chamber

At $\tau > 23$ µs, the primary shock wave comes out from the intermediate connecting tube into the duct section with the throttling insert. This shock wave, and other gasdynamic discontinuities following it, undergo here multiple reflections and interactions, so the flow pattern here becomes more complex and it is not interpreted in detail.

A flow pattern in the duct section between the prechamber and the nozzle is shown in Figure 3 for $\tau = 150$ µs. There is a shock wave located at $x \approx -0.28$ m in the duct section before the throttling grate which is reflected from the throttling grate and moves back in the direction of the prechamber. The primary shock wave has reached a cross section $x \approx -0.0134$ m somewhat before the nozzle throat, and its front has a flat shape. Distributions of flow parameters in an interval $\Delta x = -0.045...-0.012$ m are presented in Figure 3. A "smeared" contact discontinuity approximately corresponds to a segment $\Delta x = -0.032...-0.145$ m where there is a plateau pressure. Along a segment $\Delta x = -0.035...-0.32$ m, there are a downstream decrease in the flow Mach number as well as increases in the pressure and temperature. This indicates of a presence of the "smeared" inverse shock wave moving downstream more slowly than the main flow. A peak level of the pressure and temperature takes place in the flow portion between the inverse and primary shock waves. That is why this local flow pattern can be named a "shock-wave package" or a "shock-wave cork" [4]. In the flow just upstream of the shock-wave package, there are considerable inhomogeneities near the duct wall. They evidently affect the evolution of the flow in the nozzle in the process of moving the shock-wave package through it.

6. Flow in the nozzle

A flow pattern in the nozzle and distributions of the flow parameters are shown in Figure 4 for $\tau = 400$ µs. At this time, the shock-wave package has moved to an interval $\Delta x = 0.4...0.53$ m. The primary shock wave has a convex shape. The flow between the primary and inverse shock waves along the axis is moderately supersonic with $M = 1.76$-1.83, the flow just after the shock-wave package is of high Mach numbers up to $M = 7.6$.

A distinctive feature of the nozzle flow associated with moving the shock-wave package is the development of the unsteady boundary layer. After the primary shock wave, it begins to grow quickly. Its thickness is maximal near at the contact discontinuity position and then it decreases tending to reach a value corresponding to a steady supersonic flow forming in the nozzle [6]. This kind of swelling of the boundary layer in the considered flow pattern leads to a bifurcation of the inverse shock wave. As a result, an oblique shock wave arises which interacts with the inverse shock wave so the reflected shock wave emerges which falls on the boundary layer and the λ-shaped configuration of shock wave emerges.

A flow pattern in the nozzle at $\tau = 970$ µs is shown in Figure 5. At this time, the primary shock wave has reached a nozzle exit cross section ($x = 1.45$ m), and its front is almost flat. The shock-wave package extends along the nozzle axis in an interval $\Delta x = 1.27...1.45$ m. The flow between the primary and inverse shock waves is moderately supersonic with $M = 1.73$-1.79. The boundary layer swelling gets an interval $\Delta x = 0.8...1.32$ m. In the given case, this swelling leads to emergence of focusing pseudo-conical compression waves. A resultant shock wave falls onto the nozzle axis and reflects irregularly with forming a triple-shock configuration including the Mach disc and the reflected shock wave.

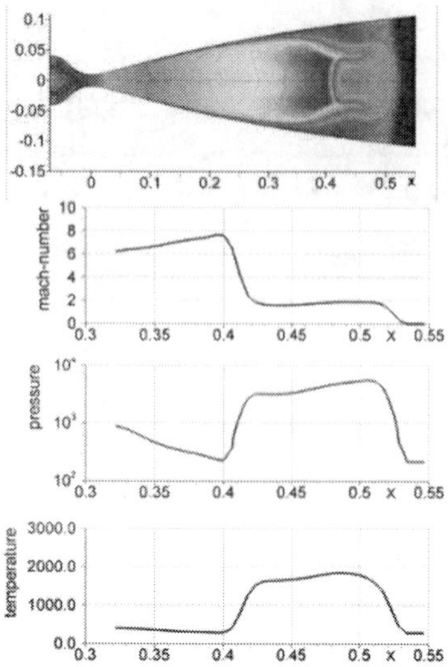

Figure 4. Flow pattern and axial distributions of flow parameters in the nozzle at $\tau = 400$ µs.

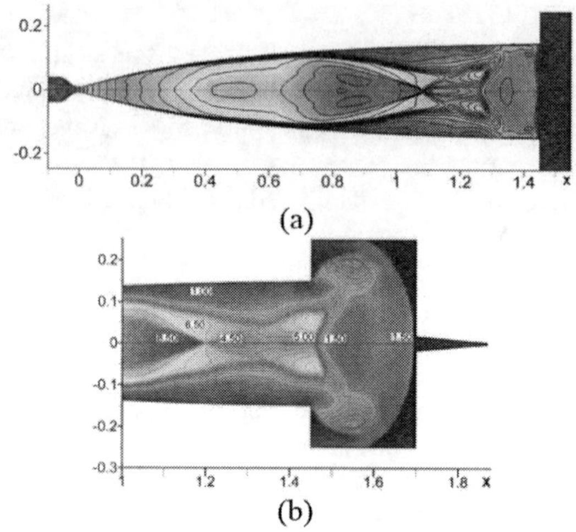

Figure 5. Flow patterns in the nozzle: (a) at $\tau = 970$ µs and (b) at $\tau = 1130$ µs.

In the given case, a size of the Mach disc is very small so "an illusory regular reflection" of shock waves from the axis is seen at $x \approx 1.07$ m.

The wave package comes out from the nozzle within ~1130 µs. The flow-field in a region at the nozzle exit for this time is shown in Figure 5. Before the exit cross section, the flow is highly non-uniform since there it contains intensive shock waves. These disturbances leave the nozzle exit at $\tau \approx 2250$ µs, then the velocity in the flow core here is enlarged up to $M = 8.5$. The process of steadying the nozzle flow begins when the swelling of the boundary layer shifts out of the nozzle exit (at $\tau \approx 2230$ µs).

7. Steadying the flow in the nozzle

Steadying the flow in the wind tunnel and in its nozzle was firstly estimated by monitoring the static pressures on the back face of the prechamber, on the side wall of the settling chamber, and at the point on the axis of the nozzle exit cross section (Figure 6). These parameters were monitored within 3.6 ms. The pressure in the prechamber decreases by ~11% within 3 ms. At the same time, the pressure in the settling chamber increases and practically stabilizes at $\tau > 2.8$ ms. The pressure in the nozzle exit cross section stabilizes on average at $\tau > 2.5$ ms, although there are its oscillations of an amplitude about 20%.

The flow parameters mass-averaged over a flow core of 0.2 m diameter were determined in various cross sections. During time period $\tau = 2.5$ - 3.6 ms, the Mach number increases by $\Delta M = 2.5\%$ to $M_{av} = 7.51$, the total pressure decreases by $\Delta p_0 = 13\%$, the total temperature decreases by $\Delta t_0 = 0.2\%$. Cross-sectional distributions of Mach number, static and total pressure and temperature are presented in Figure 7 for final monitored time $\tau \approx 3606$ µs.

8. Conclusion

The results of numerical simulations have shown the following.

The starting process of an impulse wind tunnel launches when, as it is predicted by the one-dimensional theory of the Riemann problem, the primary shock wave and contact discontinuity emerge at the diaphragm rupture. As for the considered impulse wind tunnel, these waves begin to move through the duct of variable cross-section, including the throttling insert and the settling chamber, to the nozzle. Multiple reflections and reciprocal interactions of these wave structures occur in this duct, so the flow structure is very complex and can hardly be interpreted.

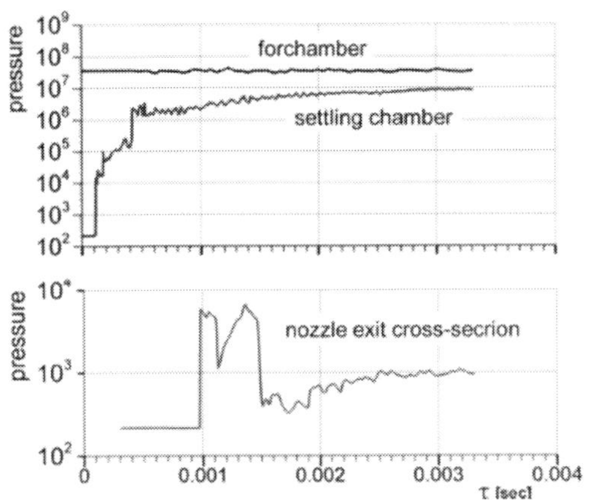

Figure 6. Varying the monitored pressures by time.

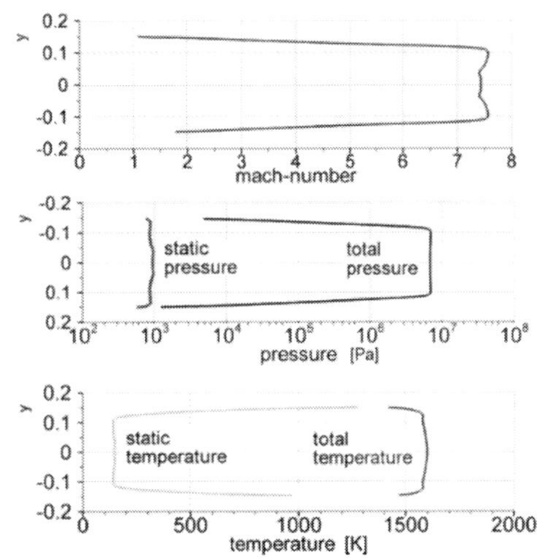

Figure 7. Distribution of flow parameters in the nozzle exit cross section at $\tau = 3606$ μs.

Despite this, further the flow in the nozzle forms where the movement of the said gas-dynamic waves is complied with that known one treated in the one-dimensional approach. The inverse shock wave subsequent to the contact discontinuity emerges, it moves downstream with a velocity lower than the flow velocity. A flow portion with very high temperature and pressure occurs between the primary and inverse shock waves which can be named as a "shock-wave package". Namely its moving through the nozzle determines the starting process of the impulse wind tunnel. Significantly, moving the shock-wave package is exposed to effects of development of the unsteady boundary layer on the nozzle wall. This is manifested in forming the current flow patterns.

In a time, when the shock-wave package comes out of the nozzle, the flow at the nozzle exit has an enlarged velocity, and then the process of steadying the flow in the wind tunnel takes place. There is no complete steadying the flow in an impulse wind tunnel in the covered time interval of the computations, only a pseudo-steadied condition was reached.

Acknowledgments
The study was performed as part of the Program of Fundamental Scientific Research of the State Academies of Sciences for the years 2013-2020 (project No. AAAA-A17-117030610137-0).

References

[1] Korolev A S, Boshenyatov B V, Drouker I G, and Zatoloka V V 1978 *Hot-shot Wind Tunnels as a Hardware for Aerodynamic Investigations* (Novosibirsk: Nauka)
[2] Zvegintsev V I 2014 *Gasdynamic Facilities of Short Duration (Part I. Facilities for Scientific Research)* (Nivosibirsk: Parallel)
[3] Fomin V M *et al.* 2016 *Experimental Methods of Shock Wave Research. Shock Wave Science and Technology Reference Library* **9** pp 315-346
[4] Gounko Y P, Mazhul I I, and Kavun N I 2017 *AIP Conf. Proc.* **1893** 030057
[5] Kutepova A I, Sidorenko A A, and Gromyko Yu V 2018 *AIP Conf. Proc.* **2027** 040069
[6] Bazhenova T V and Gvozdeva L G 1977 *Unsteady Interactions of Shock Waves* (Moscow: Nauka)

Direct Monte Carlo simulation of development of the Richtmyer-Meshkov instability on the Ar/He interface

A V Kashkovsky[1], A N Kudryavtsev[1,2] and A A Shershnev[1]

[1]Khristianovich Institute of Theoretical and Applied Mechanics, Russian Academy of Sciences, Siberian Branch, 630090 Novosibirsk, Russia
[2] Novosibirsk State University, 630090 Novosibirsk, Russia

E-mail: sasa@itam.nsc.ru

Abstract. The Richtmyer-Meshkov instability developing on the interface between helium and argon gases hit by a propagating shock wave is simulated numerically. In contrast with conventional approach based on the Navier-Stokes equations, the simulation in the current paper is performed on the molecular level using the Direct Simulation Monte Carlo method. The data averaging over short time periods is employed to decrease statistical scattering. The instability development is successfully reproduced with relatively moderate computer resources used. The time evolution of the transmitted shock wave and contact surface velocities are retrieved from numerical data as well as the instability growth rates.

1. Introduction

Instability of flows of liquids and gases is a widespread phenomenon, playing an important role in many scientific problems and technical applications. It is usually studied on the basis of the continuum approach [1, 2, 3], since it occurs, as a rule, at large Reynolds numbers and, therefore, small Knudsen numbers. However, in a number of important cases, instability can be observed for conditions when rarefaction effects can play a significant role.

One of the examples of such instabilities is the Richtmyer-Meshkov instability, which emerges under the influence of body force, the role of which in this case is played by the inertia force in non-inertial reference frame associated with a gas accelerating in a shock wave on either side of a contact surface. The shock wave moves through the contact surface from less dense to more dense medium. Natural or artificially imposed disturbances on the contact surface play the role of the initial "seed" which then grows during the instability development.

In this paper, we consider the development of two-dimensional Richtmyer-Meshkov instability at the boundary separating helium and argon for relatively low Reynolds number conditions. Numerical simulation is performed using the Direct Simulation Monte Carlo (DSMC) method. The problem statement is similar to continuum computations: at the initial time moment sinusoidal disturbance of the interface imposed, so that the initial stages of development of instability occur in the single mode regime. The resulting flow evolution reminds the flow patterns observed in experiments and continuum numerical simulations.

2. The DSMC method

The Direct Simulation Monte Carlo (DSMC) method [4] is a stochastic numerical method for solving the Boltzmann equation. In DSMC simulations, the gas flow is modeled on the basis of an ensemble of model particles, each representing a large number of real gas molecules. The simulation process is divided into two stages:

(i) free-molecular transfer of a particle according to its individual velocity

(ii) stochastic intermolecular collisions modeled in a cells of a computational grid [5, 6].

The DSMC is the most widely used technique for the numerical simulation of rarefied gas flows. The degree of rarefaction is usuallly characterized by the Knudsen number Kn, defined as $\mathrm{Kn} = \lambda/L$, where λ is the mean free path of gas molecules and L is the characteristic length scale of a flow. As the Knudsen number Kn decreases the number of model particles required for the correct DSMC simulation grows rapidly and can reach hundreds of millions and even billions of particles. However, modern supercomputers allows one to perform the DSMC simulation of the flows with Knudsen numbers as low as $\mathrm{Kn} \sim 10^{-4} \div 10^{-5}$.

3. Richtmyer–Meshkov instability

The Richtmyer-Meshkov instability (RMI) develops when a shock wave passes through an interface between two media with different densities. It is special case of the Rayleigh-Taylor instability when permanently acting body force is replaced by an impulsive action of a shock wave. This instability is widespread in natural phenomena and technical applications including astrophysics and nuclear technology. Usually it is considered on the basis of the continuum description, namely Navier-Stokes equations [7, 8, 9]. Modern computational systems, however, allows one to use kinetic approaches, including the DSMC, to simulate hydrodynamic instabilities. Kinetic approach can provide new insights to the process of an instability development [10, 11] and help in understanding of process of an instability emergence on a molecular level. Besides, many hydrodynamic instabilities can emerge in rarefied flows where the continuum approach is inapplicable.

Recently a DSMC simulation of the RMI [12] was performed using the SPARTA code [13]. However, the study was conducted for the standard atmospheric conditions in 50 μm by 400 μm domain. Approximately 10^{11} model particles and up to 1.57×10^6 cores of Sequoia, an IBM Blue Gene/Q supercomputer at Lawrence Livermore National Laboratory, were used in the computations. Using so much resources makes it difficult to run RMI multiparametric study. One of the goals of the present paper is to investigate the possibility to reproduce RMI numerically using moderate computational resources.

4. Results

A 2D plane channel with a specular solid walls (i.e. symmetry lines) was considered. The geometry of the computational domain is shown in figure 1. The left section of the channel is filled with helium, and the right one is filled with argon. The interface between these gases has a form of a half period of a sine curve with an amplitude equal to 10% of a wavelength. A shock wave with the Mach number $\mathrm{M} = 2.5$ propagates through helium from left to right. The main gasdynamic parameters are presented in table 1. 55 million of model particles and a computational grid consisting of 5000×100 cells were used in the simulation. The time step was constant and equal to $\Delta t = 10^{-7}$ s and 300,000 steps were made during the computation. Macroparameters were averaged over 1000 steps to reduce statistical fluctuation, inherent in the DSMC method. The numerical simulation was performed using 26 cores of a Intel Xeon E5-2695v4 CPU with a SMILE++ DSMC code [14, 15, 16], which is a successor to the well-known SMILE code [17].

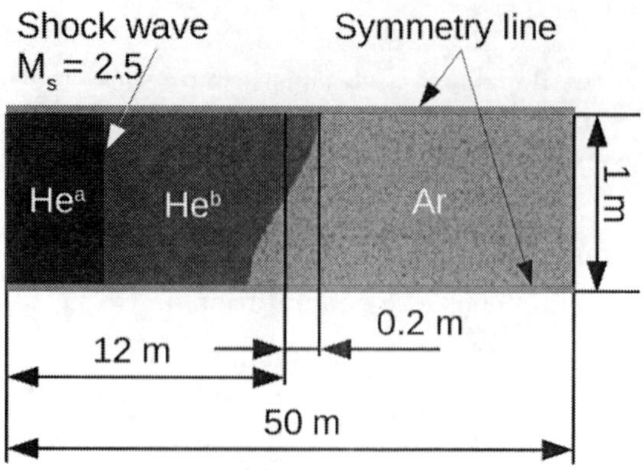

Figure 1. Computational domain geometry and initial flowfield configuration.

Table 1. Initial flowfield conditions.

	He^a	He^b	Ar
ρ, kg/m^3	2.7×10^{-5}	10^{-5}	10^{-4}
P, Pa	47	6.2	6.2
T, K	839	300	300
V, m/s	1605	0	0
Kn	6×10^{-4}	10^{-3}	4×10^{-4}

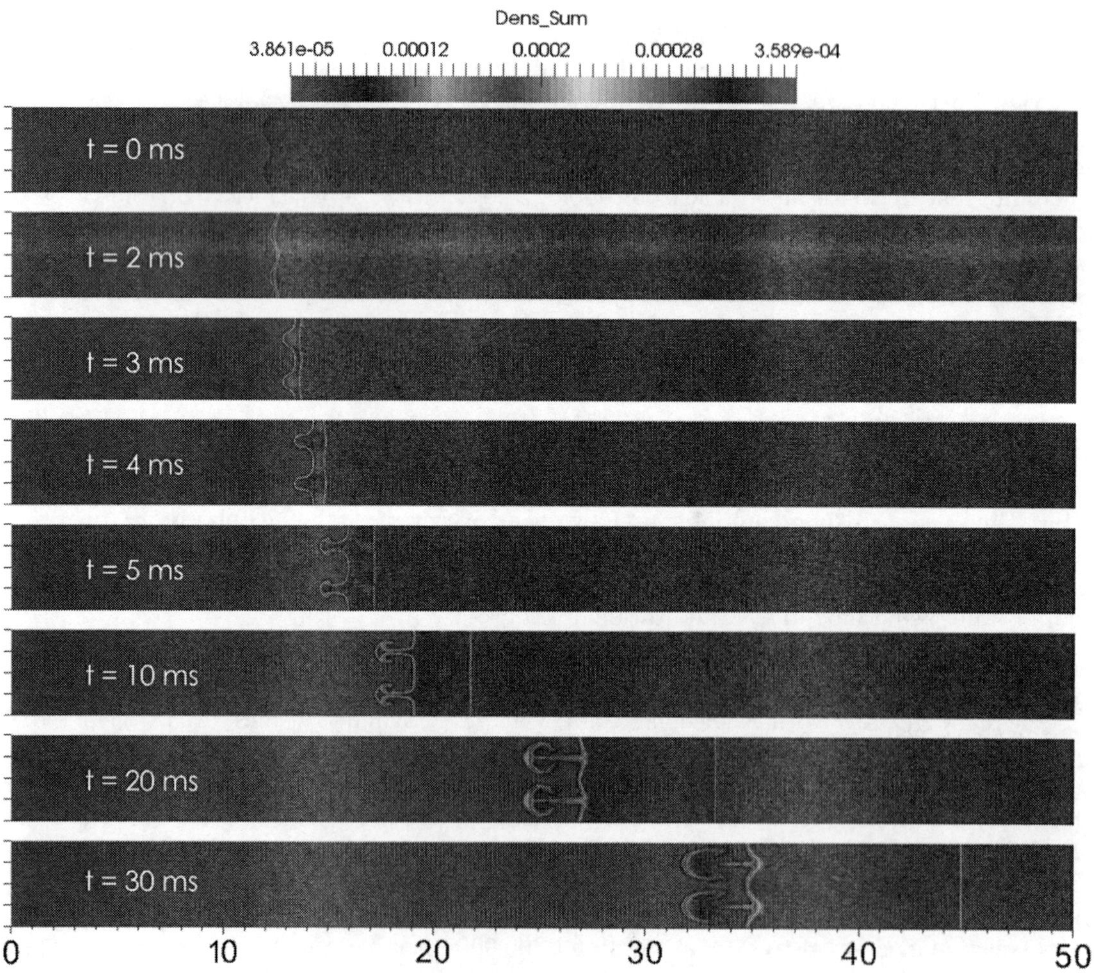

Figure 2. Density flowfields.

Figure 2 shows the density flowfields at different time moments. The computations were performed for the interface with a half-period of a sine disturbance but for better visual representation the flowfield was mirrored twice with respect to the X axis. As can be seen, the shock wave passed through the interface quickly becomes plane, while the interface itself stretches taking a mushroom-like shape. The figure 3 shows the evolution of the shock wave (SW) and contact surface (CS) positions at the Y coordinate values corresponding to the bubble and the spike. As can be seen, SW evolution is identical for both the cases, except for a short period at the beginning of the computation, while the difference in CS positions seems to be growing linearly in time. In figure 4 the difference between CS position of the bubble and the spike is shown. It is evident that instability "stretching" rate decreases and the distance between the spike and the bubble tends to a constant value. Figure 5 demonstrates the velocities of CS and SW for the spike and the bubble. Oscillations appear as a result of differentiation of the data containing stochastic fluctuations. The theoretical values of velocities of the discontinuities are calculated from the solution of the 1D Riemann problem. It can also be seen, that SW speed almost immediately reaches its theoretical value. The speed of CS spike at first is higher than the theoretical value, while the speed of CS bubble is slowly grows up its theoretical value.

Thus, the development of the RMI was simulated numerically using the DSMC method without employment of huge computational resources. It is worth mentioning that in the present computations the number of model particles in a virtual volume with a linear size equal to the local mean free path (so called λ-cell) was close to unity. As was shown in [18], one model particle in λ-cell is sufficient to keep the deterministic DSMC error low enough to ensure reasonably acurate numerical results. It seems that such a large number of particles in [12] was used to reduce statistical fluctuations. In the present work it was shown that averaging over 1000 time steps is sufficient in this kind of simulations.

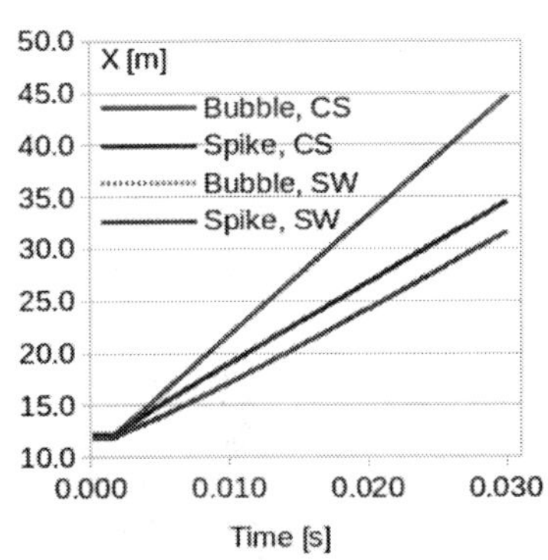

Figure 3. Time evolution of the shock wave and contact surface positions.

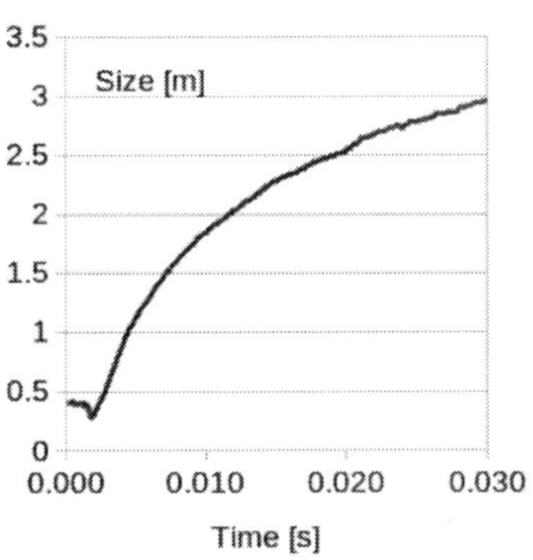

Figure 4. Distance between the bubble and the spike as a function of time.

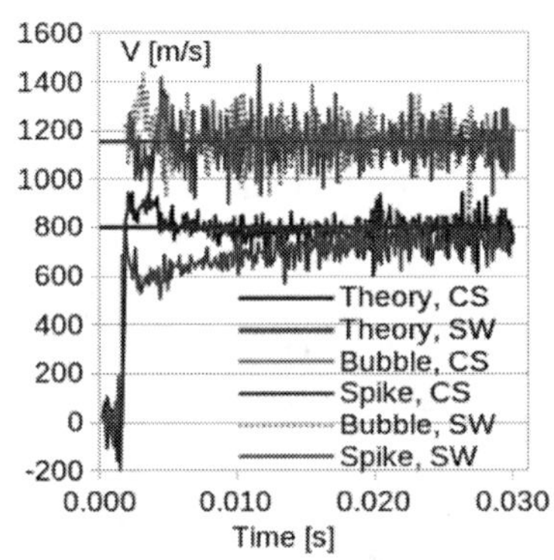

Figure 5. Shock wave and contact surface speeds as functions of time.

5. Conclusion

The DSMC method was successfully used for the numerical simulation of the Richtmyer-Meshkov instability developing at the interface between helium and argon. The instability development patterns are in good agreement with the continuum simulations and experimental data. The present work confirms that the DSMC method has reached a high level of maturity and currently allows one to simulate, on molecular level, complicated unsteady flows, previously considered only on the basis of the Navier-Stokes equations.

Acknowledgments

The current study was supported by Russian Science Foundation Grant No. 18-11-00246. This support is gratefully acknowledged. Computational resources were kindly provided by Siberian Supercomputer Center of the Institute of Computational Mathematics and Mathematical Geophysics SB RAS (sscc.ru) and Computational Center of Novosibirsk State University (nusc.nsu.ru)

References

[1] Ruev G A, Fedorov A V and Fomin V M 2004 Evolution of the diffusion mixing layer of two gases upon interaction with shock waves *J. Appl. Mech. and Techn. Phys.* **45** 328-34

[2] Ruev G A, Fedorov A V and Fomin V M 2005 Development of the Richtmyer-Meshkov instability upon interaction of a diffusion mixing layer of two gases with shock waves *J. Appl. Mech. and Techn. Phys.* **46** 307-14

[3] Ruev G A, Fedorov A V and Fomin V M 2010 Development of the Richtmyer-Meshkov instability during interaction of the diffusion mixing layer of two gases with transient and reflected shock waves *J. Appl. Mech. and Techn. Phys.* **51** 308-16

[4] Bird G A 1994 *Molecular Gas Dynamics and the Direct Simulation of Gas Flows* (Oxford: Clarendon Press)

[5] Ivanov M S and Rogasinsky S V 1988 Analysis of the numerical techniques of the direct simulation Monte Carlo method in the rarefied gas dynamics *Soviet J. Numer. Anal. Math. Modelling* **3** 453-65

[6] Ivanov M S and Rogasinsky S V 1991 Theoretical analysis of traditional and modern schemes of the DSMC method *Proc. XVII Int. Symp. on Rarefied Gas Dynamics* p 629-42

[7] Zhou Ye 2017 Rayleigh-Taylor and Richtmyer-Meshkov instability induced flow, turbulence, and mixing I *Physics Reports* **720-722** 1-136

[8] Zhou Ye 2017 Rayleigh-Taylor and Richtmyer-Meshkov instability induced flow, turbulence, and mixing II *Physics Reports* **723-725** 1-160

[9] Kudryavtsev A N and Epstein D B 2019 Richtmyer-Meshkov instability in a rarefied gas, results of continuum and kinetic numerical simulations *AIP Conf. Proc.* **2125** 020010

[10] Kashkovsky A V, Kudryavtsev A N and Shershnev A A 2019 Investigation of the Kelvin-Helmholtz instability with the DSMC method *AIP Conf. Proc.* **2125**

[11] Gallis M A, Koehler T P, Torczynski J R and Plimpton S J 2016 Direct simulation Monte Carlo investigation of the Rayleigh-Taylor instability *Phys. Rev. Fluids* **1** 043403

[12] Gallis M A, Koehler T P, Torczynski J R and Plimpton S J 2015 Direct simulation Monte Carlo investigation of the Richtmyer-Meshkov instability *Physics of Fluids* **27** 084105

[13] Gallis M A, Torczynski J R, Plimpton S J, Rader D J and Koehler T P 2014 Direct Simulation Monte Carlo: The Quest for Speed *AIP Conf. Proc.* **1628** 27

[14] Kashkovsky A V, Markelov G N and Ivanov M S 2001 An Object-Orirnted Software Design for the Direct Simulation Monte-Carlo method *AIAA Paper 2001-2895*

[15] Kashkovsky A V, Bondar Ye A, Zhukova G A, Ivanov M S and Gimelshein S F 2005 Object-oriented software design of real gas effects for the DSMC method *AIP Conf. Proc.* **762** 583-8.

[16] Ivanov M S, Kashkovsky A V, Vashchenkov P V and Bondar Ye A 2011 Parallel object-oriented software system for DSMC modeling of high-altitude aerothermodynamic problems *AIP Conf. Proc.* **1333** 211

[17] Ivanov M S, Markelov G N and Gimelshein S F 1998 Statistical simulation of reactive rarefied flows: numerical approach and applications *AIAA Paper 98-2669*

[18] Shevyrin A A, Bondar Ye A and Ivanov M S 2005 Analysis of repeated collisions in the DSMC method *AIP Conf. Proc.* **762** 565

Modeling flows of underexpanded jets in a slot space

S P Kiselev[1,2], **V P Kiselev**[1], **V N Zaikovskii**[1]

[1]Khristianovich Institute of Theoretical and Applied Mechanics SB RAS,
4/1 Institutskaya str., Novosibirsk, 63009 Russia
[2]Novosibirsk State Technical University,
20 K. Marksa av., Novosibirsk, 630092, Russia

kiselev@itam.nsc.ru, kiselevvp@itam.nsc.ru, zaikovskii@itam.nsc.ru

Abstract. In the present paper, we report on the results of a study of supersonic gas flows in a slot space. It is shown that the structure of the ejected jet exhausting into a slot ambient space substantially depends on the friction force. The latter force leads to a substantial change of the supersonic jet, which acquires, instead of a barrel-shaped, a fan-shaped form. The results of numerical simulation are consistent with experimental data.

1. Introduction

Until recently, supersonic flows in the slot space did not attract very much attention of researchers. To our knowledge, only one work [1] dealing with a supersonic flow in a radial nozzle formed by a gap between two parallel disks, has so far been published. Currently, the situation has changed after publication [2] in which it was shown that radial nozzles could be used in the cold gas spraying technology for depositing coatings onto the inner surface of pipes. In [3, 4], a pseudo-shock structure formed in a radial nozzle was studied and in [5], supersonic jets discharged into a slot space from a rectangular channel were investigated. The present paper is a continuation of the study of [5]. Here, new results concerning the structure of supersonic jets discharged from a rectangular channel into a slot space filled with quiescent air are reported.

2. Statement of the problem

In the present study, experimental and numerical modeling of the submerged gas flow both in the rectangular channel and in the slot space was carried out. A diagram of the rectangular channel and the slot space is shown in figure 1, *a* and *b*. Two disks, *1* and *2*, both of radius $r_e = 36$ mm, spaced apart by a distance *h*, are fitted onto a central rod *3* of radius $r_0 = 5$ mm. The distance between the disks was varied in various experiments and numerical calculations. A gasket with a cut was inserted in between the disks (this gasket is shown in figure 1, *a* and *b* as a shaded area). The cut made in the gasket formed a rectangular channel of width *b*, length *a*, and height *h*. The gas having temperature T_0 was supplied from chamber *4* into the channel through the slot $x_i < x < r_1$, $-b/2 < y < b/2$ under pressure p_0. In the experiments and calculations, the parameters a, b and h were varied parameters, whereas the parameters $x_i = 6$ mm and $r_1 = 9$ mm were fixed quantities. In the channel, the gas was accelerated to the speed of sound; as a result, an underexpanded supersonic jet was ejected into the slot

IOP Publishing
1404 (2019) 012110 doi:10.1088/1742-6596/1404/1/012110

space $x > r_2$. The latter space was filled with air under normal conditions, $p_\infty = 0.1$ MPa and $T_\infty = 300$ K.

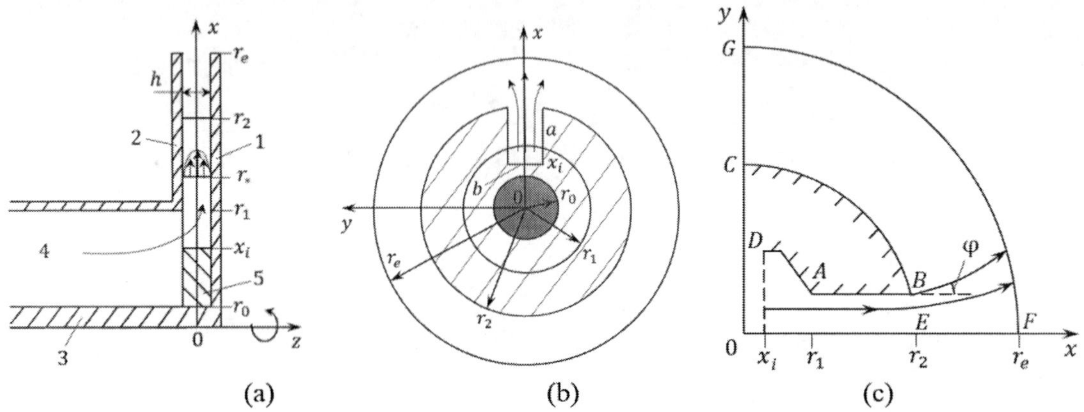

Figure 1. Diagram of the gas flow in the channel and in the slot space drawn in the planes (x,z) and (x,y) (respectively, *a* and *b*). Part *c* of the figure shows the computational domain in the plane (x,y).

The working gas was air with temperature $T_0 = 300$ K and pressure $0.9 \text{ MPa} \le p_0 \le 1.0 \text{ MPa}$. In the experiment, a flow with converging boundary layers formed in between the disks; that is why in the numerical study the gas flow in the plane (x,y) was calculated using equations averaged over the channel width [3, 4]:

$$\frac{\partial A}{\partial t}+\frac{\partial F}{\partial x}+\frac{\partial G}{\partial y}+f=0,\quad p=\rho RT,\quad e=C_VT,\quad q^2=u^2+v^2,$$

$$A=\begin{pmatrix}\rho\\ \rho u\\ \rho v\\ \rho\left(e+q^2/2\right)\end{pmatrix},\quad F=\begin{pmatrix}\rho u\\ \rho u^2+p\\ \rho uv\\ \rho u\left(e+p/\rho+\mathrm{v}^2/2\right)\end{pmatrix},\quad G=\begin{pmatrix}\rho v\\ \rho vu\\ \rho v^2+p\\ \rho v\left(e+p/\rho+q^2/2\right)\end{pmatrix},\quad f=\begin{pmatrix}0\\ \tau_x\\ \tau_y\\ 0\end{pmatrix},\qquad (1)$$

$$\tau_x = C_f\rho uq/h,\quad \tau_y = C_f\rho vq/h,\quad C_f = 12/Re + 0{,}06/(2Re)^{0{,}25},\quad \mathrm{Re} = \rho|\mathbf{v}|h/\mu.$$

In equations (1), ρ, p, e and T are the gas density, the gas pressure, the specific internal energy, and the gas temperature; u and v are the velocity components respectively along the axes x and y; q^2 is the squared velocity modulus; τ_x and τ_y are the components of the friction force acting on the gas from the side of the disks; C_f, μ and C_V are the drag coefficient, the dynamic viscosity, and the heat capacity; γ is the Poisson adiabatic index; and Re is the Reynolds number. The drag coefficient C_f involved two terms. The first term described the dependence of C_f on the Reynolds number Re for laminar flow, and the second term, the same dependence for turbulent flow. The numerical modeling was carried out in the domain shown in figure 1 *c* (for convenience in presenting the calculated and experimental data, the coordinate axes (x,y) in figure 1 *c* are turned through angle 90° compared to figure 1 *b*). The system of equations (1) was solved numerically using the modified Lax – Wendroff scheme [3]. At the lateral boundaries *DABC* (hatched ones in figure 1 *c*), the conditions of impermeability were set, and at the axis $x_i < x < r_e$, $y = 0$ and line *CG*, the conditions of symmetry were posed. The gas entered the computational domain through the inlet section $x = x_i$, at which the conditions $p_0, T_0, v=0$ were set; the fourth condition was derived from the condition of conservation of the Riemann invariant $J_- = u - 2c/(\gamma-1)$. The gas left the slot space through the boundary FG, at which the symmetry condition was specified in the case of supersonic flow with $M = q/c > 1$, and the pressure p_∞, in the case of subsonic flow with $M < 1$; the rest conditions were derived from the conditions of conservation of the Riemann invariants reaching the boundary FG [3 - 5]. Like in [3 - 5], the steady flow in the channel and in the slot space was calculated using the relaxation method.

3. Discussion of calculated and experimental data

Figure 2*a* shows the results of numerical calculation of a supersonic underexpanded jet ejected from a rectangular channel (a = 12 mm, b = 4 mm, h = 0.2 mm) into a slot space width $h = 0.2$ mm. The gas parameters in the prechamber were p_0 = 1.0 MPa and $T_0 = 300$ K. Figure 2*b* shows soot-oil visualization data in the form of a photograph taken in an experiment with two jets simultaneously discharged from channels of width b = 4 mm and b = 17 mm. The remaining parameters in the experiment were the same as those in the numerical calculation with $h = 0.2$ mm. Both in the calculation and in the experiment, the nozzle pressure ratio was $p(r_2)/p_0 = 5$, where r_2 = 20 mm. Figure 2*c* shows the dependence of non-dimensional pressure $p(x)/p_0$ on the coordinate x (along the jet centerline) obtained in two independent experiments and in numerical calculations. The scatter of experimental data permits an estimate of the experimental error, which was mainly related with the non-parallelism of the two disks.

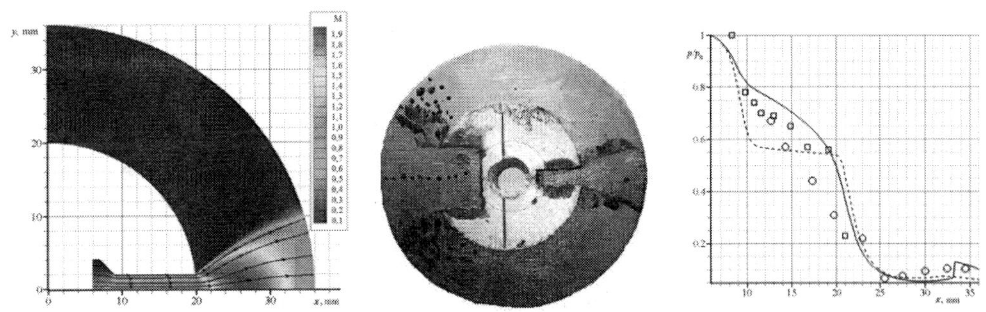

(a) (b) (c)

Figure 2. The calculated and experimental data obtained for the steady flow in the channel and in the slot space: the calculated distribution of Mach numbers $M(x, y)$ (*a*); soot-oil visualization data for the flow on the surface of the external disk (photo) (*b*); the distribution of non-dimensional pressure $p(x, y)/p_0$ along the jet axis $y = 0$ (*c*). In part *c* of the figure, the circles and rectangles show the experimental data, and the solid and dashed line, the data calculated respectively with and without allowance for the friction force.

As it is seen in figure 2*a*, a diverging supersonic jet is discharged from the channel; the boundary of this jet is inclined at an angle $\varphi = 35^\circ$ to the x-axis. A similar picture was observed in the experiment (see figure 2*b*). If the friction force were absent ($C_f = 0$), then the jet would have the barrel-shaped form shown in figure 3*b* of publication [5]. The formation of the diverging jet is a result of the action of the friction force due to the disk walls on the ejected supersonic jet. If the jet leaves the channel at point B (see figure 1*c*), there forms a centered rarefaction wave, which undergoes reflection from the centerline of the jet and extends to the jet boundary. Under the action of the friction force, the rarefaction wave gets attenuated; therefore, when this wave reaches the jet boundary, it becomes rather weak and does not disturb this boundary. From figure 2*c*, it follows that, due to the deceleration of the supersonic jet, a shock wave is formed in the vicinity of the exit from the slot space. The calculated distribution of pressure along the jet centerline is in satisfactory agreement with the pressure values that were measured in the experiments (see figure 2*c*).

The friction force that acts on the jet flow from the side of the disk walls is inversely proportional to the spacing h between the disks (see formulas (1) for the friction-force components τ_x and τ_y). A decrease in the distance between the disks leads to an increased friction force and to a change of the flow pattern in the jet. Figure 3 shows the calculated and experimental data obtained for a supersonic underexpanded jet with $p(r_2)/p_0 = 5$ ejected from a rectangular channel into a slot space with a smaller spacing between the disks, h = 0.15 mm.

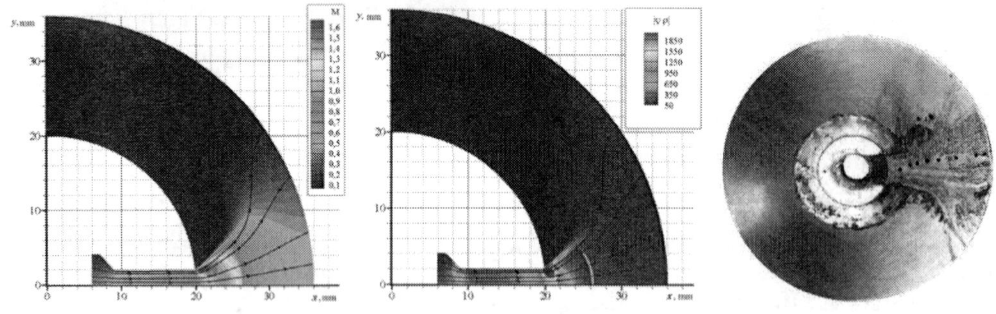

(a) (b) (c)

Figure 3. The calculated and experimental data for the steady flow obtained at $a = 12$ mm, b=7 mm, h = 0.15 mm, p_0 = 0.95 MPa, and T_0 = 300 K: the calculated distribution of Mach numbers $\mathrm{M}(x, y)$ (*a*); the calculated distribution of density gradient $|\nabla\rho(x, y)|$, kg/m^3 (*b*); and the soot-oil visualization data for the flow on the surface of the external disk (photo) (*c*).

Evidently, an increase in friction force leads to an amplification of the shock wave and to its displacement towards the channel outlet. Upstream of the shock wave, the discharged jet has a rectilinear boundary inclined to the centerline of the jet. Behind the shock, we have a subsonic flow in which the pressure increases and the jet velocity decreases. This causes an intense lateral expansion of the gas. As a result, the boundary of the jet becomes fan-shaped. The results of calculating the jet boundary (see figures 3*a* and 3*b*) are in satisfactory agreement with the soot-oil visualization obtained data (see figure 3*c*).

4. Conclusion

The paper presents results of a numerical and experimental study of supersonic underexpanded jets discharged from a rectangular channel into a slot space formed by two parallel disks. As the gas flow leaves the channel and enters the slot space, at the corner there arises a centered rarefaction wave in which the gas flow undergoes acceleration to a supersonic speed. Under the action of the friction force, the rarefaction wave rapidly decays. As a result, the rarefaction wave reflected from the jet centerline does not perturb the jet boundaries. In this case, the initial length of the jet boundary is a rectilinear streamline inclined at some angle to the jet axis. The angle of inclination is defined by the turn of the velocity vector that occurs as the streamline intersects the centered rarefaction wave. Depending on the separation between the disks, a curvilinear section of the jet, over which a most profound expansion of the jet occurs, may be adjoined to its rectilinear section. The emergence of this former section is related with the formation of a shock wave closing the zone of supersonic flow.

Behind this shock wave, the jet flow is subsonic. The subsonic stream expands intensively towards the lateral regions in which the pressure is equal to the external pressure. If the subsonic region is large enough, then a second fan-shaped section of the jet boundary can arise with the occurrence of an intensive expansion of the gas to the sides. The results of numerical calculations proved to be in satisfactory agreement with experimental data in terms of the distribution of pressure along the jet centerline and in terms of the shape of jet boundaries.

Acknowledgements
This work was supported in part by the Russian Foundation for Basic Research (grant No. 19-01-00292-a) and by the Program of Fundamental Research for the State Academies of Sciences for the years 2013-2020 (project AAAA-A17-117030610124-0).

References

[1] Moller P 1966 *J. Basic Eng. Series D* **88** 153-4
[2] Kiselev S P, Kiselev V P, Klinkov S V, Kosarev V F and Zaikovskii V N 2017 *Surface & Coating Technology* **313** 24
[3] Kiselev S P Kiselev V P and Zaikovskii V N 2018 *Shock Waves* **28** 829
[4] Kiselev S P, Kiselev V Yu, Liapidevskii V P and Zaikovskii V N 2017 *Journal of Physics: Conf. Series* **894** 012042
[5] Kiselev S P Kiselev V P and Zaikovskii V N 2019 *Journal of Physics: Conf.Series* **894** 012033

Numerical simulation of sonic boom near-field flow over 69-degree delta wing body

T Kiseleva[1,2] **and S Lukeshevich**[1]

[1] Khristianovich Institute of Theoretical and Applied Mechanics, Siberian Branch of Russian Academy of Sciences, 630090, Novosibirsk, Russia
[2] Novosibirsk State University, 630090, Novosibirsk, Russia

E-mail: bobarykina@ngs.ru

Abstract. Fast and accurate prediction of sonic boom intensity is important in performing design and optimization studies of a supersonic aircraft. The paper presents the computational mesh generation technique for sonic boom prediction in the near field of complex geometry body. Studies are done on a 69-degree delta wing body model used for testing various prediction techniques in the framework of the AIAA Sonic Boom Prediction Workshops.

1. Introduction

Many researches are developing technologies to enable future supersonic civilian aircraft to fly efficiently with reduced sonic boom (SB). The main problem of SB modelling is the large extent of the area under investigation, where the perturbed pressure levels vary by several orders of magnitude. Usually the sonic boom is studied in the near and far fields. The near field of the SB adjoins the aircraft and has a complicated flow structure with shock waves, rarefaction and compression waves; its length is about the aircraft length ($K = H/L \sim L$, where H is the height distance from the aircraft, L is its characteristic size). The length of the far field where the sonic boom parameters vary feasibly in accordance with the asymptotic law, realizes within the distances of $K \sim L^2$. Numerical simulation is the main research method, but it requires huge computational resources, the construction of complex grids to describe aircraft configurations, and, in addition, contains some assumptions. In the case of a large model with complex geometry, the time spent on meshing is significant. A reasonable alternative for this kind of simulations is a quick automatic mesh generation. At the same time you need to create more uniform meshes for improved convergence and accuracy. The challenge here is to provide a quick automatic mesh with good accuracy.

2. Numerical simulation

2.1. Numerical simulation conditions

The flow around a body is solved in a three-dimensional mode. SOLIDWORKS Flow Simulation solver is utilized for the solution process. Flow Simulation is a full Navier-Stokes solver for structured grids, however, in this study the code is run in laminar mode without taking into account friction on the model surface since viscous effects are negligible for sonic boom prediction. The calculation was carried out for a stationary flow of an ideal gas with the specific heats ratio $\gamma = 1.4$. The surfaces of the model were impenetrable, without roughness, without thermal conductivity. This does not allow simulating the boundary layer, but made it possible to correctly simulate the propagation of shock waves, compression and rarefaction waves in the near field. Initial conditions were normal corresponding to the altitude of 15000 m.

2.2. Cartesian mesh generation

The 69-degree delta wing body geometry taken from [4] is shown in figure 1. This model is used for testing various prediction techniques in the framework of the AIAA Sonic Boom Prediction Workshops.

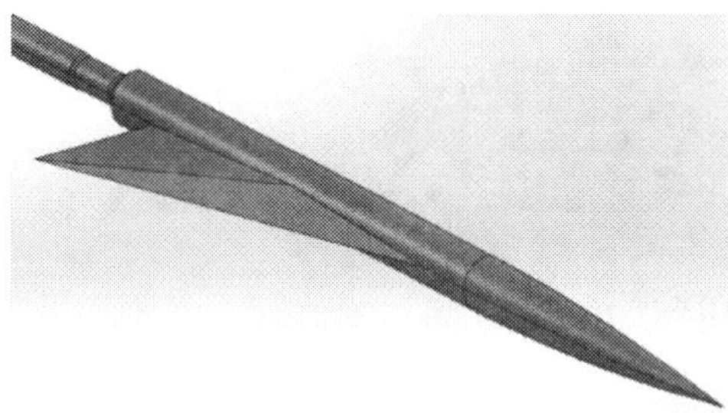

Figure 1. 69-degree delta wing body geometry.

The rectangular computational domain is shown in figure 2. The dashed line in figure 2 shows a symmetrically reflected region. One of the faces passes through the model symmetry axis. The inflow and outflow boundaries are set at Mach angles for M = 1.6. So the faces were tilted at the angle of weak shock waves [5]. The boundaries of the computational domain were located at a sufficient distance to exclude some artifacts are reflected back. Such a boundary condition is supposed to allow the disturbances to pass through the boundary. Each face of the computational domain was divided into three subregions: near the model, far from the model, and the transition between them. A Cartesian grid was used. Near the model the cell size was 0.005×0.0026×0.004 m, in the distance - 0.022×0.01×0.012 m. The grid covering up to $H/L = 4$. Here H is the distance from the model in the vertical direction, L is the length of the model.

The mesh in the symmetry plane of the model is shown in figure 2. Inside the computational domain, a flow feature-based subdomain was used, in which the cell size was refined by 4 times. The geometry of the subdomain is shown in figure 3. An additional 4-fold grid refinement was used near the surface of the model.

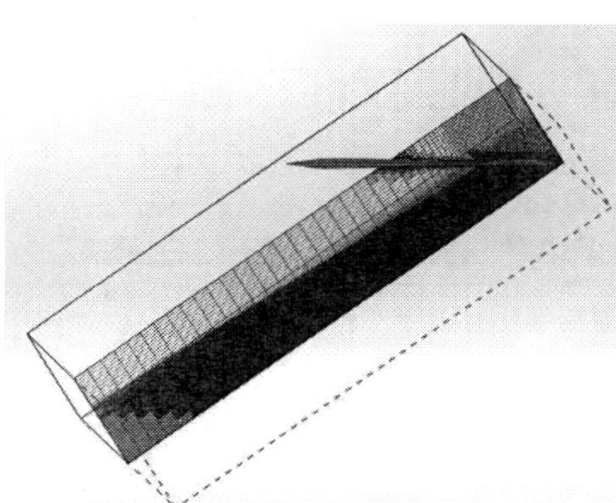

Figure 2. The computational grid in the symmetry plane of the model.

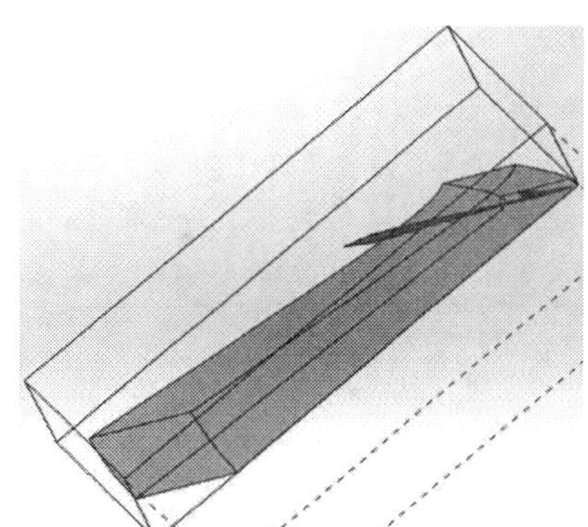

Figure 3. A subregion for grid refinement.

Journal of Physics: Conference Series **1404** (2019) 012111 doi:10.1088/1742-6596/1404/1/012111

3. Numerical results

Figure 4 shows the distribution of excess static pressure along the control line at a distance of $K = H/L = 3.6$. The x-axis, passing through the model axis, is plotted along the horizontal direction. The origin is chosen arbitrarily. On the ordinate axis is the excess (relative to the incident flow pressure P_{inf}) static pressure. The result is shown for the base grid.

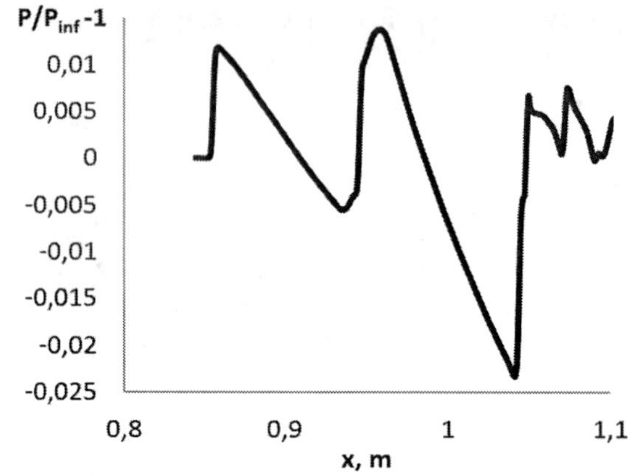

Figure 4. Distribution of excess static pressure along the control line in the near field of the model.

As can be seen, the head SW has the intensity of approximately 0.01, formed as a result of the flow around the nose of the model, followed by the rarefaction wave, which leads to a decrease in pressure to a value lower than in free stream. Next, at the point $x \approx 0.95$ the SW formed by the wing is observed. The tail shock wave ($x \approx 1.05$) closes the profile.

To reduce the computation and retain the results the cell sizes in the region remote from the model (see figure 5, region III) were changed. As shown in [5], it is acceptable to use "elongated" cells when shock-grid alignment is present. The cells with the aspect ratio of 4.4:1 were selected as base cells. Then cells length in region III was varied along the x direction. The test computations were performed for cells two times longer and two times shorter than the base cells. So, in these cases the intensity and position of the shock wave from the nose of the model did not change. Figure 6 shows the intensities of the wing SW for different refinement factors in the region III. The lines denoted as "0.5", "1" and "2" correspond to the cell sizes 0.088×0.01×0.012 m, 0.022×0.01×0.012 m and 0.011×0.01×0.012 m, respectively. It can be seen that a decrease in the cell size leads to a decrease in the intensity of the shock wave. The difference between the cases "0.5" and "2" is about 1%.

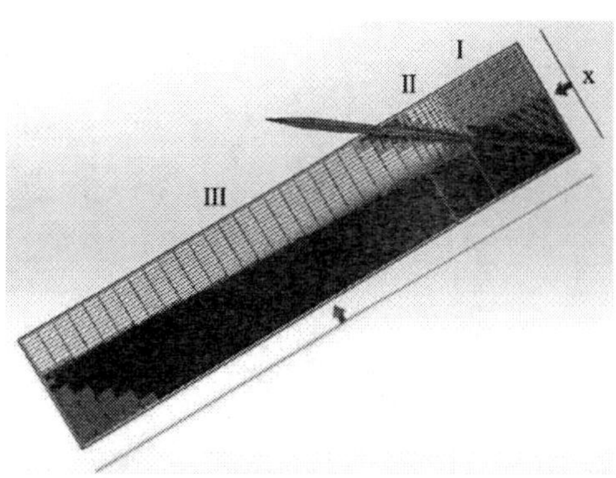

Figure 5. Grid adaptation regions.

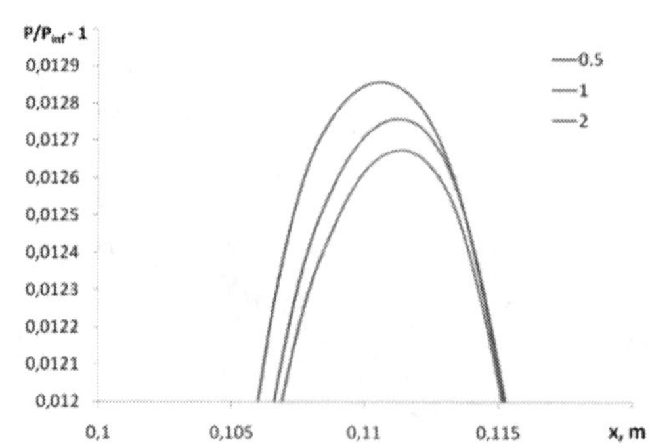

Figure 6. Shock wave from the wing.

Next, we studied the effect of the subdomain transverse size on the intensity of the shock wave on the symmetry plane. For this, the size of the region of cell refinement was varied, which was specified by the angle α (see figure 7). The figure 8 shows the results of numerical modeling for the convergence angle of refinement subdomain: 10, 20, and 30 degrees. It can be seen that the results coincide for subdomains with $\alpha = 20$, 30 degrees. For $\alpha = 10$ degrees, the SW intensity is less, due to the influence of the SW dissipating on a coarse grid in the transverse direction.

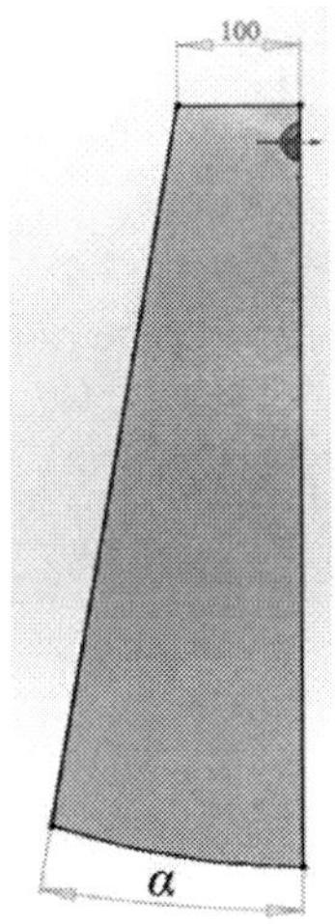

Figure 7. Grid adaptation subdomain.

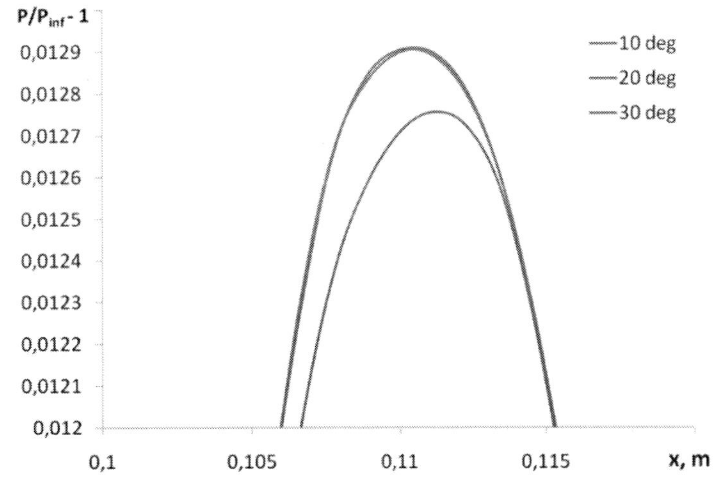

Figure 8. Shock wave from the wing.

Figure 9 shows a comparison with the results of numerical simulations obtained by researchers from Boeing, Dassault Aviation, JAXA, NASA, and ONERA [4]. The shock waves intensity in the present computation is higher than the average value, however, it is at the level of the values obtained in Boeing, Dassault Aviation and JAXA. It should also be noted that the pressure gradient on the shock waves is maximum in the calculation results of Boeing and ITAM SB RAS.

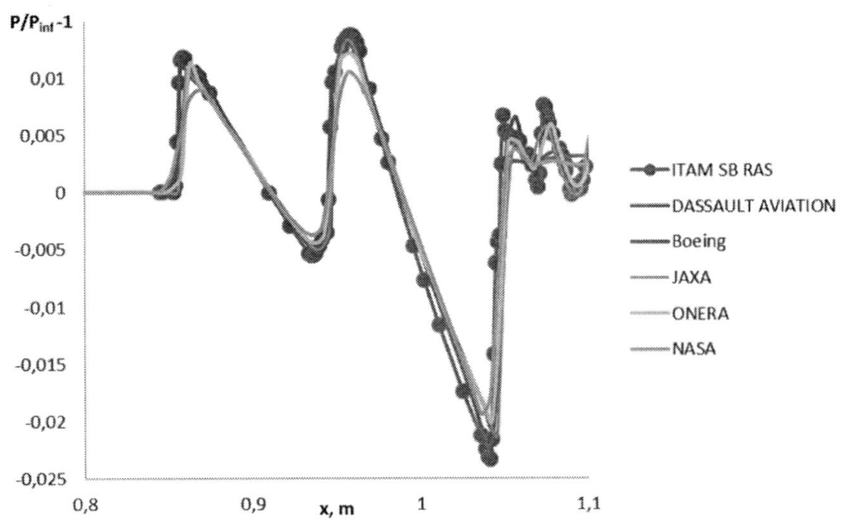

Figure 9. Comparison of the results of numerical modeling obtained by different researchers.

4. Conclusions

The computational mesh generation technique for sonic boom near field prediction of complex geometry body is presented. The rectangular computational domain consists of elongated cells and contains subdomain for a flow feature-based grid refinement. It is shown, that a decrease in the cell size of base domain leads to a decrease in the intensity of the shock wave. The difference between the results obtained with the 4 fold refinement is about 1%. Subdomain transverse size affect the shock intensity on the symmetry plane, when this size is less than $Z/L = 3$.

Studies are done on a 69-degree delta wing body model used for testing various prediction techniques in the framework of the AIAA Sonic Boom Prediction Workshops. Numerical simulation of the near-field flow of the 69-degree delta wing body model was conducted. The results have a good agreement with the results obtained by different researchers.

Acknowledgments

The research was partly carried out within the framework of the Program of Fundamental Scientific Research of the state academies of sciences in 2013-2020 (project No. AAAA- -A17-117030610121-9) and partly financially supported by the Russian Foundation for Basic Research (Grant No. 18-08-00449).

References

[1] Chernyshev S L 2011 *Sonic boom* (Moscow: Nauka Publishing) (in Russian)
[2] Zvegintsev V I and Potapkin A V 2018 *Thermophysics and Aeromechanics* **25** 321-330
[3] Kiseleva T A and Kislovskiy V A 2018 *Thermophysics and Aeromechanics* **25** 687-694
[4] Cliff S E, Durston D, Chan W M, Elmiligui A A, Moini-Yekta S, Sozer E and Jensen J C 2014 Computational and Experimental Assessment of Models for the First AIAA Sonic Boom Prediction Workshop *AIAA Paper* 2014-560.
[5] Ozcer I and Kandil O Fun3D / OptiGRID Coupling for Unstructured Grid Adaptation for Sonic Boom Problems *AIAA*-2008-61

Computational grid adaptation for the sonic boom near field modeling

T Kiseleva[1, 2] and V Temerbekov[1]

[1] Khristianovich Institute of Theoretical and Applied Mechanics SB RAS, Institutskaya 4/1, Novosibirsk, 630090, Russia
[2]National Research Novosibirsk State University, Pirogova str, 1, Novosibirsk, 630090, Russia

bobarykina@ngs.ru

Abstract. This paper discusses the numerical simulation of the 69-Degree Delta Wing Body for computing the sonic boom signature in the near field. For flow field calculation, the commercial CFD flow solver ANSYS Fluent is employed. The computational mesh in the perturbed flow regions was refined two times using the automatic gradient adaptation. The results are in agreement with the results obtained by other researchers.

1. Introduction

Sonic boom (SB) is one of the critical challenges for creating supersonic passenger aircraft [1-2]. Supersonic aircraft flying in the air generate sonic boom waves propagating on large areas, which annoys people. Finding the ways to reduce the sonic boom loudness is a major concern for the world's leading researchers.

Normally, the sonic boom is studied in the near and far fields. The near field of the SB is adjacent to the aircraft and has a complicated flow pattern with shock waves (SW), rarefaction and compression waves; its length is about the aircraft length ($K = H/L \sim L$, where H is the radial distance from the aircraft, L is its characteristic size). The far field, where the sonic boom parameters vary slightly in accordance with the asymptotic law, has the order of the distances of $K \sim L^2$. Numerical modeling is the most powerful research method, but it demands huge computational capability, requires the construction of complex grids to describe aircraft configurations, and, in addition, contains some assumptions. For modeling a real airplane model with complex geometry the time spent on meshing is important. For such simulation the compromise is a quick automatic meshing. The challenge here is the simulation of the flow in the near field of the sonic boom using mesh adaptation.

2. Test cases

2.1. Numerical simulation conditions

The flow around a body is numerically modeled in a three-dimensional formulation. The commercial CFD flow solver ANSYS Fluent is utilized for the numerical simulation. In this study the code is run in laminar mode excluding viscosity since viscous effects are negligible for sonic boom prediction. The calculation was carried out for a stationary flow of an ideal gas with the heat capacity ratio $\gamma = 1.4$. The implicit second-order scheme for temporal approximation and the AUSM flux vector splitting method with a second-order upwind scheme for spatial approximation were used. Initial conditions correspond to the flight altitude of 15 000 m and the flight velocity corresponding to the Mach number of 1.7.

2.2. Geometry of the model
The 69-Degree Delta Wing Body geometry taken from [3] is shown in figure 1. The fuselage is designed as a cylinder with a parabolic nose [4]. The wing has double-wedge 5-percent-thick sections with the ridge line located at mid-chord and is mounted on the cylindrical portion of the fuselage at the longitudinal plane of symmetry. The model has a fineness ratio (L/d) of 16.2, where L and d are the length and the diameter of the model, respectively. This model was used for testing various prediction techniques in the framework of the AIAA Sonic Boom Prediction Workshops.

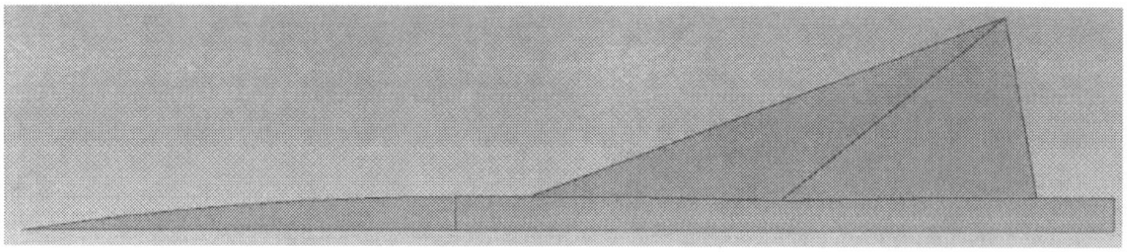

Figure 1. 69-Degree Delta Wing Body geometry.

2.3. Computational domain and mesh
The computational domain in x-y and y-z planes is shown in figure 2 and figure 3, respectively. The basic grid is generated by Fluent Meshing and contains 433 thousands cells. Due to the symmetry of the problem, only a quarter of the domain is considered. The unstructured mesh is condensed near the surface of the model. Structured mesh is aligned with the SW and the Mach angle at a distance.

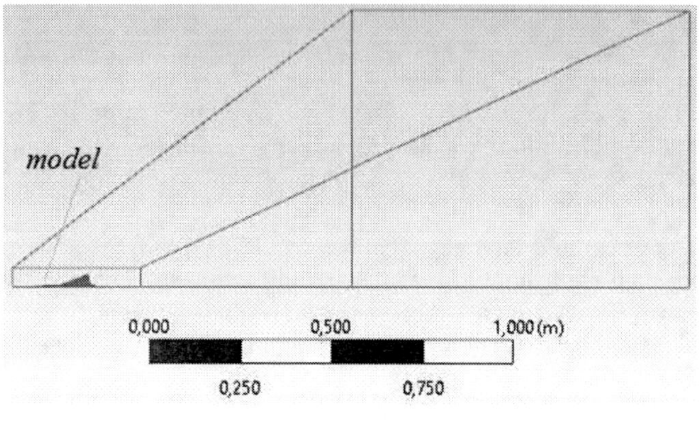

Figure 2. The computational domain (view from above).

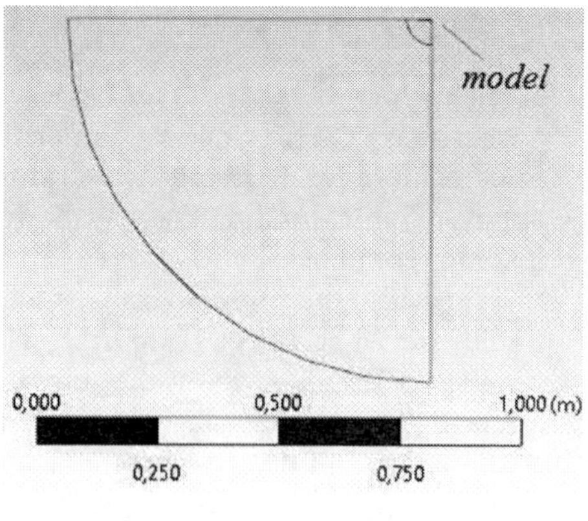

Figure 3. The computational domain (front view).

3. Numerical results
Figure 4 shows static pressure disturbance contours around the 69-Degree Delta Wing Body in x-z plane. A series of SW and rarefaction waves is observed. Based on this solution, the mesh was adapted to the flow pattern. The static pressure gradient was chosen as a criterion for mesh adaptation.

Figure 4. Static pressure disturbance contours around the 69-Degree Delta Wing Body in x-z plane.

Thus, the mesh was refined in the regions of perturbed flow, especially in shock wave fronts. The resulting computational grid in the middle region of the computational domain in the vertical symmetry plane (x-z) is shown in figure 5. After the first adaptation, the grid contained 7.5 million cells.

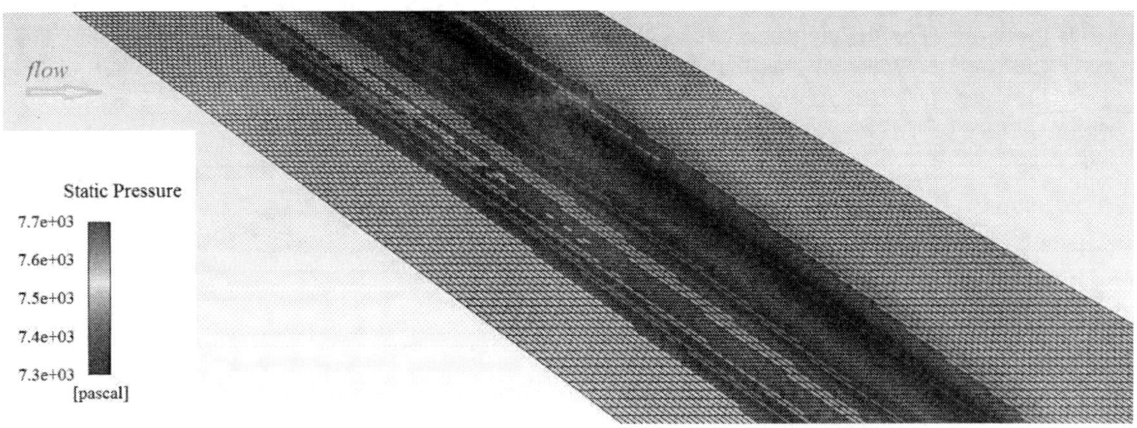

Figure 5. The computational grid after the first adaptation.

However, the use of the automatic gradient adaptation led to a significant increase in the number of cells near the surface of the model. The mesh near the nose of the model after the first adaptation is shown in figure 6. In addition, the number of cells has increased significantly in the wake area behind the aft of the model due to the large pressure gradients here (figure 7). To avoid this effect, it is necessary to divide the computational domain into subdomains (near the surface of the model and at a distance); the mesh adaptation within these subdomains should be carried out separately.

To optimize the calculations, mesh refinement and coarsening were carried out during the second gradient adaptation based on the results of the first adaptation. After the second adaptation, the grid consisted of 14.75 million cells.

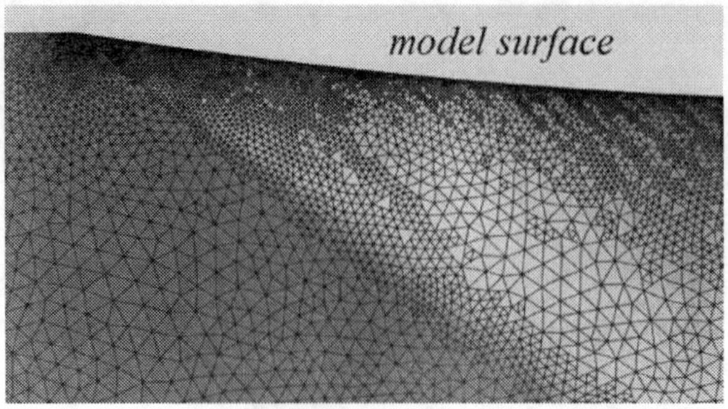

Figure 6. Mesh near the nose of the model after the first adaptation.

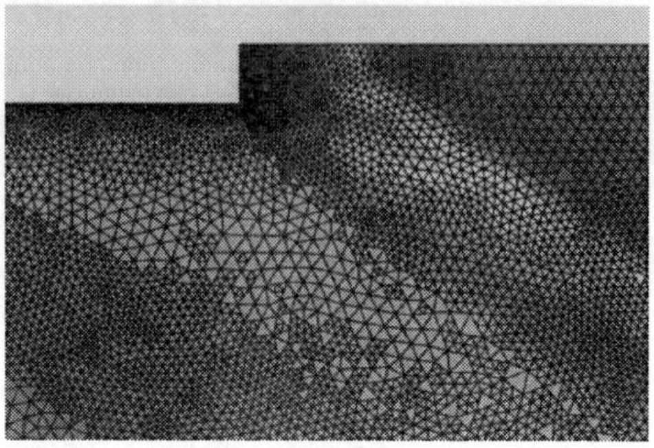

Figure 7. Mesh near the aft of the model after the first adaptation.

Figure 8 shows the distribution of excess static pressure along the control line at a distance $K = H/L = 3.6$ on the vertical symmetry plane. The x-axis, passing parallel to model axis, is normalized to L. The origin of the axis is chosen arbitrarily. On the ordinate axis is the excess (relative to the incident flow pressure – P_{inf}) static pressure. The result is shown for the base grid and for the grids produced after the first and the second adaptation. It can be seen that the results of all three calculations are almost the same. The bow SW formed as a result of the flow around the nose of the model is observed with the intensity of ~ 0.01. The bow SW is followed by a rarefaction wave, which causes a decrease in pressure to a value lower than in the incident flow. Next, a SW ($x \approx 0.8$) formed by the wing is observed. The aft SW ($x \approx 1.35$) closes the profile. A closer look (figure 9) shows that an increase in the number of cells by 17 times relative to the base grid leads to an increase in the intensity of the SW from the wing. The difference between the results is about 6%. The next step (second adaptation) gives the same SW intensity as that on the base grid with a slight increase in the pressure gradient. This is mainly due to the mismatch of the fronts of the SW and the edges of the cells [5].

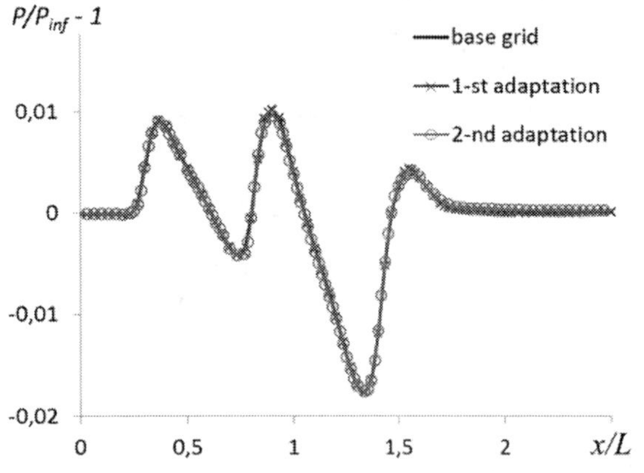

Figure 8. Distribution of excess static pressure along the control line in the near field of the model.

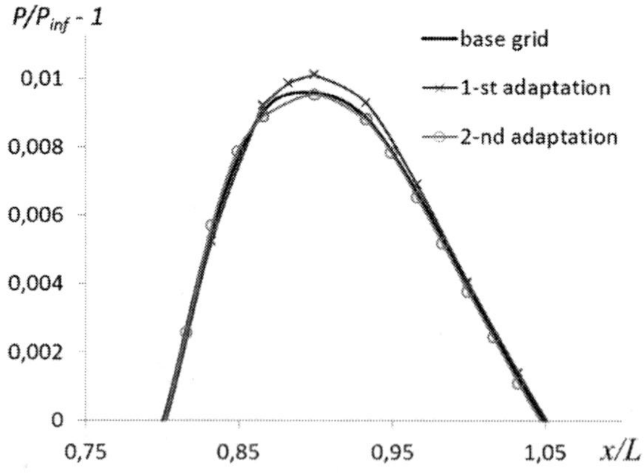

Figure 9. Shock wave from the wing.

Figure 10 shows a comparison with the results of numerical simulations obtained by researchers from Boeing, Dassault Aviation, JAXA, NASA, and ONERA [4]. The calculation results obtained at ITAM SB RAS are consistent with the results of other authors. The intensity of the shock waves is lower than the average value; however, it is at the level of the values obtained in NASA. The pressure gradient on the shock waves is lower too.

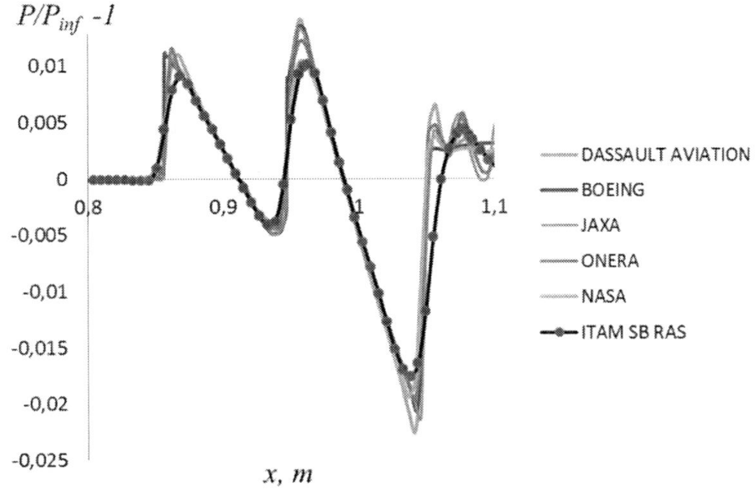

Figure 10. Comparison of the results of numerical modeling obtained by different researchers.

4. Conclusions

Numerical simulation of the flow in the near field of a sonic boom using mesh adaptation is presented. The mesh was refined in the regions of perturbed flow using the automatic gradient adaptation. This method allows refining mesh in the regions of perturbed flow, especially in shock waves. However, it led to an increase in the number of cells near the surface of the model. To avoid this effect, it is necessary to divide the computational domain into subdomains; the mesh adaptation within the subdomains should be carried out separately. The difference between the results obtaining on different meshes is about 6%.

Studies are done on a 69-Degree Delta Wing Body model used for testing various prediction techniques in the framework of the AIAA Sonic Boom Prediction Workshops. Numerical simulation of the near-field flow of the 69-Degree Delta Wing Body model was conducted. The results are in agreement with the results obtained by other researchers.

Acknowledgments

The research was carried out within the framework of the Program of Fundamental Scientific Research of the state academies of sciences in 2013-2020 (project No. AAAA-A17-117030610121-9).

References

[1] Chernyshev S L 2011 *Sonic boom* (Moscow: Nauka Publishing)
[2] Kiseleva T A and Kislovskiy V A 2018 *Thermophysics and Aeromechanics* **25** 687
[3] Cliff S E, Durston D, Chan W M, Elmiligui A A, Moini-Yekta S, Sozer E and Jensen J C 2014 *Computational and Experimental Assessment of Models for the First AIAA Sonic Boom Prediction Workshop* AIAA Paper 2014-560
[4] Hunton L W, Hicks R M and Mendoza J P 1972 *Some Effects of Wing Planform on Sonic Boom* NASA TN D-7160
[5] Ozcer I and Kandil O 2008 *Fun3D / OptiGRID Coupling for Unstructured Grid Adaptation for Sonic Boom Problems* AIAA Paper 2008-61

Numerical investigation supersonic jet near field with a longitudinal slit cutout on the expiration channel

V A Kislovskiy and A Y Melnikov

Khristianovich Institute of Theoretical and Applied Mechanics SB RAS 630090, Novosibirsk, Russia

E-mail: kislovsky_v@mail.ru

Abstract. The effects of a slit cutout length on the parameters of a supersonic flow are studied. The flow in the nozzle with a Mach number M = 2 was simulated numerically. The jet from the nozzle passed through the 30 cm long tube and then exhausted into quiescent air. Two symmetrical longitudinal slits were cut at the end of the tube. Three cases with 3, 6, and 9 cm cutouts were considered. The slit width was fixed at 1 cm for all cases. The study showed that an increase in the cutouts length leads to a slight increase in the mixing zone width at the jet boundary. However, the mixing intensity gradient was decreased. This flow case is somewhat similar to the chevrons used in some practical applications. So the cutouts can be used to reduce the intensity of the acoustic noise in the far field.

1. Introduction

The supersonic jet noise is usually generated by the high-speed jet exhausting in the ambient stationary medium. This process is one of the main sources of the acoustic noise, in situations like the technical gases exhaust at the industrial facilities. Acoustic noise from a supersonic flow can lead to a deterioration in the people health in the vicinity. Therefore, the problem of reducing acoustic noise from a supersonic jet is relevant and important.

There are many options for reducing the supersonic jet acoustic noise. The best options include screening the jet with a gas screen [1], use of sector nozzles, multi-tube nozzles, use of periodic electric discharges, injection of a liquid into the main stream, using chevrons and other active methods [2].

The jet exhausting from bypass-ratio nozzles was considered by Sayed et al. [3], where the chevrons of various configurations were used to suppress acoustic noises. Mehmed et al. [4] also used chevrons and eight micro jets as generators of longitudinal vortices. Experiments of Zapryagaev et al. [5] demonstrated the significant efficiency of water injection for changing the acoustic spectrum supersonic jets radiation. It was showed that increasing the injection rate of water jets reduces the magnitude of the main supersonic jet acoustic noise. James B et al. [6] showed that the use of the lobed mixers allows one to reduce the jet acoustic noise by 10 dB. Zapryagaev et al. [7, 8] also studied the chevrons influence on the main features of the supersonic jet flow both experimentally and numerically. Sipatov et al. [9] developed a numerical method for supersonic jet noise modeling using the ANSYS Fluent software package. Using this approach, influence of the various geometric nozzles characteristics on the acoustic radiation intensity analyzed.

The present study shows the influence of the longitudinal cutout length in tube on parameters of a supersonic flow. The study is carried out on the basis of the numerical simulation. The theoretical and analytical equations determining the jet power were shown to be valid for the values of the Mach numbers between 1.5 and 2 [10, 11].

2. Conditions of computations

Three-dimensional modeling of the supersonic jet flow into the quiescent air was performed. Modeling took into account the gas viscosity, both molecular and turbulent. For the numerical simulation, the commercial ANSYS Fluent software package was used. The general approach is to use the Reynolds-averaged Navier-Stokes equations (RANS) with a suitable turbulence model (SST k-ω) to close the system of equations.

Figure 1 shows a scheme of the computational domain indicating the main dimensions. At the inlet boundary, the static temperature and pressure drop were set. The flow passed through the profiled nozzle reaching the Mach number value M = 2. The flow enters from the nozzle into the tube 300 mm long. Several cases were considered with different lengths of symmetrically located slit cutout at the tip end l = 3 cm, 6 cm, and 9 cm. In all cases, the slit cutout had a constant width of 1 mm. The flow exhausted from the tube into the still air.

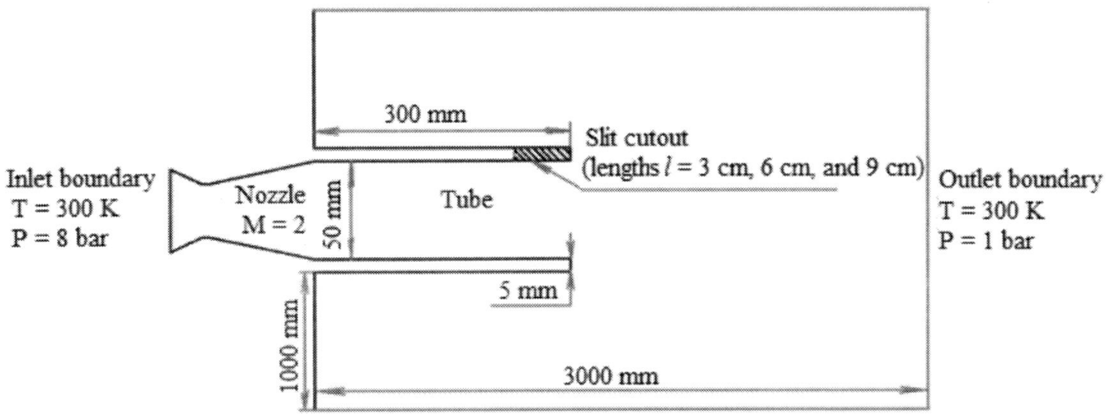

Figure 1. Scheme of the computational domain.

This type of flows require highly accurate computations, because even small changes in environmental parameters affect the flow pattern. Therefore, in the calculation, a mesh with a high density and good quality should be used. To ensure an appropriate grid structure, the area of the nozzle, tube, and jet core was divided into cubic elements 0.4 mm in size. The area near the jet core was divided into 0.8 mm elements. The maximum grid elements size at the remote boundary was 24 mm. The computational domain was divided into 12 000 000 elements, most of which were concentrated in the jet flowfield.

3. Results

Figure 2 shows the pressure contours of the jet exhausting from the tube for all cases considered. An increase in the slit cutout length at the discharge channel cut leads to a decrease in the jet plume length and to a decrease in the pressure drop intensity. The 3 cm slit cutout reduces the jet plume length by about 17%. Each subsequent increase in the length of the slit cutout by 3 cm reduced the length jet plume by about 3%.

Figure 3 shows the turbulence kinetic energy contours during the jet expiration from the tube for all cases considered. An increase in the length of the slit cutout leads to an increase in the mixing zone at the outflowing jet boundaries and to a decrease in the gradient of the mixing intensity in the near outflow field. This flow picture in the mixing zone is somewhat similar to the case of chevrons use.

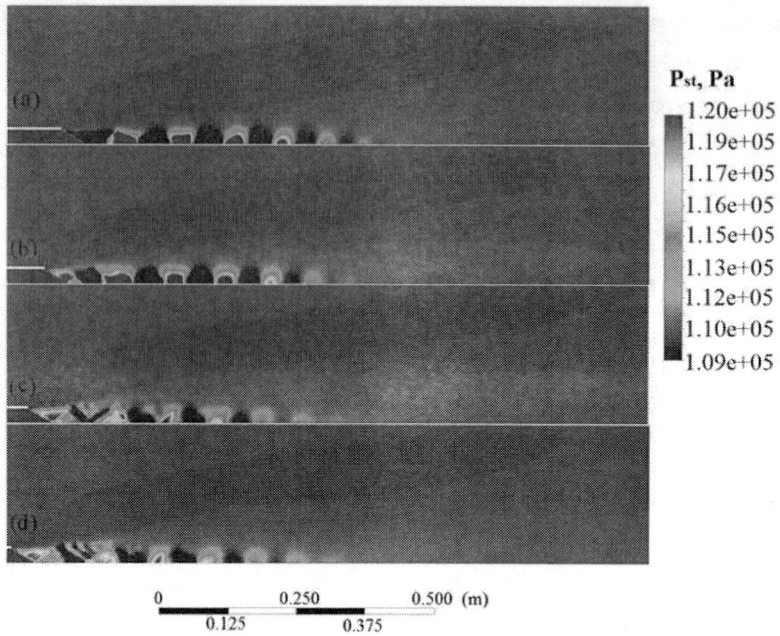

Figure 2. Pressure contours in the plane of symmetry for the case of blowing out from tube without slit cutout (a) and with slit cutout length: 3 cm (b); 6 cm (c); 9 cm (d).

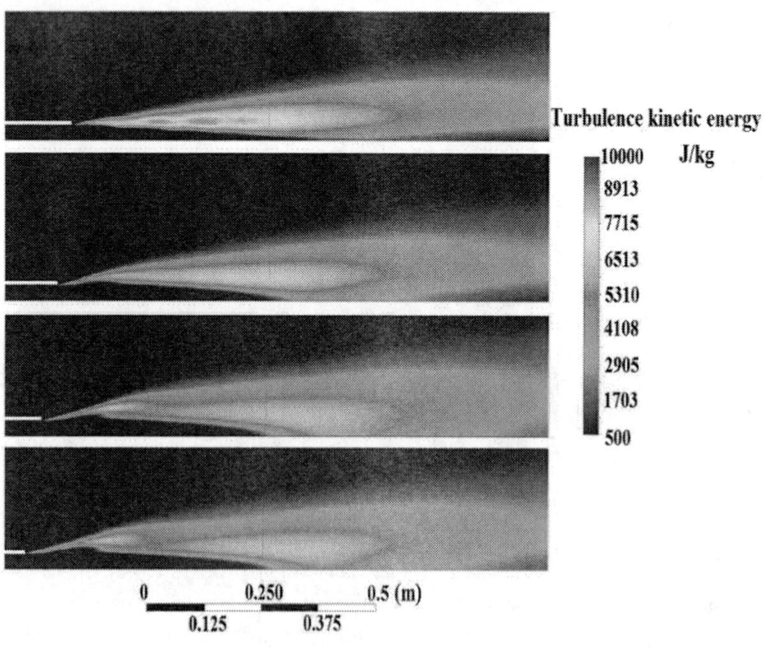

Figure 3. Turbulence kinetic energy contours in the plane of symmetry for the case of blowing out from tube without slit cutout (a) and with slit cutout length: 3 cm (b); 6 cm (c); 9 cm (d).

4. Conclusion

A study by numerical simulation showed that the use of a slit cutout on the expiration tube of a supersonic jet leads to a decrease in pressure in the outflowing jet, a decrease in the mixing zone intensity and a decrease in the jet plume length. All these factors should lead to more rapid jet attenuation in space, and, as a result, a decrease in the acoustic effect from the jet outflow. An increase in the slit cutout length leads to a more efficient attenuation of the outflowing jet, which means a more efficient reduction of acoustic noise.

Acknowledgments

The reported study was funded by RFBR according to the research project № 19-48-543003.

References

[1] Munin A G, Samokhin V F, Shipov R A et al. 1986 *Aviation Acoustics, Part 1. Noise in the Area of Subsonic Passenger Aircraft and Helicopters* (Moscow: Mechanical Engineering)
[2] Peake N and Crighton G 2000 *J. Annu. Rev. Fluid Mech.* **32** 137-64
[3] Sayed N, Mikkelsen K and Bridges J 2003 *AIAA J.* **41** 372-8
[4] Mehmet B Alkislar, Krothapalli A and Butler W G 2007 *J. Fluid Mech.* **578** 139-69
[5] Zapryagaev V I, Kiselev N P and Gubanov D A 2012 *TsAGI Science J.* **43** 481-96
[6] James B and Mark PW 2004 *Cross-stream PIV Measurements of Jets with Internal Lobed Mixers* (NASA/TM — 2004-213106)
[7] Zapryagaev V I, Kavun I N and Kiselev N P 2010 *J. of Applied Mechanics and Technical Physics.* **51** 202-10
[8] Boyko V M, Dostovalov A V, Zapryagaev V I et al. 2010 *TsAGI Science J.* **41** 187-205
[9] Sipatov A M, Usanin M V and Chuhlantseva N O 2010 *AIAA J.* **3843**
[10] Kuznetsov V M 2008 *Fundamentals of the Turbulent Jets Noise Theory* (Moscow: Fizmatlit)
[11] Krasheninnikov S Yu and Mironov A K 2008 *J. Acoustical Physics.* **54** 392-7

Journal of Physics: Conference Series **1404** (2019) 012114 doi:10.1088/1742-6596/1404/1/012114

Linear stability of supersonic boundary layer over a cooled porous surface

Alexey Kudryavtsev[1,2] **and Dmitry Khotyanovsky**[1]

[1] Khristianovich Institute of Theoretical and Applied Mechanics, Russian Academy of Sciences, Siberian Division, 630090 Novosibirsk, Russia

[2] Novosibirsk State University, 630090 Novosibirsk, Russia

E-mail: `alex@itam.nsc.ru`, `khotyanovsky@itam.nsc.ru`

Abstract. The linear stability of the boundary layer at high supersonic Mach numbers on a flat plate with a cooled porous surface is studied. As is well known, at high supersonic speeds, both vortex disturbances of the first mode and acoustic disturbances of the second mode are unstable. The more unstable disturbances of the second mode can be stabilized using sound-absorbing porous coatings. At the same time, such coatings have a destabilizing effect on the disturbances of the first mode. Surface cooling is one of the factors potentially capable of preventing such a destabilization. Based on this assumption, in the present work, the linear theory of stability is used to consider the combined effect of a sound-absorbing coating and surface cooling.

1. Introduction

The instability of the supersonic boundary layer has been studied long enough. It is well known [1, 2] that at moderate supersonic Mach numbers the instability arises due to vortical disturbances, Tollmien–Schlichting waves. In this case, the most unstable disturbances are those propagating at sufficiently large ($50 \div 60\ ^\circ$) angle to the direction of the basic flow. At higher speeds, along with unstable Tollmien–Schlichting waves (the first mode), multiple unstable acoustic-vortical high-frequency modes appear, known as Mack modes. The Mack mode with the lowest frequency is known as the second mode. For Mach numbers greater than approximately 4, the second-mode disturbances grow faster than the first-mode disturbances, and unlike the disturbances of the first mode, the fastest growing disturbances of the second mode are two-dimensional. These theoretical predictions were verified in a number of experiments with natural and artificial disturbances [3, 4, 5, 6, 7]. In general, they confirmed the presence of rapidly growing high-frequency disturbances of the second mode in hypersonic boundary layers, although they did not play a dominant role in all experiments (see, in particular, [6]).

Changes in the surface temperature and its acoustic and mechanical properties have significant effects on the stability of the flow. It is known that surface cooling stabilizes the disturbances of the first mode, while at the same time exerting a destabilizing effect on the second mode. Studies of the effect of surface permeability (porosity) on the stability of the compressible subsonic and supersonic boundary layers, started in [8, 9], showed that porosity destabilizes the first mode of disturbances. At the same time, theoretical analysis of the effect of ultrasound-absorbing porous coatings on the stability of the hypersonic boundary layer [10] led to the conclusion that they are capable of exerting a strong stabilizing effect on disturbances of the second mode. This conclusion was confirmed experimentally [10, 11, 12, 13, 14], and then numerically [15, 16, 17, 18, 19]. The calculations were carried out both with special boundary conditions simulating the effect of porosity [15, 18, 19], and with a real surface shape resolving individual pores [16, 17]. The calculations also confirmed the stabilization of disturbances of the second mode. Active studies of the suppression of instabilities using porous coatings are ongoing, in particular with the aim of choosing the optimal coating parameters [20].

As mentioned above, porous coatings have a destabilizing effect on the disturbances of the first mode. This circumstance coupled with the fact that the disturbances of the first mode remain unstable for a significantly greater spatial extent (along the surface of the body) than the disturbances of the second mode can make the first mode the most dangerous for the development of instability and transition to turbulence. It seems natural to apply surface cooling for first mode stabilization. In the works cited above one can find some examples of calculations of linear stability under the combined action of porosity and cooling, however this issue, as it seems, has not been studied in detail. This kind of study is the goal of the present work.

2. Linear stability problem formulation

We consider stability to small disturbances of the flow in the boundary layer on a semi-infinite flat plate at a zero angle of attack in a supersonic gas flow. The distributions of the mean velocity and temperature inside the boundary layer are found as self-similar solutions of the boundary layer equations.

It is assumed that the porous coating does not alter the mean flow; this assumption has been experimentally validated [12].

Small disturbances of the following form are added to the basic flow

$$q'(x, y, z, t) = \hat{q}(y)\, e^{i(\alpha x + \beta y - \omega t)}. \tag{1}$$

Here q is any of the gasdynamic quantities, x is the streamwise coordinate, y is the coordinate normal to the plate, z is the spanwise coordinate, t is time, α and β are components of the wave vector of the disturbance, ω is the disturbance frequency. We consider disturbances growing in the streamwise direction, so ω and β are real numbers, and $\alpha = \alpha_r + i\alpha_i$ is complex. The disturbance growth rate is $(-\alpha_i)$, and the phase speed along x is given by the real part of the $c = \omega/\alpha$ quantity.

The hydrodynamic stability equations can be written [21] as the system of 8 linear ordinary differential equations

$$\frac{d\hat{\mathbf{q}}}{dy} = \mathsf{G}\,\hat{\mathbf{q}}, \qquad \hat{\mathbf{q}} = (\hat{u}, d\hat{u}/dy, \hat{v}, \hat{p}, \hat{\theta}, d\hat{\theta}/dy, \hat{u}, d\hat{u}/dy), \tag{2}$$

where u, v, w are the velocity components, p is the pressure, θ is the temperature. The expressions for the entries of the matrix G dependent on the basic flow can be found, for example, in [21].

These equations should be supplemented by the conditions of boundedness of the disturbances at infinity and the boundary conditions on the plate surface. It is assumed that a large number of regularly arranged blind cylindrical holes (pores) of the radius r and the depth h are made in the plate material. The porosity n is defined as the total inlet area of all pores per unit area of the plate. For a solid surface $n = 0$, for a surface with the pores closely packed in a square lattice $n = \pi/4$. The boundary conditions on such a surface have the form

$$\hat{u}(0) = 0, \qquad \hat{v}(0) = A\,\hat{p}(0), \qquad \hat{\theta}(0) = B\,\hat{p}(0), \tag{3}$$

where coefficients A and B are proportional to n, and depend on the geometrical sizes of the holes a, h, the wave frequency ω and the basic flow parameters near the plate surface.

The boundary conditions can be derived and the coefficients A and B can be obtained considering the propagation of an acoustic wave in a cylindrical channel with a closed end [22].

The stability equations together with the boundary conditions form an eigenvalue problem with respect to the parameter α. It is solved numerically by the shooting method, setting an initial guess for the eigenvalue and refining it with Newton iterations.

3. Results

3.1. Basic flow

In the course of calculations, it was assumed that the gas is perfect, with a ratio of specific heats $\gamma = 1.4$, the Prandtl number $\mathrm{Pr} = 0.71$ and the power-law dependence of the viscosity on the temperature with an exponent of 0.75. All calculations were performed at the Mach number $\mathrm{M} = 6$.

Figure 1 shows the velocity and temperature profiles of the basic flow at various ratios of the wall temperature T_w and the temperature at the edge of the boundary layer T_e. Here, the Blasius thickness of the boundary layer δ is used as a length scale. It is also used below when determining the Reynolds number Re and in the non-dimensionalization of the depth and radius of pores.

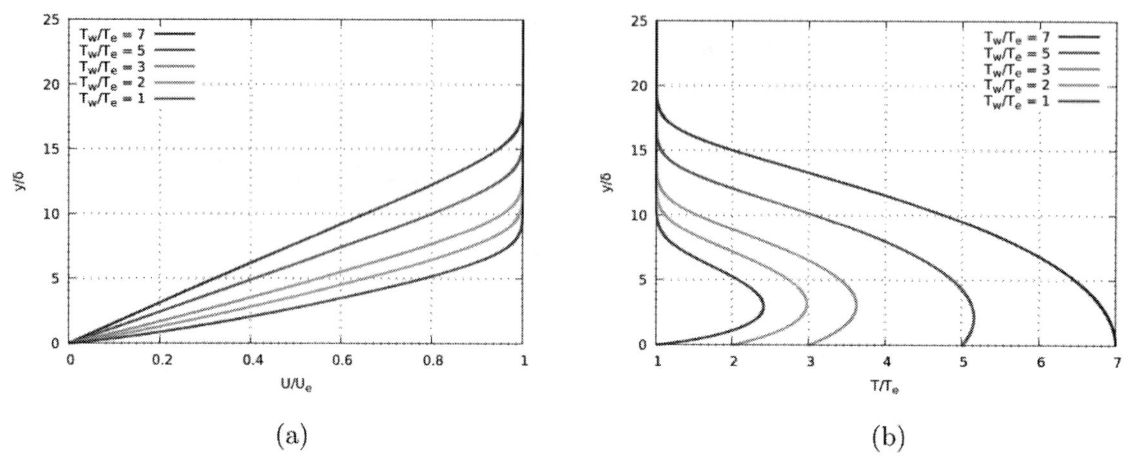

Figure 1. Velocity (a) and temperature (b) profiles of the basic flow at various temperature ratios T_w/T_e.

Wall cooling most apparently affects the temperature profile. With decreasing the wall temperature ratio from adiabatic conditions (corresponding to $T_w/T_e = 7.098$), the more and more prominent maximum appears in the temperature profile, and a strong temperature gradient emerges near the wall.

3.2. Effects of porosity parameters

The influence of the parameters characterizing the porous surface on the stability of the second mode was studied for a two-dimensional (propagation angle $\chi = \tan^{-1}(\beta/\alpha_r) = 0$) disturbance with a non-dimensional frequency $\omega = 0.14$ on nearly adiabatic ($T_w/T_e = 7$) surface at $\mathrm{Re} = 800$. A disturbance with this frequency has a growth rate close to the maximum for the given Reynolds number.

The results of calculations with varying porosity and fixed values of the pore radius and their depth, which is assumed to be quite large ($h \to \infty$) in this case, indicate that the growth rate decreases with increasing porosity, and this effect is more pronounced for large values of the pore radius. So, at $r = 0.18$ the decrease in the growth rate upon reaching the maximum porosity is 24 %, while at $r = 0.8$ it decreases by 3.26 times (Fig. 2a). When the pore depth changes from zero, the growth rate first increases slightly, then rapidly decreases and reaches a minimum at a certain value of h. With a further increase in depth, it slightly increases again and then ceases to change, reaching a certain asymptotic value (Fig. 2b). The depth at which the growth rate ceases to change depends on the radius of the pores and increases with its growth.

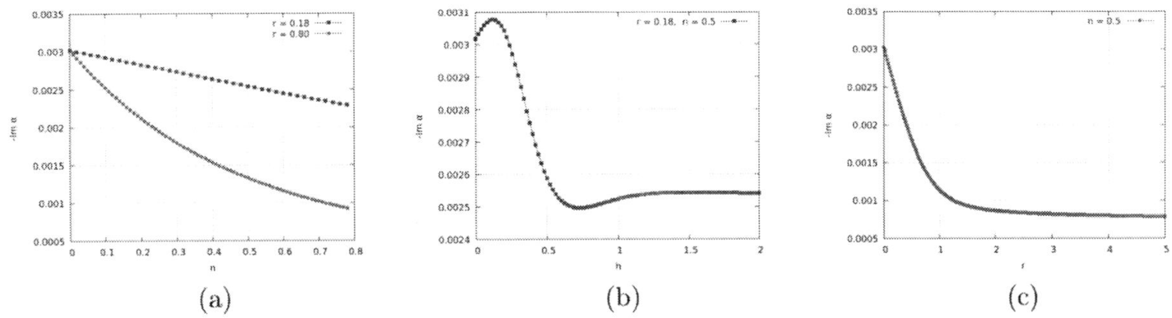

Figure 2. The effects of the porous surface parameters on the growth rate of the second mode at M = 6, $T_w/T_e = 7$, Re = 800, $\chi = 0$. Dependence of the growth rate of the disturbance with $\omega = 0.14$ on porosity n at $r = 0.18$, 0.8 and $h \to \infty$ (a), on pore depth at $n = 0.5$ and $r = 0.18$ (b), and on pore radius r at $n = 0.5$ and $h \to \infty$ (c).

An increase in the pore radius at fixed porosity and depth leads first to a rapid, then to an ever slower decrease in the growth rate (Fig. 2c).

3.3. Stability on adiabatic surface

Figure 3 shows the neutral stability curves $\alpha_i(\mathrm{Re}, \omega) = 0$ and $\alpha_i(\mathrm{Re}, F) = 0$ and the growth rates $(-\alpha_i)$ for two-dimensional disturbances in the boundary layer over the solid wall at $T_w/T_e = 7$. Here $F = \omega/\mathrm{Re}$ is the so-called frequency parameter which is convenient since, in contrast to the non-dimensional frequency ω, it remains constant for a disturbance of a given (dimensional) frequency as the boundary layer thickness grows when moving downstream.

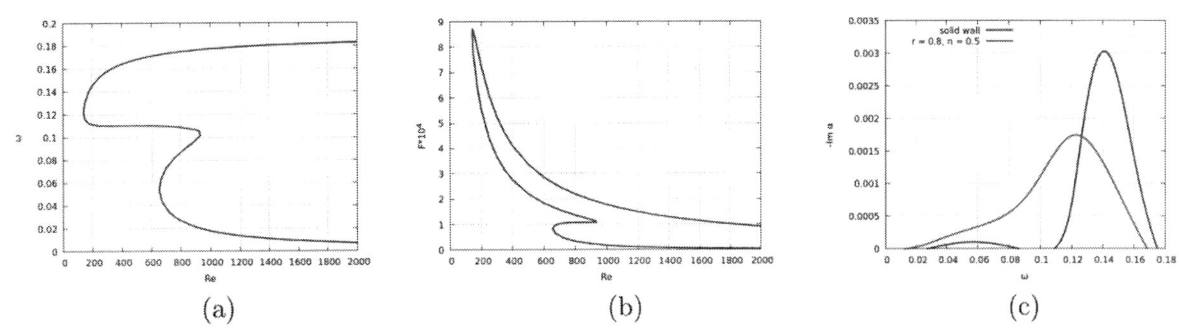

Figure 3. Neutral stability curves in the planes Re, ω (a) and Re, F (b) and growth rates of the disturbances at $Re = 800$ (c) for a supersonic boundary layer at M = 6, $T_w/T_e = 7$, $\chi = 0$.

Two parts corresponding to low-frequency disturbances of the first mode and higher-frequency disturbances of the second mode can be distinguished on the neutral curves. In a certain range of Reynolds numbers (e.g., at Re = 800), the region of unstable frequencies is divided into two separate intervals corresponding to disturbances of these two modes (Fig. 3c). The maximum growth rate of the higher-frequency disturbance of the second mode is many times, more than an order of magnitude, higher than the growth rates of the first mode. The porous surface ($n = 0.5$, $r = 0.8$, $h \to \infty$) causes merging of the two instability regions into one. In this case, the low-frequency disturbances of the first mode are destabilized: the range of unstable frequencies widens, the maximum growth rate noticeably increases. However, the porosity affects the disturbances of the second mode in the opposite way: it leads to a significant decrease in their growth rates. As a result, the maximum growth rate for all frequencies decreases almost twofold.

3.4. Stability on cooled surface

The calculations of the linear stability characteristics were also performed for a different ratio of the wall and free-stream temperatures $T_w/T_e = 1.4$, i.e., under conditions of a strongly cooled wall. The main feature of the boundary layer under such conditions is the complete stabilization of the two-dimensional disturbances of the first mode. On the neutral curves (Fig. 4), there is no portion corresponding to the first mode, the low-frequency disturbances are stable. The calculations show that with a decrease in T_w/T_e, the critical Reynolds number for the low-frequency disturbances grows rapidly and they soon become completely stable. As a result, the axis $\omega = 0$ ceases to be the asymptote of the lower branch of the neutral curve.

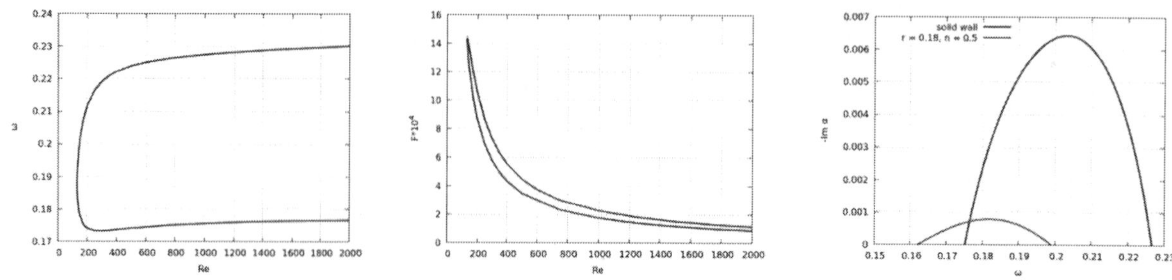

Figure 4. Neutral stability curves in the planes Re, ω (a) and Re, F (b) and growth rates of the disturbances at $Re = 800$ (c) for a supersonic boundary layer at $\mathrm{M} = 6$, $T_w/T_e = 1.4$, $\chi = 0$.

It should also be noted that the region of Reynolds numbers, at which the disturbance of the given frequency is unstable, on the cooled surface becomes very narrow (Fig. 4b). Those disturbances that become unstable close enough to the leading edge, when moving downstream will quickly leave the instability region because of the growth of the boundary layer thickness and will not have time to significantly increase their amplitude.

The disturbances of the second mode are still strongly stabilized by the porous surface. The calculations show that at $\mathrm{Re} = 800$ the maximum disturbance growth rate on a porous ($n = 0.5$, $r = 0.18$, $h \to \infty$) surface decreases by more than 8 times (Fig. 4).

4. Conclusion

The calculations of the linear stability of the supersonic boundary layer performed at the Mach number $\mathrm{M} = 6$ show that the simultaneous use of sound-absorbing porous coatings and surface cooling can be efficient for boundary layer stabilization. The sound-absorbing porous coatings suppress the growth of the disturbances of the second mode, although they destabilize the first mode to some extent. Surface cooling is a very efficient way of suppressing the disturbances of the first mode. In the future, we plan to continue this work by performing calculations for three-dimensional disturbances propagating at an angle with respect to the free stream direction.

Acknowledgments
The current investigation was supported by the Russian Foundation for Basic Research (Project No. 18-01-00866). This support is greatly appreciated.

References
[1] Mack L M 1969 *Boundary Layer Stability Theory* (JPL Doc. 900-277-REV-A, NASA CR-131501)
[2] Gaponov S A and Maslov A A *Evolution of Disturbances in Compressible Flows* (Novosibirsk: Nauka, in Russian)
[3] Demetriades A 1974 *AIAA Paper No. 1974-535*
[4] Kendall J M 1975 *AIAA J.* **13** 290
[5] Stetson K F, Thompson E R, Donaldson J C and Siler L G 1983 *AIAA Paper No. 83-1761*
[6] Vendt V, Simen M, Hanifi A 1995 *Phys. Fluids* **7** 877
[7] Maslov A A, Sidorenko A A, Shiplyuk A N 1997 *J. Appl. Mech. Techn. Phys.* **38** 64
[8] Gaponov S A 1975 *J. Appl. Mech. Techn. Phys.* **16** 95
[9] Gaponov S A 1977 *Fluid Dyn.* **12** 33
[10] Fedorov A V, Malmuth N D, Rasheed A and Hornung H G 2001 *AIAA J.* **39** 605
[11] Rasheed A, Hornung H G, Fedorov A V and Malmuth N D 2002 *AIAA J.* **40** 481
[12] Fedorov A, Shiplyuk A, Maslov A, Burov E and Malmuth N 2003 *J. Fluid Mech.* **479** 99
[13] Shiplyuk A N, Burov E V, Maslov A A and Fomin V M 2004 *J. Appl. Mech. Techn. Phys.* **45** 169
[14] Fedorov A, Kozlov V, Shiplyuk A, Maslov A and Malmuth N 2006 *AIAA J.* **44** 1866
[15] Egorov I V, Fedorov A V and Soudakov V G *J. Fluid Mech.* 2008 **601** 165
[16] Sandham N D and Lüdeke H 2009 *AIAA J.* **47** 2243
[17] Wartemann V, Lüdeke H and Sandham N D 2012 *AIAA J.* **50** 1281
[18] Wang X and Zhong X *Phys. Fluids* **24** 034105
[19] Wartemann V, Wagner A, Giese T, Eggers T, and Hanneman K 2014 *CEAS Space J.* **6** 13
[20] Lukashevich S V, Maslov A A, Shiplyuk A N, Fedorov A V and Soudakov V G 2012 *AIAA J.* **50** 1897
[21] Betchov R and Criminale W O 1967 *Stability of Parallel Flows* (N.Y./London: Academic Press)
[22] Allard J F and Atalla N 2009 *Propagation of Sound in Porous Media* (Chichester, UK: Wiley, 2nd ed.)

The influence of nozzle geometry on the gas-dynamic structure of the supersonic nonisobaric jet

S G Kundasev[1], V I Zapryagaev[1], N P Kiselev[1] and S P Rybak[2]

[1]Khristianovich Institute of Theoretical and Applied Mechanics SB RAS, 4/1, Institutskaya str., Novosibirsk, 630090, Russia
[2]Korolev Energia Rocket and Space Corporation, 4A, Lenin str., Korolev, Moscow area, 141070, Russia

kundasev.sergey@gmail.com

Abstract. The paper presents the investigation results of the flow structure of cold supersonic jets flowing out of model nozzles with a Mach number on the exit $M_a = 3.5$: an axisymmetric nozzle, a nozzle with a rectangular critical section, a beveled nozzle with a rectangular critical section. Flow schlieren images and transverse measured total pressure profiles in the flow are presented. The influence of the critical section shape and the beveled exit section on the shock-wave structure of the flow is described.

1. Introduction

In the development of promising space reentry vehicles the usage of multi-jet systems for vehicle deceleration in the final part of the landing trajectory is currently proposed. The flow structure near the reentry vehicle surface and the study of its aerodynamic characteristics during deceleration are presented in [1]. Supply conduit technical features and the requirement for an aerodynamically

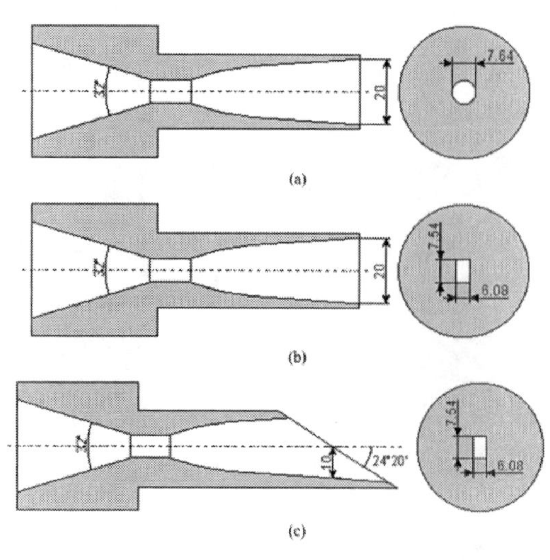

Figure 1. The geometry of the studied nozzles. (a) - nozzle 1, (b) - nozzle 2, (c) - nozzle 3

"clean" external surface of the apparatus when the nozzle section does not go beyond the vehicle profile make the goal of studying the influence of supersonic nozzles geometry on the outflowing jets gas-dynamic structure relevant.

The work was carried out at the AT-326 wind tunnel and the Vertical Jet Unit (VJU) of the ITAM SB RAS. Detailed descriptions of equipment and research methods are given in [2] and [3].

The purpose of the work is to determine the influence of the shape of the critical section of the nozzle and the beveled cut of the exit section on the gas-dynamic structure of the flow of a supersonic jet flowing out of nozzles with a geometric Mach number in the exit section $M_a = 3.5$.

Supersonic jets flowing out of three nozzles with the Mach number on the cut $M_a = 3.5$ and the output diameter $D_a = 20$ mm are studied. The nozzles geometry is shown in figure 1. To perform experiments to estimate the influence of the critical section shape on the supersonic jet, we used a pair of nozzles 1 and 2 with different geometry and equal critical section area (figure 1, (a), (b)). To estimate the effect of the beveled cut, we studied a nozzle 3 with $M_a = 3.5$, $D_a = 20$ mm, similar to nozzle 2 with the beveled cut with an angle $\theta = 24°\ 20'$, shown in figure 1 c. The research was performed for cold jets with the total pressure ratio in the prechamber of the wind tunnel to the ambient pressure $Npr = 91.5$.

2. Flow visualization

Flow schlieren visualization was carried out using a shadow device IAB-451 and a digital CCD camera with a resolution of 1400 × 1000. Figure 2 shows representative schlieren images of the studied supersonic jets initial part in the symmetry plane. Exposure time is 4 ms.

(a)

(b)

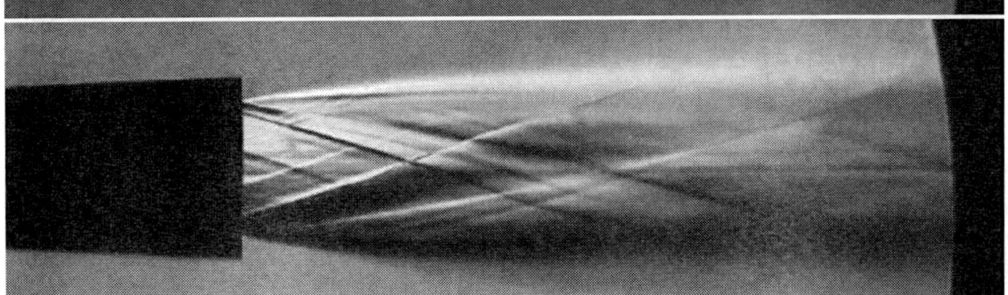

(c)

Figure 2. Schlieren images of the investigated jets. Vertical plane. $D_a = 20$ mm, $Npr = 91.5$. Exposure 4 ms. Horizontal knife. (a) - nozzle 1, (b) - nozzle 2, (c) - nozzle 3

 1404 (2019) 012115 doi:10.1088/1742-6596/1404/1/012115

The complex flow shock-wave structure is clearly registered. The groove presence in the critical section of nozzle 1 (figure 2, (a)) leads to the nozzle shock appearance. The observed additional shock is formed due to the microstep presence in each of the nozzles divergent part, leading to an abrupt expansion of the channel.

The rectangular shape of the nozzle 2 critical section (figure 2, (b)) leads to the additional shock waves formation and jet elongation along the vertical axis.

The beveled exit section of nozzle 3 (figure 2, (c)) leads to a distortion of the flow shock-wave structure and a deviation of the jet axis from the nozzle axis. For the presented mode, the deviation occurs towards the external normal to the nozzle exit.

3. Pitot pressure probe measurements

Figure 3 shows the transverse profiles of the pitot pressure in the section x/Ra = 4 in the vertical (y-var) and horizontal (z var) planes.

The recorded profiles in the perpendicular planes coincide in the case of the axisymmetric nozzle (figure 3, (a)).

For the nozzle with rectangular critical section (figure 3, (b)), the vertical profile (y-var) is elongated along the abscissa according to the outer mixing layers position, which is explained by the influence of

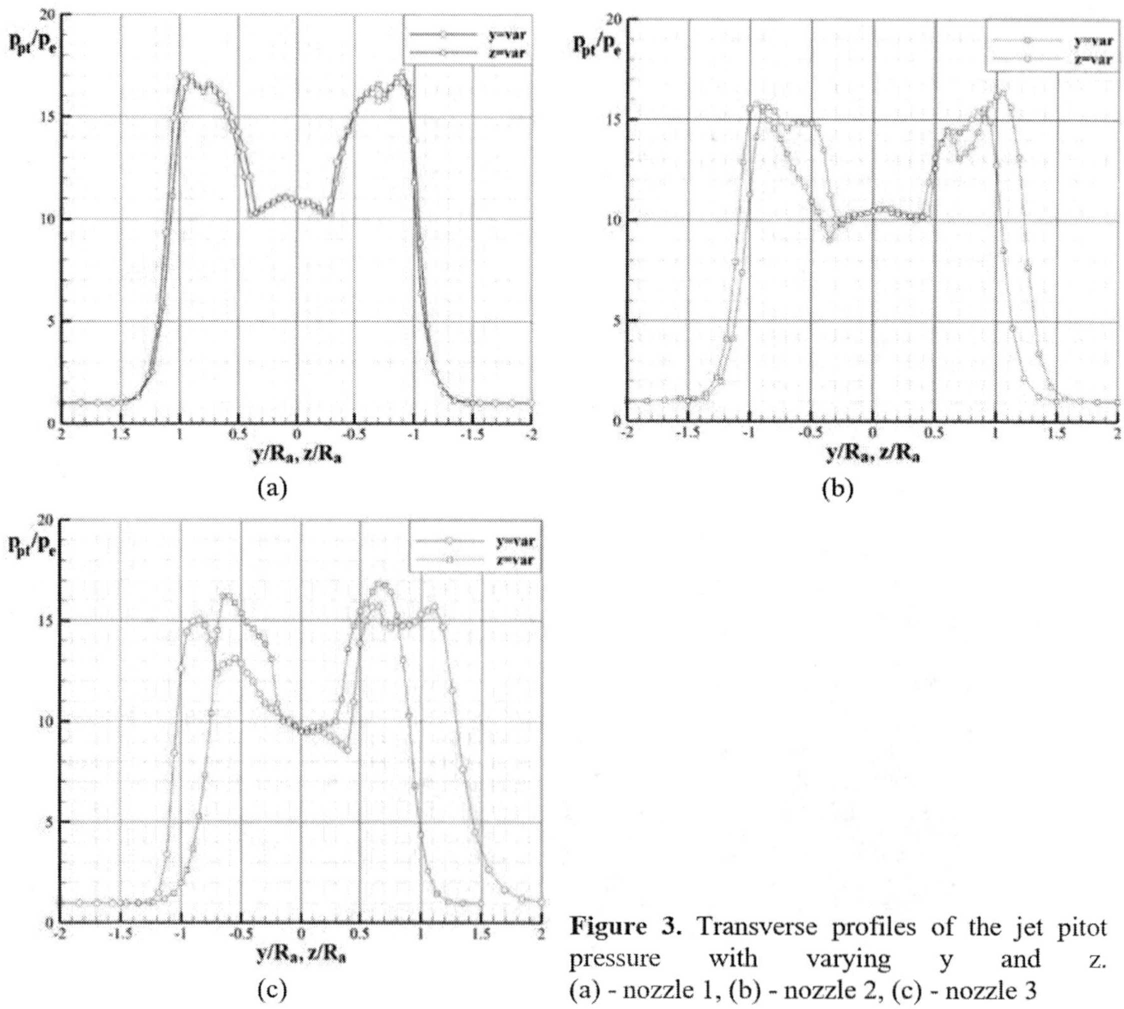

Figure 3. Transverse profiles of the jet pitot pressure with varying y and z. (a) - nozzle 1, (b) - nozzle 2, (c) - nozzle 3

the rectangular critical section. On the y axis, the pressure maximum is displaced, and the position of local pressure minima on the profile is not symmetrical with respect to the origin (y/R_a = -0.35 and y/R_a = 0.7). Such pressure distribution pattern along the y axis is caused by an inaccuracy in the nozzle 2 manufacture along the critical section long side (along y axis).

For the beveled nozzle (figure 3, (c)), the horizontal distribution (z-var) is symmetrical and qualitatively corresponds to that recorded for the case of the nozzle with the normal cut (figure 3, (b)). The vertical profile (y-var) has an asymmetric appearance with a significantly more developed external mixing layer from the side of positive y values. This distribution pattern is explained by the nozzle beveled cut, which leads to the shock waves asymmetric position in the jet flow field.

Figure 4 presents the results of a comparative analysis of the pitot pressure transverse profiles in characteristic sections of supersonic jets M_a = 3.5 Npr = 91.5.

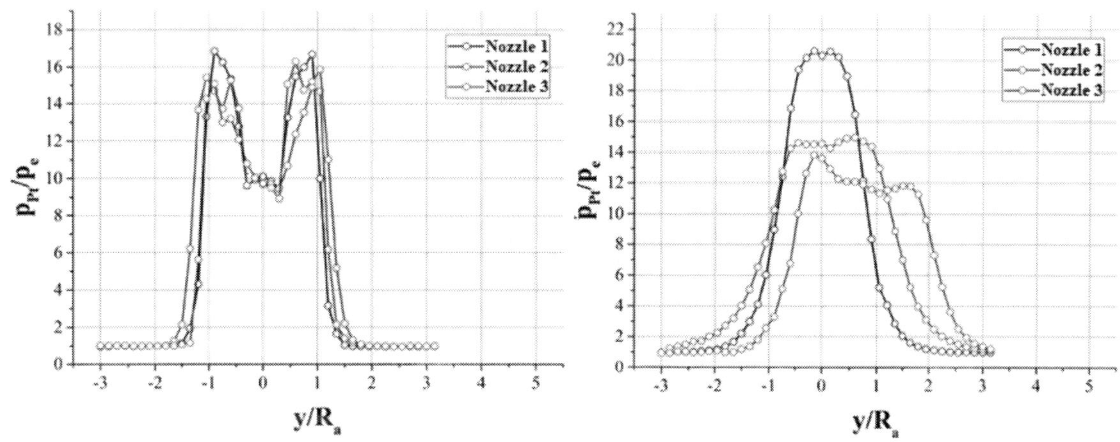

(a) (b)

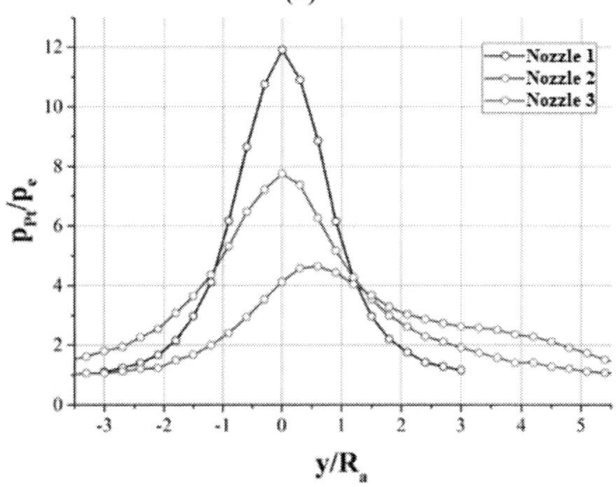

(c)

Figure 4. Transverse profiles of the pitot pressure in the symmetry plane. (a) - section x/R_a = 4, (b) - section x/R_a = 20, (c) - section x/R_a = 45

Near the nozzle exit section (figure 4, (a)), insignificant deviations in the local pressure maxima region $y/R_a = \pm 0.5$are recorded both for the pair of nozzles 1 and 2, and for 2 and 3. Pitot pressure differences in the mixing layer for positive y values between nozzles 2 and 3 are explained by the beveled cut of the nozzle 3. Differences increase with distance from the nozzle exit (figure 4, (b)). At a large distance from the nozzle exit (figure 4, (c)) for the nozzle with rectangular critical section (2) the high intensity of dissipative processes caused by a more complex shock-wave structure leads to an accelerated decrease in the measured pitot pressure. For instance, at a distance of x/R_a = 45, the difference reaches 1.5 times. For the beveled nozzle (3), the pitot pressure in the jet downstream decreases even faster due to more intensive mixing with the environment. It reaches 1.8 times at a distance of $x/R_a = 45$.

4. Conclusion

New data were obtained on the structure of supersonic jets flowing out of nozzles $M_a = 3.5$ with the rectangular critical section and the beveled exit section. It is shown that near the nozzle exit, a change in the critical section shape and the presence of the beveled exit do not significantly affect the parameters distribution in the supersonic stream. Downstream, the rectangular shape of the critical section leads to the dissipative processes intensification caused by a more complex shock-wave structure. This entails a decrease in the measured pitot pressure with increasing distance from the nozzle exit. It is shown that the beveled cut of the nozzle additionally enhances the jet energy dissipation due to more intensive mixing with the environment. This leads to a more rapid decrease in total pressure in the jet downstream.

References

[1] Dyadkin A A, Sukhorukov V P, Trashkov G A, Volkov V F, Zapryagaev V I and Kiselev N P 2014 *Proc. Congress of the International Council of the Aeronautical Sciences (St.-Petersburg)* Paper No 2014_0640

[2] Dyad'kin A A, Sukhorukov V P, Rybak S P, Zapryagaev V I, Kiselev N P, Kundasev S G, Sobolev A V and Gubanov D A 2017 *Proc. EUCASS (Milano) (Electronic Materials* DOI:10.13009/EUCASS2017-116) pp 1-9

[3] Zapryagaev V I, Kavun I N and Kundasev S G 2014 *Vestnik NGU. Ser. Fizika* **9** 75-83

Journal of Physics: Conference Series **1404** (2019) 012116 doi:10.1088/1742-6596/1404/1/012116

Essay of thermodynamic description and optimization of Ranque effect

A F Latypov

Khristianovich Institute of Theoretical and Applied Mechanics,
4/1 Institutskaya st., Novosibirsk, 630090, Russia

latypov@itam.nsc.ru

Abstract. This paper presents a thermodynamic description of Ranque effect. Including expansion and compression works into energy exchange provides a rational explanation of the observed effect. These works are performed by small portions of gas in polytropic processes. Rotational motion of the gas and presence of non-reversibility in the processes of expansion and compression are responsible for energy separation. A mathematical model with the use of the conservation laws and the polytropic expansion process was created. Optimization calculations for various values of the polytropic efficiency coefficient and the Mach number of hot gas outflow were performed. It has been obtained that the optimal relative values of exit-section area and the Mach number of inflowing high-pressure gas are invariant with respect to the values of the parameters mentioned above. Using this result, a practical model for optimizing operation process was elaborated. The efficiency coefficient of expansion work is determined during the optimization process. Using this result a practical technique of operational process has been developed.

1. Introduction

Ranque vortex tube is a cylindrical chamber with a near-axial hole with a cross-section area F_c at one tube end, and with an annular slotted opening near the wall with a cross-section area $F_{\square}$ at another one. Compressed gas with parameters $\{T_e, p_e, M_e\}$ (figure 1) is supplied into the chamber in tangential direction through the swirl nozzle with an inlet cross-section area F_e on its side wall. Ranque discovered that the temperature of gas flowing out from the near-axial hole is considerably lower than the temperature of inflow gas; and the temperature of gas flowing out from the near-axial hole is considerably higher [1]. In spite of many publications available on the effect, there is still no generally recognized physical-mathematical model.

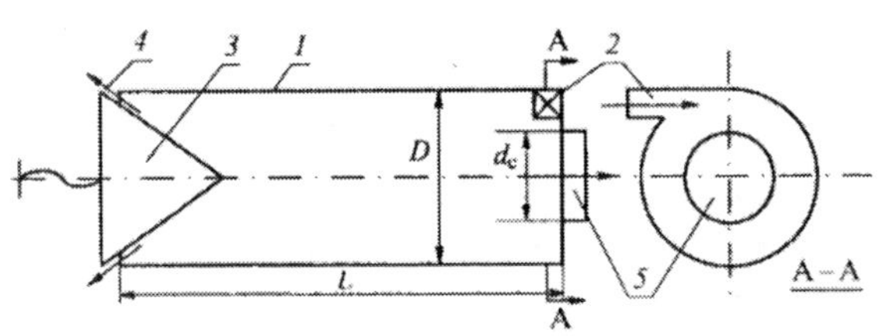

Figure 1. Basic diagram of vortex tube:
1 – smooth vortex tube,
2 – vortex generator for supplying compressed gas,
3 – throttle valve,
4 – outflow of hot gas through annular slot,
5 – diaphragm for cold gas outflow.

2. Briefly about existing approaches to describe Ranque effect

Critical reviews of significant number of works are presented in [2, 3]. The authors of the studies build stationary models of parameters distribution in the near-diaphragm section orthogonal to the rotation axis using the condition of circular motion balance, conditions of isentropic flow when establishing the link between gas parameters at various radii, and equations of ideal gas law. The models differ in distribution functions of tangential velocity depending on the radius. Unfortunately, the models are erroneous, i.e. the isentropic condition is erroneous (there is viscosity and heat exchange). Absence of analytical description of the effect using minimal experimental data stimulates conducting further theoretical and experimental research. In the present work a variant of thermodynamic description of the effect was suggested (without going into structure details of vortex flow in the tube), which also allowed us to perform the optimization procedure.

3. Physical model of Ranque effect

Let us pick out an elementary volume of gas $\sigma = \delta r \cdot \delta x \cdot r\delta\varphi$ in the cylindrical coordinate system(r, x, φ), with $\{u, v, w\}$ – axial, peripheral, and radial velocity components. Let the radial gas velocity in any elementary volume (mole of fluid [4]) be directed to the axis of rotation. When moving, gas entering the area of lower pressure expands and perfoms work, then the temperature and density in the elements decrease. The accompanying entropy increase does not allow a particle to return to the previous exergy level. Density decrease aids in further moving the particle toward the rotation axis. Part of the expansion work is spent on compressing the adjacent elementary volumes of gas. The density inside them increases causing their moving away from the rotation axis. The temperature and pressure also increase. This process is accompanied by the entropy increase, which also excludes the possibility of returning the particle to the previous exergy level. Density increase contributes in further moving the particle away from the rotation axis. As a consequence of these motions, cold gas is accumulated in the central area, and hot gas is accumulated at the periphery. As the gas temperature grows driven by compression of portions of gas, the heat flow towards the portions of gas that do the work increases. This lowers the efficiency coefficient of expansion, and, in the limit, determines the maximum temperature of the heating. Thus, the reasons of energetic gas separation during its rotation motion are the processes of expansion and compression in presence of dissipation (creating a complicated structure of interpenetrating flows). The flow is characterized by high values of relative velocities of the particles, which results in occurrence of significant noise effects.

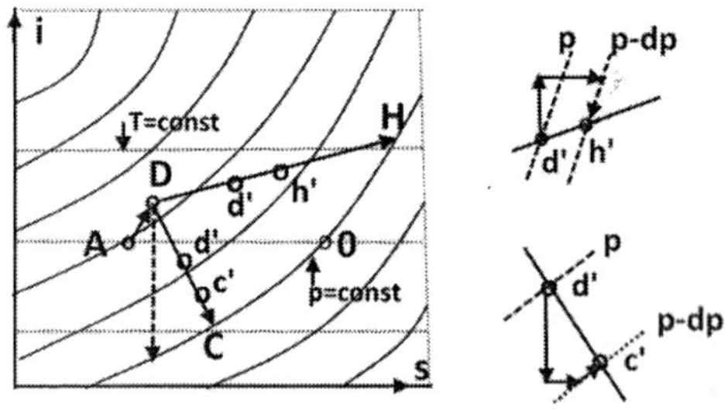

Figure 2. *i-s* – diagram of energy gas separation process.

Consider the diagram in figure 2. The point denoted as 0 corresponds to the ambient parameters. It is assumed that the pressure of outflow gases is equal to the pressure p_∞ of the ambient environment. The process of kinetic energy dissipation and flow slowdown due to viscosity and interaction with the wall is represented in the diagram by the trajectory "A–D". The trajectory "D–C" corresponds to the gas flow from the periphery to the central area. It is described by a sequence of elementary acts of isentropic expansion, isothermal throttling, and isobaric heat supply. The trajectory "D–H" corresponding to the flow of heated gas is described by the sequence of elementary acts of isentropic compression, isothermal throttling, and isobaric heat removal. The integral compression work is equivalent to the full work of expansion. Energy separation is not possible in the model of ideal gas, since only one trajectory of possible motion of the representative point issues from the point D – there is no trajectory splitting.

4. Conventions and indexes of parameters

m is the gas mass flux, V is the velocity, M is the Mach Number, Re$=\frac{V_e D}{\nu_e}$ is the Reynolds number, Pr is the Prandtl number, p is the pressure, T is the temperature, ρ is the density, i is the enthalpy, s is the entropy, $\Delta\bar{s}=\frac{\Delta s}{R}$, R is the gas constant, γ is the adiabatic index, ν is the kinematic coefficient of viscosity, $F_{ec}, F_{e\square}, F_c, F_\square$ are the cross-section areas of incoming and outgoing flows corresponding to cold and hot gas, $F_e = F_{ec} + F_{eh}$, $\alpha = m_c/m_e = F_{ce}/F_e$ is the mass ratio of cold gas, $\bar{F}_c = \frac{F_c}{F_e}, \bar{F}_\square = \frac{F_\square}{F_e}$, D is the tube diameter, $p_\infty = 1$ atm is the ambient pressure, $\bar{p}_c = \bar{p}_\square = \bar{p}_\infty = \frac{p_\infty}{p_e}$.

Indices notation is as follows e denotes entry section, c denotes cold gas, h denotes hot gas, 0 denotes stagnated flow, $\bar{x} = x/x_e$ denotes relative value of any parameter.

5. Mathematical model. Optimizing parameters

Figure 3 shows the scheme of control sections, the gas flow are indicated by the dashed line. Let us write the continuity equations for the flows of hot and cold gas of stationary process supposing that the flow parameters are uniform in control sections

$$\frac{\bar{F}_c \bar{p}_c}{\alpha\sqrt{\bar{T}_c}}\frac{\mathrm{M}_c}{\mathrm{M}_e} = 1, \frac{\bar{F}_h \bar{p}_h}{(1-\alpha)\sqrt{\bar{T}_h}}\frac{\mathrm{M}_h}{\mathrm{M}_e} = 1, \alpha = \frac{m_c}{m_e} = \frac{F_{ce}}{F_e}. \quad (1)$$

Table 1. Acceptable values of parameters to be optimized

	Minimal value	Maximal value	Location step
p_e	3.0	15.0	0.1
M_e	0.5	1.0	0.01
M_c	0.5	1.0	0.01
$\bar{F}_c$	2.0	5.0	0.01
$\bar{F}_h$	2.0	5.0	0.01

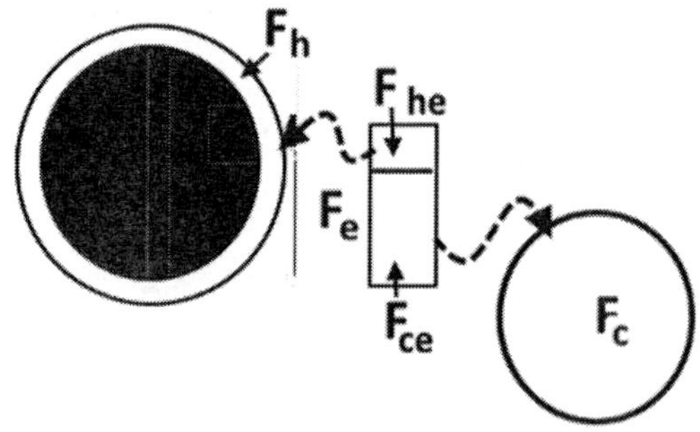

Figure 3. Scheme of control sections. Gas flows are marked by the dashed line.

It is admitted, as in gas turbines, that the dissipation energy is proportional to the change of static enthalpy $Tds = -\delta_c \cdot di$ owing to irreversible processes. Then, from the equation $Tds = du + pdv$in case $\delta_c = const$we obtain

$$\bar{p}_c = \bar{T}_c^{\frac{\gamma}{\gamma-1}(1+\delta_c)}. \quad (2)$$

Let us define the value $\eta_c = (1+\delta_c)^{-1}$ as the polytropic efficiency coefficient of gas expansion process in the Ranque tube analogically to the characteristic of gas expansion process in a turbine [5, 6]. The parameter η_c is a function of Mach number M_e, Reynolds number Re, Prandtl number Pr, specific heat ratio γ, geometrical parameters of assembly G, direction θ of the incoming flow velocity vector. It is not obvious today how to obtain theoretically integral estimation η_c. The way to obtain the experimental estimation is shown below. Energy balance equation

$$\alpha \bar{T}_c\left(1+\frac{\gamma-1}{2}\mathrm{M}_c^2\right)+(1-\alpha)\bar{T}_h\left(1+\frac{\gamma-1}{2}\mathrm{M}_h^2\right)=\left(1+\frac{\gamma-1}{2}\mathrm{M}_e^2\right). \tag{3}$$

The temperature of gas flow slowdown in control sections

$$\bar{T}_{0e}=\left(1+\frac{\gamma-1}{2}\mathrm{M}_e^2\right), \bar{T}_{0c}=\bar{T}_c\left(1+\frac{\gamma-1}{2}\mathrm{M}_c^2\right), \bar{T}_{0c\,min}=\bar{p}_c^{\frac{\gamma-1}{\gamma}}\left(1+\frac{\gamma-1}{2}\mathrm{M}_c^2\right). \tag{4}$$

Efficiency of energy separation is estimated by thermal efficiency coefficient [7]

$$\eta_t=\left(\bar{T}_{0e}-\bar{T}_{0c}\right)\left(\bar{T}_{0e}-\bar{T}_{0c\,min}\right)^{-1} \tag{5}$$

or by normalized value of refrigeration capacity [3]

$$\eta_{cp}=\left[m_c c_p\left(T_{0e}-T_{0c}\right)\right]\left[m_e c_p\left(T_{0e}-T_{0c\,min}\right)\right]^{-1}=\alpha\eta_t. \tag{6}$$

The values of parameters $\Omega=\left\{p_e, \mathrm{M}_e, \mathrm{M}_c, \bar{F}_c, \bar{F}_h\right\}$ are determined from solutions of optimization problems

$$\max_{\Omega} \eta_{cp} \tag{7}$$

for equations (1), (2), (3) with condition (7)

$$\mathrm{M}_{hmin} \le \mathrm{M}_h \le 1 \tag{8}$$

from the given values of parameters $\eta_c, \mathrm{M}_{hmin}$. The area of search of Ω parameters is defined (table 1), $\gamma = 1.4$. When solving the problem (6) the limitation (7) was considered by the method of penalty functions [8], search of the maximum was done by the method of coordinate-wise optimization [9].

Table 2. Optimal valued $\bar{F}_c, \bar{F}_h, M_e, M_c$ at various values η_c and M_{hmin}.

Parameter	Value
$\bar{F}_c$	3.1 ± 0.1
$\bar{F}_h$	2.9 ± 0.1
M_e	0.96 ± 0.04
M_c	0.86 ± 0.03

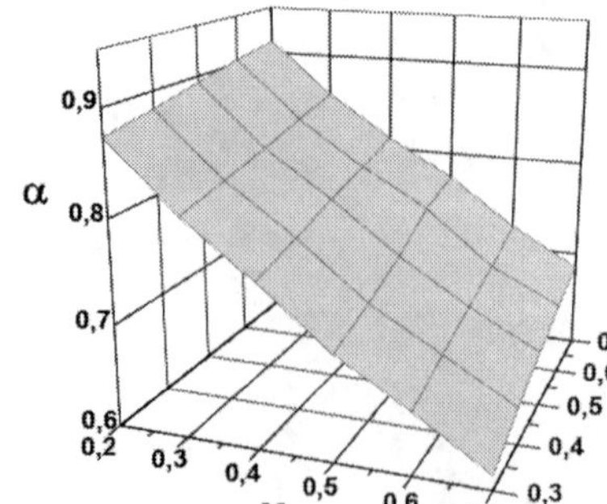

Figure 4. Optimal values of cooled gas mass ration α.

6. Calculation results

In all variant of the problem (6) $M_{hopt} = M_{hmin}$ was obtained at the point of maximum of function η_{cp}. The parameters η_c, M_{hmin} are basically determining ones in the process of energy separation: more is η_c, more is the work of expansion in case of fixed pressure fall; less is M_{hmin}, more is proportion of the gas being cooled α (figure 4) and more is the functional η_{cpopt} (figure 5). A considerable increase of the temperature of the gas being heated (figure 6) is conditioned by increase of refrigeration efficiency η_{cpopt} and decrease of heated gas flux $(1-\alpha)$. In reality, the temperature growth will be limited by increasing the outflowing heat flux toward the ambient cold gas. Figure 7 shows the optimal values of the pressure of the input gas flow p_e, figure 8 shows the optimal values of relative temperature of the gas cooled $\overline{T}_c$. The numerical results presented illustrate possible values of device characteristics.

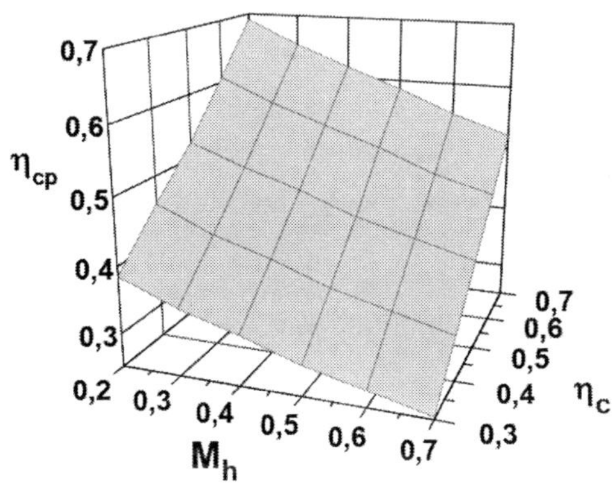

Figure 5. Optimal values of normalized value of refrigeration capacity η_{cp}.

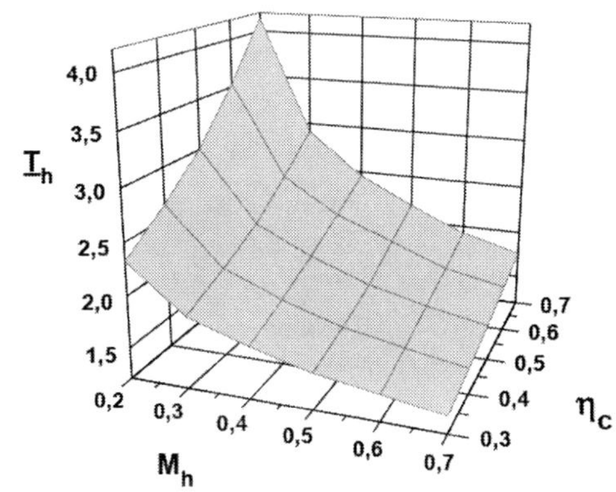

Figure 6. Optimal values of heated gas relative temperature $\overline{T}_h$.

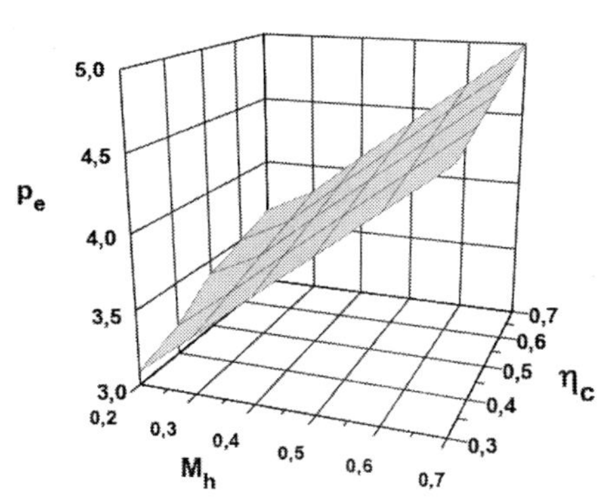

Figure 7. Optimal values of input gas flow pressure p_e.

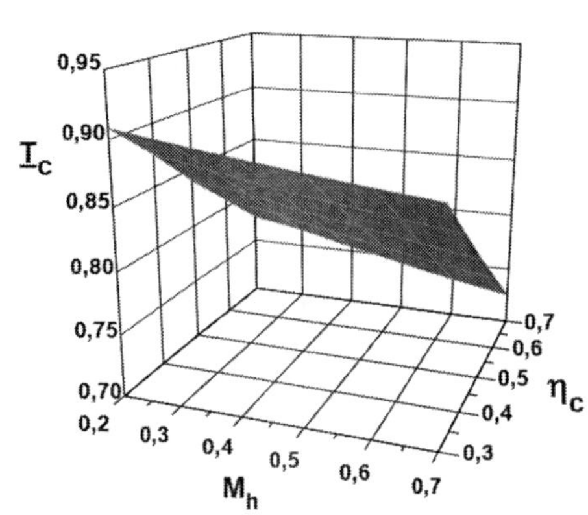

Figure. **8**. Optimal values of cooled gas relative temperature $\overline{T}_c$.

It is noteworthy that the optimal values of relative areas of pipe exit section, Mach number of incoming flow are practically independent on the values η_c, M_{hmin} (table 2). For this reason, we can accept mean values of parameters $\overline{F}_c, \overline{F}_h, M_e$ for the design implementation. The value of inlet cross-section area of the compressed gas F_e is determined by its defined flux. The diameter of diaphragm aperture is $d_c = 2\sqrt{F_e\bar{F}_c/\pi}$. In [5] a linear relation was obtained between the diaphragm aperture radius and inner diameter of the tube $d_c = \beta D, \beta \approx 0.4$ (based on the experimental results of various authors), from which we obtain the value D. Other empirical ratios were also obtained for the value d_c/D. The scatter of the values motivates using experimental technology to determine the rational value of d_c/D subject to the conditions $\bar{F}_c = const, \bar{F}_h = const$. The tube length is 9–10 calibers [10] in case of presence of a device for vortex motion damping near the throttle.

7. Practical evaluation of the optimal mode

To determine the characteristics realized, it is necessary to measure the values of the parameters $p_{0e}, p_e, T_{0e}, T_{0c}, T_{0\square}$, and then to calculate all characteristics $\bar{p}_{0e}, \bar{T}_{oe}, M_e, T_e, \alpha, M_c, T_c, M_\square, T_\square, \eta_c, \eta_t, \eta_{cp}$ according to the presented thermodynamic model. Performing this procedure for various values of pressure p_e, we can determine the optimal mode.

8. Conclusion

A thermodynamic model of Ranque effect has been created. It was obtained that the optimal values of relative areas of the tube outlet sections and the Mach numbers of incoming flow are practically independent on polytropic efficiency coefficient of expansion η_c or Mach number of hot gas outflow $M_{h\,min}$. A practical technique for optimizing the operation process has been suggested.

References

[1] Ranque G J 1933 Experiments on expantion a vortex with simultaneous exhaust of hot air and cold air *J. Phys. Radium* **4** 112–14
[2] Gutsol A F 1997 The Ranque effect *Phys. Usp.* **40** 639–58
[3] Piralishvili Sh A, Poliayev V M and Sergueyev M N 2000 *Vortex Effect* (Moscow: Energomash) 414 pp
[4] Hinze J O 1959 *Turbulence* (New York: McGraw Hill) 586 pp
[5] Abiantz V H 1965 *Theory of Gas Turbine of Ramjets* (Moscow: Mashinostroyeniye) p 312
[6] Zhiritzky G S, Lokay V I, Maksutova M K and Strunkin V A 1971 *Gas Turbines of Flying Vehicles Motors* (Moscow: Mashinostroyeniye) 620 pp
[7] Hilsh R 1947 *Review of Scientific Instruments* **18** 108-113
[8] Latypov A F 1974 *Bulletin of the SB Academy of Sciences of USSR: Technical Series* **3** 49-50
[9] Bakhvalov N S, Zhidkov N P and Kobelkov G M 1987 *Numerical Methods* (Moscow: Science) 598 pp
[10] Merkulov A P 1969 *Vortex Effect and its Application in Technic* (Moscow: Mashinostroyeniye) 184 pp

Numerical simulation of a two-component gas flow near a highway

S A Lavruk [1] and S A Valger [1,2]

[1]Khristianovich Institute of Theoretical and Applied Mechanics SD RAS, Institutskaya 4/1, 630090, Novosibirsk, Russia
[2]Novosibirsk State University of architecture and Civil Engineering(Sibstrin), Leningradskaya 113, 630008, Novosibirsk, Russia

lavruks@mail.ru

Abstract. Numerical simulation of the air flow near a highway was carried out taking into account injection of ethane through a system of low sources. Numerical model is based on full physical and mathematical models of continuum media and takes into account the phenomena of turbulent mixing of gas components in the surface boundary layer. Based on the calculation results, the fields of velocity distribution, turbulent parameters and impurity concentrations in characteristic sections near the road are obtained.

1. Introduction

One of the most important environmental criteria is the quality of the air, which contains pollutants in the gases and aerosols forms. Air pollutants, such as gaseous impurities of carbon dioxide, nitrogen oxides, sulfur dioxide, benzapyrene, aerosols with soot particles, heavy metal compounds, mineral and organic compounds are more of anthropogenic origin. According to [1], the main sources of emissions in the air of large cities are industry, transport, domestic fuel combustion and other anthropogenic sources. In contrast to industrial sources of emissions, the transfer of emissions from vehicles is local in nature near urban buildings and pedestrian zones and is characterized by a low spatial arrangement of sources relative to the level of the urban landscape. In pedestrian areas of densely built megalopolises, the composition of the air substantially depends on the quality of the transport infrastructure and the composition of the exhaust gases of vehicles.

Today, along with instruments for atmospheric monitoring and control over the level of emissions and aerosols in air, methods of mathematical and statistical analysis and methods of numerical modeling are actively used to describe and predict the parameters of transport, transformation and filtration of impurities in the lower atmosphere. An advantage of computational aerodynamics methods for solving such problems is ability to study nature of impurity transport phenomena and the influence of number of factors, such as the thermal regime of the boundary layer, turbulent mixing, filtration through buildings and green zones, on dynamics of the impurity on fairly detailed spatial scale. Thus, numerical modeling of the emission processes transfer in the surface layer of the atmosphere near the systems constructions and buildings is relevant both from the point of view of solving the fundamental problem of studying atmospheric processes, and from the point of view of solving the applied problem of predicting and improving air ecology.

Today there are a fairly large number of studies devoted to the numerical estimation of the processes of distribution and transfer of exhaust gases. In [2], a numerical simulation of air flow in the downtown of Melbourne city was performed to determine model constants and parameters of the turbulence model and further simplify the numerical model while maintaining sufficient calculation accuracy. In [3], a review of experimental work in wind tunnels aimed at studying air flows in urban areas and a description of the structure of flows and recirculation zones behind poor-flowing bodies, dangerous from the point of view of accumulation of heavy exhaust gases, is given. Detailed review of existing numerical models and schemes used to describe flows in urban canyons is given in [4]. Speaking about the numerical modeling of the processes of transporting exhaust gases, one of the most interesting works is the study [5], which analyzes the structure of the air flow taking into account the emission of exhaust gases near highways taking into account noise-suppressing barriers. The work [5] was carried out in accordance with experimental data on the estimation of flow parameters near a section corresponding to a part of the highway under conditions of a different location of this roadway relative to the main height level of the building [6]. The use of parametric models for setting the emission of exhaust gases from highways is presented in [7]. In [8, 9], the application of the RLINE model is described, which makes it possible to estimate the dispersion of pollutants from mobile sources. The authors' works [10–11] are devoted to the study of the aerodynamics of microdistricts of complex architecture and to the study of the influence of the thermal stratification of the boundary layer on the turbulent impurity transfer parameters.

2. Physical and numerical model

In this paper, we consider a model problem for studying the flow of multicomponent gas near a section of a highway. The statement of the problem corresponds to the experimental data [6]. To account for turbulent effects in the flow, we considered the Reynolds averaged Navier-Stokes equations for a two-component medium, supplemented by a *k-ω SST* turbulence model. The geometry of the computational domain is shown in Fig. 1 (a, b). The computational domain measures 140 cm wide by 245 cm length and 40 cm height (Fig. 1a). The substrate with neutral gas emission holes measures 24 cm wide and 24 cm long (Fig. 1b). To discretize the computational area, mesh of finite volumes based on tetrahedrons is constructed.

The air flow goes to the calculation area as shown in Fig. 1 (a). The profiles of the boundary conditions at the entrance to the computational area are constructed taking into account the vertical gradients of the gas-dynamic parameters. The vertical distribution of average velocity is described by a logarithmic profile [6]:

$$\frac{U}{u_*} = \frac{1}{k}\ln\left(\frac{z+d}{z_0}\right),$$

where $k=0.4$ – the von Karman constant, $u_*=0.3$ m/s – reference velocity, $d=0.006$ m, $z_0=0.0052$ m.

As injection gas is taken ethane (C_2H_6) that is used in [6] to model exhaust gases. In experimental study [6] ethane is used because with a molecular weight of 30 is only slightly heavier than air and may be regarded as neutrally buoyant Ethane is injected through circular inlets arranged linearly in accordance with Fig. 1 (b). The total rate of gas at the outlet of the injectors is 1500 cm^3/min.

A symmetry condition was established along the lateral faces of the computational domain and a soft boundary condition for static pressure was set at the outer boundaries of the computational domain.

The calculations were carried out with the second order of approximation in space. The goal was to establish the flow field in the considered computational domain. As a solver, the software ANSYS Fluent was used.

Journal of Physics: Conference Series **1404** (2019) 012117 doi:10.1088/1742-6596/1404/1/012117

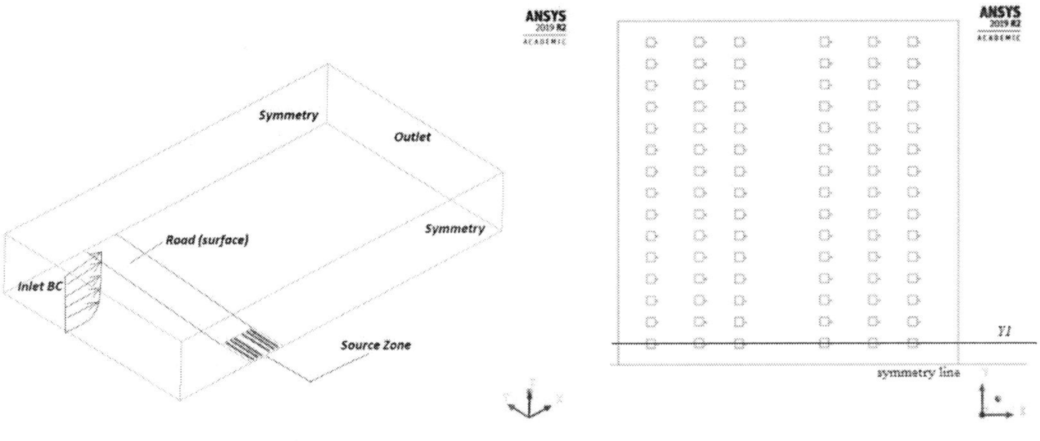

(a) (b)

Figure 1. Geometry of the computational area: scheme of the computational area (a) and scheme of the C_2H_6 emission holes substrate (b).

3. Preliminary results

Based on the calculation results, the fields of velocity distribution, turbulent parameters and impurity concentrations in characteristic sections near the road are obtained. In fig. 2 shows the velocity vectors field over the substrate in a characteristic longitudinal section *Y1*. Near the simulated highway, a compression of the flow is observed, that can be explained as interaction of the flow with the obstacle (Fig. 3). Formation of stagnant recirculation zones in the area between the rows of cars is observed.

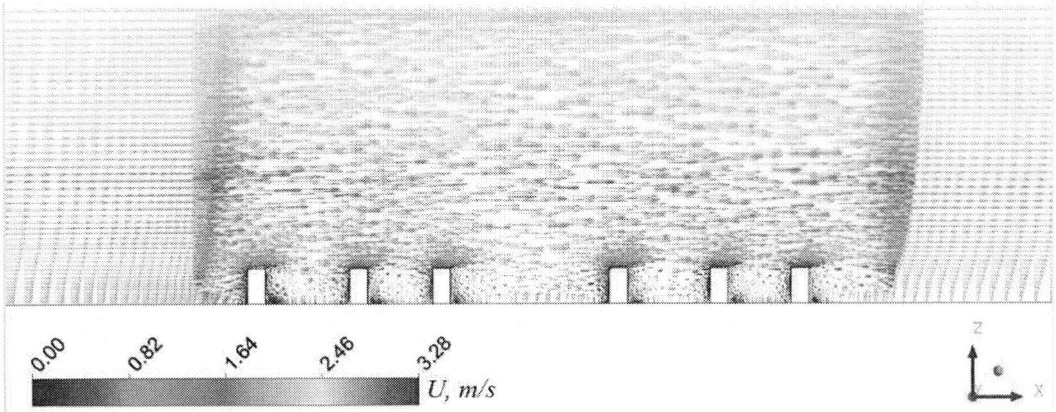

Figure 2. The velocity vectors field above the substrate in a characteristic longitudinal section *Y1*.

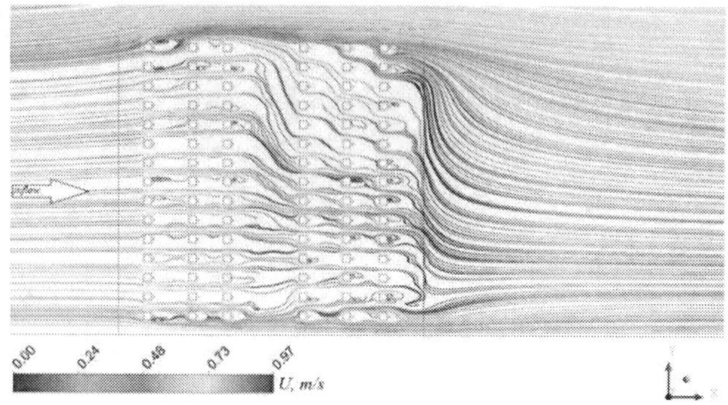

Figure 3. Flow lines near a road fragment in plan.

Figure 4 shows the field of the mass fraction of ethane *C* in a characteristic longitudinal section *Y1*. It can be seen from the figure that high concentrations of ethane are observed in recirculated stagnant zones behind bluff bodies. The transfer of impurities behind a system of bluff bodies has a highly spatial character.

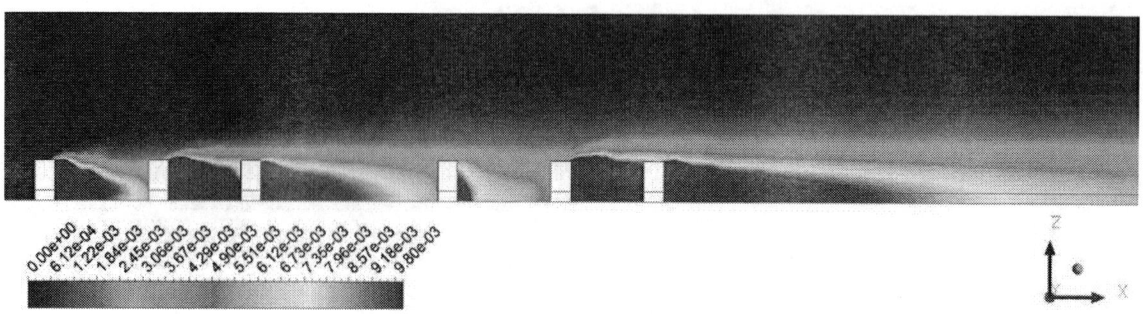

Figure 4. Field of mass fraction of ethane *C* in a characteristic vertical section *Y1*.

4. Conclusions

A numerical simulation of the air flow near the highway was carried out taking into account the injection of ethane through a system of low sources. Based on the calculation results, the distribution fields of velocity vectors, turbulent parameters and impurity concentrations in characteristic sections near the highway are obtained and the flow structure behind the system of poorly streamlined bodies is described. In further work, it is planned to clarify the parameters of the problem and conduct a quantitative comparison of the results with experimental data [6].

Acknowledgment

The work was supported by Russian Foundation for Basic Research, grant 18-08-00755 A "Numerical simulation of heat and mass transfer processes in multicomponent flows, describing mechanisms for the emission dissemination from vehicles in the vicinity of urban roads".

References

[1] Zavadskaya E K, Ivleva T P, Smirnova I V and Vorobeva I A 2009 *Air quality in the largest cities of Russia for ten years 1998-2007: Analytical review.* (Saint Peterburg: Voeikov Main Geophysical Observatory) 133 pp.

[2] Gousseau P, Blocken B, Stathopoulos T and van Heijst G J F 2015 *Comput. Fluids* **114** 151–62

[3] Ahmad K, Khare M and Chaudhry K K 2005 *J. Wind Eng. Ind. Aerodyn.* **93** 697–717
[4] Starchenko AV, Nuterman R B and Danilkin E A 2015 *Numerical Study of Turbulent Flows and Pollution Transport in Street Canyons* (Tomsk: Tomsk University Press) 252 pp
[5] Enayati Ahangar F, Heist D, Perry S and Venkatram A 2017 *Atmos. Environ.* **155** 1–10
[6] Heist D K, Perry S G and Brixey L A 2009 *Atmos. Environ.* **43** 5101–11
[7] Bäumer D, Vogel B and Fiedler F 2005 *Atmos. Environ.* **39** 5750–9
[8] Milando C W and Batterman S A 2018 *Atmos. Environ.* **182** 213–24
[9] Snyder M G, Venkatram A, Heist D K, Perry S G, Petersen W B and Isakov V 2013 *Atmos. Environ.* **77** 748–56
[10] Valger S A, Fedorova N N and Fedorov A V 2015 *Thermoph. Aeromech.* **22** 29-41
[11] Valger S and Fedorova N 2018 *AIP Conf. Proc.* **2027** 040043

NO production on the reentry spacecraft thermal protection system surface in the direct simulation Monte Carlo method

A S Litvintsev, A N Molchanova and Ye A Bondar

Khristianovich Institute of Theoretical and Applied Mechanics SB RAS,
4/1 Institutskaya str., Novosibirsk 630090, Russia

litvintsev@itam.nsc.ru

Abstract. This work is aimed at development of an approach for modeling surface chemistry with multiple reaction channels in the direct simulation Monte Carlo method. In particular, production of NO on spacecraft thermal protections system surface is considered in detail. The approach is based on the previous works of the authors where a surface chemistry molecular model based on macroscopic detailed reaction sets was developed. The new approach has been implemented in SMILE++ code and verified by comparison with theoretical values of the surface population of atomic species in thermal equilibrium. Excellent agreement between computational results and theoretical prediction has been demonstrated.

1. Introduction

Surface chemical reactions have significant impact on the heat transfer to the thermal protection system (TPS) of the descent vehicle entering the Earth atmosphere. Since these reactions are exothermic, ignoring these processes in an aerothermodynamic computational model may lead to a significant underprediction of the heat flux to the surface of the descent vehicle [1, 2]. Thus, a correctly developed surface catalytic model allows to increase the accuracy of prediction of heat flux in aerothermodynamics computations. For example, accurate prediction of heat flux to the surface reduces the weight of the heat shield and hence increases the payload.

When modeling the aerothermodynamics of descent vehicles at high altitudes (more than 80 km), one cannot rely on continuum gas models due to the significant flow nonequilibrium. The main tool for the numerical analysis of such flows is the direct simulation Monte Carlo (DSMC) method [3]. To simulate surface processes in the DSMC method, macroscopic chemical data is not enough and information on probabilities and times of processes for specific molecular species is required. In [4-6], Molchanova et al. proposed an approach for obtaining such microscopic information based on detailed macroscopic kinetic mechanisms of surface processes. One of the limitations of the proposed approach is its inability to take into account recombination reactions leading to the formation of nitric monoxide. As noted in [7, 8], such reactions can play a significant role in the kinetics of surface processes. In this work, we generalize the approach [6] to the kinetics of surface processes, including the formation of nitric monoxide. The generalized approach is implemented in the SMILE++ code [9, 10]. The approach has been applied to the Kurotaki macroscopic kinetic mechanism for SiO_2-based thermal protection system [11] that includes reactions producing NO. Verification of the approach is performed by comparing equilibrium surface fraction of the adsorbed species obtained in DSMC computations with the theoretical values.

2. Modification of the model of Molchanova et al

Processes involved in a surface chemical mechanism are typically divided into impact (such as Eley-Rideal, ER, recombination) and surface (such as Langmuir-Hinshelwood, LH, recombination) processes. NO molecules can be produced on the surface by both of the recombination processes. These reactions cannot be simulated within the model [6] because it was constructed in such a manner that only one ER and one LH process can be considered for each reagent. For example, if $N + N_s \rightarrow N_2$ recombination is modeled (which is one of the primary processes) than $N + O_s \rightarrow NO$ cannot be taken into account. Index *s* denotes here particles adsorbed at the surface. So, the modification of the approach [6] performed in the present work is primarily related to the modeling multiple ER and LH recombination channels. The modified algorithms are presented below separately for impact and surface processes. Below we focus only on the differences of the present modified approach from the approach of Molchanova et al., all the common details of the original and modified algorithms can be found in [6].

2.1. *Impact processes*

The schematic of the present modified algorithm of the impact processes is shown in figure 1. Let us consider a particle A (atom or molecule) impinging onto the panel (an elementary element) of the surface. When the particle A collides with the surface, the following events are possible: reflection, adsorption (with the probability P_{ad}), Eley-Rideal recombination between particles A and A_s (with the probability P_{ER1}), Eley-Rideal recombination between particles A and B_s (with the probability P_{ER2}), etc. All these events are incompatible and form a complete group of events. The probability of a reaction in the interaction with the surface is $P_{mult} = P_{ad} + P_{ER1} + P_{ER2}$. Comparing P_{mult} with Rn (a

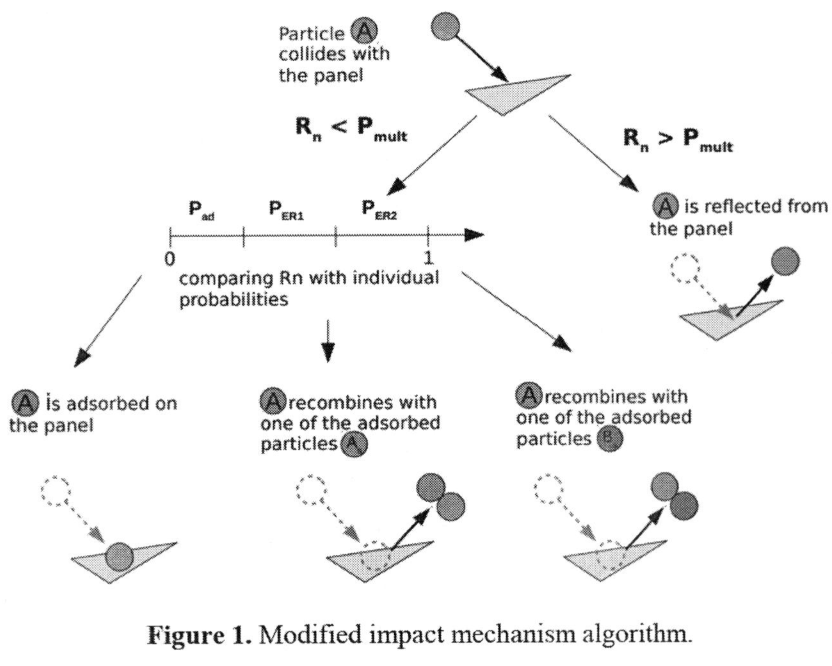

Figure 1. Modified impact mechanism algorithm.

random variable uniformly distributed on the interval [0,1]) shows whether the particle will reflect from the surface or participate in one of the surface processes. To understand which process is realized, we need to compare the position of *Rn* value on the interval [0, 1] with the position of interval visualizing the probabilities of all surface processes.

2.2. *Surface processes*

Although Molchanova et al. have already implemented the algorithm for simulating Langmuir-Hinshelwood recombination, it was shown in [12] that when this algorithm is extended in the straightforward manner to multiple reaction channels, a bias towards some of the processes arises and rates of the processes are modeled incorrectly. This behavior is due to the fact that in Molchanova et al. algorithm surface processes are modeled one by one which does not reflect the real physical situation where the reactions occur simultaneously. Such an approach works properly only for LH recombination with a single reaction channel (e.g., $N_s + N_s \rightarrow N_2$). If one includes another possible channel (e.g., $N_s + O_s \rightarrow NO$) then the rate of the reaction which is modeled first is artificially increased, and for the subsequent reactions is, on the contrary, reduced. This is due to the fact that the average time between reactions is inversely proportional to the number of reagents. Consequently, less and less of them are available to subsequent reactions, which ultimately creates the mentioned bias. A modified LH algorithm taking into account multiple reaction channels should reflect the simultaneous nature of surface reactions as accurately as possible and maintain optimal performance.

The modified LH algorithm diagram can be seen in figure 2. At the beginning of each computational iteration, all available Langmuir-Hinshelwood recombination reactions are initialized, after which their list is randomly mixed so they are modeled in the random order. For each reaction, the time equal to the computation time step Δt is allocated during which it can occur. One starts with the first reaction in the randomly shuffled reaction lists. The following steps are performed consequently:

- **IF** there are available reagents **AND** the allocated time is greater than zero than the time to the next reaction t_{reac} is modeled in the manner similar to presented in [6], the reaction is simulated, and t_{reac} is subtracted from the allocated time (**ELSE** the reaction is not modeled),
- move to the next reaction in the list.

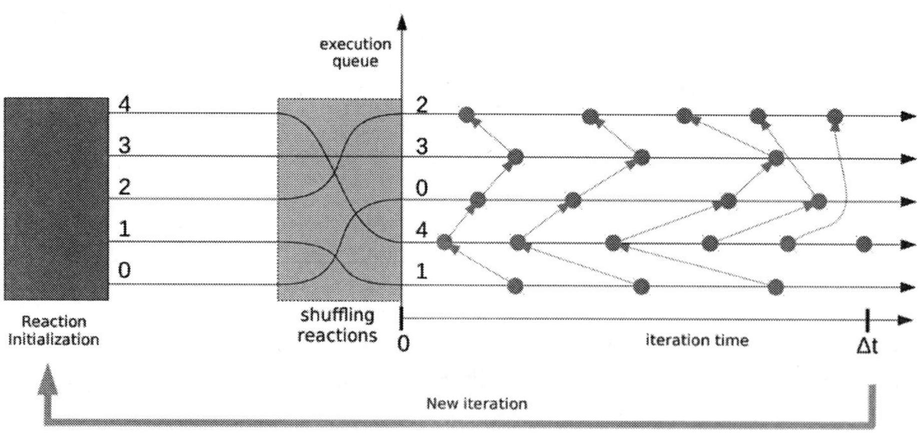

Figure 2. Schematic of the modified Langmuir-Hinshelwood recombination algorithm.

These steps are performed until there are reagents and positive allocated time for any of the reactions. The modified algorithms have been implemented into the SMILE++ software system. The results of verification are presented in the next section.

3. Numerical results

A detailed macroscopic mechanism of Kurotaki, employed in the present work includes surface reactions of adsorption, desorption, Eley-Rideal recombination, and Langmuir-Hinshelwood recombination for N and O atoms (formation of N_2, O_2 and NO).

The calculation was carried out for a stationary gas, under conditions of thermal equilibrium between the gas and the wall for different wall temperatures. The gas density was set to be $9 \cdot 10^{-6}$ kg/m^3. Gas-phase reactions were neglected.

Sampling of surface macroparameters was performed after achieving steady state in terms of surface fraction of the adsorbed atomic species N and O. This state corresponds to zero mass flux through the surface and can be calculated by solving the system of kinetic equations from the Kurotaki catalytic model. Comparison of the DSMC results with this theoretical solution was performed in order to verify implementation of the model.

The results of the calculation that was carried out for the case of adsorption and desorption reactions are shown in figure 3a. The obtained dependence is in excellent agreement with the theoretical model for all considered temperature range.

Figure 3b presents the equilibrium surface population for the case of adsorption, desorption, Eley-Rideal recombination, which also coincides well with the theoretical curve. In comparison with the adsorption-desorption case, the proportion of oxygen on the surface decreases for temperature close to 1000 K. The obtained dependence is in good agreement with the theoretical prediction too.

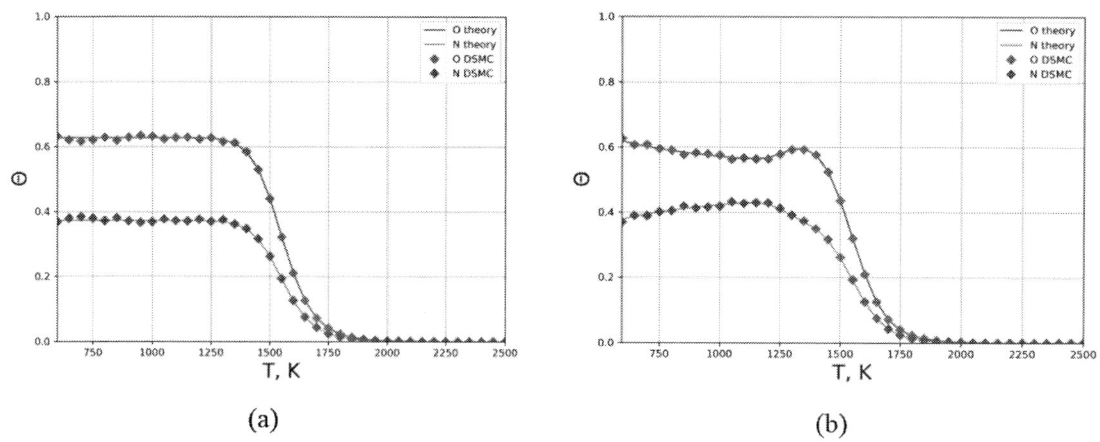

Figure 3. Comparing equilibrium surface fraction of the adsorbed species obtained in DSMC computations with the theoretical values: (a) adsorption and desorption, (b) adsorption, desorption and ER recombination.

Figure 4 shows the distribution of the fraction of nitrogen and oxygen on the surface for the case of adsorption, desorption and Langmuir-Hinshelwood recombination taken into account. It can be seen that the Langmuir-Hinshelwood recombination has a high rate, significantly reducing the concentration of both oxygen and nitrogen that their total fraction on the surface becomes less than unity at relatively low temperatures (there are some vacant sites). The obtained dependence is also in good agreement with the theoretical curve.

All numerical results demonstrate that the present DSMC model predict correct rates in thermal equilibrium for complex chemical mechanism with multiple channels including nitrogen monoxide production.

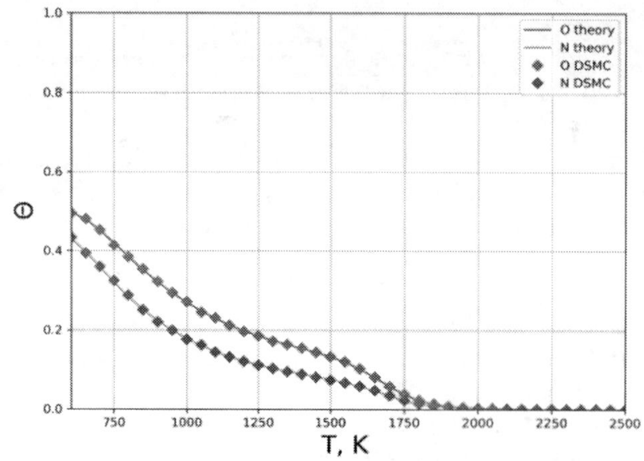

Figure 4. Comparing equilibrium surface fraction of the adsorbed species obtained in DSMC computations with the theoretical values for adsorption, desorption and Langmuir-Hinshelwood recombination.

4. Conclusion
In the present work, an approach has been developed that allows including surface chemistry with multiple reaction channels in the direct simulation Monte Carlo computations. The approach is easily generalized to any number of surface reactions and can be applied for chemical mechanism with NO production on the surface.

Modified surface chemistry algorithms have been implemented into SMILE++ code. The DSMC computations have shown excellent agreement with theoretical predictions under conditions of thermal equilibrium. The future work should address the effect of the surface reactions with NO production on high-altitude aerothermodynamics of re-entry spacecraft.

Acknowledgements
The work was supported by the Russian Science Foundation (Grant No. 17-19-01375). This support is gratefully acknowledged.

References
[1] Anderson L A 1973 *AIAA J.* **11** 649-56
[2] Kovalev V L and Kolesnikov A F 2005 *Fluid Dynamics* **40** 669-93
[3] Bird G A 1994 *Molecular Gas Dynamics and The Direct Simulation of Gas Flows* (New York: Oxford University Press Inc.)
[4] Molchanova (Shumakova) A N, Kashkovsky A V and Bondar Ye A 2014 Detailed DSMC surface chemistry model *AIP Conf. Proceedings* **1628** 131
[5] Molchanova A N, Kashkovsky A V and Bondar Ye A 2016 Effect of surface catalycity on high-altitude aerothermodynamics of reentry vehicles *AIP Conf. Proceeding* **1770** 040010
[6] Molchanova A N, Kashkovsky A V and Bondar Ye A 2018 Surface recombination in the direct simulation Monte Carlo method *Phys. Fluids* **30** 107105
[7] Pejakovic D A, Marschall J, Duan L and Martin M P 2010 Direct detection of NO produced by high-temperature surface-catalyzed atom recombination *J. of Thermophysics and Heat Transfer* **24** 603-11
[8] Weisberger J M, MacLean M G, Parker R A and DesJardin P 2013 Near-surface nitric oxide concentration measurement in the LENS-XX expansion tunnel facility *44th AIAA Thermophysics Conf.* AIAA 2013-2643

[9] Kashkovsky A V, Bondar Ye A, Zhukova G A, Ivanov M S and Gimelshein S F 2005 Object-oriented software design of real gas effects for the DSMC method *AIP Conf. Proceedings* **762** 583-8
[10] Ivanov M S, Kashkovsky A V, Vashchenkov P V and Bondar Ye A 2011 Parallel object-oriented software system for DSMC modeling of high-altitude aerothermodynamic problems *AIP Conf. Proceedings* **1333** 211-8
[11] Kurotaki T 2001 Catalytic model on SiO_2-based surface and application to real trajectory *J. of Spacecraft and Rockets* **38** 798-800
[12] Swaminathan G K, Arnaud B and Stephani K A 2018 Development of a detailed surface chemistry framework in DSMC *2018 AIAA Aerospace Sciences Meeting* 2018-0494

Journal of Physics: Conference Series **1404** (2019) 012119 doi:10.1088/1742-6596/1404/1/012119

Numerical solution of the Vlasov-Poisson equations using a semi-Lagrangian WENO scheme implemented on GPU

E A Malkov[1,2], S O Poleshkin[1], A A Shershnev[1] and A N Kudryavtsev[1,3]

[1] Khristianovich Institute of Theoretical and Applied Mechanics, Russian Academy of Sciences, Siberian Division, 630090 Novosibirsk, Russia
[2] Siberian State University of Telecommunications and Information Sciences, 630102 Novosibirsk, Russia
[3] Novosibirsk State University, 630090 Novosibirsk, Russia

E-mail: malkov@itam.nsc.ru, poleshkin@itam.nsc.ru, antony@itam.nsc.ru, alex@itam.nsc.ru

Abstract. A numerical method for solving the Vlasov–Poisson equations using a high-order semi-Lagrange conservative WENO scheme is developed. The Vlasov–Poisson equations govern evolution of the collisionless self-interacting medium and are widely used in plasma physics and astrophysics, in particular for modeling dynamics of galactic systems. The method is implemented for computations on Graphical Processing Units (GPUs). The GPU code is validated using an exact unsteady analytical solution describing nonlinear oscillations of a plane self-gravitating layer. The comparison with numerical results obtained with the serial CPU code show a significant, up to 50 times, speed-up of the computations.

1. Introduction

Rarefied gas flows in the field of a body force are governed by the kinetic equation

$$\frac{\partial f}{\partial t} + \mathbf{u} \cdot \frac{\partial f}{\partial \mathbf{r}} + \frac{\mathbf{F}}{m} \cdot \frac{\partial f}{\partial \mathbf{u}} = \frac{1}{\mathrm{Kn}} \left(\frac{\partial f}{\partial t} \right)_{coll}. \tag{1}$$

Here $f(t, \mathbf{r}, \mathbf{u})$ is the one-particle distribution function, i.e. the density, in a six-dimensional phase space, of some additive quantity such as the mass, the number of molecules, or the probability that characterizes the measure of the gas at the time moment t located at the point $\mathbf{r}$ of the physical space and having a corresponding velocity $\mathbf{u}$. The function $\mathbf{F}(t, \mathbf{r}, \mathbf{u})$ defines a body force field that, in general, is coordinated with the phase density, and Kn is the Knudsen number equal to the ratio of the molecular free-path length and the characteristic length scale of the problem.

In numerical simulations of processes in plasmas, stellar systems, and neutral gas flows in the gravitational field, when the mean free path of molecules is comparable with the relative variation of this field, and also in computations of unsteady flows in noninertial frames of reference, the question arises whether the calculation of the convective transport represented by the left-hand side of the kinetic equation (1) is correct. In the case of moderate Knudsen numbers, the excessive dissipation introduced by a numerical scheme during calculation of the convective terms can be “disguised” by the distribution function variation due to collisions. The situation is different for high Knudsen numbers (tending to infinity in the limit). Computer simulations of the dynamics of collisionless systems are particularly sensitive to the choice of the numerical scheme. Code validation is usually performed through comparisons of results obtained with two different numerical schemes. For example, the results on the dynamic evolution of a gravitating system obtained using the conservative semi-Lagrange scheme are compared with the results of N-body simulations [1, 2, 3].

In the present study, it is proposed to use the exact unsteady solutions of the Vlasov–Poisson equations [4, 5, 6, 7] for testing codes for simulating the collisionless dynamics of gravitating systems. The proposed approach is illustrated using a one-parameter family of unsteady solutions of the Vlasov–Poisson equations constructed by means of the group analysis [7] and describing nonlinear oscillations of a plane self-gravitating layer. The computations of the plane layer evolution are computationally inexpensive but demonstrate the main problems of numerical simulation of the Vlasov–Poisson equations. Moreover, the results of these computations are fairly universal for considering fundamental problems of the evolution of gravitating systems.

2. Model of unsteady plane layer

A gravitating layer can be presented as a set of parallel planes with a homogeneous surface density interacting under the action of gravity. A continuum of such particles-planes is described by the kinetic equation in a two-dimensional phase space coupled with the Poisson equation:

$$\frac{\partial f(t,u,z)}{\partial t} + u\frac{\partial f(t,u,z)}{\partial z} - \frac{\partial \Phi(t,z)}{\partial z}\frac{\partial f(t,u,z)}{\partial u} = 0, \tag{2}$$

$$\frac{\partial^2 \Phi(t,z)}{\partial z^2} = 4\pi G \int f(t,u,z)\,du. \tag{3}$$

In the case of a homogeneous spatial density and, correspondingly, a quadratic potential, there exists an exact analytical solution of system (2-3) [8].

Using the group analysis, one can obtain a family of unsteady solutions that describe finite-amplitude oscillations of a gravitating layer with the same total energy [7]. The distribution functions from this family parametrized with the parameter α are

$$f(t,z,u) = \frac{\rho_0}{\pi}\left[\alpha^2\left(z_0^2 - \frac{z^2}{a^2}\right) - a^2\left(u - \frac{\dot a}{a}z\right)^2\right]^{-1/2}, \quad a = a(t). \tag{4}$$

The homogeneous spatial density depends on time as

$$\rho(t) = \frac{\rho_0}{a(t)}. \tag{5}$$

Correspondingly, the potential is

$$\Phi(t,z) = \frac{2\pi G\rho_0}{a(t)} z^2. \tag{6}$$

The boundary of the plane layer changes as $z(t) = a(t)z_0$. Without loss of generality, it is further assumed that $4\pi G\rho_0 = 1$, $\rho_0 = 1$ and $z_0 = 1$. The function $a(t)$ satisfies the equation

$$\ddot a = \frac{\alpha^2}{a^3} - 1. \tag{7}$$

The value h of the first integral of this equation

$$h = \frac{\dot{a}^2}{2} + \frac{\alpha^2}{2a^2} + a \tag{8}$$

is taken equal to 3/2. This value corresponds to the steady layer with $a(t) \equiv 1$ and $\alpha = 1$. The distribution function with the parameter $\alpha \neq 1$ and $h = 3/2$ are unsteady, but have the same total energy. The degree of unsteadiness is determined by the so-called virial ratio, i.e. the ratio of the doubled kinetic energy to the absolute value of the potential energy

$$V_{rel} = \frac{2E_{kin}}{|W|} = \frac{\alpha^2}{a^3} \tag{9}$$

3. Numerical simulations of oscillations of a homogeneous gravitating layer

The distribution function (4) is singular at the layer boundaries. As a result, numerical simulations of the plane self-gravitating layer oscillating in accordance with this solution require some caution when specifying the initial conditions for the computations. In the present paper, the initial data are taken as the values of the exact distribution function averaged over rectangular grid cells in the 2D phase space. In this case, the initial density of layer in coordinate space is perfectly homogeneous.

The layer oscillations are simulated numerically with the semi-Lagrange conservative WENO scheme [9, 10]. In semi-Lagrange methods, the solution of the 1D advection equation on the new time level is calculated by tracing characteristics (Lagrange trajectories) backward in time from the nodes of Euler (fixed) grid to determine the origins of the characteristics on the previous time level. The function values transported along the characteristics are obtained using a high-order WENO reconstruction via the primitive function. In our case, the fifth-order reconstruction is employed to prevent spurious numerical oscillations when the solution is discontinuous.The 2D computations are performed with the second-order direction splitting scheme [11] calculating successively the Lagrange advection first in the coordinate and then in the velocity space.

The most time-consuming parts of the code are parallelized to be executed on GPUs. At the splitting stage, a separate CUDA block of threads is assigned to each of the one-dimensional transport equations with its own constant transport velocity corresponding to a grid node, and the numerical scheme is parallelized over the threads of this block. An example of implementation can be found in Appendix.

In order to change the direction of the calculations, the matrix containing the phase density values is transposed. This technique enables us to achieve 50-fold acceleration on a NVIDIA GeForce GTX 1050 GPU as compared to serial calculations on one core of the Intel Core i5 processor with a frequency of 3.3GHz.

Numerical simulations were performed with the initial conditions for the distribution function (4) specified as $a(0) = 1.2$ and $\dot{a}(0) = 0$ and, correspondingly, $\alpha^2 = 0.864$ and $V_{rel}(0) = 0.500$. The computational domain in the phase space $R_{space} \times R_{vel}$ was 7.20×4.65. The mesh in the phase space consisted of 384×384 nodes.

The time step was $\tau = 0.1$, which corresponds to $CFL = F_{max}\tau/h_u = 6.4 > 1.0$. Here F_{max} is the maximum value of the force predicted by Eq. (3) and h_u is the grid step in velocity space.

Figure 1 shows the density profiles at different time moments and the values of density at the center of the layer as a function of time, the latter ones compared with theoretical predictions.

The evolution of the phase space distribution is shown in Fig. 2. It is seen that the distribution changes its shape from circular to elliptic and then returns to the original form.

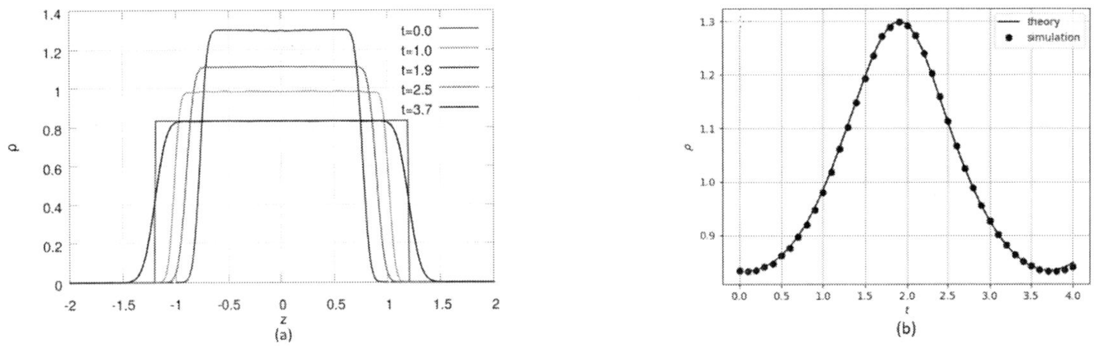

Figure 1. Evolution of density profiles of a homogeneous layer over a period of oscillations (a) and comparison of the density values at the layer center with the exact solution (b) ($V_{rel}(0) = 0.5$).

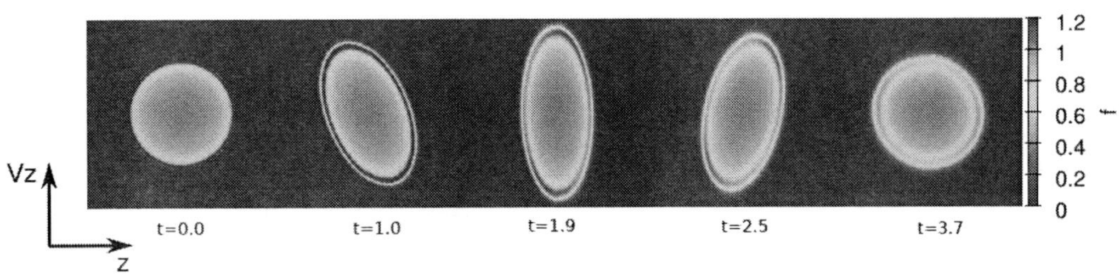

Figure 2. Phase space evolution.

4. Conclusion

A numerical code based on the semi-Lagrange conservative WENO scheme is developed for solving the Vlasov-Poisson equations which govern the evolution of a collisionless medium with long-range forces. The code is implemented on GPU using the CUDA parallel computing platform. Numerical simulations of nonlinear oscillations of a plane layer of a self-gravitating gas are performed and the results of computations are compared with the exact analytical solutions that can be constructed using the group analysis. A high accuracy and efficiency of the numerical solver is demonstrated. The 50-fold speedup of the parallel GPU code in comparison with a serial code executed on CPU is observed. It is supposed to extend the present numerical approach for solving multidimensional problems of nonlinear dynamics of self-gravitating media, in particular for studying the stability of some non-stationary analytical solutions of the Vlasov–Poisson system [4, 5, 6, 7].

Acknowledgments

The current investigation was supported by the Russian Science Foundation (Project No. 18-11-00246). This support is greatly appreciated. All computations were performed on the hybrid cluster of the Novosibirsk State University (`http://nusc.nsu.ru`).

Appendix. Program implementation

Listing 1 below shows a C++ implementation of the fifth-order WENO reconstruction used in a semi-Lagrangian numerical scheme for solving the transport equation. Calculation of the distribution function at the next time layer is based on the following instructions:

```
Df[i]=(flag)?Df[i-shift]:Df[i+shift];
Df[i]=dPhi(fP, N, i+1, xi, flag)-dPhi(fP, N, i, xi, flag);
```

where `Df` is the array of the distribution function values at the grid nodes, `fP` is the array for storage of the primitive of the distribution function, `N` is the dimension of the distribution function array, `i` is the index of the distribution function array, `h` is the grid step, `shift` and `xi` are the integer and fractional parts of the Courant number, and `flag` indicates the upstream direction. Iterations can be performed either consecutively or parallel by asynchronously calling the corresponding kernel function, which is executed on the GPU.

Listing 1.

```
#if SINGLE
#define REAL float
#else
#define REAL double
..................................................................
__host__ __device__ REAL dPhi(REAL* fP, int N, int i, REAL xi, int flag) {

  REAL U1, U2, U3, U4, U5, U6;

  if (flag) {
    U1 = fP[i - 3];
    U2 = fP[i - 2];
    U3 = fP[i - 1];
    U4 = fP[i];
    U5 = fP[i + 1];
    U6 = fP[i + 2];
  }
  else {
    U6 = fP[i - 2];
    U5 = fP[i - 1];
    U4 = fP[i];
    U3 = fP[i + 1];
    U2 = fP[i + 2];
    U1 = fP[i + 3];
  }

  REAL S1 = U4 + xi*xi*xi * (-U1 / 6. + U2 / 2. - U3 / 2. + U4 / 6.) +
    xi*xi *(-U1 / 2. + 2. * U2 - 5. * U3 / 2. + U4) +
    xi*(-U1 / 3. + 3. * U2 / 2. - 3. * U3 + 11. * U4 / 6.);

  REAL S2 = U4 + xi*xi*xi * (-U2 / 6. + U3 / 2. - U4 / 2. + U5 / 6.) +
    xi*xi * (U3 / 2. - U4 + U5 / 2.) +
    xi*(U2 / 6. - U3 + U4 / 2. + U5 / 3.);

  REAL S3 = U4 + xi*xi*xi * (-U3 / 6. + U4 / 2. - U5 / 2. + U6 / 6.) +
```

```
    xi*xi * (U3 / 2. - U4 + U5 / 2.) +
    xi*(-U3 / 3. - U4 / 2. + U5 - U6 / 6.);

  REAL C1 = (xi - 1.)*(xi - 2.) / 20.;
  REAL C2 = -(xi + 3.)*(xi - 2.) / 10.;
  REAL C3 = (xi + 3.)*(xi + 2.) / 20.;

  REAL beta1 = 4.*U1*U1 / 3. - 9.*U1*U2 + 10.*U1*U3 -
      11.*U1*U4 / 3. + 16.*U2*U2 -
      37.*U2*U3 + 14.*U2*U4 + 22.*U3*U3 -
      17.*U3*U4 + 10.*U4*U4 / 3.;

  REAL beta2 = 4.*U2*U2 / 3. - 7.*U2*U3 + 6.*U2*U4 -
      5.*U2*U5 / 3. + 10.*U3*U3 - 19.*U3*U4 +
      6.*U3*U5 + 10.*U4*U4 - 7.*U4*U5 + 4.*U5*U5 / 3.;

  REAL beta3 = 10.*U3*U3 / 3. - 17.*U3*U4 + 14.*U3*U5 -
      11.*U3*U6 / 3. + 22.*U4*U4 -
      37.*U4*U5 + 10.*U4*U6 + 16.*U5*U5 -
      9.*U5*U6 + 4.*U6*U6 / 3.;

  REAL eps = 1.0E-10;
  REAL wt1 = C1 / (eps + beta1) / (eps + beta1);
  REAL wt2 = C2 / (eps + beta2) / (eps + beta2);
  REAL wt3 = C3 / (eps + beta3) / (eps + beta3);

  REAL w1 = wt1 / (wt1 + wt2 + wt3);
  REAL w2 = wt2 / (wt1 + wt2 + wt3);
  REAL w3 = wt3 / (wt1 + wt2 + wt3);

  return w1*S1 + w2*S2 + w3*S3;
}
```

References

[1] Tanaka S and Yoshikawa K 2017 *Astrophys. J.* **849** 76
[2] De Buyl P 2010 *Vlasov dynamics of 1D models with long-range interactions* PhD thesis (Université Libre de Bruxelles)
[3] Yamaguchi Y Y 2008 *Phys. Rev.* E **78** 041114
[4] Antonov V A and Nuritdinov S N 1981 *Sov. Astron.* **25** 659
[5] Nuritdinov S N 1995 *Astron. Astrophys. Trans.* **7** 307
[6] Malkov E A 1986 *Astrophysics* **24** 221
[7] Malkov E A and Nuzhnova T N 2003 *Astron. Astrophys. Trans.* **22** 237
[8] Antonov V A 1971 *Trudy Astron. Obs. Leningrad Univ.* **28** 64
[9] Qiu J M and Shu C-W 2011 *Commun. Comput. Phys.* **10** 979
[10] Poleshkin S O, Shershnev A A, Malkov E A and Kudryavtsev A N 2018 *AIP Conf. Proc.* **2027** 040068
[11] Strang G 1968 *SIAM J. Numer. Anal.* **5** 506

Study of the flow structure in boundary layer bleed channel at the hypersonic speed of external flow

E A Obodovskaia, TA Korotaeva and M A Goldfeld

Khristianovich Institute of Theoretical and Applied Mechanics SB RAS,
4/1, Institutskaya str., Novosibirsk, 630090, Russia

katrin.nsk@ngs.ru

Abstract. The paper presents the study of the flow structure in a boundary layer bleed channel and calculation of the mass flow rate under the conditions of thick boundary layer incoming. The flow at Mach numbers of M = 3, 4, 6 and angle of attack 6° has been considered. In the framework of the problem, the Reynolds-averaged Navier-Stokes equations added with the k–ω SST turbulence model have been solved. Numerical simulation of the flow in the channel has been carried out using the Roe implicit scheme of the second order of accuracy. Pressure and temperature profiles at the outlet boundary obtained at the simulation of the flow around the inlet forebody have been used to specify the inlet boundary conditions for simulation of the flow through the boundary layer bleed hole. The pressure distribution obtained at numerical simulation of a flow in the boundary layer bleed channel has shown that there is a large high-pressure area caused by the presence of so-called "barrier shock" near the channel back wall. Thus here is a supercritical pressure drop which ensures a sonic flow in the bleed hole. This conclusion has been confirmed by the experimental data obtained for small supersonic speeds. Calculation of the mass-flow rate through exit cross-section of the cylindrical bleed channel has shown that the mass flow rate coefficient decreases 2.5 – 3 times with the Mach number increasing from 3 to 6. The mass flow coefficient also decreases by about 10% due to the change of conditions in the 3D flow before the bleed hole with the moving off from the symmetry plane.

1. Introduction

Recently, a large number of studies have been conducted in respect to the hypersonic vehicle development. The main object of the studies is a hypersonic ramjet engine that seems to be the most effective one in providing the hypersonic flight. The aim is to increase the engine efficiency, namely, to ensure the maximum air flow through the inlet, implementing the design flow regime and the total pressure loss reducing in a wide range of flight conditions.

The boundary layer separation is a serious problem that leads to instability in the inlet operation and the deterioration of its characteristics. One of the solution methods is to implement the distributed boundary layer bleeding in order to remove partly the low-pressure part of it through the perforated areas on the compression surfaces before entering the inlet channel. The boundary-layer bleed decreases the mass flow to the engine, mitigates the effects of adverse pressure gradients, and reduces the boundary-layer thickness. The method effectiveness is confirmed for supersonic inlets by the Mach number free flow of up to 2.5. For the inlets with the flight range of the Mach number from 3 to 12, this problem is much more complicated due to the significant change of the external flow conditions.

2. Problem statement

The present paper is devoted to the numerical simulation of the boundary layer bleeding through the perforated region on the ramp. The statement of the problem and the computational domain are determined by the test model and conditions studied at testing. Experiments have been carried out in the wind tunnel ITAM SO RAN T-313. Test section sizes are 0.6x0.6x2 m. Operating conditions of the wind tunnel: the Mach number range is from 2 to 7, the Reynolds number per unit length is from 8 to 54 million at temperatures of 285 – 415K. The model consists of an inlet forebody and a two-shock inlet with the boundary layer bleed (BLB) (figure 1).

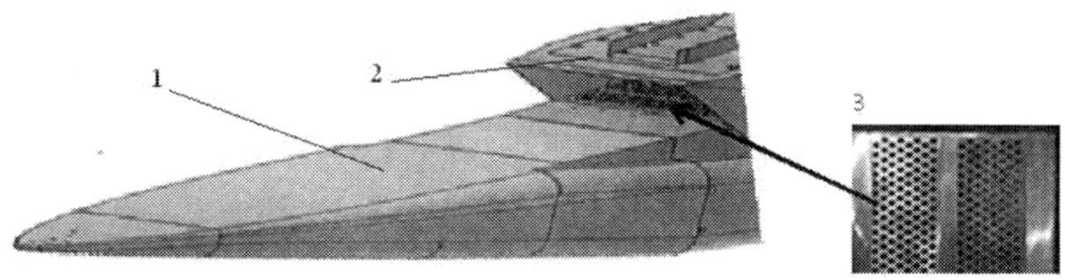

Figure 1. Experimental model: 1- hypersonic inlet forebody, 2-cowl, 3 - boundary layer bleeding surface.

Perforation on the ramp is a system of round holes with a diameter of 4 mm, spaced 13 mm apart (figure 2, a). The inclination of the orifice channels to the compression surface is of 90 °. At this stage of the study, under consideration is the single BLB channel located in the symmetry plane of the ramp and 13, 26, or 39 mm away from it (figure 2, b). Air flows through the hole in the surface and then through the outlet channel.

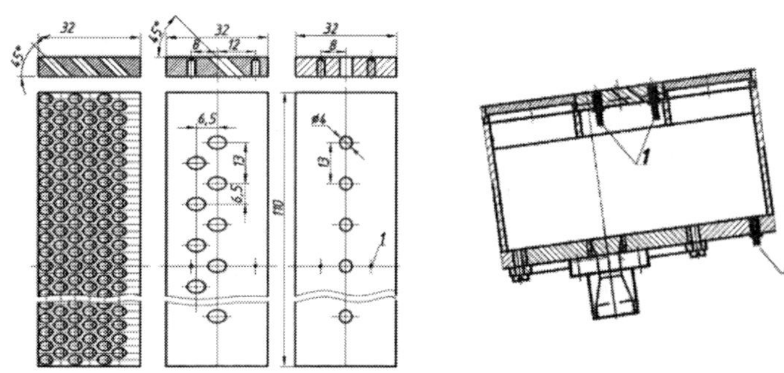

(a) (b)

Figure 2. BLB system schematic: a) General view of the porous region with multiple bleed holes, all dimensions are given in mm, b) Bleed plenum: 1- boundary layer bleed holes, 2- exit channel

The aim of the present work is to study the flow structure and define the mass flow rate characteristics in a single BLB channel at the free flow Mach numbers varying from 3 to 6 at the angle of attack equal to 6 °.

Numerical simulation of the problem has been carried out using ANSYS commercial package (the academic version). The stationary Reynolds- averaged Navier-Stokes equations have been solved for the turbulent flow. The turbulence model *k-ω SST* has been used. In the framework of the problem, a *'density-based'* solver has been exploited for an implicit second-order accuracy scheme of Roe.

The BLB characteristic definition is divided in two stages. At the first stage, the flow around the inlet forebody before the BLB holes has been modeled. At the second stage, the flow in the bleed hole is simulated. Detailed description of the first modeling stage is given in [1]. In the present paper, the main attention is paid to the second stage, i.e. to the simulation of the flow through the BLB channel.

Figure 3, a shows the computational domain and the boundary condition assignment scheme for the simulation of the flow through the BLB cylindrical channel. The domain for the flow calculating is a thick plate with a hole bounded by rectangular flow regions above and below. The type of the boundary conditions '*pressure inlet*' are set at the input boundaries, i.e. the static pressure p, the total pressure P_0 and the total temperature T_0. Above the plate (the input boundary 1), these values have been obtained at the simulation of the flow around the inlet forebody, since the plate on top, in fact, is a continuation of the compression ramp of the inlet. Under the plate (the inlet boundary 2), the conditions assigned are the same as in BLB plenum used in the experiments. The non-slip boundary condition '*wall*' with heat flax equal to zero is used on the surfaces of the plate and the hole. The boundary condition '*pressure outlet*' is set at the output borders, i.e. the static pressure p and the total temperature T_0 are assigned (figure 3, a). At the lateral boundaries, the flow symmetry condition is used. The meshes have been built using the ANSYS ICEM CFD application. The mesh for calculating the flow through the bleed hole consists of about 2.2 million elements.

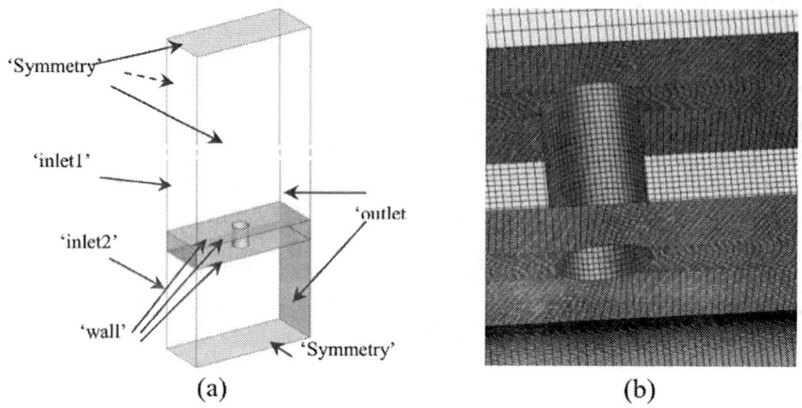

Figure 3. a) Computational domain for the flow simulation through the BLB channel and the boundary conditions schematic; b) General view of the mesh

Flow parameters are given in table 1.

Table 1. Free flow parameters.

M	p, Pa	P_0, Pa	T_0, K	α, °
3.037	10855.96	422078.2	284.5	6
4.03	6527.61	1031488.5	280	6
6.06	522.145	876461.25	376.5	6

3. Results of the flow around the inlet forebody modeling and transfer of parameter profiles

In paper [1], it is shown that a stepwise increase of the pressure takes place at the compression ramps at the flow around the inlet forebody, and the pressure increases significantly with the increasing Mach number (figure 4).

The profiles of the flow parameters behind the shock wave at the end of the last compression ramp obtained at this stage of the flow simulation have been taken as the calculation parameters and for the initial conditions assignment for the flow problem through a single BLB channel. Geometrically, the input boundary for the modeling through the BLB channel is part of the output boundary of the domain

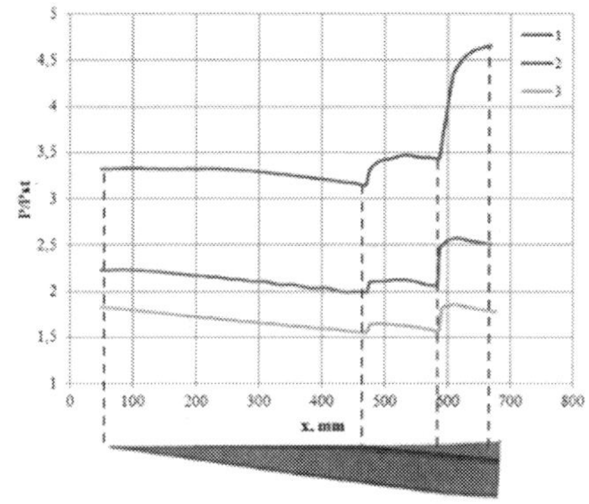

Figure 4. Pressure distribution along the plane of symmetry of the inlet forebody.

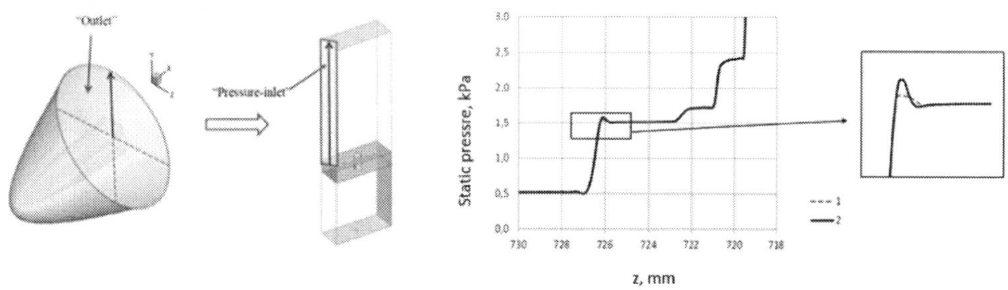

(a) (b)

Figure 5. Transfer of parameters from one computational grid to another: a) scheme; b) static pressure profile: 1 – the profile obtained from the calculation of the flow around the inlet forebody, 2 – profile after interpolation.

for simulation of the flow around the inlet forebody (figure 5, a). And yet, these are borders with different computational grids. To transfer the parameter profiles from one grid to another, cubic splines with defect 1 have been used. These splines are widely used because of their relative simplicity and the fact that they are quite sufficient to keep the continuity of the second derivative functions.

Using the known parameter values on the first grid, the parameters on the second grid have been calculated by the spline interpolation. Figure 5 presents schematically the transfer of the pressure profile as the input parameter for simulation of the BLB through the hole located in the symmetry plane of the perforated bleeding area. Figure 5, b shows the initial profile and the profile after interpolation with different location and number of nodes. One can see that there is very good agreement between the results with the exception of very small areas in the vicinity of a sharp change of parameters (figure 5, b). This phenomenon is typical for the cubic splines, but no attempts have been made to correct this defect in the work, because we assume that the parameter profiles are slightly distorted, and it cannot fundamentally change the flow flowing to the BLB channel.

4. Result of the flow in the BLB channel modeling

Figure 6 shows the pressure distribution and streamlines inside the BLB channel. The flow pattern schematic is shown in figure 6, a [2]. In figure 6, b, one can see flow pattern in the BLB channel. The arrow conventionally shows the direction of the flow flowing into the hole at the speed V.

First of all, the high degree of the flow non-uniformity in the channel should be noted in both the longitudinal and transverse directions.

The calculation results indicate a complex flow pattern with two zones with higher and lower velocities (to the left and right of the channel symmetry axis). An extensive high-pressure region is visible near the rear wall. Its appearance is caused by the presence of the so-called "barrier" shock [2]. These data also show that the "barrier" shock does not cause the flow separation on the compression ramp. This is due to the fact that the shock is inside the cylindrical channel, as is seen in figure 6. This conclusion is confirmed by the experimental data obtained for small supersonic speeds (M = 1.27) [3]. The increased-pressure area on the leeward side contributes to the reversal of the flow along the longitudinal axis of the channel and the formation of an extensive circulation region (figure 6, b), which prevents the flow increase. As a result, the main air flow passes through the area to the right of the symmetry axis, at the rear wall. It is also confirmed by the output jet structure analysis. The result is in qualitative agreement with the flow schematic presented in figure 6, a.

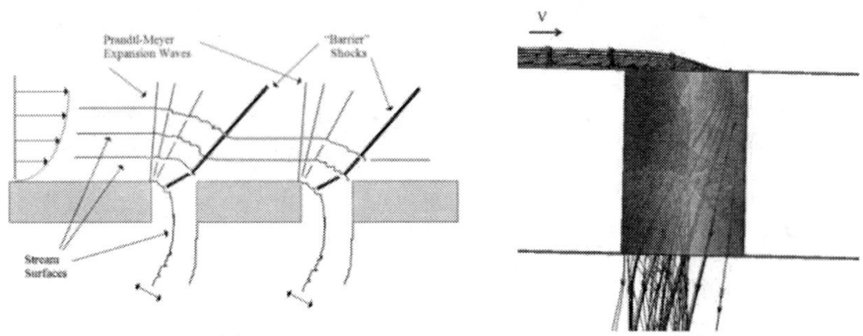

(a) (b)

Figure 6. Flow pattern in a single BLB channel at M = 4 in front of the bleeding channel: a) flow pattern schematic [2]; b) pressure distribution of and streamlines in the channel symmetry plane.

Bleed configurations are often characterized by the sonic flow coefficient (Q), which is defined as:

$$Q = \frac{w}{w_i^*}.$$

Here, the numerator is the calculated mass-flow rate at the hole exit and the denominator is the ideal mass-flow rate under the choked conditions. The perfect mass-flow rate can be defined by formula (1) at Mach number M=1 [4]:

$$w_i^* = P_{0\delta} \cdot AM\left(\frac{\gamma}{RT_{0\delta}}\right)^{\frac{1}{2}}\left(1+\frac{\gamma-1}{2}M^2\right)^{-\frac{\gamma+1}{2(\gamma-1)}}, \quad (1)$$

where A is the channel cross-sectional area, M is Mach number, $P_{0\delta}$ is the stagnation pressure, and $T_{0\delta}$ is the stagnation temperature at the boundary layer edge.

In the framework of this problem, the sonic flow coefficient has been located for the orifices in the symmetry plane of the compression ramp and at the distances of 13, 26, and 39 mm from the symmetry plane at the free flow Mach numbers from 3 to 6.

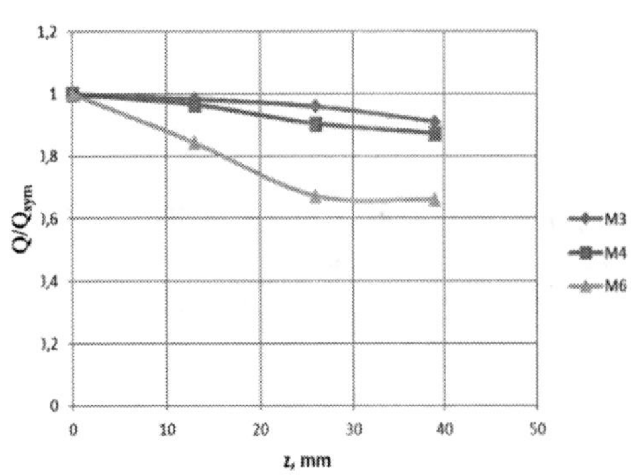

Figure 7. Normalized flow coefficient depending on the distance between a hole and the symmetry plane (M= 3, 4 and 6).

Figure 7 shows the dependence of the sonic flow coefficient on the distance from the symmetry plane for Mach numbers 3, 4 and 6. The data are normalized using the sonic flow coefficient for the orifice located in the plane of symmetry (Q_{sym}). It has been found that the sonic flow coefficient decreases more significantly as the Mach number increases. The coefficient decreases for the BLB holes located at some distance from the symmetry plane.

The coefficient decreases by 10% at the Mach number 3, and by 13% at the Mach flow equal to 4. Evident that the sonic flow coefficient decreases by about 39% for the hole located at the distance of 39 mm from the symmetry plane at the Mach number of 6.

5. Conclusions

As a result of the numerical modeling, the flow structure in a single BLB channel has been defined, and the sonic flow coefficient has been calculated versus the Mach number and the hole position relative to the symmetry plane. The main conclusions that can be drawn from the numerical simulation are the following:

- It has been found that a complex three-dimensional flow structure with separation zones and a high level of pressure non-uniformity realizes in a small cylindrical BLB channel at a supersonic external flow velocity. That leads to the BLB channel flow coefficient decreasing.
- The sonic flow coefficient through the BLB channel decreases at the free flow Mach number increasing.
- The flow coefficient decreases within the distance from the symmetry plane increasing. The higher the Mach number, the greater degree of the flow coefficient decrease.

Acknowledgments
This work has been supported by project of Russian Foundation for Basic Research No 18-31-00228

References

[1] Goldfeld M A, Korotaeva T A and Turchinovich A O 2018 *AIP Conference Proceedings* **2027** 030118 doi: 10.1063/1.5065212
[2] Rimlinger M J, Shih T I P and Chyu W J 1996 *J. of Propulsion and Power* **12** 217–24
[3] Morell A and Hamed A 2013 *AIAA Paper* 2013-0426
[4] Slater J W and Saunders J D 2010 *J. of Propulsion and Power* **26** 193-202

 IOP Publishing
1404 (2019) 012121 doi:10.1088/1742-6596/1404/1/012121

Numerical simulation of stationary flow around a wing with a subsonic leading edge at M = 2 and 2.5

A N Semenov[1, 2], **V L Kocharin** [1] **and N V Semionov**[1]

[1] Khristianovich Institute of Theoretical and Applied Mechanics 630090 Novosibirsk, Institutskaya 4/1, Russia
[2] Institute of Geology and Mineralogy, SB RAS, 630090, Novosibirsk, Ac. Koptyuga ave. 3, Russia

semenov@itam.nsc.ru

Abstract. The results of direct numerical simulation of the stationary flow above the wing with a swept angle of 72° and a subsonic leading edge are presented. The simulation was carried out at Mach numbers 2 and 2.5 and unit Reynolds numbers 9e+6 m^{-1} and 11e+6 m^{-1}.

1. Introduction

The occurrence of turbulence on the wing at supersonic flow speeds is of great interest to researchers in many countries. This is due to a practical interest since the transitional boundary layers occur on the wings of high-speed aircraft. At present, all works on numerical modeling or experiments on wings with subsonic leading edges [1-3] at supersonic speeds are related to the issue of stability and transition, and there is still no direct comparison of calculated and experimental data. One of the culprits is that the experimental data obtained at different facilities can radically differ [4]. Our work is focused on the numerical and experimental study of the development of disturbances in the boundary layer on the wing with a subsonic leading edge. Calculations are performed for a specific experimental model and completely replicate the flow parameters of the corresponding experiments. Here we present the first stage of this work, namely, the results of the numerical simulation of a stationary flow and the investigation of the parametric dependences of the flow.

2. Formulation of the problem

The equations of gas motion known as the Navier–Stokes equations, continuity, energy and state were solved using the ANSYS Fluent software package [5]:

$$\frac{\partial \rho}{\partial t} + \frac{\partial(\rho v_i)}{\partial x_i} = 0$$

$$\frac{\partial \rho v_i}{\partial t} + \frac{\partial(\rho v_j v_i)}{\partial x_j} = \frac{\partial \tau_{ij}}{\partial x_j} - \frac{\partial p}{\partial x_i}$$

$$\frac{\partial(\rho E)}{\partial t} + \frac{\partial(\rho v_j E)}{\partial x_j} = \frac{\partial}{\partial x_j}\left(k\frac{\partial T}{\partial x_j}\right) + \frac{\partial}{\partial x_j}(\tau_{ij} v_i),\ \tau_{ij} = \mu\left[\left(\frac{\partial v_i}{\partial x_j} + \frac{\partial v_j}{\partial x_i}\right) - \frac{2}{3}\frac{\partial v_k}{\partial x_k}\right]$$

Here v is velocity vector with components (u, v, w) in x, y, z direction, p, ρ, T are the pressure, density and temperature, $\mu = \mu_r \left(\frac{T}{T_r}\right)^{3/2} \frac{T_r + T_s}{T + T_s}$ is the dynamic viscosity. Air is taken as ideal gas with properties: $c_p = 1006.43$ J/(kg·K) and thermal conductivity according to the kinetic theory. Density-based solver and explicit second-order schemes were used. Flux splitting was performed using the AUSM scheme.

The experimental model of the swept wing used for numerical modeling has a lenticular profile, the sweep angle along the leading edge is 72°, the sweep angle along the trailing edge is 58°, the sweep angle along the ¼ chord is 69.5°, the central chord is 500 mm, the end chord is 200 mm, wing half-span is 200 mm, and relative thickness along the central chord is 2.8%, relative thickness along the terminal chord is 1.33%. The model is shown in figure 1. The coordinate system was set as follows: x is the streamwise coordinate, z is the spanwise coordinate, and y is the normal coordinate.

Figure 1. Wing model.

Two buffer zones were used to construct the computational grid. The first zone was obtained scaling the model to 130% and subtracting 110% from it. The second zone was obtained scaling the model to 110% and subtracting 100% from it. Next, the compound of these two zones was placed in the calculation box and the lower half was cut off, since we are interested in the flow above the wing. Free-stream conditions were set on the upper and front wall of the box, namely: at M = 2 T_s = 164 K P_s = 9200 and 11300 Pa, at M = 2.5 T_s = 130 K P_s = 5300 and 6500 Pa, which for both cases correspond to a unit Reynolds number of $9{\cdot}10^6$ m^{-1} and $11{\cdot}10^6$ m^{-1}. The exit conditions were set on the rear wall. In the lower part adiabatic wall conditions were set on the wing and symmetry conditions were set in the zones outside the wing. Zero disturbances were placed on the side walls so that they did not affect the flow above the wing. A tetrahedral mesh was used with slight condensing in the outer buffer zone toward the wing. In the area near the wing, the grid was condensed near the surface and the leading edge at the same time. As a result, we had approximately 40–50 points in the boundary layer along the entire wing along the normal coordinate, while the total number of points did not exceed 80 million. The calculation was carried out in one run. In order to make sure that we reached convergence to the stationary flow, the residual control was used at several random points near the wing surface.

3. Results

The calculation results are given for the profiles located in the middle of the span and at a distance of 80 mm from the leading edge along the streamwise coordinate. Figure 2 shows the non-dimensional streamwise velocity component, which is non-dimensionalized with its external value. It can be noted that a slight curvature of the profile occurs with an increase in the unit Reynolds number, and this effect is most noticeable at M = 2 and $Re_1 = 9{\cdot}10^6$ m^{-1}.

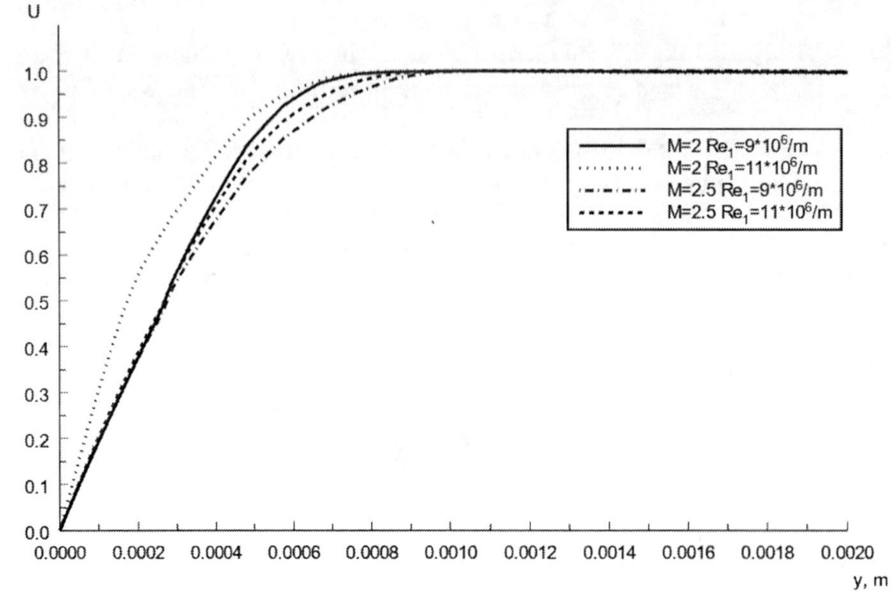

Figure 2. Streamwise velocity profiles.

The spanwise velocity component, which is non-dimensionalized with the external velocity magnitude is shown in figure 3 at the same coordinates as the streamwise velocity in figure 2. It can be seen that an increase in the Mach number and the unit Reynolds number leads to an absolute increase of spanwise velocity at the level of the critical layer ($y \approx 0.0003$ m), and to the growth of its negative value beyond the boundary layer, which is typical for three-dimensional swept wings.

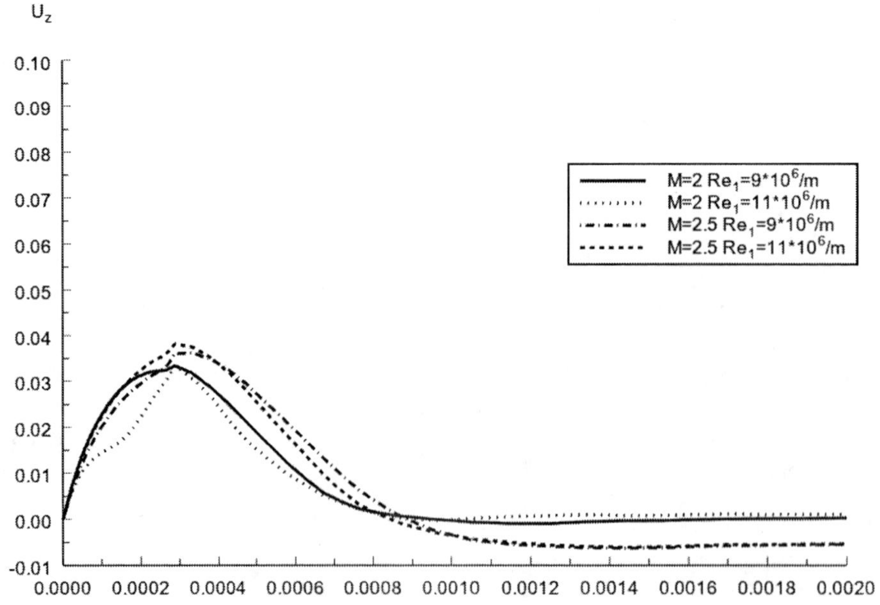

Figure 3. Spanwise velocity profiles.

If we look at the distribution of temperature that is non-dimensionalized with its external value (figure 4), then we can say that it is restored to a single value on the adiabatic wall, namely 270 K. In this case, the influence of the Reynolds number on the shape of the profile is clearly visible, the profile becomes more filled with decreasing static pressure.

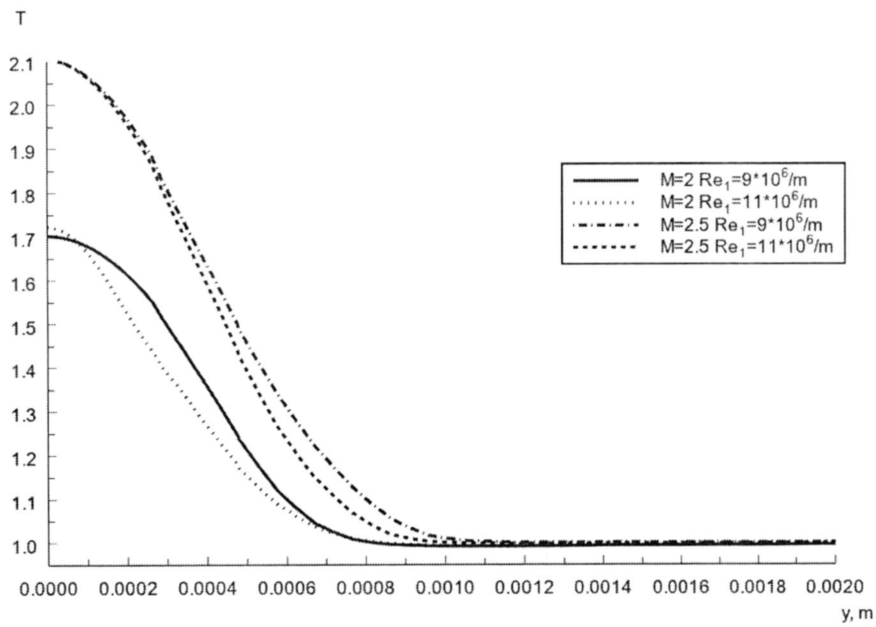

Figure 4. Temperature profiles.

It makes no sense to bring the static pressure, because at this distance from the leading edge it changes by no more than 1% of the value in the external flow, so we can say that it is constant. Therefore, based on the ideal gas equation, we can say that the density reflects the trend of the temperature distribution.

4. Conclusion

Thus, we successfully carried out the first stage of numerical simulation, we managed to solve the problems with the grid and calculate the stationary flow. Now we can expand this research by studying how the profiles are distorted when the angle of attack changes, and also try to solve the unsteady problem with the study of the development of disturbances inside the boundary layer and make a direct comparison with experiment.

Acknowledgments

The study was conducted at the Joint Access Center "Mechanics" of ITAM SB RAS. The work is supported by Russian Science Foundation (grant number №19-08-00772).

References

[1] Arnal D, Casalis G and Piot E 2008 *Proc AAAF 43ème Coll Aero Appliquée*
[2] Duan L, Choudhari M and Li F 2014 *Proceedings of the Center for Turbulence Research Summer Program*
[3] Hetsch Ti and Rist U 2004 The effect of sweep on laminar separation bubbles *IUTAM Symposium on Laminar-Turbulent Transition*
[4] Semionov N V, Yermolaev Yu G, Kosinov A D, Dryasov A D, Semenov A N and Yatskikh A A 2016 *AIP Conference Proceedings* 177
[5] ANSYS Inc. 2013 *ANSYS Fluent Theory Guide* 275 Technology Drive Canonsburg 15317 USA

Numerical simulation of jet flows in homogeneous and heterogeneous media

A K Shevchenko[1], S V Strijhak[2] and S N Yakovenko[1]

[1] Khristianovich Institute of Theoretical and Applied Mechanics, Siberian Branch of Russian Academy of Sciences, Novosibirsk 630090, Russia
[2] Ivannikov Institute for System Programming, Russian Academy of Sciences, Moscow 109004, Russia

a.shevchenko@live.com, strijhak@yandex.ru, s.yakovenko@mail.ru

Abstract. Submerged jets at different Reynolds numbers (Re) with random perturbations added to the inlet velocity profile are simulated numerically by solving the Navier–Stokes equations. The effects of Re on flow behavior and the length of laminar jet region are studied. A natural gas jet flow emitting form the sea bottom is also simulated using the URANS and VOF methods, and forces acting on an object placed within the multiphase jet are evaluated.

1. Introduction

Study of microjets is relevant [1-5] for energy issues, such as the flow control from various burner nozzles, for transport applications, where jets might be used to reduce engine noise, aircraft drag, fuel consumption. Also, multiphase jets appear [6-10] when underwater oil and gas pipelines leak or break.

The present simulations use the OpenFOAM software with the icoFoam solver for incompressible plane and rectangular jets, and the interFoam solver for a gaseous jet from the water reservoir bottom.

2. Set up of the 2D computations

To simulate a homogenous submerged jet, the unsteady continuity and Navier–Stokes equations are solved in Cartesian coordinates (x, y) at $\mathrm{Re} = u_0 h/\nu = 80, 160, 320$ where u_0 is the mean inlet velocity, h is the jet nozzle height, ν is the molecular viscosity. Cases of the top-hat, $u(y) = u_0$, and parabolic, $u(y) = 1.5\,u_0(1 - (2r/h)^2)$, inlet velocity profiles are considered. For both cases, a random noise with amplitude $A_{\mathrm{noise}} = \pm 0.001 u_0$ is added to the flow by means of the inlet profile distortion. Figure 1a shows the computational domain. Boundary conditions have been given in [1] whereas the initial conditions are: $u(x,y,t=0) = v(x,y,t=0) = 0, p(x,y,t=0) = 0$.

To get the solution independent of numerical factors, the effects of mesh density, time step, domain size, numerical schemes have been studied at the first stage. A way to achieve such a solution is as follows: for any selected parameter, results obtained by two runs are compared using the criterion: $|(u^{(1)}+u^{(2)})/(u^{(2)}{}_{\max}-u^{(2)}{}_{\min})| \leq \chi$, where $u^{(1)}$ represents "less accurate" simulation and $u^{(2)}$ is "more accurate" simulation (found, e.g. in a domain twice the size or with twice denser mesh), $u^{(2)}{}_{\max}$ and $u^{(2)}{}_{\min}$ are, respectively, the maximum and minimum of $u^{(2)}(x, y)$, and the threshold $\chi = 5\%$ is selected. Figure 1b where areas with $\chi \geq 5\%$ are marked by white shows an example of such a comparison. After achieving the difference with $\chi < 5\%$, solution is considered to be accurate enough and an examined parameter is fixed. Although this method seems to be straightforward, on practice each simulation requires it's numerical parameters to be tuned individually.

Journal of Physics: Conference Series **1404** (2019) 012122 doi:10.1088/1742-6596/1404/1/012122

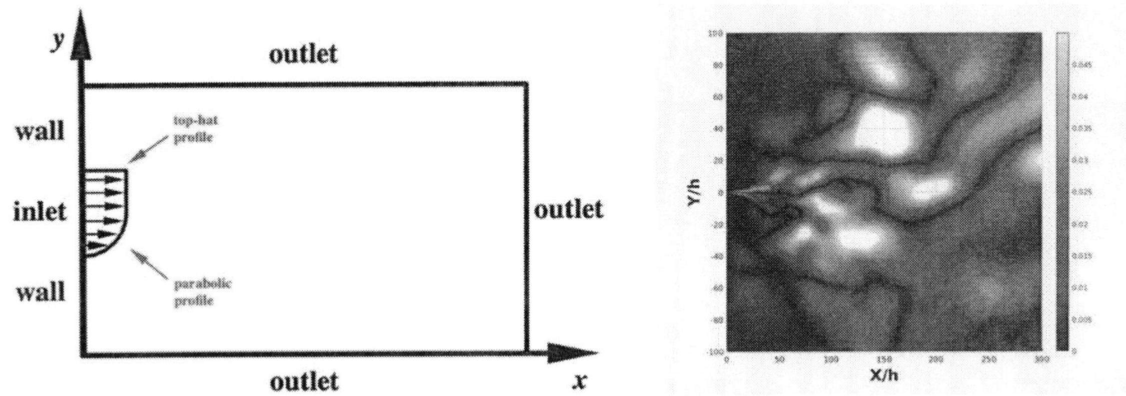

(a) (b)

Figure 1. Computational domain scheme (a), the normalized difference example for two numerical simulations at Re = 160 where the domain size effect is studied (b).

3. Results of 2D simulations of submerged incompressible jets

The computation results illustrate three jet regions. The first part near the jet nozzle is laminar; in the second region the growing asymmetric sinusoidal instability arises, which leads to the jet breakdown into an irregular flow in the third region. These results (figure 2a) agree with experiments [2, 3].

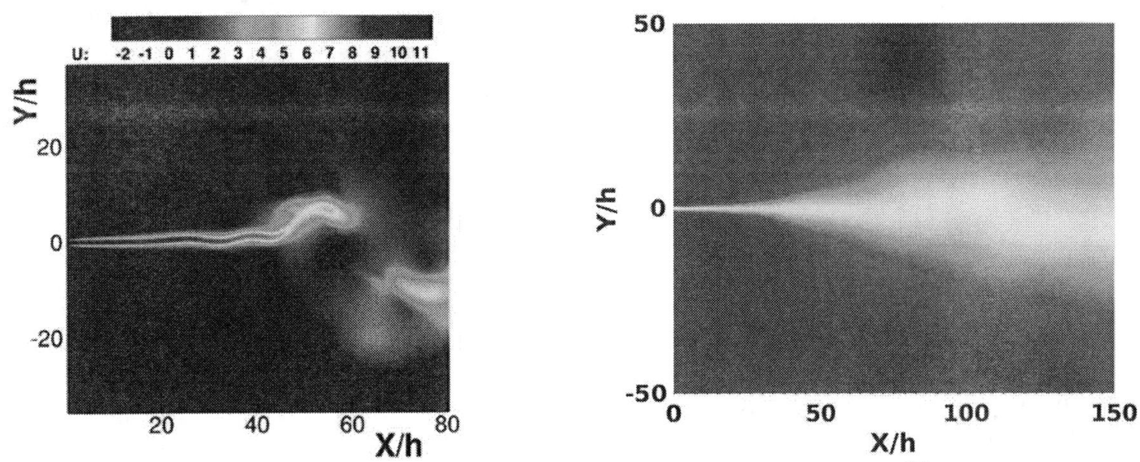

(a) (b)

Figure 2. Instantaneous flow visualized by contours of $u(x,y)$ (a), contours of $\langle u(x,y)\rangle/u_{max}(x, y = 0)$ obtained after time averaging (b) in the present studies at Re = 160 for parabolic inlet velocity profile.

The length of laminar jet region is a useful characteristic, often mentioned in the literature. In the measurements [4], the jet appeared to explode at a certain "breakdown" point. In [2], the jet was considered to be turbulent when significant disturbances of the flow started to appear. In the present work, to define the transition point, the root-mean square of horizontal velocity (U_{rms}) is calculated along the jet centerline ($y = 0$), i.e. $(U_{rms})^2 = \langle u'u'\rangle(x, y)$, where $\langle u'\rangle = u - \langle u\rangle$, and $\langle u\rangle$ is time averaged horizontal velocity (figure 2b). Time averaging is performed using the above-mentioned normalized difference criterion with χ = 5%. As seen in figure 3a, each simulation demonstrates the same behavior: the magnitude of velocity fluctuations grows, reaching the peak value (at Re = 80, there are two peaks), then decays slightly. This is in agreement with three regions, where instabilities appear in the laminar jet part, then grow until the jet breaks down into turbulence. Thus, position $x = L$ where

 1404 (2019) 012122 doi:10.1088/1742-6596/1404/1/012122

U_{rms} reaches its maximum value can be considered as the point of laminar-turbulent transition. Dashed curves represent jets with parabolic inlet velocity profile; solid ones are for top-hat jets. As one can see, top-hat jets exhibit a longer initial laminar region and smaller instability levels.

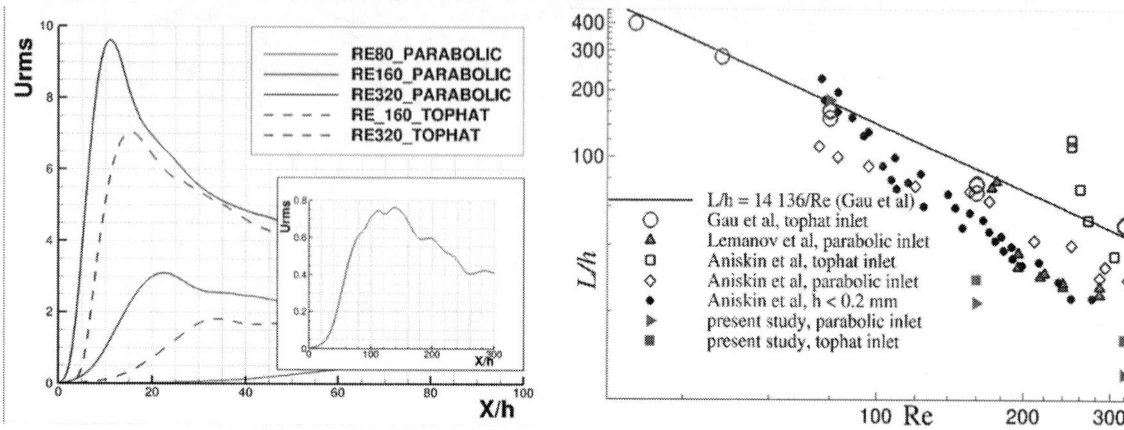

(a) (b)

Figure 3. Root-mean-square velocity values along the jet centerline (a), the length of laminar jet region in [2, 4, 5] and in the present study (b).

The computed values of L are compared with the data of measurements (figure 3b). In simulations at Re = 160, 320 for both parabolic and top-hat inlet velocity profiles, L is considerably smaller than in the laboratory experiments. A possible reason of this discrepancy is that the noise levels assigned in simulation ($A_{noise} = 0.001u_0$) are higher than in experimental jets. It can also indicate to 3D effects that exist in real jets but are not covered in two-dimensional calculations. The result where top-hat inlet jets showed the longer laminar region length than parabolic ones agrees well with the data of [5] (pricked square and rhomboid markers). Figure 4a displays the curves of the inversed maximum velocity, $u_{max}(x = y = 0)/u_{max}(x, y = 0)$, along x. The curves show an initial region with close proximity to horizontal, then growth which for Re = 80 is lengthy enough to demonstrate asymptotic approach to the power law $u_{max} \sim x^{-1/3}$ (that corresponds to laminar jets), then in third region of flow all curves follow $u_{max} \sim x^{-2/3}$ power law. Similar behavior can be seen in figure 4b, where typical thicknesses $\delta_{0.5}$ of jets are shown. Curves of $\delta_{0.5}$ basically demonstrate the same behavior as u_{max} curves, and follow x^2 power law in the far jet region. It is known that the turbulent jets should expand linearly, which means that the x^2 law expansion is an invalid result. Thus, 3D simulations are needed to correct this.

4. Results of 3D simulations of submerged incompressible jets

At this stage, the same equations as in 2D runs discussed above are solved numerically. Two cases are computed (figure 5). One is the rectangular jet at Re = 1280, where the nozzle width is equal to $20h$. The second case is the plane jet, where the nozzle width is equal to the computational domain width, and conditions at front and back boundaries (planes normal to the spanwise direction z) are cyclic.

Figure 6 shows the instability development in rectangular jet through a series of the front views on spanwise cross-sections, as well as via the side view of instantaneous velocity and the top view of averaged velocity. These views exhibit primary (sinusoidal) and secondary (periodic along the span) instabilities, and a phenomenon of axes switch in rectangular jet when a jet turns around its centerline.

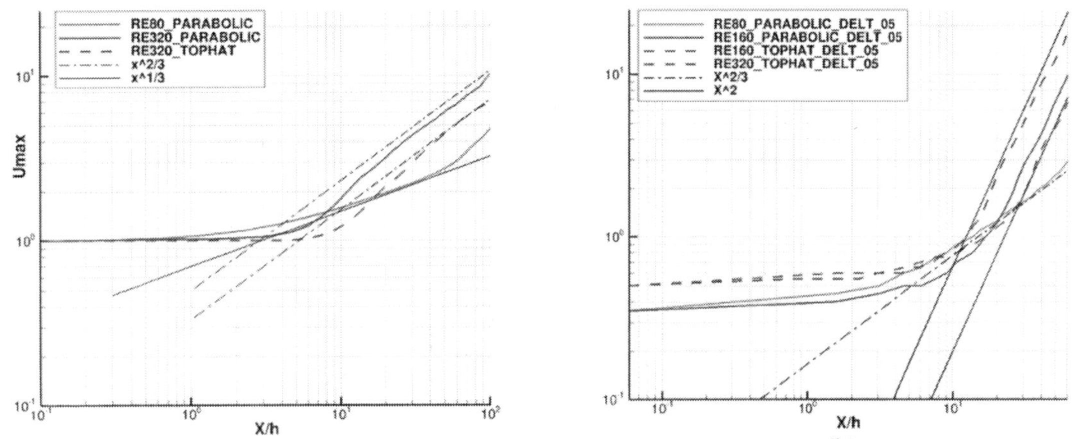

(a) (b)

Figure 4. Inversed maximum horizontal velocity $u_{max}(x = y = 0)/u_{max}(x, y = 0)$ (a), typical thickness $\delta_{0.5} = 2|y(u/u_{max} = 0.5)|$ (b) in comparison with power laws.

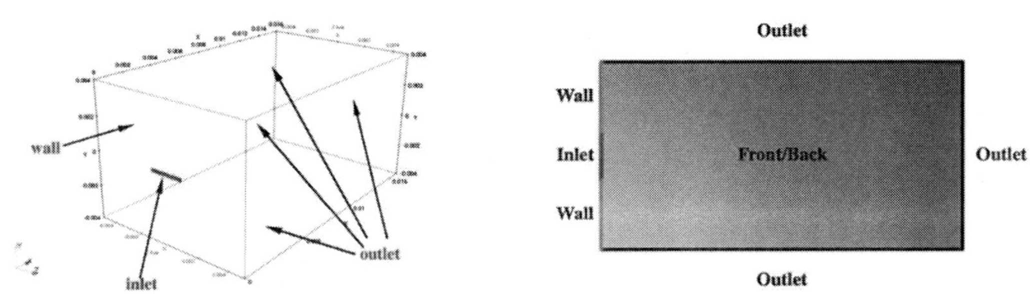

(a) (b)

Figure 5. Computational domain scheme for rectangular jet (a) and plane jet (b).

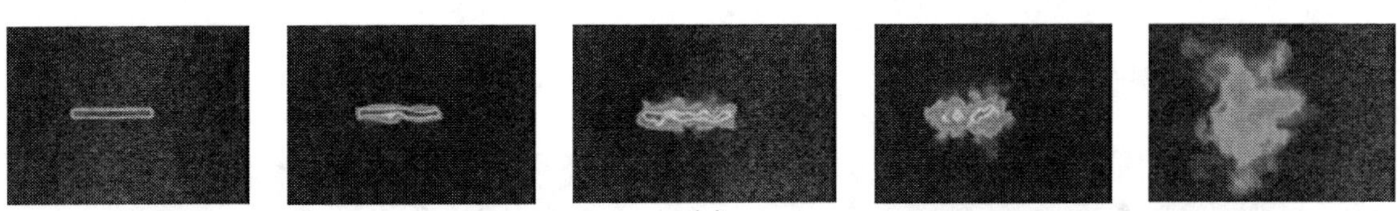

(a)

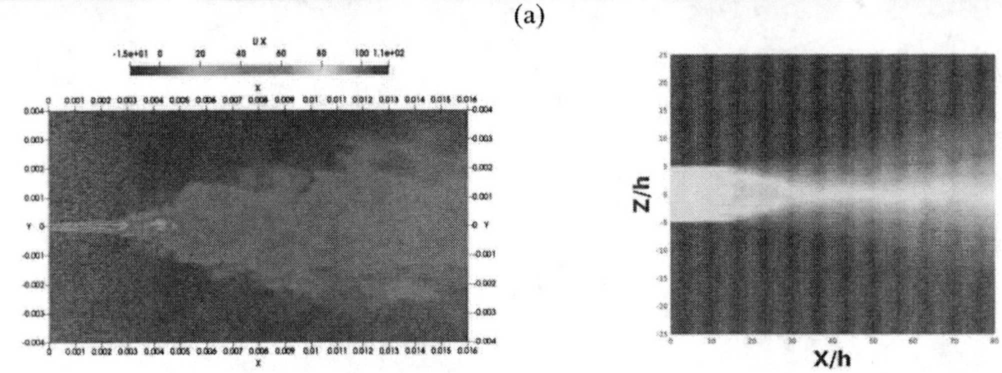

(b) (c)

Figure 6. Rectangular jet: contours of $u(x, z)$ at different x (a), contours of $u(x, y)$ at $z = 0$ (b), contours of averaged velocity $\langle u(x, z)\rangle$ at $y = 0$ (c).

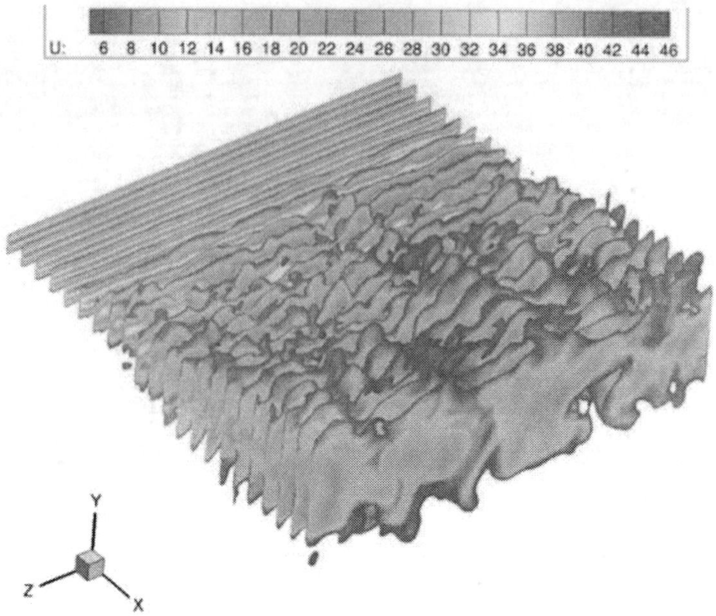

Figure 7. Flow 3D structure in the present plane-jet computations.

Figure 7 shows the plane jet structure which looks similar to that in visualizations of the wide rectangular jet in the laboratory experiments [3].

5. Simulations of multiphase jets

In the present study, computation of a gaseous jet emitting from the bottom of water reservoir of 30 m in height and 4.5 m in diameter is performed too. In the middle of domain bottom there was a nozzle of 0.3 m diameter, with the daily (24 h) discharge rate $Q = 3$ or 5×10^6 m^3 of methane through the nozzle. At distance 1 m above the bottom, a hollow cylinder of 3 m in height and 0.6 m in diameter (inner hole diameter is 0.3 m) is placed coaxially with the nozzle axis. The gas-liquid interface is resolved by means of the VOF model, turbulence is modeled using the URANS approach with the k-ω SST model. The volume fraction of gas α is found at times $t \leq 3.8$ s (figure 8), as well as velocity and pressure distributions. Flow velocity reaches about 70 m/s, maximum Reynolds number is 1.5×10^5.

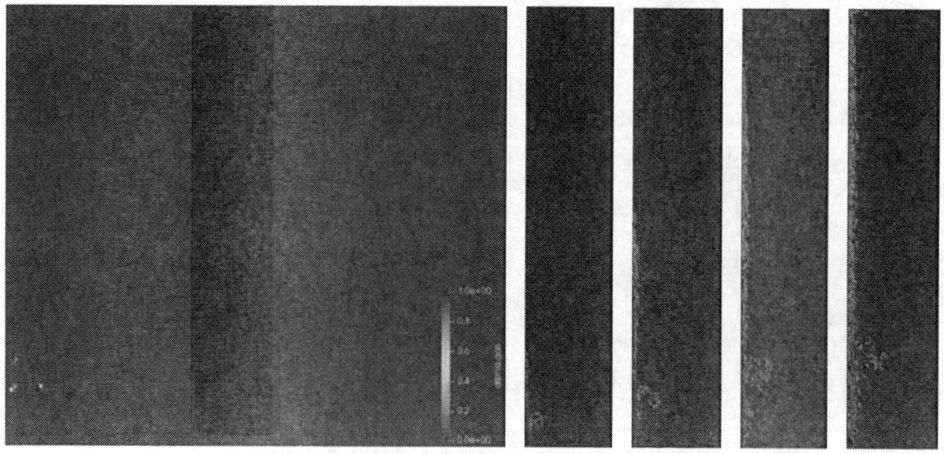

Figure 8. Volume gas fraction $\alpha = 1$ (in red) in liquid $\alpha = 0$ (in blue) at $t = 0$, 1.0, 2.0, 3.0, 3.8 s.

Acknowledgments
The study is performed within the framework of the Program of Fundamental Scientific Research of the state academies of sciences in 2013-2020 (Project No. AAAA-A17-117030610128-8).

References
[1] Shevchenko A K and Yakovenko S N 2018 *Siberian Journal of Physics* **13** 35
[2] Lemanov V V, Terekhov V I, Sharov K A and Shumeiko A A 2013 *Tech. Phys. Lett.* **39** 421
[3] Kozlov V V, Grek G R, Litvinenko Yu A, Kozlov G V and Litvinenko M V 2010 *Byull. Novosibirsk State Univ. Ser. Fiz.* **5** 28
[4] Gau C, Shen C H and Wang Z B 2009 *Phys. Fluids* **21** 1
[5] Aniskin V M, Lemanov V V, Maslov N A, Mukhin K A, Terekhov V I and Sharov K A 2015 *Tech. Phys. Lett.* **41** 46
[6] Milgram J H 1983 *J. Fluid Mech.* **133** 345
[7] Cloete S, Olsen J E and Skjetne P 2009 *Appl. Ocean Res.* **31** 220
[8] Skjetne P and Olsen J E 2012 *Prog. Comput. Fluid Dyn.* **12** 187
[9] Fraga B, Stoesser T, Lai C C K and Socolofsky S A 2016 *Ocean Modelling* **97** 27
[10] Yang D, Chen B, Socolofsky S A, Chamecki M and Meneveau C 2016 *J. Fluid Mech.* **794** 798

Comparison of modern implementations of the direct simulation Monte Carlo method

N V Shugalevskaia, A A Shevyrin and Ye A Bondar

Khristianovich Institute of Theoretical and Applied Mechanics SB RAS, Institutskaya str., 4/1, Novosibirsk, 630090, Russia

n.shugalevskaia@g.nsu.ru

Abstract. In this work, collisional schemes of the direct simulation Monte Carlo (DSMC) method are compared. The sensitivity of the schemes to the number of simulated particles is investigated. The original No-Time-Counter (NTC) scheme, proposed by G.A. Bird, its modification NTC-2007 and majorant collision frequency (MCF) scheme, proposed by M. S. Ivanov and S.V. Rogasinsky are considered. For the high-speed Couette flow with parameters corresponding to a near-continuum regime (speed ratio S is equal to 10, Knudsen number Kn is equal to 0.01) the convergence of shear stress component is analysed. Calculations with different total number of the simulated particles (in the range from 500 to 10^6) were performed. It was shown, that the deviation of NTC-2007 scheme from the reference solution is about four times larger as compared to the MCF scheme. The original NTC scheme exhibits much higher sensitivity to the number of simulated particles.

1. Introduction

The direct simulation Monte Carlo (DSMC) method is a stochastic technique to solve the Boltzmann equation numerically [1]. The main area of DSMC method application is rarefied gas flows (when the continuum approach is not applicable): the high-altitude aerodynamics of spacecraft, vacuum systems, micromechanical devices. The DSMC method is based on the gas flow representation as a finite set of the simulated particles. Each simulated particle corresponds to a large amount of real molecules. The computational domain is split into the collisional cells. At the initial time, this domain is populated with the particles with the some initial distribution over the coordinates and velocities. During one time step the following stages are performed consequently: the free-molecular transfer of the simulated particles and the binary collision of the particles located in the same cell.

There are following error types in the numerical solution calculated by the DSMC method:

1) A statistical error, which depends on the size of the sample.
2) Deterministic errors, which depends on:
   - the time discretization,
   - the spatial discretization,
   - the number of simulated particles N.

The computational resources required by the DSMC computations linearly depend on the number of particles. All modern implementations of the DSMC method are mostly different from each other in the details of collision modeling. Practical applications of the DSMC method require intensive computations on powerful computers. Therefore, the ability to carry out calculations with a small number of particles without loss of accuracy is the most attractive property of the collision scheme. In this work, only the error connected to a number of simulated particles N (particle discretization error) is investigated. The time step and cell size were small enough and the sample size was large enough for other errors being insignificant compared to the particle discretization error.

Until recently, one of the most popular DSMC collision schemes was the No-Time Counter (NTC) scheme, proposed by G.A. Bird in 1989 [2]. The NTC scheme allows getting the numerical solution of the Boltzmann equation, if the number of simulated particles in a collision cell is large enough.

Another collision scheme is the Majorant Collision Frequency (MCF) scheme, proposed by M. S. Ivanov and S. V. Rogasinsky [3]. The MCF scheme is used in the SMILE code family [4], [5], [6]. In [7] it was shown that the MCF scheme requires significantly fewer total particle number and does not depend on the particle number in the cell than conventional DSMC schemes. It means that more complicated problems can be solved with the same computational resources. The fraction of repeated collisions were introduced in [8] as indicator of the numerical solution convergences to the solution of the Boltzmann equation. The MFC scheme has been employed in the SMILE family codes for modelling various real gas effects, including energy transfer between different molecular modes, chemical reaction in the gas phase and on the vehicle surface and ionization [9] – [12].

In 2007, G.A. Bird proposed a modification of the NTC scheme [13], which, according to the author, makes the numerical modeling much more efficient. A comparison of NTC and MCF schemes was presented in [7]. However, the rigorous investigation of these schemes for classic problems of rarefied gas dynamics is still required.

The goal of this work was comparison of different schemes sensitivity to the total number of the simulated particles. Three schemes were considered:

- NTC, G.A. Bird,
- NTC-2007, G.A. Bird,
- MCF, M.S. Ivanov, S.V. Rogasinsky.

The dependence of the solution of the plane Couette flow problem on the simulated particle number is investigated. This classical problem of the rarefied gas dynamics is often used for the comparison between different DSMC schemes (see, e.g. [14]). The Couette flow problem is solved for the parameters corresponding to the near continuum high-speed gas flow.

2. Numerical results

Numerical simulation of the planar Couette flow of the monatomic dilute gas (argon) between two parallel plates was performed using different collisional schemes. This problem was considered in the following statement. Two plates are kept at constant temperature $T = 273$ K. The distance between plates is 1 m. One plate is stationary and the other one moves in its own plane with velocity of 3370 m/s. The x-axis is directed perpendicular to the walls, and the y-axis is in the direction of the motion. The gas motion occurs due to the viscous friction forces in the y-direction. The speed ratio (plate velocity divided by the most probable speed) is equal to 10, Knudsen number (ratio of the mean free path to the flow width) is equal to 0.01.

In the considered flow all macroparameters change along the x–axis. The velocity, density and temperature profiles are shown in the figure 1. In the center of the flow, a maximum of temperature and a minimum of density are observed (figures 1b and 1c).

In this work τ_{xy} component of shear stress tensor was used for the comparison of the collision schemes:

$$\tau_{xy} = \rho \overline{c_x c_y} ,$$

where c_x, c_y are the components of the particle thermal velocity vector, ρ is the density. Since $\rho = nm$, where n is the number density, m is the molecular mass, and the flow velocity in x direction is zero, τ_{xy} was estimated as $\tau_{xy} = nm\overline{uv}$, where u, v are the absolute molecule velocities in x, y directions.

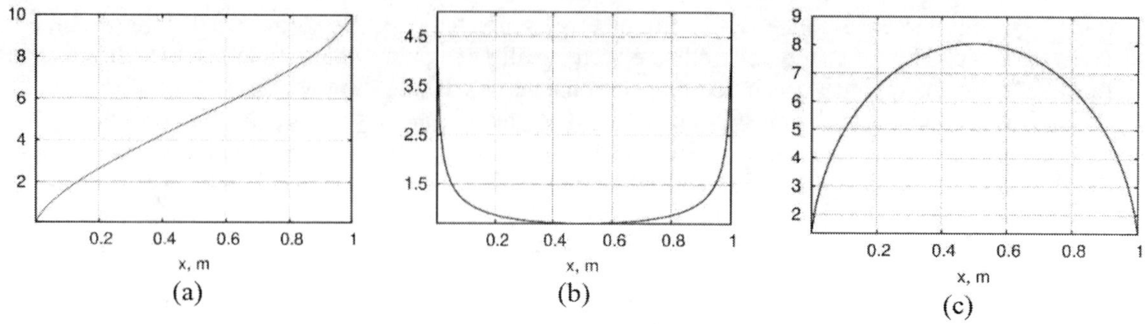

Figure 1. The non-dimensional velocity (a), density (b) and temperature (c) profiles.

The calculations were performed using the DSMC1 demonstration code written by G.A. Bird. The collisional subroutine COLLMR was modified to implement each investigated scheme. The sample size, the time and spatial discretization were taken in such a way that the corresponding errors became significantly less than the particle discretization error. The number of the cells in the computational domain was equal to 2000, and the time step was $2 \cdot 10^{-7}$ s. The calculations using each scheme were performed for N equal to 500, 1000, 2000, 4000, 6000, 8000, 10^4, $5 \cdot 10^4$, 10^5, 10^6. For $N = 10^6$, the reference result was reproduced by all three schemes, as shown in figure 2.

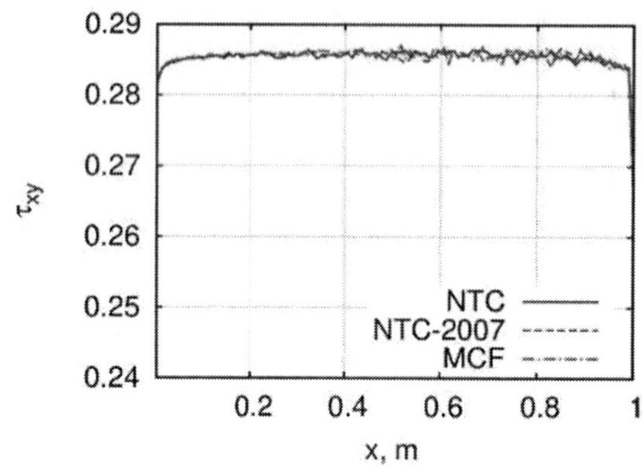

Figure 2. The reference result ($N = 10^6$) for NTC, NTC-2007 and MCF schemes.

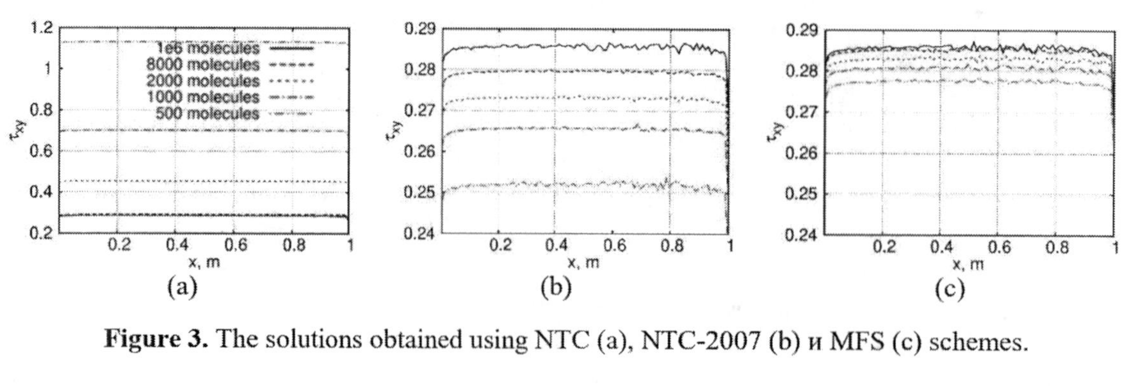

Figure 3. The solutions obtained using NTC (a), NTC-2007 (b) и MFS (c) schemes.

Journal of Physics: Conference Series **1404** (2019) 012123 doi:10.1088/1742-6596/1404/1/012123

The figure 3 shows the deviation of the solution from the reference profile as the particle number decreases for each scheme. The different lines correspond to the results of the calculations with the particle number 500, 1000, 2000, 8000, and 10^6. For the NTC scheme (figure 3a) with a decrease of the number of particles to 500 the solution deviates from the reference one severalfold. For NTC-2007, the deviation is about 12% if $N = 500$ (figure 3b). For MCF scheme, the deviation is only about 3% for $N = 500$ (figure 3c).

Since the linear convergence behavior O(1/*N*) is observed for the parameter estimation during the increase of the simulated particle number, the dependence of the estimation of τ_{xy} parameter from *N*, obtained from the calculations, may be approximated by the function:

$$\tau_{xy} = \tau^*_{xy} + CN^{-1}, \qquad (1)$$

where τ^*_{xy} is the reference value and *C* is some coefficient. Symbols on figure 4 denote the results of the calculations with different *N* (average τ_{xy} along *x*-axis), and lines represent the approximation (1). The figure 4 shows the convergence of the τ_{xy} estimation for all schemes as *N* is large enough to the reference value. However, when *N* is small, the deviation for the NTC scheme is significantly larger than for two other schemes, and the MCF scheme has smaller deviation than NTC-2007.

The smaller is absolute value of coefficient *C* in formula (1), the smaller is deviation of τ_{xy} parameter from the reference value for any fixed N. Therefore, the value of *C* can characterize the scheme sensitivity to the particle number for the flow considered. In table 1, the values of *C* normalized by the reference value τ^*_{xy} are shown.

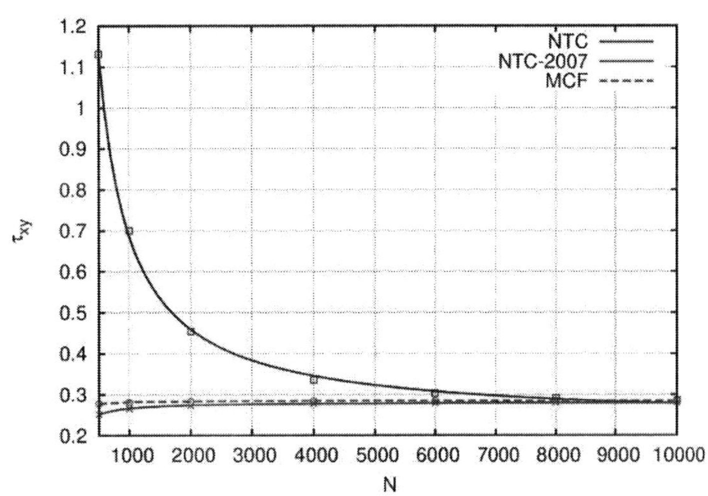

Figure 4. The estimation of the shear stress τ_{xy} value as a function of total number of the particles. Points – DSMC computations, lines – approximation.

Table 1. The values of *C* normalized by τ^*_{xy}.

Scheme	$\lvert C/\tau^*_{xy} \rvert$
NTC	1460
NTC-2007	70
MFC	18

3. Conclusions

By the test case of the high-speed Couette flow in the near-continuum regime (Kn = 0.01, S = 10), the sensitivity of three collision schemes of the DSMC method to the total number of simulated particles was studied. The NTC, NTC-2007 and MCF schemes were considered. The calculations with different simulated particle number (from 500 to 10^6) were performed. The values of the shear stress τ_{xy} was calculated using each scheme. For the case of large particle number N = 10^6, the same results for all schemes were obtained. The deviation of the computational results from the reference value for decreased total number of the simulated particles was demonstrated for the considered schemes.

The numerical results demonstrate that the MCF scheme has the lowest error growth with a decrease of the number of simulated particles. The deviation from the reference solution obtained for the MCF scheme is approximately four times lower than for the NTC-2007 scheme and 80 times lower that for the NTC.

These results have the implications for the practical applications of the DSMC method, particularly, for the high-altitude aerothermodynamics problems. The smaller sensitivity to the particle number is important in the big supercomputer computations performed with all available computational resources.

The investigation of the convergence behavior for the different flow parameters (Kn, S), and for other classical problems of the rarefied gas dynamic (e.g. the Fourier problem), and interaction of the considered deterministic error with other DSMC errors will allow to extend and generalize present results.

Acknowledgements

This work was carried out within the framework of the Program of Fundamental Scientific Research of the state academies of sciences in 2013-2020 (project No. AAAA-A17-117030610138-7).

References

[1] Bird G A 1994 *Molecular Gas Dynamics and the Direct Simulation of Gas Flow* (Oxford, England, UK: Clarendon)

[2] Bird G A 1989 *Prog. Astronaut. Aeronaut.* **117** 211–26

[3] Ivanov M and Rogasinsky S 1988 Analysis of numerical techniques of the direct simulation Monte Carlo method in the rarefied gas dynamics *Sov. J. Numer. Anal. Math. Model.* **3** 453–65

[4] Ivanov M, Kashkovsky A, Gimelshein S, Markelov G, Alekseenko A, Bondar Ye, Zhukova G, Nikiforov S and Vashchenkov P 2006 *Proc. 25th Int. Symp. on Rarefied Gas Dynamics (Saint-Petersburg)* pp 539–44

[5] Kashkovsky A, Bondar Ye, Zhukova G, Ivanov M and Gimelshein S 2004 Object-oriented software design of real gas effects for the DSMC method *AIP Conf.Proc.***762** 583–8

[6] Ivanov M, Kashkovsky A, Vashchenkov P and Bondar Ye 2010 Parallel object-oriented software system for DSMC modeling of high-altitude aerothermodynamic problems *AIP Conf. Proc.* **1333** 211–8

[7] Venkattraman A, Alexeenko A A, Gallis M A and Ivanov M S 2012 A comparative study of no-time-counter and majorant collision frequency numerical schemes in DSMC *AIP Conf. Proc.* **1501** 489

[8] Shevyrin A A, Bondar Ye A and Ivanov M S 2005 Analysis of repeated collisions in the DSMC method *AIP Conf. Proc.* **762** 565–70

[9] Bondar Ye, Shevyrin A, Chen Y, Shumakova A Kashkovsky A and Ivanov M 2013 Direct Monte Carlo simulation of high-temperature chemical reactions in air *Thermophys. Aeromech.* **20** 553–64

[10] Molchanova (Shumakova) A, Kashkovsky A and Bondar Ye 2014 A detailed DSMC surface chemistry model *AIP Conf. Proc.* **1628** 131

[11] Molchanova A Kashkovsky A and Bondar Ye 2018 Surface recombination in the direct

simulation Monte Carlo method *Phys. Fluids* **30** 107105
[12] Shevyrin A, Bondar Ye, Kalashnikov S, Khlybov V and Degtyar V G 2016 Direct simulation of rarefied high-enthalpy flow around the RAM C-II capsule *High Temp.* **54** 383-9
[13] Bird G A, Gallis M A, Torczynski J R and Rader D J 2009 Accuracy and efficiency of the sophisticated direct simulation Monte Carlo algorithm for simulating noncontinuum gas flow *Phys. Fluids* **21** 017103
[14] Russo G, Pareschi L, Trazzi S, Shevyrin A A, Bondar Ye A and Ivanov M S 2004 Plane Couette Flow Computations by TRMC and MFS Methods *AIP Conf. Proc.* **762** 577-82

Simulation of the transition to turbulence in the boundary layer on complex-shaped bodies at an angle of attack

A A Sidorenko[1], A V Boiko[1,2], K V Demyanko[1,3], S V Kirilovskiy[1], Yu M Nechepurenko[1,3] and T V Poplavskaya[1]

[1]Khristianovich Institute of Theoretical and Applied Mechanics SB RAS, 4/1, Institutskaya str., Novosibirsk, 630090, Russia
[2]University of Tyumen, 6, Volodarskogo str., Tyumen, 625003, Russia
[3]Marchuk Institute of Numerical Mathematics of RAS, 8, Gubkina str., Moscow, 119991, Russia

popla@itam.nsc.ru

Abstract. The paper deals with the numerical simulation of the laminar-turbulent transition as the gas flow moves at a nacelle at transonic speeds. The laminar-turbulent transition position in the 3D boundary layer of the nacelle is found with the aid of the laminar-turbulent transition module based on the e^N-method which involves the data of the numerical simulation of the laminar flow at the model with the gas-dynamic software package ANSYS Fluent. Performance of this module is demonstrated as the laminar-turbulent transition, caused by the Tollmien-Schlichting wave and crossflow instabilities, takes place. It is shown that the Tollmien-Schlichting wave instability is the dominating mechanism of the laminar-turbulent transition on the nacelle up to the angle of attack of 7°.

1. Introduction

It is known that the position of the laminar-turbulent transition (LTT) and its length dictate the economic efficiency of the aircraft. The concept of the natural flow laminarization (NLF or the Natural Laminar Flow) suggests that, to reduce the drag of various aircraft elements (wings, nacelles, etc.), it is necessary to design its shape in such a way to provide the maximally extended zone of the laminar flow within the Mach number and angle of attack range intrinsic for the cruising flight. This leads to the problem of simulation of the laminar-turbulent transition in the boundary layers of complex-shaped bodies.

Using of software packages which simulate the propagation of small amplitude disturbances in the boundary layer and calculate the LTT position by the criteria based on a certain threshold value of the amplitude (the N-factor method, or e^N-method) is among up-to-date approaches for the transition simulation. This method applied to aerodynamic problems is widely used due to its foundation on the physically vindicated linear theory of hydrodynamic stability. The theory of hydrodynamic stability is valid both for 2D and 3D incompressible and compressible flows at quite a low degree of free stream turbulence (Tu < 1 – 2%). Today, the e^N-method of LTT determination is not presented in the general-purpose gas-dynamic packages [1], but it can be implemented as an individually connected module – the LTT module.

In recent years, the co-authors of this work developed the LOTRAN 3.0 software package implemented in the MATLAB environment for analysis of stability and LTT in 3D boundary layers to

Journal of Physics: Conference Series **1404** (2019) 012124 doi:10.1088/1742-6596/1404/1/012124

be used with independent data on the laminar viscous compressible flow computed at an aerodynamic body with a practically acceptable accuracy by any CFD code. In this work, LOTRAN 3.0 was transformed into a corresponding LTT module integrated into the gas-dynamic package ANSYS Fluent as a plug-in to be able to estimate the transition position in 3D boundary layers of the complex-shaped bodies.

The numerical solution of 3D problems is still quite a complicated challenge. Many factors should be taken into account when determining the LTT, such as the presence of pressure gradients in different directions, surface roughness, free stream turbulence, etc. The paper deals with the effect of such aspects as the crossflow caused by the spanwise pressure gradient, and possible flow separation in the presence of the streamwise pressure gradient occurring possibly at nacelles, bodies of revolution, swept wings, etc. The development of Tollmien-Schlichting waves and stationary cross flow vortices as primary instabilities was suggested as the main possible mechanisms of the transition to turbulence. It is based on the fact that one of these mechanisms is dominating in different flow areas at ellipsoids set at a given angle of attack [2, 3]. At the same time, their relative significance is still not completely understood for the nacelles.

The main purpose of this work is to make conclusions about the dominating mechanism of the LTT in the boundary layer of a transonic gas flow at a nacelle set at an angle of attack.

2. Transonic flow at the nacelle

To create the nacelle model (NM), the profile LNT76 was designed; it provides quite an extended section of the laminar and transition boundary layer. Preliminary computations were carried out for the flow around the LNT76 profile in the axisymmetric statement at the zero angle of attack; these computations were later used to construct the 3D computational domain and numerical grid on the NM. Note that validity of the predicted LTT position based on the LOTRAN 2.0 software package aimed for LTT analysis of 2D flows with the help of e^N-method and integrated into ANSYS Fluent has been shown [4] by the good agreement of the results with experimental data [5] obtained in a transonic wind tunnel at the wing profile NASA-SR-0410 with the chord length 0.2 m.

Figure 1. Nacelle model with the numerical grid on the surface (each 4th cell).

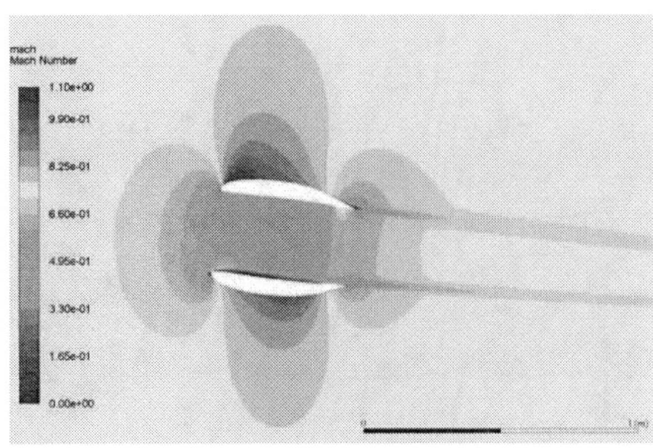

Figure 2. Velocity field in the NM plane of symmetry (the laminar mode): $M_\infty = 0.75$, $\alpha=7°$.

The 3D NM was constructed by rotation of the profile to 180° about the streamwise axis, i.e. it is a body of revolution with a duct axisymmetric about the streamwise axis (figure 1). The computational domain presents a semi-sphere with the radius of 15 m, the NM is located in the center. The sub-domain of the near-wall flow (height 10 mm) and the far field sub-domain were selected in order to have the detailed resolution of the boundary layer near the NM surface. ANSYS ICEM was used to construct the block hexahedral computational grid fine toward the NM surface in both domains. The NM has the streamwise size 500 mm. Total amount of the numerical grid cells is 10.3 million.

Figure 2 shows that the free-stream Mach number $M_\infty = 0.75$ provides the flow at the NM with the extended supersonic section and closing shock wave typical for the transonic flow mode.

The flow at the NM is numerically simulated with the ANSYS Fluent 18.0 package with the following parameters: P_∞ =1 atm, T_∞ = 300K, M_∞ = 0.5, Re_1=1.11 $\times 10^7$ 1/m , $\alpha = 0 \div 7°$ and P_∞ = 2 atm, T_∞ = 300K, M_∞ = 0.75, Re_1= 3.32$\times 10^7$ 1/m , $\alpha = 0 \div 7°$. The free-stream conditions assigned on the external boundary of the computational domain: pressure P_∞ , temperature T_∞ , Mach number M_∞ and angle of attack α, the non-slip and adiabatic wall conditionson the NM surface and the symmetry conditions in the plane of symmetry of the computational domain. In the numerical simulation, the stationary Navier-Stokes equations were solved with the aid of the solver based on density; the implicit scheme of the 2nd order of space accuracy with the Roe-FDS method of convective flow splitting was used. During the simulation of the completely laminar flow near the trailing edge of the NM, the laminar separation results in an essentially non-stationary flow. Thus, to improve the convergence and increase the computation quality, it is advisable to use the artificial turbulization deliberately downstream from the laminarized domain.

3. LTT computation module based on LOTRAN 3.0

The e^N-method is one of the most common engineering methods of LTT determination. Today, it is not presented in the general-purpose gas-dynamic packages, but it can be implemented as a plug-in for the LTT analysis. In 2013 the authors of this paper developed and registered the software package LOTRAN for the computation of the LTT position in the boundary layers of the viscous incompressible fluid flows above slightly curved surfaces [6, 7]. Then the significant work has been done to expand this software for the computation of the LTT position in compressible 2D and 3D boundary layers [8], which gave two new packages LOTRAN 2.0 and LOTRAN 3.0, respectively. They are purposed for fundamental scientific research of stability of 2D and 3D boundary layers of the viscous compressible fluid and determination of the LTT positions in them using the e^N -method. In this work, the LOTRAN 3.0 package was transformed into the LTT module for integration into the gas-dynamic package ANSYS Fluent.

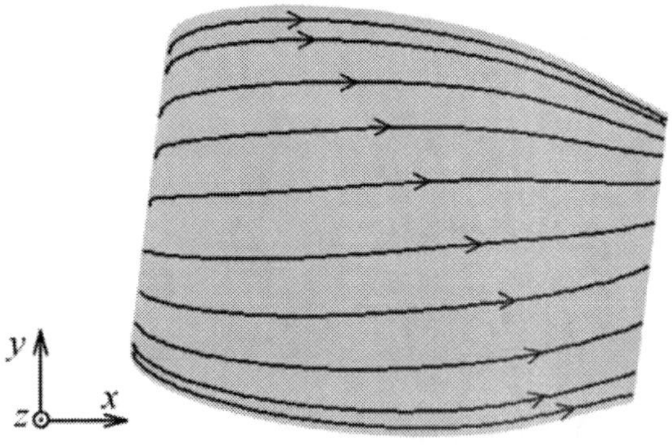

Figure 3. NM with marked streamlines (20 pcs).

The scheme of the integration of the 3D LTT module with the gas-dynamic package ANSYS Fluent [3] is the following. First, ANSYS Fluent is used to compute the laminar flow at a body with the 3D Navier-Stokes equations (see Sect. 2). Then, files with extension *.cas and *.dat are sent in the LTT module (case file contain the mesh, boundary and cell zone conditions, and solution parameters for a problem). A specially formed module of data import from ANSYS Fluent, the LTT module assimilates the obtained data and determines the boundary layer boundary. Then, the assigned number of streamlines are computed on the boundary layer boundary by the specially developed algorithm (figure 3 shows the computed streamlines on the NM). Each streamline is broken up to sections by the nods over its length; normal lines to the surface are built from these nods. Thus, the cross section forms along each computed streamline perpendicularly to the surface. Grids that are condensed near the surface are constructed on the normal lines, and the data on the main flow are interpolated into their points.

Then for each streamline the boundaries of instability regions of the most unstable disturbances and their time frequencies were computed by solving the local temporal stability problems for some given values of longitudinal wave number. The amplification factors (N-factors) of these disturbances in the instability regions were computed by solving the local spatial stability problems and then the envelope of the N-factors is computed. The e^N-method suggests that the transition to the turbulence is observed when one of the N-factors rises up to the critical (threshold) value. The empirical formula [9]: $N_c = -8.34 - 2.4\ln(\mathrm{Tu})$ is used as the threshold value of the N-factors for the LTT start position at the moderate degrees of turbulence of the free stream.

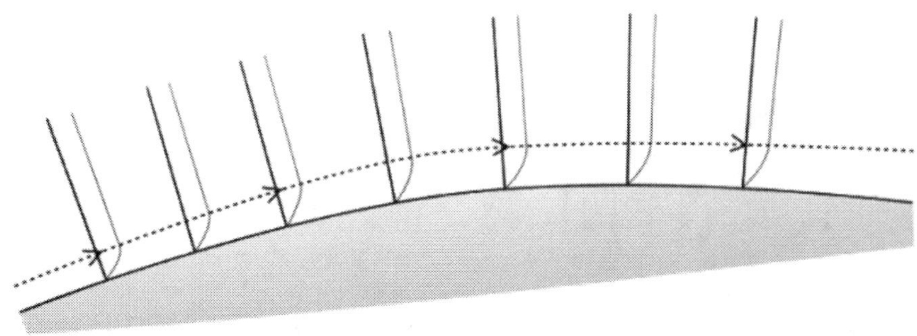

Figure 4. Schematic of the normal lines construction along each streamline followed with the interpolation of the main flow data in them.

When determining the LTT, solving 3D problems, one should take into account many factors of the flow. In this work, the NM simulation involves such aspects as the cross flow with the pressure gradient and potential separation. The 3D LTT module based on the LOTRAN 3.0 complex, has its peculiarities. First, as was said above, it is the necessity to construct 3D streamlines on the boundary layer boundary and analysis of the temporal and spatial stability along each streamline. The second feature is the presence of two types of instability: the Tollmien-Schlichting waves and cross section vortices. In [3] it was demonstrated that during the flow around the elongated ellipsoid of rotation at the angle of attack, the instability of cross flow vortices imposes the dominating effect on the LTT process. The NM also presents the model of the rotation body with the strong cross overflow. The LTT module permits investigating both types of the flow instability and find the dominating LTT mechanism.

4. Numerical results

Figures 5a, b shows the streamlines colored by the N-factors obtained with the aid of the LTT module for the case of the flow around the NM at the angle of attack $\alpha = 7°$ at $M_\infty = 0.5$ and Tu = 0.2%. The data were acquired for the LTT position (the black line) for two types of instability: the Tollmien-Schlichting wave instability and cross flow vortices instability. Figure 5c demonstrates the LT transition zones plotted with due regard to both types of instability: the laminar (blue), transition (green) and turbulent (red) flow areas. It is evident that the Tollmien-Schlichting wave instability (TS) imposes the dominating effect on the LTT process on the NM with the cross flow at $\alpha = 7°$ and $M_\infty = 0.5$.

Figure 6 presents the same LTT characteristics but for the case of NM flow around at $M_\infty = 0.75$, $Re_1 = 33.2\times10^6$ m^{-1}, $\alpha = 7°$, Tu=0.2%. It is evident that, similarly for $M_\infty = 0.5$, at $M_\infty = 0.75$ the LTT occurs earlier on the NM leeward than on the windward. It is also seen that at $M_\infty = 0.75$, the LTT comes later than at $M_\infty = 0.5$, in spite of the essential Reynolds number overrun. It means that, as the Mach number rises, the position of the transition start moves downstream on the NM. This very dependence was observed on the 2D bodies (the plate), too [10].

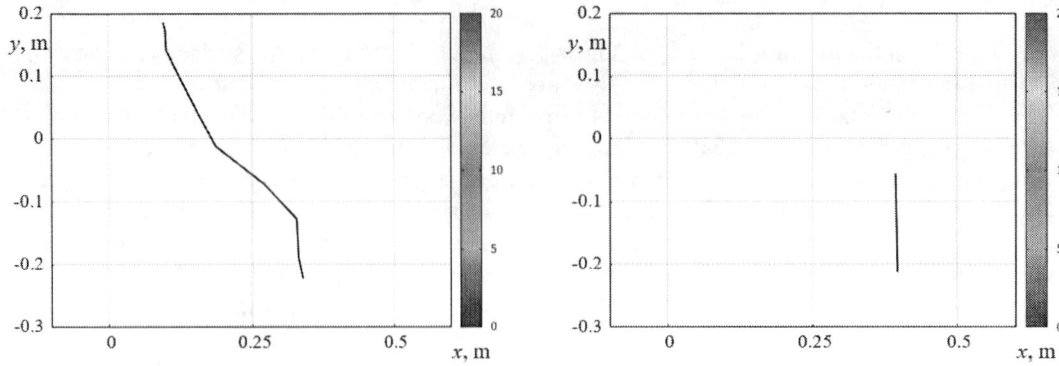

(a) (b)

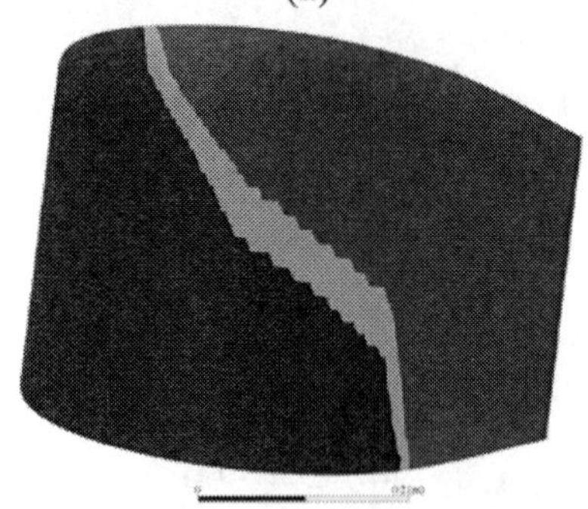

(c)

Figure 5. Streamlines colored by the *N*-factors: Tollmien-Schlichting wave instability (a); instability of cross flow vortices (b) and LT transition zone for both types of instability (c): laminar (blue), transition (green) and turbulent (red) flow areas. $M_\infty = 0.5$, $Re_1 = 11.1\times10^6$ m^{-1}, $\alpha = 7°$, Tu = 0.2%.

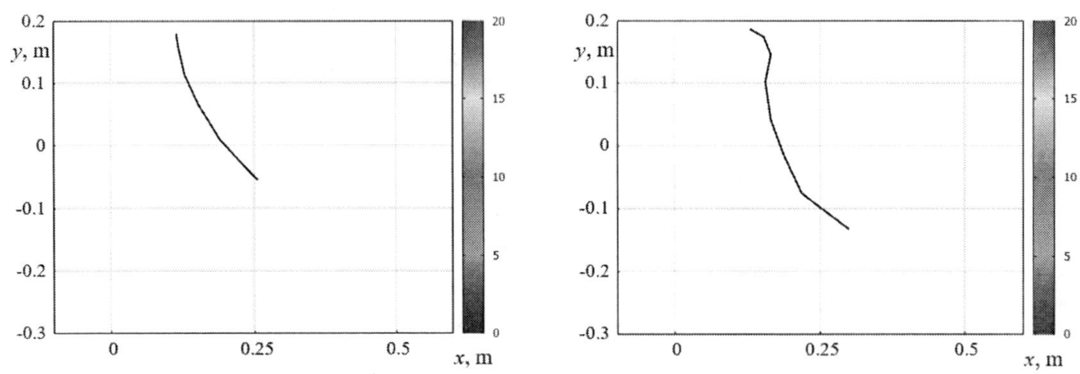

(a) (b)

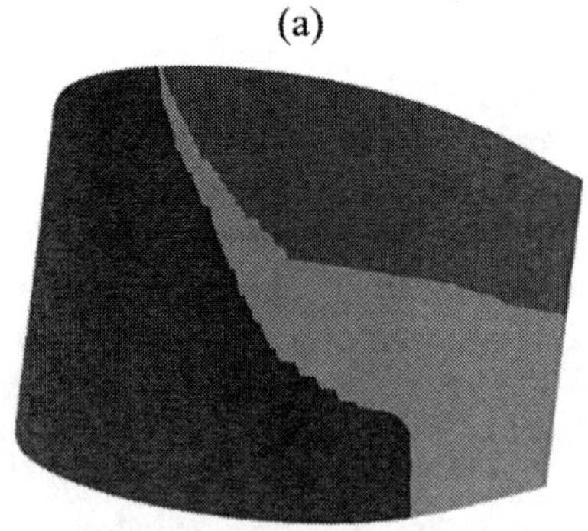

(c)

Figure 6. Streamlines colored by the *N*-factors: Tollmien-Schlichting wave instability (a); instability of cross flow vortices (b) and LT transition zone for both types of instability (c): laminar (blue), transition (green) and turbulent (red) flow areas. $M_\infty = 0.75$, $Re_1 = 33.2\times10^6$ m^{-1}, $\alpha = 7°$, Tu = 0.2%.

5. Conclusion

The laminar-turbulent flow at the nacelle is numerically simulated at the transonic velocities and low degree of turbulence of the external flow by the package ANSYS Fluent with the LTT module based of the original software package LOTRAN 3.0.

The performance of the LTT module is demonstrated during the natural LTT caused by the amplification of the Tollmien-Schlichting wave amplitudes and instability vortices of the cross flow.

It is shown that, in spite of the strong crossflow, the Tollmien-Schlichting wave instability is the dominating mechanism of the LTT up to the angle of attack of 7° on the NM at $M_\infty = 0.5$.

The data obtained on the LTT position and dominating mechanisms in the boundary layer of the nacelle flown by the transonic gas flow at the low degree of turbulence are necessary for the creation of the laminarized NM in order to reduce the drag friction on its external surface.

Acknowledgments

This work was supported by the Russian Science Foundation (Grant No. 18-19-00460).

References

[1] Boiko A V, Kirilovskiy S V, Maslov A A and Poplavskaya T V 2015 *J. Appl. Mech. Tech. Phys.* **56** 761-76
[2] Meier H U and Kreplin H P 1980 *Zeitschrift für Flugwissenschaften und Weltraumforschung* **4** 65-71
[3] Kirilovskiy S V, Boiko A V, Demyanko K V, Nechepurenko Y M, Poplavskaya T V and Sidorenko A A 2019 *AIP Conf. Proc.* **2125** 030098
[4] Poplavskaya T V, Boiko A V, Demyanko K V, Kirilovskiy S V and Nechepurenko Y M 2019 *J. of Phys.: Conf. Series* (in press)
[5] Haghiri A A, Fallahpour N, Mani M and Tadjfar M 2015 *J. Mech. Sci. Technol.* **29** 4671-79
[6] Boiko A V, Nechepurenko Y M, Zhuchkov R N and Kozelkov A S 2014 *Thermophysics and Aeromechanics* **21** 191–210
[7] Nechepurenko Y M and Boiko A V 2013 *Software official registration certificate for the* No. 2013660060 (Russia: Federal Intellectual Property Service)
[8] Boiko A V, Demyanko K V and Nechepurenko Y M 2017 *Rus. J. Num. Anal. Math. Mod.* **32** 1-12
[9] Mack L M 1977 *AGARD Conf. Proc.* AGARD-CP-224 22 (Copenhagen, Denmark) pp 1.1–1.22
[10] Boiko A V, Demyanko K V, Inozemtsev A A, Kirilovskiy S V, Nechepurenko Y M, Paduchev A P and Poplavskaya T V 2019 *Thermophysics and Aeromechanics* **26** 675-83

Journal of Physics: Conference Series **1404** (2019) 012125 doi:10.1088/1742-6596/1404/1/012125

Numerical analysis of flow structure behind a bluff body under conditions of unstable thermal stratification

S A Valger[1,2] and N N Fedorova[2,3]

[1] Novosibirsk State University of Architecture and Civil Engineering, Leningradskaya 113, Novosibirsk, 630008, Russia
[2] Khristianovich Institute of Theoretical and Applied Mechanics SB RAS, Institutskaya 4/1, Novosibirsk, 630090, Russia
[3]National Research Novosibirsk State University, Pirogova str, 1, Novosibirsk, 630090, Russia

svetlanavalger@gmail.com

Abstract. In the paper, the authors describe some results of a numerical simulation of the air flow around a bluff body mounted on a flat plate taking into account the injection of a tracer gas through a hole located in the substrate behind the body. The aim of this paper is to describe flow structure in the vicinity of a prism taking into account unstable stratification of the boundary layer which is characterized by a negative vertical temperature gradient. The results of numerical simulation for an unstable thermal state were compared with those obtained for the case of stable stratification for the same geometry configuration.

1. Introduction

Air quality is one of the most important criteria for urban ecology. High concentration of gas pollutants and aerosols injected from various anthropogenic sources into the air reduces quality of life in large agglomerations and may cause various diseases among its population. Air in large cities is often characterized by a high concentration of CO, NOx, SO2, heavy metals and their oxides, aldehydes, hydrocarbons, ammonia, atmospheric dust, etc. According to [1, 2], the main sources of gas and aerosols emission in large cities are industry and transportation system. Emissions of gases and aerosols from thermal power plants and enterprises are strictly controlled and regulated by standards at the design stage of equipment and during its operation. However, monitoring emission transfer from vehicles in an urban environment is more difficult due to a number of factors such as small spatial and temporal scales of the process and a high level of difficulty in identifying pollution sources. In addition, transport and transformation of gas pollutants and particles from high and low pollution sources depend on the characteristics of the atmospheric boundary layer significantly, as well as on interaction of the atmospheric turbulent flow with local flows, the structure of which is formed during the flow around buildings and structures.

Evaluation of concentrations for air pollutants is based nowadays on atmospheric monitoring at stationary observation posts. For example, in [3], data on microclimatic parameters of air and concentrations of dust particles PM2.5 and PM10 are delivered in real time taking into account daily dynamics. However, a small number of monitoring posts does not allow describing and evaluating the sources of formation of pollutants and its distribution and their transformation in time and space. Studies [4, 5] showed that restoring data on pollution sources based on a small number of observation posts is quite problematic. Existing pollution prediction systems (e.g., WRF-CHEM [6]) as a rule operate large spatial scales and do not allow one to consider the local transformation and transfer emission effects in an urban landscape, the scale of which is determined by hundreds of meters. Thus, predicting the transfer of pollutants from low sources in the urban environment remains a difficult task requiring solution by means of computational fluid dynamics.

In the paper, the authors describe some results of a numerical simulation of an air flow around a bluff body mounted on a flat plate, considering an injection of a tracer gas through a hole located behind the body in the substrate. This statement of the problem was proposed in the experimental work [7] and is a good test for studying processes of heat and mass transfer behind a bluff-body. The aim of this paper is to describe a flow structure in the vicinity of a prism taking into account unstable stratification of the boundary layer which is characterized by a negative vertical temperature gradient. The results of the numerical simulation for the unstable thermal state were compared with the results obtained for the case of stable stratification for the same geometry configuration. The work continues the study described in [8].

2. Problem statement and numerical methods

Figure 1 shows schematically the problem statement corresponding to [7]. A bluff-body is a prism of a square cross-section with a characteristic transverse dimension b = 0.08 m and a height H = 2b. The gas tracer (ethylene) enters the zone behind the body through a hole with a diameter of 5×10^{-3} m located at the distance $b/2$. A mass flow rate of tracer gas is 1.12×10^{-5} kg/s. The stratification of the boundary layer is specified through the set of boundary conditions described in [7]. The velocity, turbulent kinetic energy (TKE) and temperature profiles are set at the inlet boundary of the computational domain and surface temperature boundary conditions, which vary depending on the case (stable or unstable state), are set on the walls of the prism and substrate. The problem was considered in three-dimensional formulation.

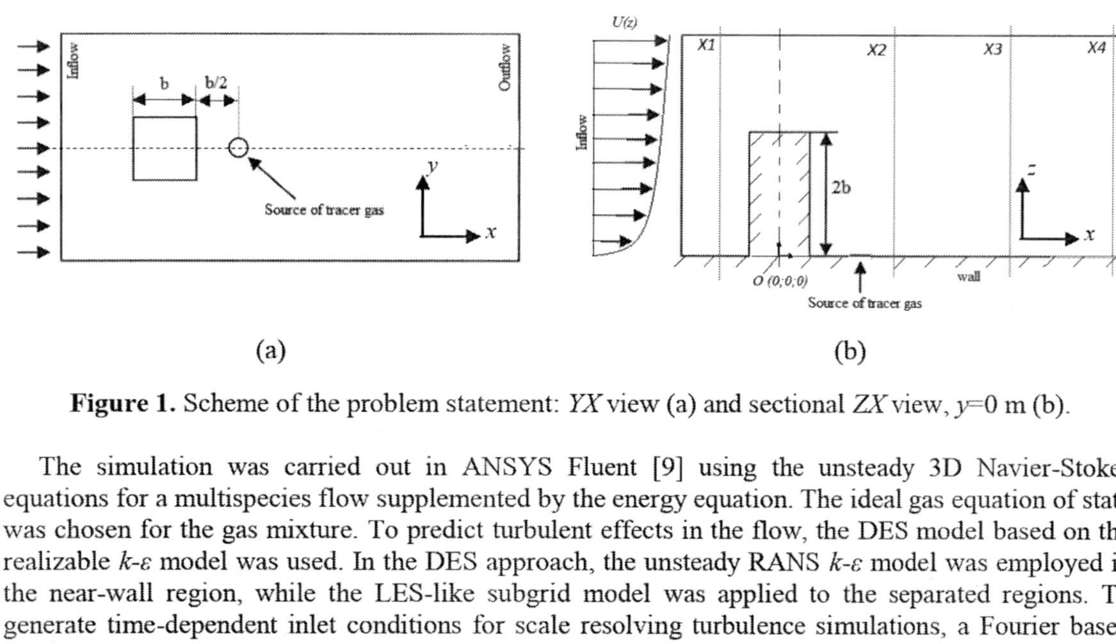

Figure 1. Scheme of the problem statement: YX view (a) and sectional ZX view, y=0 m (b).

The simulation was carried out in ANSYS Fluent [9] using the unsteady 3D Navier-Stokes equations for a multispecies flow supplemented by the energy equation. The ideal gas equation of state was chosen for the gas mixture. To predict turbulent effects in the flow, the DES model based on the realizable k-ε model was used. In the DES approach, the unsteady RANS k-ε model was employed in the near-wall region, while the LES-like subgrid model was applied to the separated regions. To generate time-dependent inlet conditions for scale resolving turbulence simulations, a Fourier based synthetic turbulence generator was used. The "no-slip" boundary conditions were used for the walls of the prism and the flat plate. At the external boundaries of the computational domain, the conditions of constant static pressure were used.

The control-volume-based method was used to convert general scalar transport equations to algebraic equations. The general grid included about 15 million cells. To resolve the flow vortex structure in the LES sub-grid model manner, the fine grid was used in a region behind the prism. Approximately 5 cells were built across the integral length scale $l_0=k^{1.5}/\varepsilon$ to resolve 80% of the turbulent kinetic energy. The $y+$ values at walls did not exceed 1. The NITA/fractional-step method with the second-order of approximation was used for the time-discretization. The bounded central-differenced scheme was used for the momentum equations and the second-order upwind schemes were used for scalars.

3. Results of numerical simulation for unstable thermal flow stratification

The instantaneous and time-averaged fields of temperature, velocity, mass concentration of the tracer gas and turbulent parameters were obtained as the results of calculation for the case of the unstable thermal stratification. Figure 2 (a) shows the time-averaged static temperature field in the central section (*ZX* plane, y = 0 m). The surface boundary layer is characterized by a negative vertical temperature gradient. The temperatures on the walls of the prism and on the substrate are significantly higher than the gas temperature at the prism height H and in the free stream flow and the temperature difference is about 34 °C. This state of stratification drives vertical movement in the atmosphere and it is a condition for the development of convection.

(a)

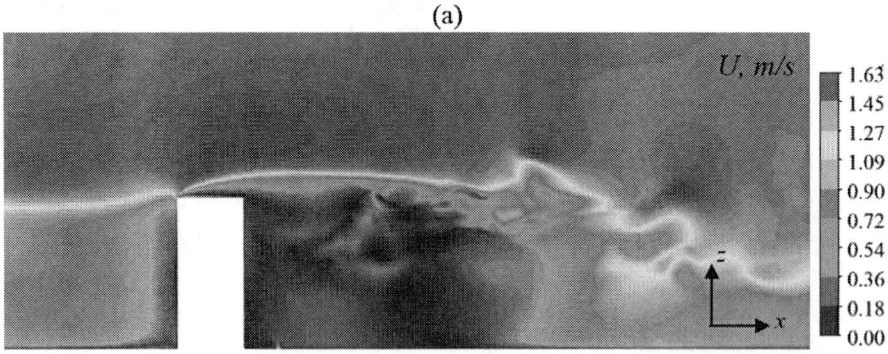

(b)

Figure 2. The mean static temperature <T> [C°] field (a) and instantaneous velocity U [m/s] (b) at the central *ZX* section (y = 0 m) for the unstable thermal state.

To evaluate the characteristics of turbulent gas mixing in the recirculation zones around the prism, the TKE profiles in the characteristic cross-sections in the vicinity of the prism were plotted and compared with those obtained for the case of stable stratification (Fig. 3, (a)). The characteristic sections *X1-4* are located at the central *ZX* plane (see Fig. 1 (b)). As Fig. 3 (a) shows, for the case of unstable stratification, the TKE level is significantly higher comparing to that for the stable stratification state. The unstable thermal state of the boundary layer leads to more intense turbulent mass transfer.

The uniformity of air-ethylene mixing can be evaluated by the uniformity index γ_a based on the area-weighted average mass concentrations of ethylene. The uniformity index γ_a [9] represents how a mass concentration of ethylene varies over the cross-section surfaces, where a value of *1* corresponds to the highest uniformity. Fig. 3 (b) shows the uniformity index values γ_a at the characteristic cross-sections *z/b* behind the prism for stable (red line) and unstable (blue line) stratification regimes.

It can be seen from Fig. 3 (b), the fields of the mass concentration of ethylene are more uniform for the case of the unstable state. A more uniform mixing of gases in the recirculation zone behind the prism in the case of unstable stratification is caused by the intensive convection transfer in the vertical direction.

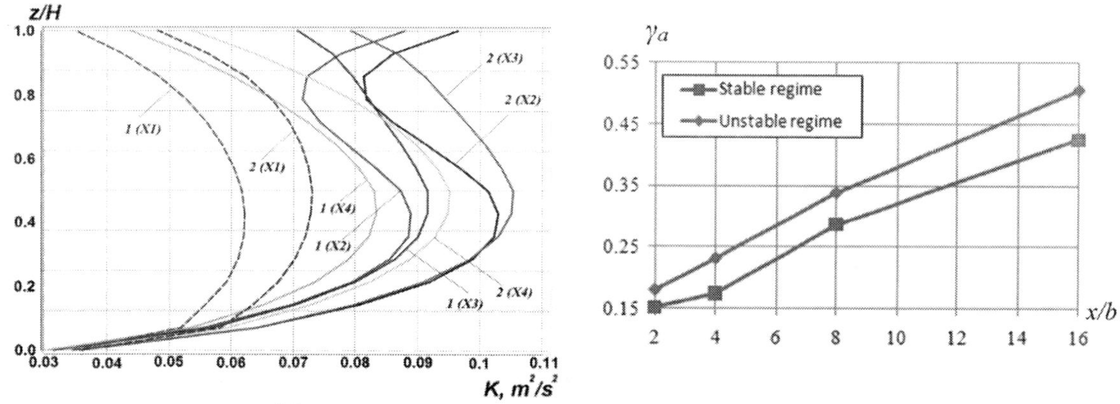

(a) (b)

Figure 3. The TKE profiles (a) for the characteristic sections *X1=-b; X2=2b; X3=4b; X4=6b* for the cases of the stable (1) and the unstable (2) thermal states; Uniformity index γ_a at the characteristic cross-sections *z/b* behind the prism for stable (red line) and unstable (blue line) stratification regimes.

4. Conclusion

The features of the unstable thermal state of the boundary layer in the vicinity of a bluff-body were reproduced. Based on the calculation results, the fields of the main gas flow parameters in characteristic sections were obtained, the flow structure and the nature of the pollutant transfer behind the body were described. It is shown that the regime of unstable stratification is accompanied by more intense turbulent mixing of gases and contributes to faster dispersion of gas emission from a low source in the zone behind the body compared to the case when the surface boundary layer is stable.

Acknowledgments

The work was supported by Russian Foundation for Basic Research, grant 18-08-00755 A.

References

[1] Bezuglaya E Yu 2009 *Air Quality in Russian Major Cities for 10 Years (1998—2007).* Analytical Review (Saint-Petersburg: FGBI MGO) 134 pp
[2] Azarov V N, Marinin N A and Jogoleva D V 2011 *Proc. Southwest State University* **38** 144-8
[3] https://cityair.io/
[4] Penenko A V and Grishina A A 2017 *Interexpo Geo-Siberia* Proc. XIII Int. Cong., 17-21 April 2017, Novosibirsk Vol 1 (Novosibirsk: SSUGT Press) 136-40
[5] Penenko V V, Penenko A V, Tsvetova E A and Gochakov A V 2019 *J. Appl. Mech. Tech. Phys.* **354** 238-46
[6] https://www2.acom.ucar.edu/wrf-chem
[7] http://www.wind.arch.t-kougei.ac.jp/info_center/pollution/Non-Isothermal_Flow.html
[8] Valger S and Fedorova N 2018 *AIP Conference Proceedings* **2027** 040043
[9] https://ansyshelp.ansys.com

Supersonic airplane engine jet influence on the intensity of tail shock wave of sonic boom waveform

V F Volkov

Khristianovich Institute of Theoretical and Applied Mechanics SB RAS,
4/1, Institutskaya str., Novosibirsk, 630090, Russia

volkov@itam.nsc.ru

Abstract. The paper presents the computational experiment results aimed at studying the influence of the aircraft propulsion system on the sonic boom parameters. As a result, at large distances from the disturbances source at jet temperature of 2400° K and velocity of 1700 m/s at the nozzle exit, the trailing shock wave intensity in the sonic boom profile was significantly reduced by 50% compared to the profile generated by the aircraft without considering the jet influence.

1. Introduction

Strict environmental requirements are imposed on supersonic passenger aircrafts to limit the sonic boom level. A preliminary analysis conducted at TsAGI [1] showed that the supersonic passenger aircraft cannot provide acceptable levels of sonic boom $\Delta P_{sb} < 5$ kg m^{-2} ($\Delta P_{sb} < 50$ *Pa*) without a noticeable deterioration in economic performance. In this regard, the implementation of supersonic flights on routes passing over the water surface or over uninhabited land is envisaged.

The solution to the question of whether flights of supersonic aircraft over populated territories will be possible depends on the sonic boom made by these aircrafts.

According to the systematic calculations of the sonic boom intensity [2-5] for various parameter values, methods of the sonic boom intensity reduction were established by changing the aircraft layout. These methods include: shifting the wing back relative to the fuselage, increasing the wing root chord, reducing the relative profile thickness, increasing the relative midship of the fuselage, etc.

The influence of the layout elements of the supersonic passenger aircraft model on the sonic boom parameters was investigated in [5]. In this paper the optimal layout is selected from the analysis of the parametric calculations results. Selected layout provides a significant reduction in the sonic boom level at large distances from the source of disturbances and without reducing aerodynamic quality. In the early works [2-5], the methods of sonic boom reduction were applicable only to aircraft airframes without taking into account the propulsion system influence. This setting does not reflect the real physical phenomenon, the influence of the propulsion system jets on the disturbed flow generated by the aircraft at large distances from the source.

2. Jet propulsion modeling

To solve this problem, the theoretical foundations of Laval nozzle calculations were used [6, 7].

- Gas constant of air $R = 287$ J kg^{-1} K^{-1}, adiabatic index $k = 1.4$.
- Pressure and temperature of air at the nozzle inlet P_0^*, T_0^*.

- Air pressure at the nozzle exit P_e.

The flow regime is determined by the comparison of the pressure ratio $\beta_i = P_i / P_0^*$ with the critical ratio

$$\beta_c = \frac{P_c}{P_0^*} = \left(\frac{2}{k+1}\right)^{\frac{k}{k-1}} \qquad (1)$$

where P_i, P_c– static pressure in i-section and in critical section, P_0^* – total pressure at the nozzle inlet, k – adiabatic index. In this study, the working fluid is air, for which $k = 1.4$ and therefore $\beta_c = 0.528$.

Calculation Procedure: find the nozzle pressure ratio if: when $\beta_e < \beta_c$ the flow regime is supersonic. To determine the speed of the jet at the nozzle exit, the pressure ratio was taken equal to $\beta_e = P_e / P_0^* = 0.05$, which provides supersonic flow regime.

$$w_e = \sqrt{\frac{2 \cdot k}{k-1} \cdot R \cdot T_0^* (1 - \beta^{\frac{k-1}{k}})}.$$

3. Calculation methods and conditions

To quantify the sonic boom impact on terrestrial objects, it is necessary to have a method of the shock wave intensity calculation at large distances from an aircraft.

The calculation method based on linear theory leads to the need to determine the shape factor of the aircraft, which makes the task difficult to solve for a real configuration because of the analytical description impossibility of the body geometry. On the other hand, the presence of engine nacelles in the aircraft design limits the applicability of linear theory. In this regard, it seems appropriate to use the flow parameters in the near zone of the model, which can be obtained experimentally or numerically, for their subsequent conversion to the far zone.

The use of numerical methods for solving the supersonic flow problems around the aircraft to determine its aerodynamic characteristics and the parameters of the sonic boom created by it at large distances is associated with some limitations.

In particular, the width of the perturbed region is much smaller than the characteristic radius of front curvature and the characteristic distance at which the parameters of the perturbed environment change significantly. Under this condition, it is advisable to calculate the propagation of small perturbations using the linear theory provisions described in [8,9]. In the calculations of perturbations at large distances from the source, the perturbed flow parameters in the near zone, determined by the numerical solution results of the flow around the model, were used as initial data.

The problem of disturbances propagation over long distances and the formation of the sonic boom wave generated by the aircraft layout was solved in two stages.

At the first stage, the problem of supersonic flow around the layout with the free-stream Mach number $M_\infty = 2.03$ was solved. Based on the results of this problem numerical solution, the aerodynamic characteristics and parameters of the perturbed flow generated by the layout in the near zone were determined on the control surface at a given distance from the body axis.

The control surface (figure 1) in the perturbed region is a cylinder with a radius $\overline{R} = R/L_f = 1.33$ whose axis coincides with the layout axis. The perturbed pressure parameters on the control surface are used as initial data to the problem solution.

The numerical solution of the flow around the layout determining the flow parameters in the perturbed region was carried out in the framework of the Navier-Stokes equations.

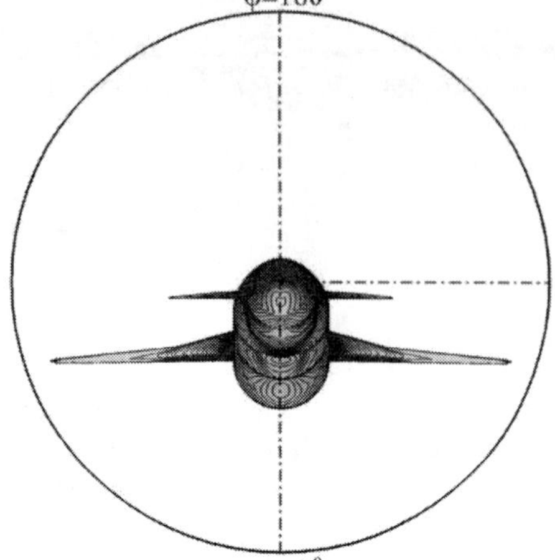

Figure 1. Control surface in the vicinity of the investigated layout.

The geometric models formation of monoplane and tandem layouts was carried out by combinations of its constituent elements joints. The necessary elements for creating layouts are the following:

- Fus1 - axisymmetric body with ogival head part with elongation $\lambda = 4.5$ and total elongation $\lambda = 15$;
- Win1 - trapezoidal wing with sweep along the leading edge $\chi = 56°$ with a flat median surface with a 3% rhomboid profile;
- Win2 - trapezoidal wing with a basal influx with sweep along the leading edge $\chi_1 = 76°$ and $\chi_2 = 56°$, with a flat median surface with a 3% rhomboid profile;
- engine nacellas.

The model geometric layout is shown in figure 2, (a). The mathematical model of supersonic flow around a tandem layout with running engines is shown in figure 2, (b).

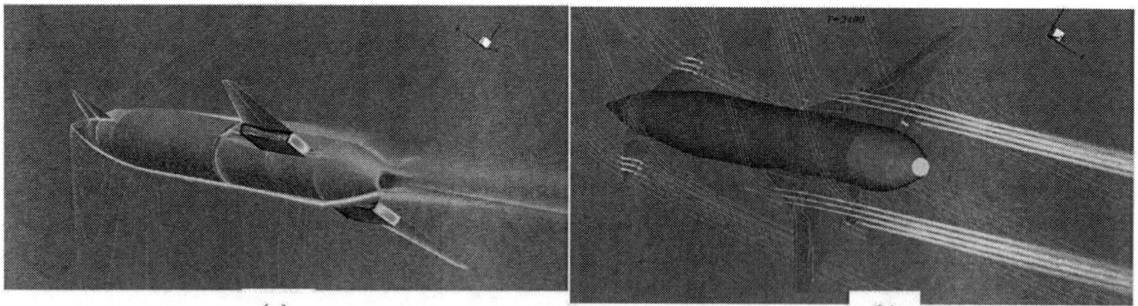

(a) (b)

Figure 2. Jet modelling: (a) - optimum tandem layout with nacelles; (b) - tandem layout with a propulsion system.

For visualization of jets and features of the parameters distribution in the disturbed region the temperature fields are presented in the calculation region.

4. Computational experiment results

To conduct the computational experiment to determine the sonic boom wave intensity in the cruise flight mode at a distance from the source of disturbances and without loss of aerodynamic quality [6], the optimal tandem configuration layout of the aircraft was used (figure 2, (a)).

Journal of Physics: Conference Series **1404** (2019) 012126 doi:10.1088/1742-6596/1404/1/012126

At a large distance from the source, the sonic boom wave profile induced by the tandem arrangement with predetermined front wing positions contains shock waves spaced along the longitudinal coordinate – the train of shock waves (figure 2, label 1). These include the leading shock wave generated by the bow of the hull with the front wing, the shock wave from the rear wing and the trailing (tail) shock wave. It should be noted that an effective reduction in the sonic boom level is achieved when the effect of the leading and following shock waves was perceived by the observer as separate pops. This condition, the so-called solvability condition of shock waves, is determined by the flight time of the distance between the leading and following shock waves, which should be $\Delta t_0 \geq 40$ ms, which corresponds to the frequency of their impact on the observer $f_0 = 25$ Hz. Otherwise, at a frequency $f > f_0$ these pops are perceived by the observer as continuous noise, which pressure level corresponds to the total overpressure of the leading and following shock waves.

A comparison of the obtained results is shown in figure 3, where the markers indicate: 1 – the sonic boom wave profile generated by the optimal tandem layout with engine nacelles [6] without taking into account the influence of the propulsion system jets; 2 - the generated sonic boom wave profile in the disturbed region with the jets of the propulsion system with a temperature at the nozzle exit $T = 1200$ K; 3 – sonic boom wave profile in the disturbed region with the jets of the propulsion system with a temperature at the nozzle exit $T = 2400$ K.

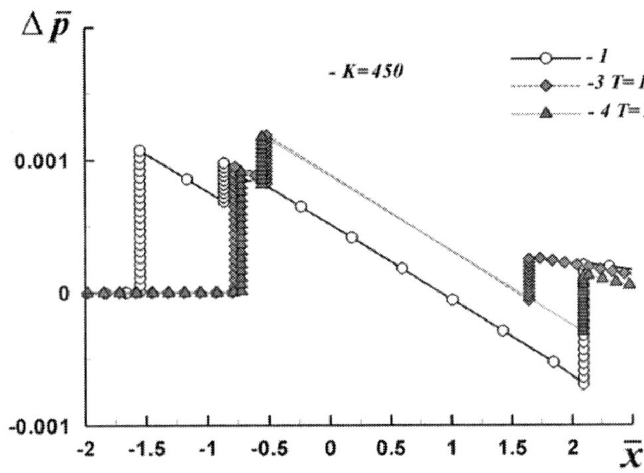

Figure 3. Sonic boom wave profiles at a distance $K = r/L_f = 450$ from the disturbances source.

It can be noted in this figure that the influence of the propulsion system jets is accompanied by significant changes in the parameters of the sonic boom wave profile (figure 3, marks 3-4). This can be attributed to: the distance between the leading and following shocks on profiles 3, 4 is significantly reduced in comparison with profile 1. This distance is much less than the distance satisfying the condition for solvability of shocks in the sonic boom wave profile, which the observer perceives as a leading pop in the N-profile of perturbed pressure waves.

The level of tail shock wave intensity of the sonic boom is reduced by more than 50%. A decrease in the intensity level of the tail shock is accompanied by a decrease in the wavelength of the sonic boom.

It should be noted that the perturbed flow parameters behind the closing shock asymptotically approach the unperturbed flow parameters.

The dimensional values of the excess pressures in the table 1 are applicable to the estimation of the sonic boom parameters created by the tandem layout with length $l_f = 40$ m, mass $G = 40$ ton in the cruise flight mode at a speed corresponding to the Mach number $M_\infty = 2.03$ at altitude $H = 18$ km.

Table 1. Shock intensities in the profile of the sonic boom wave.

$K=H/L_f=450$ $H=18\ km$ $p_k=\sqrt{p_h \cdot p_g}=27680\ Pa$					
No	$\Delta\bar{p}_h$	Δp_h [Pa]	$\Delta\bar{p}_t$	Δp_t [Pa]	dB
1	0.001075	29.7	0.000696	19.3	117.0
3	0.001191	33.0	0.000325	9.0	115.0
4	0.001191	33.0	0.000431	12.0	116.0

p_h – atmospheric pressure at flight altitude,
p_g – atmospheric pressure at earth surface,
Δp_t – tail shock intensity of sonic boom.

5. Conclusion

The results of a computational experiment aimed at studying the influence of propulsion system jets on the formation of the sonic boom wave profile generated by the supersonic passenger plane at large distances from the disturbances source (about K = 450) are presented. The data obtained are compared with the corresponding data received for the airframe with optimal tandem layout. It is shown that the jet outflow is accompanied by a significant decrease in the intensity of the closing shock in the profile of the sonic boom wave (about 50%).

References

[1] Vasiliev L E, Popov S I and Svishchev G R 1994 *Technique of the air fleet* **68** 14-7
[2] Zhilin Yu L 1967 *Tr. TsAGI* **1094** 14-22
[3] Ferri A and Ismail A 1969 *AIAA J* **7** 1538-41
[4] George A R, Seebass R 1971 *AIAA J* **9** 2091-3
[5] Volkov V F 2017 *Journal of Engineering Physics and Thermophysics* **90** 449-60
[6] Dorofeev A A 1999 *Fundamentals of the theory of thermal rocket engines (General theory of rocket engines)* (Moscow: MSTU N E Bauman)
[7] Landau L D and Lifshits E M 1986 *Theoretical Physics. Hydrodynamics* (Moscow: Nauka) (in Russian)
[8] Whitham G B 1952 *Comm. Pure Appl. Math.* **5** 301-38
[9] Ryshov O S 1961 *Prikl. Mekh. Tekh. Fiz.* **2** 15-25 (in Russian)

AUTHOR INDEX

AUTHOR INDEX

AUTHOR INDEX